Solution Key

Algebra

Structure and Method **Book 1**

Richard G. Brown
Mary P. Dolciani
Robert H. Sorgenfrey
William L. Cole

McDougal Littell/Houghton Mifflin
Evanston, Illinois
Boston Dallas Phoenix

2000 Impression

Printed in U.S.A.

ISBN-13: 978-0-395-67764-3 ISBN-10: 0-395-67764-5

10 11 12 13 – CSM – 09 08

Contents

Pages 3–4 • WRITTEN EXERCISES

A

1. $(8 - 3) + 3 = 5 + 3 = 8$ **2.** $9 + (18 - 2) = 9 + 16 = 25$

3. $5 \cdot (11 + 1) = 5 \cdot 12 = 60$ **4.** $(13 - 6) - 7 = 7 - 7 = 0$

5. $(6 + 12) \div 3 = 18 \div 3 = 6$ **6.** $6 + (12 \div 3) = 6 + 4 = 10$

7. $29 - (0 \cdot 9) = 29 - 0 = 29$ **8.** $5 - (16 \div 4) = 5 - 4 = 1$

9. $(8 \cdot 17) + (12 \cdot 17) = 136 + 204 = 340$ **10.** $(12 \cdot 11) - (2 \cdot 11) = 132 - 22 = 110$

11. $(26 + 4) \div (30 \div 2) = 30 \div 15 = 2$ **12.** $(40 \div 10) \div (1 \cdot 4) = 4 \div 4 = 1$

13. $5x = 5 \cdot 2 = 10$ **14.** $6y = 6 \cdot 3 = 18$ **15.** $xy = 2 \cdot 3 = 6$

16. $yz = 3 \cdot 4 = 12$ **17.** $(4x) + 7 = (4 \cdot 2) + 7 = 8 + 7 = 15$

18. $(3y) - 9 = (3 \cdot 3) - 9 = 9 - 9 = 0$ **19.** $(5z) - 7 = (5 \cdot 4) - 7 = 20 - 7 = 13$

20. $(8x) + 7 = (8 \cdot 2) + 7 = 16 + 7 = 23$

21. $(2x) + (2y) = (2 \cdot 2) + (2 \cdot 3) = 4 + 6 = 10$

22. $(3z) - (4x) = (3 \cdot 4) - (4 \cdot 2) = 12 - 8 = 4$

23. $8 \cdot (y + z) = 8 \cdot (3 + 4) = 8 \cdot 7 = 56$ **24.** $7 \cdot (y - x) = 7 \cdot (3 - 2) = 7 \cdot 1 = 7$

25. $\frac{1}{2} \cdot (z - x) = \frac{1}{2} \cdot (4 - 2) = \frac{1}{2} \cdot 2 = 1$ **26.** $\frac{2}{3} \cdot (y + 6) = \frac{2}{3} \cdot (3 + 6) = \frac{2}{3} \cdot 9 = 6$

27. $\frac{(y + x)}{(y - x)} = \frac{(3 + 2)}{(3 - 2)} = \frac{5}{1} = 5$ **28.** $\frac{(z + y)}{(z - y)} = \frac{(4 + 3)}{(4 - 3)} = \frac{7}{1} = 7$

29. $(4x) - 8 = (4 \cdot 2) - 8 = 8 - 8 = 0$

30. $4z - (8x) = 4 \cdot 4 - (8 \cdot 2) = 16 - 16 = 0$

31. $(5yz) - x = (5 \cdot 3 \cdot 4) - 2 = 60 - 2 = 58$

32. $8 + (9xy) = 8 + (9 \cdot 2 \cdot 3) = 8 + 54 = 62$

B

33. $lw = 25 \cdot 12 = 300$ **34.** $(2l) + (2w) = (2 \cdot 25) + (2 \cdot 12) = 50 + 24 = 74$

35. $(a + b) + c = (10 + 24) + 26 = 34 + 26 = 60$

36. $\frac{1}{2} \cdot (ab) = \frac{1}{2} \cdot (10 \cdot 24) = \frac{1}{2} \cdot 240 = 120$

37. $(1.8 \cdot C) + 32 = (1.8 \cdot 37) + 32 = 66.6 + 32 = 98.6$

38. $\left(\frac{1}{2} \cdot g\right)(t \cdot t) = \left(\frac{1}{2} \cdot 9.8\right) \cdot (16 \cdot 16) = 4.9 \cdot 256 = 1254.4$

39. $Prt = 5000 \cdot 0.125 \cdot 2 = 625 \cdot 2 = 1250$ **40.** $\frac{p \cdot c}{1000} = \frac{75 \cdot 9.5}{1000} = \frac{712.5}{1000} = 0.7125$

41. $4n = 12$; $4 \cdot 3 = 12$, so $n = 3$; 3 **42.** $8x = 16$; $8 \cdot 2 = 16$, so $x = 2$; 2

43. $y + 3 = 3 + y$; for example, $7 + 3 = 3 + 7$ and $1.5 + 3 = 3 + 1.5$; any number

44. $m + 5 = 5$; $0 + 5 = 5$, so $m = 0$; 0 **45.** $6\frac{1}{2} - y = 3 \cdot 2$; $6\frac{1}{2} - \frac{1}{2} = 6$, so $y = \frac{1}{2}$; $\frac{1}{2}$

46. $\frac{3}{4} + y = 0.75 + y$; for example, $\frac{3}{4} + 1 = 0.75 + 1$ and $\frac{3}{4} + 5 = 0.75 + 5$; any number

C **47.** $a \cdot a = 2a$; $0 \cdot 0 = 2 \cdot 0$ and $2 \cdot 2 = 2 \cdot 2$, so $a = 0$ or 2; 0, 2

48. $z \cdot 8 = z$; $0 \cdot 8 = 0$, so $z = 0$; 0

49. $(b \cdot b) + 3 = 4b$; $(1 \cdot 1) + 3 = 4 \cdot 1$ and $(3 \cdot 3) + 3 = 4 \cdot 3$, so $b = 1$ or 3; 1, 3

Page 5 • MIXED REVIEW EXERCISES

1. $(0.3)(1.2) = 0.36$ **2.** $21.25 + 8.07 = 29.32$ **3.** $1.6 \div 0.4 = 4$ **4.** $7.2 - 3.8 = 3.4$

5. $212.1 + 6.9 = 219$ **6.** $6.32 - 4.7 = 1.62$ **7.** $72.34 \times 2.1 = 151.914$

8. $33.6 \div 2.1 = 16$ **9.** $\frac{2}{3} + \frac{3}{10} = \frac{20}{30} + \frac{9}{30} = \frac{29}{30}$ **10.** $\frac{5}{\underset{1}{\cancel{7}}} \times \frac{\overset{\overset{1}{\cancel{2}}}{\cancel{14}}}{\underset{8}{\cancel{16}}} = \frac{5}{8}$ **11.** $\frac{1}{2} \div \frac{3}{8} = \frac{1}{\underset{1}{\cancel{2}}} \times \frac{\overset{4}{\cancel{8}}}{3} = \frac{4}{3}$

12. $\frac{3}{8} - \frac{1}{4} = \frac{3}{8} - \frac{2}{8} = \frac{1}{8}$ **13.** $\frac{\overset{1}{\cancel{7}}}{\underset{5}{\cancel{20}}} \times \frac{\overset{4}{\cancel{16}}}{\underset{3}{\cancel{21}}} = \frac{4}{15}$ **14.** $\frac{5}{12} - \frac{1}{15} = \frac{25}{60} - \frac{4}{60} = \frac{21}{60} = \frac{7}{20}$

15. $\frac{4}{7} + \frac{2}{3} = \frac{12}{21} + \frac{14}{21} = \frac{26}{21}$ **16.** $\frac{4}{7} \div \frac{3}{14} = \frac{4}{\underset{1}{\cancel{7}}} \times \frac{\overset{2}{\cancel{14}}}{3} = \frac{8}{3}$

Page 5 • APPLICATION

1. $p = 1008$, $t = 2$; $p \cdot t = 1008 \cdot 2 = 2016$ (watt-hours); $2016 \div 1000 = 2.016$ kW · h

2. $p = 4856$, $t = 10$; $p \cdot t = 48{,}560$ (watt-hours); $48{,}560 \div 1000 = 48.56$ kW · h

Pages 8–9 • WRITTEN EXERCISES

A **1.** **a.** $8 + 3 \cdot 4 = 8 + 12 = 20$ **b.** $(8 + 3)4 = 11 \cdot 4 = 44$

2. **a.** $9 + 5 \cdot 2 = 9 + 10 = 19$ **b.** $(9 + 5)2 = 14 \cdot 2 = 28$

3. **a.** $6 - 3 \div 3 = 6 - 1 = 5$ **b.** $(6 - 3) \div 3 = 3 \div 3 = 1$

4. **a.** $5 + 10 \div 5 = 5 + 2 = 7$ **b.** $(5 + 10) \div 5 = 15 \div 5 = 3$

5. $\frac{6 + 5 \cdot 3}{6 + 1} = \frac{6 + 15}{7} = \frac{21}{7} = 3$ **6.** $\frac{8 + 3 \cdot 2}{4 + 3} = \frac{8 + 6}{7} = \frac{14}{7} = 2$

7. $\frac{3(12 - 8)}{2 \cdot 5 - 4} = \frac{3 \cdot 4}{10 - 4} = \frac{12}{6} = 2$ **8.** $\frac{8 \cdot 5 + 2 \cdot 7}{2(7 - 4)} = \frac{40 + 14}{2 \cdot 3} = \frac{54}{6} = 9$

9. **a.** $2x + 7 = 2 \cdot 3 + 7 = 6 + 7 = 13$ **b.** $2(x + 7) = 2(3 + 7) = 2 \cdot 10 = 20$

10. **a.** $5y - 3 = 5 \cdot 4 - 3 = 20 - 3 = 17$ **b.** $5(y - 3) = 5(4 - 3) = 5 \cdot 1 = 5$

11. **a.** $18 - 4x = 18 - 4 \cdot 3 = 18 - 12 = 6$ **b.** $(18 - 4)x = (18 - 4) \cdot 3 = 14 \cdot 3 = 42$

12. **a.** $7z + 8 = 7 \cdot 5 + 8 = 35 + 8 = 43$ **b.** $7(z + 8) = 7(5 + 8) = 7 \cdot 13 = 91$

13. **a.** $xy + z = 3 \cdot 4 + 5 = 12 + 5 = 17$ **b.** $x(y + z) = 3(4 + 5) = 3 \cdot 9 = 27$

14. **a.** $zt - y = 5 \cdot 6 - 4 = 30 - 4 = 26$ **b.** $z(t - y) = 5(6 - 4) = 5 \cdot 2 = 10$

15. **a.** $4xz + 3y = 4 \cdot 3 \cdot 5 + 3 \cdot 4 = 60 + 12 = 72$

b. $4(xz + 3y) = 4(3 \cdot 5 + 3 \cdot 4) = 4(15 + 12) = 4 \cdot 27 = 108$

16. **a.** $9xyz - t = 9 \cdot 3 \cdot 4 \cdot 5 - 6 = 540 - 6 = 534$

b. $9x(zy - t) = 9 \cdot 3(5 \cdot 4 - 6) = 27(20 - 6) = 27 \cdot 14 = 378$

17. $5(3y - 4x) = 5(3 \cdot 4 - 4 \cdot 3) = 5(12 - 12) = 5 \cdot 0 = 0$

18. $6y - 2xy = 6 \cdot 4 - 2 \cdot 3 \cdot 4 = 24 - 24 = 0$

19. $xyz - 4z = 3 \cdot 4 \cdot 5 - 4 \cdot 5 = 60 - 20 = 40$

20. $(y \cdot y + z) \cdot z = (4 \cdot 4 + 5) \cdot 5 = (16 + 5) \cdot 5 = 21 \cdot 5 = 105$

21. $\frac{9x + z}{x + z} = \frac{9 \cdot 3 + 5}{3 + 5} = \frac{27 + 5}{8} = \frac{32}{8} = 4$ **22.** $\frac{4y - 2t}{y + 2t} = \frac{4 \cdot 4 - 2 \cdot 6}{4 + 2 \cdot 6} = \frac{16 - 12}{4 + 12} = \frac{4}{16} = \frac{1}{4}$

23. $\frac{10t - z}{10(t - z)} = \frac{10 \cdot 6 - 5}{10(6 - 5)} = \frac{60 - 5}{10 \cdot 1} = \frac{55}{10} = \frac{11}{2}$ or $5\frac{1}{2}$

24. $\frac{2(y + x)}{2y + x} = \frac{2(4 + 3)}{2 \cdot 4 + 3} = \frac{2 \cdot 7}{8 + 3} = \frac{14}{11}$ or $1\frac{3}{11}$

B **25.** $2[x + 4(y + z)] = 2[3 + 4(4 + 5)] = 2[3 + 4 \cdot 9] = 2[3 + 36] = 2 \cdot 39 = 78$

26. $3[z + 5(2y - x)] = 3[5 + 5(2 \cdot 4 - 3)] = 3[5 + 5(8 - 3)] = 3[5 + 5 \cdot 5] =$
$3[5 + 25] = 3 \cdot 30 = 90$

27. $[(5y + 6z) - 3t] \div y = [(5 \cdot 4 + 6 \cdot 5) - 3 \cdot 6] \div 4 = [(20 + 30) - 18] \div 4 =$
$[50 - 18] \div 4 = 32 \div 4 = 8$

28. $2t - [7z \div (y + x)] = 2 \cdot 6 - [7 \cdot 5 \div (4 + 3)] = 12 - [35 \div 7] = 12 - 5 = 7$

29. $2(a + b) = 2(7.5 + 19.5) = 2 \cdot 27 = 54$

30. $2a + b + c = 2 \cdot 20 + 16 + 48 = 40 + 16 + 48 = 104$

31. $\frac{1}{2}h(b + c) = \frac{1}{2} \cdot 12(16 + 48) = 6 \cdot 64 = 384$

32. $(\pi r)r = \left(\frac{22}{7} \cdot \frac{28}{1}\right) 28 = \left(\frac{22}{\cancel{7}_1} \cdot \frac{\cancel{28}^4}{1}\right) 28 = 88 \cdot 28 = 2464$

33. $2(r + h) + \pi r = 2(2 + 3) + 3.14 \cdot 2 = 2 \cdot 5 + 6.28 = 10 + 6.28 = 16.28$

34. $2(lw + wh + lh) = 2(14 \cdot 12 + 12 \cdot 10 + 14 \cdot 10) = 2(168 + 120 + 140) =$
$2 \cdot 428 = 856$

35. $\frac{24 \cdot 8}{12 + 4} = \frac{192}{16} = 12; \frac{12 \cdot 4}{6 + 2} = \frac{48}{8} = 6; 12 \neq 6$

36. $\frac{23 + 19}{7 \cdot 2} = \frac{42}{14} = 3; 54 \div 9 - 3 = 6 - 3 = 3; 3 = 3$

37. $2[3(12 - 7)] = 2[3 \cdot 5] = 2 \cdot 15 = 30; (5 \cdot 5) + 5 = 25 + 5 = 30; 30 = 30$

38. $3[36 \div (3 + 6)] = 3[36 \div 9] = 3 \cdot 4 = 12; 30 - [(36 \div 3) + 6] = 30 - [12 + 6] =$
$30 - 18 = 12; 12 = 12$

C **39–46.** Answers may vary; examples are given. **39.** $5 \cdot (8 - 6) \div 2 = 5$

40. $(5 \cdot 8 - 6) \div 2 = 17$ **41.** $5 \cdot (8 - 6 \div 2) = 25$ **42.** $(5 \cdot 8) - (6 \div 2) = 37$

43. $5 \cdot (2 + 8 - 4) \div 2 = 15$ **44.** $(5 \cdot 2) + 8 - (4 \div 2) = 16$

45. $[(5 \cdot 2) + 8 - 4] \div 2 = 7$ **46.** $5 \cdot 2 + [(8 - 4) \div 2] = 12$

Page 9 • MIXED REVIEW EXERCISES

1. $(10 - 4) \div 2 = 6 \div 2 = 3$ **2.** $40 \cdot 10 + 18 \cdot 2 = 400 + 36 = 436$

3. $8 \times (26 - 8) = 8 \times 18 = 144$ **4.** $7 + 15 \div 3 = 7 + 5 = 12$

5. $(28 + 4) \div (16 \div 2) = 32 \div 8 = 4$ **6.** $(6 + 5) \cdot (8 - 3) = 11 \cdot 5 = 55$

7. $x + ay = 8 + 3 \cdot 5 = 8 + 15 = 23$ **8.** $(5 - a)y = (5 - 3)5 = 2 \cdot 5 = 10$

9. $3a - 2b = 3 \cdot 3 - 2 \cdot 2 = 9 - 4 = 5$ **10.** $\frac{1}{4}x + 3 = \frac{1}{4} \cdot 8 + 3 = 2 + 3 = 5$

11. $\frac{1}{2}axy = \frac{1}{2} \cdot 3 \cdot 8 \cdot 5 = \frac{1}{2} \cdot 3 \cdot 40 = \frac{1}{2} \cdot 120 = 60$

12. $x \cdot (2a - b) = 8 \cdot (2 \cdot 3 - 2) = 8 \cdot (6 - 2) = 8 \cdot 4 = 32$

Page 9 • CHALLENGE

$2(1 + 2 + 3 + \cdots + n) = n(n + 1)$

a. for $n = 10$, $n(n + 1) = 10 \cdot 11 = 110$

b. for $n = 100$, sum $= \frac{n(n + 1)}{2} = \frac{100 \cdot 101}{2} = \frac{10{,}100}{2} = 5050$

Pages 11–12 • WRITTEN EXERCISES

A **1.**

x	$x + 5 = 9$	
0	$0 + 5 = 9$	False
1	$1 + 5 = 9$	False
2	$2 + 5 = 9$	False
3	$3 + 5 = 9$	False
4	$4 + 5 = 9$	True
5	$5 + 5 = 9$	False

∴ the solution set is {4}.

2.

x	$6 + x = 11$	
0	$6 + 0 = 11$	False
1	$6 + 1 = 11$	False
2	$6 + 2 = 11$	False
3	$6 + 3 = 11$	False
4	$6 + 4 = 11$	False
5	$6 + 5 = 11$	True

∴ the solution set is {5}.

3.

x	$x - 2 = 3$	
0	$0 - 2 = 3$	False
1	$1 - 2 = 3$	False
2	$2 - 2 = 3$	False
3	$3 - 2 = 3$	False
4	$4 - 2 = 3$	False
5	$5 - 2 = 3$	True

∴ the solution set is {5}.

4.

x	$x - 3 = 1$	
0	$0 - 3 = 1$	False
1	$1 - 3 = 1$	False
2	$2 - 3 = 1$	False
3	$3 - 3 = 1$	False
4	$4 - 3 = 1$	True
5	$5 - 3 = 1$	False

∴ the solution set is {4}.

5.

x	$7 - x = 2$	
0	$7 - 0 = 2$	False
1	$7 - 1 = 2$	False
2	$7 - 2 = 2$	False
3	$7 - 3 = 2$	False
4	$7 - 4 = 2$	False
5	$7 - 5 = 2$	True

∴ the solution set is {5}.

6.

x	$6 - x = 3$	
0	$6 - 0 = 3$	False
1	$6 - 1 = 3$	False
2	$6 - 2 = 3$	False
3	$6 - 3 = 3$	True
4	$6 - 4 = 3$	False
5	$6 - 5 = 3$	False

∴ the solution set is {3}.

7.

x	$2x = 8$	
0	$2 \cdot 0 = 8$	False
1	$2 \cdot 1 = 8$	False
2	$2 \cdot 2 = 8$	False
3	$2 \cdot 3 = 8$	False
4	$2 \cdot 4 = 8$	True
5	$2 \cdot 5 = 8$	False

∴ the solution set is {4}.

8.

x	$5x = 10$	
0	$5 \cdot 0 = 10$	False
1	$5 \cdot 1 = 10$	False
2	$5 \cdot 2 = 10$	True
3	$5 \cdot 3 = 10$	False
4	$5 \cdot 4 = 10$	False
5	$5 \cdot 5 = 10$	False

∴ the solution set is {2}.

9.

x	$8x = 16$	
0	$8 \cdot 0 = 16$	False
1	$8 \cdot 1 = 16$	False
2	$8 \cdot 2 = 16$	True
3	$8 \cdot 3 = 16$	False
4	$8 \cdot 4 = 16$	False
5	$8 \cdot 5 = 16$	False

$\therefore$ the solution set is {2}.

10.

x	$2x = 10$	
0	$2 \cdot 0 = 10$	False
1	$2 \cdot 1 = 10$	False
2	$2 \cdot 2 = 10$	False
3	$2 \cdot 3 = 10$	False
4	$2 \cdot 4 = 10$	False
5	$2 \cdot 5 = 10$	True

$\therefore$ the solution set is {5}.

11.

x	$3x = 0$	
0	$3 \cdot 0 = 0$	True
1	$3 \cdot 1 = 0$	False
2	$3 \cdot 2 = 0$	False
3	$3 \cdot 3 = 0$	False
4	$3 \cdot 4 = 0$	False
5	$3 \cdot 5 = 0$	False

$\therefore$ the solution set is {0}.

12.

x	$0 = 4x$	
0	$0 = 4 \cdot 0$	True
1	$0 = 4 \cdot 1$	False
2	$0 = 4 \cdot 2$	False
3	$0 = 4 \cdot 3$	False
4	$0 = 4 \cdot 4$	False
5	$0 = 4 \cdot 5$	False

$\therefore$ the solution set is {0}.

13.

x	$x \div 3 = 1$	
0	$0 \div 3 = 1$	False
1	$1 \div 3 = 1$	False
2	$2 \div 3 = 1$	False
3	$3 \div 3 = 1$	True
4	$4 \div 3 = 1$	False
5	$5 \div 3 = 1$	False

$\therefore$ the solution set is {3}.

14.

x	$x \div 2 = 2$	
0	$0 \div 2 = 2$	False
1	$1 \div 2 = 2$	False
2	$2 \div 2 = 2$	False
3	$3 \div 2 = 2$	False
4	$4 \div 2 = 2$	True
5	$5 \div 2 = 2$	False

$\therefore$ the solution set is {4}.

15.

x	$\frac{1}{2}x = 2$	
0	$\frac{1}{2} \cdot 0 = 2$	False
1	$\frac{1}{2} \cdot 1 = 2$	False
2	$\frac{1}{2} \cdot 2 = 2$	False
3	$\frac{1}{2} \cdot 3 = 2$	False
4	$\frac{1}{2} \cdot 4 = 2$	True
5	$\frac{1}{2} \cdot 5 = 2$	False

$\therefore$ the solution set is {4}.

16.

x	$\frac{1}{3}x = 1$	
0	$\frac{1}{3} \cdot 0 = 1$	False
1	$\frac{1}{3} \cdot 1 = 1$	False
2	$\frac{1}{3} \cdot 2 = 1$	False
3	$\frac{1}{3} \cdot 3 = 1$	True
4	$\frac{1}{3} \cdot 4 = 1$	False
5	$\frac{1}{3} \cdot 5 = 1$	False

$\therefore$ the solution set is {3}.

17.

x	$x \cdot x = 1$	
0	$0 \cdot 0 = 1$	False
1	$1 \cdot 1 = 1$	True
2	$2 \cdot 2 = 1$	False
3	$3 \cdot 3 = 1$	False
4	$4 \cdot 4 = 1$	False
5	$5 \cdot 5 = 1$	False

$\therefore$ the solution set is {1}.

18.

x	$9x = 9$	
0	$9 \cdot 0 = 9$	False
1	$9 \cdot 1 = 9$	True
2	$9 \cdot 2 = 9$	False
3	$9 \cdot 3 = 9$	False
4	$9 \cdot 4 = 9$	False
5	$9 \cdot 5 = 9$	False

$\therefore$ the solution set is {1}.

19.

x	$x - x = 0$	
0	$0 - 0 = 0$	True
1	$1 - 1 = 0$	True
2	$2 - 2 = 0$	True
3	$3 - 3 = 0$	True
4	$4 - 4 = 0$	True
5	$5 - 5 = 0$	True

$\therefore$ the solution set is {0, 1, 2, 3, 4, 5}.

20.

x	$x \cdot x = 25$	
0	$0 \cdot 0 = 25$	False
1	$1 \cdot 1 = 25$	False
2	$2 \cdot 2 = 25$	False
3	$3 \cdot 3 = 25$	False
4	$4 \cdot 4 = 25$	False
5	$5 \cdot 5 = 25$	True

$\therefore$ the solution set is {5}.

21.

Number	Eight times a number is 32.	
0	Eight times 0 is 32.	False
2	Eight times 2 is 32.	False
4	Eight times 4 is 32.	True
6	Eight times 6 is 32.	False
8	Eight times 8 is 32.	False

$\therefore$ the number is 4.

22.

Number	Twelve more than a number is 20.	
0	Twelve more than 0 is 20.	False
2	Twelve more than 2 is 20.	False
4	Twelve more than 4 is 20.	False
6	Twelve more than 6 is 20.	False
8	Twelve more than 8 is 20.	True

$\therefore$ the number is 8.

23.

Number	A number divided by four is 2.	
0	0 divided by four is 2.	False
2	2 divided by four is 2.	False
4	4 divided by four is 2.	False
6	6 divided by four is 2.	False
8	8 divided by four is 2.	True

$\therefore$ the number is 8.

24.

Number	Three less than a number is 3.	
0	Three less than 0 is 3.	False
2	Three less than 2 is 3.	False
4	Three less than 4 is 3.	False
6	Three less than 6 is 3.	True
8	Three less than 8 is 3.	False

∴ the number is 6.

B **25.**

Number	Three more than twice a number is 7.	
0	Three more than twice 0 is 7.	False
2	Three more than twice 2 is 7.	True
4	Three more than twice 4 is 7.	False
6	Three more than twice 6 is 7.	False
8	Three more than twice 8 is 7.	False

∴ the number is 2.

26.

Number	Four times a number is 6 more than the number.	
0	Four times 0 is 6 more than 0.	False
2	Four times 2 is 6 more than 2.	True
4	Four times 4 is 6 more than 4.	False
6	Four times 6 is 6 more than 6.	False
8	Four times 8 is 6 more than 8.	False

∴ the number is 2.

27.

a	$2a + 9 = 17$	
0	$2 \cdot 0 + 9 = 17$	False
1	$2 \cdot 1 + 9 = 17$	False
2	$2 \cdot 2 + 9 = 17$	False
3	$2 \cdot 3 + 9 = 17$	False
4	$2 \cdot 4 + 9 = 17$	True
5	$2 \cdot 5 + 9 = 17$	False
6	$2 \cdot 6 + 9 = 17$	False
7	$2 \cdot 7 + 9 = 17$	False
8	$2 \cdot 8 + 9 = 17$	False
9	$2 \cdot 9 + 9 = 17$	False

∴ the solution set is {4}.

28.

b	$3b - 4 = 11$	
0	$3 \cdot 0 - 4 = 11$	False
1	$3 \cdot 1 - 4 = 11$	False
2	$3 \cdot 2 - 4 = 11$	False
3	$3 \cdot 3 - 4 = 11$	False
4	$3 \cdot 4 - 4 = 11$	False
5	$3 \cdot 5 - 4 = 11$	True
6	$3 \cdot 6 - 4 = 11$	False
7	$3 \cdot 7 - 4 = 11$	False
8	$3 \cdot 8 - 4 = 11$	False
9	$3 \cdot 9 - 4 = 11$	False

∴ the solution set is {5}.

29.

c	$10 = 8c - 6$	
0	$10 = 8 \cdot 0 - 6$	False
1	$10 = 8 \cdot 1 - 6$	False
2	$10 = 8 \cdot 2 - 6$	True
3	$10 = 8 \cdot 3 - 6$	False
4	$10 = 8 \cdot 4 - 6$	False
5	$10 = 8 \cdot 5 - 6$	False
6	$10 = 8 \cdot 6 - 6$	False
7	$10 = 8 \cdot 7 - 6$	False
8	$10 = 8 \cdot 8 - 6$	False
9	$10 = 8 \cdot 9 - 6$	False

$\therefore$ the solution set is {2}.

30.

d	$13 = 6d - 5$	
0	$13 = 6 \cdot 0 - 5$	False
1	$13 = 6 \cdot 1 - 5$	False
2	$13 = 6 \cdot 2 - 5$	False
3	$13 = 6 \cdot 3 - 5$	True
4	$13 = 6 \cdot 4 - 5$	False
5	$13 = 6 \cdot 5 - 5$	False
6	$13 = 6 \cdot 6 - 5$	False
7	$13 = 6 \cdot 7 - 5$	False
8	$13 = 6 \cdot 8 - 5$	False
9	$13 = 6 \cdot 9 - 5$	False

$\therefore$ the solution set is {3}.

31.

r	$9 + 9r = 81$	
0	$9 + 9 \cdot 0 = 81$	False
1	$9 + 9 \cdot 1 = 81$	False
2	$9 + 9 \cdot 2 = 81$	False
3	$9 + 9 \cdot 3 = 81$	False
4	$9 + 9 \cdot 4 = 81$	False
5	$9 + 9 \cdot 5 = 81$	False
6	$9 + 9 \cdot 6 = 81$	False
7	$9 + 9 \cdot 7 = 81$	False
8	$9 + 9 \cdot 8 = 81$	True
9	$9 + 9 \cdot 9 = 81$	False

$\therefore$ the solution set is {8}.

32.

n	$5 + 5n = 50$	
0	$5 + 5 \cdot 0 = 50$	False
1	$5 + 5 \cdot 1 = 50$	False
2	$5 + 5 \cdot 2 = 50$	False
3	$5 + 5 \cdot 3 = 50$	False
4	$5 + 5 \cdot 4 = 50$	False
5	$5 + 5 \cdot 5 = 50$	False
6	$5 + 5 \cdot 6 = 50$	False
7	$5 + 5 \cdot 7 = 50$	False
8	$5 + 5 \cdot 8 = 50$	False
9	$5 + 5 \cdot 9 = 50$	True

$\therefore$ the solution set is {9}.

33.

x	$2x = x + 7$	
0	$2 \cdot 0 = 0 + 7$	False
1	$2 \cdot 1 = 1 + 7$	False
2	$2 \cdot 2 = 2 + 7$	False
3	$2 \cdot 3 = 3 + 7$	False
4	$2 \cdot 4 = 4 + 7$	False
5	$2 \cdot 5 = 5 + 7$	False
6	$2 \cdot 6 = 6 + 7$	False
7	$2 \cdot 7 = 7 + 7$	True
8	$2 \cdot 8 = 8 + 7$	False
9	$2 \cdot 9 = 9 + 7$	False

$\therefore$ the solution set is {7}.

34.

w	$3w = w \cdot 3$	
0	$3 \cdot 0 = 0 \cdot 3$	True
1	$3 \cdot 1 = 1 \cdot 3$	True
2	$3 \cdot 2 = 2 \cdot 3$	True
3	$3 \cdot 3 = 3 \cdot 3$	True
4	$3 \cdot 4 = 4 \cdot 3$	True
5	$3 \cdot 5 = 5 \cdot 3$	True
6	$3 \cdot 6 = 6 \cdot 3$	True
7	$3 \cdot 7 = 7 \cdot 3$	True
8	$3 \cdot 8 = 8 \cdot 3$	True
9	$3 \cdot 9 = 9 \cdot 3$	True

$\therefore$ the solution set is {0, 1, 2, 3, 4, 5, 6, 7, 8, 9}.

35.

f	$2f = f \cdot f$	
0	$2 \cdot 0 = 0 \cdot 0$	True
1	$2 \cdot 1 = 1 \cdot 1$	False
2	$2 \cdot 2 = 2 \cdot 2$	True
3	$2 \cdot 3 = 3 \cdot 3$	False
4	$2 \cdot 4 = 4 \cdot 4$	False
5	$2 \cdot 5 = 5 \cdot 5$	False
6	$2 \cdot 6 = 6 \cdot 6$	False
7	$2 \cdot 7 = 7 \cdot 7$	False
8	$2 \cdot 8 = 8 \cdot 8$	False
9	$2 \cdot 9 = 9 \cdot 9$	False

$\therefore$ the solution set is {0, 2}.

36.

y	$y(8 - y) = 0$	
0	$0(8 - 0) = 0$	True
1	$1(8 - 1) = 0$	False
2	$2(8 - 2) = 0$	False
3	$3(8 - 3) = 0$	False
4	$4(8 - 4) = 0$	False
5	$5(8 - 5) = 0$	False
6	$6(8 - 6) = 0$	False
7	$7(8 - 7) = 0$	False
8	$8(8 - 8) = 0$	True
9	$9(8 - 9) = 0$	False

$\therefore$ the solution set is {0, 8}.

37.

v	$(6 - v)(1 + v)v = 0$	
0	$(6 - 0)(1 + 0)0 = 0$	True
1	$(6 - 1)(1 + 1)1 = 0$	False
2	$(6 - 2)(1 + 2)2 = 0$	False
3	$(6 - 3)(1 + 3)3 = 0$	False
4	$(6 - 4)(1 + 4)4 = 0$	False
5	$(6 - 5)(1 + 5)5 = 0$	False
6	$(6 - 6)(1 + 6)6 = 0$	True
7	$(6 - 7)(1 + 7)7 = 0$	False
8	$(6 - 8)(1 + 8)8 = 0$	False
9	$(6 - 9)(1 + 9)9 = 0$	False

$\therefore$ the solution set is {0, 6}.

38.

k	$27k = (3k)(3k)(3k)$	
0	$27 \cdot 0 = (3 \cdot 0)(3 \cdot 0)(3 \cdot 0)$	True
1	$27 \cdot 1 = (3 \cdot 1)(3 \cdot 1)(3 \cdot 1)$	True
2	$27 \cdot 2 = (3 \cdot 2)(3 \cdot 2)(3 \cdot 2)$	False
3	$27 \cdot 3 = (3 \cdot 3)(3 \cdot 3)(3 \cdot 3)$	False
4	$27 \cdot 4 = (3 \cdot 4)(3 \cdot 4)(3 \cdot 4)$	False
5	$27 \cdot 5 = (3 \cdot 5)(3 \cdot 5)(3 \cdot 5)$	False
6	$27 \cdot 6 = (3 \cdot 6)(3 \cdot 6)(3 \cdot 6)$	False
7	$27 \cdot 7 = (3 \cdot 7)(3 \cdot 7)(3 \cdot 7)$	False
8	$27 \cdot 8 = (3 \cdot 8)(3 \cdot 8)(3 \cdot 8)$	False
9	$27 \cdot 9 = (3 \cdot 9)(3 \cdot 9)(3 \cdot 9)$	False

$\therefore$ the solution set is {0, 1}.

C **39.** **a.** More than one **b.** None **c.** One; {4}

40. **a.** None **b.** More than one **c.** None

41. **a.** One; {2} **b.** More than one **c.** None

42. **a.** More than one **b.** None **c.** One; {3}

43–46. Answers will vary; examples are given. **43.** $1 \cdot x = 3; x + 1 = 4$

44. $x + x = x; x \cdot 1 = 0$ **45.** $x - x = 0; 4x \div 2 = x + x$

46. $x(3 - x) = 0; x \cdot x \cdot x = 9x$

Page 12 • MIXED REVIEW

1. $11 \cdot 9 + 3 \cdot 11 = 99 + 33 = 132$ **2.** $7 + (15 \div 5) = 7 + 3 = 10$

3. $(17 - 8) \div 3 = 9 \div 3 = 3$ **4.** $(3 + 2 \cdot 2) \div 8 = (3 + 4) \div 8 = 7 \div 8 = \frac{7}{8}$

5. $7 - 4 \div 2 \div 2 = 7 - 2 \div 2 = 7 - 1 = 6$ **6.** $42 \div 7 \div (2 + 1) = 6 \div 3 = 2$

7. $3x + 4z = 3 \cdot 2 + 4 \cdot 4 = 6 + 16 = 22$ **8.** $6 \cdot (z - x) = 6 \cdot (4 - 2) = 6 \cdot 2 = 12$

9. $4(yz + 2) = 4(5 \cdot 4 + 2) = 4(20 + 2) = 4 \cdot 22 = 88$

10. $5xz + 3y = 5 \cdot 2 \cdot 4 + 3 \cdot 5 = 40 + 15 = 55$

11. $axz \div (2y) = 3 \cdot 2 \cdot 4 \div (2 \cdot 5) = 24 \div 10 = 2.4$

12. $2y - [4a \div (x + 1)] = 2 \cdot 5 - [4 \cdot 3 \div (2 + 1)] = 10 - [12 \div 3] = 10 - 4 = 6$

Page 13 • CALCULATOR KEY-IN

1. 0 **2.** 3.8 **3.** 0.1 **4.** 369 **5.** 333,333 **6.** 12 **7. a.** 100 **b.** 0 **c.** 37

Pages 16–18 • WRITTEN EXERCISES

A **1.** $n + 8$ **2.** $4n$ **3.** $n - 11$ **4.** $3 + n$ **5.** $\frac{1}{2}n$ **6.** $n + 10$ **7.** $\frac{17}{d}$ **8.** $\frac{n}{3}$

9. $11x$ **10.** $n - 2$ **11.** $2y + 7$ **12.** $5n - 4$ **13.** $\frac{1}{2}n - 8$ **14.** $\frac{1}{3}n + 11$

15. $5 + \frac{n}{8}$ **16.** $10(n + 9)$ **17.** i **18.** a **19.** g **20.** d **21.** h **22.** j

23. b **24.** f **25.** c **26.** e **27.** $f + 3$ **28.** $j - 9$ **29.** $m - 10$

30. $\frac{1}{2}d$ **31.** $17 - x$ **32.** $\frac{18}{y}$

B **33. a.** $s - 3$ **b.** $d + 3$ **34. a.** $r + 5$ **b.** $e - 5$ **35. a.** $g - 12$ **b.** $b + 12$

36. a. $s + 12$ **b.** $l - 12$ **37.** $t = 3, d = rt = 3r$; $3r$ mi

38. $t = h, r = 85, d = rt = 85h$; $85h$ km **39.** $n = 5, C = np = 5p$; $5p$ cents

40. $n = x, p = 75, C = np = 75x$; $75x$ cents

41.

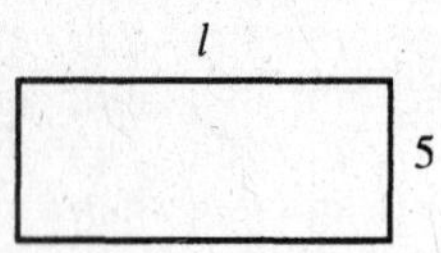

$A = l \cdot w = 5l$

42.

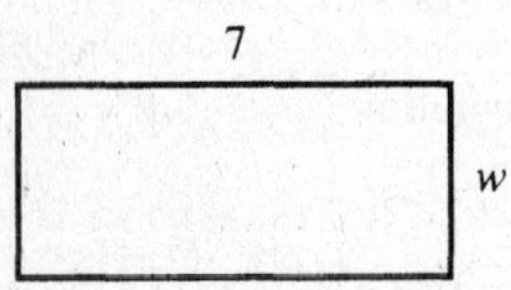

$P = 2l + 2w = 2 \cdot 7 + 2 \cdot w = 14 + 2w$

43. $C = n \cdot p$; pencil cost $= 3p$, book cost $= 2n$; Total cost $= 3p + 2n$, $(3p + 2n)$ cents

44. $C = np$; shirt cost $= 20s$, tie cost $= 12t$; Total cost $= 20s + 12t$, $(20s + 12t)$ dollars

45. Number of min $= 60 \cdot$ number of hours; $60 \cdot 3 = 180$, 180 min; $60 \cdot h = 60h$, $60h$ min

46. Number of days $= 7 \cdot$ (number of weeks); $7 \cdot 4 = 28$, 28 days; $7 \cdot w = 7w$, $7w$ days

47. Number of years = (number of months) $\div$ 12; $36 \div 12 = 3$, 3 years; $\frac{m}{12}$, $\frac{m}{12}$ years

48. Number of min = (number of seconds) $\div$ 60; $330 \div 60 = \frac{330}{60} = \frac{11}{2}$; $5\frac{1}{2}$, or 5.5 min; $\frac{s}{60}$, $\frac{s}{60}$ min

49. $10c$, $10(c - 5)$ **50.** $2r$, $2(r + 3)$ **51.** $n - 5$ **52.** $m + 8$

C **53.** Let p = calories in peach; then calories in apple $= p + 29$, calories in banana $= (p + 29) + 13 = p + 42$; total calories $= (p + 29) + 2p + (p + 42)$

54. Let w = grams of carbohydrates in a cup of walnuts; then carbohydrates in peanuts $= w + 8$, carbohydrates in almonds $= (w + 8) + 1 = w + 9$; total $= (w + 8) + w + (w + 9)$

Page 18 • MIXED REVIEW EXERCISES

1. $7x - 2y = 7 \cdot 3 - 2 \cdot 4 = 21 - 8 = 13$ **2.** $4 + 3yz = 4 + 3 \cdot 4 \cdot 5 = 4 + 60 = 64$

3. $(2x - t) \cdot 3 = (2 \cdot 3 - 2) \cdot 3 = (6 - 2) \cdot 3 = 4 \cdot 3 = 12$

4. $4z + 3x - t = 4 \cdot 5 + 3 \cdot 3 - 2 = 20 + 9 - 2 = 29 - 2 = 27$

5. $txy + 2 = 2 \cdot 3 \cdot 4 + 2 = 24 + 2 = 26$

6. $ty + z + x = 2 \cdot 4 + 5 + 3 = 8 + 5 + 3 = 13 + 3 = 16$

7.

x	$x + 4 = 8$	
0	$0 + 4 = 8$	False
1	$1 + 4 = 8$	False
2	$2 + 4 = 8$	False
3	$3 + 4 = 8$	False
4	$4 + 4 = 8$	True

∴ the solution set is {4}.

8.

x	$3x = 12$	
0	$3 \cdot 0 = 12$	False
1	$3 \cdot 1 = 12$	False
2	$3 \cdot 2 = 12$	False
3	$3 \cdot 3 = 12$	False
4	$3 \cdot 4 = 12$	True

∴ the solution set is {4}.

9.

x	$4x = 4$	
0	$4 \cdot 0 = 4$	False
1	$4 \cdot 1 = 4$	True
2	$4 \cdot 2 = 4$	False
3	$4 \cdot 3 = 4$	False
4	$4 \cdot 4 = 4$	False

∴ the solution set is {1}.

10.

x	$5x = 0$	
0	$5 \cdot 0 = 0$	True
1	$5 \cdot 1 = 0$	False
2	$5 \cdot 2 = 0$	False
3	$5 \cdot 3 = 0$	False
4	$5 \cdot 4 = 0$	False

∴ the solution set is {0}.

11.

x	$2x + 1 = 7$	
0	$2 \cdot 0 + 1 = 7$	False
1	$2 \cdot 1 + 1 = 7$	False
2	$2 \cdot 2 + 1 = 7$	False
3	$2 \cdot 3 + 1 = 7$	True
4	$2 \cdot 4 + 1 = 7$	False

∴ the solution set is {3}.

12.

x	$5 = 3x - 1$	
0	$5 = 3 \cdot 0 - 1$	False
1	$5 = 3 \cdot 1 - 1$	False
2	$5 = 3 \cdot 2 - 1$	True
3	$5 = 3 \cdot 3 - 1$	False
4	$5 = 3 \cdot 4 - 1$	False

∴ the solution set is {2}.

13.

x	$x \div 2 = 2$	
0	$0 \div 2 = 2$	False
1	$1 \div 2 = 2$	False
2	$2 \div 2 = 2$	False
3	$3 \div 2 = 2$	False
4	$4 \div 2 = 2$	True

∴ the solution set is {4}.

14.

x	$2x = x + 3$	
0	$2 \cdot 0 = 0 + 3$	False
1	$2 \cdot 1 = 1 + 3$	False
2	$2 \cdot 2 = 2 + 3$	False
3	$2 \cdot 3 = 3 + 3$	True
4	$2 \cdot 4 = 4 + 3$	False

∴ the solution set is {3}.

Pages 20–22 • WRITTEN EXERCISES

A **1.** a **2.** b **3.** a **4.** a **5.** b **6.** a **7.** b **8.** b **9.** e **10.** a **11.** b **12.** f **13.** d **14.** g **15.** h **16.** c **17.** $24 = x + 9 + 8, 24 = x + 17$ **18.** $29 = x + x + 7$ **19.** $12 = 3 + 4 + x, 12 = 7 + x$

20. $P = 2l + 2w$, $26 = 2 \cdot x + 2 \cdot 6$, $26 = 2x + 12$ **21.** $A = lw$, $40 = 8x$

22. $A = lw$, $42 = x(x + 1)$

B **23–36.** Answers may vary; examples are given.

23. a. Let s = length of a side **b.** $4s = 116$

24. a. Let e = cost of one egg **b.** $12e = 1.19$

25. a. Let c = coin's age now **b.** $c + 7 = 50$

26. a. Let r = radio station's age now **b.** $r - 9 = 13$

27. a. Let d = number of dictionaries sold **b.** $d + 12 = 120$

28. a. Let x = number of exercises solved **b.** $x + 4 = 30$

29. a. Let p = cost of whole pizza **b.** $\frac{1}{8}p = 0.95$

30. a. Let b = bridge's age now **b.** $16 = \frac{1}{4}b$

31. a. Let t = number of hours traveled **b.** $d = rt$, $r = 132$, $d = 462$, $462 = 132t$

32. a. Let w = number of tiles in width **b.** $A = l \cdot w$, $A = 928$, $l = 32$, $928 = 32w$

33. a. Let l = actual length of room **b.** $\frac{1}{100}l = 8.5$

34. a. Let c = cost of one admission **b.** $39c = 1092$

C **35. a.** Let r = number of red cars **b.** Then number of blue cars = $24 - r$; $24 - r = 2r + 3$

36. a. Let n = first number **b.** Then second number = $n - 8$; third number = $n + 4$; $n + (n - 8) + (n + 4) = 120$

Page 22 • MIXED REVIEW EXERCISES

1.

x	$3 + x = 7$	
0	$3 + 0 = 7$	False
1	$3 + 1 = 7$	False
2	$3 + 2 = 7$	False
3	$3 + 3 = 7$	False
4	$3 + 4 = 7$	True
5	$3 + 5 = 7$	False
6	$3 + 6 = 7$	False

∴ the solution set is {4}.

2.

x	$2 = x - 4$	
0	$2 = 0 - 4$	False
1	$2 = 1 - 4$	False
2	$2 = 2 - 4$	False
3	$2 = 3 - 4$	False
4	$2 = 4 - 4$	False
5	$2 = 5 - 4$	False
6	$2 = 6 - 4$	True

∴ the solution set is {6}.

3.

x	$5x = 25$	
0	$5 \cdot 0 = 25$	False
1	$5 \cdot 1 = 25$	False
2	$5 \cdot 2 = 25$	False
3	$5 \cdot 3 = 25$	False
4	$5 \cdot 4 = 25$	False
5	$5 \cdot 5 = 25$	True
6	$5 \cdot 6 = 25$	False

∴ the solution set is {5}.

4.

x	$3 = x \div 2$	
0	$3 = 0 \div 2$	False
1	$3 = 1 \div 2$	False
2	$3 = 2 \div 2$	False
3	$3 = 3 \div 2$	False
4	$3 = 4 \div 2$	False
5	$3 = 5 \div 2$	False
6	$3 = 6 \div 2$	True

∴ the solution set is {6}.

5.

x	$3x + 1 = 10$	
0	$3 \cdot 0 + 1 = 10$	False
1	$3 \cdot 1 + 1 = 10$	False
2	$3 \cdot 2 + 1 = 10$	False
3	$3 \cdot 3 + 1 = 10$	True
4	$3 \cdot 4 + 1 = 10$	False
5	$3 \cdot 5 + 1 = 10$	False
6	$3 \cdot 6 + 1 = 10$	False

∴ the solution set is {3}.

6.

x	$4x = x + 6$	
0	$4 \cdot 0 = 0 + 6$	False
1	$4 \cdot 1 = 1 + 6$	False
2	$4 \cdot 2 = 2 + 6$	True
3	$4 \cdot 3 = 3 + 6$	False
4	$4 \cdot 4 = 4 + 6$	False
5	$4 \cdot 5 = 5 + 6$	False
6	$4 \cdot 6 = 6 + 6$	False

∴ the solution set is {2}.

7.

x	$x + 5 = 2x$	
0	$0 + 5 = 2 \cdot 0$	False
1	$1 + 5 = 2 \cdot 1$	False
2	$2 + 5 = 2 \cdot 2$	False
3	$3 + 5 = 2 \cdot 3$	False
4	$4 + 5 = 2 \cdot 4$	False
5	$5 + 5 = 2 \cdot 5$	True
6	$6 + 5 = 2 \cdot 6$	False

∴ the solution set is {5}.

8.

x	$4x = x \cdot 4$	
0	$4 \cdot 0 = 0 \cdot 4$	True
1	$4 \cdot 1 = 1 \cdot 4$	True
2	$4 \cdot 2 = 2 \cdot 4$	True
3	$4 \cdot 3 = 3 \cdot 4$	True
4	$4 \cdot 4 = 4 \cdot 4$	True
5	$4 \cdot 5 = 5 \cdot 4$	True
6	$4 \cdot 6 = 6 \cdot 4$	True

∴ the solution set is {0, 1, 2, 3, 4, 5, 6}.

9. $n - 5$ **10.** $n + 5$ **11.** $\frac{5}{n}$ **12.** $2(5 + n)$

Pages 24–26 • PROBLEMS

A **1–22.** Answers may vary; examples are given.

1. Let s = Sam's amount, then $s + 5$ = Luke's amount; $s + (s + 5) = 73$

2. Let j = Jo's amount, then $2j$ = Lyn's amount; $j + 2j = 63$

3. Let w = number of women, then $w + 4$ = number of men; $w + (w + 4) = 12$

4. Let m = number of men, then $m + 20$ = number of women; $m + (m + 20) = 620$

5. Let w = Wing's distance, then $w + 3$ = Lee's distance; $w + (w + 3) = 25$

6. Let j = Jan's distance, then $3j$ = Brenda's distance; $3j = j + 24$

7. Let t = Tim's distance, then $t + 8$ = Lisa's distance; $t + 8 = 2t$

8. Let l = number of games lost, then $2l$ = number of games won; $l + 2l = 96$

9. Let c = calls made by Clark, then $c + 5$ = calls made by Shelley; $c + (c + 5) = 33$

10. Let w = number by Woody, then $w - 8$ = number by Skip; $w + (w - 8) = 20$

11. Let b = dollars spent by Barry, then $b + 2$ = dollars spent by Amanda; $b + (b + 2) = 34$

12. Let n = larger number, then $n - 7$ = smaller number; $n + (n - 7) = 33$

B **13.** Let x = length of shorter piece, then $x + 1$ = length of longer piece; $x + (x + 1) = 9$

14. Let b = building height, then $3b$ = tower height; $3b = b + 50$

15. Let w = width, then $2w$ = length; $P = 2l + 2w$, so $2(2w) + 2w = 60$

16. Let w = width, then $w + 1$ = length; $A = l \cdot w$, so $30 = (w + 1)w$

17. Let P = sum of three sides; $30 = 7 + x + (x + 1)$

18. Let x = length of each equal side; P = sum of 3 sides, $50 = x + x + 15$

C 19. Let x = number tied, then $x + 1$ = number lost and $(x + 1) + 2$ = number won; $x + (x + 1) + [(x + 1) + 2] = 12$.

20. Let h = dollars Harry has, then $2h$ = dollars Dawn has and $3(2h)$ = dollars Tina has, $h + 2h + 3(2h) = 175$

21. Let x = shortest length, then $x + 1$ = middle length and $(x + 1) + 2$ = longest length; $x + (x + 1) + [(x + 1) + 2] = 10$

22. Let x = Terri's stamps, then $4x$ = father's stamps and $7x$ = grandfather's stamps; $x + 4x + 7x = 120$

Page 26 • MIXED REVIEW EXERCISES

1.

x	$2x + 3 = 5$	
0	$2 \cdot 0 + 3 = 5$	False
1	$2 \cdot 1 + 3 = 5$	True
2	$2 \cdot 2 + 3 = 5$	False
3	$2 \cdot 3 + 3 = 5$	False
4	$2 \cdot 4 + 3 = 5$	False
5	$2 \cdot 5 + 3 = 5$	False
6	$2 \cdot 6 + 3 = 5$	False

$\therefore$ the solution set is {1}.

2.

x	$x \div 2 = 1$	
0	$0 \div 2 = 1$	False
1	$1 \div 2 = 1$	False
2	$2 \div 2 = 1$	True
3	$3 \div 2 = 1$	False
4	$4 \div 2 = 1$	False
5	$5 \div 2 = 1$	False
6	$6 \div 2 = 1$	False

$\therefore$ the solution set is {2}.

3.

x	$16 = 4x$	
0	$16 = 4 \cdot 0$	False
1	$16 = 4 \cdot 1$	False
2	$16 = 4 \cdot 2$	False
3	$16 = 4 \cdot 3$	False
4	$16 = 4 \cdot 4$	True
5	$16 = 4 \cdot 5$	False
6	$16 = 4 \cdot 6$	False

$\therefore$ the solution set is {4}.

4.

x	$3 = 2x - 5$	
0	$3 = 2 \cdot 0 - 5$	False
1	$3 = 2 \cdot 1 - 5$	False
2	$3 = 2 \cdot 2 - 5$	False
3	$3 = 2 \cdot 3 - 5$	False
4	$3 = 2 \cdot 4 - 5$	True
5	$3 = 2 \cdot 5 - 5$	False
6	$3 = 2 \cdot 6 - 5$	False

$\therefore$ the solution set is {4}.

5.

x	$6 + 6x = 36$	
0	$6 + 6 \cdot 0 = 36$	False
1	$6 + 6 \cdot 1 = 36$	False
2	$6 + 6 \cdot 2 = 36$	False
3	$6 + 6 \cdot 3 = 36$	False
4	$6 + 6 \cdot 4 = 36$	False
5	$6 + 6 \cdot 5 = 36$	True
6	$6 + 6 \cdot 6 = 36$	False

$\therefore$ the solution set is {5}.

6.

x	$7 = 2x - 3$	
0	$7 = 2 \cdot 0 - 3$	False
1	$7 = 2 \cdot 1 - 3$	False
2	$7 = 2 \cdot 2 - 3$	False
3	$7 = 2 \cdot 3 - 3$	False
4	$7 = 2 \cdot 4 - 3$	False
5	$7 = 2 \cdot 5 - 3$	True
6	$7 = 2 \cdot 6 - 3$	False

$\therefore$ the solution set is {5}.

7.

x	$3x = x + 4$	
0	$3 \cdot 0 = 0 + 4$	False
1	$3 \cdot 1 = 1 + 4$	False
2	$3 \cdot 2 = 2 + 4$	True
3	$3 \cdot 3 = 3 + 4$	False
4	$3 \cdot 4 = 4 + 4$	False
5	$3 \cdot 5 = 5 + 4$	False
6	$3 \cdot 6 = 6 + 4$	False

$\therefore$ the solution set is {2}.

8.

x	$x \cdot x = 16$	
0	$0 \cdot 0 = 16$	False
1	$1 \cdot 1 = 16$	False
2	$2 \cdot 2 = 16$	False
3	$3 \cdot 3 = 16$	False
4	$4 \cdot 4 = 16$	True
5	$5 \cdot 5 = 16$	False
6	$6 \cdot 6 = 16$	False

$\therefore$ the solution set is {4}.

9. $\frac{1}{3}x$ **10.** $4x - 2$ **11.** $2x + 5$ **12.** $\frac{1}{4}x - 4$

Page 26 • READING ALGEBRA

1. Twice as much milk as apple juice and three times as much milk as fruit punch was served; a total of 660 cartons was served.
2. The number of cartons of apple juice.
3. No; the amount of apple juice served was $\frac{1}{2}x$, not $2x$, and the amount of punch served was $\frac{1}{3}x$, not $3x$.

Pages 28–29 • PROBLEMS

A **1.** *Step 1* The unknowns are the age of oil painting and the age of watercolor painting.

Step 2 Let x = age of watercolor, then $x + 16$ = age of oil.

Step 3 $x + 16 = 3x$

Step 4 Replace x in turn by 4, 8, and 12.

x	$x + 16 = 3x$	
4	$4 + 16 = 3 \cdot 4$	False
8	$8 + 16 = 3 \cdot 8$	True
12	$12 + 16 = 3 \cdot 12$	False

Watercolor's age: $x = 8$ Oil painting's age: $x + 16 = 8 + 16 = 24$

Step 5 An oil painting is 16 years older than a watercolor. $24 \stackrel{?}{=} 8 + 16$ ✓
Oil is three times older than watercolor. $24 \stackrel{?}{=} 3 \cdot 8$ ✓

$\therefore$ Watercolor is 8 years old, and oil is 24 years old.

2. *Step 1* The unknowns are the ages of the auditorium and the gym.

Step 2 Let x = age of auditorium, then $x - 21$ = age of gym.

Step 3 $x - 21 = \frac{1}{4} \cdot x$

(*Continued*)

Step 4 Replace x in turn by 26, 27, and 28.

x	$x - 21 = \frac{1}{4} \cdot x$	
26	$26 - 21 = \frac{1}{4} \cdot 26$	False
27	$27 - 21 = \frac{1}{4} \cdot 27$	False
28	$28 - 21 = \frac{1}{4} \cdot 28$	True

Age of auditorium: $x = 28$ Age of gym: $x - 21 = 28 - 21 = 7$

Step 5 The gym is 21 years newer than the auditorium. $7 \stackrel{?}{=} 28 - 21$ ✓

The gym is one fourth as old as the auditorium. $7 \stackrel{?}{=} \frac{1}{4} \cdot 28$ ✓

∴ Auditorium is 28 years old, and gym is 7 years old.

3. *Step 1* The unknowns are two numbers.

Step 2 Let x = smaller number, then $x + 57$ = larger number.

Step 3 $x + (x + 57) = 185$

Step 4 Replace x in turn by 54, 64, and 74.

x	$x + (x + 57) = 185$	
54	$54 + (54 + 57) = 185$	False
64	$64 + (64 + 57) = 185$	True
74	$74 + (74 + 57) = 185$	False

Smaller number: $x = 64$ Larger number: $x + 57 = 64 + 57 = 121$

Step 5 Two numbers differ by 57. $121 - 64 \stackrel{?}{=} 57$ ✓
The sum of two numbers is 185. $121 + 64 \stackrel{?}{=} 185$ ✓

∴ The numbers are 64 and 121.

4. *Step 1* The unknowns are two numbers.

Step 2 Let x = smaller number, then $4x$ = larger number.

Step 3 $4x = x + 87$

Step 4 Replace x in turn by 29, 30, and 33.

x	$4x = x + 87$	
29	$4 \cdot 29 = 29 + 87$	True
30	$4 \cdot 30 = 30 + 87$	False
33	$4 \cdot 33 = 33 + 87$	False

Smaller number: $x = 29$ Larger number: $4x = 4 \cdot 29 = 116$

Step 5 One number is four times another number. $4 \cdot 29 \stackrel{?}{=} 116$ ✓
The larger number is 87 more than the smaller number. $116 \stackrel{?}{=} 29 + 87$ ✓

∴ The numbers are 29 and 116.

5. *Step 1* The unknowns are distances traveled by a car and a bus.

Step 2 Let x = car's distance, then $x + 318$ = bus' distance.

(*Continued*)

Step 3 $x = \frac{1}{3}(x + 318)$

Step 4 Replace x in turn by 147, 151, and 159.

x	$x = \frac{1}{3}(x + 318)$	
147	$147 = \frac{1}{3}(147 + 318)$	False
151	$151 = \frac{1}{3}(151 + 318)$	False
159	$159 = \frac{1}{3}(159 + 318)$	True

Car's distance: $x = 159$ Bus' distance: $x + 318 = 159 + 318 = 477$

Step 5 A bus went 318 km farther than a car. $159 + 318 \stackrel{?}{=} 477$ ✓

The car went one third as far as the bus. $159 \stackrel{?}{=} \frac{1}{3}(477)$ ✓

∴ The car traveled 159 km, and the bus traveled 477 km.

6. *Step 1* The unknowns are the weights of a clown and a trapeze artist.

Step 2 Let x = weight of trapeze artist, then $x + 60$ = weight of clown.

Step 3 $x = \frac{2}{3}(x + 60)$

Step 4 Replace x in turn by 110, 120, and 125.

x	$x = \frac{2}{3}(x + 60)$	
110	$110 = \frac{2}{3}(110 + 60)$	False
120	$120 = \frac{2}{3}(120 + 60)$	True
125	$125 = \frac{2}{3}(125 + 60)$	False

Weight of trapeze artist: $x = 120$
Weight of clown: $x + 60 = 120 + 60 = 180$

Step 5 A clown weighs 60 lb more than a trapeze artist. $180 \stackrel{?}{=} 120 + 60$ ✓
The trapeze artist weighs two thirds as much as the clown. $120 \stackrel{?}{=} \frac{2}{3} \cdot 180$ ✓

∴ The clown weighs 180 lb, and the trapeze artist weighs 120 lb.

7. *Step 1* The unknowns are the number of Democrats and Republicans in the Senate.

Step 2 Let x = number of Republicans, then $x + 12$ = number of Democrats.

Step 3 $x + (x + 12) = 100$

(*Continued*)

Step 4 Replace x in turn by 38, 42, and 44.

x	$x + (x + 12) = 100$	
38	$38 + (38 + 12) = 100$	False
42	$42 + (42 + 12) = 100$	False
44	$44 + (44 + 12) = 100$	True

Number of Republicans: $x = 44$
Number of Democrats: $x + 12 = 44 + 12 = 56$

Step 5 The U.S. Senate has 100 members, all Democrats or Republicans. $100 \stackrel{?}{=} 44 + 56$ ✓
There were 12 more Democrats than Republicans. $56 \stackrel{?}{=} 44 + 12$ ✓

∴ There were 44 Republicans and 56 Democrats.

8. *Step 1* The unknowns are the number of boys and girls in the 9th grade class.

Step 2 Let x = number of boys, then $x + 17$ = number of girls.

Step 3 $x + (x + 17) = 431$

Step 4 Replace x in turn by 191, 202, and 207.

x	$x + (x + 17) = 431$	
191	$191 + (191 + 17) = 431$	False
202	$202 + (202 + 17) = 431$	False
207	$207 + (207 + 17) = 431$	True

Number of boys: $x = 207$ Number of girls: $x + 17 = 207 + 17 = 224$

Step 5 The ninth grade class has 17 more girls than boys. $224 \stackrel{?}{=} 207 + 17$ ✓
There are 431 students in all. $224 + 207 \stackrel{?}{=} 431$ ✓

∴ There are 207 boys and 224 girls in the 9th grade.

9. *Step 1* The unknowns are the amounts of money held by Ramon and Elena.

Step 2 Let x = Ramon's amount, then $1\frac{1}{2} \cdot x$ = Elena's amount.

Step 3 $x + 1\frac{1}{2} \cdot x = 225$; $x + \frac{3}{2}x = 225$

Step 4 Replace x in turn by 80, 90, and 95.

x	$x + \frac{3}{2}x = 225$	
80	$80 + \frac{3}{2} \cdot 80 = 225$	False
90	$90 + \frac{3}{2} \cdot 90 = 225$	True
95	$95 + \frac{3}{2} \cdot 95 = 225$	False

Ramon's amount: $x = 90$ Elena's amount: $\frac{3}{2}x = \frac{3}{2} \cdot 90 = 135$

(Continued)

Step 5 Elena has one and a half times as much money as Ramon. $135 = \frac{3}{2} \cdot 90$ ✓

Together they have \$225. $90 + 135 \stackrel{?}{=} 225$ ✓

∴ Ramon has \$90, and Elena has \$135.

10. *Step 1* The unknowns are the ages of Southern State and Western State.

Step 2 Let x = Southern's age, then $x + 18$ = Western's age.

Step 3 $x + 18 = 2\frac{1}{2} \cdot x$; $x + 18 = \frac{5}{2}x$

Step 4 Replace x in turn by 12, 14, and 18.

x	$x + 18 = \frac{5}{2}x$	
12	$12 + 18 = \frac{5}{2} \cdot 12$	True
14	$14 + 18 = \frac{5}{2} \cdot 14$	False
18	$18 + 18 = \frac{5}{2} \cdot 18$	False

Southern's age: $x = 12$ Western's age: $x + 18 = 12 + 18 = 30$

Step 5 Western State College is 18 years older than Southern State. $30 \stackrel{?}{=} 12 + 18$ ✓

Western is $2\frac{1}{2}$ times as old as Southern. $30 = \frac{5}{2} \cdot 12$ ✓

∴ Southern State is 12 years old, and Western State is 30 years old.

B

11. *Step 1* The unknown is the height of the school.

Step 2 Let x = school's height, then $\frac{3}{4}x$ = height of flagpole.

Step 3 $x - \frac{3}{4}x = 4.5$

Step 4 Replace x in turn by 18, 20, and 24.

x	$x - \frac{3}{4}x = 4.5$	
18	$18 - \frac{3}{4} \cdot 18 = 4.5$	True
20	$20 - \frac{3}{4} \cdot 20 = 4.5$	False
24	$24 - \frac{3}{4} \cdot 24 = 4.5$	False

School's height: $x = 18$

Step 5 The height of the flagpole is three fourths the height of the school. $13.5 \stackrel{?}{=} \frac{3}{4} \cdot 18$ ✓

The difference in their heights is 4.5 m. $18 - 13.5 \stackrel{?}{=} 4.5$ ✓

∴ The height of the school is 18 m.

12. *Step 1* The unknowns are the distances jogged by Leah and Barb.

Step 2 Let x = Leah's distance, then $x + 1.5$ = Barb's distance.

Step 3 $x + (x + 1.5) = 10.5$

Step 4 Replace x in turn by 4.5, 4.6, and 4.7.

x	$x + (x + 1.5) = 10.5$	
4.5	$4.5 + (4.5 + 1.5) = 10.5$	True
4.6	$4.6 + (4.6 + 1.5) = 10.5$	False
4.7	$4.7 + (4.7 + 1.5) = 10.5$	False

Leah's distance: $x = 4.5$ Barb's distance: $x + 1.5 = 4.5 + 1.5 = 6$

Step 5 They were 10.5 km apart. $4.5 + 6 \stackrel{?}{=} 10.5$ ✓
Barb had traveled 1.5 km farther than Leah. $6 \stackrel{?}{=} 4.5 + 1.5$ ✓

∴ Leah jogged 4.5 km, and Barb jogged 6 km.

13. *Step 1* The unknowns are the length and width of the rectangle.

Step 2 Let x = width, then $x + 12$ = length.

Step 3 $P = 2l + 2w$; $68 = 2(x + 12) + 2x$

Step 4 Replace x in turn by 10, 11, and 12.

x	$68 = 2(x + 12) + 2x$	
10	$68 = 2(10 + 12) + 2 \cdot 10$	False
11	$68 = 2(11 + 12) + 2 \cdot 11$	True
12	$68 = 2(12 + 12) + 2 \cdot 12$	False

Width of rectangle: $x = 11$ Length of rectangle: $x + 12 = 11 + 12 = 23$

Step 5 A rectangle is 12 m longer than it is wide. $11 + 12 \stackrel{?}{=} 23$ ✓
Its perimeter is 68 m. $2(23) + 2(11) \stackrel{?}{=} 68$ ✓

∴ The width is 11 m, and the length is 23 m.

14. *Step 1* The unknowns are the length and width of the rectangle.

Step 2 Let x = width, then $3\frac{1}{2} \cdot x$ = length.

Step 3 $P = 2l + 2w$; $108 = 2\left(3\frac{1}{2} \cdot x\right) + 2x$; $108 = 2\left(\frac{7}{2}x\right) + 2x$

Step 4 Replace x in turn by 12, 18, and 24.

x	$2\left(\frac{7}{2} \cdot x\right) + 2x = 108$	
12	$2\left(\frac{7}{2} \cdot 12\right) + 2 \cdot 12 = 108$	True
18	$2\left(\frac{7}{2} \cdot 18\right) + 2 \cdot 18 = 108$	False
24	$2\left(\frac{7}{2} \cdot 24\right) + 2 \cdot 24 = 108$	False

Width of rectangle: $x = 12$ Length of rectangle: $\frac{7}{2} \cdot x = \frac{7}{2} \cdot 12 = 42$

(Continued)

Step 5 The length of a rectangle is $3\frac{1}{2}$ times the width. $42 \stackrel{?}{=} \frac{7}{2} \cdot 12$ ✓

Its perimeter is 108 cm. $2(12) + 2(42) \stackrel{?}{=} 108$ ✓

∴ The width is 12 cm, and the length is 42 cm.

15. *Step 1* The unknowns are the width and length of the rectangle

Step 2 Let x = width, then length = $x + 15$.

Step 3 $A = l \cdot w$; $324 = (x + 15)x$

Step 4 Replace x in turn by 10, 12, and 16.

x	$324 = (x + 15)x$	
10	$324 = (10 + 15)10$	False
12	$324 = (12 + 15)12$	True
16	$324 = (16 + 15)16$	False

Width of rectangle: $x = 12$ Length of rectangle: $x + 15 = 12 + 15 = 27$

Step 5 A rectangle is 15 ft longer than it is wide. $12 + 15 \stackrel{?}{=} 27$ ✓
Its area is 324 ft^2. $27 \cdot 12 \stackrel{?}{=} 324$ ✓

∴ The width is 12 ft, and the length is 27 ft.

C **16.** *Step 1* The unknowns are the dimensions of a square and a rectangle.

Step 2 Let x = width of square and rectangle, then the length of the rectangle = $x + 5$.

Step 3 Rectangle's area = $l \cdot w = (x + 5)x$; square's area = $x \cdot x$
Total area = rectangle's area + square's area; $133 = (x + 5)x + x \cdot x$

Step 4 Replace x in turn by 6, 7, and 8.

x	$133 = (x + 5)x + x \cdot x$	
6	$133 = (6 + 5)6 + 6 \cdot 6$	False
7	$133 = (7 + 5)7 + 7 \cdot 7$	True
8	$133 = (8 + 5)8 + 8 \cdot 8$	False

Width of square and rectangle: $x = 7$
Length of rectangle: $x + 5 = 7 + 5 = 12$

Step 5 The rectangle is 5 m longer than the square. $7 + 5 \stackrel{?}{=} 12$ ✓
The total area of the square and the rectangle is 133.
$133 \stackrel{?}{=} (7 + 5)7 + 7 \cdot 7$ ✓

∴ The rectangle's width is 7 m, and its length is 12 m. The square's width is 7 m, and its length is 7 m.

17. *Step 1* The unknowns are the weights of Luis, Carla, and Rita.

Step 2 Let x = Luis's weight, then $x - 5$ = Carla's and $x - 5 - 2 = x - 7$ = Rita's.

Step 3 $333 = x + (x - 5) + (x - 7)$

Step 4 Replace x in turn by 110, 115, and 120.

x	$333 = x + (x - 5) + (x - 7)$	
110	$333 = 110 + (110 - 5) + (110 - 7)$	False
115	$333 = 115 + (115 - 5) + (115 - 7)$	True
120	$333 = 120 + (120 - 5) + (120 - 7)$	False

(Continued)

Luis's weight: $x = 115$ Carla's weight: $x - 5 = 115 - 5 = 110$
Rita's weight: $x - 7 = 115 - 7 = 108$

Step 5 Luis weighs 5 lb more than Carla. $115 \stackrel{?}{=} 110 + 5$ ✓
Carla weighs 2 lb more than Rita. $110 \stackrel{?}{=} 108 + 2$ ✓
Their weights total 333 lb. $115 + 110 + 108 \stackrel{?}{=} 333$ ✓

∴ Luis weighs 115 lb, Carla weighs 110 lb, and Rita weighs 108 lb.

18. *Step 1* The unknowns are the amounts of money held by Tony, Alicia, and Ralph.

Step 2 Let x = Ralph's amount, then $x - 16$ = Alicia's amount and $2(x - 16)$ = Tony's amount.

Step 3 $x + (x - 16) + 2(x - 16) = 200$

Step 4 Replace x in turn by 59, 62, and 63.

x	$x + (x - 16) + 2(x - 16) = 200$	
59	$59 + (59 - 16) + 2(59 - 16) = 200$	False
62	$62 + (62 - 16) + 2(62 - 16) = 200$	True
63	$63 + (63 - 16) + 2(63 - 16) = 200$	False

Ralph's amount: $x = 62$ Alicia's amount: $x - 16 = 62 - 16 = 46$
Tony's amount: $2(x - 16) = 2(62 - 16) = 2(46) = 92$

Step 5 Tony has twice as much money as Alicia. $92 \stackrel{?}{=} 2 \cdot 46$ ✓
Alicia has \$16 less than Ralph. $46 \stackrel{?}{=} 62 - 16$ ✓
Together they have \$200. $62 + 46 + 92 \stackrel{?}{=} 200$ ✓

∴ Ralph has \$62, Alicia has \$46, and Tony has \$92.

Page 29 • MIXED REVIEW EXERCISES

1. $\frac{7 \cdot 2 + 6 \cdot 6}{10 - 6 + 1} = \frac{14 + 36}{4 + 1} = \frac{50}{5} = 10$

2. $(30 - 4 + 4 \div 2) \div (21 \div 3) = (30 - 4 + 2) \div 7 = 28 \div 7 = 4$

3. $30 - (12 \div 3 + 2) = 30 - (4 + 2) = 30 - 6 = 24$ **4.** $6 \cdot 17 + 17 \cdot 3 = 102 + 51 = 153$

5. $6x = 36$ **6.** $6 = x - 2$ **7.** $2x + 2 = 10$ **8.** $\frac{1}{3}x = 6$

Pages 33–35 • WRITTEN EXERCISES

A **1.** $^{-}5$; five steps up; 5 **2.** 2; two rooms to the left; $^{-}2$

3. 190; 190 m below sea level; $^{-}190$ **4.** $^{-}9$; nine degrees above freezing (0°C); 9

5. 18; a loss of \$18; $^{-}18$ **6.** $^{-}4$; four gains; 4 **7.** $^{-}15$; 15 km east; 15

8. 41; latitude of 41° south; $^{-}41$ **9.** 85; payments of \$85; $^{-}85$

10. $^{-}25$; a bank deposit of \$25; 25 **11.** $^{-}1$; one foot above ground; 1

12. 3; 3 penalty points; $^{-}3$ **13.** $^{-}50$; a gain of \$50; 50 **14.** $^{-}10$; 10 lb gained; 10

15. $6 > {}^{-}9$ **16.** ${}^{-}11 < {}^{-}1$ **17.** ${}^{-}8 > {}^{-}10$ **18.** $10 > 7$ **19.** $6 < 6.5$

20. $0 > {}^{-}0.3$ **21.** ${}^{-}13 < 0$ **22.** $\frac{1}{8} < \frac{1}{7}$ **23.** S, F **24.** M, V **25.** N, R

26. A, G **27.** Q, A, U **28.** H, D, B **29.** M, T, P **30.** S, Z, Y

31. ${}^{-}1 < 3, 3 > {}^{-}1$ **32.** ${}^{-}3 < 0, 0 > {}^{-}3$ **33.** $0 < 2, 2 > 0$ **34.** ${}^{-}2 < 2, 2 > {}^{-}2$

35. $^-6 < {^-2}, {^-2} > {^-6}$ **36.** $^-0.5 < 0.5, 0.5 > {^-0.5}$ **37.** $0 < 5, 5 > 0$

38. $^-1.5 < 2.5, 2.5 > {^-1.5}$ **39.** $^-4 < 0$ **40.** $0 > {^-5}$ **41.** $6 < 5 + 4$

42. $8 - 7 > {^-1}$ **43.** $0 > {^-1}$ **44.** $0 \times 0 < 1$ **45.** $^-3.1 > {^-3.2}$ **46.** $-\frac{2}{5} < -\frac{1}{5}$

B **47.** 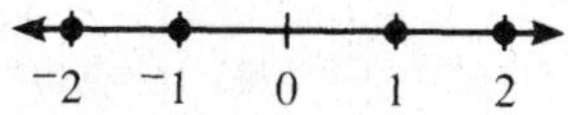 $^-2, {^-1}, 1, 2$

48. 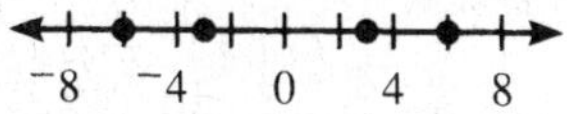 $^-6, {^-3}, 3, 6$

49. 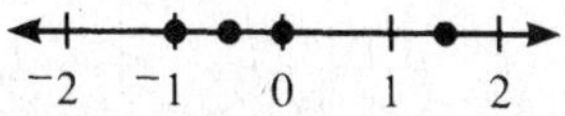 $^-1, {^-0.5}, 0, 1.5$

50.  $^-2, {^-1.5}, {^-1}, 0$

51. 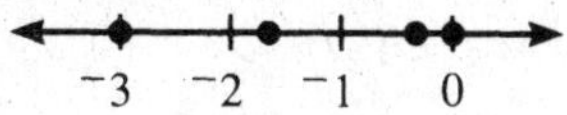 $^-3, {^-1\frac{2}{3}}, -\frac{1}{3}, 0$

52. 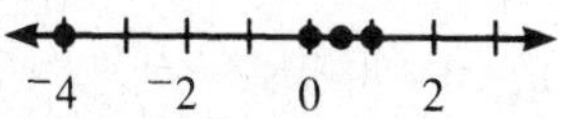 $^-4, 0, \frac{1}{2}, 1$

53. 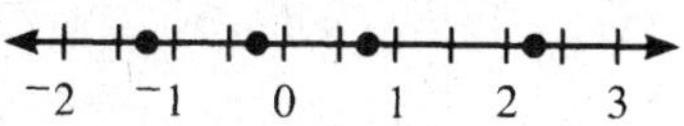 $^-1\frac{1}{4}, -\frac{1}{4}, \frac{3}{4}, 2\frac{1}{4}$

54. 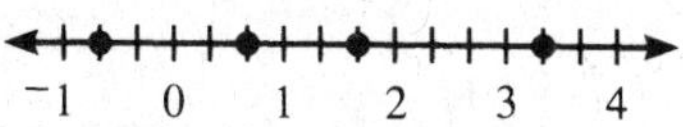 $-\frac{2}{3}, \frac{2}{3}, \frac{5}{3}, \frac{10}{3}$

55. Cora, Freida, Stu, Janelle **56.** Nick, Pete, Jack, Mona

C **57.** 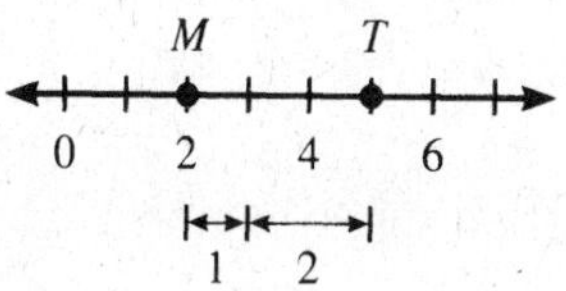

The point with coordinate 3 is 1 unit from *M* and 2 units from *T*; 3.

58. 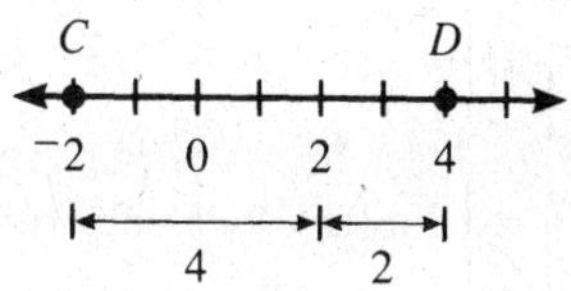

The point with coordinate 2 is 4 units from *C* and 2 units from *D*; 2.

Page 35 • MIXED REVIEW EXERCISES

1. $3ab - 2x = 3 \cdot 2 \cdot 5 - 2 \cdot 3 = 30 - 6 = 24$

2. $4x(y - b) = 4 \cdot 3 \cdot (7 - 5) = 12 \cdot 2 = 24$

3. $a \cdot (3y \div 7) = 2 \cdot (3 \cdot 7 \div 7) = 2(21 \div 7) = 2 \cdot 3 = 6$

4. $\frac{1}{5}(4c - 1) = \frac{1}{5}(4 \cdot 4 - 1) = \frac{1}{5}(16 - 1) = \frac{1}{5} \cdot 15 = 3$

5. $3y - (2a + c \div 2) = 3 \cdot 7 - (2 \cdot 2 + 4 \div 2) = 21 - (4 + 2) = 21 - 6 = 15$

6. $\left(\frac{1}{2}ac\right) + (5 - x)b = \left(\frac{1}{2} \cdot 2 \cdot 4\right) + (5 - 3)5 = 4 + 2 \cdot 5 = 4 + 10 = 14$ **7.** $30 - x = 12$

8. $x + 4 = 2$ **9.** $6x = 2$ **10.** $\frac{1}{4}x = 1$

Page 35 • CHALLENGE

1. 21, 28, 36, 45, 55

2. $n = 1$: $\frac{1 \cdot 2}{2} = 1$; $n = 2$: $\frac{2 \cdot 3}{2} = 3$; $n = 3$: $\frac{3 \cdot 4}{2} = 6$; $n = 4$: $\frac{4 \cdot 5}{2} = 10$; $n = 5$: $\frac{5 \cdot 6}{2} = 15$;
$n = 6$: $\frac{6 \cdot 7}{2} = 21$; $n = 7$: $\frac{7 \cdot 8}{2} = 28$; $n = 8$: $\frac{8 \cdot 9}{2} = 36$; $n = 9$: $\frac{9 \cdot 10}{2} = 45$; $n = 10$:
$\frac{10 \cdot 11}{2} = 55$

Pages 38–39 • WRITTEN EXERCISES

A **1.** $-(7 + 5) = -12$ **2.** $-(8 - 2) = -6$ **3.** $-(13 - 13) = 0$ **4.** $-(0 + 0) = 0$

5. $[-(-9)] + 10 = 9 + 10 = 19$ **6.** $[-(-7)] + 1 = 7 + 1 = 8$

7. $6 + [-(-2)] = 6 + 2 = 8$ **8.** $3 - [-(-1)] = 3 - 1 = 2$

9. $8 + |-3| = 8 + 3 = 11$ **10.** $|-11| + 4 = 11 + 4 = 15$

11. $|-8| + |6| = 8 + 6 = 14$ **12.** $|2| + |-9| = 2 + 9 = 11$

13. $\left|-\frac{3}{2}\right| + |0| = \frac{3}{2} + 0 = \frac{3}{2}$ **14.** $|-1| - |0| = 1 - 0 = 1$

15. $|-0.7| + |-3.3| = 0.7 + 3.3 = 4$ **16.** $|-2.8| + |2.8| = 2.8 + 2.8 = 5.6$

17. $\left|\frac{1}{2}\right| + \left|-\frac{1}{2}\right| = \frac{1}{2} + \frac{1}{2} = 1$ **18.** $\left|-\frac{3}{4}\right| - \left|\frac{1}{4}\right| = \frac{3}{4} - \frac{1}{4} = \frac{2}{4} = \frac{1}{2}$

19. $|6| - |6| = 6 - 6 = 0$ **20.** $|6| - |-6| = 6 - 6 = 0$

21. $-(-8) = 8$; $(-8) = -8$; $8 > -8$; $>$ **22.** $-(-2) = 2$; $|-3| = 3$; $2 < 3$; $<$

23. $|-8| = 8$; $|-10| = 10$; $8 < 10$; $<$ **24.** $|-15| = 15$; $|-6| = 6$; $15 > 6$; $>$

25. $-|-8| = -8$; $-8 = -8$; $=$ **26.** $-|-3| = -3$; $-3 = -3$; $=$ **27.** $|-5| > 2$

28. $4 < |-10|$ **29.** $-(-2) > -(-1)$ **30.** $-8 < -4$

B **31.** $|n| = 0$, looking for number(s) 0 units from origin; 0, {0}

32. $|p| = 2$, looking for number(s) 2 units from origin; 2 and -2, $\{2, -2\}$

33. $|t| = \frac{1}{2}$, looking for number(s) $\frac{1}{2}$ units from origin; $\frac{1}{2}$ and $-\frac{1}{2}$, $\left\{\frac{1}{2}, -\frac{1}{2}\right\}$

34. $|z| = 0.3$, looking for number(s) 0.3 units from origin; 0.3 and -0.3, $\{0.3, -0.3\}$

35. $|a| = -2$, no solution; $|a|$ is positive or 0, so it cannot be -2.

36. $|b| = -9$, no solution; $|b|$ is positive or 0, so it cannot be -9.

37. $|-q| = 1$, looking for number(s) 1 unit from origin; 1 and -1, $\{1, -1\}$

38. $|-x| = 5$, looking for number(s) 5 units from origin; 5 and -5, $\{5, -5\}$

39. $|a| + |b| = |1.5| + |-2| = 1.5 + 2 = 3.5$

40. $a + |-b| = 1.5 + |-(-2)| = 1.5 + 2 = 3.5$

41. $4a - |b| - |c| = 4 \cdot 1.5 - |-2| - |-1.7| = 6 - 2 - 1.7 = 2.3$

42. $2|b| - |a| + |c| = 2|-2| - |1.5| + |-1.7| = 2 \cdot 2 - 1.5 + 1.7 =$
$4 - 1.5 + 1.7 = 4.2$

43. $(a + 8.5) - [(-b) + |c|] = (1.5 + 8.5) - [-(-2) + |-1.7|] = 10 - [2 + 1.7] =$
$10 - 3.7 = 6.3$

44. $(10.5 - a) - [|c| + (-b)] = (10.5 - 1.5) - [|-1.7| + 2] = 9 - [1.7 + 2] = 9 - 3.7 = 5.3$

C **45.** (a) is false. 0 is a real number and $|0| = 0$, which is not a positive number.

Page 39 • MIXED REVIEW EXERCISES

1. $6 - (2 + 3) = 6 - 5 = 1$ **2.** $6 - 2 + 3 = 4 + 3 = 7$ **3.** $6 \div (2 + 1) = 6 \div 3 = 2$
4. $(4 - 2) \cdot (5 - 1) = 2 \cdot 4 = 8$ **5.** $7 - 2 \cdot 3 = 7 - 6 = 1$
6. $15 - 6 \div (2 + 1) = 15 - 6 \div 3 = 15 - 2 = 13$ **7.** $-3, 3$ **8.** $-50, 50$
9. $1000, -1000$ **10.** $0.5, -0.5$

Page 41 • CHAPTER REVIEW

1. $36 \div (9 \cdot 4) = 36 \div 36 = 1$; d **2.** $(x + y) - 3z = (7 + 8) - 3 \cdot 0 = 15 - 0 = 15$; c
3. $\frac{3x - y}{3(x - y)} = \frac{3 \cdot 5 - 3}{3(5 - 3)} = \frac{15 - 3}{3 \cdot 2} = \frac{12}{6} = 2$; b **4.** $9 = 3c - 6, c = 5$; d **5.** $7n - 3$; c
6. $9(n - 4) = 27$; b **7.** $A = l \cdot w; 28 = 7x$; c **8.** $(m + 1) + m = 5$; d
9. *Step 1* The unknowns are Wilt's and Anita's weight.

Step 2 Let x = Anita's weight, then $1\frac{1}{2} \cdot x$ = Wilt's weight.

Step 3 $x + \left(1\frac{1}{2} \cdot x\right) = 250; x + \left(\frac{3}{2}x\right) = 250$

Step 4 Replace x in turn by 90, 100, 105, and 150.

x	$x + \left(\frac{3}{2} \cdot x\right) = 250$	
90	$90 + \left(\frac{3}{2} \cdot 90\right) = 250$	False
100	$100 + \left(\frac{3}{2} \cdot 100\right) = 250$	True
105	$105 + \left(\frac{3}{2} \cdot 105\right) = 250$	False
150	$150 + \left(\frac{3}{2} \cdot 150\right) = 250$	False

Anita's weight: $x = 100$ Wilt's weight: $\frac{3}{2} \cdot x = \frac{3}{2} \cdot 100 = 150$

Step 5 Wilt weighs $1\frac{1}{2}$ times as much as Anita. $150 \stackrel{?}{=} 1\frac{1}{2} \cdot 100$ ✓

Together they weigh 250 lb. $100 + 150 \stackrel{?}{=} 250$ ✓

∴ Anita's weight is 100 lb and Wilt's weight is 150 lb; b

10. d **11.** d **12.** c

Page 42 • CHAPTER TEST

1. $5 + (15 \div 5) = 5 + 3 = 8$ **2.** $(4 + 12) \cdot 3 = 16 \cdot 3 = 48$
3. $(64 \div 4) + 2 = 16 + 2 = 18$ **4.** $(12 - 7) \cdot 9 = 5 \cdot 9 = 45$

5. $7x(y - z) = 7 \cdot 5 \cdot (8 - 6) = 35 \cdot 2 = 70$ **6.** $C = 2\pi r$, $C = 2 \cdot \frac{22}{7} \cdot 49 = 308$

7.

x	$56 = 14 + 7x$	
0	$56 = 14 + 7 \cdot 0$	False
1	$56 = 14 + 7 \cdot 1$	False
2	$56 = 14 + 7 \cdot 2$	False
3	$56 = 14 + 7 \cdot 3$	False
4	$56 = 14 + 7 \cdot 4$	False
5	$56 = 14 + 7 \cdot 5$	False
6	$56 = 14 + 7 \cdot 6$	True

$\therefore$ The solution set is {6}.

8.

x	$x \cdot x = 0$	
0	$0 \cdot 0 = 0$	True
1	$1 \cdot 1 = 0$	False
2	$2 \cdot 2 = 0$	False
3	$3 \cdot 3 = 0$	False
4	$4 \cdot 4 = 0$	False
5	$5 \cdot 5 = 0$	False
6	$6 \cdot 6 = 0$	False

$\therefore$ The solution set is {0}.

9.

x	$\frac{1}{2}x = \frac{x}{2}$	
0	$\frac{1}{2} \cdot 0 = \frac{0}{2}$	True
1	$\frac{1}{2} \cdot 1 = \frac{1}{2}$	True
2	$\frac{1}{2} \cdot 2 = \frac{2}{2}$	True
3	$\frac{1}{2} \cdot 3 = \frac{3}{2}$	True
4	$\frac{1}{2} \cdot 4 = \frac{4}{2}$	True
5	$\frac{1}{2} \cdot 5 = \frac{5}{2}$	True
6	$\frac{1}{2} \cdot 6 = \frac{6}{2}$	True

$\therefore$ The solution set is {0, 1, 2, 3, 4, 5, 6}.

10. $2b + 3$ **11.** P = sum of side lengths; $18 = 6 + 8 + x$, $18 = 14 + x$

12. Let a = Amy's amount, then $a + 5$ = Jon's amount; $a + (a + 5) = 29$

13. *Step 1* The unknowns are the width and the length of the corral.

Step 2 Let w = width, then $w + 20$ = length.

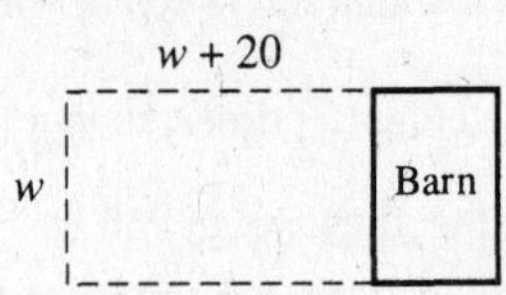

Step 3 $w + 2(w + 20) = 130$

Step 4 Replace w in turn by 25, 30, and 35.

w	$w + 2(w + 20) = 130$	
25	$25 + 2(25 + 20) = 130$	False
30	$30 + 2(30 + 20) = 130$	True
35	$35 + 2(35 + 20) = 130$	False

width: $w = 30$ length: $w + 20 = 30 + 20 = 50$

(Continued)

Step 5 The length is 20 more than the width. $50 \stackrel{?}{=} 20 + 30$

$50 = 50$ ✓

Twice the length plus the width is 130. $2 \cdot 50 + 30 = 130$

$130 = 130$ ✓

$\therefore$ The width is 30 m and the length is 50 m.

14. 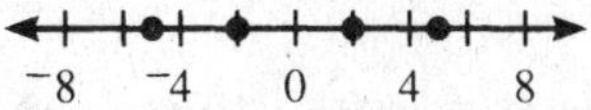 $-5, -2, 2, 5$

15. 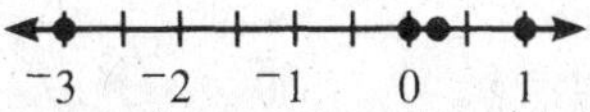 $-3, 0, \frac{1}{4}, 1$

16. $|-2.1| + 3.2 = 2.1 + 3.2 = 5.3$ **17.** $-(-2.75) = 2.75$

18. $-|4 + 1.5| = -|5.5| = -5.5$ **19.** $-(-5) = 5; 5 > -4; >$

20. $-|-10| = -10; |8| = 8; -10 < 8; <$

Page 43 • MAINTAINING SKILLS

1. $\begin{array}{r} 0.0056 \\ 2.3 \\ 18.232 \\ +\ 9.41 \\ \hline 29.9476 \end{array}$ **2.** $\begin{array}{r} 42.31 \\ 8.79 \\ +\ 13.26 \\ \hline 64.36 \end{array}$ **3.** $\begin{array}{r} 22.6 \\ 153.3 \\ +\ 201.8 \\ \hline 377.7 \end{array}$ **4.** $\begin{array}{r} 27 \\ 6.25 \\ 108.1 \\ +\ 35.72 \\ \hline 177.07 \end{array}$ **5.** $\begin{array}{r} 318 \\ \times\ 5.2 \\ \hline 636 \\ 1590 \\ \hline 1653.6 \end{array}$ **6.** $\begin{array}{r} 208.2 \\ \times\ 10.3 \\ \hline 6246 \\ 20820 \\ \hline 2144.46 \end{array}$

7. $\begin{array}{r} 7.51 \\ \times\ 2.2 \\ \hline 1502 \\ 1502 \\ \hline 16.522 \end{array}$ **8.** $\begin{array}{r} 0.876 \\ \times\ 0.09 \\ \hline 0.07884 \end{array}$ **9.** $\begin{array}{r} 824.2 \\ \times\ 1.2 \\ \hline 16484 \\ 8242 \\ \hline 989.04 \end{array}$ **10.** $\begin{array}{r} 0.222 \\ \times\ 11.1 \\ \hline 222 \\ 222 \\ 222 \\ \hline 2.4642 \end{array}$ **11.** $\begin{array}{r} 35.83 \\ +\ 9.96 \\ \hline 45.79 \end{array}$ **12.** $\begin{array}{r} 20.05 \\ +\ 8.87 \\ \hline 28.92 \end{array}$

13. $\frac{5}{8} = \frac{1}{1} \times \frac{5}{8} = \frac{5}{8}; \frac{3}{4} = \frac{3}{4} \times \frac{2}{2} = \frac{6}{8}$ **14.** $\frac{2}{3} = \frac{2}{3} \times \frac{5}{5} = \frac{10}{15}; \frac{3}{5} = \frac{3}{5} \times \frac{3}{3} = \frac{9}{15}$

15. $\frac{2}{7} = \frac{2}{7} \times \frac{3}{3} = \frac{6}{21}; \frac{2}{3} = \frac{2}{3} \times \frac{7}{7} = \frac{14}{21}$ **16.** $\frac{5}{12} = \frac{5}{12}; \frac{3}{2} = \frac{3}{2} \times \frac{6}{6} = \frac{18}{12}$

17. $\frac{3}{5} + \frac{4}{9} = \frac{3}{5} \times \frac{9}{9} + \frac{4}{9} \times \frac{5}{5} = \frac{27}{45} + \frac{20}{45} = \frac{47}{45}$ **18.** $\frac{2}{7} + \frac{1}{4} = \frac{2}{7} \times \frac{4}{4} + \frac{1}{4} \times \frac{7}{7} = \frac{8}{28} + \frac{7}{28} = \frac{15}{28}$

19. $\frac{3}{8} + \frac{5}{9} = \frac{3}{8} \times \frac{9}{9} + \frac{5}{9} \times \frac{8}{8} = \frac{27}{72} + \frac{40}{72} = \frac{67}{72}$ **20.** $\frac{7}{10} + \frac{2}{5} = \frac{7}{10} + \frac{2}{5} \times \frac{2}{2} = \frac{7}{10} + \frac{4}{10} = \frac{11}{10}$

21. $\frac{5}{13} \times \frac{3}{4} = \frac{15}{52}$ **22.** $\frac{5}{9} \times \frac{3}{5} = \frac{\overset{1}{\cancel{5}}}{\underset{3}{\cancel{9}}} \times \frac{\overset{1}{\cancel{3}}}{\underset{1}{\cancel{5}}} = \frac{1}{3}$ **23.** $\frac{10}{11} \times \frac{11}{12} = \frac{\overset{5}{\cancel{10}}}{\underset{1}{\cancel{11}}} \times \frac{\overset{1}{\cancel{11}}}{\underset{6}{\cancel{12}}} = \frac{5}{6}$

24. $\frac{5}{6} \times \frac{15}{16} = \frac{5}{\underset{2}{\cancel{6}}} \times \frac{\overset{5}{\cancel{15}}}{16} = \frac{25}{32}$ **25.** $\frac{2}{3} + \frac{3}{7} = \frac{2}{3} \times \frac{7}{7} + \frac{3}{7} \times \frac{3}{3} = \frac{14}{21} + \frac{9}{21} = \frac{23}{21}$

26. $\frac{9}{10} \times \frac{5}{3} = \frac{\overset{3}{\cancel{9}}}{\underset{2}{\cancel{10}}} \times \frac{\overset{1}{\cancel{5}}}{\underset{1}{\cancel{3}}} = \frac{3}{2}$ **27.** $\frac{7}{8} \times \frac{14}{16} = \frac{7}{8} \times \frac{\overset{7}{\cancel{14}}}{\underset{8}{\cancel{16}}} = \frac{49}{64}$

28. $\frac{9}{13} + \frac{2}{5} = \frac{9}{13} \times \frac{5}{5} + \frac{2}{5} \times \frac{13}{13} = \frac{45}{65} + \frac{26}{65} = \frac{71}{65}$

Pages 47–48 • WRITTEN EXERCISES

A

1. $275 + 52 + 25 + 8 = (275 + 25) + (52 + 8) = 300 + 60 = 360$

2. $803 + 26 + 47 + 24 = (803 + 47) + (26 + 24) = 850 + 50 = 900$

3. $2 \cdot 21 \cdot 5 \cdot 3 = (2 \cdot 5) \cdot (21 \cdot 3) = 10 \cdot 63 = 630$

4. $50 \cdot 3 \cdot 3 \cdot 20 = (50 \cdot 20) \cdot (3 \cdot 3) = 1000 \cdot 9 = 9000$

5. $25 \cdot 74 \cdot 2 \cdot 2 = 25 \cdot (2 \cdot 2) \cdot 74 = (25 \cdot 4) \cdot 74 = 100 \cdot 74 = 7400$

6. $8 \cdot 17 \cdot 9 \cdot 25 = (8 \cdot 25) \cdot (17 \cdot 9) = 200 \cdot 153 = 30{,}600$

7. $6\frac{1}{2} + 4\frac{1}{3} + 1\frac{1}{2} + \frac{2}{3} = \left(6\frac{1}{2} + 1\frac{1}{2}\right) + \left(4\frac{1}{3} + \frac{2}{3}\right) = 8 + 5 = 13$

8. $56\frac{7}{8} + \frac{3}{5} + \frac{1}{8} + 4\frac{2}{5} = \left(56\frac{7}{8} + \frac{1}{8}\right) + \left(\frac{3}{5} + 4\frac{2}{5}\right) = 57 + 5 = 62$

9. $0.1 + 1.8 + 5.9 + 0.2 = (0.1 + 5.9) + (1.8 + 0.2) = 6 + 2 = 8$

10. $4.75 + 2.95 + 1.05 + 10.25 = (4.75 + 10.25) + (2.95 + 1.05) = 15 + 4 = 19$

11. $3 + 7y + 4 = 7y + (3 + 4) = 7y + 7$ **12.** $8 + 9z + 4 = 9z + (8 + 4) = 9z + 12$

13. $5 + 2x + 1 = 2x + (5 + 1) = 2x + 6$

14. $5 + 3w + 2 = 3w + (5 + 2) = 3w + 7$ **15.** $2(5a) = (2 \cdot 5)a = 10a$

16. $3(4n) = (3 \cdot 4)n = 12n$ **17.** $(7y)(5z) = (7 \cdot 5)(yz) = 35yz$

18. $(6m)(4n) = (6 \cdot 4)(mn) = 24mn$ **19.** $(8x)(2y)(3z) = (8 \cdot 2 \cdot 3)(xyz) = 48xyz$

20. $(3p)(4q)(6r) = (3 \cdot 4 \cdot 6)(pqr) = 72pqr$

21. $a + 3 + b + 4 = a + b + (3 + 4) = a + b + 7$

22. $7 + x + y + 4 = x + y + (7 + 4) = x + y + 11$

23. $4a + 5 + 3n + 1 = 4a + 3n + (5 + 1) = 4a + 3n + 6$

24. $9m + 2 + 7n + 3 = 9m + 7n + (2 + 3) = 9m + 7n + 5$

B

25. $5 + 7x + 3 + 4y + z = 7x + 4y + z + (5 + 3) = 7x + 4y + z + 8$

26. $8p + 4 + 3q + 87 + 96 = 8p + 3q + (4 + 96) + 87 = 8p + 3q + (100 + 87) = 8p + 3q + 187$

27. $(25b)(25c)(4d)(4e) = (25 \cdot 4)(25 \cdot 4)(bcde) = (100 \cdot 100)(bcde) = 10{,}000bcde$

28. $(5a)(5b)(5c)(2d) = (5 \cdot 5)(5 \cdot 2)(abcd) = (25 \cdot 10)(abcd) = 250abcd$

29. Commutative property of multiplication **30.** Associative property of addition

31. Commutative property of addition **32.** Symmetric property of equality

33. Associative property of multiplication

34. Transitive property of equality or substitution property

35. **a.** Commutative property of addition **b.** Associative property of addition
c. Substitution (of 70 for $57 + 13$) **d.** Substitution (of 95 for $70 + 25$)

36. **a.** Commutative property of multiplication
b. Associative property of multiplication **c.** Substitution (of 200 for $25 \cdot 8$)
d. Substitution (of $100 \cdot 2$ for 200) **e.** Associative property of multiplication

f. Substitution (of 74 for $2 \cdot 37$) **g.** Substitution (of 7400 for $100 \cdot 74$)

C **37. a.** $(7 - 4) - 2 = 3 - 2 = 1; 7 - (4 - 2) = 7 - 2 = 5$ **b.** no

38. a. $(48 \div 8) \div 2 = 6 \div 2 = 3; 48 \div (8 \div 2) = 48 \div 4 = 12$ **b.** no

39. No; example: $10 \div 2 = 5$ but $2 \div 10 = \frac{2}{10} = \frac{1}{5}$

40. a. not closed **b.** closed **c.** not closed **d.** closed

Page 48 • MIXED REVIEW EXERCISES

1. $\frac{2a - y}{4y - z} = \frac{2 \cdot 3 - 2}{4 \cdot 2 - 6} = \frac{6 - 2}{8 - 6} = \frac{4}{2} = 2$ **2.** $5z(x - y) = 5 \cdot 6(5 - 2) = 30 \cdot 3 = 90$

3. $\frac{3a + x}{4z - (y + 1)} = \frac{3 \cdot 3 + 5}{4 \cdot 6 - (2 + 1)} = \frac{9 + 5}{24 - 3} = \frac{14}{21} = \frac{2}{3}$

4. $|a| + |x| - b = |-4| + |-3| - 2 = 4 + 3 - 2 = 7 - 2 = 5$

5. $2|a| + 7|x| = 2|-4| + 7|-3| = 2 \cdot 4 + 7 \cdot 3 = 8 + 21 = 29$

6. $2|x| - |b| + |a| = 2|-3| - |2| + |-4| = 2 \cdot 3 - 2 + 4 = 6 - 2 + 4 = 4 + 4 = 8$

7. $10|x| - 3|b| = 10|-3| - 3|2| = 10 \cdot 3 - 3 \cdot 2 = 30 - 6 = 24$

8. $2|b| + 3|a| = 2|2| + 3|-4| = 2 \cdot 2 + 3 \cdot 4 = 4 + 12 = 16$

9. $|-x - 1| - |b - 1| = |-(-3) - 1| - |2 - 1| = |3 - 1| - |1| = |2| - 1 = 2 - 1 = 1$

Page 48 • CHALLENGE

There can be no more than 9 dimes since 10 or more dimes would give a dollar or more (too much) change. There can be no more than 13 nickels or 13 pennies since there must be at least one of each kind of coin with a total of 15 coins. Furthermore, the number of pennies must be a multiple of 5 in order to be combined with nickels and dimes to make a dollar's worth of change. Thus, try combinations that include either 5 or 10 pennies. If 5 pennies are used with any number of dimes, there must be an *odd* number of nickels. If 10 pennies are used, there must be an *even* number of nickels. Remember that you only need to try combinations where the total *value* of the coins is one dollar. These combinations are shown below.

p	n	d	$p + n + d$
5	1	9	15 ✓
5	3	8	16
5	5	7	17
5	7	6	18
5	9	5	19
5	11	4	20
5	13	3	21

p	n	d	$p + n + d$
10	2	8	20
10	4	7	21
10	6	6	22
10	8	5	23
10	10	4	24
10	12	3	25

The only combination that works is 5 pennies, 1 nickel, and 9 dimes.

Pages 52–53 • WRITTEN EXERCISES

A **1.** $(-4 + 8) + 7 = [-4 + (4 + 4)] + 7 = (-4 + 4) + (4 + 7) = 0 + 11 = 11$

2. $(-5 + 8) + 2 = [-5 + (5 + 3)] + 2 = (-5 + 5) + (3 + 2) = 0 + 5 = 5$

3. $(-9 + 11) + (-2) = [-9 + (9 + 2)] + (-2) = (-9 + 9) + [2 + (-2)] = 0 + 0 = 0$

4. $(-3 + 5) + (-2) = [-3 + (3 + 2)] + (-2) = (-3 + 3) + [2 + (-2)] = 0 + 0 = 0$

5. $[3 + (-10)] + 6 = [3 + (-3) + (-7)] + 6 = [3 + (-3)] + (-7 + 6) = 0 + [-1 + (-6)] + 6 = -1 + (-6 + 6) = -1 + 0 = -1$

6. $[5 + (-15)] + 7 = 5 + [-5 + (-10)] + 7 = [5 + (-5)] + (-10 + 7) = 0 + [-3 + (-7)] + 7 = -3 + (-7 + 7) = -3 + 0 = -3$

7. $[-9 + (-8)] + 9 = [-8 + (-9)] + 9 = -8 + (-9 + 9) = -8 + 0 = -8$

8. $[-5 + (-23)] + 5 = [-23 + (-5)] + 5 = -23 + (-5 + 5) = -23 + 0 = -23$

9. $22 + [6 + (-11)] = (11 + 11) + (-11 + 6) = 11 + [11 + (-11)] + 6 = 11 + 0 + 6 = 17$

10. $32 + [8 + (-16)] = (16 + 16) + (-16 + 8) = 16 + [16 + (-16)] + 8 = 16 + 0 + 8 = 24$

11. $-26 + [-2 + (-8)] = -26 + [-(2 + 8)] = -26 + (-10) = -(26 + 10) = -36$

12. $-5 + [-12 + (-18)] = -5 + [-(12 + 18)] = -5 + (-30) = -(5 + 30) = -35$

13. $[37 + (-7)] + [1 + (-1)] = [(30 + 7) + (-7)] + 0 = 30 + [7 + (-7)] = 30 + 0 = 30$

14. $(-3 + 3) + [8 + (-12)] = 0 + 8 + [-8 + (-4)] = [8 + (-8)] + (-4) = 0 + (-4) = -4$

15. $[0 + (-8)] + [-7 + (-23)] = -8 + [-(7 + 23)] = -8 + (-30) = -(8 + 30) = -38$

16. $(-4 + 4) + [18 + (-8)] = 0 + (10 + 8) + (-8) = 10 + [8 + (-8)] = 10 + 0 = 10$

17. $-3 + (-4) + (-6) = -(3 + 4) + (-6) = -7 + (-6) = -(7 + 6) = -13$

18. $-5 + (-7) + (-9) = -(5 + 7) + (-9) = -12 + (-9) = -(12 + 9) = -21$

19. $-4 + (-10) + 9 + (-6) = [-(4 + 10)] + [(3 + 6) + (-6)] = -14 + 3 + [6 + (-6)] = [-11 + (-3)] + 3 + 0 = -11 + (-3 + 3) = -11 + 0 = -11$

20. $-17 + 12 + (-4) + (-13) = 12 + (-17) + [-(4 + 13)] = 12 + (-17) + (-17) = 12 + [-(17 + 17)] = 12 + (-34) = 12 + [-12 + (-22)] = [12 + (-12)] + (-22) = 0 + (-22) = -22$

21. $-3.2 + (-4.6) + 5.4 = -(3.2 + 4.6) + 5.4 = -7.8 + 5.4 = [-2.4 + (-5.4)] + 5.4 = -2.4 + (-5.4 + 5.4) = -2.4 + 0 = -2.4$

22. $5.9 + (-3.2) + (-7.8) = 5.9 + [-(3.2 + 7.8)] = 5.9 + (-11) = 5.9 + [-5.9 + (-5.1)] = [5.9 + (-5.9)] + (-5.1) = 0 + (-5.1) = -5.1$

23. $-\frac{3}{2} + 5 + \left(-\frac{7}{2}\right) = 5 + \left[-\left(\frac{3}{2} + \frac{7}{2}\right)\right] = 5 + (-5) = 0$

24. $-\frac{9}{5} + (-3) + \left(-\frac{6}{5}\right) = -\left(\frac{9}{5} + \frac{6}{5}\right) + (-3) = -3 + (-3) = -(3 + 3) = -6$

25. $3 + x + (-8) = x + 3 + (-8) = x + (-5)$ **26.** $y + (-2) + 7 = y + 5$

27. $2n + 5 + (-5) = 2n$ **28.** $8 + 3n + (-12) = 3n + 8 + (-12) = 3n + (-4)$

29. $13 + 5n + (-21) = 5n + 13 + (-21) = 5n + (-8)$

30. $-7 + (-z) + 12 = -z + (-7) + 12 = -z + 5$

B **31.** $3a + (-7) + 4b + (-6) = 3a + 4b + (-7) + (-6) = 3a + 4b + (-13)$

32. $-2 + 4n + 9 + (-7) = 4n + [-2 + 9 + (-7)] = 4n + 0 = 4n$

33. $7x + (-3) + (-y) + (-7) = 7x + (-y) + (-3) + (-7) = 7x + (-y) + (-10)$

34. $-8 + 7c + 11 + (-3) = 7c + (-8) + 11 + (-3) = 7c + 3 + (-3) = 7c + 0 = 7c$

C **35.** $b + [a + (-b)] = b + [-b + a] = [b + (-b)] + a = 0 + a = a$

36. $-m + (k + m) = -m + (m + k) = (-m + m) + k = 0 + k = k$

37. $-m + [-h + (m + h)] = -m + [(m + h) + (-h)] = (-m + m) + [h + (-h)] = 0 + 0 = 0$

38. $[a + (-b)] + [b + (-a)] = a + (-b + b) + (-a) = a + 0 + (-a) = a + (-a) = 0$

39. a. $(a + b) + [(-a) + (-b)] = [a + (-a)] + [b + (-b)]$ (Comm. and assoc. properties of add.)
$= 0 + 0$ (Prop. of opposites)
$= 0$ (Identity prop. of add.)

b. The property of opposites

c. Since the opposite of a number is unique and $-(a + b)$ and $[(-a) + (-b)]$ both have the sum of zero when added to $(a + b)$, $-(a + b) = (-a) + (-b)$.

Page 53 • MIXED REVIEW EXERCISES

1. $7 + 21 \div 7 = 7 + 3 = 10$ **2.** $|-6| + |3| = 6 + 3 = 9$ **3.** $6 - [-(-2)] = 6 - 2 = 4$

4. $\frac{4 + (3 \cdot 2)}{2} = \frac{4 + 6}{2} = \frac{10}{2} = 5$ **5.** $\left|\frac{1}{3}\right| + \left|-\frac{1}{3}\right| = \frac{1}{3} + \frac{1}{3} = \frac{2}{3}$

6. $\left|-\frac{2}{3}\right| - \left|\frac{1}{3}\right| = \frac{2}{3} - \frac{1}{3} = \frac{1}{3}$

7. $63 + 25 + 17 + 15 = (63 + 17) + (25 + 15) = 80 + 40 = 120$

8. $25 \cdot 5 \cdot 20 = 25 \cdot 100 = 2500$ **9.** $(4a)(2b)(3c) = (4 \cdot 2 \cdot 3)(abc) = 24abc$

10. $5\frac{2}{3} + 3\frac{1}{4} + 2\frac{1}{3} + 6\frac{3}{4} = \left(5\frac{2}{3} + 2\frac{1}{3}\right) + \left(3\frac{1}{4} + 6\frac{3}{4}\right) = 8 + 10 = 18$

11. $0.7 + 2.4 + 6.3 = (0.7 + 6.3) + 2.4 = 7 + 2.4 = 9.4$

12. $2 \cdot 17 \cdot 5 = (2 \cdot 5)17 = 10 \cdot 17 = 170$

Page 53 • CHALLENGE

1. The next two square numbers are 36 and 49.

2. The square numbers from 4 to 100 are: 4, 9, 16, 25, 36, 49, 64, 81, and 100. The triangular numbers are: 1, 3, 6, 10, 15, 21, 28, 36, 45, 55, ... To verify: $4 = 1 + 3$; $9 = 3 + 6$; $16 = 6 + 10$; $25 = 10 + 15$; $36 = 15 + 21$; $49 = 21 + 28$; $64 = 28 + 36$; $81 = 36 + 45$; $100 = 45 + 55$

3.

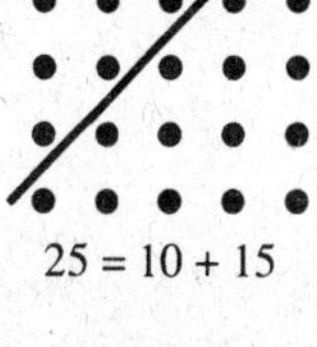

$25 = 10 + 15$

Pages 55–56 • WRITTEN EXERCISES

A

1. $\begin{array}{r} -7 \\ 6 \\ \hline -1 \end{array}$ **2.** $\begin{array}{r} -5 \\ -6 \\ \hline -11 \end{array}$ **3.** $\begin{array}{r} -34 \\ 75 \\ \hline 41 \end{array}$ **4.** $\begin{array}{r} 58 \\ -72 \\ \hline -14 \end{array}$ **5.** $\begin{array}{r} 148 \\ -72 \\ -73 \\ -31 \\ \hline -28 \end{array}$ **6.** $\begin{array}{r} -173 \\ 412 \\ -58 \\ -93 \\ \hline 88 \end{array}$

7. $-12 + 7 + (-14) + 29 = 36 + (-26) = 10$

8. $-16 + (-9) + 8 + 25 = -25 + 33 = 8$

9. $109 + (-56) + (-91) + 26 = 135 + (-147) = -12$

10. $-206 + (-75) + 153 + 37 = -281 + 190 = -91$

11. $-[24 + (-5)] + [-(-2 + 6)] = -19 + (-4) = -23$

12. $[-7 + (-1)] + [-(-7 + 1)] = -8 + 6 = -2$

13. $3.7 + 4.2 + (-2.3) + 0 + 6.4 + 12.8 = -2.3 + 27.1 = 24.8$

14. $-7.2 + 11.4 + (-8.1) + (-9.7) + 0.6 = -25 + 12 = -13$

15. $27 + 43 + (-14) + (-57) + 5 + (-36) + (-14) = 75 + (-121) = -46$

16. $46 + (-33) + 18 + 0 + (-93) + (-2) + (-34) = 64 + (-162) = -98$

17. $-\frac{1}{3} + \left(-1\frac{2}{3}\right) + 2 = -2 + 2 = 0$ **18.** $3 + \left(-\frac{5}{2}\right) + \left(-\frac{7}{2}\right) = 3 + (-6) = -3$

19. $-1\frac{3}{4} + 2\frac{1}{4} = \frac{1}{2}$ **20.** $-3\frac{2}{5} + \left(-1\frac{4}{5}\right) = -5\frac{1}{5}$ **21.** $-\frac{7}{8} + \left(-\frac{11}{8}\right) = -\frac{18}{8} = -\frac{9}{4}$

22. $\frac{16}{3} + \left(-\frac{10}{3}\right) = \frac{6}{3} = 2$

23. $-3 + x + (-5) + 4 = x + (-3) + (-5) + 4 = x + (-8) + 4 = x + (-4)$

24. $2 + (-9) + (-y) + (-11) = -y + 2 + (-9) + (-11) = -y + 2 + (-20) =$
$-y + (-18)$

25. $-4 + a + 4 + (-a) = a + (-a) + (-4) + 4 = 0 + 0 = 0$

26. $21 + 4b + (-17) + (-12) = 4b + 21 + (-17) + (-12) = 4b + 21 + (-29) =$
$4b + (-8)$

27. $-[8 + (-5)] + (-c) + 3 = -c + (-3) + 3 = -c + 0 = -c$

28. $-(-9) + 2y + (-9) + 6 = 2y + 9 + (-9) + 6 = 2y + 0 + 6 = 2y + 6$

29. $x + 5 + [(-x) + (-14)] = x + (-x) + 5 + (-14) = 0 + (-9) = -9$

30. $4x + [8 + (-5) + (-7)] = 4x + [8 + (-12)] = 4x + (-4)$

B

31. $x + y + (-1) = -3 + 6 + (-1) = -4 + 6 = 2$

32. $-21 + x + z = -21 + (-3) + (-4) = -28$

33. $-15 + (-x) + y = -15 + [-(-3)] + 6 = -15 + 3 + 6 = -15 + 9 = -6$

34. $-z + (-8) + y = -(-4) + (-8) + 6 = 4 + (-8) + 6 = 10 + (-8) = 2$

35. $1 + (-y) + z = 1 + (-6) + (-4) = 1 + (-10) = -9$

36. $-y + (-11) + x = -6 + (-11) + (-3) = -20$

37. $|x + y + z| = |(-3) + 6 + (-4)| = |-7 + 6| = |-1| = 1$

38. $|x + (-y) + z| = |(-3) + (-6) + (-4)| = |-13| = 13$

39. $-|x + z + (-8)| = -|-3 + (-4) + (-8)| = -|-15| = -15$

C 40. $-b + [-a + (a + b)] = -b + (-a + a) + b = -b + 0 + b = -b + b = 0$

41. $-(-x + y) + y = x + (-y) + y = x + 0 = x$

42. $r + s + [-(r + s + t)] = r + s + (-r) + (-s) + (-t) = [r + (-r)] + [s + (-s)] + (-t) = 0 + 0 + (-t) = -t$

43. $a + [-(a + b)] = a + (-a) + (-b) = 0 + (-b) = -b$

Pages 56–58 • PROBLEMS

A 1. **a.** 18, −7, 9 **b.** $18 + (-7) + 9 = 20$ **c.** The elevator was at the 20th floor.

2. **a.** −230, 95, −120 **b.** $-230 + 95 + (-120) = -255$
c. The submarine's depth was 255 m below the surface of the ocean.

3. **a.** 784, 96, −18, −44 **b.** $784 + 96 + (-18) + (-44) = 818$
c. The new balance is $818.

4. **a.** −3, 15, 8, −9, −4 **b.** $-3 + 15 + 8 + (-9) + (-4) = 7$
c. The net yardage was a gain of 7 yd.

5. **a.** 698, −250, −1684, −1316, 3700
b. $698 + (-250) + (-1684) + (-1316) + 3700 = 1148$
c. The Drama Club had $1148 in its treasury after the play.

6. **a.** 32, −5, −3, 4, 2, 0 **b.** $32 + (-5) + (-3) + 4 + 2 + 0 = 30$
c. The closing price was $30 per share.

7. **a.** 32, −27, 28, −31, 12, −18, 16, −25
b. $32 + (-27) + 28 + (-31) + 12 + (-18) + 16 + (-25) = -13$
c. There were 13 fewer guests at the hotel on the fourth day than there were on the first.

8. **a.** −14,250, 0, 18,180, 29,470 **b.** $-14{,}250 + 0 + 18{,}180 + 29{,}470 = 33{,}400$
c. The restaurant's net gain was $33,400 over the four-year period.

9. **a.** −5, 3, 8, −12 **b.** $-5 + 8 = 3$; $3 + (-12) = -9$
c. The truck was 3 blocks east and 9 blocks south of its starting point.

10. **a.** 32, −8, 5, −4, 6 **b.** $32 + (-8) + 5 + (-4) + 6 = 31$
c. The temperature was 31°C after the storm.

B 11. **a.** −86, 41, 28, −37 **b.** $-86 + 41 + 28 + (-37) = -54$
c. The helicopter was flying at an altitude of 54 m below sea level.

12. **a.** −156, −71, 29 **b.** $-156 + (-71) + 29 = -198$
c. The flare would rise to an altitude of 198 ft below sea level.

13. **a.** 1, 1.5, −2 **b.** $1 + 1.5 - 2 = 0.5$
c. The time in Bay View was 12:30 P.M. when Coretta landed.

14. **a.** 135, −7 **b.** $135 + (-7) = 128$
c. The passenger is traveling at 128 km/h with respect to the ground.

C 15. $x - 175 - 287 + 50 + 50 - 11.50 = -495.25$, $x + 100 - 473.50 = -495.25$, $x - 373.50 = -495.25$, $x = -121.75$; Wes owed $121.75 before making purchases.

16. $x + 375 - 240 + 93 - 164 - 157 = 54$, $x + 468 - 561 = 54$, $x - 93 = 54$, $x = 147$; there were 147 L of water in the tank at the beginning.

Page 58 • MIXED REVIEW EXERCISES

1. $5 - 9 \div 3 = 5 - 3 = 2$ **2.** $3 \cdot 5 \cdot 4 \cdot 20 = (3 \cdot 4) \cdot (5 \cdot 20) = 12 \cdot 100 = 1200$

3. $(6 - 4 \div 2) \cdot 3 + 1 = (6 - 2) \cdot 3 + 1 = 4 \cdot 3 + 1 = 12 + 1 = 13$

4. $|-7| - 6 = 7 - 6 = 1$ **5.** $|-1.2| + 1.2 = 1.2 + 1.2 = 2.4$

6. $|-15| - |-6| = 15 - 6 = 9$ **7.** $\frac{7 \cdot 6 + 3 \cdot 7}{(5 + 2)} = \frac{7(6 + 3)}{(5 + 2)} = \frac{7(9)}{7} = \frac{63}{7} = 9$

8. $6\frac{1}{7} + 5\frac{2}{3} + 8\frac{6}{7} = 6\frac{1}{7} + 8\frac{6}{7} + 5\frac{2}{3} = 15 + 5\frac{2}{3} = 20\frac{2}{3}$

9. $3.2 + 1.0 + 4.8 = 3.2 + 4.8 + 1.0 = 8 + 1 = 9$ **10.** $[15 + (-5)] + 6 = 10 + 6 = 16$

11. $(-8 + 4) + (-6) = -4 + (-6) = -10$ **12.** $-3 + (-9) + 6 + (-5) = -12 + 1 = -11$

Page 58 • CALCULATOR KEY-IN

1. -19 **2.** -574 **3.** -391 **4.** -93.4 **5.** -8.1 **6.** -273.7

Pages 61–62 • WRITTEN EXERCISES

A **1.** $25 - 213 = 25 + (-213) = -188$ **2.** $154 - 281 = 154 + (-281) = -127$

3. $39 - (-32) = 39 + 32 = 71$ **4.** $47 - (-49) = 47 + 49 = 96$

5. $-19 - (-3) = -19 + 3 = -16$ **6.** $-25 - (-9) = -25 + 9 = -16$

7. $-2.8 - 4.4 = -2.8 + (-4.4) = -7.2$ **8.** $-5.1 - 6.7 = -5.1 + (-6.7) = -11.8$

9. $174 - (-24) = 174 + 24 = 198$ **10.** $-106 - (-95) = -106 + 95 = -11$

11. $1.91 - (-1.03) = 1.91 + 1.03 = 2.94$ **12.** $2.95 - (-2.55) = 2.95 + 2.55 = 5.5$

13. $-21 - 6 = -21 + (-6) = -27$ **14.** $-6 - (-32) = -6 + 32 = 26$

15. $-1 - 24 = -1 + (-24) = -25$ **16.** $-3 - 15 = -3 + (-15) = -18$

17. $132 - (72 - 61) = 132 + (-72) + 61 = 60 + 61 = 121$

18. $275 - (80 - 65) = 275 + (-80) + 65 = 195 + 65 = 260$

19. $234 - (56 - 87) = 234 + (-56) + 87 = 178 + 87 = 265$

20. $193 - (30 - 75) = 193 + (-30) + 75 = 163 + 75 = 238$

21. $(22 - 33) - (55 - 66) = 22 + (-33) + (-55) + 66 = -11 + (-55) + 66 = -66 + 66 = 0$

22. $(42 - 50) - (73 - 60) = 42 + (-50) + (-73) + 60 = (42 + 60) + [-50 + (-73)] = 102 + (-123) = -21$

23. $(3 - 8) - (-15 + 19) = 3 + (-8) + 15 + (-19) = (3 + 15) + [-8 + (-19)] = 18 + (-27) = -9$

24. $(42 - 33) - (-7 + 12) = 42 + (-33) + 7 + (-12) = (42 + 7) + [-33 + (-12)] = 49 + (-45) = 4$

25. $106 - 492 + 776 = 106 + (-492) + 776 = -386 + 776 = 390$

26. $910 - 939 - 201 = 910 + (-939) + (-201) = -29 + (-201) = -230$

27. $12 - (-9) - [5 - (-4)] = 12 + 9 + (-5) + (-4) = (12 + 9) + [-5 + (-4)] = 21 + (-9) = 12$

28. $-15 - 6 - [-7 - (-10)] = -15 + (-6) + 7 + (-10) = [-15 + (-6) + (-10)] + 7 = -31 + 7 = -24$

29. $4 - 5 + 8 - 17 + 31 = 4 + 8 + 31 - (5 + 17) = 43 - 22 = 21$

30. $18 - 14 + 15 + 7 - 26 = (18 + 15 + 7) - (14 + 26) = 40 - 40 = 0$

31. $-6 - 19 + 4 - 8 + 20 = (4 + 20) - (6 + 19 + 8) = 24 - 33 = 24 + (-33) = -9$

32. $-11 - 43 + 1 - 9 + 30 = (1 + 30) - (11 + 43 + 9) = 31 - 63 = 31 + (-63) = -32$

33. $[81 - (-6)] - (-29) = 81 + 6 + 29 = 87 + 29 = 116$

34. $-54 - (-37 - 15) = -54 + 37 + 15 = -54 + 52 = -2$

35. $-46 - [-23 - (-61)] = -46 + 23 + (-61) = -23 + (-61) = -84$

36. $(-59 - 12) - 72 = -59 + (-12) + (-72) = -71 + (-72) = -143$

37. $-(x - 7) = -x + 7$ **38.** $-(4 - y) = -4 + y$ or $y - 4$

39. $x - (x + 5) = x - x - 5 = -5$ **40.** $8 - (y + 8) = 8 - y - 8 = -y$

41. $5 - (b - 7) = 5 - b + 7 = -b + 12$ **42.** $n - (4 + n) = n - 4 - n = -4$

B **43.** $z - (8 + y) = z - 8 - y = z - y - 8$ **44.** $-(-7 + y) = 7 - y$ or $-y + 7$

45. $(7 - z) - 5 = -z + 7 - 5 = -z + 2$ **46.** $-(x - 4) = -x + 4$

47. $7 + y - (7 - y) - y = 7 + y - 7 + y - y = 7 - 7 + y + y - y = 0 + y + 0 = y$

48. $s - (-3) - [s + (-3)] - 3 = s + 3 - s + 3 - 3 = s - s + 3 + 3 - 3 = 0 + 3 + 0 = 3$

49. $h + 8 - (-9 + h) = h + 8 + 9 - h = h - h + 8 + 9 = 0 + 17 = 17$

50. $-(10 - k) - (k - 12) = -10 + k - k + 12 = -10 + 12 + k - k = 2 + 0 = 2$

51. $(\pi - 17) - (\pi + 2) = \pi - 17 - \pi - 2 = -17 + (-2) + \pi - \pi = -19 + 0 = -19$

52. $(11 - 2\pi) - (7 - 2\pi) = 11 - 2\pi - 7 + 2\pi = 11 - 7 - 2\pi + 2\pi = 4 + 0 = 4$

53. $c - |a - b| = -1 - |-4 - 5| = -1 - |-9| = -1 - 9 = -1 + (-9) = -10$

54. $b - c - |a| = 5 - (-1) - |-4| = 5 + 1 - 4 = 6 - 4 = 2$

55. $|c - a| - b = |-1 - (-4)| - 5 = |-1 + 4| - 5 = |3| - 5 = 3 - 5 = 3 + (-5) = -2$

56. $|b - c| - a = |5 - (-1)| - (-4) = |5 + 1| + 4 = |6| + 4 = 6 + 4 = 10$

57. $a - |c| - (|a| - b) = -4 - |-1| - (|-4| - 5) = -4 - 1 - (4 - 5) = -4 - 1 - 4 + 5 = -4 + (-1) + (-4) + 5 = -9 + 5 = -4$

58. $|c| - b - (|a| - c) = |-1| - 5 - [|-4| - (-1)] = 1 - 5 - (4 + 1) = 1 - 5 - 5 = 1 + (-5) + (-5) = 1 + (-10) = -9$

59. $(a - |c|) - |a| - b = (-4 - |-1|) - |-4| - 5 = (-4 - 1) - 4 - 5 = (-5) + (-4) + (-5) = -14$

60. $(|c| - b) - |a| - c = (|-1| - 5) - |-4| - (-1) = (1 - 5) - 4 + 1 = 1 + (-5) + (-4) + 1 = (1 + 1) + [-5 + (-4)] = 2 + (-9) = -7$

C **61.** **a.** (a) Definition of subtraction (b) Commutative property of addition
(c) Associative property of addition (d) Property of opposites
(e) Identity property of addition **b.** a

62. $b - a + (a - b) = b + (-a) + [a + (-b)] = b + [-a + a] + (-b) = b + 0 + (-b) = b + (-b) = 0$; since the sum of $b - a$ and $a - b$ is 0, then $b - a$ must be the unique opposite of $a - b$.

Pages 62–63 • PROBLEMS

A

1. $141.5 - 48.6 = 92.9$; Earth is 92.9 millions of miles from the sun.
2. $58 - 146.3 = 58 + (-146.3) = -88.3$; the lowest recorded temperature is 88.3° below 0°C.
3. $-42 - (-70) = -42 + 70 = 28$; Virgil was 28 years old.
4. $-497 - (-582) = -497 + 582 = 85$; Pythagoras was 85 years old.
5. $-157.2 - (-153.4) = -3.8$; krypton's melting point is 3.8°C below its boiling point.
6. $-17 - (-8) = -17 + 8 = -9$; the temperature fell 9°C.
7. $-39 - 52 = -39 + (-52) = -91$; she rode 91 blocks west.
8. $-53 - 126 = -53 + (-126) = -179$; the ball traveled 179 yd.
9. $4392 - (-86) = 4392 + 86 = 4478$; the difference in altitude is 4478 m.
10. $163.068 - 164.592 = 163.068 + (-164.592) = -1.524$; the altitude of New Orleans is 1.524 m below sea level.

B

11. $475 - 396 = 79$; it weighed 79 lb on the moon.
12. Alonzo: $22 - 15 - 3 + 5 = (22 + 5) - (15 + 3) = 27 - 18 = 9$
Brad: $29 - 18 - 7 + 15 = (29 + 15) - (18 + 7) = 44 - 25 = 19$
Caryl: $2 + 3 + 18 + 11 = 34$
Diane: $13 - 11 - 5 + 7 = (13 + 7) - (11 + 5) = 20 - 16 = 4$
Alonzo will have \$9, Brad will have \$19, Caryl will have \$34, and Diane will have \$4.

Page 63 • MIXED REVIEW EXERCISES

1. $|-9| + |7| = 9 + 7 = 16$ **2.** $18 \cdot 3 \cdot 5 \cdot 2 = 54 \cdot 10 = 540$

3. $4 + 7x + 3y + 5 = 7x + 3y + 4 + 5 = 7x + 3y + 9$ **4.** $\left|-\frac{2}{5}\right| - \left|\frac{1}{5}\right| = \frac{2}{5} - \frac{1}{5} = \frac{1}{5}$

5. $-\frac{5}{4} + \left(-\frac{7}{4}\right) = -\frac{12}{4} = -3$ **6.** $1\frac{1}{3} + \left(-3\frac{2}{3}\right) = -2\frac{1}{3}$ **7.** $[6 + (-12)] + 9 = -6 + 9 = 3$

8. $4.2 - 0.5 + (-3.2) = 4.2 - (0.5 + 3.2) = 4.2 - 3.7 = 0.5$

9. $-6 + [-8 + (-4)] = -6 + (-12) = -18$

10. $-3.2 + 74 + (-2.8) = 74 - (3.2 + 2.8) = 74 - 6 = 68$

11. $-41 + (-28) + 32 + 49 = (32 + 49) - (41 + 28) = 81 - 69 = 12$

12. $3 + (-7) + (-12) + (-x) = -x + 3 + (-19) = -x + (-16) = -x - 16$

Page 64 • READING ALGEBRA

1. Add and subtract polynomials, multiply monomials
2. Monomial, constant monomial, constant, polynomial, binomials, trinomials, coefficient, numerical coefficient, similar or like terms, simplified, simplest form, degree of a variable in a monomial, degree of a monomial, degree of a polynomial
3. An expression that is either a numeral, a variable, or the product of a numeral and one or more variables; a sum of monomials
4. Glossary; page 1 **5.** Index and/or Table of Contents

Pages 67–69 • WRITTEN EXERCISES

A

1. $8(30 + 1) = (8 \cdot 30) + (8 \cdot 1) = 240 + 8 = 248$

2. $7(60 + 2) = (7 \cdot 60) + (7 \cdot 2) = 420 + 14 = 434$

3. $5(70 - 2) = (5 \cdot 70) - (5 \cdot 2) = 350 - 10 = 340$

4. $9(20 - 1) = (9 \cdot 20) - (9 \cdot 1) = 180 - 9 = 171$

5. $4(50 - 3) = (4 \cdot 50) - (4 \cdot 3) = 200 - 12 = 188$

6. $6(80 + 5) = (6 \cdot 80) + (6 \cdot 5) = 480 + 30 = 510$

7. $4 \cdot 52 = 4(50 + 2) = (4 \cdot 50) + (4 \cdot 2) = 200 + 8 = 208$

8. $6 \cdot 64 = 6(60 + 4) = (6 \cdot 60) + (6 \cdot 4) = 360 + 24 = 384$

9. $4(9.5) = 4(9 + 0.5) = (4 \cdot 9) + (4 \cdot 0.5) = 36 + 2 = 38$

10. $6(9.9) = 6(9 + 0.9) = (6 \cdot 9) + (6 \cdot 0.9) = 54 + 5.4 = 59.4$

11. $14\left(1\frac{1}{7}\right) = 14\left(1 + \frac{1}{7}\right) = (14 \cdot 1) + \left(14 \cdot \frac{1}{7}\right) = 14 + 2 = 16$

12. $15\left(4\frac{3}{5}\right) = 15\left(4 + \frac{3}{5}\right) = (15 \cdot 4) + \left(15 \cdot \frac{3}{5}\right) = 60 + \frac{45}{5} = 60 + 9 = 69$

13. $30 \cdot 18 + 70 \cdot 18 = (30 + 70)18 = (100)18 = 1800$

14. $11 \cdot 43 + 89 \cdot 43 = (11 + 89)43 = (100)43 = 4300$

15. $(13 \cdot 27) - (13 \cdot 27) = 13(27 - 27) = 13 \cdot 0 = 0$

16. $(64 \cdot 81) + (36 \cdot 81) = (64 + 36)81 = (100)81 = 8100$

17. $(0.75)(32) + (0.25)(32) = (0.75 + 0.25)32 = 1(32) = 32$

18. $(3.6)(25) - (1.6)(25) = (3.6 - 1.6)25 = (2)25 = 50$ **19.** $5a + 2a = (5 + 2)a = 7a$

20. $7x + 6x = (7 + 6)x = 13x$ **21.** $13y - 4y = (13 - 4)y = 9y$

22. $2n + (-8)n = [2 + (-8)]n = (-6)n$ **23.** $(-3)p + 7p = (-3 + 7)p = 4p$

24. $(-8)n - 7n = (-8 - 7)n = [-8 + (-7)]n = (-15)n$

25. $3(x + 2) = 3 \cdot x + 3 \cdot 2 = 3x + 6$ **26.** $5(a + 7) = 5 \cdot a + 5 \cdot 7 = 5a + 35$

27. $4(n - 2) = 4 \cdot n - 4 \cdot 2 = 4n - 8$ **28.** $8(b - 5) = 8 \cdot b - 8 \cdot 5 = 8b - 40$

29. $6(3n + 2) = 6 \cdot 3n + 6 \cdot 2 = 18n + 12$ **30.** $7(5n + 3) = 7 \cdot 5n + 7 \cdot 3 = 35n + 21$

31. $3(j - k) = 3 \cdot j - 3 \cdot k = 3j - 3k$ **32.** $2(3x - y) = 2 \cdot 3x - 2 \cdot y = 6x - 2y$

33. $(3n + 7)2 = 3n \cdot 2 + 7 \cdot 2 = 6n + 14$ **34.** $(4x + 3)3 = 4x \cdot 3 + 3 \cdot 3 = 12x + 9$

35. $(2x + 3y)5 = 2x \cdot 5 + 3y \cdot 5 = 10x + 15y$

36. $(6m + 7n)2 = 6m \cdot 2 + 7n \cdot 2 = 12m + 14n$ **37.** $3a + 7 + 5a = 8a + 7$

38. $6n + 1 + 3n = 9n + 1$ **39.** $8n - 5 + 4n = 12n - 5$ **40.** $3p + 8 - 3p = 8$

41. $8y - 7y + 4 = y + 4$ **42.** $y + 9z - 6z = y + 3z$

43. $2a + b + 5a + 3b = 7a + 4b$ **44.** $4c + d + 3c + 2d = 7c + 3d$

45. $3s + 5t - t - 2s = s + 4t$ **46.** $5d - 3e + 4e - d = 4d + e$

47. $7x + 5y - x - 5y = 6x$ **48.** $6g - 3f + 8f - g = 5f + 5g$

B

49. $4(x + 3) + 3 = 4 \cdot x + 4 \cdot 3 + 3 = 4x + 12 + 3 = 4x + 15$

50. $3(n - 1) + 4 = 3 \cdot n - 3 \cdot 1 + 4 = 3n - 3 + 4 = 3n + 1$

51. $3(a - 7) - 4 = 3 \cdot a - 3 \cdot 7 - 4 = 3a - 21 - 4 = 3a - 25$

52. $2(n - 8) + 9 = 2 \cdot n - 2 \cdot 8 + 9 = 2n - 16 + 9 = 2n - 7$

53. $7 + 2(5 - x) = 7 + 2 \cdot 5 - 2 \cdot x = 7 + 10 - 2x = 17 - 2x = -2x + 17$

54. $8 + 3(4 - y) = 8 + 3 \cdot 4 - 3 \cdot y = 8 + 12 - 3y = 20 - 3y = -3y + 20$

55. $2 + (x + 3)5 = 2 + x \cdot 5 + 3 \cdot 5 = 2 + 5x + 15 = 5x + 17$

56. $(y - 5)6 + 15 = y \cdot 6 - 5 \cdot 6 + 15 = 6y - 30 + 15 = 6y - 15$

57. $5(y + 3) + 7y = 5 \cdot y + 5 \cdot 3 + 7y = 5y + 15 + 7y = 12y + 15$

58. $9(n - 3) + 4n = 9 \cdot n - 9 \cdot 3 + 4n = 9n - 27 + 4n = 13n - 27$

59. $3(2x + y + 1) = 3 \cdot 2x + 3 \cdot y + 3 \cdot 1 = 6x + 3y + 3$

60. $5(7y - 3z + 4) = 5 \cdot 7y - 5 \cdot 3z + 5 \cdot 4 = 35y - 15z + 20$

61. $9(a + b) + 4(3a + 2b) = 9 \cdot a + 9 \cdot b + 4 \cdot 3a + 4 \cdot 2b = 9a + 9b + 12a + 8b =$
$21a + 17b$

62. $8(k + m) + 15(2k + 5m) = 8 \cdot k + 8 \cdot m + 15 \cdot 2k + 15 \cdot 5m =$
$8k + 8m + 30k + 75m = 38k + 83m$

63. $6(r + 5) + 9(r - 2) - 4r = 6 \cdot r + 6 \cdot 5 + 9 \cdot r - 9 \cdot 2 - 4r =$
$6r + 30 + 9r - 18 - 4r = 11r + 12$

64. $4(n + 7) + 5(n - 3) - 2n = 4 \cdot n + 4 \cdot 7 + 5 \cdot n - 5 \cdot 3 - 2n =$
$4n + 28 + 5n - 15 - 2n = 7n + 13$

65. $7(c + 2d + 8) + 3(9c - 2) = 7 \cdot c + 7 \cdot 2d + 7 \cdot 8 + 3 \cdot 9c - 3 \cdot 2 =$
$7c + 14d + 56 + 27c - 6 = 34c + 14d + 50$

66. $4(5x + 3y + 6) + 14(2y - 1) = 4 \cdot 5x + 4 \cdot 3y + 4 \cdot 6 + 14 \cdot 2y - 14 \cdot 1 =$
$20x + 12y + 24 + 28y - 14 = 20x + 40y + 10$

67. $5(c + d) + 2(3c + d) = 5 \cdot c + 5 \cdot d + 2 \cdot 3c + 2 \cdot d = 5c + 5d + 6c + 2d =$
$11c + 7d$

68. $2(11 + x) + 3(x - 7) = 2 \cdot 11 + 2 \cdot x + 3 \cdot x - 3 \cdot 7 = 22 + 2x + 3x - 21 =$
$5x + 1$

69. $[8 + (-5 + 15y)] + \frac{1}{2}(12y - 8) = [8 + (-5)] + 15y + \left(\frac{1}{2} \cdot 12y\right) - \left(\frac{1}{2} \cdot 8\right) =$
$3 + 15y + 6y - 4 = 21y - 1$

70. $[6 + 3(a + b)] + (b - 5) = 6 + 3 \cdot a + 3 \cdot b + b - 5 = 6 + 3a + 3b + b - 5 =$
$3a + 4b + 1$

C **71.** *Step 1* The unknown is a number.

Step 2 Let n = the number.

Step 3 $(n + 17)4 = n + 77$; $4n + 68 = n + 77$

Step 4

n	$4n + 68 = n + 77$	
2	$4 \cdot 2 + 68 = 2 + 77$	False
3	$4 \cdot 3 + 68 = 3 + 77$	True
4	$4 \cdot 4 + 68 = 4 + 77$	False

$\therefore$ the solution set is $\{3\}$.

Step 5 $(3 + 17)4 \stackrel{?}{=} 3 + 77$, $20 \cdot 4 \stackrel{?}{=} 80$, $80 = 80$ $\checkmark$; $\therefore n = 3$

72. *Step 1* The unknown is a number. *Step 2* Let k = the number.

Step 3 $3(2k - 5) = 4k + 1, 6k - 15 = 4k + 1$

Step 4

k	$6k - 15 = 4k + 1$	
8	$6 \cdot 8 - 15 = 4 \cdot 8 + 1$	True
9	$6 \cdot 9 - 15 = 4 \cdot 9 + 1$	False
10	$6 \cdot 10 - 15 = 4 \cdot 10 + 1$	False

Step 5 $3(16 - 5) \stackrel{?}{=} 32 + 1, 3 \cdot 11 \stackrel{?}{=} 33, 33 = 33 \checkmark \therefore n = 8$

73. $8[5x + 7(3 + 4x)] - 16x - 8 = 8(5x + 21 + 28x) - 16x - 8 = 40x + 168 + 224x - 16x - 8 = 248x + 160$

74. $-24 + 3[4y + 2(5y - 8)] - 9y = -24 + 3(4y + 10y - 16) - 9y = -24 + 12y + 30y - 48 - 9y = 33y - 72$

75. $12(3n + 2p) + 11[n + 3(2n - p - 3)] = 36n + 24p + 11(n + 6n - 3p - 9) = 36n + 24p + 11n + 66n - 33p - 99 = 113n - 9p - 99$

76. $9[7(3a + 2b - 4) + 12(a - 2)] + 3(5a - 8b) = 9(21a + 14b - 28 + 12a - 24) + 15a - 24b = 189a + 126b - 252 + 108a - 216 + 15a - 24b = 312a + 102b - 468$

77. $3[4y - 2(x + 3) + (2 - y)] = 3(4y - 2x - 6 + 2 - y) = 12y - 6x - 18 + 6 - 3y = -6x + 9y - 12$

78. $9[4(2n + r) - 11r] + 3(r - n) = 9(8n + 4r - 11r) + 3r - 3n = 72n + 36r - 99r + 3r - 3n = 69n - 60r$

Page 69 • MIXED REVIEW EXERCISES

1. $5x + 2y - c = 5 \cdot 4 + 2 \cdot 5 - 2 = 20 + 10 - 2 = 28$

2. $(c \cdot c + x) \div 2 = (2 \cdot 2 + 4) \div 2 = (4 + 4) \div 2 = 8 \div 2 = 4$

3. $2y - (4x \div c) = 2 \cdot 5 - (4 \cdot 4 \div 2) = 10 - (16 \div 2) = 10 - 8 = 2$

4. $|a| + |b| + (-c) = |-3| + |-2| + (-2) = 3 + 2 + (-2) = 3$

5. $c + |-y| + |b| = 2 + |-5| + |-2| = 2 + 5 + 2 = 9$

6. $3|a| - |b| = 3|-3| - |-2| = 3 \cdot 3 - 2 = 9 - 2 = 7$

7. $-(-b + x) + y = -[-(-2) + 4] + 5 = -(2 + 4) + 5 = -6 + 5 = -1$

8. $b + y + (-2) = -2 + 5 + (-2) = 1$ **9.** $a + x + (-y) = -3 + 4 + (-5) = -4$

10. $x - (a - b) = 4 - [-3 - (-2)] = 4 + 3 - 2 = 5$

11. $-a - b + y = -(-3) - (-2) + 5 = 3 + 2 + 5 = 10$

12. $|b + a| - y = |-2 + (-3)| - 5 = |-5| - 5 = 5 - 5 = 0$

Page 69 • CHALLENGE

1–3.

2. a.				1	4	2	8	5	7		
b.						2	8	5	7	1	4
c.					4	2	8	5	7	1	
d.		5	7	1	4	2	8				
3. a.			7	1	4	2	8	5			
b.	8	5	7	1	4	2					

Pages 72–73 • WRITTEN EXERCISES

A **1.** $(-37)(-2) = 74$ **2.** $23(-5) = -115$ **3.** $(-4)(10)(-12) = (-40)(-12) = 480$
4. $(-6)(-9)(20) = 54(20) = 1080$ **5.** $(-3)(-7)(-4) = 21(-4) = -84$
6. $(-2)(-8)(-4) = 16(-4) = -64$ **7.** $(-17)(-18)(0) = 0$ **8.** $54(-47)(0) = 0$
9. $5(-3)(-10)(-2) = (-15)20 = -300$ **10.** $(-4)(25)(-2)(-3) = (-100)6 = -600$
11. $(-6)(-1)(-7)(-10) = 6 \cdot 70 = 420$ **12.** $(-9)(-5)(-1)(-3) = 45 \cdot 3 = 135$
13. $(-2a)(-3b) = 6ab$ **14.** $(-5x)(-3y) = 15xy$ **15.** $2p(-7q) = -14pq$
16. $(-4e)(5t) = -20et$ **17.** $(-7a)(-5c)4 = 35ac \cdot 4 = 140ac$
18. $(-x)(-3y)(-z) = 3xy(-z) = -3xyz$ **19.** $-2(x - 3y) = -2x + 6y$
20. $-5(c + 2d) = -5c - 10d$ **21.** $-7(3m + 4n) = -21m - 28n$
22. $-9(-5y - 8) = 45y + 72$ **23.** $(5x - 3)(-7) = -35x + 21$
24. $(-4 + 7y)(-2) = 8 - 14y = -14y + 8$ **25.** $(-1)(a + b - 3) = -a - b + 3$
26. $(-1)(2n - y - 5) = -2n + y + 5$ **27.** $(-1)(a - b - c) = -a + b + c$
28. $(-1)(-x - y + 7) = x + y - 7$ **29.** $-[(-2)(a + b)] = -(-2a - 2b) = 2a + 2b$
30. $-[(-3)(x - y)] = -(-3x + 3y) = 3x - 3y$ **31.** $5x - 7x + 8 + 2x = 8$
32. $3y - 7 - 8y + 5 = -5y - 2$ **33.** $14p - 7c - 9c + 11p = -16c + 25p$
34. $-y - 3z + 4y - 9z - y = 2y - 12z$

B **35.** $3.4y + 1.6s - (-1.9y) - 3.6s = 1.6s - 3.6s + 3.4y + 1.9y = -2s + 5.3y$
36. $-0.8c + 4.1h - (-3.2c) - 0.1h = -0.8c + 3.2c + 4.1h - 0.1h = 2.4c + 4h$
37. $-8x - 7x - 5x = -20x$ **38.** $-34n + 18n - 4n = -20n$
39. $(-8)(-19) - 7(-19) - 5(-19) = (-8 - 7 - 5)(-19) = (-20)(-19) = 380$
40. $-34 \cdot 9 + 18 \cdot 9 - 4 \cdot 9 = (-34 + 18 - 4)9 = (-20)9 = -180$
41. $88(-57) + 13(-57) + (-1)(-57) = (88 + 13 - 1)(-57) = (100)(-57) = -5700$
42. $-63 \cdot 81 + 56 \cdot 81 - 13 \cdot 81 = (-63 + 56 - 13)(81) = (-20)(81) = -1620$
43. $-m + \frac{2}{3}m - \frac{1}{2}n + \frac{5}{2}n = -\frac{1}{3}m + \frac{4}{2}n = -\frac{1}{3}m + 2n$
44. $-5a - \left(-\frac{1}{2}b\right) + 3\frac{1}{2}a - \frac{1}{2}b = -5a + 3\frac{1}{2}a + \frac{1}{2}b - \frac{1}{2}b = -1\frac{1}{2}a$
45. $7 - 3(r + s) = 7 - 3r - 3s = -3r - 3s + 7$
46. $2 + 6(m - n) = 2 + 6m - 6n = 6m - 6n + 2$
47. $2(x + 5y) + (-3)(7x - y) = 2x + 10y - 21x + 3y = -19x + 13y$
48. $8(t - u) + 5(2t - 3u) = 8t - 8u + 10t - 15u = 18t - 23u$
49. $-2(2q + w) - 7(w - q) = -4q - 2w - 7w + 7q = 3q - 9w$
50. $-3(7c + d) - 2(10d - c) = -21c - 3d - 20d + 2c = -19c - 23d$
51. $-4(-e + 3f) - 3[e + (-5f)] = 4e - 12f - 3e + 15f = e + 3f$
52. $-6[v + (-9w)] + (-5)(3v - w) = -6v + 54w - 15v + 5w = -21v + 59w$

C **53.** $2[-7(r + 2s) - r] - 3(s + 2r) = 2(-7r - 14s - r) - 3s - 6r =$
$-14r - 28s - 2r - 3s - 6r = -22r - 31s$

54. $4[2(-5x + y) - y] - 10(y - 4x) = 4(-10x + 2y - y) - 10y + 40x =$
$-40x + 8y - 4y - 10y + 40x = -6y$

55. $-15 + (-3)[2(g - 7) - 2(1 - g)] = -15 + (-3)(2g - 14 - 2 + 2g) =$
$-15 - 6g + 42 + 6 - 6g = -12g + 33$

56. $-50 + (-2)[3(1 - f) - 3(-2 + f)] = -50 + (-2)(3 - 3f + 6 - 3f) =$
$-50 - 6 + 6f - 12 + 6f = 12f - 68$

57. $500[p - (2p + 1)] = 500(p - 2p - 1) = 500p - 1000p - 500 = -500p - 500 =$
$-(500p + 500)$; the total value of the stock had gone down $(500p + 500)$ points.

58. Cost $= 12 \cdot 12 \cdot 11 = 1584$; sales $= x(11 + 15) + (144 - x)(11 - 4)$;
income $= 26x + 7(144 - x) - 1584 = 26x + 1008 - 7x - 1584 = 19x - 576$; the store's total income from the sale was $(19x - 576)$ dollars.

59. **a.** Identity property of multiplication
b. Distributive property (of mult. with respect to add.) **c.** Property of opposites
d. Multiplication property of 0

60. **a.** Definition of subtraction **b.** Distributive property (of mult. with respect to add.)
c. Property of opposites in products **d.** Definition of subtraction

Page 73 • MIXED REVIEW EXERCISES

1. $4x = 44$ **2.** $n + 2 = 32$ **3.** $\frac{1}{2}x = 12$ **4.** $2x + 7 = 11$

5. $120 - (14 - 6) = 120 - 8 = 112$ **6.** $191 - (9 - 12) = 191 - (-3) = 191 + 3 = 194$

7. $3 + (-2) + (-y) + 11 = 1 + (-y) + 11 = -y + 12$

8. $4(30 + 2) = (4 \cdot 30) + (4 \cdot 2) = 120 + 8 = 128$ **9.** $3n + (-7n) = [3 + (-7)]n = -4n$

10. $4(n + 2) + 8 = 4n + 4 \cdot 2 + 8 = 4n + 8 + 8 = 4n + 16$

Page 74 • APPLICATION

1. 1 L = 1000 mL, so 480 mL $= \frac{480}{1000}$ L = 0.48 L; \$2.40 ÷ 0.48 = \$5.00 per liter

2. 315 mL = 0.315 L; \$0.63 ÷ .315 = \$2.00 per liter

3. 600 mL = 0.6 L; \$1.98 ÷ 0.6 = \$3.30 per liter

4. 1440 mL = 1.44 L; \$1.44 ÷ 1.44 = \$1.00 per liter

Pages 77–78 • WRITTEN EXERCISES

A **1.** Let x = the first integer; then $x + 1$ = the second integer; $x + (x + 1) = 43$

2. Let x = first integer; then $x + 1$ and $x + 2$ are the second and third integers;
$x + (x + 1) + (x + 2) = 69$

3. Let x = first integer; then $x + 1$, $x + 2$, and $x + 3$ are the second, third, and fourth integers; $x + (x + 1) + (x + 2) + (x + 3) = -106$

4. Let x = first integer; then $x + 1$, $x + 2$, and $x + 3$ are the second, third, and fourth integers; $x + (x + 1) + (x + 2) + (x + 3) = -42$

5. Let x = first odd integer; then $x + 2$ and $x + 4$ are the second and third odd integers; $x + (x + 2) + (x + 4) = 75$

6. Let x = the first odd integer; then $x + 2$ and $x + 4$ are the second and third odd integers; $x + (x + 2) + (x + 4) = 147$

7. Let x = first even integer; then $x + 2$ = second even integer; $x(x + 2) = 168$

8. Let x = first odd integer; then $x + 2$ = second odd integer; $x(x + 2) = 195$

9. Let x = first even integer; then $x + 2$, $x + 4$, and $x + 6$ are the second, third, and fourth even integers; $x + (x + 2) + (x + 4) + (x + 6) = -100$

10. Let x = smaller even integer; then $x + 2$ is the greater even integer; $x + 2 = 2x - 6$

11. Let x = age of youngest child; then $x + 2$, $x + 4$, and $x + 6$ are the ages of the other three cousins; $x + (x + 2) + (x + 4) + (x + 6) = 36$

12. Let x = smaller even integer; then $x + 2$ = greater even integer; $x = \frac{1}{2}(x + 2) + 5$

B **13.** Let x = first odd integer; then $x + 2$ and $x + 4$ are the second and third odd integers; $x + (x + 2) + (x + 4) = x + 40$, $3x + 6 = x + 40$; $x = 17$; the three odd integers are 17, 19, and 21.

14. Let x = first even integer; then $x + 2$ and $x + 4$ are the second and third even integers; $x + (x + 2) + (x + 4) = (x + 4) + 30$, $3x + 6 = x + 34$; $x = 14$; the three even integers are 14, 16, and 18.

15. Let x = first integer; then $x + 1$ = second integer; $x(x + 1) = 5[x + (x + 1)] - 5$; $5[x + (x + 1)] - 5 = 5(2x + 1) - 5 = 10x + 5 - 5 = 10x$; $x(x + 1) = 10x$; $x = 0$ or $x = 9$; either the two integers are 0 and 1 or they are 9 and 10.

16. Let x = first odd integer; then $x + 2$ = second odd integer; $x(x + 2) = 6[x + (x + 2)] - 1$; $x(x + 2) = 6(2x + 2) - 1 = 12x + 12 - 1$; $x(x + 2) = 12x + 11$; $x = -1$ or $x = 11$; either the two integers are -1 and 1 or they are 11 and 13.

C **17.** Let x = the length of the shortest rope. Then $x + 3$ and $x + 6$ are the other two lengths of rope. If each length of rope were 4 inches shorter, their lengths would be $x - 4$, $(x + 3) - 4$, and $(x + 6) - 4$; $(x - 4) + [(x + 3) - 4] + [(x + 6) - 4] = 42$.

18. Let x = Joe's weight in kg; then Jim's weight is $x + 7$ and Jack's weight is $x + 14$. If they each were 5 kg lighter, their weights would be $x - 5$, $(x + 7) - 5$, and $(x + 14) - 5$; $(x - 5) + [(x + 7) - 5] + [(x + 14) - 5] = 195$.

Page 78 • MIXED REVIEW EXERCISES

1. $(40 - 7) - (55 - 20) = 33 - 35 = 33 + (-35) = -2$

2. $-5.3 + 2.1 - 1.7 = -3.2 + (-1.7) = -4.9$

3. $-4 + 2c + (-1) + 8 = 2c + (-5) + 8 = 2c + 3$ **4.** $\frac{6}{5} + \left(-\frac{9}{5}\right) = -\frac{3}{5}$

5. $3\frac{1}{3} + 12 + 2\frac{2}{3} = 12 + 6 = 18$

6. $6\left(\frac{2}{5}\right) - 5\left(\frac{2}{5}\right) + 3\left(\frac{1}{5}\right) = (6 - 5)\left(\frac{2}{5}\right) + \frac{3}{5} = 1\left(\frac{2}{5}\right) + \frac{3}{5} = \frac{2}{5} + \frac{3}{5} = \frac{5}{5} = 1$

7. $-(11 - x) - (x - 13) = -11 + x - x + 13 = 2 + 0 = 2$

8. $15a - 3a + 7a = 12a + 7a = 19a$ **9.** $12 + 5y + 6 + (-2) = 5y + 16$

10. $(-1)(x + y - z) = -x - y + z$ **11.** $-[-6(a - b)] = -(-6a + 6b) = 6a - 6b$

12. $3(-5 + y) = -15 + 3y = 3y - 15$

Pages 81–82 • WRITTEN EXERCISES

A **1.** $\frac{1}{5}(-20) = -4$ **2.** $-\frac{1}{12}(48) = -4$ **3.** $-1000\left(\frac{1}{100}\right) = -10$

4. $(-70)\left(\frac{1}{7}\right) = -10$ **5.** $96\left(-\frac{1}{8}\right)\left(-\frac{1}{12}\right) = 96\left(\frac{1}{96}\right) = 1$

6. $-63\left(-\frac{1}{3}\right)\left(-\frac{1}{21}\right) = -63\left(\frac{1}{63}\right) = -1$ **7.** $\frac{1}{-3}(36)\left(\frac{1}{4}\right) = -\frac{1}{12}(36) = -3$

8. $-150\left(\frac{1}{2}\right)\left(-\frac{1}{3}\right) = -150\left(-\frac{1}{6}\right) = 25$ **9.** $4r\left(-\frac{1}{4}\right) = 4\left(-\frac{1}{4}\right)(r) = -1 \cdot r = -r$

10. $33p\left(-\frac{1}{11}\right) = 33\left(-\frac{1}{11}\right)p = -3p$ **11.** $\frac{1}{x}(5x) = \frac{1}{x}(x)(5) = 1 \cdot 5 = 5$

12. $7t\left(\frac{1}{t}\right) = 7 \cdot t\left(\frac{1}{t}\right) = 7 \cdot 1 = 7$ **13.** $12xy\left(\frac{1}{3}\right) = 12\left(\frac{1}{3}\right)(xy) = 4xy$

14. $15mn\left(\frac{1}{5}\right) = 15\left(\frac{1}{5}\right)(mn) = 3mn$ **15.** $8ab\left(-\frac{1}{4}\right) = 8\left(-\frac{1}{4}\right)(ab) = -2ab$

16. $9cd\left(\frac{1}{-3}\right) = 9\left(-\frac{1}{3}\right)(cd) = -3cd$ **17.** $(-4pg)\left(\frac{1}{-2}\right) = -4\left(-\frac{1}{2}\right)(pg) = 2pg$

18. $(-36ac)\left(\frac{1}{-9}\right) = -36\left(-\frac{1}{9}\right)(ac) = 4ac$

B **19.** $\frac{1}{2}(-16a + 20) = \frac{1}{2}(-16a) + \frac{1}{2}(20) = \frac{1}{2}(-16)a + 10 = -8a + 10$

20. $\frac{1}{3}(18b - 39) = \frac{1}{3}(18b) - \frac{1}{3}(39) = \frac{1}{3}(18)b - 13 = 6b - 13$

21. $-\frac{1}{5}(-45c + 10d) = -\frac{1}{5}(-45c) + \left(-\frac{1}{5}\right)(10d) = -\frac{1}{5}(-45)c + \left(-\frac{1}{5}\right)(10)d = 9c - 2d$

22. $-\frac{1}{8}(56g - 72h) = -\frac{1}{8}(56g) - \left(-\frac{1}{8}\right)(72h) = -\frac{1}{8}(56)g - \left(-\frac{1}{8}\right)(72)h =$
$-7g - (-9h) = -7g + 9h$

23. $(-42m - 91k)\left(-\frac{1}{7}\right) = -42m\left(-\frac{1}{7}\right) - (91k)\left(-\frac{1}{7}\right) = -42\left(-\frac{1}{7}\right)m - (91)\left(-\frac{1}{7}\right)k =$
$6m - (-13k) = 6m + 13k$

24. $(-39n - 52p)\left(-\frac{1}{13}\right) = -39n\left(-\frac{1}{13}\right) - (52p)\left(-\frac{1}{13}\right) = -39\left(-\frac{1}{13}\right)n - (52)\left(-\frac{1}{13}\right)p =$
$3n - (-4p) = 3n + 4p$

25. $\frac{1}{2}(8u + 10v) - \frac{1}{3}(15u - 3v) = \frac{1}{2}(8u) + \frac{1}{2}(10v) - \left[\frac{1}{3}(15u) - \frac{1}{3}(3v)\right] =$
$\frac{1}{2}(8)u + \frac{1}{2}(10)v - \left[\frac{1}{3}(15)u - \frac{1}{3}(3)v\right] = 4u + 5v - (5u - v) = 4u + 5v - 5u + v =$
$-u + 6v$

26. $\frac{1}{5}(-5a + 20b) - \frac{1}{2}(2b - 6a) = \frac{1}{5}(-5a) + \frac{1}{5}(20b) - \left[\frac{1}{2}(2b) - \frac{1}{2}(6a)\right] =$
$\frac{1}{5}(-5)a + \frac{1}{5}(20)b - \left[\frac{1}{2}(2)b - \frac{1}{2}(6)a\right] = -a + 4b - (b - 3a) =$
$-a + 4b - b + 3a = 2a + 3b$

27. $6\left(\frac{1}{3}x - \frac{1}{2}y\right) + 42\left(-\frac{1}{3}y - \frac{1}{7}x\right) = 6\left(\frac{1}{3}x\right) - 6\left(\frac{1}{2}y\right) + 42\left(-\frac{1}{3}y\right) - 42\left(\frac{1}{7}x\right) =$

$6\left(\frac{1}{3}\right)x - 6\left(\frac{1}{2}\right)y + 42\left(-\frac{1}{3}\right)y - 42\left(\frac{1}{7}\right)x = 2x - 3y - 14y - 6x = -4x - 17y$

28. $-8\left(-\frac{1}{8}p + q\right) + \frac{1}{9}(63p - 9q) = -8\left(-\frac{1}{8}p\right) + (-8)(q) + \frac{1}{9}(63p) - \frac{1}{9}(9q) =$

$-8\left(-\frac{1}{8}\right)p - 8q + \frac{1}{9}(63)p - \frac{1}{9}(9)q = p - 8q + 7p - q = 8p - 9q$

29. $-\frac{1}{8}(48m - 16) - \frac{1}{4}(84m + 8) = -\frac{1}{8}(48m) - \left(-\frac{1}{8}\right)(16) - \left[\frac{1}{4}(84m) + \frac{1}{4}(8)\right] =$

$-\frac{1}{8}(48)m - (-2) - \left[\frac{1}{4}(84)m + 2\right] = -6m + 2 - (21m + 2) =$

$-6m + 2 - 21m - 2 = -27m$

30. $-5\left(4 - \frac{1}{2}n\right) + \frac{1}{16}(-32n + 8) = -5(4) - (-5)\left(\frac{1}{2}n\right) + \frac{1}{16}(-32n) + \frac{1}{16}(8) =$

$-20 - (-5)\left(\frac{1}{2}\right)n + \frac{1}{16}(-32)n + \frac{1}{2} = -20 + \frac{5}{2}n - 2n + \frac{1}{2} = \frac{1}{2}n - \frac{39}{2}$

31. $-\frac{1}{12}(6r + 4s) + 7\left(\frac{1}{21}s - \frac{1}{14}r\right) = -\frac{1}{12}(6r) + \left(-\frac{1}{12}\right)(4s) + 7\left(\frac{1}{21}s\right) - 7\left(\frac{1}{14}r\right) =$

$-\frac{1}{12}(6)r + \left(-\frac{1}{12}\right)(4)s + 7\left(\frac{1}{21}\right)s - 7\left(\frac{1}{14}\right)r = -\frac{1}{2}r - \frac{1}{3}s + \frac{1}{3}s - \frac{1}{2}r = -1r = -r$

32. $-\frac{1}{20}(5z - 4w) - 6\left(-\frac{1}{30}w - \frac{1}{24}z\right) = -\frac{1}{20}(5z) - \left(-\frac{1}{20}\cdot 4w\right) -$

$\left[6\left(-\frac{1}{30}w\right) + 6\left(-\frac{1}{24}z\right)\right] = -\frac{1}{4}z + \frac{1}{5}w - \left(-\frac{1}{5}w - \frac{1}{4}z\right) =$

$-\frac{1}{4}z + \frac{1}{5}w + \frac{1}{5}w + \frac{1}{4}z = \frac{2}{5}w$

33. $-3\left[\frac{1}{4}(12n + 1) - \frac{1}{4}\right] + 10n = -3\left[\frac{1}{4}(12n) + \frac{1}{4}(1) - \frac{1}{4}\right] + 10n =$

$-3\left(3n + \frac{1}{4} - \frac{1}{4}\right) + 10n = -3(3n) + 10n = -9n + 10n = 1n = n$

34. $3s + \left(-\frac{1}{2}\right)\left[6 + 24\left(-\frac{1}{3} + \frac{1}{4}s\right)\right] = 3s + \left(-\frac{1}{2}\right)\left[6 + 24\left(-\frac{1}{3}\right) + 24\left(\frac{1}{4}s\right)\right] =$

$3s + \left(-\frac{1}{2}\right)(6 - 8 + 6s) = 3s + \left(-\frac{1}{2}\right)(-2 + 6s) = 3s + \left(-\frac{1}{2}\right)(-2) + \left(-\frac{1}{2}\right)(6s) =$

$3s + 1 - 3s = 1$

C **35.** *Step 1* The unknown is a number.

Step 2 Let x = the number; then its reciprocal is $\frac{1}{x}$.

Step 3 $x + \frac{1}{x} = \frac{13}{6}$

(Continued)

Step 4

x	$x + \frac{1}{x} = \frac{13}{6}$	
$\frac{2}{3}$	$\frac{2}{3} + \frac{3}{2} = \frac{13}{6}; \frac{13}{6} = \frac{13}{6}$	True
$\frac{3}{4}$	$\frac{3}{4} + \frac{4}{3} = \frac{13}{6}; \frac{25}{12} = \frac{13}{6}$	False
$\frac{5}{4}$	$\frac{5}{4} + \frac{4}{5} = \frac{13}{6}; \frac{41}{20} = \frac{13}{6}$	False
$\frac{3}{2}$	$\frac{3}{2} + \frac{2}{3} = \frac{13}{6}; \frac{13}{6} = \frac{13}{6}$	True

$x = \frac{2}{3}$ or $x = \frac{3}{2}$

Step 5 $\frac{2}{3} + \frac{3}{2} = \frac{13}{6}, \frac{3}{2} + \frac{2}{3} = \frac{13}{6}; \therefore x = \frac{2}{3}$ or $x = \frac{3}{2}$

36. *Step 1* The unknown is a number.

Step 2 Let x = the number; then its reciprocal is $\frac{1}{x}$.

Step 3 $x = \frac{1}{x} + 2\frac{1}{10}$

Step 4

x	$x = \frac{1}{x} + 2\frac{1}{10}$	
$\frac{2}{5}$	$\frac{2}{5} = \frac{5}{2} + 2\frac{1}{10}; \frac{2}{5} = \frac{23}{5}$	False
$\frac{3}{5}$	$\frac{3}{5} = \frac{5}{3} + 2\frac{1}{10}; \frac{3}{5} = \frac{113}{30}$	False
$\frac{5}{4}$	$\frac{5}{4} = \frac{4}{5} + 2\frac{1}{10}; \frac{5}{4} = \frac{29}{10}$	False
$\frac{5}{2}$	$\frac{5}{2} = \frac{2}{5} + 2\frac{1}{10}; \frac{5}{2} = \frac{5}{2}$	True

Step 5 $\frac{5}{2} = \frac{2}{5} + 2\frac{1}{10} = \frac{4}{10} + \frac{21}{10} = \frac{25}{10} = \frac{5}{2}; \therefore x = \frac{5}{2}$

Page 82 • MIXED REVIEW EXERCISES

1. $5x + 4 = 24$ **2.** $x - 15 = 250$ **3.** $x + (x + 1) = 67$ **4.** $x(x + 1) = 42$

5. $(-12)(-5)(-2) = -12(10) = -120$ **6.** $-36(25)(-2) = -36(-50) = 1800$

7. $3(-7)(-2) = 3 \cdot 14 = 42$ **8.** $-7(5a - 2d) = -7(5a) - (-7)(2d) = -35a + 14d$

9. $-4(2 + x) - 3(x - 2) = -4 \cdot 2 + (-4)(x) - [3 \cdot x - 3(2)] = -8 - 4x - (3x - 6) =$ $-8 - 4x - 3x + 6 = -7x - 2$

10. $8(x - 1) + 5(2 - x) = 8 \cdot x - 8 \cdot 1 + 5 \cdot 2 - 5 \cdot x = 8x - 8 + 10 - 5x = 3x + 2$

Page 82 • CALCULATOR KEY-IN

1. 16 **2.** −0.03125 **3.** 0.00032 **4.** 15,625 **5.** the number

6. a. Answers may vary in the last decimal places. Answers given here were obtained using a calculator that rounds to obtain the final digit and that has an eight-digit display.

a	b	$\frac{1}{ab}$	$\frac{1}{a}\cdot\frac{1}{b}$
4	16	0.015625	0.015625
−32	−0.5	0.0625	0.0625
0.234	0.654	6.5344102	6.5344102
555	222	0.00000812	0.00000812

b. $\frac{1}{ab} = \frac{1}{a}\cdot\frac{1}{b}$

Pages 85–86 • WRITTEN EXERCISES

A **1.** $-48 \div 6 = -48 \cdot \frac{1}{6} = -8$ **2.** $-64 \div 4 = -64 \cdot \frac{1}{4} = -16$

3. $-12 \div (-3) = -12\left(-\frac{1}{3}\right) = 4$ **4.** $-24 \div (-8) = -24\left(-\frac{1}{8}\right) = 3$

5. $8 \div \left(-\frac{1}{2}\right) = 8(-2) = -16$ **6.** $12 \div \left(-\frac{1}{3}\right) = 12(-3) = -36$

7. $0 \div \frac{3}{4} = 0 \cdot \frac{4}{3} = 0$ **8.** $-6 \div \left(-\frac{1}{4}\right) = -6(-4) = 24$ **9.** $\frac{-6}{-\frac{1}{6}} = -6(-6) = 36$

10. $\frac{-8}{-\frac{1}{3}} = -8(-3) = 24$ **11.** $\frac{-7}{\frac{1}{5}} = -7 \cdot 5 = -35$ **12.** $\frac{9}{-\frac{1}{9}} = 9(-9) = -81$

13. $3 \cdot \frac{x}{3} = 3 \cdot x \cdot \frac{1}{3} = 3 \cdot \frac{1}{3} \cdot x = 1 \cdot x = x$ **14.** $\frac{-w}{4} \cdot 4 = -w \cdot \frac{1}{4} \cdot 4 = -w \cdot 1 = -w$

15. $(-4)\left(\frac{-y}{2}\right) = -4(-y)\left(\frac{1}{2}\right) = -4 \cdot \frac{1}{2}(-y) = -2(-y) = 2y$

16. $(-8)\left(\frac{x}{-2}\right) = (-8)(x)\left(-\frac{1}{2}\right) = (-8)\left(-\frac{1}{2}\right)(x) = 4x$ **17.** $\frac{156a}{12} = 156 \cdot \frac{1}{12} \cdot a = 13a$

18. $\frac{384b}{-8} = 384\left(-\frac{1}{8}\right)(b) = -48b$ **19.** $\frac{-54x}{6} = -54x\left(\frac{1}{6}\right) = -54\left(\frac{1}{6}\right)x = -9x$

20. $\frac{-144y}{-24} = -144y\left(-\frac{1}{24}\right) = -144\left(-\frac{1}{24}\right)y = 6y$

21. $\frac{-12 + 4 + (-11) + (-7)}{4} = \frac{4 - 30}{4} = \frac{-26}{4} = -\frac{13}{2}$

22. $\frac{18 + (-21) + (-7) + 2}{4} = \frac{20 - 28}{4} = \frac{-8}{4} = -2$

23. $\frac{18 + (-17) + (-22) + 16 + 0}{5} = \frac{34 - 39}{5} = \frac{-5}{5} = -1$

24. $\frac{18 + (-17) + (-22) + 16}{4} = \frac{34 - 39}{4} = \frac{-5}{4}$, or $-\frac{5}{4}$

B **25.** $\frac{b + 5c}{3 - d} = \frac{-1 + 5 \cdot 2}{3 - 6} = \frac{-1 + 10}{-3} = \frac{9}{-3} = -3$

26. $\frac{bcd}{(1 - b)a} = \frac{(-1)(2)(6)}{[1 - (-1)](-3)} = \frac{-2 \cdot 6}{2(-3)} = \frac{-12}{-6} = 2$

27. $\frac{abc}{(1 - c)d} = \frac{(-3)(-1)(2)}{(1 - 2)6} = \frac{3 \cdot 2}{-1 \cdot 6} = \frac{6}{-6} = -1$

28. $\frac{7a + d}{a + c} = \frac{7(-3) + 6}{-3 + 2} = \frac{-21 + 6}{-1} = \frac{-15}{-1} = 15$

29. $\frac{a - 2d}{b + 3c} = \frac{-3 - 2 \cdot 6}{-1 + 3 \cdot 2} = \frac{-3 - 12}{-1 + 6} = \frac{-15}{5} = -3$

30. $\frac{4a - d}{abc} = \frac{4(-3) - 6}{(-3)(-1)(2)} = \frac{-12 - 6}{3 \cdot 2} = \frac{-18}{6} = -3$

31. $\frac{a - 3b}{bcd} = \frac{-3 - 3(-1)}{(-1)(2)(6)} = \frac{-3 + 3}{-2 \cdot 6} = \frac{0}{-12} = 0$

32. $\frac{abc - d}{abc} = \frac{(-3)(-1)(2) - 6}{(-3)(-1)(2)} = \frac{3 \cdot 2 - 6}{3 \cdot 2} = \frac{6 - 6}{6} = \frac{0}{6} = 0$

33. $\frac{(c + a)(c - a)}{b + d} = \frac{[2 + (-3)][2 - (-3)]}{-1 + 6} = \frac{(-1)(5)}{5} = \frac{-5}{5} = -1$

C **34. a.** Definition of division **b.** Distributive property (of mult. with respect to add.) **c.** Definition of division

35. $\frac{a - b}{c} = (a - b) \cdot \frac{1}{c} = \left(a \cdot \frac{1}{c}\right) - \left(b \cdot \frac{1}{c}\right) = \frac{a}{c} - \frac{b}{c}$

Page 86 • MIXED REVIEW EXERCISES

1.

x	$x + 3 = 8$	
0	$0 + 3 = 8$	False
1	$1 + 3 = 8$	False
2	$2 + 3 = 8$	False
3	$3 + 3 = 8$	False
4	$4 + 3 = 8$	False
5	$5 + 3 = 8$	True
6	$6 + 3 = 8$	False

∴ the solution set is {5}.

2.

x	$\frac{1}{3}x = 2$	
0	$\frac{1}{3} \cdot 0 = 2$	False
1	$\frac{1}{3} \cdot 1 = 2$	False
2	$\frac{1}{3} \cdot 2 = 2$	False
3	$\frac{1}{3} \cdot 3 = 2$	False
4	$\frac{1}{3} \cdot 4 = 2$	False
5	$\frac{1}{3} \cdot 5 = 2$	False
6	$\frac{1}{3} \cdot 6 = 2$	True

∴ the solution set is {6}.

3.

x	$x - 1 = 2$	
0	$0 - 1 = 2$	False
1	$1 - 1 = 2$	False
2	$2 - 1 = 2$	False
3	$3 - 1 = 2$	True
4	$4 - 1 = 2$	False
5	$5 - 1 = 2$	False
6	$6 - 1 = 2$	False

∴ the solution set is {3}.

4.

x	$3x = 6$	
0	$3 \cdot 0 = 6$	False
1	$3 \cdot 1 = 6$	False
2	$3 \cdot 2 = 6$	True
3	$3 \cdot 3 = 6$	False
4	$3 \cdot 4 = 6$	False
5	$3 \cdot 5 = 6$	False
6	$3 \cdot 6 = 6$	False

$\therefore$ the solution set is $\{2\}$.

5.

x	$2x + 1 = 5$	
0	$2 \cdot 0 + 1 = 5$	False
1	$2 \cdot 1 + 1 = 5$	False
2	$2 \cdot 2 + 1 = 5$	True
3	$2 \cdot 3 + 1 = 5$	False
4	$2 \cdot 4 + 1 = 5$	False
5	$2 \cdot 5 + 1 = 5$	False
6	$2 \cdot 6 + 1 = 5$	False

$\therefore$ the solution set is $\{2\}$.

6.

x	$x \div 4 = 1$	
0	$0 \div 4 = 1$	False
1	$1 \div 4 = 1$	False
2	$2 \div 4 = 1$	False
3	$3 \div 4 = 1$	False
4	$4 \div 4 = 1$	True
5	$5 \div 4 = 1$	False
6	$6 \div 4 = 1$	False

$\therefore$ the solution set is $\{4\}$.

7.

n	$\frac{1}{5}n = 1$	
0	$\frac{1}{5} \cdot 0 = 1$	False
1	$\frac{1}{5} \cdot 1 = 1$	False
2	$\frac{1}{5} \cdot 2 = 1$	False
3	$\frac{1}{5} \cdot 3 = 1$	False
4	$\frac{1}{5} \cdot 4 = 1$	False
5	$\frac{1}{5} \cdot 5 = 1$	True

$\therefore$ the solution set is $\{5\}$.

8.

y	$5y - 1 = 14$	
0	$5 \cdot 0 - 1 = 14$	False
1	$5 \cdot 1 - 1 = 14$	False
2	$5 \cdot 2 - 1 = 14$	False
3	$5 \cdot 3 - 1 = 14$	True
4	$5 \cdot 4 - 1 = 14$	False
5	$5 \cdot 5 - 1 = 14$	False

$\therefore$ the solution set is $\{3\}$.

9.

x	$x + 2 = 5$	
0	$0 + 2 = 5$	False
1	$1 + 2 = 5$	False
2	$2 + 2 = 5$	False
3	$3 + 2 = 5$	True
4	$4 + 2 = 5$	False
5	$5 + 2 = 5$	False

$\therefore$ the solution set is $\{3\}$.

10.

x	$2x = 4$	
0	$2 \cdot 0 = 4$	False
1	$2 \cdot 1 = 4$	False
2	$2 \cdot 2 = 4$	True
3	$2 \cdot 3 = 4$	False
4	$2 \cdot 4 = 4$	False
5	$2 \cdot 5 = 4$	False

$\therefore$ the solution set is $\{2\}$.

11.

x	$x \cdot x \cdot x = 1$	
0	$0 \cdot 0 \cdot 0 = 1$	False
1	$1 \cdot 1 \cdot 1 = 1$	True
2	$2 \cdot 2 \cdot 2 = 1$	False
3	$3 \cdot 3 \cdot 3 = 1$	False
4	$4 \cdot 4 \cdot 4 = 1$	False
5	$5 \cdot 5 \cdot 5 = 1$	False

$\therefore$ the solution set is $\{1\}$.

12.

n	$5n = n \cdot 5$	
0	$5 \cdot 0 = 0 \cdot 5$	True
1	$5 \cdot 1 = 1 \cdot 5$	True
2	$5 \cdot 2 = 2 \cdot 5$	True
3	$5 \cdot 3 = 3 \cdot 5$	True
4	$5 \cdot 4 = 4 \cdot 5$	True
5	$5 \cdot 5 = 5 \cdot 5$	True

$\therefore$ the solution set is $\{0, 1, 2, 3, 4, 5\}$.

Page 87 • COMPUTER KEY-IN

1. −6 **2.** 385 **3.** 17.62 **4.** 147.89 **5.** −1 **6.** 38.5 **7.** 3.524 **8.** 29.578

Page 89 • CHAPTER REVIEW

1. $125 + 62 + 75 + 38 = (125 + 75) + (62 + 38) = 200 + 100 = 300$; d
2. $(16p)(5q) = (16 \cdot 5)(p \cdot q) = 80pq$; b
3. $13 - 5x + (-17) = -5x + (13 - 17) = -5x - 4$; b
4. $33 + [7 + (-12)] = 33 + (-5) = 28$; c
5. $[-6 + (-1)] + [-(-6 + 1)] = -7 + [-(-5)] = -7 + 5 = -2$; b
6. $10 + 3 - 5 + 2 = (10 + 3 + 2) - 5 = 15 - 5 = 10$; a
7. $x + 9 - (x - 7) = x + 9 - x + 7 = (x - x) + (9 + 7) = 0 + 16 = 16$; b
8. $(23 \cdot 32) - (13 \cdot 32) = (23 - 13) \cdot 32 = 10 \cdot 32 = 320$; c
9. $3 + 7(r - 4) = 3 + 7r - 28 = 7r + (3 - 28) = 7r - 25$; c
10. $(-2)(6)(-1)(8) = (-2)(-1)(6)(8) = 2 \cdot 48 = 96$; a
11. $2(3c - 2d) - 4(c - 3d) = 6c - 4d - 4c + 12d = (6c - 4c) + (-4d + 12d) = 2c + 8d$; d
12. b **13.** $(-63st)\left(-\frac{1}{9}\right) = (-63)\left(-\frac{1}{9}\right)st = 7st$; c
14. $12\left(\frac{1}{4}x + \frac{1}{3}\right) - \frac{1}{2}(12x - 6) = 3x + 4 - 6x + 3 = (3x - 6x) + (4 + 3) = -3x + 7$; a
15. d **16.** $\frac{a + 15b}{c} = \frac{(-6) + 15 \cdot 9}{-3} = \frac{(-6) + 135}{-3} = \frac{129}{-3} = -43$; a

Page 90 • CHAPTER TEST

1. $16 + p + q + 5 = p + q + 16 + 5 = p + q + 21$
2. $25 \cdot 12 \cdot 4 = (25 \cdot 4)12 = 100 \cdot 12 = 1200$
3. $-5 + [-9 + (-9)] = -5 + (-18) = -23$ **4.** $3n - 6 + (-9) = 3n - 15$
5. $\left(-1\frac{3}{5}\right) + \frac{2}{5} + 1 = -\frac{8}{5} + \frac{2}{5} + \frac{5}{5} = \frac{-8 + 7}{5} = -\frac{1}{5}$
6. $-7 + 5 - 4 = -7 - 4 + 5 = -11 + 5 = -6$
7. $42.51 - 1.20 - 1.95 - 4.36 - 26.00 - 1.20 + 5.00 = (42.51 + 5.00) - (1.20 + 1.95 + 4.36 + 26.00 + 1.20) = 47.51 - 34.71 = 12.80$; Samantha had \$12.80 at day's end.
8. $(25 - 31) - (-6 + 11) = -6 - 5 = -11$
9. $x - (-8) - [x + (-8)] = x + 8 - x + 8 = 16$
10. $2\frac{2}{3} + 1\frac{2}{3} = \frac{8}{3} + \frac{5}{3} = \frac{13}{3} = 4\frac{1}{3}$; Kara had been gone $4\frac{1}{3}$ hours.
11. $23(0.25) - 7(0.25) = (23 - 7)(0.25) = 16(0.25) = 4$
12. $(6x + 3)4 = 6x \cdot 4 + 3 \cdot 4 = 24x + 12$
13. $5(b - 1) + 8 = 5 \cdot b - 5 \cdot 1 + 8 = 5b - 5 + 8 = 5b + 3$
14. $7(2c - 4d + 6) = 7 \cdot 2c - 7 \cdot 4d + 7 \cdot 6 = 14c - 28d + 42$ **15.** $-16(-3) = 48$
16. $(-11 + 11)19 = 0 \cdot 19 = 0$ **17.** $(-9)(8)(-1)(-3) = (-9)(-1)(8)(-3) = 9(-24) = -216$
18. $7x - y + 2x + 3y = (7x + 2x) + (-y + 3y) = 9x + 2y$
19. Let $x =$ first even integer; then $x + 2$ and $x + 4$ are the second and third even integers; $x + (x + 2) + (x + 4) = x + 30$

20. -1 **21.** $\frac{7}{5}$ **22.** $\left(-\frac{1}{17}\right)(85)\left(\frac{1}{5}\right) = \left(-\frac{1}{17}\right)(17) = -1$

23. $-\frac{1}{7}(-56m + 49n) = -\frac{1}{7}(-56m) + \left(-\frac{1}{7}\right)(49n) = 8m - 7n$

24. $\frac{-13}{\frac{1}{6}} = -13 \cdot 6 = -78$ **25.** $\frac{343w}{-7} = 343w\left(-\frac{1}{7}\right) = 343\left(-\frac{1}{7}\right)w = -49w$

Page 91 • CUMULATIVE REVIEW

1. $(56 \div 7) - (26 \div 13) = 8 - 2 = 6$ **2.** $\frac{12 + 72}{6 + 8} = \frac{84}{14} = 6$

3. $|-12| - |-6| = 12 - 6 = 6$ **4.** $-|20| \div |-4| = -20 \div 4 = -5$

5. $38 + [(-3) + 16] = 38 + 13 = 51$ **6.** $3[27 \div (12 \div 4)] = 3(27 \div 3) = 3 \cdot 9 = 27$

7. $2 + 10 \cdot 15 \div 5 = 2 + 150 \div 5 = 2 + 30 = 32$

8. $2\frac{5}{9} + 7\frac{1}{8} + 8\frac{4}{9} = 2\frac{5}{9} + 8\frac{4}{9} + 7\frac{1}{8} = 11 + 7\frac{1}{8} = 18\frac{1}{8}$

9. $-120 - (-17) = -120 + 17 = -103$

10. $6(2x + 3) - 3(7 - 3x) = 12x + 18 - 21 + 9x = 21x - 3$

11. $3(x + 2y) + 5(2x + y) = 3x + 6y + 10x + 5y = 13x + 11y$

12. $(-9)(-5)(4) - 9(5) = 180 - 45 = 135$ **13.** $-18 - 4 - [(-6) + 12] = -22 - 6 = -28$

14. $\frac{1}{12}(-552xy) \div (-23) = -46xy\left(-\frac{1}{23}\right) = 2xy$

15. $\frac{6}{7}(4a - 3b) - \frac{3}{7}(8a - 6b) = \frac{24}{7}a - \frac{18}{7}b - \frac{24}{7}a + \frac{18}{7}b = 0$

16. $-z + (-x + 7) = -5 + [-(-3) + 7] = -5 + (3 + 7) = -5 + 10 = 5$

17. $x - |y| + |z| = -3 - |4| + |5| = -3 - 4 + 5 = -7 + 5 = -2$

18. $\frac{(x + 6)(z - y)}{(x + y)} = \frac{(-3 + 6)(5 - 4)}{-3 + 4} = \frac{3 \cdot 1}{1} = 3$ **19.** $-3, -2, -\frac{1}{2}, 0, \frac{1}{2}, 4$

20. $-5, -\frac{2}{3}, -\frac{1}{3}, |-3|, |5|$ **21.**

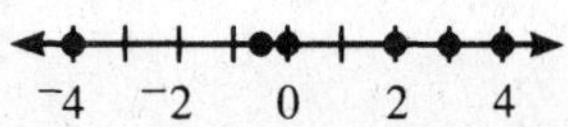

22. $\{7, -7\}$ **23.** $\{10, -10\}$ **24.** $\{3, -3\}$ **25.** $\{-12\}$ **26.** $\{9\}$ **27.** $\{0\}$

28.

x	$4x + 2 = 14$	
-5	$4(-5) + 2 = 14$	False
-3	$4(-3) + 2 = 14$	False
0	$4 \cdot 0 + 2 = 14$	False
3	$4 \cdot 3 + 2 = 14$	True
5	$4 \cdot 5 + 2 = 14$	False

$\therefore$ the solution set is $\{3\}$.

29.

x	$\frac{1}{3}x - 2 = -3$	
-5	$\frac{1}{3}(-5) - 2 = -3$	False
-3	$\frac{1}{3}(-3) - 2 = -3$	True
0	$\frac{1}{3} \cdot 0 - 2 = -3$	False
3	$\frac{1}{3} \cdot 3 - 2 = -3$	False
5	$\frac{1}{3} \cdot 5 - 2 = -3$	False

$\therefore$ the solution set is $\{-3\}$.

30.

x	$6 = 2x - 4$	
-5	$6 = 2(-5) - 4$	False
-3	$6 = 2(-3) - 4$	False
0	$6 = 2 \cdot 0 - 4$	False
3	$6 = 2 \cdot 3 - 4$	False
5	$6 = 2 \cdot 5 - 4$	True

$\therefore$ the solution set is $\{5\}$.

31. $7y + 5$ **32.** $[(x + (-a) - 2] - 3y$ **33.** $-x = 8 - 4$

34. $x + (x + 1) = 2(x + 1) - 1$ **35.** $4 + 7 - 8 = 11 - 8 = 3$; he got out on the third.

Page 92 • MAINTAINING SKILLS

1. 11.32 **2.** 566.241 **3.** 291.82 **4.** 662.07 **5.** 2.5996 **6.** 7.55 **7.** 0.757 **8.** 783

9. 55 **10.** 0.045 **11.** 662 **12.** 453 **13.** 0.0025 **14.** 47.34 **15.** 192 **16.** 3400

17. $\frac{14}{49} = \frac{2 \cdot 7}{7 \cdot 7} = \frac{2}{7}$ **18.** $\frac{21}{24} = \frac{7 \cdot 3}{8 \cdot 3} = \frac{7}{8}$ **19.** $\frac{39}{52} = \frac{3 \cdot 13}{4 \cdot 13} = \frac{3}{4}$ **20.** $\frac{27}{54} = \frac{1 \cdot 27}{2 \cdot 27} = \frac{1}{2}$

21. $\frac{63}{81} = \frac{7 \cdot 9}{9 \cdot 9} = \frac{7}{9}$ **22.** $\frac{27}{33} = \frac{9 \cdot 3}{11 \cdot 3} = \frac{9}{11}$ **23.** $\frac{2}{3} - \frac{1}{6} = \frac{2}{3} \cdot \frac{2}{2} - \frac{1}{6} = \frac{4}{6} - \frac{1}{6} = \frac{4 - 1}{6} = \frac{3}{6} = \frac{1}{2}$

24. $\frac{5}{7} - \frac{2}{3} = \frac{5}{7} \cdot \frac{3}{3} - \frac{2}{3} \cdot \frac{7}{7} = \frac{15}{21} - \frac{14}{21} = \frac{15 - 14}{21} = \frac{1}{21}$

25. $\frac{3}{4} - \frac{15}{21} = \frac{3}{4} \cdot \frac{21}{21} - \frac{15}{21} \cdot \frac{4}{4} = \frac{63}{84} - \frac{60}{84} = \frac{63 - 60}{84} = \frac{3}{84} = \frac{3 \cdot 1}{3 \cdot 28} = \frac{1}{28}$

26. $\frac{3}{4} - \frac{5}{12} = \frac{3}{4} \cdot \frac{3}{3} - \frac{5}{12} = \frac{9}{12} - \frac{5}{12} = \frac{9 - 5}{12} = \frac{4}{12} = \frac{4 \cdot 1}{4 \cdot 3} = \frac{1}{3}$

27. $\frac{4}{5} - \frac{2}{11} = \frac{4}{5} \cdot \frac{11}{11} - \frac{2}{11} \cdot \frac{5}{5} = \frac{44}{55} - \frac{10}{55} = \frac{34}{55}$

28. $\frac{3}{7} - \frac{11}{28} = \frac{3}{7} \cdot \frac{4}{4} - \frac{11}{28} = \frac{12}{28} - \frac{11}{28} = \frac{12 - 11}{28} = \frac{1}{28}$ **29.** $\frac{5}{6} \div \frac{5}{2} = \frac{5}{6} \cdot \frac{2}{5} = \frac{\overset{1}{\cancel{5}} \cdot \overset{1}{\cancel{2}}}{\underset{3}{\cancel{6}} \cdot \underset{1}{\cancel{5}}} = \frac{1}{3}$

30. $\frac{12}{21} \div \frac{8}{14} = \frac{12}{21} \cdot \frac{14}{8} = \frac{\overset{\overset{1}{\cancel{4}}}{\cancel{12}} \cdot \overset{\overset{1}{\cancel{2}}}{\cancel{14}}}{\underset{\underset{1}{\cancel{7}}}{\cancel{21}} \cdot \underset{\underset{1}{\cancel{2}}}{\cancel{8}}} = \frac{1}{1} = 1$ **31.** $\frac{21}{25} \div \frac{9}{20} = \frac{21}{25} \cdot \frac{20}{9} = \frac{\overset{7}{\cancel{21}} \cdot \overset{4}{\cancel{20}}}{\underset{5}{\cancel{25}} \cdot \underset{3}{\cancel{9}}} = \frac{28}{15}$

32. $\frac{10}{7} \div \frac{7}{10} = \frac{10}{7} \cdot \frac{10}{7} = \frac{10 \cdot 10}{7 \cdot 7} = \frac{100}{49}$ **33.** $\frac{24}{25} \div \frac{5}{6} = \frac{24}{25} \cdot \frac{6}{5} = \frac{24 \cdot 6}{25 \cdot 5} = \frac{144}{125}$

34. $\frac{24}{25} \div \frac{6}{5} = \frac{24}{25} \cdot \frac{5}{6} = \frac{\overset{4}{\cancel{24}} \cdot \overset{1}{\cancel{5}}}{\underset{5}{\cancel{25}} \cdot \underset{1}{\cancel{6}}} = \frac{4}{5}$

Page 93 • PREPARING FOR COLLEGE ENTRANCE EXAMS

1. D **2.** D **3.** A **4.** C **5.** C **6.** A **7.** C

Pages 97–98 • WRITTEN EXERCISES

A

1. $x - 7 = 13, x - 7 + 7 = 13 + 7, x = 20; \{20\}$

2. $y - 9 = 17, y - 9 + 9 = 17 + 9, y = 26; \{26\}$

3. $z + 8 = 31, z + 8 - 8 = 31 - 8, z = 23; \{23\}$

4. $x + 15 = 27, x + 15 - 15 = 27 - 15, x = 12; \{12\}$

5. $-52 + m = 84, -52 + 52 + m = 84 + 52, m = 136; \{136\}$

6. $-49 + n = 63, -49 + 49 + n = 63 + 49, n = 112; \{112\}$

7. $t - 25 = -18, t - 25 + 25 = -18 + 25, t = 7; \{7\}$

8. $x - 26 = 18, x - 26 + 26 = 18 + 26, x = 44; \{44\}$

9. $p + 18 = -32, p + 18 - 18 = -32 - 18, p = -50; \{-50\}$

10. $y + 32 = -45, y + 32 - 32 = -45 - 32, y = -77; \{-77\}$

11. $0 = 38 + k, -38 + 0 = -38 + 38 + k, -38 = k; \{-38\}$

12. $0 = z - 14, 0 + 14 = z - 14 + 14, 14 = z; \{14\}$

13. $-19 + a = 23, -19 + 19 + a = 23 + 19, a = 42; \{42\}$

14. $-32 + b = 82, -32 + 32 + b = 82 + 32, b = 114; \{114\}$

15. $c + 9 = 5, c + 9 - 9 = 5 - 9, c = -4; \{-4\}$

16. $x - 8 = 25, x - 8 + 8 = 25 + 8, x = 33; \{33\}$

17. $f + 7 = 9 - 2, f + 7 - 7 = 9 - 2 - 7, f = 0; \{0\}$

18. $g - 6 = 14 - 8, g - 6 + 6 = 14 - 8 + 6, g = 12; \{12\}$

19. $z - 57 = -67, z - 57 + 57 = -67 + 57, z = -10; \{-10\}$

20. $x - 97 = -105, x - 97 + 97 = -105 + 97, x = -8; \{-8\}$

21. $-0.7 + k = -1.7, -0.7 + 0.7 + k = -1.7 + 0.7, k = -1; \{-1\}$

22. $-1.8 + h = -3.8, -1.8 + 1.8 + h = -3.8 + 1.8, h = -2; \{-2\}$

23. $4.5 = x + 1.6, 4.5 - 1.6 = x + 1.6 - 1.6, 2.9 = x; \{2.9\}$

24. $3.9 = y + 1.2, 3.9 - 1.2 = y + 1.2 - 1.2, 2.7 = y; \{2.7\}$

B

25. $-x + 6 = 4, -x + 6 - 6 = 4 - 6, -x = -2, x = 2; \{2\}$

26. $-y + 5 = 17, -y + 5 - 5 = 17 - 5, -y = 12, y = -12; \{-12\}$

27. $21 - x = 28, 21 - 21 - x = 28 - 21, -x = 7, x = -7; \{-7\}$

28. $9 - y = 16, 9 - 9 - y = 16 - 9, -y = 7, y = -7; \{-7\}$

29. $8 = -x + 18, 8 - 18 = -x + 18 - 18, -10 = -x, 10 = x; \{10\}$

30. $11 = 32 - y, 11 - 32 = 32 - 32 - y, -21 = -y, 21 = y; \{21\}$

31. $-8 - y = 9, -8 + 8 - y = 9 + 8, -y = 17, y = -17; \{-17\}$

32. $7 = -12 + e, 7 + 12 = -12 + 12 + e, 19 = e; \{19\}$

33. $13 = -y + 8, 13 - 8 = -y + 8 - 8, 5 = -y, -5 = y; \{-5\}$

34. $(c + 2) + 8 = 4, c + 10 = 4, c + 10 - 10 = 4 - 10, c = -6; \{-6\}$

35. $(r + 4) + 2 = 1, r + 6 = 1, r + 6 - 6 = 1 - 6, r = -5; \{-5\}$

36. $2 = 10 + (x - 2), 2 = 8 + x, 2 - 8 = 8 - 8 + x, -6 = x; \{-6\}$

37. $8 = 16 + (y - 1), 8 = 15 + y, 8 - 15 = 15 - 15 + y, -7 = y; \{-7\}$

38. $-2 + (1 + p) = 5, -1 + p = 5, -1 + 1 + p = 5 + 1, p = 6; \{6\}$

39. $-3 + (1 + n) = 9, -2 + n = 9, -2 + 2 + n = 9 + 2, n = 11; \{11\}$

40. $(a - 3) + 19 = 125, a + 16 = 125, a + 16 - 16 = 125 - 16, a = 109; \{109\}$

41. $(b - 6) + 14 = 100, b + 8 = 100, b + 8 - 8 = 100 - 8, b = 92; \{92\}$

42. $4 - (1 + x) = 5, 4 - 1 - x = 5, 3 - x = 5, 3 - 3 - x = 5 - 3, -x = 2, x = -2;$ $\{-2\}$

43. $2 - (3 + y) = 6, 2 - 3 - y = 6, -1 - y = 6, -1 + 1 - y = 6 + 1, -y = 7,$ $y = -7; \{-7\}$

44. $1 = -2 - (4 - w), 1 = -2 - 4 + w, 1 = -6 + w, 1 + 6 = -6 + 6 + w, 7 = w; \{7\}$

45. $11 = 7 - (1 - q), 11 = 7 - 1 + q, 11 = 6 + q, 11 - 6 = 6 - 6 + q, 5 = q; \{5\}$

C

46. $|y| - 2 = 8, |y| - 2 + 2 = 8 + 2, |y| = 10, y = 10$ or $y = -10; \{10, -10\}$

47. $|z| + 10 = 28, |z| + 10 - 10 = 28 - 10, |z| = 18, z = 18$ or $z = -18; \{18, -18\}$

48. $-7 + |s| = 0, -7 + 7 + |s| = 0 + 7, |s| = 7, s = 7$ or $s = -7; \{7, -7\}$

49. $6 + |t| = 14, 6 - 6 + |t| = 14 - 6, |t| = 8, t = 8$ or $t = -8; \{8, -8\}$

50. $|x| + (-2) = 4, |x| - 2 + 2 = 4 + 2, |x| = 6, x = 6$ or $x = -6; \{6, -6\}$

51. $|y| + (-1) = 1, |y| - 1 + 1 = 1 + 1, |y| = 2, y = 2$ or $y = -2; \{2, -2\}$

52. $0 = -5 + |r|, 0 + 5 = -5 + 5 + |r|, 5 = |r|, r = 5$ or $r = -5; \{5, -5\}$

53. $2 = 6 + |t|, 2 - 6 = 6 - 6 + |t|, -4 = |t|$; no solution

54. $-(|a| - 9) = 1, -|a| + 9 = 1, -|a| + 9 - 9 = 1 - 9, -|a| = -8, |a| = 8, a = 8$ or $a = -8; \{8, -8\}$

55. $-(|x| + 2) = -6, -|x| - 2 = -6, -|x| - 2 + 2 = -6 + 2, -|x| = -4, |x| = 4,$ $x = 4$ or $x = -4; \{4, -4\}$

56. $4 - (2 - |n|) = 2, 4 - 2 + |n| = 2, 2 + |n| = 2, 2 - 2 + |n| = 2 - 2, |n| = 0,$ $n = 0; \{0\}$

57. $7 - (3 - |m|) = 8, 7 - 3 + |m| = 8, 4 + |m| = 8, 4 - 4 + |m| = 8 - 4, |m| = 4,$ $m = 4$ or $m = -4; \{4, -4\}$

58. $9 - (|s| + 7) = 4, 9 - |s| - 7 = 4, 2 - |s| = 4, 2 - 2 - |s| = 4 - 2, -|s| = 2,$ $|s| = -2$; no solution

59. $-3 + (15 - |a|) = 12, 12 - |a| = 12, 12 - 12 - |a| = 12 - 12, -|a| = 0, |a| = 0,$ $a = 0; \{0\}$

60. $(|e| - |-8|) + 15 = 7, |e| - 8 + 15 = 7, |e| + 7 = 7, |e| + 7 - 7 = 7 - 7,$ $|e| = 0, e = 0; \{0\}$

Pages 98–100 • PROBLEMS

A

1. $n + 51 = -12, n = -63$; the number is -63

2. $n - 22 = -7, n = 15$; the number is 15

3. $n + 28 = 7, n = -21$; the number is -21

4. $n - 8 = -21, n = -13$; the number is -13

5. $n - (-8) = 84, n + 8 = 84, n = 76$; the number is 76

6. $n - (-15) = -29$, $n + 15 = -29$, $n = -44$; the number is -44

7. *Step 1* Find how fast a giraffe can run. *Step 2* Let g = giraffe's speed.

Step 3 $50 = g + 18$ *Step 4* $50 - 18 = g + 18 - 18$, $32 = g$

Step 5 A giraffe can run 32 mi/h.

8. *Step 1* Find Corita's previous time. *Step 2* Let t = previous time.

Step 3 $56.8 = t - 1.3$ *Step 4* $56.8 + 1.3 = t - 1.3 + 1.3$, $58.1 = t$

Step 5 Corita's previous time was 58.1 s.

9. *Step 1* Find the temperature at 6 A.M. *Step 2* Let t = temperature at 6 A.M.

Step 3 $t + 25 = 18$ *Step 4* $t + 25 - 25 = 18 - 25$, $t = -7$

Step 5 The temperature at 6 A.M. was −7°C.

10. *Step 1* Find the temperature at 4 P.M. *Step 2* Let t = temperature at 4 P.M.

Step 3 $t - 17 = -11$ *Step 4* $t - 17 + 17 = -11 + 17$, $t = 6$

Step 5 The temperature at 4 P.M. was 6°F.

11. *Step 1* Find the cost of the sandwich. *Step 2* Let s = cost of the sandwich.

Step 3 $1.15 + 1.15 + s = 4.75$

Step 4 $2.30 + s = 4.75$, $2.30 - 2.30 + s = 4.75 - 2.30$, $s = 2.45$

Step 5 The sandwich cost $2.45.

12. *Step 1* Find the number of sheets put in the desk.

Step 2 Let s = number of sheets put in the desk. *Step 3* $5(6) + s = 45$

Step 4 $30 + s = 45$, $30 - 30 + s = 45 - 30$, $s = 15$

Step 5 Ruth put 15 sheets in her desk.

B 13. *Step 1* Find the number of workers at the beginning of the year.

Step 2 Let b = number of workers at the beginning of the year.

Step 3 $b + 130 - 27 - 59 = 498$

Step 4 $b + 44 = 498$, $b + 44 - 44 = 498 - 44$, $b = 454$

Step 5 There were 454 workers at the beginning of the year.

14. *Step 1* Find the value of the investor's stocks when the market opened.

Step 2 Let v = value of stocks when market opened.

Step 3 $v - 2500 + 1700 = 52{,}400$

Step 4 $v - 800 = 52{,}400$, $v - 800 + 800 = 52{,}400 + 800$, $v = 53{,}200$

Step 5 The investor's stock holdings were worth $53,200 when the market opened.

15. *Step 1* Find how many degrees Nairobi is east of New York City.

Step 2 Let n = number of degrees east from New York City to Nairobi.

Step 3 $n = |37| + |-74|$ *Step 4* $n = 37 + 74$, $n = 111$

Step 5 Nairobi is 111° east of New York City.

16. *Step 1* Find how many degrees Nairobi is south of New York City.

Step 2 Let n = number of degrees south from New York City to Nairobi.

Step 3 $n = |41| + |-1|$ *Step 4* $n = 41 + 1$, $n = 42$

Step 5 Nairobi is 42° south of New York City.

17. *Step 1* Find the value of the coupon. *Step 2* Let c = value of coupon.

Step 3 $1.79 + (1.79 - c) = 3.23$

Step 4 $3.58 - c = 3.23, 3.58 - 3.58 - c = 3.23 - 3.58, -c = -0.35, c = 0.35$

Step 5 Gino used a coupon worth \$.35.

18. *Step 1* Find how much the store reduced the price for the sale.

Step 2 Let d = amount of discount for sale. *Step 3* $4.69 + (4.69 + d) = 10.64$

Step 4 $9.38 + d = 10.64, 9.38 - 9.38 + d = 10.64 - 9.38, d = 1.26$

Step 5 The store reduced the price by \$1.26 for the sale.

C **19.** *Step 1* Find the distance between the two teams when the second team found seaweed.

Step 2 Let d = distance between Team 1 and Team 2 at that point.

Step 3 $d = |5 - 3| + |-12|$ *Step 4* $d = 2 + 12, d = 14$

Step 5 The two teams were 14 km apart when Team 2 found seaweed.

20. *Step 1* Find how far Rick will have to drive to get 28082 on his odometer.

Step 2 Let d = distance Rick must drive. *Step 3* $d = 28082 - 27972$

Step 4 $d = 110$

Step 5 Rick will have to drive 110 mi to get 28082 on his odometer.

Pages 100–101 • MIXED REVIEW EXERCISES

1. $a - |c - b| = 4 - |-7 - (-6)| = 4 - |-7 + 6| = 4 - |-1| = 4 - 1 = 3$

2. $(|b| - a) - (|d| - a) = (|-6| - 4) - (|5| - 4) = (6 - 4) - (5 - 4) = 2 - 1 = 1$

3. $2|a| - (-a) = 2|4| - (-4) = (2)(4) + 4 = 8 + 4 = 12$

4. $\frac{b - 2c}{b + 2d} = \frac{-6 - (2)(-7)}{-6 + (2)(5)} = \frac{-6 - (-14)}{-6 + 10} = \frac{-6 + 14}{4} = \frac{8}{4} = 2$

5. $\frac{4d + c + 1}{abc} = \frac{(4)(5) + (-7) + 1}{(4)(-6)(-7)} = \frac{20 + (-6)}{168} = \frac{14}{168} = \frac{1}{12}$ **6.** $\frac{3ab}{c + d} = \frac{3(4)(-6)}{-7 + 5} = \frac{-72}{-2} = 36$

7. $(-6)(-7)(10) = 420$ **8.** $(-9 \cdot 12) + (-9 \cdot 18) = -108 + (-162) = -270$

9. $148a \div (-37) = -4a$ **10.** $96\left(-\frac{1}{24}\right)\left(-\frac{1}{4}\right) = \frac{(96)(-1)(-1)}{(24)(4)} = \frac{96}{96} = 1$

11. $-\frac{12b}{5} \cdot (-5) = \frac{(-12)(-5)b}{5} = \frac{60b}{5} = 12b$

12. $\frac{1}{7}(21a - 7b) - \frac{1}{3}(12b - 6a) = \frac{21a}{7} - \frac{7b}{7} - \frac{12b}{3} + \frac{6a}{3} = 3a - b - 4b + 2a = 5a - 5b$

Page 101 • APPLICATION

1. a. Total paid = 480 + 24(115) = 480 + 2760 = \$3240 **b.** Interest = 3240 − 3000 = \$240

2. a. Total paid = 451 + 36(154) = 451 + 5544 = \$5995 **b.** Interest = 5995 − 5500 = \$495

3. a. Total paid = 498 + 48(95) = 498 + 4560 = \$5058 **b.** Interest = 5058 − 4500 = \$558

4. a. Total paid = 620 + 48(220) = 620 + 10,560 = \$11,180

b. Interest = 11,180 − 10,000 = \$1180

Page 104 • WRITTEN EXERCISES

A **1.** $4x = 44, \frac{4x}{4} = \frac{44}{4}, x = 11; \{11\}$ **2.** $5y = 65, \frac{5y}{5} = \frac{65}{5}, y = 13; \{13\}$

3. $6y = -18, \frac{6y}{6} = \frac{-18}{6}, y = -3; \{-3\}$ **4.** $3t = -27, \frac{3t}{3} = \frac{-27}{3}, t = -9; \{-9\}$

5. $-8c = 72, \frac{-8c}{-8} = \frac{72}{-8}, c = -9; \{-9\}$ **6.** $-6p = 42, \frac{-6p}{-6} = \frac{42}{-6}, p = -7; \{-7\}$

7. $-12a = -36, \frac{-12a}{-12} = \frac{-36}{-12}, a = 3; \{3\}$ **8.** $-9z = -63, \frac{-9z}{-9} = \frac{-63}{-9}, z = 7; \{7\}$

9. $\frac{1}{2}x = 12, 2\left(\frac{1}{2}x\right) = 2(12), x = 24; \{24\}$ **10.** $\frac{1}{3}y = 18, 3\left(\frac{1}{3}y\right) = 3(18), y = 54; \{54\}$

11. $\frac{x}{5} = -7, 5\left(\frac{x}{5}\right) = 5(-7), x = -35; \{-35\}$

12. $\frac{c}{4} = -9, 4\left(\frac{c}{4}\right) = 4(-9), c = -36; \{-36\}$

13. $-\frac{1}{8}b = 8, -8\left(-\frac{1}{8}b\right) = -8(8), b = -64; \{-64\}$

14. $-\frac{1}{5}t = 17, -5\left(-\frac{1}{5}t\right) = -5(17), t = -85; \{-85\}$

15. $\frac{2}{3}x = 12, \frac{3}{2}\left(\frac{2}{3}x\right) = \frac{3}{2}(12), x = 18; \{18\}$ **16.** $\frac{5}{2}y = 10, \frac{2}{5}\left(\frac{5}{2}y\right) = \frac{2}{5}(10), y = 4; \{4\}$

17. $\frac{3}{4}d = -60, \frac{4}{3}\left(\frac{3}{4}d\right) = \frac{4}{3}(-60), d = -80; \{-80\}$

18. $\frac{5}{8}c = -20, \frac{8}{5}\left(\frac{5}{8}c\right) = \frac{8}{5}(-20), c = -32, \{-32\}$

19. $-\frac{2}{3}t = 22, -\frac{3}{2}\left(-\frac{2}{3}t\right) = -\frac{3}{2}(22), t = -33; \{-33\}$

20. $-\frac{2}{11}p = 14, -\frac{11}{2}\left(-\frac{2}{11}p\right) = -\frac{11}{2}(14), p = -77; \{-77\}$

21. $600 = -25x, \frac{600}{-25} = \frac{-25x}{-25}, -24 = x; \{-24\}$

22. $-324 = -18c, \frac{-324}{-18} = \frac{-18c}{-18}, 18 = c; \{18\}$ **23.** $-17d = 0, \frac{-17d}{-17} = \frac{0}{-17}, d = 0; \{0\}$

24. $252 = -14y, \frac{252}{-14} = \frac{-14y}{-14}, -18 = y; \{-18\}$

25. $99 = -\frac{11}{5}x, -\frac{5}{11}(99) = -\frac{5}{11}\left(-\frac{11}{5}x\right), -45 = x; \{-45\}$

26. $24 = -\frac{3}{7}t, -\frac{7}{3}(24) = -\frac{7}{3}\left(-\frac{3}{7}t\right), -56 = t; \{-56\}$

27. $-6 = \frac{x}{3}, 3(-6) = 3\left(\frac{x}{3}\right), -18 = x; \{-18\}$ **28.** $21 = \frac{c}{7}, 7(21) = 7\left(\frac{c}{7}\right), 147 = c; \{147\}$

B **29.** $\frac{1}{3}y = 2\frac{1}{3}, \frac{1}{3}y = \frac{7}{3}, 3\left(\frac{1}{3}y\right) = 3\left(\frac{7}{3}\right), y = 7; \{7\}$

30. $\frac{1}{2}t = 3\frac{1}{2}, \frac{1}{2}t = \frac{7}{2}, 2\left(\frac{1}{2}t\right) = 2\left(\frac{7}{2}\right), t = 7; \{7\}$

31. $-1 = 2.5z, \frac{-1}{2.5} = \frac{2.5z}{2.5}, -0.4 = z; \{-0.4\}$ **32.** $0 = -4.5y, \frac{0}{-4.5} = \frac{-4.5y}{-4.5}, 0 = y; \{0\}$

33. $\frac{2}{5}a = 6\frac{2}{5}, \frac{2}{5}a = \frac{32}{5}, \frac{5}{2}\left(\frac{2}{5}a\right) = \frac{5}{2}\left(\frac{32}{5}\right), a = 16; \{16\}$

34. $\frac{3}{2}b = -14\frac{1}{2}, \frac{3}{2}b = -\frac{29}{2}, \frac{2}{3}\left(\frac{3}{2}b\right) = \frac{2}{3}\left(-\frac{29}{2}\right), b = -\frac{29}{3}; \left\{-\frac{29}{3}\right\}$

35. $-\frac{y}{3} = 3\frac{2}{3}, -\frac{y}{3} = \frac{11}{3}, -3\left(-\frac{y}{3}\right) = -3\left(\frac{11}{3}\right), y = -11; \{-11\}$

36. $-\frac{x}{2} = 11\frac{1}{2}, -\frac{x}{2} = \frac{23}{2}, -2\left(-\frac{x}{2}\right) = -2\left(\frac{23}{2}\right), x = -23; \{-23\}$

37. $-\frac{y}{12} = \frac{1}{4}, -12\left(-\frac{y}{12}\right) = -12\left(\frac{1}{4}\right), y = -3; \{-3\}$

38. $-\frac{n}{4} = -\frac{3}{2}, -4\left(-\frac{n}{4}\right) = -4\left(-\frac{3}{2}\right), n = 6; \{6\}$

39. $0 = -\frac{x}{5}, -5(0) = -5\left(-\frac{x}{5}\right), 0 = x; \{0\}$

40. $-\frac{5}{3}x = 10, -\frac{3}{5}\left(-\frac{5}{3}x\right) = -\frac{3}{5}(10), x = -6; \{-6\}$

C **41.** $2|x| = 18, \frac{2|x|}{2} = \frac{18}{2}, |x| = 9, x = 9$ or $x = -9; \{9, -9\}$

42. $3|y| = 33, \frac{3|y|}{3} = \frac{33}{3}, |y| = 11, y = 11$ or $y = -11; \{11, -11\}$

43. $-7|t| = 42, \frac{-7|t|}{-7} = \frac{42}{-7}, |t| = -6$; no solution

44. $-32 = 8|k|, \frac{-32}{8} = \frac{8|k|}{8}, -4 = |k|$; no solution

45. $\frac{|x|}{5} = 2, 5\left(\frac{|x|}{5}\right) = 5(2), |x| = 10, x = 10$ or $x = -10; \{10, -10\}$

46. $\frac{|a|}{8} = 4, 8\left(\frac{|a|}{8}\right) = 8(4), |a| = 32, a = 32$ or $a = -32; \{32, -32\}$

47. $3 = \frac{|n|}{7}, 7(3) = 7\left(\frac{|n|}{7}\right), 21 = |n|, n = 21$ or $n = -21; \{21, -21\}$

48. $6 = \frac{|m|}{2}, 2(6) = 2\left(\frac{|m|}{2}\right), 12 = |m|, m = 12$ or $m = -12; \{12, -12\}$

49. $\frac{5|c|}{3} = 10, \frac{3}{5}\left(\frac{5|c|}{3}\right) = \frac{3}{5}(10), |c| = 6, c = 6$ or $c = -6; \{6, -6\}$

50. $\frac{4}{7}|x| = 16, \frac{7}{4}\left(\frac{4}{7}|x|\right) = \frac{7}{4}(16), |x| = 28, x = 28$ or $x = -28; \{28, -28\}$

51. $7 - \frac{3|a|}{2} = 1, -7 + 7 - \frac{3|a|}{2} = -7 + 1, -\frac{3|a|}{2} = -6, -\frac{2}{3}\left(-\frac{3|a|}{2}\right) = -\frac{2}{3}(-6),$
$|a| = 4, a = 4$ or $a = -4; \{4, -4\}$

52. $10 - \frac{4}{3}|b| = -2, -10 + 10 - \frac{4}{3}|b| = -10 - 2, -\frac{4}{3}|b| = -12, -\frac{3}{4}\left(-\frac{4}{3}|b|\right) =$
$-\frac{3}{4}(-12), |b| = 9, b = 9$ or $b = -9; \{9, -9\}$

Pages 105–106 • PROBLEMS

A **1.** Let n = the number. $5n = -375$, $\frac{5n}{5} = \frac{-375}{5}$, $n = -75$; The number is -75.

2. Let n = the number. $-9n = -108$, $\frac{-9n}{-9} = \frac{-108}{-9}$, $n = 12$; The number is 12.

3. Let n = the number. $\frac{1}{3}n = -7$, $3\left(\frac{1}{3}n\right) = 3(-7)$, $n = -21$; The number is -21.

4. Let n = the number. $\frac{3}{4}n = 21$, $\frac{4}{3}\left(\frac{3}{4}n\right) = \frac{4}{3}(21)$, $n = 28$; The number is 28.

5. Let s = number of seniors. $120 = \frac{1}{3}s$, $3(120) = 3\left(\frac{1}{3}s\right)$, $360 = s$; 360 seniors.

6. Let s = number of students at Lincoln High School. $225 = \frac{3}{8}s$, $\frac{8}{3}(225) = \frac{8}{3}\left(\frac{3}{8}s\right)$, $600 = s$; There are 600 students at the school.

7. Let s = length of one side of the parking lot. $4s = 784$, $\frac{4s}{4} = \frac{784}{4}$, $s = 196$; Each side of the parking lot is 196 m long.

8. Let s = length of each side of the Pentagon building. $5s = 1$, $\frac{5s}{5} = \frac{1}{5}$, $s = \frac{1}{5}$; Each side of the Pentagon building is $\frac{1}{5}$ mi long.

9. Let p = price of the whole pizza. $\frac{3}{8}p = 2.70$, $\frac{8}{3}\left(\frac{3}{8}p\right) = \frac{8}{3}(2.70)$, $p = 7.20$; The price of the whole pizza was \$7.20.

10. Let q = amount paid for a whole quiche. $\frac{1}{8}q = 2.50$, $8\left(\frac{1}{8}q\right) = 8(2.50)$, $q = 20$; The restaurant receives \$20 for the whole quiche.

11. Let l = the number of games lost. $3l = 21$, $\frac{1}{3}(3l) = \frac{1}{3}(21)$, $l = 7$; The Eagles lost 7 games.

12. Let c = the number of color monitors. $5c = 40$, $\frac{5c}{5} = \frac{40}{5}$, $c = 8$; The school has 8 color monitors.

13. Let a = number of apples. $0.2a = 50$, $\frac{0.2a}{0.2} = \frac{50}{0.2}$, $a = 250$; There are 250 apples in the shipment.

14. Let t = number of hours the bulb burned. $0.075t = 3.3$, $\frac{0.075t}{0.075} = \frac{3.3}{0.075}$, $t = 44$; The bulb was left burning for 44 h.

B **15.** Let h = number of hard-cover books. $16.50h = 30(4.95)$, $16.50h = 148.50$, $\frac{16.50h}{16.50} = \frac{148.50}{16.50}$, $h = 9$; The dealer must sell 9 hard-cover books.

16. Let p = selling price of the home. $\frac{6}{100}p = 10{,}725$, $\frac{100}{6}\left(\frac{6}{100}p\right) = \frac{100}{6}(10{,}725)$, $p = 178{,}750$; The home sold for \$178,750.

17. Let w = wages earned by the worker. $\frac{24}{100}w = 5100$, $\frac{100}{24}\left(\frac{24}{100}w\right) = \frac{100}{24}(5100)$, $w = 21{,}250$; The worker earned \$21,250.

18. Let w = number of kilograms of sea water. $35w = 4.2$, $\frac{35w}{35} = \frac{4.2}{35}$, $w = 0.12$, $0.12 \times 1000 = 120$; There are 120 g of sea water.

C 19. Let r = rate Raul was driving in mi/s. $18r = \frac{1}{4}$, $\frac{1}{18}(18r) = \frac{1}{18}\left(\frac{1}{4}\right)$, $r = \frac{1}{72}$; Raul was driving at $\frac{1}{72}$ mi/s; $\frac{1}{72} \times 60 \times 60 = 50$; Raul was driving at 50 mi/h.

20. Let r = rate the truck was traveling in mi/s. $10r = \frac{1}{5}$, $\frac{10r}{10} = \frac{\left(\frac{1}{5}\right)}{10}$, $r = \frac{1}{50}$; The truck was traveling at $\frac{1}{50}$ mi/s; $\frac{1}{50} \times 60 \times 60 = 72$; The truck was traveling at 72 mi/h.

Page 106 • MIXED REVIEW EXERCISES

1. $7a - 2b = 7(2) - 2(-3) = 14 + 6 = 20$
2. $(3a - 2b)c = [3(2) - 2(-3)]4 = (6 + 6)4 = (12)4 = 48$
3. $|b| + |c| - (-c) = |-3| + |4| - (-4) = 3 + 4 + 4 = 11$
4. $|a| - |b + c| = |2| - |-3 + 4| = 2 - |1| = 2 - 1 = 1$
5. $\frac{-(5ab)}{2c} = \frac{-[5(2)(-3)]}{2(4)} = \frac{30}{8} = \frac{15}{4}$ 6. $\frac{2 + a}{c} = \frac{2 + 2}{4} = \frac{4}{4} = 1$ 7. $7a + 6 + 9a = 16a + 6$
8. $5n - 9 + 9 = 5n$ 9. $10p - p + 2 = 9p + 2$
10. $-4(m + 2) = -4m - 4(2) = -4m - 8$ 11. $(x + 7)8 = 8x + 56$
12. $3(2y - 5) = 6y - 15$

Page 106 • CALCULATOR KEY-IN

1. 0.75 2. −0.625 3. −0.2 4. 0.15 5. 0.025 6. −2.75 7. −0.24 8. 0.875
9. 0.96875 10. −0.671875 11. 0.75 12. −0.625 13. −0.2 14. 0.15 15. 0.025
16. −2.75 17. −0.24 18. 0.875 19. 0.96875
20. −0.671875; yes, except for round-off error on some calculators.

Pages 109–110 • WRITTEN EXERCISES

A 1. $2x - 1 = 11$, $2x = 12$, $x = 6$; $\{6\}$ 2. $3y - 8 = 16$, $3y = 24$, $y = 8$; $\{8\}$
3. $4n + 9 = -3$, $4n = -12$, $n = -3$; $\{-3\}$
4. $5c + 7 = -28$, $5c = -35$, $c = -7$; $\{-7\}$
5. $-2x + 5 = 19$, $-2x = 14$, $x = -7$; $\{-7\}$
6. $-8y - 11 = 13$, $-8y = 24$, $y = -3$; $\{-3\}$

7. $\frac{1}{2}x + 7 = 6, \frac{1}{2}x = -1, x = -2; \{-2\}$ **8.** $\frac{2}{3}p - 7 = 17, \frac{2}{3}p = 24, p = 36; \{36\}$

9. $\frac{2x}{3} = 8, x = 12; \{12\}$ **10.** $\frac{4y}{5} = 28, y = 35; \{35\}$

11. $\frac{x + 5}{3} = 7, x + 5 = 21, x = 16; \{16\}$ **12.** $\frac{z - 5}{4} = 8, z - 5 = 32, z = 37; \{37\}$

13. $\frac{1 - x}{2} = 7, 1 - x = 14, -x = 13, x = -13; \{-13\}$

14. $\frac{5 - n}{3} = 4, 5 - n = 12, -n = 7, n = -7; \{-7\}$

15. $7x - 4x = 54, 3x = 54, x = 18; \{18\}$ **16.** $3y - 7y = 28, -4y = 28, y = -7; \{-7\}$

17. $-3n - 5n = 0, -8n = 0, n = 0; \{0\}$ **18.** $2a - 11a = -27, -9a = -27, a = 3; \{3\}$

19. $y - 7 + 4y = 13, 5y - 7 = 13, 5y = 20, y = 4; \{4\}$

20. $2x + 5 - 7x = 15, -5x + 5 = 15, -5x = 10, x = -2; \{-2\}$

21. $0 = n - 15 - 4n, 0 = -3n - 15, 15 = -3n, -5 = n; \{-5\}$

22. $0 = p + 18 + 5p, 0 = 6p + 18, -18 = 6p, -3 = p; \{-3\}$

23. $2(x - 4) = 22, x - 4 = 11, x = 15; \{15\}$

24. $3(y - 7) = 27, y - 7 = 9, y = 16; \{16\}$ **25.** $35 = 5(n + 2), 7 = n + 2, 5 = n; \{5\}$

26. $20 = 4(x + 3), 5 = x + 3, 2 = x; \{2\}$

27. $y + 5 - 4y = -10, -3y + 5 = -10, -3y = -15, y = 5; \{5\}$

28. $4w - 3w + 2w = 24, 3w = 24, w = 8; \{8\}$

29. $15 = 8x - 5 + 2x, 15 = 10x - 5, 20 = 10x, x = 2; \{2\}$

30. $32 = 2n - 3n + 5n, 32 = 4n, 8 = n; \{8\}$

B **31.** $-\frac{1}{2}(x + 4) = 16, x + 4 = -32, x = -36; \{-36\}$

32. $\frac{3}{5}(x + 2) = 12, x + 2 = 20, x = 18; \{18\}$

33. $21 = -\frac{3}{2}(x - 2), -14 = x - 2, -12 = x; \{-12\}$

34. $66 = -\frac{6}{5}(s + 3), -55 = s + 3, -58 = s; \{-58\}$

35. $3(a - 5) + 19 = -2, 3a - 15 + 19 = -2, 3a + 4 = -2, 3a = -6, a = -2; \{-2\}$

36. $2(b + 8) - 9 = 5, 2b + 16 - 9 = 5, 2b + 7 = 5, 2b = -2, b = -1; \{-1\}$

37. $-3 = 4(k + 7) - 15, -3 = 4k + 28 - 15, -3 = 4k + 13, -16 = 4k, -4 = k; \{-4\}$

38. $3 = 7(h - 2) + 17, 3 = 7h - 14 + 17, 3 = 7h + 3, 0 = 7h, 0 = h; \{0\}$

39. $4c + 3(c - 2) = -34, 4c + 3c - 6 = -34, 7c - 6 = -34, 7c = -28, c = -4; \{-4\}$

40. $d + 4(d + 6) = -1, d + 4d + 24 = -1, 5d + 24 = -1, 5d = -25, d = -5; \{-5\}$

41. $\frac{2x - 1}{3} = 5, 2x - 1 = 15, 2x = 16, x = 8; \{8\}$

42. $\frac{4y + 3}{7} = 9, 4y + 3 = 63, 4y = 60, y = 15; \{15\}$

43. $0 = \frac{8 - 2x}{5}, 0 = 8 - 2x, -8 = -2x, 4 = x; \{4\}$

44. $7 = \frac{4 + 9y}{7}$, $49 = 4 + 9y$, $45 = 9y$, $5 = y$; $\{5\}$

45. $1 - \frac{3}{4}(v + 2) = -5$, $1 - \frac{3}{4}v - \frac{3}{2} = -5$, $-\frac{3}{4}v - \frac{1}{2} = -5$, $-\frac{3}{4}v = -\frac{9}{2}$, $v = 6$; $\{6\}$

46. $9 - \frac{4}{5}(u - 3) = 1$, $9 - \frac{4}{5}u + \frac{12}{5} = 1$, $-\frac{4}{5}u + \frac{57}{5} = 1$, $-\frac{4}{5}u = -\frac{52}{5}$, $u = 13$, $\{13\}$

47. $-9 - 3(2q - 1) = -18$, $-9 - 6q + 3 = -18$, $-6q - 6 = -18$, $-6q = -12$, $q = 2$; $\{2\}$

48. $-10 + 4(3p + 10) = 18$, $-10 + 12p + 40 = 18$, $12p + 30 = 18$, $12p = -12$, $p = -1$; $\{-1\}$

49. $-2 = 4(s + 8) - 3s$, $-2 = 4s + 32 - 3s$, $-2 = s + 32$, $s = -34$; $\{-34\}$

50. $-7 = 3(t - 5) - t$, $-7 = 3t - 15 - t$, $-7 = 2t - 15$, $8 = 2t$, $4 = t$; $\{4\}$

51. $(x - 13) - (x - 5) + 2x = 0$, $x - 13 - x + 5 + 2x = 0$, $2x - 8 = 0$, $2x = 8$, $x = 4$; $\{4\}$

52. $(5 - y) + (6 - y) - (5 - y) = 0$, $5 - y + 6 - y - 5 + y = 0$, $-y + 6 = 0$, $-y = -6$, $y = 6$; $\{6\}$

53. $b - (1 - 2b) + (b - 3) = -4$, $b - 1 + 2b + b - 3 = -4$, $4b - 4 = -4$, $4b = 0$, $b = 0$; $\{0\}$

54. $(c + 3) - 2c - (1 - 3c) = 2$, $c + 3 - 2c - 1 + 3c = 2$, $2c + 2 = 2$, $2c = 0$, $c = 0$; $\{0\}$

C **55.** $5m - 3[7 - (1 - 2m)] = 0$, $5m - 21 + 3(1 - 2m) = 0$, $5m - 21 + 3 - 6m = 0$, $-m - 18 = 0$, $-m = 18$, $m = -18$; $\{-18\}$

56. $\frac{1}{5}[4(k + 2) - (3 - k)] = 4$, $\frac{4}{5}(k + 2) - \frac{1}{5}(3 - k) = 4$, $\frac{4}{5}k + \frac{8}{5} - \frac{3}{5} + \frac{1}{5}k = 4$, $k + 1 = 4$, $k = 3$; $\{3\}$

57. $5(g - 7) + 2[g - 3(g - 5)] = 0$, $5g - 35 + 2g - 6(g - 5) = 0$, $5g - 35 + 2g - 6g + 30 = 0$, $g - 5 = 0$, $g = 5$; $\{5\}$

58. $7n + 2[3(1 - n) - 2(1 + n)] = 14$, $7n + 6(1 - n) - 4(1 + n) = 14$, $7n + 6 - 6n - 4 - 4n = 14$, $-3n + 2 = 14$, $-3n = 12$, $n = -4$; $\{-4\}$

59. $3|n| - (2|n| - 2) = 9$, $3|n| - 2|n| + 2 = 9$, $|n| + 2 = 9$, $|n| = 7$, $n = 7$ or $n = -7$; $\{7, -7\}$

60. $(9|x| + 3) - 5|x| - 3 = 12$, $9|x| + 3 - 5|x| - 3 = 12$, $4|x| = 12$, $|x| = 3$, $x = 3$ or $x = -3$; $\{3, -3\}$

Page 110 • MIXED REVIEW EXERCISES

1. $\frac{1}{7}x = -23$, $x = -161$; $\{-161\}$ **2.** $\frac{x}{8} = \frac{3}{4}$, $x = 6$; $\{6\}$ **3.** $\frac{1}{5}x = 3\frac{1}{5}$, $\frac{1}{5}x = \frac{16}{5}$, $x = 16$; $\{16\}$

4. $-3 + x = 1$, $x = 4$; $\{4\}$ **5.** $2x + 7 = 13$, $2x = 6$, $x = 3$; $\{3\}$

6. $51 = y + 17$, $34 = y$; $\{34\}$ **7.** $-12 + x = -20$, $x = -8$; $\{-8\}$

8. $32 - x = 36$, $-x = 4$, $x = -4$; $\{-4\}$ **9.** $-0.25x = 1$, $x = -4$; $\{-4\}$

10. $2.3 = x + 1$, $1.3 = x$; $\{1.3\}$ **11.** $0 = 3x$, $0 = x$; $\{0\}$ **12.** $18y = 360$, $y = 20$; $\{20\}$

Page 111 • COMPUTER EXERCISES

1.
```
10 PRINT "THIS PROGRAM SOLVES THE EQUATION"
20 PRINT "      AX + B = C"
30 PRINT "WHAT ARE A, B, AND C";
40 INPUT A, B, C
50 PRINT
60 LET X = (C - B)/A
70 PRINT "THE SOLUTION IS"; X
80 END
```

a. {8} **b.** {−2} **c.** {−10} **d.** {6} **e.** {36} **f.** {11}

2. An error message results from attempting to divide by zero. There is no solution to $0x + 9 = 12$. These modifications will give a correct response if A = 0:

```
51 IF A = 0 THEN 90
90 IF B = C THEN 120
100 PRINT "NO SOLUTION"
110 GOTO 140
120 PRINT "THE EQUATION IS TRUE FOR"
130 PRINT "ALL VALUES OF X"
140 END
```

Change line 80 to: 80 GOTO 140

3. These modifications will solve $A|x| + B = C$:

```
20 PRINT "A*ABS(X) + B = C"
61 IF X < 0 THEN 100
70 PRINT "THE SOLUTIONS ARE"; X; "AND"; -X
```

a. {2, −2} **b.** no solution **c.** {0.5, −0.5}

Page 111 • CALCULATOR KEY-IN

1. no **2.** yes **3.** no

Pages 113–115 • PROBLEMS

A

1. *Step 1* Find the number. *Step 2* Let n = the number. *Step 3* $38 + 2n = 124$
Step 4 $2n = 86, n = 43$ *Step 5* The number is 43.

2. *Step 1* Find the number. *Step 2* Let n = the number. *Step 3* $3n + 5 = 197$
Step 4 $3n = 192, n = 64$ *Step 5* The number is 64.

3. *Step 1* Find the number. *Step 2* Let n = the number. *Step 3* $\frac{n}{2} - 4 = 17$
Step 4 $\frac{n}{2} = 21, n = 42$ *Step 5* The number is 42.

4. *Step 1* Find the number. *Step 2* Let n = the number. *Step 3* $\frac{n}{3} - 11 = 38$
Step 4 $\frac{n}{3} = 49, n = 147$ *Step 5* The number is 147.

5. *Step 1* Find the number. *Step 2* Let n = the number. *Step 3* $\frac{2}{3}n + 4 = 22$

Step 4 $\frac{2}{3}n = 18, n = 27$ *Step 5* The number is 27.

6. *Step 1* Find the number. *Step 2* Let n = the number. *Step 3* $\frac{3}{4}n - 8 = 91$

Step 4 $\frac{3}{4}n = 99, n = 132$ *Step 5* The number is 132.

7. *Step 1* Find the three consecutive integers.

Step 2 Let n, $(n + 1)$, and $(n + 2)$ be the consecutive integers.

Step 3 $n + (n + 1) + (n + 2) = 171$

Step 4 $n + n + 1 + n + 2 = 171, 3n + 3 = 171, 3n = 168, n = 56; n + 1 = 57; n + 2 = 58$

Step 5 The consecutive integers are 56, 57, and 58.

8. *Step 1* Find the three consecutive odd integers.

Step 2 Let n, $(n + 2)$, and $(n + 4)$ be the consecutive odd integers.

Step 3 $n + (n + 2) + (n + 4) = 105$

Step 4 $n + n + 2 + n + 4 = 105, 3n + 6 = 105, 3n = 99, n = 33; n + 2 = 35; n + 4 = 37$

Step 5 The consecutive odd integers are 33, 35, and 37.

9. *Step 1* Find the four consecutive even integers.

Step 2 Let n, $(n + 2)$, $(n + 4)$, and $(n + 6)$ be the consecutive even integers.

Step 3 $n + (n + 2) + (n + 4) + (n + 6) = 244$

Step 4 $n + n + 2 + n + 4 + n + 6 = 244, 4n + 12 = 244, 4n = 232, n = 58; n + 2 = 60; n + 4 = 62; n + 6 = 64$

Step 5 The consecutive even integers are 58, 60, 62, and 64.

10. *Step 1* Find the five consecutive integers.

Step 2 Let n, $(n + 1)$, $(n + 2)$, $(n + 3)$, and $(n + 4)$ be the consecutive integers.

Step 3 $n + (n + 1) + (n + 2) + (n + 3) + (n + 4) = 195$

Step 4 $n + n + 1 + n + 2 + n + 3 + n + 4 = 195, 5n + 10 = 195, 5n = 185, n = 37; n + 1 = 38; n + 2 = 39; n + 3 = 40; n + 4 = 41$

Step 5 The consecutive integers are 37, 38, 39, 40, and 41.

11. *Step 1* Find the number of hamburgers sold with cheese and the number sold without cheese.

Step 2 Let x = number sold without cheese. *Step 3* $\frac{x}{2} + x = 495$

Step 4 $\frac{3}{2}x = 495, x = 330; \frac{x}{2} = 165$

Step 5 330 hamburgers without cheese were sold and 165 hamburgers with cheese were sold.

12. *Step 1* Find how much each tank holds.

Step 2 Let t = number of barrels the old tank holds.

Step 3 $t + (t + 350) = 3650$

Step 4 $t + t + 350 = 3650$, $2t + 350 = 3650$, $2t = 3300$, $t = 1650$; $t + 350 = 2000$

Step 5 The old tank holds 1650 barrels and the new tank holds 2000 barrels.

13. *Step 1* Find the number of weeks. *Step 2* Let w = number of weeks.

Step 3 $88 + 3.5w = 200$ *Step 4* $3.5w = 112$, $w = 32$

Step 5 It will take Brian 32 weeks.

14. *Step 1* Find how many minutes it will take to fill the tank.

Step 2 Let m = minutes to fill tank. *Step 3* $240 + 25m = 1000$

Step 4 $25m = 760$, $m = 30.4$ *Step 5* It will take 30.4 min to fill the tank.

15. *Step 1*

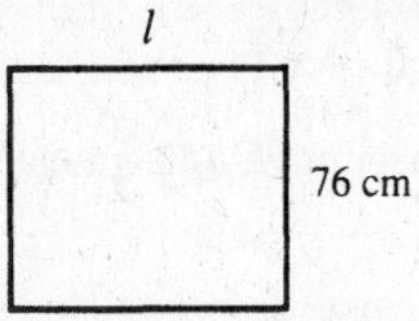

Step 2 Let l = length of rectangle. *Step 3* $2l + 2(76) = 332$

Step 4 $2l + 152 = 332$, $2l = 180$, $l = 90$ *Step 5* The rectangle is 90 cm long.

16. *Step 1*

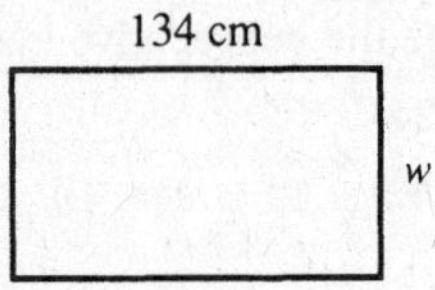

Step 2 Let w = width of rectangle. *Step 3* $2w + 2(134) = 408$

Step 4 $2w + 268 = 408$, $2w = 140$, $w = 70$ *Step 5* The rectangle is 70 cm wide.

17. *Step 1*

Step 2 Let l = length of each leg. *Step 3* $2l + 85 = 345$

Step 4 $2l = 260$, $l = 130$ *Step 5* Each leg of the triangle is 130 cm long.

18. *Step 1*

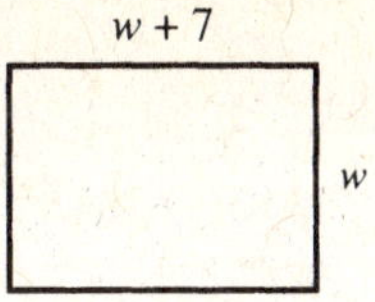

Step 2 Let w and $w + 7$ represent the width and length of the rectangle.

Step 3 $2w + 2(w + 7) = 78$

Step 4 $2w + 2w + 14 = 78$, $4w + 14 = 78$, $4w = 64$, $w = 16$; $w + 7 = 23$

Step 5 The rectangle is 16 cm wide and 23 cm long.

19. *Step 1*

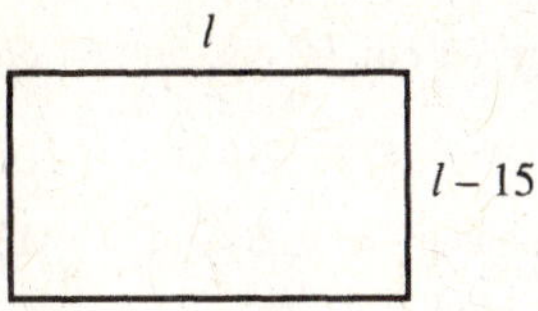

Step 2 Let l and $l - 15$ be the length and width of the rectangle.

Step 3 $2l + 2(l - 15) = 98$

Step 4 $2l + 2l - 30 = 98$, $4l - 30 = 98$, $4l = 128$, $l = 32$; $l - 15 = 17$

Step 5 The rectangle is 32 cm long and 17 cm wide.

20. *Step 1*

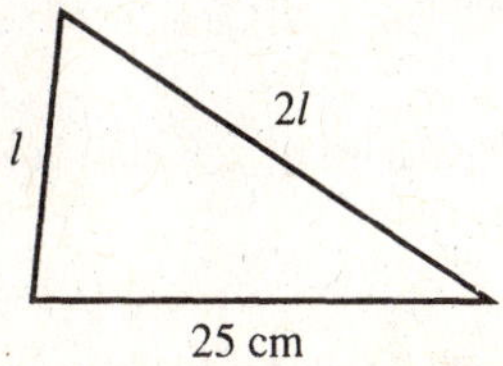

Step 2 Let l = length of shortest side and $2l$ = longest side.

Step 3 $l + 2l + 25 = 70$ *Step 4* $3l + 25 = 70$, $3l = 45$, $l = 15$; $2l = 30$

Step 5 The shortest side is 15 cm long, the longest side is 30 cm long, and the third side is 25 cm long.

21. *Step 1*

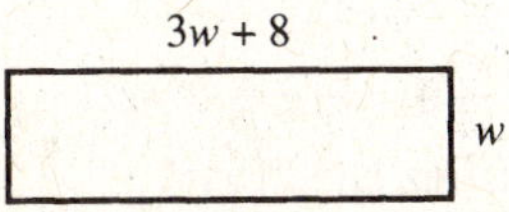

Step 2 Let w = width of the rectangle and $3w + 8$ = length.

Step 3 $2(w) + 2(3w + 8) = 128$

Step 4 $2w + 6w + 16 = 128$, $8w + 16 = 128$, $8w = 112$, $w = 14$; $3w + 8 = 50$

Step 5 The rectangle is 14 cm wide and 50 cm long.

22. *Step 1*

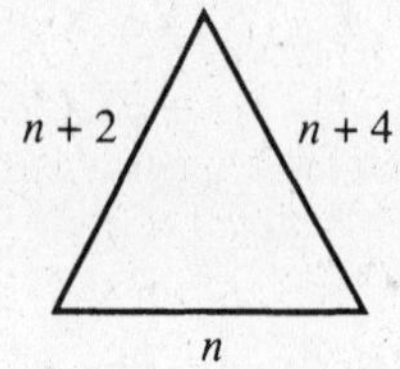

Step 2 Let n, $n + 2$, and $n + 4$ be the lengths of the sides.

Step 3 $n + (n + 2) + (n + 4) = 186$

Step 4 $n + n + 2 + n + 4 = 186$, $3n + 6 = 186$, $3n = 180$, $n = 60$; $n + 2 = 62$; $n + 4 = 64$

Step 5 The sides of the triangle measure 60 cm, 62 cm, and 64 cm.

B **23.** *Step 1* Find the measure of each angle.

Step 2 Let a = measure of $\angle A$, $\frac{a}{3}$ = measure of $\angle B$, and $a - 16$ = measure of $\angle C$.

Step 3 $a + \frac{a}{3} + (a - 16) = 180$

Step 4 $a + \frac{1}{3}a + a - 16 = 180$, $\frac{7}{3}a - 16 = 180$, $\frac{7}{3}a = 196$, $\frac{3}{7}\left(\frac{7}{3}a\right) = \frac{3}{7}(196)$, $a = 84$

Step 5 $\angle A$ measures 84°, $\angle B$ measures $\frac{84}{3} = 28°$, $\angle C$ measures $84 - 16 = 68°$

24. *Step 1* Find the measure of each angle.

Step 2 Let c = measure of $\angle C$, $c + 4$ = measure of $\angle B$, and $2(c + 4)$ = measure of $\angle A$.

Step 3 $c + (c + 4) + 2(c + 4) = 180$

Step 4 $c + c + 4 + 2c + 8 = 180$, $4c + 12 = 180$, $4c = 168$, $c = 42$; $c + 4 = 46$; $2(c + 4) = 92$

Step 5 $\angle A$ measures 92°, $\angle B$ measures 46°, and $\angle C$ measures 42°.

25. *Step 1* Find the lengths of the three sides.

Step 2 Let a = length of $\overline{AC}$, $a - 9$ = length of $\overline{AB}$, and $a + 3$ = length of $\overline{BC}$.

Step 3 $a + a - 9 + a + 3 = 48$

Step 4 $3a - 6 = 48$, $3a = 54$, $a = 18$; $a - 9 = 9$; $a + 3 = 21$

Step 5 The three sides measure 18 cm, 9 cm, and 21 cm.

26. *Step 1*

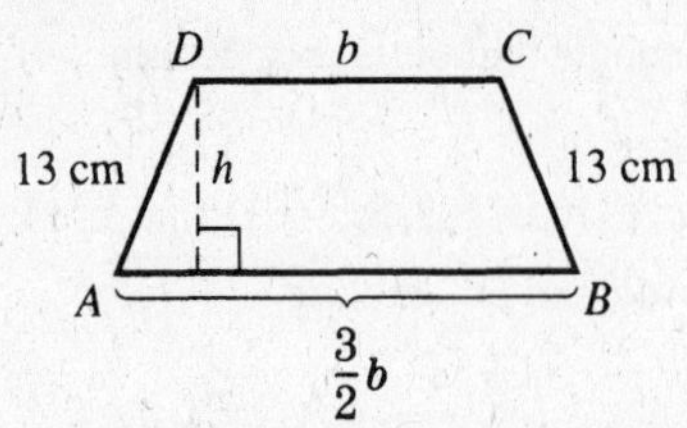

Step 2 Let b = length of shorter base and $\frac{3}{2}b$ = length of longer base.

Step 3 $13 + b + 13 + \frac{3}{2}b = 76$

(Continued)

Step 4 $\frac{5}{2}b + 26 = 76, \frac{5}{2}b = 50, b = 20; \frac{3}{2}b = 30$

Step 5 **a.** $\overline{AB}$ is 30 cm long and $\overline{DC}$ is 20 cm long.

b. area $= \frac{1}{2}(12)(20 + 30) = (6)(50) = 300 \text{ cm}^2$

27. *Step 1* Find how much money each has.

Step 2 Let r = amount Rudy has, $r + 11$ = amount Denise has, and $(r + 11) + 5$ = amount Theo has.

Step 3 $r + (r + 11) + [(r + 11) + 5] = 45$

Step 4 $r + r + 11 + r + 11 + 5 = 45, 3r + 27 = 45, 3r = 18, r = 6, r + 11 = 17, (r + 11) + 5 = 22$

Step 5 Rudy has \$6, Denise has \$17, and Theo has \$22.

28. *Step 1* Find how much money each has.

Step 2 Let l = amount Lian has, $l - 6$ = amount Nora has, and $2(l - 6)$ = amount Chandra has.

Step 3 $l + (l - 6) + 2(l - 6) = 54$

Step 4 $l + l - 6 + 2l - 12 = 54, 4l - 18 = 54, 4l = 72, l = 18; l - 6 = 12; 2(l - 6) = 24$

Step 5 Lian has \$18, Nora has \$12, and Chandra has \$24.

29. *Step 1* Find the number of bottles capped by each machine.

Step 2 Let b = number of bottles capped by B, $2b$ = number of bottles capped by A, and $2b + 500$ = number of bottles capped by C.

Step 3 $b + 2b + (2b + 500) = 40{,}000$

Step 4 $b + 2b + 2b + 500 = 40{,}000, 5b + 500 = 40{,}000, 5b = 39{,}500, b = 7900; 2b = 15{,}800; 2b + 500 = 16{,}300$

Step 5 Machine A caps 15,800 bottles, Machine B caps 7900 bottles, and Machine C caps 16,300 bottles.

30. *Step 1* Find the price of each item.

Step 2 Let a = price of apple, $\frac{3}{2}a$ = price of milk, and $a + 1.40$ = price of sandwich.

Step 3 $a + \frac{3}{2}a + (a + 1.40) = 3.50$

Step 4 $a + 1.5a + a + 1.40 = 3.50, 3.5a + 1.40 = 3.50, 3.5a = 2.10, a = 0.60;$

$\frac{3}{2}a = .90; a + 1.40 = 2.00$

Step 5 The apple costs \$.60, the milk costs \$.90, and the sandwich costs \$2.00.

C **31.** *Step 1* Find the base price of the car. *Step 2* Let b = base price.

Step 3 $b + \frac{b}{10} + \frac{b}{50} = 24{,}416$ *Step 4* $\frac{28}{25}b = 24{,}416, b = 21{,}800$

Step 5 The base price was \$21,800.

32. *Step 1* Find Keiko's second and third scores.

Step 2 Let s = score on second test and $s + 1$ = score on third test.

Step 3 $\frac{1}{3}(72 + s + s + 1) = 83$

Step 4 $72 + s + s + 1 = 249$, $2s + 73 = 249$, $2s = 176$, $s = 88$; $s + 1 = 89$

Step 5 Keiko scored 88 and 89 on her second and third tests.

33. *Step 1* Find the number. *Step 2* Let n = the number.

Step 3 $|-7 + 2n| = 23$

Step 4 $-7 + 2n = 23$ or $-7 + 2n = -23$; $2n = 30$, $n = 15$; or $2n = -16$, $n = -8$

Step 5 The number is 15 or -8.

Page 115 • MIXED REVIEW EXERCISES

1. $-5 + y = 1$, $y = 6$; $\{6\}$ **2.** $x - 1.5 = -3$, $x = -1.5$; $\{-1.5\}$

3. $y + 8 = 21$, $y = 13$; $\{13\}$ **4.** $\frac{2}{5}y = 2$, $y = 5$; $\{5\}$ **5.** $-21 = \frac{c}{3}$, $-63 = c$; $\{-63\}$

6. $-\frac{1}{5}x = 20$, $x = -100$; $\{-100\}$ **7.** $41 = y + 11$, $30 = y$; $\{30\}$

8. $x - 20 = 37$, $x = 57$; $\{57\}$ **9.** $0 = 1.75y$, $0 = y$; $\{0\}$

10. $5y + 6 = 16$, $5y = 10$, $y = 2$; $\{2\}$ **11.** $2x - 5 = 15$, $2x = 20$, $x = 10$; $\{10\}$

12. $2(a - 3) + 6 = -3$, $2a - 6 + 6 = -3$, $2a = -3$, $a = -\frac{3}{2}$; $\left\{-\frac{3}{2}\right\}$

Page 118 • WRITTEN EXERCISES

A

1. $5n = 2n + 6$, $3n = 6$, $n = 2$; $\{2\}$ **2.** $8a = 2a + 30$, $6a = 30$, $a = 5$; $\{5\}$

3. $y = 24 - 3y$, $4y = 24$, $y = 6$; $\{6\}$ **4.** $2b = 80 - 8b$, $10b = 80$, $b = 8$; $\{8\}$

5. $12n = 34 - 5n$, $17n = 34$, $n = 2$; $\{2\}$ **6.** $3x = 27 - 15x$, $18x = 27$, $x = \frac{3}{2}$; $\left\{\frac{3}{2}\right\}$

7. $30 = 8 - 2x$, $22 = -2x$, $-11 = x$; $\{-11\}$

8. $51 = 9 - 3x$, $42 = -3x$, $-14 = x$; $\{-14\}$

9. $51a - 56 = 44a$, $7a - 56 = 0$, $7a = 56$, $a = 8$; $\{8\}$

10. $39c + 78 = 33c$, $6c + 78 = 0$, $6c = -78$, $c = -13$; $\{-13\}$

11. $98 - 4b = -11b$, $98 + 7b = 0$, $7b = -98$, $b = -14$; $\{-14\}$

12. $-7a = -12a - 65$, $5a = -65$, $a = -13$; $\{-13\}$

13. $4n + 5 = 6n + 7$, $5 = 2n + 7$, $-2 = 2n$, $-1 = n$; $\{-1\}$

14. $5p - 9 = 2p + 12$, $3p - 9 = 12$, $3p = 21$, $p = 7$; $\{7\}$

15. $3p - 8 = 13 - 4p$, $7p - 8 = 13$, $7p = 21$, $p = 3$; $\{3\}$

16. $89 + x = 2 - 2x$, $89 + 3x = 2$, $3x = -87$, $x = -29$; $\{-29\}$

17. $71 - 5x = 9x - 13$, $71 = 14x - 13$, $84 = 14x$, $6 = x$; $\{6\}$

18. $5n + 1 = 5n - 1$, $1 = -1$; no solution

19. $2(x - 6) = 3x$, $2x - 12 = 3x$, $-12 = x$; $\{-12\}$

20. $4(y-6)=7y,\ 4y-24=7y,\ -24=3y,\ -8=y;\ \{-8\}$

21. $8(5-n)=2n,\ 40-8n=2n,\ 40=10n,\ 4=n;\ \{4\}$

22. $7(2-m)=3m,\ 14-7m=3m,\ 14=10m,\ \frac{14}{10}=m,\ \frac{7}{5}=m;\ \left\{\frac{7}{5}\right\}$

23. $\frac{1}{2}x+5=x,\ 5=\frac{1}{2}x,\ 10=x;\ \{10\}$ **24.** $\frac{2}{3}x-7=x,\ -7=\frac{1}{3}x,\ -21=x;\ \{-21\}$

25. $\frac{4+y}{5}=y,\ 4+y=5y,\ 4=4y,\ 1=y;\ \{1\}$

26. $\frac{x-2}{3}=x,\ x-2=3x,\ -2=2x,\ -1=x;\ \{-1\}$

27. $\frac{9-2y}{7}=y,\ 9-2y=7y,\ 9=9y,\ 1=y;\ \{1\}$

28. $\frac{6-4y}{2}=y,\ 6-4y=2y,\ 6=6y,\ 1=y;\ \{1\}$

29. $\frac{4n-28}{3}=2n,\ 4n-28=6n,\ -28=2n,\ -14=n;\ \{-14\}$

30. $\frac{23-11c}{7}=5c,\ 23-11c=35c,\ 23=46c,\ \frac{1}{2}=c;\ \left\{\frac{1}{2}\right\}$

B **31.** $\frac{1}{3}(12-6x)=4-2x,\ 4-2x=4-2x$; identity

32. $\frac{1}{4}(20-4a)=6-a,\ 5-a=6-a,\ 5=6$; no solution

33. $5(2+n)=3(n+6),\ 10+5n=3n+18,\ 10+2n=18,\ 2n=8,\ n=4;\ \{4\}$

34. $3(30+s)=4(s+19),\ 90+3s=4s+76,\ 90=s+76,\ 14=s;\ \{14\}$

35. $5u+5(1-u)=u+8,\ 5u+5-5u=u+8,\ 5=u+8,\ -3=u;\ \{-3\}$

36. $2(g-2)-4=2(g-3),\ 2g-4-4=2g-6,\ 2g-8=2g-6,\ -8=-6$; no solution

37. $3(m+5)-6=3(m+3),\ 3m+15-6=3m+9,\ 3m+9=3m+9$; identity

38. $3(2+v)-4v=v+16,\ 6+3v-4v=v+16,\ 6-v=v+16,\ 6=2v+16,\ -10=2v,\ -5=v;\ \{-5\}$

39. $3(5y+2)-y=2(y-3),\ 15y+6-y=2y-6,\ 14y+6=2y-6,\ 12y+6=-6,\ 12y=-12,\ y=-1;\ \{-1\}$

40. $4(3y-1)+13=5y+2,\ 12y-4+13=5y+2,\ 12y+9=5y+2,\ 7y+9=2,\ 7y=-7,\ y=-1;\ \{-1\}$

41. $6r-2(2-r)=4(2r-1),\ 6r-4+2r=8r-4,\ 8r-4=8r-4$; identity

42. $5x+2(1-x)=2(2x-1),\ 5x+2-2x=4x-2,\ 3x+2=4x-2,\ 2=x-2,\ 4=x;\ \{4\}$

43. $3+4(p+2)=2p+3(p+4),\ 3+4p+8=2p+3p+12,\ 4p+11=5p+12,\ 11=p+12,\ -1=p;\ \{-1\}$

44. $4(a+2)=14-2(3-2a),\ 4a+8=14-6+4a,\ 4a+8=4a+8$; identity

C **45.** $3x+2[1-3(x+2)]=2x,\ 3x+2-6(x+2)=2x,\ 3x+2-6x-12=2x,\ -3x-10=2x,\ -10=5x,\ -2=x;\ \{-2\}$

46. $2[5(w+3)-(w+1)]=3(1+w),\ 2[5w+15-w-1]=3+3w,\ 2(4w+14)=3+3w,\ 8w+28=3+3w,\ 5w+28=3,\ 5w=-25,\ w=-5;\ \{-5\}$

47. $5(2m + 3) - (1 - 2m) = 2[3(3 + 2m) - (3 - m)]$, $10m + 15 - 1 + 2m = 2(9 + 6m - 3 + m)$, $12m + 14 = 2(7m + 6)$, $12m + 14 = 14m + 12$, $14 = 2m + 12$, $2 = 2m$, $1 = m$; $\{1\}$

48. $3(r + 1) - [2(3 - 2r) - 3(3 - r)] = 2(r + 5) - 4$, $3r + 3 - (6 - 4r - 9 + 3r) = 2r + 10 - 4$, $3r + 3 - 6 + 4r + 9 - 3r = 2r + 6$, $4r + 6 = 2r + 6$, $2r + 6 = 6$, $2r = 0$, $r = 0$; $\{0\}$

Pages 118–119 • PROBLEMS

A

1. Let n = the number. $n = 96 + -n$, $2n = 96$, $n = 48$; The number is 48.

2. Let n = the number. $n = -n - 38$, $2n = -38$, $n = -19$; The number is -19.

3. Let n = the number. $9n = n + 56$, $8n = 56$, $n = 7$; The number is 7.

4. Let n = the number. $n = 3(-n) + 68$, $n = -3n + 68$, $4n = 68$, $n = 17$; The number is 17.

5. Let n = the number. $3n - 8 = 2n + 15$, $n - 8 = 15$, $n = 23$; The number is 23.

6. Let n = the number. $4n + 25 = 6n - 13$, $25 = 2n - 13$, $38 = 2n$, $19 = n$; $\{19\}$

7. Let n = the smaller number and $n + 1$ = the larger number. $n + 1 = 2n + 15$, $1 = n + 15$, $-14 = n$, $n + 1 = -13$; The numbers are -14 and -13.

8. Let n = the smaller even integer and $n + 2$ = the larger even integer. $n + 2 = 2n + 20$, $2 = n + 20$, $-18 = n$, $n + 2 = -16$; The consecutive even integers are -18 and -16.

9. Let n = baskets Cliff shot, $3n$ = baskets Lyle shot, and $n + 12$ = baskets Kyle shot. $3n = n + 12$, $2n = 12$, $n = 6$, $3n = 18$, $n + 12 = 18$; Cliff shot 6 baskets, Kyle and Lyle shot 18 baskets each.

10. Let b = mass of one steel ball. $5b = b + 100$, $4b = 100$, $b = 25$; Each steel ball has mass 25 g.

B

11. Let x = one number and $15 - x$ = other number. $3x = 5(15 - x) - 11$, $3x = 75 - 5x - 11$, $8x = 64$, $x = 8$, $15 - x = 7$; The numbers are 8 and 7.

12. Let x = smaller integer and $x + 9$ = larger integer. $5x = 3(x + 9) + 7$, $5x = 3x + 27 + 7$, $2x = 34$, $x = 17$, $x + 9 = 26$; The numbers are 17 and 26.

13. Let s = length of the shortest side, $s + 2$ = length of the next side, and $s + 4$ = length of the longest side. $s + 4 = [s + (s + 2) + (s + 4)] - 22$, $s + 4 = s + s + 2 + s + 4 - 22$, $s + 4 = 3s - 16$, $4 = 2s - 16$, $20 = 2s$, $10 = s$; $s + 4 = 14$; The longest side is 14 units long.

14.

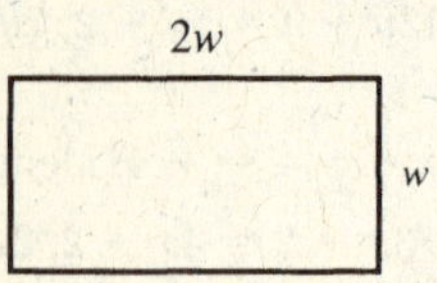

Let w = width of the rectangle. $2(w) + 2(2w) = w + 84$, $2w + 4w = w + 84$, $6w = w + 84$, $5w = 84$, $w = 16.8$; $2w = 33.6$; The rectangle is 16.8 cm wide and 33.6 cm long.

15. Let y = the number of years. $16{,}000 + 1500y = 19{,}300 + 950y$, $16{,}000 + 550y = 19{,}300$, $550y = 3300$, $y = 6$; Mei and Janet will earn the same salary in 6 years.

16. Let m = number of minutes. $550 + 75m = 1600 - 75m$, $550 + 150m = 1600$, $150m = 1050$, $m = 7$; The two tanks will hold the same amount in 7 minutes.

17. Let m = amount of money Marcia has, $2m$ = amount Eric has, and $m + 175$ = amount Laurel has. $m + 175 = m + 2m$, $m + 175 = 3m$, $175 = 2m$, $87.50 = m$; $2m = 175$; $m + 175 = 262.50$; Marcia has \$87.50, Eric has \$175, and Laurel has \$262.50.

18. Let b = weight of the boat, $b - 1500$ = weight of motor, $b - 1900$ = weight of trailer. $b + (b - 1500) = 5(b - 1900)$, $b + b - 1500 = 5b - 9500$, $2b - 1500 = 5b - 9500$, $-1500 = 3b - 9500$, $8000 = 3b$, $\frac{8000}{3} = b$; The boat weighs $2666\frac{2}{3}$ pounds.

19. Let n = smallest integer. $n + (n + 1) + (n + 2) = n + 200$, $n + n + 1 + n + 2 = n + 200$, $3n + 3 = n + 200$, $2n + 3 = 200$, $2n = 197$, $n = 98.5$; 98.5 is not an integer. $\therefore$ it is impossible.

20. Let n = smallest number. $n + (n + 2) + (n + 4) + (n + 6) = n + (n + 2) + 10$, $n + n + 2 + n + 4 + n + 6 = n + n + 2 + 10$, $4n + 12 = 2n + 12$, $2n + 12 = 12$, $2n = 0$, $n = 0$; The integers are 0, 2, 4, and 6. $\therefore$ it is possible and there is one solution.

Page 120 • MIXED REVIEW EXERCISES

1. $4 + \left(-\frac{1}{2}\right) + \left(-\frac{3}{2}\right) = \frac{8 - 1 - 3}{2} = \frac{4}{2} = 2$ **2.** $-2\frac{3}{5} + 1\frac{1}{5} = -\frac{13}{5} + \frac{6}{5} = -\frac{7}{5} = -1\frac{2}{5}$

3. $-212 - (-13) = -212 + 13 = -199$ **4.** $17x + (-3)x - 5 = 17x - 3x - 5 = 14x - 5$

5. $-5y + 6 + 20y + 12 = 15y + 18$ **6.** $7(-3)(-10)(-2) = (-21)(20) = -420$

7. $-3 - x = 9$, $-x = 12$, $x = -12$; $\{-12\}$

8. $5 - (1 + z) = 3$, $5 - 1 - z = 3$, $4 - z = 3$, $-z = -1$, $z = 1$; $\{1\}$

9. $4x = -392$, $x = -98$; $\{-98\}$ **10.** $\frac{1}{2}x = 4\frac{1}{2}$, $\frac{1}{2}x = \frac{9}{2}$, $x = 9$; $\{9\}$ **11.** $\frac{x}{7} = 6$, $x = 42$; $\{42\}$

12. $-12\frac{2}{3} = -\frac{1}{3}x$, $-\frac{38}{3} = -\frac{1}{3}x$, $38 = x$; $\{38\}$

Page 120 • COMPUTER EXERCISES

```
10 PRINT "THIS PROGRAM SOLVES EQUATIONS"
20 PRINT "OF THE FORM AX + B = CX + D"
30 PRINT "INPUT A, B, C, D"
40 INPUT A, B, C, D
50 IF A <> C THEN 110
60 IF B = D THEN 90
70 PRINT "NO SOLUTION"
80 GOTO 120
90 PRINT "IDENTITY"
100 GOTO 120
110 PRINT "THE SOLUTION IS";(D - B)/(A - C)
120 END
```

1. $\{-3\}$ **2.** 4 **3.** no solution **4.** identity **5.** $\{5\}$ **6.** $\{1.2\}$

Pages 122–125 • PROBLEMS

A 1.

	Hours per shift	× Number of shifts	= Total hours worked
6 h shift	6	$2x$	$6(2x)$
8 h shift	8	x	$8x$

$6(2x) + 8x = 280$, $12x + 8x = 280$, $20x = 280$, $x = 14$; Maria worked 14 full 8 hour shifts.

2.

	Rides per ticket	× Number of tickets sold	= Total rides
12-ride ticket	12	n	$12n$
Round-trip ticket	2	$30n$	$2(30n)$

$12n + 2(30n) = 1440$, $12n + 60n = 1440$, $72n = 1440$, $n = 20$; $30n = 600$; 600 round-trip tickets were sold.

3.

	Weight per box	× Number of boxes	= Total weight
Pecans	3	x	$3x$
Walnuts	2	$x - 24$	$2(x - 24)$

$3x + 2(x - 24) = 462$, $3x + 2x - 48 = 462$, $5x - 48 = 462$, $5x = 510$, $x = 102$; $3x = 306$; The boxes of pecans weighed 306 lb.

4.

	Movie length	× Number of movies	= Total time
Space movies	$1\frac{1}{2}$	$2a$	$1\frac{1}{2}(2a)$
Mystery movies	2	a	$2a$

$1\frac{1}{2}(2a) + 2a = 15$, $\frac{3}{2}(2a) + 2a = 15$, $3a + 2a = 15$, $5a = 15$, $a = 3$; $1\frac{1}{2}(2a) = 9$;

Joel spent 9 hours watching space adventure movies.

5.

	Calories per fruit	× Number of fruit	= Total Calories
Orange	$p + 30$	7	$7(p + 30)$
Peach	p	13	$13p$

$7(p + 30) = 13p$, $7p + 210 = 13p$, $210 = 6p$, $35 = p$; $p + 30 = 65$;
There are 65 Calories in an orange and 35 Calories in a peach.

6.

	Calories per vegetable ×	Number of vegetables =	Total Calories
Celery	x	10	$10x$
Carrot	$x + 13$	5	$5(x + 13)$

$10x + 5(x + 13) = 170$, $10x + 5x + 65 = 170$, $15x + 65 = 170$, $15x = 105$, $x = 7$; $x + 13 = 20$; A stalk of celery has 7 Calories and a carrot has 20 Calories.

7.

	Length	Width
Red rectangle	$w + 15$	w
Blue rectangle	$w + 13$	$w + 5$

$2(w + 5) + 2(w + 13) = 72$, $2w + 10 + 2w + 26 = 72$, $4w + 36 = 72$, $4w = 36$, $w = 9$; $w + 15 = 24$; $w + 5 = 14$; $w + 13 = 22$; The red rectangle is 9 cm wide and 24 cm long and the blue rectangle is 14 cm wide and 22 cm long.

8.

	Length	Width
First rectangle	$2w$	w
Second rectangle	$2w + 8$	$w - 3$

$2(w - 3) + 2(2w + 8) = 154$, $2w - 6 + 4w + 16 = 154$, $6w + 10 = 154$, $6w = 144$, $w = 24$; $2w = 48$; $2w + 8 = 56$; $w - 3 = 21$; The first rectangle is 24 cm wide and 48 cm long and the second rectangle is 21 cm wide and 56 cm long.

B 9.

	Tutoring	Pumping gas	Weekly wages
Brian	$2x$	x	$2x + x$
Sister			$\frac{1}{4}(2x)$

$2x + x = \frac{1}{4}(2x) + 150$, $3x = \frac{1}{2}x + 150$, $\frac{5}{2}x = 150$, $x = 60$, $2x = 120$; Brian earns \$120 as a tutor.

10.

	Actuary	Writer	Total income
Mona	$3x$	x	$3x + x$
Brother			$\frac{1}{2}(3x)$

$3x + x = \frac{1}{2}(3x) + 40{,}000$, $4x = \frac{3}{2}x + 40{,}000$, $\frac{5}{2}x = 40{,}000$, $x = 16{,}000$, $3x = 48{,}000$; Mona earns \$48,000 as an actuary.

11.

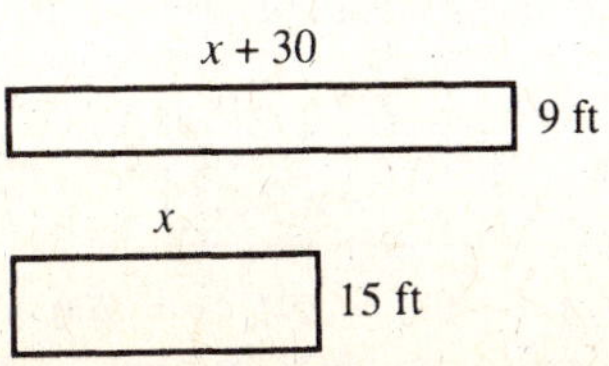

$9(x + 30) = 15x$, $9x + 270 = 15x$, $270 = 6x$, $45 = x$; $x + 30 = 75$; The first roll is 9 ft wide and 75 ft long, and the second roll is 15 ft wide and 45 ft long.

12.

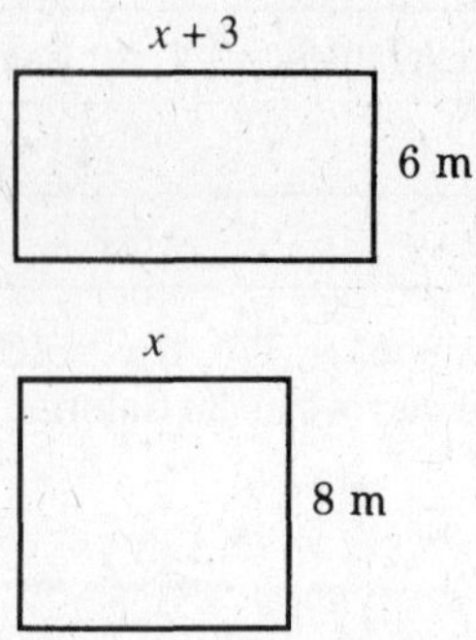

$6(x + 3) = 8x$, $6x + 18 = 8x$, $18 = 2x$, $9 = x$; $x + 3 = 12$; Leo's garden is 6 m by 12 m and Jen's garden is 8 m by 9 m.

13.

	Cars sold in March	Cars sold in April
Rodney	$2x$	$2x - 5$
Greg	x	$x + 3$

$2x - 5 = x + 3$, $x - 5 = 3$, $x = 8$; $2x = 16$; In March, Rodney sold 16 cars and Greg sold 8 cars.

14.

	First game	Second game
Maria	$3x$	$3x - 7$
Holly	x	$x + 9$

$3x - 7 = x + 9$, $2x - 7 = 9$, $2x = 16$, $x = 8$; $3x = 24$; In the first game, Maria scored 24 points and Holly scored 8 points.

15.

	Calories per cup ×	Number of cups =	Total Energy
Sunflower seeds	x	2	$2x$
Raisins	$x - 400$	3	$3(x - 400)$

$2x + 3(x - 400) = 2900$, $2x + 3x - 1200 = 2900$, $5x - 1200 = 2900$, $5x = 4100$, $x = 820$; $x - 400 = 420$; Raisins have 420 Calories per cup.

16. Let w = height of Washington monument and $w + 497$ = height of Eiffel Tower. $2(w - 58) = (w + 497) - 58$, $2w - 116 = w + 497 - 58$, $2w - 116 = w + 439$, $w - 116 = 439$, $w = 555$; $w + 497 = 1052$; The Washington Monument is 555 ft tall and the Eiffel Tower is 1052 ft tall.

17. Let n = height of Niagara Falls and $n + 750$ = height of upper Angel Falls. $16(n - 7) = (n + 750) - 7$, $16n - 112 = n + 750 - 7$, $16n - 112 = n + 743$, $15n - 112 = 743$, $15n = 855$, $n = 57$; $n + 750 = 807$; Niagara Falls are 57 m high and the upper Angel Falls are 807 m high.

18. Let s = cost of a bowl of soup.

	Cost per item ×	Number of items =	Total cost
Juice	$s - 50$	9	$9(s - 50)$
Fruit cup	$s + 50$	5	$5(s + 50)$

$9(s - 50) = 5(s + 50)$, $9s - 450 = 5s + 250$, $4s - 450 = 250$, $4s = 700$, $s = 175$; $s - 50 = 125$; $s + 50 = 225$; A carton of juice costs \$1.25, a fruit cup costs \$2.25, and a bowl of soup costs \$1.75.

19.

	Calories per serving	× Number of servings	= Total Calories
Peas	x	1	x
Carrots	$x - 45$	1	$x - 45$
Green beans	$x - 50$	3	$3(x - 50)$

$(x - 45) + 3(x - 50) = x, x - 45 + 3x - 150 = x, 4x - 195 = x, 3x - 195 = 0,$
$3x = 195, x = 65$; There are 65 Calories in one serving of peas.

C 20.

	First meet	Second meet
East High	$\frac{2}{3}w$	$\frac{2}{3}w + 7$
West High	w	$w - 7$

$w - 7 = \left(\frac{2}{3}w + 7\right) - 3, w - 7 = \frac{2}{3}w + 7 - 3, w - 7 = \frac{2}{3}w + 4, \frac{1}{3}w - 7 = 4,$

$\frac{1}{3}w = 11, w = 33; w - 7 = 26; \frac{2}{3}w = 22; \frac{2}{3}w + 7 = 29$; East High scored 22 and 29 points, and West High scored 33 and 26 points.

Page 125 • MIXED REVIEW EXERCISES

1. $25z = 600, z = 24; \{24\}$ **2.** $6 = \frac{2}{5}x, 15 = x; \{15\}$ **3.** $12z - 3z = 0, 9z = 0, z = 0; \{0\}$

4. $195 = 3x, 65 = x; \{65\}$ **5.** $7y + 3 = 24, 7y = 21, y = 3; \{3\}$

6. $-12 + 3y = -36, 3y = -24, y = -8; \{-8\}$ **7.** $5x - 2x = 21, 3x = 21, x = 7; \{7\}$

8. $4(x + 2) = 6x, 4x + 8 = 6x, 8 = 2x, 4 = x; \{4\}$

9. $7x - 11 = 2x + 44, 5x - 11 = 44, 5x = 55, x = 11; \{11\}$

10. $41 - x = -1 - 7x, 41 + 6x = -1, 6x = -42, x = -7; \{-7\}$

11. $-x = 2x - 54, 0 = 3x - 54, 54 = 3x, 18 = x; \{18\}$

12. $5(y + 2) - 3(y - 1) = -27, 5y + 10 - 3y + 3 = -27, 2y + 13 = -27, 2y = -40,$
$y = -20; \{-20\}$

Pages 127–129 • PROBLEMS

A 1.

	Number	× Price each	= Cost
Fancy	f	8	$8f$
Plain	$30 - f$	4	$4(30 - f)$

$8f + 4(30 - f) = 168,$
$8f + 120 - 4f = 168, 4f + 120 = 168,$
$4f = 48, f = 12$; 12 students bought fancy pennants.

2.

	Number ×	Price each =	Cost
Adult	$920 - s$	4	$4(920 - s)$
Student	s	2	$2s$

$4(920 - s) + 2s = 2446$, $3680 - 4s + 2s = 2446$, $3680 - 2s = 2446$, $3680 = 2446 + 2s$, $1234 = 2s$, $617 = s$; 617 student tickets were sold.

3.

	Number ×	Value of coin =	Total Value
Dimes	$40 - n$	10	$10(40 - n)$
Nickels	n	5	$5n$

$10(40 - n) + 5n = 290$, $400 - 10n + 5n = 290$, $400 - 5n = 290$, $400 = 290 + 5n$, $110 = 5n$, $22 = n$; There are 22 nickels.

4.

	Number ×	Value of coin (¢) =	Total value (¢)
Dimes	d	10	$10d$
Nickels	$52 - d$	5	$5(52 - d)$

$10d + 5(52 - d) = 450$, $10d + 260 - 5d = 450$, $5d + 260 = 450$, $5d = 190$, $d = 38$, $52 - d = 14$; There are 14 nickels.

5.

	Number ×	Price each ($) =	Cost ($)
Bought	b	1.50	$1.50b$
Sold	$b - 20$	3.00	$3(b - 20)$

$3(b - 20) - 1.5b = 15$, $3b - 60 - 1.5b = 15$, $1.5b - 60 = 15$, $1.5b = 75$, $b = 50$; Hans bought 50 programs.

6.

	Number ×	Price each (¢) =	Cost (¢)
Bought	12	b	$12b$
Sold	10	$b + 20$	$10(b + 20)$

$10(b + 20) - 12b = 100$, $10b + 200 - 12b = 100$, $-2b + 200 = 100$, $200 = 100 + 2b$, $100 = 2b$, $50 = b$, $b + 20 = 70$; Celia sold each apple for 70¢.

B **7.**

	Number ×	Value of coin (¢) =	Total Value (¢)
Nickels	$2q$	5	$5(2q)$
Quarters	q	25	$25q$

$5(2q) + 25q = 490$, $10q + 25q = 490$, $35q = 490$, $q = 14$; I have 14 quarters.

8.

	Number ×	Value of coin (¢) =	Total Value (¢)
Quarters	$d + 8$	25	$25(d + 8)$
Dimes	d	10	$10d$

$25(d + 8) + 10d = 620$, $25d + 200 + 10d = 620$, $35d + 200 = 620$, $35d = 420$, $d = 12$; I have 12 dimes.

9.

	Number ×	Price each =	Cost
Bought	60	$\frac{1}{2}s$	$\left(\frac{1}{2}s\right)60$
Sold	57	s	$57s$

$57s - \left(\frac{1}{2}s\right)60 = 1188$, $57s - 30s = 1188$, $27s = 1188$, $s = 44$; The store sold each recorder for \$44.

10.

	Number ×	Price each =	Cost
Bought	80	$2e$	$80(2e)$
Sold	60	e	$60e$

$60e - 80(2e) = -1000$, $60e - 160e = -1000$, $-100e = -1000$, $e = 10$; The employees paid \$10 for a ticket.

11.

	Hours ×	Wage (\$) per hour =	Total income (\$)
Plumber	8	$a + 4.50$	$8(a + 4.50)$
Apprentice	8	a	$8a$

$8(a + 4.5) + 8a = 372$, $8a + 36 + 8a = 372$, $16a + 36 = 372$, $16a = 336$, $a = 21$; $a + 4.50 = 25.50$; The plumber makes \$25.50 per hour and his apprentice makes \$21.00 per hour.

12.

	Hours ×	Wage (\$) per hour =	Total income (\$)
Weekdays	10	w	$10w$
Saturdays	8	$w + 2$	$8(w + 2)$

$10w + 8(w + 2) = 142$, $10w + 8w + 16 = 142$, $18w + 16 = 142$, $18w = 126$, $w = 7$; Raul makes \$7 per hour on weekdays.

13.

	Number	× Value of coin (¢) =	Total value (¢)
Dimes	d	10	$10d$
Nickels	$d + 7$	5	$5(d + 7)$
Quarters	$40 - [d + (d + 7)]$	25	$25(40 - [d + (d + 7)])$

$10d + 5(d + 7) + 25(40 - [d + (d + 7)]) = 405$, $10d + 5d + 35 + 1000 - 25[d + (d + 7)] = 405$, $15d + 1035 - 25d - 25(d + 7) = 405$, $-10d + 1035 - 25d - 175 = 405$, $-35d + 860 = 405$, $860 = 405 + 35d$, $455 = 35d$, $13 = d$; $40 - [d + (d + 7)] = 7$; Warren has 7 quarters.

14.

	Number	× Value of coin (¢) =	Total value (¢)
Nickels	n	5	$5n$
Dimes	$37 - [n + (n + 4)]$	10	$10(37 - [n + (n + 4)])$
Quarters	$n + 4$	25	$25(n + 4)$

$5n + 10(37 - [n + (n + 4)]) + 25(n + 4) = 550$, $5n + 370 - 10[n + (n + 4)] + 25n + 100 = 550$, $30n + 470 - 10n - 10(n + 4) = 550$, $20n + 470 - 10n - 40 = 550$, $10n + 430 = 550$, $10n = 120$, $n = 12$, $37 - [n + (n + 4)] = 9$; Jo has 9 dimes.

15.

	Number ×	Value of item ($) =	Total value ($)
Quarters	$2h$	0.25	$0.25(2h)$
Half dollars	h	0.50	$0.50h$
Dollars	$\frac{h}{2}$	1.00	$\frac{h}{2}$

$0.25(2h) + 0.5h + \frac{1}{2}h = 20$, $0.5h + 0.5h + 0.5h = 20$, $1.5h = 20$, $h = 13\frac{1}{3}$; It is impossible to have a fractional number of half dollars. ∴ Rory must be wrong.

16.

	Number ×	Value of coin (¢) =	Total value (¢)
Pennies	p	1	p
Nickels	$46 - p$	5	$5(46 - p)$

$p + 5(46 - p) = 100$, $p + 230 - 5p = 100$, $-4p + 230 = 100$, $-4p = -130$, $p = 32.5$; It is impossible to have 32.5 pennies. ∴ Eleanor is wrong.

C **17.**

	Number	Total Value	New Number	New Total value
Nadine	$d + 7$	$5(d + 7)$	$(d + 7) + 4$	$5(d + 7) + 40$
Delano	d	$10d$	$d - 4$	$10(d - 4)$

$5(d + 7) + 40 = 10(d - 4)$, $5d + 35 + 40 = 10d - 40$, $5d + 75 = 10d - 40$, $75 = 5d - 40$, $115 = 5d$, $23 = d$; $5(d + 7) + 10d = 380$; Nadine and Delano have $3.80 together.

18.

	Number	Total Value	New Value
Natalie	n	$5n$	$5(n - 1) + 25$
Dirk	$q + 5$	$10(q + 5)$	$10(q + 5 - 1) + 5$
Quincy	q	25	$25(q - 1) + 10$

$10(q + 5 - 1) + 5 = 25(q - 1) + 10$, $10q + 40 + 5 = 25q - 25 + 10$, $10q + 45 = 25q - 15$, $45 = 15q - 15$, $60 = 15q$, $4 = q$; $q + 5 = 9$; $10(q + 5 - 1) + 5 = 85$; $5(n - 1) + 25 = 85$, $5n - 5 + 25 = 85$, $5n + 20 = 85$, $5n = 65$, $n = 13$; Originally, Natalie had 13 nickels, Dirk had 9 dimes, and Quincy had 4 quarters.

Page 129 • MIXED REVIEW EXERCISES

1. $\frac{40 \div 5 + 3}{13 - 2} = \frac{11}{11} = 1$ **2.** $35 \div \frac{1}{6} = 35 \cdot 6 = 216$

3. $\frac{1}{3}(39y - 6) + 3 = 13y - 2 + 3 = 13y + 1$ **4.** $(-7)(5)(-1) = (-35)(-1) = 35$

5. $4(2x + 7) + 5(-x) = 8x + 28 + -5x = 3x + 28$

6. $7(x + y) + 8(2y + x) = 7x + 7y + 16y + 8x = 15x + 23y$

7. $\frac{5a + b}{x - 3} = \frac{5(4) + 5}{8 - 3} = \frac{20 + 5}{5} = \frac{25}{5} = 5$ **8.** $\frac{ab}{4x} = \frac{(4)(5)}{(4)(8)} = \frac{5}{8}$

9. $3(a + b) - x \div 2 = 3(4 + 5) - 8 \div 2 = (3)(9) - 4 = 27 - 4 = 23$

Page 129 • COMPUTER EXERCISES

1.
```
10 PRINT "NICKELS", "DIMES", "VALUE"
20 FOR I = 0 TO 12
30 PRINT I, 12 - I, 0.05 * I + 0.10 * (12 - I)
40 NEXT I
50 END
```

2.
```
10 PRINT "HOW MANY COINS"
20 INPUT K
30 PRINT
40 FOR I = 0 TO K
50 PRINT I, K - I, 0.05 * I + 0.10 * (K - I)
60 NEXT I
70 END
```

Pages 131–133 • WRITTEN EXERCISES

A **1.** (1) Reflexive prop. of = (3) Substitution principle

2. (2) Addition prop. of = (3) Assoc. prop. of add. (4) Prop. of opposites (5) Identity prop. of add.

3. (2) Multiplication prop. of = (3) Assoc. prop. of mult. (4) Prop. of reciprocals (5) Identity prop. of mult.

4. (1) Assoc. prop. of mult. (2) Prop. of reciprocals (3) Identity prop. of mult. (5) Comm. prop. of mult. (6) Comm. prop. of mult.

B **5.** (1) Prop. of opposites (2) Prop. of opposites (3) Substitution principle (4) Comm. prop. of add.

6. (1) Assoc. prop. of add. (2) Assoc. comm. props. of add. (3) Prop. of opp. (4) Identity prop. of add. (5) Prop. of opp.

7. (1) Def. of subtraction (3) Proved in Exercise 5 (4) Comm. prop. of add. (5) Def. of subtraction

8. (1) Def. of subtr. (3) Proved in Exercise 5

C **9.**

1. $a = b, c \neq 0$	1. Given
2. $a \cdot \frac{1}{c} = b \cdot \frac{1}{c}$	2. Mult. prop. of equality
3. $\frac{a}{c} = \frac{b}{c}$	3. Def. of division

10.

1. $a \neq 0$	1. Given
2. $\frac{a}{a} = a \cdot \frac{1}{a}$	2. Def. of division
3. $= 1$	3. Prop. of reciprocals

11. Since $\frac{1}{\frac{a}{b}}$ is the reciprocal of $\frac{a}{b}$, you can prove that $\frac{1}{\frac{a}{b}} = \frac{b}{a}$ by showing that $\frac{b}{a}$ is the reciprocal of $\frac{a}{b}$; that is, $\frac{a}{b} \cdot \frac{b}{a} = 1$.

1. $a \neq 0, b \neq 0$	1. Given
2. $\frac{a}{b} \cdot \frac{b}{a} = \left(a \cdot \frac{1}{b}\right)\left(b \cdot \frac{1}{a}\right)$	2. Def. of div.
3. $= \left(a \cdot \frac{1}{a}\right)\left(b \cdot \frac{1}{b}\right)$	3. Assoc. and comm. prop. of mult.
4. $= (1)(1)$	4. Prop. of reciprocals
5. $= 1$	5. Identity prop. of mult.

Page 133 • MIXED REVIEW EXERCISES

1. $12(12 - 7) \div 6 + 4 = (12)(5) \div 6 + 4 = 60 \div 6 + 4 = 10 + 4 = 14$

2. $-4(-16 + 8) = (-4)(-8) = 32$ **3.** $-7x - 2x + 12x = 3x$

4. $-\frac{1}{2}(8 + 2a) = -4 + -a = -4 - a$ **5.** $\frac{1}{2}(2b - 4) + 3 = b - 2 + 3 = b + 1$

6. $10 \div \frac{1}{2} = (10)(2) = 20$ **7.** $a(b + x) = 5(4 + 6) = (5)(10) = 50$

8. $3|x - a| = 3|6 - 5| = 3|1| = (3)(1) = 3$

9. $b - |x - a| = 4 - |6 - 5| = 4 - |1| = 4 - 1 = 3$

10. $\frac{b + 2a}{|1 - c|x} = \frac{4 + (2)(5)}{|1 - 3|6} = \frac{4 + 10}{|-2|6} = \frac{14}{(2)(6)} = \frac{14}{12} = \frac{7}{6}$

11. $\frac{5a + x + 4}{b + c} = \frac{(5)(5) + 6 + 4}{4 + 3} = \frac{25 + 10}{7} = \frac{35}{7} = 5$

12. $\frac{1}{2}(c - a) + x = \frac{1}{2}(3 - 5) + 6 = \frac{1}{2}(-2) + 6 = -1 + 6 = 5$

Page 135 • CHAPTER REVIEW

1. c **2.** d **3.** d **4.** d **5.** a **6.** b **7.** b **8.** a **9.** a **10.** a **11.** b **12.** d

Page 136 • CHAPTER TEST

1. $y + 25 = 10, y = -15; \{-15\}$ **2.** $73 = h - 13, 86 = h; \{86\}$

3. $c + 51 = 38, c = -13; \{-13\}$ **4.** $x - 38 = 12, x = 50; \{50\}$

5. $\frac{1}{13}y = 65, y = 845; \{845\}$ **6.** $-19v = -114, v = 6; \{6\}$ **7.** $-112 = 16e, -7 = e; \{-7\}$

8. $-\frac{x}{21} = 35, x = -735; \{-735\}$ **9.** $12y - 7 = 113, 12y = 120, y = 10; \{10\}$

10. $\frac{2}{3}x + 6 = 16, \frac{2}{3}x = 10, x = 15; \{15\}$

11. $\frac{3x + 90}{5} = 0, 3x + 90 = 0, 3x = -90, x = -30; \{-30\}$

12. $-\frac{7}{8}(w - 16) = 70, -\frac{7}{8}w + 14 = 70, -\frac{7}{8}w = 56, w = -64; \{-64\}$

13. *Step 1* Find the number of 2-point baskets Manuel made.

Step 2 Let x = number of two-point baskets.

Step 3 $8(1) + 2(3) + 2x = 30$

Step 4 $8 + 6 + 2x = 30,\ 14 + 2x = 30,\ 2x = 16,\ x = 8$

Step 5 Manuel made 8 two-point baskets.

14. $7(a - 6) = -3 + 6a,\ 7a - 42 = -3 + 6a,\ a - 42 = -3,\ a = 39;\ \{39\}$

15. $6(m - 1) = 6(m + 3),\ 6m - 6 = 6m + 18,\ -6 = 18$; no solution.

16. $3x + 44 = -x,\ 4x + 44 = 0,\ 4x = -44,\ x = -11$; the number is -11.

17.

	Weight	New weight
Sean	$2x + 10$	$2x + 10$
Brad	x	$x + 10$

$(2x + 10) + (x + 10) = 230,$
$2x + 10 + x + 10 = 230,\ 3x + 20 = 230,\ 3x = 210,$
$x = 70,\ 2x + 10 = 150;$
Brad weighs 70 lb and Sean weighs 150 lb.

18.

	Number ×	Value of coin (¢) =	Total value (¢)
Dimes	$5q$	10	$50q$
Quarters	q	25	$25q$

$50q + 25q = 525,\ 75q = 525,\ q = 7;$
Courtney has 7 quarters.

19. (1) Identity prop. of add. (2) Mult. prop. of = (3) Distributive prop. (5) Trans. prop. of = (6) Subtr. prop. of = (7) Comm. prop. of mult.

Page 137 • CUMULATIVE REVIEW

1. $45 - 2(16 - 6) = 45 - 32 + 12 = 25$ **2.** $\frac{56 \div 7}{16} = \frac{8}{16} = \frac{1}{2}$

3. $48 \div (9 + 3) = 48 \div 12 = 4$ **4.** $|-21| - |-14| = 21 - 14 = 7$

5. $31 - [35 \div 5] = 31 - 7 = 24$ **6.** $48 \div 3 + 7(3) = 16 + 21 = 37$

7. $-55 - (-42 + 7) = -55 - (-35) = -55 + 35 = -20$

8. $-22 + 31 + (-44) + 50 = 15$

9. $1\frac{1}{3} - 10\frac{1}{4} + 12\frac{2}{3} = \frac{4}{3} - \frac{41}{4} + \frac{38}{3} = \frac{16 - 123 + 152}{12} = \frac{45}{12} = \frac{15}{4} = 3\frac{3}{4}$

10. $2.4 + 5.1 - 6.3 = 1.2$ **11.** $9 + x - (5 - x) - 6 = 9 + x - 5 + x - 6 = 2x - 2$

12. $4(x + 3y) - 3(x - 4y) = 4x + 12y - 3x + 12y = x + 24y$

13. $-5(42)\left(-\frac{2}{5}\right)\left(-\frac{1}{3}\right) = -5\left(-\frac{2}{5}\right)(42)\left(-\frac{1}{3}\right) = (2)(-14) = -28$

14. $-5(-a - b) + 5(a + b) = 5a + 5b + 5a + 5b = 10a + 10b$ **15.** $\frac{85xy}{5y} = \frac{85x}{5} = 17x$

16. $w - |z - x| = 2 - |3 - (-3)| = 2 - |3 + 3| = 2 - |6| = 2 - 6 = -4$

17. $\frac{4y - x}{w + z} = \frac{4\left(\frac{1}{2}\right) - (-3)}{2 + 3} = \frac{2 + 3}{2 + 3} = \frac{5}{5} = 1$ **18.** $-5(x + w) = -5(-3 + 2) = -5(-1) = 5$

19. $-w + 2x + 4y = -2 + 2(-3) + 4\left(\frac{1}{2}\right) = -2 - 6 + 2 = -6$ **20.** $-\frac{1}{2}$ **21.** $-4\frac{1}{2}$

22. $|x| = 3,\ x = 3$ or $x = -3;\ \{3, -3\}$ **23.** $|y| = -2$; no solution

24. $x + 7 = 12, x = 5; \{5\}$ **25.** $z + 1 = -3, z = -4; \{-4\}$ **26.** $y + 2 = 9, y = 7; \{7\}$

27. $(x - 3) + 17 = 30, x - 3 + 17 = 30, x + 14 = 30, x = 16; \{16\}$

28. $\frac{a}{3} = -14, a = -42; \{-42\}$ **29.** $9 = \frac{1}{2}q, 18 = q; \{18\}$

30. $2x + 6 = -2, 2x = -8, x = -4; \{-4\}$ **31.** $\frac{1}{2}(2x + 4) = 2x, x + 2 = 2x, 2 = x; \{2\}$

32. $16 = \frac{3}{4}k + 1, 15 = \frac{3}{4}k, 20 = k; \{20\}$ **33.** $5(z - 3) = 40, 5z - 15 = 40, 5z = 55, z = 11; \{11\}$

34. $5y - 2 = 7y + 8, -2 = 2y + 8, -10 = 2y, -5 = y; \{-5\}$

35. $9(2 - b) = b, 18 - 9b = b, 18 = 10b, \frac{18}{10} = b, \frac{9}{5} = b; \left\{\frac{9}{5}\right\}$

36. $3(x - 4) = 6(x - 3), 3x - 12 = 6x - 18, -12 = 3x - 18, 6 = 3x, 2 = x; \{2\}$

37. Let p = price of peach. $4p + p = 1.50, 5p = 1.50, p = 0.30; 4p = 1.20$; The melon costs \$1.20 and the peach costs \$.30.

38. Let n = first consecutive integer. $n + (n + 1) + (n + 2) = 87, n + n + 1 + n + 2 = 87, 3n + 3 = 87. 3n = 84, n = 28; n + 1 = 29; n + 2 = 30$; The three consecutive integers are 28, 29, and 30.

39. Let e = number of employees. $\frac{2}{5}e = 38, e = 95$; There are 95 employees.

40. Let d = number of dimes, and $5d$ = number of nickels. $d + 5d = 30, 6d = 30, d = 5; 5d = 25; 10(5) + 5(25) = 175$; Rory has \$1.75.

Page 138 • MAINTAINING SKILLS

1. $2\frac{1}{12}$ **2.** $7\frac{1}{2}$ **3.** $5\frac{1}{5}$ **4.** $4\frac{3}{10}$ **5.** $7\frac{7}{12}$ **6.** $11\frac{6}{7}$ **7.** $9\frac{1}{4}$ **8.** $8\frac{11}{13}$ **9.** $\frac{25}{6}$ **10.** $\frac{43}{5}$

11. $\frac{25}{9}$ **12.** $\frac{51}{4}$ **13.** $\frac{47}{13}$ **14.** $\frac{52}{3}$ **15.** $\frac{119}{12}$ **16.** $\frac{87}{8}$

17. $8\frac{7}{15} + 9\frac{11}{15} = \frac{127}{15} + \frac{146}{15} = \frac{273}{15} = 18\frac{3}{15} = 18\frac{1}{5}$ **18.** $5\frac{7}{9} - 6\frac{4}{9} = \frac{52}{9} - \frac{58}{9} = -\frac{6}{9} = -\frac{2}{3}$

19. $7\frac{1}{3}\left(-\frac{1}{7}\right) = \frac{22}{3}\left(-\frac{1}{7}\right) = -\frac{22}{21} = -1\frac{1}{21}$ **20.** $4\frac{2}{5} \div 2\frac{1}{5} = \frac{22}{5} \div \frac{11}{5} = \frac{\overset{2}{\cancel{22}}}{\underset{1}{\cancel{5}}} \cdot \frac{\overset{1}{\cancel{5}}}{\underset{1}{\cancel{11}}} = 2$

21. $10\frac{1}{5} \div \frac{2}{5} = \frac{51}{5} \div \frac{2}{5} = \frac{51}{\underset{1}{\cancel{5}}} \cdot \frac{\overset{1}{\cancel{5}}}{2} = \frac{51}{2} = 25\frac{1}{2}$

22. $20\frac{5}{13} + 8\frac{4}{5} = \frac{265}{13} + \frac{44}{5} = \frac{1325}{65} + \frac{572}{65} = \frac{1897}{65} = 29\frac{12}{65}$

23. $6\frac{5}{8} - 3\frac{2}{5} = \frac{53}{8} - \frac{17}{5} = \frac{265}{40} - \frac{136}{40} = \frac{129}{40} = 3\frac{9}{40}$

24. $6\frac{3}{5} \div 2\frac{3}{4} = \frac{33}{5} \div \frac{11}{4} = \frac{\overset{3}{\cancel{33}}}{5} \cdot \frac{4}{\underset{1}{\cancel{11}}} = \frac{12}{5} = 2\frac{2}{5}$

25. $12\frac{9}{10} - 8\frac{3}{5} = \frac{129}{10} - \frac{43}{5} = \frac{129}{10} - \frac{86}{10} = \frac{43}{10} = 4\frac{3}{10}$

26. $-9\frac{1}{2} + \left(-13\frac{5}{9}\right) = -\frac{19}{2} + -\frac{122}{9} = -\frac{171}{18} - \frac{244}{18} = -\frac{415}{18} = -23\frac{1}{18}$

27. $11\frac{4}{7} + 10\frac{1}{5} = \frac{81}{7} + \frac{51}{5} = \frac{405}{35} + \frac{357}{35} = \frac{762}{35} = 21\frac{27}{35}$

28. $8\frac{5}{6} \div 6\frac{5}{8} = \frac{53}{6} \div \frac{53}{8} = \frac{\overset{1}{\cancel{53}}}{\underset{3}{\cancel{6}}} \cdot \frac{\overset{4}{\cancel{8}}}{\underset{1}{\cancel{53}}} = \frac{4}{3} = 1\frac{1}{3}$

29. $4\frac{10}{11} - 7\frac{1}{2} = \frac{54}{11} - \frac{15}{2} = \frac{108}{22} - \frac{165}{22} = -\frac{57}{22} = -2\frac{13}{22}$ **30.** $12\frac{3}{20} \cdot 1\frac{5}{9} = \frac{\overset{27}{\cancel{243}}}{\underset{10}{\cancel{20}}} \cdot \frac{\overset{7}{\cancel{14}}}{\underset{1}{\cancel{9}}} = \frac{189}{10} = 18\frac{9}{10}$

31. $9\frac{5}{9} - 8\frac{1}{4} = \frac{86}{9} - \frac{33}{4} = \frac{344}{36} - \frac{297}{36} = \frac{47}{36} = 1\frac{11}{36}$ **32.** $3\frac{9}{10} \cdot 7\frac{1}{13} = \frac{\overset{3}{\cancel{39}}}{\underset{5}{\cancel{10}}} \cdot \frac{\overset{46}{\cancel{92}}}{\underset{1}{\cancel{13}}} = \frac{138}{5} = 27\frac{3}{5}$

33. $-5\frac{2}{3} \div 3\frac{1}{2} = -\frac{17}{3} \div \frac{7}{2} = -\frac{17}{3} \cdot \frac{2}{7} = -\frac{34}{21} = -1\frac{13}{21}$ **34.** $5\frac{5}{9} \cdot 4\frac{1}{2} = \frac{\overset{25}{\cancel{50}}}{\underset{1}{\cancel{9}}} \cdot \frac{\overset{1}{\cancel{9}}}{\underset{1}{\cancel{2}}} = 25$

35. $17\frac{3}{4} + 8\frac{1}{3} = \frac{71}{4} + \frac{25}{3} = \frac{213}{12} + \frac{100}{12} = \frac{313}{12} = 26\frac{1}{12}$ **36.** $10\frac{5}{6} \cdot 3\frac{4}{5} = \frac{\overset{13}{\cancel{65}}}{6} \cdot \frac{19}{\underset{1}{\cancel{5}}} = \frac{247}{6} = 41\frac{1}{6}$

Page 139 • MIXED PROBLEM SOLVING

A

1. Let x = the number. $2x + -6 = -x + 9, 3x - 6 = 9, 3x = 15, x = 5$; The number is 5.

2. Let s = cost of softball. $22 - s = 5s - 2, 22 = 6s - 2, 24 = 6s, 4 = s; 22 - s = 18$; The softball cost \$4 and the mitt cost \$18.

3. Let t = rate of travel. $t = 450 \div 6, t = 75$; My rate of travel was 75 km/h.

4. $-35.87 + 580 - 25 - 254.09 = 265.04$; The Longs' new balance is \$265.04.

5. Let a = number of apples bought. $0.25a + 0.18(12 - a) = 2.51$, $0.25a + 2.16 - 0.18a = 2.51, 0.07a + 2.16 = 2.51, 0.07a = 0.35, a = 5$; Alice bought 5 apples and 7 oranges.

6.

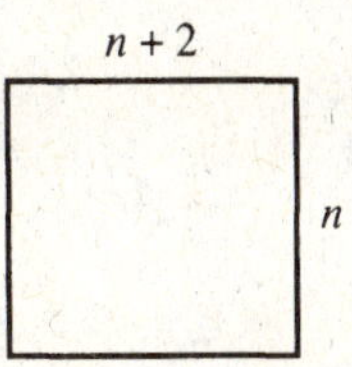

$2n + 2(n + 2) = 48, 2n + 2n + 4 = 48, 4n + 4 = 48$, $4n = 44, n = 11; n + 2 = 13$; The rectangle is 11 cm × 13 cm.

7. Let j = usual January temp. $22 = j + 27, -5 = j$; The usual January temperature is −5°C.

8. Let x = the number. $7 - x = 10, -x = 3; x = -3$; The number is −3.

9. Let n = the smallest consecutive integer. $3n = (n + 1) + (n + 2)$, $3n = n + 1 + n + 2, 3n = 2n + 3, n = 3; n + 1 = 4; n + 2 = 5$; The integers are 3, 4, and 5.

10. $357 - (-39) = 357 + 39 = 396$; The difference is 396°C.

11. Let x = number of \$10 bills. $10x + 5(x + 4) = 125, 10x + 5x + 20 = 125$, $15x + 20 = 125, 15x = 105, x = 7; x + 4 = 11$; Ruwa has 7 \$10 bills and 11 \$5 bills.

B

12. $10(c - 4) - 8c = 10c - 40 - 8c = 2c - 40$; The store's profit was $2c - 40$ dollars.

13. Let b = total zoo budget. $\frac{45}{100}b = 216{,}000, b = 480{,}000$; The zoo's total budget was \$480,000.

14. Let q = number of questions on the quiz. $q = \frac{7}{8}q + 5, \frac{1}{8}q = 5, q = 40$; There were 40 questions on the quiz.

15. Let a = number of hours Ann worked. $a + (a + 3) = 3a - 1, a + a + 3 = 3a - 1, 2a + 3 = 3a - 1, 3 = a - 1, 4 = a; a + 3 = 7$; Kim worked 7 hours.

16. Let q = number of quarters, $q - 8$ = number of nickels, and $2(q - 8)$ = number of dimes. $q + (q - 8) + 2(q - 8) = 44, q + q - 8 + 2q - 16 = 44, 4q - 24 = 44, 4q = 68, q = 17; q - 8 = 9; 2(q - 8) = 18; 25(17) + 5(9) + 10(18) = 650$; There is \$6.50 in the bank.

17. Let m = amount Miguel has. $2m + 6 = \frac{4}{3}m, \frac{2}{3}m + 6 = 0, \frac{2}{3}m = -6, m = -9$; no solution. Miguel cannot have a negative amount of money.

Pages 143–144 • WRITTEN EXERCISES

A **1.** $2(3y) = 2 \cdot 3 \cdot y = 6y$ **2.** $(3x)(5x) = 3 \cdot x \cdot 5 \cdot x = 15x^2$

3. $(2z)(4z) = 2 \cdot z \cdot 4 \cdot z = 8z^2$ **4.** $x \cdot x \cdot x \cdot x \cdot x = x^5$ **5.** $m \cdot m \cdot m = m^3$

6. $4 \cdot t \cdot t \cdot t = 4t^3$ **7.** $c \cdot c \cdot 3 \cdot c = 3c^3$ **8.** $-6 \cdot z \cdot z = -6z^2$

9. $y \cdot y \cdot (-5) = -5y^2$ **10.** $-3 \cdot x \cdot 2 \cdot x = -6x^2$ **11.** $7 \cdot n \cdot (-2) \cdot n = -14n^2$

12. $a \cdot a \cdot b \cdot b \cdot b = a^2b^3$ **13.** $c \cdot d \cdot c \cdot c = c^3d$ **14.** $m \cdot 8 \cdot m \cdot n \cdot n = 8m^2n^2$

15. $u \cdot v \cdot u \cdot u \cdot 9 = 9u^3v$ **16.** $-3 \cdot x \cdot y \cdot x = -3x^2y$ **17.** $r \cdot (-4) \cdot s \cdot s = -4rs^2$

18. $e \cdot f \cdot e \cdot g \cdot g = e^2fg^2$ **19.** $p \cdot p \cdot q \cdot q \cdot r \cdot s = p^2q^2rs$ **20.** b **21.** d **22.** a

23. c **24.** **a.** $-6^2 = -(6 \cdot 6) = -36$ **b.** $(-6)^2 = (-6)(-6) = 36$

25. **a.** $(-2)^2 = (-2)(-2) = 4$ **b.** $-2^2 = -(2 \cdot 2) = -4$

26. **a.** $2 \cdot 5^2 = 2 \cdot 5 \cdot 5 = 50$ **b.** $(2 \cdot 5)^2 = 10^2 = 100$

27. **a.** $-4^2 \cdot 3 = -(4 \cdot 4) \cdot 3 = -48$ **b.** $(-4 \cdot 3)^2 = (-12)^2 = (-12)(-12) = 144$

28. **a.** $5 - 3^4 = 5 - (3 \cdot 3 \cdot 3 \cdot 3) = 5 - 81 = -76$ **b.** $(5 - 3)^4 = 2^4 = 16$

29. **a.** $7 + 3^3 = 7 + (3 \cdot 3 \cdot 3) = 7 + 27 = 34$ **b.** $(7 + 3)^3 = 10^3 = 1000$

30. **a.** $2 \cdot 3 - 5^2 = 6 - (5 \cdot 5) = 6 - 25 = -19$

b. $2 \cdot (3 - 5)^2 = 2 \cdot (-2)^2 = 2 \cdot (-2) \cdot (-2) = 8$

31. **a.** $3 \cdot (4 - 5)^2 = 3 \cdot (-1)^2 = 3 \cdot (-1) \cdot (-1) = 3$

b. $3 \cdot 4 - 5^2 = 12 - (5 \cdot 5) = 12 - 25 = -13$

B **32.** $(1 \cdot 10^3) + (4 \cdot 10^2) + (9 \cdot 10) + 2 = (1 \cdot 1000) + (4 \cdot 100) + 90 + 2 = 1000 + 400 + 92 = 1492$

33. $(1 \cdot 10^3) + (7 \cdot 10^2) + (7 \cdot 10) + 6 = (1 \cdot 1000) + (7 \cdot 100) + 70 + 6 = 1000 + 700 + 70 + 6 = 1776$

34. $[2^3 + 3^3] \div [2^3 + (-1)^2] = (8 + 27) \div (8 + 1) = 35 \div 9 = \frac{35}{9}$

35. $[5^3 + (-3)^3] \div 7^2 = (125 - 27) \div 49 = 98 \div 48 = 2$

36. $[3^3 + (-2)^3 + (-1)^3] \div 3^2 = (27 - 8 - 1) \div 9 = 18 \div 9 = 2$

37. $(3^4 - 2^4) \div [5^3 \div (4^2 + 3^2)] = (81 - 16) \div [125 \div (16 + 9)] = 65 \div (125 \div 25) = 65 \div 5 = 13$

38. $3^2 \div (2^2 - 1) - (5^2 - 3^2) \div (-2)^3 = 9 \div (4 - 1) - (25 - 9) \div (-8) = 9 \div 3 - 16 \div (-8) = 3 - (-2) = 3 + 2 = 5$

39. $[2^2 \cdot 3^3 - 3 \cdot 2^4] \div [(2 \cdot 3)^2 - 2^4] = (4 \cdot 27 - 3 \cdot 16) \div (6^2 - 16) = (108 - 48) \div (36 - 16) = 60 \div 20 = 3$

40. **a.** $ab - a^2 = (3)(-2) - (3)^2 = -6 - 9 = -15$

b. $a(b - a)^2 = 3(-2 - 3)^2 = 3(-5)^2 = 3 \cdot 25 = 75$

41. **a.** $4 + ab^2 = 4 + 3(-2)^2 = 4 + 3 \cdot 4 = 4 + 12 = 16$

b. $(4 + ab)^2 = [4 + 3(-2)]^2 = (4 - 6)^2 = (-2)^2 = 4$

42. **a.** $(2a - b)^3 = [2 \cdot 3 - (-2)]^3 = (6 + 2)^3 = 8^3 = 512$

b. $2a - b^3 = 2 \cdot 3 - (-2)^3 = 6 - (-8) = 6 + 8 = 14$

43. **a.** $(a + 2b)^3 = [3 + 2 \cdot (-2)]^3 = (3 - 4)^3 = (-1)^3 = -1$

b. $a^3 + 2b^3 = 3^3 + 2 \cdot (-2)^3 = 27 + 2 \cdot (-8) = 27 - 16 = 11$

44. $\dfrac{(2a + b)^2}{2a + b} = \dfrac{[2 \cdot 3 + (-2)]^2}{2 \cdot 3 + (-2)} = \dfrac{(6 - 2)^2}{6 - 2} = \dfrac{4^2}{4} = \dfrac{16}{4} = 4$

45. $\dfrac{a^3 + 2b^3}{a + 2b} = \dfrac{3^3 + 2(-2)^3}{3 + 2(-2)} = \dfrac{27 + 2(-8)}{3 + (-4)} = \dfrac{27 - 16}{3 - 4} = \dfrac{11}{-1} = -11$

46. $\dfrac{a^4 + b^4}{a + b} = \dfrac{3^4 + (-2)^4}{3 + (-2)} = \dfrac{81 + 16}{3 - 2} = \dfrac{97}{1} = 97$

47. $\dfrac{4a}{ab + 4} = \dfrac{4(3)}{(3)(-2) + 4} = \dfrac{12}{-6 + 4} = \dfrac{12}{-2} = -6$

C **48.** $(x^2 + 4x + 5)(x^2 + x - 2) = [(-3)^2 + 4(-3) + 5][(-3)^2 + (-3) - 2] =$
$(9 - 12 + 5)(9 - 3 - 2) = 2 \cdot 4 = 8$

49. $(x^2 - 3x + 1)(x^2 + 2x - 8) = [(-4)^2 - 3(-4) + 1][(-4)^2 + 2(-4) - 8] =$
$(16 + 12 + 1)(16 - 8 - 8) = 29 \cdot 0 = 0$

Page 144 • MIXED REVIEW EXERCISES

1. $-7x = 56, x = -8; \{-8\}$ **2.** $3(n - 4) = 36, n - 4 = 12, n = 16; \{16\}$

3. $36 = -3x, -12 = x; \{-12\}$ **4.** $-n + 8 = 6, -n = -2, n = 2; \{2\}$

5. $x - 5 = |7 - 12|, x - 5 = |-5|, x - 5 = 5, x = 10; \{10\}$

6. $-y + 12 = 8, -y = -4, y = 4; \{4\}$ **7.** $-\frac{1}{2}(x + 2) = 3, x + 2 = -6, x = -8; \{-8\}$

8. $\frac{1}{5}x = 10, x = 50; \{50\}$ **9.** $3k = -\frac{6}{7}, k = -\frac{2}{7}; \left\{-\frac{2}{7}\right\}$

Page 145 • COMPUTER EXERCISES

1.

```
10 FOR N = 1 TO 5
20 LET X = N ↑ N
30 PRINT N; "↑"; N; "= "; X
40 NEXT N
50 END
```

Results

$1^1 = 1$
$2^2 = 4$
$3^3 = 27$
$4^4 = 256$
$5^5 = 3125$

2.

```
10 FOR N = 1 TO 5
20 LET X = 1
30 FOR I = 1 TO N
40 LET X = X * I
50 NEXT I
60 PRINT N; "FACTORIAL = "; X
70 NEXT N
80 END
```

Results

$1! = 1$
$2! = 2$
$3! = 6$
$4! = 24$
$5! = 120$

3. For values of $n > 1$, $n^n > n!$; for $n = 1$, $n^n = n! = 1$; however, for $n > 1$, n^n, which is the product of n factors, each having a value of n, is always greater than $n!$, which is the product of n factors, each except the first having a value less than n.

Page 145 • CALCULATOR KEY-IN

1. $(0.1)^2 = 0.01$ **2.** $(0.2)^2 = 0.04$ **3.** $(0.5)^2 = 0.25$ **4.** $(1.1)^2 = 1.21$

5. $(0.11)^2 = 0.0121$ **6.** $(0.02)^2 = 0.0004$ **7.** $(0.3)^3 = 0.027$ **8.** $(0.4)^3 = 0.064$

Pages 148–149 • WRITTEN EXERCISES

A **1.** $3x - 2y - x - 3y = (3 - 1)x + (-2 - 3)y = 2x - 5y$

2. $6m - 6n - 4m + n = (6 - 4)m + (-6 + 1)n = 2m - 5n$

3. $3x^2 - 2x - 2x^2 - 4x - 3 = (3 - 2)x^2 + (-2 - 4)x - 3 = x^2 - 6x - 3$

4. $n^2 - 4n - 3n^2 + 7n + 5n^2 = (1 - 3 + 5)n^2 + (-4 + 7)n = 3n^2 + 3n$

5. $a^2 + 3ab - 4ab + 3a^2 = (1 + 3)a^2 + (3 - 4)ab = 4a^2 - ab$

6. $p^2q - q^3 - 3p^2q + 4q^3 = (1 - 3)p^2q + (-1 + 4)q^3 = -2p^2q + 3q^3$

7. $r^2s - 3rs^2 + 4s^3 - 2r^2s - 3s^2 = (1 - 2)r^2s - 3rs^2 + 4s^3 - 3s^2 =$
$-r^2s - 3rs^2 - 3s^2 + 4s^3$

8. $-3x^2 + 7x^2y - x^3 + xy^2 + 4x^3 - 3x^2y = -3x^2 + (7 - 3)x^2y + (-1 + 4)x^3 +$
$xy^2 = 3x^3 - 3x^2 + 4x^2y + xy^2$

9. $\begin{array}{r} 5y - 3 \\ 2y + 9 \\ \hline 7y + 6 \end{array}$ **10.** $\begin{array}{r} 4x + 7 \\ x - 2 \\ \hline 5x + 5 \end{array}$ **11.** $\begin{array}{r} 3y + 8 \\ 2y - 5 \\ \hline 5y + 3 \end{array}$ **12.** $\begin{array}{r} 7n - 6 \\ n + 4 \\ \hline 8n - 2 \end{array}$

13. $\begin{array}{r} 2r - 3x + 5 \\ -r + 3x - 2 \\ \hline r \qquad + 3 \end{array}$ **14.** $\begin{array}{r} -2p + 4q - 7 \\ -4p - 2q + 5 \\ \hline -6p + 2q - 2 \end{array}$ **15.** $\begin{array}{r} 2x^2 - 3x - 4 \\ 3x^2 + 4x - 6 \\ \hline 5x^2 + x - 10 \end{array}$

16. $\begin{array}{r} 4 - 3n - 5n^2 \\ -2 + n - 3n^2 \\ \hline 2 - 2n - 8n^2 \end{array}$ **17.** $\begin{array}{r} 4x^2 - 3xy - 5y^2 \\ 2x^2 + xy - 3y^2 \\ \hline 6x^2 - 2xy - 8y^2 \end{array}$ **18.** $\begin{array}{r} 8p^2 - 5pq + 6q^2 \\ -2p^2 + 5pq - 4q^2 \\ \hline 6p^2 \qquad + 2q^2 \end{array}$

19. $\begin{array}{r} 3a - 7b - 5c + 2 \\ -a + 4b + c - 5 \\ 2a \qquad + 3c + 3 \\ \hline 4a - 3b - c \quad \end{array}$ **20.** $\begin{array}{r} 7x - 6y + 4z - 1 \\ -3x + 3y - 5z + 3 \\ -2x - y + z - 4 \\ \hline 2x - 4y \qquad - 2 \end{array}$ **21.** $\begin{array}{r} 5y - 3 \\ -2y - 9 \\ \hline 3y - 12 \end{array}$

22. $\begin{array}{r} 4x + 7 \\ -x + 2 \\ \hline 3x + 9 \end{array}$ **23.** $\begin{array}{r} 3y + 8 \\ -2y + 5 \\ \hline y + 13 \end{array}$ **24.** $\begin{array}{r} 7n - 6 \\ -n - 4 \\ \hline 6n - 10 \end{array}$ **25.** $\begin{array}{r} 2r - 3x + 5 \\ r - 3x + 2 \\ \hline 3r - 6x + 7 \end{array}$

26. $\begin{array}{r} -2p + 4q - 7 \\ 4p + 2q - 5 \\ \hline 2p + 6q - 12 \end{array}$ **27.** $\begin{array}{r} 2x^2 - 3x - 4 \\ -3x^2 - 4x + 6 \\ \hline -x^2 - 7x + 2 \end{array}$ **28.** $\begin{array}{r} 4 - 3n - 5n^2 \\ 2 - n + 3n^2 \\ \hline 6 - 4n - 2n^2 \end{array}$

29. $\begin{array}{r} 4x^2 - 3xy - 5y^2 \\ -2x^2 - xy + 3y^2 \\ \hline 2x^2 - 4xy - 2y^2 \end{array}$ **30.** $\begin{array}{r} 8p^2 - 5pq + 6q^2 \\ 2p^2 - 5pq + 4q^2 \\ \hline 10p^2 - 10pq + 10q^2 \end{array}$

31. $(2x - 5y + 2) + (5x + 6y - 7) = (2x + 5x) + (-5y + 6y) + (2 - 7) =$
$7x + y - 5$

32. $(2p - 7q - 4) + (3q + 2p - 1) = (2p + 2p) + (-7q + 3q) + (-4 - 1) =$
$4p - 4q - 5$

33. $(2x - 5) - (x - 2) = (2x - x) + (-5 + 2) = x - 3$

34. $(3m + 5) - (-2m + 3) = (3m + 2m) + (5 - 3) = 5m + 2$

35. $(5x - 3t - 7) - (x - 2t - 3) = (5x - x) + (-3t + 2t) + (-7 + 3) = 4x - t - 4$

36. $(a - 3b + 5) - (-a + 2b + 3) = (a + a) + (-3b - 2b) + (5 - 3) = 2a - 5b + 2$

37. $(3n^2 + 5n - 6) + (-n^2 - 3n + 3) = (3n^2 - n^2) + (5n - 3n) + (-6 + 3) = 2n^2 + 2n - 3$

38. $(y^2 + 6y - 5) + (-y^2 - 3y - 1) = (y^2 - y^2) + (6y - 3y) + (-5 - 1) = 3y - 6$

39. $(3x^2 - 4x - 2) - (-x^2 - 4x + 7) = (3x^2 + x^2) + (-4x + 4x) + (-2 - 7) = 4x^2 - 9$

40. $(y^2 - 3y - 5) - (-y^2 - 7y + 4) = (y^2 + y^2) + (-3y + 7y) + (-5 - 4) = 2y^2 + 4y - 9$

B **41.** $(u^3 - 3u^2v + 2uv^2) + (3u^2v - 2uv^2 - v^3) = u^3 + (-3u^2v + 3u^2v) + (2uv^2 - 2uv^2) - v^3 = u^3 - v^3$

42. $(2x^2y - 3xy^2 - y^3) + (2x^2y - xy^2) = (2x^2y + 2x^2y) + (-3xy^2 - xy^2) - y^3 = 4x^2y - 4xy^2 - y^3$

43. $(3a^3 - 2ab^2) - (a^3 - 4ab^2 - b^3) = (3a^3 - a^3) + (-2ab^2 + 4ab^2) + b^3 = 2a^3 + 2ab^2 + b^3$

44. $(2p^2q - 3pq^2 + q^3) - (-p^2q + q^3) = (2p^2q + p^2q) - 3pq^2 + (q^3 - q^3) = 3p^2q - 3pq^2$

45. $7x - (3x - 2) = 10, 7x - 3x + 2 = 10, 4x + 2 = 10, 4x = 8, x = 2; \{2\}$

46. $z - (4z - 5) = 8, z - 4z + 5 = 8, -3z + 5 = 8, -3z = 3, z = -1; \{-1\}$

47. $(11n - 5) - (3n - 2) = -19, 11n - 5 - 3n + 2 = -19, 8n - 3 = -19, 8n = -16, n = -2; \{-2\}$

48. $(2x + 3) - (5x - 7) = 1, 2x + 3 - 5x + 7 = 1, -3x + 10 = 1, -3x = -9, x = 3; \{3\}$

49. $(4y - 3) - (4 - y) = 3(y + 3), 4y - 3 - 4 + y = 3y + 9, 5y - 7 = 3y + 9, 2y - 7 = 9, 2y = 16, y = 8; \{8\}$

50. $3(n - 2) - 2(3 - n) = 4(n - 3), 3n - 6 - 6 + 2n = 4n - 12, 5n - 12 = 4n - 12, n - 12 = -12, n = 0; \{0\}$

51. $2 - 3x = 8(5 - x) - (x - 10), 2 - 3x = 40 - 8x - x + 10, 2 - 3x = 50 - 9x, 2 + 6x = 50, 6x = 48, x = 8; \{8\}$

52. $3(4u - 6) = 2(4u - 3) - (u - 8), 12u - 18 = 8u - 6 - u + 8, 12u - 18 = 7u + 2, 5u - 18 = 2, 5u = 20, u = 4; \{4\}$

C **53.** $(2y^2 - y + 6) - 2(y^2 - 3y + 5) = 11, 2y^2 - y + 6 - 2y^2 + 6y - 10 = 11, 5y - 4 = 11, 5y = 15, y = 3; \{3\}$

54. $y(2 - y) = 6 - (y^2 + 3y - 4), 2y - y^2 = 6 - y^2 - 3y + 4, 2y - y^2 = 10 - y^2 - 3y, 2y = 10 - 3y, 5y = 10, y = 2; \{2\}$

55. $x(3 - x) = x - (x^2 - 2x + 4), 3x - x^2 = x - x^2 + 2x - 4, 3x - x^2 = 3x - x^2 - 4, 0 = -4$; no solution

56. $3 - 2x(x - 1) = x(3 - 2x) - (x - 3), 3 - 2x^2 + 2x = 3x - 2x^2 - x + 3, 3 - 2x^2 + 2x = 2x - 2x^2 + 3, 0 = 0$; {all real numbers}

Page 150 • PROBLEMS

A **1.** $x(2x) + 1x + 2(2x) + 2(1) = 2x^2 + x + 4x + 2 = 2x^2 + 5x + 2$

2. $x(5x) + 2x + 3(5x) + 2(3) = 5x^2 + 2x + 15x + 6 = 5x^2 + 17x + 6$

3. $x(3x) + 1x + 1(3x) + 1(1) = 3x^2 + 4x + 1$

4. $x(2x) + 3x + 2(2x) + 3(2) = 2x^2 + 3x + 4x + 6 = 2x^2 + 7x + 6$

5. Let n and $n + 1$ be the consecutive integers. $n + (n + 1) = 7^2$, $2n + 1 = 49$, $2n = 48$, $n = 24$; $n + 1 = 25$. The consecutive integers are 24 and 25.

6. Let n, $n + 1$, and $n + 2$ be the consecutive integers. $n + (n + 1) + (n + 2) = 6^2$, $3n + 3 = 36$, $3n = 33$, $n = 11$; $n + 1 = 12$; $n + 2 = 13$. The consecutive integers are 11, 12, and 13.

7. Let n, $n + 1$, $n + 2$, and $n + 3$ be the consecutive integers. $n + (n + 1) + (n + 2) + (n + 3) = 2(5^3)$, $4n + 6 = 2(125)$, $4n + 6 = 250$, $4n = 244$, $n = 61$; $n + 1 = 62$; $n + 2 = 63$; $n + 3 = 64$. The consecutive integers are 61, 62, 63, and 64.

8. Let n and $n + 2$ be the consecutive odd integers. $n + (n + 2) = 4^3$, $2n + 2 = 64$, $2n = 62$, $n = 31$; $n + 2 = 33$. The consecutive odd integers are 31 and 33.

9. Let n and $n + 1$ be the consecutive integers. $n + 1 = 2n - 10$, $1 = n - 10$, $11 = n$; $n + 1 = 12$. The consecutive integers are 11 and 12.

10. Let n and $n + 2$ be the consecutive even integers. $n + 2 = 2n - 10$, $2 = n - 10$, $12 = n$; $n + 2 = 14$. The consecutive even integers are 12 and 14.

B 11. Let n, $n + 2$, and $n + 4$ be the consecutive odd integers. $2n = (n + 4) + 3$, $2n = n + 7$, $n = 7$; $n + 2 = 9$; $n + 4 = 11$. The consecutive odd integers are 7, 9, and 11.

12. Let n, $n + 1$, $n + 2$, and $n + 3$ be the consecutive integers. $(n + 2) + (n + 3) = 2[n + (n + 1)] - 17$, $2n + 5 = 2(2n + 1) - 17$, $2n + 5 = 4n + 2 - 17$, $2n + 5 = 4n - 15$, $5 = 2n - 15$, $20 = 2n$, $10 = n$; $n + 1 = 11$; $n + 2 = 12$; $n + 3 = 13$. The consecutive integers are 10, 11, 12, and 13.

13. Let n, $n + 2$, $n + 4$, and $n + 6$ be the consecutive even integers. $n + 6 = n + (n + 2)$, $n + 6 = 2n + 2$, $6 = n + 2$, $4 = n$; $n + 2 = 6$; $n + 4 = 8$; $n + 6 = 10$. The consecutive even integers are 4, 6, 8, and 10.

14. Let n, $n + 2$, $n + 4$, and $n + 6$ be the consecutive odd integers. $(n + 4) + (n + 6) = 4n$, $2n + 10 = 4n$, $10 = 2n$, $5 = n$; $n + 2 = 7$; $n + 4 = 9$; $n + 6 = 11$. The consecutive odd integers are 5, 7, 9, and 11.

Pages 150–151 • MIXED REVIEW EXERCISES

1. $-5^2 = -(5 \cdot 5) = -25$ 2. $(-5)^2 = (-5)(-5) = 25$ 3. $(2^3 + 3^2) = 8 + 9 = 17$

4. $(4 - 6)^2 = (-2)^2 = (-2)(-2) = 4$

5. $2(y + 3) - 4 = 3(5 - y)$, $2y + 6 - 4 = 15 - 3y$, $5y + 2 = 15$, $5y = 13$, $y = \frac{13}{5}$; $\left\{\frac{13}{5}\right\}$

6. $15 = 2(n + 3)$, $15 = 2n + 6$, $9 = 2n$, $\frac{9}{2} = n$; $\left\{\frac{9}{2}\right\}$

7. $5(x + 12) = 15 - 2(2x - 2)$, $5x + 60 = 15 - 4x + 4$, $5x + 60 = 19 - 4x$, $9x + 60 = 19$, $9x = -41$, $x = -\frac{41}{9}$; $\left\{-\frac{41}{9}\right\}$

8. $-\frac{4}{3}(n + 5) = 10$, $n + 5 = -\frac{30}{4}$, $n = -\frac{25}{2}$; $\left\{-\frac{25}{2}\right\}$

9. $c - 3 = |2 - 9|$, $c - 3 = |-7|$, $c - 3 = 7$, $c = 10$; $\{10\}$

10. $\frac{2}{5}(2y + 1) = y - 3$, $\frac{4}{5}y + \frac{2}{5} = y - 3$, $\frac{2}{5} = \frac{1}{5}y - 3$, $\frac{17}{5} = \frac{1}{5}y$, $17 = y$; $\{17\}$

Page 151 • COMPUTER EXERCISES

```
10 PRINT "THIS PROGRAM WILL ADD TWO POLYNOMIALS"
20 PRINT "ENTER THE DEGREE OF EACH POLYNOMIAL"
30 PRINT "(SMALLER DEGREE FIRST)"
40 PRINT "FIRST POLYNOMIAL";
50 INPUT M
60 DIM P(M + 1)
70 PRINT "SECOND POLYNOMIAL";
80 INPUT N
90 DIM R(N + 1)
100 PRINT "ENTER THE COEFFICIENTS OF THE POLYNOMIALS"
110 PRINT "IN ASCENDING ORDER"
120 PRINT "FIRST POLYNOMIAL";
130 FOR I = 1 TO M + 1
140 INPUT P(I)
150 NEXT I
160 PRINT "SECOND POLYNOMIAL";
170 FOR I = 1 TO N + 1
180 INPUT R(I)
190 NEXT I
200 PRINT
210 PRINT "SUM: "; P(1) + R(1);
220 FOR I = 2 TO M + 1
230 PRINT "+"; P(I) + R(I); "X↑"; I - 1;
240 NEXT I
250 FOR I = M + 2 TO N + 1
260 PRINT "+"; R(I); "X↑"; I - 1;
270 NEXT I
280 PRINT
290 END
```

1. $4 + 11x + 3x^2$ **2.** $-5 - 4x^2 + x^4$ **3.** $4x + 9x^2 + 5x^3$ **4.** $4 + 6x + x^2 + x^3$

Pages 153–154 • WRITTEN EXERCISES

A **1.** $n^3 \cdot n^5 = n^{3+5} = n^8$ **2.** $a^2 \cdot a^2 = a^{2+2} = a^4$ **3.** $x^3 \cdot x^4 \cdot x^2 = x^{3+4+2} = x^9$

4. $n^2 \cdot n^2 \cdot n = n^{2+2+1} = n^5$ **5.** $(2x^2)(5x^5) = (2 \cdot 5)(x^2 \cdot x^5) = 10x^7$

6. $(5a^5)(6a^6) = (5 \cdot 6)(a^5 \cdot a^6) = 30a^{11}$ **7.** $(m^2n)(mn^4) = (m^2 \cdot m)(n \cdot n^4) = m^3n^5$

8. $(y^3z)(y^2z^3) = (y^3 \cdot y^2)(z \cdot z^3) = y^5z^4$

9. $(2ab)(3ab^5) = (2 \cdot 3)(a \cdot a)(b \cdot b^5) = 6a^2b^6$

10. $(5x^2y)(3x^3y^4) = (5 \cdot 3)(x^2 \cdot x^3)(y \cdot y^4) = 15x^5y^5$

11. $(4x^5)(-3x^2) = (4)(-3)(x^5 \cdot x^2) = -12x^7$

12. $(5y^3)(-2y^4) = (5)(-2)(y^3 \cdot y^4) = -10y^7$

13. $(-3xy^3)(-2x^3y) = (-3)(-2)(x \cdot x^3)(y^3 \cdot y) = 6x^4y^4$

14. $(3r^2s^3)(-5r^3s) = (3)(-5)(r^2 \cdot r^3)(s^3 \cdot s) = -15r^5s^4$

15. $(5a^2b^3c)(2ab^4c^2) = (5 \cdot 2)(a^2 \cdot a)(b^3 \cdot b^4)(c \cdot c^2) = 10a^3b^7c^3$

16. $(3y^3z)(4y^4z^2) = (3 \cdot 4)(y^3 \cdot y^4)(z \cdot z^2) = 12y^7z^3$

17. $(2p^2q)(3pq)(4q) = (2 \cdot 3 \cdot 4)(p^2 \cdot p)(q \cdot q \cdot q) = 24p^3q^3$

18. $(ab^2)(5a^2b^3)(3a^3) = (5 \cdot 3)(a \cdot a^2 \cdot a^3)(b^2 \cdot b^3) = 15a^6b^5$

19. $(-x^2y^3)(3xy^2)(-2x^3y) = (-1)(3)(-2)(x^2 \cdot x \cdot x^3)(y^3 \cdot y^2 \cdot y) = 6x^6y^6$

20. $(-r^2s)(-3rs^3)(-s^2) = (-1)(-3)(-1)(r^2 \cdot r)(s \cdot s^3 \cdot s^2) = -3r^3s^6$

21. $\left(\frac{2}{3}t^4\right)\left(\frac{3}{2}t^2\right) = \left(\frac{2}{3} \cdot \frac{3}{2}\right)(t^4 \cdot t^2) = t^6$ **22.** $\left(\frac{2}{7}a^2\right)(21a^5) = \left(\frac{2}{7} \cdot 21\right)(a^2 \cdot a^5) = 6a^7$

23. $\frac{15a^3b}{2} \cdot \frac{8ab^2}{10} = \left(\frac{15}{2} \cdot \frac{8}{10}\right)(a^3 \cdot a)(b \cdot b^2) = 6a^4b^3$

24. $\frac{4h^3k^2}{7} \cdot \frac{21hk^5}{2} = \left(\frac{4}{7} \cdot \frac{21}{2}\right)(h^3 \cdot h)(k^2 \cdot k^5) = 6h^4k^7$

25. $(3x^3)\left(\frac{1}{6}x^2\right)(8x) = \left(3 \cdot \frac{1}{6} \cdot 8\right)(x^3 \cdot x^2 \cdot x) = 4x^6$

26. $(8c^2)(-d)\left(-\frac{1}{4}cd^2\right) = (8)(-1)\left(-\frac{1}{4}\right)(c^2 \cdot c)(d \cdot d^2) = 2c^3d^3$

27. $(3p^3q)\left(-\frac{5}{6}q^3\right)(-p^4) = (3)\left(-\frac{5}{6}\right)(-1)(p^3 \cdot p^4)(q \cdot q^3) = \frac{5}{2}p^7q^4$

28. $(-a^3b)(-a^2b^2)(-ab^3) = (-1)(-1)(-1)(a^3 \cdot a^2 \cdot a)(b \cdot b^2 \cdot b^3) = -a^6b^6$

29. $(4xy)(2xy^3)(-2y^2) = (4)(2)(-2)(x \cdot x)(y \cdot y^3 \cdot y^2) = -16x^2y^6$

30. $(5b^4)(-3a^2b)(-a^3) = (5)(-3)(-1)(a^2 \cdot a^3)(b^4 \cdot b) = 15a^5b^5$

B **31.** $(5x^2)(2x^3) + (3x)(4x^4) = (5 \cdot 2)(x^2 \cdot x^3) + (3 \cdot 4)(x \cdot x^4) = 10x^5 + 12x^5 = 22x^5$

32. $(2y)(4y^3) + (3y^2)(5y^2) = (2 \cdot 4)(y \cdot y^3) + (3 \cdot 5)(y^2 \cdot y^2) = 8y^4 + 15y^4 = 23y^4$

33. $(3a^5)(5a^3) - (6a^2)(a^6) = (3 \cdot 5)(a^5 \cdot a^3) - 6(a^2 \cdot a^6) = 15a^8 - 6a^8 = 9a^8$

34. $(6s^5)(2s^2) - (3s^4)(4s^3) = (6 \cdot 2)(s^5 \cdot s^2) - (3 \cdot 4)(s^4 \cdot s^3) = 12s^7 - 12s^7 = 0$

35. Perimeter $= 2(2ab) + 2(3ab) = 4ab + 6ab = 10ab$; Area $= (2ab)(3ab) = (2 \cdot 3)(a \cdot a)(b \cdot b) = 6a^2b^2$

36. Perimeter $= 3x + 3x + x + x + 2x + 4x = 14x$; Area $= (2x)(x) + (3x)(3x) = 2x^2 + 9x^2 = 11x^2$

37. Perimeter $= 2(4a) + 10a = 8a + 10a = 18a$; Area $= (2)(4a)(a) + (2a)(a) = 8a^2 + 2a^2 = 10a^2$

38. Surface area $= (2)(3a)(a) + 2(2a)(a) + 2(2a)(3a) = 6a^2 + 4a^2 + 12a^2 = 22a^2$

39. Surface area $= 12(x)(3x) + 2[2(x)(x) + (x)(3x)] = 36x^2 + 2(2x^2 + 3x^2) = 36x^2 + 2(5x^2) = 36x^2 + 10x^2 = 46x^2$

40. Surface area $= 4(a)(3a) + 3(2a)(3a) + (4a)(3a) + 2[2(a)(a) + (4a)(a)] = 12a^2 + 18a^2 + 12a^2 + 2(2a^2 + 4a^2) = 42a^2 + 2(6a^2) = 42a^2 + 12a^2 = 54a^2$

C **41.** $a^m \cdot a^m = a^{m+m} = a^{2m}$ **42.** $x^{3n} \cdot x^n = x^{3n+n} = x^{4n}$ **43.** $3^p \cdot 3^q = 3^{p+q}$

44. $5^x \cdot 5^4 = 5^{x+4}$ **45.** $4^2 \cdot 4^{x-2} = 4^{2+(x-2)} = 4^x$

46. $(-2)(-2)^{x-2} = (-2)^{1+(x-2)} = (-2)^{x-1}$ **47.** $a^x \cdot a^3 = a^{x+3}$ **48.** $x^2 \cdot x^n = x^{2+n}$

49. $2^{x-1} \cdot 2^{x+4} = 2^{(x-1)+(x+4)} = 2^{2x+3}$ **50.** $5^{2n} \cdot 5^{n+2} \cdot 5^n = 5^{2n+(n+2)+n} = 5^{4n+2}$

51. $(nx^5)(5x^5) = 5nx^{5+5} = 5nx^{10}$ **52.** $(3t^k)(kt^3) = 3kt^{k+3}$

Page 154 • MIXED REVIEW EXERCISES

1. $4 + 2^2 = 4 + 4 = 8$ **2.** $(4 + 2)^2 = 6^2 = 36$

3. $3p^2 + 4q^2 - 2p^2q - q^2 = 3p^2 + 3q^2 - 2p^2q$ **4.** $3 \cdot 7^2 = 3 \cdot 49 = 147$

5. $(3 \cdot 7)^2 = 21^2 = 441$ **6.** $3x^2 - 4x + 5 + 6x + 4x^2 = 7x^2 + 2x + 5$

7. $4(y + 3) = 3y, 4y + 12 = 3y, y + 12 = 0, y = -12; \{-12\}$

8. $15z = 30 + 10z, 5z = 30, z = 6; \{6\}$ **9.** $7n - 5 = 2n, 5n - 5 = 0, 5n = 5, n = 1; \{1\}$

10. $\frac{n}{5} + 3 = 6, \frac{n}{5} = 3, n = 15; \{15\}$ **11.** $2(x - 4) = 6, 2x - 8 = 6, 2x = 14, x = 7; \{7\}$

12. $\frac{y}{2} - 1 = 3, \frac{y}{2} = 4, y = 8; \{8\}$

Pages 156–157 • WRITTEN EXERCISES

A **1. a.** $3x^3 = 3 \cdot 3^3 = 3 \cdot 27 = 81$ **b.** $(3x)^3 = (3 \cdot 3)^3 = 9^3 = 729$

c. $3^3 \cdot x^3 = 27 \cdot 3^3 = 27 \cdot 27 = 729$

2. a. $5y^2 = 5 \cdot 2^2 = 5 \cdot 4 = 20$ **b.** $(5y)^2 = (5 \cdot 2)^2 = 10^2 = 100$

c. $5^2y^2 = 25 \cdot 2^2 = 25 \cdot 4 = 100$

3. a. $xy^2 = 3 \cdot 2^2 = 3 \cdot 4 = 12$ **b.** $x^2y^2 = 3^2 \cdot 2^2 = 9 \cdot 4 = 36$

c. $(xy)^2 = (3 \cdot 2)^2 = 6^2 = 36$

4. a. $xy^3 = 3 \cdot 2^3 = 3 \cdot 8 = 24$ **b.** $(xy)^3 = (3 \cdot 2)^3 = 6^3 = 216$

c. $x^3y^3 = 3^3 \cdot 2^3 = 27 \cdot 8 = 216$

5. a. $c^5 \cdot c^2 = c^{5+2} = c^7$ **b.** $(c^2)^5 = c^{2 \cdot 5} = c^{10}$ **c.** $(c^5)^2 = c^{5 \cdot 2} = c^{10}$

6. a. $x^4 \cdot x^7 = x^{4+7} = x^{11}$ **b.** $(x^4)^7 = x^{4 \cdot 7} = x^{28}$ **c.** $(x^7)^4 = x^{7 \cdot 4} = x^{28}$

7. a. $(-5a^4)^3 = (-5)^3 \cdot a^{4 \cdot 3} = -125a^{12}$ **b.** $-(5a^4)^3 = -5^3a^{4 \cdot 3} = -125a^{12}$

c. $-5(a^4)^3 = -5a^{4 \cdot 3} = -5a^{12}$

8. a. $(-2k^5)^6 = (-2)^6 \cdot (k^5)^6 = 64k^{5 \cdot 6} = 64k^{30}$ **b.** $-(2k^5)^6 = -2^6k^{5 \cdot 6} = -64k^{30}$

c. $-2(k^5)^6 = -2k^{5 \cdot 6} = -2k^{30}$

9. $(7a)^2 = 7^2a^2 = 49a^2$ **10.** $(-2t)^5 = (-2)^5t^5 = -32t^5$

11. $(-4c)^3 = (-4)^3c^3 = -64c^3$ **12.** $(5x)^3 = 5^3x^3 = 125x^3$

13. $(4k^2)^3 = 4^3k^{2 \cdot 3} = 64k^6$ **14.** $(6x^3)^2 = 6^2x^{3 \cdot 2} = 36x^6$

15. $(-3y^3)^3 = (-3)^3y^{3 \cdot 3} = -27y^9$ **16.** $(-2t^4)^4 = (-2)^4t^{4 \cdot 4} = 16t^{16}$

17. $(3a^2b)^3 = 3^3a^{2 \cdot 3}b^3 = 27a^6b^3$ **18.** $(2x^2y)^5 = 2^5x^{2 \cdot 5}y^5 = 32x^{10}y^5$

19. $(2r^3s^4)^4 = 2^4r^{3 \cdot 4}s^{4 \cdot 4} = 16r^{12}s^{16}$ **20.** $(5p^3q^4)^3 = 5^3p^{3 \cdot 3}q^{4 \cdot 3} = 125p^9q^{12}$

B **21.** $(2x)^2(2x)^4 = (2^2x^2)(2^4x^4) = 2^6x^6 = 64x^6$

22. $(3c)^2(3c)^3 = (3c)^{2+3} = (3c)^5 = 3^5c^5 = 243c^5$

23. $(10b)^3(10b)^3 = (10b)^6 = 10^6b^6 = 1{,}000{,}000b^6$

24. $\left(\frac{1}{2}y\right)^2(2y)^5 = \left(\frac{1}{2}\right)^2 y^2 2^5 y^5 = \frac{1}{4} \cdot 32 \cdot y^7 = 8y^7$

25. $(3a^2b)^3(2a^3b) = 3^3a^6b^3 \cdot 2a^3b = (27 \cdot 2)a^9b^4 = 54a^9b^4$

26. $(2x^2y^3)^4(-xy^2)^2 = 2^4x^8y^{12} \cdot (-x)^2y^4 = 16(x^8 \cdot x^2)y^{16} = 16x^{10}y^{16}$

27. $\left(\frac{1}{2}p^2q\right)^3(2pq^2)^4 = \left(\frac{1}{2}\right)^3 p^6q^3 \cdot 2^4p^4q^8 = \left(\frac{1}{8}\right)16p^{10}q^{11} = 2p^{10}q^{11}$

28. $\left(\frac{1}{10}x^3y\right)^3(10y)^4 = \left(\frac{1}{10}\right)^3 x^9y^3 \cdot 10^4y^4 = \left(\frac{1}{1{,}000}\right)10{,}000x^9y^7 = 10x^9y^7$

29. $[(2x^2)^2]^2 = (2^2x^4)^2 = 2^4x^8 = 16x^8$ **30.** $[(-x^3)^2]^3 = (x^6)^3 = x^{18}$

31. $(3x^2y)^3(2xy) = 3^3x^6y^3 \cdot 2xy = (27 \cdot 2)x^7y^4 = 54x^7y^4$

32. $(5a^2b)^2(5b)^3 = 5^2a^4b^2 \cdot 5^3b^3 = 5^5a^4b^5 = 3125a^4b^5$

33. a. $(3x^3)^2 + (2x^2)^3 = 3^2x^6 + 2^3x^6 = 9x^6 + 8x^6 = 17x^6$

b. $(3x^3)^2 \cdot (2x^2)^3 = 9x^6 \cdot 8x^6 = 72x^{12}$

34. a. $(a^2)^6 + (-2a^4)^3 = a^{12} + (-2)^3a^{12} = a^{12} - 8a^{12} = -7a^{12}$

b. $(a^2)^6 \cdot (-2a^4)^3 = (a^{12})(-8a^{12}) = -8a^{24}$

35. a. $a^2(ab^3)^2 + b^2(a^2b^2)^2 = a^2a^2b^6 + b^2a^4b^4 = a^4b^6 + a^4b^6 = 2a^4b^6$

b. $a^2(ab^3)^2 \cdot b^2(a^2b^2)^2 = a^2a^2b^6b^2a^4b^4 = a^8b^{12}$

36. a. $p(-pq)^4 + p^3(2pq^2)^2 = p(-p)^4q^4 + p^32^2p^2q^4 = pp^4q^4 + 4p^5q^4 =$
$p^5q^4 + 4p^5q^4 = 5p^5q^4$

b. $p(-pq)^4 \cdot p^3(2pq^2)^2 = p(-p)^4q^4p^32^2p^2q^4 = 4p^{10}q^8$

37. a. $a(-ab^2)^3 + (2a^2b^3)^2 = a(-a)^3b^6 + 2^2a^4b^6 = -a^4b^6 + 4a^4b^6 = 3a^4b^6$

b. $a(-ab^2)^3 \cdot (2a^2b^3)^2 = a(-a)^3b^62^2a^4b^6 = -4a^8b^{12}$

38. a. $(2x)^3(xy)^2 + (2xy)^2(-x)^3 = 2^3x^3x^2y^2 + 2^2x^2y^2(-x^3) = 8x^5y^2 - 4x^5y^2 = 4x^5y^2$

b. $(2x)^3(xy)^2 \cdot (2xy)^2(-x)^3 = 2^3x^3x^2y^22^2x^2y^2(-x^3) = -32x^{10}y^4$

C **39.** $(x^n)^2 = x^{2n}$ **40.** $(a^x)^3 = a^{3x}$ **41.** $a^x \cdot a^3 = a^{x+3}$ **42.** $x^n \cdot x^2 = x^{n+2}$

43. $x^n \cdot x^n = x^{2n}$ **44.** $a^x \cdot a^x \cdot a^x = a^{3x}$ **45.** $z^x \cdot z^x \cdot z^x = z^{3x}$

46. $(a^x)^y(a^y)^x = a^{xy}a^{xy} = a^{2xy}$ **47.** $(2r^n)^3(3r^n)^2 = 2^3r^{3n}3^2r^{2n} = 72r^{5n}$

48. $(2x^n)^3(x^n)^5 = 2^3x^{3n}x^{5n} = 8x^{8n}$ **49.** $(3x^n)^2(x^2)^n = 3^2x^{2n}x^{2n} = 9x^{4n}$

50. $(t^m)^n \cdot 3(t^n)^m = t^{mn}3t^{mn} = 3t^{2mn}$

51. a. $2x \cdot 2x \cdot 2x = 2^3x^3 = 8x^3$; $4x \cdot 4x \cdot 4x = 4^3x^3 = 64x^3$

b. The cube on the right has the larger volume.

c. The bigger cube is 8 times as large as the smaller cube.

52. a. $(3^3)^3 = 3^9$; $3^{(3^3)} = 3^{27}$; $3^{(3^3)}$ is larger. **b.** $3^{(3^3)}$ is 3^{18} times as large as $(3^3)^3$.

53. $16^x \cdot (4^x)^2 = 16^x \cdot 4^{2x} = (2^4)^x \cdot (2^2)^{2x} = 2^{4x} \cdot 2^{4x} = 2^{8x} = (2^x)^8$

Page 157 • MIXED REVIEW EXERCISES

1. $(3a^2b)(2ab)(4b) = (3 \cdot 2 \cdot 4)(a^2 \cdot a)(b \cdot b \cdot b) = 24a^3b^3$

2. $(-x^2y)(2xy)(-4y) = 8x^3y^3$ **3.** $(2x^2y^3)^4 = 2^4x^8y^{12} = 16x^8y^{12}$

4. $\left(\frac{1}{2}t^3\right)\left(\frac{2}{3}t^2\right) = \frac{1}{3}t^5$ **5.** $6c - 3a - 2c + a = 4c - 2a$

6. $(3x + 4y + 2) + (2x + y) = 5x + 5y + 2$

7. $2 \cdot 3^3 + 2 \cdot 3 = 2 \cdot 27 + 6 = 54 + 6 = 60$

8. $-5^2 \cdot 3 = -25 \cdot 3 = -75$ **9.** $(4^3 + 6^2) \div 5^2 = (64 + 36) \div 25 = 100 \div 25 = 4$

Pages 159–160 • WRITTEN EXERCISES

A **1.** $4(x - 3) = 4x - (4 \cdot 3) = 4x - 12$ **2.** $-3(y + 3) = -3y + (-3)(3) = -3y - 9$
3. $c(c - 2) = c^2 - 2c$ **4.** $a(4 - 2a) = 4a - a(2a) = 4a - 2a^2$
5. $3y(y + 5) = (3y)y + (3y)5 = 3y^2 + 15y$
6. $4x(2x - 3) = (4x)(2x) - 4x(3) = 8x^2 - 12x$
7. $-2r(7 - 3r) = (-2r)7 - (-2r)(3r) = -14r + 6r^2$
8. $-z(6 - 5z) = (-z)6 - (-z)(5z) = -6z + 5z^2$
9. $\begin{array}{r} 3y^2 - y - 5 \\ 2y \\ \hline 6y^3 - 2y^2 - 10y \end{array}$ **10.** $\begin{array}{r} 2k^2 - 3k - 7 \\ -3k \\ \hline -6k^3 + 9k^2 + 21k \end{array}$ **11.** $\begin{array}{r} a^2b - 3ab^2 + 5b^3 \\ -2ab \\ \hline -2a^3b^2 + 6a^2b^3 - 10ab^4 \end{array}$
12. $\begin{array}{r} 2p^3 - 5p^2q - 3pq^2 \\ 3q^2 \\ \hline 6p^3q^2 - 15p^2q^3 - 9pq^4 \end{array}$
13. $3x(x^2 - 2x + 4) = (3x)(x^2) - (3x)(2x) + (3x)(4) = 3x^3 - 6x^2 + 12x$
14. $4t(2t^2 - t - 5) = (4t)(2t^2) - (4t)t - (4t)5 = 8t^3 - 4t^2 - 20t$
15. $pq^2(p^2 - 3pq - 4q^2) = (pq^2)p^2 - (pq^2)(3pq) - (pq^2)(4q^2) = p^3q^2 - 3p^2q^3 - 4pq^4$
16. $2x^2y(2x^2 - 3xy + y^2) = (2x^2y)(2x^2) - (2x^2y)(3xy) + (2x^2y)(y^2) =$ $4x^4y - 6x^3y^2 + 2x^2y^3$
17. $\frac{1}{3}x^2(6x^2 - 9xy - 3y^2) = \left(\frac{1}{3}x^2\right)(6x^2) - \left(\frac{1}{3}x^2\right)(9xy) - \left(\frac{1}{3}x^2\right)(3y^2) =$ $2x^4 - 3x^3y - x^2y^2$
18. $\frac{1}{2}s^2t(4t^2 - 10st + 6s^2) = \left(\frac{1}{2}s^2t\right)(4t^2) - \left(\frac{1}{2}s^2t\right)(10st) + \left(\frac{1}{2}s^2t\right)(6s^2) = 2s^2t^3 -$ $5s^3t^2 + 3s^4t$
19. $2x(x - 3) + x(5 - x) = 2x(x) - 2x(3) + x(5) - x(x) = 2x^2 - 6x + 5x - x^2 =$ $x^2 - x$
20. $3x(5 - 2x) + 6x(x - 2) = 3x(5) - 3x(2x) + 6x(x) - 6x(2) = 15x - 6x^2 +$ $6x^2 - 12x = 3x$
21. $6r^2(2r - 1) - 3r(4r^2 - 5r) = 6r^2(2r) - 6r^2 + (-3r)(4r^2) - (-3r)(5r) = 12r^3 -$ $6r^2 - 12r^3 + 15r^2 = 9r^2$
22. $4y(2y^2 - 3y) - 3y^2(y - 4) = 4y(2y^2) - 4y(3y) + (-3y^2)y - (-3y^2)4 =$ $8y^3 - 12y^2 - 3y^3 + 12y^2 = 5y^3$
23. $-[6y - 3(5y - 4)] = -(6y - 15y + 12) = -(-9y + 12) = 9y - 12$
24. $8n - 2[n - 2(3 - n)] = 8n - 2(n - 6 + 2n) = 8n - 2(3n - 6) = 8n - 6n +$ $12 = 2n + 12$
25. $a[2a - 3(1 - a)] + 5(a - a^2) = a(2a - 3 + 3a) + 5a - 5a^2 = a(5a - 3) +$ $5a - 5a^2 = 5a^2 - 3a + 5a - 5a^2 = 2a$
26. $2x[3x - 2(3 + x)] - x^2 = 2x(3x - 6 - 2x) - x^2 = 2x(x - 6) - x^2 = 2x^2 -$ $12x - x^2 = x^2 - 12x$
27. $2(x - 3) + 5 = 7, 2x - 6 + 5 = 7, 2x - 1 = 7, 2x = 8, x = 4; \{4\}$
28. $3(z + 3) - 7 = 8, 3z + 9 - 7 = 8, 3z + 2 = 8, 3z = 6, z = 2; \{2\}$
29. $15 = 3(x - 1) + 2(4 - x), 15 = 3x - 3 + 8 - 2x, 15 = x + 5, 10 = x; \{10\}$
30. $0 = 3(1 - 2t) - 5(2 - t), 0 = 3 - 6t - 10 + 5t, 0 = -7 - t, 7 = -t, -7 = t; \{-7\}$

B **31.** $y(2 - 3y) + 3(y^2 - 4) = 0, 2y - 3y^2 + 3y^2 - 12 = 0, 2y - 12 = 0, 2y = 12, y = 6;$ $\{6\}$

32. $\frac{1}{2}(12 - 4s^2) - 2s(1 - s) = 8, 6 - 2s^2 - 2s + 2s^2 = 8, 6 - 2s = 8, 6 = 8 + 2s,$ $-2 = 2s, -1 = s; \{-1\}$

33. $2(n - 3) + n(2 - n) = 2 - n^2, 2n - 6 + 2n - n^2 = 2 - n^2, 4n - 6 - n^2 = 2 - n^2, 4n - 6 = 2, 4n = 8, n = 2; \{2\}$

34. $\frac{1}{2}(6c + 4) - 2\left(c + \frac{5}{2}\right) = \frac{2}{3}(9 - 3c), 3c + 2 - 2c - 5 = 6 - 2c, c - 3 = 6 - 2c,$ $3c - 3 = 6, 3c = 9, c = 3; \{3\}$

35. $\frac{3}{2}(4x - 6) - x(3 - x) = \frac{1}{3}x(3x + 6), 6x - 9 - 3x + x^2 = x^2 + 2x, 3x - 9 +$ $x^2 = x^2 + 2x, 3x - 9 = 2x, x - 9 = 0, x = 9; \{9\}$

36. $4(y - 7) - 2y(1 - 3y) = 6y^2, 4y - 28 - 2y + 6y^2 = 6y^2, 2y - 28 + 6y^2 = 6y^2,$ $2y - 28 = 0, 2y = 28, y = 14; \{14\}$

37. $r^2 - [4 - r(3 - r)] = 2(2r - 3), r^2 - (4 - 3r + r^2) = 4r - 6, r^2 - 4 + 3r - r^2 = 4r - 6, -4 + 3r = 4r - 6, -4 = r - 6, 2 = r; \{2\}$

38. $1 = 2x(x - 2) - [5 - 2x(1 - x)], 1 = 2x^2 - 4x - (5 - 2x + 2x^2), 1 = 2x^2 - 4x - 5 + 2x - 2x^2, 1 = -2x - 5, 6 = -2x, -3 = x; \{-3\}$

39. **a.** $x(2x) + yx = 2x^2 + xy$ **b.** $x(2x + y) = 2x^2 + xy$
c. Distributive property

40. $x(2x) + 5x = 2x^2 + 5x$ **41.** $(x + 9)(3x) - 4x = 3x^2 + 27x - 4x = 3x^2 + 23x$

42. $4(x + 8) + (2x + 4)x = 4x + 32 + 2x^2 + 4x = 2x^2 + 8x + 32$

C **43.** Surface area $= 2x(2x) + 2x(x) + 2x(2x) + 2x(2x) + 4x(2x) + 2x(3x) + 2[2x(3x) + 2x(2x)] = 4x^2 + 2x^2 + 4x^2 + 4x^2 + 8x^2 + 6x^2 + 2(6x^2 + 4x^2) = 28x^2 + 2(10x^2) = 28x^2 + 20x^2 = 48x^2$; Volume $= 2x(2x)(3x) + 2x(2x)(2x) = 12x^3 + 8x^3 = 20x^3$

44. Surface area $= 2[5x(3x) - 2(x)(x)] + 2(3x)(x) + 2(5x)(x) + 8(x)(x) = 2(15x^2 - 2x^2) + 6x^2 + 10x^2 + 8x^2 = 2(13x^2) + 24x^2 = 26x^2 + 24x^2 = 50x^2$; Volume $= 5x(3x)x - 2(x)(x)(x) = 15x^3 - 2x^3 = 13x^3$

Page 160 • MIXED REVIEW EXERCISES

1. $(3x^2y)^3 = 3^3x^6y^3 = 27x^6y^3$ **2.** $(-5r^3y^4)^2 = (-5)^2r^6y^8 = 25r^6y^8$

3. $(-3n)^3 = (-3)^3n^3 = -27n^3$

4. $(2a^3y)^3(3a^2y^2)^2 = 2^3a^9y^33^2a^4y^4 = (8 \cdot 9)a^{13}y^7 = 72a^{13}y^7$

5. $(2x^2)(3x^3) + (x^4)(4x) = 6x^5 + 4x^5 = 10x^5$

6. $(5n^4)n^2 - n^3(2n^3) = 5n^6 - 2n^6 = 3n^6$

7. $(7p - 3q + 5) + (3q + 3p) = 10p + 5$

8. $(2x + y - 3) - (y + x - 7) = 2x + y - 3 - y - x + 7 = x + 4$

9. $(5y^3)^2(3x^2y)^3 = 5^2y^63^3x^6y^3 = (25 \cdot 27)x^6y^9 = 675x^6y^9$

Pages 162–164 • WRITTEN EXERCISES

A

1. $$\begin{array}{r} 3x - 5 \\ 2x + 1 \\ \hline 6x^2 - 10x \phantom{{}-5} \\ 3x - 5 \\ \hline 6x^2 - 7x - 5 \end{array}$$

2. $$\begin{array}{r} 5t - 1 \\ 2t - 3 \\ \hline 10t^2 - 2t \phantom{{}+3} \\ -15t + 3 \\ \hline 10t^2 - 17t + 3 \end{array}$$

3. $$\begin{array}{r} a^2 - 3a + 4 \\ 2a + 3 \\ \hline 2a^3 - 6a^2 + 8a \phantom{{}+12} \\ 3a^2 - 9a + 12 \\ \hline 2a^3 - 3a^2 - a + 12 \end{array}$$

4. $$\begin{array}{r} r^2 + 2r - 5 \\ 3r - 2 \\ \hline 3r^3 + 6r^2 - 15r \phantom{{}+10} \\ -2r^2 - 4r + 10 \\ \hline 3r^3 + 4r^2 - 19r + 10 \end{array}$$

5. $$\begin{array}{r} 3x - 2y \\ 2x + 5y \\ \hline 6x^2 - 4xy \phantom{{}-10y^2} \\ 15xy - 10y^2 \\ \hline 6x^2 + 11xy - 10y^2 \end{array}$$

6. $$\begin{array}{r} 2a - 5b \\ a - 3b \\ \hline 2a^2 - 5ab \phantom{{}+15b^2} \\ -6ab + 15b^2 \\ \hline 2a^2 - 11ab + 15b^2 \end{array}$$

7. $$\begin{array}{r} c^2 + 2cd - 3d^2 \\ c - 2d \\ \hline c^3 + 2c^2d - 3cd^2 \phantom{{}+6d^3} \\ -2c^2d - 4cd^2 + 6d^3 \\ \hline c^3 - 7cd^2 + 6d^3 \end{array}$$

8. $$\begin{array}{r} 2x^2 - 3xy - y^2 \\ 2x + y \\ \hline 4x^3 - 6x^2y - 2xy^2 \phantom{{}-y^3} \\ 2x^2y - 3xy^2 - y^3 \\ \hline 4x^3 - 4x^2y - 5xy^2 - y^3 \end{array}$$

9. $(y + 3)(y + 2) = y^2 + 2y + 3y + 6 = y^2 + 5y + 6$

10. $(n + 7)(n + 5) = n^2 + 5n + 7n + 35 = n^2 + 12n + 35$

11. $(a + 4)(a - 1) = a^2 - a + 4a - 4 = a^2 + 3a - 4$

12. $(r - 3)(r + 6) = r^2 + 6r - 3r - 18 = r^2 + 3r - 18$

13. $(2x - 1)(x - 5) = 2x^2 - 10x - x + 5 = 2x^2 - 11x + 5$

14. $(3a - 2)(a - 3) = 3a^2 - 9a - 2a + 6 = 3a^2 - 11a + 6$

15. $(3z - 2)(2z + 3) = 6z^2 + 9z - 4z - 6 = 6z^2 + 5z - 6$

16. $(5k + 2)(2k - 3) = 10k^2 - 15k + 4k - 6 = 10k^2 - 11k - 6$

17. $(4s - 5)(4s + 5) = 16s^2 + 20s - 20s - 25 = 16s^2 - 25$

18. $(3x + 7)(3x - 7) = 9x^2 - 21x + 21x - 49 = 9x^2 - 49$

19. $(a + 2)(a^2 + 3a + 5) = a^3 + 3a^2 + 5a + 2a^2 + 6a + 10 = a^3 + 5a^2 + 11a + 10$

20. $(x + 3)(x^2 + 2x + 4) = x^3 + 2x^2 + 4x + 3x^2 + 6x + 12 = x^3 + 5x^2 + 10x + 12$

21. $(m - 1)(m^2 + 2m + 6) = m^3 + 2m^2 + 6m - m^2 - 2m - 6 = m^3 + m^2 + 4m - 6$

22. $(y - 5)(y^2 - 3y - 7) = y^3 - 3y^2 - 7y - 5y^2 + 15y + 35 = y^3 - 8y^2 + 8y + 35$

23. $(2x - 1)(x^2 - x + 3) = 2x^3 - 2x^2 + 6x - x^2 + x - 3 = 2x^3 - 3x^2 + 7x - 3$

24. $(3s + 2)(2s^2 - 2s + 1) = 6s^3 - 6s^2 + 3s + 4s^2 - 4s + 2 = 6s^3 - 2s^2 - s + 2$

25. $(3z - 2)(3z^2 - z + 4) = 9z^3 - 3z^2 + 12z - 6z^2 + 2z - 8 = 9z^3 - 9z^2 + 14z - 8$

26. $(2n - 5)(2n^2 - 3n - 2) = 4n^3 - 6n^2 - 4n - 10n^2 + 15n + 10 =$
$4n^3 - 16n^2 + 11n + 10$

27. a. $x^2 + 7x + 3x + 21 = x^2 + 10x + 21$

b. $(x + 3)(x + 7) = x^2 + 7x + 3x + 21 = x^2 + 10x + 21$ **c.** equal

28. a. $ac + bc + ad + bd$ **b.** $(c + d)(a + b) = ac + bc + ad + bd$ **c.** equal

For Exs. 29–36, form of answers may vary as long as correct terms are combined.

B

29. $(2 + y)(y^2 - 2y + 3) = 2y^2 - 4y + 6 + y^3 - 2y^2 + 3y = y^3 - y + 6$

30. $(5 + x)(x^2 - 5x + 4) = 5x^2 - 25x + 20 + x^3 - 5x^2 + 4x = x^3 - 21x + 20$

31. $(2y - 3)(2y - 4 + y^2) = 4y^2 - 8y + 2y^3 - 6y + 12 - 3y^2 = 2y^3 + y^2 - 14y + 12$

32. $(3y - 4)(y - 2y^2 + 6) = 3y^2 - 6y^3 + 18y - 4y + 8y^2 - 24 =$
$-6y^3 + 11y^2 + 14y - 24$

33. $(2 - 3x)(x^2 - 2x + 5) = 2x^2 - 4x + 10 - 3x^3 + 6x^2 - 15x =$
$-3x^3 + 8x^2 - 19x + 10$

34. $(1 - 2a)(a^2 - 4 + 3a) = a^2 - 4 + 3a - 2a^3 + 8a - 6a^2 = -2a^3 - 5a^2 + 11a - 4$

35. $(2r - s)(s^2 + 4r^2 - 4rs) = 2rs^2 + 8r^3 - 8r^2s - s^3 - 4r^2s + 4rs^2 =$
$8r^3 - 12r^2s + 6rs^2 - s^3$

36. $(y - 2x)(2x^2 + y^2 - 3xy) = 2x^2y + y^3 - 3xy^2 - 4x^3 - 2xy^2 + 6x^2y =$
$-4x^3 + 8x^2y - 5xy^2 + y^3$

37. $(x + 2)(x - 5) = (x - 1)(x - 3), x^2 - 3x - 10 = x^2 - 4x + 3, -3x - 10 =$
$-4x + 3, x - 10 = 3, x = 13; \{13\}$

38. $(x - 3)(x + 7) - (x + 1)(x + 5) = 0, x^2 + 4x - 21 - (x^2 + 6x + 5) = 0,$
$x^2 + 4x - 21 - x^2 - 6x - 5 = 0, -2x - 26 = 0, -2x = 26, x = -13; \{-13\}$

39. $(2x - 5)(x - 4) + 2x(1 - x) = 0, 2x^2 - 13x + 20 + 2x - 2x^2 = 0,$

$-11x + 20 = 0, -11x = -20, x = \frac{20}{11}; \left\{\frac{20}{11}\right\}$

40. $(3x + 5)(2x - 3) = (x - 1)(6x + 5), 6x^2 + x - 15 = 6x^2 - x - 5,$
$x - 15 = -x - 5, 2x - 15 = -5, 2x = 10, x = 5; \{5\}$

41. $(x - 3)(x^2 - 2x + 6) = x(x^2 - 5x + 9), x^3 - 2x^2 + 6x - 3x^2 + 6x - 18 =$
$x^3 - 5x^2 + 9x, x^3 - 5x^2 + 12x - 18 = x^3 - 5x^2 + 9x, -5x^2 + 12x - 18 =$
$-5x^2 + 9x, 12x - 18 = 9x, -18 = -3x, 6 = x; \{6\}$

42. $(2n - 3)(n^2 + 3n - 2) = (n - 1)(2n^2 + 5n - 4),$
$2n^3 + 6n^2 - 4n - 3n^2 - 9n + 6 = 2n^3 + 5n^2 - 4n - 2n^2 - 5n + 4,$
$2n^3 + 3n^2 - 13n + 6 = 2n^3 + 3n^2 - 9n + 4, 3n^2 - 13n + 6 =$

$3n^2 - 9n + 4, -13n + 6 = -9n + 4, 6 = 4n + 4, 2 = 4n, \frac{1}{2} = n; \left\{\frac{1}{2}\right\}$

43. $(x^2 - 3x + 5)^2 = (x^2 - 3x + 5)(x^2 - 3x + 5) = x^4 - 3x^3 + 5x^2 -$
$3x^3 + 9x^2 - 15x + 5x^2 - 15x + 25 = x^4 - 6x^3 + 19x^2 - 30x + 25$

44. $(n^2 + 2n - 1)^2 = (n^2 + 2n - 1)(n^2 + 2n - 1) = n^4 + 2n^3 - n^2 + 2n^3 +$
$4n^2 - 2n - n^2 - 2n + 1 = n^4 + 4n^3 + 2n^2 - 4n + 1$

45. $(x + 5)^3 = (x + 5)(x + 5)(x + 5) = (x + 5)(x^2 + 10x + 25) =$
$x^3 + 10x^2 + 25x + 5x^2 + 50x + 125 = x^3 + 15x^2 + 75x + 125$

46. $(a - 3)^3 = (a - 3)(a - 3)(a - 3) = (a - 3)(a^2 - 6a + 9) =$
$a^3 - 6a^2 + 9a - 3a^2 + 18a - 27 = a^3 - 9a^2 + 27a - 27$

47. $(2 - y)^4 = (2 - y)(2 - y)^3 = (2 - y)(8 - 12y + 6y^2 - y^3) =$
$16 - 24y + 12y^2 - 2y^3 - 8y + 12y^2 - 6y^3 + y^4 = 16 - 32y + 24y^2 - 8y^3 + y^4$

48. $(x^2 + y^2) - (2x - y)(x - 2y) = x^2 + y^2 - (2x^2 - 4xy - xy + 2y^2) =$
$x^2 + y^2 - (2x^2 - 5xy + 2y^2) = x^2 + y^2 - 2x^2 + 5xy - 2y^2 = -x^2 + 5xy - y^2$

49. $(x - 3)(x^2 + x - 2) = x^3 + x^2 - 2x - 3x^2 - 3x + 6 = x^3 - 2x^2 - 5x + 6$

C **50.** **a.** (1) $(x + 1)(x^2 - x + 1) = x^3 - x^2 + x + x^2 - x + 1 = x^3 + 1$
(2) $(x + 1)(x^3 - x^2 + x - 1) = x^4 - x^3 + x^2 - x + x^3 - x^2 + x - 1 = x^4 - 1$
(3) $(x + 1)(x^4 - x^3 + x^2 - x + 1) = x^5 - x^4 + x^3 - x^2 + x + x^4 - x^3 +$
$x^2 - x + 1 = x^5 + 1$

b. (4) $(x + 1)(x^5 - x^4 + x^3 - x^2 + x - 1) = x^6 - 1$
(5) $(x + 1)(x^6 - x^5 + x^4 - x^3 + x^2 - x + 1) = x^7 + 1$

Page 164 • MIXED REVIEW EXERCISES

1. $3(x - 1) = 6, 3x - 3 = 6, 3x = 9, x = 3; \{3\}$

2. $4(x + 2) - 3 = 21, 4x + 8 - 3 = 21, 4x + 5 = 21, 4x = 16, x = 4; \{4\}$

3. $6(2a + 2) = 3(a + 10), 12a + 12 = 3a + 30, 9a + 12 = 30, 9a = 18, a = 2; \{2\}$

4. $w + |-x| + -y = 3 + |-(-1)| + -4 = 3 + 1 - 4 = 0$

5. $w + x + 3 = 3 + (-1) + 3 = 5$ **6.** $w - |x - y| = 3 - |-1 - 4| = 3 - |-5| = 3 - 5 = -2$

7. $(w + y)^2 = (3 + 4)^2 = 7^2 = 49$ **8.** $(-x^8)x^2 = (-(-1)^8)(-1)^2 = (-1)(1) = -1$

9. $xw^3 = (-1)(3^3) = (-1)(27) = -27$

Page 166 • WRITTEN EXERCISES

A **1.** $C = 2\pi r, \frac{C}{2\pi} = r$ **2.** $F = ma, \frac{F}{m} = a; m \neq 0$ **3.** $s = \frac{v}{r}, rs = v, r = \frac{v}{s}; s \neq 0$

4. $d = \frac{m}{v}, dv = m, v = \frac{m}{d}; d \neq 0$ **5.** $I = Prt, \frac{I}{rt} = P; r \neq 0, t \neq 0$

6. $A = P + Prt, A - P = Prt, \frac{A - P}{Pr} = t; P \neq 0, r \neq 0$

7. $A = 2a^2 + 4ah, A - 2a^2 = 4ah, \frac{A - 2a^2}{4a} = h; a \neq 0$

8. $s = vt + 16t^2, s - 16t^2 = vt, \frac{s - 16t^2}{t} = v; t \neq 0$

9. $A = \frac{1}{2}h(a + b), \frac{2A}{a + b} = h; a + b \neq 0$ **10.** $S = \frac{n}{2}(a + l), \frac{2S}{a + l} = n; a + l \neq 0$

11. $p = 2(l + w), p = 2l + 2w, p - 2l = 2w, \frac{p - 2l}{2} = w$

12. $A = P(1 + rt), A = P + Prt, A - P = Prt, \frac{A - P}{Pt} = r; t \neq 0, P \neq 0$

13. $m = \frac{x + y}{2}, 2m = x + y, 2m - x = y$ **14.** $a = \frac{v - u}{t}, at = v - u, at + u = v$

B **15.** $S = \frac{n}{2}(a + l), S = \frac{an}{2} + \frac{ln}{2}, S - \frac{nl}{2} = \frac{an}{2}, \frac{2}{n}\left(S - \frac{nl}{2}\right) = a, \frac{2S - nl}{n} = a; n \neq 0$

16. $C = \frac{5}{9}(F - 32), \frac{9}{5}C = F - 32, \frac{9}{5}C + 32 = F$

17. $a = \frac{v - u}{t}, at = v - u, t = \frac{v - u}{a}; a \neq 0$

18. $m = \frac{x + y + z}{3}, 3m = x + y + z, 3m - x - z = y$

19. $v^2 = u^2 + 2as, v^2 - u^2 = 2as, \frac{v^2 - u^2}{2a} = s; a \neq 0$

20. $s = \frac{n}{2}(a + l), \frac{2s}{n} = a + l, \frac{2s}{n} - a = l; n \neq 0$

21. $S = \frac{a}{a - r}$, $S(a - r) = a$, $Sa - Sr = a$, $-Sr = a - Sa$, $r = -\frac{a - Sa}{S}$, $r = \frac{Sa - a}{S}$; $S \neq 0$

22. $l = a + (n - 1)d$, $l = a + nd - d$, $l - a + d = nd$, $\frac{l - a + d}{d} = n$; $d \neq 0$

23. $F = \frac{fg}{f + g - d}$, $F(f + g - d) = fg$, $Ff + Fg - Fd = fg$, $Ff + Fg = fg + Fd$,

$Ff + Fg - fg = Fd$, $\frac{Ff + Fg - fg}{F} = d$; $F \neq 0$

24. $S = \frac{a - rl}{l - r}$, $S(l - r) = a - rl$, $S - Sr - a = -rl$, $\frac{S - Sr - a}{-r} = l$; $r \neq 0$

25. $a = \frac{180(n - 2)}{n}$, $an = 180(n - 2)$, $an = 180n - 360$, $an - 180n = -360$,

$(a - 180)n = -360$, $n = -\frac{360}{a - 180}$; $a \neq 180$

26. $S = \frac{r}{1 - r}$, $S(1 - r) = r$, $S - Sr = r$, $S = r + Sr$, $S = (1 + S)r$, $\frac{S}{1 + S} = r$; $S \neq -1$

C **27.** $r = \frac{ab}{a + b}$, $r(a + b) = ab$, $ra + rb = ab$, $ra = ab - rb$, $ra - ab = -rb$,

$(r - b)a = -rb$, $a = -\frac{rb}{r - b}$, $a = \frac{rb}{b - r}$; $r \neq b$

28. $F = \frac{fg}{f + g - d}$, $F(f + g - d) = fg$, $Ff + Fg - Fd = fg$, $Ff = fg - Fg + Fd$,

$Ff - fg = -Fg + Fd$, $(F - g)f = Fd - Fg$, $f = \frac{Fd - Fg}{F - g}$; $F \neq g$

29. $C = K\left(\frac{Rr}{R - r}\right)$, $C(R - r) = KRr$, $CR - Cr = KRr$, $-Cr = KRr - CR$,

$-Cr = (Kr - C)R$, $-\frac{Cr}{Kr - C} = R$, $\frac{Cr}{C - Kr} = R$; $C \neq Kr$

Page 166 • MIXED REVIEW EXERCISES

1. $(y - 5)(y + 3) = y^2 + 3y - 5y - 15 = y^2 - 2y - 15$

2. $(3n - 2)(2n - 4) = 6n^2 - 12n - 4n + 8 = 6n^2 - 16n + 8$

3. $a[2a - 4(2 + a)] = a[2a - 8 - 4a] = a(-2a - 8) = -2a^2 - 8a$

4. $xy(2x + 3y) = 2x^2y + 3xy^2$ **5.** $4x(x^2 - 3x + 2) = 4x^3 - 12x^2 + 8x$

6. $(-3x^3)^3 = (-3)^3x^{3\cdot3} = -27x^9$ **7.** $n^2 \cdot n^2 \cdot n^2 = n^{2+2+2} = n^6$

8. $(3a^2)^2 \cdot 4a^3b = 3^2a^{2\cdot2} \cdot 4a^3b = (9 \cdot 4)(a^4 \cdot a^3)b = 36a^7b$

Pages 170–171 • PROBLEMS

A **1.** $500t + 700(t - 1) = 2100$, $500t + 700t - 700 = 2100$, $1200t - 700 = 2100$,

$1200t = 2800$, $t = 2\frac{1}{3}$; The aircraft passed each other at 2:20 P.M.

2. $48t = 60\left(t - \frac{1}{3}\right)$, $48t = 60t - 20$, $48t + 20 = 60t$, $20 = 12t$, $\frac{5}{3} = t$; The Garcias

overtook the Smiths after the Smiths had been driving for $1\frac{2}{3}$ h, at 9:40 A.M.

3. $4t = 6(3 - t)$, $4t = 18 - 6t$, $10t = 18$, $t = 1.8$; It took Kwan 1 h 48 min to hike up the hill.

4. $2r + r - 10 = 152$, $3r - 10 = 152$, $3r = 162$, $r = 54$; Jenny's original speed was 54 mi/h.

5.

	Rate ×	Time =	Distance
Jet flying east	$r + 50$	2	$2(r + 50)$
Jet flying west	r	2	$2r$

$2(r + 50) + 2r = 2500$, $2r + 100 + 2r = 2500$, $4r + 100 = 2500$, $r = 2400$, $r = 600$; $r + 50 = 650$; One jet flew east at 650 km/h, and the other flew west at 600 km/h.

6.

	Rate ×	Time =	Distance
Joe	6	$\frac{1}{2}$	3
Ken	r	$\frac{1}{3}$	$\frac{1}{3}r$

$3 = \frac{1}{3}r$, $9 = r$; Ken has to run 9 mi/h.

7.

	Rate ×	Time =	Distance
Andrew	12	t	$12t$
Stacey	16	$t - \frac{1}{2}$	$16\left(t - \frac{1}{2}\right)$

$12t + 16\left(t - \frac{1}{2}\right) = 20$,

$12t + 16t - 8 = 20$,
$28t - 8 = 20$, $28t = 28$, $t = 1$;
Andrew and Stacey met at 10:30 A.M.

8.

	Rate ×	Time =	Distance
From Boston	525	$t + \frac{2}{3}$	$525\left(t + \frac{2}{3}\right)$
To Boston	600	t	$600t$

$525\left(t + \frac{2}{3}\right) = 600t$, $525t + 350 = 600t$, $350 = 75t$, $\frac{14}{3} = t$; $600t = 600 \cdot \frac{14}{3} = 2800$;
Boston and Los Angeles are 2800 mi apart.

9.

	Rate ×	Time =	Distance
City	r	1	r
Highway	$2r$	4	$(2r)4$

$r + (2r)4 = 387$, $r + 8r = 387$,
$9r = 387$, $r = 43$;
The bus's city speed was 43 km/h.

10.

	Rate ×	Time =	Distance
From Abbott	r	2	$2r$
To Abbott	$r + 6$	1.5	$1.5(r + 6)$

$2r = 1.5(r + 6)$, $2r = 1.5r + 9$, $.5r = 9$, $r = 18$; $2r = 36$; Abbott and Benson are 36 km apart.

B **11.**

	Rate	× Time	= Distance
Up	r	2.5	$2.5r$
Down	$r + 3$	1.5	$1.5(r + 3)$

$2.5r = 1.5(r + 3)$, $2.5r = 1.5r + 4.5$, $r = 4.5$; $2.5r = 2.5(4.5) = 11.25$; Jerry biked 11.25 km each way or a total of 22.5 km.

12.

	Rate	× Time	= Distance
Jan	7.5	t	$7.5t$
Mary	8.0	t	$8.0t$

$7.5t = 8.0t - 25$, $7.5t + 25 = 8.0t$, $25 = .5t$, $50 = t$; $8.0t = 8.0 \cdot 50 = 400$; The track is 400 m long.

13.

	Rate	× Time	= Distance
Without lunch	66	t	$66t$
With lunch	90	$t - \frac{2}{3}$	$90\left(t - \frac{2}{3}\right)$

$66t = 90\left(t - \frac{2}{3}\right)$, $66t = 90t - 60$, $66t + 60 = 90t$, $60 = 24t$, $2.5 = t$; $66t = 66 \cdot 2.5 = 165$; Alton is 165 km away.

14.

	Rate	× Time	= Distance
Without wind	r	$\frac{2}{3}$	$\frac{2}{3}r$
With wind	$2r$	$\frac{4}{3}$	$\frac{4}{3}(2r)$

$\frac{2}{3}r + \frac{4}{3}(2r) = 160$, $\frac{2}{3}r + \frac{8}{3}r = 160$, $\frac{10}{3}r = 160$, $r = 48$; $\frac{2}{3}r = \frac{2}{3} \cdot 48 = 32$; The plane flew 32 mi in the first 40 min.

15.

	Rate	× Time	= Distance
Bad weather	16	4	64
Good weather	r	6	$6r$

Since the ship must average 22 knots for 10 hours, the distance must be 220 nautical miles. $64 + 6r = 220$, $6r = 156$, $r = 26$; The ship must maintain an average speed of 26 knots for the rest of its trip.

16.

	Rate	× Time	= Distance
First lap	8.5	t	$8.5t$
Second lap	8.0	$99 - t$	$8.0(99 - t)$

$8.5t = 8.0(99 - t)$, $8.5t = 792 - 8t$, $16.5t = 792$, $t = 48$; $99 - t = 51$; Jamie ran the first lap in 48 s and the second lap in 51 s.

C **17.**

	Rate	× Time	= Distance
Car A	r	t	rt
Car B	s	t	st

$rt + p = st$, $p = st - rt$, $p = (s - r)t$, $\frac{p}{s - r} = t$; Car B will overtake Car A in $\frac{p}{s - r}$ hours.

18.

	Rate	× Time	= Distance
Car A	r	t	rt
Car B	s	$t - q$	$s(t - q)$

$rt = s(t - q)$, $rt = st - sq$, $rt + sq = st$, $sq = st - rt$, $sq = (s - r)t$, $\frac{sq}{s - r} = t$; Car B will overtake Car A in $\frac{sq}{s - r}$ hours.

19.

	Rate	× Time	= Distance
Car A	r	t	rt
Car B	s	t	st

$rt + st = d$, $(r + s)t = d$, $t = \frac{d}{r + s}$; Car A and Car B will meet in $\frac{d}{r + s}$ hours.

20.

	Rate	× Time	= Distance
Car A	r	t	rt
Car B	s	t	st

After t hours, Car A and Car B will be $(rt + st + d)$ km apart.

Page 171 • MIXED REVIEW EXERCISES

1. $64 = -8x$, $-8 = x$; $\{-8\}$ **2.** $4(x + 2) = 32$, $4x + 8 = 32$, $4x = 24$, $x = 6$; $\{6\}$

3. $(x - 6)(x + 9) = (x + 6)(x - 2)$, $x^2 + 3x - 54 = x^2 + 4x - 12$, $3x - 54 = 4x - 12$, $-54 = x - 12$, $-42 = x$; $\{-42\}$

4. $-5x = -\frac{10}{18}$, $x = \frac{1}{9}$; $\left\{\frac{1}{9}\right\}$

5. $x - 7 = |2 - 10|$, $x - 7 = |-8|$, $x - 7 = 8$, $x = 15$; $\{15\}$

6. $(x - 4)(x + 3) = (x - 6)(x + 4)$, $x^2 - x - 12 = x^2 - 2x - 24$, $-x - 12 = -2x - 24$, $x - 12 = -24$, $x = -12$; $\{-12\}$

7. $\frac{bx - 6}{2} = 3$, $bx - 6 = 6$, $bx = 12$; $x = \frac{12}{b}$, $b \neq 0$

8. $a = 3x + 5$, $a - 5 = 3x$, $\frac{a - 5}{3} = x$

Pages 173–174 • PROBLEMS

A 1.

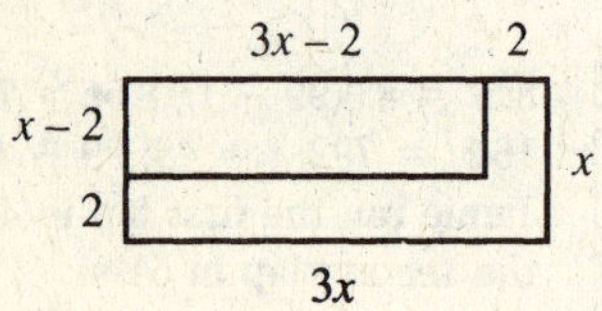

$(3x)x - (x - 2)(3x - 2) = 36$, $3x^2 - (3x^2 - 8x + 4) = 36$, $3x^2 - 3x^2 + 8x - 4 = 36$, $8x - 4 = 36$, $8x = 40$, $x = 5$; $3x = 15$; The original rectangle was 5 cm wide and 15 cm long.

2.

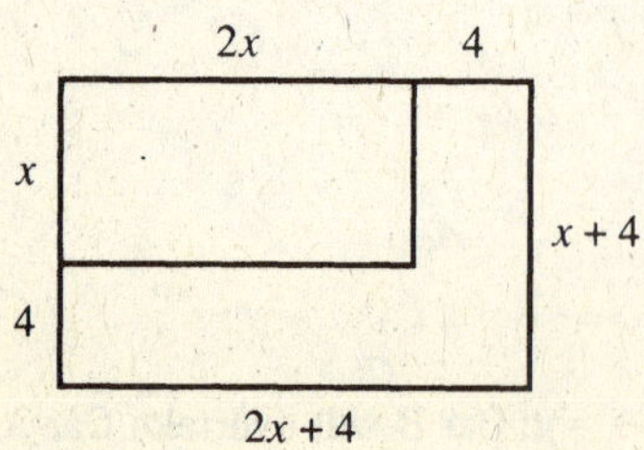

$(2x + 4)(x + 4) - (2x)x = 88$, $2x^2 + 12x + 16 - 2x^2 = 88$, $12x + 16 = 88$, $12x = 72$, $x = 6$; $2x = 12$; The original rectangle was 6 m wide and 12 m long.

3. Let x be the width of the rectangle. $(3x + 5)(x + 5) - (3x)x = 265$, $3x^2 + 20x + 25 - 3x^2 = 265$, $20x + 25 = 265$, $20x = 240$, $x = 12$; $3x = 36$; The pool is 12 m wide and 36 m long.

4. Let x be the width of the poster. $(x + 10)(x + 35) - x(x + 25) = 1350$, $x^2 + 45x + 350 - x^2 - 25x = 1350$, $20x + 350 = 1350$, $20x = 1000$, $x = 50$; $x + 25 = 75$; The poster is 50 cm wide and 75 cm long.

5. Let x be the width of the patio. $(x + 3)(2x + 3) - x(2x) = 54$, $2x^2 + 9x + 9 - 2x^2 = 54$, $9x + 9 = 54$, $9x = 45$, $x = 5$; $2x = 10$; The patio is 5 m wide and 10 m long.

6. Let x be the length of one side of the square room. $x^2 = (x + 5)(x - 4)$, $x^2 = x^2 + x - 20$, $0 = x - 20$, $20 = x$; $x^2 = 400$; Each room is 400 ft^2 in area.

B **7.** Let y be the width of the entire park. $y(y + 5) - y(y - 5) = 330$, $y^2 + 5y - y^2 + 5y = 330$, $10y = 330$, $y = 33$; $y(y - 5) = 33(33 - 5) = 33 \cdot 28 = 924$; The area of the lawn is 924 m^2.

8. Let x be the length of a side of the original lot. $x^2 - (x - 3)(x - 5) = 185$, $x^2 - (x^2 - 8x + 15) = 185$, $x^2 - x^2 + 8x - 15 = 185$, $8x - 15 = 185$, $8x = 200$, $x = 25$; $x - 3 = 22$; $x - 5 = 20$; The new lot is 20 m wide and 22 m long.

9. $\pi(r + w)^2 - \pi r^2 = \pi(r^2 + 2rw + w^2) - \pi r^2 = \pi r^2 + 2\pi rw + \pi w^2 - \pi r^2 = 2\pi rw + \pi w^2$; The area of the shaded area is $2\pi rw + \pi w^2$.

10. $2\pi r \cdot 2 + \pi(2)^2 = 176$, $4\pi r + 4\pi = 176$, $4\pi r = 176 - 4\pi$, $r = \frac{176 - 4\pi}{4\pi}$,

$r = \frac{176}{4\pi} - 1$, $r = \frac{176}{4\left(\frac{22}{7}\right)} - 1$, $r = 13$; The radius is 13 cm.

11. $2\pi r \cdot 3 + \pi(3)^2 = 198$, $6\pi r + 9\pi = 198$, $6\pi r = 198 - 9\pi$, $r = \frac{198}{6\pi} - \frac{3}{2}$,

$r = \frac{198}{6\left(\frac{22}{7}\right)} - \frac{3}{2}$, $r = 9$; The radius of the pool is 9 m.

C **12.** The width of the soccer field is $2r$ and the length is $4r$. The area of the track is the area of the semi-circles at either end minus the area of the inner semi-circles, plus the area of the straight sections of track. $A = \pi(r + 4)^2 - \pi r^2 + 2(4 \cdot 4r) = \pi(r^2 + 8r + 16) - \pi r^2 + 32r = \pi r^2 + 8\pi r + 16\pi - \pi r^2 + 32r = 8\pi r + 16\pi + 32r$

13. a. $d = 2\pi(r + .5) + 2(4r) = 2\pi r + \pi + 8r$; $(2\pi r + \pi + 8r)$ m

b. The friend runs distance D. $D = 2\pi(r + 3.5) + 2(4r) = 2\pi r + 7\pi + 8r$; $D - d = 2\pi r + 7\pi + 8r - (2\pi r + \pi + 8r) = 2\pi r + 7\pi + 8r - 2\pi r - \pi - 8r = 6\pi$; The friend runs 6π m more.

Page 174 • MIXED REVIEW EXERCISES

1. $(-5 + 9) + (-8) = 4 - 8 = -4$ **2.** $6(c - 2) + (-3)c + 5 = 6c - 12 - 3c + 5 = 3c - 7$

3. $(4xy^3)^3 = 4^3x^3y^{3 \cdot 3} = 64x^3y^9$

4. $y - (-7) - [y + (-7)] = y + 7 - (y - 7) = y + 7 - y + 7 = 14$

5. $5^3 \cdot 2^2 = 125 \cdot 4 = 500$ **6.** $(7 \cdot 4 + 4 \cdot 5) \div (3 \cdot 2) = (28 + 20) \div 6 = 48 \div 6 = 8$

7. $-\frac{9}{4} + \frac{5}{4} + 1 = -\frac{4}{4} + 1 = -1 + 1 = 0$ **8.** $\left(\frac{1}{2}x^4\right)\left(\frac{2}{5}x^3\right) = \frac{2}{10}x^{4+3} = \frac{1}{5}x^7$

9. $-\frac{x}{11}(-143) = \frac{143}{11}x = 13x$ **10.** $(-8 + 12) + (-4) = 4 - 4 = 0$

11. $\left(\frac{1}{3}y^3\right)\left(\frac{3}{5}y\right) = \frac{1}{5}y^{3+1} = \frac{1}{5}y^4$ **12.** $(3y^2z)^4 = 3^4y^{2\cdot4}z^4 = 81y^8z^4$

Pages 176–177 • PROBLEMS

A

1. No solution; not enough information.

2. Let x be the number of quarters. $10(2x) + 25x = 630$, $20x + 25x = 630$, $45x = 630$, $x = 14$; $2x = 28$; The bank contains 28 dimes and 14 quarters.

3. No solution; not enough information.

4. Let x be the number of 25¢ stamps, and $15 - x$ be the number of 40¢ stamps. $25x + 40(15 - x) = 495$, $25x + 600 - 40x = 495$, $-15x + 600 = 495$, $-15x = -105$, $x = 7$; $15 - x = 8$; Kyle bought seven 25¢ stamps and eight 40¢ stamps.

5. Let b be Ben's weight, $b - 30$ be Ann's weight, and $2b$ be the canoe's weight. $2b - 20 = b + b - 30$, $2b - 20 = 2b - 30$, $-20 = -30$; no solution; The given facts are contradictory.

6. Let n be the first consecutive even integer. $\frac{1}{4}[n + (n + 2) + (n + 4) + (n + 6)] = 17$, $\frac{1}{4}(4n + 12) = 17$, $n + 3 = 17$, $n = 14$; The consecutive even integers are 14, 16, 18, and 20.

7. Let v be the speed of the second messenger. $1\frac{2}{3}(51) = 1 \cdot v$, $\frac{5}{3} \cdot 51 = v$, $85 = v$; The second messenger must travel 85 km/h.

8. Let m be the cost of the milk. $3m = (.20 + 2m) + m$, $3m = .20 + 3m$, $0 = .20$; No sol.; contradictory facts.

9. Let n be the first consecutive odd integer. $n + (n + 2) + (n + 4) = (n + 4) + 24$, $3n + 6 = n + 28$, $2n + 6 = 28$, $2n = 22$, $n = 11$; The consecutive odd integers are 11, 13, and 15.

10. No sol.; not enough info. **11.** No sol.; not enough info.

12. Let w be the width of the original rectangle. $w(w + 8) = (w + 3)(w + 5) - 4$, $w^2 + 8w = w^2 + 8w + 15 - 4$, $0 = 11$; No sol.; The given facts are contradictory.

B

13.

	Rate ×	Time =	Distance
Scheduled route	49	5	$49 \cdot 5$
Reduced speed	42	1.5	$42 \cdot 1.5$
Rest of trip	r	$5 - 1.5$	$r(5 - 1.5)$

Let r be the driver's speed for the rest of the trip. $49 \cdot 5 = 42 \cdot (1.5) + r(5 - 1.5)$, $245 = 63 + 3.5r$, $182 = 3.5r$, $52 = r$; The driver must drive 52 mi/h.

14. Let x = length of an edge of the smaller cube. $6(x^2) + 234 = 6(x + 3)^2$, $6x^2 + 234 = 6(x^2 + 6x + 9)$, $6x^2 + 234 = 6x^2 + 36x + 54$, $234 = 36x + 54$, $180 = 36x$, $5 = x$; $x + 3 = 8$; The larger cube has edges 8 cm long.

15.

	Number of coins	Value (each coin)	Total value
Nickels	n	5	$5n$
Dimes	$2n + 3$	10	$10(2n + 3)$
Quarters	$n - 6$	25	$25(n - 6)$

Let n be the number of nickels. $5n + 10(2n + 3) + 25(n - 6) = 860$, $5n + 20n + 30 + 25n - 150 = 860$, $50n - 120 = 860$, $50n = 980$, $n = 19.6$; Jane cannot have a fractional number of nickels; no sol.; unrealistic result.

16. Let a be the number of books Alison had at the beginning of the year. $2a + (24 - a) - 3 = 46$, $a + 21 = 46$, $a = 25$; $24 - a = -1$; Sean cannot have a negative number of books; no sol.; unrealistic result.

Pages 178–179 • CHAPTER REVIEW

1. b; $(a + b)^3$ **2.** c; $9 - 4^3 = 9 - 64 = -55$

3. b; $(xy^2 + 4x^2y - 6) + (5xy^2 - 5x^2y - 7) = 6xy^2 - x^2y - 13$

4. a; $x - (15x - 6) = 104$, $x - 15x + 6 = 104$, $-14x + 6 = 104$, $-14x = 98$, $x = -7$; $\{-7\}$

5. d; $3x^6\left(-\frac{1}{3}x^6\right) = -x^{6+6} = -x^{12}$

6. d; $(3a^4b)(5a^2b^2)(2a^3) = (3 \cdot 5 \cdot 2)(a^4 \cdot a^2 \cdot a^3)(b \cdot b^2) = 30a^9b^3$

7. d; $(-3x^2y^4)^3 = (-3)^3(x^2)^3(y^4)^3 = -27x^6y^{12}$

8. c; $9n^2\left(\frac{1}{3}n\right)^4 = 9 \cdot \left(\frac{1}{3}\right)^4 \cdot n^2 \cdot n^4 = \frac{9}{81}n^6 = \frac{1}{9}n^6$

9. d; $-6[16a - 8(2a - 2)] = -6(16a - 16a + 16) = -6 \cdot 16 = -96$

10. a; $6 - 2(n - 3) = 12$, $6 - 2n + 6 = 12$, $-2n + 12 = 12$, $-2n = 0$, $n = 0$; $\{0\}$

11. d; $(4x - 3)(x - 4) = 4x^2 - 16x - 3x + 12 = 4x^2 - 19x + 12$

12. c; $(c - 6)(c^2 + 2c + 3) = c^3 + 2c^2 + 3c - 6c^2 - 12c - 18 = c^3 - 4c^2 - 9c - 18$

13. a; $(a - b)(a^2 + ab + b^2) = a^3 + a^2b + ab^2 - a^2b - ab^2 - b^3 = a^3 - b^3$

14. c; $c + by = a$, $by = a - c$, $b = \frac{a - c}{y}$

15. b; $\frac{xy + z}{2} = a$, $xy + z = 2a$, $xy = 2a - z$, $y = \frac{2a - z}{x}$

16. b; $10t = 8(.45 - t)$, $10t = 3.6 - 8t$, $18t = 3.6$, $t = 0.2$; $10t = 10 \cdot 0.2 = 2$; Laurie ran 4 miles in all.

17. a; $(x + 1)(x + 2) - x(x + 1) = 8$, $x^2 + 3x + 2 - x^2 - x = 8$, $2x + 2 = 8$, $2x = 6$, $x = 3$; $x + 1 = 4$; The picture is 3 in. by 4 in.

18. d; $5(16 - q) + 25q = 300$, $80 - 5q + 25q = 300$, $80 + 20q = 300$, $20q = 220$, $q = 11$; Estaban has 11 quarters.

Page 180 • CHAPTER TEST

1. $x^3 + y^3$ **2.** $(ab)^3$

3. $(4^2 - 3 \cdot 1 - 3^2) \div [0 - (-2)^2] = (16 - 3 - 9) \div (0 - 4) = 4 \div (-4) = -1$

4. a. $(9n - 5) + (n + 3) = 10n - 2$ **b.** $(9n - 5) - (n + 3) = 9n - 5 - n - 3 = 8n - 8$

5. **a.** $(6x^2 - 5x - 1) + (-6x^2 + 5x + 1) = 0$

b. $(6x^2 - 5x - 1) - (-6x^2 + 5x + 1) = 6x^2 - 5x - 1 + 6x^2 - 5x - 1 = 12x^2 - 10x - 2$

6. **a.** $(x^2 + 2xy + 3y^2) + (5x^2 - xy - y^2) = 6x^2 + xy + 2y^2$

b. $(x^2 + 2xy + 3y^2) - (5x^2 - xy - y^2) = x^2 + 2xy + 3y^2 - 5x^2 + xy + y^2 = -4x^2 + 3xy + 4y^2$

7. $4y^3(-3xy) = (4)(-3)xy^{3+1} = -12xy^4$ **8.** $(12a^2)\left(\frac{2}{3}a^3\right) = \left(12 \cdot \frac{2}{3}\right)a^{2+3} = 8a^5$

9. $7^2 \cdot 7x^3 = 49 \cdot 7x^3 = 343x^3$ **10.** $5^4n^4 = 625n^4$ **11.** $(2x^3)^5 = 2^5x^{3 \cdot 5} = 32x^{15}$

12. $4n^3\left(\frac{1}{2}n\right)^4 = 4n^3\left(\frac{1}{2}\right)^4 n^4 = 4 \cdot \frac{1}{16}n^{3+4} = \frac{1}{4}n^7$ **13.** $2x(5 - 4x) = 10x - 8x^2$

14. $-3xy(7x^2 - 8xy + 9y^2) = -21x^3y + 24x^2y^2 - 27xy^3$

15. $\frac{5}{6}(12x - 6) - 4(3x - 1) = 0, 10x - 5 - 12x + 4 = 0, -2x - 1 = 0, -2x = 1, x = -\frac{1}{2};$ $\left\{-\frac{1}{2}\right\}$

16. $(3x + 2)(2x + 1) = 6x^2 + 3x + 4x + 2 = 6x^2 + 7x + 2$

17. $(c + 3)(4c^2 - 6c + 1) = 4c^3 - 6c^2 + c + 12c^2 - 18c + 3 = 4c^3 + 6c^2 - 17c + 3$

18. $F = \frac{9}{5}C + 32, F - 32 = \frac{9}{5}C, \frac{5}{9}(F - 32) = C$

19. $D = \frac{a}{2}(2t - 1), 2D = a(2t - 1), \frac{2D}{2t - 1} = a; t \neq \frac{1}{2}$

20. Let l be Lee's speed. $.5l + .5(l - 1) = 3.9, .5l + .5l - .5 = 3.9, l - .5 = 3.9, l = 4.4;$ $l - 1 = 3.4$; Lee's speed was 4.4 km/h and Jessie's speed was 3.4 km/h.

21. $x(x + 4) - (x - 2)(x + 2) = 24, x^2 + 4x - (x^2 - 4) = 24, x^2 + 4x - x^2 + 4 = 24,$ $4x + 4 = 24, 4x = 20, x = 5; x + 4 = 9$; The original rectangle was 5 cm wide and 9 cm long.

22. No solution; not enough information

Page 181 • CUMULATIVE REVIEW

1. $(6x - 3) - (4x + 2) = 6x - 3 - 4x - 2 = 2x - 5$ **2.** $\frac{1}{2}(101 - 43) = \frac{1}{2} \cdot 58 = 29$

3. $-2.2 + 3.8 - 5.6 + 4 = 0$ **4.** $15s - (2s - 9) = 15s - 2s + 9 = 13s + 9$

5. $-7 - (-20) \div 2 = -7 - (-10) = -7 + 10 = 3$

6. $-\frac{3}{5} + 4\frac{5}{8} - \frac{2}{5} + 7\frac{1}{8} = -\frac{5}{5} + 11\frac{6}{8} = -1 + 11\frac{3}{4} = 10\frac{3}{4}$

7. $3^3 + 42 \div 3 + 4 = 27 + 14 + 4 = 45$ **8.** $(9a - 5) + (4a + 7) = 13a + 2$

9. $-3|6 - 12| = -3|-6| = -3 \cdot 6 = -18$

10. $(5x - 2)(2x + 3) = 10x^2 + 15x - 4x - 6 = 10x^2 + 11x - 6$

11. $(4a^3)(5a)^2 = 4a^3 \cdot 5^2a^2 = 4 \cdot 25 \cdot a^{3+2} = 100a^5$ **12.** $5(3z^2 - 2z + 4) = 15z^2 - 10z + 20$

13. $-4x^2(3x^2 - 2x - 5) = -12x^4 + 8x^3 + 20x^2$

14. $(3y^3)(2y^2) - 2(y)(y^4) = 6y^5 - 2y^5 = 4y^5$

15. $(5 - 4x)(3 + 2x) = 15 + 10x - 12x - 8x^2 = 15 - 2x - 8x^2$

16. $w(3y + x) = \frac{1}{5}[3(-3) + (-1)] = \frac{1}{5}(-9 - 1) = \frac{1}{5}(-10) = -2$

17. $(xz - y)^5 = [(-1)2 - (-3)]^5 = (-2 + 3)^5 = 1^5 = 1$

18. $w(z - (-y)) = \frac{1}{5}(2 - [-(-3)]) = \frac{1}{5}(2 - 3) = \frac{1}{5}(-1) = -\frac{1}{5}$

19. $z(x - 2y) = 2[(-1) - 2(-3)] = 2(-1 + 6) = 2 \cdot 5 = 10$

20. $|-x| = 9, -x = 9$ or $-x = -9, x = -9$ or $x = 9$; $\{9, -9\}$

21. $|y - 1| + 4 = 0$; no sol. **22.** $5 = |x| + 5, 0 = |x|, 0 = x$; $\{0\}$

23. $c - (-4) = -8, c + 4 = -8, c = -12$; $\{-12\}$

24. $3x - 2 = x + 6, 2x - 2 = 6, 2x = 8, x = 4$; $\{4\}$ **25.** $42z = -42, z = -1$; $\{-1\}$

26. $0 = \frac{1}{3}n + 2, -\frac{1}{3}n = 2, n = -6$; $\{-6\}$ **27.** $-10 = 4m + 2, -12 = 4m, -3 = m$; $\{-3\}$

28. $\frac{1}{4}x = 20, x = 80$; $\{80\}$

29. $3(2 + x) = -4(x - 5), 6 + 3x = -4x + 20, 3x = -4x + 14, 7x = 14, x = 2$; $\{2\}$

30. $(11x - 3) - (4 + 2x) = 11, 11x - 3 - 4 - 2x = 11, 9x - 7 = 11, 9x = 18, x = 2$; $\{2\}$

31. $(2n + 9) + (5n - 4) = 6n + 9, 7n + 5 = 6n + 9, n + 5 = 9, n = 4$; $\{4\}$

32. $(4y - 2) + (4 - 2y) = 30, 2y + 2 = 30, 2y = 28, y = 14$; $\{14\}$

33. $2(c - 1) - 7 = 1, 2c - 2 - 7 = 1, 2c - 9 = 1, 2c = 10, c = 5$; $\{5\}$

34. $(2x - 3)(3x + 1) = (3x - 4)(2x + 2), 6x^2 + 2x - 9x - 3 = 6x^2 + 6x - 8x - 8,$
$6x^2 - 7x - 3 = 6x^2 - 2x - 8, -7x - 3 = -2x - 8, -5x - 3 = -8, -5x = -5,$
$x = 1$; $\{1\}$

35. $am - bn = c, am = c + bn, m = \frac{c + bn}{a}; a \neq 0$

36. $by - ax = 0, by = ax, \frac{by}{a} = x; a \neq 0$

37. Let n be the first consecutive odd integer. $\frac{1}{3}(n + n + 2) = n - 5, \frac{1}{3}(2n + 2) = n - 5,$
$2n + 2 = 3(n - 5), 2n + 2 = 3n - 15, 2 = n - 15, 17 = n; n + 2 = 19$; The consecutive odd integers are 17 and 19.

38.

	Rate	× Time	= Distance
Randy	3.6	$\frac{1}{6}$	$\frac{1}{6}(3.6)$
Amy	4.2	$\frac{1}{6}$	$\frac{1}{6}(4.2)$

$\frac{1}{6}(3.6) + \frac{1}{6}(4.2) = 0.6 + 0.7 = 1.3$;

After 10 min Randy and Amy were 1.3 km apart.

39.

	Dimes	Quarters
Jessica	d	$16 - d$
Whitney	$2d$	$\frac{1}{3}(16 - d)$

Let d be the number of dimes Jessica has. $10d + 25(16 - d) = 10(2d) + 25\left[\frac{1}{3}(16 - d)\right]$, $10d + 400 - 25d = 20d + 25\left(\frac{16}{3} - \frac{1}{3}d\right)$, $400 - 15d = 20d + \frac{400}{3} - \frac{25}{3}d$, $400 - 15d = \frac{35}{3}d + \frac{400}{3}$, $400 = \frac{80}{3}d + \frac{400}{3}$, $\frac{800}{3} = \frac{80}{3}d$, $10 = d$; $16 - d = 6$; $2d = 20$; $\frac{1}{3}(16 - d) = 2$;

Jessica has 10 dimes and 6 quarters, and Whitney has 20 dimes and 2 quarters.

40. Let x be the length of a side of the square. $(x + 4)(x + 2) = x^2 + 74$, $x^2 + 6x + 8 = x^2 + 74$, $6x + 8 = 74$, $6x = 66$, $x = 11$; $x + 4 = 15$; $x + 2 = 13$; The rectangle was 15 cm by 13 cm.

Page 182 • MAINTAINING SKILLS

1. $1921 + (-876) = 1045$ **2.** $181 + 97 - 64 = 214$

3. $(55 - 82) + (91 - 108) = -27 - 17 = -44$ **4.** $-78 - 84 - (-92) = -162 + 92 = -70$

5. $(28 - 86) - (46 - 81) = -58 - (-35) = -58 + 35 = -23$

6. $284 - (93 - 165) = 284 - (-72) = 284 + 72 = 356$

7. $35 - (58 + 62) = 35 - 120 = -85$ **8.** $-325 + (-726) + 922 = -129$

9. $\frac{7}{8} - \left(-\frac{1}{4} + \frac{1}{2}\right) = \frac{7}{8} - \left(\frac{1}{4}\right) = \frac{7}{8} - \frac{2}{8} = \frac{5}{8}$

10. $\left(\frac{3}{5} - \frac{2}{3}\right) - \frac{5}{9} = \frac{9 - 10}{15} - \frac{5}{9} = -\frac{1}{15} - \frac{5}{9} = \frac{-3 - 25}{45} = -\frac{28}{45}$

11. $17.6 - (8.05 - 9.6) = 17.6 - (-1.55) = 17.6 + 1.55 = 19.15$

12. $112.72 + (92.04 - 87.6) = 112.72 + 4.44 = 117.16$

13. $-12(-16) + 5(-24) = 192 - 120 = 72$ **14.** $27(20) - 60(48) = 540 - 2880 = -2340$

15. $-65 - (412 - 385) = -65 - 27 = -92$ **16.** $-4(-50) + 8(-25) = 200 - 200 = 0$

17. $9.25(-2.3) = -21.275$ **18.** $-6.06(-5.4) = 32.724$ **19.** $-82.05 \div (-25) = 3.282$

20. $-\frac{24}{35} \div \frac{9}{14} = -\frac{24}{35} \cdot \frac{14}{9} = -\frac{16}{15}$ **21.** $7.24 \div (-0.25) = -28.96$ **22.** $-\frac{12}{25} \cdot \left(-\frac{35}{42}\right) = \frac{2}{5}$

23. $-\frac{18}{35} \cdot \frac{49}{54} = -\frac{7}{15}$ **24.** $-\frac{15}{64} \cdot \left(-\frac{40}{27}\right) = \frac{25}{72}$

25. $-2\frac{3}{8} \div \left(-2\frac{1}{8}\right) = -\frac{19}{8} \div \left(-\frac{17}{8}\right) = -\frac{19}{8} \cdot -\frac{8}{17} = \frac{19}{17}$

26. $7 \cdot 8^2 - 6 \cdot 3 - 12 \div 2 = 7 \cdot 64 - 18 - 6 = 448 - 24 = 424$

27. $(29 + 7) \div 3^2 + 13 - 2^2 = 36 \div 9 + 13 - 4 = 4 + 9 = 13$

28. $(0.6)^2 - 1.2^2 \div 8 + 0.27 \div 0.3 = 0.36 - 1.44 \div 8 + 0.9 = 0.36 - 0.18 + 0.9 = 1.08$

29. $-2(0.35 + 0.55)^2 \div 1.8 \div 2 = -2(.9)^2 \div 1.8 \div 2 = -2(.81) \div 1.8 \div 2 = -1.62 \div 1.8 \div 2 = -0.9 \div 2 = -0.45$

30. $1\frac{4}{5} \div 9 \cdot \frac{1}{2} - \left(\frac{1}{3} + \frac{1}{2}\right)^2 = \frac{9}{5} \div 9 \cdot \frac{1}{2} - \left(\frac{5}{6}\right)^2 = \frac{1}{5} \cdot \frac{1}{2} - \frac{25}{36} = \frac{1}{10} - \frac{25}{36} = \frac{18 - 125}{180} = -\frac{107}{180}$

31. $-\frac{5}{7}\left[-\left(1\frac{1}{3} - \frac{3}{4}\right)\right] + \frac{1}{3} \div 4 = \frac{5}{7}\left(\frac{4}{3} - \frac{3}{4}\right) + \frac{1}{12} = \frac{5}{7}\left(\frac{16 - 9}{12}\right) + \frac{1}{12} = \frac{5}{7} \cdot \frac{7}{12} + \frac{1}{12} =$
$\frac{5}{12} + \frac{1}{12} = \frac{6}{12} = \frac{1}{2}$

Page 183 • PREPARING FOR COLLEGE ENTRANCE EXAMS

1. B; Let x = length of a shorter side. $2(x) + 2(2x - 1) = 28$, $2x + 4x - 2 = 28$, $6x - 2 = 28$, $6x = 30$, $x = 5$; $2x - 1 = 2 \cdot 5 - 1 = 10 - 1 = 9$; The longer sides are 9 cm long.
2. C; Let n = the smaller consecutive integer. $2(n + n + 1) = 5n - 10$, $2(2n + 1) = 5n - 10$, $4n + 2 = 5n - 10$, $2 = n - 10$, $12 = n$; $n + 1 = 13$; The greater consecutive integer is 13.
3. B; Let s = the cost of the soup. $s + 2s + s + 0.30 = 4.70$, $4s + 0.30 = 4.70$, $4s = 4.40$, $s = 1.10$; The soup cost \$1.10.
4. D
5. A; $(a + b)^2 \div (2a) - b^2 = (6 + 4)^2 \div (2 \cdot 6) - 4^2 = 10^2 \div 12 - 16 = 100 \div 12 - 16 = \frac{25}{3} - 16 = \frac{25 - 48}{3} = -\frac{23}{3}$
6. C; $(n + 2)(n + 3) = (4 - n)(12 - n)$, $n^2 + 5n + 6 = 48 - 16n + n^2$, $5n + 6 = 48 - 16n$, $21n + 6 = 48$, $21n = 42$, $n = 2$; $(2n^3)^2 = (2 \cdot 2^3)^2 = (2^4)^2 = 2^8 = 256$
7. C; $q = 1 + \frac{p}{100}$, $q - 1 = \frac{p}{100}$, $100(q - 1) = p$ **8.** D

Pages 186–187 • WRITTEN EXERCISES

A **1.** (1)(13); (−1)(−13) **2.** (1)(22); (2)(11); (−1)(−22); (−2)(−11)

3. (1)(24); (2)(12); (3)(8); (4)(6); (−1)(−24); (−2)(−12); (−3)(−8); (−4)(−6)

4. (1)(18); (2)(9); (3)(6); (−1)(−18); (−2)(−9); (−3)(−6)

5. (1)(29); (−1)(−29) **6.** (1)(49); (7)(7); (−1)(−49); (−7)(−7)

7. (1)(40); (2)(20); (4)(10); (5)(8); (−1)(−40); (−2)(−20); (−4)(−10); (−5)(−8)

8. (1)(101); (−1)(−101) **9.** (1)(−121); (11)(−11); (−1)(121)

10. (1)(−52); (2)(−26); (4)(−13); (−1)(52); (−2)(26); (−4)(13)

11. (1)(−33); (3)(−11); (−1)(33); (−3)(11)

12. (1)(−48); (2)(−24); (3)(−16); (4)(−12); (6)(−8); (−1)(48); (−2)(24); (−3)(16); (−4)(12); (−6)(8)

13. (1)(53); (−1)(−53) **14.** (1)(67); (−1)(−67)

15. (1)(26); (2)(13); (−1)(−26); (−2)(−13)

16. (1)(64); (2)(32); (4)(16); (8)(8); (−1)(−64); (−2)(−32); (−4)(−16); (−8)(−8)

17. (1)(68); (2)(34); (4)(17); (−1)(−68); (−2)(−34); (−4)(−17)

18. (1)(83); (−1)(−83) **19.** (1)(38); (2)(19); (−1)(−38); (−2)(−19)

20. (1)(74); (2)(37); (−1)(−74); (−2)(−37) **21.** $16 = 2 \cdot 8 = 2 \cdot 2 \cdot 4 = 2 \cdot 2 \cdot 2 \cdot 2 = 2^4$

22. $34 = 2 \cdot 17$ **23.** $69 = 3 \cdot 23$ **24.** $65 = 5 \cdot 13$

25. $75 = 3 \cdot 25 = 3 \cdot 5 \cdot 5 = 3 \cdot 5^2$ **26.** $88 = 2 \cdot 44 = 2 \cdot 2 \cdot 22 = 2 \cdot 2 \cdot 2 \cdot 11 = 2^3 \cdot 11$

27. $27 = 3 \cdot 9 = 3 \cdot 3 \cdot 3 = 3^3$ **28.** $54 = 2 \cdot 27 = 2 \cdot 3 \cdot 9 = 2 \cdot 3 \cdot 3 \cdot 3 = 2 \cdot 3^3$

29. $99 = 3 \cdot 33 = 3 \cdot 3 \cdot 11 = 3^2 \cdot 11$

30. $120 = 2 \cdot 60 = 2 \cdot 2 \cdot 30 = 2 \cdot 2 \cdot 2 \cdot 15 = 2 \cdot 2 \cdot 2 \cdot 3 \cdot 5 = 2^3 \cdot 3 \cdot 5$

31. $104 = 2 \cdot 52 = 2 \cdot 2 \cdot 26 = 2 \cdot 2 \cdot 2 \cdot 13 = 2^3 \cdot 13$

32. $128 = 2 \cdot 64 = 2 \cdot 2 \cdot 32 = 2 \cdot 2 \cdot 2 \cdot 16 = 2 \cdot 2 \cdot 2 \cdot 2 \cdot 8 = 2 \cdot 2 \cdot 2 \cdot 2 \cdot 2 \cdot 4 = 2 \cdot 2 \cdot 2 \cdot 2 \cdot 2 \cdot 2 \cdot 2 = 2^7$

B **33.** $125 = 5 \cdot 25 = 5 \cdot 5 \cdot 5 = 5^3$

34. $200 = 2 \cdot 100 = 2 \cdot 2 \cdot 50 = 2 \cdot 2 \cdot 2 \cdot 25 = 2 \cdot 2 \cdot 2 \cdot 5 \cdot 5 = 2^3 \cdot 5^2$

35. $450 = 2 \cdot 225 = 2 \cdot 3 \cdot 75 = 2 \cdot 3 \cdot 3 \cdot 25 = 2 \cdot 3 \cdot 3 \cdot 5 \cdot 5 = 2 \cdot 3^2 \cdot 5^2$

36. $476 = 2 \cdot 238 = 2 \cdot 2 \cdot 119 = 2 \cdot 2 \cdot 7 \cdot 17 = 2^2 \cdot 7 \cdot 17$

37. $1089 = 3 \cdot 363 = 3 \cdot 3 \cdot 121 = 3 \cdot 3 \cdot 11 \cdot 11 = 3^2 \cdot 11^2$

38. $840 = 2 \cdot 420 = 2 \cdot 2 \cdot 210 = 2 \cdot 2 \cdot 2 \cdot 105 = 2 \cdot 2 \cdot 2 \cdot 3 \cdot 35 = 2 \cdot 2 \cdot 2 \cdot 3 \cdot 5 \cdot 7 = 2^3 \cdot 3 \cdot 5 \cdot 7$

39. $782 = 2 \cdot 391 = 2 \cdot 17 \cdot 23$

40. $2310 = 2 \cdot 1155 = 2 \cdot 3 \cdot 385 = 2 \cdot 3 \cdot 5 \cdot 77 = 2 \cdot 3 \cdot 5 \cdot 7 \cdot 11$

41. $66 = 2 \cdot 3 \cdot 11$; $90 = 2 \cdot 3^2 \cdot 5$; GCF $= 2 \cdot 3 = 6$

42. $132 = 2^2 \cdot 3 \cdot 11$; $220 = 2^2 \cdot 5 \cdot 11$; GCF $= 2^2 \cdot 11 = 44$

43. $182 = 2 \cdot 7 \cdot 13$; $196 = 2^2 \cdot 7^2$; GCF $= 2 \cdot 7 = 14$

44. $132 = 2^2 \cdot 3 \cdot 11$; $242 = 2 \cdot 11^2$; GCF $= 2 \cdot 11 = 22$

45. $330 = 2 \cdot 3 \cdot 5 \cdot 11$; $945 = 3^3 \cdot 5 \cdot 7$; GCF $= 3 \cdot 5 = 15$

46. $348 = 2^2 \cdot 3 \cdot 29$; $426 = 2 \cdot 3 \cdot 71$; GCF $= 2 \cdot 3 = 6$

C **47.** $1176 = 2^3 \cdot 3 \cdot 7^2$; $1617 = 3 \cdot 7^2 \cdot 11$; GCF $= 3 \cdot 7^2 = 147$

48. $1925 = 5^2 \cdot 7 \cdot 11$; $6300 = 2^2 \cdot 3^2 \cdot 5^2 \cdot 7$; GCF $= 5^2 \cdot 7 = 175$

49. $56 = 2^3 \cdot 7$; $98 = 2 \cdot 7^2$; $126 = 2 \cdot 3^2 \cdot 7$; GCF $= 2 \cdot 7 = 14$

50. $105 = 3 \cdot 5 \cdot 7$; $126 = 2 \cdot 3^2 \cdot 7$; $210 = 2 \cdot 3 \cdot 5 \cdot 7$; GCF $= 3 \cdot 7 = 21$

51. $141 = 3 \cdot 47$; $198 = 2 \cdot 3^2 \cdot 11$; $364 = 2^2 \cdot 7 \cdot 13$; GCF $= 1$

52. $90 = 2 \cdot 3^2 \cdot 5$; $126 = 2 \cdot 3^2 \cdot 7$; $252 = 2^2 \cdot 3^2 \cdot 7$; GCF $= 2 \cdot 3^2 = 18$

Page 187 • MIXED REVIEW

1. $\frac{1}{3}(6x + 3) + 2\left(\frac{1}{2}x - 1\right) = 2x + 1 + x - 2 = 3x - 1$

2. $(5 + 4)^2 = 9^2 = 81$

3. $3^2 + (2 + 1)^2 = 9 + 3^2 = 9 + 9 = 18$

4. $3x - 4 - (2x + 6) = 3x - 4 - 2x - 6 = x - 10$

5. $3ab(2a^2)3a = (3 \cdot 2 \cdot 3)(a \cdot a^2 \cdot a)b = 18a^4b$

6. $3x^3(4y)2y = (3 \cdot 4 \cdot 2)x^3(y \cdot y) = 24x^3y^2$ **7.** $(3x)^3x = 3^3x^3 \cdot x = 27x^4$

8. $2n(3n^2 - 4n) + 8n^2 = 6n^3 - 8n^2 + 8n^2 = 6n^3$ **9.** $(-2)^5x^5 = -32x^5$

10. $x(x^2 + 3) - x^2(x - 2) = x^3 + 3x - x^3 + 2x^2 = 2x^2 + 3x$

11. $(2y + 5)(y + 3) = 2y^2 + 11y + 15$

12. $(x - 4)(3x + 4) = 3x^2 - 8x - 16$

Page 187 • COMPUTER EXERCISES

1.
```
10 INPUT "ENTER AN INTEGER: "; N
20 IF N/2 = INT(N/2) THEN PRINT N; " IS EVEN."
30 IF N/2 < > INT(N/2) THEN PRINT N; " IS ODD."
40 END
```

2.
```
10 PRINT "PROGRAM TO CHECK IF B IS A FACTOR OF A"
20 INPUT "ENTER A, B (A FIRST): "; A, B
30 IF A/B = INT(A/B) THEN PRINT B; " IS A FACTOR OF "; A
40 IF A/B <> INT(A/B) THEN PRINT B; " IS NOT A FACTOR OF "; A
50 END
```

3.
```
10 INPUT "ENTER A POSITIVE INTEGER: "; N
20 LET F = 0
30 FOR J = 2 TO N - 1
40 IF N/J = INT(N/J) THEN LET F = 1
50 NEXT J
60 IF F = 0 THEN PRINT N; " IS PRIME."
70 IF F = 1 THEN PRINT N; " IS COMPOSITE."
80 END
```

Page 188 • COMPUTER KEY-IN

1. Factors of 30: (1)(30), (2)(15), (3)(10), (5)(6), (6)(5), (10)(3), (15)(2); Factors of 31: (1)(31); Factors of 36: (1)(36), (2)(18), (3)(12), (4)(9), (6)(6), (9)(4), (12)(3), (18)(2); Factors of 119: (1)(119), (7)(17), (17)(7); Factors of 323: (1)(323), (17)(19), (19)(17)

2. Let N be a number between $W/2$ and W. $W/2 < N < W$, $W/2N < 1 < W/N$, $W/2N < 1$, $W/N < 2$; $1 < W/N < 2$; Thus W/N is not an integer and therefore N is not a factor of W.

3. `31 IF W < 0 OR W < > INT(W) THEN 30`

5. No output for 1; (1)(2), prime; (1)(11), prime; (1)(51), (3)(17), (17)(3), not prime; (1)(53), prime

6.
```
10 PRINT "THIS PROGRAM WILL PRINT ALL THE PRIME NUMBERS"
20 PRINT "LESS THAN 500"
40 FOR W = 1 TO 499
45 LET C = 0
50 FOR F = 1 TO W/2
60 LET Q = W/F
70 IF Q < > INT(Q) THEN 90
85 LET C = C + 1
90 NEXT F
95 IF C > 1 THEN 100
96 PRINT W,
100 NEXT W
110 END
```

Page 188 • CHALLENGE

$7 \cdot 7 \cdot 7 \cdot 7 \cdot 7 = 7^5 = 16{,}807$; 16,807 hekats of grain were saved.

Pages 192–193 • WRITTEN EXERCISES

A

1. $\frac{42}{63} = \frac{2 \cdot 21}{3 \cdot 21} = \frac{2}{3}$ 2. $\frac{54}{72} = \frac{3 \cdot 18}{4 \cdot 18} = \frac{3}{4}$ 3. $\frac{10^4}{10^6} = \frac{1}{10^{6-4}} = \frac{1}{10^2} = \frac{1}{100}$

4. $\frac{10^9}{10^7} = 10^{9-7} = 10^2 = 100$ 5. $\frac{12a}{3a} = \frac{12}{3} \cdot \frac{a}{a} = 4 \cdot 1 = 4$

6. $\frac{10m}{5m} = \frac{10}{5} \cdot \frac{m}{m} = 2 \cdot 1 = 2$ 7. $\frac{6 \cdot 10^5}{3 \cdot 10^3} = \frac{6}{3} \cdot \frac{10^5}{10^3} = 2 \cdot 10^{5-3} = 2 \cdot 10^2 = 2 \cdot 100 = 200$

8. $\frac{12 \cdot 10^3}{4 \cdot 10^4} = \frac{12}{4} \cdot \frac{10^3}{10^4} = 3 \cdot \frac{1}{10^{4-3}} = 3 \cdot \frac{1}{10} = \frac{3}{10}$

9. $\frac{4x^5}{8x} = \frac{4}{8} \cdot \frac{x^5}{x} = \frac{1}{2} \cdot x^4 = \frac{x^4}{2}$ 10. $\frac{3n^7}{15n^3} = \frac{3}{15} \cdot \frac{n^7}{n^3} = \frac{1}{5} \cdot n^4 = \frac{n^4}{5}$

11. $\frac{12y^3}{4xy} = \frac{12}{4} \cdot \frac{1}{x} \cdot \frac{y^3}{y} = 3 \cdot \frac{1}{x} \cdot y^2 = \frac{3y^2}{x}$ 12. $\frac{3r^2s}{9rs^2} = \frac{3}{9} \cdot \frac{r^2}{r} \cdot \frac{s}{s^2} = \frac{1}{3} \cdot r \cdot \frac{1}{s} = \frac{r}{3s}$

13. $\frac{-9x^2y^3}{12xy^2} = \frac{-9}{12} \cdot \frac{x^2}{x} \cdot \frac{y^3}{y^2} = \frac{-3}{4} \cdot x \cdot y = \frac{-3xy}{4}$, or $-\frac{3xy}{4}$

14. $\frac{-6pq^2}{-15pq} = \frac{-6}{-15} \cdot \frac{p}{p} \cdot \frac{q^2}{q} = \frac{2}{5} \cdot 1 \cdot q = \frac{2q}{5}$

15. $\frac{-28cd^3}{-21bd^2} = \frac{-28}{-21} \cdot c \cdot \frac{1}{b} \cdot \frac{d^3}{d^2} = \frac{4}{3} \cdot c \cdot \frac{1}{b} \cdot d = \frac{4cd}{3b}$

16. $\frac{42yz^4}{-48xz^3} = \frac{42}{-48} \cdot \frac{1}{x} \cdot y \cdot \frac{z^4}{z^3} = -\frac{7}{8} \cdot \frac{1}{x} \cdot y \cdot z = -\frac{7yz}{8x}$

17. $\frac{xy^2z^3}{x^3y^2z} = \frac{x}{x^3} \cdot \frac{y^2}{y^2} \cdot \frac{z^3}{z} = \frac{1}{x^2} \cdot 1 \cdot z^2 = \frac{z^2}{x^2}$

18. $\frac{r^2s^4t}{r^2st^3} = \frac{r^2}{r^2} \cdot \frac{s^4}{s} \cdot \frac{t}{t^3} = 1 \cdot s^3 \cdot \frac{1}{t^2} = \frac{s^3}{t^2}$

19. $\frac{32a^2bc^3}{20abc} = \frac{32}{20} \cdot \frac{a^2}{a} \cdot \frac{b}{b} \cdot \frac{c^3}{c} = \frac{8}{5} \cdot a \cdot 1 \cdot c^2 = \frac{8ac^2}{5}$

20. $\frac{39x^3y^2z}{52x^2y} = \frac{39}{52} \cdot \frac{x^3}{x^2} \cdot \frac{y^2}{y} \cdot z = \frac{3}{4} \cdot x \cdot y \cdot z = \frac{3xyz}{4}$

21. $\frac{(2r)^4}{2r^4} = \frac{2^4r^4}{2r^4} = \frac{2^4}{2} \cdot \frac{r^4}{r^4} = 2^3 \cdot 1 = 8$

22. $\frac{7m^2}{(7m)^2} = \frac{7m^2}{7^2m^2} = \frac{7}{7^2} \cdot \frac{m^2}{m^2} = \frac{1}{7} \cdot 1 = \frac{1}{7}$

23. $\frac{(3t^2)^3}{(3t^3)^2} = \frac{3^3t^6}{3^2t^6} = \frac{3^3}{3^2} \cdot \frac{t^6}{t^6} = 3 \cdot 1 = 3$

24. $\frac{(5a^2)^3}{(5a^3)^2} = \frac{5^3a^6}{5^2a^6} = \frac{5^3}{5^2} \cdot \frac{a^6}{a^6} = 5 \cdot 1 = 5$

25. $\frac{(2ab)^2}{2ab^2} = \frac{2^2a^2b^2}{2ab^2} = \frac{2^2}{2} \cdot \frac{a^2}{a} \cdot \frac{b^2}{b^2} = 2 \cdot a \cdot 1 = 2a$

26. $\frac{(3mn)^2}{3mn^2} = \frac{3^2m^2n^2}{3mn^2} = \frac{3^2}{3} \cdot \frac{m^2}{m} \cdot \frac{n^2}{n^2} = 3 \cdot m \cdot 1 = 3m$ **27.** $\frac{(-z)^9}{(-z)^6} = \frac{-z^9}{z^6} = -z^3$

28. $\frac{(-a)^7}{(-a)^4} = \frac{-a^7}{a^4} = -a^3$ **29.** $\frac{(-xy)^9}{xy^9} = \frac{-x^9y^9}{xy^9} = \frac{-x^9}{x} \cdot \frac{y^9}{y^9} = -x^8 \cdot 1 = -x^8$

30. $\frac{(-t^4)^3}{(-t^2)^5} = \frac{-t^{12}}{-t^{10}} = t^2$ **31.** $\frac{6t^4}{2t} = \frac{6}{2} \cdot \frac{t^4}{t} = 3 \cdot t^3 = 3t^3$

32. $\frac{12w^6}{3w^2} = \frac{12}{3} \cdot \frac{w^6}{w^2} = 4 \cdot w^4 = 4w^4$ **33.** $\frac{9a^3b^4}{3a^2b^2} = \frac{9}{3} \cdot \frac{a^3}{a^2} \cdot \frac{b^4}{b^2} = 3ab^2$

34. $\frac{18pq^3}{6pq} = \frac{18}{6} \cdot \frac{p}{p} \cdot \frac{q^3}{q} = 3 \cdot 1 \cdot q^2 = 3q^2$

35. $\frac{-35x^3y^5}{7x^2y} = \frac{-35}{7} \cdot \frac{x^3}{x^2} \cdot \frac{y^5}{y} = -5xy^4$

36. $\frac{-28r^4s^3}{-7r} = \frac{-28}{-7} \cdot \frac{r^4}{r} \cdot s^3 = 4r^3s^3$

37. $\frac{48c^5d^4}{-3c^3d^2} = \frac{48}{-3} \cdot \frac{c^5}{c^3} \cdot \frac{d^4}{d^2} = -16c^2d^2$

38. $\frac{72h^3k}{-8hk} = \frac{72}{-8} \cdot \frac{h^3}{h} \cdot \frac{k}{k} = -9 \cdot h^2 \cdot 1 = -9h^2$

B **39.** $\frac{(3a^3b^2)^3}{(3a^3b^2)^2} = 3a^3b^2$ **40.** $\frac{(2s^3t)^5}{(2s^3t)^3} = (2s^3t)^2 = 2^2s^6t^2 = 4s^6t^2$

41. $\frac{(x^2y^3)^3}{x^4y^5} = \frac{x^6y^9}{x^4y^5} = \frac{x^6}{x^4} \cdot \frac{y^9}{y^5} = x^2y^4$

42. $\frac{(2c^3d^2)^5}{(2c^2d^2)^4} = \frac{2^5c^{15}d^{10}}{2^4c^8d^8} = \frac{2^5}{2^4} \cdot \frac{c^{15}}{c^8} \cdot \frac{d^{10}}{d^8} = 2c^7d^2$

43. $\frac{36r^5s^7}{(2r^3s)(6s^4)} = \frac{36r^5s^7}{12r^3s^5} = \frac{36}{12} \cdot \frac{r^5}{r^3} \cdot \frac{s^7}{s^5} = 3r^2s^2$

44. $\dfrac{48p^5q^4}{(2pq^2)(4pq)} = \dfrac{48p^5q^4}{8p^2q^3} = 6p^3q$

45. $\dfrac{72x^5y^5}{(2x^2y)^2(3y^3)} = \dfrac{72x^5y^5}{4x^4y^23y^3} = \dfrac{72}{12} \cdot \dfrac{x^5}{x^4} \cdot \dfrac{y^5}{y^5} = 6x$

46. $\dfrac{75a^6b^5}{(ab)^3(5a)^2} = \dfrac{75a^6b^5}{a^3b^35^2a^2} = \dfrac{75a^6b^5}{25a^5b^3} = 3ab^2$

47. $48a^2bc^3 = 2^4 \cdot 3 \cdot a^2bc^3$; $72ab^3c^2 = 2^3 \cdot 3^2 \cdot ab^3c^2$; GCF $= 2^3 \cdot 3 \cdot abc^2 = 24abc^2$

48. $36x^2y^2z^2 = 2^2 \cdot 3^2 \cdot x^2y^2z^2$; $24xy^2z^3 = 2^3 \cdot 3 \cdot xy^2z^3$;
GCF $= 2^2 \cdot 3 \cdot xy^2z^2 = 12xy^2z^2$

49. $25p^2q^3 = 5^2 \cdot p^2q^3$; $15p^2q^2 = 3 \cdot 5 \cdot p^2q^2$; $35pq^4 = 5 \cdot 7 \cdot pq^4$; GCF $= 5pq^2$

50. $56r^4s^3 = 2^3 \cdot 7 \cdot r^4s^3$; $28r^3s^2 = 2^2 \cdot 7 \cdot r^3s^2$; $42r^3s = 2 \cdot 3 \cdot 7 \cdot r^3s$;
GCF $= 2 \cdot 7 \cdot r^3s = 14r^3s$

51. $(x + y)(x - y)$; $2x(x + y)$; GCF $= x + y$

52. $4p^2(p - 1) = 2^2p^2(p - 1)$; $6p(p + 1)^2 = 2 \cdot 3p(p + 1)^2$; GCF $= 2p$

53. $\dfrac{(a + b)^3}{(a + b)^2} = a + b$ **54.** $\dfrac{(w + x)^4}{(w + x)^5} = \dfrac{1}{w + x}$ **55.** $\dfrac{(x + y)(x - y)}{(x + y)^2} = \dfrac{x - y}{x + y}$

56. $\dfrac{(r - s)^2}{(r + s)(r - s)} = \dfrac{r - s}{r + s}$

C **57.** $\dfrac{136x^{n+1}}{187x^n} = \dfrac{136}{187} \cdot \dfrac{x^{n+1}}{x^n} = \dfrac{8}{11}x = \dfrac{8x}{11}$

58. $\dfrac{143(xy)^n}{117xy^n} = \dfrac{143x^ny^n}{117xy^n} = \dfrac{143}{117} \cdot x^{n-1} \cdot 1 = \dfrac{11}{9}x^{n-1} = \dfrac{11x^{n-1}}{9}$

59. $\dfrac{x^{2n+1}y^{n+1}}{(xy^2)^n} = \dfrac{x^{2n+1}y^{n+1}}{x^ny^{2n}} = x^{n+1} \cdot \dfrac{1}{y^{n-1}} = \dfrac{x^{n+1}}{y^{n-1}}$ **60.** $\dfrac{(-xy)^{2n+1}}{x^{2n}y^{n+1}} = \dfrac{-x^{2n+1}y^{2n+1}}{x^{2n}y^{n+1}} = -xy^n$

61. **a.** Def. of division **b.** Prop. of the reciprocal of a product
c. Comm. prop. of mult.; Assoc. prop. of mult. **d.** Def. of division

62. **a.** Prop. of quotients **b.** Def. of division **c.** Prop. of reciprocals
d. Identity prop. of mult.

Page 193 • MIXED REVIEW EXERCISES

1. $\frac{1}{5}(-25) = -5$ **2.** $111 \cdot \frac{1}{3} = 37$ **3.** $423 \div 9 = 47$ **4.** $6n^4\left(\frac{1}{6}n^4\right) = 6 \cdot \frac{1}{6}n^{4+4} = n^8$

5. $15 \div \left(-\frac{1}{3}\right) = 15 \cdot (-3) = -45$ **6.** $8y \cdot \frac{3}{4}y^2 = 6y^3$

7. $\dfrac{xy}{(1 - z)y} = \dfrac{4 \cdot 2}{[1 - (-3)]2} = \dfrac{8}{8} = 1$ **8.** $\dfrac{2x}{y - z} = \dfrac{2 \cdot 4}{2 - (-3)} = \dfrac{8}{5}$

9. $7y^3 + 2x = 7(2)^3 + 2 \cdot 4 = 7 \cdot 8 + 8 = 56 + 8 = 64$

Page 193 • CALCULATOR KEY-IN

1. 3; $0.230769 = \dfrac{3}{13}$ **2.** 5; $0.714285 = \dfrac{5}{7}$ **3.** 5; $0.\overline{45} = \dfrac{5}{11}$

4. 13; $0.8\overline{6} = \dfrac{13}{15}$ **5.** 26; $0.7\overline{2} = \dfrac{26}{36} = \dfrac{13}{18}$ **6.** 24; $0.5\overline{3} = \dfrac{24}{45} = \dfrac{8}{15}$

Pages 196–197 • WRITTEN EXERCISES

A **1.** $\frac{6a + 9}{3} = \frac{6a}{3} + \frac{9}{3} = 2a + 3$ **2.** $\frac{4x - 6}{2} = \frac{4x}{2} - \frac{6}{2} = 2x - 3$

3. $\frac{24t - 12}{6} = \frac{24t}{6} - \frac{12}{6} = 4t - 2$ **4.** $\frac{21c + 35}{7} = 3c + 5$

5. $\frac{9m - 18n}{9} = \frac{9m}{9} - \frac{18n}{9} = m - 2n$ **6.** $\frac{15a + 25b}{5} = \frac{15a}{5} + \frac{25b}{5} = 3a + 5b$

7. $\frac{12xy + 27y}{3y} = \frac{12xy}{3y} + \frac{27y}{3y} = 4x + 9$ **8.** $\frac{24mn - 16n}{8n} = \frac{24mn}{8n} - \frac{16n}{8n} = 3m - 2$

9. $\frac{10z^2 - 15z - 20}{5} = \frac{10z^2}{5} - \frac{15z}{5} - \frac{20}{5} = 2z^2 - 3z - 4$

10. $\frac{3x^2 - 12x - 18}{3} = \frac{3x^2}{3} - \frac{12x}{3} - \frac{18}{3} = x^2 - 4x - 6$

11. $\frac{33y^4 + 11y^3 - 44y^2}{11y} = \frac{33y^4}{11y} + \frac{11y^3}{11y} - \frac{44y^2}{11y} = 3y^3 + y^2 - 4y$

12. $\frac{4u^3 + 10u^2 - 6u}{2u} = \frac{4u^3}{2u} + \frac{10u^2}{2u} - \frac{6u}{2u} = 2u^2 + 5u - 3$

13. $\frac{8r^4 - 4r^3 - 6r^2}{-2r^2} = \frac{8r^4}{-2r^2} - \frac{4r^3}{-2r^2} - \frac{6r^2}{-2r^2} = -4r^2 + 2r + 3$

14. $\frac{9m^5 + 12m^4 - 6m^3}{-m^3} = \frac{9m^5}{-m^3} + \frac{12m^4}{-m^3} - \frac{6m^3}{-m^3} = -9m^2 - 12m + 6$

15. $\frac{pq^3 - p^3q}{pq} = \frac{pq^3}{pq} - \frac{p^3q}{pq} = q^2 - p^2$

16. $\frac{10a^2b - 15ab^2}{5ab} = \frac{10a^2b}{5ab} - \frac{15ab^2}{5ab} = 2a - 3b$

17. $\frac{x^2y - xy^2 - xy}{xy} = x - y - 1$

18. $\frac{6c^3d - 12cd^3 - 15cd}{3cd} = \frac{6c^3d}{3cd} - \frac{12cd^3}{3cd} - \frac{15cd}{3cd} = 2c^2 - 4d^2 - 5$

19. $\frac{28r^3s^2 + 42r^2s^3 - 56r^3s^3}{-7r^2s^2} = \frac{28r^3s^2}{-7r^2s^2} + \frac{42r^2s^3}{-7r^2s^2} - \frac{56r^3s^3}{-7r^2s^2} = -4r - 6s + 8rs$

20. $\frac{30p^4q - 45p^3q^2 + 15p^2q^3}{5p^2q} = \frac{30p^4q}{5p^2q} - \frac{45p^3q^2}{5p^2q} + \frac{15p^2q^3}{5p^2q} = 6p^2 - 9pq + 3q^2$

21. $65 \cdot 3 + 65 \cdot 7 = 65(3 + 7) = 65 \cdot 10 = 650$

22. $43 \cdot 13 - 43 \cdot 3 = 43(13 - 3) = 43 \cdot 10 = 430$

23. $7 \cdot 19 - 3 \cdot 19 + 6 \cdot 19 = (7 - 3 + 6)19 = 10 \cdot 19 = 190$

24. $7 \cdot 13 + 8 \cdot 13 + 5 \cdot 13 = (7 + 8 + 5)13 = 20 \cdot 13 = 260$

25. $83^2 + 83 \cdot 17 = 83(83 + 17) = 83 \cdot 100 = 8300$

26. $2 \cdot 9 + 9^2 = 9(2 + 9) = 9 \cdot 11 = 99$

27. $13^2 - 5 \cdot 13 + 2 \cdot 13 = (13 - 5 + 2)13 = 10 \cdot 13 = 130$

28. $12 \cdot 13 - 60 + 12^2 = 12 \cdot 13 - 12 \cdot 5 + 12 \cdot 12 = 12(13 - 5 + 12) = 12 \cdot 20 = 240$

29. $7^2 - 28 + 7 \cdot 17 = 7(7 - 4 + 17) = 7 \cdot 20 = 140$

30. $11^2 - 6 \cdot 11 + 5 \cdot 11 = 11(11 - 6 + 5) = 11 \cdot 10 = 110$

31. $15a - 25b + 20 = 5(3a - 5b + 4)$ **32.** $18x - 12y + 36 = 6(3x - 2y + 6)$

33. $6x^2 + 10x = 2x(3x + 5)$ **34.** $14c^3 - 21c = 7c(2c^2 - 3)$

35. $6p^2q - 9pq = 3pq(2p - 3)$ **36.** $2a^2b^2 + 10ab = 2ab(ab + 5)$

37. $7y^3 - 21y^2 - 14y = 7y(y^2 - 3y - 2)$

38. $22y^4 - 33y^3 + 11y^2 = 11y^2(2y^2 - 3y + 1)$

B **39.** $6ab^2 - 8a^2b = 2ab(3b - 4a)$ **40.** $4x^2y - 16xy^2 = 4xy(x - 4y)$

41. $-15x^2y^2 - 6xy^2 = -3xy^2(5x + 2)$ **42.** $-16x^3y - 24x^4y^3 = -8x^3y(2 + 3xy^2)$

43. $5ax^2 + 10a^2x - 15a^3 = 5a(x^2 + 2ax - 3a^2)$

44. $14p^3q^3 - 21p^2q^2 + 35pq = 7pq(2p^2q^2 - 3pq + 5)$

45. $48a^3b^2 + 72a^2b^3 = 24a^2b^2(2a + 3b)$ **46.** $77r^7s^7 - 84r^8s^4 = 7r^7s^4(11s^3 - 12r)$

47. $96wx^3y^2z^2 - 144w^3xy^2z^2 = 48wxy^2z^2(2x^2 - 3w^2)$

48. $84ab^2c^3d^4 + 126a^4b^3c^2 = 42ab^2c^2(2cd^4 + 3a^3b)$

49. $\frac{4a - 6}{2} + \frac{3a + 6}{3} = 2a - 3 + a + 2 = 3a - 1$

50. $\frac{14x - 21}{7} - \frac{10x - 25}{5} = 2x - 3 - (2x - 5) = 2x - 3 - 2x + 5 = 2$

51. $\frac{6p + 9q}{3} - \frac{7p + 21q}{7} = 2p + 3q - (p + 3q) = 2p + 3q - p - 3q = p$

52. $\frac{a^2b + 2ab^2}{ab} - \frac{2ab - b^2}{b} = a + 2b - (2a - b) = a + 2b - 2a + b = -a + 3b$

53. $\frac{x^2y - 3x^2y^2}{xy} + \frac{6xy + 9xy^2}{3y} = x - 3xy + 2x + 3xy = 3x$

54. $\frac{a^3b^4 - a^4b^3}{a^2b^3} - \frac{a^3b^2 - a^2b^3}{a^2b} = ab - a^2 - (ab - b^2) = ab - a^2 - ab + b^2 = -a^2 + b^2$

Pages 197–199 • PROBLEMS

A **1.** A = Area of rectangle − area of circle = $(2r)^2 - \pi r^2 = 4r^2 - \pi r^2 = r^2(4 - \pi)$

2. A = Area of rectangle − area of circle = $4r \cdot 2r - \pi r^2 = 8r^2 - \pi r^2 = r^2(8 - \pi)$

3. A = Area of square + area of circle = $(2r)^2 + \pi r^2 = 4r^2 + \pi r^2 = r^2(4 + \pi)$

4. A = Area of square − area of circle = $(2r)^2 - \pi r^2 = 4r^2 - \pi r^2 = r^2(4 - \pi)$

5. A = Area of square − 4 · area of circle = $(4r)^2 - 4 \cdot \pi r^2 = 16r^2 - 4\pi r^2 =$ $4r^2(4 - \pi)$

6. A = Area of square − 2 · area of circle = $(4r)^2 - 2 \cdot \pi r^2 = 16r^2 - 2\pi r^2 =$ $2r^2(8 - \pi)$

B **7.** A = Area of circle − 2 · area of square = $\pi(2r)^2 - 2 \cdot r^2 = \pi \cdot 4r^2 - 2r^2 =$ $2r^2(2\pi - 1)$

8. A = Area of circle − 4 · area of triangle = $\pi r^2 - 4 \cdot \frac{1}{2}r^2 = \pi r^2 - 2r^2 = r^2(\pi - 2)$

9. A = 2 · area of straight sections + area of large semi-circle − area of small semi-circle = $2 \cdot rn + \frac{1}{2}\pi(2r)^2 - \frac{1}{2}\pi r^2 = 2nr + \frac{1}{2}\pi \cdot 4r^2 - \frac{1}{2}\pi r^2 = 2nr + 2\pi r^2 -$ $\frac{1}{2}\pi r^2 = r\left(2n + 2\pi r - \frac{1}{2}\pi r\right) = r\left(2n + \frac{3}{2}\pi r\right)$

10. A = Area of rectangle + 2 · area of quarter-circle = $(2x)(y) + 2 \cdot \frac{1}{4}\pi(2x)^2 =$

$2xy + \frac{1}{2}\pi \cdot 4x^2 = 2xy + 2\pi x^2 = 2x(y + \pi x)$

11. A = 2 · area of straight section + area of large circle − area of small circle = $2 \cdot 2r \cdot r + \pi(2r)^2 - \pi r^2 = 4r^2 + 4\pi r^2 - \pi r^2 = r^2(4 + 4\pi - \pi) = r^2(4 + 3\pi)$

Page 199 • MIXED REVIEW EXERCISES

1. $7n^3\left(\frac{1}{7}n^3\right) = \left(7 \cdot \frac{1}{7}\right)(n^3 \cdot n^3) = n^6$ **2.** $9x^2\left(\frac{2}{3}x^3\right) = \left(9 \cdot \frac{2}{3}\right)(x^2 \cdot x^3) = 6x^5$

3. $4a^2 - 7a^2c + 2a^2 - 3ac^2 = 6a^2 - 7a^2c - 3ac^2$ **4.** $\frac{3x^3y}{6xy^2} = \frac{x^2}{2y}$

5. $28 \div \left(-\frac{1}{4}\right) = 28 \cdot (-4) = -112$ **6.** $\frac{(2a^3)^3}{a^4} = \frac{2^3 \cdot a^9}{a^4} = 8a^5$

7. $(4a)^4 = 4^4 \cdot a^4 = 256a^4$

8. $5(2^2 - 1) + 3^3 \cdot 2 = 5(4 - 1) + 27 \cdot 2 = 5 \cdot 3 + 54 = 15 + 54 = 69$

9. $(x - 2)(x^2 + 2x + 5) = x^3 + 2x^2 + 5x - 2x^2 - 4x - 10 = x^3 + x - 10$

Pages 201–202 • WRITTEN EXERCISES

A

1. $(x + 5)(x + 8) = x^2 + 13x + 40$ **2.** $(x + 7)(x + 6) = x^2 + 13x + 42$

3. $(y - 4)(y - 3) = y^2 - 7y + 12$ **4.** $(c - 9)(c - 6) = c^2 - 15c + 54$

5. $(z + 4)(z - 7) = z^2 - 3z - 28$ **6.** $(y + 6)(y - 2) = y^2 + 4y - 12$

7. $(n - 3)(n + 7) = n^2 + 4n - 21$ **8.** $(u - 10)(u + 9) = u^2 - u - 90$

9. $(3 + z)(2 + z) = 6 + 5z + z^2$ **10.** $(4 - x)(1 - x) = 4 - 5x + x^2$

11. $(2y + 5)(y + 2) = 2y^2 + 9y + 10$ **12.** $(r + 4)(2r + 3) = 2r^2 + 11r + 12$

13. $(7x - 1)(x + 7) = 7x^2 + 48x - 7$ **14.** $(4k - 1)(k + 4) = 4k^2 + 15k - 4$

15. $(2y + 1)(3y + 2) = 6y^2 + 7y + 2$ **16.** $(2n + 1)(5n + 2) = 10n^2 + 9n + 2$

17. $(2 - 3s)(1 - 2s) = 2 - 7s + 6s^2$ **18.** $(3 - 2r)(2 - 3r) = 6 - 13r + 6r^2$

19. $(3h - 5)(2h + 1) = 6h^2 - 7h - 5$ **20.** $(3x + 2)(2x - 3) = 6x^2 - 5x - 6$

21. $(5n + 4)(4n - 5) = 20n^2 - 9n - 20$

B

22. $(a + 2b)(a - b) = a^2 + (-ab + 2ab) - 2b^2 = a^2 + ab - 2b^2$

23. $(3x - y)(x - 2y) = 3x^2 + (-6xy - xy) + 2y^2 = 3x^2 - 7xy + 2y^2$

24. $(2r - s)(3r + 2s) = 6r^2 + (4rs - 3rs) - 2s^2 = 6r^2 + rs - 2s^2$

25. $(4h - k)(2h + 3k) = 8h^2 + (12hk - 2hk) - 3k^2 = 8h^2 + 10hk - 3k^2$

26. $(2x + 5y)(2x - 3y) = 4x^2 + (-6xy + 10xy) - 15y^2 = 4x^2 + 4xy - 15y^2$

27. $(7a - 2b)(5a - 3b) = 35a^2 + (-21ab - 10ab) + 6b^2 = 35a^2 - 31ab + 6b^2$

28. $(x^2 - 4x)(3x^2 + 2x) = 3x^4 + (2x^3 - 12x^3) - 8x^2 = 3x^4 - 10x^3 - 8x^2$

29. $(a^2 + 3b)(3a^2 - b) = 3a^4 + (-a^2b + 9a^2b) - 3b^2 = 3a^4 + 8a^2b - 3b^2$

30. $(p^2 - q^2)(p^2 + 3q^2) = p^4 + (3p^2q^2 - p^2q^2) - 3q^4 = p^4 + 2p^2q^2 - 3q^4$

31. $(p^3 - 4q^3)(p^3 + 3q^3) = p^6 + (3p^3q^3 - 4p^3q^3) - 12q^6 = p^6 - p^3q^3 - 12q^6$

32. $(y^4 - 3y^2)(y^2 + 2) = y^6 + (2y^4 - 3y^4) - 6y^2 = y^6 - y^4 - 6y^2$

33. $(x^4 + x^2y^2)(4x^2 - y^2) = 4x^6 + (-x^4y^2 + 4x^4y^2) - x^2y^4 = 4x^6 + 3x^4y^2 - x^2y^4$

34. $y(2y - 1)(y + 4) = y[2y^2 + (8y - y) - 4] = y(2y^2 + 7y - 4) = 2y^3 + 7y^2 - 4y$

35. $y(4y + 3)(y - 2) = y[4y^2 + (-8y + 3y) - 6] = y(4y^2 - 5y - 6) =$ $4y^3 - 5y^2 - 6y$

36. $y^2(1 - 2y)(2 + 3y) = y^2[2 + (3y - 4y) - 6y^2] = y^2(2 - y - 6y^2) =$ $2y^2 - y^3 - 6y^4 = -6y^4 - y^3 + 2y^2$

37. $x^2(4 - x)(2 - 3x) = x^2[8 + (-12x - 2x) + 3x^2] = x^2(8 - 14x + 3x^2) =$ $8x^2 - 14x^3 + 3x^4 = 3x^4 - 14x^3 + 8x^2$

38. $(x - 2)(x - 3) = (x - 7)(x + 3), x^2 - 5x + 6 = x^2 - 4x - 21, -5x + 6 =$ $-4x - 21, -x + 6 = -21, -x = -27, x = 27; \{27\}$

39. $(y + 4)(y - 3) = (y - 2)(y + 5), y^2 + y - 12 = y^2 + 3y - 10, y - 12 =$ $3y - 10, -12 = 2y - 10, -2 = 2y, -1 = y; \{-1\}$

40. $(2n + 5)(3n - 4) = (n + 2)(6n - 7), 6n^2 + 7n - 20 = 6n^2 + 5n - 14,$ $7n - 20 = 5n - 14, 2n - 20 = -14, 2n = 6, n = 3; \{3\}$

41. $(2x - 1)(8x + 3) = (4x + 5)(4x - 5), 16x^2 - 2x - 3 = 16x^2 - 25, -2x - 3 =$ $-25, -2x = -22, x = 11; \{11\}$

42. $(n + 3)(2n + 3) = (n + 2)^2 + (n - 2)^2, 2n^2 + 9n + 9 = n^2 + 4n + 4 +$ $n^2 - 4n + 4, 2n^2 + 9n + 9 = 2n^2 + 8, 9n + 9 = 8, 9n = -1, n = -\frac{1}{9}; \left\{-\frac{1}{9}\right\}$

43. $(2x - 3)(x + 3) = (x - 3)^2 + (x + 3)^2, 2x^2 + 3x - 9 = x^2 - 6x +$ $9 + x^2 + 6x + 9, 2x^2 + 3x - 9 = 2x^2 + 18, 3x - 9 = 18, 3x = 27, x = 9; \{9\}$

44. $(ax + b)(cx + d) = acx^2 + adx + bcx + bd = acx^2 + (ad + bc)x + bd$

45. $(px + q)(2x + 5) = 6x^2 + 11x + r, 2px^2 + (5px + 2qx) + 5q = 6x^2 + 11x + r,$ $2px^2 + (5p + 2q)x + 5q = 6x^2 + 11x + r$; comparing the coefficients of the quadratic terms: $2p = 6, p = 3$; comparing the coeff. of the linear terms: $5p + 2q = 11, 5 \cdot 3 + 2q = 11, 15 + 2q = 11, 2q = -4, q = -2$; comparing the constant terms: $5q = r, 5 \cdot (-2) = r, -10 = r$

46. $(px + 2)(3x + q) = rx^2 + x - 2, 3px^2 + 6x + pqx + 2q = rx^2 + x - 2,$ $3px^2 + (6 + pq)x + 2q = rx^2 + x - 2$; comparing the constant terms: $2q = -2,$ $q = -1$; comparing the coeff. of the linear terms: $6 + pq = 1, 6 + p \cdot (-1) = 1,$ $6 - p = 1, 6 = 1 + p, 5 = p$; comparing the quadratic terms: $3p = r, 3 \cdot 5 = r,$ $15 = r$

C 47. $(x^n - y^n)(2x^n + 3y^n) = 2x^{2n} + 3x^ny^n - 2x^ny^n - 3y^{2n} = 2x^{2n} + x^ny^n - 3y^{2n}$

48. $(2x^n - y^n)(2x^n + y^n) = 4x^{2n} + 2x^ny^n - 2x^ny^n - y^{2n} = 4x^{2n} - y^{2n}$

49. Let $2n + 1$ be any odd integer. $(2n + 1)^2 = (2n)^2 + (2n + 2n) + 1 = 4n^2 + 4n +$ $1 = 2(2n^2 + 2n) + 1$; since $2n^2 + 2n$ is an integer, $2(2n^2 + 2n) + 1$ is an odd integer.

Page 203 • MIXED REVIEW EXERCISES

1. $(2x^3y)(-4xy^3) = -8x^4y^4$ 2. $(5x^2y^5)^3 = 5^3(x^2)^3(y^5)^3 = 125x^6y^{15}$

3. $(6n + 3)(2n^2 + 3n - 1) = 12n^3 + 18n^2 - 6n + 6n^2 + 9n - 3 =$ $12n^3 + 24n^2 + 3n - 3$

4. $\frac{10r^2 + 25r - 30}{5} = \frac{10r^2}{5} + \frac{25r}{5} - \frac{30}{5} = 2r^2 + 5r - 6$

5. $\frac{(5y)^4}{5y} = \frac{5^4 \cdot y^4}{5y} = 5^3 \cdot y^3 = 125y^3$ 6. $\frac{15 - 9x - 3x^2}{3} = \frac{15}{3} - \frac{9x}{3} - \frac{3x^2}{3} = 5 - 3x - x^2$

7. $m = 42 - 2m, 3m = 42, m = 14; \{14\}$

8. $4x - (2x + 5) = 5, 4x - 2x - 5 = 5, 2x - 5 = 5, 2x = 10, x = 5; \{5\}$

9. $4(n + 3) = 3(5 + n), 4n + 12 = 15 + 3n, n + 12 = 15, n = 3; \{3\}$

10. $6y + 3 = 63, 6y = 60, y = 10; \{10\}$

11. $2(x + 1) - 3 = 9, 2x + 2 - 3 = 9, 2x - 1 = 9, 2x = 10, x = 5; \{5\}$

12. $3(y - 3) + 7 = 10, 3y - 9 + 7 = 10, 3y - 2 = 10, 3y = 12, y = 4; \{4\}$

Page 203 • COMPUTER EXERCISES

```
10 PRINT "THIS PROGRAM MULTIPLIES TWO BINOMIALS"
20 PRINT "OF THE FORM (AX + B)(CX + D)"
30 PRINT "INPUT A, B, C, D"
40 INPUT A, B, C, D
50 PRINT "PRODUCT = "; (A * C); " * X ^ 2 + ";
60 PRINT (A * D  +  B * C); " * X  +  "; D * B
70 END
```

1. $x^2 + 8x + 15$ **2.** $2x^2 - 7x - 4$ **3.** $8x^2 - 14x - 15$ **4.** $4x^2 + 28x + 49$
5. $9x^2 - 4$ **6.** $36x^2 - 84x + 49$ **7.** $50x^2 + 40x + 8$ **8.** $72x^2 + 17x - 72$
9. $144x^2 - 100$

Page 203 • CALCULATOR KEY-IN

1. 44 **2.** 119 **3.** 11 **4.** −6.75 **5.** 600 **6.** 7560 **7.** 728 **8.** −2.76
9. 17.5

Pages 206–207 • WRITTEN EXERCISES

A **1.** $(y - 7)(y + 7) = y^2 - 49$ **2.** $(m + 8)(m - 8) = m^2 - 64$

3. $(4 + x)(4 - x) = 16 - x^2$ **4.** $(9 - w)(9 + w) = 81 - w^2$

5. $(5y - 2)(5y + 2) = (5y)^2 - 4 = 25y^2 - 4$

6. $(8x - 11)(8x + 11) = (8x)^2 - 11^2 = 64x^2 - 121$

7. $(1 + 3a)(1 - 3a) = 1 - (3a)^2 = 1 - 9a^2$

8. $(7 + 2a)(7 - 2a) = 49 - (2a)^2 = 49 - 4a^2$

9. $(3x + 2y)(3x - 2y) = (3x)^2 - (2y)^2 = 9x^2 - 4y^2$

10. $(4w + 6z)(4w - 6z) = (4w)^2 - (6z)^2 = 16w^2 - 36z^2$

11. $(4s + 5t)(4s - 5t) = (4s)^2 - (5t)^2 = 16s^2 - 25t^2$

12. $(7p + 5q)(7p - 5q) = (7p)^2 - (5q)^2 = 49p^2 - 25q^2$

13. $(x^2 - 9y)(x^2 + 9y) = (x^2)^2 - (9y)^2 = x^4 - 81y^2$

14. $(2x + n^2)(2x - n^2) = (2x)^2 - (n^2)^2 = 4x^2 - n^4$

15. $(2r^2 + 7s^2)(2r^2 - 7s^2) = (2r^2)^2 - (7s^2)^2 = 4r^4 - 49s^4$

16. $(3m^2 - 8n^2)(3m^2 + 8n^2) = (3m^2)^2 - (8n^2)^2 = 9m^4 - 64n^4$

17. $(ab - c^2)(ab + c^2) = (ab)^2 - (c^2)^2 = a^2b^2 - c^4$

18. $(xy + 3z)(xy - 3z) = (xy)^2 - (3z)^2 = x^2y^2 - 9z^2$

19. $38 \cdot 42 = (40 - 2)(40 + 2) = 1600 - 4 = 1596$

20. $53 \cdot 47 = (50 + 3)(50 - 3) = 2500 - 9 = 2491$

21. $87 \cdot 93 = (90 - 3)(90 + 3) = 8100 - 9 = 8091$

22. $49 \cdot 51 = (50 - 1)(50 + 1) = 2500 - 1 = 2499$

23. $91 \cdot 89 = (90 + 1)(90 - 1) = 8100 - 1 = 8099$

24. $102 \cdot 98 = (100 + 2)(100 - 2) = 10{,}000 - 4 = 9{,}996$

25. $74 \cdot 66 = (70 + 4)(70 - 4) = 4900 - 16 = 4884$

26. $25 \cdot 35 = (30 - 5)(30 + 5) = 900 - 25 = 875$

27. $b^2 - 36 = b^2 - 6^2 = (b + 6)(b - 6)$ **28.** $m^2 - 25 = m^2 - 5^2 = (m + 5)(m - 5)$

29. $4c^2 - 81 = (2c)^2 - 9^2 = (2c + 9)(2c - 9)$

30. $9a^2 - 100 = (3a)^2 - 10^2 = (3a + 10)(3a - 10)$

31. $25z^2 - 1 = (5z)^2 - 1^2 = (5z + 1)(5z - 1)$

32. $16x^2 - 9 = (4x)^2 - 3^2 = (4x + 3)(4x - 3)$

33. $169u^2 - 225 = (13u)^2 - 15^2 = (13u + 15)(13u - 15)$

34. $81n^2 - 121 = (9n)^2 - 11^2 = (9n + 11)(9n - 11)$

35. $1 - 9a^2 = 1^2 - (3a)^2 = (1 + 3a)(1 - 3a)$

36. $144 - y^2 = 12^2 - y^2 = (12 + y)(12 - y)$

37. $49a^2 - 9b^2 = (7a)^2 - (3b)^2 = (7a + 3b)(7a - 3b)$

38. $64u^2 - 25v^2 = (8u)^2 - (5v)^2 = (8u + 5v)(8u - 5v)$

B **39.** $16 - c^4 = 4^2 - (c^2)^2 = (4 + c^2)(4 - c^2) = (4 + c^2)(2 + c)(2 - c)$

40. $625x^4 - 1 = (25x^2)^2 - 1^2 = (25x^2 + 1)(25x^2 - 1) = (25x^2 + 1)(5x + 1)(5x - 1)$

41. $u^4 - 81v^4 = (u^2)^2 - (9v^2)^2 = (u^2 + 9v^2)(u^2 - 9v^2) = (u^2 + 9v^2)(u + 3v)(u - 3v)$

42. $81s^4 - 16t^8 = (9s^2)^2 - (4t^4)^2 = (9s^2 + 4t^4)(9s^2 - 4t^4) =$
$(9s^2 + 4t^4)(3s + 2t^2)(3s - 2t^2)$

43. $x^8 - y^8 = (x^4)^2 - (y^4)^2 = (x^4 + y^4)(x^4 - y^4) = (x^4 + y^4)(x^2 + y^2)(x^2 - y^2) =$
$(x^4 + y^4)(x^2 + y^2)(x + y)(x - y)$

44. $m^{16} - 1 = (m^8)^2 - 1^2 = (m^8 + 1)(m^8 - 1) = (m^8 + 1)(m^4 + 1)(m^4 - 1) =$
$(m^8 + 1)(m^4 + 1)(m^2 + 1)(m^2 - 1) = (m^8 + 1)(m^4 + 1)(m^2 + 1)(m + 1)(m - 1)$

45. $5x^3 - 20x = 5x(x^2 - 4) = 5x(x + 2)(x - 2)$

46. $2a^4 - 18a^2 = 2a^2(a^2 - 9) = 2a^2(a + 3)(a - 3)$

47. $36a^2 - 16a^4 = 4a^2(9 - 4a^2) = 4a^2(3 + 2a)(3 - 2a)$

48. $50r^8 - 32r^2 = 2r^2(25r^6 - 16) = 2r^2(5r^3 + 4)(5r^3 - 4)$

49. $uv^3 - u^3v = uv(v^2 - u^2) = uv(v + u)(v - u)$

50. $27a^3b - 12ab = 3ab(9a^2 - 4) = 3ab(3a + 2)(3a - 2)$

51. $2a^5 - 162a = 2a(a^4 - 81) = 2a(a^2 + 9)(a^2 - 9) = 2a(a^2 + 9)(a + 3)(a - 3)$

52. $16x^5y^2 - xy^6 = xy^2(16x^4 - y^4) = xy^2(4x^2 + y^2)(4x^2 - y^2) =$
$xy^2(4x^2 + y^2)(2x + y)(2x - y)$

53. $(x + 4)^2 - x^2 = [(x + 4) + x][(x + 4) - x] = (2x + 4)(4)$

54. $t^2 - (t - 1)^2 = [t + (t - 1)][t - (t - 1)] = (2t - 1)(t - t + 1) = (2t - 1)(1) =$
$2t - 1$

55. $(s + 2)^2 - (s - 2)^2 = [(s + 2) + (s - 2)][(s + 2) - (s - 2)] =$
$(2s)(s + 2 - s + 2) = (2s)(4) = 8s$

56. $9(a + 1)^2 - 4(a - 1)^2 = [3(a + 1)]^2 - [2(a - 1)]^2 = [3(a + 1) + 2(a - 1)]$
$[3(a + 1) - 2(a - 1)] = (3a + 3 + 2a - 2)(3a + 3 - 2a + 2) = (5a + 1)(a + 5)$

57. $a^{2n} - b^{2n} = (a^n)^2 - (b^n)^2 = (a^n + b^n)(a^n - b^n)$

58. $x^{2n} - 25 = (x^n)^2 - 5^2 = (x^n + 5)(x^n - 5)$

59. $u^{4n} - 4v^{2n} = (u^{2n})^2 - (2v^n)^2 = (u^{2n} + 2v^n)(u^{2n} - 2v^n)$

60. $45r^{2n} - 5s^{4n} = 5(9r^{2n} - s^{4n}) = 5[(3r^n)^2 - (s^{2n})^2] = 5(3r^n + s^{2n})(3r^n - s^{2n})$

61. $x^{4n} - y^{4n} = (x^{2n})^2 - (y^{2n})^2 = (x^{2n} + y^{2n})(x^{2n} - y^{2n}) =$
$(x^{2n} + y^{2n})(x^n + y^n)(x^n - y^n)$

62. $a^{4n} - 81b^{4n} = (a^{2n})^2 - (9b^{2n})^2 = (a^{2n} + 9b^{2n})(a^{2n} - 9b^{2n}) =$
$(a^{2n} + 9b^{2n})(a^n + 3b^n)(a^n - 3b^n)$

63. $rt^{4n} - 16r = r(t^{4n} - 16) = r[(t^{2n})^2 - 4^2] = r(t^{2n} + 4)(t^n - 2)(t^n + 2)$

64. $x^2 - x^2y^{4n} = x^2(1 - y^{4n}) = x^2[1^2 - (y^{2n})^2] = x^2(1 + y^{2n})(1 - y^{2n}) =$
$x^2(1 + y^{2n})(1 + y^n)(1 - y^n)$

C **65.** Let a and b be the numbers.
$(a + b)^2 - (a - b)^2 = [(a + b) + (a - b)][(a + b) - (a - b)] = (2a)(2b) = 4ab$

66. Let n be the first integer.
$|(n + 1)^2 - n^2| = |[(n + 1) + n][(n + 1) - n]| = |(2n + 1)(1)| = |2n + 1| =$
$|n + (n + 1)|$

Page 207 • MIXED REVIEW EXERCISES

1. $3z(z - 4) + 4z(z + 5) = 3z^2 - 12z + 4z^2 + 20z = 7z^2 + 8z$

2. $(x + 5)(x - 7) = x^2 - 2x - 35$

3. $-2(m + 3) - 5m(m - 2) = -2m - 6 - 5m^2 + 10m = -5m^2 + 8m - 6$

4. $\frac{25a^4b}{5a^2} = 5a^2b$ **5.** $\frac{21a + 3}{3} = 7a + 1$ **6.** $\frac{24n^3x}{6n^2x} = 4n$

7. $(a + 1)(2a - 3) = (a)(2a) + (a)(-3) + (1)(2a) + (1)(-3) =$
$2a^2 - 3a + 2a - 3 = 2a^2 - a - 3$

8. $5b(b - 2)(3b + 1) = (5b^2 - 10b)(3b + 1) = 15b^3 + 5b^2 - 30b^2 - 10b =$
$15b^3 - 25b^2 - 10b$

9. $(3x)^2\left(\frac{1}{3}\right)^2x = 9x^2 \cdot \frac{1}{9} \cdot x = x^3$ **10.** $\frac{12y^3 + 36y^2 - 6y}{6y} = 2y^2 + 6y - 1$

11. $\frac{28x^2 + 42x - 14}{14} = 2x^2 + 3x - 1$ **12.** $\frac{18x^2y^3z}{x^2y^2z} = 18y$

Pages 210–212 • WRITTEN EXERCISES

A **1.** $(n + 5)^2 = n^2 + 2(5n) + 25 = n^2 + 10n + 25$

2. $(z + 8)^2 = z^2 + 2(8z) + 64 = z^2 + 16z + 64$

3. $(a - 9)^2 = a^2 - 2(9a) + 81 = a^2 - 18a + 81$

4. $(p - 10)^2 = p^2 - 2(10p) + 100 = p^2 - 20p + 100$

5. $(4u - 1)^2 = (4u)^2 - 2(4u) + 1 = 16u^2 - 8u + 1$

6. $(6c - 1)^2 = (6c)^2 - 2(6c) + 1 = 36c^2 - 12c + 1$

7. $(5n - 4)^2 = (5n)^2 - 2(5n)(4) + 16 = 25n^2 - 40n + 16$

8. $(3y - 8)^2 = (3y)^2 - 2(3y)(8) + 64 = 9y^2 - 48y + 64$

9. $(2r + 9s)^2 = (2r)^2 + 2(2r)(9s) + (9s)^2 = 4r^2 + 36rs + 81s^2$

10. $(4u + 7v)^2 = (4u)^2 + 2(4u)(7v) + (7v)^2 = 16u^2 + 56uv + 49v^2$

11. $(5p - 6q)^2 = (5p)^2 - 2(5p)(6q) + (6q)^2 = 25p^2 - 60pq + 36q^2$

12. $(5a - 8z)^2 = (5a)^2 - 2(5a)(8z) + (8z)^2 = 25a^2 - 80az + 64z^2$

13. $(mn + 2)^2 = (mn)^2 + 2(mn)(2) + 4 = m^2n^2 + 4mn + 4$

14. $(pq - 4)^2 = (pq)^2 - 2(4pq) + 16 = p^2q^2 - 8pq + 16$

15. $(2ab + c^2)^2 = (2ab)^2 + 2(2abc^2) + (c^2)^2 = 4a^2b^2 + 4abc^2 + c^4$

16. $(-6rs + s^2)^2 = (-6rs)^2 + 2(-6rs)(s^2) + (s^2)^2 = 36r^2s^2 - 12rs^3 + s^4$

17. $(-4m^2 - 3n)^2 = (-4m^2)^2 - 2(-4m^2)(3n) + (3n)^2 = 16m^4 + 24m^2n + 9n^2$

18. $(-13u + v^2)^2 = (-13u)^2 + 2(-13u)(v^2) + (v^2)^2 = 169u^2 - 26uv^2 + v^4$

19. $(9p^3 + 10)^2 = (9p^3)^2 + 2(9p^3)(10) + 100 = 81p^6 + 180p^3 + 100$

20. $(-11t^2 - 2)^2 = (-11t^2)^2 - 2(-11t^2)(2) + (2)^2 = 121t^4 + 44t^2 + 4$

21. $y^2 + 6y + 9 = (y + 3)^2$ **22.** $x^2 - 4x + 4 = (x - 2)^2$

23. $p^2 - 14p + 49 = (p - 7)^2$ **24.** $a^2 + 16a + 64 = (a + 8)^2$

25. $121 - 22u + u^2 = (11 - u)^2$ **26.** $144 + 12y + y^2$: not a perfect square

27. $4x^2 + 9 + 12x = 4x^2 + 12x + 9 = (2x + 3)^2$

28. $9 + 16c^2 - 24c = 16c^2 - 24c + 9 = (4c - 3)^2$

29. $25x^2 - 15xy + 36y^2$: not a perfect square

30. $49a^2 + 28ab + 4b^2 = (7a + 2b)^2$ **31.** $4s^2 - 36st + 81t^2 = (2s - 9t)^2$

32. $9u^2 + 30uv + 100v^2$: not a perfect square

33. $8x^2 + 8x + 2 = 2(4x^2 + 4x + 1) = 2(2x + 1)^2$

34. $3a^2 - 18a + 27 = 3(a^2 - 6a + 9) = 3(a - 3)^2$

35. $9 - 72m + 144m^2 = 9(1 - 8m + 16m^2) = 9(1 - 4m)^2$

36. $125u^2 - 50u + 5 = 5(25u^2 - 10u + 1) = 5(5u - 1)^2$

B **37.** $x^5 + 2x^4 + x^3 = x^3(x^2 + 2x + 1) = x^3(x + 1)^2$

38. $y^4 - 14y^3 + 49y^2 = y^2(y^2 - 14y + 49) = y^2(y - 7)^2$

39. $a^2b + 6ab^2 + 9b^3 = b(a^2 + 6ab + 9b^2) = b(a + 3b)^2$

40. $8u^3 - 24u^2v + 18uv^2 = 2u(4u^2 - 12uv + 9v^2) = 2u(2u - 3v)^2$

41. $36p^4 - 48p^3 + 16p^2 = 4p^2(9p^2 - 12p + 4) = 4p^2(3p - 2)^2$

42. $3x^8 + 48x^5 + 192x^2 = 3x^2(x^6 + 16x^3 + 64) = 3x^2(x^3 + 8)^2$

43. $u^2 - 2u + 1 - v^2 = (u - 1)^2 - v^2 = [(u - 1) + v][(u - 1) - v] =$
$(u - 1 + v)(u - 1 - v)$

44. $p^2 + 4p + 4 - q^2 = (p^2 + 4p + 4) - q^2 = (p + 2)^2 - q^2 =$
$[(p + 2) + q][(p + 2) - q] = (p + 2 + q)(p + 2 - q)$

45. $a^2 - b^2 + 6b - 9 = a^2 - (b^2 - 6b + 9) = a^2 - (b - 3)^2 =$
$[a + (b - 3)][a - (b - 3)] = (a + b - 3)(a - b + 3)$

46. $p^2 - q^2 - 4q - 4 = p^2 - (q^2 + 4q + 4) = p^2 - (q + 2)^2 =$
$[p + (q + 2)][p - (q + 2)] = (p + q + 2)(p - q - 2)$

47. $x^6 + 10x^3 + 25 = (x^3 + 5)^2$ **48.** $4 - 4y^2 + y^4 = (2 - y^2)^2$

49. $p^2q^2 - 12pq + 36 = (pq - 6)^2$ **50.** $a^4 + 2a^2b^4 + b^8 = (a^2 + b^4)^2$

51. $121 - 33n^2 + 9n^4$: not a perfect square

52. $121c^4 - 264c^2 + 144 = (11c^2 - 12)^2$

53. $(x + 1)^2 - 2(x + 1) + 1 = [(x + 1) - 1]^2 = (x + 1 - 1)^2 = x^2$

54. $(x + 1)^2 + 2(x + 1) + 1 = [(x + 1) + 1]^2 = (x + 2)^2$

55. $x^4 - 8x^2 + 16 = (x^2 - 4)^2 = [(x + 2)(x - 2)]^2 = (x + 2)^2(x - 2)^2$

56. $a^4 - 18a^2 + 81 = (a^2 - 9)^2 = [(a + 3)(a - 3)]^2 = (a + 3)^2(a - 3)^2$

57. **a.** $(2x - 3y)^2 = 4x^2 - 12xy + 9y^2$; $(3y - 2x)^2 = 9y^2 - 12xy + 4x^2$

b. $(2x - 3y)$ and $(3y - 2x)$ are opposites, and the squares of opposites are equal.

58. $(x + 2)^2 - (x - 3)^2 = 35$, $x^2 + 4x + 4 - (x^2 - 6x + 9) = 35$, $x^2 + 4x + 4 - x^2 + 6x - 9 = 35$, $10x - 5 = 35$, $10x = 40$, $x = 4$; $\{4\}$

59. $(2x + 5)^2 = (2x + 3)^2$, $4x^2 + 20x + 25 = 4x^2 + 12x + 9$, $20x + 25 = 12x + 9$, $8x + 25 = 9$, $8x = -16$, $x = -2$; $\{-2\}$

60. $(3x + 2)^2 + (4x - 3)^2 = (5x - 1)^2$, $9x^2 + 12x + 4 + 16x^2 - 24x + 9 = 25x^2 - 10x + 1$, $25x^2 - 12x + 13 = 25x^2 - 10x + 1$, $-12x + 13 = -10x + 1$, $13 = 2x + 1$, $12 = 2x$, $6 = x$; $\{6\}$

61. $(x + 2)^2 - (x - 2)^2 = (x - 1)^2 - (x - 3)^2$, $x^2 + 4x + 4 - (x^2 - 4x + 4) = x^2 - 2x + 1 - (x^2 - 6x + 9)$, $x^2 + 4x + 4 - x^2 + 4x - 4 = x^2 - 2x + 1 - x^2 + 6x - 9$, $8x = 4x - 8$, $4x = -8$, $x = -2$; $\{-2\}$

62.

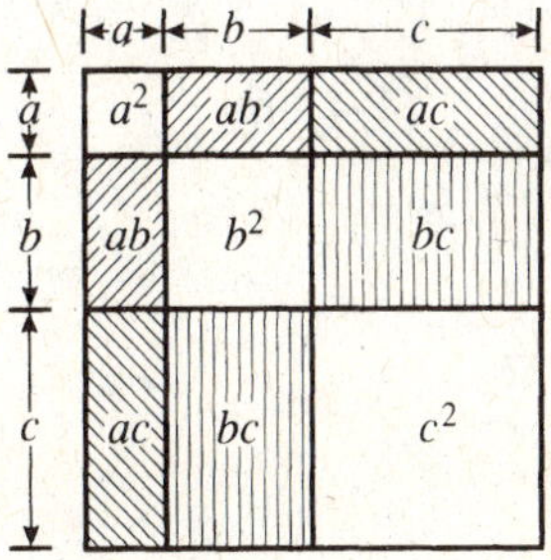

$(a + b + c)^2$

$$(a + b + c)^2 = [(a + b) + c]^2 = (a + b)^2 + 2(a + b)c + c^2 = a^2 + 2ab + b^2 + 2ac + 2bc + c^2$$

63.

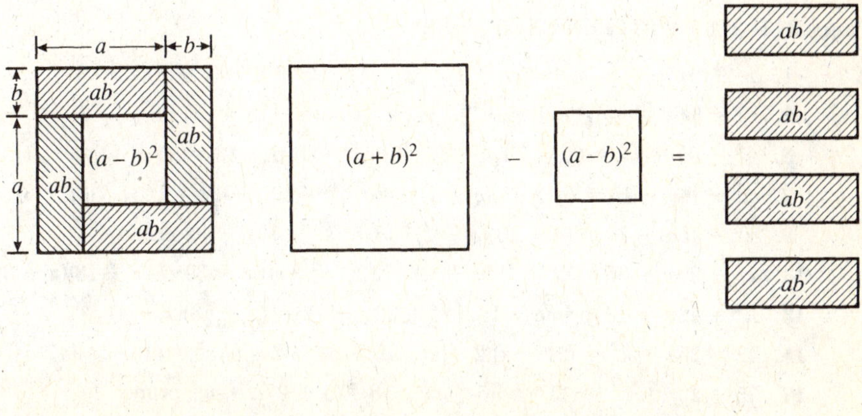

64.

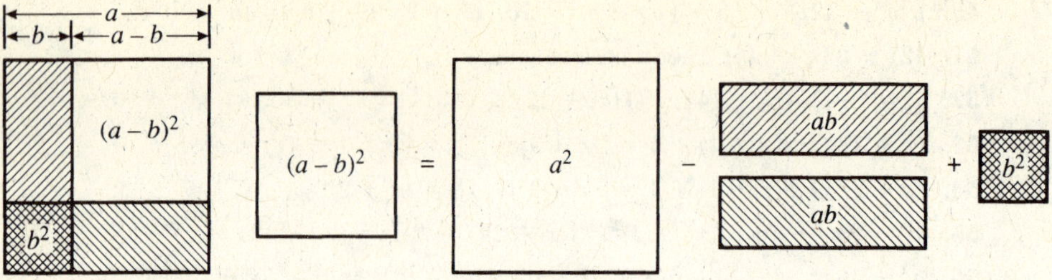

C **65.** Let x be the length of a side of the larger garden. Since the perimeter of the larger garden is 12 m greater than the smaller garden, $x - 3$ is the length of a side of the smaller garden. $x^2 = (x - 3)^2 + 105$, $x^2 = x^2 - 6x + 9 + 105$, $0 = -6x + 114$, $6x = 114$, $x = 19$; The larger garden is 19 m by 19 m.

66. **a.** $25^2 = 625$; $55^2 = 3025$; $95^2 = 9025$

b. $(10n + 5)^2 = (10n)^2 + 2(10n)(5) + 25 = 100n^2 + 100n + 25 = 100n(n + 1) + 25$

Page 212 • MIXED REVIEW EXERCISES

1. $x + y + (-8) = 4 + 2 + (-8) = 6 - 8 = -2$

2. $x - |y| - |3| = 4 - |2| - |3| = 4 - 2 - 3 = -1$

3. $9 + xy^2 = 9 + 4(2)^2 = 9 + 4 \cdot 4 = 9 + 16 = 25$

4. $(9 + xy)^2 = (9 + 4 \cdot 2)^2 = (9 + 8)^2 = 17^2 = 289$

5. $(y)^3(-y)^6 = 2^3(-2)^6 = 8 \cdot 64 = 512$

6. $(x^2y^2)^2 = (4^2 \cdot 2^2)^2 = (16 \cdot 4)^2 = 64^2 = 4096$

7. $(4s + 3)(4s - 3) = (4s)^2 - 3^2 = 16s^2 - 9$

8. $(x + 9)(x + 3) = x^2 + 3x + 9x + 27 = x^2 + 12x + 27$

9. $(5 - 3)^3 = 2^3 = 8$ **10.** $5 - 3^3 = 5 - 27 = -22$

11. $\frac{(a^6)^2}{(a^5)^3} = \frac{a^{12}}{a^{15}} = \frac{1}{a^3}$ **12.** $\frac{(2xy)^2}{4xy^2} = \frac{4x^2y^2}{4xy^2} = x$

Pages 215–216 • WRITTEN EXERCISES

A **1.** $x^2 + 5x + 4 = (x + 4)(x + 1)$ **2.** $z^2 + 9z + 8 = (z + 8)(z + 1)$

3. $r^2 - 6r + 8 = (r - 4)(r - 2)$ **4.** $c^2 - 10c + 16 = (c - 8)(c - 2)$

5. $y^2 - 9y + 14 = (y - 7)(y - 2)$ **6.** $p^2 - 14p + 13 = (p - 13)(p - 1)$

7. $q^2 + 16q + 15 = (q + 15)(q + 1)$ **8.** $n^2 + 10n + 21 = (n + 7)(n + 3)$

9. $a^2 - 15a + 26 = (a - 13)(a - 2)$ **10.** $s^2 - 12s + 40$: prime

11. $x^2 + 20x + 36 = (x + 18)(x + 2)$ **12.** $z^2 + 16z + 39 = (z + 13)(z + 3)$

13. $u^2 + 12u + 28$: prime **14.** $x^2 - 22x + 72 = (x - 18)(x - 4)$

15. $42 - 23k + k^2 = (21 - k)(2 - k)$ **16.** $64 - 20s + s^2 = (16 - s)(4 - s)$

17. $75 + 20r + r^2 = (15 + r)(5 + r)$ **18.** $75 + 27u + u^2$: prime

19. $p^2 + 19pq + 34q^2 = (p + 17q)(p + 2q)$

20. $a^2 + 10ab + 24b^2 = (a + 6b)(a + 4b)$

21. $c^2 - 16cd + 48d^2 = (c - 12d)(c - 4d)$ **22.** $x^2 - 15xy + 72y^2$: prime

23. $u^2 - 50uv + 49v^2 = (u - 49v)(u - v)$ 24. $h^2 - 14hk + 49k^2 = (h - 7k)^2$

25. $x^2 - 16xy + 45y^2$: prime 26. $m^2 + 20mn + 51n^2 = (m + 17n)(m + 3n)$

27. $a^2 + 17ab + 52b^2 = (a + 13b)(a + 4b)$ 28. $p^2 + 20pq + 50q^2$: prime

29. $r^2 - 15rs + 54s^2 = (r - 9s)(r - 6s)$ 30. $a^2 - 12ab + 27b^2 = (a - 9b)(a - 3b)$

B 31. $y^2 + 20yz + 91z^2 = (y + 13z)(y + 7z)$

32. $w^2 + 20wm + 96m^2 = (w + 12m)(w + 8m)$

33. $124 - 35y + y^2 = (31 - y)(4 - y)$ 34. $108 - 24y + y^2 = (18 - y)(6 - y)$

35. $112a^2 - 22ab + b^2 = (14a - b)(8a - b)$

36. $117x^2 - 22xy + y^2 = (13x - y)(9x - y)$

37. 14 may be factored as (1)(14), (2)(7), (−1)(−14), (−2)(−7); k may be 15, 9, −15, −9

38. 10 may be factored as (1)(10), (2)(5), (−1)(−10), (−2)(−5); k may be 11, 7, −11, −7

39. 12 may be factored as (1)(12), (2)(6), (3)(4), (−1)(−12), (−2)(−6), (−3)(−4); k may be 13, 8, 7, −13, −8, −7

40. 18 may be factored as (1)(18), (2)(9), (3)(6), (−1)(−18), (−2)(−9), (−3)(−6); k may be 19, 11, 9, −19, −11, −9

41. 9 may be factored as (1)(9), (3)(3), (−1)(−9), (−3)(−3); k may be 10, 6, −10, −6

42. 20 may be factored as (1)(20), (2)(10), (4)(5), (−1)(−20), (−2)(−10), (−4)(−5); k may be 21, 12, 9, −21, −12, −9

43. 6 may be the sum of 1 and 5, 2 and 4, 3 and 3; k may be 5, 8, 9

44. 7 may be the sum of 1 and 6, 2 and 5, 3 and 4; k may be 6, 10, 12

45. 8 may be the sum of 1 and 7, 2 and 6, 3 and 5, 4 and 4; k may be 7, 12, 15, 16

46. 9 may be the sum of 1 and 8, 2 and 7, 3 and 6, 4 and 5; k may be 8, 14, 18, 20

C 47. $(y + 2)^2 - 6(y + 2) + 5 = [(y + 2) - 1][(y + 2) - 5] = (y + 1)(y - 3)$

48. $(t + 3)^2 + 8(t + 3) + 15 = [(t + 3) + 3][(t + 3) + 5] = (t + 6)(t + 8)$

49. $(y + 3)^2 + 6(y + 3) + 9 = [(y + 3) + 3]^2 = (y + 6)^2$

50. $z^6 - 17z^4 + 16z^2 = z^2(z^4 - 17z^2 + 16) = z^2(z^2 - 16)(z^2 - 1) =$
$z^2(z + 4)(z - 4)(z + 1)(z - 1)$

51. $x^4 - 5x^2 + 4 = (x^2 - 4)(x^2 - 1) = (x + 2)(x - 2)(x + 1)(x - 1)$

52. $r^4 - 29r^2 + 100 = (r^2 - 25)(r^2 - 4) = (r + 5)(r - 5)(r + 2)(r - 2)$

53. $t^5 - 20t^3 + 64t = t(t^4 - 20t^2 + 64) = t(t^2 - 16)(t^2 - 4) =$
$t(t + 4)(t - 4)(t + 2)(t - 2)$

54. $(a - 4)^2 + 5(a - 4)(a + 2) + 6(a + 2)^2 =$
$[(a - 4) + 3(a + 2)][(a - 4) + 2(a + 2)] = (a - 4 + 3a + 6)(a - 4 + 2a + 4) =$
$(4a + 2)(3a) = (2a + 1)(2)(3a) = (2a + 1)6a = 6a(2a + 1)$

55. $a^{2n} - 30a^nb^{2n} + 209b^{4n} = (a^n - 19b^{2n})(a^n - 11b^{2n})$

56. $p^{4n} - 30p^{2n}q^n + 221q^{2n} = (p^{2n} - 17q^n)(p^{2n} - 13q^n)$

Page 216 • MIXED REVIEW EXERCISES

1. $-13 + x = -9, x = 4$; {4} 2. $d + (-5) = -6, d = -1$; {−1}

3. $-15 + b = 8, b = 23$; {23} 4. $n + 2 = |3 - 6|, n + 2 = |-3|, n + 2 = 3, n = 1$; {1}

5. $19m = 76, m = 4$; {4} 6. $3p + 18 = -72, 3p = -90, p = -30$; {−30}

7. $-\frac{1}{2}x = 12, x = -24; \{-24\}$ **8.** $\frac{r}{3} - 4 = 5, \frac{r}{3} = 9, r = 27; \{27\}$

9. $-21x = 252, x = -12; \{-12\}$ **10.** $(5y + 7)(5y - 7) = (5y)^2 - 7^2 = 25y^2 - 49$

11. $(2xy^3)^3 = 2^3 \cdot x^3y^9 = 8x^3y^9$ **12.** $(2x^2)^6 = 2^6 \cdot x^{12} = 64x^{12}$

Pages 218–219 • WRITTEN EXERCISES

A **1.** $y^2 + 5y - 6 = (y + 6)(y - 1)$ **2.** $v^2 - 3v - 4 = (v + 1)(v - 4)$

3. $x^2 - 6x - 16 = (x + 2)(x - 8)$ **4.** $x^2 + 2x - 8 = (x + 4)(x - 2)$

5. $c^2 - 4c - 12 = (c + 2)(c - 6)$ **6.** $u^2 - 10u - 9$: prime

7. $n^2 + 2n - 6$: prime **8.** $a^2 - 5a - 24 = (a + 3)(a - 8)$

9. $b^2 - 13b - 30 = (b + 2)(b - 15)$ **10.** $p^2 + 7p - 18 = (p + 9)(p - 2)$

11. $y^2 + 12y - 36$: prime **12.** $y^2 - 4y - 32 = (y + 4)(y - 8)$

13. $x^2 - 25x - 54 = (x + 2)(x - 27)$ **14.** $t^2 - 16t - 40$: prime

15. $y^2 - 21y - 72 = (y + 3)(y - 24)$ **16.** $z^2 + z - 72 = (z + 9)(z - 8)$

17. $a^2 - ab - 42b^2 = (a + 6b)(a - 7b)$

18. $r^2 - 20rs - 44s^2 = (r + 2s)(r - 22s)$ **19.** $u^2 + 9uv - 70v^2 = (u + 14v)(u - 5v)$

20. $x^2 - 2xy - 63y^2 = (x + 7y)(x - 9y)$

21. $h^2 - 25hk - 54k^2 = (h + 2k)(h - 27k)$

22. $m^2 + mn - 56n^2 = (m + 8n)(m - 7n)$

23. $p^2 - 16pq - 36q^2 = (p + 2q)(p - 18q)$

24. $a^2 - 13ab - 48b^2 = (a + 3b)(a - 16b)$

B **25.** $1 - 2n - 48n^2 = (1 + 6n)(1 - 8n)$ **26.** $1 + 15c - 34c^2 = (1 + 17c)(1 - 2c)$

27. $x^2 - 10xy - 75y^2 = (x + 5y)(x - 15y)$

28. $a^2 + 5ab - 84b^2 = (a + 12b)(a - 7b)$

29. $1 + 11pq - 80p^2q^2 = (1 + 16pq)(1 - 5pq)$

30. $1 - 15mn - 100m^2n^2 = (1 + 5mn)(1 - 20mn)$

31. $p^2 + 2p - 360 = (p + 20)(p - 18)$ **32.** $n^2 + 9n - 400 = (n + 25)(n - 16)$

33. $-380 + x + x^2 = (20 + x)(-19 + x)$

34. $-800 - 20a + a^2 = (20 + a)(-40 + a)$

35. -28 may be factored as $(1)(-28)$, $(2)(-14)$, $(4)(-7)$, $(-1)(28)$, $(-2)(14)$, $(-4)(7)$; k may be $-27, -12, -3, 27, 12, 3$

36. -20 may be factored as $(1)(-20)$, $(2)(-10)$, $(4)(-5)$, $(-1)(20)$, $(-2)(10)$, $(-4)(5)$; k may be $-19, -8, -1, 19, 8, 1$

37. -35 may be factored as $(1)(-35)$, $(5)(-7)$, $(-1)(35)$, $(-5)(7)$; k may be $-34, -2, 34, 2$

38. -36 may be factored as $(1)(-36)$, $(2)(-18)$, $(3)(-12)$, $(4)(-9)$, $(6)(-6)$, $(-1)(36)$, $(-2)(18)$, $(-3)(12)$, $(-4)(9)$; k may be $-35, -16, -9, -5, 0, 35, 16, 9, 5$

39–44. Answers may vary.

39. $-2 = -3 + 1, k = -3; -2 = -4 + 2, k = -8$

40. $4 = 5 - 1, k = -5; 4 = 6 - 2, k = -12$

41. $5 = 6 - 1, k = -6; 5 = 7 - 2, k = -14$

42. $-7 = 1 - 8, k = -8; -7 = 2 - 9, k = -18$

43. $4 = 5 - 1, k = -5; 4 = 6 - 2, k = -12$

44. $-6 = -7 + 1, k = -7; -6 = 2 - 8, k = -16$

C 45. $x^4 - 3x^2 - 4 = (x^2 - 4)(x^2 + 1) = (x + 2)(x - 2)(x^2 + 1)$

46. $t^4 - 7t^2 - 18 = (t^2 - 9)(t^2 + 2) = (t + 3)(t - 3)(t^2 + 2)$

47. $x^4 - 15x^2y^2 - 16y^4 = (x^2 - 16y^2)(x^2 + y^2) = (x + 4y)(x - 4y)(x^2 + y^2)$

48. $(x + 2)^2 - 4(x + 2) - 21 = [(x + 2) - 7][(x + 2) + 3] = (x - 5)(x + 5)$

49. $(y + 3)^2 + 5(y + 3) - 24 = [(y + 3) - 3][(y + 3) + 8] = y(y + 11)$

50. $(p + q)^2 - 2(p + q) - 15 = [(p + q) - 5][(p + q) + 3]$

51. $(a + b)^2 - (a + b) - 2 = [(a + b) - 2][(a + b) + 1]$

52. $(p + q)^2 - 2r(p + q) - 15r^2 = [(p + q) - 5r][(p + q) + 3r]$

53. $(a + b)^2 - c(a + b) - 2c^2 = [(a + b) - 2c][(a + b) + c]$

54. $(a + b)^4 - (a - b)^4 = [(a + b)^2 + (a - b)^2][(a + b)^2 - (a - b)^2] =$
$[(a^2 + 2ab + b^2) + (a^2 - 2ab + b^2)][(a^2 + 2ab + b^2) - (a^2 - 2ab + b^2)] =$
$(2a^2 + 2b^2)(a^2 + 2ab + b^2 - a^2 + 2ab - b^2) = 2(a^2 + b^2)(4ab) = 8ab(a^2 + b^2)$

55. $x^{2n} - 4x^ny^{2n} - 221y^{4n} = (x^n + 13y^{2n})(x^n - 17y^{2n})$

56. $x^{4n} - 4x^{2n}y^n - 252y^{2n} = (x^{2n} + 14y^n)(x^{2n} - 18y^n)$

Page 219 • MIXED REVIEW EXERCISES

1. $(9x^2y)(3xy^2)(2x^2) = (9 \cdot 3 \cdot 2)(x^2 \cdot x \cdot x^2)(y \cdot y^2) = 54x^5y^3$

2. $(3x - 4)(2x + 3) = 6x^2 + 9x - 8x - 12 = 6x^2 + x - 12$

3. $-7x(3x^2 - 2x + 4) = -21x^3 + 14x^2 - 28x$

4. $(3x - 4)^2 = (3x)^2 - 2(3x)(4) + 4^2 = 9x^2 - 24x + 16$

5. $(7x^5y^2)^3 = 7^3 \cdot x^{15}y^6 = 343x^{15}y^6$ 6. $5y(2y^2 + 3y + 5) = 10y^3 + 15y^2 + 25y$

7. $\frac{5(xy)^6}{10(xy)^3} = \frac{5x^6y^6}{10x^3y^3} = \frac{x^3y^3}{2}$ 8. $\frac{-4ab}{-12ab^3} = \frac{1}{3b^2}$ 9. $\frac{(-n)^6}{-n^{10}} = \frac{n^6}{-n^{10}} = -\frac{1}{n^4}$

10. $(n + 3p)^2 = n^2 + 2(n)(3p) + (3p)^2 = n^2 + 6np + 9p^2$

11. $(a - 6)(5a + 2) = 5a^2 + 2a - 30a - 12 = 5a^2 - 28a - 12$

12. $(2y + 7)^2 = (2y)^2 + 2(2y)(7) + 7^2 = 4y^2 + 28y + 49$

13. $15m - 21n + 9 = 3(5m - 7n + 3)$ 14. $121k^2 - 81 = (11k + 9)(11k - 9)$

15. $a^2 + 18a + 81 = (a + 9)^2$ 16. $a^2 - 13ab + 42b^2 = (a - 7b)(a - 6b)$

17. $16x^2 + 24x = 8x(2x + 3)$ 18. $64 - n^2 = (8 + n)(8 - n)$

19. $u^2 - 10u + 25 = (u - 5)^2$ 20. $44 + 15y + y^2 = (11 + y)(4 + y)$

21. $7a^2b^3 - 14ab = 7ab(ab^2 - 2)$ 22. $49w^4 - 16x^2 = (7w^2 + 4x)(7w^2 - 4x)$

23. $4m^2 + 20m + 24 = 4(m^2 + 5m + 6) = 4(m + 2)(m + 3)$

24. $c^2 - 11c - 26 = (c + 2)(c - 13)$ 25. $9x^2 - 24xy + 16y^2 = (3x - 4y)^2$

26. $56 - 15z + z^2 = (8 - z)(7 - z)$ 27. $x^2 - 1 = (x + 1)(x - 1)$

28. $a^2 + 13a - 68 = (a + 17)(a - 4)$

29. $25w^6 - 144x^6 = (5w^3 + 12x^3)(5w^3 - 12x^3)$ 30. $25a^2 + 20ab + 4b^2 = (5a + 2b)^2$

Pages 222–223 • WRITTEN EXERCISES

A **1.** $3x^2 + 7x + 2 = (3x + 1)(x + 2)$ **2.** $2x^2 + 5x + 3 = (2x + 3)(x + 1)$

3. $3c^2 - 8c + 5 = (3c - 5)(c - 1)$ **4.** $2x^2 - 15x + 7 = (2x - 1)(x - 7)$

5. $5y^2 + 4y - 1 = (5y - 1)(y + 1)$ **6.** $3a^2 + 4a - 4 = (3a - 2)(a + 2)$

7. $5u^2 - 6u - 2$: prime **8.** $3r^2 - 2r - 5 = (3r - 5)(r + 1)$

9. $7x^2 + 8x + 1 = (7x + 1)(x + 1)$ **10.** $2p^2 + 7p + 3 = (2p + 1)(p + 3)$

11. $5x^2 - 17x + 6 = (5x - 2)(x - 3)$ **12.** $7m^2 - 9m + 2 = (7m - 2)(m - 1)$

13. $3p^2 + 7p - 6 = (3p - 2)(p + 3)$ **14.** $4c^2 + 4c - 3 = (2c + 3)(2c - 1)$

15. $4y^2 - y - 3 = (y - 1)(4y + 3)$ **16.** $6a^2 - 5a - 2$: prime

17. $5 + 7x - 6x^2 = (5 - 3x)(1 + 2x)$ **18.** $9 + 6k - 8k^2 = (3 + 4k)(3 - 2k)$

19. $1 - 5b - 8b^2$: prime **20.** $7 - 12s - 4s^2 = (7 + 2s)(1 - 2s)$

21. $3m^2 + 11mn + 6n^2 = (3m + 2n)(m + 3n)$

22. $2p^2 - 7pq + 6q^2 = (2p - 3q)(p - 2q)$

23. $2x^2 + xy - 3y^2 = (2x + 3y)(x - y)$ **24.** $5a^2 - 2ab - 7b^2 = (5a - 7b)(a + b)$

B **25.** $9m^2 - 25mn - 6n^2 = (9m + 2n)(m - 3n)$

26. $6h^2 + 17hk + 10k^2 = (6h + 5k)(h + 2k)$

27. $6r^2 - 11rp + 5p^2 = (6r - 5p)(r - p)$

28. $4x^2 + 16xy - 9y^2 = (2x + 9y)(2x - y)$

29. $21c^2 + 4c - 12 = (7c + 6)(3c - 2)$ **30.** $18z^2 + 19z - 12 = (2z + 3)(9z - 4)$

31. $6 + 7a - 20a^2 = (2 + 5a)(3 - 4a)$ **32.** $8 + 45r - 18r^2 = (8 - 3r)(1 + 6r)$

33. $32n^2 - 4n - 15 = (4n - 3)(8n + 5)$ **34.** $33u^2 - u - 14 = (3u - 2)(11u + 7)$

35. $21c^2 + 22c - 24 = (3c - 2)(7c + 12)$ **36.** $35y^2 + 2y - 24 = (7y + 6)(5y - 4)$

C **37.** $2(a + 2)^2 + 5(a + 2) - 3 = [2(a + 2) - 1][(a + 2) + 3] = (2a + 4 - 1)(a + 5) =$
$(2a + 3)(a + 5)$

38. $2(x - 1)^2 - 9(x - 1) - 5 = [2(x - 1) + 1][(x - 1) - 5] = (2x - 2 + 1)(x - 6) =$
$(2x - 1)(x - 6)$

39. $2(a + 2b)^2 + 5(a + 2b)c - 3c^2 = [2(a + 2b) - c][(a + 2b) + 3c] =$
$(2a + 4b - c)(a + 2b + 3c)$

40. $2(x - y)^2 - 9(x - y)z - 5z^2 = [2(x - y) + z][(x - y) - 5z] =$
$(2x - 2y + z)(x - y - 5z)$

41. $4x^4 - 17x^2 + 4 = (4x^2 - 1)(x^2 - 4) = (2x + 1)(2x - 1)(x + 2)(x - 2)$

42. $2x^4 - 15x^2 - 27 = (2x^2 + 3)(x^2 - 9) = (2x^2 + 3)(x + 3)(x - 3)$

43. $(y^2 + 3y - 1)^2 - 9 = [(y^2 + 3y - 1) + 3][(y^2 + 3y - 1) - 3] =$
$(y^2 + 3y + 2)(y^2 + 3y - 4) = (y + 2)(y + 1)(y + 4)(y - 1)$

44. $(a^2 - 4a - 1)^2 - 16 = [(a^2 - 4a - 1) + 4][(a^2 - 4a - 1) - 4] =$
$(a^2 - 4a + 3)(a^2 - 4a - 5) = (a - 1)(a - 3)(a - 5)(a + 1)$

45. $(15x^2 - 14x + 3)(6x^2 + 19x - 7)(10x^2 + 29x - 21) =$
$(5x - 3)(3x - 1)(2x + 7)(3x - 1)(2x + 7)(5x - 3) =$
$(5x - 3)^2(3x - 1)^2(2x + 7)^2 = [(5x - 3)(3x - 1)(2x + 7)]^2 =$
$[(15x^2 - 14x + 3)(2x + 7)]^2 = (30x^3 + 77x^2 - 92x + 21)^2$

46. $90a^{8n+1}b^2 - 25a^{4n+1}b^{2n+2} - 240ab^{4n+2} = 5ab^2(18a^{8n} - 5a^{4n}b^{2n} - 48b^{4n}) =$
$5ab^2(9a^{4n} - 16b^{2n})(2a^{4n} + 3b^{2n}) = 5ab^2(3a^{2n} + 4b^n)(3a^{2n} - 4b^n)(2a^{4n} + 3b^{2n})$

Page 223 • MIXED REVIEW EXERCISES

1. $x^2 - 225 = (x + 15)(x - 15)$ **2.** $x^2 - 9x + 20 = (x - 4)(x - 5)$

3. $r^2 - 5r - 14 = (r - 7)(r + 2)$ **4.** $c^2 - 6c + 9 = (c - 3)^2$

5. $9y^2 - 289x^2 = (3y + 17x)(3y - 17x)$ **6.** $4a^4 - 49 = (2a^2 + 7)(2a^2 - 7)$

7. $y^2 + 15y + 56 = (y + 7)(y + 8)$ **8.** $p^2 + 12p + 36 = (p + 6)^2$

9. $16y^2 + 24y + 9 = (4y + 3)^2$ **10.** $m^2 - m - 72 = (m - 9)(m + 8)$

11. $n^2 + 15n + 36 = (n + 3)(n + 12)$ **12.** $b^2 - 2b - 24 = (b - 6)(b + 4)$

Page 223 • CHALLENGE

1. $2^{64} = ([([(2^2)^2]^2)^2]^2)^2$ You must press the squaring button six times.

2. $2^{64} - 1 = (2^{32} + 1)(2^{32} - 1) = (2^{32} + 1)(2^{16} + 1)(2^{16} - 1) =$
$(2^{32} + 1)(2^{16} + 1)(2^8 + 1)(2^8 - 1) = (2^{32} + 1)(2^{16} + 1)(2^8 + 1)(2^4 + 1)(2^4 - 1) =$
$(2^{32} + 1)(2^{16} + 1)(2^8 + 1)(2^4 + 1)(2^2 + 1)(2^2 - 1) = (2^{32} + 1)(2^{16} + 1)(2^8 + 1)(2^4 + 1) \cdot$
$(2^2 + 1)(2 + 1)(2 - 1)$; $(2 + 1) = 3$; $(2^2 + 1) = 4 + 1 = 5$; $(2^4 + 1) = 16 + 1 = 17$;
3, 5, and 17 are factors of $2^{64} - 1$.

Pages 225–226 • WRITTEN EXERCISES

A

1. $3(x + y) + z(x + y) = (3 + z)(x + y)$ **2.** $7(r - s) + t(r - s) = (7 + t)(r - s)$

3. $e(f - g) - 4(f - g) = (e - 4)(f - g)$ **4.** $w(x - y) - 8(x - y) = (w - 8)(x - y)$

5. $7(r - s) + t(s - r) = 7(r - s) + t[-(r - s)] = 7(r - s) - t(r - s) = (7 - t)(r - s)$

6. $7(m - n) + p(n - m) = 7(m - n) + p[-(m - n)] = 7(m - n) - p(m - n) =$
$(7 - p)(m - n)$

7. $2a(a + 3) - (3 + a) = 2a(a + 3) - (a + 3) = (2a - 1)(a + 3)$

8. $u(v - 2) + 2(2 - v) = u(v - 2) - 2(v - 2) = (u - 2)(v - 2)$

9. $2x(x - y) + y(y - x) = 2x(x - y) - y(x - y) = (2x - y)(x - y)$

10. $3p(2q - p) - 2q(p - 2q) = 3p(2q - p) + 2q(2q - p) = (3p + 2q)(2q - p)$

11. $2u(u - 2v) + v(u - 2v) + (u - 2v) = (2u + v + 1)(u - 2v)$

12. $a(a - b) + 4b(a - b) - a(a - b) = (a + 4b - a)(a - b) = 4b(a - b)$

13. $x(2w - 3v + u) - (2w - 3v + u) = (x - 1)(2w - 3v + u)$

14. $r(r - s - 2t) + s(r - s - 2t) = (r + s)(r - s - 2t)$

15. $(s^2 - 2ps + 2s) - (2s - 4p + 4) = s(s - 2p + 2) - 2(s - 2p + 2) =$
$(s - 2)(s - 2p + 2)$

16. $(x^2 - xy + x) - (y - x - 1) = x(x - y + 1) + (-y + x + 1) = x(x - y + 1) +$
$(x - y + 1) = (x + 1)(x - y + 1)$

17. $(3t - 3st) + (rs - r) = 3t(1 - s) + r(s - 1) = 3t(1 - s) - r(1 - s) =$
$(3t - r)(1 - s)$

18. $(9p - 3pq) + (2nq - 6n) = 3p(3 - q) + 2n(q - 3) = 3p(3 - q) - 2n(3 - q) =$
$(3p - 2n)(3 - q)$

19. $(12x^2 - 8xy) - 5(3xz - 2yz) = 4x(3x - 2y) - 5z(3x - 2y) = (4x - 5z)(3x - 2y)$

20. $(p^2 - 2pq) - 2(2qr - pr) = p(p - 2q) - 2r(2q - p) = p(p - 2q) + 2r(p - 2q) =$
$(p + 2r)(p - 2q)$

21. $3a + ab + 3c + bc = a(3 + b) + c(3 + b) = (a + c)(3 + b)$

22. $rs + 5r + st + 5t = r(s + 5) + t(s + 5) = (r + t)(s + 5)$

23. $x^2 - 2x + xy - 2y = x(x - 2) + y(x - 2) = (x + y)(x - 2)$

24. $u^2 - 2u + uv - 2v = u(u - 2) + v(u - 2) = (u + v)(u - 2)$

25. $h^2 - hk + hr - kr = h(h - k) + r(h - k) = (h + r)(h - k)$

26. $x^2 - 2xy + 4xz - 8yz = x(x - 2y) + 4z(x - 2y) = (x + 4z)(x - 2y)$

27. $p^3 - 2p^2 + 4p - 8 = p^2(p - 2) + 4(p - 2) = (p^2 + 4)(p - 2)$

28. $3a^3 + a^2 + 6a + 2 = a^2(3a + 1) + 2(3a + 1) = (a^2 + 2)(3a + 1)$

29. $p^2 - 2pq + pr - 2qr = p(p - 2q) + r(p - 2q) = (p + r)(p - 2q)$

30. $u^2 - 3uv - 6uw + 18vw = u(u - 3v) + 6w(-u + 3v) =$
$u(u - 3v) - 6w(u - 3v) = (u - 6w)(u - 3v)$

31. $3hk - 2k - 12h + 8 = k(3h - 2) + 4(-3h + 2) = k(3h - 2) - 4(3h - 2) =$
$(k - 4)(3h - 2)$

32. $3ab - b - 4 + 12a = 3ab - b + 12a - 4 = b(3a - 1) + 4(3a - 1) =$
$(b + 4)(3a - 1)$

33. $4z^3 - 6z^2 - 6z + 9 = 2z^2(2z - 3) + 3(-2z + 3) = 2z^2(2z - 3) - 3(2z - 3) =$
$(2z^2 - 3)(2z - 3)$

34. $3u^3 - u^2 - 9u + 3 = u^2(3u - 1) + 3(-3u + 1) = u^2(3u - 1) - 3(3u - 1) =$
$(u^2 - 3)(3u - 1)$

35. $(h^2k^2 + 4k^2) + (h^2k + 4k) = k^2(h^2 + 4) + k(h^2 + 4) = (k^2 + k)(h^2 + 4) =$
$k(k + 1)(h^2 + 4)$

36. $(a^2b^2 + 2a^2) - (2ab^2 + 4a) = a^2(b^2 + 2) - 2a(b^2 + 2) = (a^2 - 2a)(b^2 + 2) =$
$a(a - 2)(b^2 + 2)$

37. $x^3 - 3x^2 - x + 3 = x^2(x - 3) + (-x + 3) = x^2(x - 3) - (x - 3) =$
$(x^2 - 1)(x - 3) = (x + 1)(x - 1)(x - 3)$

38. $n^3 + 2n^2 - 4n - 8 = n^2(n + 2) + 4(-n - 2) = n^2(n + 2) - 4(n + 2) =$
$(n^2 - 4)(n + 2) = (n + 2)(n - 2)(n + 2) = (n + 2)^2(n - 2)$

39. $x^2 - (y - z)^2 = [x + (y - z)][x - (y - z)] = (x + y - z)(x - y + z)$

40. $(a + 2b)^2 - 9c^2 = [(a + 2b) + 3c][(a + 2b) - 3c] = (a + 2b + 3c)(a + 2b - 3c)$

41. $(u - 2v)^2 - 4w^2 = [(u - 2v) + 2w][(u - 2v) - 2w] =$
$(u - 2v + 2w)(u - 2v - 2w)$

42. $4p^2 - (q - 2r)^2 = [2p + (q - 2r)][2p - (q - 2r)] = (2p + q - 2r)(2p - q + 2r)$

B **43.** $(a + 2b)^2 - (2b + c)^2 = [(a + 2b) + (2b + c)][(a + 2b) - (2b + c)] =$
$(a + 4b + c)(a + 2b - 2b - c) = (a + 4b + c)(a - c)$

44. $4(x + y)^2 - (2y - z)^2 = [2(x + y) + (2y - z)][2(x + y) - (2y - z)] =$
$(2x + 2y + 2y - z)(2x + 2y - 2y + z) = (2x + 4y - z)(2x + z)$

45. $a^2 + 4a + 4 - b^2 = (a + 2)^2 - b^2 = [(a + 2) + b][(a + 2) - b] =$
$(a + 2 + b)(a + 2 - b)$

46. $x^2 - 2xy + y^2 - 4 = (x - y)^2 - 4 = [(x - y) + 2][(x - y) - 2] =$
$(x - y + 2)(x - y - 2)$

47. $u^2 - v^2 + 2v - 1 = u^2 - (v^2 - 2v + 1) = u^2 - (v - 1)^2 =$
$[u + (v - 1)][u - (v - 1)] = (u + v - 1)(u - v + 1)$

48. $m^2 - n^2 - 2m + 1 = (m^2 - 2m + 1) - n^2 = (m - 1)^2 - n^2 =$
$[(m - 1) + n][(m - 1) - n] = (m - 1 + n)(m - 1 - n)$

49. $h^2 - 4k^2 - 4h + 4 = (h^2 - 4h + 4) - 4k^2 = (h - 2)^2 - (2k)^2 =$
$[(h - 2) + 2k][(h - 2) - 2k] = (h - 2 + 2k)(h - 2 - 2k)$

50. $a^2 - b^2 - 2a + 1 = (a^2 - 2a + 1) - b^2 = (a - 1)^2 - b^2 =$
$[(a - 1) + b][(a - 1) - b] = (a - 1 + b)(a - 1 - b)$

51. $p^2 - q^2 + r^2 - 2pr = (p^2 - 2pr + r^2) - q^2 = (p - r)^2 - q^2 =$
$[(p - r) + q][(p - r) - q] = (p - r + q)(p - r - q)$

52. $4s^2 - 4t^2 + 4s + 1 = (4s^2 + 4s + 1) - 4t^2 = (2s + 1)^2 - (2t)^2 =$
$[(2s + 1) + 2t][(2s + 1) - 2t] = (2s + 1 + 2t)(2s + 1 - 2t)$

53. $x^2 - 4y^2 + 4z^2 - 4xz = (x^2 - 4xz + 4z^2) - 4y^2 = (x - 2z)^2 - (2y)^2 =$
$[(x - 2z) + 2y][(x - 2z) - 2y] = (x - 2z + 2y)(x - 2z - 2y)$

54. $m^2 - 9n^2 + 9 - 6m = (m^2 - 6m + 9) - 9n^2 = (m - 3)^2 - (3n)^2 =$
$[(m - 3) + 3n][(m - 3) - 3n] = (m - 3 + 3n)(m - 3 - 3n)$

55. $a^2 + b^2 + 2ab + 2a + 2b = (a^2 + 2ab + b^2) + (2a + 2b) = (a + b)^2 + 2(a + b) =$
$(a + b)[(a + b) + 2] = (a + b)(a + b + 2)$

56. $p^2 - q^2 - 2p + 2q = (p^2 - q^2) - (2p - 2q) = (p + q)(p - q) - 2(p - q) =$
$(p - q)[(p + q) - 2] = (p - q)(p + q - 2)$

57. $x^4 - y^4 - 4x^2 + 4 = (x^4 - 4x^2 + 4) - y^4 = (x^2 - 2)^2 - (y^2)^2 =$
$[(x^2 - 2) + y^2][(x^2 - 2) - y^2] = (x^2 - 2 + y^2)(x^2 - 2 - y^2)$

58. $a^4 + b^4 - c^4 + 2a^2b^2 = (a^4 + 2a^2b^2 + b^4) - c^4 = (a^2 + b^2)^2 - (c^2)^2 =$
$[(a^2 + b^2) + c^2][(a^2 + b^2) - c^2] = (a^2 + b^2 + c^2)(a^2 + b^2 - c^2)$

59. $p^2 + q^2 - r^2 - 2pq + 2r - 1 = (p^2 - 2pq + q^2) - (r^2 - 2r + 1) =$
$(p - q)^2 - (r - 1)^2 = [(p - q) + (r - 1)][(p - q) - (r - 1)] =$
$(p - q + r - 1)(p - q - r + 1)$

60. $h^2 - 4k^2 + 4h - 8k = [h^2 - (2k)^2] + (4h - 8k) = (h + 2k)(h - 2k) + 4(h - 2k) =$
$[(h + 2k) + 4](h - 2k) = (h + 2k + 4)(h - 2k)$

C **61.** $x^4 + 4 = x^4 + 4x^2 + 4 - 4x^2 = (x^2 + 2)^2 - (2x)^2 =$
$[(x^2 + 2) + 2x][(x^2 + 2) - 2x] = (x^2 + 2x + 2)(x^2 - 2x + 2)$

62. **a.** $64x^4 + 1 = (8x^2)^2 + 1 = (8x^2)^2 + 16x^2 + 1 - 16x^2 = (8x^2 + 1)^2 - (4x)^2 =$
$[(8x^2 + 1) + 4x][(8x^2 + 1) - 4x] = (8x^2 + 4x + 1)(8x^2 - 4x + 1)$

b. $x^4 + 4a^4 = (x^2)^2 + (2a^2)^2 = (x^2)^2 + 2(x^2)(2a^2) + (2a^2)^2 - 4x^2a^2 =$
$(x^2 + 2a^2)^2 - (2ax)^2 = [(x^2 + 2a^2) + 2ax][(x^2 + 2a^2) - 2ax] =$
$(x^2 + 2ax + 2a^2)(x^2 - 2ax + 2a^2)$

63. $a^{2n+1} + b^{2n+1} + a^{2n}b^{2n} + ab = (a^{2n+1} + ab) + (b^{2n+1} + a^{2n}b^{2n}) =$
$a(a^{2n} + b) + b^{2n}(b + a^{2n}) = (a + b^{2n})(a^{2n} + b)$

Page 226 • MIXED REVIEW EXERCISES

1. $-12 + x = -29, x = -17; \{-17\}$ **2.** $-n + 10 = 2, -n = -8, n = 8; \{8\}$

3. $18 + x = 32, x = 14; \{14\}$ **4.** $16 = 1 + 3z, 15 = 3z, 5 = z; \{5\}$

5. $10m - 6m = 36, 4m = 36, m = 9; \{9\}$

6. $5n - 2n + 8 = 9, 3n + 8 = 9, 3n = 1, n = \frac{1}{3}; \left\{\frac{1}{3}\right\}$

7. $14x = 700, x = 50; \{50\}$ **8.** $-13n = 156, n = -12; \{-12\}$

9. $9b = 108, b = 12; \{12\}$ **10.** $10n = 2n - 24, 8n = -24, n = -3; \{-3\}$

11. $19m = 55 + 14m, 5m = 55, m = 11; \{11\}$

12. $10y + 6 = 4(19 - y), 10y + 6 = 76 - 4y, 14y + 6 = 76, 14y = 70, y = 5; \{5\}$

Pages 228–229 • WRITTEN EXERCISES

A

1. $6a^2 - 9ab - 15b^2 = 3(2a^2 - 3ab - 5b^2) = 3(2a - 5b)(a + b)$

2. $18x - 8x^3 = 2x(9 - 4x^2) = 2x(3 + 2x)(3 - 2x)$

3. $15r^3 + 20r^2s - 20rs^2 = 5r(3r^2 + 4rs - 4s^2) = 5r(3r - 2s)(r + 2s)$

4. $6ab + 9a^2b - 15a^3b = 3ab(2 + 3a - 5a^2) = 3ab(2 + 5a)(1 - a)$

5. $4(z - 4)^2 - 16 = 4[(z - 4)^2 - 4] = 4[(z - 4) + 2][(z - 4) - 2] = 4(z - 2)(z - 6)$

6. $12x^2y - 36xy^2 + 27y^3 = 3y(4x^2 - 12xy + 9y^2) = 3y(2x - 3y)^2$

7. $5a^2 + 10ab + 5b^2 = 5(a^2 + 2ab + b^2) = 5(a + b)(a + b) = 5(a + b)^2$

8. $6c^2 + 18cd + 12d^2 = 6(c^2 + 3cd + 2d^2) = 6(c + 2d)(c + d)$

9. $4m^3 - m = m(4m^2 - 1) = m(2m + 1)(2m - 1)$

10. $3xy^2 - 27x^3 = 3x(y^2 - 9x^2) = 3x(y + 3x)(y - 3x)$

11. $y^4 - 2y^2 - y^3 = y^2(y^2 - 2 - y) = y^2(y^2 - y - 2) = y^2(y - 2)(y + 1)$

12. $-n^4 - 3n^2 - 2n^3 = -n^2(n^2 + 3 + 2n) = -n^2(n^2 + 2n + 3)$

13. $x^2 - xy - x + y = x(x - y) - (x - y) = (x - 1)(x - y)$

14. $-m^2 + mn + 2m - 2n = -m(m - n) + 2(m - n) = (2 - m)(m - n)$

15. $-41a + 10 + 21a^2 = 21a^2 - 41a + 10 = (7a - 2)(3a - 5)$

16. $80 - 120p + 45p^2 = 5(16 - 24p + 9p^2) = 5(4 - 3p)^2$

17. $a^3 - 2a^2b + 3a^2 - 6ab = a^2(a - 2b) + 3a(a - 2b) = (a^2 + 3a)(a - 2b) =$
$a(a + 3)(a - 2b)$

18. $8p^3q - 18pq^3 = 2pq(4p^2 - 9q^2) = 2pq(2p + 3q)(2p - 3q)$

19. $6u^2v - 11u^2v^2 - 10u^2v^3 = u^2v(6 - 11v - 10v^2) = u^2v(2 - 5v)(3 + 2v)$

20. $180x^2y - 108xy^2 - 75x^3 = -3x(-60xy + 36y^2 + 25x^2) =$
$-3x(25x^2 - 60xy + 36y^2) = -3x(5x - 6y)^2$

21. $k(k + 1)(k + 2) - 3k(k + 1) = k(k + 1)[(k + 2) - 3] = k(k + 1)(k - 1)$

22. $n(n^2 - 1) + n(n - 1) = n(n + 1)(n - 1) + n(n - 1) = n(n - 1)[(n + 1) + 1] =$
$n(n - 1)(n + 2)$

23. $2u^5 - 7u^3 - 4u = u(2u^4 - 7u^2 - 4) = u(2u^2 + 1)(u^2 - 4) =$
$u(2u^2 + 1)(u + 2)(u - 2)$

24. $81a + 18a^3 + a^5 = a(81 + 18a^2 + a^4) = a(9 + a^2)^2$

25. $r^2 - 6r - 9s^2 + 9 = (r^2 - 6r + 9) - 9s^2 = (r - 3)^2 - (3s)^2 =$
$[(r - 3) + 3s][(r - 3) - 3s] = (r - 3 + 3s)(r - 3 - 3s)$

26. $x^2 - 4y^2 - 4x + 4 = (x^2 - 4x + 4) - 4y^2 = (x - 2)^2 - (2y)^2 =$
$[(x - 2) + 2y][(x - 2) - 2y] = (x - 2 + 2y)(x - 2 - 2y)$

27. $u^2 - 4v^2 + 3u - 6v = (u^2 - 4v^2) + (3u - 6v) = (u + 2v)(u - 2v) + 3(u - 2v) =$
$(u - 2v)[(u + 2v) + 3] = (u - 2v)(u + 2v + 3)$

28. $a^2 - b^2 + ac - bc = (a^2 - b^2) + (ac - bc) = (a + b)(a - b) + c(a - b) =$
$(a - b)[(a + b) + c] = (a - b)(a + b + c)$

29. $p^2 - 1 - 4q^2 - 4q = p^2 - (1 + 4q^2 + 4q) = p^2 - (4q^2 + 4q + 1) =$
$p^2 - (2q + 1)^2 = [p + (2q + 1)][p - (2q + 1)] = (p + 2q + 1)(p - 2q - 1)$

30. $x^2 - 2x - 4y^2 - 4y = (x^2 - 4y^2) - (2x + 4y) = (x + 2y)(x - 2y) - 2(x + 2y) =$
$(x + 2y)[(x - 2y) - 2](x + 2y)(x - 2y - 2)$

B **31.** $100 + 4x^2 - 16y^2 - 40x = 4(25 + x^2 - 4y^2 - 10x) = 4[(x^2 - 10x + 25) - 4y^2] =$
$4[(x - 5)^2 - (2y)^2] = 4[(x - 5) + 2y][(x - 5) - 2y] = 4(x - 5 + 2y)(x - 5 - 2y)$

32. $16x^2 + 16y - y^2 - 64 = 16x^2 - (y^2 - 16y + 64) = 16x^2 - (y - 8)^2 =$
$[4x + (y - 8)][4x - (y - 8)] = (4x + y - 8)(4x - y + 8)$

33. $a^4 - b^4 = (a^2 + b^2)(a^2 - b^2) = (a^2 + b^2)(a + b)(a - b)$

34. $m^8 - n^8 = (m^4 + n^4)(m^4 - n^4) = (m^4 + n^4)(m^2 + n^2)(m^2 - n^2) =$
$(m^4 + n^4)(m^2 + n^2)(m + n)(m - n)$

35. $2pq + 2pr + q^2 - r^2 = 2p(q + r) + (q + r)(q - r) =$
$[2p + (q - r)](q + r) = (2p + q - r)(q + r)$

36. $8a^3 + 4a^2b - 2ab^2 - b^3 = 4a^2(2a + b) - b^2(2a + b) = (2a + b)(4a^2 - b^2) =$
$(2a + b)(2a + b)(2a - b) = (2a + b)^2(2a - b)$

37. $(a + b)^2 - (a - c)^2 = [(a + b) + (a - c)][(a + b) - (a - c)] = (2a + b - c)(b + c)$

38. $3x^5 + 15x^3 - 108x = 3x(x^4 + 5x^2 - 36) = 3x(x^2 - 4)(x^2 + 9) =$
$3x(x + 2)(x - 2)(x^2 + 9)$

39. $x^3 - x^2y - xy^2 + y^3 = x^2(x - y) - y^2(x - y) = (x - y)(x^2 - y^2) =$
$(x - y)(x + y)(x - y) = (x + y)(x - y)^2$

40. $4 - 4x^2 - 4y^2 + 8xy = 4(1 - x^2 - y^2 + 2xy) = 4[1 - (x^2 - 2xy + y^2)] =$
$4[1 - (x - y)^2] = 4[1 + (x - y)][1 - (x - y)] = 4(1 + x - y)(1 - x + y)$

41. $a(a + 2)(a - 3) - 8(a - 3) = (a - 3)[a(a + 2) - 8] = (a - 3)(a^2 + 2a - 8) =$
$(a - 3)(a + 4)(a - 2)$

42. $x(x + 1)(x - 4) + 4(x + 1) = (x + 1)[x(x - 4) + 4] = (x + 1)(x^2 - 4x + 4) =$
$(x + 1)(x - 2)^2$

43. $16c^{16} - 16 = 16(c^{16} - 1) = 16(c^8 + 1)(c^8 - 1) = 16(c^8 + 1)(c^4 + 1)(c^4 - 1) =$
$16(c^8 + 1)(c^4 + 1)(c^2 + 1)(c^2 - 1) = 16(c^8 + 1)(c^4 + 1)(c^2 + 1)(c + 1)(c - 1)$

44. $(u - v)^3 + v - u = (u - v)^3 - (u - v) = (u - v)[(u - v)^2 - 1] =$
$(u - v)[(u - v) + 1][(u - v) - 1] = (u - v)(u - v + 1)(u - v - 1)$

45. $a(a^2 - 9) - 2(a + 3)^2 = a(a + 3)(a - 3) - 2(a + 3)^2 =$
$(a + 3)[a(a - 3) - 2(a + 3)] = (a + 3)(a^2 - 3a - 2a - 6) =$
$(a + 3)(a^2 - 5a - 6) = (a + 3)(a - 6)(a + 1)$

46. $(x - 2)(x^2 - 1) - 6x - 6 = (x - 2)(x + 1)(x - 1) - 6(x + 1) = (x + 1)$
$[(x - 2)(x - 1) - 6] = (x + 1)(x^2 - 3x + 2 - 6) = (x + 1)(x^2 - 3x - 4) =$
$(x + 1)(x - 4)(x + 1) = (x + 1)^2(x - 4)$

47. $9u^2 - 9v^2 - 36w^2 + 36vw = 9(u^2 - v^2 - 4w^2 + 4vw) =$
$9[u^2 - (v^2 - 4vw + 4w^2)] = 9[u^2 - (v - 2w)^2] =$
$9[u + (v - 2w)][u - (v - 2w)] = 9(u + v - 2w)(u - v + 2w)$

48. $x^4 - x^2 + 4x - 4 = x^4 - (x^2 - 4x + 4) = x^4 - (x - 2)^2 =$
$[x^2 + (x - 2)][x^2 - (x - 2)] = (x^2 + x - 2)(x^2 - x + 2) =$
$(x + 2)(x - 1)(x^2 - x + 2)$

C **49.** $x^2(x + 2) - x(x + 2) - 12(x + 2) = (x + 2)(x^2 - x - 12) = (x + 2)$
$(x + 3)(x - 4)$

50. $(a + b)^3 - 6(a + b)^2 - 7(a + b) = (a + b)[(a + b)^2 - 6(a + b) - 7] =$
$(a + b)[(a + b) - 7][(a + b) + 1] = (a + b)(a + b - 7)(a + b + 1)$

51. $t^4 - 10t^2 + 9 = (t^2 - 9)(t^2 - 1) = (t + 3)(t - 3)(t + 1)(t - 1)$

52. $16t^4 - 8t^2 + 1 = (4t^2 - 1)^2 = [(2t + 1)(2t - 1)]^2 = (2t + 1)^2(2t - 1)^2$

53. $a^2 + b^2 - c^2 - d^2 - 2ab + 2cd = (a^2 - 2ab + b^2) - (c^2 - 2cd + d^2) = (a - b)^2 - (c - d)^2 = [(a - b) + (c - d)][(a - b) - (c - d)] = (a - b + c - d)(a - b - c + d)$

54. $(u^2 - v^2)^2 - w^2(u + v)^2 = [(u + v)(u - v)]^2 - w^2(u + v)^2 = (u + v)^2 [(u - v)^2 - w^2] = (u + v)^2[(u - v) + w][(u - v) - w] = (u + v)^2(u - v + w)(u - v - w)$

55. $x^4 + x^2 + 1 = x^4 + 2x^2 + 1 - x^2 = (x^2 + 1)^2 - x^2 = [(x^2 + 1) + x][(x^2 + 1) - x] = (x^2 + x + 1)(x^2 - x + 1)$

56. $a^4 + a^2b^2 + b^4 = a^4 + 2a^2b^2 + b^4 - a^2b^2 = (a^2 + b^2)^2 - (ab)^2 = [(a^2 + b^2) + ab][(a^2 + b^2) - ab] = (a^2 + ab + b^2)(a^2 - ab + b^2)$

57. $a^3 + b^3 = a^3 + a^2b - a^2b - ab^2 + ab^2 + b^3 = a^2(a + b) - ab(a + b) + b^2(a + b) = (a + b)(a^2 - ab + b^2)$

58. $a^3 - b^3 = a^3 - a^2b + a^2b - ab^2 + ab^2 - b^3 = a^2(a - b) + ab(a - b) + b^2(a - b) = (a - b)(a^2 + ab + b^2)$

Page 229 • MIXED REVIEW EXERCISES

1. $\left(-\frac{1}{4}\right)\left(\frac{1}{5}\right)(40) = -\frac{40}{20} = -2$ **2.** $\frac{1}{8}(56) = 7$ **3.** $-\frac{1}{9}(72)\left(-\frac{1}{8}\right) = \frac{72}{72} = 1$

4. $\frac{140b}{7} = 20b$ **5.** $52 \div \left(\frac{1}{13}\right) = 52 \cdot 13 = 676$ **6.** $625 \div (-5) = -125$

7. $x^2 - 12x + 35 = (x - 7)(x - 5)$ **8.** $x^2 + 3x - 28 = (x + 7)(x - 4)$

9. $x^2 - x - 2 = (x - 2)(x + 1)$ **10.** $2n^2 + 19n + 9 = (2n + 1)(n + 9)$

11. $3x^2 + 11x + 10 = (3x + 5)(x + 2)$

12. $(3x - 12) - 2n(4 - x) = 3(x - 4) + 2n(x - 4) = (3 + 2n)(x - 4)$

Pages 232–233 • WRITTEN EXERCISES

A **1.** $(y + 5)(y - 7) = 0; y + 5 = 0, y = -5; y - 7 = 0, y = 7; \{-5, 7\}$

2. $(n + 1)(n + 9) = 0; n + 1 = 0, n = -1; n + 9 = 0, n = -9; \{-1, -9\}$

3. $15n(n + 15) = 0; 15n = 0, n = 0; n + 15 = 0, n = -15; \{0, -15\}$

4. $2x(x - 20) = 0; 2x = 0, x = 0; x - 20 = 0, x = 20; \{0, 20\}$

5. $(2t - 3)(3t - 2) = 0; 2t - 3 = 0, 2t = 3, t = \frac{3}{2}; 3t - 2 = 0, 3t = 2, t = \frac{2}{3}; \left\{\frac{3}{2}, \frac{2}{3}\right\}$

6. $(2u + 7)(3u - 1) = 0; 2u + 7 = 0, 2u = -7, u = -\frac{7}{2}; 3u - 1 = 0, 3u = 1, u = \frac{1}{3}; \left\{-\frac{7}{2}, \frac{1}{3}\right\}$

7. $3x(2x + 1)(2x + 5) = 0; 3x = 0, x = 0; 2x + 1 = 0, 2x = -1, x = -\frac{1}{2}; 2x + 5 = 0, 2x = -5, x = -\frac{5}{2}; \left\{0, -\frac{1}{2}, -\frac{5}{2}\right\}$

8. $n(5n - 2)(2n + 5) = 0; n = 0; 5n - 2 = 0, 5n = 2, n = \frac{2}{5}; 2n + 5 = 0, 2n = -5, n = -\frac{5}{2}; \left\{0, \frac{2}{5}, -\frac{5}{2}\right\}$

9. $y^2 - 3y + 2 = 0, (y - 2)(y - 1) = 0; y - 2 = 0, y = 2; y - 1 = 0, y = 1; \{2, 1\}$

10. $p^2 - p - 6 = 0, (p - 3)(p + 2) = 0; p - 3 = 0, p = 3; p + 2 = 0, p = -2; \{3, -2\}$

11. $0 = x^2 + 14x + 48, 0 = (x + 6)(x + 8); x + 6 = 0, x = -6; x + 8 = 0, x = -8;$ $\{-6, -8\}$

12. $0 = k^2 - 12k + 35, 0 = (k - 5)(k - 7); k - 5 = 0, k = 5; k - 7 = 0, k = 7; \{5, 7\}$

13. $m^2 - 36 = 16m, m^2 - 16m - 36 = 0, (m - 18)(m + 2) = 0; m - 18 = 0, m = 18;$ $m + 2 = 0, m = -2; \{18, -2\}$

14. $r^2 + 9 = 10r, r^2 - 10r + 9 = 0, (r - 9)(r - 1) = 0; r - 9 = 0, r = 9; r - 1 = 0,$ $r = 1; \{9, 1\}$

15. $s^2 = 4s + 32, s^2 - 4s - 32 = 0, (s - 8)(s + 4) = 0; s - 8 = 0, s = 8; s + 4 = 0,$ $s = -4; \{8, -4\}$

16. $x^2 = 20x - 100, x^2 - 20x + 100 = 0, (x - 10)^2 = 0; x - 10 = 0, x = 10; \{10\}$

17. $y^2 = 16y, y^2 - 16y = 0, y(y - 16) = 0; y = 0; y - 16 = 0, y = 16; \{0, 16\}$

18. $9k^2 = 4k, 9k^2 - 4k = 0, k(9k - 4) = 0; k = 0; 9k - 4 = 0, 9k = 4, k = \frac{4}{9}; \left\{0, \frac{4}{9}\right\}$

19. $4x^2 - 9 = 0, (2x + 3)(2x - 3) = 0; 2x + 3 = 0, 2x = -3, x = -\frac{3}{2}; 2x - 3 = 0,$ $2x = 3, x = \frac{3}{2}; \left\{-\frac{3}{2}, \frac{3}{2}\right\}$

20. $25m^2 - 16 = 0, (5m + 4)(5m - 4) = 0; 5m + 4 = 0, 5m = -4, m = -\frac{4}{5};$ $5m - 4 = 0, 5m = 4, m = \frac{4}{5}; \left\{-\frac{4}{5}, \frac{4}{5}\right\}$

21. $6n^2 + n = 2, 6n^2 + n - 2 = 0, (2n - 1)(3n + 2) = 0; 2n - 1 = 0, 2n = 1, n = \frac{1}{2};$ $3n + 2 = 0, 3n = -2, n = -\frac{2}{3}; \left\{\frac{1}{2}, -\frac{2}{3}\right\}$

22. $3x^2 + x = 2, 3x^2 + x - 2 = 0, (3x - 2)(x + 1) = 0; 3x - 2 = 0, 3x = 2, x = \frac{2}{3};$ $x + 1 = 0, x = -1; \left\{\frac{2}{3}, -1\right\}$

23. $4s - 4s^2 = 1, 0 = 4s^2 - 4s + 1, 0 = (2s - 1)^2; 2s - 1 = 0, 2s = 1, s = \frac{1}{2}; \left\{\frac{1}{2}\right\}$

24. $r - 6r^2 = -1, 0 = 6r^2 - r - 1, 0 = (3r + 1)(2r - 1); 3r + 1 = 0, 3r = -1,$ $r = -\frac{1}{3}; 2r - 1 = 0, 2r = 1, r = \frac{1}{2}; \left\{-\frac{1}{3}, \frac{1}{2}\right\}$

25. $7x^2 = 18x - 11, 7x^2 - 18x + 11 = 0, (7x - 11)(x - 1) = 0; 7x - 11 = 0, 7x = 11,$ $x = \frac{11}{7}; x - 1 = 0, x = 1; \left\{\frac{11}{7}, 1\right\}$

26. $2y^2 = 25y + 13, 2y^2 - 25y - 13 = 0, (2y + 1)(y - 13) = 0; 2y + 1 = 0, 2y = -1,$ $y = -\frac{1}{2}; y - 13 = 0, y = 13; \left\{-\frac{1}{2}, 13\right\}$

27. $8u^3 - 2u^2 = 0, 2u^2(4u - 1) = 0; 2u^2 = 0, u = 0; 4u - 1 = 0, 4u = 1, u = \frac{1}{4};$
$\left\{0, \frac{1}{4}\right\}$

28. $10u^3 - 5u^2 = 0, 5u^2(2u - 1) = 0; 5u^2 = 0, u^2 = 0, u = 0; 2u - 1 = 0, 2u = 1,$
$u = \frac{1}{2}; \left\{0, \frac{1}{2}\right\}$

29. $0 = 4y^3 - 2y^2, 0 = 2y^2(2y - 1); 2y^2 = 0, y^2 = 0, y = 0; 2y - 1 = 0, 2y = 1,$
$y = \frac{1}{2}; \left\{0, \frac{1}{2}\right\}$

30. $0 = 10x^3 - 15x^2, 0 = 5x^2(2x - 3); 5x^2 = 0, x^2 = 0, x = 0; 2x - 3 = 0, 2x = 3,$
$x = \frac{3}{2}; \left\{0, \frac{3}{2}\right\}$

31. $8y^2 - 9y + 1 = 0, (8y - 1)(y - 1) = 0; 8y - 1 = 0, 8y = 1, y = \frac{1}{8}; y - 1 = 0,$
$y = 1; \left\{\frac{1}{8}, 1\right\}$

32. $6h^2 + 17h + 12 = 0, (2h + 3)(3h + 4) = 0; 2h + 3 = 0, 2h = -3, h = -\frac{3}{2};$
$3h + 4 = 0, 3h = -4, h = -\frac{4}{3}; \left\{-\frac{3}{2}, -\frac{4}{3}\right\}$

33. $15u^2 - 14u = 49, 15u^2 - 14u - 49 = 0, (3u - 7)(5u + 7) = 0; 3u - 7 = 0, 3u = 7,$
$u = \frac{7}{3}; 5u + 7 = 0, 5u = -7, u = -\frac{7}{5}; \left\{\frac{7}{3}, -\frac{7}{5}\right\}$

34. $25x^2 - 90x = -81, 25x^2 - 90x + 81 = 0, (5x - 9)^2 = 0; 5x - 9 = 0, 5x = 9,$
$x = \frac{9}{5}; \left\{\frac{9}{5}\right\}$

35. $4p^2 + 121 = 44p, 4p^2 - 44p + 121 = 0, (2p - 11)^2 = 0; 2p - 11 = 0, 2p = 11,$
$p = \frac{11}{2}; \left\{\frac{11}{2}\right\}$

36. $6c^2 - 72 = 11c, 6c^2 - 11c - 72 = 0, (2c - 9)(3c + 8) = 0; 2c - 9 = 0, 2c = 9,$
$c = \frac{9}{2}; 3c + 8 = 0, 3c = -8, c = -\frac{8}{3}; \left\{\frac{9}{2}, -\frac{8}{3}\right\}$

B **37.** $4x^3 - 12x^2 + 8x = 0, 4x(x^2 - 3x + 2) = 0, 4x(x - 2)(x - 1) = 0; 4x = 0, x = 0;$
$x - 2 = 0, x = 2; x - 1 = 0, x = 1; \{0, 1, 2\}$

38. $2n^3 - 30n^2 + 100n = 0, 2n(n^2 - 15n + 50) = 0, 2n(n - 5)(n - 10) = 0; 2n = 0,$
$n = 0; n - 5 = 0, n = 5; n - 10 = 0, n = 10; \{0, 5, 10\}$

39. $9x^3 + 9x = 30x^2, 9x^3 - 30x^2 + 9x = 0, 3x(3x^2 - 10x + 3) = 0,$
$3x(3x - 1)(x - 3) = 0; 3x = 0, x = 0; 3x - 1 = 0, 3x = 1, x = \frac{1}{3}; x - 3 = 0,$
$x = 3; \left\{0, \frac{1}{3}, 3\right\}$

40. $9x^3 + 25x = 30x^2, 9x^3 - 30x^2 + 25x = 0, x(9x^2 - 30x + 25) = 0, x(3x - 5)^2 = 0;$

$x = 0; 3x - 5 = 0, 3x = 5, x = \frac{5}{3}; \left\{0, \frac{5}{3}\right\}$

41. $y^4 - 10y^2 + 9 = 0, (y^2 - 9)(y^2 - 1) = 0, (y + 3)(y - 3)(y + 1)(y - 1) = 0;$
$y + 3 = 0, y = -3; y - 3 = 0, y = 3; y + 1 = 0, y = -1; y - 1 = 0, y = 1;$
$\{-3, 3, -1, 1\}$

42. $u^5 - 13u^3 + 36u = 0, u(u^4 - 13u^2 + 36) = 0, u(u^2 - 9)(u^2 - 4) = 0,$
$u(u + 3)(u - 3)(u + 2)(u - 2) = 0; u = 0; u + 3 = 0, u = -3; u - 3 = 0, u = 3;$
$u + 2 = 0, u = -2; u - 2 = 0, u = 2; \{0, -3, 3, -2, 2\}$

43. $(z + 1)(z - 5) = 16, z^2 - 4z - 5 = 16, z^2 - 4z - 21 = 0, (z - 7)(z + 3) = 0;$
$z - 7 = 0, z = 7; z + 3 = 0, z = -3; \{7, -3\}$

44. $(2t - 5)(t - 1) = 2, 2t^2 - 7t + 5 = 2, 2t^2 - 7t + 3 = 0, (2t - 1)(t - 3) = 0;$

$2t - 1 = 0, 2t = 1, t = \frac{1}{2}; t - 3 = 0, t = 3; \left\{\frac{1}{2}, 3\right\}$

45. $(x - 2)(x + 3) = 6, x^2 + x - 6 = 6, x^2 + x - 12 = 0, (x + 4)(x - 3) = 0;$
$x + 4 = 0, x = -4; x - 3 = 0, x = 3; \{-4, 3\}$

46. $(a - 5)(a - 2) = 28, a^2 - 7a + 10 = 28, a^2 - 7a - 18 = 0, (a - 9)(a + 2) = 0;$
$a - 9 = 0, a = 9; a + 2 = 0, a = -2; \{9, -2\}$

47. $x(x - 6) = 4(x - 4), x^2 - 6x = 4x - 16, x^2 - 10x + 16 = 0, (x - 2)(x - 8) = 0;$
$x - 2 = 0, x = 2; x - 8 = 0; x = 8; \{2, 8\}$

48. $3(m + 2) = m(m - 2), 3m + 6 = m^2 - 2m, 0 = m^2 - 5m - 6, 0 = (m - 6)$
$(m + 1); m - 6 = 0, m = 6; m + 1 = 0, m = -1; \{6, -1\}$

49. $(x - 2)(x + 3) = 0, x^2 + x - 6 = 0$ **50.** $(x + 1)(x - 5) = 0, x^2 - 4x - 5 = 0$

51. $\left(x - \frac{5}{2}\right)(x + 2) = 0, 2\left(x - \frac{5}{2}\right)(x + 2) = 0, (2x - 5)(x + 2) = 0, 2x^2 - x - 10 = 0$

52. $\left(x + \frac{1}{3}\right)(x + 1) = 0, 3\left(x + \frac{1}{3}\right)(x + 1) = 0, (3x + 1)(x + 1) = 0, 3x^2 + 4x + 1 = 0$

53. $\left(x - \frac{1}{3}\right)\left(x - \frac{3}{2}\right) = 0, 3\left(x - \frac{1}{3}\right) \cdot 2\left(x - \frac{3}{2}\right) = 0, (3x - 1)(2x - 3) = 0,$

$6x^2 - 11x + 3 = 0$

54. $\left(x + \frac{5}{2}\right)\left(x - \frac{2}{5}\right) = 0, 2\left(x + \frac{5}{2}\right) \cdot 5\left(x - \frac{2}{5}\right) = 0, (2x + 5)(5x - 2) = 0,$

$10x^2 + 21x - 10 = 0$

C **55.** **b.** Prop. of reciprocals **c.** Mult. prop. of equality **d.** Mult. prop. of zero
e. Assoc. prop. of mult. **f.** Prop. of reciprocals **g.** Identity prop. of mult.

Page 233 • MIXED REVIEW EXERCISES

1. $(x - y)^4 = (2 - 4)^4 = (-2)^4 = 16$ **2.** $x^4 \cdot y^2 = 2^4 \cdot 4^2 = 16 \cdot 16 = 256$

3. $5x^3 = 5(2)^3 = 5 \cdot 8 = 40$ **4.** $(5x)^3 = 5^3(2)^3 = 125 \cdot 8 = 1000$

5. $4x + y^2 = 4 \cdot 2 + 4^2 = 8 + 16 = 24$

6. $4x^2 + y = 4 \cdot 2^2 + 4 = 4 \cdot 4 + 4 = 16 + 4 = 20$

7. $4(x + y)^2 = 4(2 + 4)^2 = 4 \cdot 6^2 = 4 \cdot 36 = 144$ **8.** $(yx)^2 = (4 \cdot 2)^2 = 8^2 = 64$

9. $y^2x^2 = 4^2 \cdot 2^2 = 16 \cdot 4 = 64$ **10.** $(3x^3y)(-2xy^5) = -6x^4y^6$

11. $(9a)^3 = 9^3 \cdot a^3 = 729a^3$ **12.** $-5(x + 2) = -5x - 10$

Pages 235–238 • PROBLEMS

A

1. Let x be the number. $x + x^2 = 56, x^2 + x - 56 = 0, (x + 8)(x - 7) = 0; x + 8 = 0, x = -8; x - 7 = 0, x = 7$; The number is either -8 or 7.

2. Let x be the number. $x^2 - x = 72, x^2 - x - 72 = 0, (x - 9)(x + 8) = 0; x - 9 = 0, x = 9; x + 8 = 0, x = -8$; The number is either 9 or -8.

3. Let x be the number. $x^2 - 30 = x, x^2 - x - 30 = 0, (x - 6)(x + 5) = 0; x - 6 = 0, x = 6; x + 5 = 0, x = -5$ (not a pos. number); The number is 6.

4. Let x be the number. $x^2 - 42 = x, x^2 - x - 42 = 0, (x - 7)(x + 6) = 0; x - 7 = 0, x = 7$ (not a negative number); $x + 6 = 0, x = -6$; The number is -6.

5. Let n be the first integer. $n(n + 1) = 90, n^2 + n = 90, n^2 + n - 90 = 0, (n + 10)(n - 9) = 0; n + 10 = 0, n = -10; n - 9 = 0, n = 9$ (not a neg. integer); The integers are -10 and -9.

6. Let n be the first integer. $n(n + 2) = 143, n^2 + 2n = 143, n^2 + 2n - 143 = 0, (n + 13)(n - 11) = 0; n + 13 = 0, n = -13$ (not a pos. integer); $n - 11 = 0, n = 11$; The integers are 11 and 13.

7. Let n be the first integer. $n^2 + (n + 2)^2 = 340, n^2 + n^2 + 4n + 4 = 340, 2n^2 + 4n + 4 = 340, 2n^2 + 4n - 336 = 0, 2(n^2 + 2n - 168) = 0, n^2 + 2n - 168 = 0, (n + 14)(n - 12) = 0; n + 14 = 0, n = -14$ (not a pos. integer); $n - 12 = 0, n = 12$; The integers are 12 and 14.

8. Let n be the first integer. $n^2 + (n + 2)^2 = 100, n^2 + n^2 + 4n + 4 = 100, 2n^2 + 4n + 4 = 100, 2n^2 + 4n - 96 = 0, 2(n^2 + 2n - 48) = 0, n^2 + 2n - 48 = 0, (n + 8)(n - 6) = 0; n + 8 = 0, n = -8; n - 6 = 0, n = 6$ (not a neg. integer); The integers are -8 and -6.

9. Let w be the width of the rectangle. $w(w + 8) = 105, w^2 + 8w = 105, w^2 + 8w - 105 = 0, (w + 15)(w - 7) = 0; w + 15 = 0, w = -15$ (width cannot be neg.); $w - 7 = 0, w = 7; w + 8 = 15$; The rectangle is 7 cm by 15 cm.

10. Let w be the width of the rectangle. $w(2w - 6) = 108, 2w^2 - 6w = 108, 2w^2 - 6w - 108 = 0, 2(w^2 - 3w - 54) = 0, w^2 - 3w - 54 = 0, (w - 9)(w + 6) = 0; w - 9 = 0, w = 9; 2w - 6 = 18 - 6 = 12; w + 6 = 0, w = -6$ (width cannot be neg.); The rectangle is 9 cm by 12 cm.

11. Let w be the width of the rectangle. Let l be the length of the rectangle.

$2w + 2l = 46, 2l = 46 - 2w, l = \frac{1}{2}(46 - 2w), l = 23 - w; w(23 - w) = 126,$

$23w - w^2 = 126, 0 = w^2 - 23w + 126, 0 = (w - 9)(w - 14); w - 9 = 0, w = 9; 23 - w = 23 - 9 = 14; w - 14 = 0, w = 14; 23 - w = 23 - 14 = 9$; The rectangle is 9 m by 14 m.

12. Let w be the width and l the length of the rectangle. $2w + 2l = 42, 2l = 42 - 2w,$

$l = \frac{1}{2}(42 - 2w), l = 21 - w; w(21 - w) = 104, 21w - w^2 = 104,$

$0 = w^2 - 21w + 104, 0 = (w - 8)(w - 13); w - 8 = 0, w = 8; 21 - w = 21 - 8 = 13; w - 13 = 0, w = 13; 21 - w = 21 - 13 = 8$; The rectangle is 8 m by 13 m.

13. Let x be one of the numbers and $25 - x$ be the other number.
$x^2 + (25 - x)^2 = 313$, $x^2 + 625 - 50x + x^2 = 313$, $2x^2 - 50x + 625 = 313$, $2x^2 - 50x + 312 = 0$, $2(x^2 - 25x + 156) = 0$, $x^2 - 25x + 156 = 0$, $(x - 12)(x - 13) = 0$; $x - 12 = 0$, $x = 12$; $25 - x = 25 - 12 = 13$; $x - 13 = 0$, $x = 13$; $25 - x = 25 - 13 = 12$; The numbers are 12 and 13.

14. Let x be the smaller number and $x + 5$ be the larger number. $x^2 + (x + 5)^2 = 233$, $x^2 + x^2 + 10x + 25 = 233$, $2x^2 + 10x + 25 = 233$, $2x^2 + 10x - 208 = 0$, $2(x^2 + 5x - 104) = 0$, $x^2 + 5x - 104 = 0$, $(x + 13)(x - 8) = 0$; $x + 13 = 0$, $x = -13$ (not a pos. number); $x - 8 = 0$, $x = 8$; $x + 5 = 8 + 5 = 13$; The numbers are 8 and 13.

15. Let x be the amount each dimension was reduced. $(20 - x)(23 - x) = 20 \cdot 23 - 120$, $460 - 43x + x^2 = 460 - 120$, $460 - 43x + x^2 = 340$, $x^2 - 43x + 120 = 0$, $(x - 40)(x - 3) = 0$; $x - 40 = 0$, $x = 40$; $20 - x = 20 - 40 = -20$ (width cannot be neg.); $x - 3 = 0$, $x = 3$; $20 - x = 20 - 3 = 17$; $23 - x = 23 - 3 = 20$. The new rectangle is 17 cm by 20 cm.

16. Let w be the width and $2w$ be the length of the original rectangle.
$(2w + 4)(w - 3) = 600$, $2w^2 - 2w - 12 = 600$, $2(w^2 - w - 6) = 600$, $w^2 - w - 6 = 300$, $w^2 - w - 306 = 0$, $(w + 17)(w - 18) = 0$; $w + 17 = 0$, $w = -17$ (width cannot be neg.); $w - 18 = 0$, $w = 18$; $2w + 4 = 2 \cdot 18 + 4 = 36 + 4 = 40$; $w - 3 = 18 - 3 = 15$; The new rectangle is 40 m by 15 m.

B

17. Let t be the time when the ball is 19.6 m high. $19.6 = (24.5)t - 4.9t^2$, $4.9t^2 - 24.5t + 19.6 = 0$, $4.9(t^2 - 5t + 4) = 0$, $t^2 - 5t + 4 = 0$, $(t - 4)(t - 1) = 0$; $t - 4 = 0$, $t = 4$; $t - 1 = 0$, $t = 1$; The ball reaches 19.6 m at 1 s and 4 s.

18. Let t be the time when the rocket hits the ground. $0 = 1960t - 4.9t^2$, $0 = 4.9t(400 - t)$; $4.9t = 0$, $t = 0$ (when the rocket leaves the ground); $400 - t = 0$, $400 = t$; The rocket hits the ground after 400 s or $6\frac{2}{3}$ min.

19. Let t be the elapsed time. $0 = 120t - 16t^2$, $0 = 8t(15 - 2t)$; $8t = 0$, $t = 0$ (when the ball was hit); $15 - 2t = 0$, $15 = 2t$, $7.5 = t$; The catcher caught the ball after 7.5 s.

20. Let t be the time until Kathy caught the apple. $40 = 56t - 16t^2$, $16t^2 - 56t + 40 = 0$, $8(2t^2 - 7t + 5) = 0$, $2t^2 - 7t + 5 = 0$, $(t - 1)(2t - 5) = 0$; $t - 1 = 0$, $t = 1$ (on the way up); $2t - 5 = 0$, $2t = 5$, $t = 2.5$ (on the way down); The apple was in the air for 2.5 s.

21. Let t be the elapsed time from the firing of the flare. $1960 = 245t - 4.9t^2$, $4.9t^2 - 245t + 1960 = 0$, $4.9(t^2 - 50t + 400) = 0$, $t^2 - 50t + 400 = 0$, $(t - 10)(t - 40) = 0$; $t - 10 = 0$, $t = 10$ (on the way up); $t - 40 = 0$, $t = 40$ (on the way down); $40 - 10 = 30$; The flare falls past the balloon 30 s after it was fired past it.

22. Let t be elapsed time. $-98 = 39.2t - 4.9t^2$, $4.9t^2 - 39.2t - 98 = 0$, $4.9(t^2 - 8t - 20) = 0$, $t^2 - 8t - 20 = 0$, $(t + 2)(t - 10) = 0$; $t + 2 = 0$, $t = -2$ (time cannot be neg. here); $t - 10 = 0$, $t = 10$; The ball lands after 10 s.

23. **a.** Let t be the elapsed time. $400 = 160t - 16t^2$, $16t^2 - 160t + 400 = 0$, $16(t^2 - 10t + 25) = 0$, $t^2 - 10t + 25 = 0$, $(t - 5)^2 = 0$; $t - 5 = 0$, $t = 5$; The projectile hits 400 ft after 5 s.

b. There is only one time when the rocket is 400 ft high.

24. Let x be the width of the border. $(12 + 2x)(4 + x) = 2 \cdot 4 \cdot 12$, $48 + 20x + 2x^2 = 96$, $2x^2 + 20x - 48 = 0$, $2(x^2 + 10x - 24) = 0$, $x^2 + 10x - 24 = 0$, $(x + 12)(x - 2) = 0$; $x + 12 = 0$, $x = -12$ (width cannot be neg.); $x - 2 = 0$, $x = 2$; The width of the border should be 2 m.

25. Let x be the length of one of the equal sides and $20 - 2x$ be the length of the third side. $x(20 - 2x) = 48$, $20x - 2x^2 = 48$, $0 = 2x^2 - 20x + 48$, $0 = 2(x^2 - 10x + 24)$, $0 = x^2 - 10x + 24$, $0 = (x - 6)(x - 4)$; $x - 6 = 0$, $x = 6$; $20 - 2x = 20 - 2 \cdot 6 = 20 - 12 = 8$; $x - 4 = 0$, $x = 4$; $20 - 2x = 20 - 2 \cdot 4 = 20 - 8 = 12$; The pen could have dimensions of 6 m by 8 m or 4 m by 12 m.

26. Let x be the width of each path. $40x + 30x - x^2 = 325$, $70x - x^2 = 325$, $0 = x^2 - 70x + 325$, $0 = (x - 65)(x - 5)$; $x - 65 = 0$, $x = 65$ (too wide); $x - 5 = 0$, $x = 5$; The width of each path is 5 m.

27. Let x be the length of a side of the square bottom. $4(5x) + 2x^2 = 192$, $20x + 2x^2 = 192$, $2x^2 + 20x - 192 = 0$, $2(x^2 + 10x - 96) = 0$, $x^2 + 10x - 96 = 0$, $(x + 16)(x - 6) = 0$; $x + 16 = 0$, $x = -16$ (length cannot be neg.); $x - 6 = 0$, $x = 6$; $x \cdot x \cdot 5 = 6^2 \cdot 5 = 36 \cdot 5 = 180$; The volume of the box is 180 cm^3.

28. Let x be the width of the top. $2 \cdot (w)(2w) + 2 \cdot (4 \cdot w) + 2 \cdot (4 \cdot 2w) = 220$, $4w^2 + 8w + 16w = 220$, $4w^2 + 24w = 220$, $4w^2 + 24w - 220 = 0$, $4(w^2 + 6w - 55) = 0$, $w^2 + 6w - 55 = 0$, $(w + 11)(w - 5) = 0$; $w + 11 = 0$, $w = -11$ (width cannot be neg.); $w - 5 = 0$, $w = 5$; $2w = 10$; $w \cdot 2w \cdot 4 = 5 \cdot 10 \cdot 4 = 200$; The box has a volume of 200 ft^3.

29. Let x be the width of the path. $(50 - 2x)(120 - 2x) = 2 \cdot (50x) + 2 \cdot x \cdot (120 - 2x)$, $6000 - 340x + 4x^2 = 100x + 2x(120 - 2x)$, $6000 - 340x + 4x^2 = 100x + 240x - 4x^2$, $6000 - 340x + 4x^2 = 340x - 4x^2$, $8x^2 - 680x + 6000 = 0$, $8(x^2 - 85x + 750) = 0$, $x^2 - 85x + 750 = 0$, $(x - 75)(x - 10) = 0$; $x - 75 = 0$, $x = 75$; $50 - 2x = 50 - 2 \cdot 75 = 50 - 150 = -100$ (width cannot be neg.); $x - 10 = 0$, $x = 10$; $50 - 2x = 50 - 2 \cdot 10 = 50 - 20 = 30$; $120 - 2x = 120 - 2 \cdot 10 = 120 - 20 = 100$; The lawn is 30 m by 100 m.

C

30. Let x be the length of a shorter side and let y be the length of a longer side. $3x + 2y = 160$, $2y = 160 - 3x$, $y = \frac{1}{2}(160 - 3x)$, $y = 80 - \frac{3}{2}x$; $x\left(80 - \frac{3}{2}x\right) = 1000$, $80x - \frac{3}{2}x^2 = 1000$, $0 = \frac{3}{2}x^2 - 80x + 1000$, $0 = 2\left(\frac{3}{2}x^2 - 80x + 1000\right)$, $0 = 3x^2 - 160x + 2000$, $0 = (3x - 100)(x - 20)$; $3x - 100 = 0$, $3x = 100$, $x = \frac{100}{3}$; $y = 80 - \frac{3}{2}x = 80 - \frac{3}{2}\left(\frac{100}{3}\right) = 80 - 50 = 30$ (but $30 < \frac{100}{3}$ and x was the *shorter* side); $x - 20 = 0$, $x = 20$; $80 - \frac{3}{2}x = 80 - \frac{3}{2}(20) = 80 - 30 = 50$; The corral is 20 yd by 50 yd.

31. Let x be the length of each edge of the smaller cube. $(x + 2)^3 - x^3 = 98$, $(x + 2)(x^2 + 4x + 4) - x^3 = 98$, $x^3 + 4x^2 + 4x + 2x^2 + 8x + 8 - x^3 = 98$, $6x^2 + 12x + 8 = 98$, $6x^2 + 12x - 90 = 0$, $6(x^2 + 2x - 15) = 0$, $x^2 + 2x - 15 = 0$, $(x + 5)(x - 3) = 0$; $x + 5 = 0$, $x = -5$ (length cannot be neg.); $x - 3 = 0$, $x = 3$; $x + 2 = 5$; The smaller cube had sides of 3 cm each and the larger cube had sides of 5 cm each.

32. Let w be the width and $w + 10$ the length of the sheet. $(w - 2 \cdot 5)[(w + 10) - 2 \cdot 5)] \cdot 5 = 6000$, $(w - 10)(w + 10 - 10) \cdot 5 = 6000$, $(w - 10)5w = 6000$, $5w^2 - 50w = 6000$, $5w^2 - 50w - 6000 = 0$, $5(w^2 - 10w - 1200) = 0$, $w^2 - 10w - 1200 = 0$, $(w + 30)(w - 40) = 0$; $w + 30 = 0$, $w = -30$ (width cannot be neg.); $w - 40 = 0$, $w = 40$; $w + 10 = 40 + 10 = 50$; The original sheet was 40 cm by 50 cm.

Page 238 • MIXED REVIEW EXERCISES

1. $(8a^2b)(2ab^2) = 2 \cdot 8 \cdot a^{2+1}b^{1+2} = 16a^3b^3$ **2.** $(5a^2)^3 = 5^3 \cdot a^{2 \cdot 3} = 125a^6$

3. $3a(4 - 2b) = 12a - 6ab$ **4.** $(6r)\left(\frac{1}{3}rs^2\right) = \left(6 \cdot \frac{1}{3}\right)(r \cdot r)s^2 = 2r^2s^2$

5. $\left(\frac{1}{8}\right)(16n - 24p) = 2n - 3p$ **6.** $(-28x - 14y)\left(-\frac{1}{7}\right) = 4x + 2y$

7. $(3a + 2)(2a^2 + 5 - 7a) = 6a^3 + 15a - 21a^2 + 4a^2 + 10 - 14a = 6a^3 - 17a^2 + a + 10$

8. $(3b^2y)^2 = 3^2 \cdot b^4y^2 = 9b^4y^2$ **9.** $6x(x^2 - 8) = 6x^3 - 48x$

10. $-28 + 6m + 10m^2 = 2(5m^2 + 3m - 14) = 2(5m - 7)(m + 2)$

11. $36a^3 - 9ab^2 = 9a(4a^2 - b^2) = 9a(2a + b)(2a - b)$

12. $21n^2 + 22n - 8 = (7n - 2)(3n + 4)$

13. $y^4 - y^3 - 12y^2 = y^2(y^2 - y - 12) = y^2(y - 4)(y + 3)$

14. $15m^2 + 26mn + 8n^2 = (5m + 2n)(3m + 4n)$

15. $3 + 10x^2 - 17x = 10x^2 - 17x + 3 = (5x - 1)(2x - 3)$

Page 239 • EXTRA

1. $(x + y)(x^2 - xy + y^2) = x^3 - x^2y + xy^2 + x^2y - xy^2 + y^3 = x^3 + y^3$

2. $(x - y)(x^2 + xy + y^2) = x^3 + x^2y + xy^2 - x^2y - xy^2 - y^3 = x^3 - y^3$

3. $m^3 + 8 = m^3 + 2^3 = (m + 2)(m^2 - 2m + 4)$

4. $a^3 - 64 = a^3 - 4^3 = (a - 4)(a^2 + 4a + 16)$

5. $n^3 + 125 = n^3 + 5^3 = (n + 5)(n^2 - 5n + 25)$

6. $27 - 216y^3 = 27(1 - 8y^3) = 27[1 - (2y)^3] = 27(1 - 2y)(1 + 2y + 4y^2)$

7. a. $w^6 - 1 = (w^2)^3 - 1^3 = (w^2 - 1)(w^4 + w^2 + 1) = (w + 1)(w - 1)(w^4 + w^2 + 1)$

b. $w^6 - 1 = (w^3)^2 - 1^2 = (w^3 + 1)(w^3 - 1) =$
$(w + 1)(w^2 - w + 1)(w - 1)(w^2 + w + 1) =$
$(w - 1)(w + 1)(w^2 + w + 1)(w^2 - w + 1)$

c. $w^4 + w^2 + 1 = (w^4 + 2w^2 + 1) - w^2 = (w^2 + 1)^2 - w^2 =$
$[(w^2 + 1) - w][(w^2 + 1) + w] = (w^2 - w + 1)(w^2 + w + 1)$

Pages 240–241 • CHAPTER REVIEW

1. c **2.** d **3.** c; $14 = 2 \cdot 7$, $42 = 2 \cdot 3 \cdot 7$; GCF $= 2 \cdot 7 = 14$

4. d; $\frac{(-5m)^2}{-5m^2} = \frac{(-5)^2m^2}{-5m^2} = \frac{25m^2}{-5m^2} = -5$ **5.** c; $\frac{-105x^3y^6}{7xy^2} = -15x^2y^4$

6. c; $\frac{12c^2 - 30c + 6}{6} = \frac{12c^2}{6} - \frac{30c}{6} + \frac{6}{6} = 2c^2 - 5c + 1$ **7.** d

8. c; $(k - 1)(k - 1) = k^2 - k - k + 1 = k^2 - 2k + 1$

9. d; $(5m + 4n)(m + 4n) = 5m^2 + 20mn + 4mn + 16n^2 = 5m^2 + 24mn + 16n^2$

10. b; $(2m - 3n)(2m + 3n) = (2m)^2 - (3n)^2 = 4m^2 - 9n^2$

11. b; $49 - x^4 = (7 + x^2)(7 - x^2)$

12. d; $(7r - 3s)^2 = 49r^2 - 2(7r)(3s) + 9s^2 = 49r^2 - 42rs + 9s^2$ **13.** b **14.** a **15.** c

16. c **17.** b **18.** a **19.** b

20. b; $3(x - 2) - 4x(2 - x) = 3(x - 2) + 4x(x - 2) = (3 + 4x)(x - 2)$

21. c; $2x^3y - 50xy = 2xy(x^2 - 25) = 2xy(x + 5)(x - 5)$

22. b; $m^2 - 9n^2 + 2m - 6n = (m^2 - 9n^2) + (2m - 6n) = (m + 3n)(m - 3n) + 2(m - 3n) = (m + 3n + 2)(m - 3n)$

23. a; $5a = 0$, $a = 0$; $3a - 1 = 0$, $3a = 1$, $a = \frac{1}{3}$; $2a + 4 = 0$, $2a = -4$, $a = -2$; $\left\{0, \frac{1}{3}, -2\right\}$

24. d; Let n be the first integer. $(n + 1)^2 + (n + 2)^2 = 61$, $n^2 + 2n + 1 + n^2 + 4n + 4 = 61$, $2n^2 + 6n + 5 = 61$, $2n^2 + 6n - 56 = 0$, $2(n^2 + 3n - 28) = 0$, $n^2 + 3n - 28 = 0$, $(n + 7)(n - 4) = 0$; $n + 7 = 0$, $n = -7$; $n - 4 = 0$, $n = 4$; The integers are $-7, -6, -5, -4$; or $4, 5, 6, 7$.

Page 242 • CHAPTER TEST

1. -87 may be factored as $(1)(-87)$, $(3)(-29)$, $(-1)(87)$, $(-3)(29)$

2. 91 may be factored as $(1)(91)$, $(7)(13)$, $(-1)(-91)$, $(-7)(-13)$ **3.** $420 = 2^2 \cdot 3 \cdot 5 \cdot 7$

4. $168 = 2^3 \cdot 3 \cdot 7$ **5.** $\frac{-70m^5}{-42mn^7} = \frac{5m^4}{3n^7}$ **6.** $\frac{(-3x)^4}{-39x} = \frac{(-3)^4x^4}{-39x} = \frac{81x^4}{-39x} = -\frac{27x^3}{13}$

7. $\frac{49ab^2 - 56ab^8}{7ab^2} = \frac{49ab^2}{7ab^2} - \frac{56ab^8}{7ab^2} = 7 - 8b^6$

8. $\frac{-65r^6 + 78r^4 - 52r^2}{-13r^2} = \frac{-65r^6}{-13r^2} + \frac{78r^4}{-13r^2} - \frac{52r^2}{-13r^2} = 5r^4 - 6r^2 + 4$

9. $97 \times 16 - 97 \times 6 = 97(16 - 6) = 97 \cdot 10 = 970$

10. $82^2 + 82 \cdot 18 = 82(82 + 18) = 82 \cdot 100 = 8200$

11. $(5m - 1)(6m - 5) = 30m^2 - 25m - 6m + 5 = 30m^2 - 31m + 5$

12. $(7x - y)(8x + 9y) = 56x^2 + 63xy - 8xy - 9y^2 = 56x^2 + 55xy - 9y^2$

13. $(7 - 8x)(7 + 8x) = 7^2 - (8x)^2 = 49 - 64x^2$

14. $(c^4 + c^2)(c^4 - c^2) = (c^4)^2 - (c^2)^2 = c^8 - c^4$

15. $(x - 9)^2 = x^2 - 18x + 81$

16. $(4m - 6n)^2 = (4m)^2 - 2(4m)(6n) + (6n)^2 = 16m^2 - 48mn + 36n^2$

17. $n^2 + 16k - 64$: not a perfect square

18. $16x^2 - 8x + 1 = (4x - 1)^2$ **19.** $a^2 - 9ab + 81b^2$: not a perfect square

20. $b^2 - 3b + 2 = (b - 2)(b - 1)$ **21.** $x^2 - 2x + 4$: prime

22. $a^2 - 6ab + 8b^2 = (a - 4b)(a - 2b)$ **23.** $a^2 - 6a - 40 = (a - 10)(a + 4)$

24. $z^2 + z - 3$: prime **25.** $x^2 + 22xy - 48y^2 = (x + 24y)(x - 2y)$

26. $4a^2 - a - 5 = (4a - 5)(a + 1)$ **27.** $6y^2 + y - 15 = (3y + 5)(2y - 3)$

28. $7 - 23r + 6r^2 = (7 - 2r)(1 - 3r)$

29. $5(x - y) + z(y - x) = 5(x - y) - z(x - y) = (5 - z)(x - y)$

30. $ax + 2x + a + 2 = x(a + 2) + (a + 2) = (x + 1)(a + 2)$

31. $x^4 - 1 = (x^2 + 1)(x^2 - 1) = (x^2 + 1)(x + 1)(x - 1)$

32. $x^2y - y^3 = y(x^2 - y^2) = y(x + y)(x - y)$

33. $9m^3 - 63m^2 + 108m = 9m(m^2 - 7m + 12) = 9m(m - 4)(m - 3)$

34. $a^2 + a + ab + b = a(a + 1) + b(a + 1) = (a + b)(a + 1)$

35. $3x^2 - 41x = -60, 3x^2 - 41x + 60 = 0, (3x - 5)(x - 12) = 0; 3x - 5 = 0, 3x = 5,$

$x = \frac{5}{3}; x - 12 = 0, x = 12; \left\{\frac{5}{3}, 12\right\}$

36. $5m^2 = 85m, 5m^2 - 85m = 0, 5m(m - 17) = 0; 5m = 0; m = 0; m - 17 = 0, m = 17;$ $\{0, 17\}$

37. $9x^2 = 1, 9x^2 - 1 = 0, (3x + 1)(3x - 1) = 0; 3x + 1 = 0, 3x = -1, x = -\frac{1}{3}; 3x - 1 = 0,$

$3x = 1, x = \frac{1}{3}; \left\{-\frac{1}{3}, \frac{1}{3}\right\}$

38. Let w be the width and $2w + 3$ be the length of the rectangle. $w(2w + 3) = 90,$ $2w^2 + 3w = 90, 2w^2 + 3w - 90 = 0, (2w + 15)(w - 6) = 0; 2w + 15 = 0, 2w = -15,$

$w = -\frac{15}{2}$ (width cannot be neg.); $w - 6 = 0, w = 6; 2w + 3 = 2(6) + 3 = 12 + 3 = 15;$

The rectangle is 6 cm by 15 cm.

Page 243 • CUMULATIVE REVIEW

1. $-3.3 + (-27.3 + 10.6) = -3.3 + (-16.7) = -20$

2. $-\frac{3}{5}\left(-\frac{9}{17}\right) + \frac{8}{5}\left(-\frac{9}{17}\right) = \left(-\frac{9}{17}\right)\left(-\frac{3}{5} + \frac{8}{5}\right) = -\frac{9}{17} \cdot 1 = -\frac{9}{17}$

3. $\frac{-168a}{24} = -7a$ **4.** $(-4)^2 \div 2 + 2 - 8 = 16 \div 2 + 2 - 8 = 8 + 2 - 8 = 2$

5. $(2z - 3) + (3z - 4) = 5z - 7$

6. $(3m - 4) - (4m - 5) = 3m - 4 - 4m + 5 = -m + 1$

7. $(2a^2b)^3(3a^2b)^3 = 2^3 \cdot a^6b^3 \cdot 3^3a^6b^3 = 8 \cdot 27 \cdot a^{12}b^6 = 216a^{12}b^6$

8. $\frac{-45r^3st^2}{25rst} = \frac{-9r^2t}{5}$, or $-\frac{9r^2t}{5}$

9. $\frac{27x^3y^2 - 9x^2y + 8xy}{9xy} = \frac{27x^3y^2}{9xy} - \frac{9x^2y}{9xy} + \frac{8xy}{9xy} = 3x^2y - x + \frac{8}{9}$

10. $(6rs - 7t)(6rs + 7t) = (6rs)^2 - (7t)^2 = 36r^2s^2 - 49t^2$

11. $(7a + 4)^2 = 49a^2 + 56a + 16$

12. $-t(3t + 5)(2t - 5) = -t(6t^2 - 15t + 10t - 25) = -t(6t^2 - 5t - 25) = -6t^3 + 5t^2 + 25t$

13. $\frac{(z - 2w) - x}{-y + 1} = \frac{\left[(-3) - 2\left(-\frac{1}{2}\right)\right] - 2}{-(0) + 1} = \frac{(-3 + 1) - 2}{1} = -2 - 2 = -4$

14. $\frac{1}{w}(xy - z) = \frac{1}{\left(-\frac{1}{2}\right)}[2 \cdot 0 - (-3)] = -2(0 + 3) = -2 \cdot 3 = -6$

15. $(2x + z)^x = [2 \cdot 2 + (-3)]^2 = (4 - 3)^2 = 1^2 = 1$

16. $90 = 2 \cdot 3^2 \cdot 5, 756 = 2^2 \cdot 3^3 \cdot 7$; GCF $= 2 \cdot 3^2 = 18$

17. $6p^3 - 2p^3r^2 + 8p^2r^3st = 2p^2(3p - pr^2 + 4r^3st)$

18. $32a^2b - 8b = 8b(4a^2 - 1) = 8b(2a + 1)(2a - 1)$

19. $4a^2 + 12ab + 9b^2 = (2a + 3b)^2$ **20.** $x^2 + 15x + 26 = (x + 13)(x + 2)$

21. $m^2 - 9m + 18 = (m - 6)(m - 3)$ **22.** $k^2 - k - 42 = (k - 7)(k + 6)$

23. $6y^2 + 13y - 5 = (3y - 1)(2y + 5)$

24. $x^2 + 4xy + 4y^2 - 16 = (x + 2y)^2 - 16 = [(x + 2y) + 4][(x + 2y) - 4] =$
$(x + 2y + 4)(x + 2y - 4)$

25. $a^3 + a^2b - ab^2 - b^3 = a^2(a + b) - b^2(a + b) = (a + b)(a^2 - b^2) =$
$(a + b)(a + b)(a - b) = (a + b)^2(a - b)$

26. $9c - 3 = 24, 9c = 27, c = 3; \{3\}$ **27.** $|a| - 5 = 3, |a| = 8, a = 8$ or $a = -8; \{8, -8\}$

28. $\frac{2}{3}m = 18, m = 18 \cdot \frac{3}{2}, m = 27; \{27\}$

29. $7(x - 1) = 4x + 5, 7x - 7 = 4x + 5, 3x - 7 = 5, 3x = 12, x = 4; \{4\}$

30. $\frac{1}{5}n = -2, n = -10; \{-10\}$ **31.** $10 - x = -2, 10 = x - 2, 12 = x; \{12\}$

32. $3m - 2 = \frac{1}{2}(8m + 6) - (m + 5), 3m - 2 = 4m + 3 - m - 5, 3m - 2 = 3m - 2;$
identity

33. $2y^2 - 32 = 0, 2(y^2 - 16) = 0, y^2 - 16 = 0, (y + 4)(y - 4) = 0; y + 4 = 0, y = -4;$
$y - 4 = 0, y = 4; \{-4, 4\}$

34. $x^2 - 6x + 15 = 6, x^2 - 6x + 9 = 0, (x - 3)^2 = 0; x - 3 = 0, x = 3; \{3\}$

35. $(x + 7)(x + 1) = (x + 2)^2 + 5x, x^2 + 8x + 7 = x^2 + 4x + 4 + 5x, x^2 + 8x + 7 =$
$x^2 + 9x + 4, 8x + 7 = 9x + 4, 7 = x + 4, 3 = x; \{3\}$

36. $8b^2 - 10b = 3, 8b^2 - 10b - 3 = 0, (4b + 1)(2b - 3) = 0; 4b + 1 = 0, 4b = -1, b = -\frac{1}{4};$
$2b - 3 = 0, 2b = 3, b = \frac{3}{2}; \left\{-\frac{1}{4}, \frac{3}{2}\right\}$

37. $x^3 - 9x^2 + 20x = 0, x(x^2 - 9x + 20) = 0, x(x - 5)(x - 4) = 0; x = 0; x - 5 = 0, x = 5;$
$x - 4 = 0, x = 4; \{0, 5, 4\}$

38. Let n be the number of nickels. $20 - n = \frac{2}{3}n, 20 = \frac{5}{3}n, 20 \cdot \frac{3}{5} = n, 12 = n; 20 - n =$
$20 - 12 = 8$; Marvin has 12 nickels and 8 dimes.

39. Let r be the reduced speed limit. $r = \frac{d}{t}, r = \frac{42}{28 + 14}, r = \frac{42}{42}, r = 1$; The reduced speed limit is 1 km/min or 60 km/h.

40. Let n be the first integer. $n^2 + (n + 1)^2 = 8n + 9, n^2 + n^2 + 2n + 1 = 8n + 9,$
$2n^2 + 2n + 1 = 8n + 9, 2n^2 - 6n - 8 = 0, 2(n^2 - 3n - 4) = 0, n^2 - 3n - 4 = 0,$
$(n - 4)(n + 1) = 0; n - 4 = 0, n = 4; n + 1 = 5; n + 1 = 0, n = -1; n + 1 = 0;$
The two integers are 4, 5 or -1, 0.

41. Let w be the width of the rectangle. $w(3w + 5) = 22, 3w^2 + 5w = 22,$
$3w^2 + 5w - 22 = 0, (3w + 11)(w - 2) = 0; 3w + 11 = 0, 3w = -11, w = -\frac{11}{3}$ (width cannot be neg.); $w - 2 = 0, w = 2; 3w + 5 = 3 \cdot 2 + 5 = 6 + 5 = 11$; The rectangle is 2 cm by 11 cm.

Page 244 • MAINTAINING SKILLS

1. *Step 1* Find the length and width of the rectangle.
Step 2 Let w = the width. Then the length is $2w + 2$.

(continued)

Step 3 $w(2w + 2) = 40$

Step 4 $2w^2 + 2w = 40, 2w^2 + 2w - 40 = 0, 2(w^2 + w - 20) = 0, w^2 + w - 20 = 0,$ $(w + 5)(w - 4) = 0; w + 5 = 0, w = -5$ (width cannot be neg.); $w - 4 = 0,$ $w = 4; 2w + 2 = 2 \cdot 4 + 2 = 8 + 2 = 10$

Step 5 The rectangle is 4 cm by 10 cm.

2. *Step 1* Find three consecutive integers.

Step 2 Let n = the first integer. Then $(n + 1)$ and $(n + 2)$ are the next two integers.

Step 3 $(n + 2)^2 = n^2 + (n + 1)^2 - 32$

Step 4 $n^2 + 4n + 4 = n^2 + n^2 + 2n + 1 - 32, n^2 + 4n + 4 = 2n^2 + 2n - 31,$ $4n + 4 = n^2 + 2n - 31, 4 = n^2 - 2n - 31, 0 = n^2 - 2n - 35, 0 =$ $(n - 7)(n + 5); n - 7 = 0, n = 7; n + 1 = 8; n + 2 = 9; n + 5 = 0, n = -5;$ $n + 1 = -4; n + 2 = -3$

Step 5 The integers are 7, 8, 9; or −5, −4, −3.

3. *Step 1* Find the dimensions of the bottom of the box.

Step 2 Let w = width of the bottom. Then $w + 5$ = length of the bottom.

Step 3 $w(w + 5) \cdot 4 = 264$

Step 4 $4w^2 + 20w = 264, 4w^2 + 20w - 264 = 0, 4(w^2 + 5w - 66) = 0, w^2 + 5w - 66 =$ $0, (w + 11)(w - 6) = 0; w + 11 = 0, w = -11$ (width cannot be neg.); $w - 6 = 0,$ $w = 6; w + 5 = 6 + 5 = 11$

Step 5 The bottom of the box is 6 cm by 11 cm.

4. *Step 1* Find the dimensions of the painting.

Step 2 Let w = width of the painting. Then $w + 6$ = length of the painting.

Step 3 $(w + 4)[(w + 6) + 4] = 352$

Step 4 $(w + 4)(w + 10) = 352, w^2 + 14w + 40 = 352, w^2 + 14w - 312 = 0, (w + 26)$ $(w - 12) = 0; w + 26 = 0, w = -26$ (width cannot be neg.); $w - 12 = 0, w = 12;$ $w + 6 = 12 + 6 = 18$

Step 5 The painting is 12 cm by 18 cm.

5. *Step 1* Find the time (or times) when the ball is opposite a roof top.

Step 2 Let t = elapsed time.

Step 3 $14.7 = 19.6t - 4.9t^2$

Step 4 $4.9t^2 - 19.6t + 14.7 = 0, 4.9(t^2 - 4t + 3) = 0, t^2 - 4t + 3 = 0, (t - 3)(t - 1) = 0;$ $t - 3 = 0, t = 3; t - 1 = 0, t = 1$

Step 5 The ball is opposite the roof top after 1 s and 3 s.

Page 245 • MIXED PROBLEM SOLVING REVIEW

A

1. Let l = hours he worked at the library. Then $25 - l$ = hours he mowed lawns. $4l + 8(25 - l) = 164, 4l + 200 - 8l = 164, 200 - 4l = 164, 36 - 4l = 0, 36 = 4l,$ $9 = l$; Paul worked 9 hours at the library.

2. Let x = atomic number of oxygen. $\left(3\frac{1}{2}\right)x = 28, \frac{7}{2}x = 28, x = 28 \cdot \frac{2}{7}, x = 8$; The atomic number of oxygen is 8.

3. Let m = number of km Maureen jogged. $(m - 4) + [(m + 3) - 4] = 15$, $m - 4 + m + 3 - 4 = 15$, $2m - 5 = 15$, $2m = 20$, $m = 10$; $m + 3 = 13$; Maureen jogged 10 km and her sister jogged 13 km.

4. Let h = highest temperature. $h - (-44.4) = 74.6$, $h + 44.4 = 74.6$, $h = 30.2$; The highest temperature is 30.2°C.

5. Let x be one of the numbers. $x^2 + (5 - x)^2 = 53$, $x^2 + 25 - 10x + x^2 = 53$, $2x^2 - 10x + 25 = 53$, $2x^2 - 10x - 28 = 0$, $2(x^2 - 5x - 14) = 0$, $x^2 - 5x - 14 = 0$, $(x - 7)(x + 2) = 0$; $x - 7 = 0$, $x = 7$; $5 - x = 5 - 7 = -2$; $x + 2 = 0$, $x = -2$; $5 - x = 5 - (-2) = 7$; The numbers are -2 and 7.

6. Let w = cost of a can of wall paint. $7(w + 4) = 9w$, $7w + 28 = 9w$, $28 = 2w$, $14 = w$; $w + 4 = 18$; Wall paint costs \$14 and house paint costs \$18 a can.

7. Let l = length of the rectangle. $\left(\frac{2}{3}l - 2\right)(l - 2) = \left(\frac{2}{3}l\right)(l) - 36$, $\frac{2}{3}l^2 - \frac{4}{3}l - 2l + 4 = \frac{2}{3}l^2 - 36$, $\frac{2}{3}l^2 - \frac{10}{3}l + 4 = \frac{2}{3}l^2 - 36$, $-\frac{10}{3}l + 4 = -36$, $-\frac{10}{3}l = -40$, $l = (-40)\left(-\frac{3}{10}\right)$, $l = 12$; $\frac{2}{3}l = \frac{2}{3} \cdot 12 = 8$; The original dimensions of the rectangle were 8 cm by 12 cm.

8. Let q be the number of quarters. $25q + 10(30 - q) = 550$, $25q + 300 - 10q = 550$, $15q + 300 = 550$, $15q = 250$, $q = 16\frac{2}{3}$; (It is impossible to have a fractional number of quarters.) There is no solution; Unrealistic result.

9. $-64 - 37 - 28 + 70 = -59$; The submarine's new depth is 59 m below sea level.

10. Let x = Ross' balance at the beginning of the month. $x + 837.26 - 709.74 - 4 + 3.70 = 528.01$, $x + 127.22 = 528.01$, $x = 400.79$; Ross' balance at the beginning of the month was \$400.79.

B 11. Let w = width of singles court. Then $3w - 3$ = length of singles court. $(w + 9)(3w - 3) = 2808$, $3w^2 - 3w + 27w - 27 = 2808$, $3w^2 + 24w - 27 = 2808$, $3w^2 + 24w - 2835 = 0$, $3(w^2 + 8w - 945) = 0$, $w^2 + 8w - 945 = 0$, $(w + 35)(w - 27) = 0$; $w + 35 = 0$, $w = -35$ (width cannot be negative.); $w - 27 = 0$, $w = 27$; $3w - 3 = 3 \cdot 27 - 3 = 81 - 3 = 78$; $w(3w - 3) = 27 \cdot 78 = 2106$; The area of a singles court is 2106 ft².

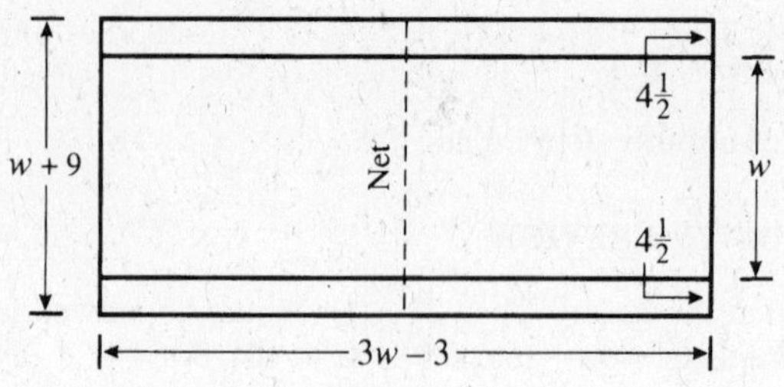

12.

	Rate × Time = Distance		
Mark	r	$1\frac{1}{2}$	$\frac{3}{2}r$
Carole	$r - 2$	1	$r - 2$

$\frac{3}{2}r + (r - 2) = 208, \frac{5}{2}r - 2 = 208, \frac{5}{2}r = 210, r = 210 \cdot \frac{2}{5} = 84; r - 2 = 82;$

Mark was driving 84 km/h and Carole was driving 82 km/h.

13. No solution; not enough information.

14. Let n be the first integer. $\frac{1}{2}[n + (n + 2) + (n + 4)] = (n + 4) - 7,$

$\frac{1}{2}(3n + 6) = n - 3, \frac{3}{2}n + 3 = n - 3, \frac{1}{2}n + 3 = -3, \frac{1}{2}n = -6, n = -12;$

$n + 2 = -10; n + 4 = -8;$ The three integers are -12, -10, and -8.

CHAPTER 6 • Fractions

Pages 248–250 • WRITTEN EXERCISES

A **1.** $\frac{2x + 2y}{4} = \frac{2(x + y)}{4} = \frac{x + y}{2}$ **2.** $\frac{12m - 15n}{9} = \frac{3(4m - 5n)}{9} = \frac{4m - 5n}{3}$

3. $\frac{5a - 10}{a - 2} = \frac{5(a - 2)}{a - 2} = 5, (a \neq 2)$ **4.** $\frac{4n + 16}{n + 4} = \frac{4(n + 4)}{n + 4} = 4, (n \neq -4)$

5. $\frac{3n + 1}{9n + 3} = \frac{3n + 1}{3(3n + 1)} = \frac{1}{3}, \left(n \neq -\frac{1}{3}\right)$ **6.** $\frac{8x - 8y}{8x + 8y} = \frac{8(x - y)}{8(x + y)} = \frac{x - y}{x + y}, (x \neq -y)$

7. $\frac{2x - 4}{x^2 - 4} = \frac{2(x - 2)}{(x + 2)(x - 2)} = \frac{2}{x + 2}, (x \neq -2, 2)$

8. $\frac{6y + 30}{y^2 - 25} = \frac{6(y + 5)}{(y - 5)(y + 5)} = \frac{6}{y - 5}, (y \neq 5, -5)$

9. $\frac{2xy}{x^2y - y^2x} = \frac{2xy}{xy(x - y)} = \frac{2}{x - y}, (x \neq 0, y \neq 0, x \neq y)$

10. $\frac{4p^2 - 8p}{4p^3} = \frac{4p(p - 2)}{4p^3} = \frac{p - 2}{p^2}, (p \neq 0)$ **11.** $\frac{(x + 4)(2x + 1)}{(1 + 2x)(x - 3)} = \frac{x + 4}{x - 3}, \left(x \neq -\frac{1}{2}, 3\right)$

12. $\frac{(x - 5)(2 + 7x)}{(5 - x)(7x + 2)} = \frac{-(5 - x)}{(5 - x)} = -1, \left(x \neq 5, -\frac{2}{7}\right)$

13. $\frac{a^2 + 8a + 16}{16 - a^2} = \frac{(a + 4)^2}{(4 + a)(4 - a)} = \frac{a + 4}{4 - a}, (a \neq 4, -4)$

14. $\frac{25 - b^2}{b^2 + 12b + 35} = \frac{(5 + b)(5 - b)}{(b + 5)(b + 7)} = \frac{5 - b}{b + 7}, (b \neq -5, -7)$

15. $\frac{2 - y}{y^2 - 4y + 4} = \frac{2 - y}{(y - 2)^2} = \frac{-(y - 2)}{(y - 2)^2} = -\frac{1}{y - 2}, (y \neq 2)$

16. $\frac{(a - 5)^2}{25 - a^2} = \frac{(a - 5)^2}{(5 - a)(5 + a)} = \frac{(a - 5)^2}{-(a - 5)(5 + a)} = -\frac{a - 5}{5 + a} = \frac{5 - a}{5 + a}, (a \neq -5, 5)$

17. $\frac{4n^2 + 144}{6n + 36} = \frac{4(n^2 + 36)}{6(n + 6)} = \frac{2(n^2 + 36)}{3(n + 6)}, (n \neq -6)$

18. $\frac{25c + 15d}{50c^2 + 30d^2} = \frac{5(5c + 3d)}{10(5c^2 + 3d^2)} = \frac{5c + 3d}{2(5c^2 + 3d^2)}$, ($c$ and d not both zero)

19. $\frac{2n^2 - 5n - 3}{4n^2 - 8n - 5} = \frac{(2n + 1)(n - 3)}{(2n + 1)(2n - 5)} = \frac{n - 3}{2n - 5}, \left(n \neq -\frac{1}{2}, \frac{5}{2}\right)$

20. $\frac{2w^2 + w - 6}{2w + 4} = \frac{(2w - 3)(w + 2)}{2(w + 2)} = \frac{2w - 3}{2}, (w \neq -2)$

21. $\frac{3x^2 + 6x}{6x^2 + 7x - 10} = \frac{3x(x + 2)}{(6x - 5)(x + 2)} = \frac{3x}{6x - 5}, \left(x \neq \frac{5}{6}, -2\right)$

22. $\frac{4b^2 - 5b - 6}{8b^2 + 6b} = \frac{(4b + 3)(b - 2)}{2b(4b + 3)} = \frac{b - 2}{2b}, \left(b \neq 0, -\frac{3}{4}\right)$

23. $\frac{10 - 3a - a^2}{a^2 - 4} = \frac{(5 + a)(2 - a)}{(a + 2)(a - 2)} = \frac{-(5 + a)(a - 2)}{(a + 2)(a - 2)} = -\frac{5 + a}{a + 2}, (a \neq 2, -2)$

24. $\frac{2y^2 - 9y + 4}{8y - 2y^2} = \frac{(2y - 1)(y - 4)}{2y(4 - y)} = \frac{-(2y - 1)(4 - y)}{2y(4 - y)} = \frac{1 - 2y}{2y}, (y \neq 0, 4)$

25. $\frac{x^2 + xy}{x^2 - xy} = \frac{x(x + y)}{x(x - y)} = \frac{x + y}{x - y}, (x \neq 0, x \neq y)$

26. $\frac{2ab + 2ac + 4a^2}{4b + 4c + 8a} = \frac{2a(b + c + 2a)}{4(b + c + 2a)} = \frac{a}{2}, (b \neq -2a - c)$

27. $\frac{r^2 - s^2}{4r^2 + rs - 5s^2} = \frac{(r + s)(r - s)}{(4r + 5s)(r - s)} = \frac{r + s}{4r + 5s}, \left(r \neq -\frac{5s}{4}, r \neq s\right)$

28. $\frac{3t^2 - 4tv - 7v^2}{t^2 - v^2} = \frac{(3t - 7v)(t + v)}{(t + v)(t - v)} = \frac{3t - 7v}{t - v}, (t \neq v, -v)$

B **29.** $cx + dx = c^2 - d^2, (c + d)x = c^2 - d^2, x = \frac{c^2 - d^2}{c + d}, x = \frac{(c + d)(c - d)}{c + d}, x = c - d,$ $(c \neq -d)$

30. $ax + bx = a^2 + 2ab + b^2, (a + b)x = (a + b)^2, x = \frac{(a + b)^2}{(a + b)}, x = a + b, (a \neq -b)$

31. $abx - b = ax - 1, abx - ax - b = -1, abx - ax = b - 1, (ab - a)x = b - 1,$ $x = \frac{b - 1}{ab - a}, x = \frac{b - 1}{a(b - 1)}, x = \frac{1}{a}, (b \neq 1, a \neq 0)$

32. $3ax + 6 = a^2x + 2a, 3ax - a^2x + 6 = 2a, 3ax - a^2x = 2a - 6,$ $(3a - a^2)x = 2a - 6, x = \frac{2(a - 3)}{a(3 - a)}, x = \frac{-2(3 - a)}{a(3 - a)}, x = -\frac{2}{a}, (a \neq 0, 3)$

33. $5kx - x = 25k^2 - 10k + 1, (5k - 1)x = (5k - 1)^2, x = \frac{(5k - 1)^2}{5k - 1}, x = 5k - 1,$ $\left(k \neq \frac{1}{5}\right)$

34. $4x - 4 = b^2 + 5b - bx, 4x + bx - 4 = b^2 + 5b, 4x + bx = b^2 + 5b + 4,$ $(4 + b)x = (b + 1)(b + 4), x = \frac{(b + 1)(b + 4)}{4 + b}, x = b + 1, (b \neq -4)$

35. $2cx + 3dx = 4c^2 + 12cd + 9d^2, (2c + 3d)x = (2c + 3d)^2, x = \frac{(2c + 3d)^2}{2c + 3d},$ $x = 2c + 3d, \left(c \neq -\frac{3d}{2}\right)$

36. $2x + 5k = 6k^2 - 3kx - 6, 2x + 3kx + 5k = 6k^2 - 6, 2x + 3kx = 6k^2 - 5k - 6,$ $(2 + 3k)x = (3k + 2)(2k - 3), x = \frac{(3k + 2)(2k - 3)}{(2 + 3k)}, x = 2k - 3, \left(k \neq -\frac{2}{3}\right)$

37. $a(x - a) + 6(x + 6) = 0, ax - a^2 + 6x + 36 = 0, ax + 6x = a^2 - 36,$ $(a + 6)x = (a + 6)(a - 6), x = \frac{(a + 6)(a - 6)}{(a + 6)}, x = a - 6, (a \neq -6)$

38. $2n(x - n) = x - 5n + 2, 2nx - 2n^2 = x - 5n + 2, 2nx - x - 2n^2 = -5n + 2,$ $2nx - x = 2n^2 - 5n + 2, (2n - 1)x = (2n - 1)(n - 2), x = \frac{(2n - 1)(n - 2)}{2n - 1},$ $x = n - 2, \left(n \neq \frac{1}{2}\right)$

39. The second answer is incorrect because the values $x = 4$ and $y = 2$ make the denominator in the original fraction zero.

40. Let $x = 2$ and $y = 3$; then $\frac{x^2 - 4y^2}{x^2 - 2xy} = \frac{2^2 - 4 \cdot 3^2}{2^2 - 2(2)(3)} = 4$. However, Donna's solution gives $\frac{2y}{x} = \frac{2(3)}{2} = 3$; $3 \neq 4$. $\frac{x^2 - 4y^2}{x^2 - 2xy} = \frac{(x + 2y)(x - 2y)}{x(x - 2y)} = \frac{x + 2y}{x}$, $(x \neq 0, 2y)$

41. $\frac{25x^2 - 36y^2}{10x^2 + 3xy - 18y^2} = \frac{(5x + 6y)(5x - 6y)}{(2x + 3y)(5x - 6y)} = \frac{5x + 6y}{2x + 3y}, \left(x \neq \frac{-3y}{2}, \frac{6y}{5}\right)$

42. $\frac{12x^2 - 5xy - 2y^2}{4y^2 - 4xy - 3x^2} = \frac{(3x - 2y)(4x + y)}{(2y + x)(2y - 3x)} = \frac{-(2y - 3x)(4x + y)}{(2y + x)(2y - 3x)} = -\frac{4x + y}{2y + x}$, $\left(x \neq -2y, \frac{2y}{3}\right)$

43. $\frac{6a^3 + 10a^2}{36a^3 - 100a} = \frac{2a^2(3a + 5)}{4a(9a^2 - 25)} = \frac{2a^2(3a + 5)}{4a(3a + 5)(3a - 5)} = \frac{a}{2(3a - 5)}, \left(a \neq \frac{5}{3}, -\frac{5}{3}, 0\right)$

44. $\frac{8a^2 + 6ab - 5b^2}{16a^2 - 25b^2} = \frac{(4a + 5b)(2a - b)}{(4a + 5b)(4a - 5b)} = \frac{2a - b}{4a - 5b}, \left(a \neq \frac{5b}{4}, -\frac{5b}{4}\right)$

45. $\frac{x(a + b) - 2(a + b)}{2 - x} = \frac{(x - 2)(a + b)}{2 - x} = \frac{-(2 - x)(a + b)}{2 - x} = -(a + b) = -a - b, (x \neq 2)$

46. $\frac{w^4 - 1}{w^4(w + 1) + (w + 1)} = \frac{(w^2 + 1)(w^2 - 1)}{(w^4 + 1)(w + 1)} = \frac{(w^2 + 1)(w - 1)(w + 1)}{(w^4 + 1)(w + 1)} =$ $\frac{(w^2 + 1)(w - 1)}{w^4 + 1}, (w \neq -1)$

C **47.** $\frac{2x^3 - 13x^2 + 15x}{15x - 7x^2 - 2x^3} = \frac{x(2x^2 - 13x + 15)}{x(15 - 7x - 2x^2)} = \frac{(2x - 3)(x - 5)}{(3 - 2x)(5 + x)} = \frac{-(3 - 2x)(x - 5)}{(3 - 2x)(5 + x)} =$ $\frac{-(x - 5)}{5 + x} = \frac{5 - x}{5 + x}, \left(x \neq \frac{3}{2}, -5, 0\right)$

48. $\frac{x^4 - 10x^2 + 9}{3 - 2x - x^2} = \frac{(x^2 - 9)(x^2 - 1)}{(3 + x)(1 - x)} = \frac{(x + 3)(x - 3)(x + 1)(x - 1)}{(3 + x)(1 - x)} =$ $\frac{-(x + 3)(x - 3)(x + 1)(1 - x)}{(3 + x)(1 - x)} = -(x - 3)(x + 1) = (3 - x)(x + 1), (x \neq -3, 1)$

49. $\frac{4a^2 - b^2 - 2a + b}{(2a + b)^2 - 1} = \frac{(2a + b)(2a - b) - (2a - b)}{[(2a + b) + 1][(2a + b) - 1]} = \frac{(2a - b)[(2a + b) - 1]}{(2a + b + 1)(2a + b - 1)} =$ $\frac{2a - b}{2a + b + 1}, (b \neq -(2a + 1), 1 - 2a)$

50. $\frac{x^2 + 2x + 1 - y^2}{(x + 1)^2 + 2y(x + 1) + y^2} = \frac{(x + 1)^2 - y^2}{[(x + 1) + y]^2} = \frac{[(x + 1) + y][(x + 1) - y]}{(x + 1 + y)^2} =$ $\frac{x + 1 - y}{x + 1 + y}, (x \neq -(1 + y))$

51. $\frac{x^2 - 4}{x^2 - 4x + 4} = \frac{(x + 2)(x - 2)}{(x - 2)^2}$, $(x \neq 2)$; the fraction is zero when $x = -2$.

52. $\frac{3x^2 + x}{2x - x^3} = \frac{x(3x + 1)}{x(2 - x^2)}$, $(x \neq 0, x^2 \neq 2)$; the fraction is zero when $x = -\frac{1}{3}$.

53. $\frac{x^2 - 2x - 15}{x^2 + 3x - 40} = \frac{(x - 5)(x + 3)}{(x - 5)(x + 8)}$, $(x \neq 5, -8)$; the fraction is zero when $x = -3$.

54. $\frac{x^4 - x^2}{x^3 + x^2 - 2x} = \frac{x^2(x^2 - 1)}{x(x^2 + x - 2)} = \frac{x^2(x + 1)(x - 1)}{x(x + 2)(x - 1)}$, $(x \neq 0, -2, 1)$; the fraction is zero when $x = -1$.

Page 250 • MIXED REVIEW EXERCISES

1. $12\left(\frac{1}{3}u + \frac{1}{4}v\right) = 4u + 3v$ **2.** $(-42n + 28p)\left(-\frac{1}{4}\right) = \frac{21n}{2} - 7p$ **3.** $\frac{15a^7b^8}{30a^2b^3} = \frac{a^5b^5}{2}$

4. $\frac{(-3y)^5}{(y^2)^5} = \frac{(-3)^5y^5}{y^{10}} = -\frac{243}{y^5}$ **5.** $\frac{3x^5 + 6x^3 - 12x^2}{x^2} = 3x^3 + 6x - 12$

6. $(-20)(-7)(-4)(-5) = 2800$ **7.** $-y + 15 = 9, 15 = 9 + y, 6 = y; \{6\}$

8. $68 = -\frac{n}{5}, -340 = n; \{-340\}$ **9.** $5p + 8 = -47, 5p = -55, p = -11; \{-11\}$

10. $5(x + 2) + 2 = 27, 5x + 10 + 2 = 27, 5x + 12 = 27, 5x = 15, x = 3; \{3\}$

11. $9y - (7y + 5) = 11, 9y - 7y - 5 = 11, 2y - 5 = 11, 2y = 16, y = 8; \{8\}$

12. $(3n - 6) - (5 - 3n) = 7, 3n - 6 - 5 + 3n = 7, 6n - 11 = 7, 6n = 18, n = 3; \{3\}$

Page 250 • COMPUTER EXERCISES

1. a.
```
10 FOR X = 10 TO 50 STEP 10
20 LET Y = X / (X ↑ 2 + 1)
30 PRINT Y
40 NEXT X
50 END
```

1) `.09900991; .0498753117; .0332963374; .0249843848; .0199920032`

2) Change 20 to `LET Y = (X ↑ 2) / (X ↑ 2 + 1)`
`.99009901; .997506234; .998890122; .99937539; .99960016`

3) Change 20 to `LET Y = (X ↑ 2) / (X + 1)`
`9.0909091; 19.0476191; 29.0322581; 39.0243903; 49.0196079`

b. In (1), the values decrease toward 0.
In (2), the values increase toward 1.
In (3), the values increase without bound.

2. a.
```
10 FOR J = 1 TO 5
20 READ X
30 LET Y = (X ^ 5 + 1) / (X ^ 4 - X ^ 3 + X ^ 2 - X + 1)
40 PRINT Y
45 NEXT J
50 DATA 2, 13, 22, 50, 99
60 END
3; 14; 23; 51; 100
```

b. $\frac{x^5 + 1}{x^4 - x^3 + x^2 - x + 1} = x + 1$

Pages 253–254 • WRITTEN EXERCISES

A **1.** $\frac{\overset{1}{\cancel{4}}}{\underset{1}{\cancel{7}}} \cdot \frac{\overset{3}{\cancel{21}}}{\underset{2}{\cancel{8}}} = \frac{3}{2}$ **2.** $\frac{\overset{1}{\cancel{4}}}{\underset{3}{\cancel{9}}} \cdot \frac{\overset{1}{\cancel{3}}}{\underset{4}{\cancel{16}}} = \frac{1}{12}$ **3.** $\frac{\overset{5}{\cancel{15}}}{\underset{1}{\cancel{4}}} \cdot \frac{\overset{2}{\cancel{8}}}{\underset{3}{\cancel{9}}} = \frac{10}{3}$ **4.** $\frac{\overset{1}{\cancel{7}}}{\underset{1}{-\cancel{2}}} \cdot \frac{\overset{5}{\cancel{10}}}{\underset{4}{\cancel{28}}} = -\frac{5}{4}$ **5.** $\frac{\overset{1}{\cancel{3}}}{\cancel{5}} \cdot \frac{\cancel{5}}{\cancel{7}} \cdot \frac{\cancel{7}}{\underset{3}{\cancel{9}}} = \frac{1}{3}$

6. $\frac{9}{5} \cdot \frac{2}{3} \cdot \frac{15}{18} = \frac{18}{15} \cdot \frac{15}{18} = 1$ **7.** $\left(-\frac{5}{2}\right)^2 \cdot \frac{8}{5} = \frac{\overset{5}{\cancel{25}}}{\underset{1}{\cancel{4}}} \cdot \frac{\overset{2}{\cancel{8}}}{\underset{1}{\cancel{5}}} = 10$

8. $(-2)^3 \cdot \frac{35}{16} = -\overset{1}{\cancel{8}} \cdot \frac{35}{\underset{2}{\cancel{16}}} = -\frac{35}{2}$ **9.** $\frac{6}{x^2} \cdot \frac{x^3}{3} = 2x$ **10.** $\frac{5y}{2} \cdot \frac{4}{15y} = \frac{2}{3}$

11. $\frac{a}{b} \cdot \frac{b}{c} \cdot \frac{c}{d} = \frac{a}{d}$ **12.** $\frac{4}{x^2} \cdot \frac{7x}{8} = \frac{7}{2x}$ **13.** $\frac{4w}{v} \cdot \frac{v^3}{2w^2} = \frac{2v^2}{w}$ **14.** $\frac{6a}{11b^4} \cdot \frac{22b}{3a^3} = \frac{4}{a^2b^3}$

15. $\frac{4d^2e}{9ef} \cdot \frac{f^2}{6d} = \frac{2df}{27}$ **16.** $\frac{2rs^2}{3t} \cdot \frac{9t^2}{4rs} = \frac{3st}{2}$ **17.** $\left(\frac{a}{6}\right)^2 = \frac{a^2}{6^2} = \frac{a^2}{36}$

18. $\left(\frac{c}{5}\right)^3 = \frac{c^3}{5^3} = \frac{c^3}{125}$ **19.** $\left(\frac{2n}{7}\right)^2 = \frac{(2n)^2}{7^2} = \frac{4n^2}{49}$ **20.** $\left(\frac{4x}{3}\right)^2 = \frac{(4x)^2}{3^2} = \frac{16x^2}{9}$

21. $\left(\frac{2a}{5b^3}\right)^2 = \frac{(2a)^2}{(5b^3)^2} = \frac{4a^2}{25b^6}$ **22.** $\left(\frac{4m}{7n^2}\right)^2 = \frac{(4m)^2}{(7n^2)^2} = \frac{16m^2}{49n^4}$

23. $\left(\frac{-x^2}{10}\right)^4 = \frac{(-x^2)^4}{10^4} = \frac{x^8}{10{,}000}$ **24.** $-\left(\frac{5b^4}{6}\right)^2 = -\frac{(5b^4)^2}{6^2} = -\frac{25b^8}{36}$

25. $\left(\frac{a}{b}\right)^2 \cdot \frac{b}{a} = \frac{a^2}{b^2} \cdot \frac{b}{a} = \frac{a}{b}$ **26.** $\left(\frac{3x}{y}\right)^3 \cdot \frac{y^2}{9} = \frac{(3x)^3}{y^3} \cdot \frac{y^2}{9} = \frac{27x^3}{y^3} \cdot \frac{y^2}{9} = \frac{3x^3}{y}$

27. $\left(-\frac{x}{4y}\right)^2 \cdot \left(-\frac{4y}{x}\right) = \frac{x^2}{16y^2} \cdot \left(-\frac{4y}{x}\right) = -\frac{x}{4y}$

28. $\left(\frac{3z}{y}\right)^3 \cdot \frac{2yz}{15} = \frac{(3z)^3}{y^3} \cdot \frac{2yz}{15} = \frac{27z^3}{y^3} \cdot \frac{2yz}{15} = \frac{18z^4}{5y^2}$

29. Area $= \left(\frac{2x}{7}\right)^2 = \frac{4x^2}{49}$; The area is $\frac{4x^2}{49}$ in.^{2}.

30. Volume $= \left(\frac{4n}{5}\right)^3 = \frac{64n^3}{125}$; The volume is $\frac{64n^3}{125}$ in.^{3}.

31. Area $= \frac{1}{2}\left(\frac{3x}{4}\right)\left(\frac{8}{9x}\right) = \frac{1}{3}$; The area is $\frac{1}{3}$ cm^2.

32. Distance $= \left(\frac{7t}{60}\right)\left(\frac{80r}{9}\right) = \frac{28rt}{27}$; You have gone $\frac{28rt}{27}$ mi.

B **33.** Total $= \left(\frac{4y}{3}\right)\left(\frac{d}{12}\right) = \frac{yd}{9}$; The total cost is $\frac{yd}{9}$ dollars.

34. Total $= 12n \cdot \left(\frac{2c}{3} \cdot \frac{1}{100}\right) = \frac{2cn}{25}$; The total cost is $\frac{2cn}{25}$ dollars.

35. $\frac{c+2}{c^2} \cdot \frac{3c}{c^2-4} = \frac{(c+2)(3c)}{c^2(c+2)(c-2)} = \frac{3}{c(c-2)}$

36. $\frac{x^2-1}{16x} \cdot \frac{4x^2}{5x+5} = \frac{(x+1)(x-1)(4x^2)}{16x \cdot 5(x+1)} = \frac{(x-1)x}{20}$

37. $\frac{a^2-x^2}{a^2} \cdot \frac{a}{3x-3a} = \frac{(a+x)(a-x)a}{a^2 \cdot 3(x-a)} = \frac{-(a+x)(x-a)}{3a(x-a)} = -\frac{a+x}{3a}$

38. $\frac{3r-rt}{6r^2t} \cdot \frac{3}{9-t^2} = \frac{r(3-t)}{2r^2t(3+t)(3-t)} = \frac{1}{2rt(3+t)}$

39. $(4b^2 - 3b) \cdot \dfrac{5b^2}{40b^3 - 30b^2} = b(4b - 3) \cdot \dfrac{5b^2}{10b^2(4b - 3)} = \dfrac{b}{2}$

40. $\dfrac{2y}{3y^2 + 15y} \cdot (3y^3 - 75y) = \dfrac{2y}{3y(y + 5)} \cdot 3y(y^2 - 25) = \dfrac{2y(y + 5)(y - 5)}{(y + 5)} = 2y(y - 5)$

41. $\dfrac{10x^4}{6x - 12} \cdot \dfrac{x^2 - x - 2}{4x} = \dfrac{5x^3(x - 2)(x + 1)}{6(x - 2) \cdot 2} = \dfrac{5x^3(x + 1)}{12}$

42. $\dfrac{2x^2 + 5x - 3}{x + 2} \cdot \dfrac{9x + 18}{1 - 2x} = \dfrac{(2x - 1)(x + 3) \cdot 9(x + 2)}{(x + 2) \cdot -(2x - 1)} = -9(x + 3)$

43. $\dfrac{n^2 - 3n + 2}{n^2 + 3n + 2} \cdot \dfrac{8n + 8}{4n + 8} = \dfrac{(n - 1)(n - 2) \cdot 8(n + 1)}{(n + 1)(n + 2) \cdot 4(n + 2)} = \dfrac{2(n - 1)(n - 2)}{(n + 2)^2}$

44. $\dfrac{x^2 + 4x - 21}{x^2 - 6x - 16} \cdot \dfrac{x^2 - 8x + 15}{x^2 + 9x + 14} = \dfrac{(x + 7)(x - 3)(x - 3)(x - 5)}{(x - 8)(x + 2)(x + 7)(x + 2)} = \dfrac{(x - 3)^2(x - 5)}{(x - 8)(x + 2)^2}$

45. $\dfrac{t^2 - 2t - 8}{4 - t^2} \cdot \dfrac{t^2 - 5t + 6}{t^2 - t - 12} = \dfrac{(t - 4)(t + 2)(t - 2)(t - 3)}{(2 + t)(2 - t)(t - 4)(t + 3)} = \dfrac{-(2 - t)(t - 3)}{(2 - t)(t + 3)} = \dfrac{3 - t}{t + 3}$

46. $\dfrac{a^2 - 7a + 6}{2a^3 - 3a^2} \cdot \dfrac{4a^3 - 9a}{a^2 - 10a + 24} = \dfrac{(a - 1)(a - 6) \cdot a(4a^2 - 9)}{a^2(2a - 3)(a - 4)(a - 6)} =$

$\dfrac{(a - 1)(2a + 3)(2a - 3)}{a(2a - 3)(a - 4)} = \dfrac{(a - 1)(2a + 3)}{a(a - 4)}$

47. $\dfrac{3x^2 - 14x - 5}{8x^2 - 12x} \cdot \dfrac{8x^2 - 18}{3x^2 + 4x + 1} = \dfrac{(3x + 1)(x - 5) \cdot 2(4x^2 - 9)}{4x(2x - 3)(3x + 1)(x + 1)} =$

$\dfrac{(x - 5)(2x + 3)(2x - 3)}{2x(2x - 3)(x + 1)} = \dfrac{(x - 5)(2x + 3)}{2x(x + 1)}$

48. $\dfrac{25 - 16y^2}{6y^3 - 36y^2} \cdot \dfrac{10y + 8y^2}{10 - 3y - 4y^2} = \dfrac{(5 + 4y)(5 - 4y) \cdot 2y(5 + 4y)}{6y^2(y - 6)(5 - 4y)(2 + y)} = \dfrac{(5 + 4y)^2}{3y(y - 6)(2 + y)}$

49. $\dfrac{3d^2 - 9d + 6}{2d^2 - 10d + 12} \cdot \dfrac{6 - 2d}{3 - 3d} = \dfrac{3(d^2 - 3d + 2) \cdot 2(3 - d)}{2(d^2 - 5d + 6) \cdot 3(1 - d)} = \dfrac{(d - 1)(d - 2)(3 - d)}{(d - 2)(d - 3)(1 - d)} =$

$\dfrac{-(d - 1)(d - 2)(d - 3)}{-(d - 2)(d - 3)(d - 1)} = 1$

50. $\dfrac{4c^2 - 8c - 5}{12c^2 + 46c + 40} \cdot \dfrac{20 + 8c}{5 - 2c} = \dfrac{(2c - 5)(2c + 1) \cdot 4(5 + 2c)}{2(6c^2 + 23c + 20) \cdot -(2c - 5)} = \dfrac{2(2c + 1)(5 + 2c)}{-(2c + 5)(3c + 4)} =$

$-\dfrac{2(2c + 1)}{3c + 4}$

51. $\left(\dfrac{x - 2}{2}\right)^2 \cdot \left(\dfrac{2}{x - 2}\right)^3 = \dfrac{(x - 2)^2}{4} \cdot \dfrac{8}{(x - 2)^3} = \dfrac{2}{x - 2}$

52. $\left(\dfrac{2n - 1}{3}\right)^4 \cdot \left(\dfrac{9}{2n - 1}\right)^2 = \dfrac{(2n - 1)^4}{81} \cdot \dfrac{81}{(2n - 1)^2} = (2n - 1)^2$

53. $\left(\dfrac{x - y}{x + y}\right)^2 \cdot \dfrac{x^2 + y^2}{x^2 - y^2} = \dfrac{(x - y)^2}{(x + y)^2} \cdot \dfrac{x^2 + y^2}{(x + y)(x - y)} = \dfrac{(x - y)(x^2 + y^2)}{(x + y)^3}$

54. $\dfrac{4a^2 - b^2}{4c^2 - d^2} \cdot \left(\dfrac{2c - d}{2a + b}\right)^2 = \dfrac{(2a + b)(2a - b)(2c - d)^2}{(2c + d)(2c - d)(2a + b)^2} = \dfrac{(2a - b)(2c - d)}{(2c + d)(2a + b)}$

C **55.** $\dfrac{9x^2 - 1}{6x^2} \cdot \dfrac{12x^2 + 12x}{(1 - 3x)^2} \cdot \dfrac{6 - 18x}{3x^2 + 4x + 1} = \dfrac{(3x + 1)(3x - 1) \cdot 12x(x + 1) \cdot 6(1 - 3x)}{6x^2(1 - 3x)^2(3x + 1)(x + 1)} =$

$\dfrac{-12(1 - 3x)}{x(1 - 3x)} = -\dfrac{12}{x}$

56. $\frac{4n^2-4}{1+n^2}\cdot\frac{1-n}{2n}\cdot\frac{1-2n^2+n^4}{2+2n}=\frac{4(n^2-1)(1-n)(1-n^2)^2}{(1+n^2)\cdot 2n\cdot 2(1+n)}=$

$\frac{(n+1)(n-1)(1-n)[(1-n)(1+n)]^2}{n(1+n^2)(1+n)}=\frac{-(1-n)(1-n)(1-n)^2(1+n)^2}{n(1+n^2)}=$

$-\frac{(1-n)^4(1+n)^2}{n(1+n^2)}$

57. $\frac{x^3+3x^2-4x-12}{2x^2-18}\cdot\frac{x^3-3x^2+3x-9}{3x^3-12x}=\frac{x^2(x+3)-4(x+3)}{2(x^2-9)}\cdot\frac{x^2(x-3)+3(x-3)}{3x(x^2-4)}=$

$\frac{(x^2-4)(x+3)(x^2+3)(x-3)}{2(x^2-9)\cdot 3x(x^2-4)}=\frac{(x+3)(x^2+3)(x-3)}{6x(x+3)(x-3)}=\frac{x^2+3}{6x}$

58. $\frac{a^2-(b-c)^2}{2a-2b+2c}\cdot\frac{6a-6b+6c}{b^2-(a-c)^2}=\frac{[a+(b-c)][a-(b-c)]\cdot 6(a-b+c)}{2(a-b+c)[b+(a-c)][b-(a-c)]}=$

$\frac{3(a+b-c)(a-b+c)}{(b+a-c)(b-a+c)}=\frac{3(a-b+c)}{b-a+c}$

59. $\frac{2x^2-2x}{3x^3}\cdot\frac{9}{x^2+3x+2}=\frac{2x(x-1)}{3x^3}\cdot\frac{9}{(x+1)(x+2)}=\frac{18x(x-1)}{3x^3(x+1)(x+2)},$

$\frac{6(x-1)}{x^2(x+2)(x+1)}$, $(x\neq 0,-1,-2)$; The fraction is zero when $x=1$.

Page 254 • MIXED REVIEW EXERCISES

1. $a^2+14a+45=(a+9)(a+5)$ **2.** $x^2-7x+10=(x-2)(x-5)$

3. $16x^4-81=(4x^2+9)(4x^2-9)=(4x^2+9)(2x+3)(2x-3)$

4. $2x^2+5x+1$ is prime **5.** $625y^2-4z^2=(25y+2z)(25y-2z)$

6. $64+16c+c^2=(8+c)^2$

7. $xy+3y-4xz-12z=y(x+3)-4z(x+3)=(y-4z)(x+3)$

8. $9x^2-12x+4=(3x-2)^2$ **9.** $3x^2+14x-5=(3x-1)(x+5)$

10. $x^4+7x^2-8x^3=x^2(x^2-8x+7)=x^2(x-7)(x-1)$

11. $n^2+5n-14=(n+7)(n-2)$ **12.** $y^2-5y-24=(y-8)(y+3)$

Page 254 • CHALLENGE

The "proof" requires a step of dividing both sides of the equation by $(r-s)$. But $r=s$ and thus we would be dividing by zero.

Pages 256–257 • WRITTEN EXERCISES

A **1.** $\frac{6}{5}\div\frac{9}{10}=\frac{\overset{2}{\cancel{6}}}{\underset{1}{\cancel{5}}}\cdot\frac{\overset{2}{\cancel{10}}}{\underset{3}{\cancel{9}}}=\frac{4}{3}$ **2.** $\frac{2}{3}\div\frac{5}{9}=\frac{2}{\underset{1}{\cancel{3}}}\cdot\frac{\overset{3}{\cancel{9}}}{5}=\frac{6}{5}$ **3.** $\frac{a}{6}\div\frac{a}{3}=\frac{a}{6}\cdot\frac{3}{a}=\frac{1}{2}$

4. $\frac{3x}{5}\div\frac{x}{15}=\frac{3x}{5}\cdot\frac{15}{x}=9$ **5.** $\frac{x}{y^2}\div\frac{x^2}{y}=\frac{x}{y^2}\cdot\frac{y}{x^2}=\frac{1}{xy}$ **6.** $\frac{3n^2}{5}\div\frac{9n}{20}=\frac{3n^2}{5}\cdot\frac{20}{9n}=\frac{4n}{3}$

7. $\frac{ab}{6}\div\frac{b}{a}=\frac{ab}{6}\cdot\frac{a}{b}=\frac{a^2}{6}$ **8.** $\frac{c}{2d}\div\frac{d}{8c}=\frac{c}{2d}\cdot\frac{8c}{d}=\frac{4c^2}{d^2}$ **9.** $\frac{3x^2}{4y}\div\frac{xy}{18}=\frac{3x^2}{4y}\cdot\frac{18}{xy}=\frac{27x}{2y^2}$

10. $\frac{2n}{3m^2}\div\frac{1}{12mn}=\frac{2n}{3m^2}\cdot\frac{12mn}{1}=\frac{8n^2}{m}$ **11.** $\frac{xy^2}{3}\div xy=\frac{xy^2}{3}\cdot\frac{1}{xy}=\frac{y}{3}$

12. $\frac{9a^2}{2b} \div 6ab = \frac{9a^2}{2b} \cdot \frac{1}{6ab} = \frac{3a}{4b^2}$ **13.** $1 \div \left(\frac{3x}{5}\right)^2 = 1 \div \frac{9x^2}{25} = 1 \cdot \frac{25}{9x^2} = \frac{25}{9x^2}$

14. $4 \div \left(\frac{2}{n}\right)^3 = 4 \div \frac{8}{n^3} = 4 \cdot \frac{n^3}{8} = \frac{n^3}{2}$ **15.** $\frac{2 + 2b}{6} \div \frac{1 + b}{9} = \frac{2(1 + b)}{6} \cdot \frac{9}{1 + b} = 3$

16. $\frac{6n - 2}{6n} \div \frac{3n - 1}{27} = \frac{2(3n - 1)}{6n} \cdot \frac{27}{(3n - 1)} = \frac{9}{n}$

17. $\frac{x^2 - 1}{2} \div \frac{x + 1}{16} = \frac{(x + 1)(x - 1)}{2} \cdot \frac{16}{x + 1} = 8(x - 1)$

18. $\frac{m^2 - 9}{2m} \div \frac{m - 3}{10} = \frac{(m + 3)(m - 3)}{2m} \cdot \frac{10}{m - 3} = \frac{5(m + 3)}{m}$

19. $\frac{1}{x - y} \div \frac{1}{y - x} = \frac{1}{x - y} \cdot y - x = \frac{1}{x - y} \cdot -(x - y) = -1$

20. $\frac{1}{8 - 2a} \div \frac{5}{3a - 12} = \frac{1}{2(4 - a)} \cdot \frac{3(a - 4)}{5} = \frac{-3(4 - a)}{10(4 - a)} = -\frac{3}{10}$

21. $\frac{2x + 2y}{x^2} \div \frac{x^2 - y^2}{4x} = \frac{2(x + y)}{x^2} \cdot \frac{4x}{(x + y)(x - y)} = \frac{8}{x(x - y)}$

22. $\frac{3}{n^2 - 9} \div \frac{3n - 9}{n + 3} = \frac{3}{(n + 3)(n - 3)} \cdot \frac{n + 3}{3(n - 3)} = \frac{1}{(n - 3)^2}$

23. $\frac{w^2 - 4}{w^2 - 1} \div \frac{w^2 - w - 2}{w^2 + w - 2} = \frac{(w + 2)(w - 2)}{(w + 1)(w - 1)} \cdot \frac{(w + 2)(w - 1)}{(w - 2)(w + 1)} = \frac{(w + 2)^2}{(w + 1)^2}$

24. $\frac{x^2 - x - 20}{5x - 25} \div \frac{x^2 + 4x - 5}{x^2 - 25} = \frac{(x - 5)(x + 4)}{5(x - 5)} \cdot \frac{(x + 5)(x - 5)}{(x + 5)(x - 1)} = \frac{(x + 4)(x - 5)}{5(x - 1)}$

B **25.** $\frac{4x^2 - 25}{x^2 - 16} \div \frac{12x + 30}{2x^2 + 8x} = \frac{(2x + 5)(2x - 5)}{(x + 4)(x - 4)} \cdot \frac{2x(x + 4)}{6(2x + 5)} = \frac{x(2x - 5)}{3(x - 4)}$

26. $\frac{2x - y}{2y - x} \div \frac{4x^2 - y^2}{4y^2 - x^2} = \frac{2x - y}{2y - x} \cdot \frac{(2y + x)(2y - x)}{(2x + y)(2x - y)} = \frac{2y + x}{2x + y}$

27. $\frac{c^4 - d^4}{5c - 5d} \div \frac{c^2 + d^2}{5} = \frac{(c^2 + d^2)(c^2 - d^2)}{5(c - d)} \cdot \frac{5}{c^2 + d^2} = \frac{(c + d)(c - d)}{c - d} = c + d$

28. $\frac{8 - 2p^4}{3p^4} \div \frac{2 + p^2}{6p^2} = \frac{2(4 - p^4)}{3p^4} \cdot \frac{6p^2}{2 + p^2} = \frac{4(2 + p^2)(2 - p^2)}{p^2(2 + p^2)} = \frac{4(2 - p^2)}{p^2}$

29. $\frac{r^2 - t^2}{r^2 + t^2} \div (r + t) = \frac{(r + t)(r - t)}{r^2 + t^2} \cdot \frac{1}{r + t} = \frac{r - t}{r^2 + t^2}$

30. $\frac{3 - 3n}{n^2 + 2n - 3} \div (2n - 2) = \frac{3(1 - n)}{(n + 3)(n - 1)} \cdot \frac{1}{2(n - 1)} = \frac{-3(n - 1)}{2(n + 3)(n - 1)^2} =$

$-\frac{3}{2(n + 3)(n - 1)}$

31. $\frac{6 + x - x^2}{x^2 - 13x + 42} \div \frac{2x^2 - 5x - 3}{2x^2 - 13x - 7} = \frac{(2 + x)(3 - x)}{(x - 7)(x - 6)} \cdot \frac{(2x + 1)(x - 7)}{(2x + 1)(x - 3)} =$

$\frac{-(2 + x)(x - 3)}{(x - 6)(x - 3)} = -\frac{2 + x}{x - 6}$

32. $\frac{2x^2 - 11x + 12}{27 - 18x} \div \frac{x^3 + 4x^2}{6x^3 - 96x} = \frac{(2x - 3)(x - 4)}{9(3 - 2x)} \cdot \frac{6x(x^2 - 16)}{x^2(x + 4)} =$

$\frac{-(3 - 2x)(x - 4) \cdot 2(x + 4)(x - 4)}{3x(3 - 2x)(x + 4)} = -\frac{2(x - 4)^2}{3x}$

33. **a.** $\frac{3}{8} \cdot \left(\frac{2}{3} \div \frac{1}{4}\right) = \frac{3}{8} \cdot \left(\frac{2}{3} \cdot 4\right) = \frac{3}{8} \cdot \frac{8}{3} = 1$ **b.** $\frac{\not{3}}{\underset{4}{\not{8}}} \cdot \frac{\overset{1}{\not{2}}}{\not{3}} \div \frac{1}{4} = \frac{1}{4} \div \frac{1}{4} = \frac{1}{4} \cdot 4 = 1$

34. a. $\frac{1}{2} \div \frac{1}{5} \cdot \frac{3}{4} = \frac{1}{2} \cdot 5 \cdot \frac{3}{4} = \frac{15}{8}$ **b.** $\frac{1}{2} \div \left(\frac{1}{5} \cdot \frac{3}{4}\right) = \frac{1}{2} \div \left(\frac{3}{20}\right) = \frac{1}{\cancel{2}_{1}} \cdot \frac{\cancel{20}^{10}}{3} = \frac{10}{3}$

35. a. $\frac{r}{5} \div \frac{t}{r} \cdot \frac{5}{t^2} = \frac{r}{5} \cdot \frac{r}{t} \cdot \frac{5}{t^2} = \frac{r^2}{t^3}$ **b.** $\frac{x}{y^2} \cdot \frac{2}{x^2} \div \frac{x}{y} = \frac{2}{xy^2} \cdot \frac{y}{x} = \frac{2}{x^2y}$

36. a. $\left(\frac{n}{2}\right)^2 \div \frac{n}{4} \cdot \frac{n}{3} = \frac{n^2}{4} \cdot \frac{4}{n} \cdot \frac{n}{3} = \frac{n^2}{3}$ **b.** $(2c)^3 \cdot \frac{3c}{d} \div \frac{6c^3}{d} = 8c^3 \cdot \frac{3c}{d} \cdot \frac{d}{6c^3} = 4c$

37. $\frac{c-d}{c+2d} \cdot \frac{2d+c}{d+c} \div \frac{d-c}{d+c} = \frac{c-d}{d+c} \cdot \frac{d+c}{d-c} = \frac{-(d-c)}{d-c} = -1$

38. $\frac{r^2}{r^2-s^2} \cdot \frac{r-s}{r+s} \div \left(\frac{r}{r+s}\right)^2 = \frac{r^2}{(r+s)(r-s)} \cdot \frac{r-s}{r+s} \div \frac{r^2}{(r+s)^2} = \frac{r^2}{(r+s)^2} \cdot \frac{(r+s)^2}{r^2} = 1$

39. $\frac{2q-6}{3} \div \left(\frac{3-q}{6}\right)^2 = \frac{2(q-3)}{3} \div \frac{(3-q)^2}{36} = \frac{2(q-3)}{3} \cdot \frac{36}{(3-q)^2} = \frac{-24(3-q)}{(3-q)^2} = -\frac{24}{3-q}$

40. $\left(\frac{2p-7}{9}\right)^3 \div \left(\frac{7-2p}{3}\right)^4 = \frac{(2p-7)^3}{729} \div \frac{(7-2p)^4}{81} = \frac{(2p-7)^3}{729} \cdot \frac{81}{(2p-7)^4} = \frac{1}{9(2p-7)}$

41. $\frac{1}{ab} \div \frac{1}{ab^2} \div \frac{1}{a^2b} = \frac{1}{ab} \cdot ab^2 \cdot a^2b = a^2b^2$

42. $\frac{6y}{6y-14} \div \frac{21}{9y-21} \div \frac{y^2}{35} = \frac{6y}{2(3y-7)} \cdot \frac{3(3y-7)}{21} \cdot \frac{35}{y^2} = \frac{15}{y}$

C **43.** $\frac{x^2-2x}{x^2-3x-4} \cdot \frac{x^2-25}{x^2-4x-5} \div \frac{x^2+5x}{5x^2+10x+5} =$

$\frac{x(x-2)}{(x-4)(x+1)} \cdot \frac{(x+5)(x-5)}{(x-5)(x+1)} \cdot \frac{5(x^2+2x+1)}{x(x+5)} = \frac{5(x-2)(x+1)^2}{(x-4)(x+1)^2} = \frac{5(x-2)}{x-4}$

44. $\frac{b^2+6b-7}{6b^2-7b-20} \cdot \frac{2b^2+b-15}{b^2+2b-3} \div \frac{b^2+5b-14}{3b^2-2b-8} =$

$\frac{(b+7)(b-1)}{(3b+4)(2b-5)} \cdot \frac{(2b-5)(b+3)}{(b+3)(b-1)} \cdot \frac{(3b+4)(b-2)}{(b+7)(b-2)} = 1$

45. $\frac{2d+2c-cd-c^2}{2+d} \div \frac{d^2-c^2}{2+c} \cdot \frac{c-d}{c^2-4} = \frac{2(d+c)-c(d+c)}{2+d} \div \frac{(d+c)(d-c)}{2+c} \cdot \frac{c-d}{(c+2)(c-2)} =$

$\frac{(2-c)(d+c)}{2+d} \cdot \frac{2+c}{(d+c)(d-c)} \cdot \frac{-(d-c)}{(c+2)(c-2)} = \frac{-(2-c)}{(2+d)(c-2)} = \frac{c-2}{(2+d)(c-2)} = \frac{1}{2+d}$

46. $\frac{x^2+2xy+y^2-16}{16x^4-16y^4} \div \frac{x+y-4}{4x^2+4y^2} \cdot \frac{x+y}{x+y+4} = \frac{(x+y)^2-16}{16(x^4-y^4)} \cdot \frac{4(x^2+y^2)}{x+y-4} \cdot \frac{x+y}{x+y+4} =$

$\frac{(x+y+4)(x+y-4)}{16(x^2+y^2)(x^2-y^2)} \cdot \frac{4(x^2+y^2)}{x+y-4} \cdot \frac{x+y}{x+y+4} = \frac{x+y}{4(x+y)(x-y)} = \frac{1}{4(x-y)}$

Page 258 • MIXED REVIEW EXERCISES

1. $4k = 5k - 13, 0 = k - 13, 13 = k; \{13\}$ **2.** $4p + 20 = 48, 4p = 28, p = 7; \{7\}$

3. $(5b-2) - (3-2b) = 9, 5b - 2 - 3 + 2b = 9, 7b - 5 = 9, 7b = 14, b = 2; \{2\}$

4. $\frac{1}{2}(6x-2) = 5, 3x - 1 = 5, 3x = 6, x = 2; \{2\}$

5. $2n^3 - 8n = 0, 2n(n^2-4) = 0, 2n(n+2)(n-2) = 0; 2n = 0, n = 0; n + 2 = 0, n = -2;$
$n - 2 = 0, n = 2; \{0, -2, 2\}$

6. $3x^2 + x = 4, 3x^2 + x - 4 = 0, (3x+4)(x-1) = 0; 3x + 4 = 0, 3x = -4, x = -\frac{4}{3};$

$x - 1 = 0, x = 1; \left\{-\frac{4}{3}, 1\right\}$

7. $256 = 2^8$ **8.** $156 = 2^2 \cdot 3 \cdot 13$ **9.** $120 = 2^3 \cdot 3 \cdot 5$ **10.** $1350 = 2 \cdot 3^3 \cdot 5^2$

Page 258 • CALCULATOR KEY-IN

1. 0.9 **2.** 4 **3.** 68.333333 **4.** -4.5 **5.** -5.75 **6.** -4.2727273

Pages 261–262 • WRITTEN EXERCISES

A **1.** $\frac{3}{4} = \frac{3 \cdot 7}{4 \cdot 7} = \frac{21}{28}$ **2.** $\frac{4}{9} = \frac{4 \cdot 3}{9 \cdot 3} = \frac{12}{27}$ **3.** $\frac{5x}{3} = \frac{5x \cdot 6}{3 \cdot 6} = \frac{30x}{18}$ **4.** $\frac{2a}{17} = \frac{2a \cdot 3}{17 \cdot 3} = \frac{6a}{51}$

5. $\frac{x-3}{4} = \frac{(x-3) \cdot 3}{4 \cdot 3} = \frac{3(x-3)}{12}$ **6.** $\frac{2n-5}{5} = \frac{(2n-5) \cdot 5}{5 \cdot 5} = \frac{5(2n-5)}{25}$

7. $\frac{6}{15x} = \frac{6 \cdot 2x}{15x \cdot 2x} = \frac{12x}{30x^2}$ **8.** $\frac{3}{4a} = \frac{3 \cdot 4a^2}{4a \cdot 4a^2} = \frac{12a^2}{16a^3}$ **9.** $\frac{x}{2y} = \frac{x \cdot 5x}{2y \cdot 5x} = \frac{5x^2}{10xy}$

10. $\frac{5m}{2n} = \frac{5m \cdot 4mn}{2n \cdot 4mn} = \frac{20m^2n}{8mn^2}$ **11.** $\frac{7}{(n-3)} = \frac{7(n+2)}{(n-3)(n+2)}$ **12.** $\frac{4}{(x+1)} = \frac{4(x-1)}{(x+1)(x-1)}$

13. $\frac{5}{2n-3} = \frac{5(2n-3)}{(2n-3)^2}$ **14.** $\frac{2y}{3y-2} = \frac{2y(3y-2)}{(3y-2)^2}$ **15.** $\frac{4}{a-1} = \frac{4 \cdot 2}{(a-1)2} = \frac{8}{2a-2}$

16. $\frac{2}{3x-1} = \frac{2 \cdot 3}{(3x-1)3} = \frac{6}{9x-3}$ **17.** $\frac{2}{x+1} = \frac{2(x-1)}{(x+1)(x-1)} = \frac{2(x-1)}{x^2-1}$

18. $\frac{5}{n-3} = \frac{5(n+3)}{(n-3)(n+3)} = \frac{5(n+3)}{n^2-9}$ **19.** $\frac{3}{4-y} = \frac{3y}{(4-y)y} = \frac{3y}{4y-y^2}$

20. $\frac{2x}{3-x} = \frac{2x \cdot x}{(3-x)x} = \frac{2x^2}{3x-x^2}$ **21.** $6 = 2 \cdot 3$, $9 = 3^2$; LCD is $2 \cdot 3^2 = 18$

22. $8 = 2^3$, $5 = 5$; LCD is $2^3 \cdot 5 = 40$ **23.** $2 = 2$, $6 = 2 \cdot 3$, $5 = 5$; LCD is $2 \cdot 3 \cdot 5 = 30$

24. $3 = 3$, $9 = 3^2$, $4 = 2^2$; LCD is $2^2 \cdot 3^2 = 36$

25. $8 = 2^3$, $12 = 2^2 \cdot 3$; LCD is $2^3 \cdot 3 = 24$

26. $20 = 2^2 \cdot 5$, $15 = 3 \cdot 5$; LCD is $2^2 \cdot 3 \cdot 5 = 60$

27. $4t = 2^2t$, $16rt^2 = 2^4rt^2$; LCD is $2^4rt^2 = 16rt^2$ **28.** LCD is x^2y

29. LCD is $(x+1)(x-2)$ **30.** LCD is $(b+5)(b-5)$

31. $m + 2 = m + 2$, $m^2 - 4 = (m+2)(m-2)$; LCD is $(m+2)(m-2)$

32. $a - 1 = a - 1$, $a^2 - 3a + 2 = (a-1)(a-2)$; LCD is $(a-1)(a-2)$

B **33.** $\frac{x}{2x^2y}, \frac{6y}{2x^2y}$ **34.** $\frac{m}{3m^2n^2}, \frac{6n}{3m^2n^2}$ **35.** $\frac{55y}{30x^2y^3}, \frac{24x}{30x^2y^3}$ **36.** $\frac{3b}{36a^2b^2}, \frac{10a^2}{36a^2b^2}$

37. $\frac{20}{4(x-3)}, \frac{7}{4(x-3)}$ **38.** $\frac{20}{15(x-2y)}, \frac{3}{15(x-2y)}$ **39.** $\frac{6(x-3)}{(x-3)^2}, \frac{4x}{(x-3)^2}$

40. $\frac{9}{(2n+1)^2}, -\frac{2n(2n+1)}{(2n+1)^2}$ **41.** $\frac{3y(y+2)}{2(y-2)(y+2)}, \frac{2}{2(y-2)(y+2)}$

42. $\frac{3(c+6)}{c(c-6)(c+6)}, \frac{5(c-6)}{c(c-6)(c+6)}$ **43.** $\frac{x(x+3)}{(x-3)(x+2)(x+3)}, \frac{9(x+2)}{(x-3)(x+2)(x+3)}$

44. $\frac{2(a+5)}{(a+5)^2(a-2)}, \frac{4a(a-2)}{(a+5)^2(a-2)}$

C **45.** **a.** $4! = 1 \cdot 2 \cdot 3 \cdot 4 = 24$; $5! = 1 \cdot 2 \cdot 3 \cdot 4 \cdot 5 = 120$; $6! = 1 \cdot 2 \cdot 3 \cdot 4 \cdot 5 \cdot 6 = 720$

b. $4! = 24 = 2^3 \cdot 3$, $5! = 120 = 2^3 \cdot 3 \cdot 5$, $6! = 720 = 2^4 \cdot 3^2 \cdot 5$; LCD is $2^4 \cdot 3^2 \cdot 5 = 720$

c. $n! = (1 \cdot 2 \cdot \ldots \cdot n)$, $(n+1)! = [1 \cdot 2 \cdot \ldots \cdot n \cdot (n+1)] = (n!)(n+1)$; LCD is $(n+1)!$

Page 262 • MIXED REVIEW EXERCISES

1. $4n - 8q + 16 = 4(n - 2q + 4)$ **2.** $3x^2 - 3 = 3(x^2 - 1) = 3(x + 1)(x - 1)$

3. $x^2 - 11x + 18 = (x - 9)(x - 2)$ **4.** $x^2 - 5x - 24 = (x - 8)(x + 3)$

5. $2x^2 - x - 3 = (2x - 3)(x + 1)$ **6.** $x^2 + 16x + 39 = (x + 13)(x + 3)$

7. $x^2 + 3x - 28 = (x + 7)(x - 4)$ **8.** $x^2 + 22x + 121 = (x + 11)^2$

9. $n^2 - 9n = n(n - 9)$ **10.** $7 = 2p - 4$ **11.** $n - \frac{1}{2} = 5\frac{1}{4}$ **12.** $\frac{2}{3}k = 16$

Page 263 • COMPUTER KEY-IN

1. LCD (7, 21) = 21 × 1 = 21 **2.** LCD (21, 7) = 7 × 3 = 21

3. LCD (24, 36) = 36 × 2 = 72 **4.** LCD (36, 24) = 24 × 3 = 72

5. LCD (13, 15) = 15 × 13 = 195 **6.** LCD (15, 13) = 13 × 15 = 195

7–12. Answers given here show the *additional* results when Line 55 is added.

7. $\frac{21}{7} = 3$ **8.** $\frac{7}{21} = 0.3333333, \frac{14}{21} = 0.6666666, \frac{21}{21} = 1$ **9.** $\frac{36}{24} = 1.5, \frac{72}{24} = 3$

10. $\frac{24}{36} = 0.6666666, \frac{48}{36} = 1.3333333, \frac{72}{36} = 2$

11. $\frac{15}{13} = 1.1538461, \frac{30}{13} = 2.3076923, \frac{45}{13} = 3.4615384, \frac{60}{13} = 4.6153846, \frac{75}{13} = 5.7692307,$

$\frac{90}{13} = 6.9230769, \frac{105}{13} = 8.076923, \frac{120}{13} = 9.2307692, \frac{135}{13} = 10.384615, \frac{150}{13} = 11.538461,$

$\frac{165}{13} = 12.692307, \frac{180}{13} = 13.846153, \frac{195}{13} = 15$

12. $\frac{13}{15} = 0.8666666, \frac{26}{15} = 1.7333333, \frac{39}{15} = 2.6, \frac{52}{15} = 3.4666666, \frac{65}{15} = 4.3333333, \frac{78}{15} = 5.2,$

$\frac{91}{15} = 6.0666666, \frac{104}{15} = 6.9333333, \frac{117}{15} = 7.8, \frac{130}{15} = 8.6666666, \frac{143}{15} = 9.5333333, \frac{156}{15} = 10.4,$

$\frac{169}{15} = 11.266666, \frac{182}{15} = 12.133333, \frac{195}{15} = 13$

13. It requires fewer steps to calculate the LCD if you enter the smaller denominator first.

14. Run the program using 27 and 30. The result is 270. Then run the program using 36 and 270. The result is 540 which is the LCD for 27, 36, and 30.

Page 263 • CHALLENGE

First, find how long the fly flies. Let t = time before horses meet. $5.5t + 4.5t = 10$, $10t = 10$, $t = 1$. The horses meet after 1 hour. The fly flies at 30 km/h for 1 hour. Therefore it travels 30 km.

Pages 267–268 • WRITTEN EXERCISES

A **1.** $\frac{x}{12} + \frac{3x}{12} = \frac{4x}{12} = \frac{x}{3}$ **2.** $\frac{7x}{15} - \frac{4x}{15} = \frac{3x}{15} = \frac{x}{5}$ **3.** $\frac{2}{n} - \frac{6}{n} = \frac{-4}{n}$, or $-\frac{4}{n}$

4. $\frac{5}{3x} + \frac{9}{3x} = \frac{14}{3x}$ **5.** $\frac{2x}{3} - \frac{x - 1}{3} = \frac{2x - (x - 1)}{3} = \frac{2x - x + 1}{3} = \frac{x + 1}{3}$

6. $\frac{3a}{10} - \frac{5a+1}{10} = \frac{3a-(5a+1)}{10} = \frac{3a-5a-1}{10} = \frac{-2a-1}{10}$, or $-\frac{2a+1}{10}$

7. $\frac{y-3}{4} - \frac{3y-5}{4} = \frac{y-3-(3y-5)}{4} = \frac{y-3-3y+5}{4} = \frac{2-2y}{4} = \frac{2(1-y)}{4} = \frac{1-y}{2}$

8. $\frac{x+3}{4} - \frac{9x+11}{4} = \frac{x+3-(9x+11)}{4} = \frac{x+3-9x-11}{4} = \frac{-8x-8}{4} = \frac{-8(x+1)}{4} =$
$-2(x+1) = -2x-2$

9. $\frac{2}{x+3} + \frac{5}{x+3} = \frac{2+5}{x+3} = \frac{7}{x+3}$

10. $\frac{9}{y-6} - \frac{3}{y-6} = \frac{6}{y-6}$ **11.** $\frac{x}{x-2} + \frac{1}{x-2} = \frac{x+1}{x-2}$

12. $\frac{n}{n-2} - \frac{3n-1}{n-2} = \frac{n-(3n-1)}{n-2} = \frac{n-3n+1}{n-2} = \frac{-2n+1}{n-2}$

13. $\frac{3}{x-5} + \frac{2}{5-x} = \frac{3}{x-5} + \frac{2}{-(x-5)} = \frac{3}{x-5} - \frac{2}{x-5} = \frac{1}{x-5}$

14. $\frac{5}{4y-1} - \frac{3}{1-4y} = \frac{5}{4y-1} - \frac{3}{-(4y-1)} = \frac{5}{4y-1} + \frac{3}{4y-1} = \frac{8}{4y-1}$

15. $\frac{5c}{c-d} + \frac{5d}{d-c} = \frac{5c}{c-d} - \frac{5d}{c-d} = \frac{5c-5d}{c-d} = \frac{5(c-d)}{c-d} = 5$

16. $\frac{6}{3-2x} - \frac{4x}{2x-3} = \frac{6}{3-2x} + \frac{4x}{3-2x} = \frac{6+4x}{3-2x}$, or $\frac{2(3+2x)}{3-2x}$

17. $\frac{2}{x} + \frac{3}{x^2} = \frac{2x}{x^2} + \frac{3}{x^2} = \frac{2x+3}{x^2}$ **18.** $\frac{8}{ab} - \frac{6}{ab^2} = \frac{8b}{ab^2} - \frac{6}{ab^2} = \frac{8b-6}{ab^2}$, or $\frac{2(4b-3)}{ab^2}$

19. $\frac{1}{2x^2} - \frac{5}{4x} = \frac{2}{4x^2} - \frac{5x}{4x^2} = \frac{2-5x}{4x^2}$ **20.** $\frac{4}{3xy} + \frac{3}{2x^2} = \frac{4 \cdot 2x}{6x^2y} + \frac{3 \cdot 3y}{6x^2y} = \frac{8x+9y}{6x^2y}$

21. $\frac{4}{5(x+1)} + \frac{x}{x+1} = \frac{4}{5(x+1)} + \frac{5x}{5(x+1)} = \frac{4+5x}{5(x+1)}$

22. $\frac{2x}{x-3} - \frac{3}{2(x-3)} = \frac{4x}{2(x-3)} - \frac{3}{2(x-3)} = \frac{4x-3}{2(x-3)}$

23. $\frac{1+2x}{4} + \frac{1+2x}{6} = \frac{3(1+2x)}{12} + \frac{2(1+2x)}{12} = \frac{3+6x+2+4x}{12} = \frac{10x+5}{12}$, or $\frac{5(2x+1)}{12}$

24. $\frac{5b+2}{8} + \frac{3b-4}{10} = \frac{5(5b+2)}{40} + \frac{4(3b-4)}{40} = \frac{25b+10+12b-16}{40} = \frac{37b-6}{40}$

25. $\frac{3y-2}{6} - \frac{y-3}{9} = \frac{3(3y-2)}{18} - \frac{2(y-3)}{18} = \frac{9y-6-2y+6}{18} = \frac{7y}{18}$

26. $\frac{a-6}{15} + \frac{4-a}{10} = \frac{2(a-6)}{30} + \frac{3(4-a)}{30} = \frac{2a-12+12-3a}{30} = \frac{-a}{30}$, or $-\frac{a}{30}$

27. $\frac{5n-2}{12} - \frac{3(n-3)}{8} = \frac{2(5n-2)}{24} - \frac{9(n-3)}{24} = \frac{10n-4-9n+27}{24} = \frac{n+23}{24}$

28. $\frac{3(a-b)}{20} - \frac{5(a+b)}{12} = \frac{9(a-b)}{60} - \frac{25(a+b)}{60} = \frac{9a-9b-25a-25b}{60} = \frac{-16a-34b}{60} =$
$\frac{-2(8a+17b)}{60} = -\frac{8a+17b}{30}$

29. $\frac{4x+3}{3} - \frac{7x}{4} + \frac{x-3}{6} = \frac{4(4x+3)}{12} - \frac{3 \cdot 7x}{12} + \frac{2(x-3)}{12} = \frac{16x+12-21x+2x-6}{12} =$
$\frac{-3x+6}{12} = \frac{3(-x+2)}{12} = \frac{-x+2}{4}$, or $-\frac{x-2}{4}$

30. $\frac{2m}{5} - \frac{m+4}{4} + \frac{2-m}{10} = \frac{4 \cdot 2m}{20} - \frac{5(m+4)}{20} + \frac{2(2-m)}{20} = \frac{8m - 5m - 20 + 4 - 2m}{20} = \frac{m-16}{20}$

B

31. $P = 2\left(\frac{4a}{5}\right) + 2\left(\frac{2a}{3}\right) = \frac{8a}{5} + \frac{4a}{3} = \frac{24a}{15} + \frac{20a}{15} = \frac{44a}{15}$

32. $P = 2\left(\frac{3a+6}{4}\right) + 2\left(\frac{2a-1}{6}\right) = \frac{3a+6}{2} + \frac{2a-1}{3} = \frac{3(3a+6)}{6} + \frac{2(2a-1)}{6} = \frac{9a + 18 + 4a - 2}{6} = \frac{13a+16}{6}$

33. $P = b + \frac{a}{2} + \frac{a}{3} + b + \left(\frac{a}{2} + \frac{a}{3}\right) = \frac{6b}{6} + \frac{3a}{6} + \frac{2a}{6} + \frac{6b}{6} + \frac{3a}{6} + \frac{2a}{6} = \frac{6b + 3a + 2a + 6b + 3a + 2a}{6} = \frac{10a + 12b}{6} = \frac{2(5a+6b)}{6} = \frac{5a+6b}{3}$

34. $P = 4\left(\frac{3x}{4}\right) + 4\left(\frac{x}{6}\right) + 4\left(\frac{x}{4}\right) = 3x + \frac{2x}{3} + x = \frac{9x + 2x + 3x}{3} = \frac{14x}{3}$

35. $\frac{1}{x-1} + \frac{1}{x} = \frac{x}{x(x-1)} + \frac{x-1}{x(x-1)} = \frac{x + x - 1}{x(x-1)} = \frac{2x-1}{x(x-1)}$

36. $\frac{3}{y-6} - \frac{1}{y} = \frac{3y}{y(y-6)} - \frac{y-6}{y(y-6)} = \frac{3y - y + 6}{y(y-6)} = \frac{2y+6}{y(y-6)}$, or $\frac{2(y+3)}{y(y-6)}$

37. $\frac{2}{x-3} + \frac{4}{x+3} = \frac{2(x+3)}{(x-3)(x+3)} + \frac{4(x-3)}{(x-3)(x+3)} = \frac{2x + 6 + 4x - 12}{(x-3)(x+3)} = \frac{6x-6}{(x-3)(x+3)}$, or $\frac{6(x-1)}{(x-3)(x+3)}$

38. $\frac{3}{x+4} - \frac{4}{x-2} = \frac{3(x-2)}{(x+4)(x-2)} - \frac{4(x+4)}{(x+4)(x-2)} = \frac{3x - 6 - 4x - 16}{(x+4)(x-2)} = \frac{-x-22}{(x+4)(x-2)}$, or $-\frac{x+22}{(x+4)(x-2)}$

39. $\frac{a+1}{a} - \frac{a}{a+1} = \frac{(a+1)^2}{a(a+1)} - \frac{a^2}{a(a+1)} = \frac{a^2 + 2a + 1 - a^2}{a(a+1)} = \frac{2a+1}{a(a+1)}$

40. $\frac{x}{x+y} + \frac{y}{x-y} = \frac{x(x-y)}{(x+y)(x-y)} + \frac{y(x+y)}{(x+y)(x-y)} = \frac{x^2 - xy + xy + y^2}{(x+y)(x-y)} = \frac{x^2+y^2}{(x+y)(x-y)}$

41. $\frac{x}{x^2-1} + \frac{4}{x+1} = \frac{x}{(x+1)(x-1)} + \frac{4(x-1)}{(x+1)(x-1)} = \frac{x + 4x - 4}{(x+1)(x-1)} = \frac{5x-4}{(x+1)(x-1)}$

42. $\frac{2y}{y^2-25} - \frac{y}{y-5} = \frac{2y}{(y+5)(y-5)} - \frac{y(y+5)}{(y+5)(y-5)} = \frac{2y - y^2 - 5y}{(y+5)(y-5)} = \frac{-y^2 - 3y}{(y+5)(y-5)} = -\frac{3y + y^2}{(y+5)(y-5)}$

43. $\frac{2m}{2m-1} + \frac{1}{1-2m} = \frac{2m}{2m-1} - \frac{1}{2m-1} = \frac{2m-1}{2m-1} = 1$

44. $\frac{3a}{a-2b} + \frac{6b}{2b-a} = \frac{3a}{a-2b} - \frac{6b}{a-2b} = \frac{3a-6b}{a-2b} = \frac{3(a-2b)}{a-2b} = 3$

45. $\frac{d+2}{d^2-1} - \frac{3}{2d+2} = \frac{d+2}{(d+1)(d-1)} - \frac{3}{2(d+1)} = \frac{2(d+2)}{2(d+1)(d-1)} - \frac{3(d-1)}{2(d+1)(d-1)} = \frac{2d + 4 - 3d + 3}{2(d+1)(d-1)} = \frac{7-d}{2(d+1)(d-1)}$

46. $\frac{2n}{n^3 - 5n^2} + \frac{2}{n^2 + 5n} = \frac{2n}{n^2(n-5)} + \frac{2}{n(n+5)} = \frac{2}{n(n-5)} + \frac{2}{n(n+5)} =$
$\frac{2(n+5)}{n(n-5)(n+5)} + \frac{2(n-5)}{n(n-5)(n+5)} = \frac{2n + 10 + 2n - 10}{n(n-5)(n+5)} = \frac{4n}{n(n-5)(n+5)} =$
$\frac{4}{(n-5)(n+5)}$

47. $\frac{a}{ab - b^2} + \frac{b}{ab - a^2} = \frac{a}{b(a-b)} + \frac{b}{a(b-a)} = \frac{a}{b(a-b)} - \frac{b}{a(a-b)} = \frac{a^2}{ab(a-b)} -$
$\frac{b^2}{ab(a-b)} = \frac{a^2 - b^2}{ab(a-b)} = \frac{(a+b)(a-b)}{ab(a-b)} = \frac{a+b}{ab}$

48. $\frac{x}{x - x^2} - \frac{1}{x - x^3} = \frac{x}{x(1-x)} - \frac{1}{x(1-x)(1+x)} = \frac{x(1+x)}{x(1-x)(1+x)} - \frac{1}{x(1-x)(1+x)} =$
$\frac{x^2 + x - 1}{x(1-x)(1+x)}$

49. $\frac{n}{n^2 + 4n} - \frac{n}{(n+4)^2} = \frac{n}{n(n+4)} - \frac{n}{(n+4)^2} = \frac{n(n+4)}{n(n+4)^2} - \frac{n^2}{n(n+4)^2} = \frac{n^2 + 4n - n^2}{n(n+4)^2} =$
$\frac{4n}{n(n+4)^2} = \frac{4}{(n+4)^2}$

50. $\frac{1}{a^2 + 4a + 4} + \frac{1}{a^2 - 4} = \frac{1}{(a+2)^2} + \frac{1}{(a+2)(a-2)} = \frac{a-2}{(a+2)^2(a-2)} + \frac{a+2}{(a+2)^2(a-2)} =$
$\frac{a - 2 + a + 2}{(a+2)^2(a-2)} = \frac{2a}{(a+2)^2(a-2)}$

51. $\frac{x - 11}{x^2 - 9} - \frac{x - 7}{x^2 - 3x} = \frac{x - 11}{(x+3)(x-3)} - \frac{x-7}{(x-3)x} = \frac{x(x-11)}{x(x+3)(x-3)} - \frac{(x-7)(x+3)}{x(x+3)(x-3)} =$
$\frac{x^2 - 11x - (x^2 - 4x - 21)}{x(x+3)(x-3)} = \frac{x^2 - 11x - x^2 + 4x + 21}{x(x+3)(x-3)} = \frac{21 - 7x}{x(x+3)(x-3)} =$
$\frac{-7(x-3)}{x(x+3)(x-3)} = \frac{-7}{x(x+3)}$, or $-\frac{7}{x(x+3)}$

52. $\frac{c - 2}{c^2 + c - 2} - \frac{c + 2}{c^2 - 3c + 2} = \frac{c-2}{(c+2)(c-1)} - \frac{c+2}{(c-2)(c-1)} = \frac{(c-2)^2}{(c+2)(c-2)(c-1)} -$
$\frac{(c+2)^2}{(c+2)(c-2)(c-1)} = \frac{c^2 - 4c + 4 - (c^2 + 4c + 4)}{(c+2)(c-2)(c-1)} = \frac{c^2 - 4c + 4 - c^2 - 4c - 4}{(c+2)(c-2)(c-1)} =$
$\frac{-8c}{(c+2)(c-2)(c-1)}$, or $-\frac{8c}{(c+2)(c-2)(c-1)}$

C **53.** $\frac{x^2 + 1}{x^2 - 1} + \frac{1}{x+1} + \frac{1}{x-1} = \frac{x^2 + 1}{(x+1)(x-1)} + \frac{x-1}{(x+1)(x-1)} + \frac{x+1}{(x+1)(x-1)} =$
$\frac{x^2 + 1 + x - 1 + x + 1}{(x+1)(x-1)} = \frac{x^2 + 2x + 1}{(x+1)(x-1)} = \frac{(x+1)^2}{(x+1)(x-1)} = \frac{x+1}{x-1}$

54. $\frac{x}{2x - 1} + \frac{x - 1}{2x + 1} - \frac{2x}{4x^2 - 1} = \frac{x(2x+1)}{(2x-1)(2x+1)} + \frac{(x-1)(2x-1)}{(2x-1)(2x+1)} -$
$\frac{2x}{(2x-1)(2x+1)} = \frac{2x^2 + x + 2x^2 - 3x + 1 - 2x}{(2x-1)(2x+1)} = \frac{4x^2 - 4x + 1}{(2x-1)(2x+1)} =$
$\frac{(2x-1)^2}{(2x-1)(2x+1)} = \frac{2x-1}{2x+1}$

55. $\dfrac{a+2}{a^2+5a+6} - \dfrac{2+a}{4-a^2} + \dfrac{2-a}{a^2+a-6} = \dfrac{a+2}{(a+2)(a+3)} - \dfrac{2+a}{(2+a)(2-a)} +$
$\dfrac{2-a}{(a+3)(a-2)} = \dfrac{(a+2)(a-2)}{(a+2)(a+3)(a-2)} - \dfrac{(2+a)\cdot -(a+3)}{(a+2)(a+3)(a-2)} + \dfrac{(2-a)(a+2)}{(a+2)(a+3)(a-2)} =$
$\dfrac{a^2-4+a^2+5a+6+4-a^2}{(a+2)(a+3)(a-2)} = \dfrac{a^2+5a+6}{(a+2)(a+3)(a-2)} = \dfrac{(a+2)(a+3)}{(a+2)(a+3)(a-2)} = \dfrac{1}{a-2}$

56. $\dfrac{x-3}{2x+6} - \dfrac{x+3}{3x-9} - \dfrac{5x^2+27}{6x^2-54} = \dfrac{x-3}{2(x+3)} - \dfrac{x+3}{3(x-3)} - \dfrac{5x^2+27}{6(x^2-9)} = \dfrac{(x-3)\cdot 3(x-3)}{6(x+3)(x-3)} -$
$\dfrac{(x+3)\cdot 2(x+3)}{6(x+3)(x-3)} - \dfrac{5x^2+27}{6(x+3)(x-3)} = \dfrac{3(x^2-6x+9)-2(x^2+6x+9)-(5x^2+27)}{6(x+3)(x-3)} =$
$\dfrac{3x^2-18x+27-2x^2-12x-18-5x^2-27}{6(x+3)(x-3)} = \dfrac{-4x^2-30x-18}{6(x+3)(x-3)} = \dfrac{-2(2x^2+15x+9)}{6(x+3)(x-3)} =$
$-\dfrac{2x^2+15x+9}{3(x+3)(x-3)}$

57. $\dfrac{b+1}{(b-1)^2} + \dfrac{2-2b}{(b-1)^3} + \dfrac{1}{b-1} = \dfrac{(b+1)(b-1)}{(b-1)^3} + \dfrac{2-2b}{(b-1)^3} + \dfrac{(b-1)^2}{(b-1)^3} =$
$\dfrac{b^2-1+2-2b+b^2-2b+1}{(b-1)^3} = \dfrac{2b^2-4b+2}{(b-1)^3} = \dfrac{2(b^2-2b+1)}{(b-1)^3} = \dfrac{2(b-1)^2}{(b-1)^3} = \dfrac{2}{b-1}$

58. $\dfrac{4}{c^2-4cd} - \dfrac{1}{cd-4d^2} - \dfrac{2}{cd} = \dfrac{4}{c(c-4d)} - \dfrac{1}{d(c-4d)} - \dfrac{2}{cd} = \dfrac{4d}{cd(c-4d)} - \dfrac{c}{cd(c-4d)} -$
$\dfrac{2(c-4d)}{cd(c-4d)} = \dfrac{4d-c-2c+8d}{cd(c-4d)} = \dfrac{12d-3c}{cd(c-4d)} = \dfrac{-3(c-4d)}{cd(c-4d)} = -\dfrac{3}{cd}$

Page 268 • MIXED REVIEW EXERCISES

1. $-8^2\cdot 3 = -64\cdot 3 = -192$ **2.** $(4\cdot 6-13)^2 = (24-13)^2 = 11^2 = 121$

3. $(-4x^2)^3 = -64x^6$ **4.** $5y(y-4)+2y(y+11) = 5y^2-20y+2y^2+22y = 7y^2+2y$

5. $n^2(n+8)-(2n^2-6)n = n^3+8n^2-2n^3+6n = -n^3+8n^2+6n$

6. $(5x^2y)(3x^3y^2)(6y^3) = 5\cdot 3\cdot 6\cdot x^2\cdot x^3\cdot y\cdot y^2\cdot y^3 = 90x^5y^6$

7. $\left(\dfrac{12x^2y^2}{5}\right)\left(\dfrac{-10xy^2}{3}\right) = \dfrac{-120x^3y^4}{15} = -8x^3y^4$ **8.** $-\dfrac{1}{6}(-42x+12y) = 7x-2y$

9. $(3n-5p+2)-(-n+6p+1) = 3n-5p+2+n-6p-1 = 4n-11p+1$

Pages 271–272 • WRITTEN EXERCISES

A

1. $4\dfrac{1}{5} = \dfrac{4}{1}+\dfrac{1}{5} = \dfrac{20}{5}+\dfrac{1}{5} = \dfrac{21}{5}$ **2.** $2\dfrac{3}{8} = \dfrac{2}{1}+\dfrac{3}{8} = \dfrac{16}{8}+\dfrac{3}{8} = \dfrac{19}{8}$

3. $8+\dfrac{1}{x} = \dfrac{8x}{x}+\dfrac{1}{x} = \dfrac{8x+1}{x}$ **4.** $3+\dfrac{7}{a} = \dfrac{3a}{a}+\dfrac{7}{a} = \dfrac{3a+7}{a}$

5. $3a-\dfrac{2}{a} = \dfrac{3a^2}{a}-\dfrac{2}{a} = \dfrac{3a^2-2}{a}$ **6.** $5x-\dfrac{3}{x} = \dfrac{5x^2}{x}-\dfrac{3}{x} = \dfrac{5x^2-3}{x}$

7. $\dfrac{a}{b}-3 = \dfrac{a}{b}-\dfrac{3b}{b} = \dfrac{a-3b}{b}$ **8.** $2-\dfrac{c}{d} = \dfrac{2d}{d}-\dfrac{c}{d} = \dfrac{2d-c}{d}$

9. $5-\dfrac{4}{x+2} = \dfrac{5(x+2)}{x+2}-\dfrac{4}{x+2} = \dfrac{5x+10-4}{x+2} = \dfrac{5x+6}{x+2}$

10. $8+\dfrac{y}{y-1} = \dfrac{8(y-1)}{y-1}+\dfrac{y}{y-1} = \dfrac{8y-8+y}{y-1} = \dfrac{9y-8}{y-1}$

11. $\frac{x}{x+2} + 6 = \frac{x}{x+2} + \frac{6(x+2)}{x+2} = \frac{x+6x+12}{x+2} = \frac{7x+12}{x+2}$

12. $\frac{a+3}{a} - 3 = \frac{a+3}{a} - \frac{3a}{a} = \frac{a+3-3a}{a} = \frac{-2a+3}{a}$, or $-\frac{2a-3}{a}$

13. $4 + \frac{2n+1}{n+1} = \frac{4(n+1)}{n+1} + \frac{2n+1}{n+1} = \frac{4n+4+2n+1}{n+1} = \frac{6n+5}{n+1}$

14. $2 - \frac{k-3}{k-4} = \frac{2(k-4)}{k-4} - \frac{k-3}{k-4} = \frac{2k-8-k+3}{k-4} = \frac{k-5}{k-4}$

15. $6x - \frac{x}{x+1} = \frac{6x(x+1)}{x+1} - \frac{x}{x+1} = \frac{6x^2+6x-x}{x+1} = \frac{6x^2+5x}{x+1}$, or $\frac{x(6x+5)}{x+1}$

16. $4y + \frac{y}{y-5} = \frac{4y(y-5)}{y-5} + \frac{y}{y-5} = \frac{4y^2-20y+y}{y-5} = \frac{4y^2-19y}{y-5}$, or $\frac{y(4y-19)}{y-5}$

17. $8n - \frac{2}{n+1} = \frac{8n(n+1)}{n+1} - \frac{2}{n+1} = \frac{8n^2+8n-2}{n+1}$, or $\frac{2(4n^2+4n-1)}{n+1}$

18. $5a + \frac{a-5}{2a+3} = \frac{5a(2a+3)}{2a+3} + \frac{a-5}{2a+3} = \frac{10a^2+15a+a-5}{2a+3} = \frac{10a^2+16a-5}{2a+3}$

19. $2a^2 - \frac{a-1}{2a+5} = \frac{2a^2(2a+5)}{2a+5} - \frac{a-1}{2a+5} = \frac{4a^3+10a^2-(a-1)}{2a+5} = \frac{4a^3+10a^2-a+1}{2a+5}$

20. $y^2 - \frac{3y+1}{y+2} = \frac{y^2(y+2)}{y+2} - \frac{3y+1}{y+2} = \frac{y^3+2y^2-(3y+1)}{y+2} = \frac{y^3+2y^2-3y-1}{y+2}$

B **21.** $x - \frac{8}{x+1} - \frac{6x-2}{x+1} = \frac{x(x+1)}{x+1} - \frac{8}{x+1} - \frac{6x-2}{x+1} = \frac{x^2+x-8-(6x-2)}{x+1} =$

$\frac{x^2+x-8-6x+2}{x+1} = \frac{x^2-5x-6}{x+1} = \frac{(x-6)(x+1)}{x+1} = x-6$

22. $y - \frac{4(y+1)}{y+2} - \frac{4}{y+2} = \frac{y(y+2)}{y+2} - \frac{4y+4}{y+2} - \frac{4}{y+2} = \frac{y^2+2y-4y-4-4}{y+2} =$

$\frac{y^2-2y-8}{y+2} = \frac{(y-4)(y+2)}{y+2} = y-4$

23. $\frac{b-1}{b} - \frac{3}{b-2} + 1 = \frac{(b-1)(b-2)}{b(b-2)} - \frac{3b}{b(b-2)} + \frac{b(b-2)}{b(b-2)} =$

$\frac{b^2-3b+2-3b+b^2-2b}{b(b-2)} = \frac{2b^2-8b+2}{b(b-2)}$, or $\frac{2(b^2-4b+1)}{b(b-2)}$

24. $\frac{x}{x+1} - \frac{x+1}{x} + 2 = \frac{x^2}{x(x+1)} - \frac{(x+1)^2}{x(x+1)} + \frac{2x(x+1)}{x(x+1)} =$

$\frac{x^2-(x^2+2x+1)+2x^2+2x}{x(x+1)} = \frac{x^2-x^2-2x-1+2x^2+2x}{x(x+1)} = \frac{2x^2-1}{x(x+1)}$

25. $\frac{3a}{a+1} + \frac{2}{a-1} - 1 = \frac{3a(a-1)}{(a+1)(a-1)} + \frac{2(a+1)}{(a+1)(a-1)} - \frac{(a+1)(a-1)}{(a+1)(a-1)} =$

$\frac{3a^2-3a+2a+2-(a^2-1)}{(a+1)(a-1)} = \frac{3a^2-a+2-a^2+1}{(a+1)(a-1)} = \frac{2a^2-a+3}{(a+1)(a-1)}$

26. $2 - \frac{x}{x-3} - \frac{1}{x+3} = \frac{2(x-3)(x+3)}{(x-3)(x+3)} - \frac{x(x+3)}{(x-3)(x+3)} - \frac{x-3}{(x-3)(x+3)} =$

$\frac{2(x^2-9)-x^2-3x-x+3}{(x-3)(x+3)} = \frac{2x^2-18-x^2-3x-x+3}{(x-3)(x+3)} = \frac{x^2-4x-15}{(x-3)(x+3)}$

27. $a - 1 - \frac{a^2+a-5}{a+2} = \frac{(a-1)(a+2)}{a+2} - \frac{a^2+a-5}{a+2} = \frac{a^2+a-2-a^2-a+5}{a+2} = \frac{3}{a+2}$

28. $2a + 3b - \frac{2a^2 - b^2}{2a - 3b} = \frac{(2a + 3b)(2a - 3b)}{2a - 3b} - \frac{2a^2 - b^2}{2a - 3b} = \frac{4a^2 - 9b^2 - 2a^2 + b^2}{2a - 3b} =$

$\frac{2a^2 - 8b^2}{2a - 3b}$, or $\frac{2(a^2 - 4b^2)}{2a - 3b} = \frac{2(a + 2b)(a - 2b)}{2a - 3b}$

29. $3x - \frac{x^2}{2x + 3} - 2 = \frac{3x(2x + 3)}{2x + 3} - \frac{x^2}{2x + 3} - \frac{2(2x + 3)}{2x + 3} = \frac{6x^2 + 9x - x^2 - 4x - 6}{2x + 3} =$

$\frac{5x^2 + 5x - 6}{2x + 3}$

30. $(x + 4)\left(\frac{4}{x} - 1\right) = 4 - x + \frac{16}{x} - 4 = \frac{4x}{x} - \frac{x^2}{x} + \frac{16}{x} - \frac{4x}{x} = \frac{4x - x^2 + 16 - 4x}{x} =$

$\frac{16 - x^2}{x}$, or $\frac{(4 - x)(4 + x)}{x}$

31. $\left(a + \frac{2}{a}\right)\left(a - \frac{2}{a}\right) = a^2 - \frac{4}{a^2} = \frac{a^4}{a^2} - \frac{4}{a^2} = \frac{a^4 - 4}{a^2}$, or $\frac{(a^2 + 2)(a^2 - 2)}{a^2}$

32. $\left(2x + \frac{3}{x}\right)\left(x - \frac{2}{x}\right) = 2x^2 - 4 + 3 - \frac{6}{x^2} = \frac{2x^4}{x^2} - \frac{x^2}{x^2} - \frac{6}{x^2} = \frac{2x^4 - x^2 - 6}{x^2}$, or

$\frac{(2x^2 + 3)(x^2 - 2)}{x^2}$

33. $\left(\frac{a + b}{a} - 1\right)\left(\frac{a}{b} + 1\right) = \frac{(a + b)a}{ab} + \frac{a + b}{a} - \frac{a}{b} - 1 = \frac{a^2 + ab}{ab} + \frac{b(a + b)}{ab} - \frac{a^2}{ab} - \frac{ab}{ab} =$

$\frac{a^2 + ab + ab + b^2 - a^2 - ab}{ab} = \frac{ab + b^2}{ab} = \frac{b(a + b)}{ab} = \frac{a + b}{a}$

34. $\left(y - \frac{2}{y + 1}\right)\left(1 - \frac{1}{y + 2}\right) = y - \frac{y}{y + 2} - \frac{2}{y + 1} + \frac{2}{(y + 1)(y + 2)} =$

$\frac{y(y + 1)(y + 2)}{(y + 1)(y + 2)} - \frac{y(y + 1)}{(y + 1)(y + 2)} - \frac{2(y + 2)}{(y + 1)(y + 2)} + \frac{2}{(y + 1)(y + 2)} =$

$\frac{y(y^2 + 3y + 2) - (y^2 + y) - (2y + 4) + 2}{(y + 1)(y + 2)} = \frac{y^3 + 3y^2 + 2y - y^2 - y - 2y - 4 + 2}{(y + 1)(y + 2)} =$

$\frac{y^3 + 2y^2 - y - 2}{(y + 1)(y + 2)} = \frac{y^2(y + 2) - 1(y + 2)}{(y + 1)(y + 2)} = \frac{(y^2 - 1)(y + 2)}{(y + 1)(y + 2)} = \frac{(y + 1)(y - 1)(y + 2)}{(y + 1)(y + 2)} =$

$y - 1$

35. $\left(\frac{m}{n} - \frac{n}{m}\right) \div \left(\frac{1}{m} + \frac{1}{n}\right) = \left(\frac{m^2}{nm} - \frac{n^2}{nm}\right) \div \left(\frac{n}{nm} + \frac{m}{nm}\right) = \left(\frac{m^2 - n^2}{nm}\right) \div \left(\frac{n + m}{nm}\right) =$

$\frac{m^2 - n^2}{nm} \cdot \frac{nm}{n + m} = \frac{(m + n)(m - n)}{n + m} = m - n$

36. $\left(9 - \frac{1}{x^2}\right) \div (3x - 1) = \left(\frac{9x^2}{x^2} - \frac{1}{x^2}\right) \cdot \frac{1}{3x - 1} = \frac{9x^2 - 1}{x^2} \cdot \frac{1}{3x - 1} = \frac{(3x - 1)(3x + 1)}{x^2(3x - 1)} =$

$\frac{3x + 1}{x^2}$

37. $\left(1 - \frac{2}{a}\right) \div \left(1 - \frac{4}{a^2}\right) = \left(\frac{a}{a} - \frac{2}{a}\right) \div \left(\frac{a^2}{a^2} - \frac{4}{a^2}\right) = \frac{a - 2}{a} \div \frac{a^2 - 4}{a^2} =$

$\frac{a - 2}{a} \cdot \frac{a^2}{(a + 2)(a - 2)} = \frac{a}{a + 2}$

38. $1 + \frac{2x}{2x - 1} - \frac{8x^2}{4x^2 - 1} = \frac{(2x + 1)(2x - 1)}{(2x + 1)(2x - 1)} + \frac{2x(2x + 1)}{(2x + 1)(2x - 1)} - \frac{8x^2}{(2x + 1)(2x - 1)} =$

$\frac{4x^2 - 1 + 4x^2 + 2x - 8x^2}{(2x + 1)(2x - 1)} = \frac{2x - 1}{(2x + 1)(2x - 1)} = \frac{1}{2x + 1}$

39. Amy and Don were right because each denominator is a factor of the common denominators they chose. Don used the LCD. Julie's common denominator will work because $\frac{8x}{x^2 - x}$ can be simplified to $\frac{8}{x - 1}$ before adding the fractions together.

40.

	Speed	× Time =	Distance
Slower	$\frac{200}{y}$	y	200
Faster	$\frac{200}{y} + 10$	$y - 2$	$\left(\frac{200}{y} + 10\right)(y - 2)$

$\left(\frac{200}{y} + 10\right)(y - 2) = 200 - \frac{400}{y} + 10y - 20 = 180 + 10y - \frac{400}{y} = \frac{(180 + 10y)y}{y} - \frac{400}{y} = \frac{180y + 10y^2 - 400}{y}$; Jan could have gone $\frac{10y^2 + 180y - 400}{y}$ km.

41.

	Number ×	Cost =	Total
Bought	n	$\frac{40}{n}$	40
Sold	$n - 2$	$\frac{40}{n} + 1$	$(n - 2)\left(\frac{40}{n} + 1\right)$

$(n - 2)\left(\frac{40}{n} + 1\right) = (n - 2)\left(\frac{40 + n}{n}\right) = \frac{40n - 80 + n^2 - 2n}{n} = \frac{n^2 + 38n - 80}{n}$, Ted received $\frac{n^2 + 38n - 80}{n}$ dollars for the film.

C **42.** $\left(1 - \frac{b^2 + c^2 - a^2}{2bc}\right) \div \left(1 - \frac{a^2 + b^2 - c^2}{2ab}\right) = \left(\frac{2bc}{2bc} - \frac{b^2 + c^2 - a^2}{2bc}\right) \div \left(\frac{2ab}{2ab} - \frac{a^2 + b^2 - c^2}{2ab}\right) = \frac{2bc - (b^2 + c^2 - a^2)}{2bc} \div \frac{2ab - (a^2 + b^2 - c^2)}{2ab} =$
$\frac{2bc - b^2 - c^2 + a^2}{2bc} \cdot \frac{2ab}{2ab - a^2 - b^2 + c^2} = \frac{a[a^2 - (b^2 - 2bc + c^2)]}{c[c^2 - (a^2 - 2ac + b^2)]} = \frac{a[a^2 - (b - c)^2]}{c[c^2 - (a - b)^2]} =$
$\frac{a(a + b - c)(a - b + c)}{c(c + a - b)(c - a + b)} = \frac{a(a + b - c)}{c(c - a + b)}$

43. $\left(2 - \frac{n}{n + 1} + \frac{n}{1 - n}\right) \div \left(\frac{1}{n - 1} - \frac{1}{n + 1}\right) =$
$\left(\frac{2(n + 1)(1 - n)}{(n + 1)(1 - n)} - \frac{n(1 - n)}{(n + 1)(1 - n)} + \frac{n(n + 1)}{(n + 1)(1 - n)}\right) \div \left(\frac{n + 1}{(n - 1)(n + 1)} - \frac{n - 1}{(n - 1)(n + 1)}\right) = \left(\frac{2(1 - n^2) - n + n^2 + n^2 + n}{(n + 1)(1 - n)}\right) \div$
$\left(\frac{n + 1 - n + 1}{(n - 1)(n + 1)}\right) = \frac{2 - 2n^2 + 2n^2}{(n + 1)(1 - n)} \cdot \frac{(n - 1)(n + 1)}{2} = \frac{2(n - 1)}{2(1 - n)} = \frac{n - 1}{-(n - 1)} = -1$

44. $\dfrac{A}{x+2}+\dfrac{B}{x-2}=\dfrac{4}{x^2-4}, \dfrac{A(x-2)}{(x+2)(x-2)}+\dfrac{B(x+2)}{(x+2)(x-2)}=\dfrac{4}{(x+2)(x-2)},$
$A(x-2)+B(x+2)=4, Ax-2A+Bx+2B=4, (A+B)x+2(B-A)=4;$
$A+B=0, A=-B; 2(B-A)=4, B-A=2, B-(-B)=2, 2B=2, B=1;$
$A=-1$

45. $\dfrac{C}{x-2}+\dfrac{D}{x+1}=\dfrac{6x}{x^2-x-2}, \dfrac{C(x+1)}{(x-2)(x+1)}+\dfrac{D(x-2)}{(x-2)(x+1)}=\dfrac{6x}{(x-2)(x+1)},$
$\dfrac{Cx+C+Dx-2D}{(x-2)(x+1)}=\dfrac{6x}{(x-2)(x+1)}, Cx+C+Dx-2D=6x,$
$(C+D)x+C-2D=6x; C+D=6, C=6-D; C-2D=0, 6-D-2D=0,$
$6-3D=0, 6=3D, 2=D; C=6-2=4$

46. $\left(1-\frac{1}{2}\right)\left(1-\frac{1}{3}\right)\left(1-\frac{1}{4}\right)\left(1-\frac{1}{5}\right)\cdot\ldots\cdot\left(1-\frac{1}{n}\right)=$
$\left(\frac{2}{2}-\frac{1}{2}\right)\left(\frac{3}{3}-\frac{1}{3}\right)\left(\frac{4}{4}-\frac{1}{4}\right)\left(\frac{5}{5}-\frac{1}{5}\right)\cdot\ldots\cdot\left(\frac{n}{n}-\frac{1}{n}\right)=\frac{1}{\cancel{2}}\cdot\frac{\cancel{2}}{\cancel{3}}\cdot\frac{\cancel{3}}{\cancel{4}}\cdot\frac{\cancel{4}}{\cancel{5}}\cdot\ldots\cdot\frac{(n-1)}{n}=\frac{1}{n}$

Page 273 • MIXED REVIEW EXERCISES

1. $\dfrac{5a+5b}{5a+10}=\dfrac{5(a+b)}{5(a+2)}=\dfrac{a+b}{a+2}$ **2.** $\dfrac{a^2-9a+14}{49-a^2}=\dfrac{(a-2)(a-7)}{(7+a)(7-a)}=\dfrac{-(a-2)(7-a)}{(7+a)(7-a)}=-\dfrac{a-2}{7+a}$

3. $\dfrac{5x^2}{3y^3}\div\dfrac{10xy}{9}=\dfrac{5x^2}{3y^3}\cdot\dfrac{9}{10xy}=\dfrac{3x}{2y^4}$ **4.** $\dfrac{n^2-4}{2}\div\dfrac{n+2}{8}=\dfrac{(n+2)(n-2)}{2}\cdot\dfrac{8}{n+2}=4(n-2)$

5. $\dfrac{8}{y^5}\cdot\dfrac{y^7}{4}=2y^2$ **6.** $(-4b^2)^2=(-4)^2(b^2)^2=16b^4$ **7.** $\dfrac{1}{2xy}, \dfrac{3}{x^2}$, LCD is $2x^2y$

8. $\dfrac{5}{b^2}, \dfrac{2}{ab}$; LCD is ab^2 **9.** $\dfrac{2}{x-1}, \dfrac{3}{1+x}, \dfrac{1}{1-x^2}$; LCD is $(1-x)(1+x)$

Page 273 • COMPUTER EXERCISES

1.

```
10 SUM = 2
20 FACT = 1
30 FOR N = 2 TO 7
40 FACT = FACT * N
50 SUM = SUM + (1/FACT)
60 PRINT "STEP "; N; "SUM = "; SUM
70 NEXT N
80 END

STEP 2  SUM = 2.5
     3        2.6666667
     4        2.7083333
     5        2.7166667
     6        2.7180556
     7        2.718254
```

2. It approaches 2.7182818. (This is the irrational number *e*.)

Page 273 • CHALLENGE

Answers may vary. For example, 11, 23, and 35 satisfy all three conditions.

Pages 276–277 • WRITTEN EXERCISES

A

1.
$$\begin{array}{r}
x+3 \\
x+2\overline{)x^2+5x+6} \\
\underline{x^2+2x} \\
3x+6 \\
\underline{3x+6} \\
0
\end{array}$$

2.
$$\begin{array}{r}
x+3 \\
x-4\overline{)x^2-x-12} \\
\underline{x^2-4x} \\
3x-12 \\
\underline{3x-12} \\
0
\end{array}$$

3.
$$\begin{array}{r}
n+6 \\
n-9\overline{)n^2-3n-54} \\
\underline{n^2-9n} \\
6n-54 \\
\underline{6n-54} \\
0
\end{array}$$

4.
$$\begin{array}{r}
n-13 \\
n+2\overline{)n^2-11n-26} \\
\underline{n^2+2n} \\
-13n-26 \\
\underline{-13n-26} \\
0
\end{array}$$

5.
$$\begin{array}{r}
y-3+\dfrac{8}{y+1} \\
y+1\overline{)y^2-2y+5} \\
\underline{y^2+y} \\
-3y+5 \\
\underline{-3y-3} \\
8
\end{array}$$

6.
$$\begin{array}{r}
a-1+\dfrac{5}{a-2} \\
a-2\overline{)a^2-3a+7} \\
\underline{a^2-2a} \\
-a+7 \\
\underline{-a+2} \\
5
\end{array}$$

7.
$$\begin{array}{r}
x-\dfrac{9}{x-3} \\
x-3\overline{)x^2-3x-9} \\
\underline{x^2-3x} \\
0-9
\end{array}$$

8.
$$\begin{array}{r}
z-12+\dfrac{36}{z+4} \\
z+4\overline{)z^2-8z-12} \\
\underline{z^2+4z} \\
-12z-12 \\
\underline{-12z-48} \\
36
\end{array}$$

9.
$$\begin{array}{r}
n+4+\dfrac{28}{n-6} \\
n-6\overline{)n^2-2n+4} \\
\underline{n^2-6n} \\
4n+4 \\
\underline{4n-24} \\
28
\end{array}$$

10.
$$\begin{array}{r}
s-3+\dfrac{4}{s+4} \\
s+4\overline{)s^2+s-8} \\
\underline{s^2+4s} \\
-3s-8 \\
\underline{-3s-12} \\
4
\end{array}$$

11.
$$\begin{array}{r}
x-2+\dfrac{8}{x+2} \\
x+2\overline{)x^2+0x+4} \\
\underline{x^2+2x} \\
-2x+4 \\
\underline{-2x-4} \\
8
\end{array}$$

12.
$$\begin{array}{r}
y-9+\dfrac{72}{y+9} \\
y+9\overline{)y^2+0y-9} \\
\underline{y^2+9y} \\
-9y-9 \\
\underline{-9y-81} \\
72
\end{array}$$

13.
$$\begin{array}{r}
x+4-\dfrac{1}{3x-2} \\
3x-2\overline{)3x^2+10x-9} \\
\underline{3x^2-2x} \\
12x-9 \\
\underline{12x-8} \\
-1
\end{array}$$

14.
$$\begin{array}{r}
a-3-\dfrac{7}{2a+1} \\
2a+1\overline{)2a^2-5a-10} \\
\underline{2a^2+a} \\
-6a-10 \\
\underline{-6a-3} \\
-7
\end{array}$$

15.
$$\begin{array}{r}
2x-1+\dfrac{9}{2x+1} \\
2x+1\overline{)4x^2+0x+8} \\
\underline{4x^2+2x} \\
-2x+8 \\
\underline{-2x-1} \\
9
\end{array}$$

16.
$$\begin{array}{r}
3y+1+\dfrac{7}{3y-1} \\
3y-1\overline{)9y^2+0y+6} \\
\underline{9y^2-3y} \\
3y+6 \\
\underline{3y-1} \\
7
\end{array}$$

17. $$\begin{array}{r} n^2 - 4n + 9 - \frac{16}{n+2} \\ n+2\overline{)\,n^3 - 2n^2 + n + 2} \\ \underline{n^3 + 2n^2} \qquad\qquad \\ -4n^2 + n \qquad \\ \underline{-4n^2 - 8n} \qquad \\ 9n + 2 \\ \underline{9n + 18} \\ -16 \end{array}$$

18. $$\begin{array}{r} a^2 + 3a + 6 + \frac{10}{a-1} \\ a-1\overline{)\,a^3 + 2a^2 + 3a + 4} \\ \underline{a^3 - a^2} \qquad\qquad \\ 3a^2 + 3a \qquad \\ \underline{3a^2 - 3a} \qquad \\ 6a + 4 \\ \underline{6a - 6} \\ 10 \end{array}$$

19. $$\begin{array}{r} a^2 + 2a + 4 + \frac{16}{a-2} \\ a-2\overline{)\,a^3 + 0a^2 + 0a + 8} \\ \underline{a^3 - 2a^2} \qquad\qquad \\ 2a^2 + 0a \qquad \\ \underline{2a^2 - 4a} \qquad \\ 4a + 8 \\ \underline{4a - 8} \\ 16 \end{array}$$

20. $$\begin{array}{r} n^2 + n + 1 \\ n-1\overline{)\,n^3 + 0n^2 + 0n - 1} \\ \underline{n^3 - n^2} \qquad\qquad \\ n^2 + 0n \qquad \\ \underline{n^2 - n} \qquad \\ n - 1 \\ \underline{n - 1} \\ 0 \end{array}$$

21. $$\begin{array}{r} 4x^2 - 2x + 1 - \frac{10}{3x+1} \\ 3x+1\overline{)\,12x^3 - 2x^2 + x - 9} \\ \underline{12x^3 + 4x^2} \qquad\qquad \\ -6x^2 + x \qquad \\ \underline{-6x^2 - 2x} \qquad \\ 3x - 9 \\ \underline{3x + 1} \\ -10 \end{array}$$

22. $$\begin{array}{r} 2n^2 - 2n + 3 + \frac{12}{4n+1} \\ 4n+1\overline{)\,8n^3 - 6n^2 + 10n + 15} \\ \underline{8n^3 + 2n^2} \qquad\qquad \\ -8n^2 + 10n \qquad \\ \underline{-8n^2 - 2n} \qquad \\ 12n + 15 \\ \underline{12n + 3} \\ 12 \end{array}$$

B **23.** $$\begin{array}{r} x^3 - 2x^2 + x - 6 \\ 3x+2\overline{)\,3x^4 - 4x^3 - x^2 - 16x - 12} \\ \underline{3x^4 + 2x^3} \qquad\qquad\qquad \\ -6x^3 - x^2 \qquad\qquad \\ \underline{-6x^3 - 4x^2} \qquad\qquad \\ 3x^2 - 16x \qquad \\ \underline{3x^2 + 2x} \qquad \\ -18x - 12 \\ \underline{-18x - 12} \\ 0 \end{array}$$

24. $$\begin{array}{r} 3x^3 + x^2 - 5x + 2 \\ 2x-1\overline{)\,6x^4 - x^3 - 11x^2 + 9x - 2} \\ \underline{6x^4 - 3x^3} \qquad\qquad\qquad \\ 2x^3 - 11x^2 \qquad\qquad \\ \underline{2x^3 - x^2} \qquad\qquad \\ -10x^2 + 9x \qquad \\ \underline{-10x^2 + 5x} \qquad \\ 4x - 2 \\ \underline{4x - 2} \\ 0 \end{array}$$

25. $$\begin{array}{r} n^3 - 1 \\ 2n-1\overline{)\,2n^4 - n^3 + 0n^2 - 2n + 1} \\ \underline{2n^4 - n^3} \qquad\qquad\qquad \\ 0 \quad + 0 \quad -2n + 1 \\ \underline{-2n + 1} \\ 0 \end{array}$$

26. $$\begin{array}{r} 2y^3 + 3 \\ 4y+5\overline{)\,8y^4 + 10y^3 + 0y^2 + 12y + 15} \\ \underline{8y^4 + 10y^3} \qquad\qquad\qquad \\ 0 \quad + 0 \quad + 12y + 15 \\ \underline{12y + 15} \\ 0 \end{array}$$

27.

$$\begin{array}{r} z^3 - 2z^2 + 4z - 8 + \frac{32}{z+2} \\ z + 2\overline{)z^4 + 0z^3 + 0z^2 + 0z + 16} \\ \underline{z^4 + 2z^3} \qquad\qquad\qquad\quad \\ -2z^3 + 0z^2 \qquad\qquad \\ \underline{-2z^3 - 4z^2} \qquad\qquad \\ 4z^2 + 0z \qquad \\ \underline{4z^2 + 8z} \qquad \\ -8z + 16 \\ \underline{-8z - 16} \\ 32 \end{array}$$

28.

$$\begin{array}{r} b^3 + 3b^2 + 9b + 27 + \frac{165}{b-3} \\ b - 3\overline{)b^4 + 0b^3 + 0b^2 + 0b + 84} \\ \underline{b^4 - 3b^3} \qquad\qquad\qquad\quad \\ 3b^3 + 0b^2 \qquad\qquad \\ \underline{3b^3 - 9b^2} \qquad\qquad \\ 9b^2 + 0b \qquad \\ \underline{9b^2 - 27b} \qquad \\ 27b + 84 \\ \underline{27b - 81} \\ 165 \end{array}$$

29.

$$\begin{array}{r} n^2 - n + 1 \qquad\qquad \\ n^2 + 2\overline{)n^4 - n^3 + 3n^2 - 2n + 2} \\ \underline{n^4 \qquad + 2n^2} \qquad\qquad \\ -n^3 + n^2 - 2n \qquad \\ \underline{-n^3 \qquad - 2n} \qquad \\ n^2 \qquad + 2 \\ \underline{n^2 \qquad + 2} \\ 0 \end{array}$$

30.

$$\begin{array}{r} x - 2 \\ x^2 + 5x + 6\overline{)x^3 + 3x^2 - 4x - 12} \\ \underline{x^3 + 5x^2 + 6x} \qquad \\ -2x^2 - 10x - 12 \\ \underline{-2x^2 - 10x - 12} \\ 0 \end{array}$$

31. Height $= \dfrac{\text{Volume}}{\text{length} \cdot \text{width}} = \dfrac{12n^3 + 8n^2 - 3n - 2}{(2n + 1)(2n - 1)} = \dfrac{12n^3 + 8n^2 - 3n - 2}{4n^2 - 1}$

$$\begin{array}{r} 3n + 2 \qquad\qquad\qquad \\ 4n^2 - 1\overline{)12n^3 + 8n^2 - 3n - 2} \\ \underline{12n^3 \qquad - 3n} \qquad \\ 8n^2 \qquad - 2 \\ \underline{8n^2 \qquad - 2} \\ 0 \end{array}$$

The height of the solid is $3n + 2$.

32. a.

$$\begin{array}{r} a^3 + 2a^2 + 5a + 10 \\ a - 2\overline{)a^4 + 0a^3 + a^2 + 0a - 20} \\ \underline{a^4 - 2a^3} \qquad\qquad\qquad \\ 2a^3 + a^2 \qquad\qquad \\ \underline{2a^3 - 4a^2} \qquad\qquad \\ 5a^2 + 0a \qquad \\ \underline{5a^2 - 10a} \qquad \\ 10a - 20 \\ \underline{10a - 20} \\ 0 \end{array}$$

b.

$$\frac{a^4 + a^2 - 20}{a - 2} = \frac{(a^2 + 5)(a^2 - 4)}{a - 2} = \frac{(a^2 + 5)(a + 2)(a - 2)}{a - 2} = (a^2 + 5)(a + 2)$$

c. $(a^2 + 5)(a + 2) = a^3 + 2a^2 + 5a + 10$

33.

$$\begin{array}{r} 2n^2 - 6n + 4 \\ n + 3\overline{)2n^3 + 0n^2 - 14n + 12} \\ \underline{2n^3 + 6n^2} \qquad\qquad\quad \\ -6n^2 - 14n \qquad \\ \underline{-6n^2 - 18n} \qquad \\ 4n + 12 \\ \underline{4n + 12} \\ 0 \end{array}$$

$$\begin{aligned} 2n^3 - 14n + 12 &= (n + 3)(2n^2 - 6n + 4) \\ &= (n + 3)(2n - 2)(n - 2) \\ &= 2(n + 3)(n - 1)(n - 2) \end{aligned}$$

34. $$\begin{array}{rl} 2x^2 - 7x - 15 & \\ 2x + 1\overline{)4x^3 - 12x^2 - 37x - 15} & \\ \underline{4x^3 + 2x^2} & \\ -14x^2 - 37x & \\ \underline{-14x^2 - 7x} & \\ -30x - 15 & \\ \underline{-30x - 15} & \\ 0 & \end{array}$$

$4x^3 - 12x^2 - 37x - 15 =$
$(2x + 1)(2x^2 - 7x - 15) =$
$(2x + 1)(2x + 3)(x - 5)$

C **35.** $$\begin{array}{r} 4x - 3 \\ x - 3\overline{)4x^2 - 15x + k} \\ \underline{4x^2 - 12x} \\ -3x + k \\ \underline{-3x + 9} \\ k - 9 \end{array}$$

If $x - 3$ is a factor, then $k - 9 = 0$, $k = 9$.

36. $$\begin{array}{r} 2x^2 - 2x - 3 \\ 2x - 1\overline{)4x^3 - 6x^2 - 4x + k} \\ \underline{4x^3 - 2x^2} \\ -4x^2 - 4x \\ \underline{-4x^2 + 2x} \\ -6x + k \\ \underline{-6x + 3} \\ k - 3 \end{array}$$

If $2x - 1$ is a factor, then $k - 3 = 0$, $k = 3$.

37. $$\begin{array}{r} y^2 - 2y + (k + 6) \\ y + 3\overline{)y^3 + y^2 + ky + 3} \\ \underline{y^3 + 3y^2} \\ -2y^2 + ky \\ \underline{-2y^2 - 6y} \\ (k + 6)y + 3 \\ \underline{(k + 6)y + 3(k + 6)} \\ 3 - 3(k + 6) \end{array}$$

If $y + 3$ is a factor, then $3 - 3(k + 6) = 0$,
$3 = 3(k + 6)$, $1 = k + 6$, $-5 = k$

38. $$\begin{array}{r} 4x^3 + 5x^2 - 2x + 1 \\ x - 1\overline{)4x^4 + x^3 - 7x^2 + 3x + k} \\ \underline{4x^4 - 4x^3} \\ 5x^3 - 7x^2 \\ \underline{5x^3 - 5x^2} \\ -2x^2 + 3x \\ \underline{-2x^2 + 2x} \\ x + k \\ \underline{x - 1} \\ k + 1 \end{array}$$

$k + 1 = 5$, $k = 4$

Page 277 • MIXED REVIEW EXERCISES

1. $\frac{x + 4}{2} + \frac{2x - 1}{2} = \frac{x + 4 + 2x - 1}{2} = \frac{3x + 3}{2}$, or $\frac{3(x + 1)}{2}$

2. $\frac{a^2}{a + 3} - \frac{9}{a + 3} = \frac{a^2 - 9}{a + 3} = \frac{(a + 3)(a - 3)}{a + 3} = a - 3$ **3.** $\frac{2}{y} + \frac{1}{3} = \frac{2 \cdot 3}{3y} + \frac{y}{3y} = \frac{6 + y}{3y}$

4. $\frac{5c + 1}{6c} + \frac{3}{2c} = \frac{5c + 1}{6c} + \frac{3 \cdot 3}{6c} = \frac{5c + 1 + 9}{6c} = \frac{5c + 10}{6c}$, or $\frac{5(c + 2)}{6c}$

5. $\dfrac{2z+1}{6} - \dfrac{3z-5}{9} = \dfrac{3(2z+1)}{18} - \dfrac{2(3z-5)}{18} = \dfrac{6z+3-6z+10}{18} = \dfrac{13}{18}$

6. $\dfrac{x}{x^2-25} - \dfrac{1}{2x+10} = \dfrac{x}{(x+5)(x-5)} - \dfrac{1}{2(x+5)} = \dfrac{2x}{2(x+5)(x-5)} - \dfrac{(x-5)}{2(x+5)(x-5)} =$

$\dfrac{2x-x+5}{2(x+5)(x-5)} = \dfrac{x+5}{2(x+5)(x-5)} = \dfrac{1}{2(x-5)}$

7. $x + \dfrac{2}{x} = \dfrac{x^2}{x} + \dfrac{2}{x} = \dfrac{x^2+2}{x}$ **8.** $4 + \dfrac{n}{n+1} = \dfrac{4(n+1)}{n+1} + \dfrac{n}{n+1} = \dfrac{4n+4+n}{n+1} = \dfrac{5n+4}{n+1}$

9. $y + 3 + \dfrac{2y-1}{y-2} = \dfrac{(y+3)(y-2)}{y-2} + \dfrac{2y-1}{y-2} = \dfrac{y^2+y-6+2y-1}{y-2} = \dfrac{y^2+3y-7}{y-2}$

Pages 278–279 • EXTRA

A **1.** $\dfrac{\frac{m}{8}}{\frac{5m}{8}} = \dfrac{\frac{m}{8}\cdot 8}{\frac{5m}{8}\cdot 8} = \dfrac{m}{5m} = \dfrac{1}{5}$ **2.** $\dfrac{\frac{3a}{4}}{\frac{15a}{12}} = \dfrac{\frac{3a}{4}\cdot 12}{\frac{15a}{12}\cdot 12} = \dfrac{9a}{15a} = \dfrac{3}{5}$

3. $\dfrac{\frac{u}{v^2}}{\frac{u}{v}} = \dfrac{\frac{u}{v^2}\cdot v^2}{\frac{u}{v}\cdot v^2} = \dfrac{u}{uv} = \dfrac{1}{v}$ **4.** $\dfrac{6e}{\frac{3}{7e}} = \dfrac{6e\cdot 7e}{\frac{3}{7e}\cdot 7e} = \dfrac{42e^2}{3} = 14e^2$

5. $\dfrac{\frac{1}{4}+\frac{1}{8}}{\frac{1}{4}-\frac{1}{8}} = \dfrac{\left(\frac{1}{4}+\frac{1}{8}\right)8}{\left(\frac{1}{4}-\frac{1}{8}\right)8} = \dfrac{2+1}{2-1} = \dfrac{3}{1} = 3$

6. $\dfrac{\frac{1}{6}+\frac{1}{3}}{\frac{1}{2}+\frac{1}{5}} = \dfrac{\left(\frac{1}{6}+\frac{1}{3}\right)30}{\left(\frac{1}{2}+\frac{1}{5}\right)30} = \dfrac{5+10}{15+6} = \dfrac{15}{21} = \dfrac{5}{7}$

7. $\dfrac{\frac{5}{6}-\frac{5}{7}}{\frac{5}{6}+\frac{5}{7}} = \dfrac{\left(\frac{5}{6}-\frac{5}{7}\right)42}{\left(\frac{5}{6}+\frac{5}{7}\right)42} = \dfrac{35-30}{35+30} = \dfrac{5}{65} = \dfrac{1}{13}$

8. $\dfrac{\frac{3}{x}}{\frac{1}{x}-\frac{1}{3x}} = \dfrac{\frac{3}{x}\cdot 3x}{\left(\frac{1}{x}-\frac{1}{3x}\right)3x} = \dfrac{9}{3-1} = \dfrac{9}{2}$ **9.** $\dfrac{\frac{9c^2}{5d^4}}{\frac{3c^2}{10d^3}} = \dfrac{\frac{9c^2}{5d^4}\cdot 10d^4}{\frac{3c^2}{10d^3}\cdot 10d^4} = \dfrac{18c^2}{3c^2d} = \dfrac{6}{d}$

10. $\dfrac{\frac{n^2-25}{n}}{n-5} = \dfrac{\frac{n^2-25}{n}\cdot n}{(n-5)\cdot n} = \dfrac{n^2-25}{n(n-5)} = \dfrac{(n+5)(n-5)}{n(n-5)} = \dfrac{n+5}{n}$

11. $\dfrac{\frac{r}{s}+2}{1-\frac{r}{s}} = \dfrac{\left(\frac{r}{s}+2\right)s}{\left(1-\frac{r}{s}\right)s} = \dfrac{r+2s}{s-r}$

12. $\dfrac{\frac{2w}{3}+2}{\frac{5w}{3}-\frac{15}{w}} = \dfrac{\left(\frac{2w}{3}+2\right)3w}{\left(\frac{5w}{3}-\frac{15}{w}\right)3w} = \dfrac{2w^2+6w}{5w^2-45} = \dfrac{2w(w+3)}{5(w^2-9)} = \dfrac{2w(w+3)}{5(w+3)(w-3)} = \dfrac{2w}{5(w-3)}$

B

13. $$\frac{z-\frac{5z}{z+5}}{z+\frac{5z}{z-5}}=\frac{\left(z-\frac{5z}{z+5}\right)(z+5)(z-5)}{\left(z+\frac{5z}{z-5}\right)(z+5)(z-5)}=\frac{z(z+5)(z-5)-5z(z-5)}{z(z+5)(z-5)+5z(z+5)}=$$
$$\frac{z(z^2-25)-5z^2+25z}{z(z^2-25)+5z^2+25z}=\frac{z^3-25z-5z^2+25z}{z^3-25z+5z^2+25z}=\frac{z^3-5z^2}{z^3+5z^2}=\frac{z^2(z-5)}{z^2(z+5)}=\frac{z-5}{z+5}$$

14. $$\frac{1-\frac{6}{s^2+2}}{\frac{4s+2}{s^2+2}+1}=\frac{\left(1-\frac{6}{s^2+2}\right)(s^2+2)}{\left(\frac{4s+2}{s^2+2}+1\right)(s^2+2)}=\frac{s^2+2-6}{4s+2+s^2+2}=\frac{s^2-4}{s^2+4s+4}=$$
$$\frac{(s+2)(s-2)}{(s+2)^2}=\frac{s-2}{s+2}$$

15. $$\frac{k+\frac{k-3}{k+1}}{k-\frac{2}{k+1}}=\frac{\left(k+\frac{k-3}{k+1}\right)(k+1)}{\left(k-\frac{2}{k+1}\right)(k+1)}=\frac{k(k+1)+k-3}{k(k+1)-2}=\frac{k^2+k+k-3}{k^2+k-2}=$$
$$\frac{k^2+2k-3}{k^2+k-2}=\frac{(k+3)(k-1)}{(k+2)(k-1)}=\frac{k+3}{k+2}$$

16. $$\frac{\frac{1}{c}-\frac{1}{3-3c}}{\frac{1}{1-c}-\frac{3}{c}}=\frac{\left(\frac{1}{c}-\frac{1}{3(1-c)}\right)3c(1-c)}{\left(\frac{1}{1-c}-\frac{3}{c}\right)3c(1-c)}=\frac{3(1-c)-c}{3c-9(1-c)}=\frac{3-3c-c}{3c-9+9c}=$$
$$\frac{3-4c}{12c-9}=\frac{-(4c-3)}{3(4c-3)}=-\frac{1}{3}$$

17. $$\frac{\frac{e}{e-f}-\frac{f}{e+f}}{\frac{f}{e-f}+\frac{e}{e+f}}=\frac{\left(\frac{e}{e-f}-\frac{f}{e+f}\right)(e-f)(e+f)}{\left(\frac{f}{e-f}+\frac{e}{e+f}\right)(e-f)(e+f)}=\frac{e(e+f)-f(e-f)}{f(e+f)+e(e-f)}=$$
$$\frac{e^2+ef-ef+f^2}{ef+f^2+e^2-ef}=\frac{e^2+f^2}{e^2+f^2}=1$$

18. $$\frac{1-2u}{1+\frac{1}{2u}}\div\frac{1+2u}{1-\frac{1}{2u}}=\frac{(1-2u)2u}{\left(1+\frac{1}{2u}\right)2u}\div\frac{(1+2u)2u}{\left(1-\frac{1}{2u}\right)2u}=\frac{(1-2u)2u}{2u+1}\div\frac{(1+2u)2u}{2u-1}=$$
$$\frac{(1-2u)2u}{2u+1}\cdot\frac{2u-1}{(1+2u)2u}=\frac{-(2u-1)^2}{(2u+1)^2}=-\frac{(2u-1)^2}{(2u+1)^2}$$

19. $$\frac{\frac{2}{z+1}-2}{\frac{2-z}{z^2-1}-2}=\frac{\left(\frac{2}{z+1}-2\right)(z+1)(z-1)}{\left(\frac{2-z}{(z+1)(z-1)}-2\right)(z+1)(z-1)}=\frac{2(z-1)-2(z+1)(z-1)}{2-z-2(z+1)(z-1)}=$$
$$\frac{2z-2-2(z^2-1)}{2-z-2(z^2-1)}=\frac{2z-2-2z^2+2}{2-z-2z^2+2}=\frac{2z-2z^2}{4-z-2z^2}=\frac{-2z(z-1)}{-(2z^2+z-4)}=\frac{2z(z-1)}{2z^2+z-4}$$

20. $$\frac{\frac{m-n}{m+n}+\frac{n}{m}}{\frac{m}{n}-\frac{m-n}{m+n}}=\frac{\left(\frac{m-n}{m+n}+\frac{n}{m}\right)(m+n)mn}{\left(\frac{m}{n}-\frac{m-n}{m+n}\right)(m+n)mn}=\frac{mn(m-n)+n^2(m+n)}{m^2(m+n)-mn(m-n)}=$$
$$\frac{m^2n-n^2m+n^2m+n^3}{m^3+m^2n-m^2n+mn^2}=\frac{m^2n+n^3}{m^3+mn^2}=\frac{n(m^2+n^2)}{m(m^2+n^2)}=\frac{n}{m}$$

21. $x = \dfrac{y-1}{y+1} = \dfrac{\dfrac{1}{1-z} - 1}{\dfrac{1}{1-z} + 1} = \dfrac{\left(\dfrac{1}{1-z} - 1\right)(1-z)}{\left(\dfrac{1}{1-z} + 1\right)(1-z)} = \dfrac{1-(1-z)}{1+(1-z)} = \dfrac{1-1+z}{2-z} = \dfrac{z}{2-z}$

22. $a = \dfrac{b-c}{1+bc} = \dfrac{\dfrac{1}{z-1} - \dfrac{1}{z+1}}{1 + \dfrac{1}{z-1}\cdot\dfrac{1}{z+1}} = \dfrac{\left(\dfrac{1}{z-1} - \dfrac{1}{z+1}\right)(z-1)(z+1)}{\left(1 + \dfrac{1}{(z-1)(z+1)}\right)(z-1)(z+1)} =$

$\dfrac{z+1-(z-1)}{(z-1)(z+1)+1} = \dfrac{z+1-z+1}{z^2-1+1} = \dfrac{2}{z^2}$

23. Average speed $= \dfrac{\text{total distance}}{\text{total time}} = \dfrac{2d}{\dfrac{d}{50} + \dfrac{d}{30}} = \dfrac{(2d)150}{\left(\dfrac{d}{50} + \dfrac{d}{30}\right)150} = \dfrac{300d}{3d + 5d} =$

$\dfrac{300d}{8d} = 37.5$; The average speed is 37.5 km/h.

24. Average speed $= \dfrac{\text{total distance}}{\text{total time}} = \dfrac{12+9}{\dfrac{12}{x} + \dfrac{9}{2x}} = \dfrac{21\cdot 2x}{\left(\dfrac{12}{x} + \dfrac{9}{2x}\right)2x} = \dfrac{42x}{24+9} = \dfrac{42x}{33} =$

$\dfrac{14x}{11}$; Her average speed is $\dfrac{14x}{11}$ km/h.

25.

	Number	× Price	= Total Price
First Price	n	$\dfrac{50}{n}$	50
Second Price	$\dfrac{50}{\dfrac{50}{n} - 10}$	$\dfrac{50}{n} - 10$	50

$\dfrac{50}{\dfrac{50}{n} - 10} = \dfrac{50n}{\left(\dfrac{50}{n} - 10\right)n} = \dfrac{50n}{50 - 10n} = \dfrac{10(5n)}{10(5-n)} = \dfrac{5n}{5-n}$; After the price is decreased, $\dfrac{5n}{5-n}$ items may be purchased for 50 cents.

26. $a = \dfrac{1-c}{1+c} = \dfrac{1 - \dfrac{1+b}{1-b}}{1 + \dfrac{1+b}{1-b}} = \dfrac{\left(1 - \dfrac{1+b}{1-b}\right)(1-b)}{\left(1 + \dfrac{1+b}{1-b}\right)(1-b)} = \dfrac{1-b-(1+b)}{1-b+1+b} =$

$\dfrac{1-b-1-b}{2} = \dfrac{-2b}{2} = -b$; $a = -b$, $a + b = 0$

27. $\left(\dfrac{w}{6} - \dfrac{6}{w}\right) \div \left(\dfrac{6}{w} - 4 + \dfrac{w}{2}\right) = \left(\dfrac{w^2}{6w} - \dfrac{36}{6w}\right) \div \left(\dfrac{12}{2w} - \dfrac{8w}{2w} + \dfrac{w^2}{2w}\right) = \dfrac{w^2 - 36}{6w} \div$

$\dfrac{w^2 - 8w + 12}{2w} = \dfrac{(w+6)(w-6)}{6w} \div \dfrac{(w-6)(w-2)}{2w} = \dfrac{(w+6)(w-6)}{6w}\cdot\dfrac{2w}{(w-6)(w-2)} =$

$\dfrac{w+6}{3(w-2)}$

28. $\left(\frac{x}{4-x^2}+\frac{1}{x-2}\right)\div\left(1-\frac{2}{2+x}\right)=\left(\frac{x}{(2-x)(2+x)}+\frac{-1}{(2-x)}\right)\div\left(\frac{2+x}{2+x}-\frac{2}{2+x}\right)=$
$\left(\frac{x}{(2-x)(2+x)}-\frac{(2+x)}{(2-x)(2+x)}\right)\div\frac{2+x-2}{2+x}=\frac{x-2-x}{(2-x)(2+x)}\cdot\frac{2+x}{x}=\frac{-2}{x(2-x)}=$
$\frac{2}{x(x-2)}$

29. $\left(\frac{1}{n-n^2}-\frac{1}{n^2+n}\right)\div\left(\frac{1}{n^2+1}-\frac{1}{n^2-1}\right)=\left(\frac{1}{n(1-n)}-\frac{1}{n(n+1)}\right)\div$
$\left(\frac{1}{n^2+1}-\frac{1}{(n+1)(n-1)}\right)=\left(\frac{n+1}{n(1-n)(n+1)}-\frac{1-n}{n(1-n)(n+1)}\right)\div$
$\left(\frac{(n+1)(n-1)}{(n^2+1)(n+1)(n-1)}-\frac{n^2+1}{(n^2+1)(n+1)(n-1)}\right)=\frac{n+1-(1-n)}{n(1-n)(1+n)}\div$
$\frac{n^2-1-(n^2+1)}{(n^2+1)(n+1)(n-1)}=\frac{n+1-1+n}{n(1-n)(1+n)}\cdot\frac{(n^2+1)(n+1)(n-1)}{n^2-1-n^2-1}=$
$\frac{2n}{-n(n-1)(n+1)}\cdot\frac{(n^2+1)(n+1)(n-1)}{-2}=n^2+1$

30. $\left(\frac{c^2+d^2}{cd}-2\right)\div\left(\frac{4c^2-4d^2}{2cd}\right)=\left(\frac{c^2+d^2}{cd}-\frac{2cd}{cd}\right)\div\frac{4(c+d)(c-d)}{2cd}=$
$\frac{c^2+d^2-2cd}{cd}\cdot\frac{cd}{2(c+d)(c-d)}=\frac{(c-d)^2}{2(c+d)(c-d)}=\frac{c-d}{2(c+d)}$

31. $\left(\frac{9e^2-5}{e-1}-\frac{1}{6}\right)\div\left(1-\frac{e-9e^2}{e-1}\right)=\left(\frac{(9e^2-5)6}{6(e-1)}-\frac{e-1}{6(e-1)}\right)\div\left(\frac{e-1}{e-1}-\frac{e-9e^2}{e-1}\right)=$
$\frac{54e^2-30-e+1}{6(e-1)}\div\left(\frac{e-1-e+9e^2}{e-1}\right)=\frac{54e^2-e-29}{6(e-1)}\div\frac{9e^2-1}{e-1}=$
$\frac{54e^2-e-29}{6(e-1)}\cdot\frac{e-1}{(3e+1)(3e-1)}=\frac{54e^2-e-29}{6(3e+1)(3e-1)}$

32. $\left(\frac{u^2+v^2}{u^2-v^2}\right)\div\left(\frac{u-v}{u+v}-\frac{u+v}{u-v}\right)=\frac{u^2+v^2}{(u+v)(u-v)}\div\left(\frac{(u-v)^2}{(u+v)(u-v)}-\frac{(u+v)^2}{(u+v)(u-v)}\right)=$
$\frac{u^2+v^2}{(u+v)(u-v)}\div\frac{u^2-2uv+v^2-(u^2+2uv+v^2)}{(u+v)(u-v)}=$
$\frac{u^2+v^2}{(u+v)(u-v)}\cdot\frac{(u+v)(u-v)}{u^2-2uv+v^2-u^2-2uv-v^2}=\frac{u^2+v^2}{-4uv}=-\frac{u^2+v^2}{4uv}$

C **33.** $\frac{r}{2}\left(\frac{r^2-s^2}{r^2s+rs^2}\right)\left(\frac{1}{r-s}\right)\left(\frac{1}{r+s}\right)\div\frac{1}{r+s}=\frac{r}{2}\left(\frac{(r+s)(r-s)}{rs(r+s)}\right)\left(\frac{1}{r-s}\right)\left(\frac{1}{r+s}\right)\cdot(r+s)=\frac{1}{2s}$

34. $\left(\frac{4x^2-y^2}{3xy}\right)\div\left(\frac{2x^2-y^2}{xy}+1\right)=\frac{(2x+y)(2x-y)}{3xy}\div\left(\frac{2x^2-y^2}{xy}+\frac{xy}{xy}\right)=$
$\frac{(2x+y)(2x-y)}{3xy}\div\frac{2x^2+xy-y^2}{xy}=\frac{(2x+y)(2x-y)}{3xy}\cdot\frac{xy}{(2x-y)(x+y)}=\frac{2x+y}{3(x+y)}$

35. $1-\frac{1}{1-\frac{1}{c-2}}=1-\frac{1}{\frac{c-2}{c-2}-\frac{1}{c-2}}=1-\frac{1}{\frac{c-2-1}{c-2}}=1-\frac{1}{\frac{c-3}{c-2}}=$
$1-\frac{c-2}{c-3}=\frac{c-3}{c-3}-\frac{c-2}{c-3}=\frac{c-3-(c-2)}{c-3}=\frac{c-3-c+2}{c-3}=-\frac{1}{c-3}=\frac{1}{3-c}$

36. $$2 + \frac{1}{1 + \frac{2}{a + \frac{1}{a}}} = 2 + \frac{1}{1 + \frac{2}{\frac{a^2 + 1}{a}}} = 2 + \frac{1}{1 + \frac{2a}{a^2 + 1}} =$$

$$2 + \frac{1}{\frac{a^2 + 1 + 2a}{a^2 + 1}} = 2 + \frac{a^2 + 1}{a^2 + 2a + 1} = \frac{2(a^2 + 2a + 1) + a^2 + 1}{a^2 + 2a + 1} =$$

$$\frac{2a^2 + 4a + 2 + a^2 + 1}{(a + 1)^2} = \frac{3a^2 + 4a + 3}{(a + 1)^2}$$

Pages 280–281 • CHAPTER REVIEW

1. b; $\frac{9x^2 - 9}{1 - x^2} = \frac{9(x^2 - 1)}{(1 + x)(1 - x)} = \frac{9(x + 1)(x - 1)}{(1 + x)(1 - x)} = \frac{9(x - 1)}{-(x - 1)} = -9$

2. b; $\frac{15xy}{10x^2y - y^2x} = \frac{15xy}{xy(10x - y)} = \frac{15}{10x - y}$ **3.** a; $\left(-\frac{2}{5}\right)^3\left(-\frac{25}{4}\right) = -\frac{\overset{2}{\cancel{8}}}{\underset{5}{\cancel{125}}} \cdot -\frac{\overset{1}{\cancel{25}}}{\underset{1}{\cancel{4}}} = \frac{2}{5}$

4. b; $\frac{2ab}{15} \cdot \frac{25}{3ab^2} = \frac{10}{9b}$ **5.** b; $5ab^2 \div \frac{10a}{b} = 5ab^2 \cdot \frac{b}{10a} = \frac{b^3}{2}$

6. a; $\frac{x^2 - 36}{6x + 36} \div (6 - x) = \frac{(x + 6)(x - 6)}{6(x + 6)} \cdot \frac{1}{6 - x} = \frac{-(6 - x)}{6(6 - x)} = -\frac{1}{6}$

7. d; $\frac{2}{x - 3} \cdot (x^2 - 6x + 9) = \frac{2(x - 3)^2}{x - 3} = 2(x - 3)$

8. c; $\frac{4n}{9n - 6} = \frac{4n}{3(3n - 2)}; \frac{2n}{15(3n - 2)^2}$; LCD is $15(3n - 2)^2$.

9. c; $\frac{3x}{x - 2} + \frac{6}{2 - x} = \frac{3x}{x - 2} - \frac{6}{x - 2} = \frac{3x - 6}{x - 2} = \frac{3(x - 2)}{x - 2} = 3$

10. b; $\frac{n - 9}{36} - \frac{n - 35}{108} = \frac{3(n - 9)}{108} - \frac{n - 35}{108} = \frac{3n - 27 - n + 35}{108} = \frac{2n + 8}{108} = \frac{2(n + 4)}{108} = \frac{n + 4}{54}$

11. c; $7 + \frac{x + 2}{x - 2} = \frac{7(x - 2)}{x - 2} + \frac{x + 2}{x - 2} = \frac{7x - 14 + x + 2}{x - 2} = \frac{8x - 12}{x - 2}$

12. a; $y + 3 + \frac{1}{y - 3} = \frac{(y + 3)(y - 3)}{y - 3} + \frac{1}{y - 3} = \frac{y^2 - 9 + 1}{y - 3} = \frac{y^2 - 8}{y - 3}$

13. c;
$$\begin{array}{r}
x^2 - x + 1 \\
x - 2\,\overline{)\,x^3 - 3x^2 + 3x + 4} \\
\underline{x^3 - 2x^2} \\
-x^2 + 3x \\
\underline{-x^2 + 2x} \\
x + 4 \\
\underline{x - 2} \\
6
\end{array}$$

14. b;
$$\begin{array}{r}
9x^2 + 6x + 4 + \frac{16}{3x - 2} \\
3x - 2\,\overline{)\,27x^3 + 0x^2 + 0x + 8} \\
\underline{27x^3 - 18x^2} \\
18x^2 + 0x \\
\underline{18x^2 - 12x} \\
12x + 8 \\
\underline{12x - 8} \\
16
\end{array}$$

Page 282 • CHAPTER TEST

1. $\frac{5x + 35}{x^2 - 49} = \frac{5(x + 7)}{(x + 7)(x - 7)} = \frac{5}{x - 7}; x \neq 7, -7$

2. $\frac{64 - n^2}{n^2 - 4n - 32} = \frac{(8 + n)(8 - n)}{(n - 8)(n + 4)} = \frac{-(8 + n)(n - 8)}{(n - 8)(n + 4)} = -\frac{8 + n}{n + 4}; n \neq 8, -4$

3. $\frac{3x^2 - 6x - 24}{3x^2 + 2x - 8} = \frac{3(x^2 - 2x - 8)}{(3x - 4)(x + 2)} = \frac{3(x - 4)(x + 2)}{(3x - 4)(x + 2)} = \frac{3(x - 4)}{3x - 4}; x \neq \frac{4}{3}, x \neq -2$

4. $\frac{15y^2 - 30y - 45}{5y^2 + 10y - 15} = \frac{15(y^2 - 2y - 3)}{5(y^2 + 2y - 3)} = \frac{3(y-3)(y+1)}{(y+3)(y-1)}$; $y \neq -3, 1$

5. $\left(-\frac{n^3}{7}\right)^2 = \frac{n^6}{49}$ **6.** $\frac{(3b)^2}{5} \cdot \frac{b^3}{5} = \frac{9b^2 \cdot b^3}{25} = \frac{9b^5}{25}$ **7.** $\left(\frac{3a}{b}\right)^2 \cdot \frac{7ab}{54} = \frac{9a^2}{b^2} \cdot \frac{7ab}{54} = \frac{7a^3}{6b}$

8. $\frac{9}{11} \div \frac{11}{9} = \frac{9}{11} \cdot \frac{9}{11} = \frac{81}{121}$ **9.** $\frac{4}{7} \div \frac{4}{7} = \frac{4}{7} \cdot \frac{7}{4} = 1$ **10.** $\frac{5x^2}{4y^2} \div 20xy = \frac{5x^2}{4y^2} \cdot \frac{1}{20xy} = \frac{x}{16y^3}$

11. $18 \div \left(\frac{3n}{2}\right)^3 = 18 \div \frac{27n^3}{8} = 18 \cdot \frac{8}{27n^3} = \frac{16}{3n^3}$

12. $\frac{6a + 36}{6a} \div \frac{a^2 - 36}{a^2} = \frac{6(a+6)}{6a} \cdot \frac{a^2}{(a+6)(a-6)} = \frac{a}{a-6}$

13. $\frac{y}{2x^3} \div \left(\frac{y}{2x}\right)^2 = \frac{y}{2x^3} \div \frac{y^2}{4x^2} = \frac{y}{2x^3} \cdot \frac{4x^2}{y^2} = \frac{2}{xy}$ **14.** $\frac{7n}{16m} = \frac{7n \cdot 2mn}{16m \cdot 2mn} = \frac{14mn^2}{32m^2n}$

15. $\frac{3}{x+5} = \frac{3(x-5)}{(x+5)(x-5)}$ **16.** $\frac{9xy^2}{24x^2y^2}, \frac{10x^2}{24x^2y^2}, \frac{20y}{24x^2y^2}$ **17.** $\frac{2(x-4)}{30}, \frac{3(x+2)}{30}$

18. $\frac{x}{x-9} + \frac{1}{x-9} - \frac{19-x}{x-9} = \frac{x + 1 - 19 + x}{x-9} = \frac{2x - 18}{x-9} = \frac{2(x-9)}{x-9} = 2$

19. $\frac{x-1}{3} + \frac{3-2x}{6} = \frac{2(x-1)}{6} + \frac{3-2x}{6} = \frac{2x - 2 + 3 - 2x}{6} = \frac{1}{6}$

20. $\frac{6n+3}{n-5} - \frac{4n+9}{5-n} = \frac{6n+3}{n-5} + \frac{4n+9}{n-5} = \frac{6n + 3 + 4n + 9}{n-5} = \frac{10n + 12}{n-5} = \frac{2(5n+6)}{n-5}$

21. $\frac{2}{y^2 - 2y} - \frac{3}{y^2 - y - 2} = \frac{2}{y(y-2)} - \frac{3}{(y-2)(y+1)} = \frac{2(y+1)}{y(y-2)(y+1)} - \frac{3y}{y(y-2)(y+1)} =$

$\frac{2y + 2 - 3y}{y(y-2)(y+1)} = \frac{2-y}{y(y-2)(y+1)} = -\frac{(y-2)}{y(y-2)(y+1)} = -\frac{1}{y(y+1)}$

22. $12 - \frac{n}{5} = \frac{60}{5} - \frac{n}{5} = \frac{60-n}{5}$

23. $2 + \frac{6}{y-7} = \frac{2(y-7)}{y-7} + \frac{6}{y-7} = \frac{2y - 14 + 6}{y-7} = \frac{2y-8}{y-7} = \frac{2(y-4)}{y-7}$

24. $4x - \frac{x+1}{x-1} = \frac{4x(x-1)}{x-1} - \frac{x+1}{x-1} = \frac{4x^2 - 4x - x - 1}{x-1} = \frac{4x^2 - 5x - 1}{x-1}$

25. $\frac{x}{x+2} + \frac{2}{x-2} + 1 = \frac{x(x-2)}{(x+2)(x-2)} + \frac{2(x+2)}{(x+2)(x-2)} + \frac{(x+2)(x-2)}{(x+2)(x-2)} =$

$\frac{x^2 - 2x + 2x + 4 + x^2 - 4}{(x+2)(x-2)} = \frac{2x^2}{(x+2)(x-2)}$

26.
$$\begin{array}{r|l} & n - 8 + \frac{5}{n-5} \\ n - 5 & n^2 - 13n + 45 \\ & \underline{n^2 - 5n} \\ & -8n + 45 \\ & \underline{-8n + 40} \\ & 5 \end{array}$$

27.
$$\begin{array}{r|l} & x^2 - x - 2 \\ 2x + 1 & 2x^3 - x^2 - 5x - 2 \\ & \underline{2x^3 + x^2} \\ & -2x^2 - 5x \\ & \underline{-2x^2 - x} \\ & -4x - 2 \\ & \underline{-4x - 2} \\ & 0 \end{array}$$

Page 283 • CUMULATIVE REVIEW

1. $(0.3 + 0.7)[(-0.5)^2] = 1(0.25) = 0.25$ **2.** $\frac{45x^3y^2z^5}{-30(xyz)^2} = -\frac{3xz^3}{2}$ **3.** $(4x^3yz^2)^3 = 64x^9y^3z^6$

4. $(8a - 9b) + (5a + 6b) = (8a + 5a) + (-9b + 6b) = 13a + (-3b) = 13a - 3b$

5. $(5x - 3)(4x + 3) = 20x^2 + 15x - 12x - 9 = 20x^2 + 3x - 9$

6. $-5y(2y^3 - 7y^2 + 1) = -10y^4 + 35y^3 - 5y$

7. $(-13t + 6s) - (4t + 9s) = (-13t - 4t) + (6s - 9s) = -17t + (-3s) = -17t - 3s$

8. $(-2b + 7c)(-2b + 7c) = 4b^2 - 14bc - 14bc + 49c^2 = 4b^2 - 28bc + 49c^2$

9. $(4t^2s - 9)(4t^2s + 9) = 16t^4s^2 - 81$

10. $\frac{a + b}{c} - cd = \frac{(-1) + 1}{-2} - (-2)(3) = 0 - (-6) = 6$

11. $(a - b)^2 \div (c + d)^2 = [(-1) - 1]^2 \div [(-2) + 3]^2 = (-2)^2 \div (1)^2 = 4$

12. $(a^2 + b^2) \div c^2 + d = [(-1)^2 + (1)^2] \div (-2)^2 + 3 = (1 + 1) \div 4 + 3 = 2 \div 4 + 3 =$
$\frac{1}{2} + 3 = 3\frac{1}{2}$

13. $6a^3b + 5a^3b^2 - 3a^2b^2 = a^2b(6a + 5ab - 3b)$ **14.** $49x^2 + 14x + 1 = (7x + 1)^2$

15. $8t^3 - 56t^2 + 98t = 2t(4t^2 - 28t + 49) = 2t(2t - 7)^2$

16. $n^4 + 8n^3 + 15n^2 = n^2(n^2 + 8n + 15) = n^2(n + 3)(n + 5)$ **17.** prime

18. $t^2 - 13t + 22 = (t - 11)(t - 2)$

19. $y^3 + 4y^2 - 32y = y(y^2 + 4y - 32) = y(y + 8)(y - 4)$

20. $5c^2 + 8c - 4 = (5c - 2)(c + 2)$

21. $4x^2 - 1 + 2x - 1 = 4x^2 + 2x - 2 = 2(2x^2 + x - 1) = 2(2x - 1)(x + 1)$

22. $6y - 3 = 27; 6y = 30; y = 5; \{5\}$ **23.** $\frac{1}{4}a - 3 = 9; \frac{1}{4}a = 12; a = 4 \cdot 12; a = 48; \{48\}$

24. $6 - \frac{3}{4}d = -3; -\frac{3}{4}d = -9; d = -9\left(-\frac{4}{3}\right); d = \frac{36}{3} = 12; \{12\}$

25. $(n + 2)^2 = (n + 4)(n - 2); n^2 + 4n + 4 = n^2 + 2n - 8;$
$n^2 + 4n + 4 - (n^2 + 2n - 8) = 0; 2n + 12 = 0; 2n = -12; n = -6; \{-6\}$

26. $10m^2 - m^3 = 25m; -m^3 + 10m^2 - 25m = 0; -m(m^2 - 10m + 25) = 0;$
$-m(m - 5)^2 = 0; m = 0 \text{ or } m = 5; \{0, 5\}$

27. $16c^2 - 9 = 0; (4c - 3)(4c + 3) = 0; 4c - 3 = 0 \text{ or } 4c + 3 = 0; c = \frac{3}{4} \text{ or } c = -\frac{3}{4}; \left\{-\frac{3}{4}, \frac{3}{4}\right\}$

28. $t^2 + 12t + 15 = -5; t^2 + 12t + 20 = 0; (t + 10)(t + 2) = 0; t = -10 \text{ or } t = -2;$
$\{-2, -10\}$

29. $x^2 - 6x = 7; x^2 - 6x - 7 = 0; (x - 7)(x + 1) = 0; x = 7 \text{ or } x = -1; \{7, -1\}$

30. $6y^3 + 13y^2 - 5y = 0; y(6y^2 + 13y - 5) = 0; y(3y - 1)(2y + 5) = 0; y = 0 \text{ or } 3y - 1 = 0$
$\text{or } 2y + 5 = 0; y = 0 \text{ or } y = \frac{1}{3} \text{ or } y = -\frac{5}{2}; \left\{0, \frac{1}{3}, -\frac{5}{2}\right\}$

31. $\frac{x^2 + 5x + 4}{2x^2 + 3x + 1} \cdot \frac{1 + 2x}{x^2 - 16} = \frac{(x + 4)(x + 1)}{(2x + 1)(x + 1)} \cdot \frac{(1 + 2x)}{(x - 4)(x + 4)} = \frac{1}{x - 4}$

32. $\frac{1}{4x - 8} + \frac{x}{x^2 - 3x + 2} = \frac{1}{4(x - 2)} + \frac{x}{(x - 2)(x - 1)} = \frac{(x - 1) + 4x}{4(x - 2)(x - 1)} = \frac{5x - 1}{4(x - 2)(x - 1)}$

33. $\dfrac{a^2 - b^2}{(cd)^3} \div \dfrac{a + b}{c^2d} = \dfrac{(a - b)(a + b)}{c^3d^3} \cdot \dfrac{c^2d}{a + b} = \dfrac{a - b}{cd^2}$

34. $\dfrac{3}{a^2 + 6a + 9} - \dfrac{a + 1}{a^2 - 9} = \dfrac{3}{(a + 3)^2} - \dfrac{a + 1}{(a - 3)(a + 3)} = \dfrac{3(a - 3) - (a + 1)(a + 3)}{(a + 3)^2(a - 3)} =$

$\dfrac{3a - 9 - (a^2 + 4a + 3)}{(a + 3)^2(a - 3)} = \dfrac{-a^2 - a - 12}{(a + 3)^2(a - 3)}$

35.
$$\begin{array}{r} 2y^2 + 3y - 8 + \frac{12}{5y - 4} \\ 5y - 4\overline{)10y^3 + 7y^2 - 52y + 44} \\ \underline{10y^3 - 8y^2} \\ 15y^2 - 52y \\ \underline{15y^2 - 12y} \\ -40y + 44 \\ \underline{-40y + 32} \\ 12 \end{array}$$

36. $x - \dfrac{x + 1}{x - 5} = \dfrac{x(x - 5) - (x + 1)}{x - 5} = \dfrac{x^2 - 5x - x - 1}{x - 5} = \dfrac{x^2 - 6x - 1}{x - 5}$

37. Let x = number of 6-hour shifts. Then $4x$ = number of 8-hour shifts. $6x + 8(4x) = 38$; $6x + 32x = 38$; $38x = 38$; $x = 1$; $4x = 4$; four 8-hour shifts.

38. Let x = number of quarters. Then $24 - x$ = number of half dollars. The value, in cents, of his money is $25(x) + 50(24 - x)$. So $25x + 50(24 - x) + 200 = \frac{1}{2}(25x) + 2[50(24 - x)]$; $50x + 100(24 - x) + 400 = 25x + 200(24 - x)$; $50x + 2400 - 100x + 400 = 25x + 4800 - 200x$; $125x = 2000$; $x = 16$; $y = 24 - 16$; $y = 8$; 16 quarters and 8 half dollars; $(25)(16) = 400$ and $(50)(8) = 400$; \$4 + \$4 = \$8.

39.

	Rate	× Time	= Distance
Bicycle	$r + 8$	$\frac{5}{12}$	$\frac{5}{12}(r + 8)$
Walking	r	$\frac{5}{4}$	$\frac{5}{4}r$

$\frac{5}{12}(r + 8) = \frac{5}{4}r$; $5(r + 8) = 3(5r)$;
$5r + 40 = 15r$; $40 = 10r$; $r = 4$;
$(4 \text{ km/h})\left(\frac{5}{4} \text{ h}\right) = 5 \text{ km}$

40. $x - y = 3$; $x^2 + y^2 = 65$; substitute $x = y + 3$: $(y + 3)^2 + y^2 = 65$; $y^2 + 6y + 9 + y^2 = 65$; $2y^2 + 6y - 56 = 0$; $2(y^2 + 3y - 28) = 0$; $2(y + 7)(y - 4) = 0$; $y = -7$ or $y = 4$; if $y = -7$ then $x - (-7) = 3$ or $x = -4$; if $y = 4$ then $x - 4 = 3$ or $x = 7$; 4 and 7, −4 and −7

Page 284 • MAINTAINING SKILLS

1. $25\overline{)23.00}$ = 0.92; $\frac{23}{25} = 0.92$ **2.** $20\overline{)35.00}$ = 1.75; $\frac{35}{20} = 1.75$ **3.** $4\overline{)10.00}$ = 2.50; $\frac{10}{4} = 2.5$

4. $50\overline{)49.00}$ = 0.98; $\frac{49}{50} = 0.98$ **5.** $25\overline{)81.00}$ = 3.24; $\frac{81}{25} = 3.24$ **6.** $3\overline{)5.00}$ = $1.6\overline{6}$; $\frac{5}{3} \approx 1.67$

7. $30\overline{)51.00}$ = 1.70; $\frac{51}{30} = 1.7$ **8.** $8\overline{)7.000}$ = 0.875; $\frac{7}{8} \approx 0.88$ **9.** $6\overline{)14.00}$ = $2.3\overline{3}$; $\frac{14}{6} \approx 2.33$

10. $24\overline{)41.000}$ with quotient 1.708; $\frac{41}{24} \approx 1.71$ **11.** $22\overline{)15.000}$ with quotient 0.681; $\frac{15}{22} \approx 0.68$ **12.** $17\overline{)21.000}$ with quotient 1.235; $\frac{21}{17} \approx 1.24$

13. $62\% = \frac{62}{100} = \frac{31}{50}$ **14.** $12\% = \frac{12}{100} = \frac{3}{25}$ **15.** $85\% = \frac{85}{100} = \frac{17}{20}$

16. $0.5\% = \frac{0.5}{100} = \frac{5}{1000} = \frac{1}{200}$ **17.** $0.03\% = \frac{0.03}{100} = \frac{3}{10{,}000}$ **18.** $9.2\% = \frac{9.2}{100} = \frac{92}{1000} = \frac{23}{250}$

19. $0.91 = \frac{91}{100} = 91\%$ **20.** $0.07 = \frac{7}{100} = 7\%$ **21.** $0.8 = \frac{8}{10} = \frac{80}{100} = 80\%$

22. $12 = \frac{1200}{100} = 1200\%$ **23.** $0.032 = \frac{32}{1000} = \frac{3.2}{100} = 3.2\%$ **24.** $1.23 = \frac{123}{100} = 123\%$

25. 32% of 85 is $0.32 \times 85 = 27.2$ **26.** 12% of 80 is $0.12 \times 80 = 9.6$

27. 0.2% of 40 is $0.002 \times 40 = 0.08$ **28.** 15.6% of 50 is $0.156 \times 50 = 7.8$

29. 130% of 40 is $1.30 \times 40 = 52$ **30.** 312% of 20 is $3.12 \times 20 = 62.4$

31. $0.30z = 21; z = 70$ **32.** $0.15m = 6; m = 40$ **33.** $0.05y = 0.6; y = 12$

34. $0.24t = 108; t = 450$ **35.** $\frac{p}{100}(50) = 30; p = 60$ **36.** $\frac{a}{100}(45) = 18; a = 40$

37. $\frac{t}{100}(105) = 21; t = 20$ **38.** $28 = \frac{n}{100}(112); 2800 = 112n; n = 25$

39. $51 = \frac{x}{100}(150); 5100 = 150x; x = 34$

Page 285 • PREPARING FOR COLLEGE ENTRANCE EXAMS

1. 24 has 8 pairs of integral factors: 2 and 12, −2 and −12, 3 and 8, −3 and −8, 4 and 6, −4 and −6, 24 and 1, −24 and −1. Therefore there are 8 integral values of k for which $x^2 + kx + 24$ is factorable; E
2. $4x^2 - 26x + 13$ is prime; B
3. $10x^3 - 7x^2 - 12x = 0; x(10x^2 - 7x - 12) = 0; x(5x + 4)(2x - 3) = 0; x = 0,$ $5x - 4 = 0, 2x - 3 = 0; x = 0, x = \frac{4}{5}, x = \frac{3}{2}; 0, \frac{4}{5}, \frac{3}{2}$; D
4. 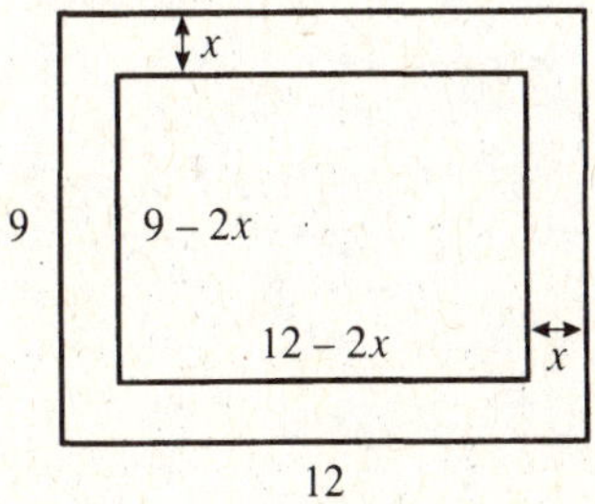

Area of flower bed $= \frac{1}{2}$(Area of garden)

Let x = the width of the border.

Then $(9 - 2x)(12 - 2x) = \frac{1}{2}(12)(9); 108 - 18x - 24x + 4x^2 = \frac{1}{2}(108);$

$108 - 42x + 4x^2 = 54; 54 - 42x + 4x^2 = 0; 27 - 21x + 2x^2 = 0; (2x - 3)(x - 9) = 0;$

$x = \frac{3}{2}$ or $x = 9$; the border can't be 9 ft wide since the garden is only 9 ft wide; use $x = \frac{3}{2}$;

width of flower bed $= 9 - 2x = 9 - 2\left(\frac{3}{2}\right) = 9 - 3 = 6$ ft; length of flower bed =

(Continued)

$12 - 2x = 12 - 2\left(\frac{3}{2}\right) = 12 - 3 = 9$ ft; perimeter of flower bed $= 2(6) + 2(9) =$

$12 + 18 = 30$ ft; B

5. $\frac{4t + v}{t + v} - \frac{4t^2 - v^2}{t^2 - v^2} = \frac{(4t + v)(t - v) - (4t^2 - v^2)}{t^2 - v^2} = \frac{4t^2 - 4tv + tv - v^2 - 4t^2 + v^2}{t^2 - v^2} =$

$\frac{-3tv}{t^2 - v^2} = \frac{3tv}{v^2 - t^2}$; D

6. Divide each binomial into $6x^3 + 29x^2 - 7x - 10$.

I. $$\begin{array}{r|l} & 2x^2 \\ \hline 3x - 1 & 6x^3 + 29x^2 - 7x - 10 \\ & \underline{6x^3 - 2x^2} \\ & 31x^2 - 7x - 10;\ \text{no} \end{array}$$

II. $$\begin{array}{r|l} & 3x^2 \\ \hline 2x + 2 & 6x^3 + 29x^2 - 7x - 10 \\ & \underline{6x^3 + 6x^2} \\ & 23x^2 - 7x - 10;\ \text{no} \end{array}$$

III. $$\begin{array}{r|l} & 6x^2 - x - 2 \\ \hline x + 5 & 6x^3 + 29x^2 - 7x - 10 \\ & \underline{6x^3 + 30x^2} \\ & -x^2 - 7x \\ & \underline{-x^2 - 5x} \\ & -2x - 10 \\ & \underline{-2x - 10} \\ & 0;\ \text{yes} \end{array}$$

$\therefore$ III only is a factor; C

7. $\frac{(3n - 5)^4}{(2n + 1)^2} \div \frac{(5 - 3n)^4}{2n^2 + 7n + 3} = \frac{(3n - 5)^4}{(2n + 1)^2} \cdot \frac{2n^2 + 7n + 3}{(5 - 3n)^4} = \frac{(3n - 5)^4}{(2n + 1)^2} \cdot \frac{(2n + 1)(n + 3)}{(3n - 5)^4(-1)^4} = \frac{n + 3}{2n + 1}$; B

CHAPTER 7 • Applying Fractions

Pages 289–290 • WRITTEN EXERCISES

A **1.** $14 : 21 = 2 : 3$ **2.** $55 : 33 = 5 : 3$ **3.** $24x : 8x = 3 : 1$ **4.** $6y : 9y = 2 : 3$

5. $\frac{s^2}{(3s)^2} = \frac{s^2}{(3)^2(s)^2} = \frac{1}{9}$ **6.** $\frac{27m^5}{45m^2} = \frac{3m^3}{5}$ **7.** $\frac{64a^3b}{16ab^4} = \frac{4a^2}{b^3}$ **8.** $\frac{72rs^5}{12r^2s^2} = \frac{6s^3}{r}$

9. $20 \text{ min} : 2 \text{ h} = \frac{20 \text{ min}}{2 \text{ h}} = \frac{20 \text{ min}}{120 \text{ min}} = \frac{1}{6} = 1 : 6$

10. $4 \text{ h} : 45 \text{ min} = \frac{4 \text{ h}}{45 \text{ min}} = \frac{240 \text{ min}}{45 \text{ min}} = \frac{16}{3} = 16 : 3$

11. $6 \text{ m} : 120 \text{ cm} = \frac{6 \text{ m}}{120 \text{ cm}} = \frac{600 \text{ cm}}{120 \text{ cm}} = \frac{5}{1} = 5 : 1$

12. $18 \text{ cm} : 1.8 \text{ m} = \frac{18 \text{ cm}}{1.8 \text{ m}} = \frac{18 \text{ cm}}{180 \text{ cm}} = \frac{1}{10} = 1 : 10$

13. $6 \text{ wk} : 3 \text{ d} = \frac{6 \text{ wk}}{3 \text{ d}} = \frac{42 \text{ d}}{3 \text{ d}} = \frac{14}{1} = 14 : 1$ **14.** $9 \text{ d} : 3 \text{ wk} = \frac{9 \text{ d}}{3 \text{ wk}} = \frac{9 \text{ d}}{21 \text{ d}} = \frac{3}{7} = 3 : 7$

15. $5 \text{ km} : 450 \text{ cm} = \frac{5 \text{ km}}{450 \text{ cm}} = \frac{500{,}000 \text{ cm}}{450 \text{ cm}} = \frac{10{,}000}{9} = 10{,}000 : 9$

16. $200 \text{ cm} : 8 \text{ km} = \frac{200 \text{ cm}}{8 \text{ km}} = \frac{2 \text{ m}}{8000 \text{ m}} = \frac{1}{4000} = 1 : 4000$

17. $1 \text{ lb} : 7 \text{ oz} = \frac{1 \text{ lb}}{7 \text{ oz}} = \frac{16 \text{ oz}}{7 \text{ oz}} = \frac{16}{7} = 16 : 7$

18. $13 \text{ oz} : 2 \text{ lb} = \frac{13 \text{ oz}}{2 \text{ lb}} = \frac{13 \text{ oz}}{32 \text{ oz}} = \frac{13}{32} = 13 : 32$

19. $150 \text{ g} : 3 \text{ kg} = \frac{150 \text{ g}}{3 \text{ kg}} = \frac{150 \text{ g}}{3000 \text{ g}} = \frac{1}{20} = 1 : 20$

20. $2 \text{ kg} : 90 \text{ g} = \frac{2 \text{ kg}}{90 \text{ g}} = \frac{2000 \text{ g}}{90 \text{ g}} = \frac{200}{9} = 200 : 9$

21. $\frac{\text{school days}}{\text{non-school days}} = \frac{180}{365 - 180} = \frac{180}{185} = \frac{36}{37} = 36 : 37$

22. $\frac{\text{students}}{\text{teachers}} = \frac{2592}{144} = 18 = 18 : 1$ **23.** $\frac{\text{new airplanes}}{\text{old airplanes}} = \frac{720 - 240}{240} = \frac{480}{240} = 2 = 2 : 1$

24. $\frac{\text{raisins}}{\text{nuts}} = \frac{3}{1\frac{1}{4}} = \frac{3}{1.25} = \frac{300}{125} = \frac{12}{5} = 12 : 5$

25. **a.** $\frac{\text{seniors taking math}}{\text{seniors in school}} = \frac{105}{270} = \frac{7}{18} = 7 : 18$

b. $\frac{\text{seniors taking math}}{\text{seniors not taking math}} = \frac{105}{270 - 105} = \frac{105}{165} = \frac{7}{11} = 7 : 11$

26. **a.** $\frac{\text{fiction books}}{\text{nonfiction books}} = \frac{1050}{1890} = \frac{5}{9} = 5 : 9$ **b.** $\frac{\text{nonfiction books}}{\text{fiction books}} = \frac{1890}{1050} = \frac{9}{5} = 9 : 5$

27. **a.** $\frac{\text{perimeter of first}}{\text{perimeter of second}} = \frac{2(8 + 6)}{2(10 + 9)} = \frac{14}{19} = 14 : 19$

b. $\frac{\text{area of first}}{\text{area of second}} = \frac{8 \cdot 6}{10 \cdot 9} = \frac{8}{15} = 8 : 15$

28. a. $\frac{\text{first perimeter}}{\text{second perimeter}} = \frac{30}{30} = 1 = 1 : 1$ **b.** $\frac{\text{first area}}{\text{second area}} = \frac{12 \cdot 3}{10 \cdot 5} = \frac{18}{25} = 18 : 25$

29. a. $\frac{\text{first perimeter}}{\text{second perimeter}} = \frac{4 \cdot 60 \text{ cm}}{4 \cdot 1 \text{ m}} = \frac{60 \text{ cm}}{100 \text{ cm}} = \frac{3}{5} = 3 : 5$

b. $\frac{\text{first area}}{\text{second area}} = \frac{(60 \text{ cm})^2}{(1 \text{ m})^2} = \frac{(60 \text{ cm})^2}{(100 \text{ cm})^2} = \frac{3600 \text{ cm}^2}{10{,}000 \text{ cm}^2} = \frac{9}{25} = 9 : 25$

30. a. $\frac{\text{first perimeter}}{\text{second perimeter}} = \frac{4 \cdot 24 \text{ in.}}{4 \cdot 2 \text{ yd}} = \frac{24 \text{ in.}}{72 \text{ in.}} = \frac{1}{3} = 1 : 3$

b. $\frac{\text{first area}}{\text{second area}} = \frac{(24 \text{ in.})^2}{(2 \text{ yd})^2} = \frac{(24 \text{ in.})^2}{(72 \text{ in.})^2} = \frac{576 \text{ in.}^2}{5184 \text{ in.}^2} = \frac{1}{9} = 1 : 9$

B **31.** $8x = 5y; x = \frac{5}{8}y; \frac{x}{y} = \frac{5}{8}$, or $5 : 8$ **32.** $7x = 4y; x = \frac{4}{7}y; \frac{x}{y} = \frac{4}{7}$, or $4 : 7$

33. $14x = 12y; x = \frac{12}{14}y; \frac{x}{y} = \frac{6}{7}$, or $6 : 7$ **34.** $10x = 26y; x = \frac{26}{10}y; \frac{x}{y} = \frac{13}{5}$, or $13 : 5$

35. $kx = 2y; x = \frac{2}{k}y; \frac{x}{y} = \frac{2}{k}$, or $2 : k$ **36.** $3x = ky; x = \frac{k}{3}y; \frac{x}{y} = \frac{k}{3}$, or $k : 3$

37. $4(x + y) = 8(x - y); 4x + 4y = 8x - 8y; 4x - 8x + 4y = -8y; -4x = -8y - 4y;$

$-4x = -12y; x = 3y; \frac{x}{y} = \frac{3}{1}$, or $3 : 1$

38. $8(2x - 3y) = 6(x + y); 16x - 24y = 6x + 6y; 16x - 6x - 24y = 6y;$

$10x = 6y + 24y; 10x = 30y; x = 3y; \frac{x}{y} = \frac{3}{1}$, or $3 : 1$

39. $ax + by = ay + bx; ax - bx + by = ay; (a - b)x = ay - by; (a - b)x = (a - b)y;$

$x = y; \frac{x}{y} = \frac{1}{1}$, or $1 : 1$

40. $cx - ay = aby - bcx; cx + bcx - ay = aby; (c + bc)x = aby + ay; (c + bc)x =$

$(ab + a)y; x = \frac{(ab + a)}{(c + bc)}y; \frac{x}{y} = \frac{a(b + 1)}{c(b + 1)}; \frac{x}{y} = \frac{a}{c}$, or $a : c$

41. $ax - a^2y = bx - b^2y; ax - bx - a^2y = -b^2y; (a - b)x = a^2y - b^2y; (a - b)x =$

$(a^2 - b^2)y; x = \frac{a^2 - b^2}{a - b}y; \frac{x}{y} = \frac{(a + b)(a - b)}{a - b}; \frac{x}{y} = \frac{a + b}{1}$, or $(a + b) : 1$

42. $r^2x = sy + ry + s^2x; r^2x - s^2x = sy + ry; (r^2 - s^2)x = (s + r)y; x = \frac{r + s}{r^2 - s^2}y;$

$\frac{x}{y} = \frac{r + s}{(r + s)(r - s)}; \frac{x}{y} = \frac{1}{r - s}$, or $1 : (r - s)$

C **43.** $x^2 + 2y^2 = 2xy + y^2; x^2 + 2y^2 - 2xy - y^2 = 0; x^2 - 2xy + y^2 = 0;$

$(x - y)^2 = 0; x - y = 0; x = y; \frac{x}{y} = \frac{1}{1}$, or $1 : 1$

44. $2(x^2 + y^2) = 5xy; 2x^2 + 2y^2 = 5xy; 2x^2 - 5xy + 2y^2 = 0; (2x - y)(x - 2y) = 0;$

$2x - y = 0, 2x = y, x = \frac{1}{2}y, \frac{x}{y} = \frac{1}{2}$, or $1 : 2$; or $x - 2y = 0, x = 2y, \frac{x}{y} = \frac{2}{1}$, or $2 : 1$

45. $\frac{x}{y} + 1 = \frac{x+y}{x}$; $\left(\frac{x}{y} + 1\right)xy = \left(\frac{x+y}{x}\right)xy$; $x^2 + xy = (x + y)y$; $x^2 + xy = xy + y^2$; $x^2 + xy - xy - y^2 = 0$; $x^2 - y^2 = 0$; $(x + y)(x - y) = 0$; $x + y = 0$, $x = -y$, $\frac{x}{y} = \frac{-1}{1}$, or $-1 : 1$, or $1 : -1$; $x - y = 0$, $x = y$, $\frac{x}{y} = \frac{1}{1}$, or $1 : 1$

Pages 291–292 • PROBLEMS

A

1. Let $4x$ and $5x$ be the numbers. $4x + 5x = 45$; $9x = 45$; $x = 5$; $4x = 20$; $5x = 25$; The numbers are 20 and 25.
2. Let $3x$ and $7x$ be the numbers. $3x + 7x = 50$; $10x = 50$; $x = 5$; $3x = 15$; $7x = 35$; The numbers are 15 and 35.
3. Let $9x$ be the number of players and x be the number of coaches. $9x + x = 180$; $10x = 180$; $x = 18$; $9x = 162$; There are 162 players.
4. Let $17x$ be the number of men and $18x$ be the number of women. $17x + 18x = 700$; $35x = 700$; $x = 20$; $18x = 360$; There were 360 women.
5. Let $7x$ be the length and $5x$ be the width. $2(7x + 5x) = 96$; $2(12x) = 96$; $12x = 48$; $x = 4$; $7x = 28$; $5x = 20$; The rectangle is 28 cm by 20 cm.
6. Let $9x$ be the length and $8x$ be the width. $2(9x + 8x) = 68$; $2(17x) = 68$; $17x = 34$; $x = 2$; $9x = 18$; $8x = 16$; The rectangle is 18 ft by 16 ft.
7. Let $3x$, $5x$, and $6x$ be the sides of the triangle. $3x + 5x + 6x = 21$; $14x = 21$; $x = \frac{3}{2}$; $3x = \frac{9}{2}$; $5x = \frac{15}{2}$; $6x = 9$; The sides of the triangle measure $4\frac{1}{2}$ cm, $7\frac{1}{2}$ cm, and 9 cm.
8. Let x, $2x$, and $3x$ be the measures of the angles. $x + 2x + 3x = 180$; $6x = 180$; $x = 30$; $2x = 60$; $3x = 90$; The angles measure 30°, 60°, and 90°.
9. Let $3x$, $6x$, and $8x$ be the number of cubic meters of each ingredient. $3x + 6x + 8x = 850$; $17x = 850$; $x = 50$; $8x = 400$; 400 m^3 of gravel is needed.
10. Let $8x$, $3x$, and x be the number of cubic meters of each metal. $8x + 3x + x = 420$; $12x = 420$; $x = 35$; $8x = 280$; $3x = 105$; 280 m^3 of iron, 105 m^3 of zinc, and 35 m^3 of tungsten are needed.

B

11.

	Rate	× Time =	Distance
Ling	$r + 18$	1.5 h	$(r + 18)(1.5)$
Eddie	r	1.5 h	$1.5r$

$\frac{(r+18)(1.5)}{1.5r} = \frac{5}{2}$, $\frac{r+18}{r} = \frac{5}{2}$, $r + 18 = \frac{5}{2}r$, $18 = \frac{5}{2}r - r$, $18 = \frac{3}{2}r$, $12 = r$; $r + 18 = 30$; Ling drives 30 mi/h and Eddie rides 12 mi/h.

12.

	Rate ×	Time =	Distance
Aldo	$6x$	$\frac{1}{3}$ h	$2x$
José	$5x$	$\frac{1}{2}$ h	$\frac{5}{2}x$

$\frac{5}{2}x - 2x = 2$, $\frac{1}{2}x = 2$, $x = 4$; $6x = 24$; $5x = 20$; Aldo is cycling at 24 km/h and José is cycling at 20 km/h.

13.

	Number	× Value	= Total
Dimes	$3x$	10¢	$30x$
Nickels	$2x$	5¢	$10x$

$30x + 10x = 560$, $40x = 560$, $x = 14$; $3x = 42$; $2x = 28$; There are 42 dimes and 28 nickels.

14.

	Number	× Value	= Total
Nickels	$3x$	5¢	$15x$
Dimes	$8x$	10¢	$80x$
Quarters	$10x$	25¢	$250x$

$15x + 80x + 250x = 690$, $345x = 690$, $x = 2$; $3x = 6$; $8x = 16$; $10x = 20$; There are 6 nickels, 16 dimes, and 20 quarters.

15.

	Number	× Price	= Cost
Pencils	$5x$	1	$5x$
Pens	$9x$	0.25	$2.25x$

$5x + 2.25x = 290$, $7.25x = 290$, $x = 40$; $5x = 200$; 200 pencils were purchased.

16.

	Number	× Price	= Cost
Sunglasses	$7x$	6	$42x$
Visors	$4x$	4.5	$18x$

$42x + 18x = 720$, $60x = 720$, $x = 12$; $7x = 84$; 84 pairs of sunglasses were purchased.

C **17.** $\frac{x+y}{x-y} = \frac{7}{5}$, $x + y = \frac{7}{5}(x - y)$, $5(x + y) = 7(x - y)$, $5x + 5y = 7x - 7y$, $5x - 7x + 5y = -7y$, $-2x = -7y - 5y$, $-2x = -12y$, $x = 6y$; $x = 6$, $y = 1$; $x = 12$, $y = 2$; $x = 18$, $y = 3$; The possible pairs of numbers are 6 and 1, 12 and 2, and 18 and 3.

18. The LCD of 5 and 8 is 40, so convert each ratio: $\frac{7}{5} = \frac{56}{40}, \frac{8}{9} = \frac{40}{45}$. Now let $56x$, $40x$, and $45x$ be the number of each size of car. $56x + 40x + 45x = 2820$, $141x = 2820$, $x = 20$; $45x = 900$; There are 900 full-size cars.

19. Let x and y be the numbers. $\frac{x+y}{x-y} = \frac{3}{2}$, $2(x + y) = 3(x - y)$, $2x + 2y = 3x - 3y$, $5y = x$; $\frac{x-y}{xy} = \frac{2}{5}$, $5(x - y) = 2xy$; $5(5y - y) = 2(5y)y$, $5(4y) = 10y^2$, $20y = 10y^2$, $0 = 10y^2 - 20y$, $0 = y(y - 2)$; $y = 0, 2$ ($y \neq 0$ because xy would then equal zero and cannot be the denominator of a ratio); $x = 5y = 10$; The two numbers are 10 and 2.

Page 292 • MIXED REVIEW EXERCISES

1. $4x = 24 - 2x$, $6x = 24$, $x = 4$; $\{4\}$

2. $4(x - 7) + x = 2$, $4x - 28 + x = 2$, $5x - 28 = 2$, $5x = 30$, $x = 6$; $\{6\}$

3. $5(4 + n) = 2(9 + 2n)$, $20 + 5n = 18 + 4n$, $20 + n = 18$, $n = -2$; $\{-2\}$

4. $\frac{z+4}{3} = -12$, $z + 4 = -36$, $z = -40$; $\{-40\}$

5. $\frac{16-2y}{2} = 3y$, $16 - 2y = 6y$, $16 = 8y$, $2 = y$; $\{2\}$ **6.** $-\frac{c}{9} = 4$, $c = -36$; $\{-36\}$

7. $(r+2)(r-5) = 0$; $r = -2, 5$; $\{-2, 5\}$

8. $3x^2 + 12x - 36 = 0$, $3(x^2 + 4x - 12) = 0$, $(x+6)(x-2) = 0$; $x = -6, 2$; $\{-6, 2\}$

9. $3(x-2) = 4(x-3)$, $3x - 6 = 4x - 12$, $-6 = x - 12$, $6 = x$; $\{6\}$

10. $\frac{3b+1}{2c} + \frac{b}{c} = \frac{3b+1}{2c} + \frac{2b}{2c} = \frac{3b+1+2b}{2c} = \frac{5b+1}{2c}$ **11.** $3x + \frac{5}{x} = \frac{3x^2}{x} + \frac{5}{x} = \frac{3x^2+5}{x}$

12. $\frac{3a}{4} + \frac{4a+1}{4} = \frac{3a+4a+1}{4} = \frac{7a+1}{4}$

Page 292 • CHALLENGE

The scientist could fill two 5 L containers and then pour back 8 L. $2 \cdot 5 - 8 = 10 - 8 = 2$.

Pages 295–296 • WRITTEN EXERCISES

A **1.** $\frac{x}{5} = \frac{3}{4}$; $4x = 5 \cdot 3$; $4x = 15$; $x = \frac{15}{4}$; $\left\{\frac{15}{4}\right\}$ **2.** $\frac{y}{3} = \frac{4}{7}$; $7y = 12$; $y = \frac{12}{7}$; $\left\{\frac{12}{7}\right\}$

3. $\frac{8}{5} = \frac{4}{3x}$; $24x = 20$; $x = \frac{20}{24}$; $x = \frac{5}{6}$; $\left\{\frac{5}{6}\right\}$ **4.** $\frac{6n}{5} = \frac{3}{1}$; $6n = 15$; $n = \frac{15}{6}$; $n = \frac{5}{2}$; $\left\{\frac{5}{2}\right\}$

5. $\frac{3n}{7} = \frac{4}{5}$; $15n = 28$; $n = \frac{28}{15}$; $\left\{\frac{28}{15}\right\}$ **6.** $\frac{8}{5k} = \frac{2}{5}$; $40 = 10k$; $4 = k$; $\{4\}$

7. $\frac{7}{6} = \frac{2a}{3}$; $21 = 12a$; $\frac{21}{12} = a$; $\frac{7}{4} = a$; $\left\{\frac{7}{4}\right\}$ **8.** $\frac{3}{8} = \frac{9}{4k}$; $12k = 72$; $k = 6$; $\{6\}$

9. $\frac{18x}{13} = \frac{36}{39}$; $\frac{18x}{13} = \frac{12}{13}$; $18x = 12$; $x = \frac{12}{18}$; $x = \frac{2}{3}$; $\left\{\frac{2}{3}\right\}$

10. $\frac{15x}{64} = \frac{45}{32}$; $480x = 2880$; $x = 6$; $\{6\}$ **11.** $\frac{24}{25} = \frac{6y}{7}$; $168 = 150y$; $\frac{168}{150} = y$; $\frac{28}{25} = y$; $\left\{\frac{28}{25}\right\}$

12. $\frac{10}{21} = \frac{5}{16a}$; $160a = 105$; $a = \frac{105}{160}$; $a = \frac{21}{32}$; $\left\{\frac{21}{32}\right\}$

13. $\frac{15a}{36} = \frac{45}{12}$; $\frac{5a}{12} = \frac{15}{4}$; $20a = 180$; $a = 9$; $\{9\}$ **14.** $\frac{81}{64} = \frac{27n}{40}$; $3240 = 1728n$; $\frac{15}{8} = n$; $\left\{\frac{15}{8}\right\}$

15. $\frac{12t}{-7} = \frac{30}{14}$; $\frac{12t}{-7} = \frac{15}{7}$; $12t = -15$; $t = -\frac{15}{12}$; $t = -\frac{5}{4}$; $\left\{-\frac{5}{4}\right\}$

16. $\frac{-15}{25} = \frac{4x}{-15}$; $\frac{-3}{5} = \frac{4x}{-15}$; $45 = 20x$; $\frac{45}{20} = x$; $\frac{9}{4} = x$; $\left\{\frac{9}{4}\right\}$

17. $\frac{4}{x} = 5$; $\frac{4}{x} = \frac{5}{1}$; $4 = 5x$; $\frac{4}{5} = x$; $\left\{\frac{4}{5}\right\}$ **18.** $\frac{5}{2n} = 3$; $\frac{5}{2n} = \frac{3}{1}$; $5 = 6n$; $\frac{5}{6} = n$; $\left\{\frac{5}{6}\right\}$

19. $-4 = \frac{8b}{5}$; $\frac{-4}{1} = \frac{8b}{5}$; $-20 = 8b$; $\frac{-20}{8} = b$; $-\frac{5}{2} = b$; $\left\{-\frac{5}{2}\right\}$

20. $4 = \frac{20}{-9w}$; $\frac{4}{1} = \frac{20}{-9w}$; $-36w = 20$; $w = \frac{20}{-36}$; $w = -\frac{5}{9}$; $\left\{-\frac{5}{9}\right\}$

21. $\frac{x-5}{4} = \frac{3}{2}$; $2(x-5) = 12$; $2x - 10 = 12$; $2x = 22$; $x = 11$; $\{11\}$

22. $\frac{y-3}{8} = \frac{3}{4}$; $4(y - 3) = 24$; $y - 3 = 6$; $y = 9$; $\{9\}$

23. $2 = \frac{6 + 4n}{7}$; $14 = 6 + 4n$; $8 = 4n$; $2 = n$; $\{2\}$

24. $5 = \frac{4 - 3y}{5}$; $25 = 4 - 3y$; $3y + 25 = 4$; $3y = -21$; $y = -7$; $\{-7\}$

25. $\frac{5n - 3}{4} = \frac{5n + 3}{6}$; $6(5n - 3) = 4(5n + 3)$; $30n - 18 = 20n + 12$; $10n - 18 = 12$;

$10n = 30$; $n = 3$; $\{3\}$

26. $\frac{6x - 2}{7} = \frac{5x + 7}{8}$; $8(6x - 2) = 7(5x + 7)$; $48x - 16 = 35x + 49$; $13x - 16 = 49$;

$13x = 65$; $x = 5$; $\{5\}$

27. $\frac{3 + 2y}{8} = \frac{1 - y}{5}$; $5(3 + 2y) = 8(1 - y)$; $15 + 10y = 8 - 8y$; $15 + 18y = 8$; $18y = -7$;

$y = -\frac{7}{18}$; $\left\{-\frac{7}{18}\right\}$

28. $\frac{2n - 9}{7} = \frac{3 - n}{4}$; $4(2n - 9) = 7(3 - n)$; $8n - 36 = 21 - 7n$; $15n - 36 = 21$;

$15n = 57$; $n = \frac{57}{15}$; $n = \frac{19}{5}$; $\left\{\frac{19}{5}\right\}$

29. $\frac{4x + 5}{5} = \frac{2x + 7}{7}$; $7(4x + 5) = 5(2x + 7)$; $28x + 35 = 10x + 35$; $18x + 35 = 35$;

$18x = 0$; $x = 0$; $\{0\}$

30. $\frac{4x + 12}{8} = \frac{x + 3}{2}$; $2(4x + 12) = 8(x + 3)$; $4x + 12 = 4(x + 3)$; $4x + 12 = 4x + 12$;

identity

B **31.** $\frac{4(2x - 11)}{13} = -(x + 5)$; $4(2x - 11) = -13(x + 5)$; $8x - 44 = -13x - 65$;

$21x - 44 = -65$; $21x = -21$; $x = -1$; $\{-1\}$

32. $2x - 2 = \frac{5(x - 3)}{11}$; $11(2x - 2) = 5(x - 3)$; $22x - 22 = 5x - 15$; $17x - 22 = -15$;

$17x = 7$; $x = \frac{7}{17}$; $\left\{\frac{7}{17}\right\}$

33. $\frac{3 + 2x}{3 - 2x} = -3$; $3 + 2x = -3(3 - 2x)$; $3 + 2x = -9 + 6x$; $12 + 2x = 6x$; $12 = 4x$;

$3 = x$; $\{3\}$

34. $-5 = \frac{4 - 7x}{4 + 7x}$; $-5(4 + 7x) = 4 - 7x$; $-20 - 35x = 4 - 7x$; $-24 - 35x = -7x$;

$-24 = 28x$; $-\frac{24}{28} = x$; $-\frac{6}{7} = x$; $\left\{-\frac{6}{7}\right\}$

35. $\frac{15}{x + 7} = \frac{45}{x + 21}$; $15(x + 21) = 45(x + 7)$; $x + 21 = 3(x + 7)$; $x + 21 = 3x + 21$;

$-2x + 21 = 21$; $-2x = 0$; $x = 0$; $\{0\}$

36. $\frac{36}{2x-3} = \frac{72}{2x+3}$; $36(2x + 3) = 72(2x - 3)$; $2x + 3 = 2(2x - 3)$; $2x + 3 = 4x - 6$; $3 = 2x - 6$; $9 = 2x$; $\frac{9}{2} = x$; $\left\{\frac{9}{2}\right\}$

37. $\frac{3x + 2y}{5} = \frac{4x - 7y}{6}$; $6(3x + 2y) = 5(4x - 7y)$; $18x + 12y = 20x - 35y$; $47y + 18x = 20x$; $47y = 2x$; $\frac{47}{2}y = x$; $\frac{47}{2} = \frac{x}{y}$

38. $\frac{9x + 3y}{4} = \frac{3x + 9y}{5}$; $5(9x + 3y) = 4(3x + 9y)$; $45x + 15y = 12x + 36y$; $33x + 15y = 36y$; $33x = 21y$; $x = \frac{21}{33}y$; $\frac{x}{y} = \frac{7}{11}$

39. $\frac{x + y}{x - y} = \frac{3}{2}$; $2(x + y) = 3(x - y)$; $2x + 2y = 3x - 3y$; $2y = x - 3y$; $5y = x$, $5 = \frac{x}{y}$; $\frac{5}{1} = \frac{x}{y}$

40. $\frac{4}{5} = \frac{x - y}{x + y}$; $4(x + y) = 5(x - y)$; $4x + 4y = 5x - 5y$; $4y = x - 5y$; $9y = x$; $9 = \frac{x}{y}$; $\frac{9}{1} = \frac{x}{y}$

41. $\frac{d}{c} = \frac{x - y}{x + y}$; $d(x + y) = c(x - y)$; $dx + dy = cx - cy$; $dx - cx + dy = -cy$; $(d - c)x = -dy - cy$; $(d - c)x = -(d + c)y$; $x = -\frac{d + c}{d - c}y$; $\frac{x}{y} = -\frac{d + c}{d - c} = \frac{d + c}{c - d}$

42. $\frac{x + y}{x - y} = \frac{a}{b}$; $b(x + y) = a(x - y)$; $bx + by = ax - ay$; $bx - ax + by = -ay$; $(b - a)x = -by - ay$; $(b - a)x = -(b + a)y$; $x = -\frac{b + a}{b - a}y$; $\frac{x}{y} = -\frac{b + a}{b - a} = \frac{b + a}{a - b}$

43. $\frac{cy}{d - c} = \frac{dx}{d^2 - c^2}$; $cy(d^2 - c^2) = dx(d - c)$; $c(d + c)(d - c)y = d(d - c)x$; $\frac{c(d + c)(d - c)}{d(d - c)}y = x$; $\frac{c(d + c)}{d} = \frac{x}{y}$

44. $\frac{b - a}{ay} = \frac{b^2 - a^2}{bx}$; $bx(b - a) = ay(b^2 - a^2)$; $x = \frac{a(b + a)(b - a)}{b(b - a)}y$; $\frac{x}{y} = \frac{a(b + a)}{b}$

45. $\frac{p}{t} = \frac{P}{T}$; $pT = tP$; $\frac{pT}{t} = P$; $T = \frac{tP}{p}$

46. a and c; the product of the means is equal to the product of the extremes.

C **47.** $\frac{5}{x} = \frac{x}{125}$; $625 = x^2$; $25 = x$; 25 **48.** $\frac{4}{x} = \frac{x}{64}$; $256 = x^2$; $16 = x$; 16

49. $\frac{3}{2} : x = x : \frac{2}{27}$; $\frac{3}{2} \cdot \frac{2}{27} = x^2$; $\frac{1}{9} = x^2$; $\frac{1}{3} = x$; $\frac{1}{3}$

50. $\frac{4}{5} : x = x : \frac{5}{16}$; $\frac{4}{5} \cdot \frac{5}{16} = x^2$; $\frac{1}{4} = x^2$; $\frac{1}{2} = x$; $\frac{1}{2}$

Pages 296–297 • PROBLEMS

1–13. Methods of estimation may vary.

A **1.** Let x be the cost of the oranges. 6 oranges cost \$.99; 10 oranges cost \$$x$. Since 10 oranges is less than twice 6 oranges, which cost about \$1.00, estimate that the answer is less than \$2.00. $\frac{6}{99} = \frac{10}{x}$; $6x = 990$; $x = \frac{990}{6}$; $x = 165$. Since $165 < 200$, the answer is reasonable; 10 oranges cost \$1.65.

2. Let x be the cost of 8 cans. Since 3 cans cost about 90¢, the cost per can is about 30¢. Estimate that 8 cans cost about \$2.40. $\frac{3}{87} = \frac{8}{x}$; $3x = 696$; $x = 232$. Since 232 is about 240, the answer is reasonable; 8 cans cost \$2.32.

3. Let x be the number of miles Maria can drive in 5 h. Maria drove 111 mi in 3 h; she can drive x mi in 5 h. Since she drove about 120 mi in 3 h, Maria drove about 40 mi/h. Estimate that she can drive about 200 mi in 5 h. $\frac{111}{3} = \frac{x}{5}$; $555 = 3x$; $185 = x$. Since 185 is about 200, the answer is reasonable; Maria can drive 185 mi in 5 h.

4. Let x be the number of liters needed to go 300 km. The truck uses 8 L to go 120 km; x L to go 300 km. Since $2 \cdot 120 < 300 < 3 \cdot 120$, estimate that $16 < x < 24$. $\frac{8}{120} = \frac{x}{300}$; $2400 = 120x$; $20 = x$. Since $16 < 20 < 24$, the answer is reasonable; the truck uses 20 L of gasoline.

5. Let x be the cost of the second car. An \$11,800 car has a tax of \$767; an \$$x$ car has a tax of \$637. Since the first car costs about $12{,}000 = 15 \cdot 800$, estimate that the second car costs about $15 \cdot 600 = 9000$. $\frac{11{,}800}{767} = \frac{x}{637}$; $7{,}516{,}600 = 767x$; $9800 = x$. Since 9800 is about 9000, the answer is reasonable; the second car costs \$9800.

6. Let x be the number of dozens of muffins made from 900 g of flour. 600 g makes $2\frac{1}{2}$ dozen; 900 g makes x dozen. Since $900 = 1\frac{1}{2} \cdot 600$, estimate that the answer is more than $1\frac{1}{2} \cdot 2$, or 3 dozen muffins. $\frac{600}{2\frac{1}{2}} = \frac{900}{x}$; $600x = \frac{5}{2} \cdot 900$; $600x = 2250$; $x = 3\frac{3}{4}$. Since $3\frac{3}{4} > 3$, the answer is reasonable; $3\frac{3}{4}$ dozen, or 45 muffins can be made with 900 g of flour.

7. Let x be the amount to be invested. \$4000 earns \$210 interest; \$$x$ earns \$336 interest. Since about \$200 is earned on \$4000 and $4000 = 20 \cdot 200$, estimate that an investment of about $20 \cdot 300 = 6000$ is needed to earn \$336 in interest. $\frac{4000}{210} = \frac{x}{336}$, $1{,}344{,}000 = 210x$; $6400 = x$. Since $6400 \approx 6000$, the answer is reasonable. You need to invest \$6400 to earn \$336 in interest.

8. Let x be the number of dentists recommending the brand. 3 out of 4 dentists recommend the brand; x out of 92 recommend the brand. Since 3 dentists is $\frac{3}{4}$ of 4 dentists, estimate that the answer is less than $\frac{3}{4}(100)$, or 75 dentists. $\frac{3}{4} = \frac{x}{92}$; $276 = 4x$; $69 = x$. Since $69 < 75$, the answer is reasonable. 69 dentists recommend the brand.

B **9.** Let x be the estimated number of homes with computers. 48 homes out of 360 have computers; x out of 18,000 homes have computers. Since about 50 homes out of 400, or $\frac{1}{8}$, have computers, estimate that the answer is about $\frac{1}{8}$ of 20,000, or 2500. $\frac{48}{360} = \frac{x}{18{,}000}$; $864{,}000 = 360x$; $2400 = x$. Since 2400 is about 2500, the answer is reasonable. Approximately 2400 homes have personal computers.

10. Let x be the total number of acres needed to yield 660 bushels. 25 acres yield 550 bushels; x acres yield 660 bushels. Since 660 is $\frac{6}{5} \cdot 550$, expect x to be $\frac{6}{5} \cdot 25 = 30$. $\frac{550}{25} = \frac{660}{x}$; $550x = 25 \cdot 660$; $x = \frac{16{,}500}{550} = 30$. Since $30 = 30$, this is reasonable. The number of additional acres needed is $30 - 25 = 5$ acres.

11. a. Let w be the new width. Old dimensions are 20 cm by 15 cm; new dimensions are 28 cm by x cm. Since 15 is $\frac{3}{4}$ of 20, expect that the answer will be $\frac{3}{4}$ of 28, or 21. $\frac{20}{15} = \frac{28}{w}$; $20w = 420$; $w = 21$. Since $21 = 21$, the answer is reasonable. The width becomes 21 cm.

b. $\frac{\text{new length}}{\text{old length}} = \frac{28}{20} = \frac{7}{5}$; $\frac{\text{new perimeter}}{\text{old perimeter}} = \frac{\cancel{2}(28 + 21)}{\cancel{2}(20 + 15)} = \frac{49}{35} = \frac{7}{5}$; $\frac{\text{new area}}{\text{old area}} = \frac{\overset{7}{\cancel{28}} \cdot \overset{7}{\cancel{21}}}{\underset{5}{\cancel{20}} \cdot \underset{5}{\cancel{15}}} = \frac{49}{25}$

12. Let a be the area of Wyoming in km^2. 1 cm represents 10 km, so 1 cm^2 represents 100 km^2; (44.5)(59.1) cm^2 represents $100x$ km^2. Since (44.5)(59.1) is about $(40)(60) = 2400$, estimate that the answer is about 100(2400), or 240,000 km^2. $\frac{1}{100} = \frac{(44.5)(59.1)}{x}$; $x = 100(2629.95) = 262{,}995$. Since 262,995 is close to 240,000, the answer is reasonable. The area of Wyoming is 262,995 km^2.

C **13.** First estimate the weight of each board. The mahogany board weighs about $34\left(\frac{1}{2}\right)\left(\frac{1}{12}\right)(6)$, or $8\frac{1}{2}$ lb; the pine board weighs about $\left(23\frac{1}{2}\right)\left(\frac{1}{4}\right)\left(\frac{1}{12}\right)(8)$, or about 4 lb; the mahogany board weighs more by estimation. Let m be the weight of the mahogany board. $\frac{33.94}{12^3} = \frac{m}{5\frac{1}{2} \cdot 1 \cdot (6 \cdot 12)}$; $\frac{33.94}{1728} = \frac{m}{396}$; $13{,}440.24 = 1728m$; $m \approx 7.8$.

Let p be the weight of the pine board. $\frac{23.45}{12^3} = \frac{p}{3\frac{1}{2} \cdot 1\frac{1}{2} \cdot (8 \cdot 12)}$; $\frac{23.45}{1728} = \frac{p}{504}$;

$11{,}818.8 = 1728p$; $p \approx 6.8$. The mahogany board weighs more, which agrees with the estimate; the answer is reasonable.

Page 297 • MIXED REVIEW EXERCISES

1. $\frac{1}{4x^2y}, \frac{5}{xy}$; LCD is $4x^2y$ **2.** $\frac{w+4}{4}, \frac{2w-6}{12}$; LCD is 12 **3.** $\frac{4}{9}, \frac{7}{12}$; LCD is 36

4. $\frac{1}{6}, \frac{1}{5}, \frac{2}{12}$; LCD is 60 **5.** $\frac{3}{x-3}, \frac{6}{x+3}$; LCD is $(x-3)(x+3)$ or $x^2 - 9$

6. $\frac{x}{3y}, \frac{x+1}{2}$; LCD is $6y$ **7.** $\frac{6}{2(x-1)} + \frac{4}{x-1} = \frac{6}{2(x-1)} + \frac{8}{2(x-1)} = \frac{14}{2(x-1)} = \frac{7}{x-1}$

8. $\frac{2r}{3} + \frac{r+1}{12} = \frac{8r}{12} + \frac{r+1}{12} = \frac{9r+1}{12}$

9. $\frac{3a}{10} + \frac{4a+1}{8} = \frac{12a}{40} + \frac{5(4a+1)}{40} = \frac{12a+20a+5}{40} = \frac{32a+5}{40}$

10. $|-7.4| - |2.6| = 7.4 - 2.6 = 4.8$ **11.** $-|3.2 + 1.1| = -|4.3| = -4.3$

12. $9 + 3 \cdot 5 = 9 + 15 = 24$

Pages 299–300 • WRITTEN EXERCISES

A **1.** $\frac{x}{3} + \frac{x}{2} = 1$; $6\left(\frac{x}{3} + \frac{x}{2}\right) = 6(1)$; $2x + 3x = 6$; $5x = 6$; $x = \frac{6}{5}$; $\left\{\frac{6}{5}\right\}$

2. $\frac{3y}{8} + \frac{y}{4} = 10$; $8\left(\frac{3y}{8} + \frac{y}{4}\right) = 8(10)$; $3y + 2y = 80$; $5y = 80$; $y = 16$; $\{16\}$

3. $\frac{3n}{10} - \frac{n}{5} = \frac{1}{2}$; $10\left(\frac{3n}{10} - \frac{n}{5}\right) = \frac{10}{2}$; $3n - 2n = 5$; $n = 5$; $\{5\}$

4. $\frac{2a}{3} - \frac{3a}{2} = -\frac{5}{6}$; $6\left(\frac{2a}{3} - \frac{3a}{2}\right) = 6\left(-\frac{5}{6}\right)$; $4a - 9a = -5$; $-5a = -5$; $a = 1$; $\{1\}$

5. $\frac{w}{3} + \frac{w}{4} = \frac{7}{4}$; $12\left(\frac{w}{3} + \frac{w}{4}\right) = 12\left(\frac{7}{4}\right)$; $4w + 3w = 21$; $7w = 21$; $w = 3$; $\{3\}$

6. $\frac{3c}{10} + \frac{c}{5} = \frac{3}{2}$; $10\left(\frac{3c}{10} + \frac{c}{5}\right) = 10\left(\frac{3}{2}\right)$; $3c + 2c = 15$; $5c = 15$; $c = 3$; $\{3\}$

7. $\frac{3a}{5} - \frac{a}{2} = \frac{1}{20}$; $20\left(\frac{3a}{5} - \frac{a}{2}\right) = 20\left(\frac{1}{20}\right)$; $12a - 10a = 1$; $2a = 1$; $a = \frac{1}{2}$; $\left\{\frac{1}{2}\right\}$

8. $\frac{7m}{8} - \frac{3m}{5} = \frac{11}{2}$; $40\left(\frac{7m}{8} - \frac{3m}{5}\right) = 40\left(\frac{11}{2}\right)$; $35m - 24m = 220$; $11m = 220$; $m = 20$; $\{20\}$

9. $\frac{x}{5} - \frac{x+4}{7} = 0$; $35\left(\frac{x}{5} - \frac{x+4}{7}\right) = 35(0)$; $7x - 5(x + 4) = 0$; $7x - 5x - 20 = 0$;
$2x - 20 = 0$; $2x = 20$; $x = 10$; $\{10\}$

10. $\frac{x}{6} - \frac{x+3}{5} = 1$; $30\left(\frac{x}{6} - \frac{x+3}{5}\right) = 30(1)$; $5x - 6(x + 3) = 30$; $5x - 6x - 18 = 30$;
$-x - 18 = 30$; $-x = 48$; $x = -48$; $\{-48\}$

11. $\frac{4y+1}{3} - \frac{2y+1}{5} = \frac{3}{5}$; $15\left(\frac{4y+1}{3} - \frac{2y+1}{5}\right) = 15\left(\frac{3}{5}\right)$; $5(4y + 1) - 3(2y + 1) = 9$;
$20y + 5 - 6y - 3 = 9$; $14y + 2 = 9$; $14y = 7$; $y = \frac{1}{2}$; $\left\{\frac{1}{2}\right\}$

12. $\frac{x+2}{15} - \frac{x-3}{5} = \frac{2}{3}$, $15\left(\frac{x+2}{15} - \frac{x-3}{5}\right) = 15\left(\frac{2}{3}\right)$; $x + 2 - 3(x - 3) = 10$;
$x + 2 - 3x + 9 = 10$; $-2x + 11 = 10$; $-2x = -1$; $x = \frac{1}{2}$; $\left\{\frac{1}{2}\right\}$

13. $3u - \frac{u}{5} = 1 + \frac{8u}{10}$; $10\left(3u - \frac{u}{5}\right) = 10\left(1 + \frac{8u}{10}\right)$; $30u - 2u = 10 + 8u$; $28u = 10 + 8u$;
$20u = 10$; $u = \frac{1}{2}$; $\left\{\frac{1}{2}\right\}$

14. $x - \frac{3x}{2} = \frac{3x}{4} - \frac{5}{6}$; $12\left(x - \frac{3x}{2}\right) = 12\left(\frac{3x}{4} - \frac{5}{6}\right)$; $12x - 18x = 9x - 10$;
$-6x = 9x - 10$; $-15x = -10$; $x = \frac{-10}{-15}$; $x = \frac{2}{3}$; $\left\{\frac{2}{3}\right\}$

15. $0 = 3m - \frac{3m-7}{8}$; $8(0) = 8\left(3m - \frac{3m-7}{8}\right)$; $0 = 24m - (3m - 7)$;
$0 = 24m - 3m + 7$; $0 = 21m + 7$; $-7 = 21m$; $-\frac{1}{3} = m$; $\left\{-\frac{1}{3}\right\}$

16. $0 = 2x + \frac{2x-8}{7}$; $7(0) = 7\left(2x + \frac{2x-8}{7}\right)$; $0 = 14x + 2x - 8$; $0 = 16x - 8$; $8 = 16x$;
$\frac{1}{2} = x$; $\left\{\frac{1}{2}\right\}$

B **17.** $\frac{n+3}{3} - \frac{n}{4} = \frac{n-2}{5}$; $60\left(\frac{n+3}{3} - \frac{n}{4}\right) = 60\left(\frac{n-2}{5}\right)$; $20(n + 3) - 15n = 12(n - 2)$;
$20n + 60 - 15n = 12n - 24$; $5n + 60 = 12n - 24$; $-7n + 60 = -24$; $-7n = -84$;
$n = 12$; $\{12\}$

18. $\frac{x+4}{3} - \frac{x}{7} = \frac{x+7}{5}$; $105\left(\frac{x+4}{3} - \frac{x}{7}\right) = 105\left(\frac{x+7}{5}\right)$; $35(x + 4) - 15x = 21(x + 7)$;
$35x + 140 - 15x = 21x + 147$; $20x + 140 = 21x + 147$; $140 = x + 147$; $-7 = x$;
$\{-7\}$

19. $\frac{1}{2}(x + 4) - \frac{2}{3}(x - 1) = 3$; $6\left[\frac{1}{2}(x + 4) - \frac{2}{3}(x - 1)\right] = 6(3)$; $3(x + 4) - 4(x - 1) =$
18; $3x + 12 - 4x + 4 = 18$; $-x + 16 = 18$; $-x = 2$; $x = -2$; $\{-2\}$

20. $\frac{2}{3}(x - 1) - \frac{1}{5}(2x - 3) = 1$; $15\left[\frac{2}{3}(x - 1) - \frac{1}{5}(2x - 3)\right] = 15(1)$;
$10(x - 1) - 3(2x - 3) = 15$; $10x - 10 - 6x + 9 = 15$; $4x - 1 = 15$; $4x = 16$;
$x = 4$; $\{4\}$

21. $0 = \frac{1}{2}(n + 3) - \frac{1}{4}(n + 4); 4(0) = 4\left[\frac{1}{2}(n + 3) - \frac{1}{4}(n + 4)\right]; 0 = 2(n + 3) - (n + 4);$

$0 = 2n + 6 - n - 4; 0 = n + 2; -2 = n; \{-2\}$

22. $1 = \frac{1}{3}(x + 6) - \frac{1}{6}(9 - x); 6(1) = 6\left[\frac{1}{3}(x + 6) - \frac{1}{6}(9 - x)\right]; 6 = 2(x + 6) - (9 - x);$

$6 = 2x + 12 - 9 + x; 6 = 3x + 3; 3 = 3x; 1 = x; \{1\}$

23. $\frac{6b - 4}{3} - 2 = \frac{18 - 4b}{3} + b; 3\left(\frac{6b - 4}{3} - 2\right) = 3\left(\frac{18 - 4b}{3} + b\right); 6b - 4 - 6 =$

$18 - 4b + 3b; 6b - 10 = 18 - b; 7b - 10 = 18; 7b = 28; b = 4; \{4\}$

24. $h - \frac{h + 1}{4} = \frac{2h + 3}{5} + 1; 20\left(h - \frac{h + 1}{4}\right) = 20\left(\frac{2h + 3}{5} + 1\right); 20h - 5(h + 1) =$

$4(2h + 3) + 20; 20h - 5h - 5 = 8h + 12 + 20; 15h - 5 = 8h + 32; 7h - 5 = 32;$

$7h = 37; h = \frac{37}{7}; \left\{\frac{37}{7}\right\}$

25. $\frac{1}{4}\left(y - \frac{1}{3}\right) - \frac{1}{6}(y - 3) = \frac{2}{3}; 12\left[\frac{1}{4}\left(y - \frac{1}{3}\right) - \frac{1}{6}(y - 3)\right] = 12\left(\frac{2}{3}\right); 3\left(y - \frac{1}{3}\right) -$

$2(y - 3) = 8; 3y - 1 - 2y + 6 = 8; y + 5 = 8; y = 3; \{3\}$

26. $\frac{2}{3}(x - 1) - \frac{1}{5}(x - 2) = \frac{x + 2}{3}; 15\left[\frac{2}{3}(x - 1) - \frac{1}{5}(x - 2)\right] = 15\left(\frac{x + 2}{3}\right);$

$10(x - 1) - 3(x - 2) = 5(x + 2); 10x - 10 - 3x + 6 = 5x + 10;$
$7x - 4 = 5x + 10; 2x - 4 = 10; 2x = 14; x = 7; \{7\}$

C **27.** $\frac{1}{2}(2x + 5) - x = \frac{3}{4}\left(\frac{6x + 7}{3}\right); 12\left[\frac{1}{2}(2x + 5) - x\right] = 12\left[\frac{3}{4}\left(\frac{6x + 7}{3}\right)\right]; 6(2x + 5) -$

$12x = 9\left(\frac{6x + 7}{3}\right); 12x + 30 - 12x = 3(6x + 7); 30 = 18x + 21; 9 = 18x; \frac{1}{2} = x; \left\{\frac{1}{2}\right\}$

28. $\frac{2}{5}\left(x + \frac{3}{2}\right) - \frac{5}{6}(2x + 1) = -\frac{44}{15}; 30\left[\frac{2}{5}\left(x + \frac{3}{2}\right) - \frac{5}{6}(2x + 1)\right] = 30\left(-\frac{44}{15}\right);$

$12\left(x + \frac{3}{2}\right) - 25(2x + 1) = -88; 12x + 18 - 50x - 25 = -88; -38x - 7 = -88;$

$-38x = -81; x = \frac{81}{38}; \left\{\frac{81}{38}\right\}$

29. $\frac{7}{3}(2x + 3) + \frac{3}{4}\left(\frac{x}{5} - \frac{15}{2}\right) = \frac{11}{8}; 120\left[\frac{7}{3}(2x + 3) + \frac{3}{4}\left(\frac{x}{5} - \frac{15}{2}\right)\right] = 120\left(\frac{11}{8}\right);$

$280(2x + 3) + 90\left(\frac{x}{5} - \frac{15}{2}\right) = 165; 560x + 840 + 18x - 675 = 165;$

$578x + 165 = 165; 578x = 0; x = 0; \{0\}$

30. $\frac{3}{4}(x - 2) - \frac{2}{3}\left(x - \frac{1}{2}\right) = \frac{x + 1}{2}; 12\left[\frac{3}{4}(x - 2) - \frac{2}{3}\left(x - \frac{1}{2}\right)\right] = 12\left(\frac{x + 1}{2}\right);$

$9(x - 2) - 8\left(x - \frac{1}{2}\right) = 6(x + 1); 9x - 18 - 8x + 4 = 6x + 6; x - 14 = 6x + 6;$

$-5x - 14 = 6; -5x = 20; x = -4; \{-4\}$

31. $\frac{x + 3a}{10} - \frac{x - a}{2} = \frac{x}{5}$; $10\left[\frac{x + 3a}{10} - \frac{x - a}{2}\right] = 10\left(\frac{x}{5}\right)$; $x + 3a - 5(x - a) = 2x$;

$x + 3a - 5x + 5a = 2x$; $-4x + 8a = 2x$; $-6x + 8a = 0$; $-6x = -8a$; $x = \frac{-8a}{-6}$;

$x = \frac{4a}{3}$; $\left\{\frac{4a}{3}\right\}$

32. $\frac{x + k}{4} - \frac{2x - 3k}{2} = \frac{k + 5x}{6}$; $12\left(\frac{x + k}{4} - \frac{2x - 3k}{2}\right) = 12\left(\frac{k + 5x}{6}\right)$;

$3(x + k) - 6(2x - 3k) = 2(k + 5x)$; $3x + 3k - 12x + 18k = 2k + 10x$;
$-9x + 21k = 2k + 10x$; $-9x + 19k = 10x$; $19k = 19x$; $k = x$; $\{k\}$

33. $\frac{m^2 + 10m + 25}{2x} = m^2 + 5m$; $2x(m^2 + 5m) = m^2 + 10m + 25$; $x = \frac{m^2 + 10m + 25}{2(m^2 + 5m)}$;

$x = \frac{(m + 5)^2}{2m(m + 5)}$; $x = \frac{m + 5}{2m}$; $\left\{\frac{m + 5}{2m}\right\}$

34. $\frac{a^2 + 6a + 9}{x} = a^2 + 3a$; $x(a^2 + 3a) = a^2 + 6a + 9$; $x = \frac{(a + 3)^2}{a(a + 3)}$; $x = \frac{a + 3}{a}$; $\left\{\frac{a + 3}{a}\right\}$

Pages 300–301 • PROBLEMS

A

1. Let x be the number. $\frac{x}{4} = \frac{x}{5} + 2$; $20\left(\frac{x}{4}\right) = 20\left(\frac{x}{5} + 2\right)$; $5x = 4x + 40$; $x = 40$; The number is 40.

2. Let x be the number. $\frac{x}{8} = \frac{x}{3} - 10$; $24\left(\frac{x}{8}\right) = 24\left(\frac{x}{3} - 10\right)$; $3x = 8x - 240$; $-5x = -240$; $x = 48$; The number is 48.

3. Let $5x$ and $2x$ be the numbers. $\frac{5x + 2x}{2} = 10\frac{1}{2}$; $\frac{7x}{2} = \frac{21}{2}$; $7x = 21$; $x = 3$; $5x = 15$; $2x = 6$; The numbers are 15 and 6.

4. Let $3x$, $5x$, and $6x$ be the numbers. $\frac{3x + 5x + 6x}{4} = 1\frac{1}{2} + 3x$; $\frac{14x}{4} = \frac{3}{2} + 3x$;

   $\frac{7}{2}x - 3x = \frac{3}{2}$; $2\left(\frac{7}{2}x - 3x\right) = 2\left(\frac{3}{2}\right)$; $7x - 6x = 3$; $x = 3$; $3x = 9$; $5x = 15$; $6x = 18$;

   The numbers are 9, 15, and 18.

5. Let l be the length and $\frac{4}{5}l$ be the width of the poster. $2\left(l + \frac{4}{5}l\right) = 36$; $l + \frac{4}{5}l = 18$;

   $5\left(l + \frac{4}{5}l\right) = 5(18)$; $5l + 4l = 90$; $9l = 90$; $l = 10$; $\frac{4}{5}l = \frac{4}{5}(10) = 8$; The poster is 8 ft by 10 ft.

6. Let w be the width and $\frac{3}{2}w$ be the length of the garden. $2\left(w + \frac{3}{2}w\right) = 50$;

   $2w + 3w = 50$; $5w = 50$; $w = 10$; $\frac{3}{2}w = \frac{3}{2}(10) = 15$; The garden is 15 m by 10 m.

7. Let x be the amount of allowance Scott receives. $x - \frac{x}{9} - \frac{2}{5}x = 2.2$;

   $45\left(x - \frac{x}{9} - \frac{2}{5}x\right) = 45(2.2)$; $45x - 5x - 18x = 99$; $22x = 99$; $x = 4.5$;

   Scott receives \$4.50 for an allowance.

8. Let x be Terri's salary. $x - \frac{3}{8}x - \frac{1}{3}x = 294$; $24\left(x - \frac{3}{8}x - \frac{1}{3}x\right) = 24(294)$;

$24x - 9x - 8x = 7056$; $7x = 7056$; $x = 1008$; Terri receives \$1008 a month.

9. Let w be the width and $w + 9$ be the length of the rectangle. $w = \frac{1}{7}[2(w + w + 9)]$;

$w = \frac{2}{7}(2w + 9)$; $7(w) = 7\left[\frac{2}{7}(2w + 9)\right]$; $7w = 2(2w + 9)$; $7w = 4w + 18$; $3w = 18$;

$w = 6$; $w + 9 = 15$; The rectangle is 15 cm by 6 cm.

10. Let x, $x + 1$, and $x + 2$ be the lengths of the sides. $\frac{1}{2}(x + x + 1 + x + 2) =$

$x + 2 + 14$; $\frac{1}{2}(3x + 3) = x + 16$; $3x + 3 = 2(x + 16)$; $3x + 3 = 2x + 32$;

$x + 3 = 32$; $x = 29$; $x + 1 = 30$; $x + 2 = 31$; $29 + 30 + 31 = 90$; The triangle's perimeter is 90.

B 11. Let d be the length of the trail. $\frac{d}{3} + \frac{1}{2} + \frac{d}{4} = 5\frac{1}{6}$; $12\left(\frac{d}{3} + \frac{1}{2} + \frac{d}{4}\right) = \left(\frac{31}{6}\right)12$;

$4d + 6 + 3d = 62$; $7d + 6 = 62$; $7d = 56$; $d = 8$; The trail was 8 km one way.

12. Let d be the distance. $\frac{d}{6} + \frac{3}{4} + \frac{d}{48} = \frac{3}{2}$; $48\left(\frac{d}{6} + \frac{3}{4} + \frac{d}{48}\right) = 48\left(\frac{3}{2}\right)$;

$8d + 36 + d = 72$; $9d + 36 = 72$; $9d = 36$; $d = 4$; Consuela's home is 4 km from the fitness center.

13. Let x be the number of coins. $\frac{x}{8} \cdot 25 + \frac{7x}{8} \cdot 5 = 360$; $8\left(\frac{25x}{8} + \frac{35x}{8}\right) = 8(360)$;

$25x + 35x = 2880$; $60x = 2880$; $x = 48$; $\frac{x}{8} = 6$; $\frac{7x}{8} = 42$; Al has 6 quarters and 42 nickels.

14. Let x be the number of coins. $\frac{2}{3}x \cdot 5 + \frac{1}{4}x \cdot 10 + \left(x - \frac{2}{3}x - \frac{1}{4}x\right)25 = 475$;

$12\left(\frac{10x}{3} + \frac{10x}{4} + 25x - \frac{50}{3}x - \frac{25}{4}x\right) = 12(475)$; $40x + 30x + 300x - 200x - 75x =$

5700; $95x = 5700$; $x = 60$; $\frac{2}{3}x = 40$; $\frac{1}{4}x = 15$; $x - \frac{2}{3}x - \frac{1}{4}x = 60 - 40 - 15 = 5$;

There are 40 nickels, 15 dimes, and 5 quarters.

C 15. Let x be the number of apples Hannah bought. Profit = Sales − Cost;

$75 = \frac{3}{4}x \cdot 45 - x \cdot \frac{95}{3}$; $75 = \frac{135x}{4} - \frac{95x}{3}$; $12(75) = 12\left(\frac{135x}{4} - \frac{95x}{3}\right)$;

$900 = 405x - 380x$; $900 = 25x$; $36 = x$; $\frac{1}{4}x = 9$; Hannah kept 9 apples.

16. Let x be the length of Diophantus' life. $\frac{x}{6} + \frac{x}{12} + \frac{x}{7} + 5 + \frac{x}{2} + 4 = x$;

$84\left(\frac{x}{6} + \frac{x}{12} + \frac{x}{7} + 5 + \frac{x}{2} + 4\right) = 84x$; $14x + 7x + 12x + 420 + 42x + 336 = 84x$;

$75x + 756 = 84x$; $756 = 9x$; $84 = x$; Diophantus lived 84 years.

Page 301 • MIXED REVIEW EXERCISES

1. 8 days : 4 wk $= \frac{8 \text{ days}}{4 \text{ wk}} = \frac{8 \text{ days}}{4 \cdot 7 \text{ days}} = \frac{8}{28} = \frac{2}{7}$, or 2 : 7 **2.** $15x : 90x = \frac{15x}{90x} = \frac{1}{6}$, or 1 : 6

3. $16 : 12 = \frac{16}{12} = \frac{4}{3}$, or 4 : 3 **4.** $\frac{16mn^2}{40mn} = \frac{2n}{5}$ **5.** $\frac{48x^2y}{36x^3y^2} = \frac{4}{3xy}$ **6.** $\frac{21a^4}{35ab^2} = \frac{3a^3}{5b^2}$

7. $\frac{7}{2n} = \frac{3}{6}$; $42 = 6n$; $7 = n$; $\{7\}$

8. $\frac{x+2}{7} = \frac{5}{3}$; $3(x+2) = 35$; $3x + 6 = 35$; $3x = 29$; $x = \frac{29}{3}$; $\left\{\frac{29}{3}\right\}$

9. $\frac{2a+3}{5} = \frac{a+8}{4}$; $4(2a+3) = 5(a+8)$; $8a + 12 = 5a + 40$; $3a + 12 = 40$; $3a = 28$; $a = \frac{28}{3}$; $\left\{\frac{28}{3}\right\}$

10. $3x - 2 = 14$; $3x = 16$; $x = \frac{16}{3}$; $\left\{\frac{16}{3}\right\}$ **11.** $|x| = 5$, $x = 5$ or -5; $\{5, -5\}$

12. $8x + 3 = 9x + 1$; $3 = x + 1$; $2 = x$; $\{2\}$

Page 303 • APPLICATION

1. $\frac{39¢}{1 \cancel{\text{dozen}}} \times 2 \cancel{\text{dozen}} = 78¢$; ¢ **2.** $\frac{45 \text{ mi}}{1 \cancel{\text{h}}} \cdot 2 \cancel{\text{h}} = 90$ mi; mi **3.** $9 \text{ ft} \times 12 \text{ ft} = 108 \text{ ft}^2$; ft^2

4. $\frac{60 \text{ mi}}{1 \text{ h}} \div \frac{60 \text{ min}}{1 \text{ h}} = \frac{\overset{1}{\cancel{60}} \text{ mi}}{1 \cancel{\text{h}}} \times \frac{1 \cancel{\text{h}}}{\underset{1}{\cancel{60}} \text{ min}} = 1$ mi/min; mi/min

5. $\frac{55 \text{ mi}}{1 \text{ h}} \times \frac{5280 \text{ ft}}{1 \text{ mi}} \div \frac{60 \text{ min}}{1 \text{ h}} = \frac{\overset{11}{\cancel{55}} \cancel{\text{mi}}}{1 \cancel{\text{h}}} \times \frac{\overset{440}{\cancel{5280}} \text{ ft}}{1 \cancel{\text{mi}}} \times \frac{1 \cancel{\text{h}}}{\underset{\underset{1}{\cancel{12}}}{\cancel{60}} \text{ min}} = 4840$ ft/min; ft/min

6. a. $\frac{36 \text{ mi}}{1 \text{ gal}} \times \frac{15 \text{ gal}}{1 \text{ tank}} = 540$ mi/tank **b.** $\frac{540 \text{ mi}}{1 \text{ tank}} \div \frac{48 \text{ mi}}{1 \text{ h}} = \frac{540 \cancel{\text{mi}}}{1 \text{ tank}} \times \frac{1 \text{ h}}{48 \cancel{\text{mi}}} = 11\frac{1}{4}$ h/tank

7. $\frac{102 \cancel{\text{kW} \cdot \text{h}}}{1 \cancel{\text{mo}}} \times \frac{1 \cancel{\text{mo}}}{30 \text{ days}} \times \frac{\$0.125}{1 \cancel{\text{kW} \cdot \text{h}}} = \$.425$/day **8.** $\frac{\$1}{2 \cdot 2.5 \cancel{\text{oz}}} \times \frac{16 \cancel{\text{oz}}}{1 \text{ lb}} = \$3.20$/lb

Pages 305–306 • WRITTEN EXERCISES

A **1.** $\frac{1}{4} + \frac{2}{x} = \frac{3}{4}$, $4x\left(\frac{1}{4} + \frac{2}{x}\right) = 4x\left(\frac{3}{4}\right)$; $x + 8 = 3x$; $8 = 2x$; $4 = x$; $\{4\}$

2. $\frac{3}{x} + \frac{1}{2} = 1$, $2x\left(\frac{3}{x} + \frac{1}{2}\right) = 2x$; $6 + x = 2x$; $6 = x$; $\{6\}$

3. $\frac{4}{y} - \frac{1}{4} = \frac{1}{12}$; $12y\left(\frac{4}{y} - \frac{1}{4}\right) = 12y\left(\frac{1}{12}\right)$; $48 - 3y = y$; $48 = 4y$; $12 = y$; $\{12\}$

4. $\frac{9}{2a} - \frac{5}{a} = -\frac{1}{4}$; $4a\left(\frac{9}{2a} - \frac{5}{a}\right) = 4a\left(-\frac{1}{4}\right)$; $18 - 20 = -a$; $-2 = -a$; $2 = a$; $\{2\}$

5. $\frac{n}{n-2} = \frac{6}{5}$; $5n = 6(n-2)$; $5n = 6n - 12$; $-n = -12$; $n = 12$; $\{12\}$

6. $\frac{1+b}{1-b} = \frac{4}{3}$; $3(1+b) = 4(1-b)$; $3 + 3b = 4 - 4b$; $3 + 7b = 4$; $7b = 1$; $b = \frac{1}{7}$; $\left\{\frac{1}{7}\right\}$

7. $\frac{6}{n} - \frac{1}{4} = \frac{8}{2n}$; $4n\left(\frac{6}{n} - \frac{1}{4}\right) = 4n\left(\frac{8}{2n}\right)$; $24 - n = 16$; $-n = -8$; $n = 8$; $\{8\}$

8. $\frac{3}{2a} - \frac{5}{6} = \frac{2}{3a}$; $6a\left(\frac{3}{2a} - \frac{5}{6}\right) = 6a\left(\frac{2}{3a}\right)$; $9 - 5a = 4$; $-5a = -5$; $a = 1$; $\{1\}$

9. $\frac{6 - x}{4 - x} = \frac{3}{5}$; $5(6 - x) = 3(4 - x)$; $30 - 5x = 12 - 3x$; $18 - 5x = -3x$; $18 = 2x$; $9 = x$; $\{9\}$

10. $\frac{5 - m}{7 - m} = \frac{2}{3}$; $3(5 - m) = 2(7 - m)$; $15 - 3m = 14 - 2m$; $1 - 3m = -2m$; $1 = m$; $\{1\}$

11. $1 = \frac{x + 5}{x - 2}$; $x - 2 = x + 5$; $-2 = 5$; No solution.

12. $2 = \frac{x - 4}{x - 2}$; $2x - 4 = x - 4$; $x - 4 = -4$; $x = 0$; $\{0\}$

13. $\frac{3y - 6}{y - 2} = 2$; $3y - 6 = 2(y - 2)$; $3y - 6 = 2y - 4$; $y - 6 = -4$; $y = 2$; but y cannot equal 2; No solution.

14. $\frac{2x - 4}{x - 2} = \frac{1}{3}$; $3(2x - 4) = x - 2$; $6x - 12 = x - 2$; $5x - 12 = -2$; $5x = 10$; $x = 2$; but x cannot equal 2; No solution.

15. $\frac{1}{2} = \frac{3 - 2m}{4m - 6}$; $4m - 6 = 2(3 - 2m)$; $4m - 6 = 6 - 4m$; $8m - 6 = 6$; $8m = 12$; $m = \frac{3}{2}$; but m cannot equal $\frac{3}{2}$; No solution.

16. $\frac{1}{4} = \frac{2w + 1}{2w + 4}$; $2w + 4 = 4(2w + 1)$; $2w + 4 = 8w + 4$; $2w = 8w$; $0 = 6w$, $0 = w$, $\{0\}$

17. $\frac{3y - 6}{y - 2} = 3$; $3y - 6 = 3(y - 2)$; $3y - 6 = 3y - 6$; {all real numbers except 2}

18. $\frac{y - 2}{4y - 8} = \frac{1}{4}$; $4(y - 2) = 4y - 8$; $4y - 8 = 4y - 8$; {all real numbers except 2}

19. $\frac{2 + c}{3c} = \frac{1}{c}$; $c(2 + c) = 3c$; $2c + c^2 = 3c$; $c^2 - c = 0$; $c(c - 1) = 0$; $c = 0, 1$; but c cannot equal 0; $\{1\}$

20. $\frac{a + 1}{6} = \frac{2}{a}$; $a(a + 1) = 2 \cdot 6$; $a^2 + a = 12$; $a^2 + a - 12 = 0$; $(a + 4)(a - 3) = 0$; $a = -4, 3$; $\{-4, 3\}$

B

21. $\frac{3}{2a + 1} = \frac{3}{2a - 1}$; $3(2a - 1) = 3(2a + 1)$; $2a - 1 = 2a + 1$; $-1 = 1$; No solution.

22. $\frac{5}{3 + 4x} = \frac{4}{4x + 3}$; $5(4x + 3) = 4(3 + 4x)$; $20x + 15 = 12 + 16x$; $4x + 15 = 12$; $4x = -3$; $x = -\frac{3}{4}$; but x cannot equal $-\frac{3}{4}$; No solution.

23. $\frac{1}{x} - \frac{2x}{x + 1} = 0$; $x(x + 1)\left(\frac{1}{x} - \frac{2x}{x + 1}\right) = 0$; $x + 1 - 2x^2 = 0$; $2x^2 - x - 1 = 0$; $(2x + 1)(x - 1) = 0$; $x = -\frac{1}{2}, 1$; $\left\{-\frac{1}{2}, 1\right\}$

24. $\frac{1}{z - 3} + \frac{3}{9 - 3z} = 0$; $3(z - 3)\left(\frac{1}{z - 3} + \frac{-3}{3(z - 3)}\right) = 0$; $3 - 3 = 0$; {all real numbers except 3}

25. $\frac{1}{y-3} = \frac{6}{y^2-9}$; $y^2 - 9 = 6(y-3)$; $y^2 - 9 = 6y - 18$; $y^2 - 6y + 9 = 0$;
$(y-3)^2 = 0$; $y = 3$; but y cannot equal 3; No solution.

26. $\frac{4}{a^2-4a} = \frac{-1}{4-a}$; $a(a-4)\left(\frac{4}{a(a-4)}\right) = a(a-4)\left(\frac{1}{a-4}\right)$; $4 = a$; but a cannot equal 4; No solution.

27. $\frac{4}{x+1} - \frac{1}{x} = 1$; $x(x+1)\left(\frac{4}{x+1} - \frac{1}{x}\right) = x(x+1)$; $4x - (x+1) = x(x+1)$;
$4x - x - 1 = x^2 + x$; $3x - 1 = x^2 + x$; $0 = x^2 - 2x + 1$; $0 = (x-1)^2$; $x = 1$; $\{1\}$

28. $\frac{3}{x-1} + \frac{2}{x} = 4$; $x(x-1)\left(\frac{3}{x-1} + \frac{2}{x}\right) = 4x(x-1)$; $3x + 2(x-1) = 4x^2 - 4x$;
$3x + 2x - 2 = 4x^2 - 4x$; $5x - 2 = 4x^2 - 4x$; $0 = 4x^2 - 9x + 2$;
$0 = (4x-1)(x-2)$; $x = \frac{1}{4}, 2$; $\left\{\frac{1}{4}, 2\right\}$

29. $1 = \frac{3}{y+2} + \frac{1}{y-2}$; $(y+2)(y-2) = (y+2)(y-2)\left(\frac{3}{y+2} + \frac{1}{y-2}\right)$;
$y^2 - 4 = 3(y-2) + y + 2$; $y^2 - 4 = 3y - 6 + y + 2$; $y^2 - 4 = 4y - 4$;
$y^2 - 4y = 0$; $y(y-4) = 0$; $y = 0, 4$; $\{0, 4\}$

30. $-2 = \frac{1}{1+x} - \frac{3}{1-x}$; $-2(1+x)(1-x) = (1+x)(1-x)\left(\frac{1}{1+x} - \frac{3}{1-x}\right)$;
$-2(1-x^2) = 1 - x - 3(1+x)$; $-2 + 2x^2 = 1 - x - 3 - 3x$; $-2 + 2x^2 =$
$-2 - 4x$; $2x^2 + 4x = 0$; $2x(x+2) = 0$; $x = 0, -2$; $\{0, -2\}$

31. $\frac{x}{x+1} + \frac{x+1}{x} = \frac{5}{2}$; $2x(x+1)\left(\frac{x}{x+1} + \frac{x+1}{x}\right) = 2x(x+1)\left(\frac{5}{2}\right)$; $2x^2 +$
$2(x+1)^2 = 5x(x+1)$; $2x^2 + 2(x^2 + 2x + 1) = 5x^2 + 5x$; $2x^2 + 2x^2 + 4x +$
$2 = 5x^2 + 5x$; $4x^2 + 4x + 2 = 5x^2 + 5x$; $0 = x^2 + x - 2$; $0 = (x+2)(x-1)$;
$x = -2, 1$; $\{-2, 1\}$

32. $\frac{y+1}{y-1} - \frac{y-1}{y+1} = \frac{8}{3}$; $3(y-1)(y+1)\left(\frac{y+1}{y-1} - \frac{y-1}{y+1}\right) = 3(y-1)(y+1)\left(\frac{8}{3}\right)$;
$3(y+1)^2 - 3(y-1)^2 = 8(y^2-1)$; $3(y^2 + 2y + 1) - 3(y^2 - 2y + 1) = 8y^2 - 8$;
$3y^2 + 6y + 3 - 3y^2 + 6y - 3 = 8y^2 - 8$; $12y = 8y^2 - 8$; $0 = 8y^2 - 12y - 8$;
$0 = 4(2y^2 - 3y - 2)$; $0 = (2y+1)(y-2)$; $y = -\frac{1}{2}, 2$; $\left\{-\frac{1}{2}, 2\right\}$

33. $\frac{n-2}{n} - \frac{n-3}{n-6} = \frac{1}{n}$; $n(n-6)\left(\frac{n-2}{n} - \frac{n-3}{n-6}\right) = n(n-6)\left(\frac{1}{n}\right)$; $(n-6)(n-2) -$
$n(n-3) = n - 6$; $n^2 - 8n + 12 - n^2 + 3n = n - 6$; $12 - 5n = n - 6$;
$18 - 5n = n$; $18 = 6n$; $3 = n$; $\{3\}$

34. $\frac{5}{y-5} - \frac{3}{y+5} = \frac{2}{y}$; $y(y-5)(y+5)\left(\frac{5}{y-5} - \frac{3}{y+5}\right) = y(y-5)(y+5)\left(\frac{2}{y}\right)$;
$5y(y+5) - 3y(y-5) = 2(y-5)(y+5)$; $5y^2 + 25y - 3y^2 + 15y = 2(y^2 - 25)$;
$2y^2 + 40y = 2y^2 - 50$; $40y = -50$; $y = -\frac{5}{4}$; $\left\{-\frac{5}{4}\right\}$

35. $\frac{3}{x-4} + \frac{2}{x^2-16} = 0$; $(x+4)(x-4)\left(\frac{3}{x-4} + \frac{2}{(x+4)(x-4)}\right) = 0$; $3(x+4) + 2 = 0$;
$3x + 12 + 2 = 0$; $3x + 14 = 0$; $3x = -14$; $x = -\frac{14}{3}$; $\left\{-\frac{14}{3}\right\}$

36. $\frac{2}{c^2 - 2c} + \frac{1}{2 - c} = 1$; $c(c - 2)\left(\frac{2}{c(c - 2)} - \frac{1}{c - 2}\right) = c(c - 2)$; $2 - c = c^2 - 2c$;
$0 = c^2 - c - 2$; $0 = (c - 2)(c + 1)$; $c = 2, -1$; but c cannot equal 2; $\{-1\}$

37. $\frac{1}{x - 4} + \frac{2}{x^2 - 16} = \frac{3}{x + 4}$; $(x + 4)(x - 4)\left(\frac{1}{x - 4} + \frac{2}{x^2 - 16}\right) = \left(\frac{3}{x + 4}\right)(x + 4)(x - 4)$;
$x + 4 + 2 = 3(x - 4)$; $x + 6 = 3x - 12$; $6 = 2x - 12$; $18 = 2x$; $9 = x$; $\{9\}$

38. $\frac{3a}{a - 1} - \frac{4}{a + 1} = \frac{4}{a^2 - 1}$; $(a - 1)(a + 1)\left(\frac{3a}{a - 1} - \frac{4}{a + 1}\right) =$
$(a - 1)(a + 1)\frac{4}{(a + 1)(a - 1)}$; $3a(a + 1) - 4(a - 1) = 4$; $3a^2 + 3a - 4a + 4 = 4$;
$3a^2 - a + 4 = 4$; $3a^2 - a = 0$; $a(3a - 1) = 0$; $a = 0, \frac{1}{3}$; $\left\{0, \frac{1}{3}\right\}$

C **39.** $\frac{3}{2x} - \frac{x}{4x^2 - 1} = \frac{7}{4x + 2}$; $2x(2x + 1)(2x - 1)\left(\frac{3}{2x} - \frac{x}{(2x + 1)(2x - 1)}\right) =$
$2x(2x + 1)(2x - 1)\left(\frac{7}{2(2x + 1)}\right)$; $3(2x + 1)(2x - 1) - 2x^2 = 7x(2x - 1)$;
$3(4x^2 - 1) - 2x^2 = 14x^2 - 7x$; $12x^2 - 3 - 2x^2 = 14x^2 - 7x$; $10x^2 - 3 =$
$14x^2 - 7x$; $0 = 4x^2 - 7x + 3$; $0 = (4x - 3)(x - 1)$; $x = \frac{3}{4}, 1$; $\left\{\frac{3}{4}, 1\right\}$

40. $\frac{x - 2}{x^2 - x - 6} = \frac{1}{x^2 - 4} + \frac{3}{2x + 4}$; $2(x - 3)(x + 2)(x - 2)\left(\frac{x - 2}{(x - 3)(x + 2)}\right) =$
$2(x - 3)(x + 2)(x - 2)\left(\frac{1}{(x + 2)(x - 2)} + \frac{3}{2(x + 2)}\right)$; $2(x - 2)^2 = 2(x - 3) +$
$3(x - 3)(x - 2)$; $2(x^2 - 4x + 4) = 2x - 6 + 3(x^2 - 5x + 6)$; $2x^2 - 8x + 8 =$
$2x - 6 + 3x^2 - 15x + 18$; $2x^2 - 8x + 8 = 3x^2 - 13x + 12$; $0 = x^2 - 5x + 4$;
$0 = (x - 4)(x - 1)$; $x = 4, 1$; $\{1, 4\}$

Pages 306–308 • PROBLEMS

A **1.** Let x be the number. $x + \frac{1}{x} = \frac{25}{12}$; $12x\left(x + \frac{1}{x}\right) = \frac{25}{12} \cdot 12x$; $12x^2 + 12 = 25x$;
$12x^2 - 25x + 12 = 0$; $(4x - 3)(3x - 4) = 0$; $x = \frac{3}{4}, \frac{4}{3}$; The number is $\frac{3}{4}$ or $\frac{4}{3}$.

2. Let x be the number. $x + \frac{1}{x} = \frac{13}{6}$; $6x\left(x + \frac{1}{x}\right) = 6x \cdot \frac{13}{6}$; $6x^2 + 6 = 13x$;
$6x^2 - 13x + 6 = 0$; $(3x - 2)(2x - 3) = 0$; $x = \frac{2}{3}, \frac{3}{2}$; The number is $\frac{2}{3}$ or $\frac{3}{2}$.

3. Let n, $n + 2$ be the two integers. $\frac{1}{n} + \frac{1}{n + 2} = \frac{8}{15}$; $15n(n + 2)\left(\frac{1}{n} + \frac{1}{n + 2}\right) =$
$15n(n + 2)\left(\frac{8}{15}\right)$; $15(n + 2) + 15n = 8n(n + 2)$; $15n + 30 + 15n = 8n^2 + 16n$;
$30n + 30 = 8n^2 + 16n$; $0 = 8n^2 - 14n - 30$; $0 = 4n^2 - 7n - 15$;
$0 = (4n + 5)(n - 3)$; $n = \frac{5}{4}, 3$; $\frac{5}{4}$ is not an odd integer; The integers are 3 and 5.

4. Let n and $n + 2$ be the two integers. $\frac{1}{n} + \frac{1}{n+2} = \frac{11}{60}$; $60n(n+2)\left(\frac{1}{n} + \frac{1}{n+2}\right) =$ $60n(n+2)\left(\frac{11}{60}\right)$; $60(n+2) + 60n = 11n(n+2)$; $60n + 120 + 60n = 11n^2 + 22n$; $120n + 120 = 11n^2 + 22n$; $0 = 11n^2 - 98n - 120$; $0 = (11n + 12)(n - 10)$; $n = -\frac{12}{11}, 10$; $-\frac{12}{11}$ is not an even integer; The integers are 10 and 12.

5. Let d be the denominator of the fraction. $\frac{d+1}{d} - \frac{1}{4} = \frac{d+3}{d+2}$; $4d(d+2)\left(\frac{d+1}{d} - \frac{1}{4}\right) = 4d(d+2)\left(\frac{d+3}{d+2}\right)$; $4(d+2)(d+1) - d(d+2) =$ $4d(d+3)$; $4(d^2 + 3d + 2) - d^2 - 2d = 4d^2 + 12d$; $4d^2 + 12d + 8 - d^2 - 2d =$ $4d^2 + 12d$; $3d^2 + 10d + 8 = 4d^2 + 12d$; $0 = d^2 + 2d - 8$; $0 = (d+4)(d-2)$; $d = -4, 2$; if $d = -4$, $\frac{1+d}{d} = \frac{1-4}{-4} = \frac{-3}{-4} = \frac{3}{4}$, but the numerator must be 1 more than the denominator; The fraction is $\frac{3}{2}$.

6. Let d be the denominator. $\frac{d-1}{d} + \frac{1}{8} = \frac{d+3}{d+4}$; $8d(d+4)\left(\frac{d-1}{d} + \frac{1}{8}\right) =$ $8d(d+4)\left(\frac{d+3}{d+4}\right)$; $8(d+4)(d-1) + d(d+4) = 8d(d+3)$; $8(d^2 + 3d - 4) +$ $d^2 + 4d = 8d^2 + 24d$; $8d^2 + 24d - 32 + d^2 + 4d = 8d^2 + 24d$; $9d^2 + 28d - 32 =$ $8d^2 + 24d$; $d^2 + 4d - 32 = 0$; $(d+8)(d-4) = 0$; $d = -8, 4$; if $d = -8$, $\frac{d-1}{d} =$ $\frac{-8-1}{-8} = \frac{-9}{-8} = \frac{9}{8}$, but the numerator must be 1 less than the denominator; The fraction is $\frac{3}{4}$.

B

7. Let x and $10 - x$ be the two numbers. $\frac{1}{x} + \frac{1}{10-x} = \frac{5}{12}$; $12x(10-x)\left(\frac{1}{x} + \frac{1}{10-x}\right) = 12x(10-x)\left(\frac{5}{12}\right)$; $12(10-x) + 12x = 5x(10-x)$; $120 - 12x + 12x = 50x - 5x^2$; $120 = 50x - 5x^2$; $5x^2 - 50x + 120 = 0$; $x^2 - 10 + 24 = 0$; $(x-6)(x-4) = 0$; $x = 6, 4$; The numbers are 6 and 4.

8. Let x and $x + 11$ be the two numbers. $\frac{x+11}{x} = 2 + \frac{4}{x}$; $x\left(\frac{x+11}{x}\right) = x\left(2 + \frac{4}{x}\right)$; $x + 11 = 2x + 4$; $11 = x + 4$; $7 = x$; The numbers are 7 and 18.

9.

	Rate ×	Time	= Distance
Up	x	$\frac{15}{x}$	15
Down	$x + 1$	$\frac{15}{x} - 0.5$	15

$(x + 1)\left(\frac{15}{x} - 0.5\right) = 15$; $10x(x + 1)\left(\frac{15}{x} - 0.5\right) = 10x(15)$; $(x + 1)(150 - 5x) = 150x$;
$145x - 5x^2 + 150 = 150x$; $0 = 5x^2 + 5x - 150$; $0 = x^2 + x - 30$;
$0 = (x - 5)(x + 6)$; $x - 5 = 0$ or $x + 6 = 0$; $x = 5$ or -6 (reject); $\frac{15}{x} = \frac{15}{5} = 3$;
$\frac{15}{x} - 0.5 = \frac{15}{5} - 0.5 = 2.5$; $3 + 2.5 = 5.5$; It took Tina 5.5 h.

10.

	Rate ×	Time	= Distance
Slower	x	$\frac{30}{x}$	30
Faster	$x + 10$	$\frac{30}{x} - \frac{1}{10}$	30

$(x + 10)\left(\frac{30}{x} - \frac{1}{10}\right) = 30$; $10x(x + 10)\left(\frac{30}{x} - \frac{1}{10}\right) = 10x(30)$; $(x + 10)(300 - x) =$
$300x$; $290x - x^2 + 3000 = 300x$; $0 = x^2 + 10x - 3000$; $0 = (x - 50)(x + 60)$;
$x - 50 = 0$ or $x + 60 = 0$; $x = 50$ or -60 (reject); $x + 10 = 50 + 10 = 60$; Jacqui's new speed is 60 mi/h.

11.

	Number ×	Price	= Cost
Planned	x	$\frac{180}{x}$	180
Actual	$x - 6$	$\frac{180}{x} + 1.50$	180

$(x - 6)\left(\frac{180}{x} + 1.50\right) = 180$; $10x(x - 6)\left(\frac{180}{x} + 1.50\right) = 10x(180)$;
$(x - 6)(1800 + 15x) = 1800x$; $15x^2 + 1710x - 10{,}800 = 1800x$;
$15x^2 - 90x - 10{,}800 = 0$; $x^2 - 6x - 720 = 0$; $(x - 30)(x + 24) = 0$;
$x - 30 = 0$ or $x + 24 = 0$; $x = 30$ or -24 (reject); $x - 6 = 30 - 6 = 24$;
24 people went on the trip.

12.

	Number ×	Price	= Cost
Planned	x	$\frac{75}{x}$	75
Actual	$x + 5$	$\frac{75}{x} - 0.50$	75

$(x + 5)\left(\frac{75}{x} - 0.50\right) = 75$; $10x(x + 5)\left(\frac{75}{x} - 0.50\right) = 10x(75)$; $(x + 5)(750 - 5x) = 750x$; $725x - 5x^2 + 3750 = 750x$; $0 = 5x^2 + 25x - 3750$; $0 = x^2 + 5x - 750$; $(x - 25)(x + 30) = 0$; $x - 25 = 0$ or $x + 30 = 0$; $x = 25$ or -30 (reject); $x + 5 = 25 + 5 = 30$; 30 people attended the party.

13.

	Rate ×	Time =	Distance
Canoe	x	$\frac{3}{x}$	3
Motor	$2x$	$\frac{3}{x} - \frac{1}{4}$	3

$2x\left(\frac{3}{x} - \frac{1}{4}\right) = 3$; $2(2x)\left(\frac{3}{x} - \frac{1}{4}\right) = 2(3)$; $12 - x = 6$; $-x = -6$; $x = 6$; $2x = 2(6) = 12$; The motorboat goes 12 km/h.

14.

	Rate ×	Hours	= Total
Plumber	x	$\frac{600}{x}$	600
Apprentice	$x - 3$	$\frac{600}{x} + 10$	600

$(x - 3)\left(\frac{600}{x} + 10\right) = 600$; $x(x - 3)\left(\frac{600}{x} + 10\right) = 600x$; $(x - 3)(600 + 10x) = 600x$; $10x^2 + 570x - 1800 = 600x$; $10x^2 - 30x - 1800 = 0$; $x^2 - 3x - 180 = 0$; $(x + 12)(x - 15) = 0$; $x + 12 = 0$ or $x - 15 = 0$; $x = -12$ (reject) or 15; The plumber makes \$15 per hour.

Page 308 • MIXED REVIEW EXERCISES

1. $\frac{2a}{5} + \frac{3a}{10} = 14$; $50\left(\frac{2a}{5} + \frac{3a}{10}\right) = 50 \cdot 14$; $20a + 15a = 700$; $35a = 700$; $a = 20$; $\{20\}$

2. $\frac{x}{5} - \frac{x}{3} = 4$; $15\left(\frac{x}{5} - \frac{x}{3}\right) = 15 \cdot 4$; $3x - 5x = 60$; $-2x = 60$; $x = -30$; $\{-30\}$

3. $\frac{1}{3}(y + 3) + \frac{1}{4}(y + 1) = 3$; $12\left[\frac{1}{3}(y + 3) + \frac{1}{4}(y + 1)\right] = 12 \cdot 3$; $4(y + 3) + 3(y + 1) = 36$; $4y + 12 + 3y + 3 = 36$; $7y + 15 = 36$; $7y = 21$; $y = 3$; $\{3\}$

4. $\frac{3}{7} = \frac{5a}{14}$; $3 \cdot 14 = 7 \cdot 5a$; $42 = 35a$; $\frac{6}{5} = a$; $\left\{\frac{6}{5}\right\}$

5. $\frac{-9}{7t} = \frac{3}{28}$; $-9 \cdot 28 = 7t \cdot 3$; $-252 = 21t$; $-12 = t$; $\{-12\}$ **6.** $\frac{z}{6} = \frac{12}{5}$; $5z = 72$; $z = \frac{72}{5}$; $\left\{\frac{72}{5}\right\}$

7. $(7 - 2)^3 = 5^3 = 125$

8. $4x^2(2x^2 - 7 + 3x) = 8x^4 - 28x^2 + 12x^3 = 8x^4 + 12x^3 - 28x^2$ **9.** $6 \cdot 2^3 = 6 \cdot 8 = 48$

10. $(4n^2 - n) + (9 + n^2) = 5n^2 - n + 9$ **11.** $(5z^3)(2y^2z) = 10y^2z^4$

12. $(3pq^2)^3 = 3^3p^3(q^2)^3 = 27p^3q^6$

Pages 312–313 • WRITTEN EXERCISES

A **1.** $25\% = \frac{25}{100} = \frac{1}{4}$ **2.** $40\% = \frac{40}{100} = \frac{2}{5}$ **3.** $37\frac{1}{2}\% = \frac{37\frac{1}{2}}{100} = \frac{75}{2} \cdot \frac{1}{100} = \frac{3}{8}$

4. $85\% = \frac{85}{100} = \frac{17}{20}$ **5.** $60\% = \frac{60}{100} = \frac{3}{5}$ **6.** $66\frac{2}{3}\% = \frac{66\frac{2}{3}}{100} = \frac{200}{3} \cdot \frac{1}{100} = \frac{2}{3}$

7. $72\% = \frac{72}{100} = \frac{18}{25}$ **8.** $16\% = \frac{16}{100} = \frac{4}{25}$ **9.** $4\% = \frac{4}{100} = \frac{1}{25}$

10. $2\frac{1}{2}\% = \frac{2\frac{1}{2}}{100} = \frac{5}{2} \cdot \frac{1}{100} = \frac{1}{40}$ **11.** $360\% = \frac{360}{100} = \frac{18}{5}$ **12.** $150\% = \frac{150}{100} = \frac{3}{2}$

13. 32% of 300 = $0.32 \cdot 300 = 96$ **14.** 6% of 145 = $0.06 \cdot 145 = 8.7$

15. 1% of 3200 = $0.01 \cdot 3200 = 32$ **16.** $6\frac{1}{4}\%$ of 500 = $0.0625 \cdot 500 = 31.25$

17. $\frac{1}{2}\%$ of 12 = $0.005 \cdot 12 = 0.06$ **18.** 9.2% of 180 = $0.092 \cdot 180 = 16.56$

19. 8.1% of 250 = $0.081 \cdot 250 = 20.25$ **20.** 110% of 30 = $1.10 \cdot 30 = 33$

21. $8 = 0.20x$; $40 = x$; 8 is 20% of 40. **22.** $15 = 0.25x$; $60 = x$; 15 is 25% of 60.

23. $21 = 0.06x$; $350 = x$; 21 is 6% of 350. **24.** $99 = 1.80x$; $55 = x$; 99 is 180% of 55.

25. $4.2 = 0.75x$; $5.6 = x$; 4.2 is 75% of 5.6. **26.** $32.4 = 0.45x$; $72 = x$; 32.4 is 45% of 72.

27. $\frac{x}{100} \cdot 45 = 7.2$; $x = 16$; 7.2 is 16% of 45.

28. $\frac{x}{100} \cdot 7.2 = 45$; $x = 625$; 45 is 625% of 7.2. **29.** $\frac{x}{100} \cdot 150 = 3$; $x = 2$; 3 is 2% of 150.

30. $\frac{x}{100} \cdot 3 = 150$; $x = 5000$; 150 is 5000% of 3.

31. $\frac{x}{100} \cdot 250 = 50$; $x = 20$; 50 is 20% of 250.

32. $\frac{x}{100} \cdot 50 = 250$; $x = 500$; 250 is 500% of 50.

	Original Price	% Discount	Price Decrease	New Price
33.	\$12	25%	\$3	\$9
34.	\$18	20%	\$3.60	\$14.40
35.	\$75	10%	\$7.50	\$67.50
36.	\$240	15%	\$36	\$204

		Original Price	% Markup	Price Increase	New Price
B	37.	\$44	25%	\$11	\$55
	38.	\$85	20%	\$17	\$102
	39.	\$160	30%	\$48	\$208
	40.	\$25	6%	\$1.50	\$26.50

41. $1.8x = 36$; $10(1.8x) = 36 \cdot 10$; $18x = 360$; $x = 20$; $\{20\}$

42. $0.06x = 120$; $100(0.06x) = 100 \cdot 120$; $6x = 12{,}000$; $x = 2{,}000$; $\{2{,}000\}$

43. $0.05t = 2.3t - 6.75$; $100(0.05t) = 100(2.3t - 6.75)$; $5t = 230t - 675$; $0 = 225t - 675$; $675 = 225t$; $3 = t$; $\{3\}$

44. $0.21x = 1.2 - 0.15x$; $100(0.21x) = 100(1.2 - 0.15x)$; $21x = 120 - 15x$; $36x = 120$; $x = \frac{10}{3}$; $\left\{\frac{10}{3}\right\}$

45. $41 = 0.75y + 1.3y$; $100(41) = 100(0.75y + 1.3y)$; $4100 = 75y + 130y$; $4100 = 205y$; $20 = y$; $\{20\}$

46. $0.05a + 0.5a = 4.4$; $100(0.05a + 0.5a) = 100 \cdot 4.4$; $5a + 50a = 440$; $55a = 440$; $a = 8$; $\{8\}$

47. $2 + 0.08x = 0.2x + 8$; $100(2 + 0.08x) = 100(0.2x + 8)$; $200 + 8x = 20x + 800$; $-600 + 8x = 20x$; $-600 = 12x$; $-50 = x$; $\{-50\}$

48. $0.47x - 20 = 0.09x + 3.25$; $100(0.47x - 20) = 100(0.09x + 3.25)$; $47x - 2000 = 9x + 325$; $38x - 2000 = 325$; $38x = 2325$; $x = \frac{2325}{38}$; $\left\{\frac{2325}{38}\right\}$

49. $90 = x + 0.04(180 - x)$; $100 \cdot 90 = 100[x + 0.04(180 - x)]$; $9000 = 100x + 4(180 - x)$; $9000 = 100x + 720 - 4x$; $9000 = 96x + 720$; $8280 = 96x$; $86.25 = x$

50. $0.4x + 0.24(x - 5) = 0.08$; $100[0.4x + 0.24(x - 5)] = 100 \cdot 0.08$; $40x + 24(x - 5) = 8$; $40x + 24x - 120 = 8$; $64x - 120 = 8$; $64x = 128$; $x = 2$; $\{2\}$

51. $0.02 = 3.685x - 0.075(100 - x)$; $1000 \cdot 0.02 = 1000[3.685x - 0.075(100 - x)]$; $20 = 3685x - 75(100 - x)$; $20 = 3685x - 7500 + 75x$; $20 = 3760x - 7500$; $7520 = 3760x$; $2 = x$; $\{2\}$

52. $0.05x + 0.065(5000 - x) = 295$; $1000[0.05x + 0.065(5000 - x)] = 1000 \cdot 295$; $50x + 65(5000 - x) = 295{,}000$; $50x + 325{,}000 - 65x = 295{,}000$; $325{,}000 - 15x = 295{,}000$; $325{,}000 = 295{,}000 + 15x$; $30{,}000 = 15x$; $2000 = x$; $\{2000\}$

53. $0.06(1000 - x) + 0.05x = 700$; $100[0.06(1000 - x) + 0.05x] = 100 \cdot 700$; $6(1000 - x) + 5x = 70{,}000$; $6000 - 6x + 5x = 70{,}000$; $6000 - x = 70{,}000$; $6000 = 70{,}000 + x$; $-64{,}000 = x$; $\{-64{,}000\}$

54. $0.1x - 0.01(10 - x) + 0.01(100 - x) = 19$; $100[0.1x - 0.01(10 - x) + 0.01(100 - x)] = 100 \cdot 19$; $10x - (10 - x) + (100 - x) = 1900$; $10x - 10 + x + 100 - x = 1900$; $10x + 90 = 1900$; $10x = 1810$; $x = 181$; $\{181\}$

55. $\frac{0.2x - 0.1}{5} = 3(x - 0.5)$; $50\left(\frac{0.2x - 0.1}{5}\right) = 50(3)(x - 0.5)$; $2x - 1 = 150x - 75$; $-1 = 148x - 75$; $74 = 148x$; $0.5 = x$; $\{0.5\}$

56. $5(x - 0.4) = \frac{0.5x - 0.5}{2}$; $20(5)(x - 0.4) = 20\left(\frac{0.5x - 0.5}{2}\right)$; $100x - 40 = 5x - 5$;

$95x - 40 = -5$; $95x = 35$; $x = \frac{7}{19}$; $\left\{\frac{7}{19}\right\}$

	Amount invested ×	Annual interest rate =	Simple interest
57.	$400	7.5%	$30
58.	$60,000	9.5%	$5700
59.	$1000	10.5%	$105
60.	$2500	11%	$275
61.	$850	16%	$136
62.	$12,000	7.25%	$870

Pages 313–314 • PROBLEMS

A

1. $\frac{30}{36} = \frac{x}{100}$; $3000 = 36x$; $83\frac{1}{3} = x$; 30 is $83\frac{1}{3}$% of 36.

2. 10.5% of 85,000 = $0.105 \cdot 85{,}000 = 8925$; Laura makes $8925.

3. 22% of x = 18,700; $0.22 \cdot x = 18{,}700$; $x = 85{,}000$; The goal is $85,000.

4. 15% of x = 6; $0.15x = 6$; $x = 40$; The original price was $40.

5. 2% of x = 190; $0.02x = 190$; $x = 9500$; The car cost $9500.

6. 3% of x = 720; $0.03x = 720$; $x = 24{,}000$; Molly's yearly salary is $24,000.

7. Total tax = 3% + $1\frac{1}{2}$% + 1% = $5\frac{1}{2}$%; $5\frac{1}{2}$% of 340 = $0.055 \cdot 340 = 18.7$; Sondra will pay $18.70 tax.

8. Bonds: $11\frac{1}{2}$% of 4000 = $0.115 \cdot 4000 = 460$; Bank: 9.1% of 5000 = $0.091 \cdot 5000 = 455$;

The bonds yield more interest than the bank accounts. They yield $5 more each year.

B

9. Amount of discount = $120 - 96 = 24$; $\frac{24}{120} = \frac{x}{100}$; $2400 = 120x$; $20 = x$; The coat was discounted by 20%.

10. 85 + 6% of 85 = $85 + 0.06 \cdot 85 = 1.06 \cdot 85 = 90.1$; The new price will be $90.10.

11. Sales price = 8400 + 20% of 8400 = $8400 + 0.2 \cdot 8400 = (1.2)(8400) = 10{,}080$;
Sales tax = 5% of 10,080 = $0.05 \cdot 10{,}080 = 504$; Total cost = $10{,}080 + 504 = 10{,}584$;
You have to pay $10,584 for the car.

12. Let x be the rural population. $x + 2.25x = 221{,}000{,}000$; $100x + 225x = 22{,}100{,}000{,}000$; $325x = 22{,}100{,}000{,}000$; $x = 68{,}000{,}000$; $2.25x = 153{,}000{,}000$; The rural population is 68,000,000 and the urban population is 153,000,000.

13. $\frac{5}{x + 5} = \frac{1}{100}$; $500 = x + 5$; $495 = x$; The fertilizer should be mixed with 495 cm^3 of water.

14. Let x be the price of a compact disc. $2\left(66\frac{2}{3}\% \text{ of } x\right) + x = 28$; $2\left(\frac{66\frac{2}{3}}{100} \cdot x\right) + x = 28$;

$\frac{133\frac{1}{3}}{100}x + x = 28$; $\left(\frac{133\frac{1}{3}}{100} + 1\right)x = 28$; $\left(\frac{133\frac{1}{3} + 100}{100}\right)x = 28$; $100\left(\frac{233\frac{1}{3}}{100}\right)x =$

$28 \cdot 100$; $233\frac{1}{3}x = 2800$; $\frac{700}{3}x = 2800$; $x = 2800 \cdot \frac{3}{700}$; $x = 12$; $66\frac{2}{3}\%$ of $12 =$

$\frac{66\frac{2}{3}}{100} \cdot 12 = 8$; Lucinda paid \$8 for a record album.

C 15. Let x be the stock's price Monday morning. Monday evening's price: $x + 10\%$ of $x = x + 0.1x = 1.1x$; Tuesday evening's price: $1.1x - 10\%$ of $1.1x = 59.40$;

$1.1x - 0.1 \cdot 1.1x = 59.40$; $(1.1 - 0.11)x = 59.40$; $0.99x = 59.40$; $x = \frac{59.40}{0.99}$; $x = 60$;

The price was \$60 per share on Monday morning.

Page 314 • MIXED REVIEW EXERCISES

1. $3a^2b + 6a = 3a(ab + 2)$ 2. $3y^3 + 6y^2 + 3y = 3y(y^2 + 2y + 1) = 3y(y + 1)^2$
3. $m^2 + 5m + 6 = (m + 2)(m + 3)$ 4. $x^2 - 9y^2 = (x + 3y)(x - 3y)$
5. $5p^2q^3 + 10p^2q - 15pq^2 = 5pq(pq^2 + 2p - 3q)$ 6. $4a^2 + a - 3 = (4a - 3)(a + 1)$
7. $a(a + b) + 4(a + b) = (a + 4)(a + b)$ 8. $x^2 + 6x + 9 = (x + 3)^2$
9. $t^2 - 2t - 15 = (t - 5)(t + 3)$

Pages 317–318 • WRITTEN EXERCISES

A 1. $\frac{33 - 30}{30} = \frac{x}{100}$; $\frac{3}{30} = \frac{x}{100}$; $300 = 30x$; $10 = x$; Percent increase $= 10\%$

2. $\frac{x}{100} = \frac{15 - 12}{12}$; $\frac{x}{100} = \frac{3}{12}$; $12x = 300$; $x = 25$; Percent increase $= 25\%$

3. $\frac{x}{100} = \frac{300 - 250}{250}$; $\frac{x}{100} = \frac{50}{250}$; $250x = 5000$; $x = 20$; Percent increase $= 20\%$

4. $\frac{x}{100} = \frac{2.1 - 1.8}{1.8}$; $\frac{x}{100} = \frac{0.3}{1.8}$; $1.8x = 30$; $x = 16\frac{2}{3}$; Percent increase $= 16\frac{2}{3}\%$

5. $\frac{x}{100} = \frac{22.5 - 20}{20}$; $\frac{x}{100} = \frac{2.5}{20}$; $20x = 250$; $x = 12.5$; Percent increase $= 12.5\%$

6. $\frac{x}{100} = \frac{253 - 220}{220}$; $\frac{x}{100} = \frac{33}{220}$; $220x = 3300$; $x = 15$; Percent increase $= 15\%$

7. $\frac{x - 6}{6} = \frac{5}{100}$; $100(x - 6) = 30$; $100x - 600 = 30$; $100x = 630$; $x = 6.3$;

New price = \$6.30

8. $\frac{x - 2.8}{2.8} = \frac{20}{100}$; $100(x - 2.8) = 2.8 \cdot 20$; $100x - 280 = 56$; $100x = 336$; $x = 3.36$;

New price = \$3.36

9. $\frac{25}{100} = \frac{8 - x}{x}$; $25x = 100(8 - x)$; $25x = 800 - 100x$; $125x = 800$; $x = 6.4$;
Original price = \$6.40

10. $\frac{10}{100} = \frac{22 - x}{x}$; $10x = 100(22 - x)$; $10x = 2200 - 100x$; $110x = 2200$; $x = 20$;
Original price = \$20

11. $\frac{12}{100} = \frac{72.8 - x}{x}$; $12x = 100(72.8 - x)$; $12x = 7280 - 100x$; $112x = 7280$; $x = 65$;
Original price = \$65

12. $\frac{3}{100} = \frac{566.5 - x}{x}$; $3x = 100(566.5 - x)$; $3x = 56{,}650 - 100x$; $103x = 56{,}650$; $x = 550$;
Original price = \$550

13. $\frac{x}{100} = \frac{25 - 20}{25}$; $\frac{x}{100} = \frac{5}{25}$; $25x = 500$; $x = 20$; Percent decrease = 20%

14. $\frac{x}{100} = \frac{36 - 27}{36}$; $\frac{x}{100} = \frac{9}{36}$; $36x = 900$; $x = 25$; Percent decrease = 25%

15. $\frac{x}{100} = \frac{2 - 1.65}{2}$; $\frac{x}{100} = \frac{0.35}{2}$; $2x = 35$; $x = 17.5$; Percent decrease = 17.5%

16. $\frac{x}{100} = \frac{120 - 108}{120}$; $\frac{x}{100} = \frac{12}{120}$; $120x = 1200$; $x = 10$; Percent decrease = 10%

17. $\frac{32 - x}{32} = \frac{20}{100}$; $100(32 - x) = 32 \cdot 20$; $3200 - 100x = 640$; $-100x = -2560$; $x = 25.6$;
New price = \$25.60

18. $\frac{2.2 - x}{2.2} = \frac{15}{100}$; $100(2.2 - x) = 2.2 \cdot 15$; $220 - 100x = 33$; $-100x = -187$; $x = 1.87$;
New price = \$1.87

19. $\frac{12.5}{100} = \frac{x - 68.25}{x}$; $12.5x = 100(x - 68.25)$; $12.5x = 100x - 6825$; $10(12.5x) =$
$10(100x - 6825)$; $125x = 1000x - 68{,}250$; $0 = 875x - 68{,}250$; $68{,}250 = 875x$;
$78 = x$; Original price = \$78

20. $\frac{5}{100} = \frac{x - 2.09}{x}$; $5x = 100(x - 2.09)$; $5x = 100x - 209$; $209 = 95x$; $2.2 = x$;
Original price = \$2.20

21. $\frac{15}{100} = \frac{x - 59.5}{x}$; $15x = 100(x - 59.5)$; $15x = 100x - 5950$; $5950 = 85x$; $70 = x$;
Original price = \$70

Pages 318–319 • PROBLEMS

A **1.** Let x be the percent of increase. $\frac{x}{100} = \frac{1926 - 1800}{1800}$; $\frac{x}{100} = \frac{126}{1800}$; $1800x = 12{,}600$;
$x = 7$; The student population increased by 7%.

2. Let x be the percent of increase. $\frac{x}{100} = \frac{162{,}000 - 150{,}000}{150{,}000}$; $\frac{x}{100} = \frac{12{,}000}{150{,}000}$;
$150{,}000x = 1{,}200{,}000$; $x = 8$; The value of the Golds' home increased 8%.

3. Let x be the percent of discount. $\frac{x}{100} = \frac{200 - 166}{200}$; $\frac{x}{100} = \frac{34}{200}$; $200x = 3400$; $x = 17$; The coat was discounted 17%.

4. Let x be the percent loss. $\frac{x}{100} = \frac{28 - 24.5}{28}$; $\frac{x}{100} = \frac{3.5}{28}$; $28x = 350$; $x = 12.5$; Elena's stock lost 12.5%.

5. Let x be the sales tax rate. $\frac{x}{100} = \frac{5.83 - 5.5}{5.5}$; $\frac{x}{100} = \frac{.33}{5.5}$; $5.5x = 33$; $x = 6$; The sales tax rate is 6%.

6. Let x be the percent of decrease. $\frac{x}{100} = \frac{3340 - 2430}{3340}$; $\frac{x}{100} = \frac{910}{3340}$; $3340x = 91{,}000$; $x \approx 27\frac{1}{4}$; The number of paid subscribers decreased by approx. $27\frac{1}{4}$%.

7. Let x be the price of the automobile. $\frac{6}{100} = \frac{11{,}448 - x}{x}$; $6x = 100(11{,}448 - x)$; $6x = 1{,}144{,}800 - 100x$; $106x = 1{,}144{,}800$; $x = 10{,}800$; The price of the automobile was \$10,800.

8. Let x be the selling price of the house. 12% of $x = 21{,}600$; $0.12x = 21{,}600$; $x = 180{,}000$; The selling price of the house was \$180,000.

9. Let x be the original price. $\frac{15}{100} = \frac{x - 35.7}{x}$; $15x = 100(x - 35.7)$; $15x = 100x - 3570$; $3570 = 85x$; $42 = x$; The original price of the shoes was \$42.

10. Let x be the number of students with licenses last year. $\frac{24}{100} = \frac{558 - x}{x}$; $24x = 100(558 - x)$; $24x = 55{,}800 - 100x$; $124x = 55{,}800$; $x = 450$; 450 students had driver's licenses last year.

B

11. Let x be the amount invested in stocks and $6000 - x$ be the amount invested in bonds. 6% of x + 11% of $(6000 - x) = 580$; $0.06x + 0.11(6000 - x) = 580$; $0.06x + 660 - 0.11x = 580$; $100(0.06x + 660 - 0.11x) = 100 \cdot 580$; $6x + 66{,}000 - 11x = 58{,}000$; $66{,}000 - 5x = 58{,}000$; $66{,}000 = 58{,}000 + 5x$; $8{,}000 = 5x$; $1600 = x$; $6000 - x = 6000 - 1600 = 4400$; Yolanda invests \$1600 in stocks and \$4400 in bonds.

12. Let x be the amount invested in bank certificates. 5.5% of x + 11% of $(4000 - x) = 352$; $0.055x + 0.11(4000 - x) = 352$; $0.055x + 440 - 0.11x = 352$; $1000(0.055x + 440 - 0.11x) = 1000 \cdot 352$; $55x + 440{,}000 - 110x = 352{,}000$; $440{,}000 - 55x = 352{,}000$; $440{,}000 = 352{,}000 + 55x$; $88{,}000 = 55x$; $1600 = x$; Craig invested \$1600 in bank certificates.

13. Let x be the amount invested in bonds. 8% of x + 9% of $(30{,}000 - x) = 2500$; $0.08x + 0.09(30{,}000 - x) = 2500$; $100(0.08x + 2700 - 0.09x) = 100 \cdot 2500$; $8x + 270{,}000 - 9x = 250{,}000$; $270{,}000 - x = 250{,}000$; $270{,}000 = 250{,}000 + x$; $20{,}000 = x$; The treasurer should invest \$20,000 in bonds.

14. Let x be the amount Bruce invested. $10\frac{1}{2}$% of $x + 75 = 7\frac{1}{2}$% of $(4000 + x)$; $0.105x + 75 = 0.075(4000 + x)$; $1000(0.105x + 75) = 1000(300 + 0.075x)$; $105x + 75{,}000 = 300{,}000 + 75x$; $30x + 75{,}000 = 300{,}000$; $30x = 225{,}000$; $x = 7500$; $x + 4000 = 11{,}500$; Bruce invested \$7,500 and Maya invested \$11,500.

15. Let x be the amount invested in bonds. 7.2% of x = 6% of $(x + 2000)$; $0.072x = 0.06(x + 2000)$; $1000(0.072x) = 1000(0.06x + 120)$; $72x = 60x + 120{,}000$; $12x = 120{,}000$; $x = 10{,}000$; total interest = 2(7.2% of 10,000) = $2(0.072)(10{,}000) = 1440$; Jesse received $1440 in interest.

16. Let x be the amount Roberta invested. 12% of $\frac{1}{2}x$ + 11% of $\frac{1}{3}x$ + 9% of $\left(x - \frac{1}{2}x - \frac{1}{3}x\right) = 1340$; $0.12 \cdot \frac{1}{2}x + 0.11 \cdot \frac{1}{3}x + 0.09\left(1 - \frac{1}{2} - \frac{1}{3}\right)x = 1340$; $6\left[0.12 \cdot \frac{1}{2}x + 0.11 \cdot \frac{1}{3}x + 0.09\left(1 - \frac{1}{2} - \frac{1}{3}\right)x\right] = 6 \cdot 1340$; $0.36x + 0.22x + (0.54 - 0.27 - 0.18)x = 8040$; $0.36x + 0.22x + 0.09x = 8040$; $0.67x = 8040$; $100(0.67x) = 100(8040)$; $67x = 804{,}000$; $x = 12{,}000$; Roberta invested $12,000.

Pages 319–320 • COMPUTER EXERCISES

1.

```
10 LET P = 100
20 PRINT "YEAR", "AMOUNT"
30 PRINT
40 FOR I = 1 TO 10
50 LET P = P + P * 0.06
60 PRINT I, P
70 NEXT I
80 END
```

YEAR	AMOUNT
1	106
2	112.36
3	119.1016
4	126.2477
5	133.82256
6	141.85191
7	150.36303
8	159.38481
9	168.9479
10	179.08477

2.

```
10 PRINT "HOW MANY TIMES";
15 INPUT N
20 FOR I = 1 TO N
30 PRINT "DATA FOR T = "I;
35 INPUT S(I)
40 NEXT I
50 FOR K = 2 TO N
60 P(K) = (S(K) - S(K - 1))/S(K - 1)
70 PRINT K - 1 "TO" K, 100 * P(K) "PERCENT"
80 PRINT
90 NEXT K
100 END
```

a. −3.38, −3.5, −3.63, −3.76; no **b.** 41, 40.99, 41.05, 41.01; yes

Page 320 • MIXED REVIEW EXERCISES

1. $\frac{3x + 1}{x - 2} = 7$; $7(x - 2) = 3x + 1$; $7x - 14 = 3x + 1$; $4x - 14 = 1$; $4x = 15$; $x = \frac{15}{4}$; $\left\{\frac{15}{4}\right\}$

2. $\frac{a + 3}{4} = \frac{1}{a}$; $a(a + 3) = 4$; $a^2 + 3a = 4$; $a^2 + 3a - 4 = 0$; $(a + 4)(a - 1) = 0$; $a = -4, 1$; $\{-4, 1\}$

3. $\frac{5}{n} = \frac{13}{6n}$; $30n = 13n$; $17n = 0$; $n = 0$; but n cannot equal 0; no solution.

4. $1.4x = 28; 10(1.4x) = 10 \cdot 28; 14x = 280; x = 20; \{20\}$

5. $m^2 + 3m + 2 = 0; (m + 2)(m + 1) = 0; m = -2, -1; \{-2, -1\}$

6. $0.5x + 5.6 = 1.2x; 10(0.5x + 5.6) = 10 \cdot 1.2x; 5x + 56 = 12x; 56 = 7x; 8 = x; \{8\}$

7. $(z + 3)9 = 9z + 27$ **8.** $-4(x - 3y) = -4x + 12y$ **9.** $9a + 7 - 2a + b = 7a + b + 7$

10. $16t\left(-\frac{1}{2}\right) = -8t$ **11.** $(-5b)(-6c) = 30bc$ **12.** $\frac{1}{7}(-21xz) = -3xz$

Pages 324–325 • PROBLEMS

A **1.**

	Number of kg	× Price per kg	= Total Cost
Cashews	x	8	$8x$
Pecans	$8 - x$	7	$7(8 - x)$
Mixture	8	7.25	58

$8x + 7(8 - x) = 58; 8x + 56 - 7x = 58; x + 56 = 58; x = 2; 8 - x = 8 - 2 = 6;$
There should be 2 kg of cashews and 6 kg of pecans.

2.

	Total amount	× % acid	= Amount of acid
Original solution	40	50%	20
Water added	x	0%	0
New solution	$40 + x$	10%	$0.10(40 + x)$

$20 + 0 = 0.1(40 + x); 20 \cdot 10 = 10[0.1(40 + x)]; 200 = 40 + x; 160 = x;$ He should add 160 mL of water.

3.

	Total amount	× % juice	= Amount of juice
Original drink	800	15%	120
Juice added	x	100%	x
New drink	$800 + x$	20%	$0.20(800 + x)$

$120 + x = 0.2(800 + x); 10(120 + x) = 10[0.2(800 + x)]; 1200 + 10x = 2(800 + x);$
$1200 + 10x = 1600 + 2x; 1200 + 8x = 1600; 8x = 400; x = 50;$ 50 mL of juice must be added.

4.

	Total amount	× % acid	= Amount of acid
1st Solution	12	45%	5.4
2nd Solution	8	70%	5.6
Mixture	20	x	$20x$

$5.4 + 5.6 = 20x; 11 = 20x; 0.55 = x;$ The mixture is 55% acid.

5.

	Number of lb × Cost per lb = Total cost		
Egg noodles	5	0.80	4
Spinach noodles	2	1.50	3
Mixture	7	x	$7x$

$4 + 3 = 7x$; $7 = 7x$; $1 = x$; The cost of the mixture will be \$1.00/lb.

6.

	Rate × Time = Distance		
1st part of trip	85	2	170
2nd part of trip	95	3	285
Entire trip	x	5	$5x$

$170 + 285 = 5x$; $455 = 5x$; $91 = x$; Susan averaged 91 km/h.

7.

	Amount invested × Rate = Interest		
Investment A	2000	$5\frac{1}{4}\%$	105
Investment B	3000	$8\frac{1}{2}\%$	255
Total Investments	5000	x	$5000x$

$105 + 255 = 5000x$; $360 = 5000x$; $0.072 = x$; Sam is earning 7.2% on his investments.

8.

	Number of coins × Value per coin = Total value		
Dimes	x	10	$10x$
Nickels	$50 - x$	5	$5(50 - x)$
Collection	50	—	430

$10x + 5(50 - x) = 430$; $10x + 250 - 5x = 430$; $5x + 250 = 430$; $5x = 180$; $x = 36$; $50 - x = 14$; Gina has 36 dimes and 14 nickels.

9.

	Total amount × % acid = Amount of acid		
Old solution	50	30%	15
Water	x	0%	0
New solution	$50 + x$	20%	$0.20(50 + x)$

$15 + 0 = 0.2(50 + x)$; $15 \cdot 10 = 10[0.2(50 + x)]$; $150 = 2(50 + x)$; $150 = 100 + 2x$; $50 = 2x$; $25 = x$; 25 L of water should be added.

10.

	Total amount	× % iodine	= Amount of iodine
Old solution	60	15%	9
Water	x	0%	0
New solution	$60 + x$	10%	$0.10(60 + x)$

$9 + 0 = 0.1(60 + x)$; $9 \cdot 10 = 10[0.1(60 + x)]$; $90 = 60 + x$; $30 = x$; 30 mL of water must be added.

11.

	Total amount	× % thyme	= Amount of thyme
Old mixture	12	25%	3
Thyme	x	100%	x
New mixture	$12 + x$	40%	$0.40(12 + x)$

$3 + x = 0.4(12 + x)$; $10(3 + x) = 10[0.4(12 + x)]$; $30 + 10x = 48 + 4x$; $30 + 6x = 48$; $6x = 18$; $x = 3$; 3 g of thyme should be added.

12.

	Number of kg	× Cost per kg	= Total cost
1st kind	x	5	$5x$
2nd kind	$40 - x$	5.8	$5.8(40 - x)$
Mixture	40	5.5	220

$5x + 5.8(40 - x) = 220$; $5x + 232 - 5.8x = 220$; $-0.8x + 232 = 220$; $-0.8x = -12$; $10(-0.8x) = 10(-12)$; $-8x = -120$; $x = 15$; $40 - x = 25$; The mixture should be 15 kg of the first kind of nuts and 25 kg of the second kind.

13.

	Number of kg	× Cost per kg	= Total cost
Apples	x	6	$6x$
Apricots	$20 - x$	8	$8(20 - x)$
Mixture	20	7.2	144

$6x + 8(20 - x) = 144$; $6x + 160 - 8x = 144$; $160 - 2x = 144$; $-2x = -16$; $x = 8$; $20 - x = 12$; The mixture should contain 8 kg dried apples and 12 kg dried apricots.

14.

	Number of L	× Cost per L	= Total cost
Apple	8	.45	3.6
Cranberry	10	1.08	10.8
Mixture	18	x	$18x$

$3.6 + 10.8 = 18x$; $14.4 = 18x$; $144 = 180x$; $x = 0.8$; He should charge 80¢/L for the mixture.

15.

	Rate × Time = Distance		
1st part of trip	80	2	160
2nd part of trip	x	1	x
Entire trip	75	3	225

$160 + x = 225$; $x = 65$; You have to drive 65 km/h.

16.

	Rate × Time = Wages		
Regular	x	40	$40x$
Overtime	$1\frac{1}{2}x$	4	$6x$
Total	—	44	552

$40x + 6x = 552$; $46x = 552$; $x = 12$; Sylvia normally earns $12/h.

B **17.**

	Total amount × % acid = Amount of acid		
Pure acid	x	100%	x
Water	$16 - x$	0%	0
Mixture	16	30%	4.8

$x = 4.8$; $16 - x = 16 - 4.8 = 11.2$; 4.8 L of acid should be mixed with 11.2 L water.

18.

	Total amount × % salt = Amount of salt		
Old solution	10	40%	4
Water	x	0%	0
New solution	$10 - x$	50%	$0.50(10 - x)$

$4 = 0.5(10 - x)$; $4 = 5 - 0.5x$; $40 = 10(5 - 0.5x)$; $40 = 50 - 5x$; $-10 = -5x$; $2 = x$; 2 L of water must be evaporated.

19.

	Total amount × % salt = Amount of salt		
Old solution	20	30%	6
Water	x	0%	0
New solution	$20 - x$	50%	$0.50(20 - x)$

$6 = 0.5(20 - x)$; $60 = 5(20 - x)$; $60 = 100 - 5x$; $-40 = -5x$; $8 = x$; 8 L of water must be evaporated.

20.

	Number of kg × Cost per kg = Total cost		
Old mixture	100	4	400
Cheaper nuts	x	3.2	$3.2x$
New mixture	$100 + x$	3.4	$3.4(100 + x)$

$400 + 3.2x = 3.4(100 + x)$; $10(400 + 3.2x) = 10[3.4(100 + x)]$; $4000 + 32x = 3400 + 34x$; $600 + 32x = 34x$; $600 = 2x$; $300 = x$; She should add 300 kg of the cheaper nuts.

21.

	Amount invested ×	Interest rate =	Interest earned
Bonds	x	12%	$0.12x$
Certificates	$21{,}000 - x$	$5\frac{1}{4}\%$	$0.0525(21{,}000 - x)$
Total	21,000	—	2250

$0.12x + 0.0525(21{,}000 - x) = 2250$;
$10{,}000[0.12x + 0.0525(21{,}000 - x)] = 10{,}000 \cdot 2250$;
$1200x + 525(21{,}000 - x) = 22{,}500{,}000$; $1200x + 11{,}025{,}000 - 525x = 22{,}500{,}000$;
$675x + 11{,}025{,}000 = 22{,}500{,}000$; $675x = 11{,}475{,}000$; $x = 17{,}000$; $21{,}000 - x =$ $21{,}000 - 17{,}000 = 4000$; \$17,000 was invested in bonds and \$4000 was invested in certificates of deposit.

22.

	Number of coins ×	Value of coin =	Total value
Dimes	x	0.1	$0.1x$
Nickels	$x + 12$	0.05	$0.05(x + 12)$
Quarters	$50 - x - (x + 12)$	0.25	$0.25(38 - 2x)$
Total	50	—	5.20

$0.1x + 0.05(x + 12) + 0.25(38 - 2x) = 5.2$;
$100[0.1x + 0.05(x + 12) + 0.25(38 - 2x)] = 5.2 \cdot 100$;
$10x + 5(x + 12) + 25(38 - 2x) = 520$; $10x + 5x + 60 + 950 - 50x = 520$;
$1010 - 35x = 520$; $-35x = -490$; $x = 14$; $x + 12 = 14 + 12 = 26$;
$50 - 14 - 26 = 10$; There are 14 dimes, 26 nickels, and 10 quarters.

23. a.

	Rate ×	Time =	Distance
1st part of trip	30	2	60
2nd part of trip	20	2	40
Whole trip	x	4	$4x$

$60 + 40 = 4x$; $100 = 4x$; $x = 25$; Your average speed is 25 km/h.

b.

	Rate ×	Time =	Distance
1st part of trip	30	2	60
2nd part of trip	20	3	60
Whole trip	x	5	$5x$

$60 + 60 = 5x$; $120 = 5x$; $24 = x$; Your average speed is 24 km/h.

C **24.** Let $3x$ be the number of nickels, $8x$ the number of dimes, and x the number of quarters. $3x \cdot 5 + 8x \cdot 10 + x \cdot 25 = (3x + 8x + x)10$;
$15x + 80x + 25x = 12x \cdot 10$; $120x = 120x$; This is true for all values of x.

25.

	Number of lb	× Price per lb	= Total Cost
Bananas	$3x$	1.17	$3.51x$
Apricots	x	3	$3x$
Dates	x	2.3	$2.3x$
Mixture	$5x$	$\frac{8.81x}{5x}$	$8.81x$

$\frac{8.81x}{5x} = 1.762$;

The cost per pound of the mixture is about \$1.76.

Page 325 • MIXED REVIEW EXERCISES

1. 9% of 60 + 0.4% of 230 = $0.09 \cdot 60 + 0.004 \cdot 230 = 5.4 + 0.92 = 6.32$

2. $\frac{x}{100} \cdot 45 = 18; x = \frac{1800}{45}; x = 40$; 18 is 40% of 45.

3. $\frac{x}{100} \cdot 160 = 12; x = \frac{1200}{160}; x = 7.5$; 12 is 7.5% of 160.

4. $16 = 0.125x; 128 = x$; 16 is $12\frac{1}{2}$% of 128. **5.** $|-5| + y = |-5| + 4 = 5 + 4 = 9$

6. $\frac{7 - 2b}{3 + b} = \frac{7 - 2 \cdot 3}{3 + 3} = \frac{7 - 6}{6} = \frac{1}{6}$ **7.** $\frac{1}{7}(9x + y) = \frac{1}{7}(9 \cdot 5 + 4) = \frac{1}{7}(45 + 4) = \frac{1}{7}(49) = 7$

8. $xy^2 = 5 \cdot 4^2 = 5 \cdot 16 = 80$ **9.** $2a + 5b = 2 \cdot 2 + 5 \cdot 3 = 4 + 15 = 19$

10. $(x - b)^2 = (5 - 3)^2 = 2^2 = 4$

Pages 328–330 • PROBLEMS

A **1.** Rita works at a rate of $\frac{1}{3}$ job/h. She can do $\frac{2}{3}$ job in 2 h. She can do $\frac{x}{3}$ job in x h.

2. Franklin's work rate is $\frac{1}{6}$. Mike's rate is $\frac{1}{4}$. $2\left(\frac{1}{6} + \frac{1}{4}\right) = \frac{1}{3} + \frac{1}{2} = \frac{2}{6} + \frac{3}{6} = \frac{5}{6}$; In 2 h they can do $\frac{5}{6}$ job. $y\left(\frac{1}{6} + \frac{1}{4}\right) = y\left(\frac{2 + 3}{12}\right) = \frac{5}{12}y$ or $\frac{5y}{12}$; In y h they can do $\frac{5y}{12}$ job.

3. Jake's work rate is $\frac{1}{10}$. Maura's rate is $\frac{1}{8}$. Together their rate is $\frac{1}{10} + \frac{1}{8} = \frac{4 + 5}{40} = \frac{9}{40}$; $2.5 \cdot \frac{9}{40} = \frac{9}{16}$; In 2.5 h they can do $\frac{9}{16}$ job; In h hours they can do $\frac{9h}{40}$ job.

4. The first pipe drains at a rate of $\frac{1}{6}$. The second pipe's rate is $\frac{1}{3}$. Together their drain rate is $\frac{1}{6} + \frac{1}{3} = \frac{1 + 2}{6} = \frac{3}{6} = \frac{1}{2}$; $2 \cdot \frac{1}{2} = 1$; The whole job is completed in 2 h. In x h they can do $\frac{x}{2}$ job.

5.

	Work rate	× Time	= Work done
Abby	$\frac{1}{2}$	x	$\frac{x}{2}$
Carla	$\frac{1}{3}$	x	$\frac{x}{3}$

$\frac{x}{2} + \frac{x}{3} = 1$; $6\left(\frac{x}{2} + \frac{x}{3}\right) = 6$; $3x + 2x = 6$; $5x = 6$; $x = 1\frac{1}{5}$; It will take them $1\frac{1}{5}$ h to mow the lawn.

6.

	Work rate	× Time	= Work done
Phil	$\frac{1}{12}$	3	$\frac{1}{4}$
Rick	$\frac{1}{10}$	$x + 3$	$\frac{x + 3}{10}$

$\frac{1}{4} + \frac{x + 3}{10} = 1$; $20\left(\frac{1}{4} + \frac{x + 3}{10}\right) = 20$; $5 + 2(x + 3) = 20$; $5 + 2x + 6 = 20$; $2x + 11 = 20$; $2x = 9$; $x = 4\frac{1}{2}$; It will take Rick $4\frac{1}{2}$ h.

7.

	Work rate	× Time	= Work done
Chuck	$\frac{1}{40}$	30	$\frac{3}{4}$
Joan	$\frac{1}{x}$	10	$\frac{10}{x}$

$\frac{3}{4} + \frac{10}{x} = 1$; $4x\left(\frac{3}{4} + \frac{10}{x}\right) = 4x$; $3x + 40 = 4x$; $40 = x$; It would take Joan 40 min to shovel the driveway.

8.

	Work rate	× Time	= Work done
Brett	$\frac{1}{50}$	25	$\frac{1}{2}$
Angela	$\frac{1}{x}$	15	$\frac{15}{x}$

$\frac{1}{2} + \frac{15}{x} = 1$; $2x\left(\frac{1}{2} + \frac{15}{x}\right) = 2x$; $x + 30 = 2x$; $30 = x$; It would take Angela 30 min.

9.

	Work rate	× Time	= Work done
Sally	$\frac{1}{15}$	x	$\frac{x}{15}$
Lisa	$\frac{1}{25}$	x	$\frac{x}{25}$

$\frac{x}{15} + \frac{x}{25} = 1$; $75\left(\frac{x}{15} + \frac{x}{25}\right) = 75$; $5x + 3x = 75$; $8x = 75$; $x = 9\frac{3}{8}$; It will take them $9\frac{3}{8}$ min.

10.

	Work rate	× Time	= Work done
Gary	1	x	x
Dana	$\frac{1}{1.5}$	x	$\frac{x}{1.5}$

$x + \frac{x}{1.5} = 1$; $1.5\left(x + \frac{x}{1.5}\right) = 1.5$; $1.5x + x = 1.5$; $2.5x = 1.5$; $x = 0.6$; It will take them 0.6 h or 36 min.

11.

	Work rate	× Time	= Work done
Contractor	$\frac{1}{20}$	x	$\frac{x}{20}$
Assistant	$\frac{1}{30}$	x	$\frac{x}{30}$

$\frac{x}{20} + \frac{x}{30} = 1$; $60\left(\frac{x}{20} + \frac{x}{30}\right) = 60$; $3x + 2x = 60$; $5x = 60$; $x = 12$; It will take them 12 h.

12.

	Work rate	× Time	= Work done
Stan	$\frac{1}{24}$	15	$\frac{5}{8}$
Brother	$\frac{1}{x}$	15	$\frac{15}{x}$

$\frac{5}{8} + \frac{15}{x} = 1$; $8x\left(\frac{5}{8} + \frac{15}{x}\right) = 8x$; $5x + 120 = 8x$; $120 = 3x$; $40 = x$; Stan's brother takes 40 min.

B 13.

	Work rate	× Time	= Work done
1st machine	$\frac{1}{2x}$	3	$\frac{3}{2x}$
2nd machine	$\frac{1}{x}$	3	$\frac{3}{x}$

$\frac{3}{2x} + \frac{3}{x} = 1$; $2x\left(\frac{3}{2x} + \frac{3}{x}\right) = 2x$; $3 + 6 = 2x$; $9 = 2x$; $4.5 = x$; One machine needs 4.5 h and the other machine needs 9 h.

14.

	Work rate	× Time	= Work done
Arthur	$\frac{1}{30}$	x	$\frac{x}{30}$
Bonnie	$\frac{1}{40}$	x	$\frac{x}{40}$
Claire	$\frac{1}{60}$	x	$\frac{x}{60}$

$\frac{x}{30} + \frac{x}{40} + \frac{x}{60} = 1$; $120\left(\frac{x}{30} + \frac{x}{40} + \frac{x}{60}\right) = 120$; $4x + 3x + 2x = 120$; $9x = 120$;

$x = \frac{120}{9}$; $x = \frac{40}{3} = 13\frac{1}{3}$; It will take them $13\frac{1}{3}$ min.

15.

	Work rate	× Time	= Work done
Father	$\frac{1}{3}$	$x + 1$	$\frac{x+1}{3}$
Child	$\frac{1}{5}$	1	$\frac{1}{5}$

$\frac{x+1}{3} + \frac{1}{5} = 1$; $15\left(\frac{x+1}{3} + \frac{1}{5}\right) = 15$; $5(x + 1) + 3 = 15$; $5x + 5 + 3 = 15$;

$5x + 8 = 15$; $5x = 7$; $x = 1\frac{2}{5}$; It will take him $1\frac{2}{5}$ h to finish plowing.

16.

	Work rate ×	Time =	Work done
Phyllis	$\frac{1}{50}$	$x + 5$	$\frac{x+5}{50}$
Me	$\frac{1}{40}$	x	$\frac{x}{40}$

$\frac{x+5}{50} + \frac{x}{40} = 1$; $200\left(\frac{x+5}{50} + \frac{x}{40}\right) = 200$; $4(x + 5) + 5x = 200$; $4x + 20 + 5x = 200$; $9x + 20 = 200$; $9x = 180$; $x = 20$; It will take us 20 min.

17.

	Work rate ×	Time =	Work done
1st pump	$\frac{1}{3}$	$\frac{1}{2}$	$\frac{1}{6}$
2nd pump	$\frac{1}{5}$	$x + \frac{1}{2}$	$\frac{1}{5}\left(x + \frac{1}{2}\right)$

$\frac{1}{6} + \frac{1}{5}\left(x + \frac{1}{2}\right) = 1$; $\frac{1}{6} + \frac{1}{5}x + \frac{1}{10} = 1$; $30\left(\frac{1}{6} + \frac{1}{5}x + \frac{1}{10}\right) = 30$; $5 + 6x + 3 = 30$;

$6x + 8 = 30$; $6x = 22$; $x = \frac{11}{3} = 3\frac{2}{3}$; The slower pump ran $3\frac{2}{3}$ h longer.

18.

	Work rate ×	Time =	Work done
Pipe A	$\frac{1}{12}$	9	$\frac{3}{4}$
Pipe B	$\frac{1}{x}$	$4\frac{1}{2}$	$\frac{9}{2x}$

$\frac{3}{4} + \frac{9}{2x} = 1$; $4x\left(\frac{3}{4} + \frac{9}{2x}\right) = 4x$; $3x + 18 = 4x$; $18 = x$; Pipe B can fill the pool in 18 h.

19.

	Work rate ×	Time =	Work done
Ramona	$\frac{1}{12}$	6	$\frac{1}{2}$
Carlotta	$\frac{1}{x}$	2	$\frac{2}{x}$

$\frac{1}{2} + \frac{2}{x} = 1$; $2x\left(\frac{1}{2} + \frac{2}{x}\right) = 2x$; $x + 4 = 2x$; $4 = x$; It would have taken Carlotta 4 days to do the whole job herself.

20.

	Work rate ×	Time =	Work done
Fill pipe	$\frac{1}{4}$	x	$\frac{x}{4}$
Drain pipe	$\frac{1}{2}$	x	$\frac{x}{2}$

$\frac{x}{2} - \frac{x}{4} = \frac{1}{2}$; $4\left(\frac{x}{2} - \frac{x}{4}\right) = 4 \cdot \frac{1}{2}$; $2x - x = 2$; $x = 2$; The tank will be empty in 2 h.

21.

	Work rate ×	Time =	Work done
Nicholas	$\frac{1}{30}$	x	$\frac{x}{30}$
Marilyn	$\frac{1}{40}$	x	$\frac{x}{40}$

$\frac{x}{30} + \frac{x}{40} = 140$; $120\left(\frac{x}{30} + \frac{x}{40}\right) = 140 \cdot 120$; $4x + 3x = 16{,}800$; $7x = 16{,}800$; $x = 2400$; It will take them 2400 s or 40 min.

C **22.**

	Work rate	× Time	= Work done
Jeff	$\frac{1}{4}$	$1\frac{1}{2}$	$\frac{3}{8}$
Brenda	$\frac{1}{4}$	$1\frac{1}{2}$	$\frac{3}{8}$
Rory	$\frac{1}{x}$	$\frac{1}{2}$	$\frac{1}{2x}$

$\frac{3}{8} + \frac{3}{8} + \frac{1}{2x} = 1; \frac{6}{8} + \frac{1}{2x} = 1;$
$\frac{3}{4} + \frac{1}{2x} = 1; 4x\left(\frac{3}{4} + \frac{1}{2x}\right) = 4x;$
$3x + 2 = 4x; 2 = x;$ It would have taken Rory 2 h.

23.

	Work rate	× Time	= Work done
Largest pipe	$\frac{1}{x}$	3	$\frac{3}{x}$
Smallest pipe	$\frac{1}{3x}$	3	$\frac{1}{x}$
Other pipe	$\frac{1}{2x}$	3	$\frac{3}{2x}$

$\frac{3}{x} + \frac{1}{x} + \frac{3}{2x} = 1; 2x\left(\frac{3}{x} + \frac{1}{x} + \frac{3}{2x}\right) = 2x; 6 + 2 + 3 = 2x; 11 = 2x; 5.5 = x;$

$3x = 16.5; 2x = 11;$ The largest pipe can fill the pool in $5\frac{1}{2}$ h, the smallest pipe takes $16\frac{1}{2}$ h, and the other pipe takes 11 h.

Page 330 • MIXED REVIEW EXERCISES

1. $\frac{x}{100} = \frac{40.12 - 34}{34}; 34x = 100(40.12 - 34); 34x = 100(6.12); 34x = 612; x = 18;$ Percent increase = 18%.

2. $\frac{10}{100} = \frac{x - 130}{130}; 1300 = 100(x - 130); 1300 = 100x - 13{,}000; 14{,}300 = 100x; 143 = x;$ New price = \$143.

3. $\frac{20}{100} = \frac{26.4 - x}{x}; 20x = 100(26.4 - x); 20x = 2640 - 100x; 120x = 2640; x = 22;$ Original Price = \$22.

4. $\frac{1}{c + 3} + \frac{4}{2c + 6} = 3; \frac{1}{c + 3} + \frac{4}{2(c + 3)} = 3; \frac{1}{c + 3} + \frac{2}{c + 3} = 3; \frac{3}{c + 3} = 3; 3 = 3(c + 3);$
$3 = 3c + 9; -6 = 3c; -2 = c; \{-2\}$

5. $\frac{8}{n} = \frac{16}{7}; 56 = 16n; \frac{56}{16} = n; 3.5 = n; \{3.5\}$

6. $\frac{x + 3}{7} = \frac{x + 5}{14}; 14(x + 3) = 7(x + 5); 14x + 42 = 7x + 35; 7x + 42 = 35; 7x = -7;$
$x = -1; \{-1\}$

7. $5a - 1 = 3(a + 7)$; $5a - 1 = 3a + 21$; $2a - 1 = 21$; $2a = 22$; $a = 11$; $\{11\}$

8. $-0.6 + k = 0.8$; $k = 1.4$; $\{1.4\}$

9. $-9p = 0$; $p = 0$; $\{0\}$

Pages 333–335 • WRITTEN EXERCISES

A **1.** $5^{-1} = \frac{1}{5^1} = \frac{1}{5}$ **2.** $4^{-1} = \frac{1}{4^1} = \frac{1}{4}$ **3.** $3^{-2} = \frac{1}{3^2} = \frac{1}{9}$ **4.** $2^{-2} = \frac{1}{2^2} = \frac{1}{4}$

5. $9^{-2} = \frac{1}{9^2} = \frac{1}{81}$ **6.** $8^{-2} = \frac{1}{8^2} = \frac{1}{64}$ **7.** $3^{-3} = \frac{1}{3^3} = \frac{1}{27}$ **8.** $2^{-3} = \frac{1}{2^3} = \frac{1}{8}$

9. $2^{-4} = \frac{1}{2^4} = \frac{1}{16}$ **10.** $3^{-4} = \frac{1}{3^4} = \frac{1}{81}$ **11.** $1^{-6} = \frac{1}{1^6} = 1$ **12.** $6^{-1} = \frac{1}{6^1} = \frac{1}{6}$

13. $4^{-2} = \frac{1}{4^2} = \frac{1}{16}$ **14.** $3^7 \cdot 3^{-9} = 3^{7-9} = 3^{-2} = \frac{1}{3^2} = \frac{1}{9}$

15. $(2^{-3})^{-1} = 2^{(-3)(-1)} = 2^3 = 8$ **16.** $(4^{-1})^{-3} = 4^{(-1)(-3)} = 4^3 = 64$

17. $\frac{3}{3^{-2}} = \frac{3^1}{3^{-2}} = 3^{1-(-2)} = 3^{1+2} = 3^3 = 27$ **18.** $\frac{6^{-2}}{6^{-3}} = 6^{-2-(-3)} = 6^{-2+3} = 6^1 = 6$

19. $\frac{7^{-2} \cdot 7}{7^{-1}} = \frac{7^{-2} \cdot 7^1}{7^{-1}} = 7^{-2+1-(-1)} = 7^0 = 1$ **20.** $\frac{3^{-5} \cdot 3^3}{3^{-4}} = 3^{-5+3-(-4)} = 3^2 = 9$

21. $\left(\frac{3^{-1}}{3}\right)^2 = \frac{3^{-2}}{3^2} = 3^{-2-2} = 3^{-4} = \frac{1}{3^4} = \frac{1}{81}$ **22.** $\left(\frac{8^4}{8^{-4}}\right)^0 = 1$

23. $\frac{(2^{-3} \cdot 4^2)^{-1}}{2^{-1}} = \frac{2^3 \cdot 4^{-2}}{2^{-1}} = \frac{2^3 \cdot (2^2)^{-2}}{2^{-1}} = \frac{2^3 \cdot 2^{-4}}{2^{-1}} = 2^{3-4-(-1)} = 2^0 = 1$

24. $\left(\frac{3^0 \cdot 2}{2^{-2}}\right)^{-3} = \frac{(1 \cdot 2)^{-3}}{(2^{-2})^{-3}} = \frac{2^{-3}}{2^6} = 2^{-3-6} = 2^{-9} = \frac{1}{2^9} = \frac{1}{512}$

25. $\left(\frac{2}{3}\right)^{-4} = \frac{2^{-4}}{3^{-4}} = \frac{\frac{1}{2^4}}{\frac{1}{3^4}} = \frac{1}{2^4} \cdot \frac{3^4}{1} = \frac{3^4}{2^4} = \frac{81}{16}$

26. $\left(\frac{5}{4}\right)^{-2} = \frac{5^{-2}}{4^{-2}} = \frac{\frac{1}{5^2}}{\frac{1}{4^2}} = \frac{1}{5^2} \cdot \frac{4^2}{1} = \frac{4^2}{5^2} = \frac{16}{25}$ **27.** $\left(\frac{4}{5}\right)^{-3} = \frac{4^{-3}}{5^{-3}} = \frac{\frac{1}{4^3}}{\frac{1}{5^3}} = \frac{1}{4^3} \cdot \frac{5^3}{1} = \frac{5^3}{4^3} = \frac{125}{64}$

28. $\left(\frac{7}{6}\right)^{-1} = \frac{7^{-1}}{6^{-1}} = \frac{\frac{1}{7^1}}{\frac{1}{6^1}} = \frac{1}{7^1} \cdot \frac{6^1}{1} = \frac{6^1}{7^1} = \frac{6}{7}$

29. a. $(2 \cdot 3)^{-3} = 6^{-3} = \frac{1}{6^3} = \frac{1}{216}$ **b.** $2 \cdot 3^{-3} = 2 \cdot \frac{1}{3^3} = 2 \cdot \frac{1}{27} = \frac{2}{27}$

30. a. $(4 \cdot 5)^{-2} = 20^{-2} = \frac{1}{20^2} = \frac{1}{400}$ **b.** $4 \cdot 5^{-2} = 4 \cdot \frac{1}{5^2} = 4 \cdot \frac{1}{25} = \frac{4}{25}$

31. a. $(2^{-1} + 2^{-2})^{-1} = \left(\frac{1}{2} + \frac{1}{4}\right)^{-1} = \left(\frac{2+1}{4}\right)^{-1} = \left(\frac{3}{4}\right)^{-1} = \frac{4}{3}$

b. $(2^{-1} \cdot 2^{-2})^{-1} = (2^{-1-2})^{-1} = (2^{-3})^{-1} = 2^3 = 8$

32. a. $(2 + 3^{-1})^{-2} = \left(2 + \frac{1}{3}\right)^{-2} = \left(\frac{6+1}{3}\right)^{-2} = \left(\frac{7}{3}\right)^{-2} = \frac{7^{-2}}{3^{-2}} = \frac{1}{7^2} \cdot 3^2 = \frac{3^2}{7^2} = \frac{9}{49}$

b. $(2 \cdot 3^{-1})^{-2} = 2^{-2} \cdot 3^2 = \frac{1}{4} \cdot 9 = \frac{9}{4}$

33. $b^{-1} = \frac{1}{b}$; $\frac{1}{b}$ is the reciprocal of b; If $b = 0$, b has no reciprocal.

34. $b^2 \cdot b^0 = b^{2+0} = b^2$; since $b^2 \cdot b^0 = b^2$, b^0 must be the multiplicative identity, 1.

B **35.** $3x^{-2} = \frac{3}{x^2}$ **36.** $2n^{-3} = \frac{2}{n^3}$ **37.** $(3x)^{-2} = 3^{-2}x^{-2} = \frac{1}{3^2} \cdot \frac{1}{x^2} = \frac{1}{9x^2}$

38. $(5m)^{-2} = 5^{-2} \cdot m^{-2} = \frac{1}{5^2} \cdot \frac{1}{m^2} = \frac{1}{25m^2}$ **39.** $x^{-1}y^2 = \frac{1}{x} \cdot y^2 = \frac{y^2}{x}$

40. $a^{-2}b^{-3} = \frac{1}{a^2} \cdot \frac{1}{b^3} = \frac{1}{a^2b^3}$ **41.** $(m^{-3})^4 = m^{-3 \cdot 4} = m^{-12} = \frac{1}{m^{12}}$

42. $(x^{-2})^3 = x^{-6} = \frac{1}{x^6}$ **43.** $m^{-3} \cdot m^4 = m^{-3+4} = m^1 = m$

44. $a^5 \cdot a^{-2} = a^{5-2} = a^3$ **45.** $(3x^{-2})^3 = 3^3 \cdot x^{-6} = 27x^{-6} = \frac{27}{x^6}$

46. $(4n^{-1})^2 = 4^2 \cdot n^{-2} = \frac{16}{n^2}$ **47.** $2x^5 \cdot x^{-5} = 2 \cdot x^{5-5} = 2 \cdot x^0 = 2 \cdot 1 = 2$

48. $3x^{-2} \cdot (3x^2)^{-1} = 3x^{-2} \cdot 3^{-1} \cdot x^{-2} = 3^{1-1}x^{-2-2} = 3^0x^{-4} = \frac{1}{x^4}$

49. $(a^2 \cdot a^{-5})^2 = (a^{2-5})^2 = (a^{-3})^2 = a^{-6} = \frac{1}{a^6}$

50. $(b^5 \cdot b^{-7})^3 = (b^{5-7})^3 = (b^{-2})^3 = b^{-6} = \frac{1}{b^6}$ **51.** $\frac{y^3}{y^{-5}} = y^{3-(-5)} = y^8$

52. $\frac{u^{-4}}{u^{10}} = u^{-4-10} = u^{-14} = \frac{1}{u^{14}}$ **53.** $\frac{c^{-5}}{c^2} = \frac{1}{c^5} \cdot \frac{1}{c^2} = \frac{1}{c^7}$ **54.** $\frac{d^{-3}}{d^{-5}} = d^{-3-(-5)} = d^2$

55. $\left(\frac{a^{-6}}{a^{-2}}\right)^{-1} = (a^{-6-(-2)})^{-1} = (a^{-4})^{-1} = a^4$ **56.** $\left(\frac{b^{-3}}{b^{-6}}\right)^4 = (b^{-3-(-6)})^4 = (b^3)^4 = b^{12}$

57. $\left(\frac{n^{-3}}{n}\right)^{-2} = (n^{-3-1})^{-2} = (n^{-4})^{-2} = n^8$

58. $\left(\frac{y^5}{y^{-2}}\right)^{-3} = (y^{5-(-2)})^{-3} = (y^7)^{-3} = y^{-21} = \frac{1}{y^{21}}$

59. $3125 \times 0.008 = 5^5 \times 5^{-3} = 5^{5-3} = 5^2 = 25$

60. $0.0000128 \times 78{,}125 = 5^{-7} \times 5^7 = 5^{-7+7} = 5^0 = 1$

61. $(78{,}125)^{-1} = (5^7)^{-1} = 5^{-7} = 0.0000128$ **62.** $(0.0016)^{-2} = (5^{-4})^{-2} = 5^8 = 390{,}625$

63. $\frac{0.2}{0.00000256} = \frac{5^{-1}}{5^{-8}} = 5^{-1-(-8)} = 5^7 = 78{,}125$

64. $\frac{0.000064}{25} = \frac{5^{-6}}{5^2} = 5^{-6-2} = 5^{-8} = 0.00000256$

65. $\frac{(3125)^2}{15{,}625} = \frac{(5^5)^2}{5^6} = \frac{5^{10}}{5^6} = 5^{10-6} = 5^4 = 625$

66. $\frac{(625)^4}{78{,}125} = \frac{(5^4)^4}{5^7} = \frac{5^{16}}{5^7} = 5^{16-7} = 5^9 = 5 \cdot 5^8 = 5 \cdot 390{,}625 = 1{,}953{,}125$

67. $(0.0000128)^3(390{,}625)^2 = (5^{-7})^3(5^8)^2 = 5^{-21} \cdot 5^{16} = 5^{-21+16} = 5^{-5} = 0.00032$

C **68.** If $9^{1/2} \cdot 9^{1/2} = 9$, then $9^{1/2}$ ought to be defined as the number 3.

69. If $16^{1/2} \cdot 16^{1/2} = 16$, then $16^{1/2} = 4$. **70.** If $(8^{1/3})^3 = 8$, then $8^{1/3} = 2$.

71. For $(x^2)^{1/2} \cdot (x^2)^{1/2}$ to equal $(x^2)^{1/2+1/2}$, or x^2, define $(x^2)^{1/2}$ as x. Similarly, define $(x^3)^{1/3}$ as x.

Page 335 • PROBLEMS

A **1.** $1.09^{-4} = \frac{1}{1.09^4} \approx 0.7084252$; $1.09^{-8} = \frac{1}{1.09^8} \approx 0.5018663$; After 4 days you will have 0.7084252 kg; after 8 days you will have 0.5018663 kg.

2.

Years	0	1	2	3	5	10
Value (V)	\$12,000	\$8571.43	\$6122.45	\$4373.18	\$2231.21	\$414.86

3. a. $12(1.03)^0 = 12 \cdot 1 = 12$; There are now about 12 million people.

b. $12(1.03)^{10} \approx 12 \cdot 1.343916 = 16.126992$; There will be about 16 million people in 10 years.

c. $12(1.03)^{-10} \approx 12 \cdot 0.7440941 = 8.9291292$; Ten years ago there were about 9 million people.

4. a. $(1.08)^3 = 1.259712$; The item will cost \$1.26.

b. $(1.08)^9 \approx 1.9990045$; The item will cost \$2.00.

c. $(1.08)^{-9} = \frac{1}{(1.08)^9} \approx \frac{1}{1.9990045} \approx 0.5002489$; The item cost \$.50.

5. a. $N = 100(2.72)^{2 \cdot 2} = 100(2.72)^4 \approx 5473.6322$; At 2 P.M., the number of bacteria is about 5474.

b. $N = 100(2.72)^{2 \cdot -2} = 100(2.72)^{-4} \approx 1.8269404$; At 10 A.M., there were about 2 bacteria.

c. $N = 100(2.72)^{2 \cdot 6} = 100(2.72)^{12} \approx 16{,}399{,}358.06$; at 6 P.M., there will be about 16,399,358 bacteria.

d. $N = 100(2.72)^{2 \cdot -6} = 100(2.72)^{-12} \approx 0.000609$; At 6 A.M., there were no bacteria.

Page 335 • MIXED REVIEW EXERCISES

1. $\frac{30x^2z^2}{18x^3z} = \frac{5z}{3x}$; $x \neq 0, z \neq 0$ **2.** $\frac{y^2 + 10y + 25}{y^2 + 7y + 10} = \frac{(y + 5)^2}{(y + 5)(y + 2)} = \frac{y + 5}{y + 2}$; $y \neq -5, -2$

3. $\left(\frac{-3p}{q}\right)^2 = \frac{9p^2}{q^2}$; $q \neq 0$ **4.** $\frac{8}{5mn} - \frac{2}{m} = \frac{8}{5mn} - \frac{10n}{5mn} = \frac{8 - 10n}{5mn} = \frac{2(4 - 5n)}{5mn}$; $m \neq 0, n \neq 0$

5. $4 - \frac{2a}{a - 2} = \frac{4(a - 2)}{a - 2} - \frac{2a}{a - 2} = \frac{4a - 8 - 2a}{a - 2} = \frac{2a - 8}{a - 2} = \frac{2(a - 4)}{a - 2}$; $a \neq 2$

6. $\frac{x^2 - 9}{x^2 + 5x + 6} = \frac{(x + 3)(x - 3)}{(x + 3)(x + 2)} = \frac{x - 3}{x + 2}$; $x \neq -3, -2$

7.
$$\begin{array}{r} 4x - 14 + \frac{92}{x + 4} \\ x + 4\overline{)4x^2 + 2x + 36} \\ \underline{4x^2 + 16x} \\ -14x + 36 \\ \underline{-14x - 56} \\ 92 \end{array}$$

8. $\frac{16}{x^2 - 9} \div \frac{24}{x^2 + 8x + 15} = \frac{16}{(x + 3)(x - 3)} \cdot \frac{(x + 3)(x + 5)}{24} = \frac{2(x + 5)}{3(x - 3)}$

9.
$$\begin{array}{r} a^2 + 3a + 9 + \frac{54}{a - 3} \\ a - 3\overline{)a^3 + 0a^2 + 0a + 27} \\ \underline{a^3 - 3a^2} \qquad\qquad \\ 3a^2 + 0a \qquad \\ \underline{3a^2 - 9a} \qquad \\ 9a + 27 \\ \underline{9a - 27} \\ 54 \end{array}$$

Page 339 • WRITTEN EXERCISES

A **1.** $3.0 \times 10^8 = 300{,}000{,}000$ **2.** $1.39 \times 10^9 = 1{,}390{,}000{,}000$

3. $2.0 \times 10^{30} = 2{,}000{,}000{,}000{,}000{,}000{,}000{,}000{,}000{,}000{,}000$

4. $1.4 \times 10^6 = 1{,}400{,}000$ **5.** $1.36 \times 10^{-6} = 0.00000136$

6. $3.0 \times 10^{-10} = 0.0000000003$ **7.** $5.0 \times 10^{-17} = 0.00000000000000005$

8. $6.65 \times 10^{-24} = 0.00000000000000000000000665$ **9. a.** 10^3 **b.** 10^{-3}

10. a. 10^2 **b.** 10^{-2} **11. a.** 10^6 **b.** 10^{-6} **12. a.** 10^{-1} **b.** 10^1

13. a. 10^6 **b.** 10^{-6} **14. a.** 10^3 **b.** 10^{-3}

15. a. $2500 = 2.5 \times 10^3$ **b.** $2500 = 2 \cdot 10^3 + 5 \cdot 10^2$

16. a. $0.0123 = 1.23 \times 10^{-2}$ **b.** $0.0123 = 1 \cdot 10^{-2} + 2 \cdot 10^{-3} + 3 \cdot 10^{-4}$

17. a. $0.0000000024 = 2.4 \times 10^{-9}$ **b.** $0.0000000024 = 2 \cdot 10^{-9} + 4 \cdot 10^{-10}$

18. a. $26{,}870{,}000 = 2.687 \times 10^7$ **b.** $26{,}870{,}000 = 2 \cdot 10^7 + 6 \cdot 10^6 + 8 \cdot 10^5 + 7 \cdot 10^4$

19. a. $3{,}030{,}000 = 3.03 \times 10^6$ **b.** $3{,}030{,}000 = 3 \cdot 10^6 + 0 \cdot 10^5 + 3 \cdot 10^4$

20. a. $0.0000485 = 4.85 \times 10^{-5}$ **b.** $0.0000485 = 4 \cdot 10^{-5} + 8 \cdot 10^{-6} + 5 \cdot 10^{-7}$

21. a. $0.0000909 = 9.09 \times 10^{-5}$ **b.** $0.0000909 = 9 \cdot 10^{-5} + 0 \cdot 10^{-6} + 9 \cdot 10^{-7}$

22. a. $798{,}100{,}000 = 7.981 \times 10^8$ **b.** $798{,}100{,}000 = 7 \cdot 10^8 + 9 \cdot 10^7 + 8 \cdot 10^6 + 1 \cdot 10^5$

B **23. a.** $98.6\% = 0.986 = 9.86 \times 10^{-1}$ **b.** $0.986 = 9 \cdot 10^{-1} + 8 \cdot 10^{-2} + 6 \cdot 10^{-3}$

24. a. 2.3 million $= 2{,}300{,}000 = 2.3 \times 10^6$ **b.** $2{,}300{,}000 = 2 \cdot 10^6 + 3 \cdot 10^5$

25. a. 12 billion $= 12{,}000{,}000{,}000 = 1.2 \times 10^{10}$ **b.** $12{,}000{,}000{,}000 = 1 \cdot 10^{10} + 2 \cdot 10^9$

26. a. 200 trillion $= 200{,}000{,}000{,}000{,}000 = 2 \times 10^{14}$ **b.** $200{,}000{,}000{,}000{,}000 = 2 \cdot 10^{14}$

27. $(3 \times 10^{-8})(1.2 \times 10^4) = (3 \cdot 1.2)(10^{-8} \cdot 10^4) = 3.6 \times 10^{-4}$

28. $\frac{2.5 \times 10^{-5}}{5 \times 10^4} = \left(\frac{2.5}{5}\right)(10^{-5-4}) = 0.5 \times 10^{-9} = 5 \times 10^{-1} \times 10^{-9} = 5 \times 10^{-10}$

29. $\frac{1.08 \times 10^8}{3 \times 10^{-10}} = \left(\frac{1.08}{3}\right)\left(\frac{10^8}{10^{-10}}\right) = 0.36 \times 10^{18} = 3.6 \times 10^{-1} \times 10^{18} = 3.6 \times 10^{17}$

30. $\frac{(5 \times 10^7)(9 \times 10^{-3})}{3 \times 10^{-2}} = \left(\frac{5 \cdot 9}{3}\right)\left(\frac{10^7 \cdot 10^{-3}}{10^{-2}}\right) = 15 \times 10^{7-3-(-2)} = 1.5 \times 10^1 \times 10^6 =$
1.5×10^7

31. $\frac{(2 \times 10^5)^2}{(8 \times 10^{-3})(5 \times 10^{12})} = \frac{(2^2)(10^{10})}{(8 \cdot 5)(10^{-3} \cdot 10^{12})} = \left(\frac{4}{40}\right)\left(\frac{10^{10}}{10^9}\right) = \frac{1}{10} \cdot 10 = 1 = 1 \times 10^0$

32. $\frac{(4 \times 10^{-5})^3}{(2 \times 10^{-7})^2} = \frac{(4^3)(10^{-15})}{(2^2)(10^{-14})} = \left(\frac{64}{4}\right)\left(\frac{10^{-15}}{10^{-14}}\right) = 16 \times 10^{-1} = 1.6 = 1.6 \times 10^0$

Pages 339–340 • PROBLEMS

A **1. a.** 3.0×10^5 km/s · 60 s/min · 60 min/h · 24 h/d · 365 d/y = $3.0 \times 10^5 \cdot 6 \times 10^1 \cdot 6 \times 10^1 \cdot 2.4 \times 10^1 \cdot 3.65 \times 10^2$ km/y = $(3.0 \cdot 6 \cdot 6 \cdot 2.4 \cdot 3.65)(10^5 \cdot 10^1 \cdot 10^1 \cdot 10^1 \cdot 10^2) = 946.08 \times 10^{10} =$ $9.4608 \times 10^2 \times 10^{10} = 9.4608 \times 10^{12}$ km/y; There are about 9.5×10^{12} km in a light year.

b. 1.5×10^6 light years · 9.4608×10^{12} km/light year = $(1.5 \cdot 9.4608) \times 10^{18} =$ $14.1912 \times 10^{18} \approx 1.4 \times 10^1 \times 10^{18} = 1.4 \times 10^{19}$; The Andromeda galaxy is about 1.4×10^{19} km from Earth.

c. $\frac{4.07 \times 10^{13} \text{ km}}{9.4608 \times 10^{12} \text{ km/y}} \approx 0.430 \times 10 = 4.30$ y; It takes about 4.3 years for light to travel from Alpha Centauri to Earth.

d. $\frac{1.5 \times 10^8 \text{ km}}{3.0 \times 10^5 \text{ km/s}} = 0.50 \times 10^3 = 5.0 \times 10^{-1} \times 10^3 = 5.0 \times 10^2 = 500$ s; It takes about 500 s or $8\frac{1}{3}$ min for the sun's light to reach Earth.

2. a. $2^{50} \cdot 0.015 \approx 1.13 \times 10^{15} \cdot 0.015 = 1.13 \times 10^{15} \cdot 1.5 \times 10^{-2} = 1.695 \times 10^{13}$; The stack would be 1.695×10^{13} cm high.

b. This is higher than your room.

c. 1.695×10^{13} cm $= 1.695 \times 10^{13} \cdot 10^{-2}$ m $= 1.695 \times 10^{11}$ m > 83 m; This is higher than the tallest redwood tree.

d. 1.695×10^{11} m > 443.18 m; This is higher than the tallest building.

e. $384{,}432 = 3.84432 \times 10^5 < 1.695 \times 10^{11}$; This is higher than the moon.

f. 1.695×10^{11} m $= 1.695 \times 10^{11} \cdot 10^{-3}$ km $= 1.695 \times 10^8$ km $> 1.5 \times 10^8$ km; This is higher than the sun.

3. a. $F \cdot L = 3.0 \times 10^{10}$; $1.3 \times 10^8 \cdot L = 3.0 \times 10^{10}$; $L = \frac{3.0 \times 10^{10}}{1.3 \times 10^8}$; $L = 2.3 \times 10^2$; The wavelength is 2.3×10^2 cm.

b. $L = \frac{3.0 \times 10^{10}}{4.0 \times 10^{14}} = .75 \times 10^{-4} = 7.5 \times 10^{-5}$; The wavelength is 7.5×10^{-5} cm.

c. $F \cdot L = 3.0 \times 10^{10}$; $F = \frac{3.0 \times 10^{10}}{L}$; $F = \frac{3.0 \times 10^{10}}{1.3 \times 10^{-8}}$; $F = 2.3 \times 10^{18}$; The frequency is 2.3×10^{18} cycles per second.

d. $F = \frac{3.0 \times 10^{10}}{1 \times 10^{-11}} = 3.0 \times 10^{21}$; The frequency is 3.0×10^{21} cycles per second.

Page 340 • MIXED REVIEW EXERCISES

1. $-[y + (-21)] + 2y + 3 = -(y - 21) + 2y + 3 = -y + 21 + 2y + 3 = y + 24$

2. $(x^{-2}y^5)^3 = x^{-6}y^{15} = \frac{y^{15}}{x^6}$ **3.** $(13 - 25) - (8 - 28) = -12 - (-20) = -12 + 20 = 8$

4. $55 : 22 = \frac{55}{22} = \frac{5}{2} = 5 : 2$ **5.** $[4 + (-6)] + 7 = 4 - 6 + 7 = 5$

6. $3t + [(-6) + (-2) + 9] = 3t - 6 - 2 + 9 = 3t + 1$

7. $\left(\frac{c^{-3}}{c^3}\right)^2 = \frac{c^{-6}}{c^6} = c^{-6-6} = c^{-12} = \frac{1}{c^{12}}$

8. $-\frac{5}{2} + 5 + \left(-\frac{1}{2}\right) = -\frac{5}{2} + \frac{10}{2} - \frac{1}{2} = \frac{-5 + 10 - 1}{2} = \frac{4}{2} = 2$ **9.** $\frac{14b^3c^2}{12c^3b} = \frac{7b^2}{6c}$

Page 341 • CALCULATOR KEY-IN

Answers may vary. $99{,}999{,}999 \approx 10^8$, so $2 \times 99{,}999{,}999 \approx 2 \times 10^8$; $20 \times 99{,}999{,}999 \approx 20 \times 10^8 = 2 \times 10^9$; $400 \times 99{,}999{,}999 \approx 400 \times 10^8 = 4 \times 10^{10}$

Pages 342–343 • CHAPTER REVIEW

1. b; $\frac{5 \cdot 60}{35} = \frac{60}{7}$ **2.** c; $3x + 5x = 24$; $8x = 24$; $x = 3$; $3x = 9$

3. d; $\frac{x + 3}{44} = \frac{42}{24}$; $24(x + 3) = 42 \cdot 44$; $24(x + 3) = 1848$; $x + 3 = 77$; $x = 74$

4. c; $\frac{1.5}{175} = \frac{x}{875}$; $1.5 \cdot 875 = 175x$; $1312.5 = 175x$; $7.5 = x$

5. c; $\frac{x}{4} - \frac{x + 4}{5} = 1$; $20\left(\frac{x}{4} - \frac{x + 4}{5}\right) = 20$; $5x - 4(x + 4) = 20$; $5x - 4x - 16 = 20$; $x - 16 = 20$; $x = 36$

6. a; $\frac{1}{a - 1} + \frac{3}{3a - 1} = 0$; $(a - 1)(3a - 1)\left(\frac{1}{a - 1} + \frac{3}{3a - 1}\right) = 0$; $3a - 1 + 3(a - 1) = 0$; $3a - 1 + 3a - 3 = 0$; $6a - 4 = 0$; $6a = 4$; $a = \frac{2}{3}$

7. c; $16\frac{2}{3}\% = 16\frac{2}{3} \cdot \frac{1}{100} = \frac{50}{3} \cdot \frac{1}{100} = \frac{1}{6}$ **8.** d; 72% of $72 = 0.72 \cdot 72 = 51.84$

9. b; $0.06x = 18$; $x = 300$ **10.** c; $\frac{x}{100} = \frac{27}{36}$; $36x = 2700$; $x = 75$

11. b; $\frac{x}{100} = \frac{57}{57 + 18}$; $75x = 5700$; $x = 76$

12. a; $\frac{20}{100} = \frac{x - 6.36}{x}$; $20x = 100(x - 6.36)$; $20x = 100x - 636$; $-80x = -636$; $x = 7.95$

13. d; $10 \cdot 0.2 = (10 + x)0.05$; $2 = 0.5 + 0.05x$; $200 = 50 + 5x$; $150 = 5x$; $30 = x$

14. a; $\left(\frac{1}{5} + \frac{1}{x}\right)3 = 1$; $\frac{3}{5} + \frac{3}{x} = 1$; $5x\left(\frac{3}{5} + \frac{3}{x}\right) = 5x$; $3x + 15 = 5x$; $15 = 2x$; $7\frac{1}{2} = x$

15. d; $(2x^3)^{-2} = \frac{1}{(2x^3)^2} = \frac{1}{(2)^2(x^3)^2} = \frac{1}{4x^6}$ **16.** b; 234 million $= 234{,}000{,}000 = 2.34 \times 10^8$

Page 344 • CHAPTER TEST

1. 21 days : 35 wk $= \frac{\overset{3}{\cancel{21}}}{35 \cdot \underset{1}{\cancel{7}}} = \frac{3}{35}$, or 3 : 35 **2.** 63 cm : 0.9 cm $= \frac{63}{0.9} = \frac{630}{9} = \frac{70}{1}$, or 70 : 1

3. Let $4x$ be the time he spent writing and let x be the time he spent revising. $4x + x = \frac{5}{2}$;

$5x = \frac{5}{2}$; $x = \frac{1}{2}$; $4x = 4 \cdot \frac{1}{2} = 2$; Sean spent 2 h writing his essay.

4. $\frac{104}{x} = \frac{13}{2}$; $13x = 208$; $x = 16$; $\{16\}$ **5.** $\frac{16}{3y} = \frac{12}{5}$; $80 = 36y$; $\frac{80}{36} = y$; $\frac{20}{9} = y$; $\left\{\frac{20}{9}\right\}$

6. $\frac{3x}{10} = \frac{5x}{6}$; $18x = 50x$; $0 = 32x$; $0 = x$; $\{0\}$

7. $\frac{4a + 1}{7} = \frac{3a - 4}{3}$; $3(4a + 1) = 7(3a - 4)$; $12a + 3 = 21a - 28$; $3 = 9a - 28$; $31 = 9a$;

$\frac{31}{9} = a$; $\left\{\frac{31}{9}\right\}$

8. $2n + \frac{3n}{5} = 5 - \frac{n}{10}$; $10\left(2n + \frac{3n}{5}\right) = 10\left(5 - \frac{n}{10}\right)$; $20n + 6n = 50 - n$; $26n = 50 - n$;

$27n = 50$; $n = \frac{50}{27}$; $\left\{\frac{50}{27}\right\}$

9. $\frac{3y + 5}{2} - \frac{3y + 2}{5} = \frac{3}{10}$; $10\left(\frac{3y + 5}{2} - \frac{3y + 2}{5}\right) = 10 \cdot \frac{3}{10}$; $5(3y + 5) - 2(3y + 2) = 3$;

$15y + 25 - 6y - 4 = 3$; $9y + 21 = 3$; $9y = -18$; $y = -2$; $\{-2\}$

10. $\frac{1}{b} + \frac{3b}{5b - 2} = 0$; $b(5b - 2)\left(\frac{1}{b} + \frac{3b}{5b - 2}\right) = 0$; $5b - 2 + 3b^2 = 0$; $3b^2 + 5b - 2 = 0$;

$(3b - 1)(b + 2) = 0$; $b = \frac{1}{3}, -2$; $\left\{\frac{1}{3}, -2\right\}$

11. $\frac{2c + 3}{2c + 6} = \frac{1}{c}$; $c(2c + 3) = 2c + 6$; $2c^2 + 3c = 2c + 6$; $2c^2 + c = 6$; $2c^2 + c - 6 = 0$;

$(2c - 3)(c + 2) = 0$; $c = \frac{3}{2}, -2$; $\left\{\frac{3}{2}, -2\right\}$

12. $49\% = \frac{49}{100}$ **13.** $73\frac{1}{2}\% = \frac{147}{2} \cdot \frac{1}{100} = \frac{147}{200}$ **14.** $540\% = \frac{540}{100} = \frac{27}{5}$

15. 22% of 78 $= 0.22 \cdot 78 = 17.16$ **16.** $48 = \frac{32}{100}x$; $\frac{4800}{32} = x$; $150 = x$; 48 is 32% of 150

17. $\frac{x}{100} = \frac{315}{945}$; $945x = 31{,}500$; $x = \frac{31{,}500}{945} = \frac{100}{3} = 33\frac{1}{3}$; 315 is $33\frac{1}{3}\%$ of 945

18. $\frac{12}{100} = \frac{1344 - x}{x}$; $12x = 100(1344 - x)$; $12x = 134{,}400 - 100x$; $112x = 134{,}400$; $x = 1200$;

1200 people walked last year.

19. Let x be the number of liters of grape juice. $\frac{16}{100} = \frac{x}{14 + x}$; $16(14 + x) = 100x$;

$224 + 16x = 100x$; $224 = 84x$; $2\frac{2}{3} = x$; $2\frac{2}{3}$ L of grape juice must be added.

20. Let x be the number of hours they need to finish the job. $\frac{x}{7} + \frac{x}{3} = 1$; $21\left(\frac{x}{7} + \frac{x}{3}\right) = 21$;

$3x + 7x = 21$; $10x = 21$; $x = 2.1$; It will take them 2.1 h.

21. $3^{-5} \cdot 3^4 = 3^{-5+4} = 3^{-1} = \frac{1}{3}$ **22.** $\frac{5^0}{5^{-2}} = 5^{0-(-2)} = 5^2 = 25$

23. $(4m^2)^{-3} = \frac{1}{(4m^2)^3} = \frac{1}{(4)^3(m^2)^3} = \frac{1}{64m^6}$ **24.** $12x^{-5} = \frac{12}{x^5}$ **25.** $(a^{-3})^2 = a^{-6} = \frac{1}{a^6}$

26. $n^2 \cdot n^{-6} = n^{2+(-6)} = n^{2-6} = n^{-4} = \frac{1}{n^4}$ **27.** $0.000128 = 1.28 \times 10^{-4}$

Page 345 • CUMULATIVE REVIEW

1. $24 \div 2 + 2^2 - 4^3 \div 2^4 = 12 + 4 - (64 \div 16) = 12 + 4 - 4 = 12$

2. $\left(\frac{18}{36}\right)\left(\frac{3}{17}\right) + \left(\frac{1}{2}\right)\left(\frac{3}{17}\right) = \left(\frac{1}{2}\right)\left(\frac{3}{17}\right) + \left(\frac{1}{2}\right)\left(\frac{3}{17}\right) = 2\left(\frac{1}{2}\right)\left(\frac{3}{17}\right) = \frac{3}{17}$

3. $(-2m^3n^4)(8m^5n^3) = -16m^8n^7$

4. $x^2y^2(-3xy^3 + 4x^3y + 14) = -3x^3y^5 + 4x^5y^3 + 14x^2y^2$ **5.** $(-4a^2b^3c)^2 = 16a^4b^6c^2$

6. $(10b + 9)^2 = 100b^2 + 2 \cdot 10b \cdot 9 + 81 = 100b^2 + 180b + 81$

7. $(5t^2v - 6)^2 = (5t^2v)^2 - 2 \cdot 5t^2v \cdot 6 + 6^2 = 25t^4v^2 - 60t^2v + 36$

8. $(3bc - 4)(5bc + 6) = (3bc)(5bc) + 6 \cdot 3bc - 4 \cdot 5bc - 24 = 15b^2c^2 + 18bc - 20bc - 24 =$
$15b^2c^2 - 2bc - 24$

9. $(2x - y)(x^2 - 5xy + 3y^2) = 2x^3 - 10x^2y + 6xy^2 - x^2y + 5xy^2 - 3y^3 =$
$2x^3 - 11x^2y + 11xy^2 - 3y^3$

10. $\frac{x^2 - 12x + 36}{x - 3} \cdot \frac{x^2 - 9}{x^2 - x - 30} = \frac{(x-6)^2}{x-3} \cdot \frac{(x+3)(x-3)}{(x-6)(x+5)} = \frac{(x-6)(x+3)}{x+5}$

11. $\frac{3z - 1}{xyz^3} \div \frac{9xz^2 - x}{yz^2} = \frac{3z-1}{xyz^3} \cdot \frac{yz^2}{(3z+1)(3z-1)x} = \frac{1}{x^2z(3z+1)}$

12. $\frac{t}{t+5} + \frac{t+4}{t^2 + 2t - 15} = \frac{t}{t+5} + \frac{t+4}{(t+5)(t-3)} = \frac{t(t-3)}{(t+5)(t-3)} + \frac{t+4}{(t+5)(t-3)} =$
$\frac{t^2 - 3t + t + 4}{(t+5)(t-3)} = \frac{t^2 - 2t + 4}{(t+5)(t-3)}$

13. $\frac{12b^2 + b - 6}{3b - 2} = \frac{(3b-2)(4b+3)}{3b-2} = 4b + 3$ **14.** $\left(\frac{y^3}{y-4}\right)^3 = \frac{y^9}{(y-4)^3}$

15. $\left(\frac{x^{-7}}{x-4}\right)^{-2} = \frac{x^{14}}{(x-4)^{-2}} = x^{14}(x-4)^2$

16. $\frac{x}{100} = \frac{336}{400}$; $400x = 33{,}600$; $x = 84$; 336 is 84% of 400.

17. $\frac{x}{100} = \frac{56}{600}$; $600x = 5600$; $x = 9\frac{1}{3}$; 56 is $9\frac{1}{3}$% of 600.

18. $-18s^3t^2 + 24s^2t^3 + 36st^3 = 6st^2(-3s^2 + 4st + 6t)$ **19.** $49z^2 - 28z + 4 = (7z - 2)^2$

20. $-64x^2 + 100t^2 = 100t^2 - 64x^2 = 4(25t^2 - 16x^2) = 4(5t + 4x)(5t - 4x)$

21. $6b^2 + 27b + 27 = 3(2b^2 + 9b + 9) = 3(2b + 3)(b + 3)$

22. $18t^2 + 27t + 12 = 3(6t^2 + 9t + 4)$ **23.** $12x^2 + 47x - 4 = (12x - 1)(x + 4)$

24. $-7x + 2 = -33$; $-7x = -35$; $x = 5$; $\{5\}$

25. $\frac{2}{3}t - 8 = 12$; $3\left(\frac{2}{3}t - 8\right) = 36$; $2t - 24 = 36$; $2t = 60$; $t = 30$; $\{30\}$

26. $(n + 4)^2 = (n + 2)^2 + 4(n + 3)$; $n^2 + 8n + 16 = n^2 + 4n + 4 + 4n + 12$;
$n^2 + 8n + 16 = n^2 + 8n + 16$; Identity.

27. $4y^2 + 28y + 49 = 0$; $(2y + 7)^2 = 0$; $2y = -7$; $y = -\frac{7}{2}$; $\left\{-\frac{7}{2}\right\}$

28. $m^2 + 14 + 9m = 0$; $m^2 + 9m + 14 = 0$; $(m + 2)(m + 7) = 0$; $m = -2, -7$; $\{-2, -7\}$

29. $y^3 - 12y^2 + 11y = 0$; $y(y^2 - 12y + 11) = 0$; $y(y - 11)(y - 1) = 0$; $y = 0, 11, 1$; $\{0, 1, 11\}$

30. $0.02(700 - x) + 0.5x = 20$; $100[0.02(700 - x) + 0.5x] = 20 \cdot 100$; $2(700 - x) + 50x = 2000$; $1400 - 2x + 50x = 2000$; $1400 + 48x = 2000$; $48x = 600$; $x = 12.5$; $\{12.5\}$

31. $\frac{n-3}{n+3} + \frac{1}{n-3} = \frac{18}{n^2 - 9}$; $(n + 3)(n - 3)\left(\frac{n-3}{n+3} + \frac{1}{n-3}\right) = (n + 3)(n - 3)\left(\frac{18}{(n+3)(n-3)}\right)$; $(n - 3)^2 + n + 3 = 18$; $n^2 - 6n + 9 + n + 3 = 18$; $n^2 - 5n + 12 = 18$; $n^2 - 5n - 6 = 0$; $(n + 1)(n - 6) = 0$; $n = -1, 6$; $\{-1, 6\}$

32. 67 billion $= 67{,}000{,}000{,}000 = 6.7 \times 10^{10}$ **33.** $0.0000043 = 4.3 \times 10^{-6}$

34. $(7.3 \times 10^5)(2.0 \times 10^{-2}) = 14.6 \times 10^3 = 1.46 \times 10^1 \times 10^3 = 1.46 \times 10^4$

35. Let x and $x + 2$ be the integers. $x(x + 2) = (x + 2)^2 - 22$; $x^2 + 2x = x^2 + 4x + 4 - 22$; $2x = 4x - 18$; $0 = 2x - 18$; $18 = 2x$; $9 = x$; $x + 2 = 11$; The integers are 9 and 11.

36. Let $3x$, $4x$, and $5x$ be the numbers. $3x + 4x + 5x = 144$; $12x = 144$; $x = 12$; $3x = 36$; $4x = 48$; $5x = 60$; The numbers are 36, 48, and 60.

37. $\frac{x}{100} = \frac{425 - 340}{425}$; $425x = (425 - 340)100$; $425x = 8500$; $x = 20$; The employee's discount is 20%.

38. Let x be the number of liters of water. $20 \cdot 0.3 = (20 + x)0.1$; $6 = 2 + 0.1x$; $60 = 20 + x$; $40 = x$; 40 L of water should be added.

39. $\left(\frac{1}{2} + \frac{1}{1\frac{3}{4}}\right)x = 1$; $\left(\frac{1}{2} + \frac{1}{\frac{7}{4}}\right)x = 1$; $\left(\frac{1}{2} + \frac{4}{7}\right)x = 1$; $\left(\frac{7+8}{14}\right)x = 1$; $\frac{15}{14}x = 1$; $x = \frac{14}{15}$; It will take $\frac{14}{15}$ h or 56 min to fill the tank.

Page 346 • MAINTAINING SKILLS

1. $7a^2 + 2 - 5a + 2a^2 + 6 = 9a^2 - 5a + 8$
2. $-4c + 2d - 3 - 5d + 3c - 6 = -c - 3d - 9$
3. $(12 - 2x) + (-8x + 5x) = 12 - 2x - 3x = 12 - 5x$
4. $(3m - 2n) - (-4m + 3n) = 3m - 2n + 4m - 3n = 7m - 5n$
5. $4(x - 3z) - 8(-x + z) = 4x - 12z + 8x - 8z = 12x - 20z$
6. $-2(5y^2 - 2y) + 3(-3y + 1) = -10y^2 + 4y - 9y + 3 = -10y^2 - 5y + 3$
7. $-2(5x + 3) + 4(3x - 2y) = -10x - 6 + 12x - 8y = 2x - 8y - 6$
8. $-3(5x - y) + 5x(2y + 2) = -15x + 3y + 10xy + 10x = -5x + 10xy + 3y$
9. $\frac{3}{4}(8a - 2b) - \frac{2}{3}(9a + b) = 6a - \frac{3}{2}b - 6a - \frac{2}{3}b = -\left(\frac{3}{2} + \frac{2}{3}\right)b = -\left(\frac{9+4}{6}\right)b = -\frac{13}{6}b$
10. $\frac{2}{3}(-12j - k) + \frac{1}{6}(18j + 8k) = -8j - \frac{2}{3}k + 3j + \frac{4}{3}k = -5j + \frac{2}{3}k$
11. $(-2a^2b^2)(3ab^2c) = -6a^3b^4c$ **12.** $(4st^2)(-s^3t)(-2s^2t^2) = 8s^6t^5$
13. $(4m^2np)\left(-\frac{1}{4}m^3np^2\right) = -m^5n^2p^3$

14. $(-5a^2b^3)(2a^3b^2) + (a^4b)(6ab^4) = -10a^5b^5 + 6a^5b^5 = -4a^5b^5$

15. $(2x^2y^3)^4 = 2^4(x^2)^4(y^3)^4 = 16x^8y^{12}$ **16.** $(-11m^4n^3)^2 = (-11)^2(m^4)^2(n^3)^2 = 121m^8n^6$

17. $(r^3s^2t)^4(rst)^3 = (r^3)^4(s^2)^4(t)^4(r^3s^3t^3) = r^{12}s^8t^4 \cdot r^3s^3t^3 = r^{15}s^{11}t^7$

18. $-9(fg^3)^2(-2f^3g)^2 = -9f^2g^6 \cdot (-2)^2(f^3)^2g^2 = -9f^2g^6 \cdot 4f^6g^2 = -36f^8g^8$

19. $(2c^2d^2e)^6 + (-2c^4d^4e^2)^3 = 2^6c^{12}d^{12}e^6 + -8c^{12}d^{12}e^6 = 64c^{12}d^{12}e^6 - 8c^{12}d^{12}e^6 =$
$56c^{12}d^{12}e^6$

20. $-ab(a^3b^2)^4 + a^3b^4(-a^2b)^5 = -ab(a^3)^4(b^2)^4 + a^3b^4(-a^2)^5(b)^5 = -ab \cdot a^{12}b^8 +$
$a^3b^4 \cdot (-a^{10}b^5) = -a^{13}b^9 + (-a^{13}b^9) = -2a^{13}b^9$

21. $3ab^2(a^2 - ab + b^2) = 3a^3b^2 - 3a^2b^3 + 3ab^4$

22. $-4m^2n(3mn^2 - 2m + n^3) = -12m^3n^3 + 8m^3n - 4m^2n^4$

23. $c^2d^3(d^4 - 2d^2c + c^3) = c^2d^7 - 2c^3d^5 + c^5d^3$

24. $-xy(5x^3y^2 - 2x^2y + 7y^5) = -5x^4y^3 + 2x^3y^2 - 7xy^6$

25. $(z - 3)(z + 8) = z^2 + 8z - 3z - 24 = z^2 + 5z - 24$

26. $(2d + 7)(7d - 2) = 14d^2 - 4d + 49d - 14 = 14d^2 + 45d - 14$

27. $(y - 2)(y + 10) = y^2 + 10y - 2y - 20 = y^2 + 8y - 20$

28. $(2y - 5)(5y - 3) = 10y^2 - 6y - 25y + 15 = 10y^2 - 31y + 15$

29. $(4z + 1)(3z - 3) = 12z^2 - 12z + 3z - 3 = 12z^2 - 9z - 3$

30. $(3c + 4)(5c + 3) = 15c^2 + 9c + 20c + 12 = 15c^2 + 29c + 12$

31. $(9d - 3)(9d + 3) = (9d)^2 - 3^2 = 81d^2 - 9$

32. $(4y + 5)^2 = (4y)^2 + 2(4y)(5) + 5^2 = 16y^2 + 40y + 25$

33. $(x^2 - 6)(x^2 + 6) = (x^2)^2 - 6^2 = x^4 - 36$

34. $(3a - 4)(7a + 2) = 21a^2 + 6a - 28a - 8 = 21a^2 - 22a - 8$

35. $(y - 3)(3y^2 - 4y + 1) = 3y^3 - 4y^2 + y - 9y^2 + 12y - 3 = 3y^3 - 13y^2 + 13y - 3$

36. $(b - 7)(-2b^2 + 5b - 2) = -2b^3 + 5b^2 - 2b + 14b^2 - 35b + 14 =$
$-2b^3 + 19b^2 - 37b + 14$

Page 347 • MIXED PROBLEM SOLVING

A

1. Let x be the number of points scored by the home team. $x = \frac{194 - x}{2} + 50$;
$2x = 2\left(\frac{194 - x}{2} + 50\right)$; $2x = 194 - x + 100$; $3x = 294$; $x = 98$; $194 - x = 96$;
$98 - 96 = 2$; The home team won by 2 points.

2. Let x be the number of kg of salt to be added. $12 \cdot 0.08 + x = (12 + x) \cdot 0.2$;
$100(0.96 + x) = 100[0.2(12 + x)]$; $96 + 100x = 20(12 + x)$; $96 + 100x = 240 + 20x$;
$96 + 80x = 240$; $80x = 144$; $x = 1.8$; 1.8 kg of salt must be added.

3. $\frac{85}{100} = \frac{187}{x}$; $85x = 18{,}700$; $x = 220$; The apple weighs 220 mg.

4. Let x and $3x$ be the lengths of the two ribbons. $(x - 4) + (3x - 4) = 4$;
$4x - 8 = 4$; $4x = 12$; $x = 3$; $x - 4 = 3 - 4 = -1$; No sol.; one new ribbon would be -1 ft long.

5. Let d be the distance one way. $\frac{d}{60} + \frac{d}{6} = 2.75$; $60\left(\frac{d}{60} + \frac{d}{6}\right) = 60 \cdot 2.75$;

$d + 10d = 165$; $11d = 165$; $d = 15$; $2d = 30$; Mia traveled 30 km.

6. Let x be the amount of the less expensive beans. $6x + (16 - x)8.8 = 16 \cdot 6.7$;
$10[6x + (16 - x)8.8] = 107.2 \cdot 10$; $60x + 88(16 - x) = 1072$;
$60x + 1408 - 88x = 1072$; $1408 - 28x = 1072$; $-28x = -336$; $x = 12$; $16 - x = 4$;
12 kg of the less expensive beans were mixed with 4 kg of the more expensive beans.

7. $\frac{x}{90} + \frac{x}{90 + 20} = 1$; $\frac{x}{90} + \frac{x}{110} = 1$; $990\left(\frac{x}{90} + \frac{x}{110}\right) = 990$; $11x + 9x = 990$; $20x = 990$;

$x = 49.5$; It will take us 49.5 min.

8. $\frac{x}{100} = \frac{24 - 18.6}{24}$; $24x = (24 - 18.6)100$; $24x = 540$; $x = 22.5$; The percent of discount

is 22.5%.

9. No sol.; not enough information.

B **10.** Let x and $40 - x$ be the numbers. $\frac{x}{40 - x} = 4 + \frac{5}{40 - x}$;

$(40 - x)\left(\frac{x}{40 - x}\right) = (40 - x)\left(4 + \frac{5}{40 - x}\right)$; $x = 4(40 - x) + 5$; $x = 160 - 4x + 5$;

$5x = 165$; $x = 33$; $40 - x = 7$; The numbers are 33 and 7.

11. $\frac{x}{80} + \frac{x + 10}{60} = \frac{3x}{240} + \frac{4(x + 10)}{240} = \frac{3x + 4x + 40}{240} = \frac{7x + 40}{240}$; Miriam drove for $\frac{7x + 40}{240}$ h.

12. Let x and $x + 2$ be the integers. $x^2 + (x + 2)^2 = 6(x + x + 2) + 2$;
$x^2 + x^2 + 4x + 4 = 6(2x + 2) + 2$; $2x^2 + 4x + 4 = 12x + 12 + 2$;
$2x^2 - 8x + 4 = 14$; $2x^2 - 8x - 10 = 0$; $2(x^2 - 4x - 5) = 0$; $(x - 5)(x + 1) = 0$;
$x = 5, -1$; The integers are 5 and 7 or -1 and 1.

13. Let x be the amount invested in an account.
$0.06x + 0.11(x + 2000) = (x + x + 2000)0.095$; $1000[0.06x + 0.11(x + 2000)] =$
$1000[(2x + 2000)0.095]$; $60x + 110(x + 2000) = 95(2x + 2000)$;
$60x + 110x + 220{,}000 = 190x + 190{,}000$; $170x + 220{,}000 = 190x + 190{,}000$;
$220{,}000 = 20x + 190{,}000$; $30{,}000 = 20x$; $1500 = x$; $x + 2000 = 3500$;
$1500 + 3500 = 5000$; I invested \$5000.

14. Let w be the width of the photograph. $(w + 4)(w + 6) = 255$;
$w^2 + 10w + 24 = 255$; $w^2 + 10w - 231 = 0$; $(w + 21)(w - 11) = 0$; $w = -21$
(reject) or 11; The width of the photo is 11 cm and the length is 13 cm.

Pages 351–352 • WRITTEN EXERCISES

A

1. (3, 4): $5(3) + 2(4) = 15 + 8 = 23$, yes; (7, −6): $5(7) + 2(-6) = 35 - 12 = 23$, yes

2. (6, 7): $4(6) - 5(7) = 24 - 35 = -11 \neq 9$, no; (6, 3): $4(6) - 5(3) = 24 - 15 = 9$; yes

3. (3, −2): $3(3) - 2(-2) = 9 + 4 = 13$, yes; (5, −1): $3(5) - 2(-1) = 15 + 2 = 17 \neq 13$, no

4. (7, 0): $3(7) - 5(0) = 21 - 0 = 21$, yes; (−2, −3): $3(-2) - 5(-3) = -6 + 15 = 9 \neq 21$, no

5. (5, 1): $2(5) + 3(1) = 10 + 3 = 13$, yes; (11, −3): $2(11) + 3(-3) = 22 - 9 = 13$, yes

6. (7, 6): $5(7) - 4(6) = 35 - 24 = 11 \neq 9$, no; (−3, −6): $5(-3) - 4(-6) = -15 + 24 = 9$, yes

7. $\left(\frac{1}{3}, -\frac{5}{2}\right)$: $3\left(\frac{1}{3}\right) - 4\left(-\frac{5}{2}\right) = 1 + 10 = 11$, yes; $\left(-\frac{5}{3}, -4\right)$: $3\left(-\frac{5}{3}\right) - 4(-4) = -5 + 16 = 11$, yes

8. (0, 3): $3(0) - 2(3) = 0 - 6 = -6 \neq 6$, no; $\left(\frac{5}{3}, -\frac{1}{2}\right)$: $3\left(\frac{5}{3}\right) - 2\left(-\frac{1}{2}\right) = 5 + 1 = 6$, yes

9. (4, −1): $(4)^2 - 3(-1)^2 = 16 - 3(1) = 16 - 3 = 13 \neq 15$, no; (−4, 1): $(-4)^2 - 3(1)^2 = 16 - 3 \cdot 1 = 16 - 3 = 13 \neq 15$, no

10. (−3, 5): $2(-3)^2 + 3(5)^2 = 2 \cdot 9 + 3 \cdot 25 = 18 + 75 = 93 \neq 57$, no; (4, −5): $2(4)^2 + 3(-5)^2 = 2 \cdot 16 + 3 \cdot 25 = 32 + 75 = 107 \neq 57$, no

11. $\left(-\frac{1}{4}, -3\right)$: $4\left(-\frac{1}{4}\right)(-3) = 3 \neq -3$, no; $\left(\frac{1}{4}, 0\right)$: $4\left(\frac{1}{4}\right)(0) = 0$, yes

12. $\left(-2, -\frac{1}{2}\right)$: $(-2)^3 = 4(-2)^2\left(-\frac{1}{2}\right)$, $-8 = 4(4)\left(-\frac{1}{2}\right)$, yes; $\left(0, \frac{2}{3}\right)$: $0^3 = 4(0)^2 \cdot \frac{2}{3} = 0$, yes

13. $y = 6 - 2x$

x	y
0	6
1	4
2	2
3	0
4	neg.

Solutions: (0, 6), (1, 4), (2, 2), (3, 0)

14. $y = 7 - 5x$

x	y
0	7
1	2
2	neg.

Solutions: (0, 7), (1, 2)

15. $y = \frac{8 - 3x}{2}$

x	y
0	4
1	$\frac{5}{2}$
2	1
3	neg.

Solutions: (0, 4), (2, 1)

16. $y = \frac{10 - 2x}{5}$

x	y
0	2
1	$\frac{8}{5}$
2	$\frac{6}{5}$
3	$\frac{4}{5}$
4	$\frac{2}{5}$
5	0
6	neg.

Solutions: (0, 2), (5, 0)

17. $y = \frac{7 - 3x}{7}$

x	y
0	1
1	$\frac{4}{7}$
2	$\frac{1}{7}$
3	neg.

Solution: (0, 1)

18. $y = \frac{9 - x}{3}$

x	y
0	3
1	$\frac{8}{3}$
2	$\frac{7}{3}$
3	2
4	$\frac{5}{3}$
5	$\frac{4}{3}$
6	1
7	$\frac{2}{3}$
8	$\frac{1}{3}$
9	0
10	neg.

Solutions: (0, 3), (3, 2), (6, 1), (9, 0)

19. $y = \frac{9 - x}{2}$

x	y
0	$\frac{9}{2}$
1	4
2	$\frac{7}{2}$
3	3
4	$\frac{5}{2}$
5	2
6	$\frac{3}{2}$
7	1
8	$\frac{1}{2}$
9	0
10	neg.

Solutions: (1, 4), (3, 3), (5, 2), (7, 1), (9, 0)

20. $y = 7 - 3x$

x	y
0	7
1	4
2	1
3	neg.

Solutions: (0, 7), (1, 4), (2, 1)

B

21. $y = \frac{4}{x}$

x	y
0	undefined
1	4
2	2
3	$\frac{4}{3}$
4	1
5	less than 1

Solutions: (1, 4), (2, 2), (4, 1)

22. $y = \frac{12}{x}$

x	y
0	undefined
1	12
2	6
3	4
4	3
5	$\frac{12}{5}$
6	2
7	$\frac{12}{7}$
8	$\frac{3}{2}$
9	$\frac{4}{3}$
10	$\frac{6}{5}$
11	$\frac{12}{11}$
12	1
13	less than 1

Solutions: (1, 12), (2, 6), (3, 4), (4, 3), (6, 2), (12, 1)

23. $y = \frac{3}{x}$

x	y
0	undefined
1	3
2	$\frac{3}{2}$
3	1
4	less than 1

Solutions:
(1, 3),
(3, 1)

24. $y = \frac{6}{x}$

x	y
0	undefined
1	6
2	3
3	2
4	$\frac{3}{2}$
5	$\frac{6}{5}$
6	1
7	less than 1

Solutions:
(1, 6),
(2, 3),
(3, 2),
(6, 1)

25. $y = \frac{16}{x}$

x	y
0	undefined
1	16
2	8
3	$\frac{16}{3}$
4	4
5	$\frac{16}{5}$
6	$\frac{8}{3}$
7	$\frac{16}{7}$
8	2
9	$\frac{16}{9}$
10	$\frac{8}{5}$
11	$\frac{16}{11}$
12	$\frac{4}{3}$
13	$\frac{16}{13}$
14	$\frac{8}{7}$
15	$\frac{16}{15}$
16	1
17	less than 1

Solutions:
(1, 16),
(2, 8),
(4, 4),
(8, 2),
(16, 1)

26. $y = \frac{6}{x}$

x	y
0	undefined
1	6
2	3
3	2
4	$\frac{3}{2}$
5	$\frac{6}{5}$
6	1
7	less than 1

Solutions:
(1, 6),
(2, 3),
(3, 2),
(6, 1)

27. $y = \frac{5}{x}$

x	y
0	undefined
1	5
2	$\frac{5}{2}$
3	$\frac{5}{3}$
4	$\frac{5}{4}$
5	1
6	less than 1

Solutions: (1, 5), (5, 1)

28. $y = \frac{7}{x}$

x	y
0	undefined
1	7
2	$\frac{7}{2}$
3	$\frac{7}{3}$
4	$\frac{7}{4}$
5	$\frac{7}{5}$
6	$\frac{7}{6}$
7	1
8	less than 1

Solutions: (1, 7), (7, 1)

29. $y = \frac{5}{x + 2}$

x	y
0	$\frac{5}{2}$
1	$\frac{5}{3}$
2	$\frac{5}{4}$
3	1
4	less than 1

Solution: (3, 1)

30. $y = \frac{7}{x + 6}$

x	y
0	$\frac{7}{6}$
1	1
2	less than 1

Solution: (1, 1)

31. $y = \frac{4}{4 - x}$

x	y
0	1
1	$\frac{4}{3}$
2	2
3	4
4	undefined

Solutions: (0, 1), (2, 2), (3, 4)

32. $y = \frac{6}{3 - x}$

x	y
0	2
1	3
2	6
3	undefined

Solutions: (0, 2), (1, 3), (2, 6)

33. $y = \frac{4 - x}{x}$

x	y
0	undefined
1	3
2	1
3	$\frac{1}{3}$
4	0
5	neg.

Solutions: (1, 3), (2, 1), (4, 0)

34. $y = \frac{6 - x}{x}$

x	y
0	undefined
1	5
2	2
3	1
4	$\frac{1}{2}$
5	$\frac{1}{5}$
6	0
7	neg.

Solutions: (1, 5), (2, 2), (3, 1), (6, 0)

35. $y = \frac{9 - x}{x}$

x	y
0	undefined
1	8
2	$\frac{7}{2}$
3	2
4	$\frac{5}{4}$
5	$\frac{4}{5}$
6	$\frac{1}{2}$
7	$\frac{2}{7}$
8	$\frac{1}{8}$
9	0
10	neg.

Solutions: (1, 8), (3, 2), (9, 0)

36. $y = \frac{12 - x}{x}$

x	y
0	undefined
1	11
2	5
3	3
4	2
5	$\frac{7}{5}$
6	1
7	$\frac{5}{7}$
8	$\frac{1}{2}$
9	$\frac{1}{3}$
10	$\frac{1}{5}$
11	$\frac{1}{11}$
12	0
13	neg.

Solutions: (1, 11), (2, 5), (3, 3), (4, 2), (6, 1), (12, 0)

37. Let x be the number of small posters and y be the number of large posters. $2x + 8y = 34$; $y = \frac{17 - x}{4}$. (Since Lisa bought some of each kind, x and y must each be at least 2.) Lisa bought 5 small posters and 3 large posters or 9 small and 2 large.

x	y
2	$\frac{15}{4}$
3	$\frac{7}{2}$
4	$\frac{13}{4}$
5	3
6	$\frac{11}{4}$
7	$\frac{5}{2}$
8	$\frac{9}{4}$
9	2
10	$\frac{7}{4}$

38. Let x be the number of blue spruce trees and y be the number of green spruce. $20x + 10y = 100$; $y = 10 - 2x$. (x and y must each be at least 2.) Linda and Bill bought 2 blue spruce and 6 green, 3 blue and 4 green, or 4 blue and 2 green.

x	y
2	6
3	4
4	2
5	0

39. Let p be the number of pigs and let c be the number of chickens. $40p + 3c = 280$; $p = \frac{280 - 3c}{40} = 7 - \frac{3}{40}c$. For p to be an integer, c must be an integer divisible by 40. Ed bought 4 pigs and 40 chickens.

c	p
40	4
80	1
120	neg.

40. $9x + 2y + 5x + 2y + 4x + 4y = 60$; $18x + 8y = 60$; $8y = 60 - 18x$; $y = \frac{60 - 18x}{8}$; $y = \frac{30 - 9x}{4}$; $x = 2$ and $y = 3$.

x	y
1	5.25
2	3
3	0.75
4	neg.

C **41.** $\frac{2x - y}{2} = \frac{1 - y}{3}$; $6\left(\frac{2x - y}{2}\right) = 6\left(\frac{1 - y}{3}\right)$; $6x - 3y = 2 - 2y$; $6x = 2 + y$; $6x - 2 = y$

42. $\frac{5y + x}{4} = \frac{3x - 2y}{3}$; $12\left(\frac{5y + x}{4}\right) = 12\left(\frac{3x - 2y}{3}\right)$; $15y + 3x = 12x - 8y$; $23y + 3x = 12x$; $23y = 9x$; $y = \frac{9x}{23}$

43. $\frac{1}{x} + \frac{1}{y} = \frac{1}{2}$; $2xy\left(\frac{1}{x} + \frac{1}{y}\right) = 2xy \cdot \frac{1}{2}$; $2y + 2x = xy$; $2y - xy + 2x = 0$;

$(2 - x)y = -2x$; $y = \frac{2x}{x - 2}$

44. $5c + 4d = 79$; $4d = 79 - 5c$; $d = \frac{79 - 5c}{4}$;

the greatest value for c is 7.

c	d
1	18.5
2	17.25
3	16
4	14.75
5	13.5
6	12.25
7	11
8	9.75
9	$c > d$

45. $2a + 3b = 100$; $2a = 100 - 3b$; $a = 50 - \frac{3}{2}b$. For a to be an integer, b must be even. Since a and b are 2-digit numbers, $10 \leq a, b < 100$. There are 9 solutions. The solutions are:

a	35	32	29	26	23	20	17	14	11
b	10	12	14	16	18	20	22	24	26

Page 352 • MIXED REVIEW EXERCISES

1. $308{,}000{,}000 = 3.08 \times 10^8$ **2.** $0.000437 = 4.37 \times 10^{-4}$

3. 119 million $= 119{,}000{,}000 = 1.19 \times 10^8$ **4.** $0.0000216 = 2.16 \times 10^{-5}$

5. $49{,}000 = 4.9 \times 10^4$ **6.** $56{,}201{,}000 = 5.6201 \times 10^7$ **7.** $\frac{3n^3}{6n} = \frac{n^2}{2}$

8. $(5x)^{-3} = 5^{-3}x^{-3} = \frac{1}{125} \cdot \frac{1}{x^3} = \frac{1}{125x^3}$ **9.** $\frac{48x^4y^3}{16x^3y^2} = 3xy$

10. $\frac{b^{-4}}{b^3} = b^{-4} \cdot b^{-3} = b^{-7} = \frac{1}{b^7}$

Page 352 • COMPUTER EXERCISES

```
10 PRINT "INPUT X AND Y"
20 INPUT X, Y
30 PRINT
40 IF 3 * X - 2 * Y = 1 THEN 70
50 PRINT "NOT A SOLUTION"
60 GOTO 80
70 PRINT "A SOLUTION"
80 END
```

1. (3, 4): A SOLUTION; (4, 6): NOT A SOLUTION

In Ex. 2–6, line 40 should be changed as shown.

2. 40 IF 4 * X + Y = 10 THEN 70; (−3, 2): NOT A SOLUTION; (5, −10): A SOLUTION

3. 40 IF X * Y + Y * Y = 6 THEN 70; (1, 2): A SOLUTION; (6, 0): NOT A SOLUTION

4. 40 IF X * X + Y * Y = 25 THEN 70; (−4, −3): A SOLUTION; (0, 5): A SOLUTION

5. 40 IF X * X * Y + X * Y * Y = 0 THEN 70; (−1, −1): NOT A SOLUTION; (1, −1): A SOLUTION

6. 40 IF X * X + (X * Y) * (X * Y) = 45 THEN 70; (3, −2): A SOLUTION; (3, 2): A SOLUTION

Pages 357–358 • WRITTEN EXERCISES

A **1–12.**

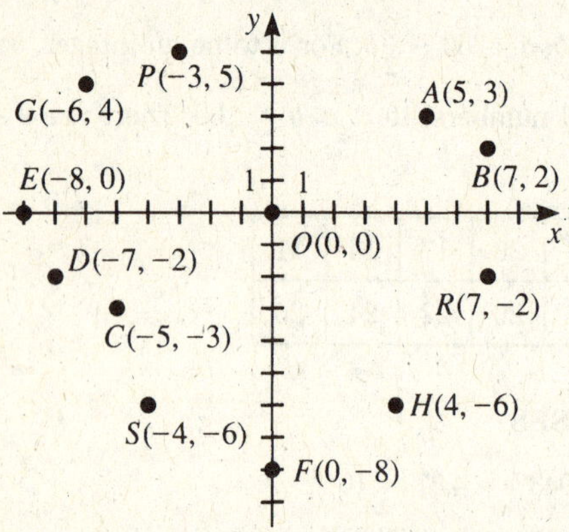

13. *E* **14.** *P* **15.** *Z, H, Y, L* **16.** *G, F, Y* **17.** *I, X, O, P* **18.** *J, O, E* **19.** *I, P, J* **20.** *X, E* **21.** *Z, O* **22.** *D, L, O* **23.** *V* **24.** *B*

25.

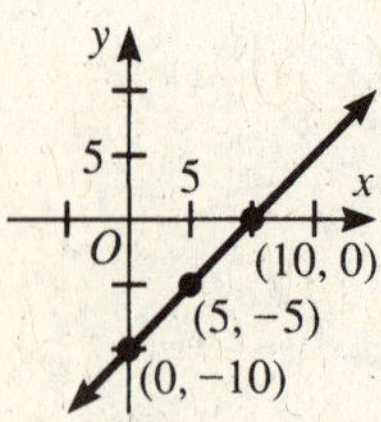

26.

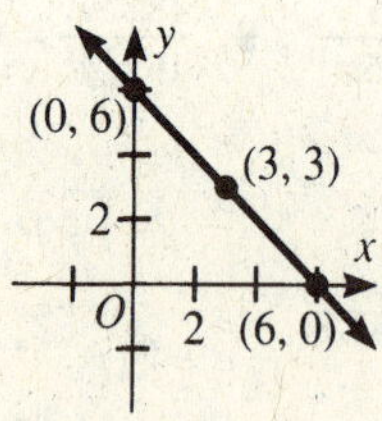

27.

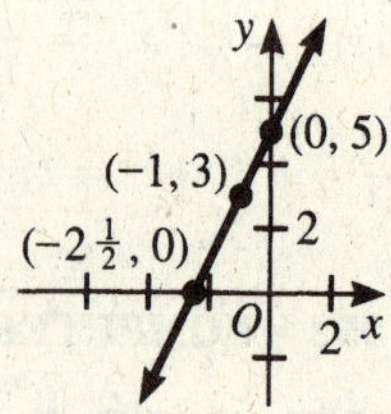

28.

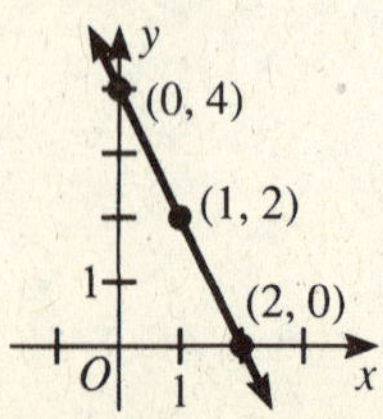

29.

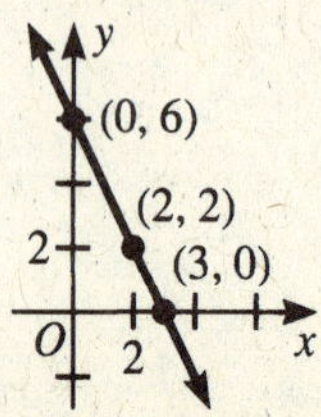

30.

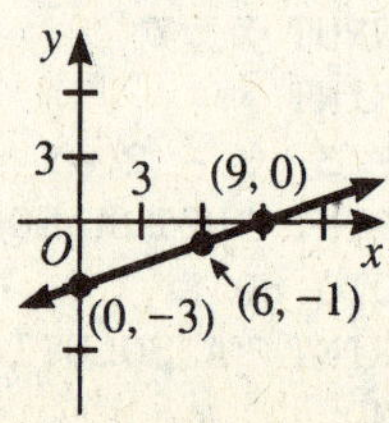

31.
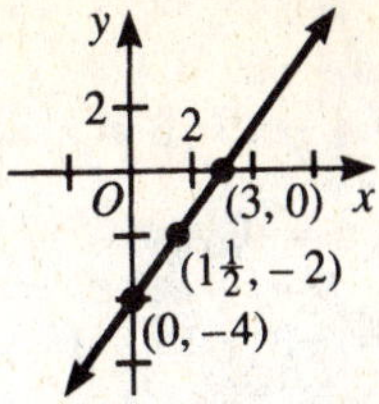

32.
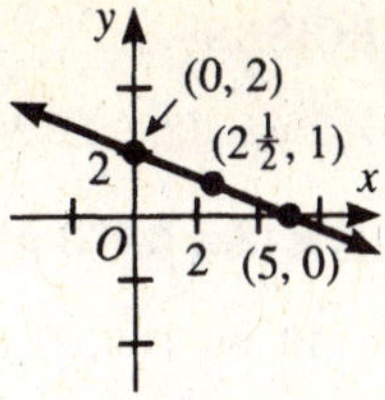

33.
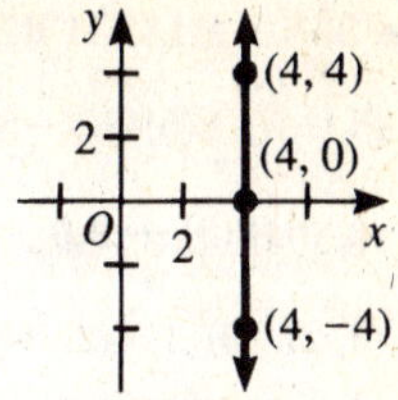

34.
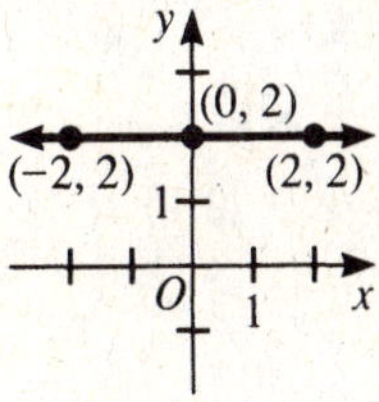

35.
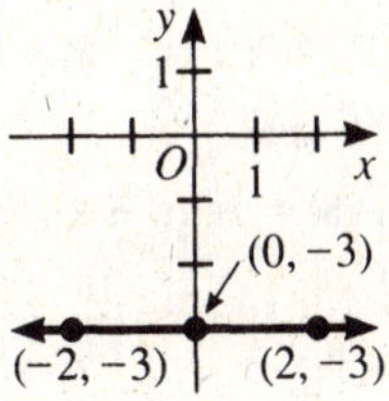

36.
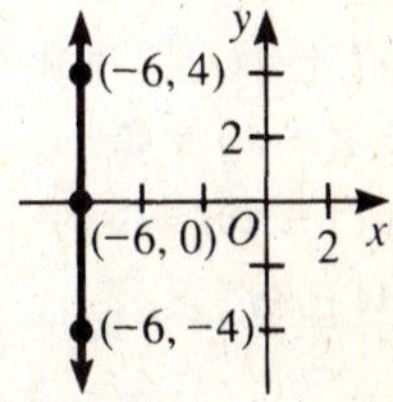

B **37.**
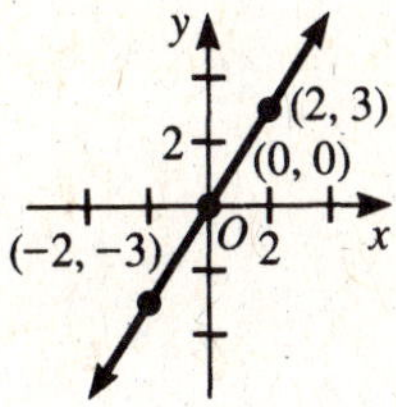

38.
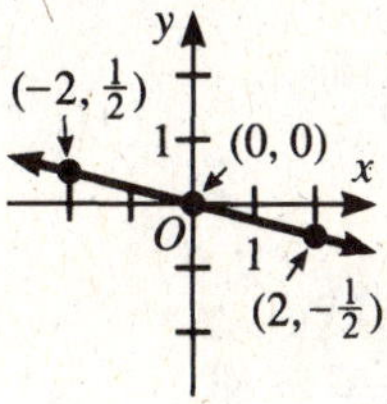

39.
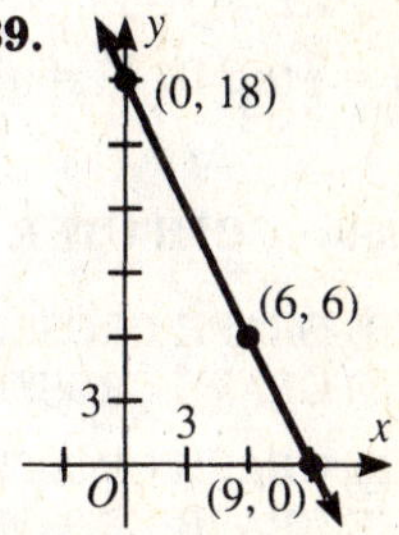

40.
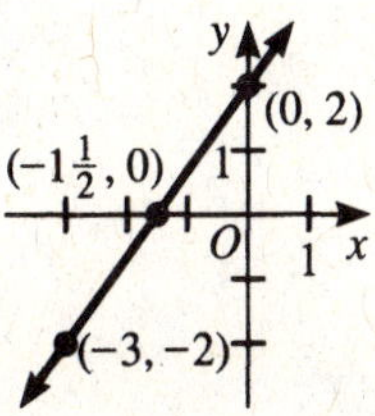

41.
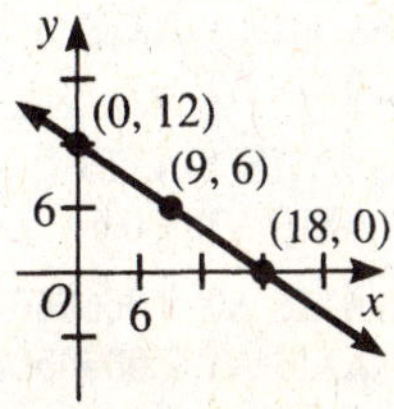

42.
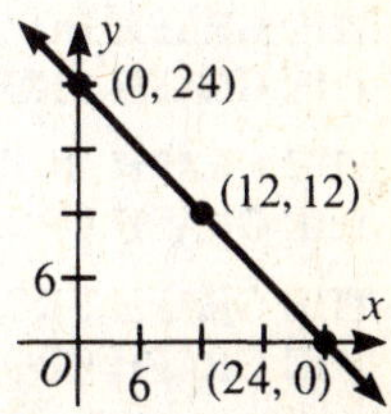

43.
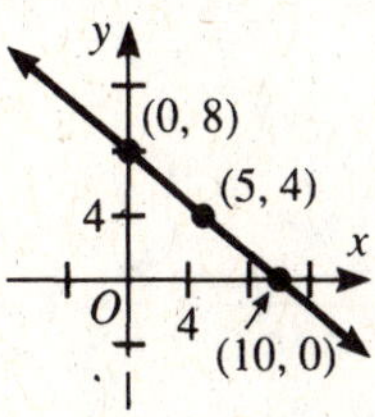

44.
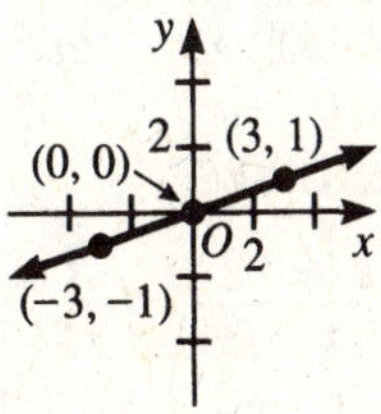

45.
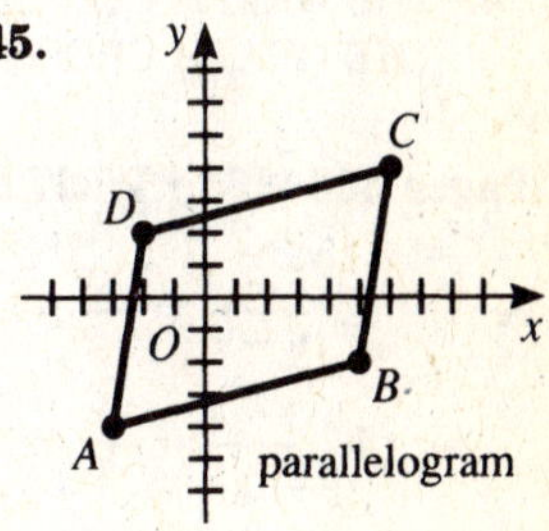

46.
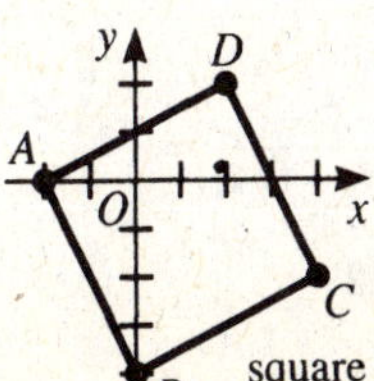

47.
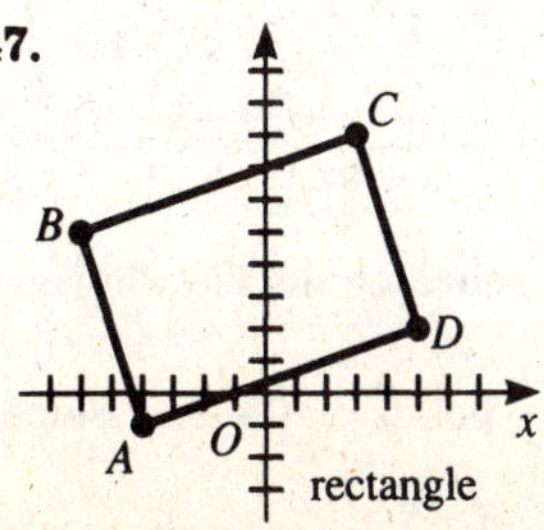

48.
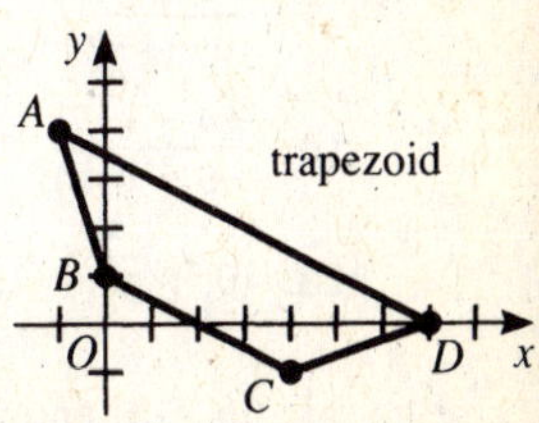

Page 358 • MIXED REVIEW EXERCISES

1. $(5, -5)$: $2(5) + (-5) = 10 - 5 = 5$, yes; $(2, -1)$: $2(2) + (-1) = 4 - 1 = 3 \neq 5$, no
2. $(3, 0)$: $3(3) + 2(0) = 9 + 0 = 9$, yes; $\left(2, \frac{3}{2}\right)$: $3(2) + 2\left(\frac{3}{2}\right) = 6 + 3 = 9$, yes
3. $(3, 2)$: $(3) + 4(2) = 3 + 8 = 11 \neq 13$, no; $(-3, 4)$: $(-3) + 4(4) = -3 + 16 = 13$, yes
4. $(4, -3)$: $5(4) + 3(-3) = 20 - 9 = 11 \neq 9$, no; $(3, -2)$: $5(3) + 3(-2) = 15 - 6 = 9$, yes
5. $x^2 + 3x + 2 = 0$; $(x + 2)(x + 1) = 0$; $x = -2, -1$; $\{-2, -1\}$
6. $-z + 13 = 2$; $-z = -11$; $z = 11$; $\{11\}$
7. $4t^2 - 20t + 24 = 0$; $4(t^2 - 5t + 6) = 0$; $(t - 2)(t - 3) = 0$; $t = 2, 3$; $\{2, 3\}$
8. $\frac{14 - 3y}{2} = 2$; $14 - 3y = 4$; $-3y = -10$; $y = \frac{10}{3}$; $\left\{\frac{10}{3}\right\}$
9. $6x + 12 = 3x - 15$; $3x + 12 = -15$; $3x = -27$; $x = -9$; $\{-9\}$
10. $12 = \frac{4}{5}(n)$; $12 \cdot \frac{5}{4} = n$; $15 = n$; $\{15\}$

Page 359 • COMPUTER KEY-IN

1. THE GRAPH CROSSES THE X-AXIS AT 3
   THE GRAPH CROSSES THE Y-AXIS AT −2
2. THE GRAPH CROSSES THE X-AXIS AT 1.5
   THE GRAPH CROSSES THE Y-AXIS AT 12
3. THE GRAPH DOES NOT CROSS THE Y-AXIS
   THE GRAPH CROSSES THE X-AXIS AT −2
4. THE GRAPH DOES NOT CROSS THE X-AXIS
   THE GRAPH CROSSES THE Y-AXIS AT −0.4
5. THE GRAPH CROSSES THE X-AXIS AT 2.6666667
   THE GRAPH CROSSES THE Y-AXIS AT 2.6666667
6. THE GRAPH CROSSES THE X-AXIS AT 1.5
   THE GRAPH CROSSES THE Y-AXIS AT −2.25

Pages 363–365 • WRITTEN EXERCISES

A 1. $\frac{5 - 4}{6 - 8} = \frac{1}{-2} = -\frac{1}{2}$ 2. $\frac{5 - 2}{-6 - (-4)} = \frac{3}{-2} = -\frac{3}{2}$ 3. $\frac{5 - 3}{-4 - (-6)} = \frac{2}{2} = 1$

4. $\frac{3 - 7}{2 - 0} = \frac{-4}{2} = -2$ 5. $\frac{5 - 3}{6 - (-5)} = \frac{2}{11}$ 6. $\frac{9 - 3}{4 - (-4)} = \frac{6}{8} = \frac{3}{4}$

7. $\frac{-1 - (-3)}{0 - (-2)} = \frac{2}{2} = 1$ 8. $\frac{0 - 3}{2 - 6} = \frac{-3}{-4} = \frac{3}{4}$ 9. $\frac{3 - 8}{1 - 4} = \frac{-5}{-3} = \frac{5}{3}$

10. $\frac{-4 - (-7)}{-6 - (-8)} = \frac{3}{2}$ 11. $\frac{-1 - (-3)}{3 - 0} = \frac{-1 + 3}{3} = \frac{2}{3}$ 12. $\frac{-7 - 7}{-5 - (-2)} = \frac{-14}{-3} = \frac{14}{3}$

13. $(0, 1)$ and $(1, 3)$ are points on the line. Slope $= \frac{3 - 1}{1 - 0} = 2$

14. $(0, -2)$ and $(1, 1)$ are points on the line. Slope $= \frac{1 - (-2)}{1 - 0} = 3$

15. $(0, 8)$, $(1, 6)$: Slope $= \frac{6 - 8}{1 - 0} = -2$ 16. $(0, 12)$, $(1, 8)$: Slope $= \frac{8 - 12}{1 - 0} = -4$

17. (0, 3), (2, 0): Slope $= \frac{0-3}{2-0} = -\frac{3}{2}$ **18.** (0, 2), (5, 0): Slope $= \frac{0-2}{5-0} = -\frac{2}{5}$

19. (0, −2), (5, 1): Slope $= \frac{1-(-2)}{5-0} = \frac{3}{5}$

20. (0, −3), (9, 0): Slope $= \frac{0-(-3)}{9-0} = \frac{3}{9} = \frac{1}{3}$ **21.** Slope = 0 **22.** Slope = 0

23. no slope **24.** no slope

25.

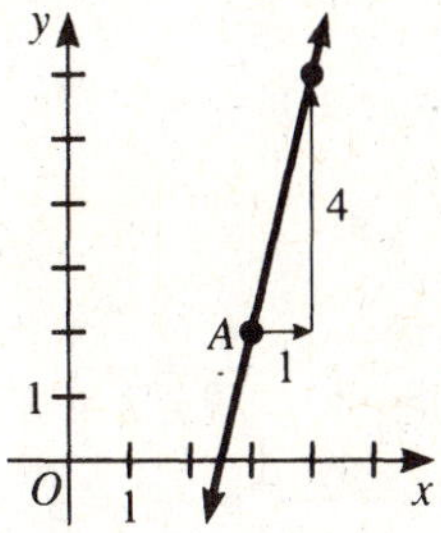

26.

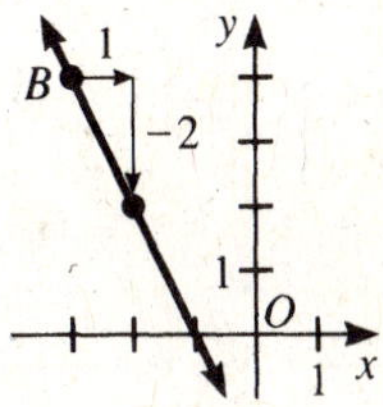

27.

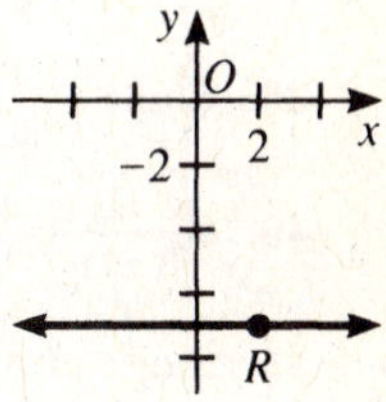

28.

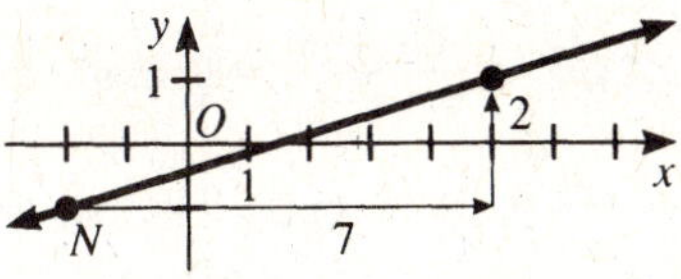

29.

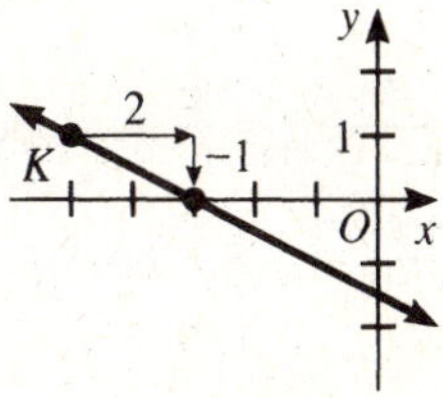

30.

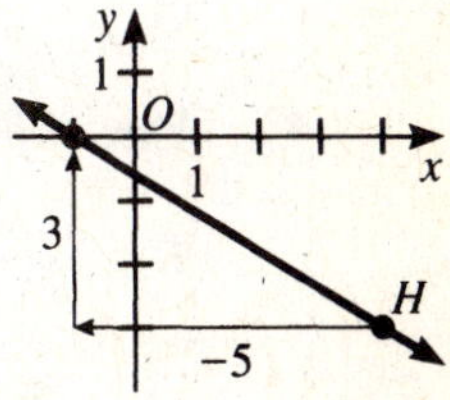

B **31.**

	x	y	
	1	4	
2	3	0	−4
2	5	−4	−4

$-\frac{4}{2} = -\frac{4}{2}$ The points are collinear.

32.

	x	y	
	8	1	
−2	6	5	4
−2	4	8	3

$\frac{4}{-2} \neq \frac{3}{-2}$ The points are not collinear.

33.

	x	y	
	−5	9	
4	−1	1	−8
2	1	−4	−5

$\frac{-8}{4} \neq \frac{-5}{2}$ The points are not collinear.

34.

	x	y	
	−3	7	
3	0	5	−2
6	6	1	−4

$\frac{-2}{3} = \frac{-4}{6}$ The points are collinear.

35.

	x	y	
	0	−1	
1	1	−2	−1
1	2	−4	−2
−3	−1	0	4

$\frac{-1}{1} \neq \frac{-2}{1} \neq \frac{4}{-3}$ The points are not collinear.

36.

	x	y	
	1	−1	
−2	−1	2	3
−2	−3	5	3
−2	−5	8	3

$\frac{3}{-2} = \frac{3}{-2} = \frac{3}{-2}$ The points are collinear.

37. Slope of $\overline{AB} = \frac{6-6}{5-(-4)} = 0$; slope of $\overline{BC} = \frac{-2-6}{-4-5} = \frac{-8}{-9} = \frac{8}{9}$; slope of $\overline{CA} = \frac{6-(-2)}{-4-(-4)} = \frac{8}{0}$, no slope

38. Slope of $\overline{MN} = \frac{2-(-3)}{3-(-2)} = \frac{5}{5} = 1$; slope of $\overline{NP} = \frac{-5-2}{10-3} = \frac{-7}{7} = -1$; slope of $\overline{PQ} = \frac{-10-(-5)}{5-10} = \frac{-5}{-5} = 1$; slope of $\overline{QM} = \frac{-3-(-10)}{-2-5} = \frac{7}{-7} = -1$

39. Slope $= \frac{4-6}{0-4} = \frac{-2}{-4} = \frac{1}{2}$; $\frac{y-4}{8-0} = \frac{1}{2}$; $\frac{y-4}{8} = \frac{1}{2}$; $y-4=4$; $y=8$

40. $\frac{y-3}{-3-1} = 1.5$; $\frac{y-3}{-4} = 1.5$; $y-3=-6$; $y=-3$

41. Slope of $\overline{AC} = \frac{5-(-5)}{3-9} = \frac{10}{-6} = -\frac{5}{3}$; slope of $\overline{AM} = \frac{5-0}{3-6} = \frac{5}{-3} = -\frac{5}{3}$; $\therefore$ M lies on $\overline{AC}$. Slope of $\overline{DB} = \frac{3-(-3)}{11-1} = \frac{6}{10} = \frac{3}{5}$; slope of $\overline{DM} = \frac{-3-0}{1-6} = \frac{-3}{-5} = \frac{3}{5}$; $\therefore$ M lies on $\overline{DB}$.

42. Slope of $\overline{PQ} = \frac{2-(-6)}{-4-(-4)} = \frac{8}{0}$, no slope. Slope of $\overline{QR} = \frac{-6-(-6)}{6-(-4)} = 0$. Slope of $\overline{PR} = \frac{2-(-6)}{-4-6} = \frac{8}{-10} = -\frac{4}{5}$; slope of $\overline{SP} = \frac{2-(-2)}{-4-1} = \frac{4}{-5} = -\frac{4}{5}$; $\therefore$ S lies on side $\overline{PR}$.

Page 365 • MIXED REVIEW EXERCISES

1. $\frac{x+3}{3} + \frac{x}{2} = 0$; $6\left(\frac{x+3}{3} + \frac{x}{2}\right) = 6 \cdot 0$; $2x + 6 + 3x = 0$; $5x = -6$; $x = -\frac{6}{5}$; $\left\{-\frac{6}{5}\right\}$

2. $-5 = \frac{9b}{4}$; $-20 = 9b$; $-\frac{20}{9} = b$; $\left\{-\frac{20}{9}\right\}$

3. $\frac{3+z}{2z} = \frac{3}{z}$; $2z\left(\frac{3+z}{2z}\right) = 2z \cdot \frac{3}{z}$; $3 + z = 6$; $z = 3$; $\{3\}$

4. $-5(y+3) = 30$; $-5y - 15 = 30$; $-5y = 45$; $y = -9$; $\{-9\}$

5. $|z| = 1.7$; $z = 1.7$ or -1.7; $\{1.7, -1.7\}$

6. $3m(m-5) = 0$; $m = 0, 5$; $\{0, 5\}$ **7.** $\frac{a+4b}{2a-b} = \frac{-3+4\cdot 4}{2(-3)-4} = \frac{-3+16}{-6-4} = \frac{13}{-10} = -\frac{13}{10}$

8. $5(x+2y) = 5(2+2\cdot 1) = 5(2+2) = 5 \cdot 4 = 20$

9. $\frac{1}{3}(4x+y) = \frac{1}{3}(4\cdot 2+1) = \frac{1}{3}(8+1) = \frac{9}{3} = 3$ **10.** $(xy^2)^2 = (2 \cdot 1^2)^2 = 2^2 = 4$

11. $(3a-2b)+7 = (3\cdot -3 - 2\cdot 4) + 7 = (-9-8)+7 = -17+7 = -10$

12. $(2y^3)^3 = (2 \cdot 1^3)^3 = 2^3 = 8$

Page 365 • CALCULATOR KEY-IN

1. 7 **2.** 10 **3.** −1.925 **4.** 1.40625 **5.** 1 **6.** −0.6

Pages 368–369 • WRITTEN EXERCISES

A 1. slope = 2; y-intercept = 1 2. slope = 3; y-intercept = −4

3. slope = $\frac{1}{2}$; y-intercept = −3 4. slope = $\frac{2}{3}$; y-intercept = 3

5. slope = $-\frac{1}{2}$; y-intercept = 6 6. slope = $-\frac{1}{3}$; y-intercept = −2

7. slope = −2; y-intercept = 8 8. slope = −3; y-intercept = 9

9. slope = 1; y-intercept = 0 10. slope = −1; y-intercept = 0

11. slope = 0; y-intercept = −4 12. slope = 0; y-intercept = 4

13. 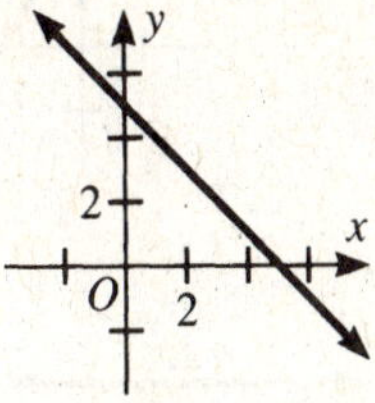

14. 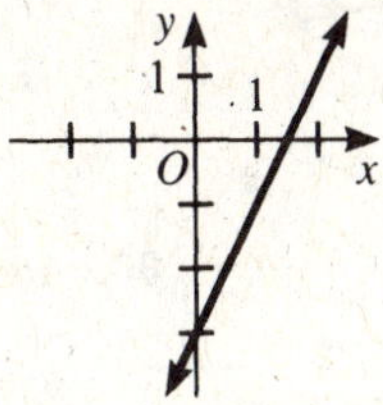

15. 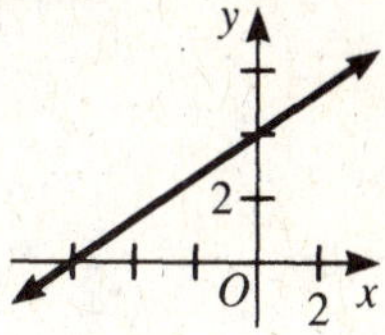

16. 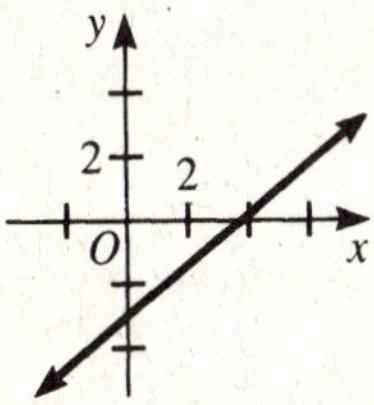

17. 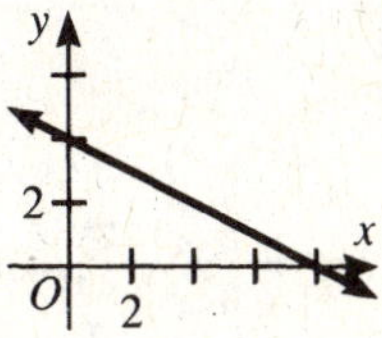

18. 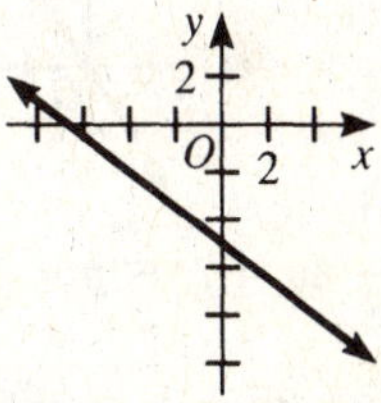

19. $x + y = 8$,
$y = -x + 8$

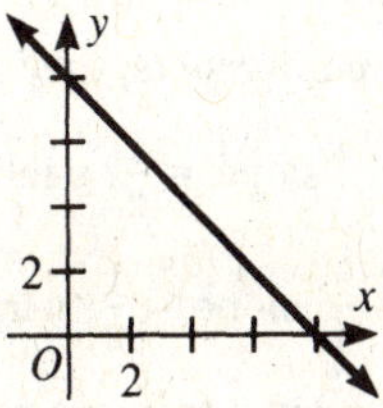

20. $2x + y = 6$,
$y = -2x + 6$

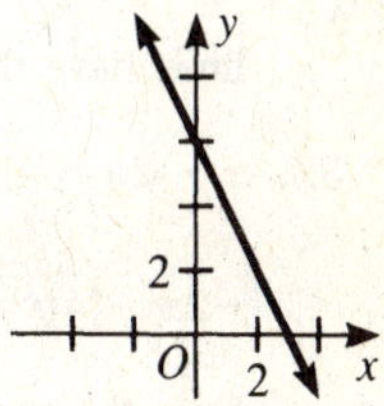

21. $3x - y = 6$,
$-y = -3x + 6$,
$y = 3x - 6$

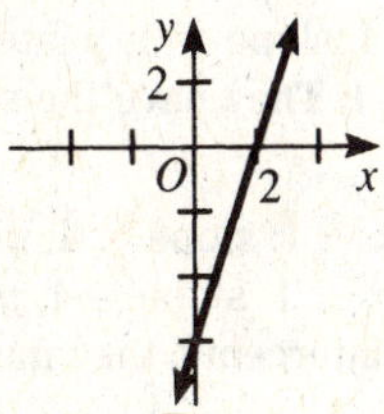

22. $2x - y = -4$,
$-y = -2x - 4$,
$y = 2x + 4$

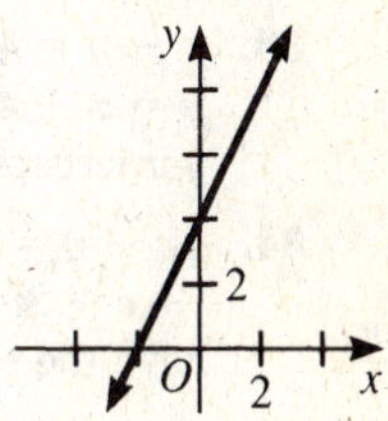

23. $3x + y = 9$,
$y = -3x + 9$

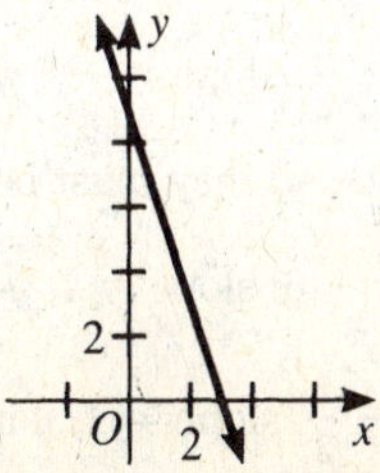

24. $2x + y = -4$,
$y = -2x - 4$

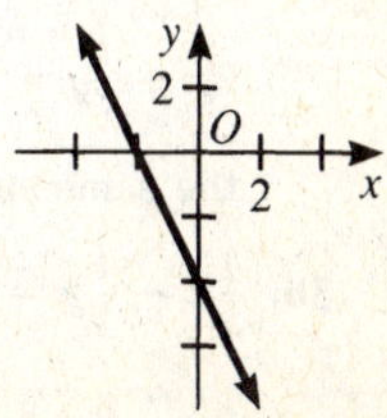

B

25. $3x - 5y = 10,$
$-5y = -3x + 10,$
$y = \frac{3}{5}x - 2$

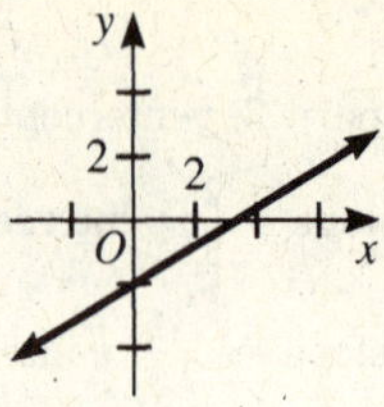

26. $4x - 3y = 9,$
$-3y = -4x + 9,$
$y = \frac{4}{3}x - 3$

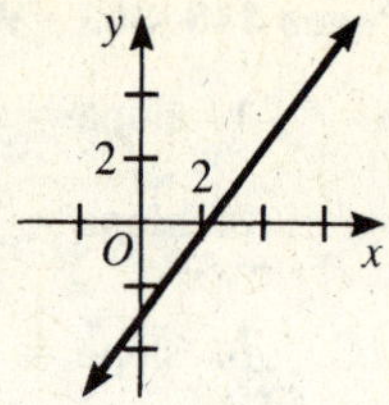

27. $x + 5y = 5,$
$5y = -x + 5,$
$y = -\frac{1}{5}x + 1$

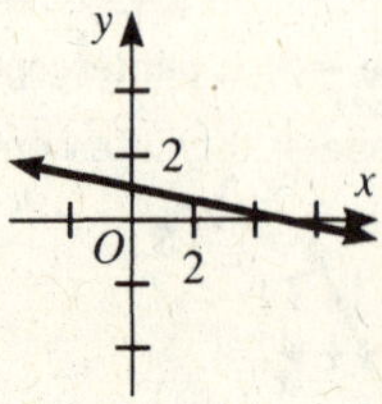

28. $6x + 4y = 8,$
$4y = -6x + 8,$
$y = -\frac{3}{2}x + 2$

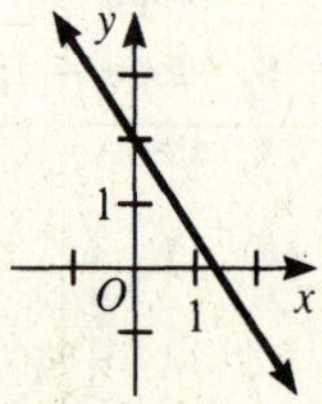

29. $3x = 2y,$
$y = \frac{3}{2}x$

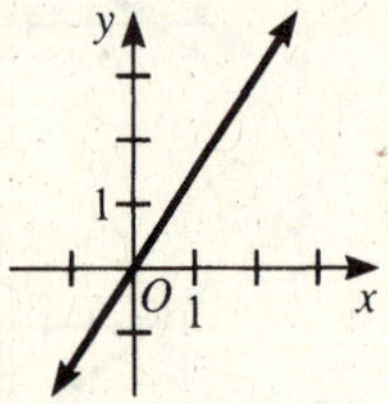

30. $4y = 3,$
$y = \frac{3}{4}$

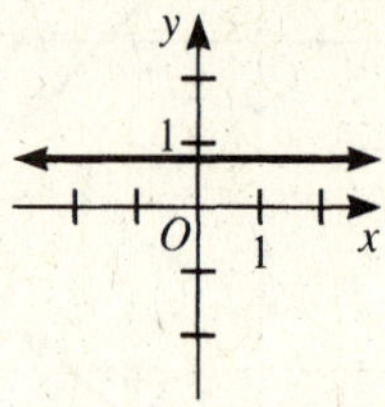

31. $2x - 3y = 7; -3y = -2x + 7; y = \frac{2}{3}x - \frac{7}{3}$: slope $= \frac{2}{3}$, y-intercept $= -\frac{7}{3}$; $-2x + 3y = 4; 3y = 2x + 4; y = \frac{2}{3}x + \frac{4}{3}$: slope $= \frac{2}{3}$, y-intercept $= \frac{4}{3}$; Since the lines have the same slope and different y-intercepts, they must be parallel.

32. $x - 2y = 1; -2y = -x + 1; y = \frac{1}{2}x - \frac{1}{2}$: slope $= \frac{1}{2}$, y-intercept $= -\frac{1}{2}$; $-2x + 4y = 9; 4y = 2x + 9; y = \frac{1}{2}x + \frac{9}{4}$: slope $= \frac{1}{2}$, y-intercept $= \frac{9}{4}$; Since they have the same slope and different y-intercepts, they are parallel.

33. $x - y = 4; -y = -x + 4; y = x - 4$: slope $= 1$, y-intercept $= -4$; $y - x = 4$; $y = x + 4$: slope $= 1$, y-intercept $= 4$; They have the same slope but different y-intercepts so they are parallel.

34. $4x - y = 1; -y = -4x + 1; y = 4x - 1$: slope $= 4$, y-intercept $= -1$; $-8x + 2y = 2; 2y = 8x + 2; y = 4x + 1$: slope $= 4$, y-intercept $= 1$; Since they have the same slope but different y-intercepts, they must be parallel.

35. $\frac{1}{2}x + \frac{1}{2}y = 5; \frac{1}{2}y = -\frac{1}{2}x + 5; y = -x + 10$: slope $= -1$, y-intercept $= 10$; $2x + 2y = 3; 2y = -2x + 3; y = -x + \frac{3}{2}$: slope $= -1$, y-intercept $= \frac{3}{2}$; They have the same slope but different y-intercepts, so they must be parallel.

36. $\frac{1}{3}x - \frac{1}{3}y = 2; -\frac{1}{3}y = -\frac{1}{3}x + 2; y = x - 6$: slope $= 1$, y-intercept $= -6$; $3x - 3y = 1; -3y = -3x + 1; y = x - \frac{1}{3}$: slope $= 1$, y-intercept $= -\frac{1}{3}$; Since they have the same slope but different y-intercepts they must be parallel.

37. y-intercept $= 7$; $y = 5x - 3$: slope $= 5$; $y = 5x + 7$

38. $y - 3x = 4$; $y = 3x + 4$: slope $= 3$; $4y + x = 36$; $4y = -x + 36$; $y = -\frac{1}{4}x + 9$: y-intercept $= 9$; $y = 3x + 9$

39. $x + y = 5$; $y = -x + 5$: slope $= -1$; $2y + px = 5$; $2y = -px + 5$; $y = -\frac{p}{2}x + \frac{5}{2}$; $-\frac{p}{2} = -1$; $p = 2$; $2y + 2x = 5$ is parallel to $x + y = 5$; $x - y = 5$; $-y = -x + 5$; $y = x - 5$: slope $= 1$; $-\frac{p}{2} = 1$; $p = -2$; $2y - 2x = 5$ is parallel to $x - y = 5$.

40. $x - 3y = 0$; $-3y = -x$; $y = \frac{1}{3}x$: slope $= \frac{1}{3}$; $dy + 3x = 2$; $dy = -3x + 2$; $y = -\frac{3}{d}x + \frac{2}{d}$; $-\frac{3}{d} = \frac{1}{3}$; $d = -9$; $-9y + 3x = 2$ is parallel to $x - 3y = 0$; $d = 0$; $3x = 2$ is parallel to the y-axis.

41. $\frac{0 - 6}{-2 - 2} = \frac{-6}{-4} = \frac{3}{2}$; The slope of the line joining B and C is $\frac{3}{2}$; $\frac{2r - 3}{r - 5} = \frac{3}{2}$; $2(2r - 3) = 3(r - 5)$; $4r - 6 = 3r - 15$; $r - 6 = -15$, $r = -9$; The line joining A and $(-9, -18)$ is parallel to the line joining B and C.

42. $\frac{0 - 3}{-2 - 5} = \frac{-3}{-7} = \frac{3}{7}$; The slope of the line joining A and C is $\frac{3}{7}$; $\frac{s - 6}{(s + 4) - 2} = \frac{3}{7}$; $7(s - 6) = 3(s + 2)$; $7s - 42 = 3s + 6$; $4s - 42 = 6$; $4s = 48$; $s = 12$; The line joining B and $(16, 12)$ is parallel to the line joining A and C.

43. $\frac{6 - 3}{2 - 5} = \frac{3}{-3} = -1$; The slope of the line joining A and B is -1; $\frac{(2t + 1) - 0}{-3t - (-2)} = -1$; $2t + 1 = -(-3t + 2)$; $2t + 1 = 3t - 2$; $1 = t - 2$; $3 = t$; The line joining C and $(-9, 7)$ is parallel to the line joining A and B.

C **44.** Let (x, y) be the navigator's position. $\frac{y - 0}{x - (-1)} = -5$; $y = -5(x + 1)$; $y = -5x - 5$; $\frac{y - (-1)}{x - 2} = 3$; $y + 1 = 3(x - 2)$; $y = 3x - 6 - 1$; $y = 3x - 7$; $-5x - 5 = 3x - 7$; $-5 = 8x - 7$; $2 = 8x$; $\frac{1}{4} = x$; $y = 3x - 7 = \frac{3}{4} - 7 = -6\frac{1}{4}$; The navigator's position is $\left(\frac{1}{4}, -6\frac{1}{4}\right)$.

45. $ax + by = c$; $by = -ax + c$; $y = -\frac{a}{b}x + \frac{c}{b}$; slope $= -\frac{a}{b}$; y-intercept $= \frac{c}{b}$

Page 369 • MIXED REVIEW EXERCISES

1. $\frac{3 - 1}{-2 - (-3)} = \frac{2}{-2 + 3} = \frac{2}{1} = 2$ **2.** $\frac{3 - 2}{4 - 1} = \frac{1}{3}$ **3.** $\frac{2 - 5}{-1 - (-2)} = \frac{-3}{-1 + 2} = \frac{-3}{1} = -3$

4. $\frac{13 - 5}{3 - 1} = \frac{8}{2} = 4$ **5.** $3s^2 + 7s + 4 = (3s + 4)(s + 1)$ **6.** $20x^2 + 15x = 5x(4x + 3)$

7. $3p^2 - 24p + 48 = 3(p^2 - 8p + 16) = 3(p - 4)^2$ **8.** $81y^2 - 16z^2 = (9y + 4z)(9y - 4z)$

9. $-3x(x + y) + 4(x + y) = (-3x + 4)(x + y)$

10. $m^2 - 9mn - 36n^2 = (m + 3n)(m - 12n)$

Page 370 • EXTRA

1. $y = \frac{3}{2}x + 7$: slope $= \frac{3}{2}$; slope of perpendicular $= -\frac{2}{3}$.

2. $y = \frac{4}{5}x - 2$: slope $= \frac{4}{5}$; slope of perpendicular $= -\frac{5}{4}$.

3. $y = -\frac{3}{8}x + 5$: slope $= -\frac{3}{8}$; slope of perpendicular $= \frac{8}{3}$

4. $2x + y = 3$, $y = -2x + 3$: slope $= -2$; slope of perpendicular $= \frac{1}{2}$

5. $3x - 5y = 110$, $-5y = -3x + 110$, $y = \frac{3}{5}x - 22$: slope $= \frac{3}{5}$; slope of perpendicular $= -\frac{5}{3}$

6. $y = -x$: slope $= -1$; slope of perpendicular $= 1$

7. $y = 1$ is a horizontal line. Its perpendicular is a vertical line, so it has no slope.

8. $x = 3$ is a vertical line. The line perpendicular to it has 0 slope.

9. $x - 3y - 8 = 0$, $-3y = -x + 8$, $y = \frac{1}{3}x - \frac{8}{3}$: slope $= \frac{1}{3}$; slope of perpendicular $= -3$

10. $3x + 6y = 8$; $6y = -3x + 8$; $y = -\frac{1}{2}x + \frac{4}{3}$: slope $= -\frac{1}{2}$; $y = 2x - 8$: slope $= 2$; $-\frac{1}{2} \cdot 2 = -1$; The two lines are perpendicular.

11. $3x + y = 7$; $y = -3x + 7$: slope $= -3$, y-intercept $= 7$; $y = -3x + 2$: slope $= -3$, y-intercept $= 2$; The lines have the same slope but different y-intercepts, so they are parallel.

12. $2x + 5y = 7$; $5y = -2x + 7$; $y = -\frac{2}{5} + \frac{7}{5}$: slope $= -\frac{2}{5}$, y-intercept $= \frac{7}{5}$; $2x + 5y = 9$; $5y = -2x + 9$; $y = -\frac{2}{5}x + \frac{9}{5}$: slope $= -\frac{2}{5}$, y-intercept $= \frac{9}{5}$; The lines have the same slope but different y-intercepts, so they are parallel.

13. $2x - 8y = 9$; $-8y = -2x + 9$; $y = \frac{1}{4}x - \frac{9}{8}$: slope $= \frac{1}{4}$; $12x + 3y = 7$; $3y = -12x + 7$; $y = -4x + \frac{7}{3}$: slope $= -4$; $\frac{1}{4} \cdot -4 = -1$; The lines are perpendicular.

14. $y = x + 5$: slope $= 1$; $y = 8 - x$; $y = -x + 8$: slope $= -1$; $1 \cdot -1 = -1$; The lines are perpendicular.

15. $4x + 6y = 9$; $6y = -4x + 9$; $y = -\frac{2}{3}x + \frac{3}{2}$: slope $= -\frac{2}{3}$, y-intercept $= \frac{3}{2}$; $2x + 3y = 5$; $3y = -2x + 5$; $y = -\frac{2}{3}x + \frac{5}{3}$: slope $= -\frac{2}{3}$, y-intercept $= \frac{5}{3}$; The lines have the same slope but different y-intercepts, so they must be parallel.

16. $y = 4 - 3x$, $y = -3x + 4$: slope $= -3$, y-intercept $= 4$; $3y - x = 12$; $3y = x + 12$; $y = \frac{1}{3}x + 4$: slope $= \frac{1}{3}$, y-intercept $= 4$; $\frac{1}{3} \cdot -3 = -1$; The lines have the same y-intercept and are perpendicular.

17. $y = \frac{2}{3}x - 12$ has slope $\frac{2}{3}$ and y-intercept -12 (at Q). The line perpendicular to it has slope $-\frac{3}{2}$ and y-intercept -12 (since it intersects the y-axis at Q). The perpendicular line is $y = -\frac{3}{2}x - 12$. At the point $P(x, 0)$: $0 = -\frac{3}{2}x - 12$; $\frac{3}{2}x = -12$; $x = -8$.

Pages 372–373 • WRITTEN EXERCISES

A **1.** $y = 2x + 5$ **2.** $y = -3x + 4$ **3.** $y = -\frac{1}{2}x + 7$ **4.** $y = \frac{2}{3}x + 9$

5. $y = \frac{1}{2}x + b$; $0 = \frac{1}{2}(4) + b$; $0 = 2 + b$; $-2 = b$; $y = \frac{1}{2}x - 2$

6. $y = -\frac{1}{3}x + b$; $0 = -\frac{1}{3}(3) + b$; $0 = -1 + b$; $1 = b$; $y = -\frac{1}{3}x + 1$

7. $y = 2x + b$; $1 = 2(-5) + b$; $1 = -10 + b$; $11 = b$; $y = 2x + 11$

8. $y = 3x + b$; $1 = 3(4) + b$; $1 = 12 + b$; $-11 = b$; $y = 3x - 11$

9. $y = -2x + b$; $5 = -2(3) + b$; $5 = -6 + b$; $11 = b$; $y = -2x + 11$

10. $y = -x + b$; $5 = -8 + b$; $13 = b$; $y = -x + 13$

11. $y = \frac{1}{2}x + b$; $-1 = \frac{1}{2}(4) + b$; $-1 = 2 + b$; $-3 = b$; $y = \frac{1}{2}x - 3$

12. $y = \frac{1}{3}x + b$; $0 = \frac{1}{3}(6) + b$; $0 = 2 + b$; $-2 = b$; $y = \frac{1}{3}x - 2$

13. $y = -\frac{3}{4}x + b$; $-3 = -\frac{3}{4}(8) + b$; $-3 = -6 + b$; $3 = b$; $y = -\frac{3}{4}x + 3$

14. $y = -\frac{2}{5}x + b$; $7 = -\frac{2}{5}(5) + b$; $7 = -2 + b$; $9 = b$; $y = -\frac{2}{5}x + 9$

15. $y = 0x + b$; $y = b$, $3 = b$; $y = 3$ **16.** $y = 0x + b$; $y = b$; $4 = b$; $y = 4$

17. $\frac{9 - 7}{1 - 0} = \frac{2}{1} = 2$; $y = 2x + b$; $7 = 2(0) + b$; $7 = b$; $y = 2x + 7$

18. $\frac{-1 - 3}{2 - 0} = \frac{-4}{2} = -2$; $y = -2x + b$; $3 = -2(0) + b$; $3 = b$; $y = -2x + 3$

19. $\frac{0 - 2}{7 - 5} = \frac{-2}{2} = -1$; $y = -x + b$; $2 = -5 + b$; $7 = b$; $y = -x + 7$

20. $\frac{-4 - 4}{3 - (-3)} = \frac{-8}{6} = -\frac{4}{3}$; $y = -\frac{4}{3}x + b$; $4 = -\frac{4}{3}(-3) + b$; $4 = 4 + b$, $0 = b$; $y = -\frac{4}{3}x$

21. $\frac{8 - 1}{1 - 8} = \frac{7}{-7} = -1$; $y = -x + b$; $1 = -8 + b$; $9 = b$; $y = -x + 9$

22. $\frac{2 - 4}{4 - (-2)} = \frac{-2}{6} = -\frac{1}{3}$; $y = -\frac{1}{3}x + b$; $4 = -\frac{1}{3}(-2) + b$; $4 = \frac{2}{3} + b$; $\frac{10}{3} = b$; $y = -\frac{1}{3}x + \frac{10}{3}$

23. $\frac{7-(-1)}{6-3} = \frac{8}{3}$; $y = \frac{8}{3}x + b$; $-1 = \frac{8}{3}(3) + b$; $-1 = 8 + b$; $-9 = b$; $y = \frac{8}{3}x - 9$

24. $\frac{-4-(-1)}{1-(-3)} = \frac{-3}{4} = -\frac{3}{4}$; $y = -\frac{3}{4}x + b$; $-1 = -\frac{3}{4}(-3) + b$; $-1 = \frac{9}{4} + b$, $-\frac{13}{4} = b$;

$y = -\frac{3}{4}x - \frac{13}{4}$

25. $\frac{4-(-1)}{1-0} = \frac{5}{1} = 5$; $y = 5x + b$; $-1 = 5(0) + b$; $-1 = b$; $y = 5x - 1$

26. $\frac{3-(-2)}{0-(-1)} = \frac{5}{1} = 5$; $y = 5x + b$; $-2 = 5(-1) + b$; $-2 = -5 + b$; $3 = b$; $y = 5x + 3$

27. $\frac{-3-0}{2-(-2)} = \frac{-3}{4} = -\frac{3}{4}$; $y = -\frac{3}{4}x + b$; $0 = -\frac{3}{4}(-2) + b$; $0 = \frac{3}{2} + b$; $-\frac{3}{2} = b$;

$y = -\frac{3}{4}x - \frac{3}{2}$

28. $\frac{5-0}{-2-3} = \frac{5}{-5} = -1$; $y = -x + b$; $0 = -3 + b$; $3 = b$; $y = -x + 3$

B **29.** $(0, -2), (5, 0)$: $\frac{0-(-2)}{5-0} = \frac{2}{5}$; $y = \frac{2}{5}x + b$; $-2 = \frac{2}{5}(0) + b$; $-2 = b$; $y = \frac{2}{5}x - 2$

30. $(0, -5), (3, 0)$: $\frac{0-(-5)}{3-0} = \frac{5}{3}$; $y = \frac{5}{3}x + b$; $-5 = \frac{5}{3}(0) + b$; $-5 = b$; $y = \frac{5}{3}x - 5$

31. $(6, 0), (0, -3)$: $\frac{-3-0}{0-6} = \frac{-3}{-6} = \frac{1}{2}$; $y = \frac{1}{2}x + b$; $0 = \frac{1}{2}(6) + b$; $0 = 3 + b$; $-3 = b$;

$y = \frac{1}{2}x - 3$

32. $(-4, 0), (0, -2)$: $\frac{-2-0}{0-(-4)} = \frac{-2}{4} = -\frac{1}{2}$; $y = -\frac{1}{2}x + b$; $0 = \frac{1}{2}(-4) + b$; $0 = 2 + b$;

$-2 = b$; $y = -\frac{1}{2}x - 2$

33. $y = 5$ **34.** $y = -1$

35. $3x - y = 4$; $-y = -3x + 4$; $y = 3x - 4$: slope $= 3$; $y = 3x + b$, $3 = 3(-1) + b$; $3 = -3 + b$; $6 = b$; $y = 3x + 6$, $-3x + y = 6$ or $3x - y = -6$

36. $x - 2y + 7 = 0$; $-2y = -x - 7$; $y = \frac{1}{2}x + \frac{7}{2}$: slope $= \frac{1}{2}$; $y = \frac{1}{2}x + b$;

$0 = \frac{1}{2}(-4) + b$; $0 = -2 + b$; $2 = b$; $y = \frac{1}{2}x + 2$; $-\frac{1}{2}x + y = 2$; $-x + 2y = 4$ or

$x - 2y = -4$

37. $x + 3y + 9 = 0$; $3y = -x - 9$; $y = -\frac{1}{3}x - 3$: y-intercept $= -3$; $(-4, -5), (0, -3)$:

$\frac{-3-(-5)}{0-(-4)} = \frac{2}{4} = \frac{1}{2}$; $y = \frac{1}{2}x - 3$; $-\frac{1}{2}x + y = -3$; $x - 2y = 6$

38. $(0, p)$: y-intercept is p; $\frac{0-1}{p-7} = -\frac{1}{p-7} = \frac{1}{7-p}$; $y = \frac{1}{7-p}x + p$; $1 = \frac{1}{7-p}(7) + p$;

$1 = \frac{7}{7-p} + p$; $7 - p = \left(\frac{7}{7-p} + p\right)(7-p)$; $7 - p = 7 + p(7-p)$; $-p = 7p - p^2$;

$0 = 8p - p^2$; $0 = p(8-p)$; $p = 8$; $y = \frac{1}{7-8}x + 8$; $y = -x + 8$; $x + y = 8$

39. **a.** The line through (2, 5) and (2, 8) is a vertical line. Since it has no slope, it cannot be written in slope-intercept form.

b. $x = 2$

40. The horizontal line is $y = 7$ and the vertical line is $x = -3$.

C **41.** $\frac{5-3}{2-(-2)} = \frac{2}{4} = \frac{1}{2}$; $y = \frac{1}{2}x + b$; $3 = \frac{1}{2}(-2) + b$; $3 = -1 + b$; $4 = b$; $y = \frac{1}{2}x + 4$;

$k = \frac{1}{2}(6) + 4$; $k = 3 + 4$; $k = 7$

42. If x-intercept $= -4$, then the line passes through the point $(-4, 0)$.

$\frac{6-0}{2-(-4)} = \frac{6}{6} = 1$; $y = x + b$; $0 = -4 + b$; $4 = b$; $y = x + 4$; $10 = p + 4$, $6 = p$

Page 373 • MIXED REVIEW EXERCISES

1. $\left(\frac{4}{3}t^3\right)(12t^2) = \left(\frac{4}{3} \cdot 12\right)(t^3 \cdot t^2) = 16t^5$ **2.** $\frac{1}{2}(4s^2 - 8st) = 2s^2 - 4st$

3. $(7pq^2)^2 = 49p^2q^4$ **4.** $(-3m^2n^4)^3 = -27m^6n^{12}$ **5.** $3 \cdot 4 - 3^2 = 12 - 9 = 3$

6. $(3a^3b)(-6ab^2) = -18a^4b^3$ **7.** $2 \cdot (8 - 4)^3 = 2 \cdot (4)^3 = 2 \cdot 64 = 128$

8. $(8x + 3y) - (3x + y) = 8x + 3y - 3x - y = 5x + 2y$

Page 377 • WRITTEN EXERCISES

A **1.**

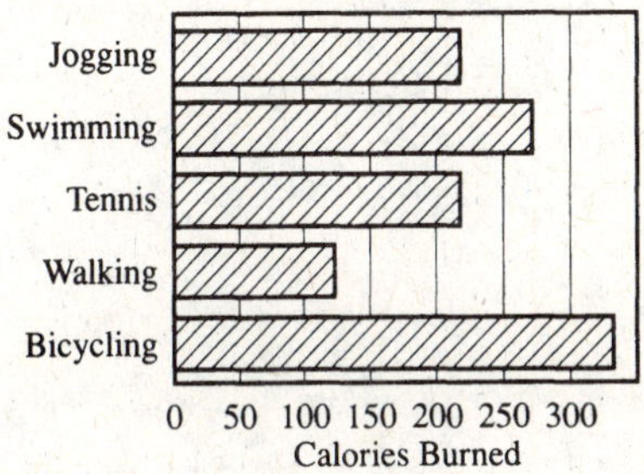

2.

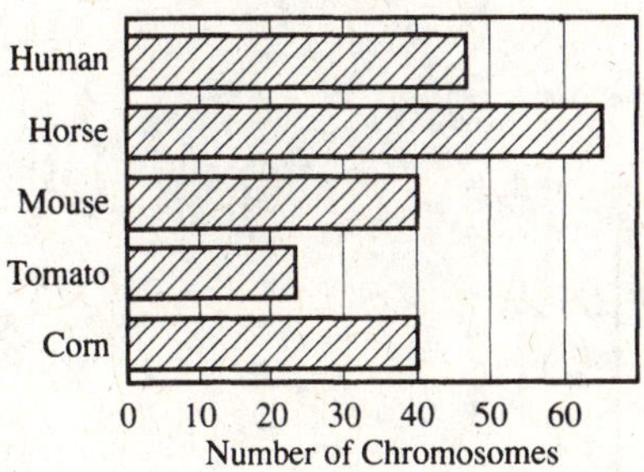

3.

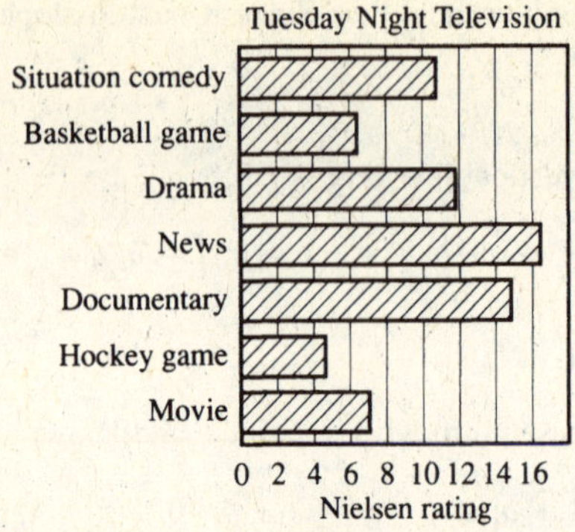

4.

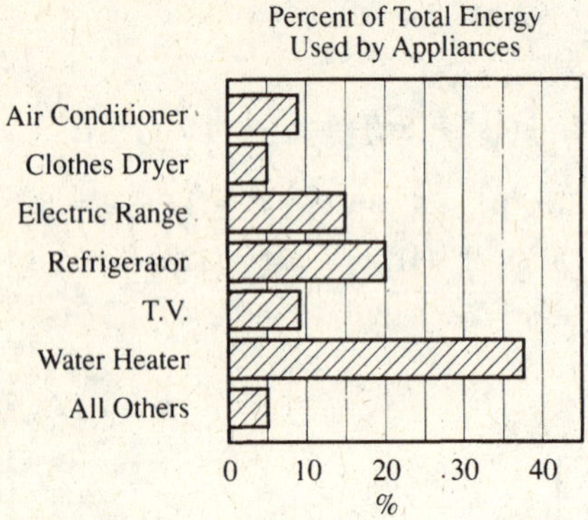

5.

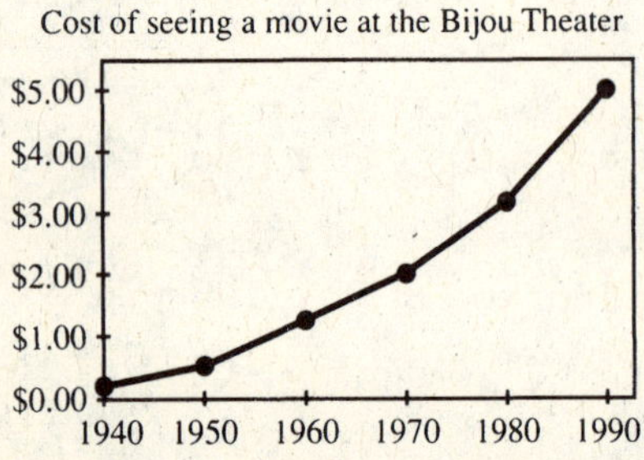

6.

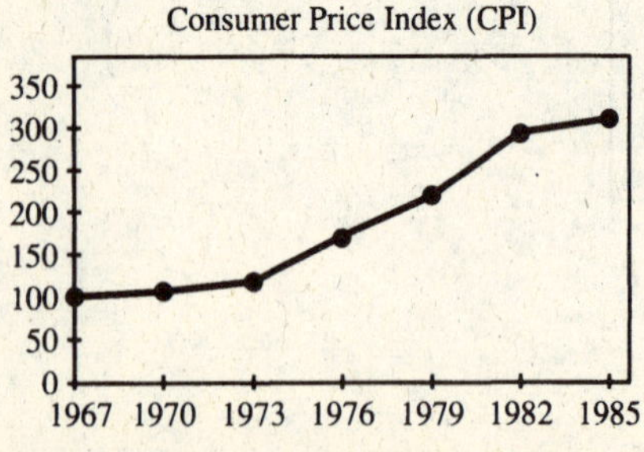

B **7.** Cost in 1955 ≈ \$.90; cost in 1975 ≈ \$2.50

8. CPI in 1980 ≈ 230; CPI in 1984 ≈ 300

9–10. Answers will vary.

9. Example: Paper carriers and number of customers on paper route

10. Example: Year and price of a pound of beef.

Page 377 • MIXED REVIEW EXERCISES

1. $\frac{6-4}{4-2} = \frac{2}{2} = 1; y = x + b; 4 = 2 + b; b = 2; y = x + 2$

2. $y = 5x + b; 7 = 5(1) + b; 7 = 5 + b; 2 = b; y = 5x + 2$

3. $y = \frac{1}{3}x - 4$

4. $\frac{-3-3}{0-(-4)} = \frac{-6}{4} = -\frac{3}{2}; y = -\frac{3}{2}x + b; 3 = -\frac{3}{2}(-4) + b; 3 = 6 + b; -3 = b; y = -\frac{3}{2}x - 3$

5.

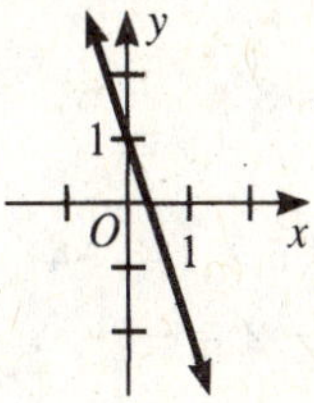

6. $x + y = 9,$
$y = -x + 9$

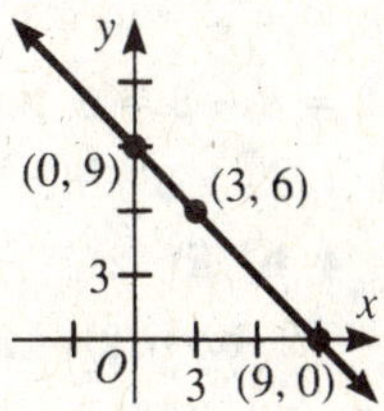

7.

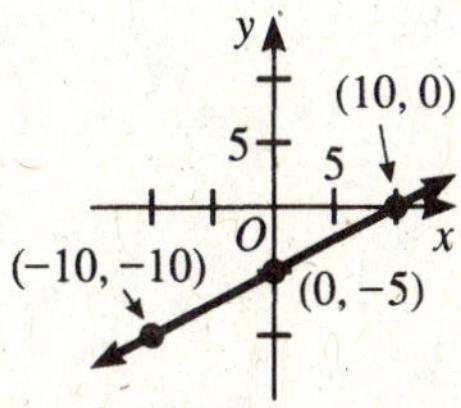

8.

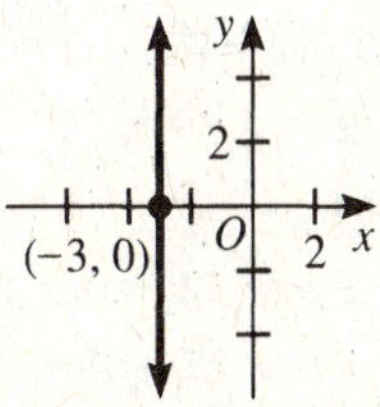

Pages 380–381 • WRITTEN EXERCISES

A

1. $g(x) = 5x + 1; g(-1) = 5 \cdot (-1) + 1 = -5 + 1 = -4; g(0) = 5 \cdot 0 + 1 = 1;$
$g(1) = 5 \cdot 1 + 1 = 5 + 1 = 6; R = \{-4, 1, 6\}$

2. $f(x) = 3x - 4; f(1) = 3 \cdot 1 - 4 = 3 - 4 = -1; f(2) = 3 \cdot 2 - 4 = 6 - 4 = 2;$
$f(3) = 3 \cdot 3 - 4 = 9 - 4 = 5; R = \{-1, 2, 5\}$

3. $s(z) = 5 - 4z; s(-2) = 5 - 4 \cdot (-2) = 5 + 8 = 13; s(0) = 5 - 4 \cdot 0 = 5 - 0 = 5;$
$s(2) = 5 - 4 \cdot 2 = 5 - 8 = -3; R = \{-3, 5, 13\}$

4. $h(y) = 1 - 2y; h(-3) = 1 - 2 \cdot (-3) = 1 + 6 = 7; h(0) = 1 - 2 \cdot 0 = 1 - 0 = 1;$
$h(1) = 1 - 2 \cdot 1 = 1 - 2 = -1; R = \{-1, 1, 7\}$

5. $G(a) = 4a^2 - 1; G(-1) = 4 \cdot (-1)^2 - 1 = 4 - 1 = 3;$
$G(0) = 4 \cdot 0^2 - 1 = 0 - 1 = -1; G(1) = 4 \cdot 1^2 - 1 = 4 - 1 = 3; R = \{-1, 3\}$

6. $H(b) = b^2 + 3; H(0) = 0^2 + 3 = 0 + 3 = 3; H(2) = 2^2 + 3 = 4 + 3 = 7;$
$H(4) = 4^2 + 3 = 16 + 3 = 19; R = \{3, 7, 19\}$

7. $F(x) = x^2 + 4x - 3; F(1) = 1^2 + 4 \cdot 1 - 3 = 1 + 4 - 3 = 2;$
$F(2) = 2^2 + 4 \cdot 2 - 3 = 4 + 8 - 3 = 9; F(4) = 4^2 + 4 \cdot 4 - 3 = 16 + 16 - 3 = 29;$
$R = \{2, 9, 29\}$

8. $M(x) = x^2 + 5x + 2; M(-1) = (-1)^2 + 5 \cdot (-1) + 2 = 1 - 5 + 2 = -2;$
$M(-2) = (-2)^2 + 5 \cdot (-2) + 2 = 4 - 10 + 2 = -4;$
$M(-4) = (-4)^2 + 5 \cdot (-4) + 2 = 16 - 20 + 2 = -2; R = \{-4, -2\}$

9. $P(z) = z^2 - 5z + 6; P(2) = 2^2 - 5 \cdot 2 + 6 = 4 - 10 + 6 = 0;$
$P(3) = 3^2 - 5 \cdot 3 + 6 = 9 - 15 + 6 = 0; P(4) = 4^2 - 5 \cdot 4 + 6 = 16 - 20 + 6 = 2;$
$R = \{0, 2\}$

10. $N(a) = a^2 - 3a - 2; N(-2) = (-2)^2 - 3 \cdot (-2) - 2 = 4 + 6 - 2 = 8;$
$N(-1) = (-1)^2 - 3 \cdot (-1) - 2 = 1 + 3 - 2 = 2;$
$N(0) = 0^2 - 3 \cdot 0 - 2 = 0 - 0 - 2 = -2; R = \{-2, 2, 8\}$

11. $q(c) = c^3 + c^2 + 3c; q(0) = 0^3 + 0^2 + 3 \cdot 0 = 0 + 0 + 0 = 0;$
$q(1) = 1^3 + 1^2 + 3 \cdot 1 = 1 + 1 + 3 = 5; q(3) = 3^3 + 3^2 + 3 \cdot 3 = 27 + 9 + 9 = 45;$
$R = \{0, 5, 45\}$

12. $K(x) = x^3 - 5x^2 + 6x; K(0) = 0^3 - 5 \cdot 0^2 + 6 \cdot 0 = 0 + 0 + 0 = 0;$
$K(2) = 2^3 - 5 \cdot 2^2 + 6 \cdot 2 = 8 - 20 + 12 = 0;$
$K(3) = 3^3 - 5 \cdot 3^2 + 6 \cdot 3 = 27 - 45 + 18 = 0; R = \{0\}$

13. **a.** $f(3) = 5 \cdot 3 - 9 = 15 - 9 = 6$ **b.** $f(-3) = 5 \cdot (-3) - 9 = -15 - 9 = -24$
c. $f(-8) = 5 \cdot (-8) - 9 = -40 - 9 = -49$

14. **a.** $p(2) = 8 - 4 \cdot 2 = 8 - 8 = 0$ **b.** $p(0) = 8 - 4 \cdot 0 = 8 - 0 = 8$
c. $p(-2) = 8 - 4 \cdot (-2) = 8 + 8 = 16$

15. **a.** $R(3) = 3 + 1 = 4$ **b.** $R(-2) = -2 + 1 = -1$ **c.** $R(-5) = -5 + 1 = -4$

16. **a.** $G(0) = 0 - 2 = -2$ **b.** $G(2) = 2 - 2 = 0$ **c.** $G(-3) = -3 - 2 = -5$

17. **a.** $h(4) = 3 \cdot 4^2 - 2 = 3 \cdot 16 - 2 = 48 - 2 = 46$
b. $h(-5) = 3 \cdot (-5)^2 - 2 = 3 \cdot 25 - 2 = 75 - 2 = 73$
c. $h(0) = 3 \cdot 0^2 - 2 = 0 - 2 = -2$

18. **a.** $k(5) = 3 \cdot 5^2 + 9 = 75 + 9 = 84$
b. $k(-2) = 3 \cdot (-2)^2 + 9 = 3 \cdot 4 + 9 = 12 + 9 = 21$
c. $k(-3) = 3 \cdot (-3)^2 + 9 = 3 \cdot 9 + 9 = 27 + 9 = 36$

19. **a.** $g(5) = 5^2 - 2 = 25 - 2 = 23$ **b.** $g(-5) = (-5)^2 - 2 = 25 - 2 = 23$
c. $g(0) = 0^2 - 2 = 0 - 2 = -2$

20. **a.** $h(2) = 3 \cdot 2^2 + 2 = 12 + 2 = 14$ **b.** $h(-2) = 3 \cdot (-2)^2 + 2 = 12 + 2 = 14$
c. $h(-4) = 3 \cdot (-4)^2 + 2 = 3 \cdot 16 + 2 = 48 + 2 = 50$

21. **a.** $R(0) = 0^3 + 8 = 0 + 8 = 8$ **b.** $R(-2) = (-2)^3 + 8 = -8 + 8 = 0$
c. $R(2) = 2^3 + 8 = 8 + 8 = 16$

22. **a.** $N(3) = 3^3 - 27 = 27 - 27 = 0$ **b.** $N(-3) = (-3)^3 - 27 = -27 - 27 = -54$
c. $N(0) = 0^3 - 27 = -27$

B 23. **a.** $f(7) = 7^2 + 3 \cdot 7 = 49 + 21 = 70$ **b.** $f(-7) = (-7)^2 + 3 \cdot (-7) = 49 - 21 = 28$
c. $f(-3) = (-3)^2 + 3 \cdot (-3) = 9 - 9 = 0$

24. **a.** $g(3) = 5 \cdot 3^2 - 7 \cdot 3 = 45 - 21 = 24$
b. $g(-3) = 5 \cdot (-3)^2 - 7 \cdot (-3) = 5 \cdot 9 + 21 = 45 + 21 = 66$
c. $g(0) = 5 \cdot 0^2 - 7 \cdot 0 = 0 - 0 = 0$

25. **a.** $P(3) = 3 - 3^3 = 3 - 27 = -24$ **b.** $P(1) = 1 - 1^3 = 1 - 1 = 0$
c. $P(-1) = -1 - (-1)^3 = -1 - (-1) = -1 + 1 = 0$

26. **a.** $m\left(-\frac{1}{2}\right) = -\frac{1}{2}\left[1 - 2 \cdot \left(-\frac{1}{2}\right)\right] = -\frac{1}{2}(1 + 1) = -\frac{1}{2} \cdot 2 = -1$

b. $m\left(\frac{1}{2}\right) = \frac{1}{2}\left(1 - 2 \cdot \frac{1}{2}\right) = \frac{1}{2}(1 - 1) = \frac{1}{2} \cdot 0 = 0$ **c.** $m(0) = 0(1 - 2 \cdot 0) = 0$

27. **a.** $Z\left(\frac{1}{3}\right) = \left|3 \cdot \frac{1}{3} - 1\right| = |1 - 1| = |0| = 0$

b. $Z(0) = |3 \cdot 0 - 1| = |0 - 1| = |-1| = 1$

c. $Z\left(\frac{2}{3}\right) = \left|3 \cdot \frac{2}{3} - 1\right| = |2 - 1| = |1| = 1$

28. **a.** $Q(-2) = |3 - 6 \cdot (-2)| = |3 + 12| = |15| = 15$

b. $Q\left(\frac{1}{2}\right) = \left|3 - 6 \cdot \frac{1}{2}\right| = |3 - 3| = |0| = 0$

c. $Q\left(-\frac{1}{2}\right) = \left|3 - 6 \cdot \left(-\frac{1}{2}\right)\right| = |3 + 3| = |6| = 6$

29. **a.** $f(0) = 3 \cdot 0 - 12 = 0 - 12 = -12$ **b.** $3x - 12 = 0; 3x = 12; x = 4; \{4\}$

30. **a.** $f(0) = 4 \cdot 0 + 7 = 0 + 7 = 7$ **b.** $4x + 7 = 0; 4x = -7; x = -\frac{7}{4}; \left\{-\frac{7}{4}\right\}$

31. **a.** $f(0) = 5 \cdot 0 - 1 = 0 - 1 = -1$ **b.** $5x - 1 = 0; 5x = 1; x = \frac{1}{5}; \left\{\frac{1}{5}\right\}$

32. **a.** $f(0) = 2 - 3 \cdot 0 = 2 - 0 = 2$ **b.** $2 - 3x = 0; 2 = 3x; \frac{2}{3} = x; \left\{\frac{2}{3}\right\}$

33. **a.** $f(0) = -\frac{1}{2} \cdot 0 + 7 = 0 + 7 = 7$ **b.** $-\frac{1}{2}x + 7 = 0; 7 = \frac{1}{2}x; 14 = x; \{14\}$

34. **a.** $f(0) = 3 - \frac{1}{2} \cdot 0 = 3 - 0 = 3$ **b.** $3 - \frac{1}{2}x = 0; 3 = \frac{1}{2}x; 6 = x; \{6\}$

35. **a.** $f(0) = 0^2 - 2 \cdot 0 - 3 = 0 - 0 - 3 = -3$

b. $x^2 - 2x - 3 = 0; (x - 3)(x + 1) = 0; x = 3$ or $x = -1; \{-1, 3\}$

36. **a.** $f(0) = 0^2 - 13 \cdot 0 + 40 = 0 - 0 + 40 = 40$

b. $x^2 - 13x + 40 = 0; (x - 5)(x - 8) = 0; x = 5$ or $x = 8; \{5, 8\}$

37. **a.** $f(0) = 0^4 - 0^2 = 0$

b. $x^4 - x^2 = 0; x^2(x^2 - 1) = 0; x^2(x + 1)(x - 1) = 0; x = 0$ or $x = -1$ or $x = 1$; $\{-1, 0, 1\}$

38. **a.** $f(0) = \frac{0 + 1}{0 - 2} = \frac{1}{-2} = -\frac{1}{2}$ **b.** $\frac{x + 1}{x - 2} = 0; x + 1 = 0; x = -1; \{-1\}$

39. **a.** $f(0) = \frac{0^2 - 1}{0^2 + 1} = \frac{-1}{1} = -1$

b. $\frac{x^2 - 1}{x^2 + 1} = 0; x^2 - 1 = 0; (x + 1)(x - 1) = 0; x = -1$ or $x = 1; \{-1, 1\}$

40. **a.** $f(0) = \frac{0^2 - 3 \cdot 0 + 2}{0 + 1} = \frac{2}{1} = 2$

b. $\frac{x^2 - 3x + 2}{x + 1} = 0; x^2 - 3x + 2 = 0; (x - 2)(x - 1) = 0; x = 2$ or $x = 1; \{1, 2\}$

41. $\frac{1}{2}g(6) = \frac{1}{2} \cdot (-6^2) = \frac{1}{2} \cdot (-36) = -18$

42. $2f(5) = 2 \cdot (3 \cdot 5 + 4) = 2 \cdot (15 + 4) = 2 \cdot 19 = 38$

43. $f(1) + g(1) = (3 \cdot 1 + 4) + (-1^2) = (3 + 4) + (-1) = 7 - 1 = 6$

44. $f(2) \cdot g(2) = (3 \cdot 2 + 4) \cdot (-2^2) = (6 + 4) \cdot (-4) = 10 \cdot (-4) = -40$

C **45.** **a.** $g(1) = 2 \cdot 1 = 2$ **b.** $f(1) = 1^2 = 1$ **c.** $g[f(1)] = g(1) = 2$

d. $f[g(1)] = f(2) = 2^2 = 4$

46. **a.** $g(-2) = 2 \cdot (-2) = -4$ **b.** $f(-2) = (-2)^2 = 4$ **c.** $g[f(-2)] = g(4) = 2 \cdot 4 = 8$

d. $f[g(-2)] = f(-4) = (-4)^2 = 16$

47. $f[g(x)] = g[f(x)]; f(2x) = g(x^2); (2x)^2 = 2 \cdot x^2; 4x^2 = 2x^2; 4x^2 - 2x^2 = 0; 2x^2 = 0;$ $x = 0$

48. $g[f(x)] = x; g(x + 1) = x; g(x) = x - 1$

Page 382 • MIXED REVIEW EXERCISES

1. $\frac{4n - 1}{3n^3} + \frac{3}{n} = \frac{4n - 1}{3n^3} + \frac{3n^2 \cdot 3}{3n^3} = \frac{4n - 1 + 9n^2}{3n^3} = \frac{9n^2 + 4n - 1}{3n^3}$

2. $7\frac{1}{4} + 3\frac{2}{3} + 2\frac{3}{4} + 5\frac{1}{3} = \frac{29}{4} + \frac{11}{3} + \frac{11}{4} + \frac{16}{3} = \frac{3 \cdot 29}{12} + \frac{4 \cdot 11}{12} + \frac{3 \cdot 11}{12} + \frac{4 \cdot 16}{12} =$

$\frac{87 + 44 + 33 + 64}{12} = \frac{228}{12} = 19$

3. $(-15)\left(\frac{y}{5}\right) = -3y$ **4.** $(-3)(3a - 2b + c) = -9a + 6b - 3c$

5. $-[9 + (-2)] + (-x) + 8 = -(7) - x + 8 = 1 - x$ **6.** $(4m - 7)3 = 12m - 21$

7. $-120\left(\frac{1}{2}\right)\left(\frac{1}{5}\right) = -12$ **8.** $\frac{3ef^2}{2e^3} \cdot \frac{4de^2}{6ef} = \frac{3f^2}{2e^2} \cdot \frac{2de}{3f} = \frac{df}{e}$

9. $\frac{x^2 - y^2}{x^2 + 2xy + y^2} = \frac{(x + y)(x - y)}{(x + y)^2} = \frac{x - y}{x + y}$

Page 382 • COMPUTER EXERCISES

```
10 DEF FNA(X) = 3 * X - 7
20 READ X
30 IF X = -100 THEN 70
40 PRINT "F("; X;") = "; FNA(X)
50 GOTO 20
60 DATA -2, 0, 1, 10, -100
70 END
```

1. F(−2) = −13; F(0) = −7; F(1) = −4; F(10) = 23

2. Change lines 10 and 60:

```
10 DEF FNA(X) = X * X + 3 * X + 2
60 DATA 0, 1, -1, -2, -100
```

F(0) = 2; F(1) = 6; F(−1) = 0; F(−2) = 0

3.

```
10 DEF FNA(X) = 1/(X * X + 1)
60 DATA -10, -2, 0, 2, 10, -100
```

F(−10) = 0.0099009; F(−2) = 0.2; F(0) = 1; F(2) = 0.2; F(10) = 0.0099009

4.
```
10 DEF FNA(X) = 4 * X - X * X * X
20 DATA -3, -2, -1, 0, 1, 2, 3, -100
```
F(−3) = 15; F(−2) = 0; F(−1) = −3; F(0) = 0; F(1) = 3; F(2) = 0; F(3) = −15

Page 382 • CHALLENGE

1. a. $\frac{1}{1\cdot2}+\frac{1}{2\cdot3}=\frac{1}{2}+\frac{1}{6}=\frac{3}{6}+\frac{1}{6}=\frac{4}{6}=\frac{2}{3}$; $\frac{1}{1\cdot2}+\frac{1}{2\cdot3}+\frac{1}{3\cdot4}=\frac{2}{3}+\frac{1}{3\cdot4}=\frac{8}{12}+\frac{1}{12}=\frac{9}{12}=\frac{3}{4}$; $\frac{1}{1\cdot2}+\frac{1}{2\cdot3}+\frac{1}{3\cdot4}+\frac{1}{4\cdot5}=\frac{3}{4}+\frac{1}{4\cdot5}=\frac{15}{20}+\frac{1}{20}=\frac{16}{20}=\frac{4}{5}$

b. $\frac{100}{101}$ **c.** $\frac{n}{n+1}$

2. a. $\frac{1}{1\cdot3}+\frac{1}{3\cdot5}=\frac{1}{3}+\frac{1}{15}=\frac{5}{15}+\frac{1}{15}=\frac{6}{15}=\frac{2}{5}$; $\frac{1}{1\cdot3}+\frac{1}{3\cdot5}+\frac{1}{5\cdot7}=\frac{2}{5}+\frac{1}{5\cdot7}=\frac{14}{35}+\frac{1}{35}=\frac{15}{35}=\frac{3}{7}$; $\frac{1}{1\cdot3}+\frac{1}{3\cdot5}+\frac{1}{5\cdot7}+\frac{1}{7\cdot9}=\frac{3}{7}+\frac{1}{7\cdot9}=\frac{27}{63}+\frac{1}{63}=\frac{28}{63}=\frac{4}{9}$

b. $\frac{50}{101}$ **c.** $\frac{n}{2n+1}$

Pages 386–387 • WRITTEN EXERCISES

A

1. $y = x - 3$

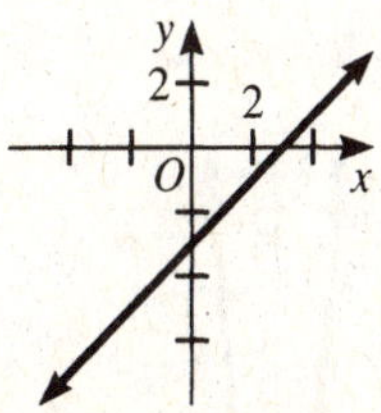

2. $y = -x + 1$

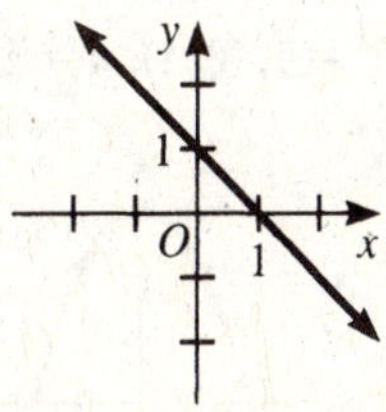

3. $y = 2 - \frac{1}{3}x$

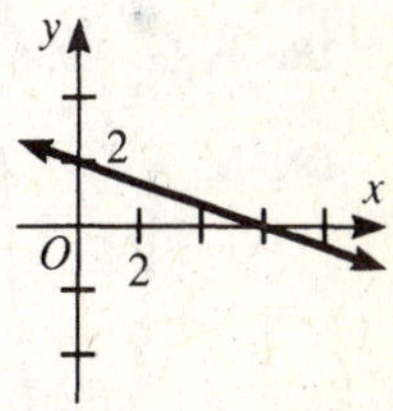

4. $y = -\frac{3}{4}x$

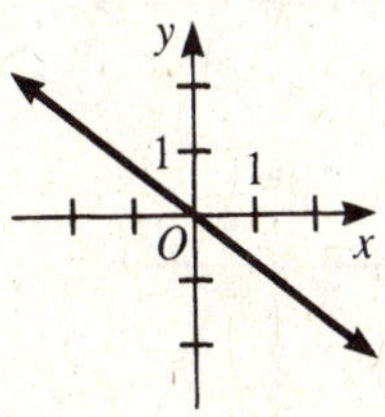

5. $y = -7$

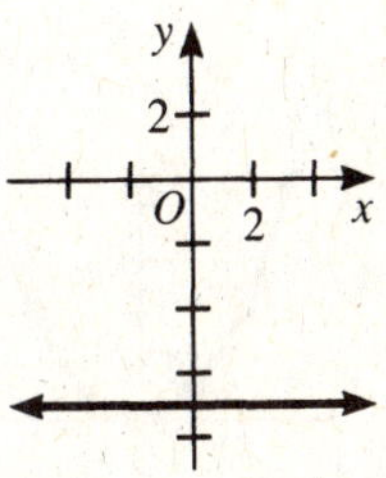

6. $y = 0$

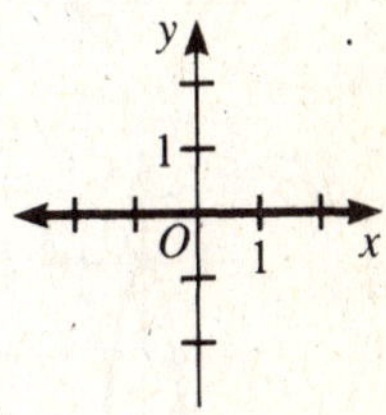

7. $-\frac{b}{2a} = 0$; $y = 0^2 - 5$; $y = -5$; vertex = (0, −5); axis of symm. is $x = 0$; least value is −5

8. $-\frac{b}{2a} = 0$; $y = 0^2 + 4$; $y = 4$; (0, 4); $x = 0$; 4

9. $-\frac{b}{2a} = -\frac{-1}{2} = \frac{1}{2}$; $y = \left(\frac{1}{2}\right)^2 - \frac{1}{2} - 6 = \frac{1}{4} - \frac{1}{2} - 6 = -6\frac{1}{4}$; $\left(\frac{1}{2}, -6\frac{1}{4}\right)$; $x = \frac{1}{2}$; $-6\frac{1}{4}$

10. $-\frac{b}{2a} = -\frac{-10}{10} = 1; y = 4 - 10 \cdot 1 + 5 \cdot 1^2 = 4 - 10 + 5 = -1; (1, -1); x = 1; -1$

11. $-\frac{b}{2a} = 0; y = 9 \cdot 0^2 - 4 = -4; (0, -4); x = 0; -4$

12. $-\frac{b}{2a} = 0; y = \frac{1}{4} \cdot 0^2 = 0; (0, 0); x = 0; 0$

13. $-\frac{b}{2a} = -\frac{-3}{2 \cdot (-1)} = -\frac{3}{2}; y = -\left(-\frac{3}{2}\right)^2 - 3\left(-\frac{3}{2}\right) = -\frac{9}{4} + \frac{9}{2} = -\frac{9}{4} + \frac{18}{4} = \frac{9}{4};$
$\left(-\frac{3}{2}, \frac{9}{4}\right); x = -\frac{3}{2}; \frac{9}{4}$

14. $-\frac{b}{2a} = -\frac{4}{2 \cdot (-1)} = 2; y = 4 \cdot 2 - 2^2 = 8 - 4 = 4; (2, 4); x = 2; 4$

15. $-\frac{b}{2a} = -\frac{-8}{2 \cdot (-1)} = -4; y = -(-4)^2 - 8 \cdot (-4) - 15 = -16 + 32 - 15 = 1; (-4, 1);$
$x = -4; y = 1$

16. $-\frac{b}{2a} = -\frac{-3}{2 \cdot (-6)} = -\frac{1}{4}; y = 1 - 3\left(-\frac{1}{4}\right) - 6\left(-\frac{1}{4}\right)^2 = 1 + \frac{3}{4} - \frac{3}{8} = \frac{8}{8} + \frac{6}{8} - \frac{3}{8} = \frac{11}{8};$
$\left(-\frac{1}{4}, \frac{11}{8}\right); x = -\frac{1}{4}; y = \frac{11}{8}$

17. $-\frac{b}{2a} = -\frac{1}{2 \cdot (-2)} = \frac{1}{4}; y = \frac{1}{4} - 2\left(\frac{1}{4}\right)^2 = \frac{1}{4} - \frac{1}{8} = \frac{1}{8}; \left(\frac{1}{4}, \frac{1}{8}\right); x = \frac{1}{4}; \frac{1}{8}$

18. $-\frac{b}{2a} = 0; y = 1 - \frac{1}{3} \cdot 0^2 = 1; (0, 1); x = 0; 1$

19.
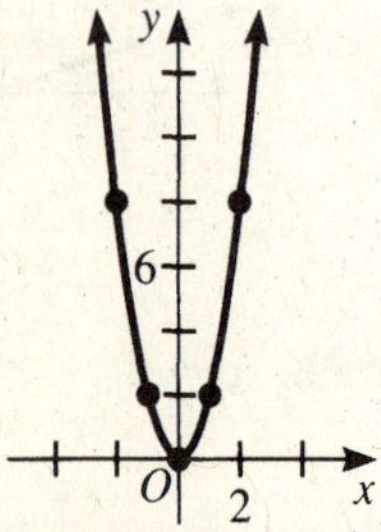

20.
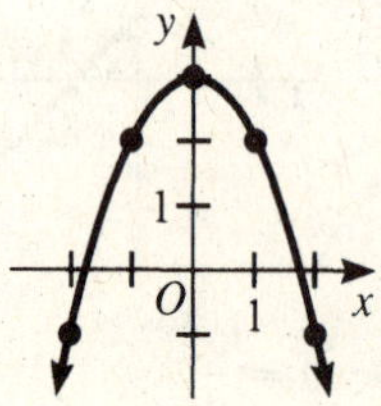

21.
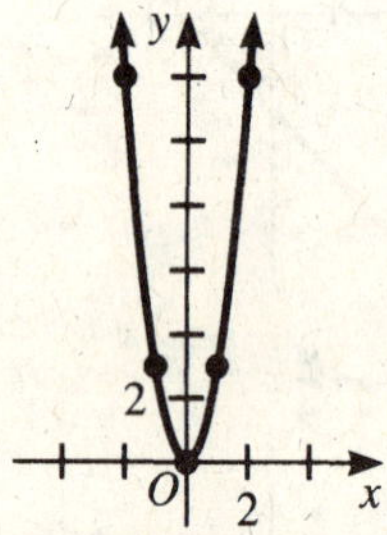

22.
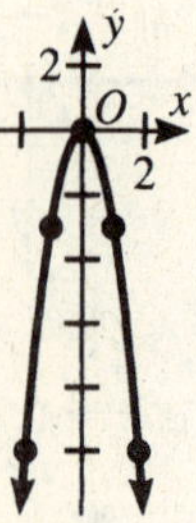

23.
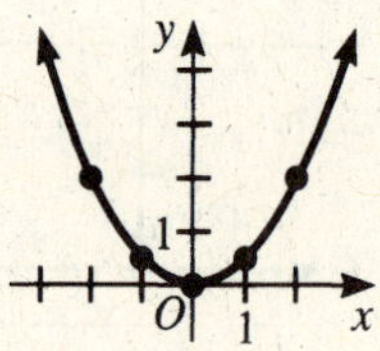

24.
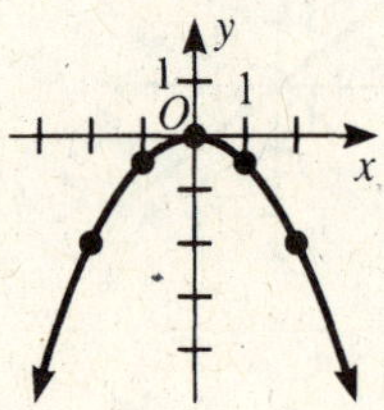

25. 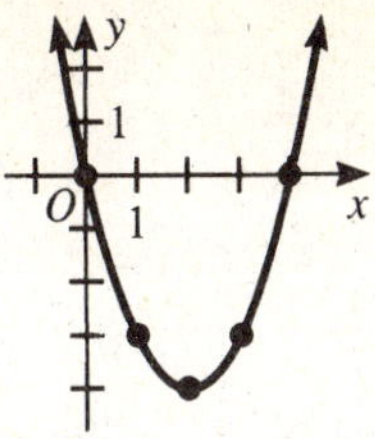 **26.** 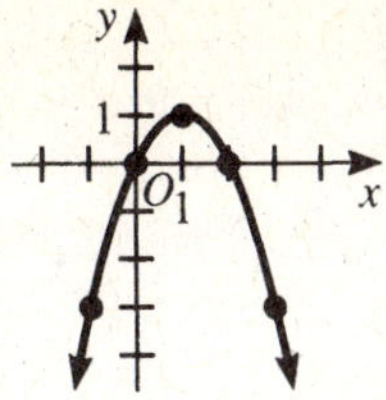 **27.** 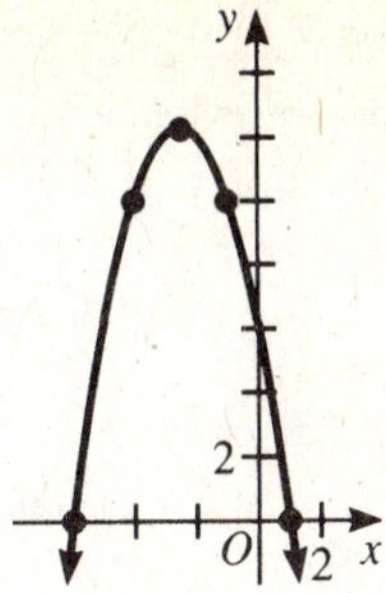

28. 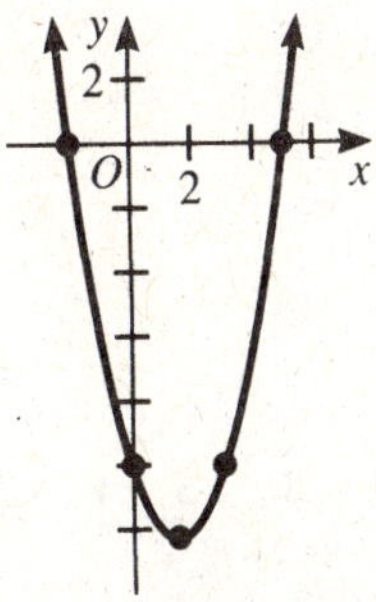 **29.** 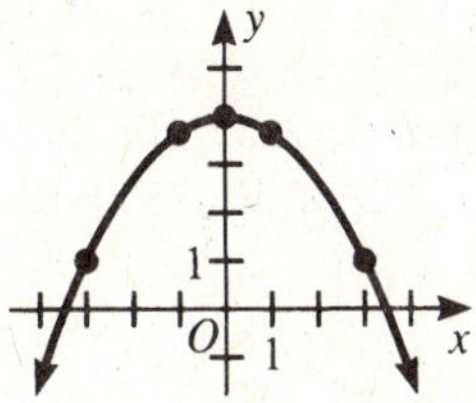 **30.** 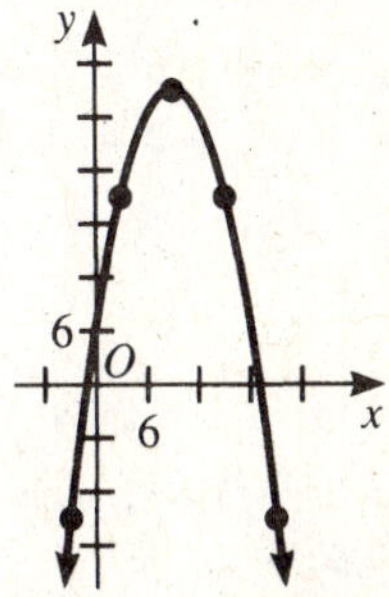

B **31. a.** 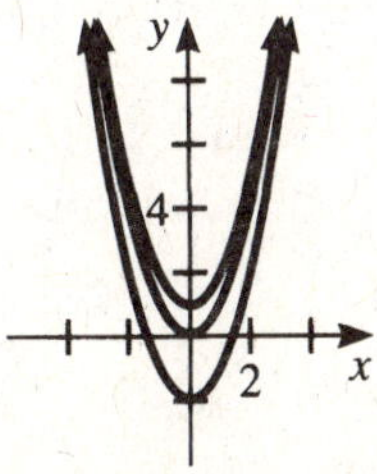

b. As c increases, the graph moves vertically upward; as c decreases, the graph moves vertically downward.

32. a. 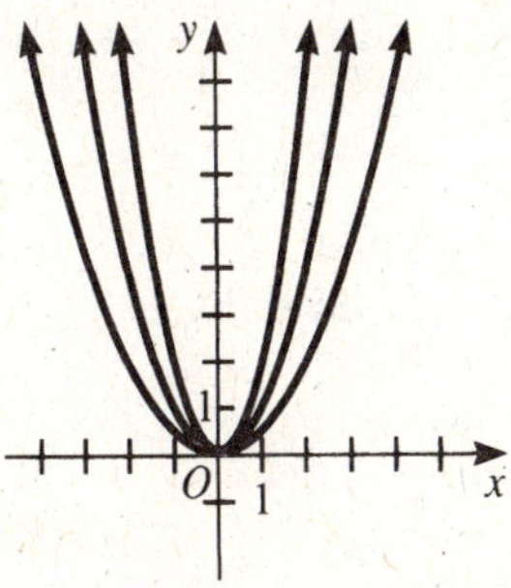

b. 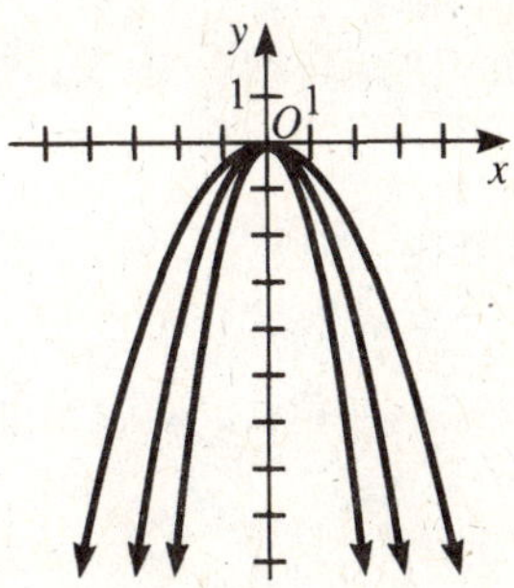

c. The parabola becomes a narrower curve.

33. a. $f(0) = 2 \cdot 0 + 10 = 10$ **b.** $2x + 10 = 0$; $2x = -10$; $x = -5$; $\{-5\}$

34. a. $f(0) = 3 \cdot 0 - 9 = -9$ **b.** $3x - 9 = 0$; $3x = 9$; $x = 3$; $\{3\}$

35. a. $f(0) = 4 \cdot 0 - 11 = -11$ **b.** $4x - 11 = 0$; $4x = 11$; $x = \frac{11}{4}$; $\left\{\frac{11}{4}\right\}$

36. a. $f(0) = 5 \cdot 0 + 8 = 8$ **b.** $5x + 8 = 0$; $5x = -8$; $x = -\frac{8}{5}$; $\left\{-\frac{8}{5}\right\}$

37. a. $f(0) = 0^2 - 6 \cdot 0 + 8 = 8$

b. $x^2 - 6x + 8 = 0$; $(x - 4)(x - 2) = 0$; $x = 4$ or $x = 2$; $\{4, 2\}$

38. a. $f(0) = 0^2 - 8 \cdot 0 + 15 = 15$

b. $x^2 - 8x + 15 = 0$; $(x - 5)(x - 3) = 0$; $x = 5$ or $x = 3$; $\{5, 3\}$

39. a. $f(0) = 0^2 + 8 \cdot 0 + 12 = 12$

b. $x^2 + 8x + 12 = 0$; $(x + 6)(x + 2) = 0$; $x = -6$ or $x = -2$; $\{-6, -2\}$

40. a. $f(0) = 0^2 + 9 \cdot 0 + 20 = 20$

b. $x^2 + 9x + 20 = 0$; $(x + 4)(x + 5) = 0$; $x = -4$ or $x = -5$; $\{-4, -5\}$

41. They are the x-intercepts. **42.** It is the y-intercept.

Page 387 • MIXED REVIEW EXERCISES

1. $H(0) = 3 \cdot 0^2 + 2 = 0 + 2 = 2$; $H(1) = 3 \cdot 1^2 + 2 = 3 + 2 = 5$;
$H(2) = 3 \cdot 2^2 + 2 = 12 + 2 = 14$; $R = \{2, 5, 14\}$

2. $f(-2) = 2 \cdot (-2) - 4 = -4 - 4 = -8$; $f(0) = 2 \cdot 0 - 4 = -4$;
$f(2) = 2 \cdot 2 - 4 = 4 - 4 = 0$; $R = \{-8, -4, 0\}$

3. $M(-1) = (-1)^3 + 7 = -1 + 7 = 6$; $M(1) = 1^3 + 7 = 1 + 7 = 8$;
$M(3) = 3^3 + 7 = 27 + 7 = 34$; $R = \{6, 8, 34\}$

4. $K(1) = 1^2 - 1 = 1 - 1 = 0$; $K(2) = 2^2 - 2 = 4 - 2 = 2$; $K(3) = 3^2 - 3 = 9 - 3 = 6$;
$R = \{0, 2, 6\}$

5. $7(x + 4)$ **6.** $x - 3$ **7.** $9x$ **8.** $\frac{1}{3}x - 5$

Page 388 • COMPUTER EXERCISES

```
10 PRINT "INPUT A,B,C ";
20 INPUT A,B,C
30 IF A = 0 THEN 130
40 X = -B/(2 * A)
50 Y = A * X * X + B * X + C
60 PRINT "VERTEX: ("; X; ", "; Y; ")"
70 PRINT "OTHER POINTS:"
80 FOR I = -3 TO 3
90 IF I = 0 THEN 110
100 PRINT "("; X + I; ", "; A * (X + I) ↑ 2 + B * (X + I) + C;")";
110 NEXT I
120 PRINT
130 END
```

1. VERTEX: (0, 0)
OTHER POINTS:
(−3, 9)(−2, 4)(−1, 1)(1, 1)(2, 4)(3, 9)

2. VERTEX: (1, 3)
OTHER POINTS:
(−2, −6)(−1, −1)(0, 2)(2, 2)(3, −1)(4, −6)

3. VERTEX: (2, −1)
OTHER POINTS: (−1, 8)(0, 3)(1, 0)(3, 0)(4, 3)(5, 8)

4. VERTEX: (2, −4)
OTHER POINTS:
(5, 5)(4, 0)(3, −3)(1, −3)(0, 0)(−1, 5)

5. VERTEX: $(-1, -5)$
OTHER POINTS:
$(-4, 13)(-3, 3)(-2, -3)(0, -3)(1, 3)(2, 13)$

6. VERTEX: $(-2.5, 12.25)$
OTHER POINTS:
$(-5.5, 3.25)(-4.5, 8.25)(-3.5, 11.25)(-1.5, 11.25)(-0.5, 8.25)(0.5, 3.25)$

Pages 389–390 • EXTRA

1. $D = \{2, 3\}; R = \{0, 3, 4, 6\}$; No. **2.** $D = \{1, 2, 3, 4\}; R = \{-1, 3, 5, 8\}$; Yes.

3. $D = \{1, 2, 3, 4\}; R = \{-3, -1, 0, 6\}$; Yes. **4.** $D = \{0, 5\}; R = \{0, 1, 2, 4\}$; No.

5. $D = \{-2, -1, 1, 2\}; R = \{1, 4\}$; Yes. **6.** $D = \{4, 9\}; R = \{-3, -2, 2, 3\}$; No.

7. $D = \{-1, 1, 2, 4\}; R = \{1, 2\}$; Yes. **8.** $D = \{0\}; R = \{-3, -1, 1, 2\}$; No.

9. $D = \{-2, 2, 4\}; R = \{-3, -1, 0, 1\}$; No.

10. $D = \{-3, -2, -1, 1, 2, 3\}; R = \{-3, -2, -1, 1, 2, 3\}$; Yes.

11. $D = \{-3, -2, -1, 0, 1, 2, 3\}; R = \{-2, -1, 0, 1\}$; Yes.

12. $D = \{-3, -2, -1, 1\}; R = \{-3, -1, 1, 2, 3\}$; No.

13.
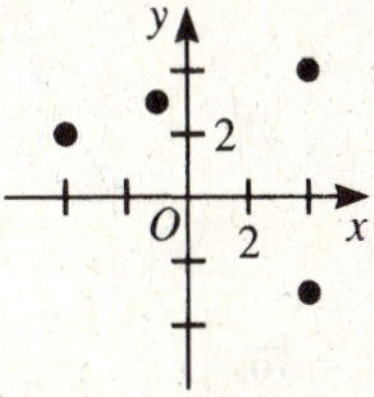

Not a function; $(4, 4)$ and $(4, -3)$ have the same first coordinate.

14.
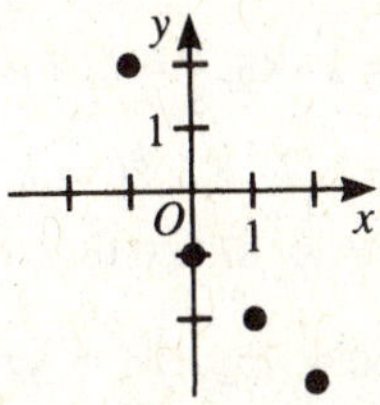

Yes it is a function.

15.
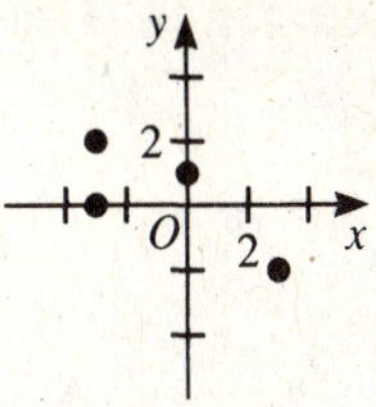

Not a function; $(-3, 0)$ and $(-3, 2)$ have the same first coordinate.

16.
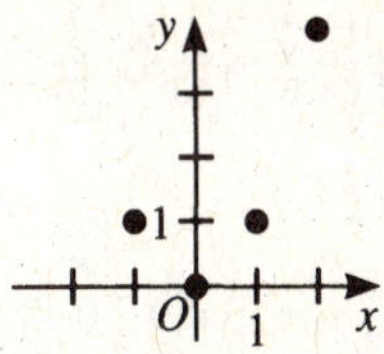

Yes, it is a function.

17.
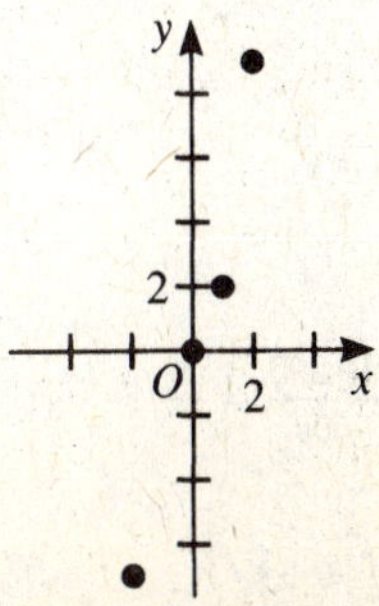

Yes, it is a function.

18.
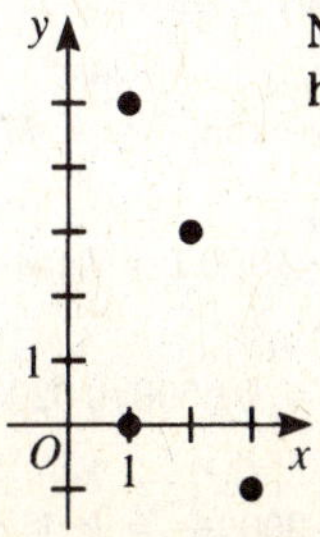

Not a function; $(1, 0)$ and $(1, 5)$ have the same first coordinate.

19. Yes. No. It is a relation because it is a set of ordered pairs. It is not a function because all nonzero x's are paired with $2y$'s; for example, (1, 1) and (1, −1) are both ordered pairs of the set.

20. Yes. No. They are a relation because they are a set of ordered pairs; they are not a function because for every x between −2 and 2, there are 2 ordered pairs belonging to the circle; for example, (0, 2) and (0, −2). The domain of the relation is the set of all real numbers between −2 and 2, including −2 and 2. The range is the same.

21. Relation, not a function (Example: (Joe Smith, 555-1212), (Joe Smith, 555-1213)).

22. Relation, not a function (Example: (Mrs. Tate, Mary), (Mrs. Tate, Melinda)).

23. Relation, function **24.** Relation, function

Pages 393–395 • WRITTEN EXERCISES

A **1.** $y = kx; 9 = k \cdot 54; \frac{1}{6} = k$ **2.** $y = kx; 6 = k \cdot 72; \frac{1}{12} = k$

3. $t = ks; -16 = k \cdot (-2); 8 = k$ **4.** $h = km; 112 = k \cdot (-16); -7 = k$

5. $W = km; 150 = k \cdot 6; 25 = k$ **6.** $P = kt; 210 = k \cdot 14; 15 = k$

7. $y = kx; 450 = k \cdot 6; 75 = k; y = 75 \cdot 10; y = 750$

8. $d = kz; 6 = k \cdot 48; \frac{1}{8} = k; d = \frac{1}{8} \cdot 20; d = 2\frac{1}{2}$

9. $h = ka; 425 = k \cdot 8.5; 50 = k; h = 50 \cdot 12; h = 600$

10. $r = kA; 14 = k \cdot 87.5; 0.16 = k; r = 0.16 \cdot 25; r = 4$

11. $\frac{9}{15} = \frac{y_2}{40}; 360 = 15y_2; 24 = y_2$ **12.** $\frac{y_1}{45} = \frac{100}{60}; 60y_1 = 4500; y_1 = 75$

13. $\frac{3}{3.6} = \frac{1}{x_2}; 3x_2 = 3.6; x_2 = 1.2$ **14.** $\frac{7}{x_1} = \frac{9}{7.65}; 53.55 = 9x_1; 5.95 = x_1$

15. $\frac{\frac{1}{6}}{\frac{1}{10}} = \frac{y_2}{\frac{2}{5}}; \frac{1}{6} \cdot \frac{2}{5} = \frac{1}{10}y_2; \frac{1}{15} = \frac{1}{10}y_2; \frac{2}{3} = y_2$

16. $\frac{y_1}{\frac{6}{5}} = \frac{\frac{1}{9}}{\frac{2}{3}}; \frac{2}{3}y_1 = \frac{1}{9} \cdot \frac{6}{5}; \frac{2}{3}y_1 = \frac{2}{15}; y_1 = \frac{2}{15} \cdot \frac{3}{2}; y_1 = \frac{1}{5}$

B **17. a.** $L = kh; 14 = k \cdot 20; 0.7 = k; L = 0.7h$ **b.** $\frac{14}{20} = \frac{L_2}{h_2}$

18. a. $H = km; 49 = k \cdot 1; 49 = k; H = 49m$ **b.** $\frac{49}{1} = \frac{H_2}{m_2}$

19. a. $M = kE; 28 = k \cdot 168; \frac{1}{6} = k; M = \frac{1}{6}E$ **b.** $\frac{28}{168} = \frac{M_2}{E_2}$

20. a. $m = kd; 1 = k \cdot 10; 0.1 = k; m = 0.1d$ **b.** $\frac{1}{10} = \frac{m_2}{d_2}$

21. a. $R = kL; 10.295 = k \cdot 500; 0.02059 = k; R = 0.02059\,L$ **b.** $\frac{10.295}{500} = \frac{R_2}{L_2}$

22. a. $V = kT; 5 = k \cdot 300; \frac{1}{60} = k; V = \frac{1}{60}T$ **b.** $\frac{5}{300} = \frac{V_2}{T_2}$

23. Since (x_1, y_1) and (x_2, y_2) are ordered pairs of a direct variation and neither is (0, 0), $\frac{x_1}{y_1}$ and $\frac{x_2}{y_2}$ are equivalent fractions, so $\frac{x_1}{y_1} = \frac{x_2}{y_2}$; $x_1y_2 = x_2y_1$; $y_2 = \frac{x_2y_1}{x_1}$; $\frac{y_2}{x_2} = \frac{y_1}{x_1}$, or $\frac{y_1}{x_1} = \frac{y_2}{x_2}$.

24. We know that $\frac{x_1}{y_1} = \frac{x_2}{y_2}$ (see Ex. 23); $x_1y_2 = x_2y_1$; $x_1 = \frac{x_2y_1}{y_2}$; $\frac{x_1}{x_2} = \frac{y_1}{y_2}$, or $\frac{y_1}{y_2} = \frac{x_1}{x_2}$.

Pages 395–396 • PROBLEMS

A **1.** Let w be the wage earned in 18 h. $\frac{100}{5} = \frac{w}{18}$; $1800 = 5w$; $360 = w$. The employee earns \$360 in 18 h.

2. Let g be the gallons of gas used on a 35-hour trip. $\frac{15}{3} = \frac{g}{35}$; $525 = 3g$; $175 = g$. The car uses 175 gal of gasoline in 35 h.

3. Let x be the payment for 1500 words. $\frac{720}{1200} = \frac{x}{1500}$; $1{,}080{,}000 = 1200x$; $900 = x$. The magazine pays \$900 for a 1500-word article.

4. Let d be the distance traveled in 7 h. $\frac{168}{4} = \frac{d}{7}$; $1176 = 4d$; $294 = d$. The truck will travel 294 mi in 7 h.

5. Let t be the time to type a 935-word essay. $\frac{275}{5} = \frac{935}{t}$; $275t = 4675$; $t = 17$. It will take 17 min to type 935 words.

6. Let x be the force when current is 65 A. $\frac{288}{32} = \frac{x}{65}$; $18{,}720 = 32x$; $585 = x$. The force is 585 V when the current is 65 A.

7. Let A be the area covered in 24 h. $\frac{52}{8} = \frac{A}{24}$; $1248 = 8A$; $156 = A$. The painter will cover 156 m^2 in 24 h.

8. Let x be the number of servings of chili made from 30 lb of beef. $\frac{110}{20} = \frac{x}{30}$; $3300 = 20x$; $165 = x$. The restaurant can make 165 servings of chili from 30 lb of ground beef.

9. Let x be the mass that stretches the spring 22 cm. $\frac{25}{10} = \frac{x}{22}$; $550 = 10x$; $55 = x$. A 55 g mass will stretch the spring 22 cm.

10. Let x be the units of chlorine needed for 3750 L of water. $\frac{5}{2500} = \frac{x}{3750}$; $18{,}750 = 2500x$; $7.5 = x$. 7.5 units of chlorine are needed for 3750 L of water.

B **11.** Let c be the degrees on thermometer C. $\frac{66.6}{180} = \frac{c}{100}$; $6660 = 180c$; $37 = c$. Thermometer C reads 37°.

12. Let d be the distance of the return trip. $\frac{216.7}{220} = \frac{453.1}{220 + d}$; $216.7(220 + d) = 99{,}682$; $47{,}674 + 216.7d = 99{,}682$; $216.7d = 52{,}008$; $d = 240$. The detour was actually $240 - 220 = 20$ mi.

13. Let w = width of the piece of land and l = length of the piece of land. $\frac{75}{1} = \frac{w}{11.5}$; $862.5 = w$; $\frac{75}{1} = \frac{l}{18.5}$; $1387.5 = l$; Area $= wl = 862.5 \cdot 1387.5 = 1{,}196{,}718.75$. The piece of land has an area of $1{,}196{,}718.75 \text{ m}^2$.

14. Let h be the height of the mast in the model. $\frac{8}{6} = \frac{h}{38}$; $304 = 6h$; $50\frac{2}{3} = h$. The model mast should be $50\frac{2}{3}$ in. or $4\frac{2}{9}$ ft tall.

C **15.** $C = k_1d$; $C = k_1(k_2r)$; $C = (k_1k_2)r$; k_1k_2 is a constant, so the circumference varies directly as the radius.

Page 396 • MIXED REVIEW EXERCISES

1. $(2p + 3)(3p + 1) = 6p^2 + 2p + 9p + 3 = 6p^2 + 11p + 3$

2. $(4x - 2)(x^2 + 3x - 6) = 4x^3 + 12x^2 - 24x - 2x^2 - 6x + 12 = 4x^3 + 10x^2 - 30x + 12$

3. $-3s(5 - 4s) = -15s + 12s^2$ **4.** $(3c + 2)(3c - 2) = (3c)^2 - 2^2 = 9c^2 - 4$

5. $(t + 2)(3t - 5) = 3t^2 - 5t + 6t - 10 = 3t^2 + t - 10$

6. $(7y - 3)(2y + 4) = 14y^2 + 28y - 6y - 12 = 14y^2 + 22y - 12$

7.

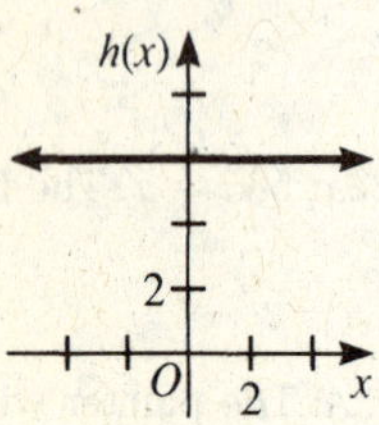

8.

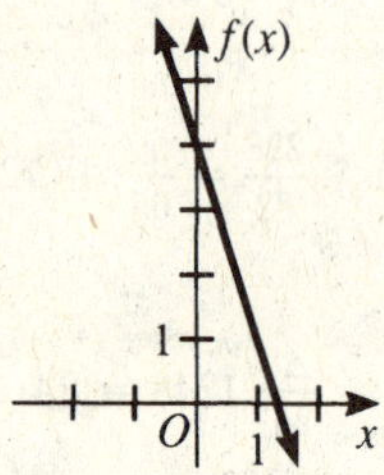

9.

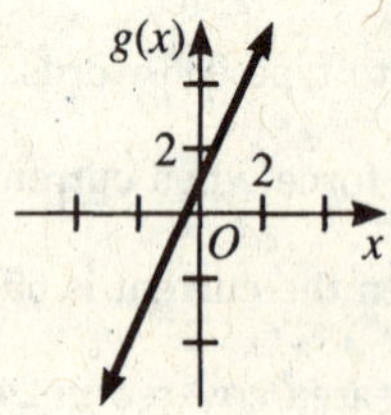

10.

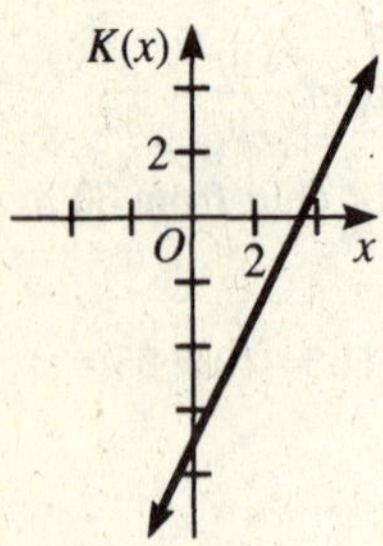

11.

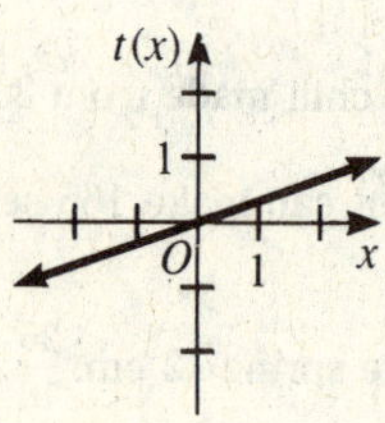

12.

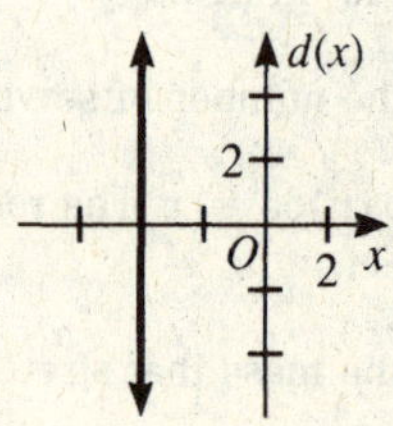

Pages 399–400 • WRITTEN EXERCISES

A **1.**

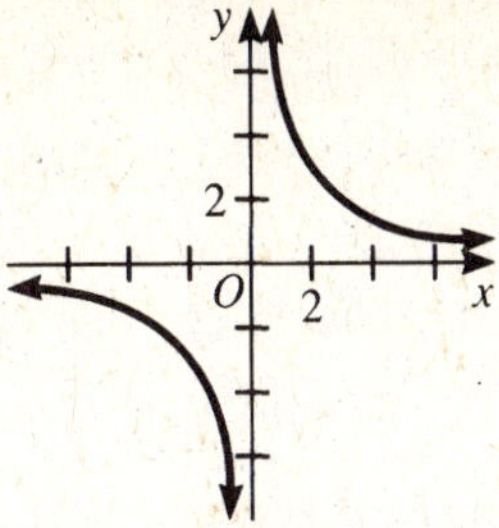

2.

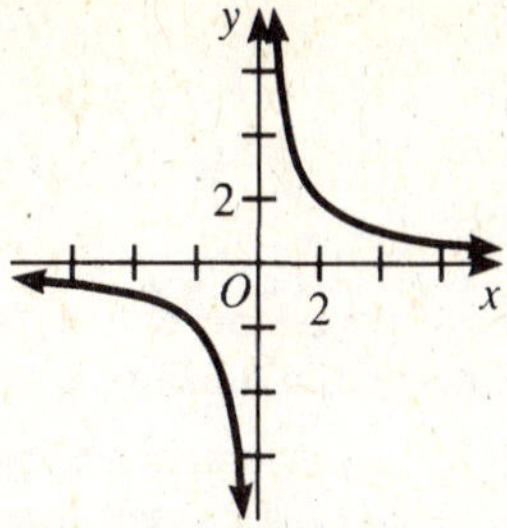

3.

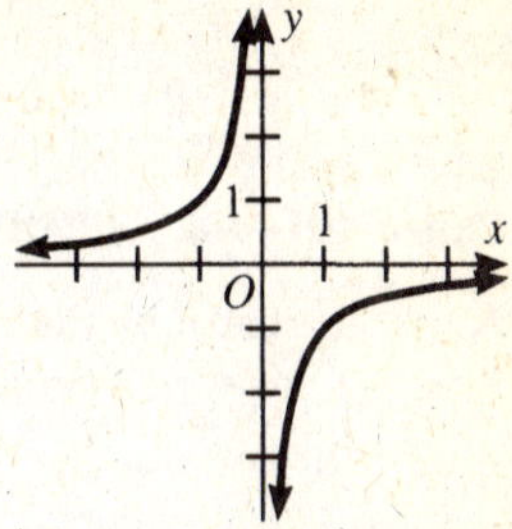

4.

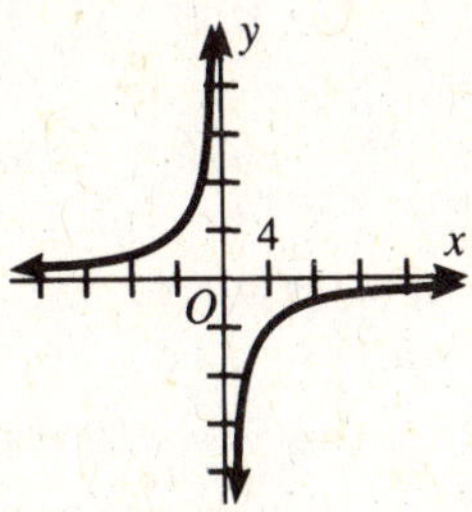

5.

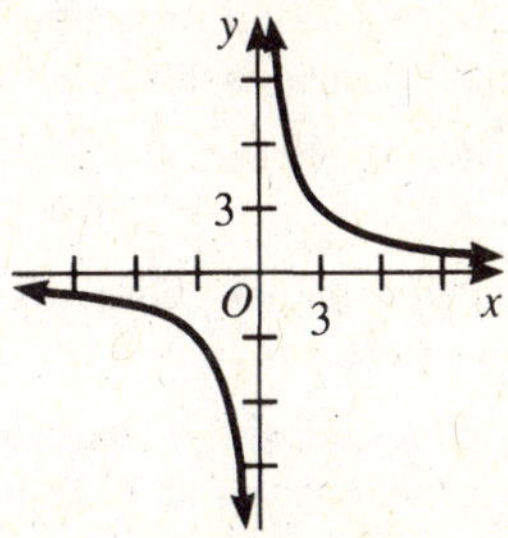

6.

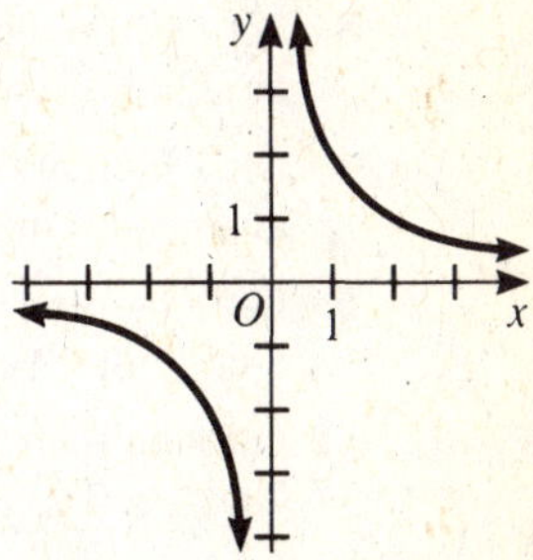

7.

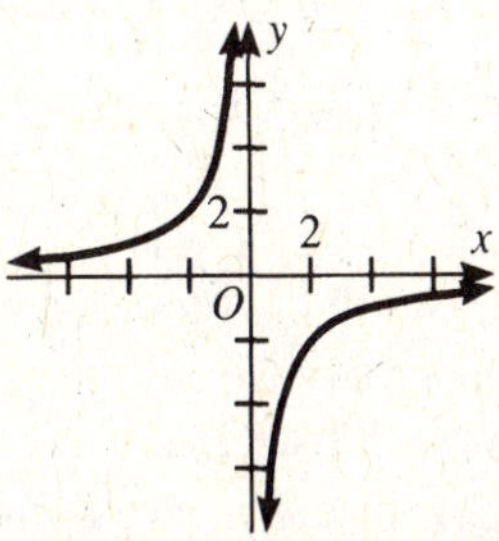

8.

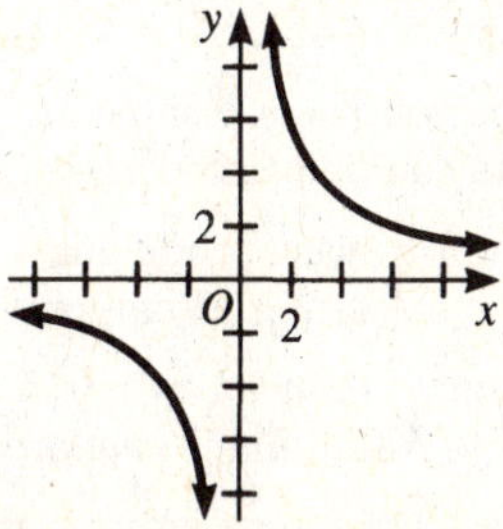

9. $m_1d_1 = m_2d_2$; $12 \cdot 30 = 9 \cdot d_2$; $360 = 9d_2$; $40 = d_2$

10. $m_1d_1 = m_2d_2$; $m_1 \cdot 5 = 60 \cdot 9$; $5m_1 = 540$; $m_1 = 108$

11. $x_1y_1 = x_2y_2$; $4 \cdot 54 = 8 \cdot y_2$; $216 = 8y_2$; $27 = y_2$

12. $x_1y_1 = x_2y_2$; $32 \cdot 9 = x_2 \cdot 12$; $288 = 12x_2$; $24 = x_2$

13. $x_1y_1 = x_2y_2$; $x_1 \cdot 19.5 = 11.7 \cdot 10.5$; $19.5x_1 = 122.85$; $x_1 = 6.3$

14. $x_1y_1 = x_2y_2$; $10 \cdot y_1 = 8 \cdot \frac{1}{4}$; $10y_1 = 2$; $y_1 = \frac{1}{5}$

B **15. a.** $lw = k$; $18 \cdot 5 = k$; $90 = k$; $lw = 90$ **b.** $\frac{18}{w_2} = \frac{l_2}{5}$

16. a. $fl = k$; $1.5 \cdot 2 = k$; $3 = k$; $fl = 3$ **b.** $\frac{1.5}{l_2} = \frac{f_2}{2}$

17. a. $fl = k$; $2.5 \cdot 0.60 = k$; $1.5 = k$; $fl = 1.5$ **b.** $\frac{2.5}{l_2} = \frac{f_2}{0.60}$

18. a. $rt = k$; $52 \cdot 3 = k$; $156 = k$; $rt = 156$ **b.** $\frac{52}{t_2} = \frac{r_2}{3}$

19. **a.** $PV = k$; $725 \cdot 465 = k$; $337{,}125 = k$; $PV = 337{,}125$ **b.** $\frac{725}{V_2} = \frac{P_2}{465}$

20. **a.** $IR = k$; $0.25 \cdot 18 = k$; $4.5 = k$; $IR = 4.5$ **b.** $\frac{0.25}{R_2} = \frac{I_2}{18}$

Pages 401–402 • PROBLEMS

A

1. Let p be the number of people needed to finish in 8 weeks. $18 \cdot 4 = 8p$; $72 = 8p$; $9 = p$. 9 people are needed to finish in 8 weeks.

2. Let p be the number of people needed to make the cost \$60 each. $3 \cdot 100 = p \cdot 60$; $300 = 60p$; $5 = p$. 5 people are needed to make the cost \$60 each.

3. Let l be the length of the second rectangle. $36 \cdot 28 = l \cdot 21$; $1008 = 21l$; $48 = l$. The length of the second rectangle is 48 cm.

4. Let r be the loser's average speed. $45 \cdot 9.6 = 48 \cdot r$; $432 = 48r$; $9 = r$. The loser averaged 9 m/s.

5. Let r be the r/min of the smaller pulley. $15 \cdot 250 = 5 \cdot r$; $3750 = 5r$; $750 = r$. The 5-cm pulley revolves at 750 r/min.

6. Let r be the rotational speed of the larger gear. $42 \cdot 1200 = 72 \cdot r$; $50{,}400 = 72r$; $700 = r$. The larger gear revolves at 700 r/min.

7. Let s be the space between chairs on a 40-chair lift. $10 \cdot 32 = 40 \cdot s$; $320 = 40s$; $8 = s$. The chairs are 8 m apart.

8. Let p be the number of plants planted 15 cm apart. $20 \cdot 75 = 15 \cdot p$; $1500 = 15p$; $100 = p$. 100 plants can be planted 15 cm apart.

9. Let l be the length of a string producing a tone of 516 Hz. $25.8 \cdot 440 = l \cdot 516$; $11{,}352 = 516l$; $22 = l$. A string 22 cm long produces a 516 Hz tone.

10. Let f be the frequency produced by a 0.55 mm string. $0.50 \cdot 440 = 0.55 \cdot f$; $220 = 0.55f$; $400 = f$. A 0.55 mm string produces a tone of 400 Hz.

11. Let d be the distance Sara must sit from the support. $106 \cdot d = 156 \cdot 6$; $106d = 936$; $d \approx 8.8$. Sara must sit about 8.8 ft from the support.

12. Let m be the mass at the other end of the bar. $350 \cdot 15 = m \cdot 175$; $5250 = 175m$; $30 = m$. A mass of 30 kg at the end of the bar will balance the boulder.

B

13. Let x be the distance of the fulcrum from the 18-kg mass. $18 \cdot x = 35(1 - x)$; $18x = 35 - 35x$; $53x = 35$; $x \approx 0.66$. The fulcrum should be placed about 0.66 m from the 18-kg mass.

14. Let x be the distance from the fulcrum to the log. $300x = 500(x - 1)$; $300x = 500x - 500$; $-200x = -500$; $x = 2.5$. The fulcrum is 2.5 m from the log.

15. Let f_1 and l_1 be the frequency and wavelength of the lower note and let f_2 and l_2 be the frequency and wavelength of the higher note. $f_1 l_1 = f_2 l_2$; $f_1 l_1 = (2f_1)l_2$; $\frac{1}{2}l_1 = l_2$. The wavelength of the higher note is $\frac{1}{2}$ the wavelength of the lower note.

16. Let f_1, l_1 and f_2, l_2 be the frequency and wavelength of the two notes. $f_1 l_1 = f_2 l_2$; $f_1 l_1 = f_2\left(\frac{3}{2}l_1\right)$; $\frac{2}{3}f_1 = f_2$. The frequency is $\frac{2}{3}$ as fast.

17. Let p be the length of the pipe. $440 \cdot 2p = 335$; $p \approx 0.4$. The pipe is about 0.4 m.

18. Let f be the frequency of the sound produced by the pipe. $f \cdot (4 \cdot 2) = 335$; $f \approx 42$. The pipe produces a sound with frequency about 42 cycles/s.

Page 402 • MIXED REVIEW EXERCISES

1. $2x + 3y = 3$; $3y = -2x + 3$; $y = -\frac{2}{3}x + 1$; $2x + 3y = 6$; $3y = -2x + 6$; $y = -\frac{2}{3}x + 2$; The lines are parallel because they have the same slope but different y-intercepts.

2. $2x + 10y = 3$; $10y = -2x + 3$; $y = -0.2x + 0.3$; $x + 5y = 2$; $5y = -x + 2$; $y = -0.2x + 0.4$; The lines are parallel because they have the same slope but different y-intercepts.

3. $x - y = 5$; $-y = -x + 5$; $y = x - 5$; $y - x = 5$; $y = x + 5$; The lines are parallel because they have the same slope but different y-intercepts.

4. $-4x + 6y = 1$; $6y = 4x + 1$; $y = \frac{2}{3}x + \frac{1}{6}$; $2x - 3y = 2$; $-3y = -2x + 2$; $y = \frac{2}{3}x - \frac{2}{3}$; The lines are parallel because they have the same slope but different y-intercepts.

5. $t = ks$; $24 = k \cdot (-4)$; $24 = -4k$; $-6 = k$

6. $y = kx$; $14 = k \cdot 70$; $\frac{1}{5} = k$ **7.** $m = kn$; $63 = k \cdot 7$; $9 = k$

Page 402 • COMPUTER EXERCISES

```
10 PRINT "HOW MANY POINTS";
20 INPUT N
30 FOR I = 1 TO N
40 PRINT "INPUT X, Y";
50 INPUT A, B
60 LET X(I) = A
70 LET Y(I) = B
80 NEXT I
90 FOR I = 2 TO N
100 IF X(I - 1) * Y(I) = Y(I - 1) * X(I) THEN 130
110 PRINT "NOT DIRECT"
120 GOTO 160
130 NEXT I
140 PRINT "DIRECT"
150 GOTO 220
160 FOR I = 2 TO N
170 IF X(I - 1) * Y(I - 1) = X(I) * Y(I) THEN 200
180 PRINT "NOT INVERSE"
190 GOTO 220
200 NEXT I
210 PRINT "INVERSE"
220 END
```

1. DIRECT **2.** NOT DIRECT **3.** NOT DIRECT
NOT INVERSE **4.** NOT INVERSE

Page 403 • CALCULATOR KEY-IN

1. 21 **2.** 19.5 **3.** 155 **4.** 7.35 **5.** 0.36 **6.** 1.8 **7.** 19.2 **8.** 38.4 **9.** 5.2 **10.** 0.6 **11.** 9 **12.** 7.2

Page 404 • READING ALGEBRA

1. Let d = number of dimes and $157 - d$ = number of quarters. $10d + 25(157 - d) = 2650$; $10d + 3925 - 25d = 2650$; $-15d + 3925 = 2650$; $-15d = -1275$; $d = 85$; $157 - d = 72$; There are 85 dimes and 72 quarters.

2. Let w = width of the rectangle and $2w + 4$ = length. $2(w + 2w + 4) = 56$; $2(3w + 4) = 56$; $6w + 8 = 56$; $6w = 48$; $w = 8$; $2w + 4 = 20$; The rectangle measures 20 cm by 8 cm.

3. Let x be the speed of the second plane. $1100 \cdot 4 + x \cdot 4 = 7600$; $4400 + 4x = 7600$; $4x = 3200$; $x = 800$; The second plane is traveling at 800 km/h.

4. Let d be the distance between the two towns. $\frac{d}{75} = \frac{d}{90} + \frac{2}{15}$; $450\left(\frac{d}{75}\right) = 450\left(\frac{d}{90} + \frac{2}{15}\right)$; $6d = 5d + 60$; $d = 60$; The towns are 60 km apart.

5. Let w = width of the rectangle and $w + 5$ = length. $w(w + 5) + 420 = 2(w + 5) \cdot 3w$; $w^2 + 5w + 420 = 6w^2 + 30w$; $0 = 5w^2 + 25w - 420$; $0 = 5(w^2 + 5w - 84)$; $0 = w^2 + 5w - 84$; $0 = (w + 12)(w - 7)$; $w = -12$ (reject) or $w = 7$; The original dimensions of the rectangle are 7 cm by 12 cm.

6. Let x be the first number and $11 - x$ be the second number. $x^2 + (11 - x)^2 = 65$; $x^2 + 121 - 22x + x^2 = 65$; $2x^2 - 22x + 56 = 0$; $2(x^2 - 11x + 28) = 0$; $x^2 - 11x + 28 = 0$; $(x - 7)(x - 4) = 0$; $x = 7$ or $x = 4$; The numbers are 7 and 4.

7. Let x be the length of the edges of the smaller cube. $x^3 = (x + 2)^3 - 152$; $x^3 = (x + 2)(x^2 + 4x + 4) - 152$; $x^3 = x^3 + 4x^2 + 4x + 2x^2 + 8x + 8 - 152$; $x^3 = x^3 + 6x^2 + 12x - 144$; $0 = 6(x^2 + 2x - 24)$; $0 = x^2 + 2x - 24$; $0 = (x + 6)(x - 4)$; $x = -6$ (reject) or $x = 4$; The smaller cube is 4 cm × 4 cm × 4 cm and the larger cube is 6 cm × 6 cm × 6 cm.

8. Let $7x$ be the number of new players and $4x$ the number of returning players. $4x + 7x = 440$; $11x = 440$; $x = 40$; $7x = 280$; There were 280 new players.

Pages 405–406 • CHAPTER REVIEW

1. a; $3(1) - 2(-1) = 3 + 2 = 5$ **2.** b; $2(0) + 3(3) = 0 + 9 = 9$; $2(3) + 3(1) = 6 + 3 = 9$

3. b **4.** c **5.** c; $\frac{6 - 4}{-4 - 4} = \frac{2}{-8} = -\frac{1}{4}$. **6.** b **7.** d **8.** a

9. c; $y = -\frac{4}{3}x + b$; $-3 = -\frac{4}{3}(12) + b$; $-3 = -16 + b$; $13 = b$; $y = -\frac{4}{3}x + 13$

10. c; slope $= \frac{0 - (-7)}{-7 - 0} = \frac{7}{-7} = -1$; y-intercept $= (0, -7)$; $y = -x - 7$; $x + y = -7$

11. d **12.** d; $G(0) = 4 - 8 \cdot 0 = 4 - 0 = 4$ **13.** c; $F(-1) = 5 - (-1)^2 = 5 - 1 = 4$

14. a; $-\frac{b}{2a} = -\frac{8}{2 \cdot 1} = -4$; $y = (-4)^2 + 8(-4) + 3 = 16 - 32 + 3 = -13$; vertex $= (-4, -13)$; least value $= -13$.

15. c; $-\frac{b}{2a} = -\frac{2}{2 \cdot (-1)} = 1$; $y = 2(1) - (1)^2 = 2 - 1 = 1$; vertex $= (1, 1)$, greatest value $= 1$.

16. a; $y = kx$; $95 = k \cdot 19$; $5 = k$ **17.** a; $75 \cdot 30 = 18x$; $2250 = 18x$; $125 = x$

Pages 407–408 • CHAPTER TEST

1. $3(-2) - 2(3) = -6 - 6 = -12$, No; $3(4) - 2(0) = 12 - 0 = 12$, Yes

2. $7(-1) + 5(-2) = -7 - 10 = -17$, No; $7(-4) + 5(-5) = -28 - 25 = -53$, No

3. $2(1) + 5(2) = 2 + 10 = 12$; $2(6) + 5(0) = 12 + 0 = 12$; The solutions are (1, 2) and (6, 0).

4. $2(0) + 3(5) = 0 + 15 = 15$; $2(3) + 3(3) = 6 + 9 = 15$; $2(6) + 3(1) = 12 + 3 = 15$; The solutions are (0, 5), (3, 3), and (6, 1).

5–8.

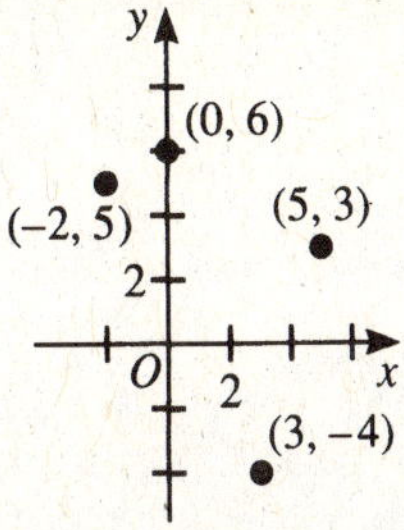

9.

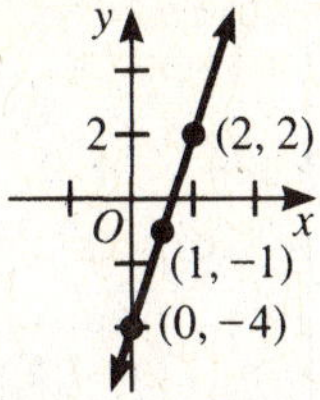

10.

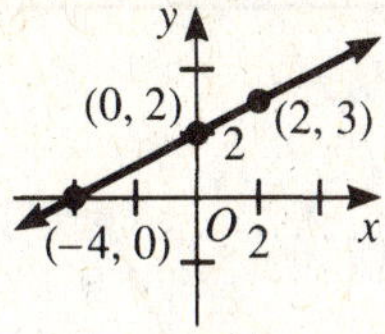

11. $m = \frac{-7 - 0}{8 - (-6)} = \frac{-7}{14} = -\frac{1}{2}$ **12.** $m = \frac{7 - 7}{8 - (-5)} = 0$ **13.** $m = \frac{-5 - 5}{9 - 9} = \frac{-10}{0}$: no slope

14. $m = \frac{-1 - 4}{3 - (-2)} = \frac{-5}{5} = -1$ **15.** $m = -3$ **16.** $x + y = 0$; $y = -x$; $m = -1$

17. $3x - 5y = 15$; $-5y = -3x + 15$; $y = \frac{3}{5}x - 3$; $m = \frac{3}{5}$

18. $2x + 4y = 1$; $4y = -2x + 1$; $y = -\frac{1}{2}x + \frac{1}{4}$; $m = -\frac{1}{2}$ **19.** slope = 3; y-intercept = 2

20. slope = $-\frac{2}{3}$; y-intercept = 4

21. slope = $-\frac{2}{3}$; y-intercept = -3; This line is parallel to the line in Ex. 20.

22. slope = $\frac{9}{4}$; y-intercept = 6

23. $5x - y = 1$,
$-y = -5x + 1$,
$y = 5x - 1$

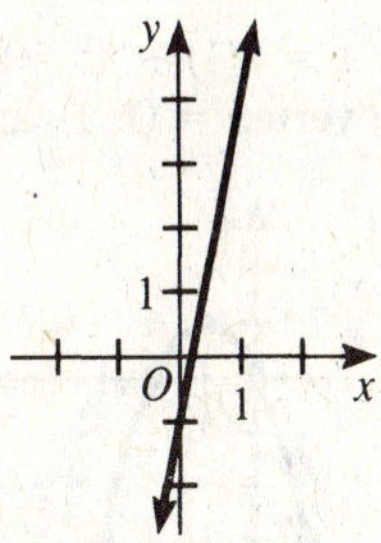

24. $-x + 3y + 6 = 0$,
$3y = x - 6$,
$y = \frac{1}{3}x - 2$

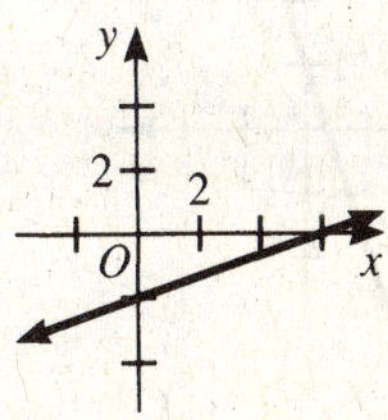

25. $2y = 7$,
$y = \frac{7}{2}$

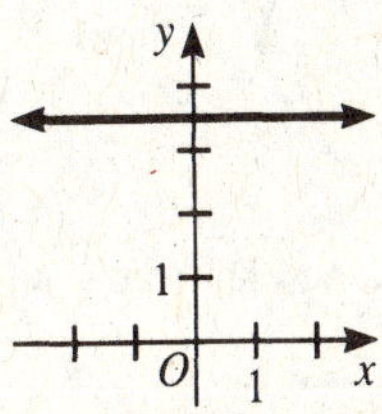

26. $x + 2y - 6 = 0$,
$2y = -x + 6$,
$y = -\frac{1}{2}x + 3$

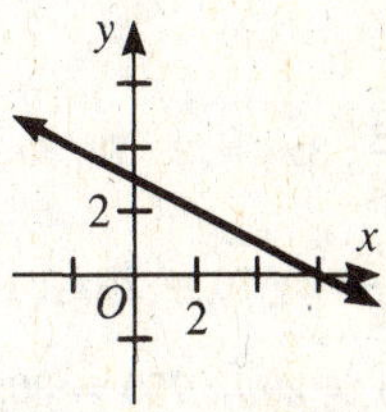

27. $y = -3x + b$; $-1 = -3(0) + b$; $-1 = b$; $y = -3x - 1$

28. $y = \frac{5}{2}x + b$; $7 = \frac{5}{2}(-6) + b$; $7 = -15 + b$; $22 = b$; $y = \frac{5}{2}x + 22$

29. $y = b$; $y = -\frac{1}{4}$ **30.** $y = \frac{3}{4}x + b$; $-4 = \frac{3}{4}(-3) + b$; $-4 = \frac{-9}{4} + b$; $-\frac{7}{4} = b$; $y = \frac{3}{4}x - \frac{7}{4}$

31. $m = \frac{10 - (-6)}{4 - (-4)} = \frac{16}{8} = 2$; $y = 2x + b$; $10 = 2(4) + b$; $10 = 8 + b$; $2 = b$; $y = 2x + 2$

32. $D = \{1960, 1965, 1970, 1975, 1980, 1985\}$; $R = \{\$29, \$41, \$66, \$156, \$345, \$359\}$

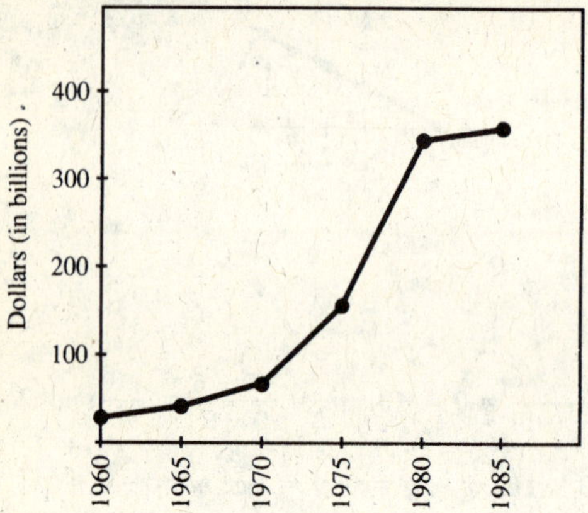

33. $F(-1) = 2(-1) - 2 = -2 - 2 = -4$; $F(0) = 2(0) - 2 = 0 - 2 = -2$;
$F(2) = 2(2) - 2 = 4 - 2 = 2$; $R = \{-4, -2, 2\}$

34. $g(-1) = (-1)^2 - 1 = 1 - 1 = 0$; $g(0) = 0^2 - 1 = -1$; $g(1) = 1^2 - 1 = 0$; $R = \{-1, 0\}$

35. **a.** $f(3) = \frac{2}{3}(3) - 1 = 2 - 1 = 1$ **b.** $f(-3) = \frac{2}{3}(-3) - 1 = -2 - 1 = -3$

36. **a.** $f(3) = \frac{2(3) - 1}{3} = \frac{6 - 1}{3} = \frac{5}{3}$ **b.** $f(-3) = \frac{2(-3) - 1}{3} = \frac{-6 - 1}{3} = -\frac{7}{3}$

37.

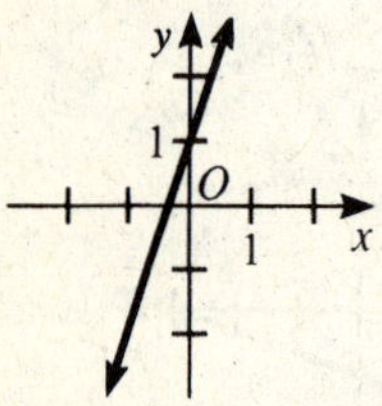

38.

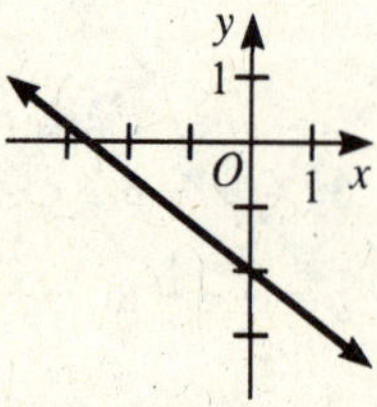

39. $-\frac{b}{2a} = -\frac{2}{2(-1)} = 1$;

$y = 2(1) - 1^2 = 2 - 1 = 1$
vertex = (1, 1); axis: $x = 1$

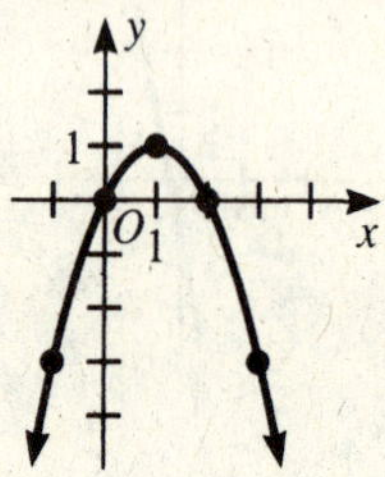

40. $\frac{3}{72} = \frac{2}{x_2}$; $3x_2 = 2 \cdot 72$; $3x_2 = 144$; $x_2 = 48$

41. $\frac{70}{x_1} = \frac{25}{2}$; $2 \cdot 70 = 25x_1$; $140 = 25x_1$; $5.6 = x_1$ **42.** $20 \cdot 5 = x_2 \cdot 50$; $100 = 50x_2$; $2 = x_2$

43. $10 \cdot 75 = 30y_2$; $750 = 30y_2$; $25 = y_2$

Page 409 • CUMULATIVE REVIEW

1. $(7p - 2r) - (-3p + 4r) = 7p - 2r + 3p - 4r = 10p - 6r$

2. $(8b - 4c)(2b + 3c) = 16b^2 + 24bc - 8bc - 12c^2 = 16b^2 + 16bc - 12c^2$

3. $3y^3z(10y^2 + 4yz - 7z^3) = 30y^5z + 12y^4z^2 - 21y^3z^4$

4. $(2x^3y^2z^4)^4 \div (4x^2yz^3)^2 = \dfrac{2^4x^{12}y^8z^{16}}{4^2x^4y^2z^6} = x^8y^6z^{10}$

5. $(3m - 5n)^2 = (3m)^2 - 2(3m)(5n) + (5n)^2 = 9m^2 - 30mn + 25n^2$

6. $(3.2 \times 10^{-8})(1.5 \times 10^{12}) = (3.2)(1.5) \times 10^{-8+12} = 4.8 \times 10^4$

7. $20x^2 + 13x - 15 = (5x - 3)(4x + 5)$ **8.** $4y^2 - 12y + 9 = (2y - 3)^2$

9. $3z^2 - 21z - 24 = 3(z^2 - 7z - 8) = 3(z - 8)(z + 1)$

10. $\dfrac{b^2 + 2b + 1}{b^2 - 2b - 3} = \dfrac{(b + 1)^2}{(b + 1)(b - 3)} = \dfrac{b + 1}{b - 3}$

11. $\dfrac{x^2 + 4x + 3}{x + 1} \cdot \dfrac{(x^2 - 1)}{2x^2 + x - 1} = \dfrac{(x + 3)(x + 1)}{x + 1} \cdot \dfrac{(x + 1)(x - 1)}{(2x - 1)(x + 1)} = \dfrac{(x + 3)(x - 1)}{2x - 1}$

12. $\dfrac{3z^2 + 8z - 3}{3z^2 + 5z - 2} \div \dfrac{z^2 + 6z + 9}{z^3 - 4z} = \dfrac{(3z - 1)(z + 3)}{(3z - 1)(z + 2)} \cdot \dfrac{z(z^2 - 4)}{(z + 3)^2} = \dfrac{z(z + 2)(z - 2)}{(z + 2)(z + 3)} = \dfrac{z(z - 2)}{z + 3}$

13. $\dfrac{r}{r - 4} - \dfrac{r}{r + 4} - 5 = \dfrac{r(r + 4)}{(r - 4)(r + 4)} - \dfrac{r(r - 4)}{(r - 4)(r + 4)} - \dfrac{5(r + 4)(r - 4)}{(r - 4)(r + 4)} =$

$\dfrac{r^2 + 4r - r^2 + 4r - 5(r^2 - 16)}{(r - 4)(r + 4)} = \dfrac{-5r^2 + 8r + 80}{(r - 4)(r + 4)}$

14. $y = -3x + b; -2 = -3(2) + b; -2 = -6 + b; 4 = b; y = -3x + 4$

15. $m = \dfrac{5 - 2}{2 - 5} = \dfrac{3}{-3} = -1; y = -x + b; 2 = -5 + b; 7 = b; y = -x + 7$

16. $2x + y = 4; y = -2x + 4;$ slope $= -2; y = -2x + b; 3 = -2(1) + b; 3 = -2 + b; 5 = b;$
$y = -2x + 5$

17. $f(-3) = (-3)^2 + 2(-3) + 1 = 9 - 6 + 1 = 4; f(-2) = (-2)^2 + 2(-2) + 1 =$
$4 - 4 + 1 = 1; f(-1) = (-1)^2 + 2(-1) + 1 = 1 - 2 + 1 = 0; f(0) = 0^2 + 2(0) + 1 = 1;$
$f(1) = 1^2 + 2(1) + 1 = 1 + 2 + 1 = 4; f(2) = 2^2 + 2(2) + 1 = 4 + 4 + 1 = 9;$
$f(3) = 3^2 + 2(3) + 1 = 9 + 6 + 1 = 16; R = \{0, 1, 4, 9, 16\}$

18. $3x + 5y = 10,$
$5y = -3x + 10,$
$y = -\dfrac{3}{5}x + 2$

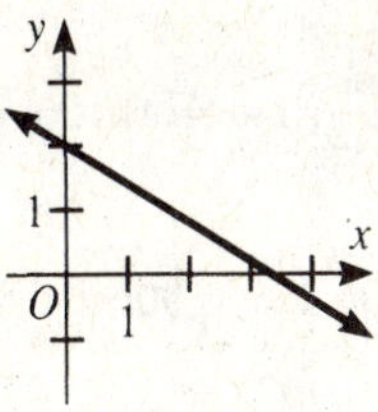

19.

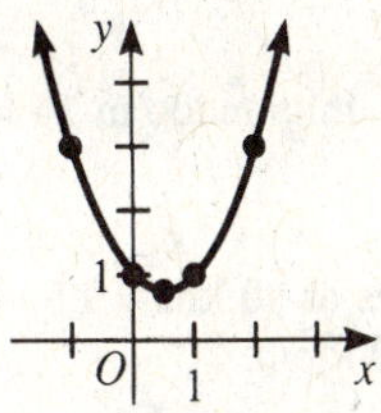

20. $-\dfrac{b}{2a} = -\dfrac{3}{2(2)} = -\dfrac{3}{4}; y = 2\left(-\dfrac{3}{4}\right)^2 + 3 \cdot \left(-\dfrac{3}{4}\right) + 4 = 2\left(\dfrac{9}{16}\right) - \dfrac{9}{4} + 4 = \dfrac{9}{8} - \dfrac{9}{4} + 4 =$

$\dfrac{9 - 18 + 32}{8} = \dfrac{23}{8};$ vertex $= \left(-\dfrac{3}{4}, \dfrac{23}{8}\right);$ axis: $x = -\dfrac{3}{4}$

21. $4(x + 3) = 3(3x - 7); 4x + 12 = 9x - 21; 12 = 5x - 21; 33 = 5x; \dfrac{33}{5} = x; \left\{\dfrac{33}{5}\right\}$

22. $2|x| + 1 = 5; 2|x| = 4; |x| = 2; x = 2$ or $x = -2; \{2, -2\}$

23. $\dfrac{y}{100} \cdot 50 = 27; \dfrac{y}{2} = 27; y = 54;$ 54% of 50 is 27; $\{54\}$

24. $9a^2 - 12a + 4 = 0; (3a - 2)^2 = 0; 3a - 2 = 0; 3a = 2; a = \dfrac{2}{3}; \left\{\dfrac{2}{3}\right\}$

25. $49t^2 - 36 = 0$; $(7t + 6)(7t - 6) = 0$; $7t + 6 = 0$; $7t = -6$; $t = -\frac{6}{7}$; $7t - 6 = 0$; $7t = 6$; $t = \frac{6}{7}$; $\left\{-\frac{6}{7}, \frac{6}{7}\right\}$

26. $10t^2 + t = 3$; $10t^2 + t - 3 = 0$; $(5t + 3)(2t - 1) = 0$; $5t + 3 = 0$; $5t = -3$; $t = -\frac{3}{5}$; $2t - 1 = 0$; $2t = 1$; $t = \frac{1}{2}$; $\left\{-\frac{3}{5}, \frac{1}{2}\right\}$

27. $\frac{1}{x + 1} + 1 = x + 2$; $(x + 1)\left(\frac{1}{x + 1} + 1\right) = (x + 1)(x + 2)$; $1 + x + 1 = x^2 + 3x + 2$; $x + 2 = x^2 + 3x + 2$; $0 = x^2 + 2x$; $0 = x(x + 2)$; $x = 0$ or $x = -2$; $\{0, -2\}$

28. $\frac{a + 2}{a - 1} = \frac{a + 1}{a + 4}$; $(a + 2)(a + 4) = (a + 1)(a - 1)$; $a^2 + 6a + 8 = a^2 - 1$; $6a + 8 = -1$; $6a = -9$; $a = -\frac{3}{2}$; $\left\{-\frac{3}{2}\right\}$

29. $\frac{2z}{z - 3} - \frac{1}{z + 1} = -1$; $(z + 1)(z - 3)\left(\frac{2z}{z - 3} - \frac{1}{z + 1}\right) = -1(z + 1)(z - 3)$;
$2z(z + 1) - (z - 3) = -(z^2 - 2z - 3)$; $2z^2 + 2z - z + 3 = -z^2 + 2z + 3$;
$2z^2 + z + 3 = -z^2 + 2z + 3$; $3z^2 - z = 0$; $z(3z - 1) = 0$; $z = 0$; $3z - 1 = 0$; $3z = 1$;
$z = \frac{1}{3}$; $\left\{0, \frac{1}{3}\right\}$

30. $m\%$ of $60 = 48$; $\frac{m}{100} \cdot 60 = 48$; $\frac{3m}{5} = 48$; $m = 80$; 80% of $60 = 48$; $\{80\}$

31. $2(x + 7) = 5(3x - 5)$; $2x + 14 = 15x - 25$; $14 = 13x - 25$; $39 = 13x$; $3 = x$; $\{3\}$

32. $\frac{1}{3}(m + 2) = m - 4$; $\frac{m}{3} + \frac{2}{3} = m - 4$; $3\left(\frac{m}{3} + \frac{2}{3}\right) = 3(m - 4)$; $m + 2 = 3m - 12$; $2 = 2m - 12$; $14 = 2m$; $7 = m$; $\{7\}$

33. Let n be the number of units made in 14 h. $\frac{10}{4} = \frac{n}{14}$; $140 = 4n$; $35 = n$; 35 units are manufactured in 14 h.

34. Let t be the time it takes at 90 km/h. $1.5 \cdot 84 = 90t$; $126 = 90t$; $1.4 = t$; It takes 1.4 h to travel the distance at 90 km/h.

Page 410 • MAINTAINING SKILLS

1. $25b^2c^3 + 15b^5c^4 = 5b^2c^3(5 + 3b^3c)$ **2.** $12m^3 - 15mn^2 - 8n^2$; prime

3. $9u^5v + 36u^4v^2 - 15u^4v^3 = 3u^4v(3u + 12v - 5v^2)$

4. $-vx^3 + 40x^2v^3 - v^6 = v(-x^3 + 40x^2v^2 - v^5)$

5. $-24x^7y^5 + 32x^6y^3 - 8x^2y^4 = 8x^2y^3(-3x^5y^2 + 4x^4 - y)$

6. $20m^6n^6 - 4m^6n^5 + 24m^5n^7 = 4m^5n^5(5mn - m + 6n^2)$

7. $z^3 - 121z = z(z^2 - 121) = z(z + 11)(z - 11)$

8. $27bc^2 - 12b = 3b(9c^2 - 4) = 3b(3c + 2)(3c - 2)$ **9.** $-x^6 + 49 = (7 + x^3)(7 - x^3)$

10. $9x^2 - 12x + 4 = (3x - 2)^2$ **11.** $25m^4 + 9$; prime

12. $1 - 16y^8 = (1 + 4y^4)(1 - 4y^4) = (1 + 4y^4)(1 + 2y^2)(1 - 2y^2)$

13. $16a^2 - 40a + 25 = (4a - 5)^2$ **14.** $4n^2 - 16n + 16 = 4(n^2 - 4n + 4) = 4(n - 2)^2$

15. $25x^2 + 30x + 9 = (5x + 3)^2$ **16.** $z^2 + 8z - 16$; prime

17. $n^2 + 11n + 18 = (n + 2)(n + 9)$ **18.** $y^2 - 9y + 18 = (y - 3)(y - 6)$

19. $-c^2 + 5cd + 14d^2 = -(c^2 - 5cd - 14d^2) = -(c - 7d)(c + 2d) = (7d - c)(c + 2d)$

20. $2d^2 + 18d - 72 = 2(d^2 + 9d - 36) = 2(d + 12)(d - 3)$

21. $x^2 + 10x + 21 = (x + 7)(x + 3)$

22. $4c^2 - 36c + 32 = 4(c^2 - 9c + 8) = 4(c - 8)(c - 1)$

23. $n^2 - 5np + 6p^2 = (n - 2p)(n - 3p)$ **24.** $7f + f^2 - 30 = (f + 10)(f - 3)$

25. $ry - 2sy + 3r - 6s = y(r - 2s) + 3(r - 2s) = (y + 3)(r - 2s)$

26. $2xy - y^2 + 6x - 3y = y(2x - y) + 3(2x - y) = (y + 3)(2x - y)$

27. $s^2 - 4t^2 - 12t - 9 = s^2 - (4t^2 + 12t + 9) = s^2 - (2t + 3)^2 =$
$[s + (2t + 3)][s - (2t + 3)] = (s + 2t + 3)(s - 2t - 3)$

28. $x^2 - 10x + 25 - y^2 = (x - 5)^2 - y^2 = (x - 5 + y)(x - 5 - y)$

29. $16a^2 - 9b^2 + 30b - 25 = 16a^2 - (9b^2 - 30b + 25) = 16a^2 - (3b - 5)^2 =$
$[4a + (3b - 5)][4a - (3b - 5)] = (4a + 3b - 5)(4a - 3b + 5)$

30. $m^2t - 5m^2 + 5t - 25 = m^2(t - 5) + 5(t - 5) = (m^2 + 5)(t - 5)$

31. $3b^2 + 2b - 5 = (3b + 5)(b - 1)$ **32.** $10n^2 + 3n - 1 = (5n - 1)(2n + 1)$

33. $6m^2 - 8m - 8 = 2(3m^2 - 4m - 4) = 2(3m + 2)(m - 2)$

34. $6a^2 + 7a - 3 = (2a + 3)(3a - 1)$

35. $25z^3 + 15z^2 + 2z = z(25z^2 + 15z + 2) = z(5z + 2)(5z + 1)$

36. $14 + 15y - 9y^2 = (2 + 3y)(7 - 3y)$

37. $-7y^2 - 20y + 3 = -(7y^2 + 20y - 3) = -(7y - 1)(y + 3) = (1 - 7y)(y + 3)$

38. $22n + 8n^2 - 6 = 2(4n^2 + 11n - 3) = 2(4n - 1)(n + 3)$

39. $12b^2 - 14b - 10 = 2(6b^2 - 7b - 5) = 2(3b - 5)(2b + 1)$

Page 411 • PREPARING FOR COLLEGE ENTRANCE EXAMS

1. A; Let x be the amount invested in bonds. $0.10x + 0.06(15{,}000 - x) + 160 = 0.10(2x) + 0.06(15{,}000 - 2x)$; $100[0.10x + 0.06(15{,}000 - x) + 160] = 100[0.10(2x) + 0.06(15{,}000 - 2x)]$; $10x + 90{,}000 - 6x + 16{,}000 = 20x + 90{,}000 - 12x$; $4x + 106{,}000 = 8x + 90{,}000$; $16{,}000 = 4x$; $4{,}000 = x$

2. A; $y = -\frac{1}{4}x + b$; $3 = -\frac{1}{4}(-4) + b$; $3 = 1 + b$; $2 = b$; $y = -\frac{1}{4}x + 2$; I: $-\frac{1}{4}(8) + 2 = -2 + 2 = 0$; II: $-\frac{1}{4}(-1) + 2 = \frac{1}{4} + 2 \neq -3$; III: $-\frac{1}{4}(-12) + 2 = 3 + 2 \neq -13$; I only.

3. D; $xy = 24$ is a hyperbola; but if x and y are sides of a rectangle, only positive values are allowed. So, the graph is one branch of a hyperbola.

4. A; Let x be the number of hours it takes to drain the tank. $\left(\frac{1}{x} - \frac{1}{8}\right)24 = 1$; $\frac{24}{x} - 3 = 1$; $x\left(\frac{24}{x} - 3\right) = x$; $24 - 3x = x$; $24 = 4x$; $6 = x$; If only the drain is open, it would take 6 h to drain the tank.

5. B; Let x be the number of minutes gained each day. $\frac{0.4}{15} = \frac{x}{1440}$; $576 = 15x$; $38.4 = x$; The watch gained 38.4 min a day.

6. A; $m_1 = \frac{-1-2}{-4-2} = \frac{-3}{-6} = \frac{1}{2}$; $y = \frac{1}{2}x + b_1$; $2 = \frac{1}{2}(2) + b_1$; $2 = 1 + b_1$; $1 = b_1$; $y = \frac{1}{2}x + 1$; $m_2 = \frac{3-6}{-3-6} = \frac{-3}{-9} = \frac{1}{3}$; $y = \frac{1}{3}x + 1$; $-\frac{1}{3}x + y = 1$; $-3\left(-\frac{1}{3}x + y\right) = -3$; $x - 3y = -3$

Pages 415–416 • WRITTEN EXERCISES

A **1.**

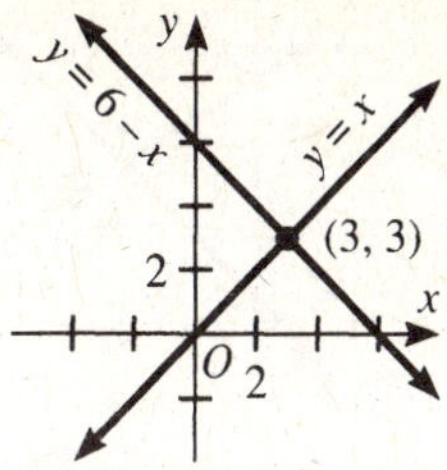

2.

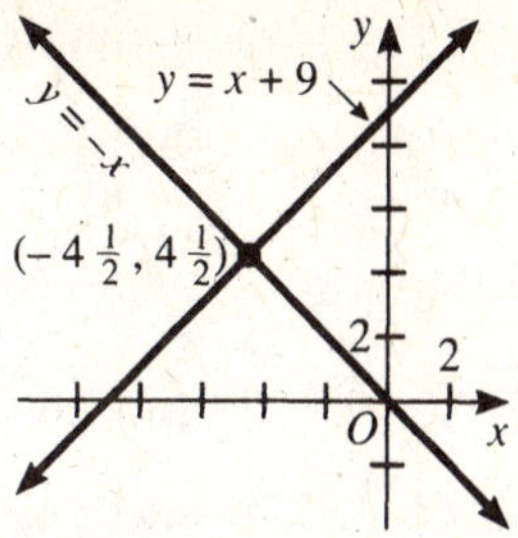

3.

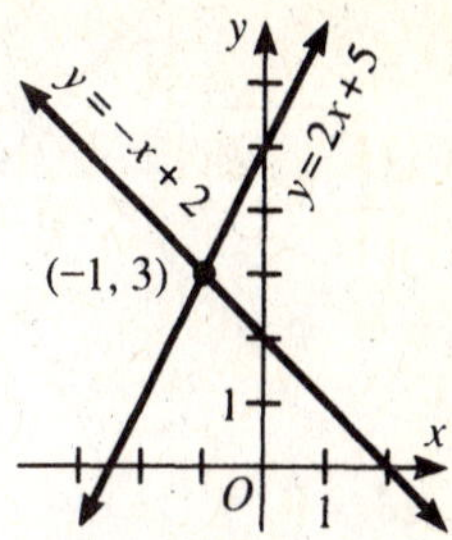

4.

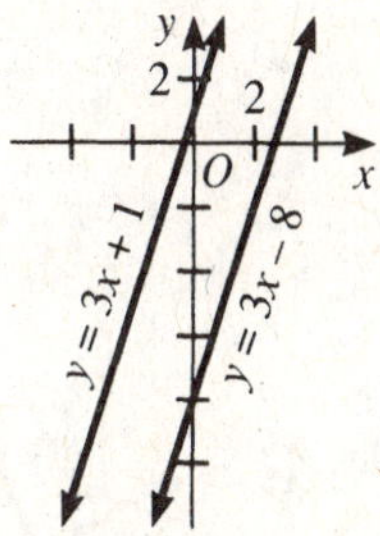

no solution

5.

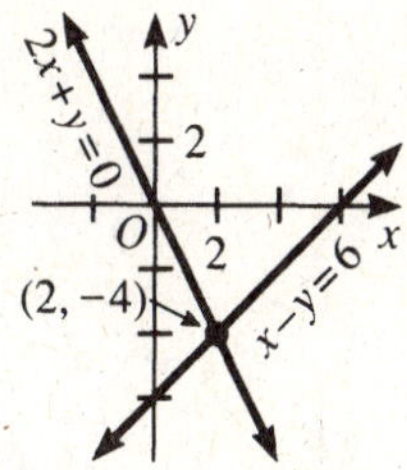

6.

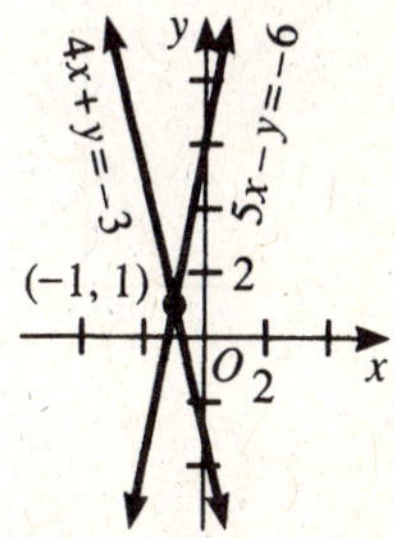

7.

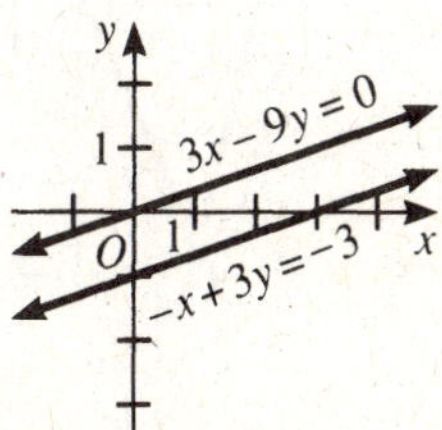

no solution

8.

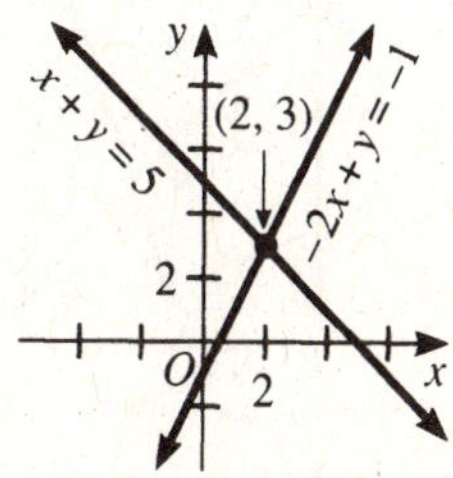

9.

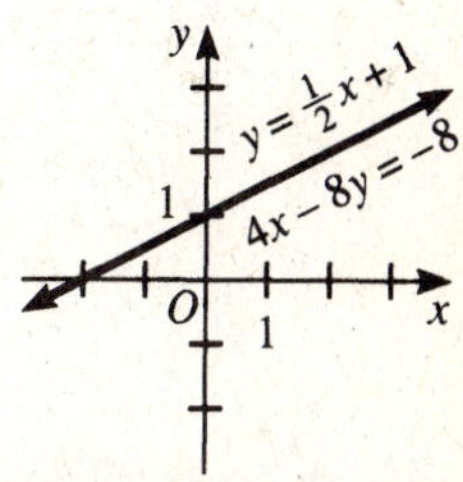

infinitely many

10.

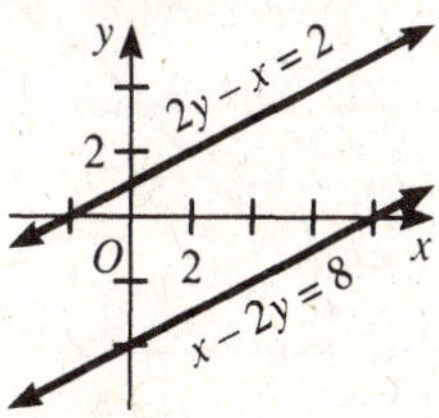

no solution

11.

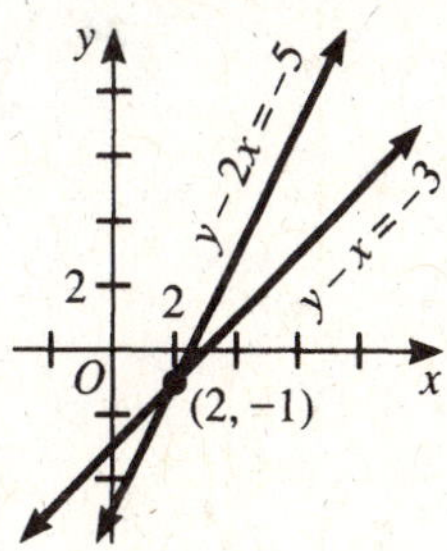

12.

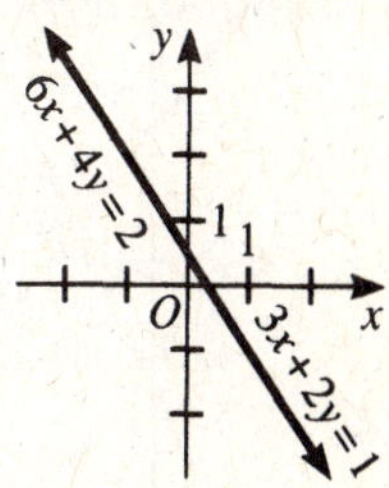

infinitely many

B **13.**

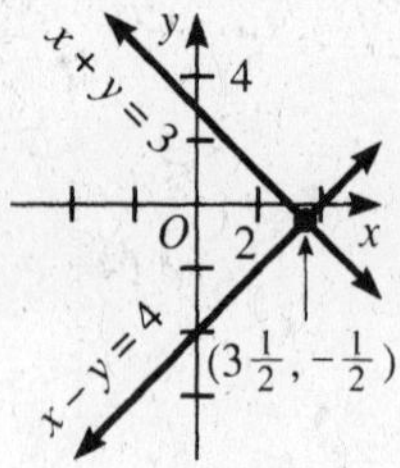

14.

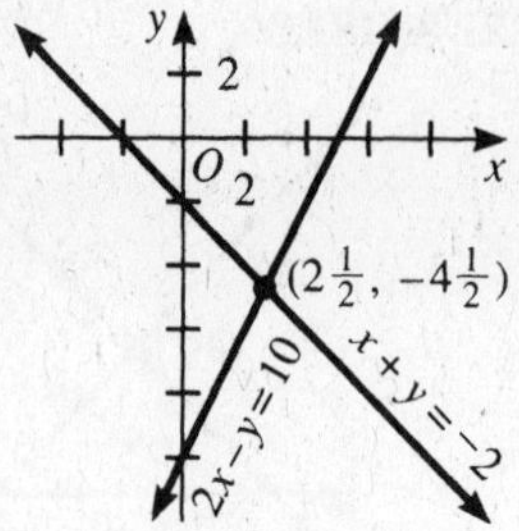

15.

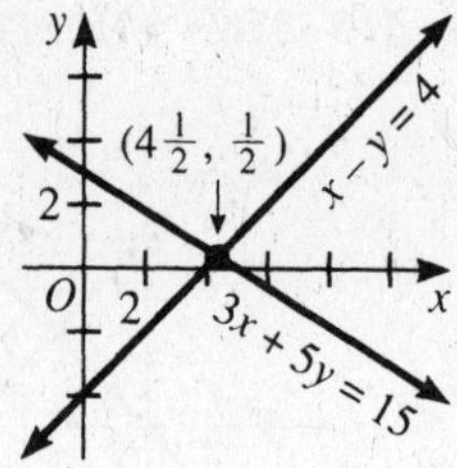

16.

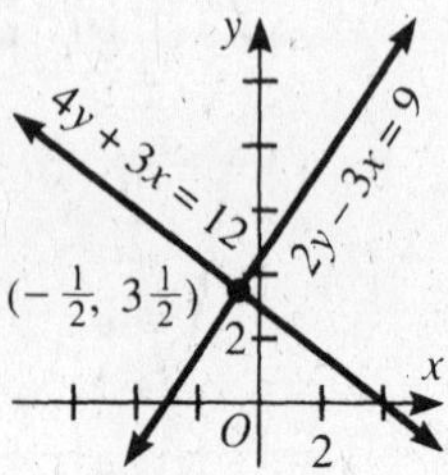

17. $(-2, 4)$ and $(2, 4)$

18. $(1, -1)$ and $(3, 3)$

19.

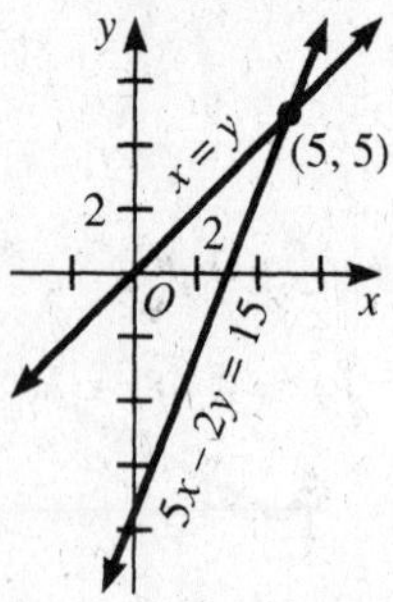

20.

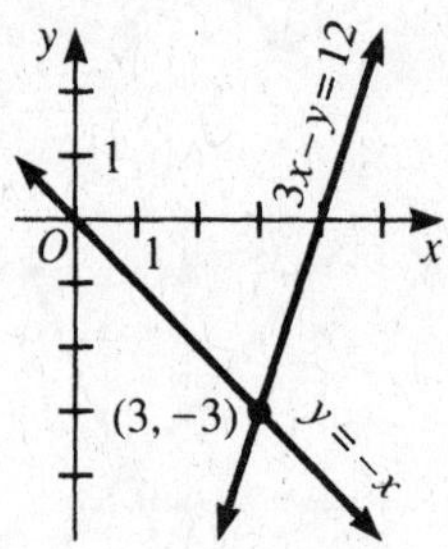

21.

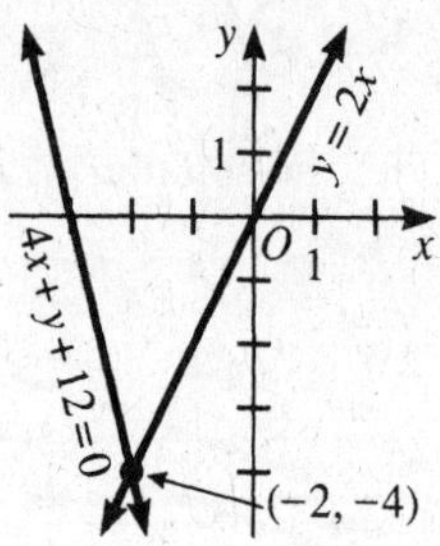

22. $A = \frac{1}{2}bh = \frac{1}{2}(8)(3) = 12$ sq. units

C **23.**

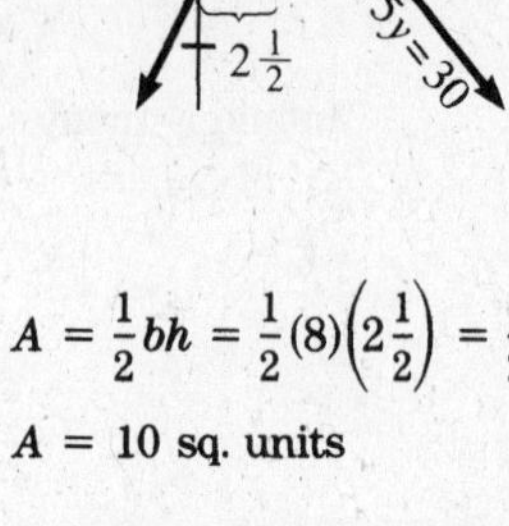

24.

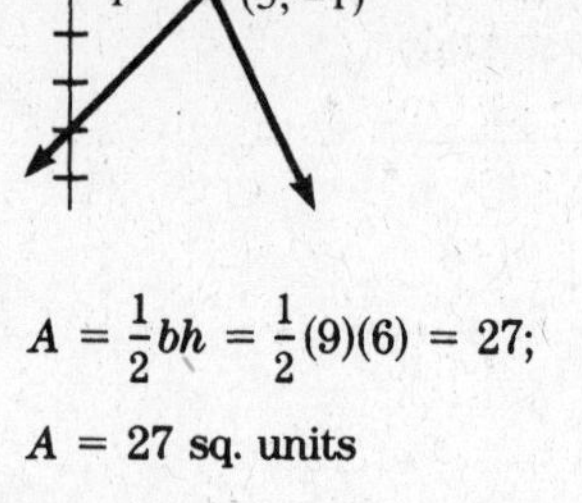

$A = \frac{1}{2}bh = \frac{1}{2}(8)\left(2\frac{1}{2}\right) = \frac{1}{2}(8)\left(\frac{5}{2}\right) = 10;$

$A = 10$ sq. units

$A = \frac{1}{2}bh = \frac{1}{2}(9)(6) = 27;$

$A = 27$ sq. units

Page 416 • MIXED REVIEW EXERCISES

1. $\dfrac{15s^3t}{3s^2t^3} = \dfrac{5s}{t^2}$ **2.** $(a^{-3}b^2)^4 = a^{-12}b^8 = \dfrac{b^8}{a^{12}}$ **3.** $\dfrac{14m^6n}{35m^4n^3} = \dfrac{2m^2}{5n^2}$ **4.** $(x^2y)^{-2} = x^{-4}y^{-2} = \dfrac{1}{x^4y^2}$

5. $p^5p^{-2} = p^3$ **6.** $\dfrac{x^4y^3}{x^{-3}y} = x^7y^2$

Page 416 • COMPUTER EXERCISES

```
10 PRINT "INPUT A, B, C";
20 INPUT A, B, C
30 PRINT "INPUT D, E, F";
40 INPUT D, E, F
50 PRINT
60 FOR X = 0 TO 5
70 LET Y1 = (C - A * X)/B
80 LET Y2 = (F - D * X)/E
90 PRINT "("; X; ", "; Y1; ")"; TAB(20) "("; X; ", "; Y2; ")";
100 IF Y1 <> Y2 THEN 120
110 PRINT "SOLUTION";
120 PRINT
130 NEXT X
140 END
```

1.

(0, −1.286)	(0, 0)	
(1, −1)	(1, −1)	solution
(2, −0.714)	(2, −2)	
(3, −0.429)	(3, −3)	
(4, −0.143)	(4, −4)	
(5, 0.143)	(5, −5)	

2. Change line 60: 60 FOR X = −3 TO 1

(−3, 1.333)	(−3, 0.875)	
(−2, 1)	(−2, 1)	solution
(−1, 0.667)	(−1, 1.125)	
(0, 0.333)	(0, 1.25)	
(1, 0)	(1, 1.375)	

3. Change line 60: 60 FOR X = 0 TO 4

(0, 0)	(0, 2)
(1, 1.333)	(1, 0)
(2, 2.667)	(2, −2)
(3, 4)	(3, −4)
(4, 5.333)	(4, −6)

4. Change line 60: 60 FOR X = −2 TO 2

(−2, 0)	(−2, −3)	
(−1, 0.75)	(−1, −1.5)	
(0, 1.5)	(0, 0)	
(1, 2.25)	(1, 1.5)	
(2, 3)	(2, 3)	solution

Pages 419–420 • WRITTEN EXERCISES

A

1. $x + (6x) = 28$; $7x = 28$; $x = 4$; $y = 6 \cdot 4 = 24$; (4, 24)

2. $5x - (2x) = 30$; $3x = 30$; $x = 10$; $y = 2 \cdot 10 = 20$; (10, 20)

3. $(3b) - b = 12$; $2b = 12$; $b = 6$; $a = 3 \cdot 6 = 18$; (18, 6)

4. $3(4n) - 2n = 20$; $12n - 2n = 20$; $10n = 20$; $n = 2$; $m = 4 \cdot 2 = 8$; (8, 2)

5. $2t + (t + 2) = 17$; $3t + 2 = 17$; $3t = 15$; $t = 5$; $s = 5 + 2 = 7$; (7, 5)

6. $(3d - 4) + d = 16$; $4d - 4 = 16$; $4d = 20$; $d = 5$; $c = 3 \cdot 5 - 4 = 15 - 4 = 11$; (11, 5)

7. $2x + 3(3x + 1) = 25$; $2x + 9x + 3 = 25$; $11x = 22$; $x = 2$; $y = 3 \cdot 2 + 1 = 6 + 1 = 7$; (2, 7)

8. $3(b - 1) = 2b - 6$; $3b - 3 = 2b - 6$; $b = -3$; $a = -3 - 1 = -4$; (−4, −3)

9. $f + 4h = 19; f = 19 - 4h; 4(19 - 4h) - 3h = 0; 76 - 16h - 3h = 0; 76 - 19h = 0;$ $76 = 19h; 4 = h; f + 4 \cdot 4 = 19; f + 16 = 19; f = 3;$ (3, 4)

10. $m - 4n = 6; m = 4n + 6; 3n + 5(4n + 6) = 7; 3n + 20n + 30 = 7; 23n = -23;$ $n = -1; m - 4(-1) = 6; m + 4 = 6; m = 2;$ (2, −1)

11. $2a - b = 17; -b = -2a + 17; b = 2a - 17; 3a + 4(2a - 17) = -13;$ $3a + 8a - 68 = -13; 11a = 55; a = 5; 2 \cdot 5 - b = 17; 10 - b = 17; -b = 7;$ $b = -7;$ (5, −7)

12. $3y - x = -9; -x = -3y - 9; x = 3y + 9; 2y + 5(3y + 9) = 11; 2y + 15y + 45 =$ $11; 17y = -34; y = -2; x = 3(-2) + 9 = -6 + 9 = 3;$ (3, −2)

13. $r + 5 = 6s; r = 6s - 5; 2(6s - 5) + 3s = 0; 12s - 10 + 3s = 0; 15s = 10; s = \frac{2}{3};$

$r = 6\left(\frac{2}{3}\right) - 5 = 4 - 5 = -1; \left(-1, \frac{2}{3}\right)$

14. $x - 2 = -4y; x = 2 - 4y; 3(2 - 4y) + 2y = 11; 6 - 12y + 2y = 11; 6 - 10y = 11;$

$-10y = 5; y = -\frac{1}{2}; x = 2 - 4\left(-\frac{1}{2}\right) = 2 + 2 = 4; \left(4, -\frac{1}{2}\right)$

15. $2\left(\frac{3}{5}b\right) - b = 1; \frac{6}{5}b - b = 1; \frac{1}{5}b = 1; b = 5; a = \frac{3}{5}(5) = 3;$ (3, 5)

16. $3\left(\frac{4}{5}y\right) + 2y = 550; \frac{12}{5}y + 2y = 550; \left(\frac{12 + 10}{5}\right)y = 550; \frac{22}{5}y = 550; y = 125;$

$x = \frac{4}{5}(125) = 100;$ (100, 125)

17. $c - d = 8; c = d + 8; \frac{d + 8}{5} = d + 4; 5\left(\frac{d + 8}{5}\right) = 5(d + 4); d + 8 = 5d + 20;$

$8 = 4d + 20; -12 = 4d; -3 = d; c = (-3) + 8 = 5;$ (5, −3)

18. $x + y + 7 = 0; y = -x - 7; \frac{-x - 7}{2} - x = 1; -x - 7 - 2x = 2; -3x = 9; x = -3;$

$y = -(-3) - 7 = -4;$ (−3, −4)

19. $2u + v = 9; v = 9 - 2u; \frac{u}{2} - \frac{1}{2} = -(9 - 2u); 2\left(\frac{u}{2} - \frac{1}{2}\right) = 2(2u - 9);$

$u - 1 = 4u - 18; -1 = 3u - 18; 17 = 3u; \frac{17}{3} = u; v = 9 - 2\left(\frac{17}{3}\right) = 9 - \frac{34}{3} =$

$\frac{27 - 34}{3} = -\frac{7}{3}; \left(\frac{17}{3}, -\frac{7}{3}\right)$

20. $\frac{m}{3} + \frac{n}{3} = 2; 3\left(\frac{m}{3} + \frac{n}{3}\right) = 3 \cdot 2; m + n = 6; m = 6 - n; 2(6 - n) + 3n = 10;$

$12 - 2n + 3n = 10; 12 + n = 10; n = -2; m = 6 - (-2) = 6 + 2 = 8;$ (8, −2)

21. $\frac{y}{2} + \frac{x}{2} = 7; 2\left(\frac{y}{2} + \frac{x}{2}\right) = 2 \cdot 7; y + x = 14; y = 14 - x; 3(14 - x) + 2x = 48;$

$42 - 3x + 2x = 48; 42 - x = 48; 42 = 48 + x; -6 = x; y = 14 - (-6) = 20;$ (−6, 20)

22. $3b - 2a = 4; -2a = 4 - 3b; a = -\frac{1}{2}(4 - 3b); a = -2 + \frac{3}{2}b; \frac{b}{2} - \frac{2}{3}\left(-2 + \frac{3}{2}b\right) = 1;$

$\frac{b}{2} + \frac{4}{3} - b = 1; \frac{4}{3} - \frac{b}{2} = 1; 6\left(\frac{4}{3} - \frac{b}{2}\right) = 6; 8 - 3b = 6; 8 = 3b + 6; 2 = 3b; \frac{2}{3} = b;$

$a = -2 + \frac{3}{2}\left(\frac{2}{3}\right) = -2 + 1 = -1; \left(-1, \frac{2}{3}\right)$

23. $\frac{r}{2} + \frac{s}{3} = 1; \frac{r}{2} = 1 - \frac{s}{3}; r = 2\left(1 - \frac{s}{3}\right); r = 2 - \frac{2}{3}s; \frac{\left(2 - \frac{2}{3}s\right)}{4} + \frac{2s}{3} = -1;$

$12\left(\frac{2 - \frac{2}{3}s}{4} + \frac{2s}{3}\right) = -12; 3\left(2 - \frac{2}{3}s\right) + 8s = -12; 6 - 2s + 8s = -12;$

$6 + 6s = -12; 6s = -18, s = -3; r = 2 - \frac{2}{3}(-3); r = 2 + 2 = 4; (4, -3)$

24. $\frac{c}{2} - \frac{d}{5} = -4; \frac{c}{2} = \frac{d}{5} - 4; c = 2\left(\frac{d}{5} - 4\right) = \frac{2}{5}d - 8; \frac{2}{3}\left(\frac{2}{5}d - 8\right) - \frac{3d}{5} = -7;$

$\frac{4}{15}d - \frac{16}{3} - \frac{3d}{5} = -7; 15\left(\frac{4}{15}d - \frac{16}{3} - \frac{3d}{5}\right) = 15(-7); 4d - 80 - 9d = -105;$

$-5d - 80 = -105; -5d = -25; d = 5; c = \frac{2}{5}(5) - 8; c = 2 - 8 = -6; (-6, 5)$

B

25. $x + y = 1000; x = 1000 - y; 0.05(1000 - y) + 0.06y = 57; 50 - 0.05y + 0.06y = 57; 100(50 + 0.01y) = 5700; 5000 + y = 5700; y = 700; x = 1000 - (700) = 300;$ $(300, 700)$

26. $a + b = 5000; a = 5000 - b; 0.08(5000 - b) = 0.06b - 20; 400 - 0.08b = 0.06b - 20; 400 = 0.14b - 20; 420 = 0.14b; 3000 = b; a = 5000 - (3000) = 2000;$ $(2000, 3000)$

27. $\frac{x + y}{2} = 10; x + y = 20; x = 20 - y; 2(20 - y) = 3y; 40 - 2y = 3y; 40 = 5y; 8 = y;$

$x = 20 - (8) = 12; (12, 8)$

28. $\frac{d - e}{4} = 1; d - e = 4; d = 4 + e; \frac{3(4 + e) + e}{4} = \frac{(4 + e) + 1}{2}; 3(4 + e) + e =$

$2[(4 + e) + 1]; 12 + 3e + e = 2(e + 5); 12 + 4e = 2e + 10; 12 + 2e = 10; 2e = -2;$ $e = -1; d = 4 + (-1) = 3; (3, -1)$

29. $\frac{5p + q}{2} = p; 5p + q = 2p; q = -3p; \frac{3p - (-3p)}{4} = p + 1; 3p + 3p = 4(p + 1);$

$6p = 4p + 4; 2p = 4; p = 2; q = -3(2) = -6; (2, -6)$

30. $2x - \frac{y + 2}{2} = 22; 2\left(2x - \frac{y + 2}{2}\right) = 2 \cdot 22; 4x - (y + 2) = 44; 4x - y - 2 = 44;$

$4x - 2 = y + 44; 4x - 46 = y; 3x + \frac{(4x - 46) + 2}{3} = 7; 3\left(3x + \frac{4x - 44}{3}\right) = 3 \cdot 7,$

$9x + 4x - 44 = 21; 13x - 44 = 21; 13x = 65, x = 5; y = 4(5) - 46 = 20 - 46 = -26; (5, -26)$

31. $5y - 2x = 3; 5y = 2x + 3; y = \frac{2}{5}x + \frac{3}{5}; 6x = 15\left(\frac{2}{5}x + \frac{3}{5}\right) - 1; 6x = 6x + 9 - 1;$

$0 = 8 \leftarrow$ False; There are no solutions to this system.

32. $8 + 2x = y$; $3(8 + 2x) - 6x = 24$; $24 + 6x - 6x = 24$; $24 = 24$; There are an infinite number of solutions. For example, (0, 8), (1, 10) and (2, 12)

33. $2x = 14y$; $x = 7y$; $\frac{(7y) - y}{3} = \frac{(7y) + y}{4}$; $\frac{6y}{3} = \frac{8y}{4}$; $2y = 2y$; There are an infinite number of solutions. For example, (7, 1), (14, 2), and (0, 0)

34. a.

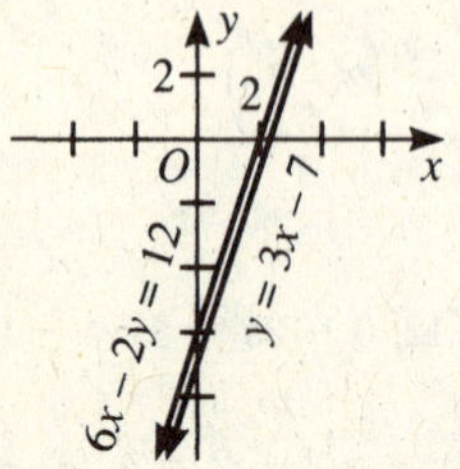

b. $6x - 2(3x - 7) = 12$; $6x - 6x + 14 = 12$; $14 = 12 \leftarrow$ False. There is no solution to this system.

35. a.

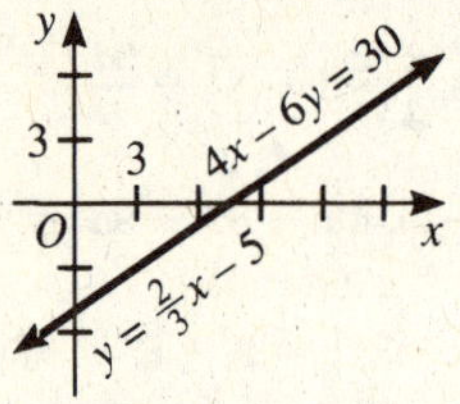

b. $4x - 6\left(\frac{2}{3}x - 5\right) = 30$; $4x - 4x + 30 = 30$; $30 = 30$; There are infinitely many solutions to this system.

36. a.

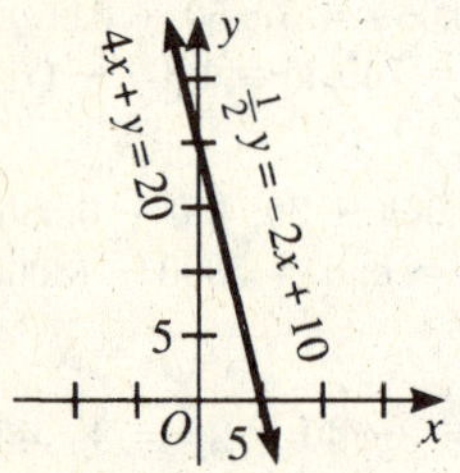

b. $4x + y = 20$; $y = 20 - 4x$; $\frac{1}{2}(20 - 4x) = -2x + 10$; $10 - 2x = -2x + 10$; $10 = 10$; There are infinitely many solutions.

37. $a(-2) + 4(3) = 6$; $-2a + 12 = 6$; $-2a = -6$; $a = 3$; $(-2) + b(3) = -8$; $-2 + 3b = -8$; $3b = -6$; $b = -2$

38. $a(1) + b(4) = 13$; $a + 4b = 13$; $a = 13 - 4b$; $a(1) - b(4) = -3$; $a - 4b = -3$; $(13 - 4b) - 4b = -3$; $13 - 8b = -3$; $-8b = -16$; $b = 2$; $a = 13 - 4(2) = 13 - 8 = 5$; $a = 5$ and $b = 2$.

C **39.** $y = 3x$; $z = 5x$; $x + 3x + 5x = 180$; $9x = 180$; $x = 20$; $y = 3(20) = 60$; $z = 5(20) = 100$; (20, 60, 100)

40. $a = 2c - 5$; $b = 3c - 5$; $(2c - 5) + (3c - 5) + c = 62$; $6c - 10 = 62$; $6c = 72$; $c = 12$; $a = 2(12) - 5 = 24 - 5 = 19$; $b = 3(12) - 5 = 36 - 5 = 31$; (19, 31, 12)

41. $x - y = 1$; $x - 1 = y$; $x - z = 2$; $x - 2 = z$; $x + (x - 1) + 2(x - 2) = 1$; $x + x - 1 + 2x - 4 = 1$; $4x - 5 = 1$; $4x = 6$; $x = \frac{3}{2}$; $y = \frac{3}{2} - 1 = \frac{3}{2} - \frac{2}{2} = \frac{1}{2}$; $z = \frac{3}{2} - 2 = \frac{3}{2} - \frac{4}{2} = -\frac{1}{2}$; $\left(\frac{3}{2}, \frac{1}{2}, -\frac{1}{2}\right)$

Page 420 • MIXED REVIEW EXERCISES

1. 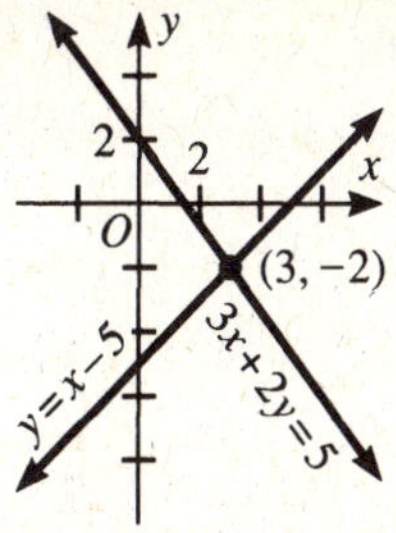

2. 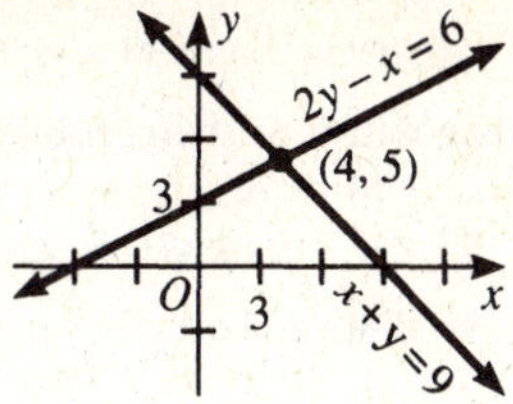

3. 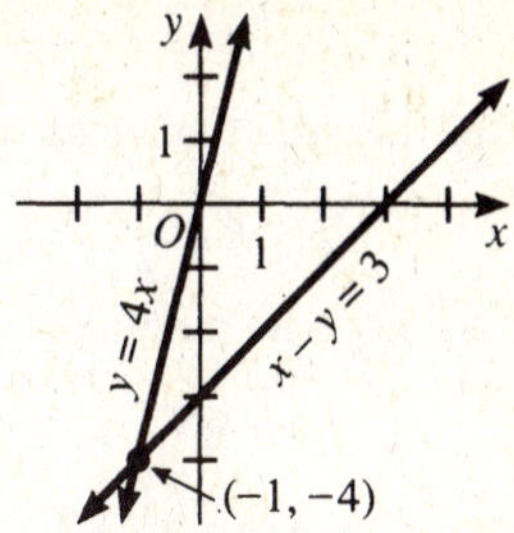

4. 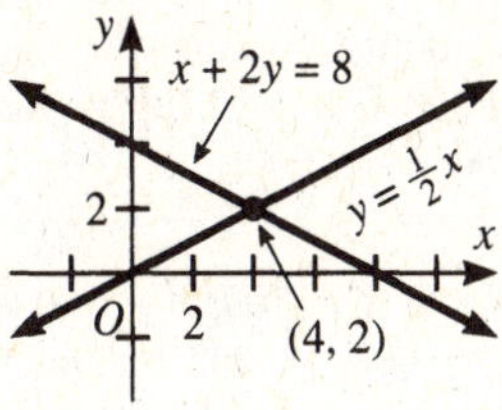

5. 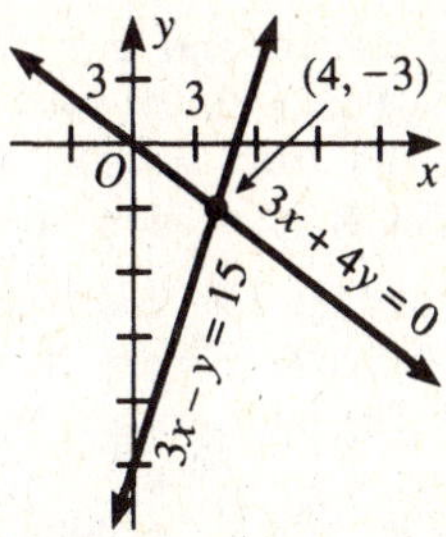

6. 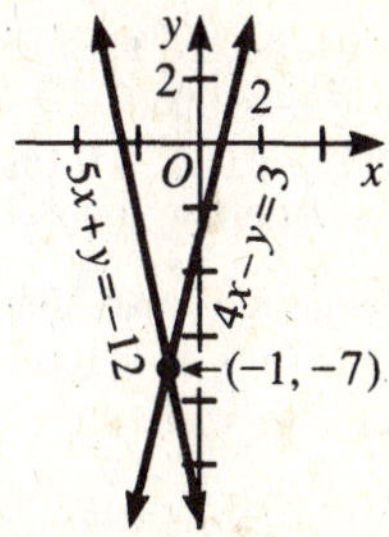

7. $y = -4x + b;\ -7 = -4(2) + b;\ -7 = -8 + b;\ 1 = b;\ y = -4x + 1$

8. $y = \frac{2}{3}x + b;\ 6 = \frac{2}{3}(6) + b;\ 6 = 4 + b;\ 2 = b;\ y = \frac{2}{3}x + 2$ 9. $y = -\frac{1}{2}x - 6$

10. $m = \frac{-1 - 8}{0 - 3} = \frac{-9}{-3} = 3;\ y = 3x + b;\ -1 = 3(0) + b;\ -1 = b;\ y = 3x - 1$

11. $m = \frac{-7 - (-4)}{5 - 2} = \frac{-7 + 4}{3} = \frac{-3}{3} = -1;\ y = -x + b;\ -4 = -(2) + b;\ -2 = b;\ y = -x - 2$

12. $y = 2$

Page 424 • PROBLEMS

A

1. $n + d = 30;\ n = 30 - d;\ 5n + 10d = 240;\ 5(30 - d) + 10d = 240;$ $150 - 5d + 10d = 240;\ 150 + 5d = 240;\ 5d = 90;\ d = 18;\ n = 30 - (18) = 12;$ Sam has 12 nickels.

2. $d + q = 24;\ d = 24 - q;\ 10d + 25q = 360;\ 10(24 - q) + 25q = 360;$ $240 - 10q + 25q = 360;\ 240 + 15q = 360;\ 15q = 120;\ q = 8;\ d = 24 - (8) = 16;$ Kelley has 8 quarters.

3. $q = d + 8;\ 10d + 25q = 550;\ 10d + 25(d + 8) = 550;\ 10d + 25d + 200 = 550;$ $35d = 350;\ d = 10;\ q = (10) + 8 = 18;$ Bruce has 18 quarters.

4. $d = q;\ 25q = 10d + 180;\ 25q = 10(q) + 180;\ 15q = 180;\ q = 12;\ d = 12;$ Luis and Julia each have 12 coins.

5. Let D be the amount Dick contributed and let C be the amount Connie paid. $D + C = 128;\ D = 128 - C;\ D = C + 36;\ (128 - C) = C + 36;\ 128 = 2C + 36;$ $92 = 2C;\ 46 = C;\ D = 128 - (46) = 82;$ Dick paid \$82 and Connie paid \$46.

6. Let A be Annette's score and let J be June's score. $A + J = 425;\ A = 425 - J;$ $J = A - 25;\ J = (425 - J) - 25;\ 2J = 400;\ J = 200;\ A = 425 - (200) = 225;$ Annette's score is 225 and June's score is 200.

7. Let S be the amount Steve has and let T be the amount Tracy has. $S = 2T + 3$; $S + T = 57$; $(2T + 3) + T = 57$; $3T + 3 = 57$; $3T = 54$; $T = 18$; $S = 2(18) + 3 = 36 + 3 = 39$; Steve has \$39 and Tracy has \$18.

8. Let l be the length and w the width of the rectangle. $l = 3w - 5$; $2(l + w) = 70$; $2[(3w - 5) + w] = 70$; $2(4w - 5) = 70$; $8w - 10 = 70$; $8w = 80$; $w = 10$; $l = 3(10) - 5 = 30 - 5 = 25$; The rectangle measures 10 cm by 25 cm.

9. Let c be the time spent on commercials and let p be the programming time. $c + p = 20$; $\frac{c}{p} = \frac{1}{4}$, $4c = p$; $c + p = 20$, $c + 4c = 20$, $5c = 20$, $c = 4$; The station spends 4 h each day broadcasting commercials.

10. Let n be the amount invested in notes and let b be the amount in bonds. $n + b = 5000$; $n = 5000 - b$; $0.08n + 0.10b = 480$; $0.08(5000 - b) + 0.10b = 480$; $400 - 0.08b + 0.10b = 480$; $400 + 0.02b = 480$; $0.02b = 80$; $b = 4000$; $n = 5000 - 4000 = 1000$; \$1000 is invested in Treasury notes.

B

11. Let x be the smaller number and let y be the larger one. $x + y = 100$; $x = 100 - y$; $5x = y + 8$; $5(100 - y) = y + 8$; $500 - 5y = y + 8$; $500 = 6y + 8$; $492 = 6y$; $82 = y$; $x = 100 - (82) = 18$; The numbers are 18 and 82.

12. Let x be the first number and let y be the second number. $x = \frac{1}{2}y + 12$; $x + y = 60$; $\left(\frac{1}{2}y + 12\right) + y = 60$; $\frac{3}{2}y + 12 = 60$; $\frac{3}{2}y = 48$; $y = 32$; $x = \frac{1}{2}(32) + 12 = 16 + 12 = 28$; The numbers are 28 and 32.

13. Let h be the cost of a hamburger and m be the cost of a carton of milk. $2h + m = 5$; $m = 5 - 2h$; $h + 2m = 4.6$; $h + 2(5 - 2h) = 4.6$; $h + 10 - 4h = 4.6$; $10 - 3h = 4.6$; $-3h = -5.4$; $h = 1.8$; $m = 5 - 2(1.8) = 5 - 3.6 = 1.4$; A carton of milk costs \$1.40.

14. Let p be the price of a pen and m be the price of a mechanical pencil. $6p + m = 9$; $m = 9 - 6p$; $4p + 2m = 8$; $4p + 2(9 - 6p) = 8$; $4p + 18 - 12p = 8$; $18 - 8p = 8$; $-8p = -10$; $p = 1.25$; $m = 9 - 6(1.25) = 9 - 7.5 = 1.5$; A pen costs \$1.25 and a mechanical pencil costs \$1.50.

15. Let s be the amount invested in stocks and let b be the amount in bonds. $s + b = 9000$; $s = 9000 - b$; $0.05s + 0.08b = 525$; $0.05(9000 - b) + 0.08b = 525$; $100(450 - 0.05b + 0.08b) = 100 \cdot 525$; $45{,}000 - 5b + 8b = 52{,}500$; $45{,}000 + 3b = 52{,}500$; $3b = 7500$, $b = 2500$; Mr. Lister invested \$2500 in bonds.

16. Let x be the amount invested at 6% and let y be the amount invested at 5%. $0.06x + 0.05y = 234.50$; $100(0.06x + 0.05y) = 100 \cdot 234.50$; $6x + 5y = 23{,}450$; $5y = 23{,}450 - 6x$; $y = 4690 - \frac{6}{5}x$; $0.06y + 0.05x = 234.50 + 5.10$; $0.06\left(4690 - \frac{6}{5}x\right) + 0.05x = 239.60$; $100\left[0.06\left(4690 - \frac{6}{5}x\right) + 0.05x\right] = 100 \cdot 239.60$; $6\left(4690 - \frac{6}{5}x\right) + 5x = 23{,}960$; $28{,}140 - 7.2x + 5x = 23{,}960$; $-2.2x = -4180$; $x = 1900$; $y = 4690 - \frac{6}{5}(1900) = 4690 - 2280 = 2410$; Ms. Bowen invested \$1900 at 6% and \$2410 at 5%.

17. Let c be the speed of the car and let b be the speed of the bike. $c = 4b$; $c = b + 60$; $(4b) = b + 60$; $3b = 60$; $b = 20$; $c = 4(20) = 80$; The car travels at 80 km/h and the bike travels at 20 km/h.

18.

	Amount	× Price/kg	= Total
Cashews	c	8	$8c$
Brazil nuts	b	10	$10b$
Mixture	12	8.50	102

$c + b = 12$; $c = 12 - b$; $8c + 10b = 102$; $8(12 - b) + 10b = 102$; $96 - 8b + 10b = 102$; $96 + 2b = 102$; $2b = 6$; $b = 3$; $c = 12 - (3) = 9$; 9 kg of cashews are needed.

19. Let x be the smaller number and let y be the larger number. $y = x + 16$; $\frac{2}{3}y - x = \frac{1}{4}(x + y)$; $\frac{2}{3}(x + 16) - x = \frac{1}{4}[x + (x + 16)]$; $12\left[\frac{2}{3}(x + 16) - x\right] = 12 \cdot \frac{1}{4}(2x + 16)$; $8(x + 16) - 12x = 6x + 48$; $8x + 128 - 12x = 6x + 48$; $128 - 4x = 6x + 48$; $128 = 10x + 48$; $80 = 10x$; $8 = x$; $y = (8) + 16 = 24$; The numbers are 8 and 24.

20. Let b be Bill's biking speed and let w be his walking speed. $\frac{4}{b} = \frac{1}{3}$, $b = 12$; $\frac{3}{b} + \frac{1}{w} = \frac{1}{2}$; $\frac{3}{(12)} + \frac{1}{w} = \frac{1}{2}$; $\frac{1}{4} + \frac{1}{w} = \frac{1}{2}$; $4w\left(\frac{1}{4} + \frac{1}{w}\right) = 4w \cdot \frac{1}{2}$; $w + 4 = 2w$; $4 = w$; Bill's biking speed is 12 mi/h and his walking speed is 4 mi/h.

C **21.** Let n be the number of nickels, d the number of dimes, and q the number of quarters. $n + d + q = 30$; $n = 30 - d - q$; $5n + 10d + 25q = 460$; $5(30 - d - q) + 10d + 25q = 460$; $150 - 5d - 5q + 10d + 25q = 460$; $150 + 5d + 20q = 460$; $5d + 20q = 310$; $5d = 310 - 20q$; $d = 62 - 4q$; $d = q + 2$; $(62 - 4q) = q + 2$; $62 = 5q + 2$; $60 = 5q$; $12 = q$; $d = (12) + 2 = 14$; $n = 30 - (14) - (12) = 4$; I have 4 nickels, 14 dimes, and 12 quarters.

22. Let c = Carl's amount, d = Diane's amount, and e = Ed's amount. $c = \frac{1}{2}d$; $e = d - 2$; $c + d + e = 46$; $\left(\frac{1}{2}d\right) + d + (d - 2) = 46$; $\frac{5}{2}d - 2 = 46$; $\frac{5}{2}d = 48$; $d = 19.2$; $c = \frac{1}{2}(19.2) = 9.6$; $e = (19.2) - 2 = 17.2$; Carl has \$9.60, Diane has \$19.20, and Ed has \$17.20.

Page 425 • MIXED REVIEW EXERCISES

1. $\frac{1}{2}s + 8 = 3$; $2\left(\frac{1}{2}s + 8\right) = 6$; $s + 16 = 6$; $s = -10$; $\{-10\}$

2. $\frac{1}{4}y = 2\frac{1}{4}$; $4\left(\frac{1}{4}y\right) = 4 \cdot \left(\frac{9}{4}\right)$; $y = 9$; $\{9\}$ **3.** $\frac{x + 2}{3} = 11$; $x + 2 = 33$; $x = 31$; $\{31\}$

4. $4(a + 3) = 12 - 2(a - 18)$; $4a + 12 = 12 - 2a + 36$; $6a + 12 = 48$; $6a = 36$; $a = 6$; $\{6\}$

5. $-12 + n = 5$; $n = 17$; $\{17\}$ **6.** $3x + 14 = x + 4$; $2x = -10$; $x = -5$; $\{-5\}$

7. $2x + 10 = 4\left(\frac{1}{2}x - \frac{5}{2}\right)$; $2x + 10 = 2x - 10$; $10 = -10 \leftarrow$ False; The system has no solution.

8. $b - 2a = -1; b = 2a - 1; 3(2a - 1) + a = 18; 6a - 3 + a = 18; 7a - 3 = 18; 7a = 21;$ $a = 3; b = 2(3) - 1 = 6 - 1 = 5; (3, 5)$

9. $\frac{y-1}{3} = x; y - 1 = 3x; (3x - 2) - 1 = 3x; 3x - 3 = 3x, -3 = 0$ ← False; The system has no solution.

10. $x + (x - 3) = 5; 2x - 3 = 5; 2x = 8; x = 4; y = (4) - 3 = 1; (4, 1)$

11. $2c + d = -7; d = -2c - 7; 3c - 4(-2c - 7) = 6; 3c + 8c + 28 = 6; 11c = -22; c = -2;$ $d = -2(-2) - 7 = 4 - 7 = -3; (-2, -3)$

12. $2p + (3p) = 15; 5p = 15; p = 3; q = 3(3) = 9; (3, 9)$

Page 425 • CHALLENGE

All; (Answers may vary); $23 = 44 \div \sqrt{4} + 4 \div 4; 24 = \sqrt{4} \cdot 4 \cdot 4 - 4 - 4;$
$25 = (\sqrt{4} + \sqrt{4} + 4 \div 4)^{\sqrt{4}}; 26 = 44 \div \sqrt{4} + \sqrt{4} + \sqrt{4}; 27 = 44 \div 4 + 4 \cdot 4;$
$28 = 4 \cdot 4 + 4 \cdot 4 - 4; 29 = (4 + 4 \div 4)^{\sqrt{4}} + 4$

Pages 427–428 • WRITTEN EXERCISES

A

1. $2x = 10; x = 5; (5) + y = 7; y = 2; (5, 2)$

2. $2a = 12; a = 6; (6) + b = 5; b = -1; (6, -1)$

3. $8n = 24; n = 3; 3(3) - 2t = 16; 9 - 2t = 16; -2t = 7; t = -\frac{7}{2}; \left(3, -\frac{7}{2}\right)$

4. $-3d = 12; d = -4; 3c + 5(-4) = 17; 3c - 20 = 17; 3c = 37; c = \frac{37}{3}; \left(\frac{37}{3}, -4\right)$

5. $7n = 14; n = 2; 12(2) + 3m = 18; 24 + 3m = 18; 3m = -6; m = -2; (-2, 2)$

6. $-4a = -28; a = 7; 3(7) - 5b = 31; 21 - 5b = 31; -5b = 10; b = -2; (7, -2)$

7. $-4q = 16; q = -4; 6p - 7(-4) = 28; 6p + 28 = 28; 6p = 0; p = 0; (0, -4)$

8. $2a = 10; a = 5; 4(5) - 7b = 13; 20 - 7b = 13; -7b = -7; b = 1; (5, 1)$

9. $7c = 21; c = 3; 2(3) + 3d = 0; 6 + 3d = 0; 3d = -6; d = -2; (3, -2)$

10. $18y = 54; y = 3; -3x + 5(3) = 45; -3x + 15 = 45; -3x = 30; x = -10; (-10, 3)$

11. $27p = 18; p = \frac{2}{3}; 12\left(\frac{2}{3}\right) - 18q = 14; 8 - 18q = 14; -18q = 6; q = -\frac{1}{3}; \left(\frac{2}{3}, -\frac{1}{3}\right)$

12. $-5z = 3; z = -\frac{3}{5}; 9y + 5\left(-\frac{3}{5}\right) = 1; 9y - 3 = 1; 9y = 4; y = \frac{4}{9}; \left(\frac{4}{9}, -\frac{3}{5}\right)$

13. $54x = 18; x = \frac{1}{3}; 21\left(\frac{1}{3}\right) - 16y = -1; 7 - 16y = -1; -16y = -8; y = \frac{1}{2}; \left(\frac{1}{3}, \frac{1}{2}\right)$

14. $55h = 11; h = \frac{1}{5}; -28g + 15\left(\frac{1}{5}\right) = -18; -28g + 3 = -18; -28g = -21; g = \frac{3}{4};$ $\left(\frac{3}{4}, \frac{1}{5}\right)$

15. $-0.02y = -4; y = 200; 0.02x + 0.03(200) = 9; 0.02x + 6 = 9; 0.02x = 3; x = 150;$ $(150, 200)$

16. $0.08a = 24; a = 300; 0.07(300) - 0.03b = 12; 21 - 0.03b = 12; -0.03b = -9;$ $b = 300; (300, 300)$

17. $3x = 12; x = 4; \frac{1}{2}(4) + \frac{1}{3}y = 4; 2 + \frac{1}{3}y = 4; \frac{1}{3}y = 2; y = 6; (4, 6)$

18. $-\frac{1}{2}x = 10; x = -20; \frac{3}{4}(-20) - \frac{1}{6}y = -12; -15 - \frac{1}{6}y = -12; -\frac{1}{6}y = 3; y = -18;$ $(-20, -18)$

B **19.** $(6b + 3) + 2b = 5; 8b + 3 = 5; 8b = 2; b = \frac{1}{4}; a = 6\left(\frac{1}{4}\right) + 3 = \frac{3}{2} + \frac{6}{2} = \frac{9}{2}; \left(\frac{9}{2}, \frac{1}{4}\right)$

20. $x - 5y = 2; x = 5y + 2; 2(5y + 2) + y = 4; 10y + 4 + y = 4; 11y = 0; y = 0;$ $x = 5(0) + 2 = 2; (2, 0)$

21. $x + 6y = 2; x = 2 - 6y; 4[(2 - 6y) - 2y] = 8; 4(2 - 8y) = 8; 2 - 8y = 2; -8y = 0;$ $y = 0; x = 2 - 6(0) = 2; (2, 0)$

22. $4a - 8b = 8; 2a + 8b = -8; 6a = 0; a = 0; 4(0) - 8b = 8; -8b = 8; b = -1; (0, -1)$

23. $\frac{1}{2}(2 - 6m) - m = 1; 1 - 3m - m = 1; -4m = 0; m = 0; n = 2 - 6(0) = 2; (0, 2)$

24. $a + b + 7 = 0; a = -b - 7; \frac{1}{3}(-b - 7) - \frac{2}{3}b = 1; -\frac{1}{3}b - \frac{7}{3} - \frac{2}{3}b = 1; -b - \frac{7}{3} = 1;$ $-b = \frac{10}{3}; b = -\frac{10}{3}; a = -\left(-\frac{10}{3}\right) - 7 = \frac{10}{3} - \frac{21}{3} = -\frac{11}{3}; \left(-\frac{11}{3}, -\frac{10}{3}\right)$

25. $x + 6\left(\frac{2}{3}x\right) = 30; x + 4x = 30; 5x = 30; x = 6; y = \frac{2}{3}(6) = 4; (6, 4)$

26. $a + 2b = 0; a = -2b; \frac{5}{6}(-2b) - \frac{b}{3} = 6; -\frac{5}{3}b - \frac{b}{3} = 6; -2b = 6; b = -3;$ $a = -2(-3) = 6; (6, -3)$

27. $3\left(\frac{p+1}{15}\right) + 1 = \frac{p}{4}; \frac{p+1}{5} + 1 = \frac{p}{4}; 20\left(\frac{p+1}{5} + 1\right) = 20\left(\frac{p}{4}\right); 4(p + 1) + 20 = 5p;$ $4p + 4 + 20 = 5p; 24 = p; n = \frac{(24) + 1}{15} = \frac{25}{15} = \frac{5}{3}; (n, p) = \left(\frac{5}{3}, 24\right)$

28. **a.** $\left(4\frac{1}{2}, -1\frac{1}{2}\right)$ or $\left(4\frac{1}{2}, -1\right)$

b. $9x = 40; x = \frac{40}{9} = 4\frac{4}{9}; 7\left(\frac{40}{9}\right) + 5y = 25; 9\left(\frac{280}{9} + 5y\right) = 9 \cdot 25; 280 + 45y = 225;$ $45y = -55; y = -\frac{11}{9} = -1\frac{2}{9}; \left(4\frac{4}{9}, -1\frac{2}{9}\right)$

29. $a(3) + b(-2) = 18; 3a - 2b = 18; a(3) - b(-2) = 6; 3a + 2b = 6; 6a = 24; a = 4;$ $3(4) - 2b = 18; 12 - 2b = 18; -2b = 6; b = -3$

C **30.** $-2x = -8; x = 4; 5(4) - 3y = 35; 20 - 3y = 35; -3y = 15; y = -5$; The lines intersect at $(4, -5)$. $4(4) - a(-5) = 61; 16 + 5a = 61; 5a = 45; a = 9$

31. $\frac{x - 2y}{x + 2y} = \frac{1}{3}; 3(x - 2y) = x + 2y; 3x - 6y = x + 2y; 2x = 8y; y = \frac{1}{4}x; \frac{x - y}{x + y} = \frac{3}{5};$ $5(x - y) = 3(x + y); 5x - 5y = 3x + 3y; 2x = 8y; \frac{1}{4}x = y$; Both graphs are the line $y = \frac{1}{4}x$.

Pages 428–429 • PROBLEMS

A **1.** Let x and y be the numbers. $x + y = 21; x - y = 5; 2x = 26; x = 13;$ $(13) + y = 21, y = 8$; The numbers are 13 and 8.

2. Let x and y be the numbers. $x + y = 64$; $y = 64 - x$; $2x = y - 10$; $2x = (64 - x) - 10$; $2x = 54 - x$; $3x = 54$; $x = 18$; $y = 64 - (18) = 46$; The numbers are 18 and 46.

3. Let g = number of girls and b = number of boys. $g + b = 812$; $b = 812 - g$; $g = b + 36$; $g = (812 - g) + 36$; $g = 848 - g$; $2g = 848$; $g = 424$; There are 424 girls.

4. Let p = price of a pizza and s = price of a sandwich. $3p + 4s = 34$; $3p + 7s = 41.50$; $-3s = -7.50$; $s = 2.50$; $3p + 4(2.50) = 34$; $3p + 10 = 34$; $3p = 24$; $p = 8$; A pizza costs \$8.00.

B 5. Let s = number of students and t = number of teachers. $s = 26t$; $s + t = 891$; $(26t) + t = 891$; $27t = 891$; $t = 33$; $s = 26(33) = 858$; There are 858 students at Lincoln High.

6. Let b = number of bull's eyes and m = number of misses. $b + m = 30$; $m = 30 - b$; $5b - 10m = -90$; $5b - 10(30 - b) = -90$; $5b - 300 + 10b = -90$; $15b = 210$; $b = 14$; Yolanda made 14 bull's eyes.

7. Let d = number of dogs and c = number of chickens. $d + c = 60$; $d = 60 - c$; $4d + 2c = 148$; $4(60 - c) + 2c = 148$; $240 - 4c + 2c = 148$; $240 - 2c = 148$; $-2c = -92$; $c = 46$; $d = 60 - (46) = 14$; My uncle has 14 dogs and 46 chickens.

8. Let x be the number of lighter cars and let y be the number of heavier cars. $x + y = 18$; $x = 18 - y$; $3000x + 5000y = 30(2000)$; $3000(18 - y) + 5000y = 60{,}000$; $54{,}000 - 3000y + 5000y = 60{,}000$; $2000y = 6000$; $y = 3$; $x = 18 - (3) = 15$; The shipment contained 15 lighter cars and 3 heavier ones.

C 9. Let t be the amount Tom has in cents, and let m be the amount Maria has in cents. $t - 30 = m + 30$; $t = m + 60$; $(t - 30) + 50 = 2(m + 30 - 50)$; $t + 20 = 2(m - 20)$; $t + 20 = 2m - 40$; $-t + 2m = 60$; $\begin{array}{r} t - m = 60 \\ \underline{-t + 2m = 60} \\ m = 120 \end{array}$; $t - 30 = 120 + 30$; $t = 180$; Maria has \$1.20 and Tom has \$1.80.

10. Let s = Peggy's swimming speed and r = her running speed. $\frac{3}{r} = \frac{18}{60}$; $180 = 18r$; $10 = r$; $\frac{2}{r} + \frac{1}{s} = \frac{42}{60}$; $\frac{2}{(10)} + \frac{1}{s} = \frac{7}{10}$; $10s\left(\frac{2}{10} + \frac{1}{s}\right) = 10s \cdot \frac{7}{10}$; $2s + 10 = 7s$; $10 = 5s$; $s = 2$; Peggy's swimming rate is 2 mi/h.

Page 429 • MIXED REVIEW EXERCISES

1. $7x^3 + 5x^2 - 3x + 6x^2 = 7x^3 + 11x^2 - 3x$ 2. $3 \cdot 4^2 = 3 \cdot 16 = 48$

3. $(1 \cdot 10^3) + (3 \cdot 10^2) + (4 \cdot 10) = 1000 + 300 + 40 = 1340$

4. $-2[3n - 2(n - 1)] = -2(3n - 2n + 2) = -2(n + 2) = -2n - 4$

5. $(6x^4y^3)\left(\frac{3}{2}xy^2\right) = 9x^5y^5$ 6. $(3a^3)^2 = 3^2a^6 = 9a^6$ 7. $(-2p^2q)^3 = (-2)^3p^6q^3 = -8p^6q^3$

8. $3x[4x + 3(2 - x)] = 3x(4x + 6 - 3x) = 3x(x + 6) = 3x^2 + 18x$

9. $(6ab)(-3a^2b)(2a^3b^2) = -36a^6b^4$ 10. $\left(-\frac{1}{15}\right)(105)\left(\frac{1}{7}\right) = -1$ 11. $\frac{-8}{\frac{1}{3}} = -8 \cdot 3 = -24$

12. $\frac{1}{6}(-54m + 36n) = -9m + 6n$

Pages 432–433 • WRITTEN EXERCISES

A

1.
$$\begin{array}{l} 2x + y = 8 \rightarrow 4x + 2y = 16 \\ 3x - 2y = 5 \rightarrow \underline{3x - 2y = 5} \\ \qquad\qquad\quad 7x = 21;\ x = 3;\ 2(3) + y = 8;\ 6 + y = 8;\ y = 2;\ (3, 2) \end{array}$$

2.
$$\begin{array}{l} 3a + 5b = 3 \rightarrow 3a + 5b = 3 \\ a + 2b = 13 \rightarrow \underline{3a + 6b = 39} \\ \qquad\qquad\quad -b = -36;\ b = 36;\ a + 2(36) = 13;\ a + 72 = 13;\ a = -59; \\ \qquad\qquad\quad (-59, 36) \end{array}$$

3.
$$\begin{array}{l} 3x - 2t = 4 \rightarrow 3x - 2t = 4 \\ 2x + t = 5 \rightarrow \underline{4x + 2t = 10} \\ \qquad\qquad\quad 7x = 14;\ x = 2;\ 2(2) + t = 5;\ t = 1;\ (t, x) = (1, 2) \end{array}$$

4.
$$\begin{array}{l} 4x - 3y = 7 \rightarrow 4x - 3y = 7 \\ 5x - y = 6 \rightarrow \underline{15x - 3y = 18} \\ \qquad\qquad\quad -11x = -11;\ x = 1;\ 4(1) - 3y = 7;\ -3y = 3;\ y = -1; \\ \qquad\qquad\quad (1, -1) \end{array}$$

5.
$$\begin{array}{l} 4p - q = 6 \rightarrow 12p - 3q = 18 \\ 2p - 3q = 8 \rightarrow \underline{2p - 3q = 8} \\ \qquad\qquad\quad 10p = 10;\ p = 1;\ 4(1) - q = 6;\ -q = 2;\ q = -2;\ (1, -2) \end{array}$$

6.
$$\begin{array}{l} 4x - 3y = 8 \rightarrow 4x - 3y = 8 \\ 2x + y = 14 \rightarrow \underline{6x + 3y = 42} \\ \qquad\qquad\quad 10x = 50;\ x = 5;\ 2(5) + y = 14;\ y = 4;\ (5, 4) \end{array}$$

7.
$$\begin{array}{l} 5p - 2q = 1 \rightarrow 25p - 10q = 5 \\ 4p + 5q = 47 \rightarrow \underline{8p + 10q = 94} \\ \qquad\qquad\quad 33p = 99;\ p = 3;\ 5(3) - 2q = 1;\ -2q = -14;\ q = 7;\ (3, 7) \end{array}$$

8.
$$\begin{array}{l} 2c - 3d = -1 \rightarrow 6c - 9d = -3 \\ 3c - 4d = -3 \rightarrow \underline{6c - 8d = -6} \\ \qquad\qquad\quad -d = 3;\ d = -3;\ 2c - 3(-3) = -1;\ 2c + 9 = -1; \\ \qquad\qquad\quad 2c = -10;\ c = -5;\ (-5, -3) \end{array}$$

9.
$$\begin{array}{l} 3r - 2s = 15 \rightarrow 9r - 6s = 45 \\ 7r - 3s = 15 \rightarrow \underline{14r - 6s = 30} \\ \qquad\qquad\quad -5r = 15;\ r = -3;\ 3(-3) - 2s = 15;\ -9 - 2s = 15; \\ \qquad\qquad\quad -2s = 24;\ s = -12;\ (-3, -12) \end{array}$$

10.
$$\begin{array}{l} 3c - 8d = 7 \rightarrow 3c - 8d = 7 \\ c + 2d = -7 \rightarrow \underline{3c + 6d = -21} \\ \qquad\qquad\quad -14d = 28;\ d = -2;\ c + 2(-2) = -7;\ c = -3;\ (-3, -2) \end{array}$$

11.
$$\begin{array}{l} 3a + b = 4 \rightarrow 3a + b = 4 \\ a - 2b = 6 \rightarrow \underline{3a - 6b = 18} \\ \qquad\qquad\quad 7b = -14;\ b = -2;\ a - 2(-2) = 6;\ a + 4 = 6;\ a = 2; \\ \qquad\qquad\quad (2, -2) \end{array}$$

12.
$$\begin{array}{l} x + y = 7 \rightarrow 2x + 2y = 14 \\ 3x - 2y = 11 \rightarrow \underline{3x - 2y = 11} \\ \qquad\qquad\quad 5x = 25;\ x = 5;\ (5) + y = 7;\ y = 2;\ (5, 2) \end{array}$$

13.
$$\begin{array}{l} 4x + 5t = 22 \rightarrow 4x + 5t = 22 \\ 5x - t = 13 \rightarrow \underline{25x - 5t = 65} \\ \qquad\qquad\quad 29x = 87;\ x = 3;\ 5(3) - t = 13;\ 15 - t = 13;\ t = 2; \\ \qquad\qquad\quad (t, x) = (2, 3) \end{array}$$

14. $2n + 5a = 14 \rightarrow 6n + 15a = 42$
$6n + 7a = 10 \rightarrow \underline{6n + 7a = 10}$
$8a = 32$; $a = 4$; $2n + 5(4) = 14$; $2n + 20 = 14$; $2n = -6$; $n = -3$; $(a, n) = (4, -3)$

15. $3p + 4q = 4 \rightarrow 3p + 4q = 4$
$5p + 2q = 16 \rightarrow \underline{10p + 4q = 32}$
$-7p = -28$; $p = 4$; $3(4) + 4q = 4$; $4q = -8$; $q = -2$; $(4, -2)$

16. $3t - 8z = 34 \rightarrow 3t - 8z = 34$
$7t + 4z = -34 \rightarrow \underline{14t + 8z = -68}$
$17t = -34$; $t = -2$; $3(-2) - 8z = 34$; $-6 - 8z = 34$; $-8z = 40$; $z = -5$; $(t, z) = (-2, -5)$

17. $2c - 7d = 41 \rightarrow 6c - 21d = 123$
$6c + 5d = -7 \rightarrow \underline{6c + 5d = -7}$
$-26d = 130$; $d = -5$; $2c - 7(-5) = 41$; $2c + 35 = 41$; $2c = 6$; $c = 3$; $(3, -5)$

18. $4r + 9s = 23 \rightarrow 4r + 9s = 23$
$-7r + 3s = -34 \rightarrow \underline{-21r + 9s = -102}$
$25r = 125$; $r = 5$; $4(5) + 9s = 23$; $9s = 3$; $s = \frac{1}{3}$; $\left(5, \frac{1}{3}\right)$

19. $4b + 13c = -24 \rightarrow 12b + 39c = -72$
$12b - 5c = 16 \rightarrow \underline{12b - 5c = 16}$
$44c = -88$; $c = -2$; $4b + 13(-2) = -24$; $4b - 26 = -24$; $4b = 2$; $b = \frac{1}{2}$; $\left(\frac{1}{2}, -2\right)$

20. $18a - 5b = 17 \rightarrow 36a - 10b = 34$
$6a + 10b = -6 \rightarrow \underline{6a + 10b = -6}$
$42a = 28$; $a = \frac{2}{3}$; $6\left(\frac{2}{3}\right) + 10b = -6$; $4 + 10b = -6$; $10b = -10$; $b = -1$; $\left(\frac{2}{3}, -1\right)$

21. $3p + 8q = 8 \rightarrow 3p + 8q = 8$
$5p - 2q = 21 \rightarrow \underline{20p - 8q = 84}$
$23p = 92$; $p = 4$; $3(4) + 8q = 8$; $8q = -4$; $q = -\frac{1}{2}$; $\left(4, -\frac{1}{2}\right)$

22. $4x + 15t = 10 \rightarrow 8x + 30t = 20$
$3x + 10t = 5 \rightarrow \underline{9x + 30t = 15}$
$-x = 5$; $x = -5$; $4(-5) + 15t = 10$; $-20 + 15t = 10$; $15t = 30$; $t = 2$; $(t, x) = (2, -5)$

23. $6n + 8c - 4 = 0 \rightarrow 18n + 24c - 12 = 0$
$9n + 10c - 7 = 0 \rightarrow \underline{18n + 20c - 14 = 0}$
$4c + 2 = 0$; $4c = -2$; $c = -\frac{1}{2}$; $6n + 8\left(-\frac{1}{2}\right) - 4 = 0$; $6n - 4 - 4 = 0$; $6n = 8$; $n = \frac{4}{3}$; $(c, n) = \left(-\frac{1}{2}, \frac{4}{3}\right)$

24. $6z - 5t + 10 = 0 \rightarrow 12z - 10t + 20 = 0$
$4z - 7t + 25 = 0 \rightarrow \underline{12z - 21t + 75 = 0}$
$11t - 55 = 0;\ 11t = 55;\ t = 5;\ 6z - 5(5) + 10 = 0;$
$6z - 25 + 10 = 0;\ 6z = 15;\ z = \frac{5}{2};$
$(t, z) = \left(5, \frac{5}{2}\right)$

B **25.** $6n + 4c = 5 \rightarrow 18n + 12c = 15$
$9n + 6c = 8 \rightarrow \underline{18n + 12c = 16}$
$0 = -1$; False. The system has no solution.

26. $8a - 10b = \frac{1}{3} \rightarrow 24a - 30b = 1 \rightarrow 120a - 150b = 5$
$20a - 25b = \frac{5}{6} \rightarrow 120a - 150b = 5 \rightarrow \underline{120a - 150b = 5}$
$0 = 0$ Infinitely many solutions.

27. $0.4x + 1.5y = -5.7 \rightarrow 4x + 15y = -57$
$-0.2x + 0.8y = -1.8 \rightarrow \underline{-4x + 16y = -36}$
$31y = -93;\ y = -3;\ 0.4x + 1.5(-3) = -5.7;$
$0.4x - 4.5 = -5.7;\ 4x - 45 = -57;$
$4x = -12;\ x = -3;\ (-3, -3)$

28. $0.9x - 2.1y = 12.3 \rightarrow 27x - 63y = 369$
$4.6x - 6.3y = 40.7 \rightarrow \underline{46x - 63y = 407}$
$-19x = -38;\ x = 2;\ 0.9(2) - 2.1y = 12.3;$
$1.8 - 2.1y = 12.3;\ 18 - 21y = 123;$
$-21y = 105;\ y = -5;\ (2, -5)$

29. $0.3x + 0.5y = 31 \rightarrow 3x + 5y = 310$
$0.2x - 0.1y = -1 \rightarrow \underline{10x - 5y = -50}$
$13x = 260;\ x = 20;\ 0.3(20) + 0.5y = 31;\ 6 + 0.5y = 31;$
$0.5y = 25;\ y = 50;\ (20, 50)$

30. $0.05x + 0.06y = 215 \rightarrow 5x + 6y = 21{,}500$
$x + y = 400 \rightarrow \underline{5x + 5y = 2000}$
$y = 19{,}500;\ x + 19{,}500 = 400;\ x = -19{,}100;$
$(-19{,}100;\ 19{,}500)$

31. $\frac{a}{4} + \frac{b}{3} = 2 \rightarrow 3a + 4b = 24 \rightarrow 3a + 4b = 24$
$\frac{a}{2} - b = -1 \rightarrow a - 2b = -2 \rightarrow \underline{2a - 4b = -4}$
$5a = 20;\ a = 4;\ \frac{(4)}{4} + \frac{b}{3} = 2;\ 1 + \frac{b}{3} = 2;$
$\frac{b}{3} = 1;\ b = 3;\ (4, 3)$

32. $\frac{x}{2} - y = 9 \rightarrow x - 2y = 18 \rightarrow x - 2y = 18$
$x + \frac{y}{2} = 8 \rightarrow 2x + y = 16 \rightarrow \underline{4x + 2y = 32}$
$5x = 50;\ x = 10;\ \frac{(10)}{2} - y = 9;\ 5 - y = 9;$
$-y = 4;\ y = -4;\ (10, -4)$

33. $\frac{a}{6} + \frac{b}{4} = \frac{3}{2} \rightarrow 2a + 3b = 18$

$\frac{2a}{3} = \frac{b}{2} \quad \rightarrow \underline{4a - 3b = 0}$

$6a = 18;\ a = 3;\ \frac{2(3)}{3} = \frac{b}{2};\ 2 = \frac{b}{2};\ 4 = b;\ (3, 4)$

34. $\frac{p}{8} + \frac{q}{6} = 1 \rightarrow 3p + 4q = 24$

$\frac{p}{q} = -\frac{2}{3} \quad \rightarrow \underline{3p + 2q = 0}$

$2q = 24;\ q = 12;\ \frac{p}{(12)} = -\frac{2}{3};\ p = -8;\ (-8, 12)$

35. $\frac{x}{2} + \frac{y}{3} = -4 \rightarrow 3x + 2y = -24 \rightarrow 3x + 2y = -24$

$\frac{x}{5} + \frac{y}{5} = -2 \rightarrow x + y = -10 \rightarrow \underline{2x + 2y = -20}$

$x = -4;\ \frac{x}{5} + \frac{y}{5} = -2;\ \frac{-4}{5} + \frac{y}{5} = -2;$

$-4 + y = -10;\ y = -6;$
$(-4, -6)$

C **36.** $3x - 4y = 3 \rightarrow 6x - 8y = 6$

$6x + 6y = 13 \rightarrow \underline{6x + 6y = 13}$

$-14y = -7;\ y = \frac{1}{2};\ 6x + 6\left(\frac{1}{2}\right) = 13;\ 6x + 3 = 13;\ 6x = 10,$

$x = \frac{5}{3}$; The first two lines intersect at $\left(\frac{5}{3}, \frac{1}{2}\right)$.

$3x - 4y = 3 \rightarrow 3x - 4y = 3$

$9x + 2y = 16 \rightarrow \underline{18x + 4y = 32}$

$21x = 35,\ x = \frac{5}{3};\ 3\left(\frac{5}{3}\right) - 4y = 3,\ 5 - 4y = 3,\ -4y = -2,$

$y = \frac{1}{2}$; The three lines intersect at $\left(\frac{5}{3}, \frac{1}{2}\right)$.

37. $a(8) + b(-3) = 25 \rightarrow 8a - 3b = 25 \rightarrow 24a - 9b = 75$

$3a(8) - 5b(-3) = 3 \rightarrow 24a + 15b = 3 \rightarrow \underline{24a + 15b = 3}$

$-24b = 72;\ b = -3;$

$8a - 3(-3) = 25;$
$8a + 9 = 25;\ 8a = 16;$
$a = 2;\ b = -3$

38. $\frac{1}{x} + \frac{1}{y} = 1;\ a + b = 1;\ a = 1 - b;\ \frac{3}{x} - \frac{2}{y} = 8;\ 3a - 2b = 8;\ 3(1 - b) - 2b = 8;$

$3 - 3b - 2b = 8;\ 3 - 5b = 8;\ -5b = 5;\ b = -1;\ a = 1 - (-1) = 2;\ \frac{1}{x} = 2;\ x = \frac{1}{2};$

$\frac{1}{y} = -1,\ y = -1;\ \left(\frac{1}{2}, -1\right)$

39. $\frac{1}{x} + \frac{1}{y} = 5;\ a + b = 5;\ a = 5 - b;\ \frac{3}{x} - \frac{5}{y} = -9;\ 3a - 5b = -9;\ 3(5 - b) - 5b = -9;$

$15 - 3b - 5b = -9;\ -8b = -24;\ b = 3;\ a = 5 - (3) = 2;\ \frac{1}{x} = 2;\ x = \frac{1}{2};\ \frac{1}{y} = 3;\ y = \frac{1}{3};$

$\left(\frac{1}{2}, \frac{1}{3}\right)$

40. $\frac{8}{x} + \frac{15}{y} = 33 \rightarrow 8a + 15b = 33$

$\frac{4}{x} - \frac{35}{y} = -43 \rightarrow 8a - 70b = -86$

$85b = 119$; $b = 1.4$; $8a + 15(1.4) = 33$; $8a + 21 = 33$;

$8a = 12$; $a = 1.5$; $\frac{1}{x} = 1.5$; $x = \frac{2}{3}$; $\frac{1}{y} = 1.4$; $y = \frac{5}{7}$;

$\left(\frac{2}{3}, \frac{5}{7}\right)$

41. $ax + y = 5 \rightarrow 2ax + 2y = 10$

$3ax - 2y = 0 \rightarrow 3ax - 2y = 0$

$5ax = 10$; $x = \frac{2}{a}$; $a\left(\frac{2}{a}\right) + y = 5$; $2 + y = 5$; $y = 3$; $\left(\frac{2}{a}, 3\right)$

42. $ax + by = 1 \rightarrow 2ax + 2by = 2$

$3ax - 2by = 4 \rightarrow 3ax - 2by = 4$

$5ax = 6$; $x = \frac{6}{5a}$; $a\left(\frac{6}{5a}\right) + by = 1$; $\frac{6}{5} + by = 1$; $by = -\frac{1}{5}$;

$y = -\frac{1}{5b}$; $\left(\frac{6}{5a}, -\frac{1}{5b}\right)$

43. $ax - by = 1 \rightarrow a^2x - aby = a$

$bx + ay = 1 \rightarrow b^2x + aby = b$

$(a^2 + b^2)x = a + b$; $x = \frac{a + b}{a^2 + b^2}$; $a\left(\frac{a + b}{a^2 + b^2}\right) - by = 1$;

$(a^2 + b^2)\left(\frac{a(a + b)}{a^2 + b^2} - by\right) = a^2 + b^2$;

$a^2 + ab - (a^2 + b^2)by = a^2 + b^2$;

$-(a^2 + b^2)by = a^2 + b^2 - a^2 - ab$;

$y = \frac{b^2 - ab}{-b(a^2 + b^2)}$; $y = \frac{-b(a - b)}{-b(a^2 + b^2)}$;

$y = \frac{a - b}{a^2 + b^2}$; $\left(\frac{a + b}{a^2 + b^2}, \frac{a - b}{a^2 + b^2}\right)$

44. $x + y + z = 7$, $2x - y + z = 4$ → $-x + 2y = 3$

$2x - y + z = 4$, $3x + 2y + z = 11$ → $-x - 3y = -7$

$-x + 2y = 3$, $-x - 3y = -7$ → $5y = 10$; $y = 2$; $-x + 2(2) = 3$;

$-x = -1$; $x = 1$;

$(1) + (2) + z = 7$;

$z = 4$; $(1, 2, 4)$

45. $x + y + z = 4$, $2x - y + z = 0$ → $3x + 2z = 4$

$x + y + z = 4$, $x - y + 2z = -3$ → $2x + 3z = 1$

$3x + 2z = 4 \rightarrow 6x + 4z = 8$

$2x + 3z = 1 \rightarrow 6x + 9z = 3$

$-5z = 5$;

$z = -1$; $3x + 2(-1) = 4$; $3x - 2 = 4$; $3x = 6$; $x = 2$; $(2) + y + (-1) = 4$;
$y + 1 = 4$; $y = 3$; $(2, 3, -1)$

46. $6x + 4y - z = -7 \rightarrow 12x + 8y - 2z = -14$
$-5x + 6y + 2z = 14 \rightarrow \underline{-5x + 6y + 2z = 14}$
$7x + 14y = 0$

$6x + 4y - z = -7 \rightarrow 18x + 12y - 3z = -21$ $\quad 7x + 14y = 0$
$3x - 2y - 3z = -13 \rightarrow \underline{3x - 2y - 3z = -13}$ $\quad \underline{15x + 14y = -8}$
$15x + 14y = -8$ $\quad -8x = 8;$

$x = -1; 7(-1) + 14y = 0; 14y = 7; y = \frac{1}{2}; 6(-1) + 4\left(\frac{1}{2}\right) - z = -7;$

$-6 + 2 - z = -7; -4 - z = -7; -z = -3; z = 3; \left(-1, \frac{1}{2}, 3\right)$

Page 433 • MIXED PRACTICE EXERCISES

A **1.**

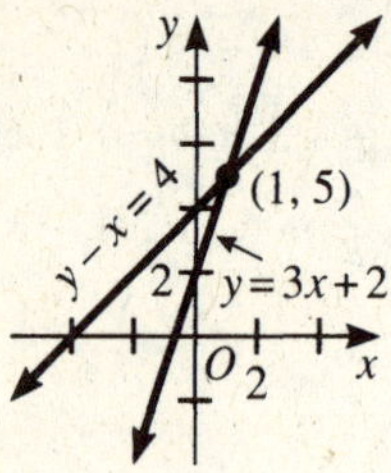

2.

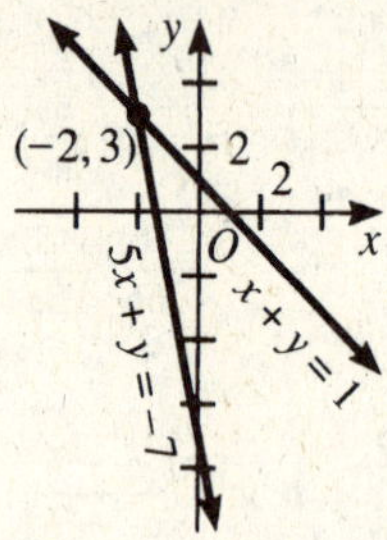

3.

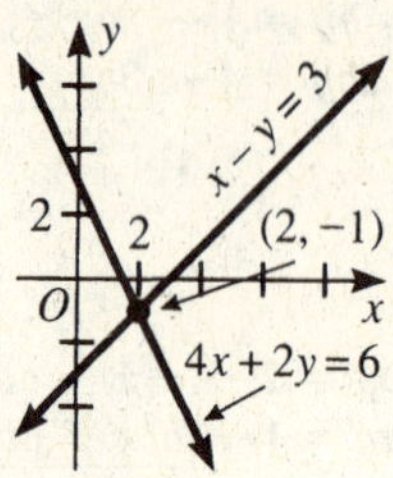

4. $(3b) - 5b = 16; -2b = 16; b = -8; a = 3(-8) = -24; (-24, -8)$

5. $8c - d = -3; -d = -8c - 3; d = 8c + 3; 4c + 5(8c + 3) = 15;$
$4c + 40c + 15 = 15; 44c = 0; c = 0; d = 8(0) + 3 = 3; (0, 3)$

6. $3p - q = 12; -q = -3p + 12; q = 3p - 12; 9p = 2(3p - 12) - 6;$
$9p = 6p - 24 - 6; 3p = -30; p = -10; q = 3(-10) - 12 = -30 - 12 = -42;$
$(-10, -42)$

7. $2a + 3b = -1$
$\underline{a - 3b = 4}$
$3a = 3; a = 1; (1) - 3b = 4; -3b = 3; b = -1; (1, -1)$

8. $5x - 9y = -3 \rightarrow 5x - 9y = -3$
$4x - 3y = 6 \rightarrow \underline{12x - 9y = 18}$
$-7x = -21; x = 3; 5(3) - 9y = -3; 15 - 9y = -3;$
$-9y = -18; y = 2; (3, 2)$

9. $2p + 3q + 1 = 0 \rightarrow 6p + 9q = -3$
$3p + 5q + 2 = 0 \rightarrow \underline{6p + 10q = -4}$
$-q = 1; q = -1; 2p + 3(-1) + 1 = 0; 2p - 2 = 0;$
$2p = 2; p = 1; (1, -1)$

B **10.** $2x + (x + 2) = 11; 3x + 2 = 11; 3x = 9; x = 3; y = (3) + 2 = 5; (3, 5)$

11. $x + y = 9$
$\underline{x - 3y = -3}$
$4y = 12; y = 3; x + (3) = 9; x = 6; (6, 3)$

12. $3x - 2y = 1$
$\underline{-3x + 4y = 7}$
$2y = 8; y = 4; 3x - 2(4) = 1; 3x - 8 = 1; 3x = 9; x = 3; (3, 4)$

13. $3x + 5y = 14 \rightarrow 3x + 5y = 14$
$2x - y = -1 \rightarrow \underline{10x - 5y = -5}$
$13x = 9; x = \frac{9}{13}; 2\left(\frac{9}{13}\right) - y = -1; -y = -1 - \frac{18}{13};$
$y = \frac{31}{13}; \left(\frac{9}{13}, \frac{31}{13}\right)$

14. $7 + a = -3b; a = -3b - 7; 2(-3b - 7) - 4b = 6; -6b - 14 - 4b = 6; -10b = 20;$
$b = -2; a = -3(-2) - 7 = 6 - 7 = -1; (-1, -2)$

15. $r - s = 4 \rightarrow r - s = 4$
$r - 6 = 2s - 12 \rightarrow \underline{r - 2s = -6}$
$s = 10; r - (10) = 4; r = 14; (14, 10)$

16. $a - 2b = 10 \rightarrow a - 2b = 10$
$a + b = 2b + 12 \rightarrow \underline{a - b = 12}$
$-b = -2; b = 2; a - 2(2) = 10; a - 4 = 10; a = 14;$
$(14, 2)$

17. $t + u = 11; u = 11 - t; (10t + u) - (10u + t) = 27; 10t + u - 10u - t = 27;$
$9t - 9u = 27; 9t - 9(11 - t) = 27; 9t - 99 + 9t = 27; 18t = 126; t = 7;$
$u = 11 - (7) = 4; (7, 4)$

18. $u - t = 5; u = t + 5; 10t + u = 3(t + u); 10t + u = 3t + 3u; 7t - 2u = 0;$
$7t - 2(t + 5) = 0; 7t - 2t - 10 = 0; 5t = 10; t = 2; u = (2) + 5 = 7; (t, u) = (2, 7)$

19. $4x + 3y = 1 \rightarrow 8x + 6y = 2$
$6x - 2y = 21 \rightarrow \underline{18x - 6y = 63}$
$26x = 65; x = \frac{5}{2}; 4\left(\frac{5}{2}\right) + 3y = 1; 10 + 3y = 1; 3y = -9;$
$y = -3; \left(\frac{5}{2}, -3\right)$

20. $3a + 4b = -25 \rightarrow 9a + 12b = -75$
$2a - 3b = 6 \rightarrow \underline{8a - 12b = 24}$
$17a = -51; a = -3; 3(-3) + 4b = -25; -9 + 4b = -25;$
$4b = -16; b = -4; (-3, -4)$

21. $5n - 2m = 1 \rightarrow 25n - 10m = 5$
$4n + 5m = 47 \rightarrow \underline{8n + 10m = 94}$
$33n = 99; n = 3; 5(3) - 2m = 1; 15 - 2m = 1;$
$-2m = -14; m = 7; (7, 3)$

22. $x + y = 6000; x = 6000 - y; 0.04x - 0.06y = 40; 4x - 6y = 4000;$
$4(6000 - y) - 6y = 4000; 24{,}000 - 4y - 6y = 4000; -10y = -20{,}000; y = 2000;$
$x = 6000 - (2000) = 4000; (4000, 2000)$

23. $2.4 = 0.3x + 0.4y \rightarrow 3x + 4y = 24 \rightarrow 9x + 12y = 72$
$5x = 2 + 6y \rightarrow 5x - 6y = 2 \rightarrow \underline{10x - 12y = 4}$
$19x = 76; x = 4; 3(4) + 4y = 24;$
$12 + 4y = 24; 4y = 12;$
$y = 3; (4, 3)$

24. $3a + 2b = 4 \rightarrow 3a + 2b = 4$
$\frac{1}{3}(2a + b) = 1 \rightarrow \underline{4a + 2b = 6}$
$-a = -2; a = 2; 3(2) + 2b = 4; 6 + 2b = 4; 2b = -2;$
$b = -1; (2, -1)$

25. $3(a - b) = -9; a - b = -3; a = b - 3; \frac{1}{3}(3a - 2b) = -3; 3a - 2b = -9;$

$3(b - 3) - 2b = -9; 3b - 9 - 2b = -9; b = 0; a = (0) - 3 = -3; (-3, 0)$

26. $\frac{5c}{4} + d = \frac{11}{2} \rightarrow 5c + 4d = 22 \rightarrow \quad 5c + 4d = 22$

$c + \frac{d}{3} = 3 \rightarrow 3c + d = 9 \rightarrow \quad \underline{12c + 4d = 36}$

$-7c = -14; c = 2; (2) + \frac{d}{3} = 3; \frac{d}{3} = 1; d = 3;$

$(2, 3)$

27. $2x - \frac{5}{2}y = 13 \rightarrow 4x - 5y = 26$

$\frac{x}{3} + \frac{y}{3} = \frac{14}{15} \rightarrow \underline{5x + 5y = 14}$

$9x = 40; x = \frac{40}{9}; 2\left(\frac{40}{9}\right) - \frac{5}{2}y = 13; \frac{80}{9} - \frac{5}{2}y = 13;$

$160 - 45y = 234; -45y = 74; y = -\frac{74}{45}; \left(\frac{40}{9}, -\frac{74}{45}\right)$

Pages 434–436 • PROBLEMS

A

1. Let x and y be the numbers. $x + y = 25$

$\underline{x - y = 7}$

$2x = 32; x = 16; (16) + y = 25; y = 9;$ The numbers are 9 and 16.

2. Let x and y be the numbers. $x = 11y - 4; x + y = 92; (11y - 4) + y = 92; 12y - 4 = 92; 12y = 96; y = 8; x = 11(8) - 4; x = 88 - 4; x = 84;$ The numbers are 84 and 8.

3. Let c be the amount that Cory has and let s be the amount Stan has. $c = 2s + 24; c + s = 150; (2s + 24) + s = 150; 3s + 24 = 150; 3s = 126; s = 42; c = 2(42) + 24 = 84 + 24 = 108;$ Cory has \$108 and Stan has \$42.

4. Let m = Marcia's amount and s = Sue's amount. $m = 3s - 84; m + s = 132; (3s - 84) + s = 132; 4s - 84 = 132; 4s = 216; s = 54; m = 3(54) - 84 = 162 - 84 = 78;$ Marcia has \$78.

5. Let l and w be the length and width of the rectangle. $l = 2w + 5; 2(l + w) = 130; 2[(2w + 5) + w] = 130; 2(3w + 5) = 130; 3w + 5 = 65; 3w = 60; w = 20; l = 2(20) + 5 = 45;$ The area of the rectangle is $20 \cdot 45 = 900$ sq units.

6. Let l and w be the length and width of the rectangle. $l = 5w; l - 24 = w + 24; (5w) - 24 = w + 24; 4w - 24 = 24; 4w = 48; w = 12; l = 5(12) = 60;$ The rectangle is 12 cm by 60 cm.

7. Let n and d be the number of nickels and dimes.

$n + d = 50 \rightarrow 5n + 5d = 250$

$5n + 10d = 415 \rightarrow \underline{5n + 10d = 415}$

$-5d = -165; d = 33;$ Phil has 33 dimes.

8. Let d and q be the number of dimes and quarters.

$d + q = 100 \rightarrow 10d + 10q = 1000$

$10d + 25q = 2140 \rightarrow \underline{10d + 25q = 2140}$

$-15q = -1140; q = 76; d + (76) = 100; d = 24;$ Sally has 24 dimes and 76 quarters.

9. Let j = price of apple juice and s = price of salad.

$5j + 4s = 950 \rightarrow 20j + 16s = 3800$

$4j + 5s = 1030 \rightarrow 20j + 25s = 5150$

$-9s = -1350$; $s = 150$; $5j + 4(150) = 950$; $5j + 600 = 950$; $5j = 350$; $j = 70$; $70 + 150 = 220$; A glass of juice and a salad costs $2.20.

10. Let p and n be the prices of a pen and a notebook.

$3p + 2n = 825 \rightarrow 6p + 4n = 1650$

$2p + 3n = 800 \rightarrow 6p + 9n = 2400$

$-5n = -750$; $n = 150$; $3p + 2(150) = 825$; $3p + 300 = 825$; $3p = 525$; $p = 175$; $2(175) + 2(150) = 350 + 300 = 650$; Two pens and two notebooks would cost $6.50.

11. Let a and c be the number of adults and children. $a + c = 785$; $a = 785 - c$; $5a + 2c = 3280$; $5(785 - c) + 2c = 3280$; $3925 - 5c + 2c = 3280$; $-3c = -645$; $c = 215$; 215 children went to the movies.

12. Let g = cost of a grapefruit and h = cost of an orange. $6g = 12h$; $g = 2h$; $12g + 24h = 12$; $12(2h) + 24h = 12$; $24h + 24h = 12$; $48h = 12$; $h = 0.25$; $g = 2(0.25) = 0.5$; One grapefruit costs 50¢.

13.

	Amount	× % acid =	Total acid
Orig. solution	1000	40%	400
Water	w	0%	0
New solution	$1000 + w$	25%	$0.25(1000 + w)$

$0.25(1000 + w) = 400$; $250 + 0.25w = 400$; $0.25w = 150$; $w = 600$; 600 g of water must be added.

14.

	Amount	× % dye =	Total dye
Orig. solution	800	20%	160
Dye	d	100%	d
New solution	$800 + d$	50%	$0.5(800 + d)$

$0.5(800 + d) = 160 + d$; $400 + 0.5d = 160 + d$; $240 = 0.5d$; $480 = d$; 480 g of dye must be added.

15. Let A be the amount invested in stock A and let B be the amount in stock B.

$A + B = 5000 \rightarrow 5A + 5B = 25{,}000$

$0.05A + 0.06B = 280 \rightarrow 5A + 6B = 28{,}000$

$-B = -3{,}000$; $B = 3000$; $A + (3000) = 5000$; $A = 2000$; Edna invested $2000 in Stock A and $3000 in Stock B.

16. Let s be the amount invested in stocks and b the amount in bonds.

$0.1b + 0.05s = 375 \rightarrow 10b + 5s = 37{,}500$

$b + s = 6000 \rightarrow 5b + 5s = 30{,}000$

$5b = 7{,}500$; $b = 1500$; $(1500) + s = 6000$; $s = 4500$; Beatrice invested $4500 in stocks.

B **17.** Let g = cost of a can of grape juice and h = cost of a can of orange juice.

$6g + 4h = 1320 \rightarrow 24g + 16h = 5280$

$4g + 6h = 1380 \rightarrow \underline{24g + 36h = 8280}$

$-20h = -3000$; $h = 150$; $4g + 6(150) = 1380$;
$4g + 900 = 1380$; $4g = 480$; $g = 120$;
A can of orange juice is \$1.50 and a can of grape juice is \$1.20.

18. Let f = number of field goals and $38 - f$ = number of points after touchdown.
$3f + (38 - f) = 70$; $2f + 38 = 70$; $2f = 32$; $f = 16$; Joe kicked 16 field goals.

19. Let t = cost of a tape deck and r = cost of a radio. $5t + 7r = 823$; $5t = 7r + 137$;
$5t - 7r = 137$.

$5t + 7r = 823$

$\underline{5t - 7r = 137}$

$10t = 960$; $t = 96$; $5(96) + 7r = 823$; $480 + 7r = 823$; $7r = 343$; $r = 49$; The cost of a radio is \$49.

20. Let n = number of nickels and d = number of dimes.

$n + d = 45 \rightarrow 5n + 5d = 225$

$5n + 10d = 360 \rightarrow \underline{5n + 10d = 360}$

$5d = 135$; $d = 27$; $n + 27 = 45$; $n = 18$; Rebecca has 18 nickels and 27 dimes.

21.

	Amount	× Price	= Total cost
Peanuts	p	4.25	$4.25p$
Raisins	$3 - p$	3.50	$3.5(3 - p)$
Mixture	3	4.00	12

$4.25p + 3.5(3 - p) = 12$; $100(4.25p + 10.5 - 3.5p) = 1200$;
$425p + 1050 - 350p = 1200$; $75p = 150$; $p = 2$; Juanita had 2 kg of peanuts.

22.

	Amount	× Price	= Total cost
First type	x	4.30	$4.30x$
Second type	$30 - x$	3.90	$3.90(30 - x)$
Mixture	30	4.10	123

$4.30x + 3.90(30 - x) = 123$; $430x + 390(30 - x) = 12{,}300$;
$430x + 11{,}700 - 390x = 12{,}300$; $40x = 600$; $x = 15$; $30 - x = 30 - 15 = 15$; The grocer should use 15 lb of each kind of fruit.

23.

	Rate ×	Time =	Distance
Going	r	$\frac{120}{r}$	120
Returning	$\frac{1}{2}r$	$\frac{120}{\frac{1}{2}r}$	120
Whole trip		4.5	240

$\frac{120}{r} + \frac{120}{\frac{1}{2}r} = 4.5$; $\frac{1}{2}r\left(\frac{120}{r} + \frac{120}{\frac{1}{2}r}\right) = \frac{1}{2}r \cdot 4.5$; $60 + 120 = 2.25r$; $180 = 2.25r$;

$80 = r$; $\frac{1}{2}r = \frac{1}{2}(80) = 40$; The car first drove at 80 km/h and then returned at 40 km/h.

24. Let p be the time Pat would take to do the job alone. $\frac{5}{8} + \frac{2}{p} = 1$; $8p\left(\frac{5}{8} + \frac{2}{p}\right) = 8p$;

$5p + 16 = 8p$; $16 = 3p$; $5\frac{1}{3} = p$; It would take Pat $5\frac{1}{3}$ h to do the job alone.

25. Let c be the number of correctly spelled words and let $48 - c$ be the number of incorrectly spelled words. $10c - 6(48 - c) = 192$; $10c - 288 + 6c = 192$; $16c = 480$; $c = 30$; Todd spells 30 words correctly.

26. Let x = number of packed dishes and $188 - x$ = number of broken dishes. $12x - 18(188 - x) = 2016$; $12x - 3384 + 18x = 2016$; $30x = 5400$; $x = 180$; $188 - x = 188 - 180 = 8$; Elsa breaks 8 dishes.

27. If x and y are both whole numbers, $2x + 6y$ is even. But 35 is odd. Hence, $2x + 6y \neq 35$ if x and y are whole numbers.

28.

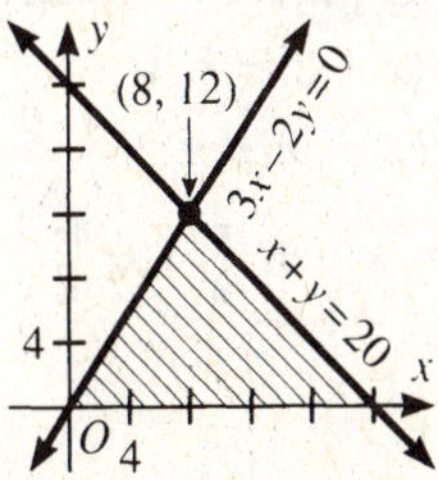

Area $= \frac{1}{2}bh = \frac{1}{2}(20)(12) = 120$ sq. units

C **29.** Let a = weight of an apple; b = banana's weight, and p = plum's weight.

$3a + b = 10p \rightarrow 3a + b = 10p$

$a + 6p = b \rightarrow \underline{a - b = -6p}$

$4a = 4p$; $a = p$; $(p) + 6p = b$; $7p = b$; 7 plums will balance one banana.

30. Let r = amount Roger has, s = Sue's amount, and t = Tim's amount.
$t + 5 = 2(s - 5)$; $t + 5 = 2s - 10$; $t = 2s - 15$; $r + s + t = 155$;
$r + s + (2s - 15) = 155$; $r + 3s - 15 = 155$; $r + 3s = 170$; $r = s + t + 5$;
$r = s + (2s - 15) + 5$; $r = 3s - 10$;

$$\begin{array}{rcl} r + 3s &=& 170 \\ r - 3s &=& -10 \\ \hline 2r &=& 160 \end{array}$$

$r = 80$; $80 = 3s - 10$; $90 = 3s$; $30 = s$; $t = 2(30) - 15 = 60 - 15 = 45$; Roger has \$80, Sue has \$30, and Tim has \$45.

31. Let l and w be the original length and width of the rectangle. $lw = (l + 12)(w - 8)$;
$lw = lw - 8l + 12w - 96$; $0 = -8l + 12w - 96$; $lw = (l + 5)(w - 4)$;
$lw = lw - 4l + 5w - 20$; $0 = -4l + 5w - 20$

$$\begin{array}{rcl} -8l + 12w = 96 &\rightarrow& -8l + 12w = 96 \\ -4l + 5w = 20 &\rightarrow& -8l + 10w = 40 \\ \hline && 2w = 56 \end{array}$$

$w = 28$; $-4l + 5(28) = 20$; $-4l + 140 = 20$; $-4l = -120$; $l = 30$; The original rectangle was 28 by 30.

32.

	Rate	× Time =	Distance
Usual	x	y	xy
Faster	$x + 15$	$y - 2$	$(x + 15)(y - 2)$
Slower	$x - 15$	$y + 3$	$(x - 15)(y + 3)$

$(x + 15)(y - 2) = xy \rightarrow xy - 2x + 15y - 30 = xy \rightarrow -2x + 15y - 30 = 0$
$(x - 15)(y + 3) = xy \rightarrow xy + 3x - 15y - 45 = xy \rightarrow \underline{3x - 15y - 45 = 0}$
$x - 75 = 0$;
$x = 75$; $3(75) - 15y = 45$; $225 - 15y = 45$; $180 = 15y$; $12 = y$; $xy = 75 \cdot 12 = 900$;
Alexandra's trip is 900 km long.

Page 436 • MIXED REVIEW EXERCISES

1. $6 - 36t + 54t^2 = 6(1 - 6t + 9t^2) = 6(1 - 3t)^2$ **2.** $12m^2n - 24mn^3 = 12mn(m - 2n^2)$
3. $25c^2 - 36d^2 = (5c - 6d)(5c + 6d)$ **4.** $x^2 + 7x + 12 = (x + 3)(x + 4)$
5. $2y^2 + 11y + 15 = (2y + 5)(y + 3)$ **6.** $p^2 - 3p - 10 = (p - 5)(p + 2)$
7. $y = kx$; $68 = k(17)$; $4 = k$ **8.** $t = ks$; $-24 = k(72)$; $-\frac{1}{3} = k$ **9.** $\frac{p}{n} = k$; $\frac{36}{54} = k$; $\frac{2}{3} = k$
10. $\frac{h}{b} = k$; $\frac{35}{5} = k$; $7 = k$

Page 436 • COMPUTER EXERCISES

1. $\begin{array}{l} Ax + By = C \rightarrow AEx + BEy = CE \\ Dx + Ey = F \rightarrow BDx + BEy = BF \end{array} \rightarrow (AE - BD)x = CE - BF$; $x = \dfrac{CE - BF}{AE - BD}$.

The original system is also equivalent to $\begin{array}{l} ADx + BDy = CD \\ ADx + AEy = AF \end{array}$, so $(AE - BD)y = AF - CD$,

$y = (AF - CD)/(AE - BD)$.

2.
```
10 PRINT "INPUT A, B, AND C";
20 INPUT A, B, C
30 PRINT "INPUT D, E, AND F";
40 INPUT D, E, F
50 PRINT
60 IF A * E <> B * D THEN 140
70 IF A * F <> D * C THEN 110
80 PRINT "THE LINES COINCIDE"
90 PRINT "INFINITELY MANY SOLUTIONS"
100 GOTO 180
110 PRINT "THE LINES ARE PARALLEL"
120 PRINT "NO SOLUTION"
130 GOTO 180
140 LET X = (C * E - B * F)/(A * E - B * D)
150 LET Y = (A * F - C * D)/(A * E - B * D)
160 PRINT "THE SOLUTION IS "
170 PRINT "("; X; ", "; Y; ")"
180 END
```

3. **a.** $(-3, -4)$ **b.** $(2, 3)$ **c.** No solution **d.** $(3, 2)$

Page 437 • CHALLENGE

Let a = cylinders, b = spheres, c = cones, and d = cubes.

$$\begin{array}{lrcl} (1) & 2c &=& 2a + b - 3d \\ (2) & c &=& -a + 2b + d \\ (3) & c &=& -a + 6b - d \\ \hline & 4c &=& 9b - 3d \end{array}$$

From equation 1, $3d = 2a + b - 2c$. Thus, $4c = 9b - 3d$ becomes $4c = 9b - (2a + b - 2c)$ or $2c = 8b - 2a$. From equation 2, $a = 2b + d - c$. Thus, $2c = 8b - 2a$ becomes $2c = 8b - 2(2b + d - c) = 8b - 4b - 2d + 2c$. This reduces to $4b = 2d$ or $2b = d$. $\therefore 4c = 9b - 6b = 3b = b + d$. One sphere and one cube will balance four cones.

Pages 440–442 • PROBLEMS

A **1.** $r + c = 15$ **2.** $r - c = 10$ **3.** $\frac{24}{r - c} = 3$ **4.** $\frac{20}{r - c} = 2$

5. Let r = rate of the motorboat in still water and c = rate of the current.

$$\begin{array}{rcl} r + c &=& 15 \\ r - c &=& 7 \\ \hline 2r &=& 22 \end{array}$$

$2r = 22$; $r = 11$; $(11) + c = 15$; $c = 4$; The rate of the current is 4 km/h.

6. Let r = rate Camille can swim in still water and c = rate of the current.

$$\begin{array}{rcl} r - c &=& 1.5 \\ r + c &=& 3.5 \\ \hline 2r &=& 5 \end{array}$$

$2r = 5$; $r = 2.5$; Camille can swim 2.5 m/s in still water.

7.

No current	Rate	× Time	= Distance
Downstream	10	3	30
Upstream	10	3	30

Total distance = 60 km
Total time = 6 h
Average speed = 10 km/h

5 km/h current

Rate	× Time	= Distance
15	2	30
5	6	30

Total distance = 60 km
Total time = 8 h
Average speed = 7.5 km/h

8.

	Rate	× Time	= Distance
Upstream	4	75	300
Downstream	6	50	300

Total distance = 600 m
Total time = 125 s
Average speed = 4.8 m/s

9.

	Rate	× Time	= Distance
No wind	200	3	600
With wind	250	2.4	600

Total distance = 1200 km
Total time = 5.4 h
Average speed = 222.2 km/h

10.

	Rate	× Time	= Distance
No wind	140	6	840
With wind	210	4	840

Total distance = 1680 mi
Total time = 10 h
Average speed = 168 mi/h

11.

	Rate	× Time	= Distance
Against wind	$r - w$	7	8400
With wind	$r + w$	6	8400

$$\begin{aligned} r - w &= 1200 \\ r + w &= 1400 \\ \hline 2r &= 2600 \end{aligned}$$

; $r = 1300$; $(1300) + w = 1400$; $w = 100$; The plane flies 1300 km/h in still air and the wind is 100 km/h.

12.

	Rate	× Time	= Distance
Downstream	$r + c$	60	480
Upstream	$r - c$	80	480

$$\begin{aligned} r + c &= 8 \\ r - c &= 6 \\ \hline 2r &= 14 \end{aligned}$$

; $r = 7$; $(7) + c = 8$; $c = 1$; The team travels 7 m/s in still water and the current is 1 m/s.

13.

	Rate	× Time	= Distance
Downstream	$r + c$	2	12
Upstream	$r - c$	3	12

$r + c = 6$
$r - c = 4$
$2r = 10$; $r = 5$; $(5) + c = 6$; $c = 1$; The sailboat's speed in still water is 5 mi/h and the current is 1 mi/h.

14.

	Rate	× Time	= Distance
Against wind	$r - w$	1.5	600
With wind	$r + w$	1	600

$r - w = 400$
$r + w = 600$
$2r = 1000$; $r = 500$; $\frac{1200}{500} = 2.4$; Total flying time with no wind would be 2.4 h.

15.

	Rate	× Time	= Distance
Against wind	$r - w$	6	4200
With wind	$r + w$	5	4200

$r - w = 700$
$r + w = 840$
$2r = 1540$; $r = 770$; $(770) + w = 840$; $w = 70$; The plane flies 770 km/h in still air and the wind is 70 km/h.

16.

	Rate	× Time	= Distance
Against wind	$r - w$	2	1080
With wind	$r + w$	1.5	1080

$r - w = 540$
$r + w = 720$
$2r = 1260$; $r = 630$; $(630) + w = 720$; $w = 90$; The speed of the plane in still air is 630 km/h and the wind speed is 90 km/h.

B 17.

	Rate	× Time	= Distance
Downstream	5	t	$5t$
Upstream	1	$3 - t$	$3 - t$

$5t = 3 - t$; $6t = 3$; $t = 0.5$; $5t = 5(0.5) = 2.5$; Len can spend 0.5 h paddling downstream and he can travel 2.5 mi downstream.

18.

	Rate	× Time =	Distance
Out	20	t	$20t$
Back	30	$4 - t$	$30(4 - t)$

$20t = 30(4 - t)$; $20t = 120 - 30t$; $50t = 120$; $t = 2.4$; $20t = 20 \cdot 2.4 = 48$; The powerboat can travel 48 mi from the marina.

19.

	Rate	× Time =	Distance
Downstream	$r + 3$	$\frac{36}{r + 3}$	36
Upstream	$r - 3$	$\frac{24}{r - 3}$	24

$\frac{36}{r + 3} = \frac{24}{r - 3}$;
$36(r - 3) = 24(r + 3)$;
$36r - 108 = 24r + 72$; $12r = 180$; $r = 15$; The boat travels 15 km/h in still water.

20.

	Rate	× Time =	Distance
Downstream	$r + 4$	$\frac{5}{r + 4}$	5
Upstream	$r - 4$	$\frac{1}{r - 4}$	1

$\frac{5}{r + 4} = \frac{1}{r - 4}$; $5(r - 4) = r + 4$; $5r - 20 = r + 4$; $4r = 24$; $r = 6$; She can paddle 6 km/h in still water.

21.

	Rate	× Time =	Distance
Upstream	$30 - c$	$\frac{45}{30 - c}$	45
Downstream	$30 + c$	$\frac{63}{30 + c}$	63

$\frac{45}{30 - c} = \frac{63}{30 + c}$;
$45(30 + c) = 63(30 - c)$;
$1350 + 45c = 1890 - 63c$;
$108c = 540$; $c = 5$; The current is 5 km/h.

22.

	Rate	× Time	= Distance
With wind	$760 + w$	$\frac{2000}{760 + w}$	2000
Against wind	$760 - w$	$\frac{1800}{760 - w}$	1800

$\frac{2000}{760 + w} = \frac{1800}{760 - w}$; $2000(760 - w) = 1800(760 + w)$; $1{,}520{,}000 - 2000w = 1{,}368{,}000 + 1800w$; $152{,}000 = 3800w$; $40 = w$; The windspeed is 40 km/h.

C 23.

	Rate	× Time	= Distance
With wind	$p + w$	$\frac{d}{p + w}$	d
Against wind	$p - w$	$\frac{d}{p - w}$	d

$$\text{Average speed} = \frac{\text{Total distance}}{\text{Total time}} = \frac{2d}{\frac{d}{p+w} + \frac{d}{p-w}} = \frac{2}{\frac{1}{p+w} + \frac{1}{p-w}} =$$

$\frac{2(p + w)(p - w)}{(p - w) + (p + w)} = \frac{2(p^2 - w^2)}{2p} = \frac{p^2 - w^2}{p}$; The average speed is $\frac{p^2 - w^2}{p}$ km/h.

Page 443 • MIXED REVIEW EXERCISES

1. $3x + 2y = 11 \rightarrow 3x + 2y = 11$
$x - y = 7 \rightarrow \underline{2x - 2y = 14}$
$5x = 25$; $x = 5$; $(5) - y = 7$; $-y = 2$, $y = -2$; $(5, -2)$

2. $5a + 4b = 3 \rightarrow 10a + 8b = 6$
$2a + 3b = 3 \rightarrow \underline{10a + 15b = 15}$
$-7b = -9$; $b = \frac{9}{7}$; $2a + 3\left(\frac{9}{7}\right) = 3$; $2a + \frac{27}{7} = 3$; $2a = -\frac{6}{7}$; $a = -\frac{3}{7}$;
$\left(-\frac{3}{7}, \frac{9}{7}\right)$

3. $7p + 2q = 10 \rightarrow 7p + 2q = 10$
$p + 3q = -4 \rightarrow \underline{7p + 21q = -28}$
$-19q = 38$; $q = -2$; $p + 3(-2) = -4$; $p + (-6) = -4$; $p = 2$;
$(2, -2)$

4. $\frac{9b^2 + b - 2}{12b + 4} = \frac{9b^2 + b - 2}{4(3b + 1)}$ **5.** $\frac{4}{x - 2} + \frac{5}{2 - x} = \frac{4}{x - 2} - \frac{5}{x - 2} = \frac{-1}{x - 2} = \frac{1}{2 - x}$

6. $(-2t^2)^2 \cdot \frac{4s}{t} = 4t^4 \cdot \frac{4s}{t} = 16st^3$

7. $\frac{4r^2 - t^2}{r^2 - t^2} \div \frac{2r - t}{5} = \frac{(2r + t)(2r - t)}{(r + t)(r - t)} \cdot \frac{5}{2r - t} = \frac{5(2r + t)}{(r + t)(r - t)}$

8. $\frac{3p + 2}{8} - \frac{2p + 1}{6} = \frac{3(3p + 2)}{24} - \frac{4(2p + 1)}{24} = \frac{9p + 6 - 8p - 4}{24} = \frac{p + 2}{24}$

9. $2a - 1 - \frac{a - 2}{a + 3} = \frac{(a + 3)(2a - 1)}{a + 3} - \frac{a - 2}{a + 3} = \frac{2a^2 - a + 6a - 3 - a + 2}{a + 3} = \frac{2a^2 + 4a - 1}{a + 3}$

Pages 447–449 • PROBLEMS

A **1.** $t + u = 15$ **2.** $h + t + u = 10$ **3.** $10t + u = 4(t + u)$

4. $100h + 10t + u = 16(h + t + u)$ **5.** $(10t + u) + 18 = 10u + t$

6. $(100h + 10t + u) - 198 = 100u + 10t + h$

7. Let t = tens digit and u = units digit. $10t + u = 4(t + u)$; $10t + u = 4t + 4u$; $6t - 3u = 0$; $2t - u = 0$; $t = u - 3$; $2(u - 3) - u = 0$; $2u - 6 - u = 0$; $u = 6$; $t = 3$; The number is 36.

8. Let t = tens digit and u = units digit. $t + u = \frac{1}{3}(10t + u)$; $3(t + u) = 10t + u$; $3t + 3u = 10t + u$; $-7t + 2u = 0$; $t + 5 = u$; $-7t + 2(t + 5) = 0$; $-7t + 2t + 10 = 0$; $-5t = -10$; $t = 2$; $(2) + 5 = u$; $7 = u$; The number is 27.

9.

	Tens	Units	Value
Original	t	u	$10t + u$
Reversed	u	t	$10u + t$

Let t = tens digit and u = units digit. $u = 2t$; $10t + u + 36 = 10u + t$; $10t + (2t) + 36 = 10(2t) + t$; $12t + 36 = 20t + t$; $36 = 9t$; $4 = t$; $u = 2(4) = 8$; The number is 48.

10.

	Tens	Units	Value
Original	t	u	$10t + u$
Reversed	u	t	$10u + t$

Let t = tens digit and u = units digit. $t + u = 8$; $t = 8 - u$; $10t + u - 54 = 10u + t$; $10(8 - u) + u - 54 = 10u + (8 - u)$; $80 - 10u + u - 54 = 10u + 8 - u$; $26 - 9u = 9u + 8$; $18 = 18u$; $1 = u$; $t = 8 - (1) = 7$; The number is 71.

11.

Age	Now	Last year
Nicole	n	$n - 1$
Pierre	p	$p - 1$

$n = p + 5$; $(n - 1) = 2(p - 1)$; $(p + 5) - 1 = 2p - 2$; $p + 4 = 2p - 2$; $6 = p$; $n = (6) + 5 = 11$, Nicole is 11 years old and Pierre is 6 years old.

12.

Age	Now	6 years ago
Cecilia	c	$c - 6$
Joe	j	$j - 6$

$c = j - 24$; $(c - 6) = \frac{1}{2}(j - 6)$; $2(c - 6) = j - 6$; $2c - 12 = j - 6$; $2c - j = 6$; $2(j - 24) - j = 6$; $2j - 48 - j = 6$; $j = 54$; $c = j - 24 = (54) - 24 = 30$; Cecilia is 30 years old and Joe is 54 years old.

13.

Age	Now	4 years from now
Steve	s	$s + 4$
Theresa	t	$t + 4$

$s = 3t$; $s + 4 = 2(t + 4)$; $(3t) + 4 = 2t + 8$; $t = 4$; $s = 3(4) = 12$; Steve is 12 years old and Theresa is 4 years old.

14.

Age	Now	4 years ago
Marion	m	$m - 4$
Les	l	$l - 4$

$m = \frac{3}{4}l$; $(m - 4) = \frac{2}{3}(l - 4)$; $3(m - 4) = 2(l - 4)$; $3m - 12 = 2l - 8$;

$3\left(\frac{3}{4}l\right) - 12 = 2l - 8$; $\frac{9}{4}l - 12 = 2l - 8$; $\frac{1}{4}l = 4$; $l = 16$; $m = \frac{3}{4}(16) = 12$; Marion is 12 years old and Les is 16 years old.

15. Let $\frac{n}{d}$ = the fraction. $d = n + 8$; $\frac{n + 3}{d + 3} = \frac{1}{2}$; $2(n + 3) = d + 3$;

$2n + 6 = (n + 8) + 3$; $2n + 6 = n + 11$; $n = 5$; $d = (5) + 8 = 13$; The fraction is $\frac{5}{13}$.

16. Let $\frac{n}{d}$ be the fraction. $d = n + 9$; $\frac{n - 3}{d + 3} = \frac{1}{4}$; $4(n - 3) = d + 3$;

$4n - 12 = (n + 9) + 3$; $4n - 12 = n + 12$; $3n = 24$; $n = 8$; $d = (8) + 9 = 17$; The fraction is $\frac{8}{17}$.

17. Let $\frac{n}{d}$ be the fraction. $\frac{n - 1}{d} = \frac{1}{2}$; $2(n - 1) = d$; $2n - 2 = d$; $\frac{n}{d + 7} = \frac{1}{3}$; $3n = d + 7$;

$3n = (2n - 2) + 7$; $3n = 2n + 5$; $n = 5$; $d = 2(5) - 2 = 10 - 2 = 8$; The fraction is $\frac{5}{8}$.

18. Let $\frac{n}{d}$ be the fraction. $\frac{n}{d + 2} = \frac{1}{2}$; $2n = d + 2$; $2n - 2 = d$; $\frac{n + 4}{d} = \frac{3}{4}$; $4(n + 4) = 3d$;

$4(n + 4) = 3(2n - 2)$; $4n + 16 = 6n - 6$; $22 = 2n$; $11 = n$;

$d = 2(11) - 2 = 22 - 2 = 20$; The fraction is $\frac{11}{20}$.

B **19.**

Age	Now	Next year	4 years ago
Lyle	l	$l + 1$	$l - 4$
Sean	s	$s + 1$	$s - 4$

$$\begin{aligned} (l + 1) = 2(s + 1) \rightarrow l + 1 = 2s + 2 \rightarrow l - 2s &= 1 \\ (l - 4) = 3(s - 4) \rightarrow l - 4 = 3s - 12 \rightarrow \underline{l - 3s} &\underline{= -8} \\ s &= 9; l - 2(9) = 1; l = 19; \end{aligned}$$

Lyle is 19 years old and Sean is 9 years old.

20.

Age	Now	Next year
Mary	m	$m + 1$
Brothers	b	$b + 1$

$m = b + 3$; $(m + 1) + 2(b + 1) = 102$; $(b + 3) + 1 + 2b + 2 = 102$; $3b + 6 = 102$; $3b = 96$; $b = 32$; $m = (32) + 3 = 35$; Mary is 35 years old and her brothers are 32 years old.

21.

Bridge	Year completed
Brooklyn	b
Golden Gate	g

$g = b + 54$; $(1983 - g) = \frac{23}{50}(1983 - b)$; $1983 - (b + 54) = \frac{23}{50}(1983 - b)$;

$50(1983 - b - 54) = 23(1983 - b)$; $50(1929 - b) = 23(1983 - b)$; $96{,}450 - 50b = 45{,}609 - 23b$; $50{,}841 = 27b$; $1883 = b$; $g = (1883) + 54 = 1937$; The Golden Gate bridge was completed in 1937 and the Brooklyn Bridge was completed in 1883.

22.

	Year built
Lincoln Memorial	l
Jefferson Memorial	j

$l = j - 21$; $1985 - l = \frac{3}{2}(1985 - j)$; $1985 - (j - 21) = \frac{3}{2}(1985 - j)$;

$2(1985 - j + 21) = 3(1985 - j)$; $2(2006 - j) = 3(1985 - j)$; $4012 - 2j = 5955 - 3j$; $j = 1943$; $l = (1943) - 21 = 1922$; The Lincoln Memorial was built in 1922 and the Jefferson Memorial was built in 1943.

23–26.

	Hundreds	Tens	Units	Value
Original	h	t	u	$100h + 10t + u$
Reversed	u	t	h	$100u + 10t + h$

23. Let h, t, and u be the hundreds, tens, and units digits. $h = 3$; $t = u + 6$; $100h + 10t + u = 40(h + t + u)$; $100(3) + 10(u + 6) + u = 40[(3) + (u + 6) + u]$; $300 + 10u + 60 + u = 40(2u + 9)$; $11u + 360 = 80u + 360$; $0 = 69u$; $0 = u$; $t = (0) + 6 = 6$; The number is 360.

24. Let h and t be the hundreds and tens digits (the units digit is 0). $h = t - 1$; $100h + 10t = 52(h + t)$; $100(t - 1) + 10t = 52[(t - 1) + t]$; $100t - 100 + 10t = 52(2t - 1)$; $110t - 100 = 104t - 52$; $6t = 48$; $t = 8$; $h = (8) - 1 = 7$; The number is 780.

25. Let h, t, and u be the hundreds, tens, and units digits. $h + t + u = 12$; $t - h = u - t$; $-h + 2t - u = 0$; $100u + 10t + h = 100h + 10t + u + 198$; $-99h + 99u = 198$;

$$\begin{aligned} h + t + u &= 12 \rightarrow & 2h + 2t + 2u &= 24 \\ -h + 2t - u &= 0 \rightarrow & -h + 2t - u &= 0 \\ & & \overline{3h \qquad + 3u} &= 24 \rightarrow 99h + 99u = 792; \end{aligned}$$

$$\begin{aligned} -99h + 99u &= 198 \\ 99h + 99u &= 792 \\ \hline 198u &= 990 \end{aligned}$$

$u = 5$; $3h + 3(5) = 24$; $h + 5 = 8$; $h = 3$; $(3) + t + (5) = 12$; $t + 8 = 12$, $t = 4$; The number is 345.

26. Let h, t, and u be the hundreds, tens, and units digits. $t = h + 1$; $h + t + u = 9$; $h + (h + 1) + u = 9$; $2h + 1 + u = 9$; $u = 8 - 2h$; $100u + 10t + h = 100h + 10t + u - 99$; $100(8 - 2h) + 10(h + 1) + h = 100h + 10(h + 1) + (8 - 2h) - 99$; $800 - 200h + 10h + 10 + h = 100h + 10h + 10 + 8 - 2h - 99$; $810 - 189h = 108h - 81$; $891 = 297h$; $3 = h$; $t = (3) + 1 = 4$; $u = 8 - 2(3) = 8 - 6 = 2$; The number is 342.

27.

Age	Now	6 years ago
Father	f	$f - 6$
Son	s	$s - 6$

Let f = father's age now and s = son's age now. $\frac{f + 4}{4} = s$; $f + 4 = 4s$;

$f - 4s = -4$; $f - 6 = \frac{15}{2}(s - 6)$; $2(f - 6) = 15(s - 6)$; $2f - 12 = 15s - 90$;

$$\begin{aligned} 2f - 15s &= -78 \rightarrow 2f - 15s = -78 \\ f - 4s &= -4 \rightarrow 2f - 8s = -8 \\ & \qquad\qquad -7s = -70;\ s = 10;\ f - 4(10) = -4;\ f = 36; \end{aligned}$$

The father is 36 years old and his son is 10 years old.

28. Let t and u be the tens and units digits of the numerator. $\frac{10t + u}{10u + t} = \frac{4}{7}$;

$7(10t + u) = 4(10u + t)$; $70t + 7u = 40u + 4t$; $66t = 33u$; $2t = u$; $\frac{10t + u + 16}{10u + t - 5} = \frac{7}{4}$;

$4(10t + u + 16) = 7(10u + t - 5)$; $40t + 4u + 64 = 70u + 7t - 35$; $33t = 66u - 99$; $t = 2u - 3$; $t = 2(2t) - 3$; $t = 4t - 3$; $3 = 3t$; $1 = t$; $u = 2(1) = 2$;

The fraction is $\frac{12}{21}$.

29. Let t and u be the tens and units digits of the denominator. $\frac{t + u}{10t + u} = \frac{1}{4}$;

$4(t + u) = 10t + u$; $4t + 4u = 10t + u$; $3u = 6t$; $u = 2t$; $\frac{t + u + 3}{10t + u + 3} = \frac{1}{3}$;

$3(t + u + 3) = 10t + u + 3$; $3t + 3u + 9 = 10t + u + 3$; $2u + 6 = 7t$;

$2(2t) + 6 = 7t$; $4t + 6 = 7t$; $6 = 3t$; $2 = t$; $u = 2(2) = 4$; The fraction is $\frac{6}{24}$.

30. Let t and u be the tens and units digits of the numerator. $\frac{10t + u - 1}{10u + t - 1} = \frac{1}{2}$;

$2(10t + u - 1) = 10u + t - 1$; $20t + 2u - 2 = 10u + t - 1$; $19t - 1 = 8u$;

$\frac{u - t}{u + t} = \frac{2}{5}$; $5(u - t) = 2(u + t)$; $5u - 5t = 2u + 2t$; $3u - 7t = 0$;

$$\begin{aligned} 19t - 8u &= 1 \rightarrow 57t - 24u = 3 \\ -7t + 3u &= 0 \rightarrow -56t + 24u = 0 \\ & \qquad\qquad t = 3;\ 3u - 7(3) = 0;\ 3u = 21;\ u = 7; \end{aligned}$$

The fraction is $\frac{37}{73}$.

C **31.**

Age	Now	In 2 years	$l - m$ years ago	2 years ago
Laura	l	$l + 2$	$l - (l - m) = m$	$l - 2$
Maria	m	$m + 2$	$m - (l - m) = 2m - l$	$m - 2$

$l = 3[m - (l - m)]$; $l = 3(2m - l)$; $l = 6m - 3l$; $4l = 6m$; $l + 2 = 2(m - 2)$; $l + 2 = 2m - 4$; $l = 2m - 6$;
$4l - 6m = 0 \rightarrow 4l - 6m = 0$
$l - 2m = -6 \rightarrow 3l - 6m = -18$
$l = 18$; $(18) - 2m = -6$; $-2m = -24$; $m = 12$; Laura is 18 years old and Maria is 12 years old.

32. Let t = tenths digit and h = hundredths digit. $t + h = 9$; $t = 9 - h$; $\frac{h}{10} + \frac{t}{100} = \frac{t}{10} + \frac{h}{100} + 0.27$; $100\left(\frac{h}{10} + \frac{t}{100}\right) = 100\left(\frac{t}{10} + \frac{h}{100} + 0.27\right)$; $10h + t = 10t + h + 27$; $10h + (9 - h) = 10(9 - h) + h + 27$; $9h + 9 = 90 - 10h + h + 27$; $9h + 9 = 117 - 9h$; $18h = 108$; $h = 6$; $t = 9 - (6) = 3$; The number is 0.36

33.

Age	Now	2 years ago	2 years from now
Cindy	c	$c - 2$	$c + 2$
Sue	s	$s - 2$	$s + 2$
Paul	p	$p - 2$	$p + 2$

$c = s + p$; $c - 2 = 4(s - 2)$; $c - 2 = 4s - 8$; $c = 4s - 6$; $c + 2 = 1.4(p + 2)$; $c + 2 = 1.4p + 2.8$; $c = 1.4p + 0.8$; $(s + p) = (4s - 6)$; $p = 3s - 6$; $(s + p) = 1.4p + 0.8$; $s = 0.4p + 0.8$; $s = 0.4(3s - 6) + 0.8$; $s = 1.2s - 2.4 + 0.8$; $0 = 0.2s - 1.6$; $1.6 = 0.2s$; $8 = s$; $p = 3(8) - 6 = 24 - 6 = 18$; $c = (8) + (18) = 26$; Cindy is 26 years old, Paul is 18 years old, and Sue is 8 years old.

34.

Age	Now	$m - 2(s + 2)$ years ago
Father	m	$2(s + 2)$
Son	s	$s - (m - 2(s + 2)) = 3s - m + 4$

$m + s = 55$; $m = 55 - s$; $m = 3(s - [m - 2(s + 2)])$; $m = 3(s - [m - 2s - 4])$; $m = 3(s - m + 2s + 4)$; $m = 3s - 3m + 6s + 12$; $4m = 9s + 12$; $4(55 - s) = 9s + 12$; $220 - 4s = 9s + 12$; $208 = 13s$; $16 = s$; $m = 55 - (16) = 39$; The father is 39 years old and his son is 16 years old.

Page 449 • MIXED REVIEW EXERCISES

1. $(27 - 12) - (4 - 7) = 27 - 12 - (-3) = 15 + 3 = 18$ **2.** $4 + 6 \div 3 = 4 + 2 = 6$
3. $|10| + 3 - |2| = 10 + 3 - 2 = 11$ **4.** $7 - (18 \div 6) = 7 - 3 = 4$
5. $[40 + (-4)] + [1 + (-3)] = 36 + (-2) = 34$ **6.** $4 \cdot (15 - 3) = 4 \cdot 12 = 48$
7. $\frac{x + 4}{3} - \frac{x}{4} = \frac{2}{3}$; $12\left(\frac{x + 4}{3} - \frac{x}{4}\right) = 12 \cdot \frac{2}{3}$; $4(x + 4) - 3x = 8$; $4x + 16 - 3x = 8$; $x = -8$; $\{-8\}$

8. $\frac{3n+5}{2} = \frac{n-1}{5}$; $5(3n+5) = 2(n-1)$; $15n + 25 = 2n - 2$; $13n = -27$; $n = -\frac{27}{13}$; $\left\{-\frac{27}{13}\right\}$

9. $\frac{4+c}{2+c} = \frac{1}{2}$; $2(4+c) = 2 + c$; $8 + 2c = 2 + c$; $c = -6$; $\{-6\}$

10. $|x| = 7.7$; $x = 7.7$ or $x = -7.7$; $\{-7.7, 7.7\}$

11. $3.3 = 70t - 0.2$; $10(3.3) = 10(70t - 0.2)$; $33 = 700t - 2$; $35 = 700t$; $0.05 = t$; $\{0.05\}$

12. $-\frac{1}{6}p = 14$; $p = -84$; $\{-84\}$

Page 451 • CHAPTER REVIEW

1. c

2. a; $2x + y = 6$; $y = 6 - 2x$; $3x - 2y = 2$; $3x - 2(6 - 2x) = 2$; $3x - 12 + 4x = 2$; $7x - 12 = 2$; $7x = 14$; $x = 2$; $y = 6 - 2(2) = 6 - 4 = 2$; (2, 2)

3. d; Let x and y be the integers. $x = y + 17$; $x + 2y = 152$; $(y + 17) + 2y = 152$; $3y + 17 = 152$; $3y = 135$; $y = 45$; $x = (45) + 17 = 62$; (62, 45)

4. d; $$\begin{array}{rl} a + 3b = & 9 \\ \underline{2a + 3b =} & \underline{15} \\ -a \qquad = & -6 \end{array}$$ $a = 6$; $(6) + 3b = 9$; $3b = 3$, $b = 1$; (6, 1)

5. d; $$\begin{array}{rl} 3s - 5t = 27 \rightarrow & 3s - 5t = 27 \\ s + 4t = -8 \rightarrow & \underline{3s + 12t = -24} \\ & -17t = 51 \end{array}$$ $t = -3$; $s + 4(-3) = -8$; $s - 12 = -8$; $s = 4$; (4, −3)

6. b; Let b = boat's speed in still water and c = speed of the current. $$\begin{array}{rl} b + c = & 37 \\ \underline{b - c =} & \underline{17} \\ 2b \qquad = & 54 \end{array}$$ $b = 27$; $(27) + c = 37$; $c = 10$; The current is 10 mi/h.

7. b; Let t and u be the tens and units digits. $t + u = 8$; $t = 8 - u$; $10t + u + 54 = 10u + t$; $10(8 - u) + u + 54 = 10u + (8 - u)$; $80 - 10u + u + 54 = 10u + 8 - u$; $134 - 9u = 9u + 8$; $126 = 18u$; $7 = u$; $t = 8 - (7) = 1$; The number is 17.

8. d; Let r = Rick's age now and b = Barbara's age now. $r - 2 = 2(b - 2) - 1$; $r - 2 = 2b - 4 - 1$; $r = 2b - 3$; $r + 4 = \frac{1}{2}(b + 4) + 8$; $r + 4 = \frac{1}{2}b + 2 + 8$; $r = \frac{1}{2}b + 6$; $(2b - 3) = \frac{1}{2}b + 6$; $\frac{3}{2}b = 9$; $b = 6$; $r = 2(6) - 3 = 12 - 3 = 9$; Rick is 9 years old.

9. d; Let $\frac{n}{d}$ be the fraction. $d = n + 7$; $\frac{n+5}{d+5} = \frac{1}{2}$; $2(n + 5) = d + 5$; $2n + 10 = d + 5$; $2n + 5 = d$; $2n + 5 = (n + 7)$; $n = 2$; $d = (2) + 7 = 9$; The fraction is $\frac{2}{9}$.

Page 452 • CHAPTER TEST

1. 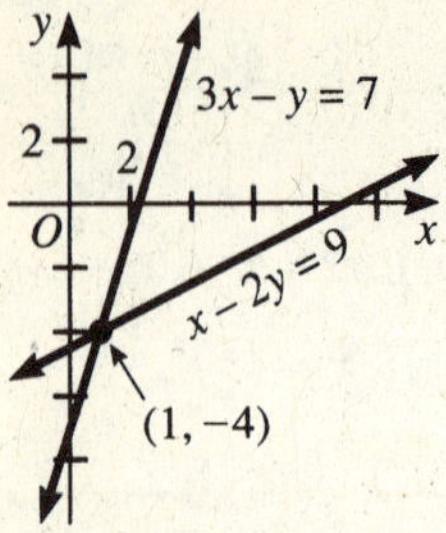

2. 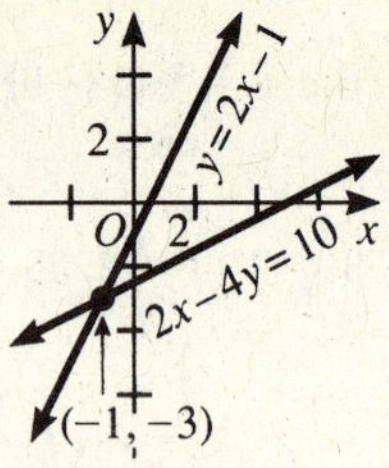

3. $8m + n = 3$; $n = 3 - 8m$; $5m + 2(3 - 8m) = -27$; $5m + 6 - 16m = -27$; $33 = 11m$; $3 = m$; $n = 3 - 8(3) = 3 - 24 = -21$; $(3, -21)$

4. $a + 2b = 1$; $a = 1 - 2b$; $3(1 - 2b) - 4b = 17$; $3 - 6b - 4b = 17$; $3 - 10b = 17$; $-10b = 14$; $b = -\frac{7}{5}$; $a = 1 - 2\left(-\frac{7}{5}\right) = \frac{5}{5} + \frac{14}{5} = \frac{19}{5}$; $\left(\frac{19}{5}, -\frac{7}{5}\right)$

5. $6x - 2y = 8$; $-2y = -6x + 8$; $y = 3x - 4$; The two equations are equivalent, thus there are infinitely many solutions.

6.
$$\begin{array}{l} y = 4x - 3 \rightarrow -4x + y = -3 \rightarrow -8x + 2y = -6 \\ 2y - 8x = -8 \rightarrow -8x + 2y = -8 \rightarrow \underline{-8x + 2y = -8} \\ \qquad\qquad\qquad\qquad\qquad\qquad\qquad 0 = 2 \end{array}$$
There is no solution.

7. Let x = number of small postcards and y = number of large postcards. $x + y = 7$; $x = 7 - y$; $20x + 30y = 180$; $20(7 - y) + 30y = 180$; $140 - 20y + 30y = 180$; $10y = 40$; $y = 4$; $x = 7 - (4) = 3$; Jason bought 3 small and 4 large postcards.

8.
$$\begin{array}{l} 4x - 5y = 0 \\ \underline{8x + 5y = -60} \\ 12x \quad = -60 \end{array}$$
$x = -5$; $4(-5) - 5y = 0$; $-20 - 5y = 0$; $-5y = 20$; $y = -4$; $(-5, -4)$

9.
$$\begin{array}{l} 10p + 4q = 2 \\ \underline{10p - 8q = 26} \\ \quad 12q = -24 \end{array}$$
$q = -2$; $10p + 4(-2) = 2$; $10p - 8 = 2$; $10p = 10$; $p = 1$; $(1, -2)$

10.
$$\begin{array}{l} -8r + s = -17 \rightarrow -24r + 3s = -51 \\ 5r - 3s = -6 \rightarrow \underline{5r - 3s = -6} \\ \qquad\qquad\qquad -19r \quad = -57 \end{array}$$
$r = 3$; $-8(3) + s = -17$; $s = 7$; $(3, 7)$

11.
$$\begin{array}{l} 2x + 5y = 16 \rightarrow 6x + 15y = 48 \\ 5x - 3y = -22 \rightarrow \underline{25x - 15y = -110} \\ \qquad\qquad\qquad 31x \quad = -62 \end{array}$$
$x = -2$; $2(-2) + 5y = 16$; $5y = 20$; $y = 4$; $(-2, 4)$

12. Let p = Gretchen's paddling speed in still water and c = current speed.
$$\begin{array}{l} p - c = 3 \\ \underline{p + c = 8} \\ 2p \quad = 11 \end{array}$$
$p = 5.5$; $(5.5) - c = 3$; $2.5 = c$; Gretchen paddles at 5.5 mi/h in still water, and the current is 2.5 mi/h.

13. Let f = Mr. Foster's age now and d = his daughter's age now. $f - 6 = 5(d - 6) + 2$; $f - 6 = 5d - 30 + 2$; $f - 6 = 5d - 28$; $f = 5d - 22$; $f + 6 = 2(d + 6) + 11$; $(5d - 22) + 6 = 2d + 12 + 11$; $5d - 16 = 2d + 23$; $3d = 39$; $d = 13$; $f = 5(13) - 22 = 65 - 22 = 43$; Mr. Foster is 43 years old.

14. Let $\frac{n}{d}$ be the fraction. $n = d - 4$; $\frac{n + 17}{d + 17} = \frac{5}{6}$; $6(n + 17) = 5(d + 17)$; $6n + 102 = 5d + 85$;

$6(d - 4) + 102 = 5d + 85$; $6d - 24 + 102 = 5d + 85$; $6d + 78 = 5d + 85$; $d = 7$;

$n = (7) - 4 = 3$; The fraction is $\frac{3}{7}$.

Page 453 • CUMULATIVE REVIEW

1. $(28p^4q^3 + 16p^3q^2 - 84p^2q^5) \div (2pq)^2 = \frac{4p^2q^2(7p^2q + 4p - 21q^3)}{4p^2q^2} = 7p^2q + 4p - 21q^3$

2. $(-2x^3yz^2)^4(x^2y^2)^3 = 16x^{12}y^4z^8 \cdot x^6y^6 = 16x^{18}y^{10}z^8$

3. $4j^4(jk^3 - 6j^2k + 7k^4) = 4j^5k^3 - 24j^6k + 28j^4k^4$

4. $(5c - 3)(2c^2 + 4c + 1) = 10c^3 + 20c^2 + 5c - 6c^2 - 12c - 3 = 10c^3 + 14c^2 - 7c - 3$

5. $(21z - 3)(2z + 4) = 42z^2 + 84z - 6z - 12 = 42z^2 + 78z - 12$

6. $(7x^2 + 3b)(7x - 3b) = 49x^3 - 21bx^2 + 21bx - 9b^2$

7. $\frac{15t^2 - 11t - 12}{3t - 4} = \frac{(3t - 4)(5t + 3)}{3t - 4} = 5t + 3$

8. $\frac{3}{2b - 1} + \frac{b^2 + 2}{1 - 4b^2} + b = \frac{-3(1 + 2b)}{(1 + 2b)(1 - 2b)} + \frac{b^2 + 2}{(1 + 2b)(1 - 2b)} + \frac{b(1 + 2b)(1 - 2b)}{(1 + 2b)(1 - 2b)} =$

$\frac{-3 - 6b + b^2 + 2 + b(1 - 4b^2)}{(1 + 2b)(1 - 2b)} = \frac{b^2 - 6b - 1 + b - 4b^3}{(1 + 2b)(1 - 2b)} = \frac{b^2 - 5b - 1 - 4b^3}{(1 + 2b)(1 - 2b)}$

9. $\frac{x^2 - 7x + 10}{x^2 - 4} \div \frac{x + 3}{x^2 - x - 6} = \frac{(x - 2)(x - 5)}{(x + 2)(x - 2)} \cdot \frac{(x + 2)(x - 3)}{x + 3} = \frac{(x - 5)(x - 3)}{x + 3}$

10.
$$\begin{array}{r} 5y - 4 + \frac{13}{2y + 7} \\ 2y + 7\overline{)10y^2 + 27y - 15} \\ \underline{10y^2 + 35y} \\ -8y - 15 \\ \underline{-8y - 28} \\ 13 \end{array}$$

11. $(1.7 \times 10^5)(0.4 \times 10^6) = 0.68 \times 10^{11} = 6.8 \times 10^{10}$

12. $\left(\frac{m^{-4}}{m^{-5}}\right)^{-3} = (m^{-4-(-5)})^{-3} = m^{-3} = \frac{1}{m^3}$ **13.** $49x^2 + 98xy + 81y^2$: prime

14. $64x^4 - 81y^4 = (8x^2 + 9y^2)(8x^2 - 9y^2)$ **15.** $30x^2 + 15x + 1$: prime

16. $28y^2 + 37y + 12 = (7y + 4)(4y + 3)$ **17.** $8y^2 - 26y + 5$: prime

18. $16y^2 - 25x^2 + 10x - 1 = 16y^2 - (25x^2 - 10x + 1) = 16y^2 - (5x - 1)^2 =$
$[4y + (5x - 1)][4y - (5x - 1)] = (4y + 5x - 1)(4y - 5x + 1)$

19. $16 - \frac{1}{2}|m| = 5$; $11 = \frac{1}{2}|m|$; $22 = |m|$; $m = 22$ or $m = -22$; $\{-22, 22\}$

20. $\frac{1}{5}(x - 6) = 34 - 3x$; $x - 6 = 5(34 - 3x)$; $x - 6 = 170 - 15x$; $16x = 176$; $x = 11$; $\{11\}$

21. $196x^2 - 16 = 0$; $4(49x^2 - 4) = 0$; $49x^2 - 4 = 0$; $(7x + 2)(7x - 2) = 0$; $7x + 2 = 0$;

$7x = -2$; $x = -\frac{2}{7}$; $7x - 2 = 0$; $7x = 2$; $x = \frac{2}{7}$; $\left\{-\frac{2}{7}, \frac{2}{7}\right\}$

22. $18t^2 + 53t - 3 = 0$; $(18t - 1)(t + 3) = 0$; $18t - 1 = 0$; $18t = 1$; $t = \frac{1}{18}$; $t + 3 = 0$;

$t = -3$; $\left\{-3, \frac{1}{18}\right\}$

23. $\frac{2x + 1}{x - 3} + \frac{x + 3}{2x + 1} = \frac{2x}{3 - x}$; $\frac{(2x + 1)(2x + 1)}{(x - 3)(2x + 1)} + \frac{(x + 3)(x - 3)}{(x - 3)(2x + 1)} = \frac{-2x(2x + 1)}{(x - 3)(2x + 1)}$;

$4x^2 + 2x + 2x + 1 + x^2 - 9 = -4x^2 - 2x$; $5x^2 + 4x - 8 = -4x^2 - 2x$;

$9x^2 + 6x - 8 = 0$; $(3x + 4)(3x - 2) = 0$; $3x + 4 = 0$; $3x = -4$; $x = -\frac{4}{3}$; $3x - 2 = 0$;

$3x = 2$; $x = \frac{2}{3}$; $\left\{-\frac{4}{3}, \frac{2}{3}\right\}$

24. $y = \frac{1}{4}x + b$; $2 = \frac{1}{4}(12) + b$; $2 = 3 + b$; $-1 = b$; $y = \frac{1}{4}x - 1$; $4y = x - 4$; $x - 4y = 4$

25. $m = \frac{-3 - 0}{2 - 1} = \frac{-3}{1} = -3$; $y = -3x + b$; $0 = -3(1) + b$; $3 = b$; $y = -3x + 3$; $3x + y = 3$

26.
$$\begin{aligned} 4x + 3y &= 12 \rightarrow & 8x + 6y &= 24 \\ 3y - 8x &= -12 \rightarrow & -8x + 3y &= -12 \\ & & \hline 9y &= 12; \ y = \tfrac{4}{3}; \ 4x + 3\left(\tfrac{4}{3}\right) = 12; \end{aligned}$$

$4x + 4 = 12$; $4x = 8$; $x = 2$; $\left(2, \frac{4}{3}\right)$

27.
$$\begin{aligned} 5x &= y + 3 \rightarrow 5x - y = 3 \rightarrow & 10x - 2y &= 6 \\ 3x + 2y &= 20 \rightarrow 3x + 2y = 20 \rightarrow & 3x + 2y &= 20 \\ & & \hline 13x \quad &= 26; \ x = 2; \ 5(2) = y + 3; \ 10 = y + 3; \end{aligned}$$

$7 = y$; $(2, 7)$

28.
$$\begin{aligned} 2x + 5y &= 9 \rightarrow & 6x + 15y &= 27 \\ 3x - 2y &= 4 \rightarrow & 6x - 4y &= 8 \\ & & \hline 19y &= 19; \ y = 1; \ 2x + 5(1) = 9; \ 2x = 4; \ x = 2; \ (2, 1) \end{aligned}$$

29. Let t and u be the tens and units digits. $t + u = 6$; $t = 6 - u$; $10u + t = 3(10t + u) + 6$; $10u + (6 - u) = 3[10(6 - u) + u] + 6$; $9u + 6 = 3(60 - 10u + u) + 6$; $9u + 6 = 3(60 - 9u) + 6$; $9u + 6 = 180 - 27u + 6$; $36u = 180$; $u = 5$; $t = 6 - (5) = 1$; The number is 15.

30.

	Rate	× Time	= Distance
Against wind	$r - w$	6	48
With wind	$r + w$	3	48

$$\begin{aligned} 6(r - w) &= 48 \rightarrow 6r - 6w = 48 \rightarrow & 6r - 6w &= 48 \\ 3(r + w) &= 48 \rightarrow 3r + 3w = 48 \rightarrow & 6r + 6w &= 96 \\ & & \hline 12r \quad &= 144; \ r = 12; \end{aligned}$$

The cyclist traveled at 12 mi/h when there was no wind.

31.

	Amount	× Cost	= Total cost
First kind	x	5	$5x$
Second kind	6	9	54
Mix	$6 + x$	6.5	$5x + 54$

$5x + 54 = 6.5(6 + x)$; $10(5x + 54) = 10[6.5(6 + x)]$; $50x + 540 = 65(6 + x)$;
$50x + 540 = 390 + 65x$; $150 = 15x$; $10 = x$; 10 kg of the cheaper nuts should be added.

Page 454 • MAINTAINING SKILLS

1. $\frac{ab}{4a - 4b} \cdot \frac{a^2 - ab}{ab} = \frac{a(a - b)}{4(a - b)} = \frac{a}{4}$

2. $\frac{2x - 14}{-5x} \cdot \frac{5x^3}{3x - 21} = \frac{2(x - 7)}{-5x} \cdot \frac{5x^3}{3(x - 7)} = -\frac{2x^2}{3}$

3. $\frac{8f + 24g}{2f - 4g} \cdot \frac{4f - 8g}{2f + 6g} = \frac{8(f + 3g)}{2(f - 2g)} \cdot \frac{4(f - 2g)}{2(f + 3g)} = 8$

4. $\frac{m + 2}{2m - 6} \cdot \frac{m^2 - 5m + 6}{2m + 4} = \frac{m + 2}{2(m - 3)} \cdot \frac{(m - 2)(m - 3)}{2(m + 2)} = \frac{m - 2}{4}$

5. $\frac{3z - 6}{5z} \cdot \frac{z^2 - z - 6}{z^2 - 4} = \frac{3(z - 2)}{5z} \cdot \frac{(z - 3)(z + 2)}{(z + 2)(z - 2)} = \frac{3(z - 3)}{5z}$

6. $\frac{x^2 - 4}{x^2 - 9} \cdot \frac{2x^2 + 6x}{x - 2} = \frac{(x + 2)(x - 2)}{(x + 3)(x - 3)} \cdot \frac{2x(x + 3)}{x - 2} = \frac{2x(x + 2)}{x - 3}$

7. $\frac{3n^2 + 2n - 1}{5n^2 - 9n - 2} \cdot \frac{10n^2 - 13n - 3}{2n^2 - n - 3} = \frac{(3n - 1)(n + 1)}{(5n + 1)(n - 2)} \cdot \frac{(5n + 1)(2n - 3)}{(2n - 3)(n + 1)} = \frac{3n - 1}{n - 2}$

8. $\frac{y^2 + 12y + 36}{36 - y^2} \cdot \frac{y^2 - 5y - 6}{2y^2 - 2y - 84} = \frac{(y + 6)^2}{(6 + y)(6 - y)} \cdot \frac{(y - 6)(y + 1)}{2(y^2 - y - 42)} =$
$-\frac{(y + 6)}{(y - 6)} \cdot \frac{(y - 6)(y + 1)}{2(y - 7)(y + 6)} = \frac{y + 1}{2(7 - y)}$

9. $\frac{10a - 10b}{ab} \div \frac{2a - 2b}{a^2b^2} = \frac{10(a - b)}{ab} \cdot \frac{a^2b^2}{2(a - b)} = 5ab$

10. $\frac{16m^2 - n^2}{6m + 3n} \div \frac{4mn - n^2}{2mn + n^2} = \frac{(4m + n)(4m - n)}{3(2m + n)} \cdot \frac{n(2m + n)}{n(4m - n)} = \frac{4m + n}{3}$

11. $\frac{2b^2 + 17b + 21}{b + 1} \div (b + 7) = \frac{(2b + 3)(b + 7)}{b + 1} \cdot \frac{1}{b + 7} = \frac{2b + 3}{b + 1}$

12. $\frac{x^2 + x}{-x^2 - 2x - 1} \div \frac{x^2 - 3x}{2x^2 - 2} = \frac{x(x + 1)}{-(x^2 + 2x + 1)} \cdot \frac{2(x^2 - 1)}{x(x - 3)} = \frac{(x + 1)}{-(x + 1)^2} \cdot \frac{2(x + 1)(x - 1)}{x - 3} = \frac{2(x - 1)}{3 - x}$

13. $\frac{c^2 - c - 6}{c^2 + 2c - 15} \div \frac{c^2 - 4c - 5}{c^2 - 25} = \frac{(c - 3)(c + 2)}{(c + 5)(c - 3)} \cdot \frac{(c + 5)(c - 5)}{(c - 5)(c + 1)} = \frac{c + 2}{c + 1}$

14. $\frac{6y^2 - y - 2}{12y^2 + 5y - 2} \div \frac{8y^2 - 6y + 1}{4y^2 - 1} = \frac{(2y + 1)(3y - 2)}{(3y + 2)(4y - 1)} \cdot \frac{(2y + 1)(2y - 1)}{(4y - 1)(2y - 1)} = \frac{(3y - 2)(2y + 1)^2}{(3y + 2)(4y - 1)^2}$

15. $\frac{r^2 - r - 20}{r^2 - 6r + 5} \cdot \frac{r^2 - 36}{r^2 - 9} \div \frac{r^2 + 7r + 12}{r^2 + 5r - 6} = \frac{(r - 5)(r + 4)}{(r - 5)(r - 1)} \cdot \frac{(r + 6)(r - 6)}{(r + 3)(r - 3)} \cdot \frac{(r + 6)(r - 1)}{(r + 3)(r + 4)} =$
$\frac{(r + 6)^2(r - 6)}{(r + 3)^2(r - 3)}$

16. $\frac{2t - 3}{r^2t} + \frac{r - t}{r^2t} = \frac{2t - 3 + r - t}{r^2t} = \frac{r + t - 3}{r^2t}$

17. $\frac{x + 2y}{x - y} + \frac{x + y}{x - y} = \frac{x + 2y + x + y}{x - y} = \frac{2x + 3y}{x - y}$

18. $\frac{3b+2}{b^3}-\frac{2b+2}{b^3}=\frac{3b+2-(2b+2)}{b^3}=\frac{3b+2-2b-2}{b^3}=\frac{b}{b^3}=\frac{1}{b^2}$

19. $\frac{w-2}{w-3}-\frac{w^2+w}{w-3}=\frac{w-2-(w^2+w)}{w-3}=\frac{w-2-w^2-w}{w-3}=\frac{2+w^2}{3-w}$

20. $z+\frac{z-1}{z+1}+1=\frac{z(z+1)+(z-1)+(z+1)}{z+1}=\frac{z^2+z+z-1+z+1}{z+1}=\frac{z^2+3z}{z+1}=\frac{z(z+3)}{z+1}$

21. $\frac{8}{v^2-1}+\frac{v}{v+1}=\frac{8+v(v-1)}{(v+1)(v-1)}=\frac{v^2-v+8}{(v+1)(v-1)}$

22. $\frac{s}{s-t}-\frac{s}{s+t}=\frac{s(s+t)-s(s-t)}{(s-t)(s+t)}=\frac{s^2+st-s^2+st}{(s-t)(s+t)}=\frac{2st}{(s-t)(s+t)}$, or $\frac{2st}{s^2-t^2}$

23. $\frac{3b}{2b-1}-b=\frac{3b-b(2b-1)}{2b-1}=\frac{3b-2b^2+b}{2b-1}=\frac{4b-2b^2}{2b-1}=\frac{2b(2-b)}{2b-1}$

24. $\frac{x+3}{x-1}+\frac{x}{x+2}=\frac{(x+3)(x+2)+x(x-1)}{(x-1)(x+2)}=\frac{x^2+5x+6+x^2-x}{(x-1)(x+2)}=\frac{2x^2+4x+6}{(x-1)(x+2)}=$

$\frac{2(x^2+2x+3)}{(x-1)(x+2)}$

25. $\frac{d-3}{2d+1}+\frac{1}{d+1}=\frac{(d-3)(d+1)+(2d+1)}{(2d+1)(d+1)}=\frac{d^2-2d-3+2d+1}{(2d+1)(d+1)}=\frac{d^2-2}{(2d+1)(d+1)}$

26. $\frac{v+4}{v+2}-\frac{v+2}{v+4}=\frac{(v+4)^2-(v+2)^2}{(v+2)(v+4)}=\frac{(v^2+8v+16)-(v^2+4v+4)}{(v+2)(v+4)}=$

$\frac{v^2+8v+16-v^2-4v-4}{(v+2)(v+4)}=\frac{4v+12}{(v+2)(v+4)}=\frac{4(v+3)}{(v+2)(v+4)}$

27. $\frac{t}{t^2-4}-\frac{5t}{4-t^2}=\frac{t+5t}{t^2-4}=\frac{6t}{(t+2)(t-2)}$, or $\frac{6t}{t^2-4}$

28. $\frac{2b}{b+2}+\frac{b+1}{b}-1=\frac{(2b)b+(b+1)(b+2)-b(b+2)}{b(b+2)}=\frac{2b^2+b^2+3b+2-b^2-2b}{b(b+2)}=$

$\frac{2b^2+b+2}{b(b+2)}$

29. $\frac{x+4}{x-4}-\frac{x-4}{x+3}+2=\frac{(x+4)(x+3)-(x-4)(x-4)+2(x-4)(x+3)}{(x-4)(x+3)}=$

$\frac{x^2+7x+12-(x^2-8x+16)+2(x^2-x-12)}{(x-4)(x+3)}=$

$\frac{x^2+7x+12-x^2+8x-16+2x^2-2x-24}{(x-4)(x+3)}=\frac{2x^2+13x-28}{(x-4)(x+3)}$

30. $\frac{3a+b}{a^2-b^2}-\frac{1}{a+b}=\frac{3a+b-(a-b)}{(a+b)(a-b)}=\frac{3a+b-a+b}{(a+b)(a-b)}=\frac{2a+2b}{(a+b)(a-b)}=\frac{2(a+b)}{(a+b)(a-b)}=$

$\frac{2}{a-b}$

Page 455 • MIXED PROBLEM SOLVING

A **1.** $\frac{x}{100}=\frac{101.3-47.2}{101.3}$; $\frac{x}{100}=\frac{54.1}{101.3}$; $101.3x=5410$; $x\approx 53.4$; The decrease in the air pressure is about 53.4%.

2. Let $11x$ = number of boys and $12x$ = number of girls. $11x+12x=460$; $23x=460$; $x=20$; $12x=240$; There are 240 girls.

3. Let n and $n + 2$ be the integers. $n^2 + (n + 2)^2 = 34$; $n^2 + n^2 + 4n + 4 = 34$; $2n^2 + 4n - 30 = 0$; $n^2 + 2n - 15 = 0$; $(n + 5)(n - 3) = 0$; $n = -5$ or 3; No solution; Contradictory facts; The integers were supposed to be even.

4. Let x be the cost of an ear of corn and let y be the cost of a cantaloupe.

$$\begin{aligned} 8x + y = 2.37 &\rightarrow 24x + 3y = 7.11 \\ 6x + 3y = 3.51 &\rightarrow \underline{\ \ 6x + 3y = 3.51} \\ &\quad\ 18x \qquad\ = 3.6 \end{aligned}$$

$x = 0.2$; $8(0.2) + y = 2.37$; $1.6 + y = 2.37$; $y = 0.77$; A cantaloupe costs 77¢.

5.

	Rate	× Time	= Distance
With wind	$r + w$	6	1800
Against wind	$r - w$	6	$\frac{2}{3}(1800)$

$$\begin{aligned} 6(r + w) = 1800 &\rightarrow 6r + 6w = 1800 \\ 6(r - w) = 1200 &\rightarrow \underline{\ 6r - 6w = 1200} \\ &\quad\ 12r \qquad = 3000 \end{aligned}$$

$r = 250$; $6(250) + 6w = 1800$; $1500 + 6w = 1800$; $6w = 300$; $w = 50$; The wind speed is 50 km/h.

6.

	Rate	× Time	= Distance
Boys	6	t	$6t$
Mother	60	$t - \frac{3}{2}$	$60\left(t - \frac{3}{2}\right)$

$6t = 60\left(t - \frac{3}{2}\right)$; $6t = 60t - 90$; $90 = 54t$; $\frac{5}{3} = t$; $\frac{5}{3} - \frac{3}{2} = \frac{10 - 9}{6} = \frac{1}{6}$; Their mother caught up with them at 10:40 A.M.

7. Let t be the amount of tax on a \$9450 car. $\frac{738}{12{,}300} = \frac{t}{9450}$; $12{,}300t = 6{,}974{,}100$; $t = 567$; The tax is \$567.

8. Let t and u be the tens and units digits. $t = 3u + 1$; $10t + u - 45 = 10u + t$; $9t - 9u = 45$; $9(3u + 1) - 9u = 45$; $27u + 9 - 9u = 45$; $18u = 36$; $u = 2$; $t = 3(2) + 1 = 6 + 1 = 7$; The number is 72.

B **9.** Let t be the time it takes them to finish. $\frac{t + \frac{1}{2}}{2} + \frac{t}{3} = 1$; $6\left(\frac{t + \frac{1}{2}}{2} + \frac{t}{3}\right) = 6$; $3\left(t + \frac{1}{2}\right) + 2t = 6$; $3t + \frac{3}{2} + 2t = 6$; $5t = \frac{9}{2}$; $t = 0.9$; It took them 54 min to finish.

10. Let w be the width and let $2w$ be the length of the base. $(2w - 7)(w + 2) = w(2w) - 32$; $2w^2 + 4w - 7w - 14 = 2w^2 - 32$; $-3w - 14 = -32$; $18 = 3w$; $6 = w$; $2w = 12$; The original dimensions of the receiver were 6 in. by 12 in.

11.

	Amount	× % salt	= Total salt
First solution	8	25%	2
Water	w	0%	0
Second solution	$8 - w$	40%	2

$0.4(8 - w) = 2$; $3.2 - 0.4w = 2$; $1.2 = 0.4w$; $3 = w$; 3 kg of water must be evaporated.

12. Let n = numerator and d = denominator of the fraction. $\frac{n}{d} = \frac{3}{4}$; $4n = 3d$; $\frac{4}{3}n = d$; $\frac{n+7}{d} = \frac{d}{n}$; $n^2 + 7n = d^2$; $n^2 + 7n = \left(\frac{4}{3}n\right)^2$; $n^2 + 7n = \frac{16}{9}n^2$; $9n^2 + 63n = 16n^2$; $0 = 7n^2 - 63n$; $0 = 7n(n - 9)$; $n = 0$ (reject) or 9; $d = \frac{4}{3}(9)$; $d = 12$; Original fraction is $\frac{9}{12}$.

13. Let x be the amount invested at $7\frac{1}{4}\%$ and let $6000 - x$ be the amount invested at 10%. $x \cdot 0.0725 + (6000 - x)0.10 = 567$; $10{,}000[0.0725x + 600 - 0.1x] = 5{,}670{,}000$; $725x + 6{,}000{,}000 - 1000x = 5{,}670{,}000$; $330{,}000 = 275x$; $x = 1200$; $6000 - x = 6000 - 1200 = 4800$. The Kwons invested \$1200 at $7\frac{1}{4}\%$ and \$4800 at 10%.

14. Let d and q be the number of dimes and quarters. $10d + 25q = 115$; $25q = 115 - 10d$; $q = \frac{115 - 10d}{25}$

d	0	1	2	3	4	5	6	7	8	9	10	11	12
q	4.6	4.2	3.8	3.4	3	2.6	2.2	1.8	1.4	1	0.6	0.2	< 0

I could have 4 dimes and 3 quarters, or 9 dimes and 1 quarter.

Pages 459–460 • WRITTEN EXERCISES

A **1.** $4 > -7$ **2.** $-5 < -3$ **3.** $-12 \le -9$ **4.** $6 \ge 2$ **5.** $2 < 3 < 3.5$

6. $-\frac{1}{2} < 0 < 1$ **7.** $-10 < -8 < 0$ **8.** $-9 < 5 < 9$ **9.** $0 < 4 < 4.6$

10. $-1.5 < -1 < 2$ **11.** $n > 10$ **12.** $n < 20$ **13.** $|-2| > 1$ **14.** $|n| \le n$

15. True **16.** True **17.** False **18.** False **19.** True **20.** True

21. False **22.** True

23. $4x < 8$
$4(-4) < 8$ True
$4(-3) < 8$ True
$4(-2) < 8$ True
$4(-1) < 8$ True
$4(0) < 8$ True
$4(1) < 8$ True
$4(2) < 8$ False
$4(3) < 8$ False
$4(4) < 8$ False
Solution: $\{-4, -3, -2, -1, 0, 1\}$

24. $3x \ge 6$
$3(-4) \ge 6$ False
$3(-3) \ge 6$ False
$3(-2) \ge 6$ False
$3(-1) \ge 6$ False
$3(0) \ge 6$ False
$3(1) \ge 6$ False
$3(2) \ge 6$ True
$3(3) \ge 6$ True
$3(4) \ge 6$ True
Solution: $\{2, 3, 4\}$

25. $-3x \le 9$
$-3(-4) \le 9$ False
$-3(-3) \le 9$ True
$-3(-2) \le 9$ True
$-3(-1) \le 9$ True
$-3(0) \le 9$ True
$-3(1) \le 9$ True
$-3(2) \le 9$ True
$-3(3) \le 9$ True
$-3(4) \le 9$ True
Solution: $\{-3, -2, -1, 0, 1, 2, 3, 4\}$

26. $x + 2 < 3$
$(-4) + 2 < 3$ True
$(-3) + 2 < 3$ True
$(-2) + 2 < 3$ True
$(-1) + 2 < 3$ True
$0 + 2 < 3$ True
$1 + 2 < 3$ False
$2 + 2 < 3$ False
$3 + 2 < 3$ False
$4 + 2 < 3$ False
Solution: $\{-4, -3, -2, -1, 0\}$

27. $-5 - x \le 1$
$-5 - (-4) \le 1$ True
$-5 - (-3) \le 1$ True
$-5 - (-2) \le 1$ True
$-5 - (-1) \le 1$ True
$-5 - 0 \le 1$ True
$-5 - 1 \le 1$ True
$-5 - 2 \le 1$ True
$-5 - 3 \le 1$ True
$-5 - 4 \le 1$ True
Solution: $\{-4, -3, -2, -1, 0, 1, 2, 3, 4\}$

28. $1 - x \ge 0$
$1 - (-4) \ge 0$ True
$1 - (-3) \ge 0$ True
$1 - (-2) \ge 0$ True
$1 - (-1) \ge 0$ True
$1 - 0 \ge 0$ True
$1 - 1 \ge 0$ True
$1 - 2 \ge 0$ False
$1 - 3 \ge 0$ False
$1 - 4 \ge 0$ False
Solution: $\{-4, -3, -2, -1, 0, 1\}$

29. $x^2 \geq 10$
$(-4)^2 \geq 10$ True
$(-3)^2 \geq 10$ False
$(-2)^2 \geq 10$ False
$(-1)^2 \geq 10$ False
$0^2 \geq 10$ False
$1^2 \geq 10$ False
$2^2 \geq 10$ False
$3^2 \geq 10$ False
$4^2 \geq 10$ True
Solution: $\{-4, 4\}$

30. $x^2 \leq 4$
$(-4)^2 \leq 4$ False
$(-3)^2 \leq 4$ False
$(-2)^2 \leq 4$ True
$(-1)^2 \leq 4$ True
$0^2 \leq 4$ True
$1^2 \leq 4$ True
$2^2 \leq 4$ True
$3^2 \leq 4$ False
$4^2 \leq 4$ False
Solution: $\{-2, -1, 0, 1, 2\}$

B

31. 

32. 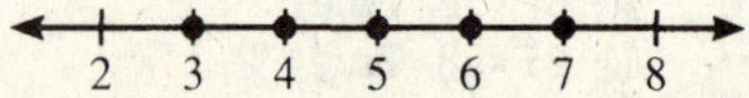

33. 

34. 

35. 

36. 

37. 

38. 

C For Exercises 39–42 answers may vary.

39. a. $(-2, -3)$ **b.** $(3, 2)$ **40. a.** $(0, 1)$ **b.** $(-1, 1)$ **41. a.** $(2, 1)$ **b.** $(2, -1)$

42. a. $(2, -3)$ **b.** $(2, 3)$

Page 460 • COMPUTER EXERCISES

1.
```
10 PRINT "SOLUTIONS"
20 FOR X = 0 TO 10
30 IF 2 * X - 5 > = 7 THEN 50
40 PRINT X;
50 NEXT X
60 END
SOLUTIONS: 0  1  2  3  4  5
```

2.
```
20 FOR X = -4 TO 4
30 IF(X * X - X) > = 6 THEN 50
SOLUTIONS: -1  0  1  2
```

3.
```
20 FOR X = 0 TO 5
30 IF ABS(4 - 5 * X) < = 13 THEN 50
SOLUTIONS: 4  5
```

4.
```
20 FOR X = -3 TO 3
30 IF (X * X + 3 * X + 4) > = 10 THEN 50
SOLUTIONS: -3  -2  -1  0  1
```

5.
```
20 FOR X = 0 TO 10
30 IF (X * X - 4 * X - 5) < = 0 THEN 50
SOLUTIONS: 6  7  8  9  10
```

Page 461 • MIXED REVIEW EXERCISES

1. $x - 5 = 13, x = 18; \{18\}$ **2.** $14 = 2(c + 1), 14 = 2c + 2, 12 = 2c, 6 = c; \{6\}$

3. $5 - 2a = 17, -2a = 12, a = -6; \{-6\}$ **4.** $\frac{s}{5} = -10, s = -50; \{-50\}$

5. $\frac{21}{y} = \frac{7}{9}, 7y = 189, y = 27; \{27\}$

6. $\frac{3x + 1}{5} = \frac{x + 12}{4}, 4(3x + 1) = 5(x + 12), 12x + 4 = 5x + 60, 7x = 56, x = 8; \{8\}$

7. $5(2 + n) = 4(n - 6), 10 + 5n = 4n - 24, n = -34; \{-34\}$

8. $(x + 4)(x + 7) = (x + 5)^2, x^2 + 11x + 28 = x^2 + 10x + 25, 11x + 28 = 10x + 25,$ $x = -3; \{-3\}$

9. $\frac{b}{5} + 8 = b, b + 40 = 5b, 40 = 4b, 10 = b; \{10\}$

10. $\frac{x}{4} = \frac{x - 3}{5}, 5x = 4(x - 3), 5x = 4x - 12, x = -12; \{-12\}$

11. $\frac{3p}{10} + \frac{p}{5} = \frac{1}{2}, 3p + 2p = 5, 5p = 5, p = 1; \{1\}$

12. $\frac{1}{3}(x + 9) = 3, x + 9 = 9, x = 0; \{0\}$

Pages 466–467 • WRITTEN EXERCISES

A **1.** $y - 2 \geq 7, y \geq 9$; d **2.** $10 < z + 8, 2 < z$; a **3.** $6p < 24, p < 4$; j

4. $18 \leq 6v, 3 \leq v$; h **5.** $-28 > -7m, 4 < m$; b **6.** $\frac{d}{2} > -10, d > -20$; g

7. $2 - g > 0, 2 > g$; e **8.** $3 \leq \frac{x}{-2}, -6 \geq x$; i

9. $b - 1 < b - 2, -1 < -2$ (no solution); f

10. $t + 2 > t + 1, 2 > 1$ (true for all reals); c

11. $n - 4 > 11, n > 15$; {the real numbers greater than 15}

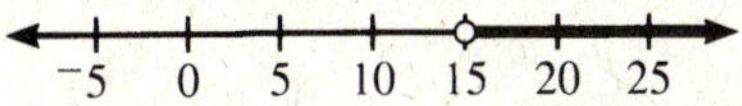

12. $a + 3 < 11, a < 8$; {the real numbers less than 8}

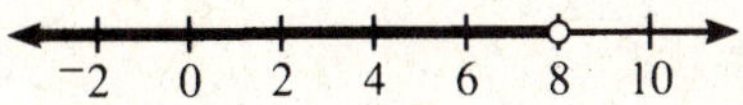

13. $-\frac{y}{2} > 4, y < -8$; {the real numbers less than -8}

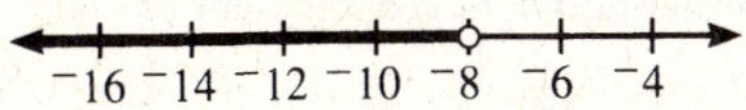

14. $4q < 12, q < 3$; {the real numbers less than 3}

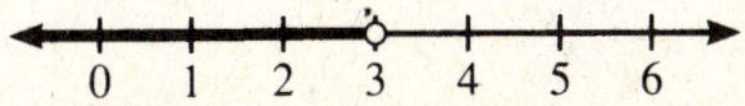

15. $36 < \frac{x}{-9}$, $-324 > x$; {the real numbers less than -324}

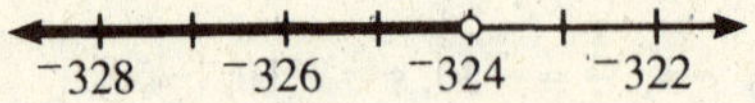

16. $-6 > \frac{s}{3}$, $-18 > s$; {the real numbers less than -18}

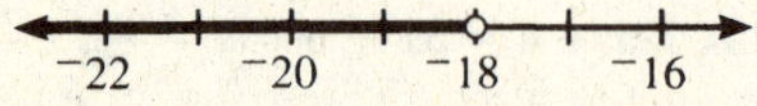

17. $-6m \geq 6$, $m \leq -1$; {the real numbers less than or equal to -1}

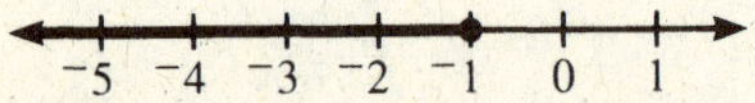

18. $-5w \leq 15$, $w \geq -3$; {the real numbers greater than or equal to -3}

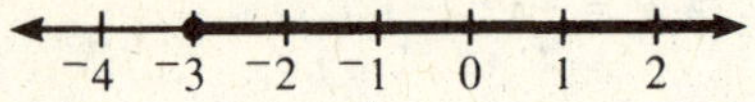

19. $\frac{x}{2} - 4 > -6$, $x - 8 > -12$, $x > -4$; {the real numbers greater than -4}

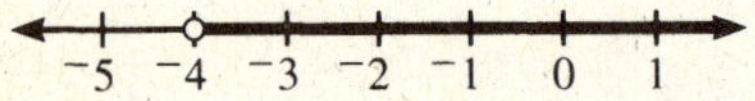

20. $2v + 1 > 7$, $2v > 6$, $v > 3$; {the real numbers greater than 3}

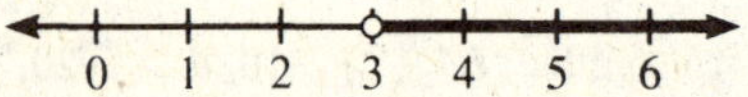

21. $-\frac{d}{3} \geq 0$, $d \leq 0$; {the real numbers less than or equal to 0}

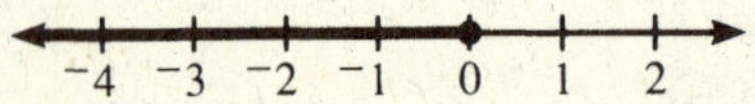

22. $7 \geq 2k - 7$, $14 \geq 2k$, $7 \geq k$; {the real numbers less than or equal to 7}

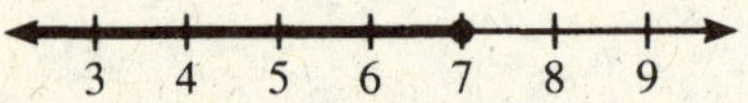

23. $\frac{3}{2}t - 7 < 2$, $\frac{3}{2}t < 9$, $t < 6$; {the real numbers less than 6}

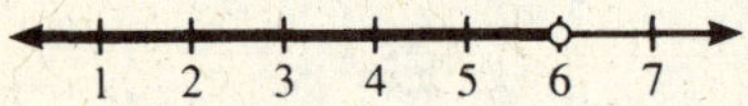

24. $-4 \geq 4 - \frac{u}{2}$, $-8 \geq -\frac{u}{2}$, $16 \leq u$; {the real numbers greater than or equal to 16}

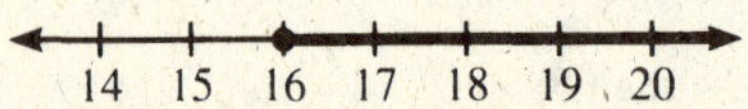

25. $-1 < 9 + \frac{2}{3}g$, $-10 < \frac{2}{3}g$, $-15 < g$; {the real numbers greater than -15}

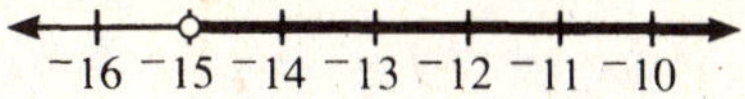

26. $5 + \frac{x}{3} \le 6$, $\frac{x}{3} \le 1$, $x \le 3$; {the real numbers less than or equal to 3}

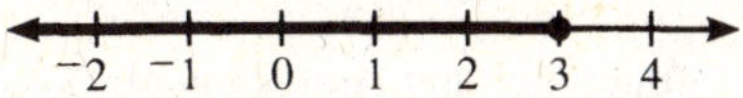

27. $12 - \frac{3}{2}c > 0$, $12 > \frac{3}{2}c$, $8 > c$; {the real numbers less than 8}

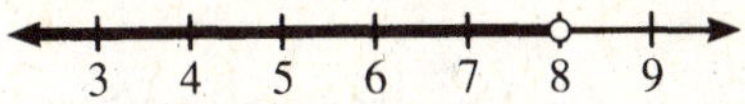

28. $0 > 6 - \frac{2}{3}d$, $\frac{2}{3}d > 6$, $d > 9$; {the real numbers greater than 9}

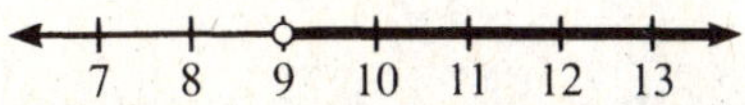

29. $5y < 4y + 6$, $y < 6$; {the real numbers less than 6}

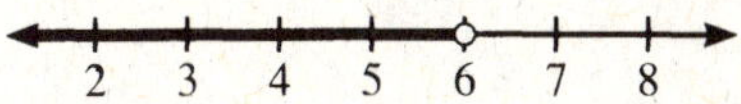

30. $3f - 4 < 2f + 5$, $f < 9$; {the real numbers less than 9}

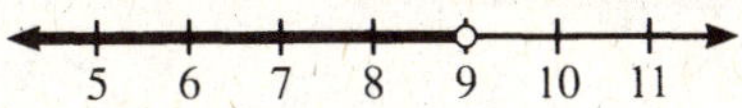

31. $4r - 5 < 5r + 7$, $-12 < r$; {the real numbers greater than -12}

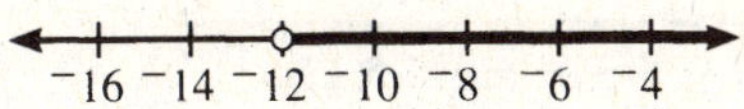

32. $8 - 2b > 4 - b$, $4 > b$; {the real numbers less than 4}

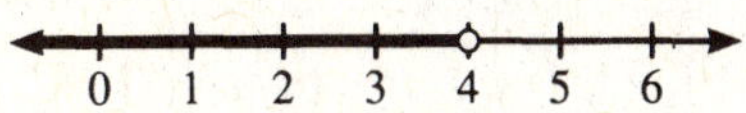

33. $2(x - 4) \le 6$, $2x - 8 \le 6$, $2x \le 14$, $x \le 7$; {the real numbers less than or equal to 7}

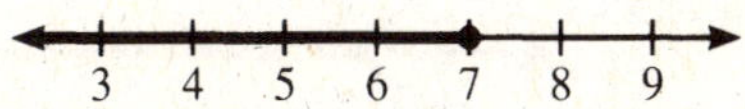

34. $8 < 2(4 - m)$, $8 < 8 - 2m$, $2m < 0$, $m < 0$; {the real numbers less than 0}

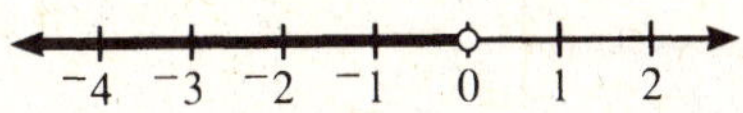

35. $-5(x + 3) < -5x + 1$, $-5x - 15 < -5x + 1$, $-15 < 1$; {the real numbers}

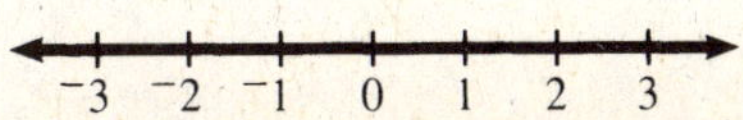

36. $7a < 3 + 7(a - 1)$, $7a < 3 + 7a - 7$, $0 < -4$; Since $0 < -4$ is false, the original inequality has no solution; there is no graph.

37. $5(1 - t) > 4(3 - t)$, $5 - 5t > 12 - 4t$, $-7 > t$; {the real numbers less than -7}

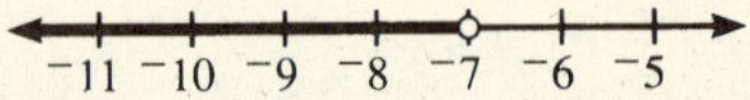

38. $4(2 - v) \geq -(v - 5)$, $8 - 4v \geq -v + 5$, $3 \geq 3v$, $1 \geq v$; {the real numbers less than or equal to 1}

39. $\frac{5}{6}r + 1 \geq \frac{4}{3}$, $5r + 6 \geq 8$, $5r \geq 2$, $r \geq \frac{2}{5}$; {the real numbers greater than or equal to $\frac{2}{5}$}

40. $\frac{3}{4} < 6 - \frac{1}{2}a$, $3 < 24 - 2a$, $2a < 21$, $a < \frac{21}{2}$, $a < 10\frac{1}{2}$; $\left\{\text{the real numbers less than } 10\frac{1}{2}\right\}$

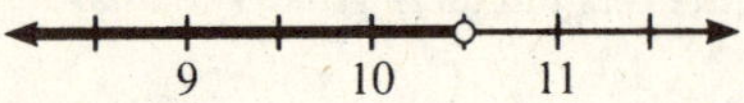

B **41.** $5(x - 1) > \frac{5}{4}x$, $20(x - 1) > 5x$, $20x - 20 > 5x$, $15x > 20$, $x > \frac{4}{3}$; $\left\{\text{the real numbers greater than } \frac{4}{3}\right\}$

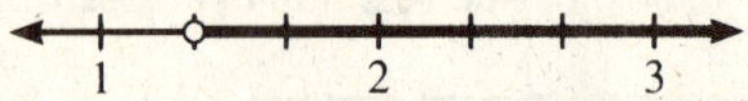

42. $\frac{1}{2}y \geq 2(y - 3)$, $y \geq 4(y - 3)$, $y \geq 4y - 12$, $12 \geq 3y$, $4 \geq y$; {the real numbers less than or equal to 4}

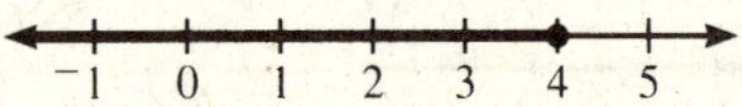

43. $n - \frac{5}{2} > \frac{3}{4}(n - 6)$, $4n - 10 > 3(n - 6)$, $4n - 10 > 3n - 18$, $n > -8$; {the real numbers greater than -8}

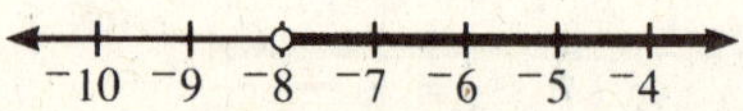

44. $3w - \frac{1}{2}(2w + 8) > 0$, $6w - (2w + 8) > 0$, $6w - 2w - 8 > 0$, $4w > 8$, $w > 2$; {the real numbers greater than 2}

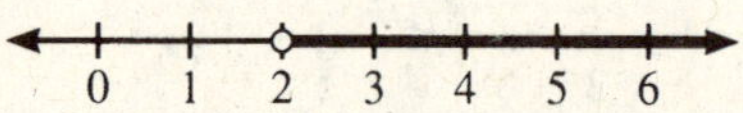

45. $5(5 - k) - 7(7 + k) < 0$, $25 - 5k - 49 - 7k < 0$, $-24 - 12k < 0$, $-24 < 12k$, $-2 < k$; {the real numbers greater than -2}

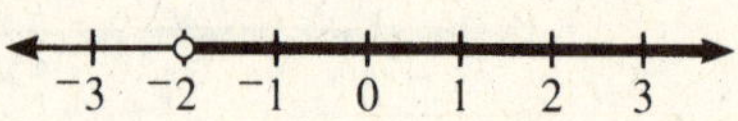

46. $5(v - 1) > 3(v + 4) - 5$, $5v - 5 > 3v + 12 - 5$, $2v > 12$, $v > 6$; {the real numbers greater than 6}

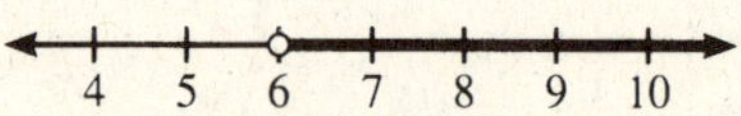

47. $4\left(r - \frac{1}{2}\right) - 3 \le 5(r - 1) + 4, 4r - 2 - 3 \le 5r - 5 + 4, 4r - 5 \le 5r - 1, -4 \le r;$ {the real numbers greater than or equal to -4}

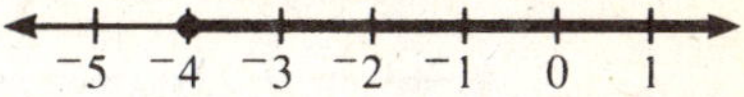

48. $4(t + 1) - 2(t - 1) \le 3 - t, 4t + 4 - 2t + 2 \le 3 - t, 2t + 6 \le 3 - t, 3t \le -3,$ $t \le -1$; {the real numbers less than or equal to -1}

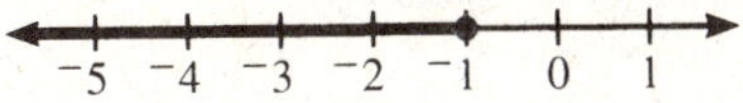

49. $5 - a + 6(a - 2) \ge 5(a + 1), 5 - a + 6a - 12 \ge 5a + 5, 5a - 7 \ge 5a + 5,$ $-7 \ge 5$; Since $-7 \ge 5$ is false, the original inequality has no solution; there is no graph.

50. $5(2b + 1) - 3(b + 1) < 7b + 5, 10b + 5 - 3b - 3 < 7b + 5, 7b + 2 < 7b + 5,$ $2 < 5$; {the real numbers}

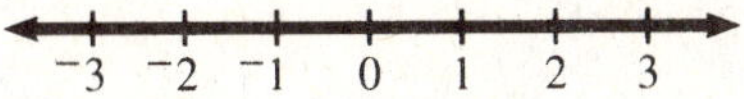

C **51.** All real numbers **52.** No real numbers **53.** All positive real numbers **54.** All negative real numbers **55.** Zero **56.** All positive real numbers **57.** All negative real numbers **58.** No real numbers **59.** True **60.** False; for example, $\left(\frac{1}{2}\right)^2 > \frac{1}{2}$ is false. **61.** True **62.** True **63.** False; for example, $\left(\frac{1}{2}\right)^2 > \frac{1}{3}$ is false. **64.** True **65.** True **66.** True

Page 467 • MIXED REVIEW EXERCISES

1. $-(-2) < |-3|, 2 < 3$; True **2.** $|-3| \ge |3|, 3 \ge 3$; True

3. $|-13| < |-12|, 13 < 12$; False

4. $5(4 + 3 \cdot 2) = 70, 5(4 + 6) = 70, 5 \cdot 10 = 70, 50 = 70$; False

5. $2x + 10 = 2(x + 5), 2x + 10 = 2x + 10$; True

6. $(x + 3)(x + 7) = x^2 + 10x, x^2 + 10x + 21 = x^2 + 10x; 21 = 0$; False

7. $3f - 4 = 2f + 5, f = 9$; {9} **8.** $0 = 2x + 5, -5 = 2x, -\frac{5}{2} = x; \left\{-\frac{5}{2}\right\}$

9. $a(a + 5) = (a - 9)(a - 7), a^2 + 5a = a^2 - 16a + 63, 5a = -16a + 63, 21a = 63,$ $a = 3$; {3}

10. $4y - 2(y + 1) = -6, 4y - 2y - 2 = -6, 2y = -4, y = -2$; {−2}

11. $p - 2(6 - p) = -p, p - 12 + 2p = -p, 3p - 12 = -p, 4p = 12, p = 3$; {3}

12. $3b + 2(b + 1) = b + 6, 3b + 2b + 2 = b + 6, 5b + 2 = b + 6, 4b = 4, b = 1$; {1}

Page 468 • READING ALGEBRA

1. a. greater than **b.** open; right **c.** $x > 15$; {the real numbers greater than 15};

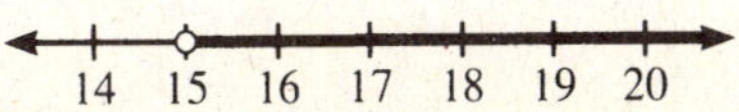

2. a. greater than **b.** open; right

c. $3y > 6$; $y > 2$; {the real numbers greater than 2};

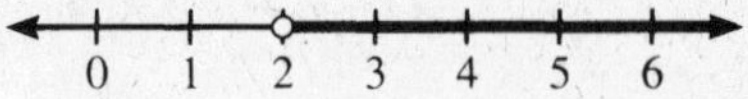

3. a. greater than or equal to **b.** closed; left

c. $8 \geq 2y$; $4 \geq y$; {the real numbers less than or equal to 4};

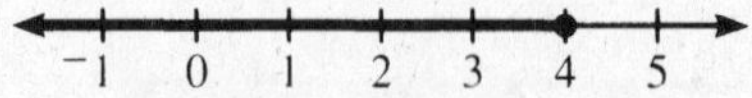

4. a. less than or equal to **b.** closed; left

c. $y \leq 12$; {the real numbers less than or equal to 12};

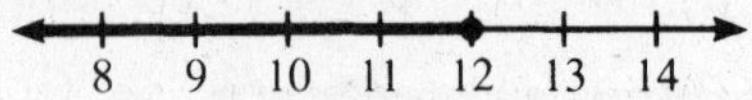

5. a. less than **b.** open; right

c. $-30 < 2m$; $-15 < m$; {the real numbers greater than -15};

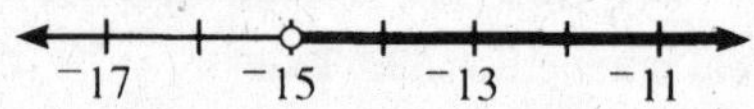

6. a. less than or equal to **b.** closed; left

c. $-12 \leq -k$; $k \leq 12$; {the real numbers less than or equal to 12};

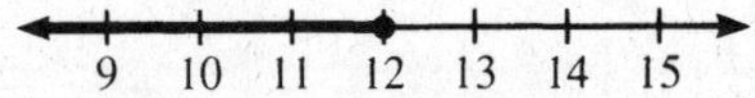

7. a. greater than **b.** open; left **c.** $18 > 3y$; $6 > y$; {the real numbers less than 6};

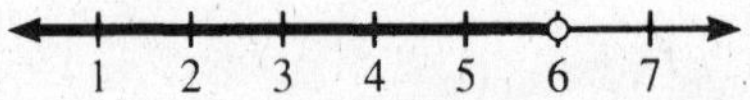

8. a. greater than or equal to **b.** closed; right

c. $3r + r \geq 5 - 1$; $4r \geq 4$; $r \geq 1$; {the real numbers greater than or equal to 1};

9. a. less than **b.** open; right

c. $2x - 4x < 5 + 3$; $-2x < 8$; $x > -4$; {the real numbers greater than -4};

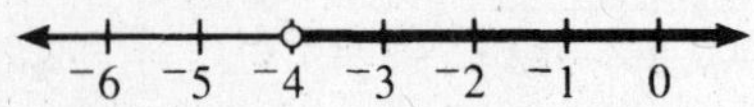

10. a. less than **b.** open; right

c. $21 - 15x < -8x - 7$; $8x - 15x < -7 - 21$; $-7x < -28$; $x > 4$; {the real numbers greater than 4};

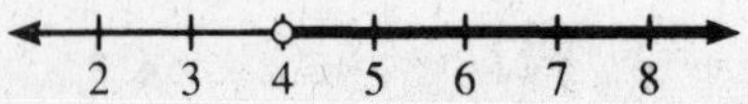

11. a. greater than **b.** open; left

c. $4(t - 2) > 5(t - 3)$; $4t - 8 > 5t - 15$; $4t - 5t > -15 + 8$; $-t > -7$; $t < 7$; {the real numbers less than 7};

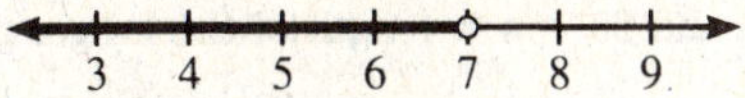

12. a. less than or equal to **b.** closed; left

c. $6(z - 5) \leq 15 + 5(7 - 2z)$; $6z - 30 \leq 15 + 35 - 10z$; $6z + 10z \leq 50 + 30$; $16z \leq 80$; $z \leq 5$; {the real numbers less than or equal to 5};

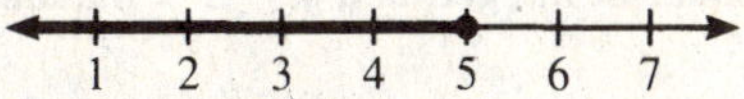

Pages 471–472 • WRITTEN EXERCISES

A

1. a. Let i = Ida's age and $i + 2$ = Ilene's age. **b.** $i + 2 < 21$

2. a. Let e = total number of envelopes. **b.** $e - 75 < 26$

3. a. Let x = number of km traveled by airplane and $0.1x$ = number of km by automobile.

b. $x + 0.1x > 3000$

4. a. Let c = number of hours Clark worked and $c + \frac{1}{2}$ = number of hours John worked.

b. $c + \left(c + \frac{1}{2}\right) \geq 2$

5. a. Let d = amount of deposit. **b.** $75 + d \geq 110$

6. a. Let w = width of the rectangle and $w + 4$ = length. **b.** $2w + 2(w + 4) \leq 28$

7. a. Let j = number of kilometers Juan ran and $\frac{j}{2} + 15$ = number of km Peter ran.

b. $\frac{j}{2} + 15 \leq 36$

8. a. Let p = amount of the monthly payment and $\frac{p}{2}$ = cost for gas and repairs

b. $p + \frac{p}{2} \geq 190$

9. a. Let m = middle integer, $m - 2$ = smallest integer, and $m + 2$ = greatest integer.

b. $(m - 2) + m + (m + 2) \geq 51$

10. a. Let g = the greater integer and $(g - 2)$ = smaller integer. **b.** $g(g - 2) \leq 255$

B

11. a. Let t = number of hours of traveling. **b.** $86t + 78t \geq 650$

12. a. Let s = the motorist's present speed. **b.** $3(s + 10) > 4s$

13. a. Let d = number of dimes, $2d$ = number of nickels, and $\frac{2d}{3}$ = number of quarters.

b. $10d + 5(2d) + 25\left(\frac{2d}{3}\right) \leq 280$

14. **a.** Let m = cost of the muffin and $125 - m$ = cost of the milk.

b. $125 - m < \frac{2}{3}m$

15. **a.** Let s = length of the shorter piece and $25 - s$ = length of the longer piece.

b. $25 - s \geq 2s + 1$

16. **a.** Let s = smaller integer and $s + 2$ = greater integer. **b.** $s + 2 \leq 5s - 50$

17. **a.** Let s = smaller integer and $s + 2$ = greater integer. **b.** $s + (s + 2) \leq \frac{s}{8} - 100$

18. **a.** Let g = greatest integer, $g - 4$ = smallest integer, and $g - 2$ = middle integer.

b. $(g - 4) + (g - 2) + g > 60 - 2(g - 4)$

C **19.** $|a + b| \leq |a| + |b|$ **20.** $m^2 + n^2 \geq 2mn$

Pages 473–474 • PROBLEMS

A **1.** Let c = number of copies originally. $c - 12 < 75$, $c < 87$; The newsdealer had fewer than 87 copies originally.

2. Let l = cost of the lot, and $7l + 1000$ = cost of the house. $l + (7l + 1000) > 89{,}000$, $8l + 1000 > 89{,}000$, $8l > 88{,}000$, $l > 11{,}000$; The lot costs more than $11,000.

3. Let m = number of miles Martha can drive. $94 + m(0.12) \leq 130$, $0.12m \leq 36$, $m \leq 300$; Martha may drive no more than 300 miles.

4. Let s = amount of Jardan's sales. $1250 + 0.05s \geq 1500$, $0.05s \geq 250$, $s \geq 5000$; Jardan's sales must be at least $5000.

5. Let n and $n + 1$ be the integers. $n + (n + 1) < 55$, $2n + 1 < 55$, $2n < 54$, $n < 27$; The greatest value for the first integer is 26. Thus, the pair of integers with the greatest sum is 26 and 27.

6. Let n and $n + 2$ be the integers. $n + (n + 2) \leq 400$, $2n + 2 \leq 400$, $2n \leq 398$, $n \leq 199$; The greatest value for the first even integer is 198. The pair with the greatest sum is 198 and 200.

7. Let t = length of time the trucks travel. $88t + 72t \geq 672$, $160t \geq 672$, $t \geq 4.2$; The trucks must travel for at least 4.2 h.

8. Let t = number of cassettes Terry has and $50 - t$ = number of cassettes Shelley has. $50 - t > \frac{2}{3}t$, $150 - 3t > 2t$, $150 > 5t$, $30 > t$; Terry has 29 cassettes at most and Shelley has at least 21.

9. Let b = number of blue marbles and $100 - b$ = number of red marbles. $100 - b \leq \frac{3}{2}b$, $200 - 2b \leq 3b$, $200 \leq 5b$, $40 \leq b$; There are at least 40 blue marbles and no more than 60 red marbles.

10. Let q = number of quarters, and $22 - q$ = number of dimes.
$25q + (22 - q)10 > 340$, $25q + 220 - 10q > 340$, $15q + 220 > 340$, $15q > 120$, $q > 8$; At least 9 of the coins are quarters, and there are at most 13 dimes.

11. Let w = width of the rectangle and $w + 4$ = length of the rectangle.
$2w + 2(w + 4) \geq 48$, $2w + 2w + 8 \geq 48$, $4w \geq 40$, $w \geq 10$; The smallest possible dimensions for the rectangle are 10 cm by 14 cm.

12. Let n and $n + 1$ be the integers. $7n < 6(n + 1)$, $7n < 6n + 6$, $n < 6$; The greatest such integers are 5 and 6.

B **13.** Let t = time trains travel. They will meet earliest when their speeds are the greatest; that is, when one travels 82 km/h and the other at 74 km/h.
$82t + 74t = 312$, $156t = 312$, $t = 2$; The earliest time will be 3 P.M.

14. Let s = Maura's present speed and $s + 3.5$ = her increased speed. $2(s + 3.5) \geq 3s$, $2s + 7 \geq 3s$, $7 \geq s$; Maura's best average speed is 7 km/h.

15. Let x = score on third exam. $\frac{1}{3}(75 + 81 + x) \geq 80$, $156 + x \geq 240$, $x \geq 84$; The student must get at least 84 on the third exam.

16. Let s = amount of Betty's sales. $15{,}600 \leq 14{,}000 + 0.08s$, $1600 \leq 0.08s$, $20{,}000 \leq s$; Betty's sales must be at least \$20,000.

17. Let h = number of hours worked. $0.75(20h) \geq 450$, $15h \geq 450$, $h \geq 30$; The mechanic must work at least 30 h.

18.

	Amount	× % silver	= Amount silver
Copper	x	0%	0
Silver	387	100%	387
Alloy	$387 + x$	$\leq$ 90%	387

$\frac{387}{387 + x} \leq 0.90$, $387 \leq 0.90(387 + x)$ [note: $387 + x$ is a pos. number],

$387 \leq 348.3 + 0.9x$, $38.7 \leq 0.9x$, $43 \leq x$; At least 43 g of copper must be used.

19.

	Rate	× Time	= Distance
Going	5.2	t	$5.2t$
Returning	7.8	$3.5 - t$	$7.8(3.5 - t)$

The distance will be greatest when the time is greatest; that is, when total time = 3.5 h. $5.2t = 7.8(3.5 - t)$, $5.2t = 27.3 - 7.8t$, $13t = 27.3$, $t = 2.1$; $5.2t = 5.2(2.1) = 10.92$; Randy's campsite is at most 10.92 km from the ranch.

20. Let w = width of the rectangle and $w + 10$ = length.
$(w + 3)(w + 13) \geq w(w + 10) + 111$, $w^2 + 16w + 39 \geq w^2 + 10w + 111$, $6w \geq 72$, $w \geq 12$; The rectangle's least possible dimensions are 12 cm by 22 cm.

C **21.** Let v = amount of vacation money and $200 + \frac{3}{5}v$ = amount spent.

$v - \left(200 + \frac{3}{5}v\right) > \frac{1}{2}v - 400$, $v - 200 - \frac{3}{5}v > \frac{1}{2}v - 400$,

$10v - 2000 - 6v > 5v - 4000$, $4v - 2000 > 5v - 4000$, $2000 > v$; They had at most \$1999 for their vacation.

22. Let n, $n + 1$, and $n + 2$ be the integers. $(n + 1)^2 - n^2 > (n + 2) + 3$, $n^2 + 2n + 1 - n^2 > n + 5$, $2n + 1 > n + 5$, $n > 4$; The smallest integers are 5, 6, and 7.

23. Let b = number of books in her original collection.

$$2 + \frac{1}{5}(b - 2) > 6 + \frac{1}{5}\left[\frac{4}{5}(b - 2) - 6\right], 25\left(2 + \frac{1}{5}b - \frac{2}{5}\right) > 25\left[6 + \frac{1}{5}\left(\frac{4}{5}b - \frac{8}{5} - 6\right)\right],$$

$$50 + 5b - 10 > 150 + 5\left(\frac{4}{5}b - \frac{8}{5} - 6\right), 40 + 5b > 150 + 4b - 8 - 30,$$

$40 + 5b > 112 + 4b$, $b > 72$; Because Fred and Joan buy a whole number of books, $b - 2$ and $\left[\frac{4}{5}(b - 2) - 6\right]$ must be divisible by 5. Checking 77, 82, 87, 92, and 97 as possible values for b shows the least possible number of books in her collection was 97.

Page 475 • MIXED REVIEW EXERCISES

1. $|s| = 7; s = 7$ or $s = -7; \{-7, 7\}$ **2.** $|2 - 5| = k, |-3| = k, 3 = k; \{3\}$
3. $|y| - 2 = 10, |y| = 12; y = 12$ or $y = -12; \{-12, 12\}$
4. $3|b| = 12, |b| = 4, b = 4$ or $b = -4; \{-4, 4\}$
5. $x = |-2 - (-6)|, x = |-2 + 6|, x = |4|, x = 4; \{4\}$
6. $p = -|7 - 10|, p = -|-3|, p = -3; \{-3\}$ **7.** $x^2 + 15x + 36 = (x + 3)(x + 12)$
8. $x^3 - 5x^2 - 24x = x(x^2 - 5x - 24) = x(x - 8)(x + 3)$
9. $49x^2 - 25 = (7x + 5)(7x - 5)$ **10.** $2y^2 + y - 3 = (2y + 3)(y - 1)$
11. $x^2 + 6xy + 9y^2 = (x + 3y)^2$
12. $16x^3 - 4x = 4x(4x^2 - 1) = 4x(2x + 1)(2x - 1)$

Page 475 • CHALLENGE

Step 4 If you divide by $3 - a$, which is negative (because $a > 3$), the inequality must be reversed so that Step 5 is $a < 3 + a$.

Pages 476–477 • EXTRA

1. {6, 9} **2.** {0, 1, 2, 5, 6, 7, 8, 9} **3.** {0, 3, 4, 5, 6, 7, 8, 9} **4.** {6, 7, 8} **5.** {6}
6. {1, 2, 5, 6, 7, 8, 9} **7.** {5, 6, 9} **8.** {0, 1, 2, 3, 4, 5, 6, 7, 8, 9} **9.** {5, 6, 7, 8}
10. $(E \cap D) \cup (E \cap F)$ or $E \cap (D \cup F)$
11. $\cup = \{-4, -3, -2, -1, 0\}; \cap = \{-3, -2\}$ **12.** $\cup = \{-1, 0, 1, 2, 3\}; \cap = \{0, 1\}$
13. $\cup = \{3, 4, 5, 6, 7, 8\}; \cap = \emptyset$; the sets are disjoint
14. $\cup = \{5, 6, 7, 8, 9\}; \cap = \{6, 8\}$ **15.** $\cup = \{2, 3, 4, 6, 8, 10, 12\}; \cap = \{4, 6, 8\}$
16. $\cup = \{-5, -4, -3, -2, -1, 0\}; \cap = \emptyset$; the sets are disjoint
17. {the real numbers between −1 and 2} **18.** {the real numbers between −1 and 3}
19. {the real numbers greater than −3} **20.** {the real numbers}
21. {the real numbers between −3 and 2} **22.** {the real numbers less than 3}
23. {the real numbers between −1 and 2} **24.** {the real numbers greater than −3}

25–26. The two diagrams for each exercise are identical.

25.

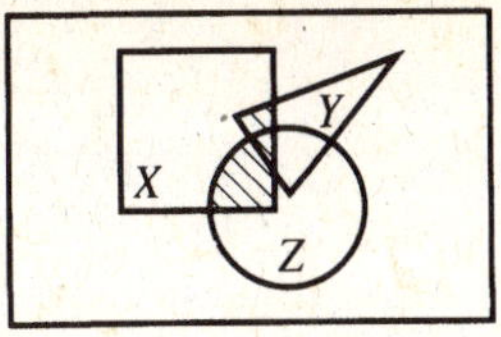

26.

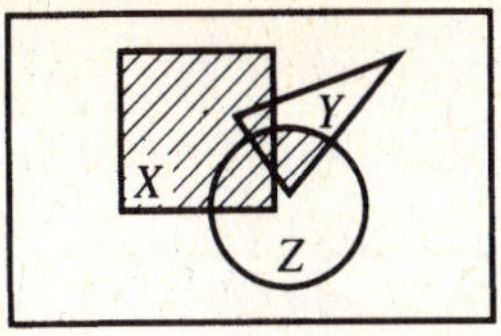

27. $X \cap (Y \cup Z) = (X \cap Y) \cup (X \cap Z)$ **28.** $X \cup (Y \cap Z) = (X \cup Y) \cap (X \cup Z)$

29.

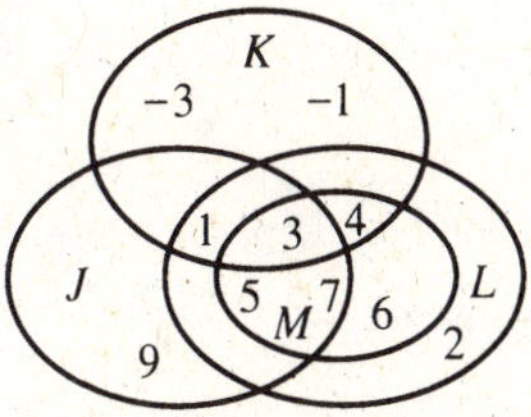

Pages 480–481 • WRITTEN EXERCISES

A **1.**

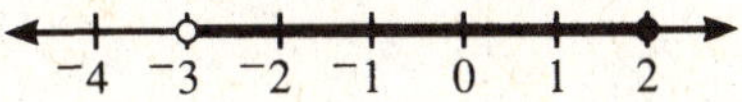

2.

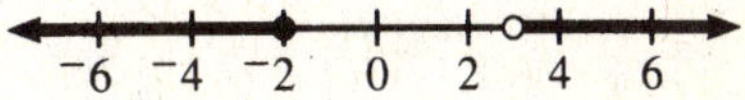

3.

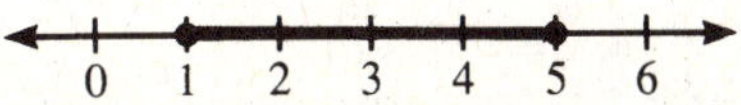

4.

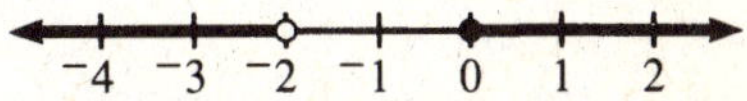

5. $-2 < a + 1 < 3$, $-3 < a < 2$; {the real numbers between -3 and 2}

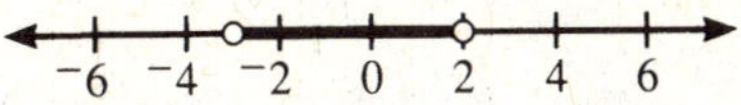

6. $-2 < y - 2 \leq 1$, $0 < y \leq 3$; {3, and the real numbers between 0 and 3}

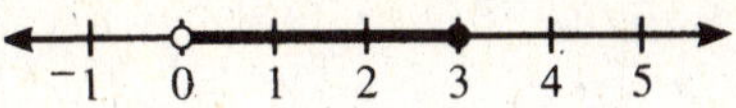

7. $-4 < -3 + d \leq 2$, $-1 < d \leq 5$; {5, and the real numbers between -1 and 5}

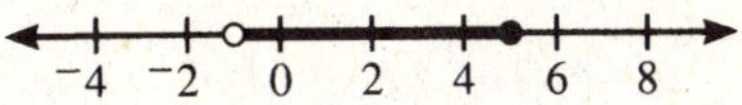

8. $-6 \leq 3 + r < 4$, $-9 \leq r < 1$; {-9, and the real numbers between -9 and 1}

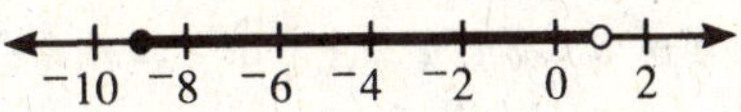

9. $-2 \leq 2a + 4 < 8$, $-6 \leq 2a < 4$, $-3 \leq a < 2$; {-3, and the real numbers between -3 and 2}

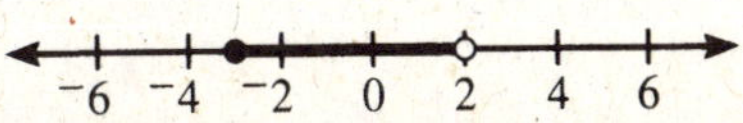

10. $-2 < 2b - 1 \le 5$, $-1 < 2b \le 6$, $-\frac{1}{2} < b \le 3$; $\left\{3\text{, and the real numbers between } -\frac{1}{2} \text{ and } 3\right\}$

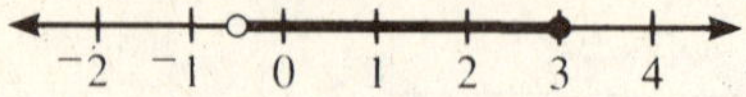

11. $-5 \le 3m + 1 < 4$, $-6 \le 3m < 3$, $-2 \le m < 1$; {−2, and the real numbers between −2 and 1}

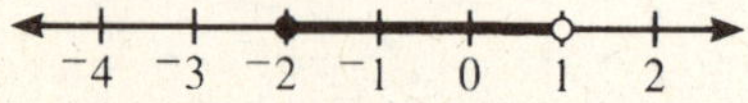

12. $-8 < 3n + 7 \le 1$, $-15 < 3n \le -6$, $-5 < n \le -2$; {−2, and the real numbers between −5 and −2}

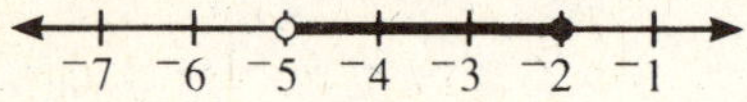

13. $y - 2 < -5$ or $y - 2 > 5$, $y < -3$ or $y > 7$; {the real numbers less than −3 or greater than 7}

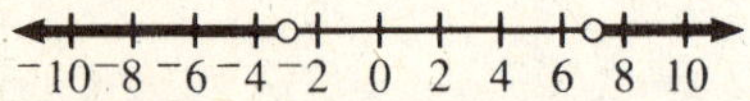

14. $h + 5 \le -2$ or $h + 5 \ge 2$, $h \le -7$ or $h \ge -3$; {−7, −3, and the real numbers less than −7 or greater than −3}

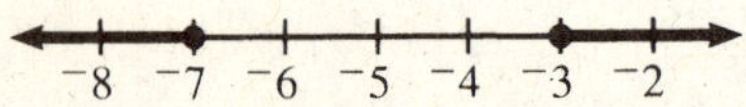

15. $2x + 1 \le -3$ or $2x + 1 \ge 3$, $2x \le -4$ or $2x \ge 2$, $x \le -2$ or $x \ge 1$; {−2, 1, and the real numbers less than −2 or greater than 1}

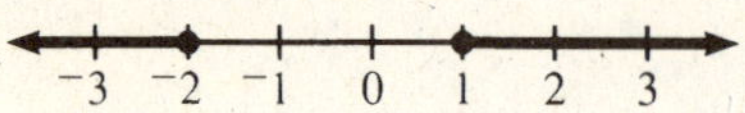

16. $1 + 2y < -9$ or $1 + 2y > 9$, $2y < -10$ or $2y > 8$, $y < -5$ or $y > 4$; {the real numbers less than −5 or greater than 4}

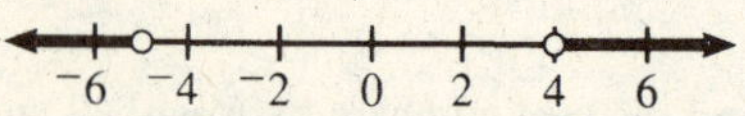

17. $3n - 1 \le -5$ or $5 \le 3n - 1$, $3n \le -4$ or $6 \le 3n$, $n \le -\frac{4}{3}$ or $2 \le n$; $\left\{-\frac{4}{3}, 2\text{, and the real numbers less than } -\frac{4}{3} \text{ or greater than } 2\right\}$

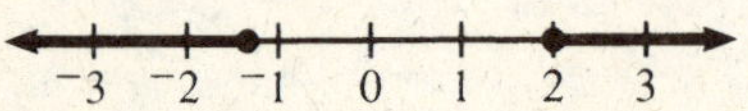

18. $2d - 5 < -7$ or $7 < 2d - 5$, $2d < -2$ or $12 < 2d$, $d < -1$ or $6 < d$; {the real numbers less than −1 or greater than 6}

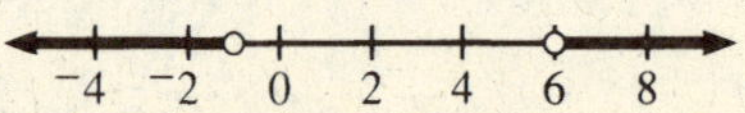

B **19.** $-6 \le -x \le 2$, $6 \ge x \ge -2$; {−2, 6, and the real numbers between −2 and 6}

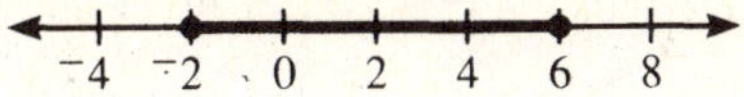

20. $-8 \le -2x \le 6$, $4 \ge x \ge -3$; {−3, 4, and the real numbers between −3 and 4}

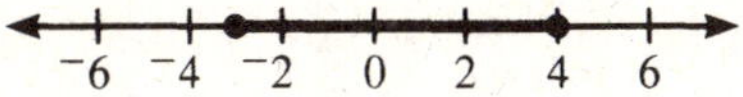

21. $-4 \le 2 - t < 3$, $-6 \le -t < 1$, $6 \ge t > -1$; {6, and the real numbers between −1 and 6}

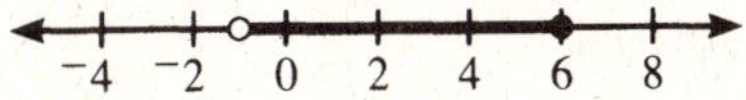

22. $-2 < 3 - y \le 1$, $-5 < -y \le -2$, $5 > y \ge 2$; {2, and the real numbers between 2 and 5}

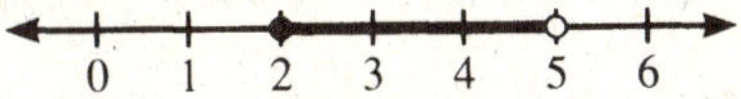

23. $-5 < 1 - 2x \le 7$, $-6 < -2x \le 6$, $3 > x \ge -3$; {−3, and the real numbers between −3 and 3}

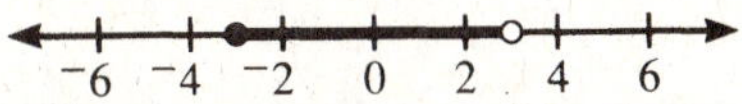

24. $-7 < -1 - 3x \le 8$, $-6 < -3x \le 9$, $2 > x \ge -3$; {−3, and the real numbers between −3 and 2}

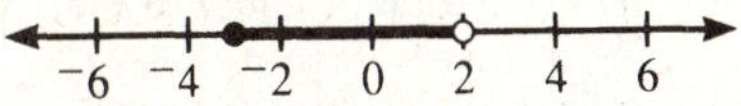

25. $-2m < 4$ and $12 + 2m < 0$, $m > -2$ and $2m < -12$, $m > -2$ and $m < -6$; no solution; There is no graph.

26. $-6r > 18$ or $12 + 3r \ge 0$, $r < -3$ or $3r \ge -12$, $r < -3$ or $r \ge -4$; {all real numbers}

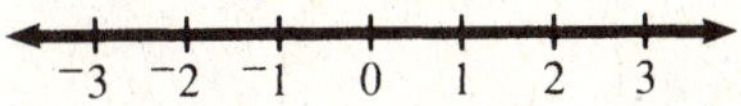

27. $-8 \le -1 - s < -3$, $-7 \le -s < -2$, $7 \ge s > 2$; {7, and the real numbers between 2 and 7}

28. $-9 < -10 - p \le -4$, $1 < -p \le 6$, $-1 > p \ge -6$; {−6, and the real numbers between −6 and −1}

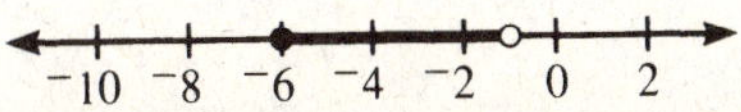

29. $5 - v > 7$ or $v - 5 > 7$, $-v > 2$ or $v > 12$, $v < -2$ or $v > 12$; {the real numbers less than -2 or greater than 12}

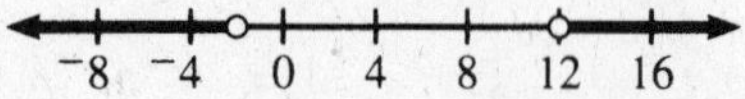

30. $t - 5 \geq 2$ or $5 - t \geq 2$, $t \geq 7$ or $-t \geq -3$, $t \geq 7$ or $t \leq 3$; {3, 7, and the real numbers less than 3 or greater than 7}

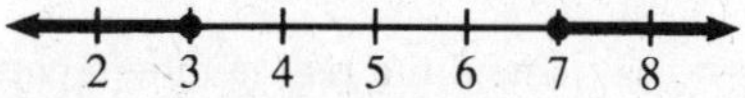

31. $5 - 2p \geq 11$ or $5 - 2p < -1$, $-2p \geq 6$ or $-2p < -6$, $p \leq -3$ or $p > 3$; {-3, and the real numbers less than -3 or greater than 3}

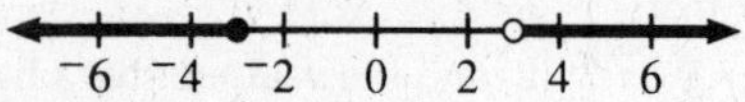

32. $7 - 3q \geq 10$ and $3q - 7 \geq 2$, $-3q \geq 3$ and $3q \geq 9$, $q \leq -1$ and $q \geq 3$; no solution; There is no graph.

33. $2d - 5 \geq -8$ and $-2d - 5 < d - 3$, $2d \geq -3$ and $-5 < 3d - 3$, $d \geq -\frac{3}{2}$ and $-2 < 3d$, $d \geq -\frac{3}{2}$ and $-\frac{2}{3} < d$; $\left\{\text{the real numbers greater than } -\frac{2}{3}\right\}$

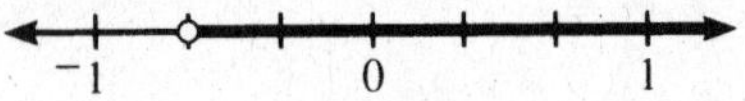

34. $9 - y \leq 3 - 2y$ and $-1 - 2y \geq -5$, $9 + y \leq 3$ and $-1 \geq -5 + 2y$, $y \leq -6$ and $4 \geq 2y$, $y \leq -6$ and $2 \geq y$; {the real numbers less than or equal to -6}

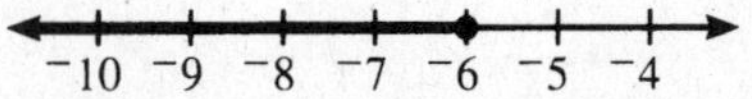

C **35.** $6 - c < 2c + 3 \leq 8 + c$, $6 - c < 2c + 3$ and $2c + 3 \leq 8 + c$, $3 < 3c$ and $c \leq 5$, $1 < c$ and $c \leq 5$, $1 < c \leq 5$; {5, and the real numbers between 1 and 5}

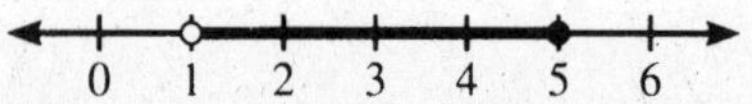

36. $5 - d \leq 3 - 2d$ or $d + 2 > 3d - 2$, $5 + d \leq 3$ or $2 > 2d - 2$, $d \leq -2$ or $4 > 2d$, $d \leq -2$ or $2 > d$; {the real numbers less than 2}

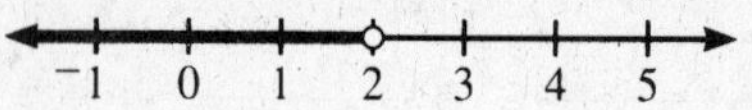

37. $2(1 - w) \geq 6$ or $4w - 5 \leq 3w - 1$, $2 - 2w \geq 6$ or $w - 5 \leq -1$, $-2w \geq 4$ or $w \leq 4$, $w \leq -2$ or $w \leq 4$; {the real numbers less than or equal to 4}

38. $1 - 4m \le 3 - 5m \le m - 3$, $1 - 4m \le 3 - 5m$ and $3 - 5m \le m - 3$, $1 + m \le 3$ and $3 - 6m \le -3$, $m \le 2$ and $-6m \le -6$, $m \le 2$ and $m \ge 1$, $1 \le m \le 2$; {1, 2, and the real numbers between 1 and 2}

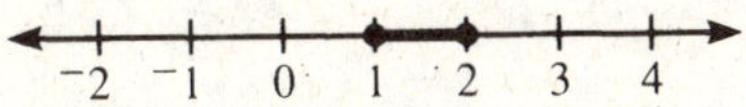

39. Answers may vary. **a.** $2 > 1$ and $3 > 2$, and $2 \cdot 3 > 1 \cdot 2$, $6 > 3$; True

b. $1 > -2$ and $2 > -3$, and $1 \cdot 2 > (-2)(-3)$, $2 > 6$; False

40. $3x - 7 \le 2(1 + x)$ and $5x - 7 \ge 23$, $3x - 7 \le 2 + 2x$ and $5x \ge 30$, $x \le 9$ and $x \ge 6$, $6 \le x \le 9$; $k - 5 \le x - 6 \le 3$, $k + 1 \le x \le 9$; $k + 1 = 6$, $k = 5$

Page 481 • COMPUTER EXERCISES

```
10 FOR X = 1 TO 9
20 FOR Y = 1 TO 9
30 IF X + Y > 10 THEN 60
40 IF X * Y < 9 THEN 60
50 PRINT "("; X; ","; Y; ")";
60 NEXT Y
70 NEXT X
80 END
(1,9)  (2,5)  (2,6)  (2,7)  (2,8)
(3,3)  (3,4)  (3,5)  (3,6)  (3,7)
(4,3)  (4,4)  (4,5)  (4,6)  (5,2)
(5,3)  (5,4)  (5,5)  (6,2)  (6,3)
(6,4)  (7,2)  (7,3)  (8,2)  (9,1)
```

Page 481 • MIXED REVIEW EXERCISES

1. Let n = distance to the net. $n \ge 10$ **2.** Let t = temperature. $t \le 78$

3. Let w = weight. $w \le 150$ **4.** Let t = length of the trip. $t \ge 3$

5. Let c = cost. $c \ge 17$ **6.** Let g = size of the gap. $g < 2$

7. Let s = her score. $s \le 340$

8. Let s = number of pairs of shoes Jane owns. $s \ge 5$

9. $|(-4) - 7| = |-11| = 11$ **10.** $|7 - (-4)| = |7 + 4| = |11| = 11$

11. $|7 - 4| = |3| = 3$ **12.** $|4 - 7| = |-3| = 3$

13. $|-4 - 4| = |-8| = 8$ **14.** $|4 - (-4)| = |8| = 8$

Pages 484–485 • WRITTEN EXERCISES

A **1.** **a.** The distance between x and 5 is 1. **b.** $x - 5 = -1$ or $x - 5 = 1$

2. **a.** The distance between y and -2 is 2. **b.** $y + 2 = -2$ or $y + 2 = 2$

3. **a.** The distance between r and -3 is greater than 5. **b.** $r + 3 < -5$ or $r + 3 > 5$

4. **a.** The distance between p and 7 is less than 4. **b.** $-4 < p - 7 < 4$

5. **a.** The distance between n and 4 is less than or equal to 1. **b.** $-1 \le n - 4 \le 1$

6. **a.** The distance between n and -6 is greater than or equal to 3.

b. $n + 6 \le -3$ or $n + 6 \ge 3$

7. $|x| < 2$ **8.** $|x| > 2$ **9.** $|x + 1| = 2$

10. $|x - 2| = 1$ **11.** $|x - 1| \geq 3$ **12.** $|x + 3| < 3$

13. $|m - 6| = 8$; $m - 6 = 8$ or $m - 6 = -8$; $m = 14$ or $m = -2$; $\{-2, 14\}$

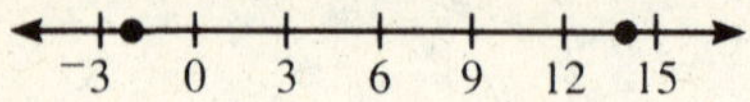

14. $|k + 7| = 3$; $k + 7 = 3$ or $k + 7 = -3$, $k = -4$ or $k = -10$; $\{-10, -4\}$

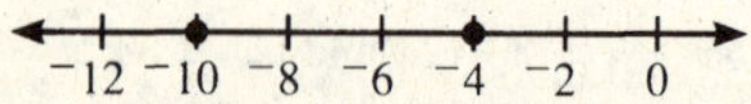

15. $|3 + x| = 5$; $3 + x = 5$ or $3 + x = -5$, $x = 2$ or $x = -8$; $\{-8, 2\}$

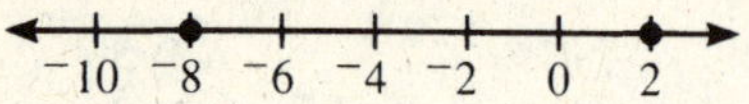

16. $|7 - y| = 4$; $7 - y = 4$ or $7 - y = -4$, $-y = -3$ or $-y = -11$, $y = 3$ or $y = 11$; $\{3, 11\}$

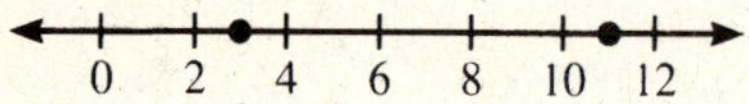

17. $|r| < 2.5$; $-2.5 < r < 2.5$; {the real numbers between -2.5 and 2.5}

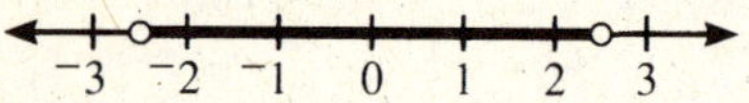

18. $|s| \geq 1.5$; $s \geq 1.5$ or $s \leq -1.5$; {-1.5, 1.5, and the real numbers less than -1.5 or greater than 1.5}

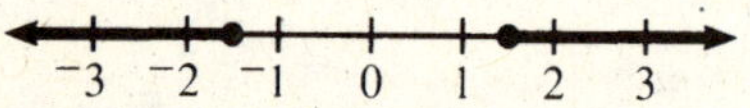

19. $|y + 5| > 8$; $y + 5 > 8$ or $y + 5 < -8$, $y > 3$ or $y < -13$; {the real numbers less than -13 or greater than 3}

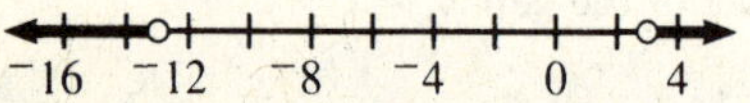

20. $|t + 4| < 10$; $-10 < t + 4 < 10$, $-14 < t < 6$; {the real numbers between -14 and 6}

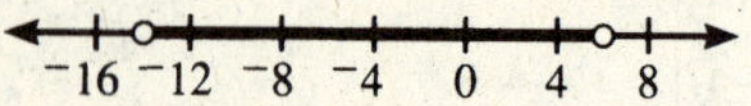

21. $|6 - p| \leq 2$; $-2 \leq 6 - p \leq 2$, $-8 \leq -p \leq -4$, $8 \geq p \geq 4$, $4 \leq p \leq 8$; {4, 8, and the real numbers between 4 and 8}

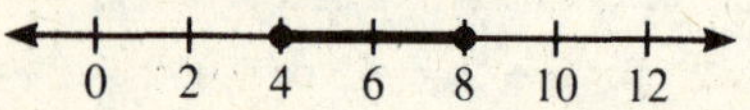

22. $|4 - v| \geq 5$; $4 - v \geq 5$ or $4 - v \leq -5$, $-v \geq 1$ or $-v \leq -9$, $v \leq -1$ or $v \geq 9$; {-1, 9, and the real numbers less than -1 or greater than 9}

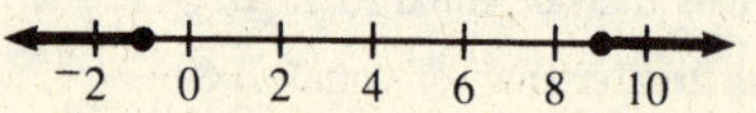

23. $|-2 - a| \geq 4$; $-2 - a \geq 4$ or $-2 - a \leq -4$, $-a \geq 6$ or $-a \leq -2$, $a \leq -6$ or $a \geq 2$; {−6, 2, and the real numbers less than −6 or greater than 2}

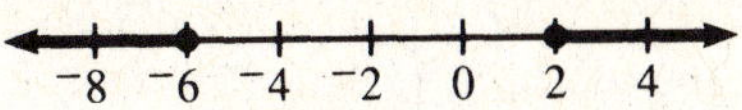

24. $|-1 - b| \leq 6$; $-6 \leq -1 - b \leq 6$, $-5 \leq -b \leq 7$, $5 \geq b \geq -7$, $-7 \leq b \leq 5$; {−7, 5, and the real numbers between −7 and 5}

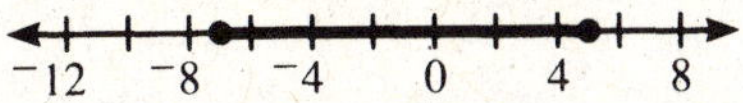

B **25.** $3|s| - 2 \geq 7$, $3|s| \geq 9$, $|s| \geq 3$; $s \geq 3$ or $s \leq -3$; {−3, 3, and the real numbers less than −3 or greater than 3}

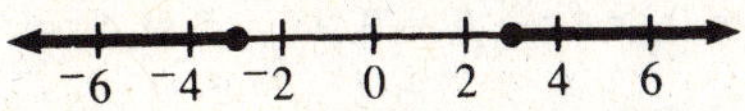

26. $4|p| - 1 < 15$, $4|p| < 16$, $|p| < 4$; $-4 < p < 4$; {the real numbers between −4 and 4}

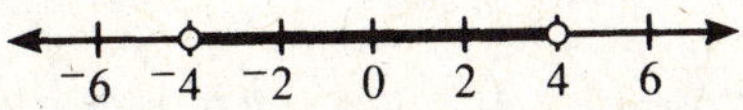

27. $|2 - z| + 3 > 8$, $|2 - z| > 5$; $2 - z > 5$ or $2 - z < -5$, $-z > 3$ or $-z < -7$, $z < -3$ or $z > 7$; {the real numbers less than −3 or greater than 7}

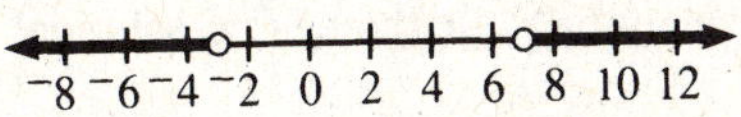

28. $|2 - q| - 3 < 1$, $|2 - q| < 4$; $-4 < 2 - q < 4$, $-6 < -q < 2$, $6 > q > -2$, $-2 < q < 6$; {the real numbers between −2 and 6}

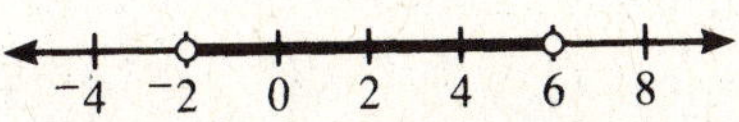

29. $4 - 3|r| \geq -7$, $-3|r| \geq -11$, $|r| \leq \frac{11}{3}$; $-\frac{11}{3} \leq r \leq \frac{11}{3}$; $\left\{-\frac{11}{3}, \frac{11}{3}\right.$, and the real numbers between $-\frac{11}{3}$ and $\left.\frac{11}{3}\right\}$

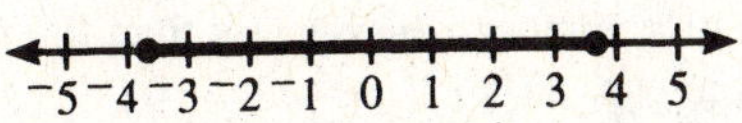

30. $4 - 3|y| < 1$, $-3|y| < -3$, $|y| > 1$, $y > 1$ or $y < -1$; {the real numbers less than −1 or greater than 1}

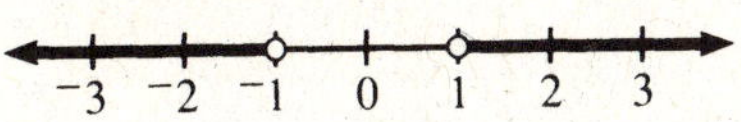

31. $6 - |2 - p| \leq 4$, $-|2 - p| \leq -2$, $|2 - p| \geq 2$; $2 - p \geq 2$ or $2 - p \leq -2$, $-p \geq 0$ or $-p \leq -4$, $p \leq 0$ or $p \geq 4$; {0, 4, and the real numbers less than 0 or greater than 4}

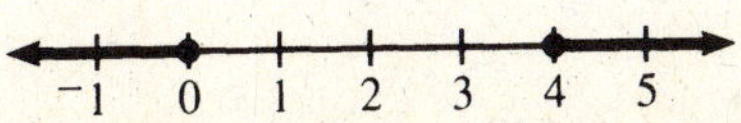

32. $8 - |1 - x| > 7, -|1 - x| > -1, |1 - x| < 1; -1 < 1 - x < 1, -2 < -x < 0, 2 > x > 0, 0 < x < 2$; {the real numbers between 0 and 2}

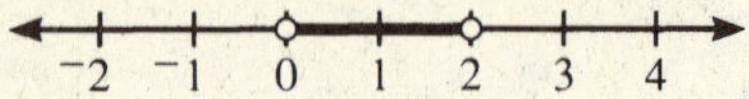

33. $2(3|a| - 1) \leq 10, 3|a| - 1 \leq 5, 3|a| \leq 6, |a| \leq 2; -2 \leq a \leq 2$; {−2, 2, and the real numbers between −2 and 2}

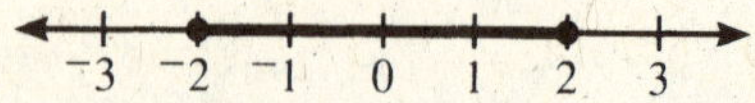

C **34.** $|x - 2| = \frac{x}{3}; x - 2 = \frac{x}{3}$ or $x - 2 = -\frac{x}{3}, 3(x - 2) = x$ or $3(x - 2) = -x$, $3x - 6 = x$ or $3x - 6 = -x, 2x - 6 = 0$ or $4x - 6 = 0, 2x = 6$ or $4x = 6, x = 3$ or $x = \frac{3}{2}; \left\{\frac{3}{2}, 3\right\}$

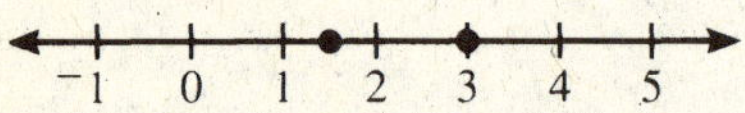

35. $|y + 1| = \frac{y}{2}; y + 1 = \frac{y}{2}$ or $y + 1 = -\frac{y}{2}, 2(y + 1) = y$ or $2(y + 1) = -y$, $2y + 2 = y$ or $2y + 2 = -y, y = -2$ or $3y = -2, y = -2$ or $y = -\frac{2}{3}$; Since both of these solutions are impossible (y cannot be < 0), there is no solution; There is no graph.

36. $|y - 4| = y - 4$ if and only if $y - 4 \geq 0$. $y \geq 4$; {4 and the real numbers greater than 4}

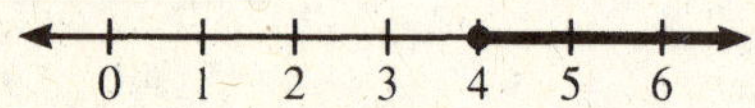

37. $|t - 4| > t - 4; t - 4 > t - 4$ or $t - 4 < -(t - 4), 0 > 0$ or $t - 4 < 4 - t, 0 > 0$ or $2t < 8, 0 > 0$ or $t < 4$; {the real numbers less than 4}

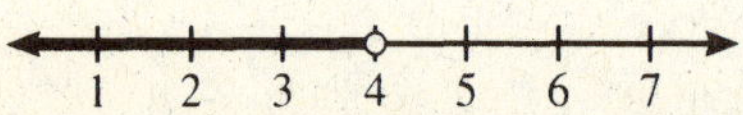

38. $|n + 5| < n + 5$; The absolute value of a number is never less than the number; There is no solution; There is no graph.

39. $|n + 5| \geq n + 5$; The absolute value of a number is always greater than or equal to the number; {the real numbers}

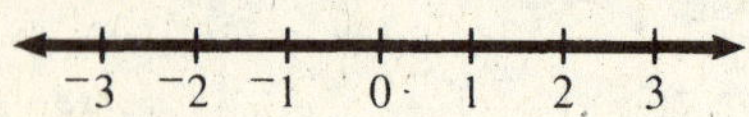

Page 485 • MIXED REVIEW EXERCISES

1. $x - 5 < 4, x < 9$; {the real numbers less than 9}

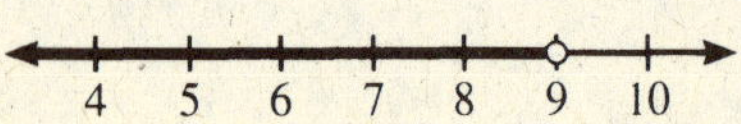

2. $\frac{x}{2} + 5 < 1, \frac{x}{2} < -4, x < -8$; {the real numbers less than -8}

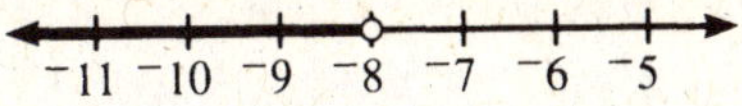

3. $10 < 5(2 + m), 10 < 10 + 5m, 0 < 5m, 0 < m$; {the real numbers greater than 0}

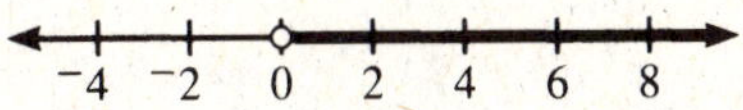

4. $-1 < x + 5 < 2, -6 < x < -3$; {the real numbers between -6 and -3}

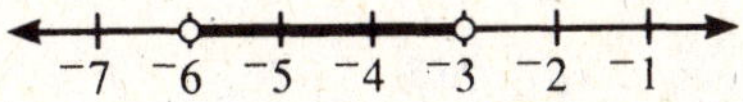

5. $h + 1 \le 10$ or $h - 5 > 7$, $h \le 9$ or $h > 12$; {9, and the real numbers less than 9 or greater than 12}

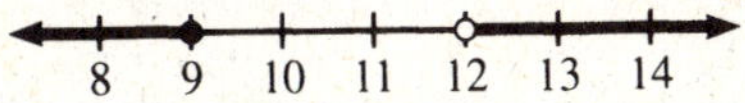

6. $3 \le -x \le 10, -3 \ge x \ge -10$; {$-10$, -3, and the real numbers between -10 and -3}

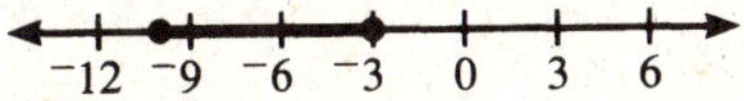

7. $\frac{20x}{3y^2} \div 5xy = \frac{20x}{3y^2} \cdot \frac{1}{5xy} = \frac{4}{3y^3}$ **8.** $\left(\frac{2a}{b}\right)^3 \cdot \left(\frac{4b^2}{3a}\right) = \frac{8a^3}{b^3} \cdot \frac{4b^2}{3a} = \frac{32a^2}{3b}$

9. $\frac{x + 4}{3} - \frac{2x}{9} = \frac{3(x + 4)}{9} - \frac{2x}{9} = \frac{3x + 12 - 2x}{9} = \frac{x + 12}{9}$

10. $3x + \frac{x}{7} = \frac{7(3x)}{7} + \frac{x}{7} = \frac{21x + x}{7} = \frac{22x}{7}$

Page 488 • WRITTEN EXERCISES

A **1.** $|3x| = 12, 3|x| = 12, |x| = 4; x = 4$ or $x = -4$; {4, -4}

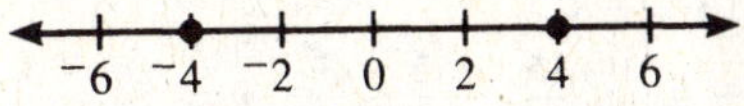

2. $|5y| = 35, 5|y| = 35, |y| = 7; y = 7$ or $y = -7$; {-7, 7}

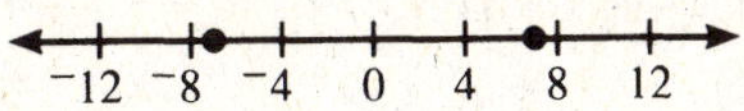

3. $\left|\frac{y}{4}\right| \ge 1, \frac{1}{4}|y| \ge 1, |y| \ge 4; y \ge 4$ or $y \le -4$; {-4, 4, and the real numbers less than -4 or greater than 4}

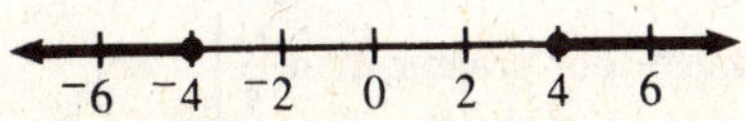

4. $\left|\frac{t}{3}\right| \le 2, \frac{1}{3}|t| \le 2, |t| \le 6; -6 \le t \le 6;$ {-6, 6, and the real numbers between -6 and 6}

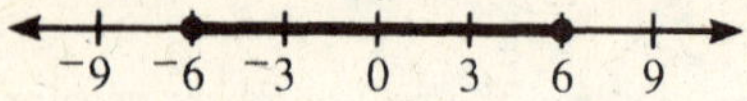

5. $|2a - 1| = 9, 2\left|a - \frac{1}{2}\right| = 9, \left|a - \frac{1}{2}\right| = \frac{9}{2}; a - \frac{1}{2} = \frac{9}{2}$ or $a - \frac{1}{2} = -\frac{9}{2}$, $a = 5$ or $a = -4$; $\{-4, 5\}$

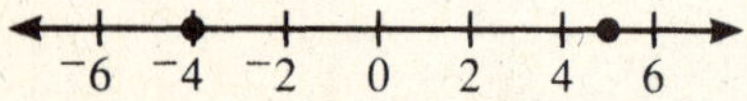

6. $|3b - 2| = 7, 3\left|b - \frac{2}{3}\right| = 7, \left|b - \frac{2}{3}\right| = \frac{7}{3}; b - \frac{2}{3} = \frac{7}{3}$ or $b - \frac{2}{3} = -\frac{7}{3}$, $b = 3$ or $b = -\frac{5}{3}$; $\left\{-\frac{5}{3}, 3\right\}$

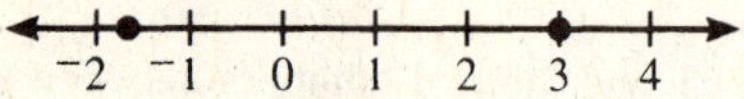

7. $|4d - 11| \le 3, 4\left|d - \frac{11}{4}\right| \le 3, \left|d - \frac{11}{4}\right| \le \frac{3}{4}, -\frac{3}{4} \le d - \frac{11}{4} \le \frac{3}{4}, 2 \le d \le \frac{7}{2}$; $\left\{2, \frac{7}{2}\right.$, and the real numbers between 2 and $\left.\frac{7}{2}\right\}$

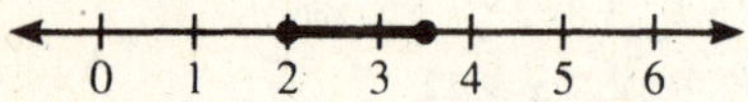

8. $|5h - 2| \ge 8, 5\left|h - \frac{2}{5}\right| \ge 8, \left|h - \frac{2}{5}\right| \ge \frac{8}{5}; h - \frac{2}{5} \ge \frac{8}{5}$ or $h - \frac{2}{5} \le -\frac{8}{5}$, $h \ge 2$ or $h \le -\frac{6}{5}$; $\left\{-\frac{6}{5}\right.$, 2, and the real numbers less than $-\frac{6}{5}$ or greater than $\left.2\right\}$

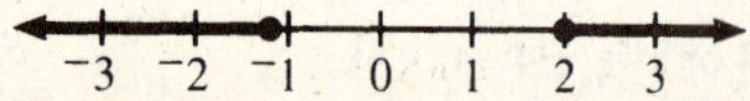

9. $|3 + 4n| \le 15, 4\left|\frac{3}{4} + n\right| \le 15, \left|\frac{3}{4} + n\right| \le \frac{15}{4}, -\frac{15}{4} \le \frac{3}{4} + n \le \frac{15}{4}, -\frac{9}{2} \le n \le 3$; $\left\{-\frac{9}{2}\right.$, 3, and the real numbers between $-\frac{9}{2}$ and $\left.3\right\}$

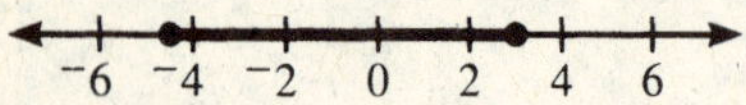

10. $|1 + 8c| > 23, 8\left|\frac{1}{8} + c\right| > 23, \left|\frac{1}{8} + c\right| > \frac{23}{8}; \frac{1}{8} + c > \frac{23}{8}$ or $\frac{1}{8} + c < -\frac{23}{8}$, $c > \frac{11}{4}$ or $c < -3$; $\left\{\text{the real numbers less than } -3 \text{ or greater than } \frac{11}{4}\right\}$

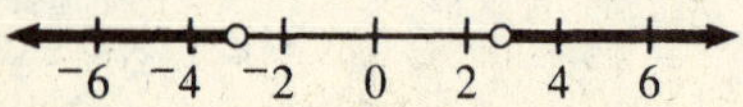

11. $\left|\frac{y}{5} - 2\right| \geq 4, \frac{1}{5}|y - 10| \geq 4, |y - 10| \geq 20; y - 10 \geq 20$ or $y - 10 \leq -20, y \geq 30$ or $y \leq -10$; {-10, 30, and the real numbers less than -10 or greater than 30}

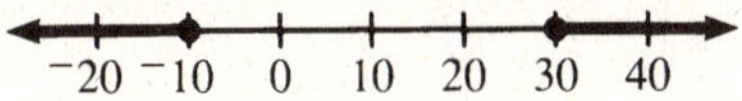

12. $\left|\frac{x}{2} - 3\right| \leq 2, \frac{1}{2}|x - 6| \leq 2, |x - 6| \leq 4, -4 \leq x - 6 \leq 4, 2 \leq x \leq 10$; {2, 10, and the real numbers between 2 and 10}

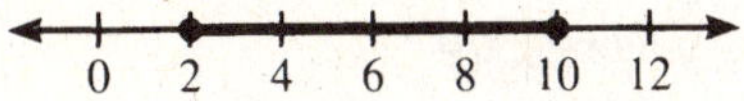

B **13.** $|1 - (3 - 2x)| < 18, |1 - 3 + 2x| < 18, |-2 + 2x| < 18, 2|x - 1| < 18,$ $|x - 1| < 9; -9 < x - 1 < 9, -8 < x < 10$; {the real numbers between -8 and 10}

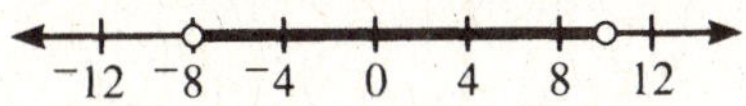

14. $|6 - (2y - 3)| \leq 9, |6 - 2y + 3| \leq 9, |9 - 2y| \leq 9, |-2| \cdot \left|y - \frac{9}{2}\right| \leq 9,$ $2\left|y - \frac{9}{2}\right| \leq 9, \left|y - \frac{9}{2}\right| \leq \frac{9}{2}; -\frac{9}{2} \leq y - \frac{9}{2} \leq \frac{9}{2}, 0 \leq y \leq 9$; {0, 9, and the real numbers between 0 and 9}

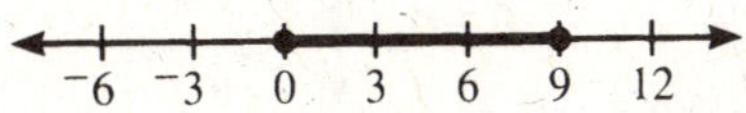

15. $4 + 3|5n + 1| = 13, 3|5n + 1| = 9, |5n + 1| = 3, 5\left|n + \frac{1}{5}\right| = 3, \left|n + \frac{1}{5}\right| = \frac{3}{5};$ $n + \frac{1}{5} = \frac{3}{5}$ or $n + \frac{1}{5} = -\frac{3}{5}, n = \frac{2}{5}$ or $n = -\frac{4}{5}; \left\{-\frac{4}{5}, \frac{2}{5}\right\}$

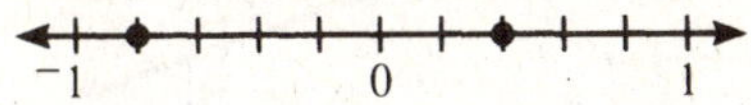

16. $2|3k - 7| + 11 = 19, 2|3k - 7| = 8, |3k - 7| = 4, 3\left|k - \frac{7}{3}\right| = 4, \left|k - \frac{7}{3}\right| = \frac{4}{3};$ $k - \frac{7}{3} = \frac{4}{3}$ or $k - \frac{7}{3} = -\frac{4}{3}, k = \frac{11}{3}$ or $k = 1; \left\{\frac{11}{3}, 1\right\}$

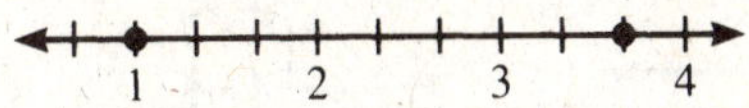

17. $5 - 4|2 - 3t| > 21, -4|2 - 3t| > 16, |2 - 3t| < -4$; The absolute value of a number cannot be < 0, so there is no solution; There is no graph.

18. $21 - 4|2 - 5w| > 13, -4|2 - 5w| > -8, |2 - 5w| < 2, \left|(-5)\left(w - \frac{2}{5}\right)\right| < 2,$ $|-5| \cdot \left|w - \frac{2}{5}\right| < 2, 5\left|w - \frac{2}{5}\right| < 2, \left|w - \frac{2}{5}\right| < \frac{2}{5}; -\frac{2}{5} < w - \frac{2}{5} < \frac{2}{5}, 0 < w < \frac{4}{5};$ $\left\{\text{the real numbers between 0 and } \frac{4}{5}\right\}$

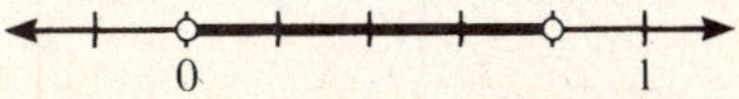

C **19.** True **20.** True **21.** False; for example, $\left|\frac{-1}{3}\right| \neq \frac{-1}{3}$; $a = -1$

22. False; for example, $|(-10) - 3| \not< |(-10) + 3|$; $a = -10$ **23.** True **24.** True

25. $\left|\frac{5}{9}(F - 32)\right| \leq 10$, $-10 \leq \frac{5}{9}(F - 32) \leq 10$, $-18 \leq F - 32 \leq 18$, $14 \leq F \leq 50$; The temperature ranged between 14°F and 50°F.

Pages 488–489 • MIXED REVIEW EXERCISES

1. slope = 4; y-intercept = 2

2. $4y = 20x - 8$, $y = 5x - 2$; slope = 5; y-intercept = −2

3. $2x + 3y + 9 = 0$, $3y = -2x - 9$, $y = -\frac{2}{3}x - 3$; slope = $-\frac{2}{3}$; y-intercept = −3

4. slope = 0; y-intercept = 9

5. $3x - y - 11 = 0$, $3x - 11 = y$; slope = 3; y-intercept = −11

6. $x = -2y + 8$, $2y = -x + 8$, $y = -\frac{1}{2}x + 4$; slope = $-\frac{1}{2}$; y-intercept = 4

7. 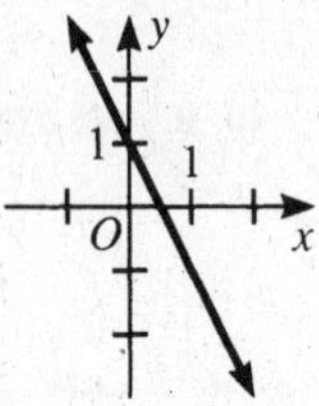

8. 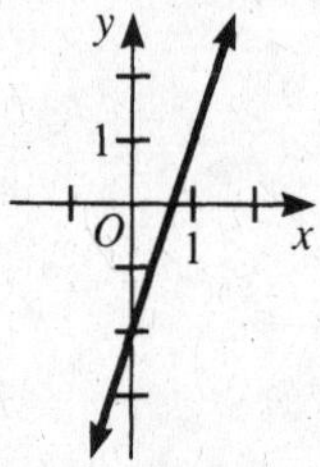

9. 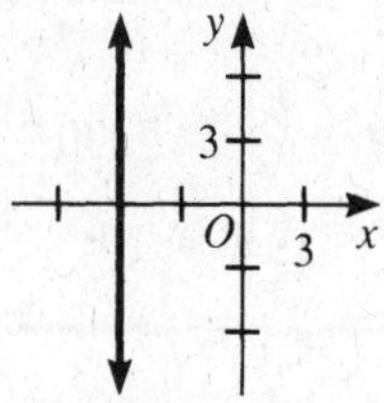

10. 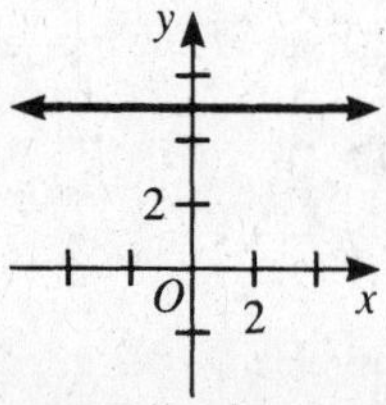

11. 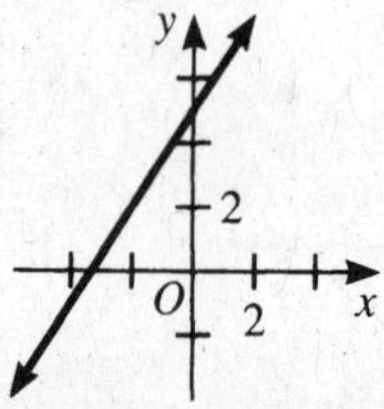

12. 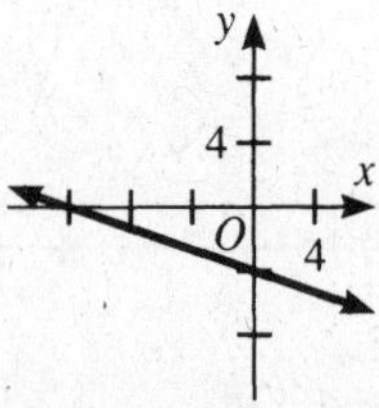

Pages 492–493 • WRITTEN EXERCISES

A **1.** 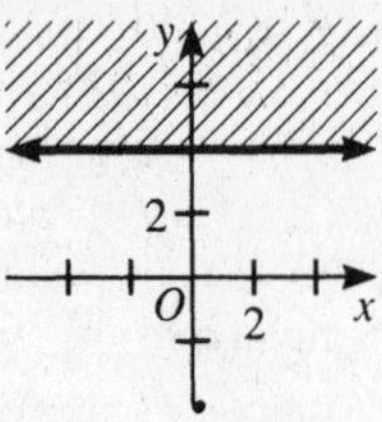

2. 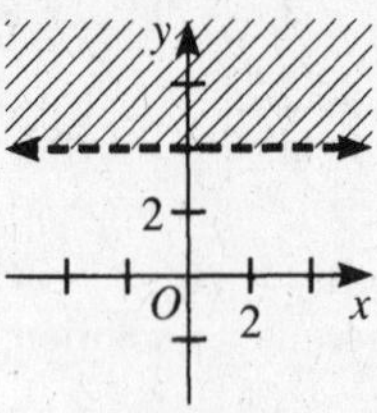

3. 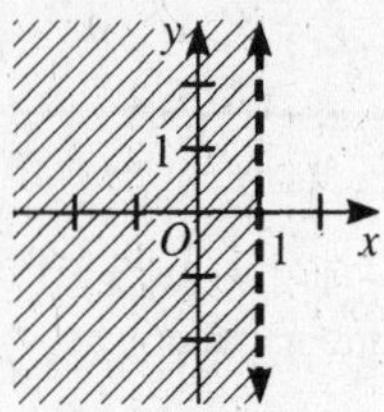

4. 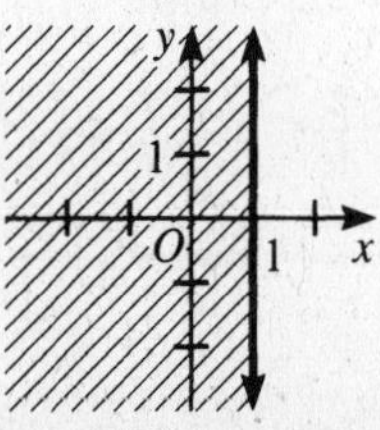

5. 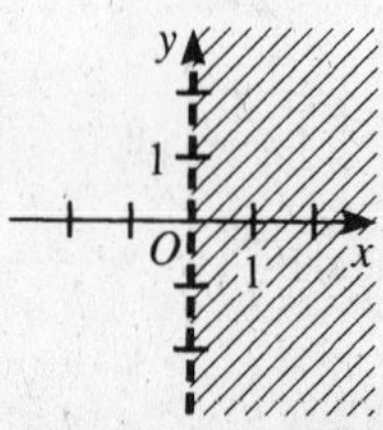

6. 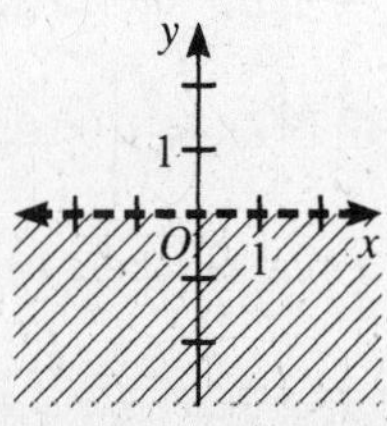

7.

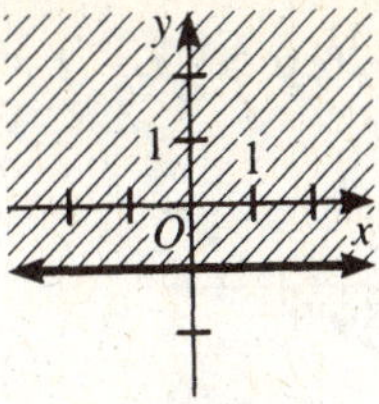

8.

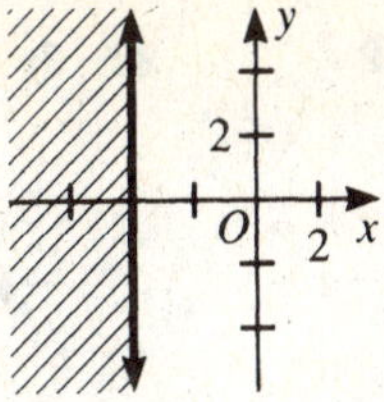

9.

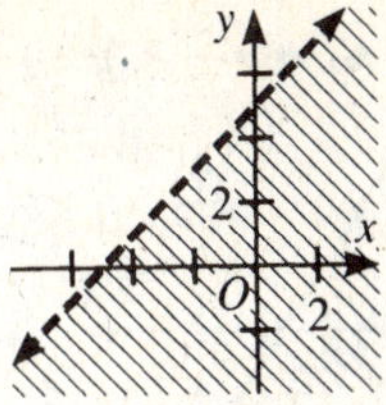

10.

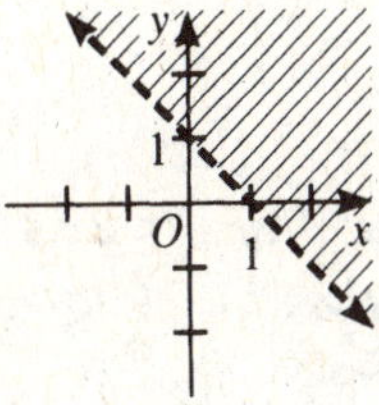

11.

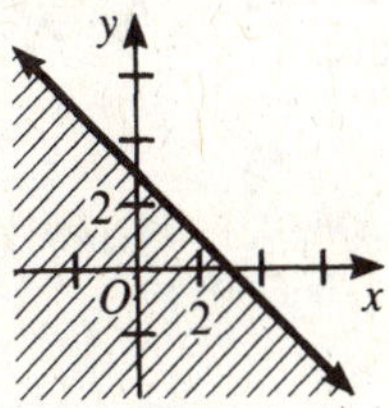

12.

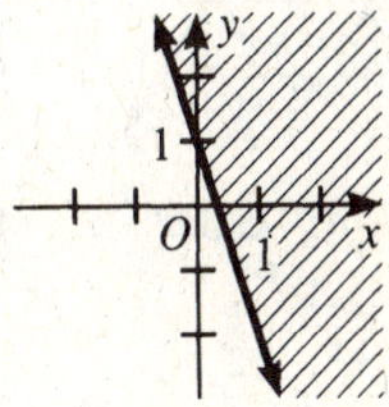

13. $x + y < 1$,
$y < 1 - x$

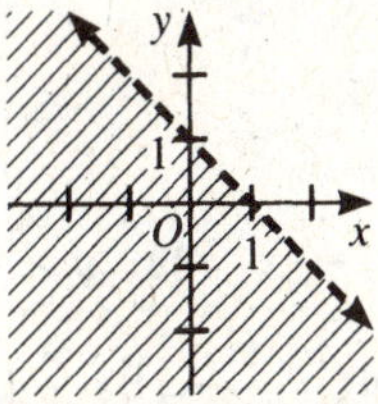

14. $x - 3 \geq y$,
$y \leq x - 3$

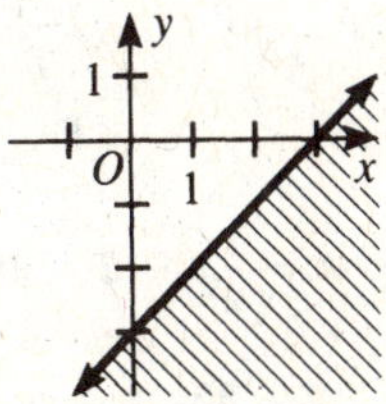

15. $x + 6 \leq 3y$,
$\frac{1}{3}x + 2 \leq y$

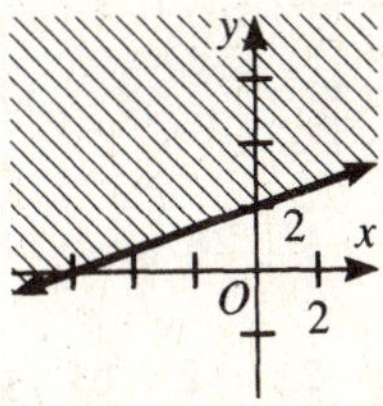

16. $y > -4 - 2x$

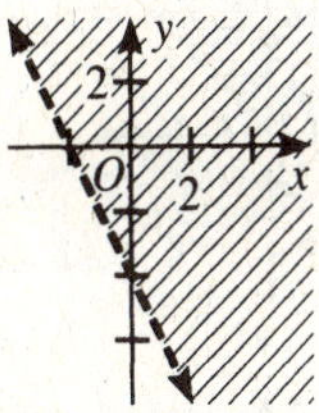

17. $-y > 6 - 4x$,
$y < 4x - 6$

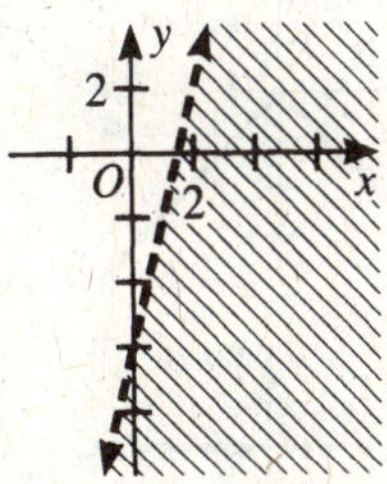

18. $y \leq 2x - 3$

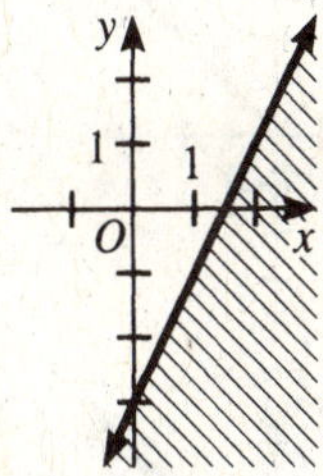

19. $-2y \geq 8 - 3x$, $y \leq \frac{3}{2}x - 4$

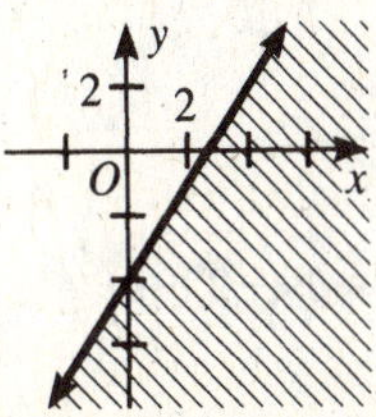

20. $3y < 2x$, $y < \frac{2}{3}x$

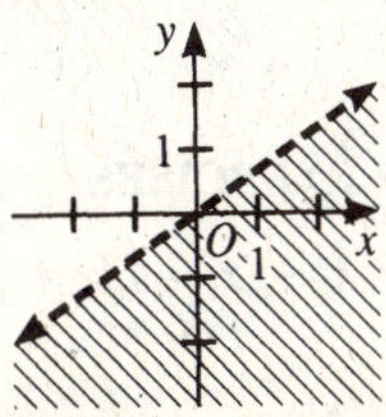

21. $4y \le -6x - 8,\ y \le -\frac{3}{2}x - 2$

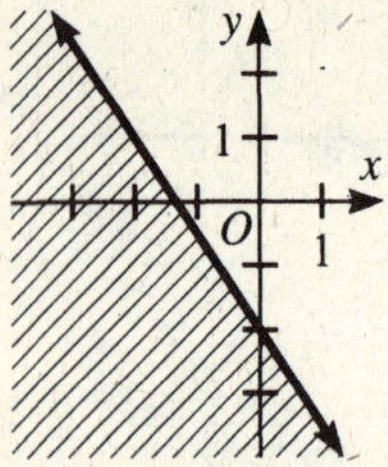

22. $3y > 2x - 6,\ y > \frac{2}{3}x - 2$

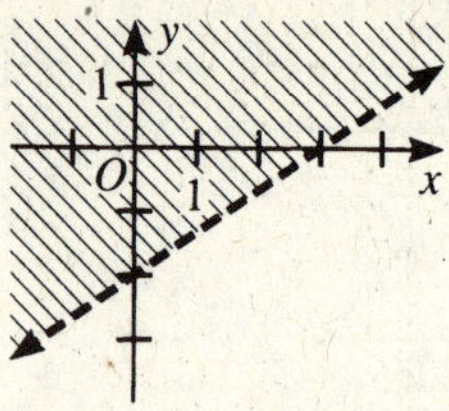

23. $4x - 4y \ge 3x + 2,\ -4y \ge -x + 2,$
$y \le \frac{1}{4}x - \frac{1}{2}$

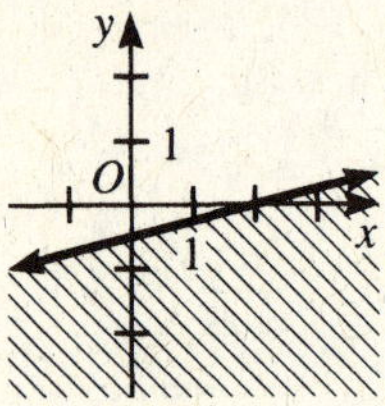

24. $5y - 8 < 2x + 4y;\ y < 2x + 8$

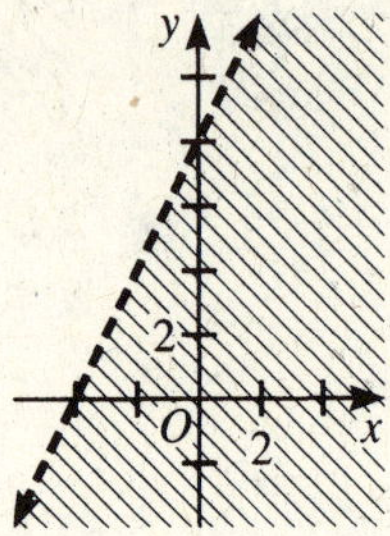

B **25.** $y > -1$ **26.** $x \le 2$ **27.** $y < -\frac{1}{2}x + 2$ **28.** $y < -\frac{3}{2}x$ **29.** $y \ge \frac{4}{3}x + 4$

30. $y < -2x - 2$ **31.** $y < \frac{1}{2}x + 2$ **32.** $y \ge x + 3$ **33.** $y \le -\frac{3}{2}x + 5$

C **34.**

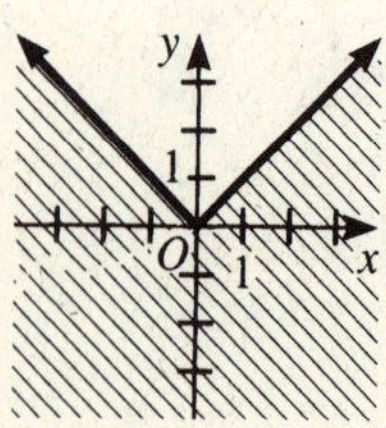

35.

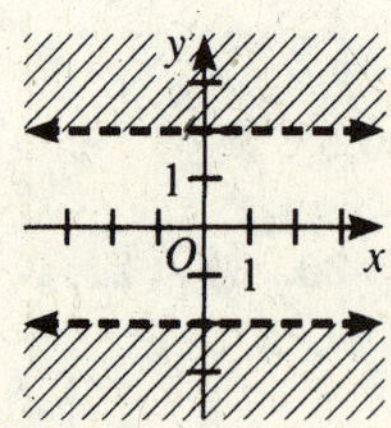

36.

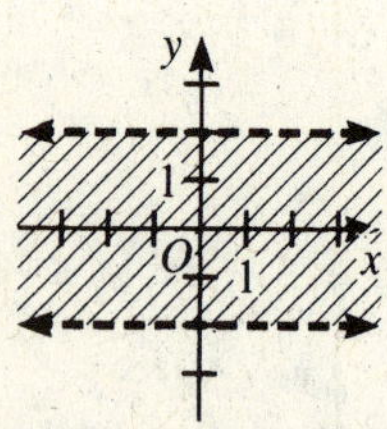

37.

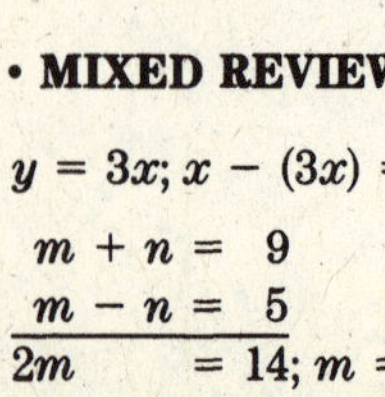

38.

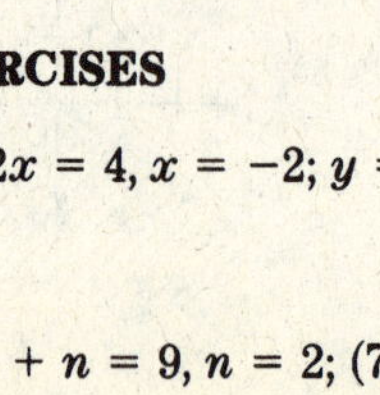

39.

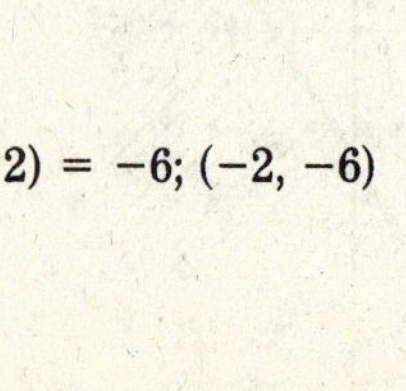

Page 494 • MIXED REVIEW EXERCISES

1. $y = 3x;\ x - (3x) = 4,\ -2x = 4,\ x = -2;\ y = 3(-2) = -6;\ (-2, -6)$

2.
$$\begin{array}{rl} m + n = & 9 \\ m - n = & 5 \\ \hline 2m \quad = & 14 \end{array}$$
$m = 7;\ (7) + n = 9,\ n = 2;\ (7, 2)$

3.
$$\begin{array}{rl} 10p + 4q = & 2 \\ 10p - 12q = & 34 \\ \hline 16q = & -32 \end{array}$$
$q = -2;\ 10p + 4(-2) = 2,\ 10p - 8 = 2,\ 10p = 10,\ p = 1;\ (1, -2)$

4. $|4p| = 20, 4|p| = 20, |p| = 5; p = 5 \text{ or } p = -5; \{-5, 5\}$

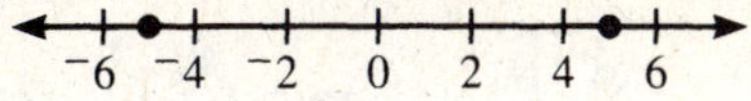

5. $|3p + 3| = 15; 3|p + 1| = 15, |p + 1| = 5, p + 1 = 5 \text{ or } p + 1 = -5; p = 4 \text{ or } p = -6; \{-6, 4\}$

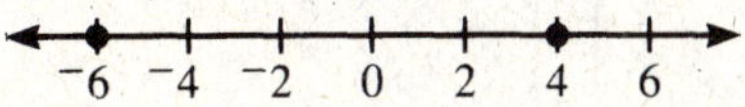

6. $|3x| < 21, 3|x| < 21, |x| < 7; -7 < x < 7;$ {the real numbers between -7 and 7}

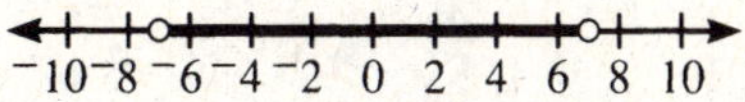

7. $|2y + 5| \geq 6, 2\left|y + \frac{5}{2}\right| \geq 6, \left|y + \frac{5}{2}\right| \geq 3, y + \frac{5}{2} \geq 3 \text{ or } y + \frac{5}{2} \leq -3; y \geq \frac{1}{2} \text{ or } y \leq -\frac{11}{2}; \left\{-\frac{11}{2}, \frac{1}{2}, \text{and the real numbers less than } -\frac{11}{2} \text{ or greater than } \frac{1}{2}\right\}$

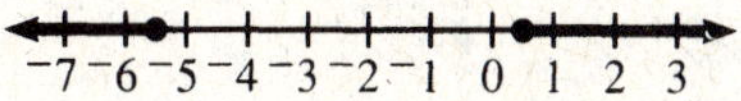

8. $-6 \leq x + 1 < 5, -7 \leq x < 4;$ {-7, and the real numbers between -7 and 4}

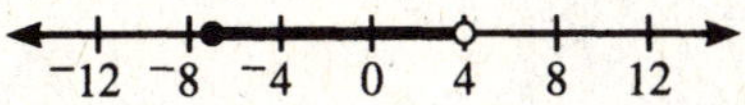

9. $2x + 5 > 7 \text{ or } 2 - x \geq 2, 2x > 2 \text{ or } 2 \geq x + 2, x > 1 \text{ or } 0 \geq x;$ {0, and the real numbers less than 0 or greater than 1}

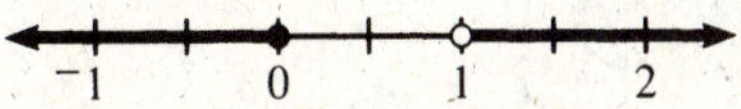

Page 494 • EXTRA

1. a–c.

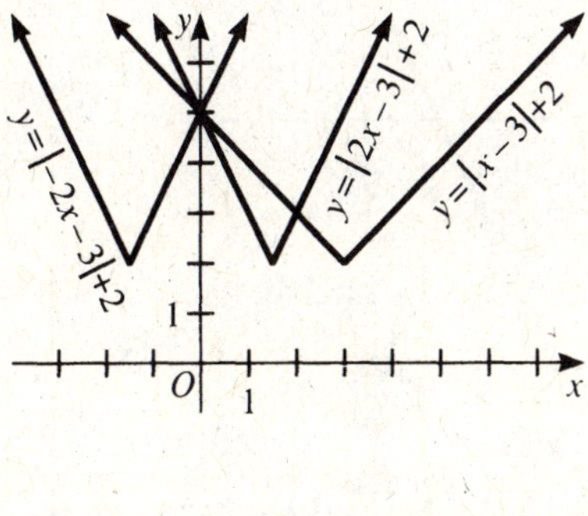

2. a–c.

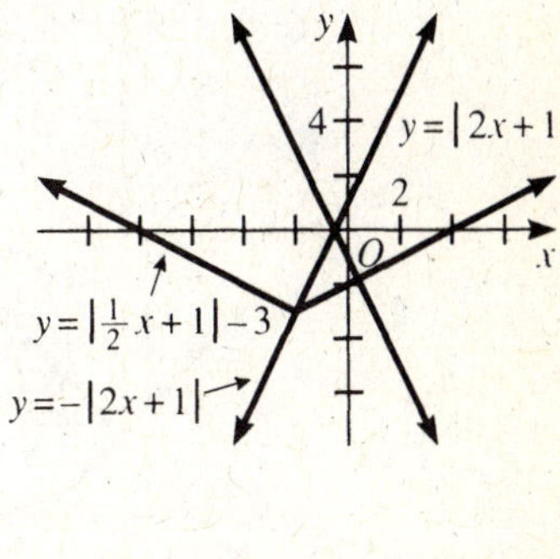

Pages 496–498 • WRITTEN EXERCISES

A

1.

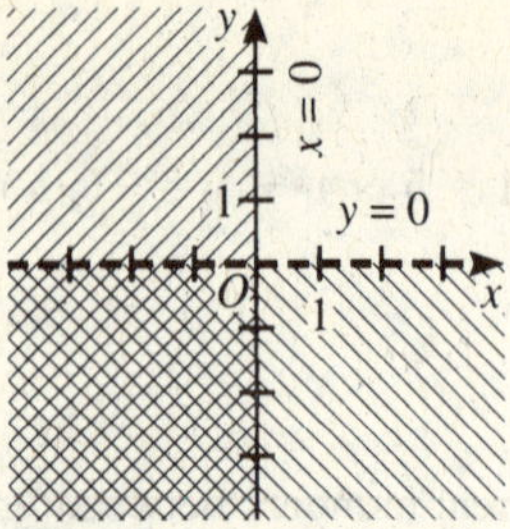

2.

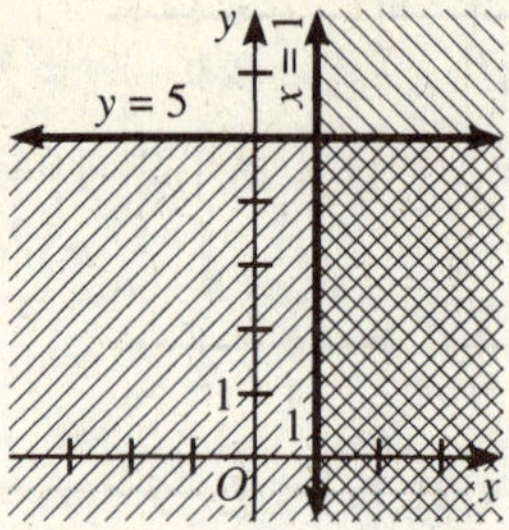

3.

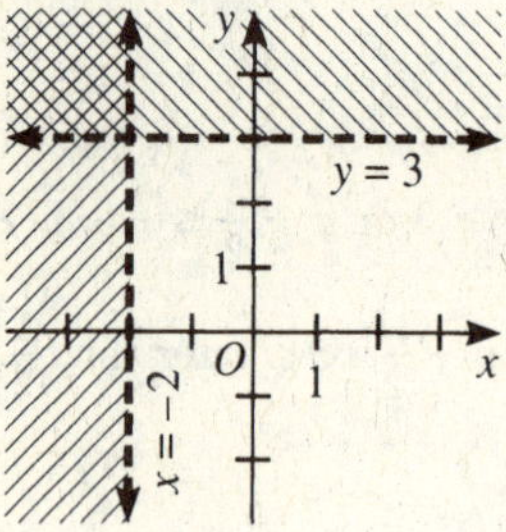

4.

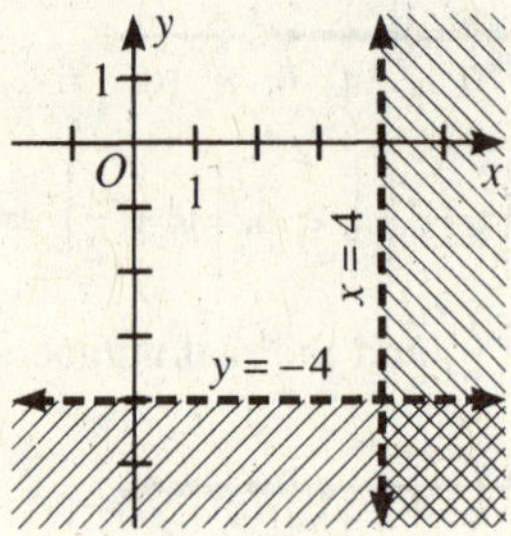

5.

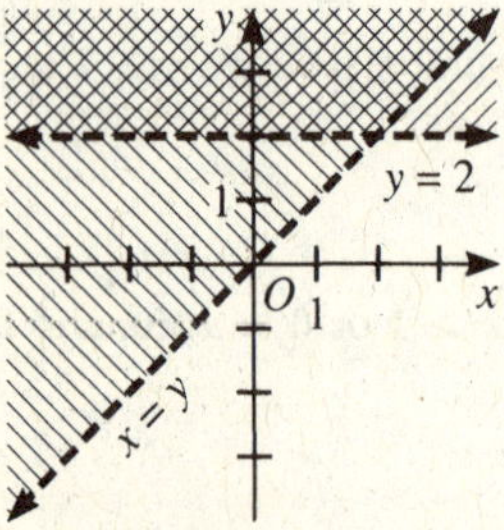

6.

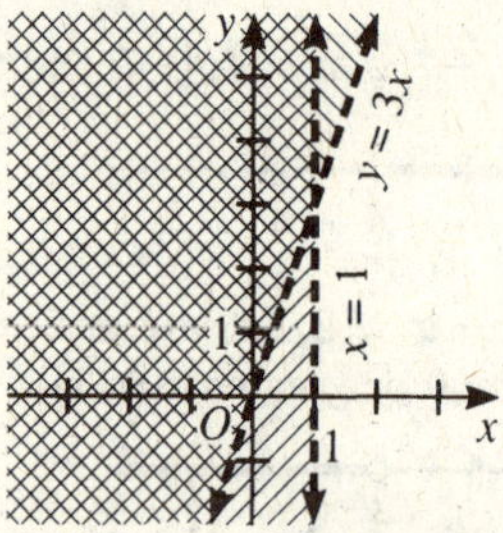

7.

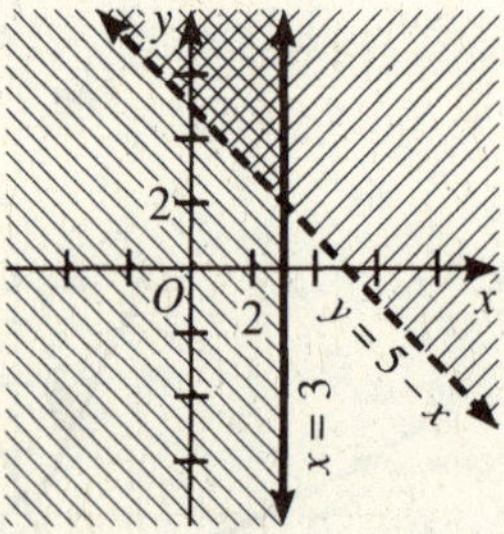

8.

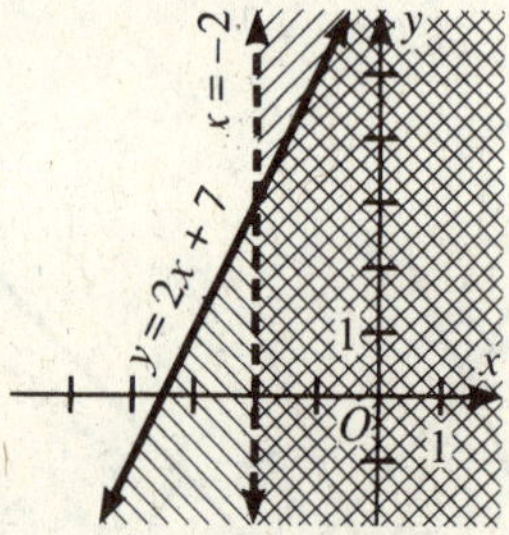

9.

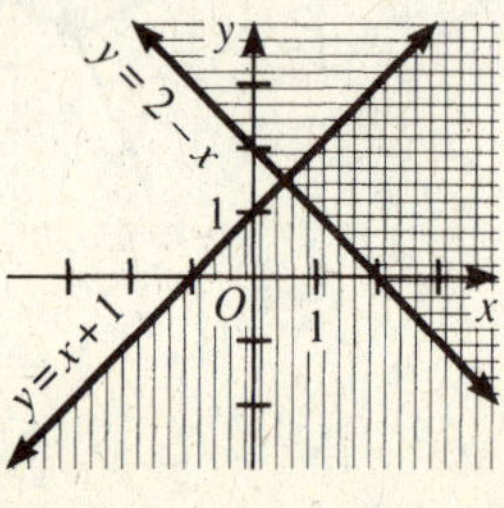

10.

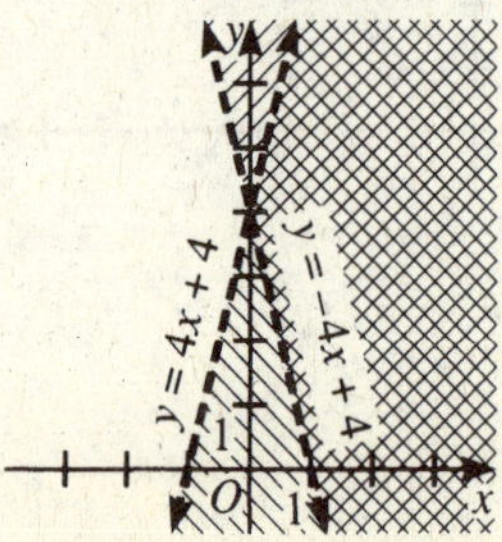

11. 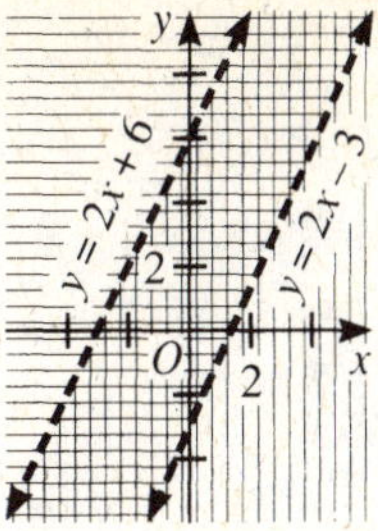

12. 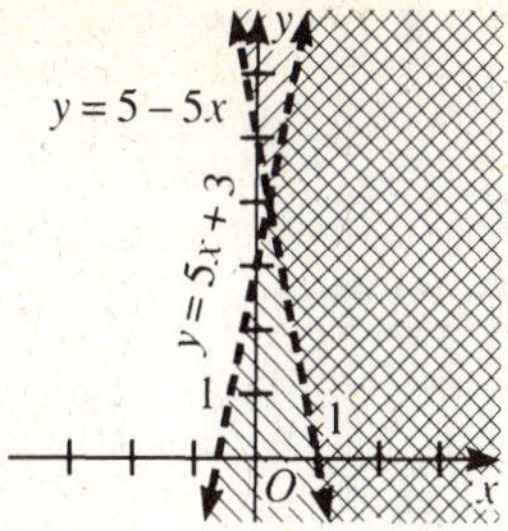

B 13. 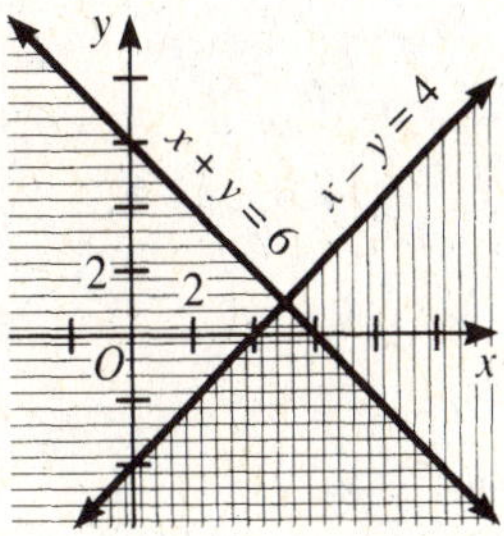

14. 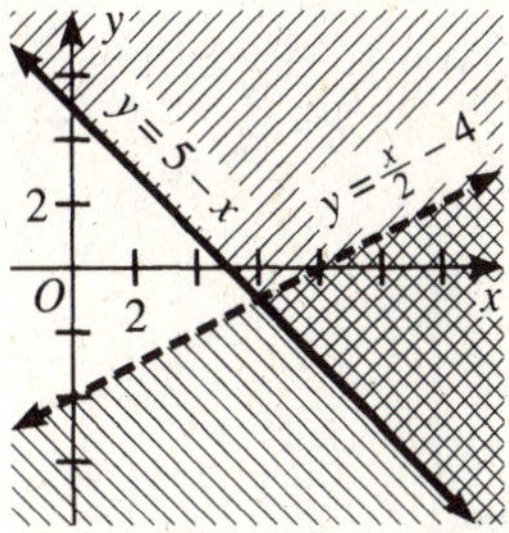

15. 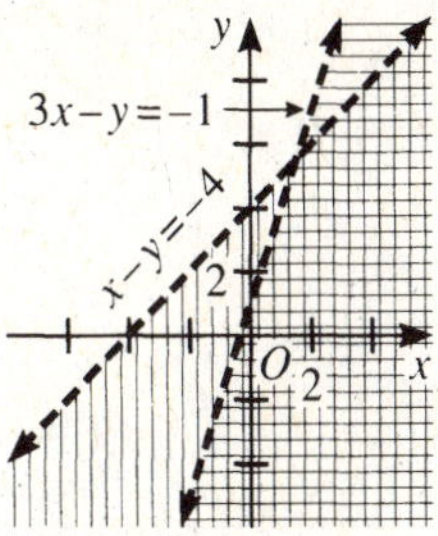

16. 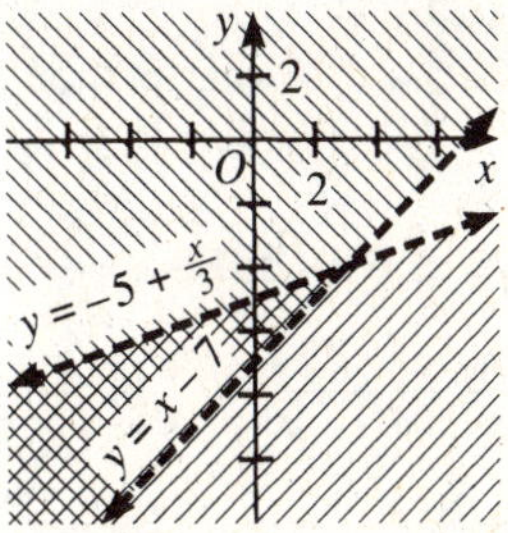

17. 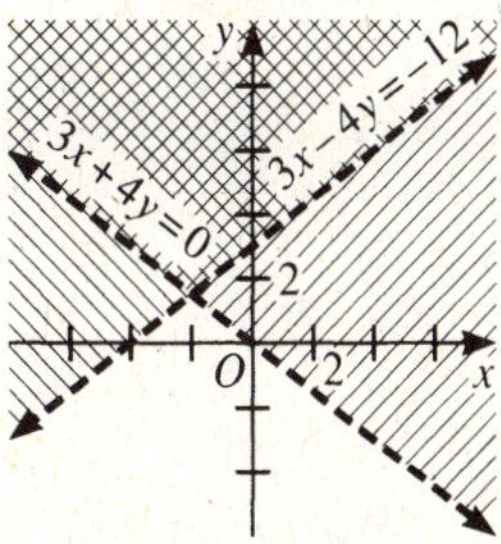

18. 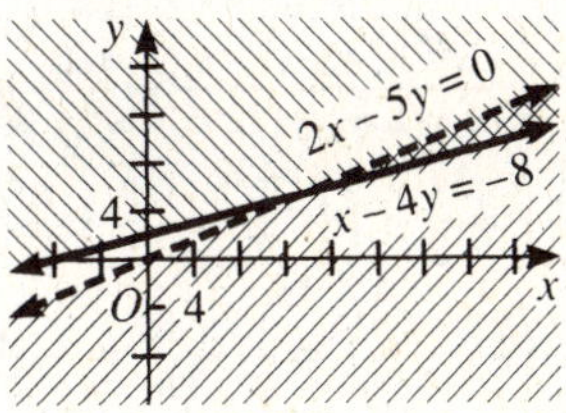

19. $y \geq -3$
$y < x - 3$

20. $x < -2$
$y \leq -x - 2$

21. $y < \frac{2}{3}x - 2$
$y < -2x + 3$

22. $y \leq x + 3$
$y \geq -2x$

C 23. $x \leq 3$
$y \leq \frac{1}{2}x + 2$
$y \geq -\frac{1}{4}x - 1$

24. $x > -2$
$x < 4$
$y \leq \frac{1}{3}x + \frac{8}{3}$
$y \geq -\frac{1}{3}x - \frac{8}{3}$

25. 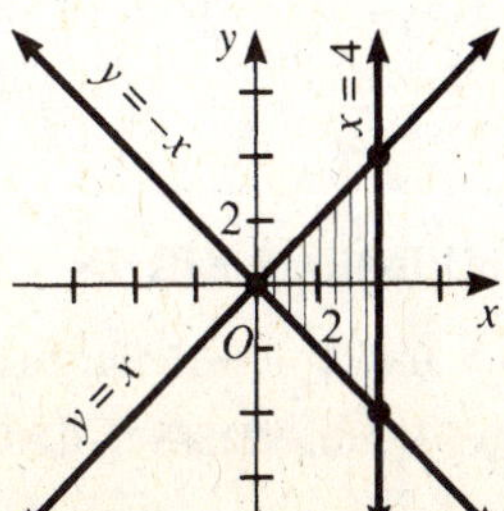

corner points: (0, 0), (4, 4), (4, −4)

26.

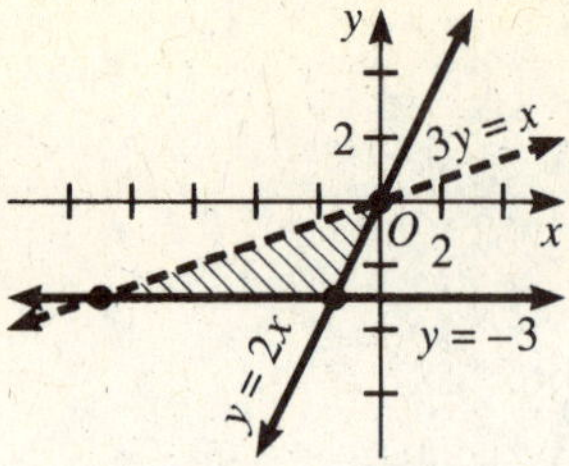

corner points:

$(-9, -3), \left(-\frac{3}{2}, -3\right), (0, 0)$

27. There is no solution to this system.

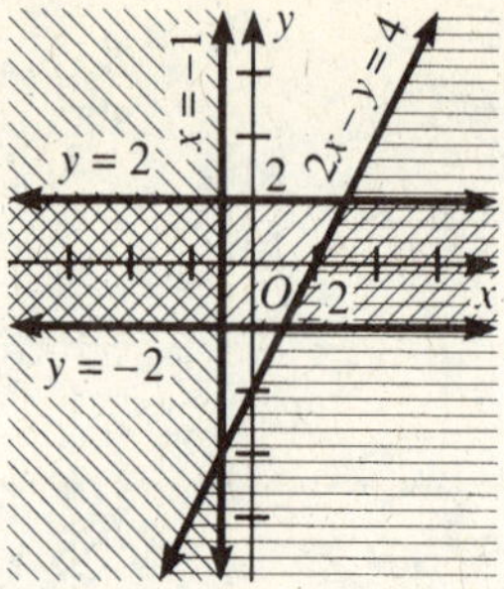

28.

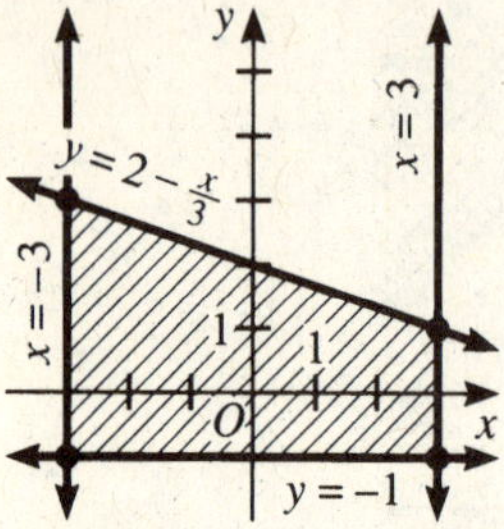

corner points:
$(-3, -3), (-3, -1), (3, -1), (3, 1)$

Page 498 • MIXED REVIEW EXERCISES

1. $\frac{5}{30}, \frac{8}{30}, \frac{18}{30}$ **2.** $\frac{12a}{36}, \frac{8}{36}, \frac{3(a+1)}{36}$ **3.** $\frac{k(k+5)}{(k+5)^2}, \frac{3k}{(k+5)^2}$

4. $\frac{5y(2y+3)}{5y(y-6)}, \frac{20(y-6)}{5y(y-6)}, \frac{y^2(y-6)}{5y(y-6)}$ **5.** $\frac{8(x-2)}{(x+2)(x-2)}, \frac{x(x+2)}{(x+2)(x-2)}$

6. $\frac{1}{(x+4)(x-4)}, \frac{-5(x+4)}{(x+4)(x-4)}, \frac{x(x-4)}{(x+4)(x-4)}$

7. $\left(\frac{3}{5}\right) + \left(\frac{1}{2} \div \frac{5}{12}\right) = \frac{3}{5} + \left(\frac{1}{2} \cdot \frac{12}{5}\right) = \frac{3}{5} + \frac{6}{5} = \frac{9}{5}$

8. $\frac{1}{2}\left(\frac{5}{12} - \frac{3}{5}\right) = \frac{1}{2}\left(\frac{25-36}{60}\right) = \frac{1}{2} \cdot \frac{-11}{60} = -\frac{11}{120}$

9. $\frac{3}{5} - \left(\frac{1}{2} + \frac{5}{12}\right) = \frac{3}{5} - \left(\frac{6+5}{12}\right) = \frac{3}{5} - \frac{11}{12} = \frac{36-55}{60} = -\frac{19}{60}$

10. $\frac{1}{2}\left(\frac{3}{5} + \frac{1}{2} + \frac{5}{12}\right) = \frac{1}{2}\left(\frac{36+30+25}{60}\right) = \frac{1}{2}\left(\frac{91}{60}\right) = \frac{91}{120}$

11. $\frac{5}{12} + \frac{1}{2}\left(\frac{1}{2} - \frac{5}{12}\right) = \frac{5}{12} + \frac{1}{2}\left(\frac{6-5}{12}\right) = \frac{5}{12} + \frac{1}{2} \cdot \frac{1}{12} = \frac{10+1}{24} = \frac{11}{24}$

12. $\frac{1}{2} - \frac{1}{2}\left(\frac{3}{5} - \frac{1}{2}\right) = \frac{1}{2} - \frac{1}{2}\left(\frac{6-5}{10}\right) = \frac{1}{2} - \frac{1}{2} \cdot \frac{1}{10} = \frac{1}{2} - \frac{1}{20} = \frac{10-1}{20} = \frac{9}{20}$

Page 498 • CALCULATOR KEY-IN

1. $0.625 > 0.45 \not> 0.4615384$; False **2.** $0.7391304 < 0.75 < 0.76$; True

3. $0.8910891 \not> 0.8921568 \not> 0.8932038$; False

Page 500 • APPLICATION

1. a.

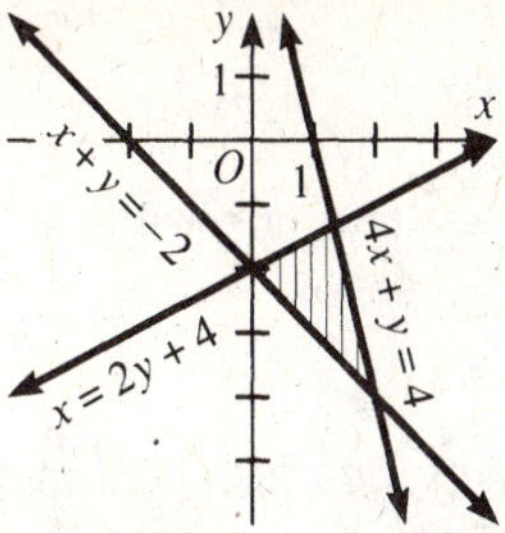

b. $\left(\frac{4}{3}, -\frac{4}{3}\right)$, (2, −4), (0, −2)

c. $4\left(\frac{4}{3}\right) + 7\left(-\frac{4}{3}\right) = \frac{16}{3} - \frac{28}{3} = -4$

$4(2) + 7(-4) = 8 - 28 = -20;$
$4(0) + 7(-2) = 0 - 14 = -14$
Maximum: −4, minimum: −20

2.

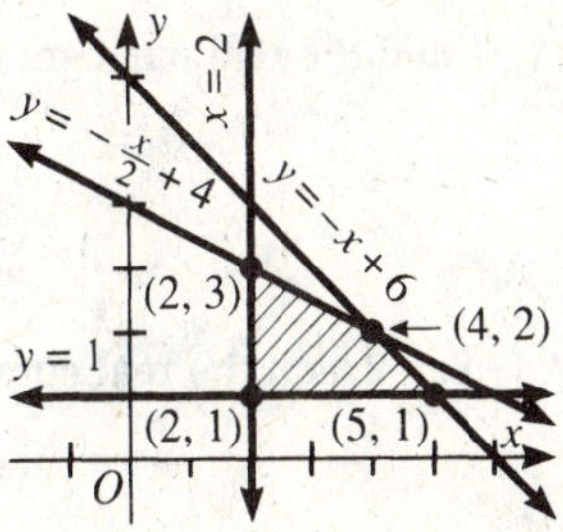

$R = x + 3y$
$R = 2 + 3 = 5$
$R = 5 + 3 = 8$
$R = 4 + 6 = 10$
$R = 2 + 9 = 11$
maximum at (2, 3)

3. a. Let c = number of acres of corn, s = number of acres of soybeans.
$6c + 12s \leq 2400$; $12c + 16s \leq 3600$; $c \geq 0$; $s \geq 0$.

b.

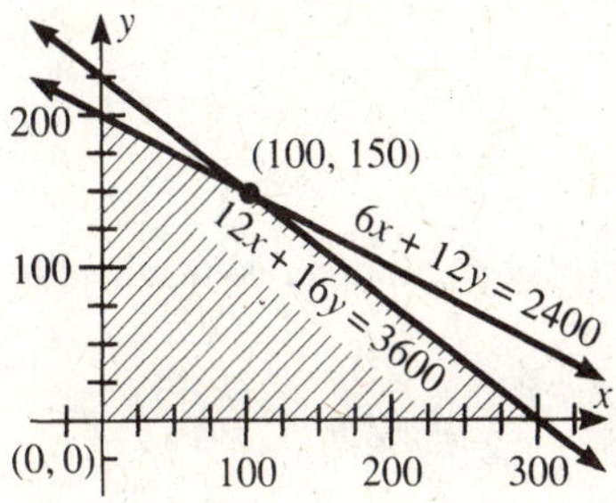

Corner points: (0, 0)
(0, 200)
(100, 150)
(300, 0)

c. $P = 24c + 48s$
$P = 24(0) + 48(0) = 0$
$P = 24(0) + 48(200) = 9600$
$P = 24(100) + 48(150) = 2400 + 7200 = 9600$
$P = 24(300) + 48(0) = 7200$
The farmer can plant 200 acres of soybeans and no corn or 150 acres of soybeans and 100 acres of corn, both for a profit of $9600.

Pages 501–502 • CHAPTER REVIEW

1. c **2.** c **3.** c; $-4x < 12, x > -3$

4. b; $3(3 + y) < 5(5 + y), 9 + 3y < 25 + 5y, -16 < 2y, -8 < y$ **5.** d **6.** a

7. b **8.** c **9.** b; $|8n - 4| < 12, 4|2n - 1| < 12, |2n - 1| < 3$

10. b; $x - 2y = 4, x - 4 = 2y, \frac{1}{2}x - 2 = y$ **11.** b; $0 < 2$ and $-5 \leq 3$

Page 502 • CHAPTER TEST

1. True
2. $(-2) + 1 = -1 < 3$; $(-1) + 1 = 0 < 3$; $(0) + 1 = 1 < 3$; $(1) + 1 = 2 < 3$; $(2) + 1 = 3 \not< 3$; $\{-2, -1, 0, 1\}$
3. $5 \geq 2x - 3$, $8 \geq 2x$, $4 \geq x$; {the real numbers less than or equal to 4}

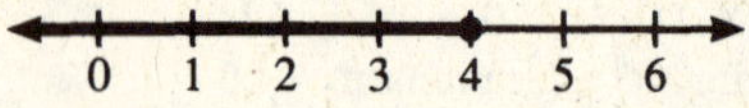

4. Let n and $n + 2$ be the two integers. $n + (n + 2) > 75$, $2n + 2 > 75$, $2n > 73$, $n > 36.5$; The pair whose sum is the least is 37 and 39.
5. $-1 \leq 2c + 1 < 5$, $-2 \leq 2c < 4$, $-1 \leq c < 2$; $\{-1$ and the real numbers between -1 and $2\}$

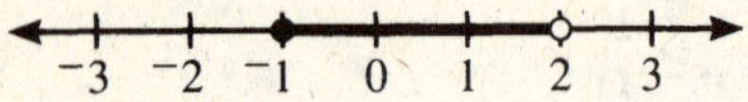

6. $k + 5 \leq -3$ or $k + 5 \geq 3$, $k \leq -8$ or $k \geq -2$; $\{-8, -2$, and the real numbers less than -8 or greater than $-2\}$

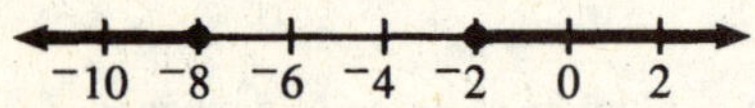

7. $|2 - y| = 3$; $2 - y = 3$ or $2 - y = -3$, $-y = 1$ or $-y = -5$, $y = -1$ or $y = 5$; $\{-1, 5\}$

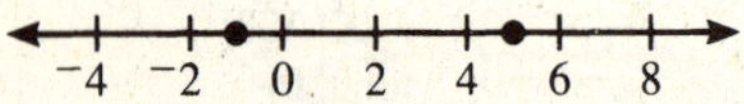

8. $|y + 1| < 12$; $-12 < y + 1 < 12$, $-13 < y < 11$; {the real numbers between -13 and 11}

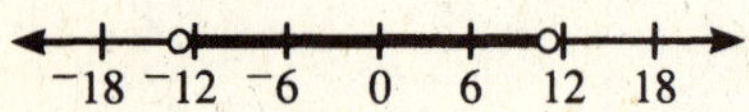

9. $|6 - 2x| = 8$, $2|3 - x| = 8$, $|3 - x| = 4$; $3 - x = 4$ or $3 - x = -4$, $-1 = x$ or $7 = x$; $\{-1, 7\}$

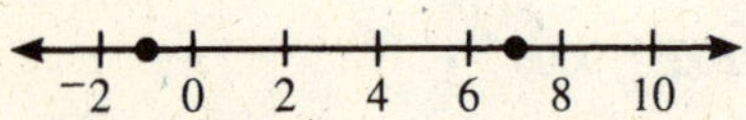

10. $|2m - 1| \leq 7$; $-7 \leq 2m - 1 \leq 7$, $-6 \leq 2m \leq 8$, $-3 \leq m \leq 4$; $\{-3, 4$, and the real numbers between -3 and $4\}$

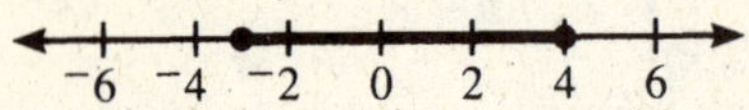

11. 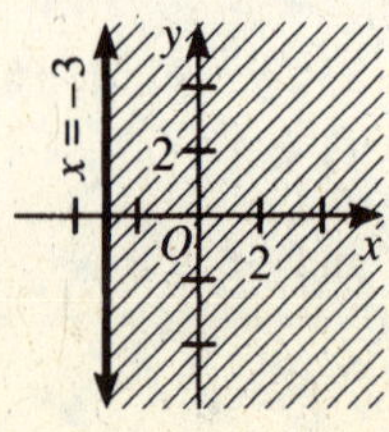

12. 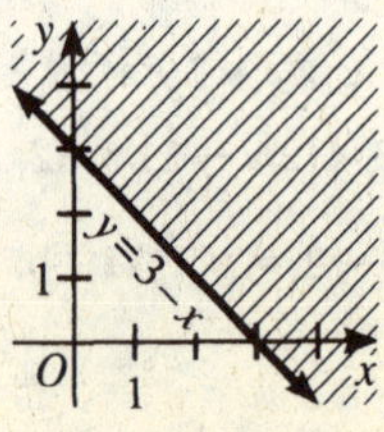

13. 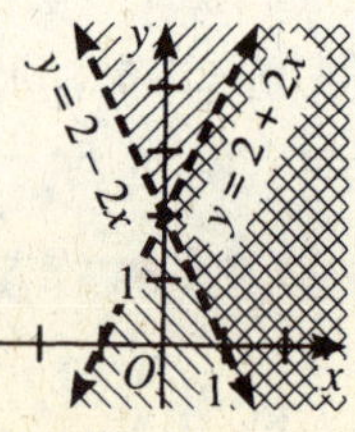

Page 503 • CUMULATIVE REVIEW

1. $(-7b + 2) + (8b - 5) = b - 3$
2. $-3x^2(4x^3 - 2x^2 - 11x) = -12x^5 + 6x^4 + 33x^3$
3. $(3x^3y^4)^5 \div (9xy^8) = 3^5x^{15}y^{20} \cdot \frac{1}{3^2xy^8} = 3^3x^{14}y^{12} = 27x^{14}y^{12}$
4. $(3x + 5y)(7x - 4y) = 21x^2 - 12xy + 35xy - 20y^2 = 21x^2 + 23xy - 20y^2$
5. $(3c^2 - 4cd + 8d^2)(c - 2d) = 3c^3 - 4c^2d + 8cd^2 - 6c^2d + 8cd^2 - 16d^3 =$
$3c^3 - 10c^2d + 16cd^2 - 16d^3$
6. $(5t - 7v)^2 = (5t)^2 - 2(5t)(7v) + (7v)^2 = 25t^2 - 70tv + 49v^2$
7. $\frac{3x^2 - 8x - 3}{9x^2 + 9x + 2} = \frac{(3x + 1)(x - 3)}{(3x + 2)(3x + 1)} = \frac{x - 3}{3x + 2}$
8. $(12m^3n^2 - 11r^3)(12m^3n^2 + 11r^3) = (12m^3n^2)^2 - (11r^3)^2 = 144m^6n^4 - 121r^6$
9. $\frac{4.8 \times 10^{25}}{2.0 \times 10^{11}} = 2.4 \times 10^{14}$
10. $\frac{z^2 + 8z + 16}{z^2 + 7z + 12} \cdot \frac{z^2 - 9}{z^2 + z - 12} = \frac{(z + 4)^2}{(z + 4)(z + 3)} \cdot \frac{(z + 3)(z - 3)}{(z + 4)(z - 3)} = 1$
11. $\frac{y^3 + 4y^2 + 3y}{y^2 + 7y + 10} \div \frac{y^2 + 3y}{y^3 + 2y^2} = \frac{y(y^2 + 4y + 3)}{(y + 5)(y + 2)} \cdot \frac{y^2(y + 2)}{y(y + 3)} = \frac{y^2(y + 3)(y + 1)}{(y + 5)(y + 3)} = \frac{y^2(y + 1)}{y + 5}$
12. $2 + \frac{4}{b - 1} - \frac{3b}{b^2 - 1} = \frac{2(b + 1)(b - 1) + 4(b + 1) - 3b}{(b + 1)(b - 1)} = \frac{2(b^2 - 1) + 4b + 4 - 3b}{(b + 1)(b - 1)} =$
$\frac{2b^2 - 2 + b + 4}{(b + 1)(b - 1)} = \frac{2b^2 + b + 2}{(b + 1)(b - 1)}$
13. $30z^2 + 240z + 96 = 6(5z^2 + 40z + 16)$
14. $81r^4 - 1 = (9r^2 + 1)(9r^2 - 1) = (9r^2 + 1)(3r + 1)(3r - 1)$
15. $16z^2 - y^2 - 10y - 25 = 16z^2 - (y^2 + 10y + 25) = (4z)^2 - (y + 5)^2 =$
$[4z + (y + 5)][4z - (y + 5)] = (4z + y + 5)(4z - y - 5)$
16. slope $= \frac{-5 - 10}{4 - (-1)} = \frac{-15}{5} = -3$; $y = -3x + b$, $10 = -3(-1) + b$, $10 = 3 + b$, $7 = b$;
$y = -3x + 7$, $3x + y = 7$
17. $y = \frac{1}{2}x + b$; $6 = \frac{1}{2}(5) + b$, $6 = \frac{5}{2} + b$, $\frac{7}{2} = b$; $y = \frac{1}{2}x + \frac{7}{2}$, $2y = x + 7$, $x - 2y = -7$
18. $|x - 3| \le 7$; $-7 \le x - 3 \le 7$, $-4 \le x \le 10$; {−4, 10, and the real numbers between −4 and 10}

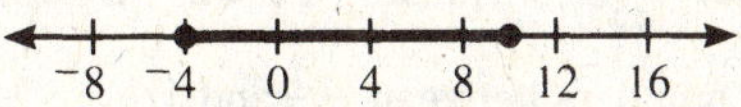

19. $4y + 5 < -7$ or $2y - 1 > 3$, $4y < -12$ or $2y > 4$, $y < -3$ or $y > 2$; {the real numbers less than −3 or greater than 2}

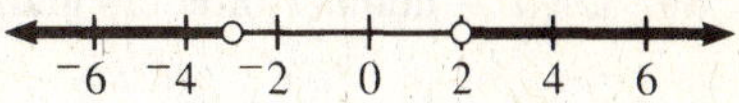

20.

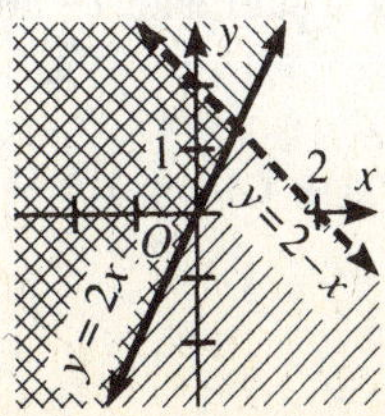

21. $|z| - 3 = 4$, $|z| = 7$; $z = 7$ or $z = -7$; {−7, 7}
22. $5q - 8 = 22$, $5q = 30$, $q = 6$; {6}

23. $|z + 1| = 10; z + 1 = -10$ or $z + 1 = 10, z = -11$ or $z = 9; \{-11, 9\}$

24. $b^2 - 8 = 1, b^2 = 9; b = 3$ or $b = -3; \{-3, 3\}$

25. $5(d - 3) = 3(2d - 5), 5d - 15 = 6d - 15, 5d = 6d, d = 0; \{0\}$

26. $(x - 2)(x - 1) = 0; x - 2 = 0$ or $x - 1 = 0, x = 2$ or $x = 1; \{1, 2\}$

27. $x^2 + 11x + 10 = 0, (x + 10)(x + 1) = 0; x + 10 = 0$ or $x + 1 = 0, x = -10$ or $x = -1; \{-10, -1\}$

28. $2b^2 + 9b - 5 = 0, (2b - 1)(b + 5) = 0; 2b - 1 = 0$ or $b + 5 = 0, 2b = 1$ or $b = -5$, $b = \frac{1}{2}$ or $b = -5; \left\{-5, \frac{1}{2}\right\}$

29. $z^3 + 3z^2 - 10z = 0, z(z^2 + 3z - 10) = 0, z(z + 5)(z - 2) = 0; z = 0$ or $z + 5 = 0$ or $z - 2 = 0, z = 0$ or $z = -5$ or $z = 2; \{-5, 0, 2\}$

30. $\frac{18}{x} = \frac{3 + 3x}{x}, x\left(\frac{18}{x}\right) = x\left(\frac{3 + 3x}{x}\right), 18 = 3 + 3x, 15 = 3x, 5 = x; \{5\}$

31. $\frac{b + 1}{b + 2} - \frac{b + 2}{b + 1} = \frac{1}{b + 2}, (b + 2)(b + 1)\left(\frac{b + 1}{b + 2} - \frac{b + 2}{b + 1}\right) = (b + 2)(b + 1)\left(\frac{1}{b + 2}\right),$
$(b + 1)^2 - (b + 2)^2 = b + 1, b^2 + 2b + 1 - (b^2 + 4b + 4) = b + 1,$
$b^2 + 2b + 1 - b^2 - 4b - 4 = b + 1, -2b - 3 = b + 1, -4 = 3b, -\frac{4}{3} = b; \left\{-\frac{4}{3}\right\}$

32. $\frac{1}{a^2 - a} = \frac{3}{a} - 1, a(a - 1)\left(\frac{1}{a(a - 1)}\right) = a(a - 1)\left(\frac{3}{a} - 1\right), 1 = 3(a - 1) - a(a - 1),$
$1 = 3a - 3 - a^2 + a, a^2 - 4a + 4 = 0, (a - 2)^2 = 0; a - 2 = 0, a = 2; \{2\}$

33. $y = 5x - 1; 2x + 3(5x - 1) = 14, 2x + 15x - 3 = 14, 17x = 17, x = 1;$
$y = 5(1) - 1, y = 4; (1, 4)$

34.
$$\begin{array}{rl} 3r + 2s = & -4 \\ \underline{4r - 2s =} & \underline{-10} \\ 7r \quad\;\; = & -14, \end{array}$$
$r = -2; 3(-2) + 2s = -4, -6 + 2s = -4, 2s = 2, s = 1; (-2, 1)$

35.
$$\begin{array}{l} 6m + 5n = 4 \rightarrow 18m + 15n = 12 \\ 4m + 3n = 2 \rightarrow \underline{20m + 15n = 10} \\ \qquad\qquad\qquad\;\; -2m \qquad\;\; = 2, \end{array}$$
$m = -1; 4(-1) + 3n = 2, -4 + 3n = 2,$
$3n = 6, n = 2; (-1, 2)$

36. $1 + 3t \leq 10, 3t \leq 9, t \leq 3$; {the real numbers less than or equal to 3}

37. $2 - b > 7, -b > 5, b < -5$; {the real numbers less than -5}

38. $2|3x - 1| + 1 < 5, 2|3x - 1| < 4, |3x - 1| < 2, -2 < 3x - 1 < 2, -1 < 3x < 3,$
$-\frac{1}{3} < x < 1$; $\left\{\text{the real numbers between } -\frac{1}{3} \text{ and } 1\right\}$

39. Let x = number of hours Mark would take working alone. $\frac{3}{5} + \frac{3}{x} = 1$,
$5x\left(\frac{3}{5} + \frac{3}{x}\right) = 5x, 3x + 15 = 5x, 15 = 2x, \frac{15}{2} = x$; Mark would take $7\frac{1}{2}$ hours working alone.

40. Let t = tens digit and $t - 1$ = units digit. $9t < 10(t - 1) + t, 9t < 10t - 10 + t,$
$10 < 2t, 5 < t$; The least such number is 65.

Page 504 • MAINTAINING SKILLS

1. $t^2 - 6t - 40 = 0$, $(t - 10)(t + 4) = 0$; $t - 10 = 0$ or $t + 4 = 0$, $t = 10$ or $t = -4$; $\{-4, 10\}$
2. $x^2 = 15x - 56$, $x^2 - 15x + 56 = 0$, $(x - 7)(x - 8) = 0$; $x - 7 = 0$ or $x - 8 = 0$, $x = 7$ or $x = 8$; $\{7, 8\}$
3. $y^2 + 12y + 36 = 0$, $(y + 6)^2 = 0$; $y + 6 = 0$, $y = -6$; $\{-6\}$
4. $n^2 + 400 = 40n$, $n^2 - 40n + 400 = 0$, $(n - 20)^2 = 0$; $n - 20 = 0$, $n = 20$; $\{20\}$
5. $5z^2 + 4z = 1$, $5z^2 + 4z - 1 = 0$, $(5z - 1)(z + 1) = 0$; $5z - 1 = 0$ or $z + 1 = 0$, $5z = 1$ or $z = -1$, $z = \frac{1}{5}$ or $z = -1$; $\left\{-1, \frac{1}{5}\right\}$
6. $10m^2 + 29m + 10 = 0$, $(5m + 2)(2m + 5) = 0$; $5m + 2 = 0$ or $2m + 5 = 0$, $5m = -2$ or $2m = -5$, $m = -\frac{2}{5}$ or $m = -\frac{5}{2}$; $\left\{-\frac{5}{2}, -\frac{2}{5}\right\}$
7. $8 + 14z = 15z^2$, $0 = 15z^2 - 14z - 8$, $0 = (5z + 2)(3z - 4)$; $5z + 2 = 0$ or $3z - 4 = 0$, $5z = -2$ or $3z = 4$, $z = -\frac{2}{5}$ or $z = \frac{4}{3}$; $\left\{-\frac{2}{5}, \frac{4}{3}\right\}$
8. $21b^2 + 25b + 6 = 0$, $(7b + 6)(3b + 1) = 0$; $7b + 6 = 0$ or $3b + 1 = 0$, $7b = -6$ or $3b = -1$, $b = -\frac{6}{7}$ or $b = -\frac{1}{3}$; $\left\{-\frac{6}{7}, -\frac{1}{3}\right\}$
9. $9d^2 + 7d = 16$, $9d^2 + 7d - 16 = 0$, $(9d + 16)(d - 1) = 0$; $9d + 16 = 0$ or $d - 1 = 0$, $9d = -16$ or $d = 1$, $d = -\frac{16}{9}$ or $d = 1$; $\left\{-\frac{16}{9}, 1\right\}$
10. $\frac{1}{x} + \frac{3}{x} = 4$ $(x \neq 0)$, $x\left(\frac{1}{x} + \frac{3}{x}\right) = 4x$, $1 + 3 = 4x$, $4 = 4x$, $1 = x$; $\{1\}$
11. $\frac{2}{3} + \frac{2}{t} = \frac{10}{3t}$ $(t \neq 0)$, $3t\left(\frac{2}{3} + \frac{2}{t}\right) = 3t\left(\frac{10}{3t}\right)$, $2t + 6 = 10$, $2t = 4$, $t = 2$; $\{2\}$
12. $\frac{5}{y} - \frac{4}{y} = \frac{1}{y}$ $(y \neq 0)$, $y\left(\frac{5}{y} - \frac{4}{y}\right) = \frac{1}{y} \cdot y$, $5 - 4 = 1$, $1 = 1$; identity; {the real numbers except 0}
13. $\frac{a + 5}{a} + \frac{2}{a^2} = 1$ $(a \neq 0)$, $a^2\left(\frac{a + 5}{a} + \frac{2}{a^2}\right) = a^2$, $a(a + 5) + 2 = a^2$, $a^2 + 5a + 2 = a^2$, $5a + 2 = 0$, $5a = -2$, $a = -\frac{2}{5}$; $\left\{-\frac{2}{5}\right\}$
14. $y - \frac{y}{y + 1} = \frac{1}{y + 1}$ $(y \neq -1)$, $(y + 1)\left(y - \frac{y}{y + 1}\right) = (y + 1)\left(\frac{1}{y + 1}\right)$, $y(y + 1) - y = 1$, $y^2 + y - y = 1$, $y^2 = 1$; $y = 1$ or $y = -1$; $\{1\}$
15. $\frac{3}{4b - 6} - \frac{1}{2b - 3} = 0$, $2(2b - 3)\left(\frac{3}{2(2b - 3)} - \frac{1}{2b - 3}\right) = 0$, $3 - 2 = 0$, $1 = 0$; There is no solution.

16. $\frac{3x}{x+1} - \frac{1}{x} = \frac{x+1}{x}$ $(x \neq 0, -1)$, $x(x+1)\left(\frac{3x}{x+1} - \frac{1}{x}\right) = x(x+1)\left(\frac{x+1}{x}\right)$, $3x^2 - (x+1) = (x+1)^2$, $3x^2 - x - 1 = x^2 + 2x + 1$, $2x^2 - 3x - 2 = 0$, $(2x+1)(x-2) = 0$; $2x + 1 = 0$ or $x - 2 = 0$, $2x = -1$ or $x = 2$, $x = -\frac{1}{2}$ or $x = 2$; $\left\{-\frac{1}{2}, 2\right\}$

17. $\frac{b-2}{b+2} + \frac{b}{b-2} = 2$ $(b \neq 2, -2)$, $(b+2)(b-2)\left(\frac{b-2}{b+2} + \frac{b}{b-2}\right) = 2(b+2)(b-2)$, $(b-2)^2 + b(b+2) = 2(b^2-4)$, $b^2 - 4b + 4 + b^2 + 2b = 2b^2 - 8$, $2b^2 - 2b + 4 = 2b^2 - 8$, $-2b + 4 = -8$, $-2b = -12$, $b = 6$; $\{6\}$

18. $\frac{1}{t} - \frac{1}{t+1} = \frac{t+2}{t^2+t}$ $(t \neq 0, -1)$, $t(t+1)\left(\frac{1}{t} - \frac{1}{t+1}\right) = t(t+1)\left(\frac{t+2}{t(t+1)}\right)$, $t + 1 - t = t + 2$, $1 = t + 2$, $-1 = t$; There is no solution.

19. $\frac{x^2+12}{x^2-4} - \frac{1}{x-2} = \frac{3}{x-2}$ $(x \neq 2, -2)$, $(x+2)(x-2)\left(\frac{x^2+12}{(x+2)(x-2)} - \frac{1}{x-2}\right) = (x+2)(x-2)\left(\frac{3}{x-2}\right)$, $x^2 + 12 - (x+2) = 3(x+2)$, $x^2 + 12 - x - 2 = 3x + 6$, $x^2 - 4x + 4 = 0$, $(x-2)^2 = 0$; $x = 2$; There is no solution.

20. $\frac{5}{d-1} - \frac{10}{d^2-1} = 3$ $(d \neq 1, -1)$, $(d+1)(d-1)\left(\frac{5}{d-1} - \frac{10}{(d+1)(d-1)}\right) = 3(d+1)(d-1)$, $5(d+1) - 10 = 3(d^2-1)$, $5d + 5 - 10 = 3d^2 - 3$, $0 = 3d^2 - 5d + 2$, $0 = (3d-2)(d-1)$; $3d - 2 = 0$ or $d - 1 = 0$, $3d = 2$ or $d = 1$, $d = \frac{2}{3}$ or $d = 1$; $\left\{\frac{2}{3}\right\}$

21. $\frac{b^2+1}{b^2-1} = \frac{b}{b-1} + \frac{2}{b+1}$ $(b \neq 1, -1)$, $(b+1)(b-1)\left(\frac{b^2+1}{b^2-1}\right) = (b+1)(b-1)\left(\frac{b}{b-1} + \frac{2}{b+1}\right)$, $b^2 + 1 = b(b+1) + 2(b-1)$, $b^2 + 1 = b^2 + b + 2b - 2$, $1 = 3b - 2$, $3 = 3b$, $1 = b$; There is no solution.

Page 505 • PREPARING FOR COLLEGE ENTRANCE EXAMS

1. D

2. B; Let w = width of the first rectangle, and $2w - 3$ = its length. $2\left(\frac{1}{w} + \frac{1}{2w-3}\right) = \frac{1}{5}[2(w + 2w - 3)]$, $5w(2w-3)\left(\frac{2}{w} + \frac{2}{2w-3}\right) = \frac{2}{5}(3w-3) \cdot 5w(2w-3)$, $10(2w-3) + 10w = 2w(6w^2 - 15w + 9)$, $20w - 30 + 10w = 12w^3 - 30w^2 + 18w$, $30w - 30 = 12w^3 - 30w^2 + 18w$, $0 = 12w^3 - 30w^2 - 12w + 30$, $0 = 2w^3 - 5w^2 - 2w + 5$, $0 = w^2(2w-5) - (2w-5)$, $0 = (w^2-1)(2w-5)$, $0 = (w+1)(w-1)(2w-5)$; $w + 1 = 0$ or $w - 1 = 0$ or $2w - 5 = 0$, $w = -1$ or $w = 1$ or $2w = 5$, $w = -1$ or $w = 1$ or $w = \frac{5}{2}$; Width cannot be neg., so $w \neq -1$. $2w - 3 = 2(1) - 3 = -1$; Length cannot be neg., so $w \neq 1$. $2\left(\frac{5}{2}\right) - 3 = 5 - 3 = 2$; $2\left(\frac{5}{2} + 2\right) = 5 + 4 = 9$; The perimeter of the first rectangle is 9 cm.

3. D; Let n, $n + 2$, and $n + 4$ be the integers. $3(n + n + 2) < 5(n + 4)$, $3(2n + 2) < 5n + 20$, $6n + 6 < 5n + 20$, $n < 14$; Solution: (2, 4, 6), (4, 6, 8), (6, 8, 10), (8, 10, 12), (10, 12, 14), and (12, 14, 16).

4. C; Let n = cost of a notebook and p = cost of a package of pencils.

$$\begin{aligned} 2n + 3p &= 7 \rightarrow 6n + 9p = 21 \\ 3n + 4p &= 10 \rightarrow \underline{6n + 8p = 20} \\ &\qquad\qquad\qquad\; p = 1 \end{aligned}$$

$p = 1$; $2n + 3(1) = 7$, $2n = 4$, $n = 2$; $n + p = 2 + 1 = 3$; It would cost $3 to buy one notebook and one package of pencils.

5. C; Let n and $n + 1$ be the integers. $n + (n + 1)^2 = 131$, $n + n^2 + 2n + 1 = 131$, $n^2 + 3n - 130 = 0$, $(n + 13)(n - 10) = 0$; $n + 13 = 0$ or $n - 10 = 0$, $n = -13$ or $n = 10$; $n + (n + 1) = 10 + 11 = 21$; The sum of the integers is 21.

6. B; $y \geq -\frac{1}{3}x + 2$, $3y \geq -x + 6$, $x + 3y \geq 6$

CHAPTER 11 • Rational and Irrational Numbers

Pages 510–511 • WRITTEN EXERCISES

A **1.** $<$ **2.** $>$ **3.** $<$ **4.** $<$ **5.** $=$ **6.** $<$ **7.** $>$ **8.** $>$ **9.** $\frac{3}{8}, \frac{5}{9}, \frac{4}{7}$

10. $\frac{2}{9}, \frac{3}{5}, \frac{5}{7}$ **11.** $-5, 5.6, \frac{107}{18}$ **12.** $-\frac{35}{8}, -3.8, -3$ **13.** $\frac{4}{15}, \frac{7}{24}, \frac{5}{16}, \frac{1}{2}$

14. $-\frac{5}{6}, -\frac{7}{9}, -\frac{3}{5}, -\frac{3}{7}$ **15.** $\frac{1}{2}\left(\frac{5}{9} + \frac{4}{7}\right) = \frac{1}{2}\left(\frac{35 + 36}{63}\right) = \frac{71}{126}$

16. $\frac{1}{2}\left(\frac{7}{11} + \frac{3}{4}\right) = \frac{1}{2}\left(\frac{28 + 33}{44}\right) = \frac{61}{88}$

17. $\frac{1}{2}\left(-\frac{4}{25} - \frac{9}{50}\right) = \frac{1}{2}\left(\frac{-8 - 9}{50}\right) = \frac{-17}{100} = -\frac{17}{100}$

18. $\frac{1}{2}\left(-\frac{5}{39} + -\frac{6}{117}\right) = \frac{1}{2}\left(-\frac{5}{39} - \frac{2}{39}\right) = \frac{1}{2}\left(-\frac{7}{39}\right) = -\frac{7}{78}$

19. $\frac{1}{2}\left(2\frac{3}{4} + -3\frac{3}{4}\right) = \frac{1}{2}\left(\frac{11}{4} - \frac{15}{4}\right) = \frac{1}{2}\left(-\frac{4}{4}\right) = -\frac{1}{2}$

20. $\frac{1}{2}\left(-5\frac{1}{5} + 8\frac{2}{3}\right) = \frac{1}{2}\left(\frac{-26}{5} + \frac{26}{3}\right) = \frac{1}{2}\left(\frac{-78 + 130}{15}\right) = \frac{52}{30} = \frac{26}{15} = 1\frac{11}{15}$

B **21.** $\frac{0}{3} < \frac{1}{3} < \frac{2}{3} < \frac{3}{3}$; the fraction increases. **22.** $\frac{0}{5} < \frac{1}{5} < \frac{2}{5} < \frac{3}{5}$; increases

23. $\frac{7}{1} > \frac{7}{2} > \frac{7}{3} > \frac{7}{4}$; decreases **24.** $\frac{1}{6} < \frac{2}{6} < \frac{3}{6} < \frac{4}{6}$; increases

25. $\frac{9}{5} > \frac{7}{5} > \frac{5}{5} > \frac{3}{5}$; decreases **26.** $\frac{14}{12} < \frac{14}{9} < \frac{14}{6} < \frac{14}{3}$; increases

27. $\frac{1}{6} + \frac{1}{3}\left(\frac{4}{5} - \frac{1}{6}\right) = \frac{1}{6} + \frac{1}{3}\left(\frac{24 - 5}{30}\right) = \frac{1}{6} + \frac{1}{3}\left(\frac{19}{30}\right) = \frac{1}{6} + \frac{19}{90} = \frac{15 + 19}{90} = \frac{34}{90} = \frac{17}{45}$

28. $\frac{5}{8} + \frac{1}{4}\left(1\frac{1}{4} - \frac{5}{8}\right) = \frac{5}{8} + \frac{1}{4}\left(\frac{5}{4} - \frac{5}{8}\right) = \frac{5}{8} + \frac{1}{4}\left(\frac{10 - 5}{8}\right) = \frac{5}{8} + \frac{1}{4}\left(\frac{5}{8}\right) = \frac{20 + 5}{32} = \frac{25}{32}$

29. $-\frac{2}{3} + \frac{1}{5}\left[\frac{3}{5} - \left(-\frac{2}{3}\right)\right] = -\frac{2}{3} + \frac{1}{5}\left(\frac{9 + 10}{15}\right) = -\frac{2}{3} + \frac{19}{75} = \frac{-50 + 19}{75} = -\frac{31}{75}$

30. $\frac{1}{2} + \frac{3}{4}\left(\frac{8}{9} - \frac{1}{2}\right) = \frac{1}{2} + \frac{3}{4}\left(\frac{16 - 9}{18}\right) = \frac{1}{2} + \frac{3}{4}\left(\frac{7}{18}\right) = \frac{1}{2} + \frac{21}{72} = \frac{36 + 21}{72} = \frac{57}{72} = \frac{19}{24}$

31. Answers may vary. For example, $\frac{1}{2}\left(\frac{1}{3} + \frac{3}{4}\right) = \frac{1}{2}\left(\frac{4 + 9}{12}\right) = \frac{1}{2}\left(\frac{13}{12}\right) = \frac{13}{24}$

32. Answers may vary. For example, $\frac{1}{2}\left(-\frac{1}{6} - \frac{1}{4}\right) = \frac{1}{2}\left(\frac{-2 - 3}{12}\right) = -\frac{5}{24}$

33. $\frac{1}{2}\left(\frac{3a}{7} + -\frac{a}{12}\right) = \frac{1}{2}\left(\frac{36a - 7a}{84}\right) = \frac{29a}{168}$

34. $-\frac{7b}{5} + \frac{2}{3}\left[-\frac{2b}{3} - \left(-\frac{7b}{5}\right)\right] = -\frac{7b}{5} + \frac{2}{3}\left(\frac{-10b + 21b}{15}\right) = -\frac{7b}{5} + \frac{2}{3}\left(\frac{11b}{15}\right) =$

$\frac{-63b + 22b}{45} = -\frac{41b}{45}$

35. **a.** 27.50 + 4.00 = 35.50 − 4.00; yes **b.** $a + \frac{1}{2}(b - a) = b - \frac{1}{2}(b - a)$, $a + \frac{1}{2}b - \frac{1}{2}a = b - \frac{1}{2}b + \frac{1}{2}a$, $\frac{1}{2}a + \frac{1}{2}b = \frac{1}{2}b + \frac{1}{2}a$, $\frac{a+b}{2} = \frac{a+b}{2}$

C **36.** 2. Multiplication rule for fractions, 3. Multiplication prop. of order, 2. Multiplication prop. of order, 3. Rule for simplifying fractions.

37. $w < z$ and $wz > 0$; or $z < 0 < w$ **38.** $w < z$

Page 511 • MIXED REVIEW EXERCISES

1. $3y + 4 \le 10$, $3y \le 6$, $y \le 2$; {the real numbers less than or equal to 2}

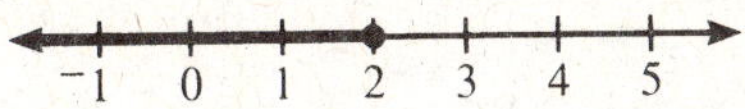

2. $|0.5 + x| < 10$; $-10 < 0.5 + x < 10$, $-10.5 < x < 9.5$; {the real numbers between −10.5 and 9.5}

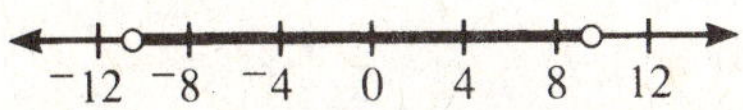

3. $10 + 6|2 - k| \ge 22$, $6|2 - k| \ge 12$, $|2 - k| \ge 2$; $2 - k \ge 2$ or $2 - k \le -2$, $-k \ge 0$ or $-k \le -4$, $k \le 0$ or $k \ge 4$; {0, 4, and the real numbers less than 0 or greater than 4}

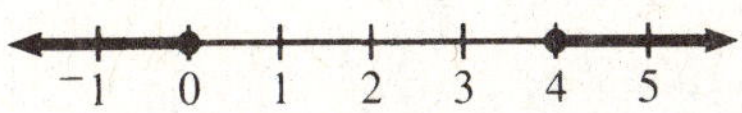

4. $|3 + z| \ge 5$; $3 + z \ge 5$ or $3 + z \le -5$, $z \ge 2$ or $z \le -8$; {2, −8, and the real numbers less than −8 or greater than 2}

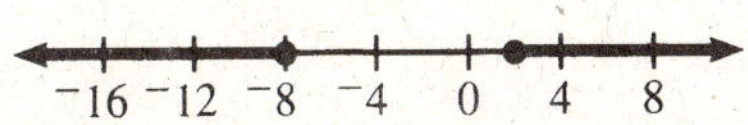

5. $4 \le 2x + 4 < 6$, $0 \le 2x < 2$, $0 \le x < 1$; {0 and the real numbers between 0 and 1}

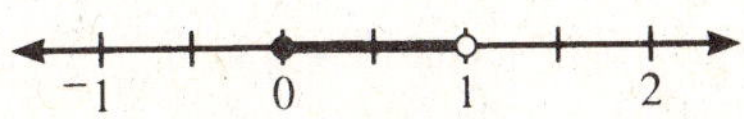

6. $6 \le 2 - 4m$, $4 \le -4m$, $-1 \ge m$; {the real numbers less than or equal to −1}

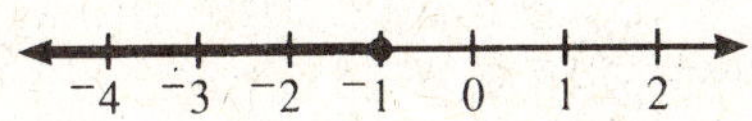

7. $35\% = \frac{35}{100} = \frac{7}{20}$ **8.** $0.6 = \frac{6}{10} = \frac{3}{5}$ **9.** $4.4 = \frac{44}{10} = \frac{22}{5}$ **10.** $3 \cdot 10^{-4} = \frac{3}{10^4} = \frac{3}{10{,}000}$

11. $62\frac{1}{2}\% = 62.5\% = \frac{62.5}{100} = \frac{625}{1000} = \frac{5}{8}$ **12.** $\frac{1}{4}\% = 0.25\% = \frac{0.25}{100} = \frac{25}{10{,}000} = \frac{1}{400}$

Pages 515–516 • WRITTEN EXERCISES

A **1.** **a.** $0.\overline{6}$ **b.** $0.0\overline{6}$ **2.** **a.** 4.5 **b.** 0.045 **3.** **a.** $-0.\overline{4}$ **b.** $-0.000\overline{4}$

4. **a.** −0.6 **b.** −0.06 **5.** 1.375 **6.** $0.91\overline{6}$ **7.** $1.\overline{36}$ **8.** $0.\overline{571428}$ **9.** $-0.3\overline{8}$

10. −0.46875 **11.** 3.45 **12.** $2.\overline{45}$ **13.** $0.\overline{27}$ **14.** −6.75 **15.** $-0.\overline{486}$

16. $0.\overline{594}$ **17.** $0.2 = \frac{2}{10} = \frac{1}{5}$ **18.** $0.66 = \frac{66}{100} = \frac{33}{50}$ **19.** $0.325 = \frac{325}{1000} = \frac{13}{40}$

20. $3.8 = \frac{38}{10} = \frac{19}{5}$ **21.** $10N = 4.\overline{4}$, $10N - N = 4.\overline{4} - 0.\overline{4}$, $9N = 4$, $N = \frac{4}{9}$

22. $100N = 115.\overline{15}$, $100N - N = 115.\overline{15} - 1.\overline{15}$, $99N = 114$, $N = \frac{114}{99} = \frac{38}{33}$

23. $10N = 2.83\overline{3}$, $10N - N = 2.83\overline{3} - 0.28\overline{3}$, $9N = 2.55$, $N = \frac{2.55}{9} = \frac{255}{900} = \frac{17}{60}$;

$-0.28\overline{3} = -\frac{17}{60}$

24. $10N = 23.9\overline{9}$, $10N - N = 23.9\overline{9} - 2.3\overline{9}$, $9N = 21.6$, $N = \frac{21.6}{9} = \frac{216}{90} = \frac{12}{5}$

25. $100N = 7.\overline{07}$, $100N - N = 7.\overline{07} - 0.\overline{07}$, $99N = 7$, $N = \frac{7}{99}$

26. $100N = 136.\overline{36}$, $100N - N = 136.\overline{36} - 1.\overline{36}$, $99N = 135$, $N = \frac{135}{99} = \frac{15}{11}$; $-1.\overline{36} = -\frac{15}{11}$

27. $10N = 23.\overline{3}$, $10N - N = 23.\overline{3} - 2.\overline{3}$, $9N = 21$, $N = \frac{21}{9} = \frac{7}{3}$; $-2.\overline{3} = -\frac{7}{3}$

28. $1{,}000{,}000N = 857{,}142.\overline{857142}$, $1{,}000{,}000N - N = 857{,}142.\overline{857142} - 0.\overline{857142}$,

$999{,}999N = 857{,}142$, $N = \frac{857{,}142}{999{,}999} = \frac{6}{7}$

B

29. $0.25 + \frac{1}{2}(0.259 - 0.25) = 0.25 + \frac{1}{2}(0.009) = 0.25 + 0.0045 = 0.2545$

30. $0.625 + \frac{1}{2}(0.634 - 0.625) = 0.625 + \frac{1}{2}(0.009) = 0.625 + 0.0045 = 0.6295$

31. $0.44 + \frac{1}{2}(0.\overline{4} - 0.44) = 0.44 + \frac{1}{2}(0.00\overline{4}) = 0.44 + 0.00\overline{2} = 0.44\overline{2}$

32. $0.77 + \frac{1}{2}(0.\overline{7} - 0.77) = 0.77 + \frac{1}{2}(0.00\overline{7}) = 0.77 + 0.003\overline{8} = 0.773\overline{8}$

33. $0.83 + \frac{1}{2}(0.8\overline{3} - 0.83) = 0.83 + \frac{1}{2}(0.00\overline{3}) = 0.83 + 0.001\overline{6} = 0.831\overline{6}$

34. $0.121 + \frac{1}{2}(0.125 - 0.121) = 0.121 + \frac{1}{2}(0.004) = 0.121 + 0.002 = 0.123$

35. $0.75 = \frac{75}{100} = \frac{3}{4}$; $\frac{3}{5} \cdot \frac{3}{4} = \frac{9}{20}$ **36.** $0.875 = \frac{875}{1000} = \frac{7}{8}$; $\frac{7}{8} \cdot \frac{5}{7} = \frac{5}{8}$

37. $0.\overline{6} = \frac{2}{3}$; $\frac{2}{3} \cdot \frac{7}{12} = \frac{7}{18}$ **38.** $0.\overline{5} = \frac{5}{9}$; $\frac{9}{20} \cdot \frac{5}{9} = \frac{1}{4}$

39. $0.\overline{407} = \frac{11}{27}$; $0.2\overline{7} = \frac{5}{18}$; $\frac{11}{27} \cdot \frac{5}{18} = \frac{55}{486}$

40. $0.3\overline{5} = \frac{16}{45}$, $1.3\overline{36} = \frac{147}{110}$, $\frac{16}{45} \cdot \frac{147}{110} = \frac{392}{825}$

C

41. a. $0.\overline{1}$, $0.\overline{5}$, $0.\overline{8}$ **b.** $0.\overline{037}$, $0.\overline{185}$, $0.\overline{296}$ **c.** $0.\overline{037} = \frac{1}{3}(0.11\overline{1})$, $0.\overline{185} = \frac{1}{3}(0.55\overline{5})$,

$0.\overline{296} = \frac{1}{3}(0.88\overline{8})$

42. a. $0.\overline{142857}, 0.\overline{857142}$

b. Each repeating block contains the same digits in the same order, but each one starts with a different digit.

c. $0.\overline{428571}, 0.\overline{571428}, 0.\overline{714285}$

43. a. $0.\overline{08}, 0.\overline{32}, 0.\overline{87}$ **b.** $1 = \frac{99}{99} = 99\left(\frac{1}{99}\right) = 99(0.\overline{01}) = 0.\overline{99} = 0.\overline{9}$

c. $10N = 9.\overline{9}, 10N - N = 9.\overline{9} - 0.\overline{9}, 9N = 9, N = 1$

Page 516 • MIXED REVIEW EXERCISES

1. $200 = 2^3 \cdot 5^2$ **2.** $98 = 2 \cdot 7^2$ **3.** $1089 = 3^2 \cdot 11^2$ **4.** $2250 = 2 \cdot 3^2 \cdot 5^3$

5. $392 = 2^3 \cdot 7^2$ **6.** $576 = 2^6 \cdot 3^2$

7. $(y + 3)(y - 4) = 0; y + 3 = 0$ or $y - 4 = 0, y = -3$ or $y = 4; \{-3, 4\}$

8. $(a + 5)^2 = 9, a^2 + 10a + 25 = 9, a^2 + 10a + 16 = 0, (a + 8)(a + 2) = 0;$ $a + 8 = 0$ or $a + 2 = 0, a = -8$ or $a = -2; \{-8, -2\}$

9. $y^2 = -36$; no solution

10. $k^3 - 16k = 0, k(k^2 - 16) = 0, k(k + 4)(k - 4) = 0; k = 0$ or $k + 4 = 0$ or $k - 4 = 0, k = 0$ or $k = -4$ or $k = 4; \{0, -4, 4\}$

11. $|x + 2| = 10; x + 2 = 10$ or $x + 2 = -10, x = 8$ or $x = -12; \{-12, 8\}$

12. $k + 4 < 16, k < 12$; {the real numbers less than 12}

Page 519 • WRITTEN EXERCISES

A **1.** 6 **2.** 8 **3.** 10 **4.** 11

5. $-\sqrt{400} = -\sqrt{4 \cdot 100} = -\sqrt{4} \cdot \sqrt{100} = -2 \cdot 10 = -20$

6. $-\sqrt{196} = -\sqrt{4 \cdot 49} = -\sqrt{4} \cdot \sqrt{49} = -2 \cdot 7 = -14$

7. $\sqrt{625} = \sqrt{5^4} = \sqrt{(5^2)^2} = 25$ **8.** $\sqrt{576} = \sqrt{64 \cdot 9} = \sqrt{64} \cdot \sqrt{9} = 8 \cdot 3 = 24$

9. $\pm\sqrt{2500} = \pm\sqrt{25 \cdot 100} = \pm\sqrt{25} \cdot \sqrt{100} = \pm 5 \cdot 10 = \pm 50$

10. $\pm\sqrt{1225} = \pm\sqrt{25 \cdot 49} = \pm\sqrt{25} \cdot \sqrt{49} = \pm 5 \cdot 7 = \pm 35$

11. $-\sqrt{\frac{81}{1600}} = -\frac{\sqrt{81}}{\sqrt{16 \cdot 100}} = -\frac{9}{\sqrt{16} \cdot \sqrt{100}} = -\frac{9}{4 \cdot 10} = -\frac{9}{40}$

12. $-\sqrt{\frac{225}{49}} = -\frac{\sqrt{225}}{\sqrt{49}} = -\frac{15}{7}$ **13.** $\pm\sqrt{\frac{121}{25}} = \pm\frac{\sqrt{121}}{\sqrt{25}} = \pm\frac{11}{5}$

14. $\pm\sqrt{\frac{1}{256}} = \pm\frac{1}{\sqrt{256}} = \pm\frac{1}{16}$ **15.** $\sqrt{\frac{144}{441}} = \frac{\sqrt{144}}{\sqrt{9 \cdot 49}} = \frac{12}{3 \cdot 7} = \frac{12}{21} = \frac{4}{7}$

16. $\sqrt{\frac{529}{576}} = \frac{\sqrt{529}}{\sqrt{9 \cdot 64}} = \frac{23}{3 \cdot 8} = \frac{23}{24}$

17. $-\sqrt{\frac{484}{100}} = -\frac{\sqrt{4 \cdot 121}}{\sqrt{100}} = -\frac{2 \cdot 11}{10} = -\frac{22}{10} = -\frac{11}{5}$

18. $-\sqrt{\frac{324}{729}} = -\frac{\sqrt{4 \cdot 81}}{\sqrt{9 \cdot 81}} = -\frac{2 \cdot 9}{3 \cdot 9} = -\frac{2}{3}$

19. $\sqrt{\frac{361}{2304}} = \frac{\sqrt{19^2}}{\sqrt{36 \cdot 64}} = \frac{19}{6 \cdot 8} = \frac{19}{48}$ **20.** $\sqrt{\frac{1156}{289}} = \frac{\sqrt{2^2 \cdot 17^2}}{\sqrt{17^2}} = \frac{2 \cdot 17}{17} = 2$

B **21.** $-\sqrt{\frac{28}{63}} = -\sqrt{\frac{4 \cdot 7}{9 \cdot 7}} = -\sqrt{\frac{4}{9}} = -\frac{2}{3}$ **22.** $-\sqrt{\frac{12}{75}} = -\sqrt{\frac{3 \cdot 4}{3 \cdot 25}} = -\sqrt{\frac{4}{25}} = -\frac{2}{5}$

23. $\sqrt{\frac{99}{44}} = \sqrt{\frac{9 \cdot 11}{4 \cdot 11}} = \sqrt{\frac{9}{4}} = \frac{3}{2}$ **24.** $\sqrt{\frac{20}{45}} = \sqrt{\frac{4 \cdot 5}{9 \cdot 5}} = \sqrt{\frac{4}{9}} = \frac{2}{3}$

25. $\pm\sqrt{\frac{175}{28}} = \pm\sqrt{\frac{7 \cdot 25}{7 \cdot 4}} = \pm\sqrt{\frac{25}{4}} = \pm\frac{5}{2}$ **26.** $\pm\sqrt{\frac{92}{207}} = \pm\sqrt{\frac{4 \cdot 23}{9 \cdot 23}} = \pm\sqrt{\frac{4}{9}} = \pm\frac{2}{3}$

27. $\sqrt{\frac{7}{175}} = \sqrt{\frac{7}{7 \cdot 25}} = \sqrt{\frac{1}{25}} = \frac{1}{5}$ **28.** $\sqrt{\frac{5}{180}} = \sqrt{\frac{5}{5 \cdot 36}} = \sqrt{\frac{1}{36}} = \frac{1}{6}$

29. $\pm\sqrt{\frac{33}{132}} = \pm\sqrt{\frac{33}{4 \cdot 33}} = \pm\sqrt{\frac{1}{4}} = \pm\frac{1}{2}$

30. $\pm\sqrt{\frac{180}{845}} = \pm\sqrt{\frac{36 \cdot 5}{169 \cdot 5}} = \pm\sqrt{\frac{36}{169}} = \pm\frac{6}{13}$ **31.** $-\sqrt{\frac{3200}{648}} = -\sqrt{\frac{400}{81}} = -\frac{20}{9}$

32. $-\sqrt{\frac{1682}{20{,}000}} = -\sqrt{\frac{2 \cdot 29^2}{2 \cdot 10{,}000}} = -\sqrt{\frac{29^2}{100^2}} = -\frac{29}{100}$ **33.** $\sqrt{0.04} = \sqrt{\frac{4}{100}} = \frac{2}{10} = 0.2$

34. $\sqrt{0.09} = \sqrt{\frac{9}{100}} = \frac{3}{10} = 0.3$ **35.** $-\sqrt{0.81} = -\sqrt{\frac{81}{100}} = -\frac{9}{10} = -0.9$

36. $-\sqrt{0.64} = -\sqrt{\frac{64}{100}} = -\frac{8}{10} = -0.8$ **37.** $\sqrt{1.21} = \sqrt{\frac{121}{100}} = \frac{11}{10} = 1.1$

38. $\sqrt{2.25} = \sqrt{\frac{225}{100}} = \frac{15}{10} = 1.5$

39. $\pm\sqrt{7.84} = \pm\sqrt{\frac{784}{100}} = \pm\frac{\sqrt{16 \cdot 49}}{10} = \pm\frac{4 \cdot 7}{10} = \pm\frac{28}{10} = \pm 2.8$

40. $\pm\sqrt{12.25} = \pm\sqrt{\frac{1225}{100}} = \pm\frac{\sqrt{25 \cdot 49}}{\sqrt{100}} = \pm\frac{5 \cdot 7}{10} = \pm\frac{35}{10} = \pm 3.5$

41. $\sqrt{0.0196} = \sqrt{\frac{196}{10{,}000}} = \sqrt{\frac{4 \cdot 49}{10{,}000}} = \frac{2 \cdot 7}{100} = \frac{14}{100} = 0.14$

42. $\sqrt{0.0289} = \sqrt{\frac{289}{10{,}000}} = \frac{17}{100} = 0.17$ **43.** $\sqrt{0.0009} = \sqrt{\frac{9}{10{,}000}} = \frac{3}{100} = 0.03$

44. $\sqrt{0.000049} = \sqrt{\frac{49}{1{,}000{,}000}} = \frac{7}{1{,}000} = 0.007$

45. $\sqrt{5^2 - 3^2} - (\sqrt{5})^2 = \sqrt{25 - 9} - 5 = \sqrt{16} - 5 = 4 - 5 = -1$

46. $\sqrt{(17)^2 - (15)^2} - (\sqrt{17})^2 = \sqrt{289 - 225} - 17 = \sqrt{64} - 17 = 8 - 17 = -9$

47. $\sqrt{(20)^2 - (16)^2} - (\sqrt{20})^2 = \sqrt{400 - 256} - 20 = \sqrt{144} - 20 = 12 - 20 = -8$

48. $\sqrt{(37)^2 - (12)^2} - (\sqrt{37})^2 = \sqrt{1369 - 144} - 37 = \sqrt{1225} - 37 = 35 - 37 = -2$

Page 520 • MIXED REVIEW EXERCISES

1. $0.125 = \frac{125}{1000} = \frac{1}{8}$ **2.** $-6.4 = -\frac{64}{10} = -\frac{32}{5}$

3. $10N = 3.\overline{3}$, $10N - N = 3.\overline{3} - 0.\overline{3}$, $9N = 3$, $N = \frac{3}{9} = \frac{1}{3}$

4. $100N = 216.\overline{16}$, $100N - N = 216.\overline{16} - 2.\overline{16}$, $99N = 214$, $N = \frac{214}{99}$

5. $\frac{1}{2}\left(\frac{4}{5} - \frac{2}{3}\right) = \frac{2}{5} - \frac{1}{3} = \frac{6-5}{15} = \frac{1}{15}$ **6.** $\frac{2}{3}\left(\frac{x}{4} - \frac{2x}{7}\right) = \frac{2}{3}\left(\frac{7x - 8x}{28}\right) = \frac{2}{3}\left(\frac{-x}{28}\right) = -\frac{x}{42}$

7. $4b^2 - 12b - 72 = 4(b^2 - 3b - 18) = 4(b - 6)(b + 3)$

8. $16t^2 - 8t - 9v^2 + 1 = 16t^2 - 8t + 1 - 9v^2 = (4t - 1)^2 - (3v)^2 =$
$[(4t - 1) + 3v][(4t - 1) - 3v] = (4t - 1 + 3v)(4t - 1 - 3v)$

9. $64k^3 - k = k(64k^2 - 1) = k(8k + 1)(8k - 1)$

10. $9w^2 - 24wy + 16y^2 = (3w - 4y)^2$ **11.** $2x^2 - 9xy - 5y^2 = (2x + y)(x - 5y)$

12. $3 - 2ab - 5a^2b^2 = (3 - 5ab)(1 + ab)$

Page 520 • CHALLENGE

$4a - 11b + 12c = 22 \rightarrow 4a - 11b + 12c = 22$
$a + 5b - 4c = 17 \rightarrow \underline{3a + 15b - 12c = 51}$
$7a + 4b = 73,\ 4b = 73 - 7a,\ b = \frac{73 - 7a}{4}$

b is a positive integer when $a = 3$ or 7

$a = 3$: $b = \frac{73 - 21}{4} = \frac{52}{4} = 13$; $(3) + 5(13) - 4c = 17$, $68 - 4c = 17$, $51 = 4c$, $12.75 = c$; but c must be an integer, so $a \neq 3$

$a = 7$: $b = \frac{73 - 49}{4} = \frac{24}{4} = 6$; $(7) + 5(6) - 4c = 17$, $37 - 4c = 17$, $20 = 4c$, $5 = c$; $(7, 6, 5)$

Pages 522–523 • WRITTEN EXERCISES

A **1.** $\sqrt{63} = \sqrt{9 \cdot 7} = \sqrt{9} \cdot \sqrt{7} = 3\sqrt{7}$ **2.** $\sqrt{28} = \sqrt{4 \cdot 7} = 2\sqrt{7}$

3. $\sqrt{98} = \sqrt{2 \cdot 49} = 7\sqrt{2}$ **4.** $\sqrt{50} = \sqrt{25 \cdot 2} = 5\sqrt{2}$

5. $\sqrt{75} = \sqrt{25 \cdot 3} = 5\sqrt{3}$ **6.** $\sqrt{24} = \sqrt{2^3 \cdot 3} = \sqrt{2^2 \cdot 2 \cdot 3} = 2\sqrt{6}$

7. $\sqrt{256} = \sqrt{16^2} = 16$ **8.** $\sqrt{120} = \sqrt{4 \cdot 30} = 2\sqrt{30}$

9. $2\sqrt{48} = 2\sqrt{16 \cdot 3} = 2 \cdot 4\sqrt{3} = 8\sqrt{3}$

10. $6\sqrt{108} = 6\sqrt{9 \cdot 4 \cdot 3} = 6 \cdot 3 \cdot 2\sqrt{3} = 36\sqrt{3}$

11. $5\sqrt{72} = 5\sqrt{9 \cdot 4 \cdot 2} = 5 \cdot 3 \cdot 2\sqrt{2} = 30\sqrt{2}$

12. $9\sqrt{90} = 9\sqrt{9 \cdot 10} = 9 \cdot 3\sqrt{10} = 27\sqrt{10}$ **13.** $\sqrt{529} = 23$ **14.** $\sqrt{324} = 18$

15. $6\sqrt{45} = 6\sqrt{9 \cdot 5} = 6 \cdot 3\sqrt{5} = 18\sqrt{5}$

16. $14\sqrt{75} = 14\sqrt{25 \cdot 3} = 14 \cdot 5\sqrt{3} = 70\sqrt{3}$ **17.** $\sqrt{361} = 19$

18. $\sqrt{864} = \sqrt{2^5 \cdot 3^3} = \sqrt{2^4 \cdot 3^2} \cdot \sqrt{2 \cdot 3} = 2^2 \cdot 3\sqrt{6} = 12\sqrt{6}$

19. $10\sqrt{125} = 10\sqrt{25 \cdot 5} = 10 \cdot 5\sqrt{5} = 50\sqrt{5}$

20. $3\sqrt{160} = 3\sqrt{16 \cdot 10} = 3 \cdot 4\sqrt{10} = 12\sqrt{10}$ **21.** $\sqrt{192} = \sqrt{64 \cdot 3} = 8\sqrt{3}$

22. $\sqrt{432} = \sqrt{9 \cdot 16 \cdot 3} = 3 \cdot 4\sqrt{3} = 12\sqrt{3}$

23. $5\sqrt{600} = 5\sqrt{100 \cdot 6} = 5 \cdot 10\sqrt{6} = 50\sqrt{6}$

24. $4\sqrt{363} = 4\sqrt{121 \cdot 3} = 4 \cdot 11\sqrt{3} = 44\sqrt{3}$

25. $6\sqrt{245} = 6\sqrt{49 \cdot 5} = 6 \cdot 7\sqrt{5} = 42\sqrt{5}$

26. $5\sqrt{567} = 5\sqrt{81 \cdot 7} = 5 \cdot 9\sqrt{7} = 45\sqrt{7}$ **27.** $\sqrt{5625} = \sqrt{25 \cdot 225} = 5 \cdot 15 = 75$

28. $\sqrt{9200} = \sqrt{100 \cdot 4 \cdot 23} = 10 \cdot 2\sqrt{23} = 20\sqrt{23}$

29. $7\sqrt{1200} = 7\sqrt{100 \cdot 4 \cdot 3} = 7 \cdot 10 \cdot 2\sqrt{3} = 140\sqrt{3}$

30. $5\sqrt{2050} = 5\sqrt{25 \cdot 82} = 5 \cdot 5\sqrt{82} = 25\sqrt{82}$

B **31.** $\sqrt{800} = \sqrt{400 \cdot 2} = 20\sqrt{2} \approx 20(1.4142) = 28.284$; $\sqrt{800} \approx 28.3$

32. $\sqrt{500} = \sqrt{100 \cdot 5} = 10\sqrt{5} \approx 10(2.2361) = 22.361$; $\sqrt{500} \approx 22.4$

33. $-\sqrt{700} = -\sqrt{100 \cdot 7} = -10\sqrt{7} \approx -10(2.6458) = -26.458$; $-\sqrt{700} \approx -26.5$

34. $-\sqrt{600} = -\sqrt{100 \cdot 6} = -10\sqrt{6} \approx -10(2.4495) = -24.495$; $-\sqrt{600} \approx -24.5$

35. $-\sqrt{5900} = -\sqrt{100 \cdot 59} = -10\sqrt{59} \approx -10(7.6811) = -76.811$; $-\sqrt{5900} \approx -76.8$

36. $-\sqrt{4800} = -\sqrt{100 \cdot 16 \cdot 3} = -10 \cdot 4\sqrt{3} = -40\sqrt{3} \approx -40(1.7321) =$
-69.284; $-\sqrt{4800} \approx -69.3$

37. $\pm\sqrt{7800} = \pm\sqrt{100 \cdot 78} = \pm 10\sqrt{78} \approx \pm 10(8.8318) = \pm 88.318$; $\pm\sqrt{7800} \approx \pm 88.3$

38. $\pm\sqrt{5600} = \pm\sqrt{100 \cdot 4 \cdot 14} = \pm 10 \cdot 2\sqrt{14} \approx \pm 20(3.7417) = \pm 74.834$;
$\pm\sqrt{5600} \approx \pm 74.8$

39. $\sqrt{68} = \sqrt{4 \cdot 17} = 2\sqrt{17} \approx 2(4.1231) = 8.2462$; $\sqrt{68} \approx 8.25$ **40.** $\sqrt{42} \approx 6.48$

41. $-\sqrt{0.5} = -\sqrt{\frac{50}{100}} = -\frac{\sqrt{25 \cdot 2}}{10} = -\frac{5\sqrt{2}}{10} = -\frac{\sqrt{2}}{2} \approx -\frac{1.4142}{2} = -0.7071$;
$-\sqrt{0.5} \approx -0.71$

42. $-\sqrt{0.3} = -\sqrt{\frac{30}{100}} = -\frac{\sqrt{30}}{10} \approx -\frac{5.4772}{10} = -0.54772$; $-\sqrt{0.3} \approx -0.55$

43. $\pm\sqrt{0.87} = \pm\sqrt{\frac{87}{100}} = \pm\frac{\sqrt{87}}{10} \approx \pm\frac{9.3274}{10} = \pm 0.93274$; $\pm\sqrt{0.87} \approx \pm 0.93$

44. $\pm\sqrt{0.73} = \pm\sqrt{\frac{73}{100}} = \pm\frac{\sqrt{73}}{10} \approx \pm\frac{8.5440}{10} = \pm 0.85440$; $\pm\sqrt{0.73} \approx \pm 0.85$

45. $-\sqrt{0.07} = -\sqrt{\frac{7}{100}} = -\frac{\sqrt{7}}{10} \approx -\frac{2.6458}{10} = -0.26458$; $-\sqrt{0.07} \approx -0.26$

46. $-\sqrt{0.08} = -\sqrt{\frac{8}{100}} = -\frac{\sqrt{4 \cdot 2}}{10} = -\frac{2\sqrt{2}}{10} \approx -\frac{1.4142}{5} = -0.28284$; $-\sqrt{0.08} \approx -0.28$

47. $\sqrt{150{,}000} = \sqrt{15 \cdot 10{,}000} = 100\sqrt{15} \approx 100(3.8730) = 387.30$; $\sqrt{150{,}000} \approx 387$

48. $\sqrt{240{,}000} = \sqrt{6 \cdot 4 \cdot 10{,}000} = 2 \cdot 100\sqrt{6} \approx 200(2.4495) = 489.9$; $\sqrt{240{,}000} \approx 490$

49. $\sqrt{420{,}000} = \sqrt{42 \cdot 10{,}000} = 100\sqrt{42} \approx 100(6.4807) = 648.07$; $\sqrt{420{,}000} \approx 648$

50. $\sqrt{580{,}000} = \sqrt{58 \cdot 10{,}000} = 100\sqrt{58} \approx 100(7.6158) = 761.58$; $\sqrt{580{,}000} \approx 762$

Page 523 • MIXED REVIEW EXERCISES

1. $\sqrt{400} = 20$ **2.** $-\sqrt{169} = -13$ **3.** $\sqrt{\frac{25}{81}} = \frac{5}{9}$ **4.** $-\sqrt{\frac{49}{225}} = -\frac{7}{15}$

5. $\sqrt{176^2} = 176$ **6.** $\sqrt{\left(\frac{3}{8}\right)^2} = \frac{3}{8}$ **7.** $(17x)^2 = 289x^2$ **8.** $(3y^4z^9)^2 = 9y^8z^{18}$

9. $(2x + 3y)^2 = 4x^2 + 12xy + 9y^2$

10. $[15(a + 2)]^2 = 225(a^2 + 4a + 4) = 225a^2 + 900a + 900$

11. $(11a^4b^{11}c)^2 = 121a^8b^{22}c^2$

12. $(6z^3 + 5y^4)(6z^3 - 5y^4) = (6z^3)^2 - (5y^4)^2 = 36z^6 - 25y^8$

Page 524 • EXTRA

1. Assume that 3 has a rational square root.
2. Then $\sqrt{3} = \frac{p}{q}$ where p and q are positive integers that have no common prime factor.
3. If $\sqrt{3} = \frac{p}{q}$, $3 = \frac{p^2}{q^2}$ where p^2 and q^2 have no common prime factor.
4. $3q^2 = p^2$, so p^2 is divisible by 3, and so is p.
5. Let $p = 3m$; then $3q^2 = (3m)^2 = 9m^2$.
6. If $3q^2 = 9m^2$, $q^2 = 3m^2$, so both q^2 and q are divisible by 3.
7. This contradicts the assumption that p and q have no common prime factor.

Page 526 • WRITTEN EXERCISES

A 1. $\sqrt{121a^2} = 11|a|$ 2. $\sqrt{100z^2} = 10|z|$ 3. $\sqrt{28x^2} = \sqrt{4x^2 \cdot 7} = 2|x|\sqrt{7}$

4. $\sqrt{32b^4} = \sqrt{16b^4 \cdot 2} = 4b^2\sqrt{2}$ 5. $-\sqrt{9c^4} = -3c^2$ 6. $-\sqrt{64x^2} = -8|x|$

7. $-\sqrt{25d^6} = -5|d^3|$ 8. $-\sqrt{16d^8} = -4d^4$

9. $\sqrt{80a^2b^2} = \sqrt{16a^2b^2 \cdot 5} = 4|ab|\sqrt{5}$ 10. $\sqrt{49a^2b^2} = 7|ab|$

11. $\sqrt{75r^3} = \sqrt{25r^2 \cdot 3r} = 5r\sqrt{3r}$ (Assuming that $\sqrt{75r^3}$ is real, then $r \geq 0$.)

12. $\sqrt{80n^6} = \sqrt{16n^6 \cdot 5} = 4|n^3|\sqrt{5}$

13. $\pm\sqrt{54x^2y^3} = \pm\sqrt{9x^2y^2 \cdot 6y} = \pm3|x|y\sqrt{6y}$ (Assuming that $\sqrt{56x^2y^3}$ is real, then $y \geq 0$.)

14. $\pm\sqrt{56r^6s^4} = \pm\sqrt{4r^6s^4 \cdot 14} = \pm2|r^3|s^2\sqrt{14}$ 15. $-\sqrt{144x^2z^2} = -12|xz|$

16. $-\sqrt{400a^6b^4} = -20|a^3|b^2$ 17. $\pm\sqrt{\frac{100f^{10}}{121}} = \pm\frac{10|f^5|}{11}$

18. $\pm\sqrt{\frac{256}{400s^{12}}} = \pm\frac{16}{20s^6} = \pm\frac{4}{5s^6}$ 19. $\sqrt{\frac{x^4y^6}{4r^2}} = \frac{x^2|y^3|}{2|r|}$ 20. $\sqrt{\frac{45m^3n^2}{5mn^4}} = \sqrt{\frac{9m^2}{n^2}} = \frac{3|m|}{|n|}$

21. $\sqrt{\frac{64g^{18}}{8100h^{20}}} = \frac{8|g^9|}{90h^{10}} = \frac{4|g^9|}{45h^{10}}$ 22. $\sqrt{\frac{324r^{50}}{49}} = \frac{18|r^{25}|}{7}$ 23. $-\sqrt{3.24x^4} = -1.8x^2$

24. $\sqrt{1.96k^6} = 1.4|k^3|$ 25. $\sqrt{y^2 - 8y + 16} = \sqrt{(y-4)^2} = |y - 4|$

26. $\sqrt{m^2 - 12m + 36} = \sqrt{(m-6)^2} = |m - 6|$ 27. $x^2 = 25; x = \pm5; \{5, -5\}$

28. $n^2 = 64; n = \pm8; \{8, -8\}$ 29. $x^2 - 4 = 0; x^2 = 4; x = \pm2; \{2, -2\}$

30. $d^2 - 49 = 0; d^2 = 49; d = \pm7; \{7, -7\}$

31. $0 = a^2 - 100; 100 = a^2; \pm10 = a; \{10, -10\}$

32. $0 = m^2 - 81; 81 = m^2; \pm9 = m; \{9, -9\}$

33. $2m^2 - 50 = 0; 2m^2 = 50; m^2 = 25; m = \pm5; \{5, -5\}$

34. $50b^2 - 450 = 0; 50b^2 = 450; b^2 = 9; b = \pm3; \{3, -3\}$

35. $81y^2 - 16 = 0; 81y^2 = 16; y^2 = \frac{16}{81}; y = \pm\frac{4}{9}; \left\{\frac{4}{9}, -\frac{4}{9}\right\}$

36. $9c^2 - 64 = 0; 9c^2 = 64; c^2 = \frac{64}{9}; c = \pm\frac{8}{3}; \left\{\frac{8}{3}, -\frac{8}{3}\right\}$

37. $0 = 81z^2 - 49$; $49 = 81z^2$; $\frac{49}{81} = z^2$; $\pm\frac{7}{9} = z$; $\left\{\frac{7}{9}, -\frac{7}{9}\right\}$

38. $0 = 80p^2 - 125$; $125 = 80p^2$; $\frac{125}{80} = p^2$; $\frac{25}{16} = p^2$; $\pm\frac{5}{4} = p$; $\left\{\frac{5}{4}, -\frac{5}{4}\right\}$

B **39.** $x^2 = 618$; $x = \pm\sqrt{618} \approx \pm 24.9$; $\{24.9, -24.9\}$

40. $a^2 = 154$; $a = \pm\sqrt{154} \approx \pm 12.4$; $\{12.4, -12.4\}$

41. $0.38 = r^2$; $r = \pm\sqrt{0.38} \approx \pm 0.6$; $\{0.6, -0.6\}$

42. $0.29 = k^2$; $k = \pm\sqrt{0.29} \approx \pm 0.5$; $\{0.5, -0.5\}$

43. $c^2 - 212 = 0$; $c^2 = 212$; $c = \pm\sqrt{212} \approx \pm 14.6$; $\{14.6, -14.6\}$

44. $w^2 - 204 = 0$; $w^2 = 204$; $w = \pm\sqrt{204} \approx \pm 14.3$; $\{14.3, -14.3\}$

45. $y^2 - 10.25 = 0$; $y^2 = 10.25$; $y = \pm\sqrt{10.25} \approx \pm 3.2$; $\{3.2, -3.2\}$

46. $n^2 - 13.08 = 0$; $n^2 = 13.08$; $n = \pm\sqrt{13.08} \approx \pm 3.6$; $\{3.6, -3.6\}$

47. $9z^2 = 513$; $z^2 = 57$; $z = \pm\sqrt{57} \approx \pm 7.5$; $\{7.5, -7.5\}$

48. $7z^2 = 133$; $z^2 = 19$; $z = \pm\sqrt{19} \approx \pm 4.4$; $\{4.4, -4.4\}$

49. $0 = 4b^2 - 0.48$; $4b^2 = 0.48$; $b^2 = 0.12$; $b = \pm\sqrt{0.12} \approx \pm 0.3$; $\{0.3, -0.3\}$

50. $0 = 6n^2 - 0.42$; $6n^2 = 0.42$; $n^2 = 0.07$; $n = \pm\sqrt{0.07} \approx \pm 0.3$; $\{0.3, -0.3\}$

C **51.** $0.13m^2 = 9.36$; $m^2 = 72$; $m = \pm\sqrt{72} \approx \pm 8.5$; $\{8.5, -8.5\}$

52. $9k^2 = 168$; $k^2 = \frac{56}{3}$; $k = \pm\sqrt{\frac{56}{3}} \approx \pm 4.3$; $\{4.3, -4.3\}$

53. $(x + 1)^2 + (x - 1)^2 = 176$; $x^2 + 2x + 1 + x^2 - 2x + 1 = 176$; $2x^2 + 2 = 176$;
$2x^2 = 174$; $x^2 = 87$; $x = \pm\sqrt{87} \approx \pm 9.3$; $\{9.3, -9.3\}$

54. $(a + 3)^2 + (a - 3)^2 = 256$; $a^2 + 6a + 9 + a^2 - 6a + 9 = 256$; $2a^2 + 18 = 256$;
$2a^2 = 238$; $a^2 = 119$; $a = \pm\sqrt{119} \approx \pm 10.9$; $\{10.9, -10.9\}$

Page 527 • PROBLEMS

A **1.** Let s = length of a side. $s^2 = 300$; $s = \sqrt{300} \approx 17.3$; The length of a side of the square is about 17.3 cm.

2. Let s = length of a side. $s^2 = 24 \cdot 30$, $s^2 = 720$; $s = \sqrt{720} \approx 26.8$; The length of a side of the square is about 26.8 cm.

3. Let a = length of the altitude and $3a$ = length of the base. $\frac{1}{2}(3a)(a) = 54$, $\frac{3}{2}a^2 = 54$, $a^2 = 36$; $a = 6$; $3a = 18$; The length of the base of the triangle is 18 m.

4. Let s = length of a side. $s^2 = \frac{1}{2}(11)(18)$, $s^2 = 99$; $s = \sqrt{99} \approx 9.9$; A side of the square is about 9.9 cm long.

5. Let r = radius of the region. $\pi r^2 = 164$, $r^2 = \frac{164}{\pi}$; $r = \sqrt{\frac{164}{\pi}} \approx 7.2$; The radius of the search region was about 7.2 km.

6. $(3s)(5s) + s^2 = 600$, $15s^2 + s^2 = 600$, $16s^2 = 600$, $s^2 = 37.5$; $s = \sqrt{37.5} \approx 6.1$; s is approximately 6.1 mm.

B 7. Let t = number of seconds the rock falls. $1587.6 = 4.9t^2$, $324 = t^2$; $t = 18$; It takes the rock 18 s to fall.

8. Let s = length of a side of the square and $\frac{1}{2}s$ = radius of the circle. $\pi\left(\frac{1}{2}s\right)^2 = 341.9$, $\pi\frac{s^2}{4} = 341.9$, $s^2 = \frac{1367.6}{\pi}$; $s = \sqrt{\frac{1367.6}{\pi}} \approx 20.9$. The square has sides about 20.9 cm long.

9. Let r_1 = radius of the old pipe and r_2 = radius of the new pipe. $\pi r_2^2 = 2(\pi r_1^2)$, $r_2^2 = 2r_1^2$, $\frac{r_2^2}{r_1^2} = 2$; $\left(\frac{r_2}{r_1}\right)^2 = 2$, $\frac{r_2}{r_1} = \frac{\sqrt{2}}{1} \approx \frac{1.4}{1}$; The ratio of the radius of the new pipe to that of the old pipe is about $\frac{1.4}{1}$.

C 10. $s = \frac{1}{2}(8 + 10 + 14) = \frac{1}{2}(32) = 16$; $\mathbf{A} = \sqrt{16(16 - 8)(16 - 10)(16 - 14)} = \sqrt{16 \cdot 8 \cdot 6 \cdot 2} = \sqrt{1536} \approx 39.2$; The area is about 39.2 cm^2.

11. $s = \frac{1}{2}(14 + 19 + 25) = \frac{1}{2} \cdot 58 = 29$; $\mathbf{A} = \sqrt{29(29 - 14)(29 - 19)(29 - 25)} = \sqrt{17{,}400} \approx 131.9$; The area is about 131.9 m^2.

12. $s = \frac{1}{2}(6 + 6 + 6) = 9$; $\mathbf{A} = \sqrt{9(9 - 6)^3} = \sqrt{243} \approx 15.6$; The area is about 15.6 cm^2.

Page 528 • MIXED REVIEW EXERCISES

1. $\pm\sqrt{250} = \pm\sqrt{25 \cdot 10} = \pm 5\sqrt{10}$ 2. $-6\sqrt{76} = -6\sqrt{4 \cdot 19} = -12\sqrt{19}$

3. $5\sqrt{396} = 5\sqrt{36 \cdot 11} = 30\sqrt{11}$ 4. $2^{-2} - 3^{-2} = \frac{1}{2^2} - \frac{1}{3^2} = \frac{1}{4} - \frac{1}{9} = \frac{9 - 4}{36} = \frac{5}{36}$

5. $6^2 \cdot 3^{-3} = \frac{6^2}{3^3} = \frac{36}{27} = \frac{4}{3}$ 6. $(4x^3)^2(-2x^5)^{-3} = \frac{16x^6}{(-2)^3x^{15}} = \frac{16}{-8x^9} = -\frac{2}{x^9}$

7. $(9)^2 + (4)^2 = 81 + 16 = 97$ 8. $(9)^2(4)^2 = 81 \cdot 16 = 1296$

9. $(9)^2 - (1)^2 = 81 - 1 = 80$

10. $\sqrt{\frac{1}{9}} = \frac{\sqrt{1}}{\sqrt{9}} = \frac{1}{3}$ 11. $\sqrt{\frac{9}{4}} = \frac{\sqrt{9}}{\sqrt{4}} = \frac{3}{2}$ 12. $(\sqrt{4})^4 = (2)^4 = 16$

Page 528 • CALCULATOR KEY-IN

1. $s = \frac{1}{2}(7 + 11 + 14) = 16$; $A = \sqrt{16(16 - 7)(16 - 11)(16 - 14)} = \sqrt{1440} \approx 37.95$; The area is about 37.95 cm^2.

2. $s = \frac{1}{2}(13 + 18 + 27) = 29$; $A = \sqrt{29(29 - 13)(29 - 18)(29 - 27)} = \sqrt{10{,}208} \approx 101.03$; The area is about 101.03 mm^2.

3. $s = \frac{1}{2}(12 + 12 + 16) = 20$; $A = \sqrt{20(20 - 12)^2(20 - 16)} = \sqrt{5120} \approx 71.55$; The area is about 71.55 m^2.

4. $s = \frac{1}{2}(48 + 64 + 84) = 98; A = \sqrt{98(98 - 48)(98 - 64)(98 - 84)} = \sqrt{2{,}332{,}400} \approx$ 1527.22; The area is about 1527.22 mm^2.

5. $s = \frac{1}{2}(2.8 + 3.9 + 5.7) = 6.2; A = \sqrt{6.2(6.2 - 2.8)(6.2 - 3.9)(6.2 - 5.7)} =$ $\sqrt{24.242} \approx 4.92$; The area is about 4.92 m^2.

6. $s = \frac{1}{2}(15.8 + 16.9 + 23.4) = 28.05$;

$A = \sqrt{28.05(28.05 - 15.8)(28.05 - 16.9)(28.05 - 23.4)} \approx 133.47$; The area is about 133.47 cm^2.

7. $s = \frac{1}{2}(9.2 + 11.8 + 17.1) = 19.05$;

$A = \sqrt{19.05(19.05 - 9.2)(19.05 - 11.8)(19.05 - 17.1)} \approx 51.51$; The area is about 51.51 cm^2.

8. $s = \frac{1}{2}(37.1 + 46.4 + 69.7) = 76.6$;

$A = \sqrt{76.6(76.6 - 37.1)(76.6 - 46.4)(76.6 - 69.7)} \approx 794.04$; The area is about 794.04 m^2.

Page 532 • WRITTEN EXERCISES

A

1. $c^2 = (10)^2 + (24)^2 = 100 + 576 = 676; c = \sqrt{676} = \sqrt{4 \cdot 169} = 2 \cdot 13 = 26$
2. $c^2 = (5)^2 + (12)^2 = 25 + 144 = 169; c = \sqrt{169} = 13$
3. $c^2 = (8)^2 + (5)^2 = 64 + 25 = 89; c = \sqrt{89} \approx 9.43$
4. $c^2 = (13)^2 + (9)^2 = 169 + 81 = 250; c = \sqrt{250} = \sqrt{25 \cdot 10} = 5\sqrt{10} \approx 15.81$
5. $c^2 = (8)^2 + (8)^2 = 64 + 64 = 128; c = \sqrt{128} = \sqrt{64 \cdot 2} = 8\sqrt{2} \approx 11.31$
6. $c^2 = (16)^2 + (8)^2 = 256 + 64 = 320; c = \sqrt{320} = \sqrt{64 \cdot 5} = 8\sqrt{5} \approx 17.89$
7. $a^2 + (21)^2 = (29)^2; a^2 = 841 - 441 = 400; a = \sqrt{400} = 20$
8. $a^2 + (11)^2 = (17)^2; a^2 = 289 - 121 = 168; a = \sqrt{168} = \sqrt{4 \cdot 42} = 2\sqrt{42} \approx 12.96$
9. $(6)^2 + b^2 = (40)^2; b^2 = 1600 - 36 = 1564; b = \sqrt{1564} = \sqrt{4 \cdot 391} =$ $2\sqrt{391} \approx 39.55$
10. $(5)^2 + b^2 = (8)^2; b^2 = 64 - 25 = 39; b = \sqrt{39} \approx 6.24$
11. $(20)^2 + (21)^2 = 400 + 441 = 841 = (29)^2$; yes
12. $(3)^2 + (9)^2 = 9 + 81 = 90 \neq (11)^2$; no
13. $(12)^2 + (16)^2 = 144 + 256 = 400 = (20)^2$; yes
14. $(16)^2 + (32)^2 = 256 + 1024 = 1280 \neq (36)^2$; no
15. $(15)^2 + (20)^2 = 225 + 400 = 625 = (25)^2$; yes
16. $(17)^2 + (34)^2 = 289 + 1156 = 1445 \neq (39)^2$; no

B

17. $(2a)^2 + (3a)^2 = 4a^2 + 9a^2 = 13a^2 \neq (4a)^2$; no
18. $(3a)^2 + (4a)^2 = 9a^2 + 16a^2 = 25a^2 = (5a)^2$; yes

19. $(8a)^2 + (15a)^2 = 64a^2 + 225a^2 = 289a^2 = (17a)^2$; yes

20. $(6a)^2 + (7a)^2 = 36a^2 + 49a^2 = 85a^2 \neq (8a)^2$; no

21. $c^2 = (12)^2 + (12)^2 = 2(144) = 288$; $c = \sqrt{288} = \sqrt{16 \cdot 9 \cdot 2} = 4 \cdot 3\sqrt{2} = 12\sqrt{2} \approx 16.97$

22. $c^2 = (15)^2 + \left(\frac{1}{5} \cdot 15\right)^2 = 225 + (3)^2 = 225 + 9 = 234$; $c = \sqrt{234} = \sqrt{9 \cdot 26} = 3\sqrt{26} \approx 15.30$

23. $c^2 = (18)^2 + \left(\frac{1}{3} \cdot 18\right)^2 = 324 + (6)^2 = 324 + 36 = 360$; $c = \sqrt{360} = \sqrt{36 \cdot 10} = 6\sqrt{10} \approx 18.97$

24. $c^2 = \left(\frac{1}{2} \cdot 14\right)^2 + (14)^2 = (7)^2 + 196 = 49 + 196 = 245$; $c = \sqrt{245} = \sqrt{49 \cdot 5} = 7\sqrt{5} \approx 15.65$

25. $c^2 = \left(\frac{4}{5} \cdot 20\right)^2 + (20)^2 = (16)^2 + 400 = 256 + 400 = 656$; $c = \sqrt{656} = \sqrt{16 \cdot 41} = 4\sqrt{41} \approx 25.61$

26. $c^2 = \left(\frac{5}{7} \cdot 28\right)^2 + (28)^2 = (20)^2 + (28)^2 = 400 + 784 = 1184$; $c = \sqrt{1184} = \sqrt{16 \cdot 74} = 4\sqrt{74} \approx 34.41$

C **27.** $a^2 + a^2 = (60)^2$, $2a^2 = 3600$, $a^2 = 1800$; $a = \sqrt{1800} = \sqrt{900 \cdot 2} = 30\sqrt{2} \approx 42.43$; $b \approx 42.43$

28. $(3b)^2 + b^2 = (20)^2$, $9b^2 + b^2 = 400$, $10b^2 = 400$, $b^2 = 40$; $b = \sqrt{40} = \sqrt{4 \cdot 10} = 2\sqrt{10} \approx 6.32$; $a \approx 3(6.32) = 18.96$

29. $\left(\frac{1}{3}b\right)^2 + b^2 = (30)^2$, $\frac{b^2}{9} + b^2 = 900$, $b^2 + 9b^2 = 8100$, $10b^2 = 8100$, $b^2 = 810$; $b = \sqrt{810} = \sqrt{81 \cdot 10} = 9\sqrt{10} \approx 28.46$; $a \approx \frac{1}{3}(28.46) \approx 9.49$

30. $\left(\frac{2}{3}b\right)^2 + b^2 = (52)^2$, $\frac{4}{9}b^2 + b^2 = 2704$, $4b^2 + 9b^2 = 24{,}336$, $13b^2 = 24{,}336$, $b^2 = 1872$; $b = \sqrt{1872} = \sqrt{144 \cdot 13} = 12\sqrt{13} \approx 43.27$; $a \approx \frac{2}{3}(43.27) \approx 28.85$

Page 532 • COMPUTER EXERCISES

```
10 PRINT "ENTER THREE POSITIVE NUMBERS"
20 PRINT "A, B, AND C, WITH C THE LARGEST"
30 INPUT A, B, C
40 IF A < = 0 THEN 10
50 IF B < = 0 THEN 10
60 IF A > C THEN 10
70 IF B > C THEN 10
80 PRINT
```

(Program continued on next page.)

```
90 IF ABS(C ↑ 2 - (A ↑ 2 + B ↑ 2)) < .0001 THEN 120
100 PRINT "NOT THE SIDES OF A RIGHT TRIANGLE"
110 GOTO 130
120 PRINT "SIDES OF A RIGHT TRIANGLE"
130 END
```

1. YES **2.** YES **3.** YES

Page 533 • PROBLEMS

A **1.** Let d = length of the diagonal. $d^2 = (33)^2 + (56)^2 = 4225$; $d = \sqrt{4225} = 65$. The diagonal of the rectangle is 65 cm long.

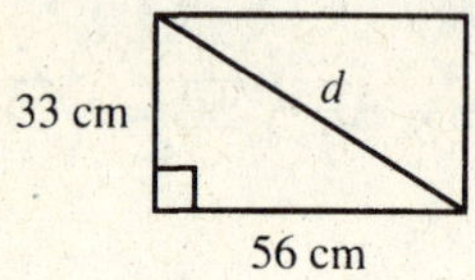

2. Let h = height of the telephone pole. $h^2 + (12)^2 = (20)^2$, $h^2 = 400 - 144 = 256$; $h = \sqrt{256} = 16$. The pole is 16 m high.

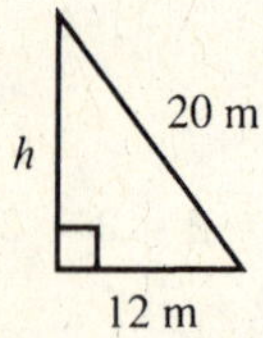

3. Let c = length from first base to third base (diagonal of the square).

$c^2 = (90)^2 + (90)^2 = 16{,}200$; $c = \sqrt{16{,}200} = \sqrt{8100 \cdot 2} = 90\sqrt{2} \approx 127.28$. The distance from first to third base is about 127.28 ft.

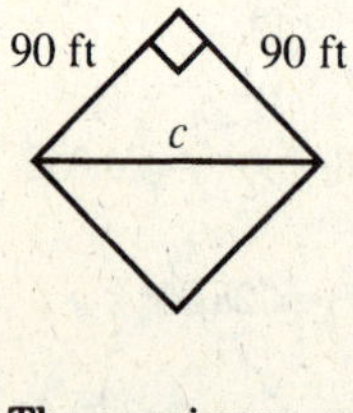

4. The maximum room in the doorway for the table top will occur at its diagonal, d.

$d^2 = (200)^2 + (90)^2 = 40{,}000 + 8100 = 48{,}100$; $d = \sqrt{48{,}100} = \sqrt{100 \cdot 481} = 10\sqrt{481} \approx 219.32$. Since the table top is only 210 cm in diameter, yes it will fit through the doorway.

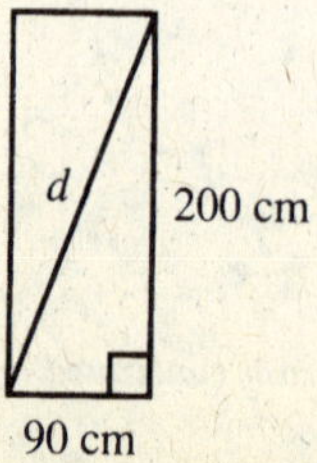

5. $a^2 = (24)^2 - \left(\frac{18}{2}\right)^2 = 576 - 9^2 = 576 - 81 = 495$; $a = \sqrt{495} = 3\sqrt{55} \approx 22.25$. The altitude is about 22.25 cm.

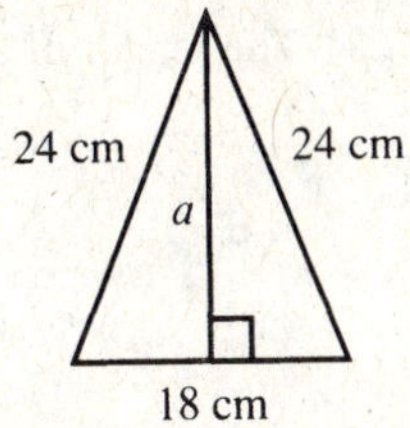

B **6.** Let n, $n + 2$, and $n + 4$ be the lengths of the sides. $n^2 + (n + 2)^2 = (n + 4)^2$, $n^2 + n^2 + 4n + 4 = n^2 + 8n + 16$, $n^2 - 4n - 12 = 0$, $(n - 6)(n + 2) = 0$; $n - 6 = 0$ or $n + 2 = 0$, $n = 6$ or $n = -2$. The sides of the triangle measure 6 ft, 8 ft, and 10 ft.

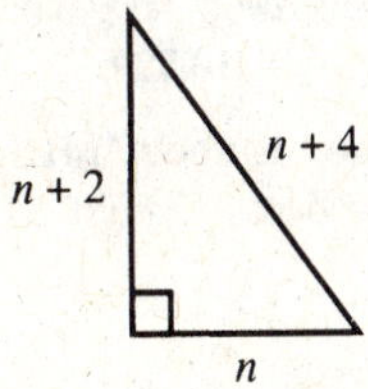

7. Let s = length of the shorter leg. Then the longer leg has length $6s + 6$, and the length of the hypotenuse is $6s + 7$. $(6s + 6)^2 + s^2 = (6s + 7)^2$, $36s^2 + 72s + 36 + s^2 = 36s^2 + 84s + 49$, $s^2 - 12s - 13 = 0$, $(s - 13)(s + 1) = 0$; $s = 13$ or $s = -1$; Perimeter $= 13 + 6(13) + 6 + 6(13) + 7 = 182$. The perimeter of the triangle is 182 cm.

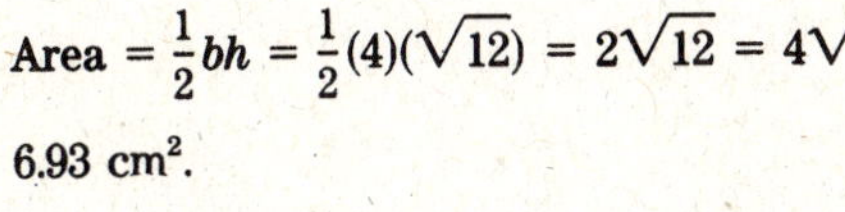

8. Let h = height of the triangle. $h^2 = (4)^2 - \left(\frac{4}{2}\right)^2 = 16 - 4 = 12$; $h = \sqrt{12}$;

Area $= \frac{1}{2}bh = \frac{1}{2}(4)(\sqrt{12}) = 2\sqrt{12} = 4\sqrt{3} \approx 6.93$. The area of the triangle is about 6.93 cm^2.

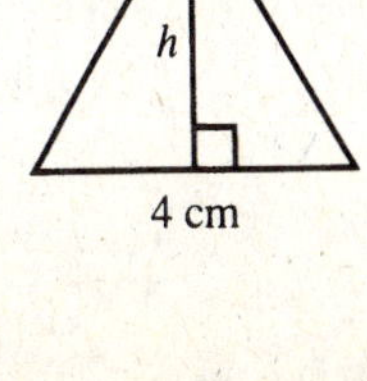

C **9.** First, find the diagonal, y, of a side: $y^2 = 2(45)^2 = 4050$; $y = \sqrt{4050} = \sqrt{25 \cdot 81 \cdot 2} = 5 \cdot 9\sqrt{2} = 45\sqrt{2}$. A diagonal, x, of the cube is the hypotenuse of the triangle formed by a side's diagonal and an edge: $x^2 = (45)^2 + (45\sqrt{2})^2 = 2025 + 4050 = 6075$; $x = \sqrt{6075} = \sqrt{25 \cdot 81 \cdot 3} = 45\sqrt{3} \approx 77.94$. The length of a diagonal of the cube is about 77.94 cm.

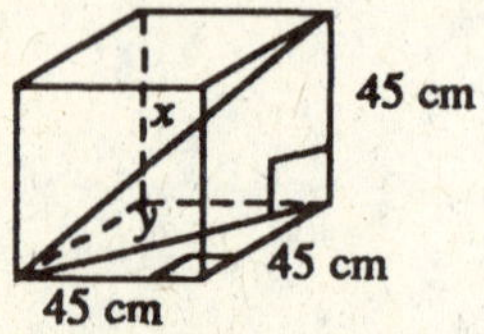

10. $x^2 + y^2$ is the largest quantity, so it must be the hypotenuse. $(2xy)^2 + (x^2 - y^2)^2 = 4x^2y^2 + x^4 - 2x^2y^2 + y^4 = x^4 + 2x^2y^2 + y^4 = (x^2 + y^2)^2$. This is a right triangle.

11. First, find the diagonal, y, of a side: $y^2 = (55)^2 + (48)^2 = 5329$; $y = \sqrt{5329} = 73$. Let x be the length of a diagonal of the box. $x^2 = (70)^2 + (73)^2 = 10{,}229$; $x = \sqrt{10{,}229} \approx 101.14$. The length of a diagonal of the box is about 101.14 cm. Since a meter stick is 100 cm long, yes it would fit in the box.

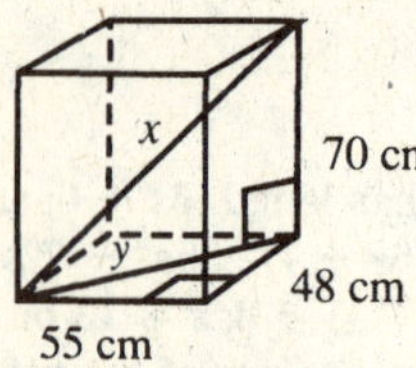

12. Let x = distance from dock before pulling rope. $x^2 + 2^2 = (5.2)^2$, $x^2 = 27.04 - 4 = 23.04$; $x = \sqrt{23.04} = \sqrt{\frac{2304}{100}} = \frac{48}{10} = 4.8$. Let y = distance from dock after pulling rope. $y^2 + 2^2 = (2.9)^2$, $y^2 = 8.41 - 4 = 4.41$; $y = \sqrt{4.41} = \sqrt{\frac{441}{100}} = \frac{21}{10} = 2.1$; Gary moved the boat 4.8 − 2.1 or 2.7 m.

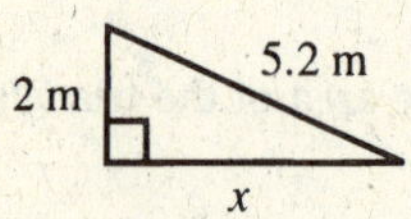

Page 534 • MIXED REVIEW EXERCISES

1. $\sqrt{16x^{16}y^2} = 4x^8|y|$ **2.** $\sqrt{c^2 - 10c + 25} = \sqrt{(c-5)^2} = |c - 5|$

3. $\sqrt{50a^5(b+4)^2} = \sqrt{25a^4(b+4)^2 \cdot 2a} = 5a^2|b+4|\sqrt{2a}$

4. $(3 \cdot 10^{-4})^2 = \left(\frac{3}{10^4}\right)^2 = \frac{3^2}{10^8} = \frac{9}{100{,}000{,}000}$

5. $(2x^{-3}y^{-5})^2 = \left(\frac{2}{x^3y^5}\right)^2 = \frac{4}{x^6y^{10}}$ **6.** $\frac{7}{8} \div \frac{8}{7} = \frac{7}{8} \cdot \frac{7}{8} = \frac{49}{64}$

7. $\frac{a-1}{6} + \frac{3-2a}{9} = \frac{3(a-1) + 2(3-2a)}{18} = \frac{3a - 3 + 6 - 4a}{18} = \frac{3-a}{18}$

8. $5y + \frac{3-y}{y-3} = 5y - \frac{y-3}{y-3} = 5y - 1$ **9.** $\frac{3r^2}{8x} \div 24rx = \frac{3r^2}{8x} \cdot \frac{1}{24rx} = \frac{r}{64x^2}$

10. $\left(-\frac{k^2}{6}\right)^3 = -\frac{k^6}{216}$

11. $\frac{2r^3 - 10r^2 - 28r}{2r - 14} = \frac{2r(r^2 - 5r - 14)}{2(r-7)} = \frac{r(r-7)(r+2)}{r-7} = r(r+2) = r^2 + 2r$

12. $\frac{(2s^2 + 9s - 5)}{5 + s} = \frac{(2s-1)(s+5)}{s+5} = 2s - 1$

Page 534 • CHALLENGE

Let r, h, and c be the number of roosters, hens, and chicks.

$5r + 3h + \frac{1}{3}c = 100 \rightarrow 15r + 9h + c = 300$

$r + h + c = 100 \rightarrow \underline{r + h + c = 100}$

$14r + 8h = 200$

$8h = 200 - 14r$, $h = 25 - \frac{7r}{4}$. r must be a multiple of 4 in order for h to be an integer, and we know that $r > 4$. Try $r = 8$: $h = 25 - \frac{7}{4}(8) = 25 - 14 = 11$; $c = 100 - 8 - 11 = 81$. Try $r = 12$: $h = 25 - \frac{7}{4}(12) = 25 - 21 = 4$; $c = 100 - 12 - 4 = 84$. Try $r = 16$: $h = 25 - \frac{7}{4}(16) = 25 - 28 = -3$; all $r \geq 16$ will result in a neg. value of h. Therefore, there are either 8 roosters, 11 hens, and 81 chicks, or 12 roosters, 4 hens, and 84 chicks.

Page 536 • EXTRA

1. $P_1P_2 = \sqrt{(4 - (-6))^2 + 0^2} = \sqrt{10^2} = 10$

2. $P_1P_2 = \sqrt{0^2 + (7 - (-9))^2} = \sqrt{16^2} = 16$

3. $P_1P_2 = \sqrt{(9-3)^2 + (7-4)^2} = \sqrt{36 + 9} = \sqrt{45} = \sqrt{9 \cdot 5} = 3\sqrt{5} \approx 6.7$

4. $P_1P_2 = \sqrt{(-4 - 10)^2 + (9 - 3)^2} = \sqrt{196 + 36} = \sqrt{232} = \sqrt{4 \cdot 58} = 2\sqrt{58} \approx 15.2$

5. $P_1P_2 = \sqrt{(-9 - (-5))^2 + (-6 - (-3))^2} = \sqrt{16 + 9} = \sqrt{25} = 5$

6. $P_1P_2 = \sqrt{(4 - 7)^2 + (-6 - (-2))^2} = \sqrt{9 + 16} = \sqrt{25} = 5$

7. $P_1P_2 = \sqrt{(5 - (-4))^2 + (2 - 6)^2} = \sqrt{81 + 16} = \sqrt{97} \approx 9.8$

8. $P_1P_2 = \sqrt{(-9 - (-4))^2 + (-5 - 7)^2} = \sqrt{25 + 144} = \sqrt{169} = 13$

9. $P_1P_2 = \sqrt{(9 - 4)^2 + (-8 - (-4))^2} = \sqrt{5^2 + 16} = \sqrt{41} \approx 6.4$

10. $P_1P_2 = \sqrt{(12 - 3)^2 + (-8 - (-7))^2} = \sqrt{81 + 1} = \sqrt{82} \approx 9.1$

11. $P_1P_2 = \sqrt{(-8 - (-2))^2 + (-5 - 1)^2} = \sqrt{36 + 36} = \sqrt{2 \cdot 36} = 6\sqrt{2} \approx 8.5$

12. $P_1P_2 = \sqrt{(-9 - 10)^2 + (3 - (-11))^2} = \sqrt{361 + 196} = \sqrt{557} \approx 23.6$

13. $MA = \sqrt{(3 - (-2))^2 + (9 - (-3))^2} = \sqrt{25 + 144} = \sqrt{169} = 13$;
$MB = \sqrt{(-7 - (-2))^2 + (-15 - (-3))^2} = \sqrt{25 + 144} = \sqrt{169} = 13$; Therefore, M is equidistant from A and B.

14. $RS = \sqrt{(3 - (-4))^2 + (6 - (-1))^2} = \sqrt{49 + 49} = \sqrt{2 \cdot 49} = 7\sqrt{2}$;
$RT = \sqrt{(2 - (-4))^2 + (7 - (-1))^2} = \sqrt{36 + 64} = \sqrt{100} = 10$;
$ST = \sqrt{(2 - 3)^2 + (7 - 6)^2} = \sqrt{1 + 1} = \sqrt{2}$; $(RS)^2 + (ST)^2 = (7\sqrt{2})^2 + (\sqrt{2})^2 =$
$98 + 2 = 100 = (RT)^2$; Therefore, it is a right triangle.

Page 536 • CHALLENGE

1. $x = 0, 2 \le y \le 7$: Endpoints are (0, 2), (0, 7); $x = 18, 4 \le y \le 9$: (18, 4), (18, 9);
$x + 3y = 21, 0 \le x \le 6$: (0, 7), (6, 5); $x - 3y = 6, 6 \le x \le 18$: (6, 0), (18, 4);
$x - 3y = -27, 3 \le x \le 15$: (3, 10), (15, 14); $5x + 3y = 117, 15 \le x \le 18$: (15, 14), (18, 9);
$x = 6, 0 \le y \le 5$: (6, 0), (6, 5); $x + 3y = 6, 0 \le x \le 6$: (0, 2), (6, 0); $x - y = -7$,
$0 \le x \le 3$: (0, 7), (3, 10); $x - 3y = -9, 6 \le x \le 18$: (6, 5), (18, 9); $5x + 3y = 45$,
$3 \le x \le 6$: (3, 10), (6, 5)

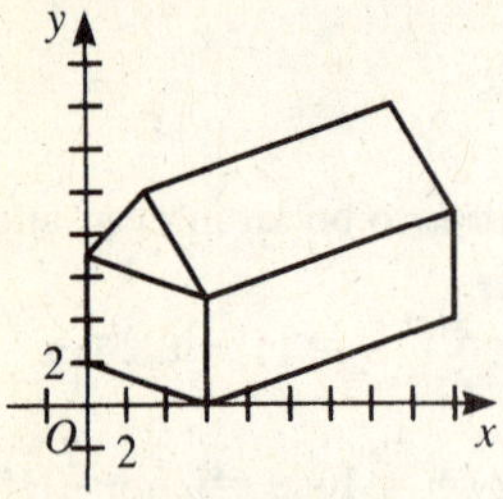

2. Answers will vary.

Pages 538–539 • WRITTEN EXERCISES

A

1. $5\sqrt{3} \cdot 2\sqrt{3} = 5 \cdot 2(\sqrt{3})^2 = 10 \cdot 3 = 30$ **2.** $4\sqrt{7} \cdot 2\sqrt{7} = 4 \cdot 2(\sqrt{7})^2 = 8 \cdot 7 = 56$

3. $\sqrt{3} \cdot \sqrt{3} \cdot \sqrt{4} = (\sqrt{3})^2 \cdot 2 = 3 \cdot 2 = 6$

4. $\sqrt{5} \cdot \sqrt{5} \cdot \sqrt{9} = (\sqrt{5})^2 \cdot 3 = 5 \cdot 3 = 15$ **5.** $2\sqrt{5} \cdot \sqrt{7} = 2\sqrt{35}$

6. $6\sqrt{2} \cdot \sqrt{5} = 6\sqrt{10}$ **7.** $\sqrt{3} \cdot \sqrt{27} = \sqrt{81} = 9$ **8.** $\sqrt{5} \cdot \sqrt{20} = \sqrt{100} = 10$

9. $\sqrt{11} \cdot \sqrt{44} = \sqrt{484} = 22$ **10.** $\sqrt{7} \cdot \sqrt{35} = \sqrt{245} = \sqrt{49 \cdot 5} = 7\sqrt{5}$

11. $6\sqrt{72} = 6\sqrt{36 \cdot 2} = 6 \cdot 6\sqrt{2} = 36\sqrt{2}$

12. $9\sqrt{242} = 9\sqrt{121 \cdot 2} = 9 \cdot 11\sqrt{2} = 99\sqrt{2}$ **13.** $\sqrt{\frac{3}{8}} \cdot \sqrt{\frac{8}{3}} = \sqrt{\frac{3}{8} \cdot \frac{8}{3}} = \sqrt{1} = 1$

14. $\sqrt{\frac{4}{9}} \cdot \sqrt{\frac{18}{4}} = \sqrt{\frac{4 \cdot 18}{9 \cdot 4}} = \sqrt{2}$ **15.** $\sqrt{\frac{8}{11} \cdot \frac{22}{32}} = \sqrt{\frac{1}{2}} = \frac{1}{\sqrt{2}} \cdot \frac{\sqrt{2}}{\sqrt{2}} = \frac{\sqrt{2}}{2}$

16. $\sqrt{\frac{7}{3}} \cdot \sqrt{\frac{3}{28}} = \sqrt{\frac{7 \cdot 3}{3 \cdot 28}} = \sqrt{\frac{1}{4}} = \frac{1}{\sqrt{4}} = \frac{1}{2}$

17. $\sqrt{\frac{3}{4}} \cdot \sqrt{\frac{8}{9}} = \sqrt{\frac{3 \cdot 8}{4 \cdot 9}} = \frac{\sqrt{24}}{\sqrt{36}} = \frac{\sqrt{4}\sqrt{6}}{6} = \frac{2\sqrt{6}}{6} = \frac{\sqrt{6}}{3}$

18. $\sqrt{\frac{4}{5}} \cdot \sqrt{\frac{10}{36}} = \sqrt{\frac{40}{180}} = \sqrt{\frac{2}{9}} = \frac{\sqrt{2}}{\sqrt{9}} = \frac{\sqrt{2}}{3}$ **19.** $\sqrt{3\frac{3}{5}} \cdot \sqrt{2\frac{1}{2}} = \sqrt{\frac{18}{5} \cdot \frac{5}{2}} = \sqrt{9} = 3$

20. $\sqrt{2\frac{2}{5}} \cdot \sqrt{1\frac{2}{3}} = \sqrt{\frac{12}{5} \cdot \frac{5}{3}} = \sqrt{4} = 2$

21. $\sqrt{\frac{10}{13}} \cdot \sqrt{\frac{1}{2}} = \sqrt{\frac{10}{26}} = \sqrt{\frac{5}{13}} = \frac{\sqrt{5}}{\sqrt{13}} \cdot \frac{\sqrt{13}}{\sqrt{13}} = \frac{\sqrt{65}}{13}$

22. $\sqrt{\frac{15}{11}} \cdot \sqrt{\frac{1}{3}} = \sqrt{\frac{15}{33}} = \sqrt{\frac{5}{11}} = \frac{\sqrt{5}}{\sqrt{11}} \cdot \frac{\sqrt{11}}{\sqrt{11}} = \frac{\sqrt{55}}{11}$

23. $\frac{6\sqrt{7}}{\sqrt{35}} = \frac{6\sqrt{7}}{\sqrt{7}\sqrt{5}} = \frac{6}{\sqrt{5}} \cdot \frac{\sqrt{5}}{\sqrt{5}} = \frac{6\sqrt{5}}{5}$

24. $\frac{5\sqrt{48}}{\sqrt{39}} = \frac{5\sqrt{16 \cdot 3}}{\sqrt{13 \cdot 3}} = \frac{5 \cdot 4\sqrt{3}}{\sqrt{13}\sqrt{3}} = \frac{20}{\sqrt{13}} \cdot \frac{\sqrt{13}}{\sqrt{13}} = \frac{20\sqrt{13}}{13}$

25. $3\sqrt{\frac{48}{9}} = 3 \cdot \frac{\sqrt{16 \cdot 3}}{\sqrt{9}} = 3 \cdot \frac{4\sqrt{3}}{3} = 4\sqrt{3}$ **26.** $7\sqrt{\frac{40}{49}} = 7 \cdot \frac{\sqrt{40}}{\sqrt{49}} = 7 \cdot \frac{\sqrt{4 \cdot 10}}{7} = 2\sqrt{10}$

27. $\frac{14\sqrt{320}}{2\sqrt{5}} = \frac{7\sqrt{64 \cdot 5}}{\sqrt{5}} = \frac{7 \cdot 8\sqrt{5}}{\sqrt{5}} = \frac{56}{1} \cdot \frac{\sqrt{5}}{\sqrt{5}} = 56 \cdot 1 = 56$

28. $\frac{15\sqrt{6}}{\sqrt{90}} = \frac{15\sqrt{6}}{\sqrt{9 \cdot 10}} = \frac{15\sqrt{6}}{3\sqrt{10}} = \frac{5\sqrt{6}}{\sqrt{10}} \cdot \frac{\sqrt{10}}{\sqrt{10}} = \frac{5\sqrt{60}}{10} = \frac{5\sqrt{4 \cdot 15}}{10} = \frac{10\sqrt{15}}{10} = \sqrt{15}$

29. $\sqrt{5}(\sqrt{5} - 2) = (\sqrt{5})^2 - 2\sqrt{5} = 5 - 2\sqrt{5}$ **30.** $\sqrt{7}(6 - \sqrt{2}) = 6\sqrt{7} - \sqrt{14}$

31. $(3\sqrt{2})(-2\sqrt{8})(3\sqrt{27}) = 3 \cdot (-2) \cdot 3\sqrt{2 \cdot 8 \cdot 27} = -18\sqrt{16 \cdot 9 \cdot 3} =$
$-18 \cdot 4 \cdot 3\sqrt{3} = -216\sqrt{3}$

32. $(3\sqrt{5})(-\sqrt{10})(\sqrt{27}) = -3\sqrt{5 \cdot 10 \cdot 27} = -3\sqrt{1350} = -3\sqrt{225 \cdot 6} =$
$-3 \cdot 15\sqrt{6} = -45\sqrt{6}$

B **33.** $(4\sqrt{a^2b})(3\sqrt{b}) = 3 \cdot 4\sqrt{a^2b \cdot b} = 12\sqrt{a^2b^2} = 12ab$

34. $(5\sqrt{mn^2})(-2\sqrt{m}) = -10\sqrt{m^2n^2} = -10mn$

35. $(-x\sqrt{x^2y})(y\sqrt{xy^2}) = -xy\sqrt{x^3y^3} = -xy\sqrt{x^2y^2 \cdot xy} = -xy \cdot xy\sqrt{xy} =$
$-x^2y^2\sqrt{xy}$

36. $(-r\sqrt{r^2s})(-s\sqrt{r^2s}) = rs\sqrt{r^4s^2} = rs \cdot r^2s = r^3s^2$

37. $\sqrt{m}(\sqrt{m^3} + 6) = \sqrt{m^4} + 6\sqrt{m} = m^2 + 6\sqrt{m}$

38. $\sqrt{x}(\sqrt{x^5} + 7) = \sqrt{x^6} + 7\sqrt{x} = x^3 + 7\sqrt{x}$

39. $(\sqrt{5x})(\sqrt{2x})(3\sqrt{10x^2}) = 3\sqrt{100x^4} = 3 \cdot 10x^2 = 30x^2$

40. $(\sqrt{3a})(\sqrt{2a})(2\sqrt{6a^2}) = 2\sqrt{36a^4} = 2 \cdot 6a^2 = 12a^2$

41. $\sqrt{32} \cdot \sqrt{2x} \cdot \sqrt{3x} = \sqrt{192x^2} = \sqrt{64x^2 \cdot 3} = 8x\sqrt{3}$

42. $\sqrt{27} \cdot \sqrt{3n} \cdot \sqrt{5n} = \sqrt{9 \cdot 3} \cdot \sqrt{15n^2} = 3\sqrt{45n^2} = 3\sqrt{9n^2 \cdot 5} = 3 \cdot 3n \cdot \sqrt{5} =$
$9n\sqrt{5}$

43. $(2\sqrt{5x})^2 = 4 \cdot 5x = 20x$ **44.** $2n(\sqrt{7n})^2 = 2n \cdot 7n = 14n^2$

45. $3x\sqrt{\frac{x}{y}} \cdot \sqrt{\frac{9x}{y}} = 3x\sqrt{\frac{9x^2}{y^2}} = 3x \cdot \frac{3x}{y} = \frac{9x^2}{y}$

46. $3q\sqrt{\frac{3q}{2r}} \cdot \sqrt{\frac{q}{r}} = 3q\sqrt{\frac{3q^2}{2r^2}} = 3q \cdot \frac{q}{r} \cdot \sqrt{\frac{3}{2}} = \frac{3q^2\sqrt{3}}{r\sqrt{2}} \cdot \frac{\sqrt{2}}{\sqrt{2}} = \frac{3q^2\sqrt{6}}{2r}$

C 47. $\sqrt{3a}(\sqrt{12a} - 2\sqrt{8a^2}) = \sqrt{3a \cdot 12a} - 2\sqrt{3a \cdot 8a^2} = \sqrt{36a^2} - 2\sqrt{24a^3} =$
$6a - 2\sqrt{4a^2 \cdot 6a} = 6a - 4a\sqrt{6a}$

48. $\sqrt{6x}(\sqrt{3x} - 4\sqrt{8x^2}) = \sqrt{18x^2} - 4\sqrt{48x^3} = \sqrt{9x^2 \cdot 2} - 4\sqrt{16x^2 \cdot 3x} =$
$3x\sqrt{2} - 4 \cdot 4x\sqrt{3x} = 3x\sqrt{2} - 16x\sqrt{3x}$

49. $3\sqrt{8m^3}(2\sqrt{2m} - 5\sqrt{8m^4}) = 6\sqrt{16m^4} - 15\sqrt{64m^7} = 6 \cdot 4m^2 - 15 \cdot 8m^3\sqrt{m} =$
$24m^2 - 120m^3\sqrt{m}$

50. $2\sqrt{6x^3}(3\sqrt{8x} - 5\sqrt{3x^4}) = 6\sqrt{48x^4} - 10\sqrt{18x^7} = 6\sqrt{16x^4 \cdot 3} - 10\sqrt{9x^6 \cdot 2x} =$
$6 \cdot 4x^2\sqrt{3} - 10 \cdot 3x^3\sqrt{2x} = 24x^2\sqrt{3} - 30x^3\sqrt{2x}$

51. $(2\sqrt{3y^3})^3 = 2^3\sqrt{3^3y^9} = 8\sqrt{9y^8 \cdot 3y} = 8 \cdot 3y^4\sqrt{3y} = 24y^4\sqrt{3y}$

52. $(5\sqrt{2x^3})^3 = 5^3\sqrt{2^3x^9} = 125\sqrt{4x^8 \cdot 2x} = 125 \cdot 2x^4\sqrt{2x} = 250x^4\sqrt{2x}$

53. $(\sqrt{10xy})^3(x\sqrt{5x^3y} - y\sqrt{10xy^3}) = \sqrt{10^3x^3y^3}(x\sqrt{5xy \cdot x^2} - y\sqrt{10xy \cdot y^2}) =$
$\sqrt{10^2x^2y^2 \cdot 10xy}(x^2\sqrt{5xy} - y^2\sqrt{10xy}) = 10xy\sqrt{10xy}(x^2\sqrt{5xy} - y^2\sqrt{10xy}) =$
$10x^3y\sqrt{50x^2y^2} - 10xy^3\sqrt{100x^2y^2} = 10x^3y\sqrt{25x^2y^2 \cdot 2} - 10xy^3 \cdot 10xy =$
$10x^3y \cdot 5xy\sqrt{2} - 100x^2y^4 = 50x^4y^2\sqrt{2} - 100x^2y^4$

54. $(\sqrt{18ab})^3(a\sqrt{3a^2b} + b\sqrt{5ab^2}) = 18ab\sqrt{18ab}(a \cdot a\sqrt{3b} + b \cdot b\sqrt{5a}) =$
$18a^3b\sqrt{54ab^2} + 18ab^3\sqrt{90a^2b} = 18a^3b\sqrt{9b^2 \cdot 6a} + 18ab^3\sqrt{9a^2 \cdot 10b} =$
$18a^3b \cdot 3b\sqrt{6a} + 18ab^3 \cdot 3a\sqrt{10b} = 54a^3b^2\sqrt{6a} + 54a^2b^3\sqrt{10b}$

55. $\frac{\sqrt{11}}{\sqrt{3}} \cdot \frac{\sqrt{11}}{\sqrt{11}} = \frac{11}{\sqrt{33}}$ 56. $\frac{\sqrt{2}}{\sqrt{5}} \cdot \frac{\sqrt{2}}{\sqrt{2}} = \frac{2}{\sqrt{10}}$ 57. $\frac{\sqrt{6}}{\sqrt{13}} \cdot \frac{\sqrt{6}}{\sqrt{6}} = \frac{6}{\sqrt{78}}$

58. $\frac{\sqrt{7}}{\sqrt{15}} \cdot \frac{\sqrt{7}}{\sqrt{7}} = \frac{7}{\sqrt{105}}$ 59. $\frac{\sqrt{19}}{\sqrt{8}} \cdot \frac{\sqrt{19}}{\sqrt{19}} = \frac{19}{\sqrt{152}} = \frac{19}{2\sqrt{38}}$

Page 539 • MIXED REVIEW EXERCISES

1. $x^2 = 169; x = \pm 13; \{-13, 13\}$
2. $2s^2 - 200 = 0, 2s^2 = 200, s^2 = 100; s = \pm 10; \{-10, 10\}$
3. $25z^2 - 1 = 15, 25z^2 = 16, z^2 = \frac{16}{25}; z = \pm\frac{4}{5}; \left\{-\frac{4}{5}, \frac{4}{5}\right\}$
4. $\frac{1}{c} + \frac{1}{3} = \frac{1}{2}, 6c\left(\frac{1}{c} + \frac{1}{3}\right) = 6c \cdot \frac{1}{2}, 6 + 2c = 3c, 6 = c; \{6\}$
5. $\frac{5}{9} = \frac{25}{y}, 5y = 225, y = 45; \{45\}$
6. $\frac{8b - 5}{5b - 4} = \frac{13}{8}, 8(8b - 5) = 13(5b - 4), 64b - 40 = 65b - 52, 12 = b; \{12\}$
7. $19x + 2(3x - 4) + 5 = 19x + 6x - 8 + 5 = 25x - 3$
8. $12a + 7 - (8a - 17) = 12a + 7 - 8a + 17 = 4a + 24$
9. $4(2b - 6) - 5(b - 4) = 8b - 24 - 5b + 20 = 3b - 4$
10. $(-5c^2d)(-4cd^4) = 20c^3d^5$

11. $-4m + 3 + 11m - 4 = 7m - 1$

12. $x(x + 2) + (x - 4)(2x + 1) = x^2 + 2x + 2x^2 + x - 8x - 4 = 3x^2 - 5x - 4$

Page 541 • WRITTEN EXERCISES

A **1.** $8\sqrt{3} - 6\sqrt{3} = (8 - 6)\sqrt{3} = 2\sqrt{3}$ **2.** $9\sqrt{5} + 4\sqrt{5} = (9 + 4)\sqrt{5} = 13\sqrt{5}$

3. $-13\sqrt{17} - 7\sqrt{17} = (-13 - 7)\sqrt{17} = -20\sqrt{17}$

4. $5\sqrt{80} - 12\sqrt{5} = 5\sqrt{16 \cdot 5} - 12\sqrt{5} = 5 \cdot 4\sqrt{5} - 12\sqrt{5} = (20 - 12)\sqrt{5} = 8\sqrt{5}$

5. $5\sqrt{3} + 2\sqrt{75} = 5\sqrt{3} + 2\sqrt{25 \cdot 3} = 5\sqrt{3} + 2 \cdot 5\sqrt{3} = (5 + 10)\sqrt{3} = 15\sqrt{3}$

6. $-2\sqrt{24} - 3\sqrt{6} = -2\sqrt{4 \cdot 6} - 3\sqrt{6} = -2 \cdot 2\sqrt{6} - 3\sqrt{6} = (-4 - 3)\sqrt{6} = -7\sqrt{6}$

7. $3\sqrt{32} - 4\sqrt{63} = 3\sqrt{16 \cdot 2} - 4\sqrt{9 \cdot 7} = 3 \cdot 4\sqrt{2} - 4 \cdot 3\sqrt{7} = 12\sqrt{2} - 12\sqrt{7}$

8. $3\sqrt{45} + 7\sqrt{36} = 3\sqrt{9 \cdot 5} + 7 \cdot 6 = 3 \cdot 3\sqrt{5} + 42 = 9\sqrt{5} + 42$

9. $5\sqrt{28} - 2\sqrt{45} = 5\sqrt{4 \cdot 7} - 2\sqrt{9 \cdot 5} = 5 \cdot 2\sqrt{7} - 2 \cdot 3\sqrt{5} = 10\sqrt{7} - 6\sqrt{5}$

10. $-4\sqrt{75} + 3\sqrt{147} = -4\sqrt{25 \cdot 3} + 3\sqrt{49 \cdot 3} = -4 \cdot 5\sqrt{3} + 3 \cdot 7\sqrt{3} =$
$-20\sqrt{3} + 21\sqrt{3} = \sqrt{3}$

11. $-11\sqrt{8} - 7\sqrt{12} = -11\sqrt{4 \cdot 2} - 7\sqrt{4 \cdot 3} = -11 \cdot 2\sqrt{2} - 7 \cdot 2\sqrt{3} =$
$-22\sqrt{2} - 14\sqrt{3}$

12. $\sqrt{150} - 5\sqrt{96} = \sqrt{25 \cdot 6} - 5\sqrt{16 \cdot 6} = 5\sqrt{6} - 5 \cdot 4\sqrt{6} = (5 - 20)\sqrt{6} = -15\sqrt{6}$

13. $9\sqrt{13} - 6\sqrt{11} + \sqrt{13} = (9 + 1)\sqrt{13} - 6\sqrt{11} = 10\sqrt{13} - 6\sqrt{11}$

14. $-4\sqrt{2} + 6\sqrt{72} - 8\sqrt{32} = -4\sqrt{2} + 6\sqrt{36 \cdot 2} - 8\sqrt{16 \cdot 2} = -4\sqrt{2} + 6 \cdot 6\sqrt{2} -$
$8 \cdot 4\sqrt{2} = (-4 + 36 - 32)\sqrt{2} = 0 \cdot \sqrt{2} = 0$

15. $5\sqrt{28} + 2\sqrt{7} - \sqrt{14} = 5\sqrt{4 \cdot 7} + 2\sqrt{7} - \sqrt{14} = 5 \cdot 2\sqrt{7} + 2\sqrt{7} - \sqrt{14} =$
$(10 + 2)\sqrt{7} - \sqrt{14} = 12\sqrt{7} - \sqrt{14}$

16. $-3\sqrt{72} + 6\sqrt{52} - 7\sqrt{128} = -3\sqrt{36 \cdot 2} + 6\sqrt{4 \cdot 13} - 7\sqrt{64 \cdot 2} = -3 \cdot 6\sqrt{2} +$
$6 \cdot 2\sqrt{13} - 7 \cdot 8\sqrt{2} = (-18 - 56)\sqrt{2} + 12\sqrt{13} = -74\sqrt{2} + 12\sqrt{13}$

17. $-\sqrt{338} - \sqrt{200} + \sqrt{162} = -\sqrt{169 \cdot 2} - \sqrt{100 \cdot 2} + \sqrt{81 \cdot 2} =$
$-13\sqrt{2} - 10\sqrt{2} + 9\sqrt{2} = (-13 - 10 + 9)\sqrt{2} = -14\sqrt{2}$

18. $4\sqrt{112} + 5\sqrt{56} - 9\sqrt{126} = 4\sqrt{16 \cdot 7} + 5\sqrt{4 \cdot 14} - 9\sqrt{9 \cdot 14} = 4 \cdot 4\sqrt{7} +$
$5 \cdot 2\sqrt{14} - 9 \cdot 3\sqrt{14} = 16\sqrt{7} + (10 - 27)\sqrt{14} = 16\sqrt{7} - 17\sqrt{14}$

B **19.** $\sqrt{55} - 7\sqrt{\frac{5}{11}} = \sqrt{55} - 7\frac{\sqrt{5}}{\sqrt{11}} \cdot \frac{\sqrt{11}}{\sqrt{11}} = \sqrt{55} - \frac{7\sqrt{55}}{11} = \frac{11\sqrt{55} - 7\sqrt{55}}{11} = \frac{4\sqrt{55}}{11}$

20. $\sqrt{3} - \sqrt{\frac{1}{3}} = \sqrt{3} - \frac{1}{\sqrt{3}} \cdot \frac{\sqrt{3}}{\sqrt{3}} = \sqrt{3} - \frac{\sqrt{3}}{3} = \frac{3\sqrt{3} - \sqrt{3}}{3} = \frac{2\sqrt{3}}{3}$

21. $3\sqrt{18} + \sqrt{\frac{2}{25}} = 3\sqrt{9 \cdot 2} + \frac{\sqrt{2}}{\sqrt{25}} = 3 \cdot 3\sqrt{2} + \frac{\sqrt{2}}{5} = \frac{9 \cdot 5\sqrt{2} + \sqrt{2}}{5} =$
$\frac{45\sqrt{2} + \sqrt{2}}{5} = \frac{46\sqrt{2}}{5}$

22. $2\sqrt{75} + \sqrt{\frac{3}{16}} = 2\sqrt{25 \cdot 3} + \frac{\sqrt{3}}{\sqrt{16}} = 2 \cdot 5\sqrt{3} + \frac{\sqrt{3}}{4} = \frac{10 \cdot 4\sqrt{3} + \sqrt{3}}{4} =$

$\frac{(40 + 1)\sqrt{3}}{4} = \frac{41\sqrt{3}}{4}$

23. $\sqrt{\frac{5}{11}} - \sqrt{\frac{11}{5}} = \frac{\sqrt{5}\sqrt{11}}{\sqrt{11}\sqrt{11}} - \frac{\sqrt{11}\sqrt{5}}{\sqrt{5}\sqrt{5}} = \frac{\sqrt{55}}{11} - \frac{\sqrt{55}}{5} = \frac{5\sqrt{55} - 11\sqrt{55}}{55} = -\frac{6\sqrt{55}}{55}$

24. $\sqrt{\frac{2}{7}} - \sqrt{\frac{7}{2}} = \frac{\sqrt{2}\sqrt{7}}{\sqrt{7}\sqrt{7}} - \frac{\sqrt{7}\sqrt{2}}{\sqrt{2}\sqrt{2}} = \frac{\sqrt{14}}{7} - \frac{\sqrt{14}}{2} = \frac{2\sqrt{14} - 7\sqrt{14}}{14} = -\frac{5\sqrt{14}}{14}$

25. $4\sqrt{\frac{5}{6}} - \sqrt{\frac{3}{10}} = \frac{4\sqrt{5}\sqrt{6}}{\sqrt{6}\sqrt{6}} - \frac{\sqrt{3}\sqrt{10}}{\sqrt{10}\sqrt{10}} = \frac{4\sqrt{30}}{6} - \frac{\sqrt{30}}{10} = \frac{20\sqrt{30} - 3\sqrt{30}}{30} = \frac{17\sqrt{30}}{30}$

26. $5\sqrt{\frac{16}{3}} - \sqrt{\frac{9}{2}} = \frac{5 \cdot 4\sqrt{3}}{\sqrt{3}\sqrt{3}} - \frac{3\sqrt{2}}{\sqrt{2}\sqrt{2}} = \frac{20\sqrt{3}}{3} - \frac{3\sqrt{2}}{2} = \frac{40\sqrt{3} - 9\sqrt{2}}{6}$

27. $3\sqrt{3} - 2\sqrt{12} + 4\sqrt{\frac{1}{3}} = 3\sqrt{3} - 2\sqrt{4 \cdot 3} + \frac{4\sqrt{3}}{\sqrt{3}\sqrt{3}} = 3\sqrt{3} - 2 \cdot 2\sqrt{3} + \frac{4\sqrt{3}}{3} =$

$3\sqrt{3} - 4\sqrt{3} + \frac{4}{3}\sqrt{3} = \frac{3 \cdot 3\sqrt{3} - 4 \cdot 3\sqrt{3} + 4\sqrt{3}}{3} = \frac{(9 - 12 + 4)\sqrt{3}}{3} = \frac{\sqrt{3}}{3}$

28. $8\sqrt{10} - 3\sqrt{40} + 5\sqrt{\frac{1}{10}} = 8\sqrt{10} - 3\sqrt{4 \cdot 10} + \frac{5\sqrt{10}}{\sqrt{10}\sqrt{10}} = 8\sqrt{10} - 3 \cdot 2\sqrt{10} +$

$\frac{5\sqrt{10}}{10} = 8\sqrt{10} - 6\sqrt{10} + \frac{\sqrt{10}}{2} = \frac{16\sqrt{10} - 12\sqrt{10} + \sqrt{10}}{2} = \frac{5\sqrt{10}}{2}$

29. $2\sqrt{\frac{7}{2}} + 4\sqrt{\frac{7}{8}} - \frac{1}{2}\sqrt{98} = \frac{2\sqrt{7}\sqrt{2}}{\sqrt{2}\sqrt{2}} + \frac{4\sqrt{7}}{\sqrt{4}\sqrt{2}} - \frac{1}{2}\sqrt{49 \cdot 2} = \frac{2\sqrt{14}}{2} + \frac{4\sqrt{7}\sqrt{2}}{2\sqrt{2}\sqrt{2}} -$

$\frac{7}{2}\sqrt{2} = \frac{2\sqrt{14}}{2} + \frac{4\sqrt{14}}{4} - \frac{7\sqrt{2}}{2} = \frac{2\sqrt{14} + 2\sqrt{14} - 7\sqrt{2}}{2} = \frac{4\sqrt{14} - 7\sqrt{2}}{2}$

30. $3\sqrt{\frac{5}{12}} + \sqrt{\frac{12}{5}} - \frac{1}{3}\sqrt{60} = \frac{3\sqrt{5}}{\sqrt{4 \cdot 3}} + \frac{\sqrt{4 \cdot 3}}{\sqrt{5}} - \frac{1}{3}\sqrt{4 \cdot 15} = \frac{3\sqrt{5}\sqrt{3}}{2\sqrt{3}\sqrt{3}} + \frac{2\sqrt{3}\sqrt{5}}{\sqrt{5}\sqrt{5}} -$

$\frac{1}{3} \cdot 2\sqrt{15} = \frac{3\sqrt{15}}{6} + \frac{2\sqrt{15}}{5} - \frac{2\sqrt{15}}{3} = \frac{\sqrt{15}}{2} + \frac{2\sqrt{15}}{5} - \frac{2\sqrt{15}}{3} =$

$\frac{15\sqrt{15} + 12\sqrt{15} - 20\sqrt{15}}{30} = \frac{7\sqrt{15}}{30}$

31. $5\sqrt{3}(\sqrt{6} + 2\sqrt{8}) = 5\sqrt{18} + 10\sqrt{24} = 5\sqrt{9 \cdot 2} + 10\sqrt{4 \cdot 6} = 5 \cdot 3\sqrt{2} +$

$10 \cdot 2\sqrt{6} = 15\sqrt{2} + 20\sqrt{6}$

32. $5\sqrt{2}(4\sqrt{8} - 2\sqrt{12}) = 20\sqrt{16} - 10\sqrt{24} = 20 \cdot 4 - 10\sqrt{4 \cdot 6} = 80 - 10 \cdot 2\sqrt{6} =$

$80 - 20\sqrt{6}$

C **33.** $2\sqrt{49x^3} - 3\sqrt{16x^5} = 2\sqrt{49x^2 \cdot x} - 3\sqrt{16x^4 \cdot x} = 2 \cdot 7x\sqrt{x} - 3 \cdot 4x^2\sqrt{x} =$

$14x\sqrt{x} - 12x^2\sqrt{x}$

34. $4\sqrt{72s^4} - 2s\sqrt{200s^2} = 4\sqrt{36s^4 \cdot 2} - 2s\sqrt{100s^2 \cdot 2} = 4 \cdot 6s^2\sqrt{2} - 2s \cdot 10s\sqrt{2} =$

$24s^2\sqrt{2} - 20s^2\sqrt{2} = (24 - 20)s^2\sqrt{2} = 4s^2\sqrt{2}$

35. $\sqrt{\frac{x^2}{16} + \frac{x^2}{25}} = \sqrt{\frac{25x^2 + 16x^2}{16 \cdot 25}} = \frac{\sqrt{41x^2}}{\sqrt{16 \cdot 25}} = \frac{x\sqrt{41}}{4 \cdot 5} = \frac{x\sqrt{41}}{20}$

36. $\sqrt{\frac{x^2}{49} - \frac{x^2}{121}} = \sqrt{\frac{121x^2 - 49x^2}{49 \cdot 121}} = \frac{\sqrt{72x^2}}{\sqrt{49 \cdot 121}} = \frac{\sqrt{36x^2 \cdot 2}}{7 \cdot 11} = \frac{6x\sqrt{2}}{77}$

37. $\sqrt{\frac{x^2}{a^2} + \frac{x^2}{b^2}} = \sqrt{\frac{b^2x^2 + a^2x^2}{a^2b^2}} = \frac{\sqrt{x^2(a^2 + b^2)}}{\sqrt{a^2b^2}} = \frac{x\sqrt{a^2 + b^2}}{ab}$

38. $\sqrt{\frac{x}{a}} - \sqrt{\frac{a}{x}} = \frac{\sqrt{x}\sqrt{a}}{\sqrt{a}\sqrt{a}} - \frac{\sqrt{a}\sqrt{x}}{\sqrt{x}\sqrt{x}} = \frac{\sqrt{ax}}{a} - \frac{\sqrt{ax}}{x} = \frac{x\sqrt{ax} - a\sqrt{ax}}{ax} = \frac{(x - a)\sqrt{ax}}{ax}$

Page 542 • MIXED REVIEW EXERCISES

1. $2y = 6x + 10, y = 3x + 5$ **2.** $5y - x + 10 = 0, 5y = x - 10, y = \frac{1}{5}x - 2$

3. $3x - y = 3, -y = -3x + 3, y = 3x - 3$ **4.** $5x + 5y = 4, 5y = -5x + 4,$
$y = -x + \frac{4}{5}$

5. $x = 2y + 6, -2y = -x + 6, y = \frac{1}{2}x - 3$ **6.** $2x - 7y = 0, -7y = -2x, y = \frac{2}{7}x$

7. $x = -y + 11, y = -x + 11$ **8.** $7 - x + 11y = 0, 11y = x - 7, y = \frac{1}{11}x - \frac{7}{11}$

9. $12x - 4 = -2y, 2y = -12x + 4, y = -6x + 2$

10. $-\frac{b}{2a} = 0; y = -4(0)^2 = 0;$ vertex: $(0, 0)$; axis: $x = 0$

11. $-\frac{b}{2a} = -\frac{(-6)}{2(1)} = 3; y = (3)^2 - 6(3) + 9 = 9 - 18 + 9 = 0;$ vertex: $(3, 0)$; axis: $x = 3$

12. $-\frac{b}{2a} = -\frac{0}{2(-2)} = 0; y = 5 - 2(0)^2 = 5;$ vertex: $(0, 5)$; axis: $x = 0$

13. $\frac{4x}{5} = \frac{10x}{9}, 45\left(\frac{4x}{5}\right) = 45\left(\frac{10x}{9}\right), 36x = 50x, x = 0; \{0\}$

14. $\frac{4}{b + 6} = \frac{-1}{2b + 3}, 4(2b + 3) = -(b + 6), 8b + 12 = -b - 6, 9b = -18, b = -2; \{-2\}$

15. $11 - \frac{10}{x} = x, x\left(11 - \frac{10}{x}\right) = x^2, 11x - 10 = x^2, 0 = x^2 - 11x + 10,$
$0 = (x - 10)(x - 1); x - 10 = 0$ or $x - 1 = 0, x = 10$ or $x = 1; \{10, 1\}$

16. $\frac{3}{t} = \frac{-7}{2t^2 - 1}, 3(2t^2 - 1) = -7t, 6t^2 - 3 = -7t, 6t^2 + 7t - 3 = 0,$
$(2t + 3)(3t - 1) = 0; 2t + 3 = 0$ or $3t - 1 = 0, 2t = -3$ or $3t = 1, t = -\frac{3}{2}$ or
$t = \frac{1}{3}; \left\{-\frac{3}{2}, \frac{1}{3}\right\}$

17. $\frac{1}{y} + \frac{2 - y}{y + 3} = \frac{1}{y + 3}, y(y + 3)\left[\frac{1}{y} + \frac{2 - y}{y + 3}\right] = y(y + 3)\left(\frac{1}{y + 3}\right), y + 3 + (2 - y)y = y,$
$y + 3 + 2y - y^2 = y, 0 = y^2 - 2y - 3, 0 = (y - 3)(y + 1); y - 3 = 0$ or
$y + 1 = 0, y = 3$ or $y = -1; \{-1, 3\}$

18. $\frac{p+2}{4p-1} = \frac{p}{4p+3}$, $(4p + 3)(p + 2) = p(4p - 1)$, $4p^2 + 11p + 6 = 4p^2 - p$,

$12p = -6$, $p = -\frac{1}{2}$; $\left\{-\frac{1}{2}\right\}$

Page 543 • EXTRA

1. Let h, t, and u be the hundreds, tens, and units digits. $100h + 10t + u = (99h + 9t) + (h + t + u)$. Since $99h + 9t$ is divisible by 3, the entire right side of the equation is divisible by 3 if and only if the sum of the digits, $h + t + u$, is divisible by 3.
2. Since 100 is divisible by 4, the entire number is divisible by 4 if and only if the last two digits are divisible by 4.
3. Let a, b, c, d, e, and f be the hundred-thousands, ten-thousands, thousands, hundreds, tens, and units digits. $100{,}000a + 10{,}000b + 1{,}000c + 100d + 10e + f = (1{,}000a + 100b + 10c + d)100 + 10e + f$. Since 100 is divisible by 4, the entire right-hand side of the equation is divisible by 4 if and only if the last two digits are divisible by 4.
4. Let a, b, c, d, e, and f be the digits. $100{,}000a + 10{,}000b + 1{,}000c + 100d + 10e + f = (100a + 10b + c)1{,}000 + 100d + 10e + f$. Since 1,000 is divisible by 8, the entire right side of the equation is divisible by 8 if and only if the last three digits are divisible by 8.
5. Let a, b, c, d, e, and f be the digits. $100{,}000a + 10{,}000b + 1{,}000c + 100d + 10e + f = 100{,}001a + 1{,}001c + 11e - (a + c + e) + 9{,}999b + 99d + (b + d + f) = 11(9091a + 909b + 91c + 9d + e) - [(a + c + e) - (b + d + f)]$. Since the first five terms on the right side of the equation are divisible by 11, the entire right side is divisible by 11 if and only if $(a + c + e) - (b + d + f)$ is divisible by 11.
6. An eleven digit number is divisible by 11 if and only if the sum of the first, third, fifth, seventh, ninth, and eleventh digits minus the sum of the second, fourth, sixth, eighth, and tenth digits is divisible by 11.

Pages 545–546 • WRITTEN EXERCISES

A

1. $(2 + \sqrt{3})(2 - \sqrt{3}) = (2)^2 - (\sqrt{3})^2 = 4 - 3 = 1$
2. $(4 + \sqrt{11})(4 - \sqrt{11}) = (4)^2 - (\sqrt{11})^2 = 16 - 11 = 5$
3. $(\sqrt{15} + 6)(\sqrt{15} - 6) = (\sqrt{15})^2 - (6)^2 = 15 - 36 = -21$
4. $(\sqrt{19} - 9)(\sqrt{19} + 9) = (\sqrt{19})^2 - (9)^2 = 19 - 81 = -62$
5. $(\sqrt{3} - \sqrt{2})(\sqrt{3} + \sqrt{2}) = (\sqrt{3})^2 - (\sqrt{2})^2 = 3 - 2 = 1$
6. $(\sqrt{15} - \sqrt{3})(\sqrt{15} + \sqrt{3}) = (\sqrt{15})^2 - (\sqrt{3})^2 = 15 - 3 = 12$
7. $(2 + \sqrt{7})^2 = (2)^2 + 2(2)(\sqrt{7}) + (\sqrt{7})^2 = 4 + 4\sqrt{7} + 7 = 11 + 4\sqrt{7}$
8. $(8 - \sqrt{6})^2 = 64 - 16\sqrt{6} + (\sqrt{6})^2 = 64 - 16\sqrt{6} + 6 = 70 - 16\sqrt{6}$
9. $(2\sqrt{2} - 3)^2 = (2\sqrt{2})^2 - 2(2\sqrt{2})(3) + 9 = 4 \cdot 2 - 12\sqrt{2} + 9 = 8 - 12\sqrt{2} + 9 = 17 - 12\sqrt{2}$
10. $(4\sqrt{10} + 3)^2 = (4\sqrt{10})^2 + 2(4\sqrt{10})(3) + 9 = 16 \cdot 10 + 24\sqrt{10} + 9 = 160 + 24\sqrt{10} + 9 = 169 + 24\sqrt{10}$

11. $(\sqrt{13} - 2\sqrt{5})^2 = (\sqrt{13})^2 - 2(\sqrt{13})(2\sqrt{5}) + (2\sqrt{5})^2 = 13 - 4\sqrt{65} + 4 \cdot 5 =$
$13 - 4\sqrt{65} + 20 = 33 - 4\sqrt{65}$

12. $(3\sqrt{7} - \sqrt{3})^2 = (3\sqrt{7})^2 - 2(3\sqrt{7})(\sqrt{3}) + (\sqrt{3})^2 = 9 \cdot 7 - 6\sqrt{21} + 3 =$
$63 - 6\sqrt{21} + 3 = 66 - 6\sqrt{21}$

13. $(2\sqrt{7} + \sqrt{3})(2\sqrt{7} - \sqrt{3}) = (2\sqrt{7})^2 - (\sqrt{3})^2 = 4 \cdot 7 - 3 = 28 - 3 = 25$

14. $(3\sqrt{5} - \sqrt{2})(3\sqrt{5} + \sqrt{2}) = (3\sqrt{5})^2 - (\sqrt{2})^2 = 9 \cdot 5 - 2 = 45 - 2 = 43$

15. $(6\sqrt{5} - \sqrt{7})(6\sqrt{5} + \sqrt{7}) = (6\sqrt{5})^2 - (\sqrt{7})^2 = 36 \cdot 5 - 7 = 180 - 7 = 173$

16. $(8\sqrt{11} + 2\sqrt{6})(8\sqrt{11} - 2\sqrt{6}) = (8\sqrt{11})^2 - (2\sqrt{6})^2 = 64 \cdot 11 - 4 \cdot 6 =$
$704 - 24 = 680$

17. $(4\sqrt{3} - 5)(2\sqrt{3} + 3) = (4\sqrt{3})(2\sqrt{3}) + 12\sqrt{3} - 10\sqrt{3} - 15 = 8 \cdot 3 + 2\sqrt{3} - 15 =$
$24 + 2\sqrt{3} - 15 = 9 + 2\sqrt{3}$

18. $(6\sqrt{2} + 4)(3\sqrt{2} - 5) = (6\sqrt{2})(3\sqrt{2}) - 30\sqrt{2} + 12\sqrt{2} - 20 =$
$18 \cdot 2 - 18\sqrt{2} - 20 = 36 - 18\sqrt{2} - 20 = 16 - 18\sqrt{2}$

B

19. $(2\sqrt{5} - 6\sqrt{7})(3\sqrt{5} + \sqrt{7}) = (2\sqrt{5})(3\sqrt{5}) + (2\sqrt{5})(\sqrt{7}) - (18\sqrt{7})(\sqrt{5}) -$
$(6\sqrt{7})(\sqrt{7}) = 6 \cdot 5 + 2\sqrt{35} - 18\sqrt{35} - 6 \cdot 7 = 30 - 16\sqrt{35} - 42 =$
$-12 - 16\sqrt{35}$

20. $(7\sqrt{13} + 2\sqrt{6})(2\sqrt{13} + 3\sqrt{6}) = (7\sqrt{13})(2\sqrt{13}) + (7\sqrt{13})(3\sqrt{6}) +$
$(2\sqrt{6})(2\sqrt{13}) + (2\sqrt{6})(3\sqrt{6}) = 14 \cdot 13 + 21\sqrt{78} + 4\sqrt{78} + 6 \cdot 6 =$
$182 + 25\sqrt{78} + 36 = 218 + 25\sqrt{78}$

21. $(4\sqrt{11} - 2\sqrt{2})(6\sqrt{11} + 8\sqrt{2}) = (4\sqrt{11})(6\sqrt{11}) + (4\sqrt{11})(8\sqrt{2}) -$
$(2\sqrt{2})(6\sqrt{11}) - (2\sqrt{2})(8\sqrt{2}) = 24 \cdot 11 + 32\sqrt{22} - 12\sqrt{22} - 16 \cdot 2 =$
$264 + 20\sqrt{22} - 32 = 232 + 20\sqrt{22}$

22. $(8\sqrt{6} - 2\sqrt{3})(2\sqrt{6} - 3\sqrt{3}) = (8\sqrt{6})(2\sqrt{6}) - (3\sqrt{3})(8\sqrt{6}) - (2\sqrt{3})(2\sqrt{6}) +$
$(2\sqrt{3})(3\sqrt{3}) = 16 \cdot 6 - 24\sqrt{18} - 4\sqrt{18} + 6 \cdot 3 = 96 - 28\sqrt{9 \cdot 2} + 18 =$
$114 - 28 \cdot 3\sqrt{2} = 114 - 84\sqrt{2}$

23. $\frac{1}{1 + \sqrt{5}} \cdot \frac{1 - \sqrt{5}}{1 - \sqrt{5}} = \frac{1 - \sqrt{5}}{1^2 - (\sqrt{5})^2} = \frac{1 - \sqrt{5}}{1 - 5} = \frac{1 - \sqrt{5}}{-4} = -\frac{1 - \sqrt{5}}{4} = \frac{-1 + \sqrt{5}}{4}$

24. $\frac{1}{2 + \sqrt{3}} \cdot \frac{2 - \sqrt{3}}{2 - \sqrt{3}} = \frac{2 - \sqrt{3}}{4 - 3} = \frac{2 - \sqrt{3}}{1} = 2 - \sqrt{3}$

25. $\frac{3}{\sqrt{5} - 2} \cdot \frac{\sqrt{5} + 2}{\sqrt{5} + 2} = \frac{3(\sqrt{5} + 2)}{5 - 4} = \frac{3(\sqrt{5} + 2)}{1} = 3\sqrt{5} + 6$

26. $\frac{1}{\sqrt{7} - 3} \cdot \frac{\sqrt{7} + 3}{\sqrt{7} + 3} = \frac{\sqrt{7} + 3}{7 - 9} = \frac{\sqrt{7} + 3}{-2} = -\frac{\sqrt{7} + 3}{2} = \frac{-3 - \sqrt{7}}{2}$

27. $\frac{1 + \sqrt{7}}{2 - \sqrt{7}} \cdot \frac{2 + \sqrt{7}}{2 + \sqrt{7}} = \frac{2 + \sqrt{7} + 2\sqrt{7} + (\sqrt{7})^2}{4 - 7} = \frac{2 + 3\sqrt{7} + 7}{-3} = -\frac{9 + 3\sqrt{7}}{3} =$
$-3 - \sqrt{7}$

28. $\frac{4 + \sqrt{5}}{3 - \sqrt{5}} \cdot \frac{3 + \sqrt{5}}{3 + \sqrt{5}} = \frac{12 + 4\sqrt{5} + 3\sqrt{5} + (\sqrt{5})^2}{9 - 5} = \frac{12 + 7\sqrt{5} + 5}{4} = \frac{17 + 7\sqrt{5}}{4}$

29. $\dfrac{\sqrt{3}-4}{\sqrt{7}+2}\cdot\dfrac{\sqrt{7}-2}{\sqrt{7}-2}=\dfrac{(\sqrt{3})(\sqrt{7})-2\sqrt{3}-4\sqrt{7}+8}{7-4}=\dfrac{\sqrt{21}-2\sqrt{3}-4\sqrt{7}+8}{3}$

30. $\dfrac{\sqrt{5}-2}{\sqrt{3}+1}\cdot\dfrac{\sqrt{3}-1}{\sqrt{3}-1}=\dfrac{\sqrt{15}-\sqrt{5}-2\sqrt{3}+2}{3-1}=\dfrac{\sqrt{15}-\sqrt{5}-2\sqrt{3}+2}{2}$

31. $\dfrac{7}{2\sqrt{5}+1}\cdot\dfrac{2\sqrt{5}-1}{2\sqrt{5}-1}=\dfrac{7(2\sqrt{5}-1)}{(2\sqrt{5})^2-1}=\dfrac{14\sqrt{5}-7}{4\cdot 5-1}=\dfrac{14\sqrt{5}-7}{20-1}=\dfrac{14\sqrt{5}-7}{19}$

32. $\dfrac{5}{3\sqrt{7}-5}\cdot\dfrac{3\sqrt{7}+5}{3\sqrt{7}+5}=\dfrac{5(3\sqrt{7}+5)}{(3\sqrt{7})^2-25}=\dfrac{15\sqrt{7}+25}{9\cdot 7-25}=\dfrac{15\sqrt{7}+25}{38}$

33. $\dfrac{4+2\sqrt{2}}{2\sqrt{5}-3}\cdot\dfrac{2\sqrt{5}+3}{2\sqrt{5}+3}=\dfrac{8\sqrt{5}+12+4\sqrt{10}+6\sqrt{2}}{4\cdot 5-9}=\dfrac{8\sqrt{5}+12+4\sqrt{10}+6\sqrt{2}}{11}$

34. $\dfrac{6-2\sqrt{3}}{3\sqrt{2}+3}\cdot\dfrac{3\sqrt{2}-3}{3\sqrt{2}-3}=\dfrac{18\sqrt{2}-18-6\sqrt{6}+6\sqrt{3}}{9\cdot 2-9}=\dfrac{3(6\sqrt{2}-6-2\sqrt{6}+2\sqrt{3})}{9}=$
$\dfrac{6\sqrt{2}-6-2\sqrt{6}+2\sqrt{3}}{3}$

35. $f(\sqrt{6})=(\sqrt{6})^2-5(\sqrt{6})-7=6-5\sqrt{6}-7=-1-5\sqrt{6}$

36. $f(\sqrt{10})=(\sqrt{10})^2-5(\sqrt{10})-7=10-5\sqrt{10}-7=3-5\sqrt{10}$

37. $f(\sqrt{2}+1)=(\sqrt{2}+1)^2-5(\sqrt{2}+1)-7=(\sqrt{2})^2+2\sqrt{2}+1-5\sqrt{2}-5-7=$
$2-11-3\sqrt{2}=-9-3\sqrt{2}$

38. $f(\sqrt{3}+2)=(\sqrt{3}+2)^2-5(\sqrt{3}+2)-7=(\sqrt{3})^2+4\sqrt{3}+4-5\sqrt{3}-10-$
$7=3-13-\sqrt{3}=-10-\sqrt{3}$

39. $f(-2+\sqrt{11})=(-2+\sqrt{11})^2-5(-2+\sqrt{11})-7=4-4\sqrt{11}+(\sqrt{11})^2+10-$
$5\sqrt{11}-7=7+11-9\sqrt{11}=18-9\sqrt{11}$

40. $f(\sqrt{7}-2)=(\sqrt{7}-2)^2-5(\sqrt{7}-2)-7=(\sqrt{7})^2-4\sqrt{7}+4-5\sqrt{7}+10-$
$7=7+7-9\sqrt{7}=14-9\sqrt{7}$

41. $(4+\sqrt{7})^2-8(4+\sqrt{7})+9=16+8\sqrt{7}+(\sqrt{7})^2-32-8\sqrt{7}+9=$
$-7+7=0$; $(4-\sqrt{7})^2-8(4-\sqrt{7})+9=16-8\sqrt{7}+(\sqrt{7})^2-32+$
$8\sqrt{7}+9=-7+7=0$; Therefore $(4+\sqrt{7})$ and $(4-\sqrt{7})$ are roots of the equation.

42. $(5+\sqrt{3})^2-10(5+\sqrt{3})+22=25+10\sqrt{3}+(\sqrt{3})^2-50-10\sqrt{3}+22=$
$-3+3=0$; $(5-\sqrt{3})^2-10(5-\sqrt{3})+22=25-10\sqrt{3}+(\sqrt{3})^2-50+$
$10\sqrt{3}+22=-3+3=0$; Therefore $(5+\sqrt{3})$ and $(5-\sqrt{3})$ are roots of the equation.

43. $3\left(\frac{2}{3}+\frac{\sqrt{7}}{3}\right)^2-4\left(\frac{2}{3}+\frac{\sqrt{7}}{3}\right)-1=3\left(\frac{4}{9}+2\left(\frac{2}{3}\right)\left(\frac{\sqrt{7}}{3}\right)+\left(\frac{\sqrt{7}}{3}\right)^2\right)-\frac{8}{3}-\frac{4\sqrt{7}}{3}-1=$
$3\left(\frac{4}{9}+\frac{4\sqrt{7}}{9}+\frac{7}{9}\right)-\frac{8}{3}-\frac{4\sqrt{7}}{3}-1=\frac{4}{3}+\frac{4\sqrt{7}}{3}+\frac{7}{3}-\frac{8}{3}-\frac{4\sqrt{7}}{3}-\frac{3}{3}=\frac{4+7-8-3}{3}+$

(Continued)

$\frac{4\sqrt{7} - 4\sqrt{7}}{3} = 0$; $3\left(\frac{2}{3} - \frac{\sqrt{7}}{3}\right)^2 - 4\left(\frac{2}{3} - \frac{\sqrt{7}}{3}\right) - 1 = 3\left(\frac{4}{9} - \frac{4\sqrt{7}}{9} + \frac{7}{9}\right) - \frac{8}{3} + \frac{4\sqrt{7}}{3} - \frac{3}{3} = \frac{4}{3} - \frac{4\sqrt{7}}{3} + \frac{7}{3} - \frac{11}{3} + \frac{4\sqrt{7}}{3} = \frac{4 + 7 - 11}{3} + \frac{4\sqrt{7} - 4\sqrt{7}}{3} = 0$; therefore they are both roots.

C **44.** $A = \frac{1}{2}bh = \frac{1}{2}\left(\frac{4\sqrt{5} - 2}{5}\right)\left(\frac{\sqrt{5} + 6}{2}\right) = \frac{4 \cdot 5 + 24\sqrt{5} - 2\sqrt{5} - 12}{20} = \frac{8 + 22\sqrt{5}}{20} = \frac{4 + 11\sqrt{5}}{10}$; the area is $\frac{4 + 11\sqrt{5}}{10}$ square units.

45. $(x + \sqrt{y})(x - \sqrt{y}) = x^2 - (\sqrt{y})^2 = x^2 - y$

46. $(x - 3\sqrt{2})^2 = x^2 - 6x\sqrt{2} + 9(\sqrt{2})^2 = x^2 - 6x\sqrt{2} + 9 \cdot 2 = x^2 - 6x\sqrt{2} + 18$

47. $(3a\sqrt{b} - c)(5a\sqrt{b} + 3c) = 15a^2(\sqrt{b})^2 + 9ac\sqrt{b} - 5ac\sqrt{b} - 3c^2 = 15a^2b + 4ac\sqrt{b} - 3c^2$

Page 546 • MIXED REVIEW EXERCISES

1. $\sqrt{18x^8} = \sqrt{9x^8 \cdot 2} = 3x^4\sqrt{2}$

2. $4\sqrt{15x} \cdot 3\sqrt{5} = 12\sqrt{75x} = 12\sqrt{25 \cdot 3x} = 12 \cdot 5\sqrt{3x} = 60\sqrt{3x}$

3. $6\sqrt{8} - 4\sqrt{2} = 6\sqrt{4 \cdot 2} - 4\sqrt{2} = 6 \cdot 2\sqrt{2} - 4\sqrt{2} = (12 - 4)\sqrt{2} = 8\sqrt{2}$

4. $4\sqrt{63} + 5\sqrt{28} = 4\sqrt{9 \cdot 7} + 5\sqrt{4 \cdot 7} = 4 \cdot 3\sqrt{7} + 5 \cdot 2\sqrt{7} = (12 + 10)\sqrt{7} = 22\sqrt{7}$

5. $\sqrt{\frac{3}{5}} \cdot \sqrt{\frac{5}{9}} = \sqrt{\frac{3}{5} \cdot \frac{5}{9}} = \sqrt{\frac{1}{3}} = \frac{1}{\sqrt{3}} \cdot \frac{\sqrt{3}}{\sqrt{3}} = \frac{\sqrt{3}}{3}$

6. $\sqrt{1\frac{5}{6}} \cdot \sqrt{4\frac{1}{6}} = \sqrt{\frac{11}{6} \cdot \frac{25}{6}} = \sqrt{\frac{25}{36} \cdot 11} = \frac{5}{6}\sqrt{11}$ or $\frac{5\sqrt{11}}{6}$

7. $(3 - 5k^2)^2 = 9 - 30k^2 + 25k^4$ **8.** $(2p + 7z)^2 = 4p^2 + 28pz + 49z^2$

9. $(6ab + x)(6ab - x) = 36a^2b^2 - x^2$

10. $7p - 3 = 6(p + 2)$, $7p - 3 = 6p + 12$, $p = 15$; $\{15\}$

11. $x^2 - 14x + 45 = 0$, $(x - 9)(x - 5) = 0$, $x - 9 = 0$ or $x - 5 = 0$, $x = 9$ or $x = 5$; $\{5, 9\}$

12. $36g^2 = 16$, $g^2 = \frac{16}{36}$, $g = \pm\sqrt{\frac{16}{36}} = \pm\frac{4}{6} = \pm\frac{2}{3}$; $\left\{-\frac{2}{3}, \frac{2}{3}\right\}$

Pages 548–549 • WRITTEN EXERCISES

A **1.** $\sqrt{x} = 3$, $x = 9$; $\{9\}$ **2.** $\sqrt{y} = 14$, $y = 196$; $\{196\}$

3. $4 = \sqrt{2x}$, $16 = 2x$, $8 = x$; $\{8\}$ **4.** $9 = \sqrt{3a}$, $81 = 3a$, $27 = a$; $\{27\}$

5. $\sqrt{8x} = \frac{2}{5}$, $8x = \frac{4}{25}$, $x = \frac{1}{50}$; $\left\{\frac{1}{50}\right\}$ **6.** $\sqrt{4n} = \frac{1}{3}$, $4n = \frac{1}{9}$, $n = \frac{1}{36}$; $\left\{\frac{1}{36}\right\}$

7. $1 = \sqrt{m} - 3, 4 = \sqrt{m}, 16 = m; \{16\}$ **8.** $7 = \sqrt{z} - 2, 9 = \sqrt{z}, 81 = z; \{81\}$

9. $\frac{2}{3} + \sqrt{b} = 1, \sqrt{b} = \frac{1}{3}, b = \frac{1}{9}; \left\{\frac{1}{9}\right\}$ **10.** $\sqrt{y} - \frac{1}{2} = 2, \sqrt{y} = \frac{5}{2}, y = \frac{25}{4}; \left\{\frac{25}{4}\right\}$

11. $3 = \sqrt{\frac{x}{2}}, 9 = \frac{x}{2}, 18 = x; \{18\}$ **12.** $8 = \sqrt{\frac{s}{5}}, 64 = \frac{s}{5}, 320 = s; \{320\}$

13. $\sqrt{x + 1} = 3, x + 1 = 9, x = 8; \{8\}$ **14.** $\sqrt{m + 5} = 1, m + 5 = 1, m = -4; \{-4\}$

15. $20 = 5\sqrt{2x}, 4 = \sqrt{2x}, 16 = 2x, 8 = x; \{8\}$

16. $5 = 2\sqrt{3x}, \frac{5}{2} = \sqrt{3x}, \frac{25}{4} = 3x, \frac{25}{12} = x; \left\{\frac{25}{12}\right\}$

17. $\sqrt{4x} + 2 = 6, \sqrt{4x} = 4, 4x = 16, x = 4; \{4\}$

18. $\sqrt{3x} + 4 = 7, \sqrt{3x} = 3, 3x = 9, x = 3; \{3\}$

19. $4 = \sqrt{8a} + 3, 1 = \sqrt{8a}, 1 = 8a, a = \frac{1}{8}; \left\{\frac{1}{8}\right\}$

20. $3 = \sqrt{4x + 1}, 9 = 4x + 1, 8 = 4x, 2 = x; \{2\}$

21. $\sqrt{5y - 2} + 3 = 9, \sqrt{5y - 2} = 6, 5y - 2 = 36, 5y = 38, y = \frac{38}{5}; \left\{\frac{38}{5}\right\}$

22. $\sqrt{5m - 5} + 6 = 7, \sqrt{5m - 5} = 1, 5m - 5 = 1, 5m = 6, m = \frac{6}{5}; \left\{\frac{6}{5}\right\}$

23. $\sqrt{x} = 3\sqrt{7}, x = 9 \cdot 7, x = 63; \{63\}$ **24.** $\sqrt{r} = 5\sqrt{2}, r = 25 \cdot 2 = 50; \{50\}$

B **25.** $8 = \sqrt{\frac{5a}{4}} - 2, 10 = \sqrt{\frac{5a}{4}}, 100 = \frac{5a}{4}, 400 = 5a, 80 = a; \{80\}$

26. $14 = \sqrt{\frac{7x}{3}} + 2, 12 = \sqrt{\frac{7x}{3}}, 144 = \frac{7x}{3}, 432 = 7x, \frac{432}{7} = x; \left\{\frac{432}{7}\right\}$

27. $\sqrt{\frac{2x + 9}{5}} = 3, \frac{2x + 9}{5} = 9, 2x + 9 = 45, 2x = 36, x = 18; \{18\}$

28. $\sqrt{\frac{2n - 4}{8}} = 2, \frac{2n - 4}{8} = 4, 2n - 4 = 32, 2n = 36, n = 18; \{18\}$

29. $4 = \sqrt{\frac{7k - 10}{9}}, 16 = \frac{7k - 10}{9}, 144 = 7k - 10, 154 = 7k, 22 = k; \{22\}$

30. $3 = \sqrt{\frac{4x - 5}{7}}, 9 = \frac{4x - 5}{7}, 63 = 4x - 5, 68 = 4x, 17 = x; \{17\}$

31. $15\sqrt{2} = 5\sqrt{t}, 225 \cdot 2 = 25t, 450 = 25t, 18 = t; \{18\}$

32. $5\sqrt{10} = 6\sqrt{m}, 25 \cdot 10 = 36 \cdot m, \frac{250}{36} = m, \frac{125}{18} = m; \left\{\frac{125}{18}\right\}$

33. $\sqrt{2a^2 - 5} = 11, 2a^2 - 5 = 121, 2a^2 = 126, a^2 = 63, a = \pm\sqrt{63} = \pm\sqrt{9 \cdot 7} = \pm 3\sqrt{7}; \{-3\sqrt{7}, 3\sqrt{7}\}$

34. $\sqrt{2m^2 - 10} = 4, 2m^2 - 10 = 16, 2m^2 = 26, m^2 = 13, m = \pm\sqrt{13}; \{-\sqrt{13}, \sqrt{13}\}$

35. $10 = 2\sqrt{3c^2 - 2}, 5 = \sqrt{3c^2 - 2}, 25 = 3c^2 - 2, 27 = 3c^2, 9 = c^2, c = \pm 3; \{3, -3\}$

36. $36 = 4\sqrt{4m^2 + 5}, 9 = \sqrt{4m^2 + 5}, 81 = 4m^2 + 5, 76 = 4m^2, 19 = m^2, m = \pm\sqrt{19}; \{-\sqrt{19}, \sqrt{19}\}$

37. $\sqrt{5b^2 - 36} = 2b$, $5b^2 - 36 = 4b^2$, $b^2 = 36$, $b = \pm 6$; $\sqrt{5(-6)^2 - 36} = 2(-6)$, $\sqrt{5 \cdot 36 - 36} = -12$, $\sqrt{144} = -12$, $12 \neq -12$; Therefore -6 is not a solution; $\{6\}$

38. $\sqrt{19x^2 - 51} = 4x$, $19x^2 - 51 = 16x^2$, $3x^2 = 51$, $x^2 = 17$, $x = \pm\sqrt{17}$; $\sqrt{19(-\sqrt{17})^2 - 51} = 4(-\sqrt{17})$, $\sqrt{19 \cdot 17 - 51} = -4\sqrt{17}$, $\sqrt{272} = -4\sqrt{17}$, $\sqrt{16 \cdot 17} = -4\sqrt{17}$, $4\sqrt{17} \neq -4\sqrt{17}$; therefore $-\sqrt{17}$ is not a solution; $\{\sqrt{17}\}$

39. $\sqrt{x^2 + 1} = 1 - x$, $x^2 + 1 = (1 - x)^2$, $x^2 + 1 = 1 - 2x + x^2$, $2x = 0$, $x = 0$; $\{0\}$

40. $\sqrt{x^2 + 9} = 3 - x$, $x^2 + 9 = (3 - x)^2$, $x^2 + 9 = 9 - 6x + x^2$, $9 = 9 - 6x$, $0 = -6x$, $x = 0$, $\{0\}$

41. $\sqrt{3a^2 - 32} = a$, $3a^2 - 32 = a^2$, $2a^2 = 32$, $a^2 = 16$, $a = \pm 4$; $\sqrt{3(-4)^2 - 32} = -4$, $\sqrt{3 \cdot 16 - 32} = -4$, $\sqrt{16} \neq -4$; therefore -4 is not a solution; $\{4\}$

42. $\sqrt{13b^2 + 33} = 4b$, $13b^2 + 33 = 16b^2$, $33 = 3b^2$, $b^2 = 11$, $b = \pm\sqrt{11}$; $\sqrt{13(-\sqrt{11})^2 + 33} = 4(-\sqrt{11})$, $\sqrt{13 \cdot 11 + 33} = -4\sqrt{11}$, $\sqrt{176} = -4\sqrt{11}$, $\sqrt{16 \cdot 11} \neq -4\sqrt{11}$; therefore $-\sqrt{11}$ is not a solution; $\{\sqrt{11}\}$

C **43.** $\sqrt{x^2 + 6x} = 4$, $x^2 + 6x = 16$, $x^2 + 6x - 16 = 0$, $(x + 8)(x - 2) = 0$; $x + 8 = 0$ or $x - 2 = 0$, $x = -8$ or $x = 2$; $\{-8, 2\}$

44. $\sqrt{a^2 + 3a} = 2$, $a^2 + 3a = 4$, $a^2 + 3a - 4 = 0$, $(a + 4)(a - 1) = 0$; $a + 4 = 0$ or $a - 1 = 0$, $a = -4$ or $a = 1$; $\{-4, 1\}$

45. $\sqrt{15x^2 - 12x} = 9x$, $15x^2 - 12x = 81x^2$, $0 = 66x^2 + 12x$, $0 = 6x(11x + 2)$; $6x = 0$ or $11x + 2 = 0$, $x = 0$ or $11x = -2$, $x = 0$ or $x = -\frac{2}{11}$; $\sqrt{15\left(-\frac{2}{11}\right)^2 - 12\left(-\frac{2}{11}\right)} = 9\left(-\frac{2}{11}\right)$, $\sqrt{\frac{15 \cdot 4}{121} + \frac{24}{11}} = -\frac{18}{11}$, $\sqrt{\frac{60 + 264}{121}} = -\frac{18}{11}$, $\sqrt{\frac{324}{121}} = -\frac{18}{11}$, $\frac{18}{11} \neq -\frac{18}{11}$; therefore $-\frac{2}{11}$ is not a solution; $\{0\}$

46. $\sqrt{20y^2 - 13y} = 5y$, $20y^2 - 13y = 25y^2$, $0 = 5y^2 + 13y$, $0 = y(5y + 13)$; $y = 0$ or $5y + 13 = 0$, $y = 0$ or $5y = -13$, $y = 0$ or $y = -\frac{13}{5}$; $\sqrt{20\left(-\frac{13}{5}\right)^2 - 13\left(-\frac{13}{5}\right)} = 5\left(-\frac{13}{5}\right)$, $\sqrt{\frac{20 \cdot 169}{25} + \frac{169}{5}} = -13$, $\sqrt{\frac{676 + 169}{5}} = -13$, $\sqrt{\frac{5 \cdot 169}{5}} = -13$, $\sqrt{169} \neq -13$; therefore $-\frac{13}{5}$ is not a solution; $\{0\}$

47. $\sqrt{x} + 6 = \sqrt{16x}$, $\sqrt{x} + 6 = 4\sqrt{x}$, $6 = 3\sqrt{x}$, $2 = \sqrt{x}$, $4 = x$; $\{4\}$

48. $3\sqrt{a} + 7 = \sqrt{16a}$, $3\sqrt{a} + 7 = 4\sqrt{a}$, $7 = \sqrt{a}$, $49 = a$; $\{49\}$

49.
$$\begin{array}{rcl} 3\sqrt{a} + 5\sqrt{b} &=& 31 \\ \underline{5\sqrt{a} - 5\sqrt{b}} &=& \underline{-15} \\ 8\sqrt{a} \qquad &=& 16 \end{array}$$
$\sqrt{a} = 2$, $a = 4$; $3\sqrt{(4)} + 5\sqrt{b} = 31$, $6 + 5\sqrt{b} = 31$, $5\sqrt{b} = 25$, $\sqrt{b} = 5$, $b = 25$; $(4, 25)$

50. $5\sqrt{x} - 2\sqrt{y} = 4\sqrt{2} \rightarrow 15\sqrt{x} - 6\sqrt{y} = 12\sqrt{2}$

$2\sqrt{x} + 3\sqrt{y} = 13\sqrt{2} \rightarrow \underline{4\sqrt{x} + 6\sqrt{y} = 26\sqrt{2}}$

$19\sqrt{x} = 38\sqrt{2}, \sqrt{x} = 2\sqrt{2}, x = 4 \cdot 2 = 8;$

$2\sqrt{(8)} + 3\sqrt{y} = 13\sqrt{2}, 3\sqrt{y} = 13\sqrt{2} - 2 \cdot 2\sqrt{2},$

$3\sqrt{y} = 9\sqrt{2}, \sqrt{y} = 3\sqrt{2}, y = 9 \cdot 2 = 18;$ (8, 18)

Pages 549–550 • PROBLEMS

A

1. $\sqrt{3x} = 15, 3x = 225, x = 75$; The number is 75.

2. $2\sqrt{x} = 22, \sqrt{x} = 11, x = 121$; The number is 121.

3. $\frac{1}{8}\sqrt{x} = 3, \sqrt{x} = 24, x = 576$; The number is 576.

4. $\sqrt{\frac{x}{8}} = 3, \frac{x}{8} = 9, x = 72$; The number is 72.

5. $\sqrt{4x + 5} = 11, 4x + 5 = 121, 4x = 116, x = 29$; The number is 29.

6. $\sqrt{3x} - 23 = 16, \sqrt{3x} = 39, 3x = 1521, x = 507$; The number is 507.

B

7. $(15) = \sqrt{\frac{V}{\pi(36)}}, 225 = \frac{V}{36\pi}, 8100\pi = V$; The volume of the cylinder is 8100π cm^3.

8. $(15) = \sqrt{\frac{2s}{9.8}}, 225 = \frac{2s}{9.8}, 2205 = 2s, 1102.5 = s$; The object falls 1102.5 m.

9. $(4.5) = \sqrt{\frac{P}{60}}, 20.25 = \frac{P}{60}, 1215 = P$; The hair dryer uses 1215 W.

C

10. Let n and $n + 2$ be the integers. $\sqrt{n(n + 2)} = 4\sqrt{39}, n(n + 2) = 16 \cdot 39,$ $n^2 + 2n - 624 = 0, (n + 26)(n - 24) = 0, n = -26$ or $n = 24$; but n cannot be neg. so $n = 24$; $n + 2 = 26$; The integers are 24 and 26.

11. $(8) = 2\pi\sqrt{\frac{l}{9.8}}, \frac{4}{\pi} = \sqrt{\frac{l}{9.8}}, \frac{16}{\pi^2} = \frac{l}{9.8}, l = \frac{16 \cdot 9.8}{(3.14)^2} \approx 15.9$; The pendulum is about 15.9 m long.

Page 550 • MIXED REVIEW EXERCISES

1. $(5 + \sqrt{6})(5 - \sqrt{6}) = 25 - 6 = 19$ **2.** $(2 + \sqrt{5})^2 = 4 + 4\sqrt{5} + 5 = 9 + 4\sqrt{5}$

3. $\frac{2}{3 + \sqrt{11}} \cdot \frac{3 - \sqrt{11}}{3 - \sqrt{11}} = \frac{2(3 - \sqrt{11})}{9 - 11} = \frac{6 - 2\sqrt{11}}{-2} = -(3 - \sqrt{11}) = \sqrt{11} - 3$

4. $\frac{2 + \sqrt{5}}{1 - \sqrt{5}} \cdot \frac{1 + \sqrt{5}}{1 + \sqrt{5}} = \frac{2 + 2\sqrt{5} + \sqrt{5} + 5}{1 - 5} = \frac{7 + 3\sqrt{5}}{-4} = \frac{-7 - 3\sqrt{5}}{4}$

5. $3\sqrt{5}(\sqrt{15} - 2\sqrt{5}) = 3\sqrt{75} - 6\sqrt{25} = 3\sqrt{25 \cdot 3} - 6 \cdot 5 = 3 \cdot 5\sqrt{3} - 30 =$ $15\sqrt{3} - 30$

6. $(2\sqrt{3} + 1)(\sqrt{3} - 4) = 6 - 8\sqrt{3} + \sqrt{3} - 4 = 2 - 7\sqrt{3}$

7. $7a^2 - 14a + 7 = 7(a^2 - 2a + 1) = 7(a - 1)^2$

8. $t^3 - 4t^2 - 45t = t(t^2 - 4t - 45) = t(t - 9)(t + 5)$

9. $6x(x + 2) + 4(x + 2) = (6x + 4)(x + 2) = 2(3x + 2)(x + 2)$

10. $y^3 + y^2 - 6y - 6 = y^2(y + 1) - 6(y + 1) = (y^2 - 6)(y + 1)$

11. $4g^5 - 100g = 4g(g^4 - 25) = 4g(g^2 + 5)(g^2 - 5)$

12. $36x^2 + 24xy + 4y^2 = 4(9x^2 + 6xy + y^2) = 4(3x + y)^2$

Page 553 • EXTRA

1. $\sqrt{6} = 6^{1/2}$ **2.** $\sqrt{10} = 10^{1/2}$ **3.** $\sqrt[3]{11} = 11^{1/3}$ **4.** $\sqrt[7]{5} = 5^{1/7}$ **5.** $\sqrt[4]{2} = 2^{1/4}$

6. $\sqrt[5]{3} = 3^{1/5}$ **7.** $\sqrt[3]{-15} = (-15)^{1/3}$ **8.** $\sqrt[4]{14} = (14)^{1/4}$ **9.** $\sqrt[6]{10} = (10)^{1/6}$

10. $\sqrt[5]{-12} = (-12)^{1/5}$ **11.** $(\sqrt[3]{4})^2 = 4^{2/3}$ **12.** $(\sqrt[6]{8})^5 = 8^{5/6}$ **13.** $(\sqrt{2})^3 = 2^{3/2}$

14. $(\sqrt{7})^5 = 7^{5/2}$ **15.** $(\sqrt[4]{3})^5 = 3^{5/4}$ **16.** $(\sqrt[7]{5})^3 = 5^{3/7}$ **17.** $64^{1/2} = \sqrt{64} = 8$

18. $8^{1/3} = \sqrt[3]{8} = 2$ **19.** $27^{1/3} = \sqrt[3]{27} = 3$ **20.** $100^{1/2} = \sqrt{100} = 10$

21. $(-64)^{1/3} = \sqrt[3]{(-64)} = -4$ **22.** $(-625)^{1/4} = \sqrt[4]{-625}$: not a real number.

23. $(-1296)^{1/4} = \sqrt[4]{-1296}$: not a real number. **24.** $(-343)^{1/3} = \sqrt[3]{-343} = -7$

25. $32^{3/5} = (\sqrt[5]{32})^3 = 2^3 = 8$ **26.** $16^{3/4} = (\sqrt[4]{16})^3 = 2^3 = 8$

27. $9^{5/2} = (\sqrt{9})^5 = 3^5 = 243$ **28.** $81^{3/4} = (\sqrt[4]{81})^3 = 3^3 = 27$

29. $(-9)^{3/2} = (\sqrt{-9})^3$: not a real number.

30. $(-1000)^{2/3} = (\sqrt[3]{-1000})^2 = (-10)^2 = 100$ **31.** $36^{3/2} = (\sqrt{36})^3 = 6^3 = 216$

32. $25^{3/2} = (\sqrt{25})^3 = 5^3 = 125$ **33.** $125^{4/3} = (\sqrt[3]{125})^4 = (5)^4 = 625$

34. $243^{4/5} = (\sqrt[5]{243})^4 = 3^4 = 81$ **35.** $5^n \cdot 5^n = 5, 5^{2n} = 5^1; 2n = 1, n = \frac{1}{2}; \left\{\frac{1}{2}\right\}$

36. $3^n \cdot 3^n \cdot 3^n \cdot 3^n = 3, 3^{4n} = 3^1; 4n = 1, n = \frac{1}{4}; \left\{\frac{1}{4}\right\}$

37. $2^n \cdot 2^n \cdot 2^n = 2, 2^{3n} = 2^1, 3n = 1, n = \frac{1}{3}; \left\{\frac{1}{3}\right\}$

38. $6^{(1/2)m} \cdot 6^{(1/2)m} = 36, 6^n = 6^2, n = 2; \{2\}$

39. $4^{(1/3)m} \cdot 4^{(1/3)m} = 16, 4^{2n/3} = 4^2; \frac{2n}{3} = 2, 2n = 6, n = 3; \{3\}$

40. $7^{(7/5)m} \cdot 7^{(3/5)m} = 343, 7^{2n} = 7^3; 2n = 3, n = \frac{3}{2}; \left\{\frac{3}{2}\right\}$

Page 554 • CALCULATOR KEY-IN

1. 2.632148 **2.** 3.7416574 **3.** 1.8707313 **4.** 1.5756979 **5.** 1.3753959

6. 1.1800684

Pages 555–556 • CHAPTER REVIEW

1. b; $\frac{1}{2} + \frac{1}{2}\left(\frac{5}{8} - \frac{1}{2}\right) = \frac{1}{2} + \frac{1}{2}\left(\frac{5 - 4}{8}\right) = \frac{1}{2} + \frac{1}{16} = \frac{8 + 1}{16} = \frac{9}{16}$

2. a; 3(15) ? 4(11), 45 > 44; $\frac{3}{4} > \frac{11}{15}$ **3.** c

4. d; $100n - n = 57.\overline{57} - 0.\overline{57}, 99n = 57, n = \frac{57}{99} = \frac{19}{33}$ **5.** b **6.** c; $\sqrt{\frac{576}{400}} = \frac{24}{20} = \frac{6}{5}$

7. b **8.** c; $\sqrt{96} = \sqrt{16 \cdot 6} = 4\sqrt{6}$ **9.** a

10. b; $5n^2 - 405 = 0, 5n^2 = 405, n^2 = 81; n = \pm\sqrt{81} = \pm 9; \{-9, 9\}$

11. c; $\sqrt{16^2 + 30^2} = \sqrt{256 + 900} = \sqrt{1156} = 34$

12. b; $15^2 + 20^2 = 225 + 400 = 625 \neq 36^2$; not a right triangle.

13. a; $\sqrt{15} \cdot \sqrt{12} = \sqrt{15 \cdot 12} = \sqrt{3 \cdot 5 \cdot 3 \cdot 4} = \sqrt{9 \cdot 20} = 3\sqrt{20} = 3\sqrt{2 \cdot 2 \cdot 5} =$
$6\sqrt{5}$

14. c; $\sqrt{16} + 3\sqrt{8} - 2\sqrt{2} = 4 + 3\sqrt{4 \cdot 2} - 2\sqrt{2} = 4 + 3 \cdot 2\sqrt{2} - 2\sqrt{2} =$
$4 + (6 - 2)\sqrt{2} = 4 + 4\sqrt{2}$

15. d; $(4 + \sqrt{7})(4 - \sqrt{7}) = 16 - 7 = 9$

16. b; $\frac{\sqrt{3}}{\sqrt{3} - 2} \cdot \frac{\sqrt{3} + 2}{\sqrt{3} + 2} = \frac{\sqrt{3}(\sqrt{3} + 2)}{3 - 4} = -(3 + 2\sqrt{3}) = -3 - 2\sqrt{3}$

17. a; $\sqrt{5x + 1} - 6 = 8, \sqrt{5x + 1} = 14, 5x + 1 = 196, 5x = 195, x = 39; \{39\}$

Page 556 • CHAPTER TEST

1. $\frac{5}{16} + \frac{1}{2}\left(\frac{7}{8} - \frac{5}{16}\right) = \frac{5}{16} + \frac{1}{2}\left(\frac{14 - 5}{16}\right) = \frac{5}{16} + \frac{9}{32} = \frac{10 + 9}{32} = \frac{19}{32}$

2. $\frac{4}{21} \cdot \frac{45}{45} = \frac{180}{945}; \frac{2}{15} \cdot \frac{63}{63} = \frac{126}{945}; \frac{5}{27} \cdot \frac{35}{35} = \frac{175}{945}$; therefore $\frac{2}{15} < \frac{5}{27} < \frac{4}{21}$ **3.** $0.58\overline{3}$

4. $10n = 1.\overline{6}, 10n - n = 1.6\overline{6} - 0.1\overline{6}, 9n = 1.5, n = \frac{1.5}{9} = \frac{15}{90} = \frac{1}{6}$ **5.** $\sqrt{\frac{225}{900}} = \frac{15}{30} = \frac{1}{2}$

6. $\sqrt{0.0049} = \sqrt{\frac{49}{10{,}000}} = \frac{7}{100} = 0.07$ **7.** $\sqrt{3400} = 10\sqrt{34} \approx 58.3$

8. $\sqrt{150} = \sqrt{25 \cdot 6} = 5\sqrt{6}$ **9.** $4\sqrt{216} = 4\sqrt{36 \cdot 6} = 4 \cdot 6\sqrt{6} = 24\sqrt{6}$

10. $\sqrt{121a^4b^{10}} = 11a^2|b^5|$ **11.** $3a^2 - 108 = 0, 3a^2 = 108, a^2 = 36, a = \pm 6; \{-6, 6\}$

12. $(8)^2 + b^2 = (19)^2, 64 + b^2 = 361, b^2 = 297, b = \sqrt{297} \approx 17.23$; The length of the other side is about 17.23 m.

13. $(14)^2 + (40)^2 = 196 + 1600 = 1796 \neq (50)^2$; Not a right triangle.

14. $\sqrt{\frac{4}{5} \cdot \frac{35}{4}} = \sqrt{7}$ **15.** $\frac{6\sqrt{18}}{18\sqrt{6}} = \frac{6\sqrt{9 \cdot 2}}{18\sqrt{6}} \cdot \frac{\sqrt{6}}{\sqrt{6}} = \frac{6 \cdot 3\sqrt{12}}{18 \cdot 6} = \frac{\sqrt{4 \cdot 3}}{6} = \frac{2\sqrt{3}}{6} = \frac{\sqrt{3}}{3}$

16. $4\sqrt{2} + \sqrt{72} = 4\sqrt{2} + \sqrt{36 \cdot 2} = 4\sqrt{2} + 6\sqrt{2} = 10\sqrt{2}$

17. $3\sqrt{24} - 4\sqrt{54} = 3\sqrt{4 \cdot 6} - 4\sqrt{9 \cdot 6} = 3 \cdot 2\sqrt{6} - 4 \cdot 3\sqrt{6} = (6 - 12)\sqrt{6} =$
$-6\sqrt{6}$

18. $(\sqrt{5} + 3)^2 = 5 + 6\sqrt{5} + 9 = 14 + 6\sqrt{5}$

19. $(\sqrt{7} - \sqrt{2})(\sqrt{7} + \sqrt{2}) = 7 - 2 = 5$

20. $\frac{3}{5 - 2\sqrt{3}} \cdot \frac{5 + 2\sqrt{3}}{5 + 2\sqrt{3}} = \frac{3(5 + 2\sqrt{3})}{25 - 4 \cdot 3} = \frac{15 + 6\sqrt{3}}{25 - 12} = \frac{15 + 6\sqrt{3}}{13}$

21. $\sqrt{3x + 2} = 4, 3x + 2 = 16, 3x = 14, x = \frac{14}{3}; \left\{\frac{14}{3}\right\}$

Page 557 • CUMULATIVE REVIEW

1. $-21 - 2 \cdot 6 \div 3 = -21 - 12 \div 3 = -21 - 4 = -25$

2. $(2ab^2)^2(3a^2 + 5ab + b^2) = 4a^2b^4(3a^2 + 5ab + b^2) = 12a^4b^4 + 20a^3b^5 + 4a^2b^6$

3. $(4a^3b + 9c^2)^2 = 16a^6b^2 + 72a^3bc^2 + 81c^4$

4. $(-3p + 2q)(12p - q) = -36p^2 + 3pq + 24pq - 2q^2 = -36p^2 + 27pq - 2q^2$

5. $(3.2 \times 10^{-4})^2 = 10.24 \times 10^{-8} = 1.024 \times 10 \times 10^{-8} = 1.024 \times 10^{-7}$

6.
$$\begin{array}{r l}
 & 3x + 4 \\
2x - 7 & \overline{)\,6x^2 - 13x - 28} \\
 & \underline{6x^2 - 21x} \\
 & \qquad 8x - 28 \\
 & \qquad \underline{8x - 28} \\
 & \qquad\qquad 0
\end{array}$$

7. $\dfrac{96 - 40x + 4x^2}{x^2 - 8x + 16} \cdot \dfrac{x^2 + 5x - 6}{2x^2 - 72} = \dfrac{4(24 - 10x + x^2)}{(x - 4)^2} \cdot \dfrac{(x - 1)(x + 6)}{2(x^2 - 36)} =$

$\dfrac{4(6 - x)(4 - x)}{(x - 4)^2} \cdot \dfrac{(x - 1)(x + 6)}{2(x + 6)(x - 6)} = \dfrac{2(x - 6)(x - 4)(x - 1)}{(x - 4)^2(x - 6)} = \dfrac{2(x - 1)}{x - 4}$ or $\dfrac{2x - 2}{x - 4}$

8. $\dfrac{3x^2 - 14x + 15}{x^2 + 10x + 25} \div \dfrac{3x^2 + 10x - 25}{x^2 - 25} = \dfrac{(3x - 5)(x - 3)}{(x + 5)^2} \cdot \dfrac{(x + 5)(x - 5)}{(3x - 5)(x + 5)} = \dfrac{(x - 3)(x - 5)}{(x + 5)^2}$

9. $\dfrac{t - 3}{t + 2} + \dfrac{1}{t} - 5 = \dfrac{t(t - 3)}{t(t + 2)} + \dfrac{t + 2}{t(t + 2)} - \dfrac{5t(t + 2)}{t(t + 2)} = \dfrac{t^2 - 3t + t + 2 - 5t^2 - 10t}{t(t + 2)} =$

$\dfrac{-4t^2 - 12t + 2}{t(t + 2)}$ or $-\dfrac{2(2t^2 + 6t - 1)}{t(t + 2)}$

10. $\sqrt{720} = \sqrt{144 \cdot 5} = 12\sqrt{5}$

11. $\sqrt{112} - \sqrt{63} = \sqrt{16 \cdot 7} - \sqrt{9 \cdot 7} = 4\sqrt{7} - 3\sqrt{7} = \sqrt{7}$

12. $2\sqrt{3}(5\sqrt{6} + 4\sqrt{2}) = 10\sqrt{18} + 8\sqrt{6} = 10\sqrt{9 \cdot 2} + 8\sqrt{6} = 10 \cdot 3\sqrt{2} + 8\sqrt{6} =$

$30\sqrt{2} + 8\sqrt{6}$

13. $\dfrac{5\sqrt{75}}{\sqrt{90}} = \dfrac{5\sqrt{25 \cdot 3}}{\sqrt{9 \cdot 10}} = \dfrac{5 \cdot 5\sqrt{3}}{3\sqrt{10}} \cdot \dfrac{\sqrt{10}}{\sqrt{10}} = \dfrac{25\sqrt{30}}{3 \cdot 10} = \dfrac{25\sqrt{30}}{30} = \dfrac{5\sqrt{30}}{6}$

14. $\dfrac{5}{\sqrt{6} + 1} \cdot \dfrac{\sqrt{6} - 1}{\sqrt{6} - 1} = \dfrac{5(\sqrt{6} - 1)}{6 - 1} = \dfrac{5(\sqrt{6} - 1)}{5} = \sqrt{6} - 1$

15. $\sqrt{\dfrac{2}{3}} + \sqrt{\dfrac{5}{6}} = \dfrac{\sqrt{2}}{\sqrt{3}} + \dfrac{\sqrt{5}}{\sqrt{6}} = \dfrac{\sqrt{2}\sqrt{3}}{\sqrt{3}\sqrt{3}} + \dfrac{\sqrt{5}\sqrt{6}}{\sqrt{6}\sqrt{6}} = \dfrac{\sqrt{6}}{3} + \dfrac{\sqrt{30}}{6} = \dfrac{2\sqrt{6} + \sqrt{30}}{6}$

16. $16y^4 - 8y^2 + 1 = (4y^2 - 1)^2 = [(2y + 1)(2y - 1)]^2 = (2y + 1)^2(2y - 1)^2$

17. $4c^2 - 4cd + d^2 - 9 = (2c - d)^2 - 3^2 = [(2c - d) + 3][(2c - d) - 3] =$
$(2c - d + 3)(2c - d - 3)$

18. $12c^2 - 17c - 5 = (4c + 1)(3c - 5)$

19. $|2 - x| \le 4$, $-4 \le 2 - x \le 4$, $-6 \le -x \le 2$, $6 \ge x \ge -2$; {6, −2, and the real numbers between −2 and 6}

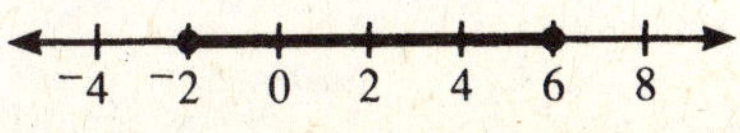

20. $2x - 3y = 3$
$-3y = -2x + 3$
$y = \frac{2}{3}x - 1$

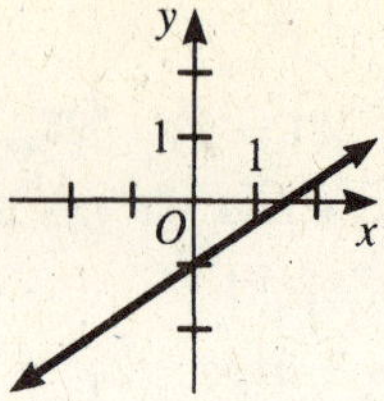

21. $2x + y < 6$
$y < -2x + 6$

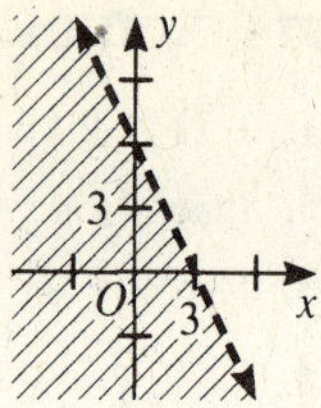

22. $m = \frac{8 - 2}{-1 - 0} = \frac{6}{-1} = -6; y = -6x + b, 2 = (-6)(0) + b, b = 2; y = -6x + 2;$
$6x + y = 2$

23. $-\frac{b}{2a} = -\frac{-4}{2} = 2; f(2) = 2^2 - 4(2) + 4 = 4 - 8 + 4 = 0$

24. $10n = 5.\overline{5}, 10n - n = 5.\overline{5} - 0.\overline{5}, 9n = 5, n = \frac{5}{9}$; therefore $0.\overline{5} = \frac{5}{9}$; $100n = 15.\overline{15}$,
$100n - n = 15.\overline{15} - 0.\overline{15}, 99n = 15, n = \frac{15}{99} = \frac{5}{33}$; therefore $0.\overline{15} = \frac{5}{33}$;
$0.\overline{5} + 0.\overline{15} = \frac{5}{9} + \frac{5}{33} = \frac{5 \cdot 11}{99} + \frac{5 \cdot 3}{99} = \frac{55 + 15}{99} = \frac{70}{99}$, or $0.\overline{70}$

25. $y < -x + 4$
$y \geq 2x$

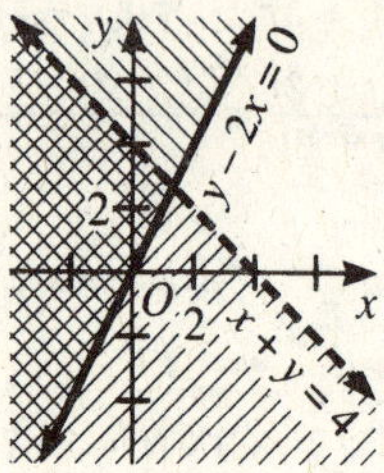

26. $2y - 3 = |-15|, 2y - 3 = 15, 2y = 18, y = 9; \{9\}$

27. $y\%$ of $75 = 12, \frac{y}{100} \cdot 75 = 12, y = \frac{12}{75} \cdot 100, y = 16; \{16\}$

28. $\frac{1}{3}y - 2 = \frac{1}{2}, \frac{1}{3}y = \frac{5}{2}, y = \frac{15}{2}; \left\{\frac{15}{2}\right\}$

29. $4d^2 - 12d + 9 = 0, (2d - 3)^2 = 0, 2d - 3 = 0, 2d = 3, d = \frac{3}{2}; \left\{\frac{3}{2}\right\}$

30. $2n^2 + 5n - 3 = 0, (2n - 1)(n + 3) = 0, 2n - 1 = 0, 2n = 1, n = \frac{1}{2}, n + 3 = 0,$
$n = -3; \left\{\frac{1}{2}, -3\right\}$

31. $\frac{1}{x} + \frac{3 - x}{x + 6} = \frac{x - 1}{2x + 12}, 2x(x + 6)\left[\frac{1}{x} + \frac{3 - x}{x + 6}\right] = 2x(x + 6)\left(\frac{x - 1}{2(x + 6)}\right),$
$2(x + 6) + 2x(3 - x) = x(x - 1), 2x + 12 + 6x - 2x^2 = x^2 - x,$
$0 = 3x^2 - 9x - 12, 0 = 3(x^2 - 3x - 4), 0 = 3(x - 4)(x + 1); x = 4$ or $x = -1;$
$\{-1, 4\}$

32. $x - 2y = 8$
$\underline{3x + 2y = 9}$
$4x = 17, x = \frac{17}{4}; \left(\frac{17}{4}\right) - 2y = 8, -2y = \frac{15}{4}, y = -\frac{15}{8}; \left(\frac{17}{4}, -\frac{15}{8}\right)$

33. $$\begin{aligned} 2x - 5y &= -1 \rightarrow 4x - 10y = -2 \\ 3x - 10y &= -1 \rightarrow \underline{3x - 10y = -1} \\ & \quad\quad\quad\quad x = -1;\ 2(-1) - 5y = -1,\ -5y = 1,\ y = -\frac{1}{5};\ \left(-1, -\frac{1}{5}\right) \end{aligned}$$

34. $$\begin{aligned} 2x + 11y &= 3 \rightarrow 10x + 55y = 15 \\ 5x + 28y &= 7 \rightarrow \underline{10x + 56y = 14} \\ & \quad\quad\quad\quad -y = 1,\ y = -1;\ 2x + 11(-1) = 3,\ 2x = 14,\ x = 7;\ (7, -1) \end{aligned}$$

35. $5 \le 2x - 1 \le 9$, $6 \le 2x \le 10$, $3 \le x \le 5$; {3, 5, and the real numbers between 3 and 5}

36. $4|y - 2| > 12$, $|y - 2| > 3$; $y - 2 < -3$ or $y - 2 > 3$, $y < -1$ or $y > 5$; {the real numbers less than -1 or greater than 5}

37. $7 - 3b < 9$, $-3b < 2$, $b > -\frac{2}{3}$; $\left\{\text{the real numbers greater than } -\frac{2}{3}\right\}$

38. $2\sqrt{t} = 3\sqrt{2}$, $\sqrt{t} = \frac{3\sqrt{2}}{2}$, $t = \frac{9 \cdot 2}{4} = \frac{9}{2}$; $\left\{\frac{9}{2}\right\}$

39. $\sqrt{2p - 1} = p$, $2p - 1 = p^2$, $0 = p^2 - 2p + 1$, $0 = (p - 1)^2$; $p - 1 = 0$, $p = 1$; {1}

40. $\sqrt{z^2 + 7} = z - 1$, $z^2 + 7 = (z - 1)^2$, $z^2 + 7 = z^2 - 2z + 1$, $2z = -6$, $z = -3$; $\sqrt{(-3)^2 + 7} = (-3) - 1$, $\sqrt{9 + 7} = -4$, $\sqrt{16} \neq -4$; There is no solution.

41. Let b = length of the third side. $(5)^2 + b^2 = (9)^2$, $b^2 = 81 - 25$, $b^2 = 56$, $b = \sqrt{56} \approx 7.48$; The third side is about 7.48 cm long.

42. Let a = admission price for adults and c = admission for children.

$$\begin{aligned} 10a + 4c &= 70 \rightarrow 10a + 4c = 70 \\ 5a + 16c &= 70 \rightarrow \underline{10a + 32c = 140} \\ & \quad\quad\quad\quad -28c = -70,\ c = \frac{70}{28} = \frac{5}{2};\ 10a + 4\left(\frac{5}{2}\right) = 70,\ 10a = 70 - 10, \end{aligned}$$

$10a = 60$, $a = 6$; The admission price is \$6.00 for adults and \$2.50 for children.

Page 558 • MAINTAINING SKILLS

1. $x + y = 1$, $x = 1 - y$; $2(1 - y) + y = 0$, $2 - 2y + y = 0$, $2 - y = 0$, $2 = y$; $x + (2) = 1$, $x = -1$; $(-1, 2)$

2. $x + y = 3$, $x = 3 - y$; $2(3 - y) - y = 6$, $6 - 2y - y = 6$, $-3y = 0$, $y = 0$; $x + (0) = 3$, $x = 3$; $(3, 0)$

3. $x + 2y = 15$, $x = 15 - 2y$; $2(15 - 2y) + y = 15$, $30 - 4y + y = 15$, $15 = 3y$, $5 = y$; $x + 2(5) = 15$, $x = 5$; $(5, 5)$

4. $$\begin{aligned} 3x + 4y &= -2 \rightarrow 9x + 12y = -6 \\ 4x + 3y &= -12 \rightarrow \underline{16x + 12y = -48} \\ & \quad\quad\quad\quad -7x = 42,\ x = -6;\ 3(-6) + 4y = -2,\ -18 + 4y = -2, \end{aligned}$$

$4y = 16$, $y = 4$; $(-6, 4)$

5. $$\begin{aligned} 2x + 3y &= 2 \rightarrow 8x + 12y = 8 \\ 8x + 9y &= 7 \rightarrow \underline{8x + 9y = 7} \\ & \quad\quad\quad\quad 3y = 1,\ y = \frac{1}{3};\ 2x + 3\left(\frac{1}{3}\right) = 2,\ 2x + 1 = 2,\ 2x = 1,\ x = \frac{1}{2}; \end{aligned}$$

$\left(\frac{1}{2}, \frac{1}{3}\right)$

6. $y = 4x; 9x + 4(4x) = 5, 9x + 16x = 5, 25x = 5, x = \frac{1}{5}; y = 4\left(\frac{1}{5}\right) = \frac{4}{5}; \left(\frac{1}{5}, \frac{4}{5}\right)$

7. $y = 4x + 5; 2x + (4x + 5) = 1, 6x + 5 = 1, 6x = -4, x = -\frac{2}{3}; y = 4\left(-\frac{2}{3}\right) + 5 = -\frac{8}{3} + \frac{15}{3} = \frac{7}{3}; \left(-\frac{2}{3}, \frac{7}{3}\right)$

8. $y - x = 20, y = 20 + x; 2x + 3(20 + x) = 10, 2x + 60 + 3x = 10, 5x = -50, x = -10; y - (-10) = 20, y + 10 = 20, y = 10; (-10, 10)$

9. $x + 2y = 13, x = 13 - 2y; 3(13 - 2y) - 4y = 9, 39 - 6y - 4y = 9, 39 - 10y = 9, 30 = 10y, y = 3; x + 2(3) = 13, x + 6 = 13, x = 7; (7, 3)$

10.
$$\begin{array}{rcl} 5x - 6y = 2 & \rightarrow & 5x - 6y = 2 \\ 4y - 5x = 12 & \rightarrow & \underline{-5x + 4y = 12} \\ & & -2y = 14 \end{array}$$
$y = -7; 5x - 6(-7) = 2, 5x + 42 = 2, 5x = -40, x = -8; (-8, -7)$

11.
$$\begin{array}{rcl} x + 10y = 2 & \rightarrow & x + 10y = 2 \\ \frac{1}{3}x + y = 3 & \rightarrow & \underline{x + 3y = 9} \\ & & 7y = -7 \end{array}$$
$y = -1; x + 10(-1) = 2, x - 10 = 2, x = 12; (12, -1)$

12.
$$\begin{array}{rcccl} \frac{4}{5}x - \frac{1}{2}y = 18 & \rightarrow & 8x - 5y = 180 & \rightarrow & 8x - 5y = 180 \\ \frac{1}{2}x - \frac{3}{4}y = 20 & \rightarrow & 2x - 3y = 80 & \rightarrow & \underline{8x - 12y = 320} \\ & & & & 7y = -140 \end{array}$$
$y = -20; \frac{1}{2}x - \frac{3}{4}(-20) = 20, \frac{1}{2}x + 15 = 20, \frac{1}{2}x = 5, x = 10; (10, -20)$

13. $y = 7x; 5x - 8(7x) = 0, 5x - 56x = 0, -51x = 0, x = 0; y = 7(0) = 0; (0, 0)$

14. $x - 2y = 5, x = 2y + 5; -2(2y + 5) + 3y = -5, -4y - 10 + 3y = -5, -10 - y = -5, -5 = y; x - 2(-5) = 5, x + 10 = 5, x = -5; (-5, -5)$

15.
$$\begin{array}{rcl} 2x + 3y = 11 & \rightarrow & 6x + 9y = 33 \\ 3x + 14y = 7 & \rightarrow & \underline{6x + 28y = 14} \\ & & -19y = 19 \end{array}$$
$y = -1; 2x + 3(-1) = 11, 2x = 14, x = 7; (7, -1)$

16.
$$\begin{array}{r} 5x + 6y = 4 \\ \underline{3x + 6y = 3} \\ 2x \quad\quad = 1 \end{array}$$
$x = \frac{1}{2}; 5\left(\frac{1}{2}\right) + 6y = 4, 6y = 4 - \frac{5}{2}, 6y = \frac{3}{2}, y = \frac{1}{4}; \left(\frac{1}{2}, \frac{1}{4}\right)$

17.
$$\begin{array}{rcl} 4x + 3y = 8 & \rightarrow & 16x + 12y = 32 \\ 5x + 4y = 13 & \rightarrow & \underline{15x + 12y = 39} \\ & & x \quad\quad = -7 \end{array}$$
$4(-7) + 3y = 8, -28 + 3y = 8, 3y = 36, y = 12; (-7, 12)$

18.
$$\begin{array}{rcl} 5x - 6y = 1 & \rightarrow & 5x - 6y = 1 \\ 2x - 3y = 7 & \rightarrow & \underline{4x - 6y = 14} \\ & & x \quad\quad = -13 \end{array}$$
$2(-13) - 3y = 7, -26 - 3y = 7, -3y = 33, y = -11; (-13, -11)$

19. $6x - 5y = 0 \rightarrow 18x - 15y = 0$
$5x - 3y = 7 \rightarrow \underline{25x - 15y = 35}$
$-7x = -35, x = 5; 6(5) - 5y = 0, 30 = 5y, 6 = y; (5, 6)$

20. $x + 8y = 16, x = 16 - 8y; 2(16 - 8y) - 7y = 9, 32 - 16y - 7y = 9, 32 - 23y = 9,$
$23 = 23y, y = 1; x + 8(1) = 16, x = 8; (8, 1)$

21. $3x - 5y = 7 \rightarrow 6x - 10y = 14$
$-2x + 7y = 10 \rightarrow \underline{-6x + 21y = 30}$
$11y = 44, y = 4; 3x - 5(4) = 7, 3x - 20 = 7, 3x = 27,$
$x = 9; (9, 4)$

Page 559 • MIXED PROBLEM SOLVING

A **1.** Let b = rate of boat, c = rate of current. $\frac{1}{3}(b + c) = 4, \frac{2}{5}(b - c) = 4;$

$b + c = 12$
$\underline{b - c = 10}$
$2b = 22, b = 11; 11 + c = 12, c = 1;$ The rate of the boat is 11 km/h in still water and the rate of the current is 1 km/h.

2. Let x = the number. $\frac{1}{2}\sqrt{2x} = 5, \sqrt{2x} = 10, 2x = 100, x = 50;$ the number is 50

3. Let h = height of the triangle. $h \cdot 15 = (18)(10), 15h = 180, h = 12;$ The height of the triangle is 12 cm.

4.

	Rate ×	Time	= Distance
First train	80	$\frac{d}{80}$	d
Second train	100	$\frac{225 - d}{100}$	$225 - d$

$\frac{d}{80} = \frac{225 - d}{100}, 5d = 4(225 - d),$
$5d = 900 - 4d, 9d = 900, d = 100;$
$80 \cdot t = 100, t = 1.25;$ The trains were 225 km apart at 1:15 P.M.

5. Let s = speed of Saturn and m = speed of Mars. $s = m - 14; s = \frac{5}{12}m;$

$\frac{5}{12}m = m - 14, 14 = \frac{7}{12}m, 24 = m; s = \frac{5}{12}(24) = 10;$ Saturn orbits at 10 km/s and Mars orbits at 24 km/s.

6. Let $\frac{n}{d}$ = the fraction. $\frac{n}{d} = \frac{3}{5}, 5n = 3d, 5n - 3d = 0; \frac{n + 3}{d + 3} = \frac{5}{8}, 8(n + 3) = 5(d + 3),$

$8n + 24 = 5d + 15, 8n - 5d = -9; 5n - 3d = 0 \rightarrow 25n - 15d = 0$
$8n - 5d = -9 \rightarrow \underline{24n - 15d = -27}$
$n = 27;$

$5(27) - 3d = 0, 135 = 3d, d = 45;$ The original fraction was $\frac{27}{45}$.

7. Let x = the usual price. $(x - 0.25x) + (x - 0.25x)0.04 = 1170,$
$0.75x + (0.75x)0.04 = 1170, 0.75x + 0.03x = 1170, 0.78x = 1170, x = 1500.$ The usual price for the computer is \$1500.

8.

	Amount ×	% acid =	Total acid
First Solution	24	18%	4.32
Water	x	0%	0
Final Solution	$24 + x$	15%	4.32

$(24 + x)0.15 = 4.32$, $3.6 + 0.15x = 4.32$, $0.15x = 0.72$, $x = 4.8$; Add 4.8 L of water.

B

9. Let x = Kris' time to complete the route alone. $\frac{18}{45} + \frac{18}{x} = 1$, $18x + 810 = 45x$, $810 = 27x$, $x = 30$; Kris takes 30 min to complete the route.

10. Let a = amount in the account and $8000 - a$ = amount in bonds. $0.06a + 0.08(8000 - a) = 0.06(8000 - a) + 0.08a + 100$, $6a + 8(8000 - a) = 6(8000 - a) + 8a + 10{,}000$, $6a + 64{,}000 - 8a = 48{,}000 - 6a + 8a + 10{,}000$, $-2a + 64{,}000 = 58{,}000 + 2a$, $6{,}000 = 4a$, $1500 = a$; $8000 - a = 6500$; Isabel invested \$1500 in the account and \$6500 in bonds.

11. Let $23x$ and $25x$ be the numbers of nickels and quarters. $5(23x) + 25(25x) = 700$; $115x + 625x = 700$; $740x = 700$; $x \approx 0.95$; There is no solution since x must be a positive integer.

12. Let l and w be the length and width of the rectangle. $\sqrt{l^2 + w^2} = l + 1$; $l = 2w + 2$; $\sqrt{(2w + 2)^2 + w^2} = (2w + 2) + 1$, $\sqrt{4w^2 + 8w + 4 + w^2} = 2w + 3$, $\sqrt{5w^2 + 8w + 4} = 2w + 3$, $5w^2 + 8w + 4 = (2w + 3)^2$, $5w^2 + 8w + 4 = 4w^2 + 12w + 9$, $w^2 - 4w - 5 = 0$, $(w - 5)(w + 1) = 0$; $w = 5$ or $w = -1$; $l = 2(5) + 2 = 12$; Perimeter $= 2(l + w) = 2(12 + 5) = 2(17) = 34$; The perimeter of the rectangle is 34 cm.

13.

	Amount ×	Price =	Total Cost
Almonds	8	x	$8x$
Raisins	4	$x + 2$	$4(x + 2)$
Mixture	12	$\frac{8x + 4(x + 2)}{12}$	$8x + 4(x + 2)$

$\frac{8x + 4(x + 2)}{12} = \frac{8x + 4x + 8}{12} = \frac{12x + 8}{12} = \frac{4(3x + 2)}{12} = \frac{3x + 2}{3}$; The mixture costs $\frac{3x + 2}{3}$ dollars per kg.

Page 563 • WRITTEN EXERCISES

A **1.** $x^2 = 64; x = \pm\sqrt{64}, x = \pm 8; \{-8, 8\}$ **2.** $(y-7)^2 = 0; y - 7 = 0; y = 7; \{7\}$

3. $t^2 = \frac{100}{169}; t = \pm\frac{10}{13}; \left\{-\frac{10}{13}, \frac{10}{13}\right\}$ **4.** $a^2 = -9$; no solution

5. $3x^2 = 108, x^2 = 36; x = \pm 6; \{-6, 6\}$

6. $6t^2 = 156, t^2 = 26, t = \pm\sqrt{26}; \{-\sqrt{26}, \sqrt{26}\}$

7. $14t^2 = 126, t^2 = 9; t = \pm 3; \{-3, 3\}$

8. $x^2 - 48 = 0, x^2 = 48; x = \pm\sqrt{48} = \pm\sqrt{16 \cdot 3} = \pm 4\sqrt{3}; \{-4\sqrt{3}, 4\sqrt{3}\}$

9. $x^2 + 32 = 0, x^2 = -32$; no solution

10. $m^2 - 54 = 0, m^2 = 54; m = \pm\sqrt{54} = \pm\sqrt{9 \cdot 6} = \pm 3\sqrt{6}; \{-3\sqrt{6}, 3\sqrt{6}\}$

11. $6x^2 - 18 = 0, 6x^2 = 18, x^2 = 3; x = \pm\sqrt{3}; \{-\sqrt{3}, \sqrt{3}\}$

12. $7m^2 - 42 = 0, 7m^2 = 42, m^2 = 6; m = \pm\sqrt{6}; \{-\sqrt{6}, \sqrt{6}\}$

13. $4y^2 + 7 = 19, 4y^2 = 12, y^2 = 3; y = \pm\sqrt{3}; \{-\sqrt{3}, \sqrt{3}\}$

14. $4r^2 - 7 = 29, 4r^2 = 36, r^2 = 9; r = \pm 3; \{-3, 3\}$

15. $3z^2 - 18 = 3, 3z^2 = 21, z^2 = 7; z = \pm\sqrt{7}; \{-\sqrt{7}, \sqrt{7}\}$

16. $(x+6)^2 = 16, x + 6 = \pm 4, x = -6 \pm 4, x = -10$ or -2; $\{-10, -2\}$

17. $(y-6)^2 = 13; y - 6 = \pm\sqrt{13}, y = 6 \pm \sqrt{13}; \{6 + \sqrt{13}, 6 - \sqrt{13}\}$

18. $(y+2)^2 = 12; y + 2 = \pm\sqrt{12}, y = -2 \pm \sqrt{4 \cdot 3}; \{-2 + 2\sqrt{3}, -2 - 2\sqrt{3}\}$

19. $(s-7)^2 = 28; s - 7 = \pm\sqrt{28}, s = 7 \pm 2\sqrt{7}; \{7 + 2\sqrt{7}, 7 - 2\sqrt{7}\}$

20. $(z-3)^2 = 32; z - 3 = \pm\sqrt{32}, z = 3 \pm 4\sqrt{2}; \{3 + 4\sqrt{2}, 3 - 4\sqrt{2}\}$

21. $2(x-5)^2 = 18, (x-5)^2 = 9; x - 5 = \pm 3, x = 5 \pm 3, x = 2$ or 8; $\{2, 8\}$

22. $5(m-8)^2 = 25, (m-8)^2 = 5; m - 8 = \pm\sqrt{5}; m = 8 \pm \sqrt{5}; \{8 + \sqrt{5}, 8 - \sqrt{5}\}$

23. $8(x+3)^2 = 56, (x+3)^2 = 7; x + 3 = \pm\sqrt{7}, x = -3 \pm \sqrt{7}; \{-3 + \sqrt{7}, -3 - \sqrt{7}\}$

24. $6(z+5)^2 = 42, (z+5)^2 = 7; z + 5 = \pm\sqrt{7}, z = -5 \pm \sqrt{7}; \{-5 + \sqrt{7}, -5 - \sqrt{7}\}$

B **25.** $x^2 + 6x + 9 = 16, (x+3)^2 = 16; x + 3 = \pm 4, x = -3 \pm 4, x = -7$ or 1; $\{-7, 1\}$

26. $x^2 - 14x + 49 = 64, (x-7)^2 = 64; x - 7 = \pm 8, x = 7 \pm 8, x = -1$ or 15; $\{-1, 15\}$

27. $r^2 - 22r + 121 = 4, (r-11)^2 = 4; r - 11 = \pm 2, r = 11 \pm 2, r = 9$ or 13; $\{9, 13\}$

28. $y^2 - 18y + 81 = 144, (y-9)^2 = 144; y - 9 = \pm 12, y = 9 \pm 12 = 21$ or -3; $\{-3, 21\}$

29. $\frac{1}{5}x^2 - \frac{5}{49} = 0, \frac{1}{5}x^2 = \frac{5}{49}, x^2 = \frac{25}{49}; x = \pm\sqrt{\frac{25}{49}} = \pm\frac{5}{7}; \left\{-\frac{5}{7}, \frac{5}{7}\right\}$

30. $\frac{1}{4}t^2 - \frac{9}{64} = 0, \frac{1}{4}t^2 = \frac{9}{64}, t^2 = \frac{9}{16}, t = \pm\sqrt{\frac{9}{16}} = \pm\frac{3}{4}; \left\{-\frac{3}{4}, \frac{3}{4}\right\}$

31. $\frac{1}{5}r^2 - 2 = \frac{5}{6}, \frac{1}{5}r^2 = \frac{12+5}{6}, \frac{1}{5}r^2 = \frac{17}{6}, r^2 = \frac{85}{6}; r = \pm\sqrt{\frac{85}{6}} = \pm\frac{\sqrt{85}\cdot\sqrt{6}}{\sqrt{6}\cdot\sqrt{6}} = \pm\frac{\sqrt{510}}{6};$
$\left\{-\frac{\sqrt{510}}{6}, \frac{\sqrt{510}}{6}\right\}$

32. $\frac{1}{6}x^2 - 4 = \frac{5}{6}, \frac{1}{6}x^2 = \frac{24+5}{6}, x^2 = 29; x = \pm\sqrt{29}; \{-\sqrt{29}, \sqrt{29}\}$

33. $t^2 + 18t + 81 = 225, (t+9)^2 = 225; t + 9 = \pm 15, t = -9 \pm 15 = -24$ or $6; \{-24, 6\}$

34. $0.49x^2 + 2 = 3.96, 49x^2 + 200 = 396, 49x^2 = 196, x^2 = 4; x = \pm 2; \{-2, 2\}$

35. $1.44z^2 - 1.36 = -0.64, 144z^2 - 136 = -64, 144z^2 = 72, z^2 = \frac{1}{2}; z = \pm\sqrt{\frac{1}{2}} =$
$\pm\frac{1\cdot\sqrt{2}}{\sqrt{2}\sqrt{2}} = \pm\frac{\sqrt{2}}{2}; \left\{-\frac{\sqrt{2}}{2}, \frac{\sqrt{2}}{2}\right\}$

36. $5(t+2)^2 = \frac{3}{5}, (t+2)^2 = \frac{3}{25}; t + 2 = \pm\sqrt{\frac{3}{25}}, t = -2 \pm \frac{\sqrt{3}}{5} = \frac{-10 \pm \sqrt{3}}{5};$
$\left\{\frac{-10+\sqrt{3}}{5}, \frac{-10-\sqrt{3}}{5}\right\}$

37. $4(x-2)^2 = \frac{1}{49}, (x-2)^2 = \frac{1}{196}; x - 2 = \pm\sqrt{\frac{1}{196}}, x = 2 \pm \frac{1}{14} = \frac{28 \pm 1}{14} = \frac{27}{14}$
or $\frac{29}{14}; \left\{\frac{27}{14}, \frac{29}{14}\right\}$

38. $\left(y - \frac{3}{7}\right)^2 = -\frac{8}{9};$ no solution

39. $\left(z - \frac{3}{5}\right)^2 = \frac{7}{16}; z - \frac{3}{5} = \pm\sqrt{\frac{7}{16}}, z = \frac{3}{5} \pm \frac{\sqrt{7}}{4} = \frac{12 \pm 5\sqrt{7}}{20}; \left\{\frac{12+5\sqrt{7}}{20}, \frac{12-5\sqrt{7}}{20}\right\}$

40. $7y^3 - 28y = 0, 7y(y^2 - 4) = 0, 7y(y+2)(y-2) = 0; y = 0$ or $y = -2$ or $y = 2;$
$\{-2, 0, 2\}$

41. $7a^3 - 175a = 0, 7a(a^2 - 25) = 0, 7a(a+5)(a-5) = 0; a = 0$ or $a = -5$ or $a = 5;$
$\{0, -5, 5\}$

42. $\frac{1}{4}t^3 - 16t = 0, t^3 - 64t = 0, t(t^2 - 64) = 0, t(t+8)(t-8) = 0; t = 0$ or $t = -8$ or
$t = 8; \{-8, 0, 8\}$

43. $4b^3 - \frac{1}{4}b = 0, 16b^3 - b = 0, b(16b^2 - 1) = 0, b(4b+1)(4b-1) = 0; b = 0$ or
$4b + 1 = 0$ or $4b - 1 = 0, b = 0$ or $4b = -1$ or $4b = 1, b = 0$ or $b = -\frac{1}{4}$ or $b = \frac{1}{4};$
$\left\{-\frac{1}{4}, 0, \frac{1}{4}\right\}$

44. $8x^3 = 392x, 8x^3 - 392x = 0, 8x(x^2 - 49) = 0, 8x(x+7)(x-7) = 0; x = 0$ or
$x = -7$ or $x = 7; \{-7, 0, 7\}$

45. $8x^3 = 512x, 8x^3 - 512x = 0, 8x(x^2 - 64) = 0, 8x(x+8)(x-8) = 0; x = 0$ or
$x = -8$ or $x = 8; \{-8, 0, 8\}$

C **46.** $3(5x-2)^2 = 27, (5x-2)^2 = 9; 5x - 2 = \pm 3, 5x = 2 \pm 3, x = \frac{2 \pm 3}{5} = \frac{5}{5}$ or $\frac{-1}{5}; \left\{1, -\frac{1}{5}\right\}$

47. $5(6x-1)^2 = 5, (6x-1)^2 = 1; 6x - 1 = \pm 1, 6x = 1 \pm 1, x = \frac{1 \pm 1}{6} = 0$ or $\frac{1}{3}; \left\{0, \frac{1}{3}\right\}$

48. $2(7x-2)^2+5=11$, $2(7x-2)^2=6$, $(7x-2)^2=3$; $7x-2=\pm\sqrt{3}$, $7x=2\pm\sqrt{3}$,

$x=\frac{2\pm\sqrt{3}}{7}$; $\left\{\frac{2+\sqrt{3}}{7}, \frac{2-\sqrt{3}}{7}\right\}$

49. **a.** two **b.** two **c.** none **d.** one

Page 563 • MIXED REVIEW EXERCISES

1. $(x-11)^2=x^2-22x+121$ **2.** $(2x+5)^2=4x^2+20x+25$

3. $(6x-7)^2=36x^2-84x+49$ **4.** $(-3c+4)^2=9c^2-24c+16$

5. $\left(x+\frac{1}{2}\right)^2=x^2+x+\frac{1}{4}$ **6.** $\left(x+\frac{1}{3}\right)^2=x^2+\frac{2}{3}x+\frac{1}{9}$

7. $\left(\frac{1}{2}x+\frac{2}{3}\right)^2=\frac{1}{4}x^2+\frac{2}{3}x+\frac{4}{9}$ **8.** $\left(\frac{1}{3}x+\frac{3}{4}\right)^2=\frac{1}{9}x^2+\frac{1}{2}x+\frac{9}{16}$

Page 566 • WRITTEN EXERCISES

A **1.** $x^2-4x=17$, $x^2-4x+(-2)^2=17+(-2)^2$, $x^2-4x+4=17+4$,

$(x-2)^2=21$; $x-2=\pm\sqrt{21}$, $x=2\pm\sqrt{21}\approx 6.6$ or -2.6; $\{-2.6, 6.6\}$

2. $y^2+8y=10$, $y^2+8y+(4)^2=10+(4)^2$, $y^2+8y+16=10+16$, $(y+4)^2=26$;

$y+4=\pm\sqrt{26}$, $y=-4\pm\sqrt{26}\approx 1.1$ or -9.1; $\{-9.1, 1.1\}$

3. $w^2+6w=-3$, $w^2+6w+(3)^2=-3+(3)^2$, $w^2+6w+9=-3+9$, $(w+3)^2=6$;

$w+3=\pm\sqrt{6}$, $w=-3\pm\sqrt{6}\approx -0.6$ or -5.4; $\{-0.6, -5.4\}$

4. $c^2+18c-175=0$, $c^2+18c=175$, $c^2+18c+(9)^2=175+(9)^2$,

$c^2+18c+81=175+81$, $(c+9)^2=256$; $c+9=\pm\sqrt{256}$, $c+9=\pm 16$,
$c=-9\pm 16$; $\{-25, 7\}$

5. $v^2-20v+19=0$, $v^2-20v=-19$, $v^2-20v+(-10)^2=-19+(-10)^2$,

$v^2-20v+100=-19+100$, $(v-10)^2=81$; $v-10=\pm\sqrt{81}$, $v-10=\pm 9$,
$v=10\pm 9$; $\{1, 19\}$

6. $z^2-6z-307=8$, $z^2-6z=315$, $z^2-6z+(-3)^2=315+(-3)^2$,

$z^2-6z+9=315+9$, $(z-3)^2=324$; $z-3=\pm\sqrt{324}$, $z=3\pm 18$; $\{-15, 21\}$

7. $y^2+18y+32=226$, $y^2+18y=194$, $y^2+18y+(9)^2=194+(9)^2$,

$y^2+18y+81=194+81$, $(y+9)^2=275$; $y+9=\pm\sqrt{275}$,

$y=-9\pm 5\sqrt{11}\approx 7.6$ or -25.6; $\{7.6, -25.6\}$

8. $2b^2+16b=4$, $b^2+8b=2$, $b^2+8b+(4)^2=2+4^2$, $b^2+8b+16=2+16$,

$(b+4)^2=18$; $b+4=\pm\sqrt{18}$, $b=-4\pm 3\sqrt{2}\approx 0.2$ or -8.2; $\{-8.2, 0.2\}$

9. $5x^2-20x=10$, $x^2-4x=2$, $x^2-4x+(-2)^2=2+(-2)^2$, $x^2-4x+4=2+4$,

$(x-2)^2=6$; $x-2=\pm\sqrt{6}$, $x=2\pm\sqrt{6}\approx 4.4$ or -0.4; $\{-0.4, 4.4\}$

10. $w^2+w=3$, $w^2+w+\left(\frac{1}{2}\right)^2=3+\left(\frac{1}{2}\right)^2$, $w^2+w+\frac{1}{4}=3+\frac{1}{4}$, $\left(w+\frac{1}{2}\right)^2=3\frac{1}{4}$;

$w+\frac{1}{2}=\pm\sqrt{\frac{13}{4}}$, $w=-\frac{1}{2}\pm\frac{\sqrt{13}}{2}=\frac{-1\pm\sqrt{13}}{2}\approx 1.3$ or -2.3; $\{1.3, -2.3\}$

11. $x^2 - 5x = 2, x^2 - 5x + \left(-\frac{5}{2}\right)^2 = 2 + \left(-\frac{5}{2}\right)^2, x^2 - 5x + \frac{25}{4} = 2 + \frac{25}{4},$

$\left(x - \frac{5}{2}\right)^2 = \frac{8}{4} + \frac{25}{4}, x - \frac{5}{2} = \pm\sqrt{\frac{33}{4}}, x = \frac{5}{2} \pm \frac{\sqrt{33}}{2} = \frac{5 \pm \sqrt{33}}{2} \approx 5.4$ or -0.4;

$\{-0.4, 5.4\}$

12. $c^2 - 7c - 2 = 4, c^2 - 7c + \left(-\frac{7}{2}\right)^2 = 6 + \left(-\frac{7}{2}\right)^2, c^2 - 7c + \frac{49}{4} = 6 + \frac{49}{4},$

$\left(c - \frac{7}{2}\right)^2 = \frac{24}{4} + \frac{49}{4}; c - \frac{7}{2} = \pm\sqrt{\frac{73}{4}}, c = \frac{7}{2} \pm \frac{\sqrt{73}}{2} = \frac{7 \pm \sqrt{73}}{2} \approx 7.8$

or -0.8; $\{-0.8, 7.8\}$

13. a. $x^2 - 12x = -35, x^2 - 12x + (-6)^2 = -35 + (-6)^2, x^2 - 12x + 36 = 1,$
$(x - 6)^2 = 1; x - 6 = \pm 1, x = 6 \pm 1; \{5, 7\}$

b. $x^2 - 12x + 35 = 0, (x - 7)(x - 5) = 0; x = 7$ or $x = 5$; $\{5, 7\}$

14. a. $y^2 + 15y = -56, y^2 + 15y + \left(\frac{15}{2}\right)^2 = -56 + \left(\frac{15}{2}\right)^2, y^2 + 15y + \frac{225}{4} =$

$-\frac{224}{4} + \frac{225}{4}, \left(y + \frac{15}{2}\right)^2 = \frac{1}{4}; y + \frac{15}{2} = \pm\sqrt{\frac{1}{4}}, y = -\frac{15}{2} \pm \frac{1}{2} = \frac{-15 \pm 1}{2}; \{-8, -7\}$

b. $y^2 + 15y + 56 = 0, (y + 7)(y + 8) = 0; y = -7$ or $y = -8$; $\{-7, -8\}$

15. a. $z^2 - 2z = 35, z^2 - 2z + (-1)^2 = 35 + (-1)^2, z^2 - 2z + 1 = 35 + 1,$

$(z - 1)^2 = 36; z - 1 = \pm\sqrt{36}, z = 1 \pm 6; \{-5, 7\}$

b. $z^2 - 2z - 35 = 0, (z - 7)(z + 5) = 0; z = 7$ or $z = -5$; $\{-5, 7\}$

16. a. $3c^2 - 7c = 6, c^2 - \frac{7}{3}c = 2, c^2 - \frac{7}{3}c + \left(-\frac{7}{6}\right)^2 = 2 + \left(-\frac{7}{6}\right)^2, c^2 - \frac{7}{3}c + \frac{49}{36} =$

$2 + \frac{49}{36}, \left(c - \frac{7}{6}\right)^2 = \frac{72}{36} + \frac{49}{36}; c - \frac{7}{6} = \pm\sqrt{\frac{121}{36}}, c = \frac{7}{6} \pm \frac{11}{6} = \frac{7 \pm 11}{6}; \left\{-\frac{2}{3}, 3\right\}$

b. $3c^2 - 7c - 6 = 0, (3c + 2)(c - 3) = 0; 3c = -2$ or $c = 3, c = -\frac{2}{3}$ or $c = 3$;

$\left\{-\frac{2}{3}, 3\right\}$

17. a. $2y^2 - 9y = -9, y^2 - \frac{9}{2}y = -\frac{9}{2}, y^2 - \frac{9}{2}y + \left(-\frac{9}{4}\right)^2 = -\frac{9}{2} + \left(-\frac{9}{4}\right)^2, y^2 - \frac{9}{2}y +$

$\frac{81}{16} = -\frac{9}{2} + \frac{81}{16}, \left(y - \frac{9}{4}\right)^2 = \frac{-72 + 81}{16}, y - \frac{9}{4} = \pm\sqrt{\frac{9}{16}}, y = \frac{9}{4} \pm \frac{3}{4} = \frac{9 \pm 3}{4}; \left\{\frac{3}{2}, 3\right\}$

b. $2y^2 - 9y + 9 = 0, (2y - 3)(y - 3) = 0; 2y = 3$ or $y = 3, y = \frac{3}{2}$ or $y = 3$; $\left\{\frac{3}{2}, 3\right\}$

18. a. $4n^2 + 12n = -5, n^2 + 3n = -\frac{5}{4}, n^2 + 3n + \left(\frac{3}{2}\right)^2 = -\frac{5}{4} + \left(\frac{3}{2}\right)^2,$

$n^2 + 3n + \frac{9}{4} = -\frac{5}{4} + \frac{9}{4}, \left(n + \frac{3}{2}\right)^2 = \frac{4}{4}; n + \frac{3}{2} = \pm\sqrt{1}, n = -\frac{3}{2} \pm 1; \left\{-\frac{5}{2}, -\frac{1}{2}\right\}$

b. $4n^2 + 12n + 5 = 0, (2n + 5)(2n + 1) = 0; 2n = -5$ or $2n = -1, n = -\frac{5}{2}$ or

$n = -\frac{1}{2}; \left\{-\frac{5}{2}, -\frac{1}{2}\right\}$

B **19.** $\frac{x^2}{4} - x = 3, x^2 - 4x = 12, x^2 - 4x + (-2)^2 = 12 + (-2)^2, x^2 - 4x + 4 = 12 + 4,$
$(x - 2)^2 = 16; x - 2 = \pm\sqrt{16}, x = 2 \pm 4; \{-2, 6\}$

20. $\frac{2y^2}{3} - y - 3 = 0, y^2 - \frac{3}{2}y = \frac{9}{2}, y^2 - \frac{3}{2}y + \left(-\frac{3}{4}\right)^2 = \frac{9}{2} + \left(-\frac{3}{4}\right)^2, y^2 - \frac{3}{2}y + \frac{9}{16} =$
$\frac{9}{2} + \frac{9}{16}, \left(y - \frac{3}{4}\right)^2 = \frac{72 + 9}{16}; y - \frac{3}{4} = \pm\sqrt{\frac{81}{16}}, y = \frac{3}{4} \pm \frac{9}{4} = \frac{3 \pm 9}{4}; \left\{-\frac{3}{2}, 3\right\}$

21. $m^2 - 3 = \frac{11m}{2}, m^2 - \frac{11m}{2} = 3, m^2 - \frac{11}{2}m + \left(-\frac{11}{4}\right)^2 = 3 + \left(-\frac{11}{4}\right)^2,$
$m^2 - \frac{11}{2}m + \frac{121}{16} = 3 + \frac{121}{16}, \left(m - \frac{11}{4}\right)^2 = \frac{48 + 121}{16}; m - \frac{11}{4} = \pm\sqrt{\frac{169}{16}},$
$m = \frac{11}{4} \pm \frac{13}{4} = \frac{11 \pm 13}{4}; \left\{-\frac{1}{2}, 6\right\}$

22. $a^2 + \frac{a}{3} = 3, a^2 + \frac{a}{3} + \left(\frac{1}{6}\right)^2 = 3 + \left(\frac{1}{6}\right)^2, \left(a + \frac{1}{6}\right)^2 = \frac{109}{36}; a + \frac{1}{6} = \pm\sqrt{\frac{109}{36}},$
$a = -\frac{1}{6} \pm \frac{\sqrt{109}}{6} = \frac{-1 \pm \sqrt{109}}{6}; \left\{\frac{-1 + \sqrt{109}}{6}, \frac{-1 - \sqrt{109}}{6}\right\}$

23. $\frac{3m^2}{4} - 3 = \frac{m}{2}, m^2 - 4 = \frac{2}{3}m, m^2 - \frac{2}{3}m = 4, m^2 - \frac{2}{3}m + \left(-\frac{1}{3}\right)^2 = 4 + \left(-\frac{1}{3}\right)^2,$
$m^2 - \frac{2}{3}m + \frac{1}{9} = 4 + \frac{1}{9}, \left(m - \frac{1}{3}\right)^2 = \frac{37}{9}; m - \frac{1}{3} = \pm\sqrt{\frac{37}{9}}, m = \frac{1}{3} \pm \frac{\sqrt{37}}{3} = \frac{1 \pm \sqrt{37}}{3};$
$\left\{\frac{1 - \sqrt{37}}{3}, \frac{1 + \sqrt{37}}{3}\right\}$

24. $\frac{y^2}{2} - \frac{y}{4} = 3, y^2 - \frac{y}{2} = 6, y^2 - \frac{1}{2}y + \left(-\frac{1}{4}\right)^2 = 6 + \left(-\frac{1}{4}\right)^2, y^2 - \frac{1}{2}y + \frac{1}{16} = 6 + \frac{1}{16},$
$\left(y - \frac{1}{4}\right)^2 = \frac{97}{16}; y - \frac{1}{4} = \pm\sqrt{\frac{97}{16}}, y = \frac{1}{4} \pm \frac{\sqrt{97}}{4} = \frac{1 \pm \sqrt{97}}{4}; \left\{\frac{1 + \sqrt{97}}{4}, \frac{1 - \sqrt{97}}{4}\right\}$

25. $x + 1 = \frac{2 + x}{4x}, 4x(x + 1) = 2 + x, 4x^2 + 4x = 2 + x, 4x^2 + 3x = 2, x^2 + \frac{3}{4}x = \frac{1}{2},$
$x^2 + \frac{3}{4}x + \left(\frac{3}{8}\right)^2 = \frac{1}{2} + \left(\frac{3}{8}\right)^2, x^2 + \frac{3}{4}x + \frac{9}{64} = \frac{1}{2} + \frac{9}{64}, \left(x + \frac{3}{8}\right)^2 = \frac{41}{64}; x + \frac{3}{8} =$
$\pm\sqrt{\frac{41}{64}}, x = -\frac{3}{8} \pm \frac{\sqrt{41}}{8} = \frac{-3 \pm \sqrt{41}}{8}; \left\{\frac{-3 + \sqrt{41}}{8}, \frac{-3 - \sqrt{41}}{8}\right\}$

26. $y - 2 = \frac{3y - 2}{2y}, 2y(y - 2) = 3y - 2, 2y^2 - 4y = 3y - 2, 2y^2 - 7y = -2,$
$y^2 - \frac{7}{2}y = -1, y^2 - \frac{7}{2}y + \left(-\frac{7}{4}\right)^2 = -1 + \left(-\frac{7}{4}\right)^2, y^2 - \frac{7}{2}y + \frac{49}{16} = -1 + \frac{49}{16},$
$\left(y - \frac{7}{4}\right)^2 = \frac{33}{16}; y - \frac{7}{4} = \pm\sqrt{\frac{33}{16}}, y = \frac{7}{4} \pm \frac{\sqrt{33}}{4} = \frac{7 \pm \sqrt{33}}{4}; \left\{\frac{7 - \sqrt{33}}{4}, \frac{7 + \sqrt{33}}{4}\right\}$

27. $9t = \frac{6t - 1}{t + 2}, 9t(t + 2) = 6t - 1, 9t^2 + 18t = 6t - 1, 9t^2 + 12t = -1, t^2 + \frac{4}{3}t = -\frac{1}{9},$
$t^2 + \frac{4}{3}t + \left(\frac{2}{3}\right)^2 = -\frac{1}{9} + \left(\frac{2}{3}\right)^2, t^2 + \frac{4}{3}t + \frac{4}{9} = -\frac{1}{9} + \frac{4}{9}, \left(t + \frac{2}{3}\right)^2 = \frac{3}{9}; t + \frac{2}{3} = \pm\sqrt{\frac{3}{9}},$
$t = -\frac{2}{3} \pm \frac{\sqrt{3}}{3} = \frac{-2 \pm \sqrt{3}}{3}; \left\{\frac{-2 + \sqrt{3}}{3}, \frac{-2 - \sqrt{3}}{3}\right\}$

C **28.** $x^2 + bx + 1 = 0, x^2 + bx = -1, x^2 + bx + \left(\frac{b}{2}\right)^2 = -1 + \left(\frac{b}{2}\right)^2, x^2 + bx + \frac{b^2}{4} =$

$-1 + \frac{b^2}{4}, \left(x + \frac{b}{2}\right)^2 = \frac{-4 + b^2}{4}; x + \frac{b}{2} = \pm\sqrt{\frac{-4 + b^2}{4}}, x = -\frac{b}{2} \pm \frac{\sqrt{-4 + b^2}}{2} =$

$\frac{-b \pm \sqrt{b^2 - 4}}{2}; \left\{\frac{-b + \sqrt{b^2 - 4}}{2}, \frac{-b - \sqrt{b^2 - 4}}{2}; b^2 \geq 4\right\}$

29. $x^2 - bx + c = 0, x^2 - bx = -c, x^2 - bx + \left(-\frac{b}{2}\right)^2 = -c + \left(-\frac{b}{2}\right)^2,$

$x^2 - bx + \frac{b^2}{4} = -c + \frac{b^2}{4}, \left(x - \frac{b}{2}\right)^2 = \frac{-4c + b^2}{4}; x - \frac{b}{2} = \pm\sqrt{\frac{b^2 - 4c}{4}},$

$x = \frac{b}{2} \pm \frac{\sqrt{b^2 - 4c}}{2} = \frac{b \pm \sqrt{b^2 - 4c}}{2}; \left\{\frac{b + \sqrt{b^2 - 4c}}{2}, \frac{b - \sqrt{b^2 - 4c}}{2}; b^2 \geq 4c\right\}$

30. $ax^2 + bx + c = 0, x^2 + \frac{b}{a}x = -\frac{c}{a}, x^2 + \frac{b}{a}x + \left(\frac{b}{2a}\right)^2 = -\frac{c}{a} + \left(\frac{b}{2a}\right)^2,$

$x^2 + \frac{b}{a}x + \frac{b^2}{4a^2} = -\frac{c}{a} + \frac{b^2}{4a^2}, \left(x + \frac{b}{2a}\right)^2 = \frac{-4ac + b^2}{4a^2}; x + \frac{b}{2a} = \pm\sqrt{\frac{b^2 - 4ac}{4a^2}},$

$x = -\frac{b}{2a} \pm \frac{\sqrt{b^2 - 4ac}}{2a} = \frac{-b \pm \sqrt{b^2 - 4ac}}{2a}; \left\{\frac{-b + \sqrt{b^2 - 4ac}}{2a}, \frac{-b - \sqrt{b^2 - 4ac}}{2a};\right.$

$\left. b^2 \geq 4ac, a \neq 0\right\}$

Page 566 • MIXED REVIEW EXERCISES

1. $\sqrt{188} = \sqrt{4 \cdot 47} = 2\sqrt{47}$ **2.** $\sqrt{2400} = \sqrt{400 \cdot 6} = 20\sqrt{6}$

3. $\sqrt{10x^4y^9} = x^2y^4\sqrt{10y}$ **4.** $\sqrt{120a^6b^7c^5} = \sqrt{4a^6b^6c^4 \cdot 30bc} = 2|a^3b^3|c^2\sqrt{30bc}$

5. $\frac{z}{4 - z} + \frac{6}{4 + z} = \frac{z(4 + z) + 6(4 - z)}{(4 - z)(4 + z)} = \frac{4z + z^2 + 24 - 6z}{(4 - z)(4 + z)} = \frac{z^2 - 2z + 24}{(4 - z)(4 + z)}$ or $\frac{z^2 - 2z + 24}{16 - z^2}$

6. $1 + \frac{b^2}{2b + 1} = \frac{2b + 1 + b^2}{2b + 1}$ or $\frac{(b + 1)^2}{2b + 1}$

7. $3(2x - 1) + (3x + 5)(5x + 3) = 6x - 3 + 15x^2 + 9x + 25x + 15 =$
$15x^2 + 40x + 12$

Page 569 • WRITTEN EXERCISES

A **1.** $x = \frac{-(-3) \pm \sqrt{(-3)^2 - 4(1)(-10)}}{2(1)} = \frac{3 \pm \sqrt{9 + 40}}{2} = \frac{3 \pm \sqrt{49}}{2} = \frac{3 \pm 7}{2}; \{-2, 5\}$

2. $s = \frac{-(-3) \pm \sqrt{(-3)^2 - 4(2)(-2)}}{2(2)} = \frac{3 \pm \sqrt{9 + 16}}{4} = \frac{3 \pm \sqrt{25}}{4} = \frac{3 \pm 5}{4}; \left\{-\frac{1}{2}, 2\right\}$

3. $z = \frac{-(-11) \pm \sqrt{(-11)^2 - 4(5)(2)}}{2(5)} = \frac{11 \pm \sqrt{121 - 40}}{10} = \frac{11 \pm \sqrt{81}}{10} = \frac{11 \pm 9}{10}; \left\{\frac{1}{5}, 2\right\}$

4. $y = \frac{-(-6) \pm \sqrt{(-6)^2 - 4(2)(-8)}}{2(2)} = \frac{6 \pm \sqrt{36 + 64}}{4} = \frac{6 \pm \sqrt{100}}{4} = \frac{6 \pm 10}{4}; \{-1, 4\}$

5. $z = \dfrac{-(-5) \pm \sqrt{(-5)^2 - 4(1)(-6)}}{2(1)} = \dfrac{5 \pm \sqrt{25 + 24}}{2} = \dfrac{5 \pm \sqrt{49}}{2} = \dfrac{5 \pm 7}{2}$; $\{-1, 6\}$

6. $m = \dfrac{-8 \pm \sqrt{(8)^2 - 4(1)(7)}}{2(1)} = \dfrac{-8 \pm \sqrt{64 - 28}}{2} = \dfrac{-8 \pm \sqrt{36}}{2} = \dfrac{-8 \pm 6}{2}$; $\{-7, -1\}$

7. $x = \dfrac{-(-6) \pm \sqrt{(-6)^2 - 4(1)(-11)}}{2(1)} = \dfrac{6 \pm \sqrt{36 + 44}}{2} = \dfrac{6 \pm \sqrt{80}}{2} = \dfrac{6 \pm 4\sqrt{5}}{2} = 3 \pm 2\sqrt{5}$;

$\{3 + 2\sqrt{5}, 3 - 2\sqrt{5}\}$ or $\{-1.5, 7.5\}$

8. $k = \dfrac{-(-3) \pm \sqrt{(-3)^2 - 4(1)(-1)}}{2(1)} = \dfrac{3 \pm \sqrt{9 + 4}}{2} = \dfrac{3 \pm \sqrt{13}}{2}$; $\left\{\dfrac{3 + \sqrt{13}}{2}, \dfrac{3 - \sqrt{13}}{2}\right\}$ or

$\{-0.3, 3.3\}$

9. $r = \dfrac{-8 \pm \sqrt{8^2 - 4(1)(5)}}{2(1)} = \dfrac{-8 \pm \sqrt{64 - 20}}{2} = \dfrac{-8 \pm \sqrt{44}}{2} = \dfrac{-8 \pm 2\sqrt{11}}{2} = -4 \pm \sqrt{11}$;

$\{-4 + \sqrt{11}, -4 - \sqrt{11}\}$ or $\{-7.3, -0.7\}$

10. $n = \dfrac{-(-6) \pm \sqrt{(-6)^2 - 4(1)(-1)}}{2(1)} = \dfrac{6 \pm \sqrt{36 + 4}}{2} = \dfrac{6 \pm \sqrt{40}}{2} = \dfrac{6 \pm 2\sqrt{10}}{2} = 3 \pm \sqrt{10}$;

$\{3 + \sqrt{10}, 3 - \sqrt{10}\}$ or $\{6.2, -0.2\}$

11. $x = \dfrac{-(2) \pm \sqrt{2^2 - 4(7)(-2)}}{2(7)} = \dfrac{-2 \pm \sqrt{4 + 56}}{14} = \dfrac{-2 \pm \sqrt{60}}{14} = \dfrac{-2 \pm 2\sqrt{15}}{14} = \dfrac{-1 \pm \sqrt{15}}{7}$;

$\left\{\dfrac{-1 + \sqrt{15}}{7}, \dfrac{-1 - \sqrt{15}}{7}\right\}$ or $\{-0.7, 0.4\}$

12. $z = \dfrac{-8 \pm \sqrt{8^2 - 4(-2)(5)}}{2(-2)} = \dfrac{-8 \pm \sqrt{64 + 40}}{-4} = \dfrac{-8 \pm \sqrt{104}}{-4} = \dfrac{-8 \pm 2\sqrt{26}}{-4} = \dfrac{4 \pm \sqrt{26}}{2}$;

$\left\{\dfrac{4 + \sqrt{26}}{2}, \dfrac{4 - \sqrt{26}}{2}\right\}$ or $\{-0.5, 4.5\}$

13. $x = \dfrac{-2 \pm \sqrt{2^2 - 4(-4)(3)}}{2(-4)} = \dfrac{-2 \pm \sqrt{4 + 48}}{-8} = \dfrac{-2 \pm \sqrt{52}}{-8} = \dfrac{-2 \pm 2\sqrt{13}}{-8} = \dfrac{1 \pm \sqrt{13}}{4}$;

$\left\{\dfrac{1 + \sqrt{13}}{4}, \dfrac{1 - \sqrt{13}}{4}\right\}$ or $\{-0.7, 1.2\}$

14. $j^2 - 6j - 13 = 0; j = \dfrac{-(-6) \pm \sqrt{(-6)^2 - 4(1)(-13)}}{2(1)} = \dfrac{6 \pm \sqrt{36 + 52}}{2} = \dfrac{6 \pm \sqrt{88}}{2} =$

$\dfrac{6 \pm 2\sqrt{22}}{2} = 3 \pm \sqrt{22}$; $\{3 + \sqrt{22}, 3 - \sqrt{22}\}$ or $\{-1.7, 7.7\}$

15. $4y^2 - 12y + 7 = 0; y = \dfrac{-(-12) \pm \sqrt{(-12)^2 - 4(4)(7)}}{2(4)} = \dfrac{12 \pm \sqrt{144 - 112}}{8} =$

$\dfrac{12 \pm \sqrt{32}}{8} = \dfrac{12 \pm 4\sqrt{2}}{8} = \dfrac{3 \pm \sqrt{2}}{2}$; $\left\{\dfrac{3 + \sqrt{2}}{2}, \dfrac{3 - \sqrt{2}}{2}\right\}$ or $\{2.2, 0.8\}$

16. $4v^2 - 10v + 5 = 0; v = \dfrac{-(-10) \pm \sqrt{(-10)^2 - 4(4)(5)}}{2(4)} = \dfrac{10 \pm \sqrt{100 - 80}}{8} =$

$\dfrac{10 \pm \sqrt{20}}{8} = \dfrac{10 \pm 2\sqrt{5}}{8} = \dfrac{5 \pm \sqrt{5}}{4}$; $\left\{\dfrac{5 + \sqrt{5}}{4}, \dfrac{5 - \sqrt{5}}{4}\right\}$ or $\{1.8, 0.7\}$

17. $3x^2 + 8x - 2 = 0; x = \frac{-8 \pm \sqrt{8^2 - 4(3)(-2)}}{2(3)} = \frac{-8 \pm \sqrt{64 + 24}}{6} = \frac{-8 \pm \sqrt{88}}{6} =$

$\frac{-8 \pm 2\sqrt{22}}{6} = \frac{-4 \pm \sqrt{22}}{3}; \left\{\frac{-4 + \sqrt{22}}{3}, \frac{-4 - \sqrt{22}}{3}\right\}$ or $\{-2.9, 0.2\}$

18. $4r^2 + 2r - 5 = 0; r = \frac{-2 \pm \sqrt{2^2 - 4(4)(-5)}}{2(4)} = \frac{-2 \pm \sqrt{4 + 80}}{8} = \frac{-2 \pm \sqrt{84}}{8} =$

$\frac{-2 \pm 2\sqrt{21}}{8} = \frac{-1 \pm \sqrt{21}}{4}; \left\{\frac{-1 + \sqrt{21}}{4}, \frac{-1 - \sqrt{21}}{4}\right\}$ or $\{-1.4, 0.9\}$

B **19.** $10a^2 + 7a - 1 = 0; a = \frac{-7 \pm \sqrt{7^2 - 4(10)(-1)}}{2(10)} = \frac{-7 \pm \sqrt{49 + 40}}{20} = \frac{-7 \pm \sqrt{89}}{20};$

$\left\{\frac{-7 + \sqrt{89}}{20}, \frac{-7 - \sqrt{89}}{20}\right\}$ or $\{-0.8, 0.1\}$

20. $300x^2 - 180x + 3 = 0; x = \frac{-(-180) \pm \sqrt{(-180)^2 - 4(300)(3)}}{2(300)} =$

$\frac{180 \pm \sqrt{32{,}400 - 3{,}600}}{600} = \frac{180 \pm \sqrt{28{,}800}}{600} = \frac{180 \pm 120\sqrt{2}}{600} = \frac{3 \pm 2\sqrt{2}}{10};$

$\left\{\frac{3 + 2\sqrt{2}}{10}, \frac{3 - 2\sqrt{2}}{10}\right\}$ or $\{0.6, 0.0\}$

21. $40r^2 - 6r - 5 = 0; r = \frac{-(-6) \pm \sqrt{(-6)^2 - 4(40)(-5)}}{2(40)} = \frac{6 \pm \sqrt{36 + 800}}{80} =$

$\frac{6 \pm \sqrt{836}}{80} = \frac{6 \pm 2\sqrt{209}}{80} = \frac{3 \pm \sqrt{209}}{40}; \left\{\frac{3 + \sqrt{209}}{40}, \frac{3 - \sqrt{209}}{40}\right\}$ or $\{-0.3, 0.4\}$

22. $6t^2 + 9t + 4 = 0; t = \frac{-9 \pm \sqrt{9^2 - 4(6)(4)}}{2(6)} = \frac{-9 \pm \sqrt{81 - 96}}{12} = \frac{-9 \pm \sqrt{-15}}{12}$; no real-number solution.

23. $12c^2 + 3c + 4 = 0; c = \frac{-3 \pm \sqrt{3^2 - 4(12)(4)}}{2(12)} = \frac{-3 \pm \sqrt{9 - 192}}{24} = \frac{-3 \pm \sqrt{-183}}{24}$; no real-number solution.

24. $x\left(x + \frac{1}{x}\right) = x\left(\frac{3}{x} + 3\right), x^2 + 1 = 3 + 3x, x^2 - 3x - 2 = 0;$

$x = \frac{-(-3) \pm \sqrt{(-3)^2 - 4(1)(-2)}}{2(1)} = \frac{3 \pm \sqrt{9 + 8}}{2} = \frac{3 \pm \sqrt{17}}{2}; \left\{\frac{3 + \sqrt{17}}{2}, \frac{3 - \sqrt{17}}{2}\right\}$ or $\{-0.6, 3.6\}$

25. $9x^2 + 2x + 4 = 0; x = \frac{-2 \pm \sqrt{2^2 - 4(9)(4)}}{2(9)} = \frac{-2 \pm \sqrt{4 - 144}}{18} = \frac{-2 \pm \sqrt{-140}}{18}$; no real-number solution.

26. $\frac{1}{3} - \frac{2}{2y + 1} = \frac{3}{y}, 3y(2y + 1)\left[\frac{1}{3} - \frac{2}{2y + 1}\right] = 3y(2y + 1)\left(\frac{3}{y}\right),$

$y(2y + 1) - 6y = 9(2y + 1), 2y^2 + y - 6y = 18y + 9, 2y^2 - 23y - 9 = 0;$

$y = \frac{-(-23) \pm \sqrt{(-23)^2 - 4(2)(-9)}}{2(2)} = \frac{23 \pm \sqrt{529 + 72}}{4} = \frac{23 \pm \sqrt{601}}{4};$

$\left\{\frac{23 + \sqrt{601}}{4}, \frac{23 - \sqrt{601}}{4}\right\}$ or $\{-0.4, 11.9\}$

27. $(x+2)(x-1)\left[\frac{3x}{x+2}-\frac{x+1}{x-1}\right]=0$, $3x(x-1)-(x+1)(x+2)=0$,

$3x^2-3x-(x^2+3x+2)=0$, $3x^2-3x-x^2-3x-2=0$, $2x^2-6x-2=0$;

$x=\frac{-(-6)\pm\sqrt{(-6)^2-4(2)(-2)}}{2(2)}=\frac{6\pm\sqrt{36+16}}{4}=\frac{6\pm\sqrt{52}}{4}=\frac{6\pm2\sqrt{13}}{4}=\frac{3\pm\sqrt{13}}{2}$;

$\left\{\frac{3+\sqrt{13}}{2},\frac{3-\sqrt{13}}{2}\right\}$ or $\{-0.3, 3.3\}$

28. $(w-1)(w+5)\left(\frac{w+3}{w-1}-\frac{7}{w+5}\right)=3(w-1)(w+5)$, $(w+3)(w+5)-7(w-1)=$

$3(w^2+4w-5)$, $w^2+8w+15-7w+7=3w^2+12w-15$,

$0=2w^2+11w-37$; $w=\frac{-11\pm\sqrt{11^2-4(2)(-37)}}{2(2)}=\frac{-11\pm\sqrt{121+296}}{4}=$

$\frac{-11\pm\sqrt{417}}{4}$; $\left\{\frac{-11+\sqrt{417}}{4},\frac{-11-\sqrt{417}}{4}\right\}$ or $\{2.4, -7.9\}$

C **29.** $\frac{-b+\sqrt{b^2-4ac}}{2a}+\frac{-b-\sqrt{b^2-4ac}}{2a}=\frac{-b+\sqrt{b^2-4ac}-b-\sqrt{b^2-4ac}}{2a}=\frac{-2b}{2a}=$

$-\frac{b}{a}$; the sum of the roots is $-\frac{b}{a}$.

30. $\left(\frac{-b+\sqrt{b^2-4ac}}{2a}\right)\left(\frac{-b-\sqrt{b^2-4ac}}{2a}\right)=\frac{(-b+\sqrt{b^2-4ac})(-b-\sqrt{b^2-4ac})}{(2a)(2a)}=$

$\frac{(-b)^2-(\sqrt{b^2-4ac})^2}{4a^2}=\frac{b^2-(b^2-4ac)}{4a^2}=\frac{4ac}{4a^2}=\frac{c}{a}$

31. $-\frac{b}{a}=(2+\sqrt{5})+(2-\sqrt{5})$, $-\frac{b}{a}=4$, $b=-4a$; $\frac{c}{a}=(2+\sqrt{5})(2-\sqrt{5})$,

$\frac{c}{a}=4-5$, $\frac{c}{a}=-1$, $c=-a$; Let $a=1$. Then, $b=-4$ and $c=-a$ or -1. The roots of the equation $x^2-4x-1=0$ are $2\pm\sqrt{5}$. (Answers may vary.)

Page 570 • MIXED REVIEW EXERCISES

1. $|x-4|\le 7$; $-7\le x-4\le 7$, $-3\le x\le 11$; {-3, 11, and the real numbers between -3 and 11}

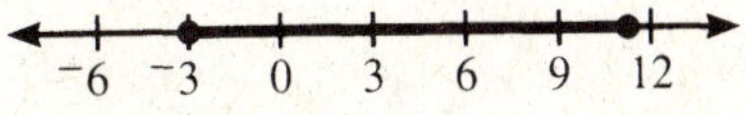

2. $2|y+7|=10$, $|y+7|=5$; $y+7=-5$ or $y+7=5$, $y=-12$ or $y=-2$; $\{-12, -2\}$

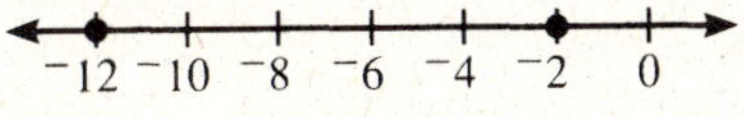

3. $|3p+5|<7$; $-7<3p+5<7$, $-12<3p<2$, $-4<p<\frac{2}{3}$; {the real numbers between -4 and $\frac{2}{3}$}

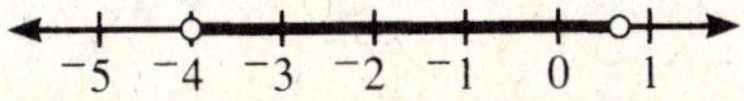

4. $0 < 2z + 1 \le 9, -1 < 2z \le 8, -\frac{1}{2} < z \le 4;$ $\left\{4, \text{ and the real numbers between } -\frac{1}{2} \text{ and } 4\right\}$

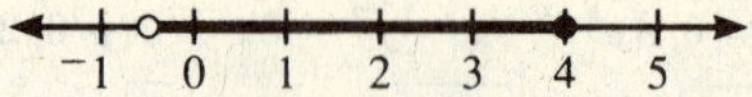

5. $\sqrt{x} = 9; x = 81; \{81\}$

6. $\sqrt{6n + 1} = 5; 6n + 1 = 25, 6n = 24, n = 4; \{4\}$

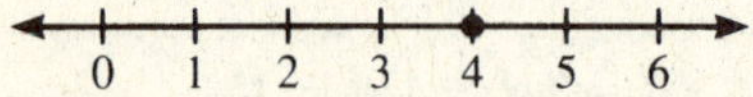

7. $3\sqrt{3p} = 18, \sqrt{3p} = 6; 3p = 36, p = 12; \{12\}$

8. $|9 + 2k| = 15; 9 + 2k = 15$ or $9 + 2k = -15, 2k = 6$ or $2k = -24, k = 3$ or $k = -12; \{-12, 3\}$

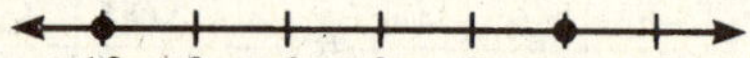

9. $4|4 - m| = 20, |4 - m| = 5; 4 - m = 5$ or $4 - m = -5, -m = 1$ or $-m = -9, m = -1$ or $m = 9; \{-1, 9\}$

10. $y^2 - 10y = -16, y^2 - 10y + (-5)^2 = -16 + (-5)^2, y^2 - 10y + 25 = -16 + 25, (y - 5)^2 = 9, y - 5 = \pm 3, y = 5 \pm 3; y = 2, y = 8; \{2, 8\}$

11. $2p^2 + 12p = 0, p^2 + 6p = 0, p^2 + 6p + 3^2 = 0 + 3^2, p^2 + 6p + 9 = 9, (p + 3)^2 = 9, p + 3 = \pm 3, p = -3 \pm 3; p = 0, p = -6; \{0, -6\}$

12. $c^2 - c = 6, c^2 - c + \left(-\frac{1}{2}\right)^2 = 6 + \left(-\frac{1}{2}\right)^2, c^2 - c + \frac{1}{4} = 6 + \frac{1}{4}, \left(c - \frac{1}{2}\right)^2 = \frac{25}{4},$

$c - \frac{1}{2} = \pm\frac{5}{2}, c = \frac{1}{2} \pm \frac{5}{2}; c = \frac{1}{2} - \frac{5}{2} = -\frac{4}{2} = -2, c = \frac{1}{2} + \frac{5}{2} = \frac{6}{2} = 3; \{-2, 3\}$

Page 570 • COMPUTER EXERCISES

1.
```
10 PRINT "TO FIND THE ROOTS OF"
20 PRINT "A * X ↑ 2 + B * X + C = 0"
30 PRINT "ENTER A, B, AND C"
40 INPUT A, B, C
50 PRINT
60 LET D = B * B - 4 * A * C
70 IF D > = 0 THEN 100
80 PRINT "NO REAL ROOTS"
90 GOTO 150
100 IF D > 0 THEN 130
110 PRINT "ROOT IS: "; -B/(2 * A)
120 GOTO 150
130 PRINT "ROOTS ARE: "; (-B + SQR(D))/(2 * A);
140 PRINT "AND"; (-B - SQR(D))/(2 * A)
150 END
```

2. ROOTS ARE: -0.5 AND -3
3. ROOTS ARE: 1 AND 0.75
4. ROOTS ARE: 2.387 AND 0.279
5. ROOTS ARE: 3.4142 AND 0.5859
6. ROOT IS: 3
7. NO REAL ROOTS

Pages 570–571 • EXTRA

1. $i^{14} = i^{12} \cdot i^2 = (i^4)^3 \cdot (-1) = 1^3(-1) = -1$
2. $i^{23} = i^{20} \cdot i^3 = (i^4)^5(-i) = 1^5(-i) = -i$ **3.** $i^{17} = i^{16} \cdot i = (i^4)^4(i) = 1^4 \cdot i = i$
4. $i^{36} = (i^4)^9 = 1^9 = 1$ **5.** $i^{19} = i^{16} \cdot i^3 = (i^4)^4(-i) = 1^4(-i) = -i$
6. $i^{27} = i^{24} \cdot i^3 = (i^4)^6(-i) = 1^6(-i) = -i$ **7.** $\sqrt{-49} = 7i$
8. $-\sqrt{-500} = -\sqrt{-1 \cdot 100 \cdot 5} = -10i\sqrt{5}$ **9.** $\sqrt{96} = 4\sqrt{6}$
10. $-\sqrt{-144} = -12i$ **11.** $\sqrt{12} \cdot \sqrt{-75} = 2\sqrt{3} \cdot 5i\sqrt{3} = 10i(\sqrt{3})^2 = 30i$
12. $\sqrt{-3} \cdot \sqrt{48} = \sqrt{-144} = 12i$ **13.** $\sqrt{18} \cdot \sqrt{-108} = 3\sqrt{2} \cdot 6i\sqrt{3} = 18i\sqrt{6}$
14. $(\sqrt{-4} \cdot \sqrt{9})^2 = (\sqrt{-4})^2(\sqrt{9})^2 = -4 \cdot 9 = -36$
15. $\sqrt{\frac{16}{-81}} = \sqrt{\frac{-16}{81}} = \frac{\sqrt{-16}}{\sqrt{81}} = \frac{4i}{9}$ **16.** $-\sqrt{\frac{-25}{64}} = -\frac{\sqrt{-25}}{\sqrt{64}} = -\frac{5i}{8}$
17. $-\sqrt{\frac{-49}{121}} = -\frac{\sqrt{-49}}{\sqrt{121}} = -\frac{7i}{11}$ **18.** $\frac{\sqrt{-150}}{-\sqrt{54}} = \frac{5i\sqrt{6}}{-3\sqrt{6}} = \frac{5i}{-3} = -\frac{5i}{3}$
19. $x^2 = -100; x = \pm\sqrt{-100} = \pm 10i; \{-10i, 10i\}$
20. $x^2 + 63 = 0, x^2 = -63; x = \pm\sqrt{-63} = \pm 3i\sqrt{7}; \{-3i\sqrt{7}, 3i\sqrt{7}\}$
21. $x^2 = -98; x = \pm\sqrt{-98} = \pm 7i\sqrt{2}; \{-7i\sqrt{2}, 7i\sqrt{2}\}$
22. $x^2 + 272 = 0, x^2 = -272; x = \pm\sqrt{-272} = \pm 4i\sqrt{17}; \{-4i\sqrt{17}, 4i\sqrt{17}\}$

Page 574 • WRITTEN EXERCISES

A 1. $(-5)^2 - 4(1)(4) = 25 - 16 = 9$; two real-number roots
2. $(-3)^2 - 4(1)(9) = 9 - 36 = -27$; none
3. $(-16)^2 - 4(1)(64) = 256 - 256 = 0$; one **4.** $(5)^2 - 4(2)(-3) = 25 + 24 = 49$; two
5. $(-12)^2 - 4(9)(4) = 144 - 144 = 0$; one
6. $(-4)^2 - 4(3)(3) = 16 - 36 = -20$; none **7.** $(20)^2 - 4(4)(25) = 400 - 400 = 0$; one
8. $(-2)^2 - 4(3)(-6) = 4 + 72 = 76$; two **9.** $(1)^2 - 4(-5)(1) = 1 + 20 = 21$; two
10. $(3)^2 - 4(-4)(-2) = 9 - 32 = -23$; none
11. $(-1.4)^2 - 4(2)(0.1) = 1.96 - 0.8 = 1.16$; two
12. $(-2)^2 - 4\left(\frac{1}{2}\right)(2) = 4 - 4 = 0$; one

B 13. **a.** $(-5)^2 - 4(1)(5) = 25 - 20 = 5$; two x-intercepts
b. Coefficient of $x^2 > 0$, so it opens up; vertex is below the x-axis.

14. a. $(3)^2 - 4(-1)(6) = 9 + 24 = 33$; two x-intercepts

b. Coefficient of $x^2 < 0$, so it opens down; vertex is above the x-axis.

15. a. $(-8)^2 - 4(1)(16) = 64 - 64 = 0$; one x-intercept

b. Vertex is on the x-axis.

16. a. $(4)^2 - 4(2)(3) = 16 - 24 = -8$; no x-intercepts

b. Coeff. of $x^2 > 0$, so it opens up; vertex is above the x-axis.

17. a. $(7)^2 - 4(-3)(2) = 49 + 24 = 73$; two x-intercepts

b. Coeff. of $x^2 < 0$, so it opens down; vertex is above the x-axis.

18. a. $(2)^2 - 4(4)(-1) = 4 + 16 = 20$; two x-intercepts

b. Coeff. of $x^2 > 0$, so it opens up; vertex is below the x-axis.

C **19.** $(12)^2 - 4(9)(k) = 0$, $144 - 36k = 0$, $144 = 36k$, $4 = k$; the equation $9x^2 + 12x + 4 = 0$ has one real-number root.

20. $(12k)^2 - 4(4)(9) = 0$, $144k^2 - 144 = 0$, $144k^2 = 144$, $k^2 = 1$, $k = \pm 1$; the equations $4x^2 + 12x + 9 = 0$ and $4x^2 - 12x + 9 = 0$ each have one real-number root.

Page 575 • MIXED REVIEW EXERCISES

1. $\frac{\sqrt{3} - 3}{\sqrt{6}} \cdot \frac{\sqrt{6}}{\sqrt{6}} = \frac{\sqrt{18} - 3\sqrt{6}}{6} = \frac{3\sqrt{2} - 3\sqrt{6}}{6} = \frac{\sqrt{2} - \sqrt{6}}{2}$

2. $\sqrt{\frac{12c^3}{5}} \cdot \sqrt{\frac{3c^5}{20}} = \sqrt{\frac{12c^3 \cdot 3c^5}{5 \cdot 20}} = \sqrt{\frac{36c^8}{100}} = \frac{6c^4}{10} = \frac{3c^4}{5}$

3. $6\sqrt{8} - 15\sqrt{2} + \sqrt{18} = 6 \cdot 2\sqrt{2} - 15\sqrt{2} + 3\sqrt{2} = (12 - 15 + 3)\sqrt{2} = 0$

4. $\frac{2x + 14}{x^3 + 8x^2 + 7x} = \frac{2(x + 7)}{x(x^2 + 8x + 7)} = \frac{2(x + 7)}{x(x + 7)(x + 1)} = \frac{2}{x(x + 1)}$ or $\frac{2}{x^2 + x}$

5. $\frac{b}{b + 1} + \frac{5}{4b + 4} = \frac{4b}{4(b + 1)} + \frac{5}{4(b + 1)} = \frac{4b + 5}{4(b + 1)}$ or $\frac{4b + 5}{4b + 4}$

6. $(6.2 \cdot 10^4)(8.1 \cdot 10^3) = 50.22 \cdot 10^7 = 5.022 \cdot 10 \cdot 10^7 = 5.022 \cdot 10^8$

7. $x = -\frac{0}{2(4)} = 0$; $y = 4(0)^2 = 0$; vertex is $(0, 0)$; axis is $x = 0$

8. $x = -\frac{10}{2(1)} = -5$; $y = (-5)^2 + 10(-5) + 25 = 25 - 50 + 25 = 0$; vertex is $(-5, 0)$; axis is $x = -5$

9. $x = -\frac{0}{2(2)} = 0$; $y = 2(0)^2 + 3 = 3$; vertex is $(0, 3)$; axis is $x = 0$

10. $x = -\frac{9}{2(-1)} = \frac{9}{2}$; $y = -\left(\frac{9}{2}\right)^2 + 9\left(\frac{9}{2}\right) - 3 = -\frac{81}{4} + \frac{81}{2} - 3 = \frac{-81 + 162 - 12}{4} = \frac{69}{4}$; vertex is $\left(\frac{9}{2}, \frac{69}{4}\right)$ or $\left(4\frac{1}{2}, 17\frac{1}{4}\right)$; axis is $x = \frac{9}{2}$ or $4\frac{1}{2}$

11. $x = -\dfrac{(-5)}{\frac{2}{3}} = 5 \cdot \dfrac{3}{2} = \dfrac{15}{2}$; $y = 6 - 5\left(\dfrac{15}{2}\right) + \dfrac{1}{3}\left(\dfrac{15}{2}\right)^2 = 6 - \dfrac{75}{2} + \dfrac{225}{12} =$

$6 - \dfrac{75}{2} + \dfrac{75}{4} = \dfrac{24 - 150 + 75}{4} = -\dfrac{51}{4}$; vertex is $\left(\dfrac{15}{2}, -\dfrac{51}{4}\right)$; axis is $x = \dfrac{15}{2}$

12. $x = -\dfrac{3}{2(-2)} = \dfrac{3}{4}$; $y = -2\left(\dfrac{3}{4}\right)^2 + 3\left(\dfrac{3}{4}\right) = -2 \cdot \dfrac{9}{16} + \dfrac{9}{4} = -\dfrac{9}{8} + \dfrac{9}{4} = \dfrac{-9 + 18}{8} = \dfrac{9}{8}$;

vertex is $\left(\dfrac{3}{4}, \dfrac{9}{8}\right)$; axis is $x = \dfrac{3}{4}$

Page 575 • COMPUTER EXERCISES

```
10 PRINT "FOR THE EQUATION A * X ↑ 2 + B * X + C = 0"
20 PRINT "ENTER A, B, AND C"
30 INPUT A, B, C
40 PRINT
50 LET D = B * B - 4 * A * C
60 PRINT "DISCRIMINANT IS: "; D
70 IF D < 0 THEN 130
80 IF D = 0 THEN 110
90 PRINT "THERE ARE TWO REAL ROOTS"
100 GOTO 140
110 PRINT "THERE IS ONE REAL ROOT"
120 GOTO 140
130 PRINT "THERE ARE NO REAL ROOTS"
140 END
```

1. DISCRIMINANT IS 17
THERE ARE TWO REAL ROOTS

2. DISCRIMINANT IS 0
THERE IS ONE REAL ROOT

3. DISCRIMINANT IS -3
THERE ARE NO REAL ROOTS

Pages 577–578 • WRITTEN EXERCISES

A

1. $x^2 + 5x - 6 = 0, (x + 6)(x - 1) = 0; x + 6 = 0$ or $x - 1 = 0, x = -6$ or $x = 1$; $\{-6, 1\}$

2. $x^2 - 2x = 1, x^2 - 2x + (-1)^2 = 1 + (-1)^2, x^2 - 2x + 1 = 1 + 1,$
$(x - 1)^2 = 2; x - 1 = \pm\sqrt{2}, x = 1 \pm \sqrt{2}; \{1 + \sqrt{2}, 1 - \sqrt{2}\}$

3. $11x^2 = 44, x^2 = 4; x = \pm 2; \{-2, 2\}$

4. $8x^2 + 11x = 0, x(8x + 11) = 0; x = 0$ or $8x + 11 = 0, x = 0$ or $8x = -11, x = 0$ or $x = -\dfrac{11}{8}$; $\left\{-\dfrac{11}{8}, 0\right\}$

5. $x^2 + 7x + 2 = 0; x = \dfrac{-7 \pm \sqrt{7^2 - 4(1)(2)}}{2(1)} = \dfrac{-7 \pm \sqrt{49 - 8}}{2} = \dfrac{-7 \pm \sqrt{41}}{2}$;
$\left\{\dfrac{-7 + \sqrt{41}}{2}, \dfrac{-7 - \sqrt{41}}{2}\right\}$

6. $3x^2 - 5x = 4, 3x^2 - 5x - 4 = 0; x = \frac{-(-5) \pm \sqrt{(-5)^2 - 4(3)(-4)}}{2 \cdot 3} =$

$\frac{5 \pm \sqrt{25 + 48}}{6} = \frac{5 \pm \sqrt{73}}{6}; \left\{\frac{5 + \sqrt{73}}{6}, \frac{5 - \sqrt{73}}{6}\right\}$

7. $x^2 + 6x = 5, x^2 + 6x + 3^2 = 5 + 3^2, (x + 3)^2 = 5 + 9; x + 3 = \pm\sqrt{14},$

$x = -3 \pm \sqrt{14}; \{-3 + \sqrt{14}, -3 - \sqrt{14}\}$

8. $2x^2 + 7x + 3 = 0, (2x + 1)(x + 3) = 0; 2x + 1 = 0$ or $x + 3 = 0, 2x = -1$ or

$x = -3, x = -\frac{1}{2}$ or $x = -3; \left\{-3, -\frac{1}{2}\right\}$

9. $5x^2 - 20x = 0, 5x(x - 4) = 0; x = 0$ or $x - 4 = 0, x = 0$ or $x = 4; \{0, 4\}$

10. $(x - 2)^2 = 7; x - 2 = \pm\sqrt{7}, x = 2 \pm \sqrt{7}; \{2 + \sqrt{7}, 2 - \sqrt{7}\}$

11. $x^2 - 8x + 1 = 3, x^2 - 8x = 2, x^2 - 8x + (-4)^2 = 2 + (-4)^2, (x - 4)^2 = 2 + 16;$

$x - 4 = \pm\sqrt{18}, x = 4 \pm 3\sqrt{2}, \{4 + 3\sqrt{2}, 4 - 3\sqrt{2}\}$

12. $4x^2 - 2x - 3 = 0; x = \frac{-(-2) \pm \sqrt{(-2)^2 - 4(4)(-3)}}{2(4)} = \frac{2 \pm \sqrt{4 + 48}}{8} = \frac{2 \pm \sqrt{52}}{8} =$

$\frac{2 \pm 2\sqrt{13}}{8} = \frac{1 \pm \sqrt{13}}{4}; \left\{\frac{1 + \sqrt{13}}{4}, \frac{1 - \sqrt{13}}{4}\right\}$

13. $5x^2 + 19x = 4, 5x^2 + 19x - 4 = 0, (5x - 1)(x + 4) = 0; 5x - 1 = 0$ or $x + 4 = 0,$

$5x = 1$ or $x = -4, x = \frac{1}{5}$ or $x = -4; \left\{-4, \frac{1}{5}\right\}$

14. $x^2 - 8x = 11, x^2 - 8x + (-4)^2 = 11 + (-4)^2, (x - 4)^2 = 11 + 16; x - 4 = \pm\sqrt{27},$

$x = 4 \pm 3\sqrt{3}; \{4 + 3\sqrt{3}, 4 - 3\sqrt{3}\}$

15. $3x^2 = 2x + 7, 3x^2 - 2x - 7 = 0; x = \frac{-(-2) \pm \sqrt{(-2)^2 - 4(3)(-7)}}{2(3)} = \frac{2 \pm \sqrt{4 + 84}}{6} =$

$\frac{2 \pm \sqrt{88}}{6} = \frac{2 \pm 2\sqrt{22}}{6} = \frac{1 \pm \sqrt{22}}{3}; \left\{\frac{1 + \sqrt{22}}{3}, \frac{1 - \sqrt{22}}{3}\right\}$

16. $\frac{2x^2}{3} + \frac{3x}{4} = 1, 12\left(\frac{2x^2}{3} + \frac{3x}{4}\right) = 12, 8x^2 + 9x = 12, 8x^2 + 9x - 12 = 0;$

$x = \frac{-9 \pm \sqrt{9^2 - 4 \cdot 8 \cdot (-12)}}{2 \cdot 8} = \frac{-9 \pm \sqrt{81 + 384}}{16} = \frac{-9 \pm \sqrt{465}}{16};$

$\left\{\frac{-9 + \sqrt{465}}{16}, \frac{-9 - \sqrt{465}}{16}\right\}$

17. $\frac{(y + 3)^2}{4} = 5, (y + 3)^2 = 20; y + 3 = \pm\sqrt{20}, y = -3 \pm 2\sqrt{5}; \{-3 + 2\sqrt{5},$

$-3 - 2\sqrt{5}\}$

18. $0.75x^2 - 0.3x + 0.03 = 0, 75x^2 - 30x + 3 = 0, 3(25x^2 - 10x + 1) = 0,$

$(5x - 1)^2 = 0; 5x - 1 = 0, 5x = 1, x = \frac{1}{5}$ or $0.2; \{0.2\}$

19. $1.4x^2 - 0.7x - 0.2 = 0, 14x^2 - 7x - 2 = 0; x = \frac{-(-7) \pm \sqrt{(-7)^2 - 4(14)(-2)}}{2(14)} =$

$\frac{7 \pm \sqrt{49 + 112}}{28} = \frac{7 \pm \sqrt{161}}{28}; \left\{\frac{7 + \sqrt{161}}{28}, \frac{7 - \sqrt{161}}{28}\right\}$

20. $\frac{1}{3x} = \frac{2x-3}{2}$, $3x(2x-3) = 2$, $6x^2 - 9x = 2$, $6x^2 - 9x - 2 = 0$;

$x = \frac{-(-9) \pm \sqrt{(-9)^2 - 4(6)(-2)}}{2(6)} = \frac{9 \pm \sqrt{81 + 48}}{12} = \frac{9 \pm \sqrt{129}}{12}$; $\left\{\frac{9 + \sqrt{129}}{12}, \frac{9 - \sqrt{129}}{12}\right\}$

21. $\frac{4x-3}{2x-3} = \frac{5x-3}{x+1}$, $(4x-3)(x+1) = (2x-3)(5x-3)$, $4x^2 + x - 3 =$

$10x^2 - 21x + 9$, $-6x^2 + 22x - 12 = 0$, $-3x^2 + 11x - 6 = 0$;

$x = \frac{-11 \pm \sqrt{121 - 4(-3)(-6)}}{2(-3)} = \frac{-11 \pm \sqrt{121 - 72}}{-6} = \frac{-11 \pm \sqrt{49}}{-6} = \frac{11 \pm 7}{6}$; $\left\{3, \frac{2}{3}\right\}$

B **22.** $5x(x-3) + 4(x+4) = 31 - 7x^2$, $5x^2 - 15x + 4x + 16 = 31 - 7x^2$,

$12x^2 - 11x - 15 = 0$; $x = \frac{-(-11) \pm \sqrt{(-11)^2 - 4(12)(-15)}}{2(12)} = \frac{11 \pm \sqrt{121 + 720}}{24} =$

$\frac{11 \pm \sqrt{841}}{24} = \frac{11 \pm 29}{24}$; $\left\{\frac{5}{3}, -\frac{3}{4}\right\}$

23. $2x(x-3) + 7(x^2-1) = 2$, $2x^2 - 6x + 7x^2 - 7 = 2$, $9x^2 - 6x - 9 = 0$,

$3x^2 - 2x - 3 = 0$; $x = \frac{-(-2) \pm \sqrt{(-2)^2 - 4(3)(-3)}}{2(3)} = \frac{2 \pm \sqrt{4 + 36}}{6} = \frac{2 \pm \sqrt{40}}{6} =$

$\frac{2 \pm 2\sqrt{10}}{6} = \frac{1 \pm \sqrt{10}}{3}$; $\left\{\frac{1 + \sqrt{10}}{3}, \frac{1 - \sqrt{10}}{3}\right\}$

24. $(x+3)^2 + 6(x+3) = 16$, $x^2 + 6x + 9 + 6x + 18 = 16$, $x^2 + 12x + 11 = 0$, $(x+11)(x+1) = 0$; $x + 11 = 0$ or $x + 1 = 0$, $x = -11$ or $x = -1$; $\{-11, -1\}$

25. $9(x-4)^2 + 4(x-4) = 0$, $9(x^2 - 8x + 16) + 4x - 16 = 0$,
$9x^2 - 72x + 144 + 4x - 16 = 0$, $9x^2 - 68x + 128 = 0$;

$x = \frac{-(-68) \pm \sqrt{(-68)^2 - 4(9)(128)}}{2(9)} = \frac{68 \pm \sqrt{4624 - 4608}}{18} = \frac{68 \pm \sqrt{16}}{18} = \frac{68 \pm 4}{18}$;

$\left\{4, \frac{32}{9}\right\}$

26. $(3x+2)(x-1) - 13 = 2x(2-x)$, $3x^2 - x - 2 - 13 = 4x - 2x^2$,

$5x^2 - 5x - 15 = 0$, $x^2 - x - 3 = 0$; $x = \frac{-(-1) \pm \sqrt{(-1)^2 - 4(1)(-3)}}{2 \cdot 1} =$

$\frac{1 \pm \sqrt{1 + 12}}{2} = \frac{1 \pm \sqrt{13}}{2}$; $\left\{\frac{1 + \sqrt{13}}{2}, \frac{1 - \sqrt{13}}{2}\right\}$

27. $(3x-5)^2 = (8x+3)^2$, $3x - 5 = \pm(8x+3)$; $3x - 5 = 8x + 3$ or $3x - 5 =$

$-(8x+3)$, $-8 = 5x$ or $3x - 5 = -8x - 3$, $x = -\frac{8}{5}$ or $11x = 2$, $x = -\frac{8}{5}$ or $x = \frac{2}{11}$;

$\left\{-\frac{8}{5}, \frac{2}{11}\right\}$

28. $9x^4 - 14x^2 + 5 = 0$, $9y^2 - 14y + 5 = 0$; $y = \frac{-(-14) \pm \sqrt{(-14)^2 - 4(9)(5)}}{2 \cdot 9} =$

$\frac{14 \pm \sqrt{196 - 180}}{18} = \frac{14 \pm \sqrt{16}}{18} = \frac{14 \pm 4}{18} = \frac{5}{9}$ or 1; $x^2 = \frac{5}{9}$ or $x^2 = 1$; $x = \pm\sqrt{\frac{5}{9}}$ or

$x = \pm 1$, $x = \pm\frac{\sqrt{5}}{3}$ or $x = \pm 1$; $\left\{-\frac{\sqrt{5}}{3}, \frac{\sqrt{5}}{3}, -1, 1\right\}$

29. $4x^4 + 21x^2 - 18 = 0, 4y^2 + 21y - 18 = 0; y = \frac{-21 \pm \sqrt{(21)^2 - 4(4)(-18)}}{2(4)} =$

$\frac{-21 \pm \sqrt{441 + 288}}{8} = \frac{-21 \pm \sqrt{729}}{8} = \frac{-21 \pm 27}{8} = \frac{3}{4}$ or -6; $x^2 = \frac{3}{4}$ or $x^2 = -6$

($x^2 = -6$ has no real solution.), $x = \pm\sqrt{\frac{3}{4}} = \pm\frac{\sqrt{3}}{2}$; $\left\{-\frac{\sqrt{3}}{2}, \frac{\sqrt{3}}{2}\right\}$

30. $3x^4 - 7x^2 = 0, x^2(3x^2 - 7) = 0; x^2 = 0$ or $3x^2 - 7 = 0, x = 0$ or $3x^2 = 7, x = 0$

or $x^2 = \frac{7}{3}, x = 0$ or $x = \pm\sqrt{\frac{7}{3}} = \pm\frac{\sqrt{7}\sqrt{3}}{\sqrt{3}\sqrt{3}} = \pm\frac{\sqrt{21}}{3}$; $\left\{-\frac{\sqrt{21}}{3}, 0, \frac{\sqrt{21}}{3}\right\}$

31. $9x^4 - 64 = 0, (3x^2 - 8)(3x^2 + 8) = 0; 3x^2 - 8 = 0$ or $3x^2 + 8 = 0, 3x^2 = 8$ or

$3x^2 = -8$ ($3x^2 = -8$ has no real solution), $x^2 = \frac{8}{3}; x = \pm\sqrt{\frac{8}{3}} = \pm\frac{2\sqrt{2}\sqrt{3}}{\sqrt{3}\sqrt{3}} =$

$\pm\frac{2\sqrt{6}}{3}$; $\left\{-\frac{2\sqrt{6}}{3}, \frac{2\sqrt{6}}{3}\right\}$

Page 578 • MIXED REVIEW EXERCISES

1. $\pm\sqrt{(8)^2 - 4(1)(-9)} = \pm\sqrt{64 + 36} = \pm\sqrt{100} = \pm 10$

2. $-\sqrt{(8)^2 + 4(-9)(1)} = -\sqrt{64 - 36} = -\sqrt{28} = -2\sqrt{7}$

3. $\sqrt{(-9)^2 - 4(1)(8)} = \sqrt{81 - 32} = \sqrt{49} = 7$

4. $\sqrt{(-9)^2 + 4(1)(8)} = \sqrt{81 + 32} = \sqrt{113}$

5. $\pm\sqrt{(1)^2 - 4(8)(-9)} = \pm\sqrt{1 + 288} = \pm\sqrt{289} = \pm 17$

6. $\sqrt{(1)^2 + 4(8)(-9)} = \sqrt{1 - 288} = \sqrt{-287}$; not a real number

7. $4x^2 - 4x - 15 = 0, (2x - 5)(2x + 3) = 0; 2x - 5 = 0$ or $2x + 3 = 0, 2x = 5$ or

$2x = -3, x = \frac{5}{2}$ or $x = -\frac{3}{2}$; $\left\{\frac{5}{2}, -\frac{3}{2}\right\}$

8. $2d^2 + 3d - 77 = 0; d = \frac{-3 \pm \sqrt{(3)^2 - 4(2)(-77)}}{2 \cdot 2} = \frac{-3 \pm \sqrt{9 + 616}}{4} =$

$\frac{-3 \pm \sqrt{625}}{4} = \frac{-3 \pm 25}{4}$; $\left\{\frac{11}{2}, -7\right\}$

9. $2(x + 4)(x + 7) = (2x + 1)(x - 9) - 3, 2(x^2 + 11x + 28) = 2x^2 - 17x - 9 - 3,$

$2x^2 + 22x + 56 = 2x^2 - 17x - 12, 39x = -68, x = -\frac{68}{39}$; $\left\{-\frac{68}{39}\right\}$

10. $c^2 - 4c - 2 = 0, c^2 - 4c = 2, c^2 - 4c + (-2)^2 = 2 + (-2)^2, (c - 2)^2 = 2 + 4,$

$c - 2 = \pm\sqrt{6}, c = 2 \pm \sqrt{6}$; $\{2 + \sqrt{6}, 2 - \sqrt{6}\}$

11. $3p^2 - 4p + 1 = 0, (3p - 1)(p - 1) = 0; 3p - 1 = 0$ or $p - 1 = 0, 3p = 1$ or $p = 1,$

$p = \frac{1}{3}$ or $p = 1$; $\left\{\frac{1}{3}, 1\right\}$

12. $12y^2 + 24y = 5, y^2 + 2y = \frac{5}{12}, y^2 + 2y + (1)^2 = \frac{5}{12} + (1)^2, (y + 1)^2 = \frac{5}{12} + 1;$

$y + 1 = \pm\sqrt{\frac{5 + 12}{12}}, y = -1 \pm \frac{\sqrt{17}}{2\sqrt{3}}\frac{\sqrt{3}}{\sqrt{3}} = -1 \pm \frac{\sqrt{51}}{2 \cdot 3} = -\frac{6}{6} \pm \frac{\sqrt{51}}{6} = \frac{-6 \pm \sqrt{51}}{6};$

$\left\{\frac{-6 + \sqrt{51}}{6}, \frac{-6 - \sqrt{51}}{6}\right\}$

Pages 580–581 • PROBLEMS

A **1.** Let l = length of the park. $l(l - 5) = 300$, $l^2 - 5l = 300$, $l^2 - 5l - 300 = 0$, $(l - 20)(l + 15) = 0$; $l - 20 = 0$ or $l + 15 = 0$, $l = 20$ or $l = -15$; $l - 5 = 20 - 5 = 15$; The park is 20 m × 15 m.

2. Let w = width of the rectangle. $w(3w) = 75$, $3w^2 = 75$, $w^2 = 25$; $w = \pm\sqrt{25} = \pm 5$; $3w = 3(5) = 15$; The rectangle is 5 cm × 15 cm.

3. Let x = the number. $x^2 + x = 56$, $x^2 + x - 56 = 0$, $(x - 7)(x + 8) = 0$; $x = 7$ or $x = -8$; The number is 7 or −8.

4. Let x = the number. $x^2 - x = 182$, $x^2 - x - 182 = 0$, $(x - 14)(x + 13) = 0$; $x = 14$ or $x = -13$; The number is 14 or −13.

5. Let a = the altitude of the triangle. $\frac{1}{2}(4a)a = 162$, $2a^2 = 162$, $a^2 = 81$;

$a = \pm\sqrt{81} = \pm 9$; The triangle has altitude 9 cm.

6. Let b = base of the triangle. $\frac{1}{2}b(b - 5) = 42$, $b(b - 5) = 84$, $b^2 - 5b - 84 = 0$,

$(b - 12)(b + 7) = 0$; $b = 12$ or $b = -7$; The base of the triangle is 12 m.

7. Let s = side of the original square. $(s + 3)^2 = 100$; $s + 3 = \pm 10$; $s + 3 = 10$ or $s + 3 = -10$, $s = 7$ or $s = -13$; The sides of the original square were 7 cm long.

8. Let x = the amount of increase in the width and length. $(12 + x)(14 + x) = 255$, $168 + 26x + x^2 = 255$, $x^2 + 26x - 87 = 0$, $(x + 29)(x - 3) = 0$; $x = -29$ or 3; $12 + x = 12 + 3 = 15$; $14 + x = 14 + 3 = 17$; The new garden measures 15 m × 17 m.

9. Let x = Olaf's rate of travel and $x + 2$ = Cindy's rate. $(x)^2 + (x + 2)^2 = (10)^2$, $x^2 + x^2 + 4x + 4 = 100$, $2x^2 + 4x + 4 = 100$, $x^2 + 2x + 2 = 50$, $x^2 + 2x = 48$, $x^2 + 2x - 48 = 0$; $(x + 8)(x - 6) = 0$; $x = -8$ or 6; $x + 2 = 6 + 2 = 8$; Cindy is traveling at 8 mi/h and Olaf is traveling at 6 mi/h.

B **10.** Let x = number of hours James needs to paint the house by himself.

$\frac{10}{x - 2} + \frac{10}{x} = 1$, $x(x - 2)\left(\frac{10}{x - 2} + \frac{10}{x}\right) = x(x - 2)$, $10x + 10(x - 2) = x^2 - 2x$,

$10x + 10x - 20 = x^2 - 2x$, $x^2 - 22x = -20$, $x^2 - 22x + (-11)^2 = -20 + (-11)^2$;

$x^2 - 22x + 121 = -20 + 121$, $(x - 11)^2 = 101$; $x - 11 = \pm\sqrt{101}$,

$x = 11 \pm \sqrt{101} \approx 21.0$ or 1.0 (1.0 is not a realistic solution because Colleen requires 2 hours fewer than James). James requires 21.0 hours to paint the house by himself.

11. Let x = number of minutes it takes for Johannes to pick a bushel. $\frac{16}{x + 4} + \frac{16}{x} = 1$,

$x(x + 4)\left(\frac{16}{x + 4} + \frac{16}{x}\right) = x(x + 4)$, $16x + 16(x + 4) = x^2 + 4x$,

$16x + 16x + 64 = x^2 + 4x$, $x^2 - 28x = 64$, $x^2 - 28x + (-14)^2 = 64 + (-14)^2$,

$(x - 14)^2 = 64 + 196$; $x - 14 = \pm\sqrt{260}$, $x = 14 \pm 2\sqrt{65} \approx 30.1$ or −2.1; $x + 4 \approx 30.1 + 4 = 34.1$; It takes Johannes 30.1 min to pick a bushel and it takes Maria 34.1 min.

12. Let x = number of students who went to the museum. $\frac{240}{x} = \frac{240}{x + 4} + 2$,

$x(x + 4)\left(\frac{240}{x}\right) = x(x + 4)\left(\frac{240}{x + 4} + 2\right)$, $240(x + 4) = 240x + 2x(x + 4)$,

$240x + 960 = 240x + 2x^2 + 8x$, $2x^2 + 8x - 960 = 0$, $x^2 + 4x - 480 = 0$,
$(x + 24)(x - 20) = 0$; $x = -24$ or 20; 20 students went to the museum.

13. Let x = number of students in the club. $\frac{72}{x} - 0.50 = \frac{72}{x + 2}$, $2x(x + 2)\left(\frac{72}{x} - \frac{1}{2}\right) =$

$2x(x + 2)\left(\frac{72}{x + 2}\right)$, $144(x + 2) - x(x + 2) = 144x$, $144x + 288 - x^2 - 2x = 144x$,

$x^2 + 2x - 288 = 0$, $(x + 18)(x - 16) = 0$; $x = -18$ or $x = 16$; There are 16 students in the club.

14. Let x = Kelly's speed. $\frac{50}{x} - 0.8 = \frac{50}{x + 2}$, $10x(x + 2)\left(\frac{50}{x} - 0.8\right) =$

$\left(\frac{50}{x + 2}\right)(10x)(x + 2)$, $500(x + 2) - 8x(x + 2) = 500x$,

$500x + 1000 - 8x^2 - 16x = 500x$, $8x^2 + 16x - 1000 = 0$, $x^2 + 2x - 125 = 0$,

$x^2 + 2x = 125$, $x^2 + 2x + (1)^2 = 125 + (1)^2$, $(x + 1)^2 = 126$; $x + 1 = \pm\sqrt{126}$,

$x = -1 \pm 3\sqrt{14} \approx 10.2$ or -12.2; $x + 2 \approx 10.2 + 2 = 12.2$; Kelly rides at 10.2 mi/h and Judith rides at 12.2 mi/h.

C **15.** Let c = time required for pipe C to empty the tank. $\frac{2}{4} + \frac{2}{c - 9} - \frac{2}{c} = 1$,

$2c(c - 9)\left(\frac{1}{2} + \frac{2}{c - 9} - \frac{2}{c}\right) = 2c(c - 9)$, $c(c - 9) + 4c - 4(c - 9) = 2c^2 - 18c$,

$c^2 - 9c + 4c - 4c + 36 = 2c^2 - 18c$, $0 = c^2 - 9c - 36$, $0 = (c - 12)(c + 3)$;
$c = 12$ or -3; Pipe C requires 12 hours to empty the tank.

Page 581 • MIXED REVIEW EXERCISES

1. $\frac{\sqrt{n}}{3} = \frac{\sqrt{5}}{1}$; $\frac{n}{9} = \frac{5}{1}$, $n = 45$; $\{45\}$

2. $\frac{\sqrt{x - 3}}{6} = \frac{3}{2}$; $\frac{x - 3}{36} = \frac{9}{4}$, $4(x - 3) = 324$, $x - 3 = 81$, $x = 84$; $\{84\}$

3. $\frac{\sqrt{5t}}{8} = \frac{1}{4}$, $\frac{5t}{64} = \frac{1}{16}$, $80t = 64$, $t = \frac{4}{5}$; $\left\{\frac{4}{5}\right\}$ **4.** $\frac{13}{7} = \frac{20}{m}$, $13m = 140$, $m = \frac{140}{13}$; $\left\{\frac{140}{13}\right\}$

5. $m^2 = 14m - 45$, $m^2 - 14m + 45 = 0$, $(m - 9)(m - 5) = 0$; $m = 9$ or 5; $\{5, 9\}$

6. $y - \frac{7}{y + 2} - \frac{3y - 1}{y + 2} = 5$, $(y + 2)\left(y - \frac{7 + 3y - 1}{y + 2}\right) = 5(y + 2)$,

$y(y + 2) - (6 + 3y) = 5y + 10$, $y^2 + 2y - 6 - 3y = 5y + 10$, $y^2 - 6y - 16 = 0$,
$(y - 8)(y + 2) = 0$; $y = 8$ or -2, but $y = -2$ makes a denominator zero; $\{8\}$

7. $\frac{1}{3} = \frac{4}{y_2}$, $y_2 = 12$ **8.** $\frac{10}{y_1} = \frac{2}{5}$, $50 = 2y_1$, $25 = y_1$ **9.** $\frac{24}{64} = \frac{x_2}{8}$, $192 = 64x_2$, $3 = x_2$

Page 582 • EXTRA

1.
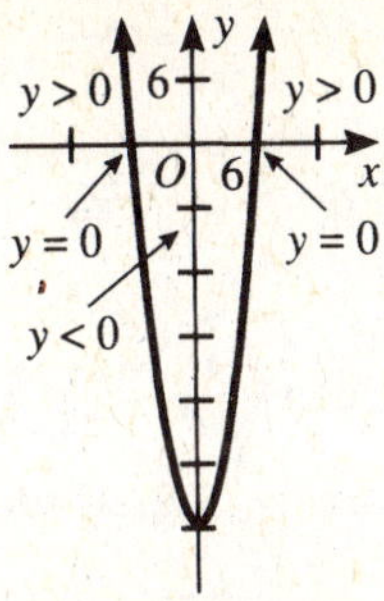

2.
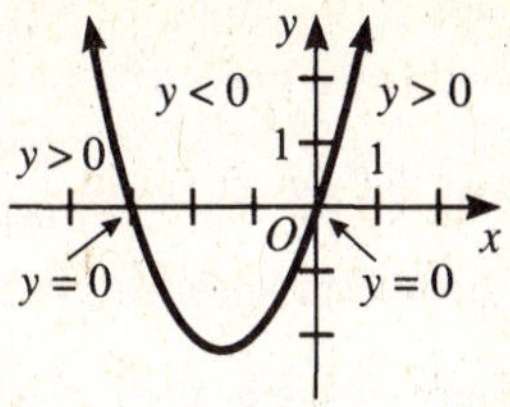

3.
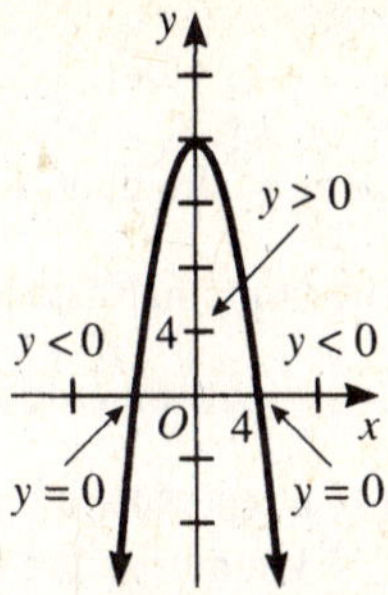

4.
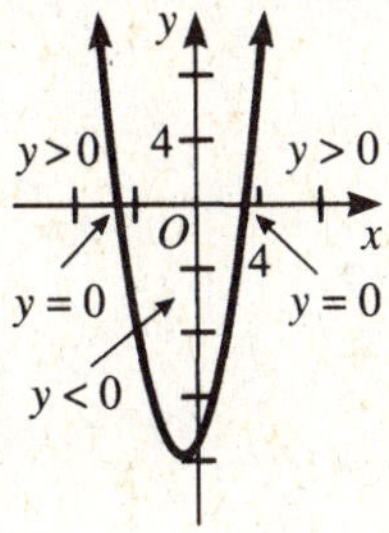

5.
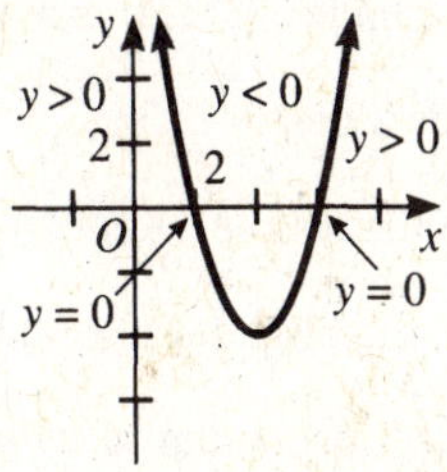

6.
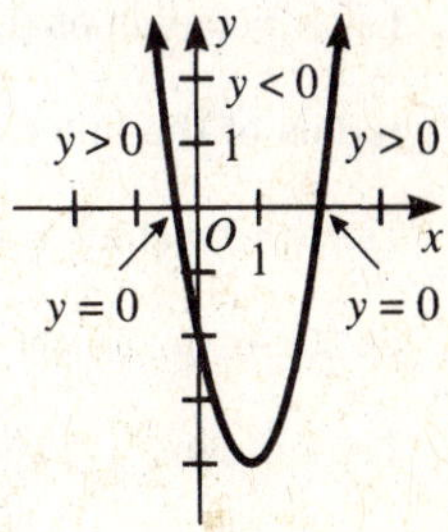

7. $\sqrt{x^2 - 3x}$ represents a real number when $x^2 - 3x \geq 0$. $x(x - 3) = 0$; $x = 0, 3$. Since the coefficient of x^2 is > 0, the parabola opens up, $\therefore x^2 - 3x \geq 0$ when $x \leq 0$ or $x \geq 3$.

8. $x^2 + 4x + 3 = 0$, $(x + 3)(x + 1) = 0$; $x = -3, -1$. Since the coefficient of x^2 is > 0, the parabola opens up, $\therefore x^2 + 4x + 3 \geq 0$ when $x \leq -3$ or $x \geq -1$.

9. $x^2 - 10x + 21 = 0$, $(x - 3)(x - 7) = 0$; $x = 3, 7$. Since the coefficient of x^2 is > 0, the parabola opens up, $\therefore x^2 - 10x + 21 \geq 0$ when $x \leq 3$ or $x \geq 7$.

Page 583 • APPLICATION

1. $720 \cdot 420 = 840V'$, $V' = \dfrac{720 \cdot 420}{840} = 360$; Its volume will be 360 cm^3.

2. $760 \cdot 2 = p'(1.9)$, $p' = \dfrac{760 \cdot 2}{1.9} = 800$; The pressure must be 800 mm.

3. $750 \cdot 500 = 625V'$, $V' = \dfrac{750 \cdot 500}{625} = 600$; The resulting volume will be 600 cm^3.

4. $760 \cdot 2 = p'(1)$, $p' = 760 \cdot 2 = 1520$; The new pressure is 1520 mm.

5. $760 \cdot 9 = 304V'$, $V' = \dfrac{760 \cdot 9}{304} = 22.5$; Its volume will be 22.5 m^3.

Pages 586–587 • PROBLEMS

A 1. Let d = diameter of the larger ball. $\dfrac{452}{12^2} = \dfrac{1017}{d^2}$, $452d^2 = 146{,}448$; $d^2 = 324$; $d = \pm 18$; The diameter of the larger ball is 18 cm.

2. Let m = mass of the larger diamond. $\frac{2700}{(1.5)^2} = \frac{4800}{m^2}$, $2700m^2 = 10{,}800$; $m^2 = 4$; $m = \pm 2$; The larger diamond has a mass of 2 carats.

3. Let d = stopping distance at the slower speed. $\frac{175}{(80)^2} = \frac{d}{(64)^2}$, $d = \frac{175 \cdot (64)^2}{(80)^2} = 112$; The stopping distance at the slower speed is 112 m.

4. Let d = distance the captain can see at the higher viewpoint. $\frac{4}{(30)^2} = \frac{10}{d^2}$, $4d^2 = 10 \cdot 900$, $d^2 = 2250$; $d = \pm\sqrt{2250} = \pm 15\sqrt{10} \approx \pm 47.4$; The captain can see 47.4 km across the ocean.

5. Let r = radius of the taller cylinder. $(6)^2(24) = r^2(96)$, $r^2 = \frac{36 \cdot 24}{96} = 9$; $r = \pm 3$; The radius of the taller cylinder is 3 cm.

6. Let b = brightness closer to the source. $(1.2)^2(25) = (1.2 - 0.4)^2(b)$, $b = \frac{(1.44)(25)}{(0.8)^2} = 56.25$; At the closer position, the brightness would measure 56.25 lm/m^2.

7. Let t = time it takes the longer pendulum to swing. $\frac{100}{(1)^2} = \frac{150}{t^2}$, $100t^2 = 150$, $t^2 = \frac{150}{100} = 1.5$; $t = \pm\sqrt{1.5} \approx \pm 1.2$; It takes 1.2 s for the longer pendulum to swing.

8. Let r = radius of the lens for the shorter time. $0.01(1)^2 = 0.0025(r)^2$, $r^2 = \frac{0.01}{0.0025} = 4$; $r = \pm 2$; The lens radius for the shorter time is 2 cm.

B **9.** Let r_1 = radius of the shorter cylinder and r_2 = radius of the taller cylinder. $3(r_1)^2 = 6(r_2)^2$, $r_1^2 = 2r_2^2$; $r_1 = \pm\sqrt{2r_2^2} = \pm\sqrt{2}(r_2)$; The radius of the shorter cylinder is $\sqrt{2}$ times as big as the radius of the taller cylinder.

10. Let d = closer distance. $(200)^2 = 16(d)^2$, $d^2 = \frac{40{,}000}{16} = 2500$; $d = \pm\sqrt{2500} = \pm 50$; The observer is now 50 m away from the source.

C **11.** Let $4x$ = radius of the first sphere and $3x$ = radius of the second sphere. $\frac{(4x)^3}{(3x)^3} = \frac{64x^3}{27x^3} = \frac{64}{27}$; The ratio of the volumes of the spheres is 64 : 27.

12. Let x_1 = surface temperature of one star and x_2 = surface temperature of the other. $\frac{x_1^4}{x_2^4} = \frac{16}{81}$, $\left(\frac{x_1}{x_2}\right)^4 = \frac{16}{81}$, $\frac{x_1}{x_2} = \pm\sqrt[4]{\frac{16}{81}} = \pm\frac{2}{3}$; The ratio of the stars' surface temperatures is 2 : 3.

Page 587 • MIXED REVIEW EXERCISES

1. $3x^2 - 2y = 1$, $3x^2 - 1 = 2y$, $y = \frac{3x^2 - 1}{2}$

2. $A = \frac{h(b + c)}{2}$, $\frac{2A}{b + c} = h$; $b + c \neq 0$, $b \neq -c$

3. $A = \frac{h(b + c)}{2}$, $\frac{2A}{h} = b + c$, $b = \frac{2A}{h} - c$ or $\frac{2A - hc}{h}$; $h \neq 0$

3. $A = \frac{h(b + c)}{2}, \frac{2A}{h} = b + c, b = \frac{2A}{h} - c$ or $\frac{2A - hc}{h}; h \neq 0$

4. $1 - x = ax, 1 = ax + x, 1 = (a + 1)x, x = \frac{1}{a + 1}; a + 1 \neq 0, a \neq -1$

5. $t = \frac{V - k^2}{g}, gt = V - k^2, gt + k^2 = V; g \neq 0$

6. $Er^2 = k, r^2 = \frac{k}{E}; r = \pm\sqrt{\frac{k}{E}}; r = \pm\frac{\sqrt{k}}{\sqrt{E}} \cdot \frac{\sqrt{E}}{\sqrt{E}}; r = \pm\frac{\sqrt{kE}}{E}; k \geq 0, E > 0$

7. $m = \frac{-7 - 1}{0 - 4} = \frac{-8}{-4} = 2; y = 2x + b, -7 = 2(0) + b, b = -7; y = 2x - 7,$

$-2x + y = -7, 2x - y = 7$

8. $m = \frac{-4 - 2}{-6 - 3} = \frac{-6}{-9} = \frac{2}{3}; y = \frac{2}{3}x + b, 2 = \frac{2}{3}(3) + b, 2 = 2 + b, b = 0; y = \frac{2}{3}x,$

$3y = 2x, 2x - 3y = 0$

9. $m = \frac{7 - 3}{-4 - 4} = \frac{4}{-8} = -\frac{1}{2}; y = -\frac{1}{2}x + b, 3 = -\frac{1}{2}(4) + b, 3 = -2 + b, 5 = b;$

$y = -\frac{1}{2}x + 5, 2y = -x + 10, x + 2y = 10$

10. $m = \frac{23 - (-1)}{7 - (-1)} = \frac{24}{8} = 3, y = 3x + b, -1 = 3(-1) + b, -1 = -3 + b, 2 = b,$

$y = 3x + 2, 3x - y = -2$

11. $m = \frac{-6 - (-4)}{-12 - (0)} = \frac{-6 + 4}{-12} = \frac{-2}{-12} = \frac{1}{6}; y = \frac{1}{6}x + b, -6 = \frac{1}{6}(-12) + b, -6 = -2 + b,$

$-4 = b; y = \frac{1}{6}x - 4, 6y = x - 24, x - 6y = 24$

12. $m = \frac{-5 - (-10)}{13 - 0} = \frac{-5 + 10}{13} = \frac{5}{13}; y = \frac{5}{13}x + b, -10 = \frac{5}{13}(0) + b, -10 = b;$

$y = \frac{5}{13}x - 10, \frac{5}{13}x - y = 10, 5x - 13y = 130$

Page 590 • WRITTEN EXERCISES

A **1.** $\frac{c_1b_1}{a_1} = \frac{c_2b_2}{a_2}, \frac{(15)(40)}{(18)} = \frac{c_2(25)}{36}, \frac{100}{3} \cdot \frac{36}{25} = c_2, c_2 = 48; \{48\}$

2. $\frac{d_1}{r_1t_1} = \frac{d_2}{r_2t_2}, \frac{45}{(15)(14)} = \frac{d_2}{(21)(8)}, \frac{3}{14} \cdot (21)(8) = d_2, d_2 = 36; \{36\}$

3. $\frac{m_1}{v_1u_1^2} = \frac{m_2}{v_2u_2^2}, \frac{9}{(15)(36)} = \frac{60}{25(u_2^2)}, 5u_2^2 = 12 \cdot 60, u_2^2 = 144; u_2 = \pm 12; \{12, -12\}$

4. $\frac{c_1h_1^2}{n_1} = \frac{c_2h_2^2}{n_2}, \frac{1 \cdot 11^2}{50} = \frac{(3.63) \cdot h_2^2}{6}, \frac{121}{50} \cdot \frac{6}{3.63} = h_2^2, h_2^2 = 4; h_2 = \pm 2; \{-2, 2\}$

5. $\frac{y_1^2x_1}{z_1} = \frac{y_2^2x_2}{z_2}, \frac{3^2 \cdot 12}{21} = \frac{y_2^2 \cdot 8}{28}, \frac{36}{7} \cdot \frac{7}{2} = y_2^2, y_2^2 = 18; y_2 = \pm\sqrt{18} = \pm 3\sqrt{2} \approx \pm 4.2;$

$\{-4.2, 4.2\}$

6. $\frac{a_1v_1^2}{r_1} = \frac{a_2v_2^2}{r_2}, \frac{15(8)^2}{27} = \frac{10v_2^2}{4.5}, \frac{5(64)}{9} \cdot \frac{4.5}{10} = v_2^2, v_2^2 = 16; v_2 = \pm 4; \{-4, 4\}$

B **7.** $\frac{x_1}{u_1v_1} = \frac{x_2}{u_2v_2}, \frac{x_1}{u_1v_1} = \frac{x_2}{(2u_1)(2v_1)}, x_1 \cdot 4u_1v_1 = x_2 \cdot u_1v_1, x_2 = 4x_1$; x is quadrupled when u and v are both doubled. $\frac{x_1}{u_1v_1} = \frac{x_2}{(2u_1)\left(\frac{1}{2}v_1\right)}, \frac{x_1}{u_1v_1} = \frac{x_2}{u_1v_1}, x_1 = x_2$; x is unchanged when u is doubled and v is halved.

8. $\frac{m_1s_1}{r_1^2} = \frac{m_2(2s_1)}{(2r_1)^2}, m_1s_1 \cdot 4r_1^2 = 2m_2s_1 \cdot r_1^2, 2m_1 = m_2$; m is doubled when both r and s are doubled. $\frac{m_1s_1}{r_1^2} = \frac{m_2(3s_1)}{(3r_1)^2}, 9m_1s_1r_1^2 = 3m_2s_1r_1^2, 3m_1 = m_2$; m is tripled when both r and s are tripled.

9. $h_1^2 = \frac{3r^2}{\pi d}; h_2^2 = \frac{3(2r)^2}{\pi\left(\frac{1}{2}d\right)} = \frac{24r^2}{\pi d}; \frac{h_1^2}{h_2^2} = \left(\frac{3r^2}{\pi d}\right) \div \left(\frac{24r^2}{\pi d}\right), \left(\frac{h_1}{h_2}\right)^2 = \frac{3r^2}{\pi d} \cdot \frac{\pi d}{24r^2} = \frac{1}{8}$;

$\frac{h_1}{h_2} = \pm\frac{1}{\sqrt{8}}, h_2 = \pm 2\sqrt{2}(h_1)$; h is increased by a factor of $2\sqrt{2}$ (≈ 2.8) when r is doubled and d is halved.

10. $t_1^2 = \frac{Fs}{r}; t_2^2 = \frac{F(3s)}{\left(\frac{1}{2}r\right)}; \frac{t_2^2}{t_1^2} = \left(\frac{F(3s)}{\left(\frac{1}{2}r\right)}\right) \div \left(\frac{Fs}{r}\right), \left(\frac{t_2}{t_1}\right)^2 = \frac{6Fs}{r} \cdot \frac{r}{Fs} = 6, \frac{t_2}{t_1} = \pm\sqrt{6}$,

$t_2 = \pm\sqrt{6}(t_1)$; t is increased by a factor of $\sqrt{6}$ (≈ 2.4) when s is tripled and r is halved.

Pages 590–592 • PROBLEMS

A **1.** Let v = volume of the shorter pyramid. $\frac{420}{(20)(63)} = \frac{v}{(8)(39)}, 104 = v$; The volume of the shorter pyramid is 104 cm^3.

2. Let h = height of the narrower jar. $\frac{588}{(10.4)(18)} = \frac{294}{(7.8)h}, 4586.4h = 55{,}036.8; h = 12$; The narrower jar is 12 cm high.

3. Let s = number of students needed. $\frac{5 \cdot 2}{280} = \frac{3s}{2100}, 840s = 21{,}000; s = 25$; 25 students are needed.

4. Let t = time it takes to mow the larger lawn. $\frac{3 \cdot 0.5}{300} = \frac{7t}{7000}, 2100t = 10{,}500; t = 5$; It will take 5 h to mow the larger lawn.

5. Let c = cost of cooking the two meals. $\frac{0.15}{\cancel{(6000)}\left(\frac{1}{6}\right)\cancel{(0.075)}} = \frac{c}{\cancel{(6000)}\left(\frac{7}{3}\right)\cancel{(0.075)}}$,

$\frac{1}{6}c = 0.35, c = 2.1$; The cost of cooking the two meals is \$2.10.

6. Let r = radius of the taller cone. $\frac{702\cancel{\pi}}{(15)(12)^2} = \frac{576\cancel{\pi}}{(27)r^2}, (702)(27)r^2 = (576)(15)(144)$,

$r^2 = \frac{2560}{39}, r = \pm\sqrt{\frac{2560}{39}} \approx \pm 8.1$; The radius of the taller cone is 8.1 cm.

7. Let t = time it takes to travel 1800 m. $\frac{250}{(20)(5)^2} = \frac{1800}{(25)t^2}$, $6250t^2 = 900{,}000$; $t^2 = 144$; $t = \pm 12$; It takes the car 12 s.

8. Let r = radius of the shorter wire. $\frac{(0.3125)(0.48)^2}{800} = \frac{(0.5)r^2}{500}$, $r^2 = 0.09$; $r = \pm 0.3$; The radius of the shorter wire is 0.3 cm.

B **9.** Let m = total mass of Capella. $\frac{(3.45)(50.0)^2}{(41.0)^3} = \frac{m(0.285)^2}{(1.51)^3}$, $m = \frac{(3.45)(50.0)^2(1.51)^3}{(41.0)^3(0.285)^2} = 5.30$; The total mass of Capella is 5.30.

10. Let g = growth factor of the smaller investment. $\frac{(1.5)(\sqrt{2000})}{\sqrt{4500}} = \frac{g\sqrt{1600}}{\sqrt{4900}}$, $g = \frac{(1.5)(20\sqrt{5})}{30\sqrt{5}} \cdot \frac{70}{40} = 1.75$; The growth factor is 1.75.

11. Let h = amount of heat lost through the second pane. $\frac{(49.5)(0.8)}{(44)(\cancel{40} \cdot \cancel{28})} = \frac{h(0.5)}{(40)\frac{1}{4}(\cancel{40} \cdot \cancel{28})}$; $h = \frac{(49.5)(0.8)(10)}{(44)(0.5)} = 18$; 18 joules of heat are lost through the second window.

12. Let w = wind speed. $\frac{20}{(4 \cdot 1.5)(12)^2} = \frac{30}{(2 \cdot 2)(w^2)}$, $80w^2 = 25{,}920$; $w^2 = 324$; $w = \pm\sqrt{324} = \pm 18$; The speed of the wind is 18 mi/h.

C **13. a.** $\frac{h_1}{r_1c_1^2} = \frac{h_2}{(2r_1)c_1^2}$, $h_2 = \frac{2r_1c_1^2}{r_1c_1^2}h_1$, $h_2 = 2h_1$; the heat is doubled.

b. $\frac{h_1}{r_1c_1^2} = \frac{h_2}{r_1(2c_1)^2}$, $h_2 = \frac{r_1(4c_1^2)}{r_1c_1^2}h_1$, $h_2 = 4h_1$; the heat is quadrupled.

c. $\frac{h_1}{r_1c_1^2} = \frac{h_2}{(2r_1)(3c_1)^2}$, $h_2 = \frac{18r_1c_1^2}{r_1c_1^2}h_1$, $h_2 = 18h_1$; the heat is increased by a factor of 18.

14. a. $\frac{p_1r_1}{(v_1)^2} = \frac{p_2r_1}{\left(\frac{1}{2}v_1\right)^2}$, $p_2 = \frac{\frac{1}{4}r_1v_1^2}{r_1v_1^2}p_1$, $p_2 = \frac{1}{4}p_1$; the power is reduced to $\frac{1}{4}$ of the original level.

b. $\frac{p_1r_1}{(v_1)^2} = \frac{p_2\frac{1}{2}r_1}{(v_1)^2}$, $p_2 = \frac{r_1v_1^2}{\frac{1}{2}r_1v_1^2}p_1$, $p_2 = 2p_1$; the power is doubled.

c. $\frac{p_1r_1}{(v_1)^2} = \frac{p_2(4r_1)}{(3v_1)^2}$, $p_2 = \frac{r_1 \cdot 9v_1^2}{4r_1v_1^2}p_1$, $p_2 = \frac{9}{4}p_1$; the power increases by a factor of $\frac{9}{4}$.

Page 592 • MIXED REVIEW EXERCISES

1.
$$\begin{aligned} 3x - 2y = 1 &\rightarrow 3x - 2y = 1 \\ x - y = 0 &\rightarrow \underline{3x - 3y = 0} \\ & \qquad y = 1;\ x - (1) = 0,\ x = 1;\ (1, 1) \end{aligned}$$

2.
$$\begin{aligned} x + y &= 10 \\ \underline{x - 2y} &\underline{= 16} \\ 3y &= -6,\ y = -2;\ x + (-2) = 10,\ x = 12;\ (12, -2) \end{aligned}$$

3. $$\begin{array}{l} 2x - 5y = 3 \\ \underline{-x + 5y = 6} \\ x \qquad = 9;\ -(9) + 5y = 6,\ 5y = 15,\ y = 3;\ (9, 3) \end{array}$$

4. $$\begin{array}{l} 5x + 3y = \ \ 5 \\ \underline{4x + 3y = 10} \\ x \qquad = -5;\ 5(-5) + 3y = 5,\ 3y = 30,\ y = 10;\ (-5, 10) \end{array}$$

5. $$\begin{array}{l} x - y = 0 \\ \underline{x + y = 4} \\ 2x \qquad = 4,\ x = 2;\ (2) + y = 4,\ y = 2;\ (2, 2) \end{array}$$

6. $$\begin{array}{l} x + \ \ y = \ \ 3 \rightarrow 3x + 3y = \ \ 9 \\ 2x + 3y = -2 \rightarrow \underline{2x + 3y = -2} \\ \qquad\qquad\qquad\ \ x \qquad = 11;\ (11) + y = 3,\ y = -8;\ (11, -8) \end{array}$$

7. $(2x + 1)(4x + 2) = 8x^2 + 4x + 4x + 2 = 8x^2 + 8x + 2$

8. $(8s + 4)(5 + 6s) = 40s + 48s^2 + 20 + 24s = 48s^2 + 64s + 20$

9. $(7t - 2y)(7t + 2y) = (7t)^2 - (2y)^2 = 49t^2 - 4y^2$

10. $(3p - 4k)^2 = (3p)^2 - 2(3p)(4k) + (4k)^2 = 9p^2 - 24pk + 16k^2$

11. $(r - 2)^2 = r^2 - 4r + 4$

12. $(xy - 2)(3xy + 4) = 3x^2y^2 + 4xy - 6xy - 8 = 3x^2y^2 - 2xy - 8$

Page 593 • APPLICATION

1. $E = \frac{160}{(4)^2} = \frac{160}{16} = 10$; the illumination is 10 lx.

2. $E = \frac{315}{(3)^2} = \frac{315}{9} = 35$; the illumination is 35 lx.

3. $24 = \frac{216}{r^2}$, $r^2 = \frac{216}{24} = 9$; $r = \pm 3$; it must be placed 3 m from the table surface.

4. $E = \frac{120}{(2)^2} = \frac{120}{4} = 30$; the illumination is 30 lx.

5. $E = \frac{125}{(2.5)^2} = \frac{125}{6.25} = 20$; the illumination is 20 lx.

Pages 595–596 • CHAPTER REVIEW

1. d; $2x^2 - 128 = 0$, $2x^2 = 128$, $x^2 = 64$; $x = \pm 8$; $\{-8, 8\}$

2. a; $(a - 25)^2 = 9$; $a - 25 = \pm 3$, $a = 25 \pm 3$, $a = 22$ or 28; $\{22, 28\}$

3. c; $x^2 - 4x = 77$, $x^2 - 4x + (-2)^2 = 77 + 4$, $(x - 2)^2 = 81$; $x - 2 = \pm 9$, $x = 2 \pm 9$, $x = -7$ or 11; $\{-7, 11\}$

4. c; $c^2 - 10c = 20$, $c^2 - 10c + (-5)^2 = 20 + 25$, $(c - 5)^2 = 45$; $c - 5 = \pm\sqrt{45}$, $c = 5 \pm 3\sqrt{5}$; $\{5 + 3\sqrt{5}, 5 - 3\sqrt{5}\}$

5. a; $z = \frac{-11 \pm \sqrt{(11)^2 - 4(5)(2)}}{2(5)} = \frac{-11 \pm \sqrt{121 - 40}}{10} = \frac{-11 \pm \sqrt{81}}{10} = \frac{-11 \pm 9}{10} = -2$ or $-\frac{1}{5}$; $\left\{-2, -\frac{1}{5}\right\}$

6. d; $y = \frac{-(-1) \pm \sqrt{(-1)^2 - 4(10)(-3)}}{2(10)} = \frac{1 \pm \sqrt{1 + 120}}{20} = \frac{1 \pm \sqrt{121}}{20} = \frac{1 \pm 11}{20} = -\frac{1}{2}$ or $\frac{3}{5}$; $\left\{-\frac{1}{2}, \frac{3}{5}\right\}$

7. b; $b^2 - 4ac = (-3)^2 - 4(2)(-2) = 9 + 16 = 25$

8. a; $b^2 - 4ac = (2)^2 - 4(1)(5) = 4 - 20 = -16$; no real roots. **9.** a

10. d; $x(x + 2) = 143$, $x^2 + 2x = 143$, $x^2 + 2x + (1)^2 = 143 + 1$, $(x + 1)^2 = 144$; $x + 1 = \pm 12$, $x = -1 \pm 12 = -13$ or 11; $x + 2 = 13$; the paper is 11 cm by 13 cm.

11. b; $y_1 x_1{}^2 = y_2\left(\frac{1}{2}x_1\right)^2$, $y_1 x_1{}^2 = y_2 \frac{1}{4} x_1{}^2$, $y_2 = 4y_1$; If x is halved, y is quadrupled.

12. d; $\frac{\frac{1}{2}(6)^2}{(2)^3} = \frac{r(12)^2}{(4)^3}$, $r = \frac{\frac{1}{2} \cdot 36 \cdot 64}{8 \cdot 144} = 1$

Page 596 • CHAPTER TEST

1. $36x^2 = 49$, $x^2 = \frac{49}{36}$; $x = \pm\sqrt{\frac{49}{36}} = \pm\frac{7}{6}$; $\left\{-\frac{7}{6}, \frac{7}{6}\right\}$

2. $3(a - 5)^2 = 21$, $(a - 5)^2 = 7$, $a - 5 = \pm\sqrt{7}$, $a = 5 \pm \sqrt{7}$; $\{5 + \sqrt{7}, 5 - \sqrt{7}\}$

3. $c^2 + 2c = 15$, $c^2 + 2c + (1)^2 = 15 + 1$, $(c + 1)^2 = 16$; $c + 1 = \pm 4$, $c = -1 \pm 4 = -5$ or 3; $\{-5, 3\}$

4. $b^2 - 4b = 10$, $b^2 - 4b + (-2)^2 = 10 + 4$, $(b - 2)^2 = 14$, $b - 2 = \pm\sqrt{14}$, $b = 2 \pm \sqrt{14}$; $\{2 + \sqrt{14}, 2 - \sqrt{14}\}$

5. $z = \frac{-(-6) \pm \sqrt{(-6)^2 - 4(3)(-9)}}{2(3)} = \frac{6 \pm \sqrt{36 + 108}}{6} = \frac{6 \pm \sqrt{144}}{6} = \frac{6 \pm 12}{6} = 1 \pm 2 =$ -1 or 3; $\{-1, 3\}$

6. $y = \frac{-6 \pm \sqrt{6^2 - 4(1)(-23)}}{2(1)} = \frac{-6 \pm \sqrt{36 + 92}}{2} = \frac{-6 \pm \sqrt{128}}{2} = \frac{-6 \pm 8\sqrt{2}}{2} =$ $-3 \pm 4\sqrt{2}$; $\{-3 + 4\sqrt{2}, -3 - 4\sqrt{2}\}$

7. $b^2 - 4ac = 3^2 - 4(2)(2) = 9 - 16 = -7$; no real roots.

8. $b^2 - 4ac = 4^2 - 4(1)(-4) = 16 + 16 = 32$; two real roots.

9. $7x^2 - 3x = 0$, $x(7x - 3) = 0$; $x = 0$ or $7x - 3 = 0$, $x = 0$ or $7x = 3$, $x = 0$ or $x = \frac{3}{7}$; $\left\{0, \frac{3}{7}\right\}$

10. $x = \frac{-(-10) \pm \sqrt{(-10)^2 - 4(3)(-4)}}{2(3)} = \frac{10 \pm \sqrt{100 + 48}}{6} = \frac{10 \pm \sqrt{148}}{6} = \frac{10 \pm 2\sqrt{37}}{6} =$ $\frac{5 \pm \sqrt{37}}{3}$; $\left\{\frac{5 + \sqrt{37}}{3}, \frac{5 - \sqrt{37}}{3}\right\}$

11. Let x = amount of increase in the width and length. $(4 + x)(7 + x) = 2(4 \cdot 7)$, $28 + 11x + x^2 = 56$, $x^2 + 11x - 28 = 0$; $x = \frac{-11 \pm \sqrt{11^2 - 4(1)(-28)}}{2(1)} =$ $\frac{-11 \pm \sqrt{121 + 112}}{2} = \frac{-11 \pm \sqrt{233}}{2} \approx 2.1$ or -13.1; $4 + (2.1) = 6.1$; $7 + (2.1) = 9.1$; The new garden measures 6.1 m by 9.1 m.

12. Let a = area of the larger square. $\frac{128}{(16)^2} = \frac{a}{(24)^2}$, $a = \frac{128 \cdot 576}{256} = 288$; The area of the larger square is 288 cm^2.

13. $\frac{50}{(25)(6)} = \frac{108}{36h}$, $1800h = 16{,}200$; $h = 9$

Pages 597–598 • CUMULATIVE REVIEW

1. $(-1)\left[(1) - \left(-\frac{5}{2}\right)\right] = -\left(1 + \frac{5}{2}\right) = -\frac{7}{2}$

2. $\left[(1) \div (3) + \left(-\frac{5}{2}\right)\right]^{(-1)} = \frac{1}{\frac{1}{3} - \frac{5}{2}} = \frac{1}{\frac{1}{3} - \frac{5}{2}} \cdot \frac{6}{6} = \frac{6}{2 - 15} = \frac{6}{-13} = -\frac{6}{13}$

3. $0 \div \left(-\frac{5}{2}\right) = 0$

4. $[(-1)(3)]^2 - [(-1)(3)]^3 = (-3)^2 - (-3)^3 = 9 - (-27) = 9 + 27 = 36$

5. $\left[(3)\left(-\frac{5}{2}\right) + (-1)(1)\right]^0 = 1$ **6.** $(3) \div \left(-\frac{5}{2}\right)^{(-1)} = 3 \div \left(-\frac{2}{5}\right) = 3 \cdot \left(-\frac{5}{2}\right) = -\frac{15}{2}$

7. $5 \cdot 58 - 7 \cdot 58 - (-58) = 58(5 - 7 + 1) = 58 \cdot (-1) = -58$

8. $6^2 \div 3^2 + 4 \div 3 = \frac{36}{9} + \frac{4}{3} = \frac{36 + 12}{9} = \frac{48}{9} = \frac{16}{3}$

9. $(6 - 13)(6 + 13) = 6^2 - 13^2 = 36 - 169 = -133$

10. $3x - (0.4x + 1.2y) = 3x - 0.4x - 1.2y = 2.6x - 1.2y$

11. $15(3x - 4) + 5 - 9c$, $45x - 60 + 5 - 9c$, $45x - 55 - 9c$

12. $\left(-\frac{1}{8}x^5y^3z\right)(-24x^2y) = 3x^7y^4z$ **13.** $-\frac{3}{5}a^3b(-20ab^2 - 15b^3) = 12a^4b^3 + 9a^3b^4$

14. $(5x^3y^2)^2(3x^4y^3)^2 = 25x^6y^4 \cdot 9x^8y^6 = 225x^{14}y^{10}$ **15.** $\frac{-72wx^4y^5z}{-96x^3y^7z} = \frac{3wx}{4y^2}$

16. $\frac{(3r^3s^4)^3}{(-9r^5s^6)^2} = \frac{27r^9s^{12}}{81r^{10}s^{12}} = \frac{1}{3r}$

17. $(4u^2 - 2u + 1)(2u + 1) = 8u^3 + 4u^2 - 4u^2 - 2u + 2u + 1 = 8u^3 + 1$

18. $a - 2(3a - 2) = a - 6a + 4 = 4 - 5a$

19. $(9z - 3)(5z + 4) = 45z^2 + 36z - 15z - 12 = 45z^2 + 21z - 12$

20. $(3a^3 - 2)^2 = (3a^3)^2 - 2(3a^3)(2) + (-2)^2 = 9a^6 - 12a^3 + 4$

21. $(8r^3t - 3v)(8r^3t + 3v) = 64r^6t^2 - 9v^2$

22. $(3.4 \times 10^{12})(0.2 \times 10^{-3}) = 0.68 \times 10^9 = 6.8 \times 10^{-1} \times 10^9 = 6.8 \times 10^8$

23. $\frac{12c^2 - 26cd + 12d^2}{12c^2 - 44cd + 24d^2} = \frac{2(6c^2 - 13cd + 6d^2)}{4(3c^2 - 11cd + 6d^2)} = \frac{2(2c - 3d)(3c - 2d)}{4(3c - 2d)(c - 3d)} = \frac{2c - 3d}{2(c - 3d)}$

24. $(18x^2 - x - 4) \div (9x + 4) = \frac{(9x + 4)(2x - 1)}{9x + 4} = 2x - 1$

25. $(27x^3 + 64) \div (3x + 4) = \frac{(3x + 4)(9x^2 - 12x + 16)}{3x + 4} = 9x^2 - 12x + 16$

26. $\frac{b^2 + b - 12}{b + 2} \cdot \frac{b^2 - 4}{b^2 - 5b - 6} = \frac{(b + 4)(b - 3)}{b + 2} \cdot \frac{(b + 2)(b - 2)}{(b - 6)(b + 1)} = \frac{(b + 4)(b - 3)(b - 2)}{(b - 6)(b + 1)}$

27. $\frac{3z^2 - 8z - 16}{z^2 + 6z + 9} \div \frac{z^2 - 3z - 4}{z + 3} = \frac{(3z + 4)(z - 4)}{(z + 3)^2} \cdot \frac{z + 3}{(z - 4)(z + 1)} = \frac{3z + 4}{(z + 3)(z + 1)}$

28. $\frac{85k^8 + 51k^4 - 17k^2}{-17k^4} = \frac{17k^2(5k^6 + 3k^2 - 1)}{-17k^4} = -\frac{5k^6 + 3k^2 - 1}{k^2}$

29. $\frac{4a - 4}{3} - \frac{5 - 2a}{3} = \frac{4a - 4 - 5 + 2a}{3} = \frac{6a - 9}{3} = 2a - 3$

30. $\frac{n + 1}{2n - 3} - \frac{4}{n + 1} + 3 = \frac{(n + 1)(n + 1)}{(2n - 3)(n + 1)} - \frac{4(2n - 3)}{(2n - 3)(n + 1)} + \frac{3(2n - 3)(n + 1)}{(2n - 3)(n + 1)} =$

$\frac{n^2 + 2n + 1 - 8n + 12 + 3(2n^2 - n - 3)}{(2n - 3)(n + 1)} = \frac{n^2 - 6n + 13 + 6n^2 - 3n - 9}{(2n - 3)(n + 1)} =$

$\frac{7n^2 - 9n + 4}{(2n - 3)(n + 1)}$

31. $\frac{3}{m + 3} - \frac{1}{2m - 6} = \frac{3 \cdot 2(m - 3)}{2(m + 3)(m - 3)} - \frac{m + 3}{2(m + 3)(m - 3)} = \frac{6m - 18 - m - 3}{2(m + 3)(m - 3)} =$

$\frac{5m - 21}{2(m + 3)(m - 3)}$

32. $-\sqrt{2.25x^6z^3} = -1.5\,|x^3|z\sqrt{z}$

33. $3\sqrt{5}(\sqrt{180} - \sqrt{81}) = 3\sqrt{5}(6\sqrt{5} - 9) = 18 \cdot 5 - 27\sqrt{5} = 90 - 27\sqrt{5}$

34. $(3\sqrt{2} + 1)^2 = (3\sqrt{2})^2 + 2(3\sqrt{2})(1) + (1)^2 = 9 \cdot 2 + 6\sqrt{2} + 1 = 18 + 6\sqrt{2} + 1 =$

$19 + 6\sqrt{2}$

35. $\sqrt{\frac{18a^7}{7}} \cdot \sqrt{\frac{112}{2a^5}} = \sqrt{\frac{18 \cdot 112a^7}{14a^5}} = \sqrt{144a^2} = 12a$

36. $\sqrt{108} - \sqrt{24} + \sqrt{6} = 6\sqrt{3} - 2\sqrt{6} + \sqrt{6} = 6\sqrt{3} - \sqrt{6}$

37. $3\sqrt{80} + \frac{2\sqrt{35}}{\sqrt{7}} = 3 \cdot 4\sqrt{5} + \frac{2\sqrt{5}\sqrt{7}}{\sqrt{7}} = 12\sqrt{5} + 2\sqrt{5} = 14\sqrt{5}$

38. $\frac{3\sqrt{5} - 4}{\sqrt{2} - 3} = \frac{3\sqrt{5} - 4}{\sqrt{2} - 3} \cdot \frac{\sqrt{2} + 3}{\sqrt{2} + 3} = \frac{3\sqrt{10} + 9\sqrt{5} - 4\sqrt{2} - 12}{2 - 9} = \frac{3\sqrt{10} + 9\sqrt{5} - 4\sqrt{2} - 12}{-7}$

39. $(2\sqrt{5} + 4\sqrt{2})(6\sqrt{5} + 4\sqrt{2}) = 12 \cdot 5 + 8\sqrt{10} + 24\sqrt{10} + 16 \cdot 2 =$

$60 + 32\sqrt{10} + 32 = 92 + 32\sqrt{10}$

40. $49r^2 + 56rs + 16s^2 = (7r + 4s)^2$ **41.** $4t^5 + 9t^3 = t^3(4t^2 + 9)$

42. $30x^2 + 91x - 30 = (3x + 10)(10x - 3)$

43. $-16z^4 - 48z^3 - 64z^2 = -16z^2(z^2 + 3z + 4)$

44. $125a^4 - 45a^2 = 5a^2(25a^2 - 9) = 5a^2(5a + 3)(5a - 3)$

45. $12a^2 - 5a - 12$: prime

46. $2a^3 - a^2b - 8a + 4b = a^2(2a - b) - 4(2a - b) = (a^2 - 4)(2a - b) =$
$(a + 2)(a - 2)(2a - b)$

47. $c^2 - 8c + 16 = (c - 4)^2$

48. $-10p^2 + 27p + 28 = -(10p^2 - 27p - 28) = -(2p - 7)(5p + 4)$

49. $\frac{1}{5}x - 2 = \frac{2}{3}$, $15\left(\frac{1}{5}x - 2\right) = 15 \cdot \frac{2}{3}$, $3x - 30 = 10$, $3x = 40$, $x = \frac{40}{3}$; $\left\{\frac{40}{3}\right\}$

50. $15 - 3(y + 1) = 14 - 5y$, $15 - 3y - 3 = 14 - 5y$, $12 - 3y = 14 - 5y$, $2y =$
2, $y = 1$; $\{1\}$

51. $7w + 4 = |-39|$, $7w = 39 - 4$, $7w = 35$, $w = 5$; $\{5\}$

52. $16 = 4.8x - 2.48$, $18.48 = 4.8x$, $x = \frac{18.48}{4.8} = 3.85$; $\{3.85\}$

53. 35% of $t = 105$, $0.35t = 105$, $t = \frac{105}{0.35} = 300$; $\{300\}$

54. $3|x| + 5 = 9$, $3|x| = 4$, $|x| = \frac{4}{3}$, $x = -\frac{4}{3}$ or $\frac{4}{3}$; $\left\{-\frac{4}{3}, \frac{4}{3}\right\}$

55. $z^2 - 9z + 18 = 0$, $(z - 6)(z - 3) = 0$; $z = 6$ or 3; $\{3, 6\}$

56. $9t^2 - 24t + 16 = 0$, $(3t - 4)^2 = 0$; $3t - 4 = 0$, $3t = 4$, $t = \frac{4}{3}$; $\left\{\frac{4}{3}\right\}$

57. $12b^2 + 5b = 2$, $12b^2 + 5b - 2 = 0$, $(3b + 2)(4b - 1) = 0$; $3b + 2 = 0$ or $4b - 1 = 0$, $3b = -2$ or $4b = 1$, $b = -\frac{2}{3}$ or $b = \frac{1}{4}$; $\left\{-\frac{2}{3}, \frac{1}{4}\right\}$

58. $(3a - 2)(4a + 1) = 5$, $12a^2 - 5a - 2 = 5$, $12a^2 - 5a - 7 = 0$, $(12a + 7)(a - 1) = 0$; $12a + 7 = 0$ or $a - 1 = 0$, $12a = -7$ or $a = 1$, $a = -\frac{7}{12}$ or $a = 1$; $\left\{-\frac{7}{12}, 1\right\}$

59.
$$\begin{array}{rcl} 3x + 4y & = & 2 \\ -5x + 4y & = & -2 \\ \hline 8x & = & 4 \end{array}$$
$x = \frac{1}{2}$; $3\left(\frac{1}{2}\right) + 4y = 2$, $4y = 2 - \frac{3}{2}$, $4y = \frac{1}{2}$, $y = \frac{1}{8}$; $\left(\frac{1}{2}, \frac{1}{8}\right)$

60.
$$\begin{array}{rcl} 2x + y = 7 & \rightarrow & 4x + 2y = 14 \\ x - 2y = 11 & \rightarrow & x - 2y = 11 \\ \hline & & 5x = 25 \end{array}$$
$x = 5$; $2(5) + y = 7$, $10 + y = 7$, $y = -3$; $(5, -3)$

61. $\frac{x - 2}{8} - \frac{2x + 1}{12} = \frac{1}{3}$, $24\left(\frac{x - 2}{8} - \frac{2x + 1}{12}\right) = 24 \cdot \frac{1}{3}$, $3(x - 2) - 2(2x + 1) = 8$, $3x - 6 - 4x - 2 = 8$, $-x - 8 = 8$, $-16 = x$; $\{-16\}$

62. $\frac{3x + 2}{x - 1} = \frac{3x + 4}{x + 1}$, $(x + 1)(3x + 2) = (3x + 4)(x - 1)$, $3x^2 + 5x + 2 = 3x^2 + x - 4$, $5x + 2 = x - 4$, $4x = -6$, $x = -\frac{3}{2}$; $\left\{-\frac{3}{2}\right\}$

63. $|t - 3| \geq 5$; $t - 3 \geq 5$ or $t - 3 \leq -5$, $t \geq 8$ or $t \leq -2$; {−2, 8, and the real numbers less than −2 or greater than 8}

64. $-4 < 3t - 5 < 13$, $1 < 3t < 18$, $\frac{1}{3} < t < 6$; {the real numbers between $\frac{1}{3}$ and 6}

65. $0 \leq 4 - 2t \leq 7$, $-4 \leq -2t \leq 3$, $2 \geq t \geq -\frac{3}{2}$; $\left\{-\frac{3}{2}\right.$, 2, and the real numbers between $-\frac{3}{2}$ and $\left.2\right\}$

66. $\frac{1}{3} - \frac{7}{2}t < -\frac{5}{6}$, $-\frac{7}{2}t < -\frac{5}{6} - \frac{1}{3}$, $-\frac{7}{2}t < \frac{-5 - 2}{6}$, $-\frac{7}{2}t < \frac{-7}{6}$, $t > \frac{-7}{6} \cdot \frac{-2}{7}$, $t > \frac{1}{3}$; {the real numbers greater than $\frac{1}{3}$}

67. $-2\sqrt{t - 1} = 6$, $\sqrt{t - 1} = -3$; no solution.

68. $\sqrt{x^2+1} = x - 1$; $x^2 + 1 = (x-1)^2$, $x^2 + 1 = x^2 - 2x + 1$, $1 = -2x + 1$,
$0 = -2x$, $x = 0$; $\sqrt{0^2+1} = 0 - 1$, $\sqrt{1} \neq -1$; no solution.

69. $\sqrt{3r+1} = 2r - 6$, $3r + 1 = (2r-6)^2$, $3r + 1 = 4r^2 - 24r + 36$,
$4r^2 - 27r + 35 = 0$; $r = \dfrac{-(-27) \pm \sqrt{(-27)^2 - 4(4)(35)}}{2 \cdot 4} = \dfrac{27 \pm \sqrt{729 - 560}}{8} =$
$\dfrac{27 \pm \sqrt{169}}{8} = \dfrac{27 \pm 13}{8} = \dfrac{14}{8}$ or $\dfrac{40}{8} = \dfrac{7}{4}$ or 5; $\sqrt{3\left(\dfrac{7}{4}\right) + 1} = 2\left(\dfrac{7}{4}\right) - 6$,
$\sqrt{\dfrac{21+4}{4}} = \dfrac{7-12}{2}$, $\sqrt{\dfrac{25}{4}} \neq -\dfrac{5}{2}$; $\sqrt{3(5)+1} = 2(5) - 6$; $\sqrt{16} = 4$; $4 = 4$; $\{5\}$

70. $\frac{1}{2}m^2 + 2m + 2 = 0$, $m^2 + 4m + 4 = 0$, $(m+2)^2 = 0$; $m + 2 = 0$, $m = -2$; $\{-2\}$

71. $3t^2 - 5t + 1 = 0$; $t = \dfrac{-(-5) \pm \sqrt{(-5)^2 - 4(3)(1)}}{2(3)} = \dfrac{5 \pm \sqrt{25-12}}{6} = \dfrac{5 \pm \sqrt{13}}{6}$;
$\left\{\dfrac{5+\sqrt{13}}{6}, \dfrac{5-\sqrt{13}}{6}\right\}$

72. $3t^2 + 2t - 7 = 0$; $t = \dfrac{-2 \pm \sqrt{2^2 - 4(3)(-7)}}{2(3)} = \dfrac{-2 \pm \sqrt{4+84}}{6} = \dfrac{-2 \pm \sqrt{88}}{6} =$
$\dfrac{-2 \pm 2\sqrt{22}}{6} = \dfrac{-1 \pm \sqrt{22}}{3}$; $\left\{\dfrac{-1+\sqrt{22}}{3}, \dfrac{-1-\sqrt{22}}{3}\right\}$

73. $|x+1| > 3$; $x + 1 > 3$ or $x + 1 < -3$, $x > 2$ or $x < -4$; {the real numbers less than −4 or greater than 2}

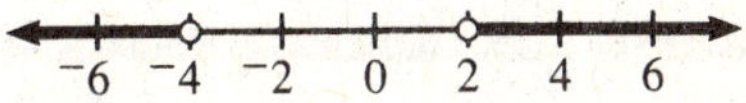

74.

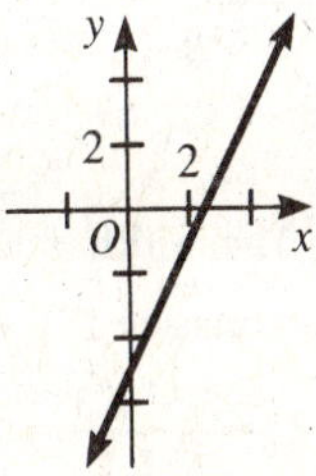

75.

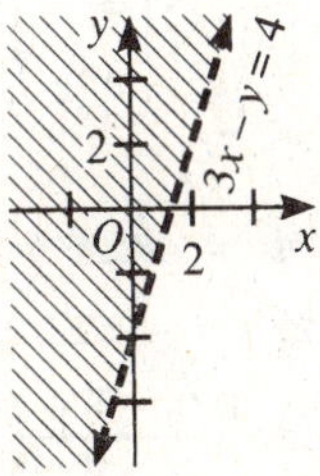

76. $m = \dfrac{-1-1}{6-(-1)} = \dfrac{-2}{6+1} = -\dfrac{2}{7}$; $y = -\dfrac{2}{7}x + b$, $1 = -\dfrac{2}{7}(-1) + b$, $1 = \dfrac{2}{7} + b$, $b = \dfrac{5}{7}$;
$y = -\dfrac{2}{7}x + \dfrac{5}{7}$, $7y = -2x + 5$, $2x + 7y = 5$

77.

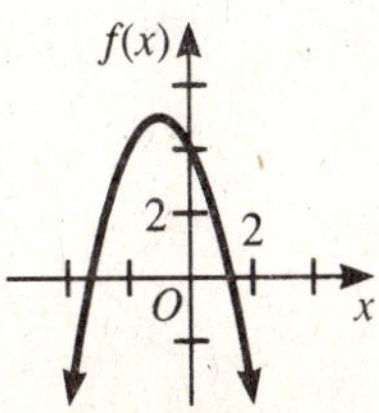

78. Let $N = 0.4\overline{25}$
$100N = 42.5\overline{25}$
$N = 0.4\overline{25}$
$99N = 42.1$
$N = \dfrac{42.1}{99} = \dfrac{421}{990}$

79.

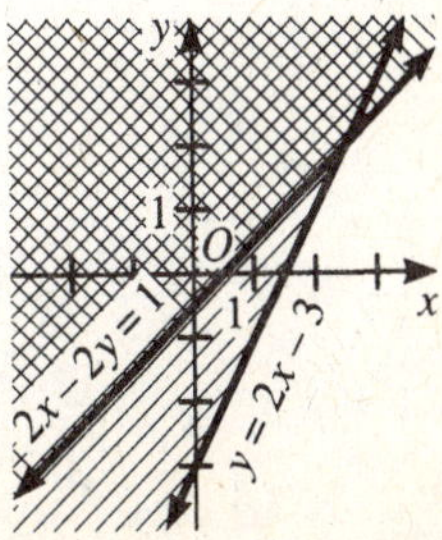

80. Let d = length of the diagonal. $d^2 = \sqrt{10^2 + 9^2} = \sqrt{181}$; $d = \pm\sqrt{181} \approx \pm 13.45$; The diagonal is 13.45 m long.

81. Let x = number of hours the assistant requires. $\left(\frac{1}{6} + \frac{1}{x}\right)4 = 1$, $6x \cdot 4\left(\frac{1}{6} + \frac{1}{x}\right) = 6x$, $4x + 24 = 6x$, $24 = 2x$, $x = 12$; The assistant requires 12 h to complete the job.

82. $\frac{10 \cdot 2^2}{8} = \frac{a \cdot 5^2}{20}$, $\frac{40}{8} = \frac{25a}{20}$, $5 = \frac{5a}{4}$, $a = 4$

83. Let e = number of hours Edward paddled. $2e + e + 1 < 10$, $3e + 1 < 10$, $3e < 9$, $e < 3$; $2e < 6$; Roberto paddled less than six hours.

84.

	Rate	× Time =	Distance
Without wind	r	t	rt
With wind	$r + 100$	$t - 1$	$(r + 100)(t - 1)$

$rt = 3000$, $r = \frac{3000}{t}$; $(r + 100)(t - 1) = 3000$, $\left(\frac{3000}{t} + 100\right)(t - 1) = 3000$,

$3000 - \frac{3000}{t} + 100t - 100 = 3000$, $t\left(-\frac{3000}{t} + 100t - 100\right) = 0$,

$-3000 + 100t^2 - 100t = 0$, $t^2 - t - 30 = 0$, $(t - 6)(t + 5) = 0$; $t = 6$ or $t = -5$;

$r = \frac{3000}{6} = 500$; The plane's speed is 500 mi/h.

Page 599 • PREPARING FOR COLLEGE ENTRANCE EXAMS

1. D; $-\frac{4}{7} + \frac{3}{5}\left[1\frac{1}{8} - \left(-\frac{4}{7}\right)\right] = -\frac{4}{7} + \frac{3}{5}\left[\frac{9}{8} + \frac{4}{7}\right] = -\frac{4}{7} + \frac{3}{5}\left[\frac{63 + 32}{56}\right] = -\frac{4}{7} + \frac{3(95)}{5(56)} =$
$-\frac{4}{7} + \frac{57}{56} = \frac{-32 + 57}{56} = \frac{25}{56}$

2. C; $(w)(4w) = 184.96$, $4w^2 = 184.96$, $w^2 = 46.24$; $w = \pm 6.8$; The width is 6.8 m.

3. B; $c^2 = 8^2 + 15^2$, $c^2 = 64 + 225 = 289$, $c = \pm 17$; The hypotenuse is 17.0 cm long.

4. A; $\frac{\sqrt{3}(\sqrt{3} + 1)}{\sqrt{3} + 2} = \frac{(3 + \sqrt{3})(\sqrt{3} - 2)}{(\sqrt{3} + 2)(\sqrt{3} - 2)} = \frac{3\sqrt{3} - 6 + 3 - 2\sqrt{3}}{3 - 4} = \frac{-3 + \sqrt{3}}{-1} = 3 - \sqrt{3}$

5. C; $x + (x + 1)^2 = 131$, $x + x^2 + 2x + 1 = 131$, $x^2 + 3x - 130 = 0$, $(x + 13)(x - 10) = 0$; $x = -13$ or 10; $x + (x + 1) = 10 + 11 = 21$

6. D **7.** E

Page 602 • WRITTEN EXERCISES

A 1. {A, B, C, D, E, F, G, H, I, J, K} **a.** {A, E, I} **b.** {B, C, D, F, G, H, J, K} **c.** {C, F, H}

2. $\{(Y, Y), (Y, G), (Y, B), (R, Y), (R, G), (R, B)\}$ **a.** $\{(Y, G), (R, G)\}$

b. $\{(Y, Y), (Y, G), (Y, B), (R, Y)\}$ **c.** $\{(Y, Y), (Y, G), (R, Y), (R, G)\}$

3. {(2, 1), (2, 4), (2, 9), (4, 1), (4, 4), (4, 9), (8, 1), (8, 4), (8, 9), (15, 1), (15, 4), (15, 9)}

a. {(4, 4)} **b.** {(15, 1), (15, 9)} **c.** {(4, 9), (8, 9), (15, 1), (15, 4), (15, 9)}

d. {(2, 1), (4, 1)}

4. {(3, 1), (3, 5), (3, 10), (3, 15), (3, 20), (6, 1), (6, 5), (6, 10), (6, 15), (6, 20), (9, 1), (9, 5), (9, 10), (9, 15), (9, 20)}

a. {(3, 1), (3, 5), (3, 15), (9, 1), (9, 5), (9, 15)}

b. {(3, 10), (3, 20), (6, 1), (6, 5), (6, 15), (9, 10), (9, 20)}

c. {(3, 10), (3, 15), (6, 10), (9, 5), (9, 10)} **d.** {(6, 15), (6, 20), (9, 10), (9, 15), (9, 20)}

B 5. O = on and F = off {(O, O, O), (O, O, F), (O, F, F), (O, F, O), (F, O, O), (F, O, F), (F, F, O), (F, F, F)}

a. {(O, O, O)} **b.** {(O, F, F), (F, O, F), (F, F, O), (F, F, F)}

c. {(O, O, F), (O, F, O), (O, F, F), (F, O, O), (F, O, F), (F, F, O), (F, F, F)}

6. {(1, 1), (1, 2), (1, 3), (2, 1), (2, 2), (2, 3), (3, 1), (3, 2), (3, 3), (4, 1), (4, 2), (4, 3)}

a. {(2, 3), (3, 2), (3, 3), (4, 1), (4, 2), (4, 3)} **b.** {(1, 2), (1, 3), (2, 3)}

c. {(1, 2), (2, 1), (3, 3), (4, 2)} **d.** {(2, 3), (3, 2), (3, 3), (4, 2), (4, 3)}

C 7. {p, n, d, (p, n), (p, d), (n, d), (p, n, d)} **a.** {d, (p, n), (p, n, d)}

b. {p, n, (p, d), (n, d)} **c.** {n, d, (p, n), (p, d), (n, d)} **d.** {p, n, (p, n, d)}

Pages 604–605 • WRITTEN EXERCISES

A 1. **a.** *Event A* = {opal, opal, opal, opal}, so $P(A) = \frac{4}{15}$

b. *Event B* = {opal, opal, opal, opal, pearl, pearl, pearl, pearl, pearl, pearl}, so

$P(B) = \frac{10}{15} = \frac{2}{3}$

c. *Event C* = {garnet, garnet, garnet, garnet, garnet}, so $P(C) = \frac{5}{15} = \frac{1}{3}$

d. *Event D* = {opal, opal, opal, opal, pearl, pearl, pearl, pearl, pearl, pearl}, so

$P(D) = \frac{10}{15} = \frac{2}{3}$

2. a. *Event A* = {5(H), 5(C), 5(D), 5(S)}, so $P(A) = \frac{4}{52} = \frac{1}{13}$

b. *Event B* = {7(H), 7(D)}, so $P(B) = \frac{2}{52} = \frac{1}{26}$

c. *Event C* = {2(C), 3(C), 4(C), 5(C), 6(C), 7(C), 8(C), 9(C), 10(C), J(C), Q(C), K(C), A(C)}, so $P(C) = \frac{13}{52} = \frac{1}{4}$

d. *Event D* = {K(H)}, so $P(D) = \frac{1}{52}$

e. *Event E* = {3(H), 3(C), 3(D), 3(S), 7(H), 7(C), 7(D), 7(S)} so $P(E) = \frac{8}{52} = \frac{2}{13}$

f. *Event F* = {8(H), 8(C), 8(D), 8(S), 9(H), 9(C), 9(D), 9(S), 10(H), 10(C), 10(D), 10(S)}, so $P(F) = \frac{12}{52} = \frac{3}{13}$

3. a. *Event A* = {1}, so $P(A) = \frac{1}{24}$

b. *Event B* = {1, 2, 3, 4, 5, 6, 7, 8, 9}, so $P(B) = \frac{9}{24} = \frac{3}{8}$

c. *Event C* = {9, 10, 11, 12, 13, 14, 15}, so $P(C) = \frac{7}{24}$

d. *Event D* = {5, 10, 15, 20}, so $P(D) = \frac{4}{24} = \frac{1}{6}$

4.

6	(6, 1)	(6, 2)	(6, 3)	(6, 4)	(6, 5)	(6, 6)
5	(5, 1)	(5, 2)	(5, 3)	(5, 4)	(5, 5)	(5, 6)
4	(4, 1)	(4, 2)	(4, 3)	(4, 4)	(4, 5)	(4, 6)
3	(3, 1)	(3, 2)	(3, 3)	(3, 4)	(3, 5)	(3, 6)
2	(2, 1)	(2, 2)	(2, 3)	(2, 4)	(2, 5)	(2, 6)
1	(1, 1)	(1, 2)	(1, 3)	(1, 4)	(1, 5)	(1, 6)
	1	2	3	4	5	6

a. $P(5) = \frac{4}{36} = \frac{1}{9}$ b. $P(\text{not } 5) = \frac{32}{36} = \frac{8}{9}$ c. $P(4 \text{ or } 7) = \frac{9}{36} = \frac{1}{4}$ d. $P(12) = \frac{1}{36}$

e. $P(13) = 0$ f. $P(2 \text{ or } 8) = \frac{6}{36} = \frac{1}{6}$ g. $P(\text{not } 6) = \frac{31}{36}$

h. $P(\text{even number}) = \frac{18}{36} = \frac{1}{2}$ i. $P(\text{odd number}) = \frac{18}{36} = \frac{1}{2}$

B 5. The sample space is {(*H*, *H*, *H*), (*H*, *H*, *T*), (*H*, *T*, *H*), (*H*, *T*, *T*), (*T*, *H*, *H*), (*T*, *H*, *T*), (*T*, *T*, *H*), (*T*, *T*, *T*)}

a. $P(3 \text{ heads}) = \frac{1}{8}$ b. $P(\text{exactly 2 tails}) = \frac{3}{8}$ c. $P(\text{at least 2 heads}) = \frac{4}{8} = \frac{1}{2}$

d. $P(\text{one or two heads}) = \frac{6}{8} = \frac{3}{4}$

6.

3	(3, 2)	(3, 4)	(3, 6)	(3, 7)
2	(2, 2)	(2, 4)	(2, 6)	(2, 7)
1	(1, 2)	(1, 4)	(1, 6)	(1, 7)
	2	4	6	7

a. $P(1, 4) = \frac{1}{12}$ **b.** $P(2, \text{not } 2) = \frac{3}{12} = \frac{1}{4}$ **c.** $P(3, 6) = \frac{1}{12}$

d. $P(\text{even number, even number}) = \frac{3}{12} = \frac{1}{4}$ **e.** $P(\text{sum is } 7) = \frac{2}{12} = \frac{1}{6}$

f. $P(\text{sum is less than } 8) = \frac{7}{12}$

C **7.** Let $R, G_1, G_2, W_1, W_2, W_3$ be the marbles. Then the sample space is $\{(R, G_1), (R, G_2), (R, W_1), (R, W_2), (R, W_3), (G_1, G_2), (G_1, W_1), (G_1, W_2), (G_1, W_3), (G_2, W_1), (G_2, W_2), (G_2, W_3), (W_1, W_2), (W_1, W_3), (W_2, W_3)\}$

a. $P(\text{1 red and 1 green}) = \frac{2}{15}$ **b.** $P(\text{1 green and 1 white}) = \frac{6}{15} = \frac{2}{5}$

c. $P(\text{neither red nor green}) = \frac{3}{15} = \frac{1}{5}$ **d.** $P(\text{neither red nor white}) = \frac{1}{15}$

e. $P(\text{two of the same color}) = \frac{4}{15}$ **f.** $P(\text{1 red and a green or a white}) = \frac{5}{15} = \frac{1}{3}$

Page 608 • WRITTEN EXERCISES

A **1. a.** $\frac{53 + 32 + 49 + 24 + 62}{5} = \frac{220}{5} = 44$ **b.** 24, 32, 49, 53, 62; median = 49

c. no mode **d.** 38

2. a. $\frac{8 + 15 + 15 + 31 + 41}{5} = \frac{110}{5} = 22$ **b.** 15 **c.** 15 **d.** 33

3. a. $\frac{17 + 22 + 29 + 32 + 32 + 40}{6} = \frac{172}{6} \approx 28.7$ **b.** $\frac{29 + 32}{2} = 30.5$ **c.** 32 **d.** 23

4. a. $\frac{9 + 9 + 9 + 11 + 13 + 19 + 19}{7} = \frac{89}{7} \approx 12.7$ **b.** 11 **c.** 9 **d.** 10

5. a. $\frac{52 + 53 + 53 + 56 + 56 + 61 + 61}{7} = \frac{392}{7} = 56$ **b.** 56 **c.** 53 and 56 and 61 **d.** 9

6. a. $\frac{49 + 50 + 63 + 66 + 72 + 78 + 81}{7} = \frac{459}{7} \approx 65.6$ **b.** 66 **c.** none **d.** 32

7. $\frac{81 + 81 + 84 + 85 + 92 + 92}{6} = \frac{515}{6} \approx 85.8$; $\frac{84 + 85}{2} = 84.5$; The mean, median, and range of David's scores are 85.8, 84.5, and 11, respectively. The modes are 81 and 92.

8. $\frac{-2 + (-1.5) + 2.5 + 5 + 6 + 16.5 + 18.5 + 20 + 24.5 + 28 + 29.5 + 33.5}{12} = \frac{180.5}{12} \approx 15.0$;

$33.5 - (-2) = 35.5$; The mean temperature is 15.0°C and the range is 35.5°C.

9. $\frac{7 + 8 + 9 + 10 + 10 + 11 + 11 + 12 + 12 + 12}{10} = 10.2$; $\frac{10 + 11}{2} = 10.5$; $12 - 7 = 5$; The mean, median, mode and range are 10.2, 10.5, 12, and 5 respectively.

10.

Score	100	98	96	95	94	90	88	85	84	83	81
Number of Students	2	1	1	1	1	4	2	1	1	1	3

Score	80	79	78	76	73	65
Number of Students	1	1	1	2	1	1

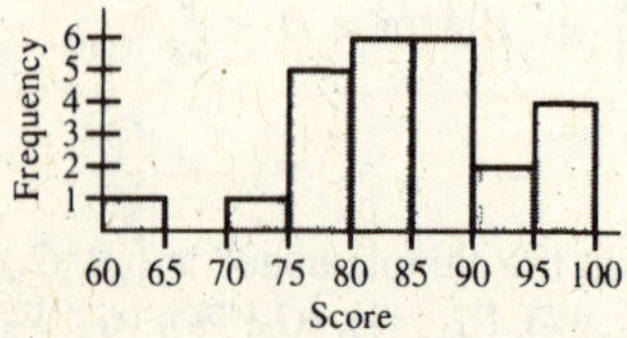

B **11.** Let p = percentage in the sixth class. $\frac{85 + 98 + 87 + 85 + 89 + p}{6} = 90$, $\frac{444 + p}{6} = 90$, $444 + p = 540$, $p = 96$; Vin needs 96% in his sixth class.

12. **a.** The mean will be increased by 10 points.
b. The mode will be increased by 10 points.
c. The median will be increased by 10 points. **d.** The range will be the same.

13. **a.** The mean is multiplied by 3. **b.** The mode is multiplied by 3.
c. The median is multiplied by 3. **d.** The range is multiplied by 3.

14. Let s = sum of the numbers. $\frac{s}{8} = 17$, $s = 136$; The sum of the numbers is 136.

C **15.** Let x = sum of the first three numbers and let y = the sum of the last nine numbers. $\frac{x}{3} = 32$, $x = 96$; $\frac{y}{9} = 40$, $y = 360$; $\frac{x + y}{12} = \frac{96 + 360}{12} = 38$; The mean of the numbers is 38.

16. Let s = score on the tenth test. $\frac{8(89) + 96 + s}{10} = 90$, $712 + 96 + s = 900$, $808 + s = 900$, $s = 92$; Felicia must score 92 on the tenth test.

Page 610 • EXTRA

1. None of the scores deviates from the mean.
2. In Group B more scores deviate from the mean. Variance of Group A $= \frac{(11 - 14)^2 + (12 - 14)^2 + 4(14 - 14)^2 + (19 - 14)^2}{7} = \frac{9 + 4 + 0 + 25}{7} \approx 5.4$; variance of Group B =

$$\frac{(11 - 15)^2 + (12 - 15)^2 + (14 - 15)^2 + (16 - 15)^2 + 2(17 - 15)^2 + (18 - 15)^2}{7} =$$

$$\frac{16 + 9 + 1 + 1 + 8 + 9}{7} \approx 6.3$$

3. a. $\dfrac{36 + 38 + 38 + 41 + 45 + 48}{6} = 41$

b. $\sigma^2 = \dfrac{(36 - 41)^2 + 2(38 - 41)^2 + (41 - 41)^2 + (45 - 41)^2 + (48 - 41)^2}{6} =$

$\dfrac{25 + 18 + 0 + 16 + 49}{6} = 18; \sigma = \sqrt{18} = 3\sqrt{2} \approx 4.2$

c.

x_i	36	38	38	41	45	48
standard score	−1.2	−0.7	−0.7	0	1.0	1.7

4. a. $\dfrac{31 - 26}{5} = 1$ **b.** $\dfrac{22 - 26}{5} = -0.8$ **c.** $\dfrac{17 - 26}{5} = -1.8$ **d.** $\dfrac{29 - 26}{5} = 0.6$

e. $\dfrac{26 - 26}{5} = 0$ **f.** $\dfrac{40 - 26}{5} = 2.8$

5. In Factory II the production rates were more nearly alike. We know this because the standard deviation is lower, thus the scores deviate less from the mean.

6. Let x = actual score on test. $\dfrac{x - 76}{5} = 3.4, x - 76 = 17, x = 93$; Your score on the test was 93.

Pages 613–614 • WRITTEN EXERCISES

A **1. a.** 51, 53, 54, 62, 63, 65, 68, 70, 71, 73, 76, 77, 79, 84, 84, 88, 90, 95

b. median $= \dfrac{71 + 73}{2} = 72$; mode = 84; range = 44

2.

Stem	Leaves
7	3, 5
8	7, 0, 6, 8
9	1
10	2, 0, 9
11	5, 4
12	7, 0, 9
13	7, 5, 4, 3
14	5

3. 4, 5, 6, 7, 7, 8, 8, 8, 9, 9, 10 **a.** 8 **b.** 6 **c.** 9 **d.** 10 **e.** 4 **f.** 6

4.

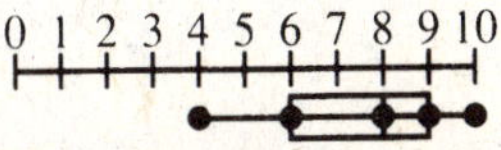

B **5.** 19, 19, 20, 21, 21, 21, 22, 22, 24, 24, 28, 28, 28, 28, 30; median = 22; first quartile = 21; third quartile = 28; highest score = 30; lowest score = 19

6. 23, 23, 24, 24, 24, 25, 26, 26, 26, 27, 29, 30, 31, 32, 38; median = 26; first quartile = 24; third quartile = 30; highest = 38; lowest = 23

7.

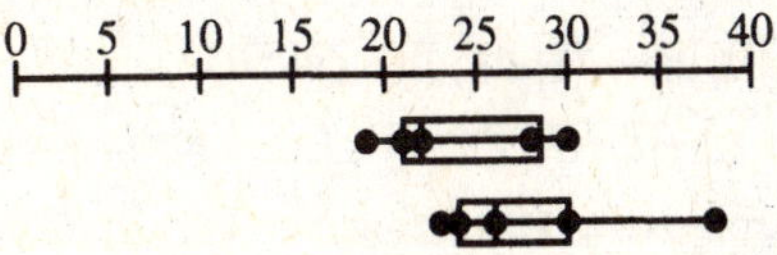

Page 615 • APPLICATION

1. The survey sample of 1000 is large enough to not be misleading.
2. This is misleading because in the fine print you see that the statistics are based on a small sample (50 doctors), but also, it is only the doctors who expressed a preference between brands! It could be a very small percentage of doctors who actually prefer VITA-TRUE, the rest not preferring any brand. Furthermore, the conclusion is based only on returned questionnaires.
3. **a.** Median is $22,000

   b. $\frac{21{,}500 + 2(21{,}750) + 21{,}800 + 3(22{,}000) + 22{,}350 + 47{,}500 + 53{,}000}{10} = 27{,}565$; The mean is $27,565.

   c. The median is a better description because the two employees who are paid considerably more make the mean much higher than the median.

   d. For the scale on the vertical axis, for example, make each unit 50 dollars (in thousands).

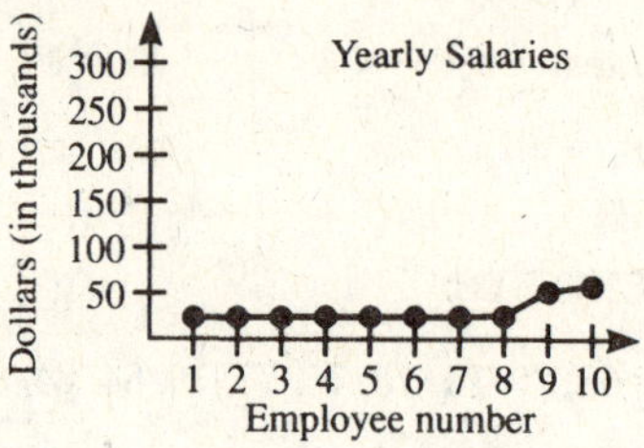

   e. For the scale on the vertical axis, for example, make each unit 5 dollars (in thousands).

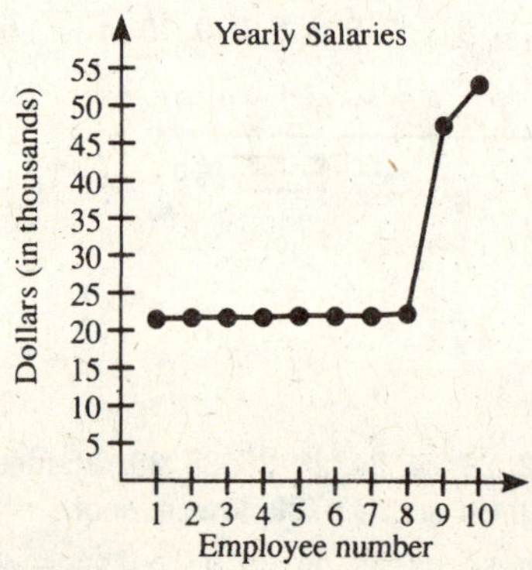

Page 618 • WRITTEN EXERCISES

A **1.** $\overline{WX}, \overline{WY}, \overline{WZ}, \overline{XY}, \overline{XZ}, \overline{YZ}$ **2.** $\overline{MR}, \overline{MP}, \overline{MN}, \overline{NP}, \overline{PR}, \overline{RQ}, \overline{QP}, \overline{NQ}$

3. $\overline{AE}, \overline{BC}, \overline{EB}, \overline{ED}, \overline{DB}, \overline{AC}, \overline{AD}, \overline{DC}$ **4.** $\overline{HF}, \overline{HK}, \overline{FK}, \overline{FG}, \overline{KG}$ **5.** acute

6. obtuse **7.** right **8.** acute **9.** obtuse **10.** straight **11.** acute **12.** right

B **13.**

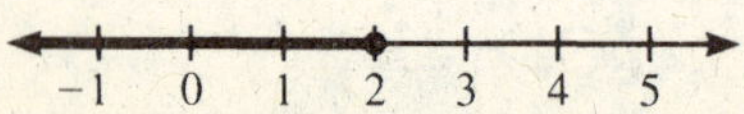

ray

14.

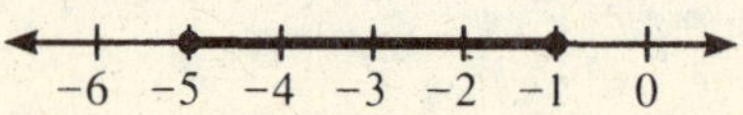

line segment

15.

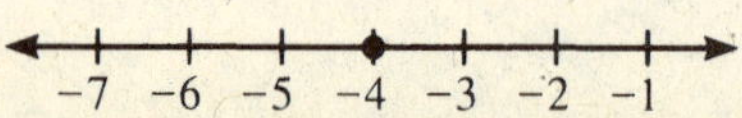

point

16. 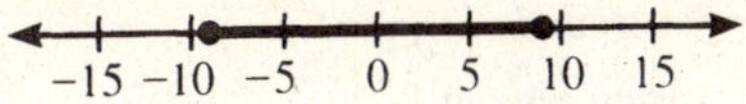

line segment

17. 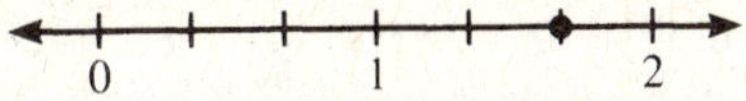

point

18. 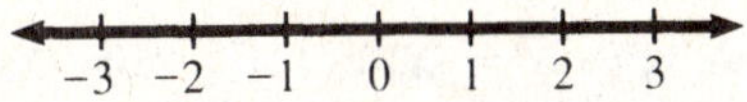

line

C 19. 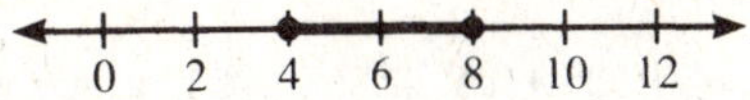

line segment

20. 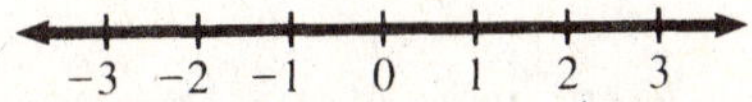

line

21. 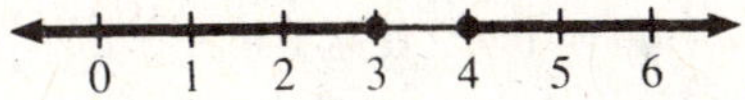

two rays

22. 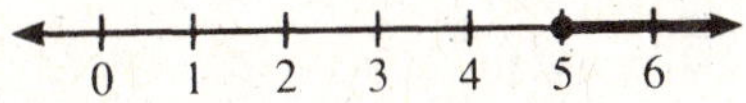

ray

Page 620 • WRITTEN EXERCISES

A **1.** $\angle EAB$, $\angle GBK$, $\angle HAD$ **2.** $\angle HAE$, $\angle DAB$, $\angle ABG$, $\angle FBK$ **3.** 30° **4.** 135°

5. Let x = measure of the smaller angle. $x = (90 - x) - 50$, $x = 40 - x$, $2x = 40$, $x = 20$; $90 - x = 70$; The two angles measure 20° and 70°.

6. Let x = measure of the smaller angle. $180 - x = 8x$, $180 = 9x$, $x = 20$; $180 - x = 160$; The two angles measure 20° and 160°.

7. Let x = the measure of the angle. $x = (180 - x) + 74$, $x = 254 - x$, $2x = 254$, $x = 127$; The angle measures 127°.

8. Let x = the measure of the angle. $x = \frac{1}{4}(90 - x)$, $4x = 90 - x$, $5x = 90$, $x = 18$; The angle measures 18°.

B **9.** Let x = the measure of the angle. $(90 - x) + (180 - x) = 144$, $270 - 2x = 144$, $126 = 2x$, $x = 63$; The angle measures 63°.

10. Let x = the measure of the angle. $(180 - x) = 10(90 - x) + 9$, $180 - x = 900 - 10x + 9$, $180 + 9x = 909$, $9x = 729$, $x = 81$; The angle measures 81°.

Page 622 • WRITTEN EXERCISES

A **1.** $180 - (28 + 59) = 180 - 87 = 93$; 93°

2. $180 - (122 + 41) = 180 - 163 = 17$; 17°

3. $180 - (38 + 52) = 180 - 90 = 90$; 90°

4. $180 - (15 + 27) = 180 - 42 = 138$; 138°

5. $180 - (90 + 24) = 180 - 114 = 66$; 66°

6. $180 - (138 + 21) = 180 - 159 = 21$; 21°

7. $9^2 + 8^2 = 81 + 64 = 145 \neq 13^2$; not a right triangle

8. $8^2 + 6^2 = 64 + 36 = 100 = 10^2$; right triangle

9. $12^2 + 12^2 = 144 + 144 = 288 \neq 16^2$; not a right triangle

10. $15^2 + 8^2 = 225 + 64 = 289 = 17^2$; right triangle

11. $24^2 + 10^2 = 576 + 100 = 676 = 26^2$; right triangle

12. $10^2 + 8^2 = 100 + 64 = 164 \neq 12^2$; not a right triangle

13. If $\angle T = 90°$, then SU is the hypotenuse. $12^2 + 9^2 = (SU)^2$, $144 + 81 = (SU)^2$, $SU = \sqrt{225} = 15$

14. If $\angle Z = 90°$, then XY is the hypotenuse. $12^2 + 5^2 = (XY)^2$, $144 + 25 = (XY)^2$, $XY = \sqrt{169} = 13$

15. If $\angle A = 90°$, then BC is the hypotenuse. $(AC)^2 + 24^2 = 30^2$, $(AC)^2 = 900 - 576$, $AC = \sqrt{324} = 18$

16. If $DE = DF$, then $\angle E = \angle F$. $180 = 50 + 2(\angle E)$, $130 = 2(\angle E)$, $\angle E = 65°$

17. If $GH = GI$, then $\angle H = \angle I = 20$. $\angle G + 40 = 180$, $\angle G = 140°$

18. $\angle N = \angle O$; $90 + 2(\angle N) = 180$, $2(\angle N) = 90$, $\angle N = 45°$; $\angle O = 45°$

B **19.** $AB = \sqrt{4^2 + 12^2} = \sqrt{16 + 144} = \sqrt{160} = 4\sqrt{10}$

20. $AB = \sqrt{3^2 + 6^2} = \sqrt{9 + 36} = \sqrt{45} = 3\sqrt{5}$

21. $10^2 = 6^2 + (BC)^2$, $(BC)^2 = 100 - 36$; $BC = \sqrt{64} = 8$

22. $41^2 = (AC)^2 + 9^2$, $(AC)^2 = 1681 - 81$; $AC = \sqrt{1600} = 40$

23. $(AC)^2 + 16^2 = 24^2$, $(AC)^2 = 576 - 256$; $AC = \sqrt{320} = 8\sqrt{5}$

24. $11^2 + (BC)^2 = 25^2$, $(BC)^2 = 625 - 121$; $BC = \sqrt{504} = 6\sqrt{14}$

C **25.** $x^2 + x^2 = (AB)^2$, $(AB)^2 = 2x^2$; $AB = \sqrt{2x^2} = x\sqrt{2}$

26. $y^2 + (AB)^2 = (2y)^2$, $(AB)^2 = 4y^2 - y^2$; $AB = \sqrt{3y^2} = y\sqrt{3}$

Page 623 • PROBLEMS

A **1.** Let x and $4x$ be the acute angles. $90 + x + 4x = 180$, $5x = 90$, $x = 18$; $4x = 72$; The acute angles measure 18° and 72°.

2. Let x = measure of each base angle. $x + x + 10x = 180$, $12x = 180$, $x = 15$; $10x = 150$; The angles measure 150°, 15°, and 15°.

3. Let x = first angle. $x + 7x + 10x = 180$, $18x = 180$, $x = 10$; $7x = 70$; $10x = 100$; The angles measure 10°, 70°, and 100°.

4. Let $4x$, $5x$, and $6x$ be the angles. $4x + 5x + 6x = 180$, $15x = 180$, $x = 12$; $4x = 48$; $5x = 60$; $6x = 72$; The angles measure 48°, 60°, and 72°.

5. Let x = first angle. $x + 3x + 2(3x) = 180$, $4x + 6x = 180$, $10x = 180$, $x = 18$; $3x = 54$; $6x = 108$; The angles measure 18°, 54°, and 108°.

6. Let x = first angle. $x + 4x + (4x - 63) = 180$, $5x + 4x - 63 = 180$, $9x = 243$, $x = 27$; $4x = 108$; $4x - 63 = 108 - 63 = 45$; The angles measure 27°, 108°, and 45°.

7. Let x = measure of each of the equal angles. $x + x + \frac{1}{2}(x + x) = 180$,

$2x + \frac{1}{2}(2x) = 180$, $2x + x = 180$, $3x = 180$, $x = 60$; The angles measure 60° each.

8. Let x = first angle. $x + 2x + (4x + 5) = 180$, $7x + 5 = 180$, $7x = 175$, $x = 25$; $2x = 50$; $4x + 5 = 100 + 5 = 105$; The angles measure 25°, 50°, and 105°.

B 9. Let x = first angle. $x + (4x + 5) + [3(4x + 5) - 27] = 180$, $5x + 5 + 12x + 15 - 27 = 180$, $17x - 7 = 180$, $17x = 187$, $x = 11$; $4x + 5 = 44 + 5 = 49$, $3(49) - 27 = 147 - 27 = 120$; The angles measure 11°, 49°, and 120°.

10. Let x = second angle. $[2(2x + 3) + 24] + x + (2x + 3) = 180$, $4x + 6 + 24 + x + 2x + 3 = 180$, $7x + 33 = 180$, $7x = 147$, $x = 21$; $2x + 3 = 42 + 3 = 45$; $2(45) + 24 = 90 + 24 = 114$; The angles measure 114°, 21°, and 45°.

11. Let x = second angle. $(2x - 2) + x + \left[\frac{1}{2}(2x - 2) + 35\right] = 180$, $3x - 2 + x - 1 + 35 = 180$, $4x + 32 = 180$, $4x = 148$, $x = 37$; $2x - 2 = 74 - 2 = 72$; $\frac{1}{2}(72) + 35 = 36 + 35 = 71$; The angles measure 37°, 72°, and 71°.

C 12. Let x = first angle. $x + 6x + [(6x + x^2) - 18] = 180$, $13x + x^2 - 18 = 180$, $x^2 + 13x - 198 = 0$, $(x - 9)(x + 22) = 0$; $x = 9$ or -22; $6x = 54$; $54 + 9^2 - 18 = 36 + 81 = 117$; The angles measure 9°, 54°, and 117°.

13. Let x = first angle. $x + [(90 - x) + 22] + [(180 - x) - 49] = 180$, $x + 112 - x + 131 - x = 180$, $243 - x = 180$, $x = 63$; $(90 - x) + 22 = 27 + 22 = 49$; $(180 - x) - 49 = 117 - 49 = 68$; The angles measure 63°, 49°, and 68°.

Pages 625–626 • WRITTEN EXERCISES

A 1. a. $\angle R$ and $\angle Y$, $\angle S$ and $\angle Z$, $\angle T$ and $\angle X$ b. $\overline{ST}$ and $\overline{XZ}$, $\overline{RT}$ and $\overline{XY}$, $\overline{RS}$ and $\overline{YZ}$ c. $\triangle RST \sim \triangle YZX$

2. $\frac{JL}{PN} = \frac{JK}{PM} = \frac{KL}{MN}$ 3. F 4. F 5. T 6. T 7. T

8. $\frac{DE}{AB} = \frac{EF}{BC} = \frac{DF}{AC}$; $\frac{9}{3} = \frac{EF}{5}$, $EF = \frac{45}{3} = 15$; $\frac{9}{3} = \frac{DF}{6}$, $DF = 18$

9. $\frac{15}{AB} = \frac{12}{8}$, $12AB = 120$, $AB = 10$; $\frac{21}{BC} = \frac{12}{8}$, $12BC = 168$, $BC = 14$

10. $\frac{DE}{24} = \frac{12}{16}$, $DE = 18$; $\frac{DF}{32} = \frac{12}{16}$, $DF = 24$ 11. $\frac{EF}{10} = \frac{14}{10}$, $EF = 14$; $\frac{DF}{15} = \frac{14}{10}$, $DF = 21$

B 12. $\frac{AB}{DE} = \frac{BC}{EF} = \frac{AC}{DF}$; $\frac{4}{6} = \frac{6}{EF}$, $4EF = 36$, $EF = 9$; $\frac{4}{6} = \frac{AC}{12}$, $AC = 8$

13. $\frac{16}{DE} = \frac{20}{15}$, $20DE = 240$, $DE = 12$; $\frac{BC}{18} = \frac{20}{15}$, $BC = 24$

14. $\frac{21}{EF} = \frac{28}{20}$, $28EF = 420$, $EF = 15$; $\frac{AB}{25} = \frac{28}{20}$, $20AB = 700$, $AB = 35$

15. $\frac{AB}{39} = \frac{32}{48}$, $AB = 26$; $\frac{18}{EF} = \frac{32}{48}$, $32EF = 864$, $EF = 27$

C 16. $\triangle RTS \sim \triangle RUT \sim \triangle TUS$

Page 626 • PROBLEMS

A 1. Let s = shortest side of the second triangle. $\frac{s}{6} = \frac{18}{12}$, $s = 9$; The shortest side of the second triangle measures 9 cm.

2. Let t = height of the tree. $\frac{5}{2} = \frac{t}{26}$, $t = 65$; The tree is 65 ft tall.

3. Let b = height of the building. $\frac{b}{27} = \frac{4}{3}$, $b = 36$; The building is 36 m high.

4. Let s = length of one of the equal sides in the smaller triangle. $\frac{s}{28} = \frac{15}{35}$, $s = 12$; $15 + 12 + 12 = 39$; The perimeter of the smaller triangle is 39 cm.

5. Let l = length of the swamp. $\frac{l}{100} = \frac{80 + 100}{80}$, $80l = 100(180)$, $l = \frac{18{,}000}{80} = 225$; The swamp is 225 m long.

B **6.** $\frac{EB}{20} = \frac{x}{4x} = \frac{1}{4}$, $EB = 5$

7. Let h = his height 10 m farther up the ramp. $\frac{h}{5 + 10} = \frac{2}{5}$, $h = \frac{2}{5} \cdot 15 = 6$; Chen would be 6 m above the ground.

8. Let x = distance from the point to the base of the building. $\frac{x}{7} = \frac{400}{8}$, $x = 350$; The point is 350 m from the base of the building.

C **9.** Let t = height of the tree. $\frac{t}{1.5} = \frac{15}{2}$, $t = 11.25$; The tree is 11.25 m tall.

Page 629 • WRITTEN EXERCISES

A **1.** $\sin A = \frac{8}{10} = \frac{4}{5}$; $\cos A = \frac{6}{10} = \frac{3}{5}$; $\tan A = \frac{8}{6} = \frac{4}{3}$; $\sin B = \frac{6}{10} = \frac{3}{5}$; $\cos B = \frac{8}{10} = \frac{4}{5}$; $\tan B = \frac{6}{8} = \frac{3}{4}$

2. $\sin A = \frac{35}{37}$; $\cos A = \frac{12}{37}$; $\tan A = \frac{35}{12}$; $\sin B = \frac{12}{37}$; $\cos B = \frac{35}{37}$; $\tan B = \frac{12}{35}$

3. $\sin A = \frac{5}{\sqrt{34}} \cdot \frac{\sqrt{34}}{\sqrt{34}} = \frac{5\sqrt{34}}{34}$; $\cos A = \frac{3}{\sqrt{34}} \cdot \frac{\sqrt{34}}{\sqrt{34}} = \frac{3\sqrt{34}}{34}$; $\tan A = \frac{5}{3}$;

$\sin B = \frac{3}{\sqrt{34}} \cdot \frac{\sqrt{34}}{\sqrt{34}} = \frac{3\sqrt{34}}{34}$; $\cos B = \frac{5}{\sqrt{34}} \cdot \frac{\sqrt{34}}{\sqrt{34}} = \frac{5\sqrt{34}}{34}$; $\tan B = \frac{3}{5}$

4. $\sin A = \frac{3\sqrt{5}}{9} = \frac{\sqrt{5}}{3}$; $\cos A = \frac{6}{9} = \frac{2}{3}$; $\tan A = \frac{3\sqrt{5}}{6} = \frac{\sqrt{5}}{2}$; $\sin B = \frac{6}{9} = \frac{2}{3}$;

$\cos B = \frac{3\sqrt{5}}{9} = \frac{\sqrt{5}}{3}$; $\tan B = \frac{6}{3\sqrt{5}} \cdot \frac{\sqrt{5}}{\sqrt{5}} = \frac{6\sqrt{5}}{15} = \frac{2\sqrt{5}}{5}$

5. Let h = length of the hypotenuse. $h = \sqrt{4^2 + 7^2} = \sqrt{16 + 49} = \sqrt{65}$;

$\sin A = \frac{4}{\sqrt{65}} \cdot \frac{\sqrt{65}}{\sqrt{65}} = \frac{4\sqrt{65}}{65}$; $\cos A = \frac{7}{\sqrt{65}} \cdot \frac{\sqrt{65}}{\sqrt{65}} = \frac{7\sqrt{65}}{65}$; $\tan A = \frac{4}{7}$;

$\sin B = \frac{7}{\sqrt{65}} = \frac{7\sqrt{65}}{65}$; $\cos B = \frac{4}{\sqrt{65}} = \frac{4\sqrt{65}}{65}$; $\tan B = \frac{7}{4}$

6. $h = \sqrt{5^2 + 6^2} = \sqrt{25 + 36} = \sqrt{61}$; $\sin A = \frac{6}{\sqrt{61}} = \frac{6\sqrt{61}}{61}$; $\cos A = \frac{5}{\sqrt{61}} = \frac{5\sqrt{61}}{61}$;

$\tan A = \frac{6}{5}$; $\sin B = \frac{5}{\sqrt{61}} = \frac{5\sqrt{61}}{61}$; $\cos B = \frac{6}{\sqrt{61}} = \frac{6\sqrt{61}}{61}$; $\tan B = \frac{5}{6}$

7. $h = \sqrt{7^2 + 7^2} = \sqrt{2 \cdot 49} = 7\sqrt{2}$; $\sin A = \cos A = \frac{7}{7\sqrt{2}} \cdot \frac{\sqrt{2}}{\sqrt{2}} = \frac{\sqrt{2}}{2}$;

$\tan A = \frac{7}{7} = 1$; $\sin B = \cos B = \frac{7}{7\sqrt{2}} = \frac{\sqrt{2}}{2}$; $\tan B = \frac{7}{7} = 1$

8. Let l = length of the other leg. $4^2 + l^2 = 8^2$, $l^2 = 64 - 16$, $l = \sqrt{48} = 4\sqrt{3}$;

$\sin A = \frac{4\sqrt{3}}{8} = \frac{\sqrt{3}}{2}$; $\cos A = \frac{4}{8} = \frac{1}{2}$; $\tan A = \frac{4\sqrt{3}}{4} = \sqrt{3}$; $\sin B = \frac{4}{8} = \frac{1}{2}$;

$\cos B = \frac{4\sqrt{3}}{8} = \frac{\sqrt{3}}{2}$; $\tan B = \frac{4}{4\sqrt{3}} = \frac{\sqrt{3}}{3}$

9. $h = \sqrt{3^2 + 11^2} = \sqrt{9 + 121} = \sqrt{130}$; $\sin A = \frac{3}{\sqrt{130}} = \frac{3\sqrt{130}}{130}$; $\cos A = \frac{11}{\sqrt{130}} =$

$\frac{11\sqrt{130}}{130}$; $\tan A = \frac{3}{11}$; $\sin B = \frac{11}{\sqrt{130}} = \frac{11\sqrt{130}}{130}$; $\cos B = \frac{3}{\sqrt{130}} = \frac{3\sqrt{130}}{130}$; $\tan B = \frac{11}{3}$

B **10.** $\sin A = \frac{a}{c}$; $\cos B = \frac{a}{c}$; $\therefore \sin A = \cos B$

11. $\cos A = \frac{b}{c}$; $\sin B = \frac{b}{c}$; $\therefore \cos A = \sin B$

12. $(\sin A)^2 + (\cos A)^2 = \left(\frac{a}{c}\right)^2 + \left(\frac{b}{c}\right)^2 = \frac{a^2}{c^2} + \frac{b^2}{c^2} = \frac{a^2 + b^2}{c^2} = \frac{c^2}{c^2} = 1$

13. $\tan A = \frac{a}{b}$; $\frac{\sin A}{\cos A} = \frac{\frac{a}{c}}{\frac{b}{c}} \cdot \frac{c}{c} = \frac{a}{b}$; $\therefore \tan A = \frac{\sin A}{\cos A}$

C **14.** From Ex. 12 we know that $(\sin R)^2 + (\cos R)^2 = 1$, $\left(\frac{3}{5}\right)^2 + (\cos R)^2 = 1$,

$(\cos R)^2 = 1 - \frac{9}{25}$, $\cos R = \sqrt{\frac{25 - 9}{25}} = \sqrt{\frac{16}{25}} = \frac{4}{5}$

15. Sine of complement of $Z = \cos Z$. Let x = side adjacent to $\angle Z$. $x^2 + 3^2 = 8^2$,

$x^2 = 64 - 9$; $x = \sqrt{55}$; $\cos Z = \frac{\sqrt{55}}{8}$

16. Let h = the hypotenuse. $h = \sqrt{11^2 + 60^2} = \sqrt{121 + 3600} = \sqrt{3721} = 61$;

$\sin X = \frac{11}{61}$; $\cos X = \frac{60}{61}$

Page 632 • WRITTEN EXERCISES

A **1.** sin (20°) = 0.3420; cos (20°) = 0.9397; tan (20°) = 0.3640

2. sin (40°) = 0.6428; cos (40°) = 0.7660; tan (40°) = 0.8391

3. sin (15°) = 0.2588; cos (15°) = 0.9659; tan (15°) = 0.2679

4. sin (39°) = 0.6293; cos (39°) = 0.7771; tan (39°) = 0.8098

5. sin (68°) = 0.9272; cos (68°) = 0.3746; tan (68°) = 2.4751

6. sin (83°) = 0.9925; cos (83°) = 0.1219; tan (83°) = 8.1443

7. sin (24°) = 0.4067; cos (24°) = 0.9135; tan (24°) = 0.4452

8. sin (59°) = 0.8572; cos (59°) = 0.5150; tan (59°) = 1.6643

9. sin (4°) = 0.0698; cos (4°) = 0.9976; tan (4°) = 0.0699

10. sin (71°) = 0.9455; cos (71°) = 0.3256; tan (71°) = 2.9042

11. sin (45°) = 0.7071; cos (45°) = 0.7071; tan (45°) = 1

12. sin (35°) = 0.5736; cos (35°) = 0.8192; tan (35°) = 0.7002

13. 78° **14.** 49° **15.** 9° **16.** 43° **17.** 57° **18.** 66° **19.** 22° **20.** 83°

21. 35° **22.** 14° **23.** 41° **24.** 8° **25.** 20° **26.** 68° **27.** 34° **28.** 73°

B **29.** $>$ **30.** $<$ **31.** $=$

C **32.** $2 \cdot \sin(23°) \cdot \cos(23°) = 2 \cdot (0.3907) \cdot (0.9205) = 0.7193 = \sin(46°)$

33. $(\cos(36°))^2 - (\sin(36°))^2 = (0.8090)^2 - (0.5878)^2 = 0.6545 - 0.3455 = 0.3090 = \cos(72°)$

Page 632 • CHALLENGE

1. No **2.** Yes **3.** Yes

Page 635 • WRITTEN EXERCISES

A **1.** $\sin(21°) = \frac{x}{50}$, $0.3584 = \frac{x}{50}$, $50(0.3584) = x$, $x \approx 18$

2. $\cos(48°) = \frac{x}{70}$, $70(0.6691) = x$, $x \approx 47$ **3.** $\tan(35°) = \frac{x}{60}$, $x = 60(0.7002) \approx 42$

4. $\cos(72°) = \frac{x}{30}$, $x = 30(0.3090) \approx 9$ **5.** $\sin(54°) = \frac{x}{20}$, $x = 20(0.8090) \approx 16$

6. $\tan(67°) = \frac{x}{80}$, $80(2.3559) = x$, $x \approx 188$

7. $\sin(39°) = \frac{BC}{53}$, $BC = 53(0.6293) \approx 33$; $\cos(39°) = \frac{AC}{53}$, $AC = 53(0.7771) \approx 41$

8. $\sin(21°) = \frac{AC}{12}$, $AC = 12(0.3584) \approx 4$; $\cos(21°) = \frac{BC}{12}$, $12(0.9336) = BC$, $BC \approx 11$

9. $\tan(80°) = \frac{48}{BC}$, $BC = \frac{48}{5.6713} \approx 8$; $\sin(80°) = \frac{48}{AB}$, $AB = \frac{48}{0.9848} \approx 49$

10. $\tan(65°) = \frac{28}{AC}$, $AC = \frac{28}{2.1445} \approx 13$; $\sin(65°) = \frac{28}{AB}$, $AB = \frac{28}{0.9063} \approx 31$

B **11.** $\tan(\angle D) = \frac{36}{48} = 0.75$; $\angle D \approx 37°$; $90 - 37 = 53$; $\angle F \approx 53°$

12. $\sin(\angle D) = \frac{25}{30} = 0.8333$; $\angle D \approx 56°$; $90 - 56 = 34$; $\angle F \approx 34°$

13. $\sin(\angle D) = \frac{42}{54} \approx 0.7778$; $\angle D \approx 51°$; $90 - 51 = 39$; $\angle F \approx 39°$

14. $\tan(\angle D) = \frac{81}{63} \approx 1.2857$; $\angle D \approx 52°$; $90 - 52 = 38$; $\angle F \approx 38°$

15. $\cos(41°) = \frac{XY}{95}$, $XY = 95(0.7547) \approx 72$; $\sin(41°) = \frac{ZY}{95}$, $ZY = 95(0.6561) \approx 62$

16. $\sin(30°) = \frac{RS}{120}$, $RS = 120(0.5) = 60$; $\cos(30°) = \frac{TS}{120}$, $TS = 120(0.8660) \approx 104$

Pages 635–636 • PROBLEMS

A **1.**

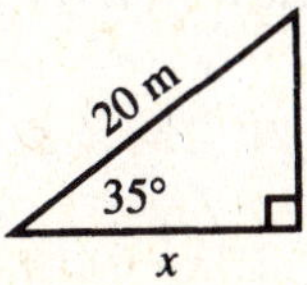

$\cos(35°) = \frac{x}{20}$, $x = 20(0.8192) \approx 16$; the ladder is 16 m from the foot of the building.

2.

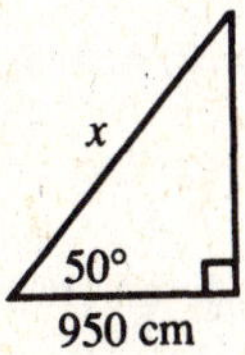

$\cos(50°) = \frac{950}{x}$, $x = \frac{950}{0.6428} \approx 1478$; The cable is 1478 cm long.

3.

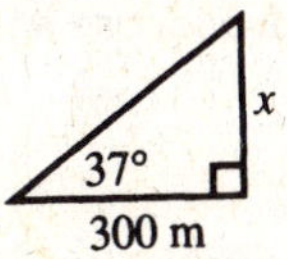

$\tan(37°) = \frac{x}{300}$, $x = 300(0.7536) \approx 226$; The cliff is 226 m high.

4.

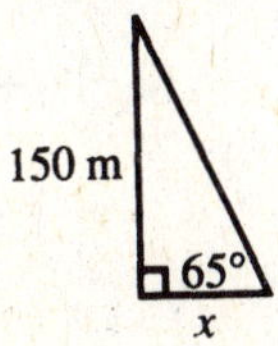

$\tan(65°) = \frac{150}{x}$, $x = \frac{150}{2.1445} \approx 70$; The boy is 70 m from the monument.

5.

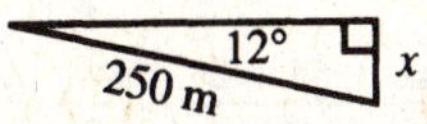

$\tan(12°) = \frac{x}{250}$, $x = 250(0.2079) \approx 52$; The porpoise will be 52 m below sea level.

6.

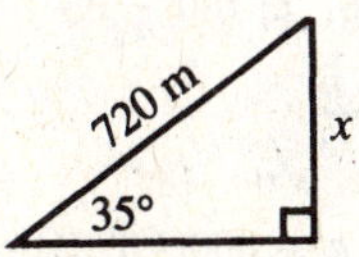

$\sin(35°) = \frac{x}{720}$, $x = 720(0.5736) \approx 413$; The jump is 413 cm high.

7.

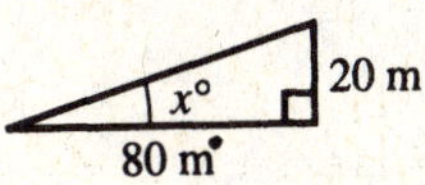

$\tan(x) = \frac{20}{80} = 0.25$; $x \approx 14°$; The angle of elevation is 14°.

8.

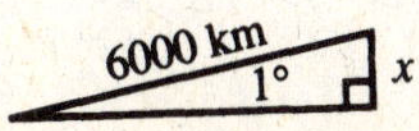

$\sin(1°) = \frac{x}{6000}$, $x = 6000(0.0175) = 105$; The plane will be 105 km away from the correct path.

B **9.** 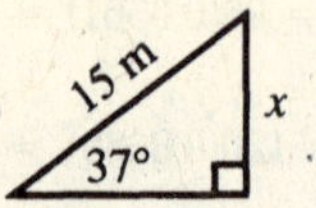

$\sin(37°) = \frac{x}{15}$, $x = 15(0.6018) \approx 9$; The escalator is 9 m high.

10. 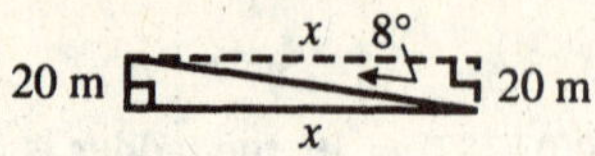

$\tan(8°) = \frac{20}{x}$, $x = \frac{20}{0.1405} \approx 142$; The bottom of the lighthouse is 142 m from the beach.

11. 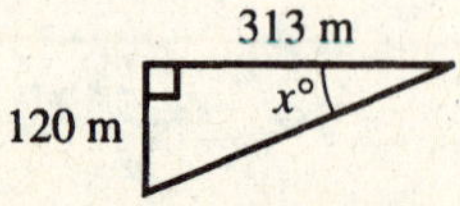

$\tan x = \frac{120}{313} = 0.3834$; $x \approx 21°$; The angle of depression will be 21°.

12. 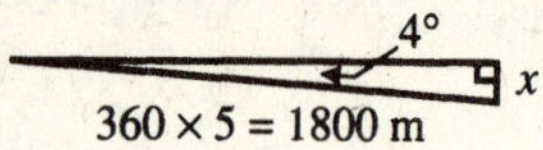

$\sin(4°) = \frac{x}{1800}$, $x = 1800(0.0698) \approx 126$; The submarine is 126 m below the surface.

C **13.** 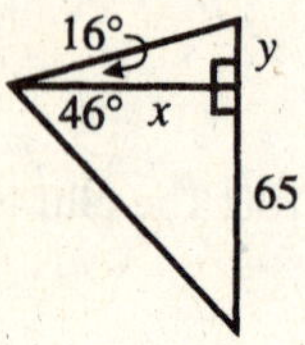

$\tan(46°) = \frac{65}{x}$, $x = \frac{65}{1.0355} \approx 63$; $\tan(16°) = \frac{y}{63}$, $y = 63(0.2867) \approx 18$; $18 + 65 = 83$; The whale was 63 m from the base of the lighthouse and the plane was flying at 83 m.

14. 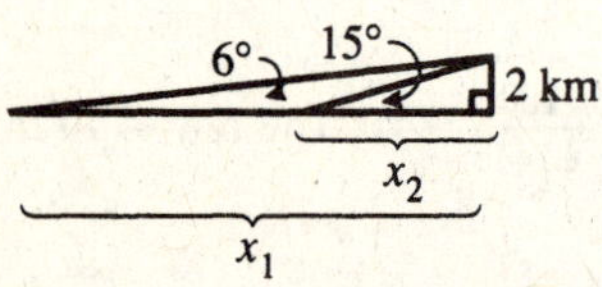

$\tan(15°) = \frac{2}{x_2}$, $x_2 = \frac{2}{0.2679} \approx 7$; $\tan(6°) = \frac{2}{x_1}$; $x_2 = \frac{2}{0.1051} \approx 19$; $x_1 - x_2 = 19 - 7 = 12$; The car has traveled 12 km.

Pages 637–638 • REVIEW

1. c **2.** d; E = {1, 3, 5}; $P(E) = \frac{3}{5}$ **3.** a; $\frac{42 + 44 + 46 + 53 + 55}{5} = \frac{240}{5} = 48$

4. c; $76 - 33 = 43$ **5.** c; $8 + 4 = 12$ **6.** d **7.** c **8.** c; $180 - 63 = 117$

9. a **10.** c **11.** d; $\frac{CA}{FD} = \frac{12}{5}$, $12FD = 5(8)$, $FD = \frac{40}{12} = \frac{10}{3}$

12. b; $\cos(A) = \frac{12}{15} = \frac{4}{5}$ **13.** c **14.** b; $\tan(47°) = \frac{h}{800}$, $h = 800(1.0724) \approx 858$

Pages 639–640 • CHAPTER 1

1. $5 + 32 = 37$ **2.** $10 \times 2 = 20$ **3.** $90 + 10 = 100$ **4.** $10 - 4 = 6$

5. $33 \times 6 = 198$ **6.** $36 \div 4 = 9$ **7.** $9 + 7 - 16 \div 4 = 9 + 7 - 4 = 12$

8. $4 + 21 - 6 = 19$ **9.** $24 - 8 + 7 = 23$ **10.** $2 \cdot 5 - 3 = 10 - 3 = 7$

11. $5 \cdot 4 + 4 \cdot 1 = 20 + 4 = 24$ **12.** $0 \cdot 5 + 3 \cdot 4 = 12$

13. $1(5 - 3) + 4 = 1 \cdot 2 + 4 = 2 + 4 = 6$ **14.** $(3 \cdot 4) \cdot (2 + 0) = 12 \cdot 2 = 24$

15. $(5 - 0)(1) + 4 = 5 \cdot 1 + 4 = 5 + 4 = 9$ **16.** $(2 + 3)(4 + 5) = (5)(9) = 45$

17. $2(0 + 5 - 1) = 2(4) = 8$ **18.** $(4 \cdot 2 - 2 \cdot 3)(5 + 1) = (8 - 6)(6) = 2 \cdot 6 = 12$

19. $\frac{2 + 4}{5 - 3} = \frac{6}{2} = 3$ **20.** $\frac{4 \cdot 5 - 5 \cdot 2}{5 - 3 \cdot 0} = \frac{20 - 10}{5 - 0} = \frac{10}{5} = 2$

21. $3(1 \cdot 2 + 5) + \frac{4}{2} = 3(2 + 5) + 2 = 3(7) + 2 = 21 + 2 = 23$

22.

x	$7 + x = 12$	
0	$7 + 0 = 12$	False
1	$7 + 1 = 12$	False
2	$7 + 2 = 12$	False
3	$7 + 3 = 12$	False
4	$7 + 4 = 12$	False
5	$7 + 5 = 12$	True
6	$7 + 6 = 12$	False

∴ the solution set is {5}.

23.

x	$x - 4 = 2$	
0	$0 - 4 = 2$	False
1	$1 - 4 = 2$	False
2	$2 - 4 = 2$	False
3	$3 - 4 = 2$	False
4	$4 - 4 = 2$	False
5	$5 - 4 = 2$	False
6	$6 - 4 = 2$	True

∴ the solution set is {6}.

24.

x	$8 - x = 3$	
0	$8 - 0 = 3$	False
1	$8 - 1 = 3$	False
2	$8 - 2 = 3$	False
3	$8 - 3 = 3$	False
4	$8 - 4 = 3$	False
5	$8 - 5 = 3$	True
6	$8 - 6 = 3$	False

∴ the solution set is {5}.

25.

x	$x - x = 0$	
0	$0 - 0 = 0$	True
1	$1 - 1 = 0$	True
2	$2 - 2 = 0$	True
3	$3 - 3 = 0$	True
4	$4 - 4 = 0$	True
5	$5 - 5 = 0$	True
6	$6 - 6 = 0$	True

∴ the solution set is {0, 1, 2, 3, 4, 5, 6}.

26.

x	$6x = 18$	
0	$6 \cdot 0 = 18$	False
1	$6 \cdot 1 = 18$	False
2	$6 \cdot 2 = 18$	False
3	$6 \cdot 3 = 18$	True
4	$6 \cdot 4 = 18$	False
5	$6 \cdot 5 = 18$	False
6	$6 \cdot 6 = 18$	False

$\therefore$ the solution set is $\{3\}$.

27.

x	$0 = 5x$	
0	$0 = 5 \cdot 0$	True
1	$0 = 5 \cdot 1$	False
2	$0 = 5 \cdot 2$	False
3	$0 = 5 \cdot 3$	False
4	$0 = 5 \cdot 4$	False
5	$0 = 5 \cdot 5$	False
6	$0 = 5 \cdot 6$	False

$\therefore$ the solution set is $\{0\}$.

28.

x	$8x = 32$	
0	$8 \cdot 0 = 32$	False
1	$8 \cdot 1 = 32$	False
2	$8 \cdot 2 = 32$	False
3	$8 \cdot 3 = 32$	False
4	$8 \cdot 4 = 32$	True
5	$8 \cdot 5 = 32$	False
6	$8 \cdot 6 = 32$	False

$\therefore$ the solution set is $\{4\}$.

29.

x	$x \cdot x = 36$	
0	$0 \cdot 0 = 36$	False
1	$1 \cdot 1 = 36$	False
2	$2 \cdot 2 = 36$	False
3	$3 \cdot 3 = 36$	False
4	$4 \cdot 4 = 36$	False
5	$5 \cdot 5 = 36$	False
6	$6 \cdot 6 = 36$	True

$\therefore$ the solution set is $\{6\}$.

30.

x	$x \cdot x = 1$	
0	$0 \cdot 0 = 1$	False
1	$1 \cdot 1 = 1$	True
2	$2 \cdot 2 = 1$	False
3	$3 \cdot 3 = 1$	False
4	$4 \cdot 4 = 1$	False
5	$5 \cdot 5 = 1$	False
6	$6 \cdot 6 = 1$	False

$\therefore$ the solution set is $\{1\}$.

31.

x	$\frac{1}{2}x = 2$	
0	$\frac{1}{2} \cdot 0 = 2$	False
1	$\frac{1}{2} \cdot 1 = 2$	False
2	$\frac{1}{2} \cdot 2 = 2$	False
3	$\frac{1}{2} \cdot 3 = 2$	False
4	$\frac{1}{2} \cdot 4 = 2$	True
5	$\frac{1}{2} \cdot 5 = 2$	False
6	$\frac{1}{2} \cdot 6 = 2$	False

$\therefore$ the solution set is $\{4\}$.

32.

x	$\frac{1}{3}x = 2$	
0	$\frac{1}{3} \cdot 0 = 2$	False
1	$\frac{1}{3} \cdot 1 = 2$	False
2	$\frac{1}{3} \cdot 2 = 2$	False
3	$\frac{1}{3} \cdot 3 = 2$	False
4	$\frac{1}{3} \cdot 4 = 2$	False
5	$\frac{1}{3} \cdot 5 = 2$	False
6	$\frac{1}{3} \cdot 6 = 2$	True

$\therefore$ the solution set is {6}.

33.

x	$x \cdot x = 5x$	
0	$0 \cdot 0 = 5 \cdot 0$	True
1	$1 \cdot 1 = 5 \cdot 1$	False
2	$2 \cdot 2 = 5 \cdot 2$	False
3	$3 \cdot 3 = 5 \cdot 3$	False
4	$4 \cdot 4 = 5 \cdot 4$	False
5	$5 \cdot 5 = 5 \cdot 5$	True
6	$6 \cdot 6 = 5 \cdot 6$	False

$\therefore$ the solution set is {0, 5}.

34.

x	$3x + 9 = 26$	
0	$3 \cdot 0 + 9 = 26$	False
1	$3 \cdot 1 + 9 = 26$	False
2	$3 \cdot 2 + 9 = 26$	False
3	$3 \cdot 3 + 9 = 26$	False
4	$3 \cdot 4 + 9 = 26$	False
5	$3 \cdot 5 + 9 = 26$	False
6	$3 \cdot 6 + 9 = 26$	False

$\therefore$ there is no solution.

35.

x	$15 = 9x - 3$	
0	$15 = 9 \cdot 0 - 3$	False
1	$15 = 9 \cdot 1 - 3$	False
2	$15 = 9 \cdot 2 - 3$	True
3	$15 = 9 \cdot 3 - 3$	False
4	$15 = 9 \cdot 4 - 3$	False
5	$15 = 9 \cdot 5 - 3$	False
6	$15 = 9 \cdot 6 - 3$	False

$\therefore$ the solution set is {2}.

36.

x	$4x = x \cdot 4$	
0	$4 \cdot 0 = 0 \cdot 4$	True
1	$4 \cdot 1 = 1 \cdot 4$	True
2	$4 \cdot 2 = 2 \cdot 4$	True
3	$4 \cdot 3 = 3 \cdot 4$	True
4	$4 \cdot 4 = 4 \cdot 4$	True
5	$4 \cdot 5 = 5 \cdot 4$	True
6	$4 \cdot 6 = 6 \cdot 4$	True

$\therefore$ the solution set is {0, 1, 2, 3, 4, 5, 6}.

37.

x	$x(9 - x) = 0$	
0	$0(9 - 0) = 0$	True
1	$1(9 - 1) = 0$	False
2	$2(9 - 2) = 0$	False
3	$3(9 - 3) = 0$	False
4	$4(9 - 4) = 0$	False
5	$5(9 - 5) = 0$	False
6	$6(9 - 6) = 0$	False

$\therefore$ the solution set is {0}.

38. $2m + 3$ **39.** $\frac{1}{2}z - 4$ **40.** $8k + 2$ **41.** $5w - 1$ **42.** $3(h + 6)$ **43.** $7x$

44. $3y$ **45.** $x - 4$ **46.** $w + 7$ **47.** $n - 5$

48. a. Let c = cost of one pencil. **b.** $12c = 139$

49. a. Let s = length of a side. **b.** $4s = 52$

50. a. Let g = number of guests at the party. **b.** $g + 5 = 34$

51. Let x = Celia's age; $(x + 4)x = 140$ **52.** Let x = width; $x(x + 5) = 176$

53. *Step 1:* Problem asks for the number of tickets each sold.

Step 2: Let n = number of tickets Holly sold. *Step 3:* $n + \left(\frac{1}{2}n - 12\right) = 114$

Step 4: $\frac{3}{2}n - 12 = 114; \frac{3}{2}(68) - 12 = 102 - 12 = 90 \neq 114;$

$\frac{3}{2}(72) - 12 = 108 - 12 = 96 \neq 114; \frac{3}{2}(84) - 12 = 126 - 12 = 114;$

$\frac{1}{2}(84) - 12 = 42 - 12 = 30$

Step 5: $84 + 30 = 114$; Holly sold 84 tickets and Cynthia sold 30.

54. *Step 1:* Problem asks how many pounds each weighs.

Step 2: Let s = Stephanie's weight. *Step 3:* $\frac{3}{4}(s + 40) = s$

Step 4: $\frac{3}{4}s + 30 = s; \frac{3}{4}(100) + 30 = 75 + 30 = 105 \neq 100;$

$\frac{3}{4}(118) + 30 = 88.5 + 30 = 118.5 \neq 118; \frac{3}{4}(120) + 30 = 90 + 30 = 120;$

$120 + 40 = 160$

Step 5: $\frac{3}{4}(160) = 120$; Stephanie weighs 120 lb and Jim weighs 160 lb.

55. 400; 400 ft below sea level; −400 **56.** −50; a deposit of $50; 50

57. −10; 10 wins; 10 **58.** 7; Seven floors down; −7

59. 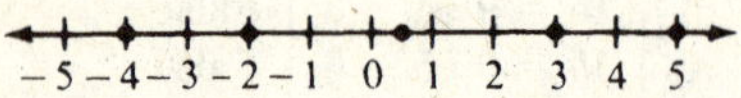

60. 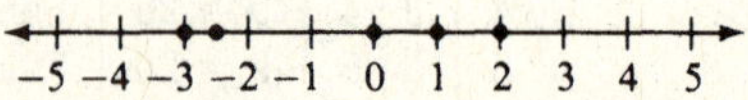

61. $-(3) = -3$ **62.** $8 + 10 = 18$ **63.** $3 + 6 = 9$ **64.** $2 + 9 = 11$

65. $3 + 0 = 3$ **66.** $6 - 6 = 0$ **67.** $3.2 + 0.8 = 4$ **68.** $4.7 + 4.7 = 9.4$ **69.** $>$

70. $<$ **71.** $<$ **72.** $>$ **73.** $<$ **74.** $<$

Pages 641–643 • CHAPTER 2

1. $(237 + 13) + (75 + 25) = 250 + 100 = 350$

2. $(456 + 44) + (29 + 21) = 500 + 50 = 550$

3. $(0.2 + 2.8) + (16.4 + 0.6) = 3 + 17 = 20$

4. $(3.75 + 1.25) + (4.85 + 3.15) = 5 + 8 = 13$

5. $\left(6\frac{3}{8} + 4\frac{5}{8}\right) + \left(1\frac{2}{7} + 3\frac{5}{7}\right) = 11 + 5 = 16$

6. $\left(25\frac{3}{4} + \frac{1}{4}\right) + \left(\frac{4}{5} + 2\frac{1}{5}\right) = 26 + 3 = 29$ **7.** $3m + 12$ **8.** $5f + 22$

9. $6w + 12$ **10.** $35u$ **11.** $88n$ **12.** $36b$ **13.** $60pqr$ **14.** $70klx$ **15.** $60hmw$

16. $4 + 9 = 13$ **17.** $3 + (-3) = 0$ **18.** $-5 + 4 = -1$ **19.** $-18 + 6 = -12$

20. $-7 + (-30) = -37$ **21.** $20 + 0 = 20$ **22.** -16 **23.** -19

24. $-10.7 + 10.7 = 0$ **25.** $5.4 - (11) = -5.6$ **26.** $21 - 3 = 18$

27. $-13 + 12 = -1$ **28.** $112 + (-97) = 15$ **29.** $-353 + 200 = -153$

30. $-16 + (-2) = -16 - 2 = -18$ **31.** $-11 + [-(-7)] = -11 + 7 = -4$

32. $-8 + (2) + [-(-5)] = -6 + 5 = -1$ **33.** $-(3) + (-5) + (-4) = -12$

34. $1 + (-2) + 3 = 2$ **35.** $|(2) + (-5) + (3)| = |0| = 0$

36. $(2) + (-3) + (-12) = 2 - 15 = -13$

37. $-|(3) + [-(-5)] + (2)| = -|5 + 5| = -|10| = -10$ **38.** -170

39. $53 + 47 = 100$ **40.** $-18 + 5 = -13$ **41.** -83 **42.** $133 - (3) = 130$

43. $186 - (-29) = 186 + 29 = 215$ **44.** $(-11) - (-11) = -11 + 11 = 0$

45. $22 - 5 = 17$ **46.** $14 + 8 - (9) = 22 - 9 = 13$

47. $-25 - (-6 + 11) = -25 - 5 = -30$ **48.** $6 + x - 6 + x - x = x$

49. $y + 4 - y + 4 - 4 = 4$ **50.** $5 + 10 = 15$ **51.** $\frac{1}{5}(24 + 16) = \frac{1}{5}(40) = 8$

52. $\frac{1}{4} \cdot 16 + \frac{1}{4} \cdot 12 = 4 + 3 = 7$ **53.** $34(0.25 + 0.75) = 34 \cdot 1 = 34$

54. $22(37 - 7) = 22(30) = 660$ **55.** $58(16 - 6) = 58 \cdot 10 = 580$

56. $m(14 + 7) = 21m$ **57.** $q(15 - 8) = 7q$ **58.** $n(53 - 110) = -57n$

59. $a(79 - 37) = 42a$ **60.** $10u + 8$ **61.** $7c + 21 + 6 = 7c + 27$

62. $26 + 4h + 12 = 4h + 38$ **63.** $8j - 32 + 17 = 8j - 15$

64. $23 + 6t - 12 = 6t + 11$ **65.** $8x + 20$ **66.** $8m + 20$

67. $2u + 5$ **68.** $2h + 20k$ **69.** $2f + 10g$ **70.** $4x + 11y$ **71.** 135 **72.** -76

73. -180 **74.** 1440 **75.** -135 **76.** 0 **77.** -480 **78.** $-3(-11) = 33$

79. $(-11)(-1) = 11$ **80.** $-6(13 + 15) = -6(28) = -168$

81. $27(-5 - 5) = 27(-10) = -270$ **82.** $-16(-1 + 11) = -16(10) = -160$

83. $-7m + 42p$ **84.** $-10u + 5h$ **85.** $-24n + 36v$ **86.** $5x + 2$ **87.** $-4 - 8t$

88. $5l + 5$ **89.** $3x + 12y - 32x + 4y = -29x + 16y$

90. $-8u - 4v + 5u - 5v = -3u - 9v$ **91.** $-6c - 2d - 15d + 3c = -3c - 17d$

92. $7e - 7f - 6e + 9f = e + 2f$

Answers for Exs. 93–98 may vary.

93. Let n, $n + 1$, and $n + 2$ be the integers. $n + (n + 1) + (n + 2) = 75$

94. Let n, $n + 2$, and $n + 4$ be the integers. $n + (n + 2) + (n + 4) = 87$

95. Let n, $n + 2$, and $n + 4$ be the integers. $n + (n + 2) + (n + 4) = 138$

96. Let n and $n + 1$ be the integers. $n(n + 1) = 156$

97. Let n and $n + 2$ be the integers. $n + 2 = 3n + 8$

98. Let n and $n + 2$ be the integers. $n = \frac{1}{2}(n + 2) - 1$ **99.** -5 **100.** -100

101. 7 **102.** 8 **103.** $\frac{-1}{\underset{1}{\cancel{20}}} \cdot \overset{4}{\cancel{80}} = -4$ **104.** $-uv$ **105.** $11xy$ **106.** $3n$ **107.** $8g$

108. $-7a + 3$ **109.** $9h - 6$ **110.** $8e - 10f$ **111.** $-6x + 9y$ **112.** $-12m - 40$

113. $5p + 10q$ **114.** -7 **115.** -8 **116.** $55 \cdot (-5) = -275$ **117.** 0

118. $-36(-6) = 216$ **119.** $8(-5) = -40$ **120.** $-12(4) = -48$ **121.** 0 **122.** $-14m$

123. $-\frac{63}{2}a$ **124.** -108 **125.** 11 **126.** c **127.** $-x$ **128.** $8w$ **129.** $5h$

Pages 643–645 • CHAPTER 3

1. $a - 13 + 13 = 17 + 13, a = 30; \{30\}$ **2.** $c + 8 - 8 = 22 - 8, c = 14; \{14\}$

3. $s - 20 + 20 = -12 + 20, s = 8; \{8\}$ **4.** $y + 14 - 14 = -33 - 14, y = -47; \{-47\}$

5. $15 + h - 15 = 0 - 15, h = -15; \{-15\}$ **6.** $0 + 13 = k - 13 + 13, 13 = k; \{13\}$

7. $f - 4 + 4 = 16 + 4, f = 20; \{20\}$ **8.** $g + 7 - 7 = 2 - 7, g = -5; \{-5\}$

9. $-x + 6 - 6 = 9 - 6, -x = 3, -(-x) = -3, x = -3; \{-3\}$

10. $23 - y - 23 = 47 - 23, -y = 24, -(-y) = -24, y = -24; \{-24\}$

11. $-5 - m + 5 = 7 + 5, -m = 12, -(-m) = -12, m = -12; \{-12\}$

12. $13 - 8 = -q + 8 - 8, 5 = -q, -5 = q; \{-5\}$

13. $e + 7 = 9, e + 7 - 7 = 9 - 7, e = 2; \{2\}$

14. $6 = n + 13, 6 - 13 = n + 13 - 13, -7 = n; \{-7\}$ **15.** $-4 + z = 8, z = 12; \{12\}$

16. $\frac{1}{13}(13u) = (338)\frac{1}{13}, u = 26; \{26\}$ **17.** $\frac{-396}{22} = \frac{22a}{22}, -18 = a; \{-18\}$

18. $\frac{-12x}{-12} = \frac{-444}{-12}, x = 37; \{37\}$ **19.** $\frac{126}{-9} = \frac{-9w}{-9}, -14 = w; \{-14\}$

20. $7\left(\frac{1}{7}t\right) = 7(13), t = 91; \{91\}$ **21.** $8\left(\frac{1}{8}h\right) = 8(-8), h = -64; \{-64\}$

22. $-4(11) = -4\left(-\frac{1}{4}v\right), -44 = v; \{-44\}$ **23.** $-5(-10) = -5\left(-\frac{1}{5}m\right), 50 = m; \{50\}$

24. $7(-42) = 7\left(\frac{n}{7}\right), -294 = n; \{-294\}$ **25.** $-4\left(-\frac{c}{4}\right) = -4(32), c = -128; \{-128\}$

26. $-27\left(-\frac{m}{27}\right) = -27(0), m = 0; \{0\}$ **27.** $m = 120; \{120\}$ **28.** $x = -\frac{1}{14}; \left\{-\frac{1}{14}\right\}$

29. $\frac{1}{6} = z; \left\{\frac{1}{6}\right\}$ **30.** $v = 4\left(\frac{11}{4}\right), v = 11; \{11\}$ **31.** $2\left(\frac{7}{2}\right) = u, u = 7; \{7\}$

32. $5k = 35, k = 7; \{7\}$ **33.** $7h = 42, h = 6; \{6\}$ **34.** $3m = -42, m = -14; \{-14\}$

35. $10n = 80, n = 8; \{8\}$ **36.** $4v = 44, v = 11; \{11\}$ **37.** $-5c = 65, c = -13; \{-13\}$

38. $\frac{n}{5} = -20, n = -100; \{-100\}$ **39.** $-\frac{x}{3} = 9, x = -27; \{-27\}$

40. $\frac{5}{6}u = -15, u = -15 \cdot \frac{6}{5}, u = -18; \{-18\}$

41. $-5x - 5 = -25, -5x = -20, x = 4; \{4\}$

42. $0 = -2y - 14, 2y = -14, y = -7; \{-7\}$ **43.** $8e = 48, e = 6; \{6\}$

44. $5k + 15 = -10, 5k = -25, k = -5; \{-5\}$

45. $-\frac{4}{3}n + 8 = 12, -\frac{4}{3}n = 4, n = -3; \{-3\}$

46. $2v + 14 - 9 = 19, 2v + 5 = 19, 2v = 14, v = 7; \{7\}$

47. *Step 1:* Problem asks for a number.

Step 2: Let n be the number.

Step 3: $37 + 3n = 67$

Step 4: $3n = 30, n = 10$

Step 5: $37 + 3(10) = 37 + 30 = 67$; The number is 10.

48. *Step 1:* Problem asks for a number.

Step 2: Let n be the number.

Step 3: $4n - 24 = -20$

Step 4: $4n = 4, n = 1$

Step 5: $4(1) - 24 = 4 - 24 = -20$; The number is 1.

49. *Step 1:* Problem asks for the width of the rectangle.

Step 2: Let w be the width.

Step 3: $2(w + 33) = 108$

Step 4: $2w + 66 = 108, 2w = 42, w = 21$

Step 5: $2(21 + 33) = 2(54) = 108$; The width of the rectangle is 21.

50. *Step 1:* Problem asks for the capacity of the large bucket.

Step 2: Let x be the capacity of the small bucket, and let $2x + 3$ be the capacity of the large bucket.

Step 3: $2x + 5(2x + 3) = 63$

Step 4: $2x + 10x + 15 = 63, 12x = 48, x = 4; 2(4) + 3 = 8 + 3 = 11$

Step 5: $2(4) + 5(11) = 8 + 55 = 63$; The large bucket holds 11 L.

51. *Step 1:* Problem asks for the lengths of the sides of the triangle.

Step 2: Let n, $n + 2$, and $n + 4$ be the sides of the triangle.

Step 3: $n + (n + 2) + (n + 4) = 18$

Step 4: $3n + 6 = 18, 3n = 12, n = 4; n + 2 = 6; n + 4 = 8$

Step 5: $4 + 6 + 8 = 18$; The sides of the triangle measure 4 m, 6 m, and 8 m.

52. *Step 1:* Problem asks for the amount in each account.

Step 2: Let x be the amount in Bruce's brother's account.

Step 3: $x + (x + 122) = 354$

Step 4: $2x + 122 = 354, 2x = 232, x = 116; x + 122 = 116 + 122 = 238$

Step 5: $116 + 238 = 354$; Bruce has \$238 in his account and his brother has \$116.

53. $2w = 14, w = 7; \{7\}$ **54.** $5x = 45, x = 9; \{9\}$ **55.** $48 = -6k, -8 = k; \{-8\}$

56. $3m + 3 = 21, 3m = 18, m = 6; \{6\}$ **57.** $27 + 4u = 3, 4u = -24, u = -6; \{-6\}$

58. $1 = -1$; no solution **59.** $2v - 16 = 6v, -16 = 4v, v = -4; \{-4\}$

60. $3x = 5x - 30, -2x = -30, x = 15; \{15\}$

61. $7y - 3 = 6y + 12, y - 3 = 12, y = 15; \{15\}$

62. $6 - 3c = 6 - 3c$; identity

63. $m - 5 = 6 - 7m, 8m - 5 = 6, 8m = 11, m = \frac{11}{8}; \left\{\frac{11}{8}\right\}$

64. $20x - 12 = 50x + 38, -12 = 30x + 38, -50 = 30x, -\frac{5}{3} = x; \left\{-\frac{5}{3}\right\}$

65. $15 + 5h = 4h + 8, 15 + h = 8, h = -7; \{-7\}$

66. $12x - 6 = 12x + 21, 6 = 21$; no solution

67. $7n - 21 = 5n - 15, 2n - 21 = -15, 2n = 6, n = 3; \{3\}$

68.

	Jay +	Wife	= Total Income
Current	$\frac{2}{3}x$	x	$\frac{5}{3}x$
After raise	$\frac{2}{3}x + 2000$	$x + 2000$	49,000

$\left(\frac{2}{3}x + 2000\right) + (x + 2000) = 49{,}000$, $\frac{5}{3}x + 4000 = 49{,}000$, $\frac{5}{3}x = 45{,}000$, $x = 27{,}000$;

$\frac{2}{3}(27{,}000) = 18{,}000$; Jay's current salary is \$18,000 and his wife's salary is \$27,000.

69. Let n, $n + 2$, and $n + 4$ be Erin's three test scores. $n + (n + 2) + (n + 4) + (n + 4 + 18) = 328$, $4n + 28 = 328$, $4n = 300$, $n = 75$; $75 + 2 = 77$; $75 + 4 = 79$; $79 + 18 = 97$; Erin's test scores were 75, 77, 79, and 97.

70.

	Now	After gain
Julius	$2x$	$2x + 10$
Twin	x	$x + 5$
Twin	x	$x + 5$
Total	$4x$	240

$2x + 10 + 2(x + 5) = 240$, $2x + 10 + 2x + 10 = 240$, $4x + 20 = 240$, $4x = 220$, $x = 55$; $2(55) = 110$; Julius weighs 110 lb and each of his twin brothers weighs 55 lb.

71.

	Width	Length	Perimeter
First Rectangle	$l - 6$	l	$2(l - 6) + 2l$
Second Rectangle	$(l - 6) + 3$	$l - 2$	54

$2(l - 3) + 2(l - 2) = 54$, $2l - 6 + 2l - 4 = 54$, $4l - 10 = 54$, $4l = 64$, $l = 16$; $16 - 6 = 10$; $10 + 3 = 13$; $16 - 2 = 14$; The first rectangle measures 16 cm by 10 cm, and the second rectangle is 13 cm by 14 cm.

72.

	Number of coins ×	Value =	Total Amount
Nickels	$3x$	5	$15x$
Dimes	x	10	$10x$

$15x + 10x = 625$, $25x = 625$, $x = 25$; $3(25) = 75$; Martha has 75 nickels.

73.

	Number of Dozens ×	Cost =	Total
Bought	x	1.50	$1.5x$
Sold	$x - 5$	2.00	$2(x - 5)$
Profit			18

$2(x - 5) - 1.5x = 18$, $2x - 10 - 1.5x = 18$, $0.5x - 10 = 18$, $0.5x = 28$, $x = 56$; Elliot bought 56 dozen flowers.

74.

	Cans	× Cost =	Total
Cat food	$x + 2$	69	$69(x + 2)$
Dog food	x	79	$79x$

$79x + 69(x + 2) = 1618$, $79x + 69x + 138 = 1618$, $148x + 138 = 1618$, $148x = 1480$, $x = 10$; $10 + 2 = 12$; Rachel bought 10 cans of dog food and 12 cans of cat food.

75.

	Hours	× Wage =	Total
After school	x	3	$3x$
Saturday	$13 - x$	4	$4(13 - x)$

$3x + 4(13 - x) = 43$, $3x + 52 - 4x = 43$, $-x + 52 = 43$, $52 = 43 + x$, $9 = x$; $13 - 9 = 4$; Victor worked 4 h on Saturday.

76. Comm. prop. of add.; assoc. prop. of add.; subst. principle

77. Subst. principle; assoc. prop. of add.; prop. of opposites; identity prop. of add.

78. Comm. prop. of add.; subst. principle; prop. of opp.; identity prop. of add.

Pages 645–646 • CHAPTER 4

1. 343 **2.** 625 **3.** -48 **4.** -1000 **5.** $7 + 25 = 32$ **6.** $4^3 = 64$
7. $6 - 32 = -26$ **8.** $11^2 = 121$ **9.** $125 \div (9 + 16) = 125 \div 25 = 5$
10. $(64 - 36) \div 7 = 28 \div 7 = 4$ **11.** $4(81 - 64) = 4(17) = 68$
12. $3(-3) + 2^2 = -9 + 4 = -5$ **13.** $[3(-3) + 2]^2 = (-9 + 2)^2 = (-7)^2 = 49$
14. $4(-3) - 2^3 = -12 - 8 = -20$ **15.** $[4(-3) - 2]^3 = (-12 - 2)^3 = (-14)^3 = -2744$
16. $7 + (-3)(4) = 7 - 12 = -5$ **17.** $[7 + (-3)(2)]^2 = (7 - 6)^2 = 1$
18. $-\frac{3(-3)}{2^2} = \frac{9}{4}$ **19.** $\left[-\frac{3(-3)}{2}\right]^2 = \left(\frac{9}{2}\right)^2 = \frac{81}{4}$ **20.** $11x + 5$ **21.** $b - 2$
22. $9m + 11$ **23.** $4t - 10$ **24.** $2l + 6$ **25.** $10x^2 - 3xy + 2y^2$
26. $-6m^2 - 3mn - 6n$ **27.** $3a^2 + 3ab - b^2$ **28.** $-3x - 11$ **29.** $5b + 10$
30. $m + 5$ **31.** $-8t - 4$ **32.** $10k - 14l + 2$ **33.** $2x^2 - xy + 4y^2$
34. $10m^2 - 3mn - 4n$ **35.** $7a^2 - 15ab + b^2$ **36.** e^{10} **37.** $8f^7$ **38.** $12c^3d^3$
39. $-10g^4h^3$ **40.** $36m^3n^4$ **41.** $60j^5k^3l^5$ **42.** $12x^6y^7$ **43.** $-a^6$ **44.** $-\frac{28u^4v^4}{3}$
45. $3^6 = 729$ **46.** 4^{2a+3} **47.** $2^{11} = 2048$ **48.** $15p^7 + 14p^7 = 29p^7$
49. $16d^{10} - 12d^{10} = 4d^{10}$ **50.** w^{10} **51.** x^{10} **52.** y^7 **53.** z^{2n} **54.** a^{3n}
55. b^{3n} **56.** c^{n+3} **57.** d^{3n} **58.** $25f^2$ **59.** g^4h^4 **60.** $36m^6$ **61.** $64m^3n^{15}$
62. $32u^{15}v^5$ **63.** $9a^{10}b^8$ **64.** $49x^8$ **65.** $-512x^{15}$ **66.** $9k^2 \cdot 81k^4 = 729k^6$
67. $-8x^9 \cdot 25x^4 = -200x^{13}$ **68.** $-16t^4 \cdot 27t^3 = -432t^7$
69. $125x^6y^3 \cdot 3xy^2 = 375x^7y^5$ **70.** $7x + 21$ **71.** $5y - 20$ **72.** $-3n + 6$
73. $-8 - 32m$ **74.** $3n^2 + 15n$ **75.** $-12t + 8t^2$ **76.** $12k^2 - 42k$
77. $-40h^2 - 15h$ **78.** $9a^3 - 27a^2 - 36a$ **79.** $-15b^4 + 10b^3 - 30b^2$
80. $2c^3 - c^2d + 3cd^2$ **81.** $5u^3v^2 - 2u^2v^3 + 4uv^4$ **82.** $m^2 + 6m + 8$

83. $n^2 + 2n - 15$ **84.** $a^2 - 13a + 42$ **85.** $5x^2 + 33x - 14$

86. $12y^2 - 10y + 2$ **87.** $30m^2 + 18m + 20m + 12 = 30m^2 + 38m + 12$

88. $u^3 + 2u^2 + 5u + 3u^2 + 6u + 15 = u^3 + 5u^2 + 11u + 15$

89. $3v^3 + 4v^2 + 7v - 3v^2 - 4v - 7 = 3v^3 + v^2 + 3v - 7$

90. $6c^3 - 3c^2 + 24c - 10c^2 + 5c - 40 = 6c^3 - 13c^2 + 29c - 40$

91. $21x^2 - 26xy + 8y^2$ **92.** $20a^2 - 27ab - 8b^2$

93. $e^3 + e^2f + ef^2 + e^2f + ef^2 + f^3 = e^3 + 2e^2f + 2ef^2 + f^3$

94. $15m^3 - 20m^2n + 5mn^2 + 3m^2n - 4mn^2 + n^3 = 15m^3 - 17m^2n + mn^2 + n^3$

95. $2A = ap; \frac{2A}{p} = a; p \neq 0$ **96.** $3V = Bh, \frac{3V}{B} = h; B \neq 0$

97. $2A = h(b_1 + b_2), \frac{2A}{h} = b_1 + b_2, \frac{2A}{h} - b_2 = b_1; h \neq 0$ **98.** $b = y - mx$

99. $\frac{A}{\pi} = r^2; r = \pm\sqrt{\frac{A}{\pi}}; A \geq 0$ **100.** $\frac{s}{180} = n - 2, \frac{s}{180} + 2 = n$

101. $F - 32 = \frac{9}{5}C, \frac{5}{9}(F - 32) = C$ **102.** $P(1 + rt) = A, P + Prt = A; 1 + rt \neq 0$

103. $Prt = I, t = \frac{I}{Pr}; P, r, t \neq 0$

104.

	Rate	× Time	= Distance
Bus 1 (N)	$x + 15$	3	$3(x + 15)$
Bus 2 (S)	x	3	$3x$

$3(x + 15) + 3x = 255, 3x + 45 + 3x = 255, 6x = 210, x = 35; (35) + 15 = 50$; The southbound bus travels at 35 mi/h and the northbound bus travels at 50 mi/h.

105.

	Rate	× Time	= Distance
Alex	36	x	$36x$
Allison	40	$x - \frac{1}{6}$	$40\left(x - \frac{1}{6}\right)$

$36x = 40\left(x - \frac{1}{6}\right), 36x = 40x - \frac{40}{6}, \frac{20}{3} = 4x, \frac{5}{3} = x; \left(\frac{5}{3}\right) - \frac{1}{6} = \frac{10 - 1}{6} = \frac{9}{6} = \frac{3}{2}$; It took Allison $1\frac{1}{2}$ h or 90 min to overtake Alex.

106.

	Rate ×	Time =	Distance
To Plainsville	800	x	$800x$
To Sky City	900	$x - \frac{1}{2}$	$900\left(x - \frac{1}{2}\right)$

$800x = 900\left(x - \frac{1}{2}\right)$, $800x = 900x - 450$, $450 = 100x$, $x = 4.5$; $800(4.5) = 3600$; The airports are 3600 km apart and the trip to Plainsville took 4.5 h.

107.

	Length	Width
Poster	$3x$	x
Outside of mat	$3x + 8$	$x + 8$

$(3x + 8)(x + 8) - (3x)(x) = 488$, $3x^2 + 24x + 8x + 64 - 3x^2 = 488$, $32x + 64 = 488$, $32x = 424$, $x = 13.25$; $3(13.25) = 39.75$; The poster measures 13.25 in. by 39.75 in.

108.

	Length	Width
Square	x	x
Rectangle	$x + 1$	$x - \frac{5}{9}$

$x^2 = (x + 1)\left(x - \frac{5}{9}\right)$, $x^2 = x^2 - \frac{5}{9}x + x - \frac{5}{9}$, $0 = \frac{4}{9}x - \frac{5}{9}$, $\frac{5}{9} = \frac{4}{9}x$, $\frac{5}{4} = x$;

$2\left(\frac{5}{4}\right)^2 = 2 \cdot \frac{25}{16} = \frac{50}{16} = \frac{25}{8} = 3\frac{1}{8}$; You will get $3\frac{1}{8}$ yd^2 altogether.

Pages 647–649 • CHAPTER 5

1. 1, 42; 2, 21; 3, 14; 6, 7; −1, −42; −2, −21; −3, −14; −6, −7

2. 1, 80; 2, 40; 4, 20; 5, 16; 8, 10; −1, −80; −2, −40; −4, −20; −5, −16; −8, −10

3. 1, 91; 7, 13; −1, −91; −7, −13

4. 1, 72; 2, 36; 3, 24; 4, 18; 6, 12; 8, 9; −1, −72; −2, −36; −3, −24; −4, −18; −6, −12; −8, −9

5. 1, 52; 2, 26; 4, 13; −1, −52; −2, −26; −4, −13

6. $2 \cdot 3 \cdot 7$ **7.** $2^4 \cdot 5$ **8.** $7 \cdot 13$ **9.** $2^3 \cdot 3^2$ **10.** $2^2 \cdot 13$

11. $126 = 2 \cdot 3^2 \cdot 7$, $168 = 2^3 \cdot 3 \cdot 7$; GCF $= 2 \cdot 3 \cdot 7 = 42$

12. $144 = 2^4 \cdot 3^2$, $84 = 2^2 \cdot 3 \cdot 7$; GCF $= 2^2 \cdot 3 = 12$

13. $65 = 5 \cdot 13$, $52 = 2^2 \cdot 13$; GCF $= 13$

14. $90 = 2 \cdot 3^2 \cdot 5$, $330 = 2 \cdot 3 \cdot 5 \cdot 11$; GCF $= 2 \cdot 3 \cdot 5 = 30$

15. $\frac{12x^5}{4x} = 3x^4$ **16.** $\frac{25m^4n}{-15mn^6} = -\frac{5m^3}{3n^5}$ **17.** $\frac{-7ab}{21ab^5} = -\frac{1}{3b^4}$ **18.** $\frac{-8(uv)^7}{-10(uv)^5} = \frac{4u^2v^2}{5}$

19. $\frac{w^8}{w^{20}} = \frac{1}{w^{12}}$ **20.** $\frac{25k^2}{5k^2} = 5$ **21.** $-\frac{27y^3}{y^6} = -\frac{27}{y^3}$ **22.** $\frac{(2c^5)(4c^3)}{(8c^2)^3} = \frac{c^2}{64}$

23. $\frac{4(3e + 2)}{4} = 3e + 2$ **24.** $\frac{3(2x - 3y + 4)}{3} = 2x - 3y + 4$

25. $\frac{x(2x^2 + 6x + 1)}{x} = 2x^2 + 6x + 1$ **26.** $\frac{-6a(-3b + 4a)}{-6a} = -3b + 4a$

27. $\frac{5m(3 - 5m - m^2)}{5m} = 3 - 5m - m^2$ **28.** $\frac{7hk^2(4h^4k - 5)}{7hk^2} = 4h^4k - 5$

29. $5(3w^2 - 2w + 1)$ **30.** $9x(x + 2)$ **31.** $7u^2(u + 2)$ **32.** $6a(2a^2 - a + 3)$

33. $3c(5c + d)$ **34.** $8mn(m - 3n)$ **35.** $x^2 + 8x + 15$ **36.** $b^2 - 6b + 8$

37. $n^2 + 4n - 21$ **38.** $e^2 - 2e - 48$ **39.** $6 + 5m + m^2$ **40.** $3f^2 + 17f + 10$

41. $8y^2 - 10y + 3$ **42.** $8z^2 - 9z - 14$ **43.** $20n^2 - 22n + 6$

44. $a(30a^2 - 38a + 12) = 30a^3 - 38a^2 + 12a$

45. $h(12h^2 + 55h + 63) = 12h^3 + 55h^2 + 63h$

46. $2x(18x^2 + 25x - 3) = 36x^3 + 50x^2 - 6x$ **47.** $k^2 - 25$ **48.** $9 - y^2$

49. $16d^2 - 64$ **50.** $w^4 - 36$ **51.** $25m^4 - n^2$ **52.** $a^2b^2 - c^4$

53. $(4e - 3)(4e + 3)$ **54.** $(6u - 5)(6u + 5)$ **55.** $(9 - f)(9 + f)$

56. $16(9a^2 - 4b^2) = 16(3a - 2b)(3a + 2b)$ **57.** $(7 - 10y)(7 + 10y)$

58. $(v^2 - w^2)(v^2 + w^2) = (v - w)(v + w)(v^2 + w^2)$ **59.** $(s^3 - 2)(s^3 + 2)$

60. $(4x^4 - 25)(4x^4 + 25) = (2x^2 - 5)(2x^2 + 5)(4x^4 + 25)$ **61.** $g^2 + 14g + 49$

62. $k^2 - 6k + 9$ **63.** $4x^2 + 24x + 36$ **64.** $25y^2 - 30y + 9$

65. $4m^2 + 12mn + 9n^2$ **66.** $49a^2 - 70ab + 25b^2$ **67.** $e^2f^2 - 16ef + 64$

68. $81f^2 - 72f + 16$ **69.** $(x - 3)^2$ **70.** $(e + 9)^2$ **71.** $(2 - 7h)^2$ **72.** $(8x + 5y)^2$

73. $(2m - 9n)^2$ **74.** $(4w + 3z)^2$ **75.** $(k + 7)(k + 1)$ **76.** $(v - 5)(v - 4)$

77. $(a - 1)^2$ **78.** $(7 + u)(5 + u)$ **79.** $(n - 4)(n - 12)$ **80.** $(w + 8)(w + 10)$

81. $(x + 6y)(x + 7y)$ **82.** $(m - 7n)(m - 3n)$ **83.** $(e - 4f)(e - 11f)$

84. $(c + 6)(c - 3)$ **85.** $(x - 7)(x + 5)$ **86.** prime **87.** $(h - 9)(h + 2)$

88. $(b + 10)(b - 3)$ **89.** $(y - 9)(y + 5)$ **90.** $(a - 3b)(a + b)$

91. $(u + 4v)(u - v)$ **92.** $(m - 5n)(m + 4n)$ **93.** $(2x + 3)(x + 4)$

94. prime **95.** $(5d + 3)(2d - 1)$ **96.** $-2(5 + 13y + 6y^2) = -2(5 + 3y)(1 + 2y)$

97. $-1(18z^2 + 39z + 7)$; prime **98.** $-2(4z^2 - 12z + 5) = -2(2z - 1)(2z - 5)$

99. $(5x + y)(3x + 2y)$ **100.** $2(4a^2 - 11ab + 6b^2) = 2(a - 2b)(4a - 3b)$

101. $(7m + 3n)(2m - n)$ **102.** $(m - 3)(8 + 5m)$

103. $(a + 2)(6a + 4) = 2(a + 2)(3a + 2)$ **104.** $(u + 1)(u - 2v)$

105. $(b^2 - 2b)(b + 1) - 3(1 + b) = (b + 1)(b^2 - 2b - 3) = (b + 1)(b - 3)(b + 1) = (b + 1)^2(b - 3)$

106. $a(a + 2) + b(a + 2) = (a + 2)(a + b)$

107. $c(7w + 3) - w(7w + 3) = (7w + 3)(c - w)$

108. $n^2(n + 1) - 6(n + 1) = (n + 1)(n^2 - 6)$

109. $64(1 - m^2) + m^4(1 - m^2) = (1 - m^2)(64 + m^4) = (1 - m)(1 + m)(64 + m^4)$

110. $2x(21x^2 + 34x + 8) = 2x(3x + 4)(7x + 2)$

111. $6y(10y^2 - 3y - 1) = 6y(5y + 1)(2y - 1)$

112. $x^3(12x^2 - 20x + 3) = x^3(6x - 1)(2x - 3)$

113. $16a^2(a^2 - 9) = 16a^2(a - 3)(a + 3)$

114. $4n(n^4 - 25) = 4n(n^2 - 5)(n^2 + 5)$ **115.** $2w^5(14w^2 - 51)$

116. $4(9m^2 + 6mn + n^2) = 4(3m + n)^2$ **117.** $-12(c^2 - 2cd + d^2) = -12(c - d)^2$

118. $-7x(x^2 - 2xy + y^2) = -7x(x - y)^2$ **119.** $a = -13, -8; \{-13, -8\}$

120. $f = 16, 27; \{16, 27\}$

121. $2x - 4 = 0$ or $3x - 5 = 0$, $2x = 4$ or $3x = 5$, $x = 2$ or $x = \frac{5}{3}; \left\{2, \frac{5}{3}\right\}$

122. $6h = 5$ or $6h = -5$, $h = \frac{5}{6}, -\frac{5}{6}; \left\{\frac{5}{6}, -\frac{5}{6}\right\}$

123. $w = 0$ or $4w = -3$, $w = 0$ or $w = -\frac{3}{4}; \left\{0, -\frac{3}{4}\right\}$

124. $m = 0$ or $2m = -7$ or $3m = 4$, $m = 0$ or $m = -\frac{7}{2}$ or $m = \frac{4}{3}; \left\{0, -\frac{7}{2}, \frac{4}{3}\right\}$

125. $(a + 6)(a + 1) = 0$; $a + 6 = 0$ or $a + 1 = 0$, $a = -6$ or $a = -1; \{-6, -1\}$

126. $q^2 - 21q + 20 = 0$, $(q - 20)(q - 1) = 0$; $q = 20$, $q = 1; \{1, 20\}$

127. $d^2 - 14d + 45 = 0$, $(d - 9)(d - 5) = 0$; $d = 9$ or $5; \{5, 9\}$

128. $(y - 9)(y + 2) = 0$, $y = 9$ or $y = -2; \{-2, 9\}$

129. $c^2 + 5c - 36 = 0$, $(c + 9)(c - 4) = 0$; $c = -9$ or $c = 4; \{-9, 4\}$

130. $h^2 + 3h - 54 = 0$, $(h + 9)(h - 6) = 0$; $h = -9$ or $h = 6; \{-9, 6\}$

131. $4z^2 + 23z - 6 = 0$, $(4z - 1)(z + 6) = 0$; $4z = 1$ or $z = -6$, $z = \frac{1}{4}$ or $-6; \left\{-6, \frac{1}{4}\right\}$

132. $3m^2 - 4m + 1 = 0$, $(3m - 1)(m - 1) = 0$; $3m = 1$ or $m = 1$, $m = \frac{1}{3}$ or $1; \left\{\frac{1}{3}, 1\right\}$

133. $2n^2 - n - 10 = 0$, $(2n - 5)(n + 2) = 0$; $2n = 5$ or $n = -2$, $n = \frac{5}{2}$ or $-2; \left\{-2, \frac{5}{2}\right\}$

134. $(e - 7)(e + 7) = 0$; $e = 7$ or $-7; \{7, -7\}$

135. $36g^2 - 16 = 0$, $4(9g^2 - 4) = 0$, $9g^2 - 4 = 0$, $(3g + 2)(3g - 2) = 0$; $3g = -2$ or $3g = 2$, $g = -\frac{2}{3}$ or $g = \frac{2}{3}; \left\{-\frac{2}{3}, \frac{2}{3}\right\}$

136. $w(w^2 - 9) = 0$, $w(w + 3)(w - 3) = 0$; $w = 0$ or $w = -3$ or $w = 3; \{-3, 0, 3\}$

137. Let x be the number. $x + x^2 = 56$, $x^2 + x - 56 = 0$, $(x + 8)(x - 7) = 0$; $x = -8, 7$; The number is -8 or 7.

138. Let n and $n + 2$ be the integers. $(n + 2)n = 143$, $n^2 + 2n - 143 = 0$, $(n + 13)(n - 11) = 0$; $n = -13$ or $n = 11$ (reject); The integers are -13 and -11.

139. Let w be the width of the rectangle. $w(2w - 5) = 88$, $2w^2 - 5w - 88 = 0$, $(2w + 11)(w - 8) = 0$; $2w = -11$ or $w = 8$, $w = -\frac{11}{2}$ or $w = 8$; $2(8) - 5 = 16 - 5 = 11$; The rectangle is 8 cm by 11 cm.

140. Let x be the first number and let $12 - x$ be the second number.
$x^2 + (12 - x)^2 = 74, x^2 + 144 - 24x + x^2 = 74, 2x^2 - 24x + 70 = 0,$
$(2x - 10)(x - 7) = 0; 2x = 10$ or $x = 7, x = 5$ or 7; The numbers are 5 and 7.

Pages 649–651 • CHAPTER 6

1. $\frac{5(m-3)}{m-3} = 5, m \neq 3$ **2.** $\frac{2a+1}{3(2a+1)} = \frac{1}{3}, a \neq -\frac{1}{2}$ **3.** $\frac{7(c-d)}{7(c+d)} = \frac{c-d}{c+d}, c \neq -d$

4. $\frac{k(6-k)}{(6-k)(6+k)} = \frac{k}{6+k}, k \neq -6, 6$ **5.** $\frac{3uv}{uv(u-v)} = \frac{3}{u-v}; u \neq 0, v \neq 0; u \neq v$

6. $\frac{8w^3}{4w(2w-3)} = \frac{2w^2}{2w-3}; w \neq 0; w \neq \frac{3}{2}$ **7.** $\frac{(x-8)(x+8)}{(x-8)(x+7)} = \frac{x+8}{x+7}; x \neq 8, x \neq -7$

8. $\frac{(e-7)^2}{(7-e)(7+e)} = \frac{(7-e)^2}{(7-e)(7+e)} = \frac{7-e}{7+e}; e \neq 7, e \neq -7$

9. $\frac{3(5m+2n)}{(5m+2n)(5m-2n)} = \frac{3}{5m-2n}, m \neq \frac{2}{5}n, m \neq -\frac{2}{5}n$

10. $\frac{a(a+b)}{a(a-b)} = \frac{a+b}{a-b}, a \neq 0, a \neq b$ **11.** $\frac{(k-3)(7k-2)}{-(7k-2)(k-3)} = -1; k \neq 3, k \neq \frac{2}{7}$

12. $\frac{(3x+5y)(x+4y)}{(3x+5y)(x-2y)} = \frac{x+4y}{x-2y}, x \neq 2y, x \neq -\frac{5y}{3}$

13. $\frac{(t-7)(t-2)}{(t-2)(t+2)} = \frac{t-7}{t+2}, t \neq 2, t \neq -2$ **14.** $\frac{(u-v)(u+v)}{(u+v)(u+v)} = \frac{u-v}{u+v}, u \neq -v$

15. $\frac{(5w-x)^5}{[-(5w-x)]^7} = \frac{(5w-x)^5}{-(5w-x)^7} = -\frac{1}{(5w-x)^2}, w \neq \frac{x}{5}$

16. $\frac{(4s-6)^2(3s-2)}{(3s-2)(4s-6)} = 4s - 6, s \neq \frac{2}{3}, s \neq \frac{3}{2}$ **17.** $\frac{5}{8} \cdot \frac{32}{15} = \frac{4}{3}$ **18.** $\frac{4}{3} \cdot \frac{3}{5} \cdot \frac{5}{7} = \frac{4}{7}$

19. $\frac{4}{25} \cdot \frac{15}{16} = \frac{3}{20}$ **20.** $-\frac{27}{8} \cdot \frac{24}{9} = -9$ **21.** $\frac{e}{f} \cdot \frac{f}{g} \cdot \frac{g}{h} = \frac{e}{h}$ **22.** $\frac{5}{w} \cdot \frac{w^2}{10} = \frac{w}{2}$

23. $\frac{8m}{3} \cdot \frac{9}{12m} = 2$ **24.** $\frac{a^2}{3b} \cdot \frac{b^2}{4a} = \frac{ab}{12}$ **25.** $\frac{14v}{12v^2} \cdot \frac{4uw^2}{7v^2} = \frac{2uw^2}{3v^3}$

26. $\frac{a+5}{a} \cdot \frac{a^2}{(a+5)(a-5)} = \frac{a}{a-5}$ **27.** $\frac{x(4-y)}{8x^2y} \cdot \frac{2}{(4-y)(4+y)} = \frac{1}{4xy(4+y)}$

28. $\frac{m+n}{m-n} \cdot \frac{(m-n)(m+n)}{3(m+n)} = \frac{m+n}{3}$ **29.** $25k^6$ **30.** $\frac{x^2}{49}$ **31.** $\frac{9x^2}{16}$ **32.** $\frac{8m^3}{27n^6}$

33. $-\frac{x^6}{125}$ **34.** $\frac{e^3}{f^3} \cdot \frac{e}{f} = \frac{e^4}{f^4}$ **35.** $\frac{64c^3}{d^3} \cdot \frac{c^2}{8} = \frac{8c^5}{d^3}$ **36.** $\frac{49a^2}{b^2} \cdot \frac{3ab}{14} = \frac{21a^3}{2b}$

37. $\frac{4}{9} \cdot \frac{3}{16} = \frac{1}{12}$ **38.** $\frac{a^2}{4} \cdot \frac{12}{a} = 3a$ **39.** $\frac{m}{3n} \cdot \frac{6}{mn} = \frac{2}{n^2}$ **40.** $\frac{8x^2}{5y} \cdot \frac{1}{4xy} = \frac{2x}{5y^2}$

41. $\frac{e+f}{5} \cdot \frac{15}{3(e+f)} = 1$ **42.** $\frac{(u+v)(u-v)}{u^2+v^2} \cdot \frac{1}{u+v} = \frac{u-v}{u^2+v^2}$

43. $\frac{5}{(n+5)(n-5)} \cdot \frac{n+5}{5(n-3)} = \frac{1}{(n-3)(n-5)}$ **44.** $\frac{4(n-3)}{4} \cdot \frac{8}{5(n-3)} = \frac{8}{5}$

45. $\frac{(x^2+y^2)(x-y)(x+y)}{2x(x+4)} \cdot \frac{(x-4)(x+4)}{x^2+y^2} = \frac{(x-y)(x+y)(x-4)}{2x}$

46. $\frac{(2a+5)(2a-5)}{6a^2} \cdot \frac{3a^4}{6(2a-5)} = \frac{(2a+5)a^2}{12}$

47. $\frac{m^2+n^2}{2(4s-5t)} \cdot \frac{2(2t-3s)}{7(m+n)} = \frac{(2t-3s)(m^2+n^2)}{7(4s-5t)(m+n)}$ **48.** $\frac{1}{3} \div \frac{2}{6} \cdot \frac{5}{7} = \frac{1}{3} \cdot \frac{6}{2} \cdot \frac{5}{7} = \frac{5}{7}$

49. $\frac{x}{7} \cdot \frac{x}{y^2} \cdot \frac{7}{y} = \frac{x^2}{y^3}$ **50.** $\frac{e^2}{3} \cdot \frac{f^2}{e} \cdot \frac{f}{e} = \frac{f^3}{3}$ **51.** $\frac{c^2}{9} \cdot \frac{9}{c} \cdot \frac{c}{3} = \frac{c^2}{3}$

52. $\frac{w^2}{16} \cdot \frac{16}{w^2} = 1$ **53.** $\frac{a-b}{a+3b} \cdot \frac{a+3b}{a+b} \cdot \frac{a+b}{b-a} = -1$ **54.** $\frac{5x \cdot 3}{11 \cdot 3} = \frac{15x}{33}$; $15x$

55. $\frac{2(h-4)}{7 \cdot 2} = \frac{2h-8}{14}$; $2h-8$ **56.** $\frac{4(3k-5)}{2 \cdot 4} = \frac{12k-20}{8}$; $12k-20$

57. $\frac{5(6m-n)}{7 \cdot 5} = \frac{30m-5n}{35}$; $30m-5n$ **58.** $\frac{c \cdot c^3}{c^3 \cdot d} = \frac{c^4}{c^3 d}$; c^4 **59.** $\frac{4s \cdot 3st}{15st^2} = \frac{12s^2t}{15st^2}$; $12s^2t$

60. $\frac{8(7d+2)}{(7d+2)^2} = \frac{56d+16}{(7d+2)^2}$; $56d+16$ **61.** $\frac{5(e+1)}{(e-1)(e+1)} = \frac{5e+5}{e^2-1}$; $5e+5$

62. $\frac{-3}{-(h-2)} = \frac{-3}{2-h}$; -3 **63.** $\frac{5z}{z(z-3)} = \frac{5z}{z^2-3z}$; $5z$ **64.** $\frac{w^2}{w(w+4)} = \frac{w^2}{w^2+4w}$; w^2

65. $\frac{3ab}{a^2b \cdot ab} = \frac{3ab}{a^3b^2}$; $3ab$ **66.** $\frac{70}{105}, \frac{84}{105}, \frac{45}{105}$ **67.** $\frac{3x-6}{48}, \frac{4x+12}{48}$ **68.** $\frac{9m-3n}{30}, \frac{6m+2n}{30}$

69. $\frac{d}{3cd^2}, \frac{12}{3cd^2}$ **70.** $\frac{7ef}{ef(e+f)}, \frac{5ef+5f^2}{ef(e+f)}, \frac{6e^2+6ef}{ef(e+f)}$

71. $\frac{2w}{3(x-3)}, \frac{1}{(x-3)(x+3)}$; $\frac{2wx+6w}{3(x-3)(x+3)}, \frac{3}{3(x-3)(x+3)}$ **72.** $\frac{8-4}{a} = \frac{4}{a}$

73. $\frac{13-5}{2x} = \frac{8}{2x} = \frac{4}{x}$ **74.** $\frac{u-3u-5}{7} = \frac{-2u-5}{7}$ **75.** $\frac{k+3-2k-7}{5} = \frac{-k-4}{5}$

76. $\frac{c+1-8+c}{c-3} = \frac{2c-7}{c-3}$ **77.** $\frac{4}{g-3} + \frac{3}{g-3} = \frac{7}{g-3}$

78. $\frac{3m}{m-n} - \frac{3n}{m-n} = \frac{3m-3n}{m-n} = \frac{3(m-n)}{m-n} = 3$ **79.** $\frac{5}{y^2} + \frac{3y}{y^2} = \frac{3y+5}{y^2}$

80. $\frac{12x}{15x^2} - \frac{1}{15x^2} = \frac{12x-1}{15x^2}$ **81.** $\frac{2}{10wx} - \frac{3x}{10wx} = \frac{2-3x}{10wx}$

82. $\frac{3k-2}{2k^3} + \frac{12k}{2k^3} = \frac{3k-2+12k}{2k^3} = \frac{15k-2}{2k^3}$ **83.** $\frac{2}{3(x+2)} + \frac{3x}{3(x+2)} = \frac{2+3x}{3(x+2)}$

84. $\frac{3(5a-4)}{18} + \frac{2(a-2)}{18} = \frac{15a-12+2a-4}{18} = \frac{17a-16}{18}$

85. $\frac{5(2h+4)}{40} - \frac{10h}{40} + \frac{4(3h-2)}{40} = \frac{10h+20-10h+12h-8}{40} = \frac{12h+12}{40} = \frac{12(h+1)}{40} = \frac{3(h+1)}{10}$

86. $\frac{(m-n)}{4} - \frac{(m+n)}{4} = \frac{m-n-m-n}{4} = \frac{-2n}{4} = \frac{-n}{2}$

87. $\frac{3}{x+2} - \frac{1}{x+3} = \frac{3(x+3)}{(x+2)(x+3)} - \frac{x+2}{(x+2)(x+3)} = \frac{3x+9-x-2}{(x+2)(x+3)} = \frac{2x+7}{(x+2)(x+3)}$

88. $\frac{3z}{(z-4)(z+4)} + \frac{z(z+4)}{(z+4)(z-4)} = \frac{3z+z^2+4z}{(z+4)(z-4)} = \frac{z(z+7)}{(z+4)(z-4)}$

89. $\frac{u}{u-4} - \frac{3}{u-4} = \frac{u-3}{u-4}$ **90.** $\frac{22}{3}$ **91.** $\frac{5n+1}{n}$ **92.** $\frac{4m^2-3}{m}$ **93.** $\frac{x+3y}{y}$

94. $\frac{6k+18-5}{k+3} = \frac{6k+13}{k+3}$ **95.** $\frac{n+7(n-2)}{n-2} = \frac{n+7n-14}{n-2} = \frac{8n-14}{n-2} = \frac{2(4n-7)}{n-2}$

96. $\frac{x+3-2x}{x} = \frac{-x+3}{x}$ **97.** $\frac{8h(h+3)-h}{h+3} = \frac{8h^2+24h-h}{h+3} = \frac{8h^2+23h}{h+3} = \frac{h(8h+23)}{h+3}$

98. $\frac{3y(2y+7)+y}{2y+7} = \frac{6y^2+21y+y}{2y+7} = \frac{6y^2+22y}{2y+7} = \frac{2y(3y+11)}{2y+7}$

99. $\dfrac{5(e^2 - 1) - (e + 3)}{e^2 - 1} = \dfrac{5e^2 - 5 - e - 3}{e^2 - 1} = \dfrac{5e^2 - e - 8}{(e^2 - 1)} = \dfrac{5e^2 - e - 8}{(e + 1)(e - 1)}$

100. $\dfrac{a(a + 3) + 5a + 3}{a + 3} = \dfrac{a^2 + 3a + 5a + 3}{a + 3} = \dfrac{a^2 + 8a + 3}{a + 3}$

101. $\dfrac{2w(w - 3) - (w + 3)}{w - 3} = \dfrac{2w^2 - 6w - w - 3}{w - 3} = \dfrac{2w^2 - 7w - 3}{w - 3}$

102. $\dfrac{n(n + 2) - 7 - (3n - 1)}{n + 2} = \dfrac{n^2 + 2n - 7 - 3n + 1}{n + 2} = \dfrac{n^2 - n - 6}{n + 2} = \dfrac{(n - 3)(n + 2)}{n + 2} = n - 3$

103. $\dfrac{v(v - u) + u(u + v) + (u + v)(v - u)}{(u + v)(v - u)} = \dfrac{v^2 - uv + u^2 + uv + v^2 - u^2}{(u + v)(v - u)} = \dfrac{2v^2}{(u + v)(v - u)}$

104. $\dfrac{x(x + 4) + x(x - 4) - 3(x - 4)(x + 4)}{(x - 4)(x + 4)} = \dfrac{x^2 + 4x + x^2 - 4x - 3(x^2 - 16)}{(x - 4)(x + 4)} = \dfrac{2x^2 - 3x^2 + 48}{(x - 4)(x + 4)} =$

$\dfrac{-x^2 + 48}{(x - 4)(x + 4)}$

105.
$$\begin{array}{r} x + 5 \\ x + 2\overline{)x^2 + 7x + 10} \\ \underline{x^2 + 2x} \\ 5x + 10 \\ \underline{5x + 10} \end{array}$$

106.
$$\begin{array}{r} y + 5 \\ y - 7\overline{)y^2 - 2y - 35} \\ \underline{y^2 - 7y} \\ 5y - 35 \\ \underline{5y - 35} \end{array}$$

107.
$$\begin{array}{r} a - 7 + \dfrac{11}{a + 2} \\ a + 2\overline{)a^2 - 5a - 3} \\ \underline{a^2 + 2a} \\ -7a - 3 \\ \underline{-7a - 14} \\ 11 \end{array}$$

108.
$$\begin{array}{r} n - 4 \\ n + 4\overline{)n^2 + 0n - 16} \\ \underline{n^2 + 4n} \\ -4n - 16 \\ \underline{-4n - 16} \end{array}$$

109.
$$\begin{array}{r} k + 1 + \dfrac{12}{k - 5} \\ k - 5\overline{)k^2 - 4k + 7} \\ \underline{k^2 - 5k} \\ k + 7 \\ \underline{k - 5} \\ 12 \end{array}$$

110.
$$\begin{array}{r} 4y + 2 + \dfrac{8}{2y - 1} \\ 2y - 1\overline{)8y^2 + 0y + 6} \\ \underline{8y^2 - 4y} \\ 4y + 6 \\ \underline{4y - 2} \\ 8 \end{array}$$

111.
$$\begin{array}{r} b^2 - b + 1 - \dfrac{2}{b + 1} \\ b + 1\overline{)b^3 + 0b^2 + 0b - 1} \\ \underline{b^3 + b^2} \\ -b^2 + 0b \\ \underline{-b^2 - b} \\ b - 1 \\ \underline{b + 1} \\ -2 \end{array}$$

112.
$$\begin{array}{r} x^2 - 3x + 9 - \dfrac{22}{x + 3} \\ x + 3\overline{)x^3 + 0x^2 + 0x + 5} \\ \underline{x^3 + 3x^2} \\ -3x^2 + 0x \\ \underline{-3x^2 - 9x} \\ 9x + 5 \\ \underline{9x + 27} \\ -22 \end{array}$$

113.
$$\begin{array}{r} w^2 + 2w + 4 \\ w - 1\overline{)w^3 + w^2 + 2w - 4} \\ \underline{w^3 - w^2} \\ 2w^2 + 2w \\ \underline{2w^2 - 2w} \\ 4w - 4 \\ \underline{4w - 4} \end{array}$$

114.
$$\begin{array}{r} u^2 + 4u + 8 \\ u - 2\overline{)u^3 + 2u^2 + 0u - 16} \\ \underline{u^3 - 2u^2} \\ 4u^2 + 0u \\ \underline{4u^2 - 8u} \\ 8u - 16 \\ \underline{8u - 16} \end{array}$$

115.
$$\begin{array}{r} n - 4 \\ 2n - 5\overline{)2n^2 - 13n + 20} \\ \underline{2n^2 - 5n} \\ -8n + 20 \\ \underline{-8n + 20} \end{array}$$

116.
$$\begin{array}{r} h - 1 \\ 7h - 2\overline{)7h^2 - 9h + 2} \\ \underline{7h^2 - 2h} \\ -7h + 2 \\ \underline{-7h + 2} \end{array}$$

117.
$$\begin{array}{r@{}l} & v^2 + 3v + 7 + \dfrac{15}{v-2} \\ v - 2 \big) & \overline{v^3 + v^2 + v + 1} \\ & \underline{v^3 - 2v^2} \\ & \quad 3v^2 + v \\ & \quad \underline{3v^2 - 6v} \\ & \qquad 7v + 1 \\ & \qquad \underline{7v - 14} \\ & \qquad\quad 15 \end{array}$$

118.
$$\begin{array}{r@{}l} & 3n^2 - 2n + 3 \\ 2n + 3 \big) & \overline{6n^3 + 5n^2 + 0n + 9} \\ & \underline{6n^3 + 9n^2} \\ & \quad -4n^2 + 0n \\ & \quad \underline{-4n^2 - 6n} \\ & \qquad\quad 6n + 9 \\ & \qquad\quad \underline{6n + 9} \end{array}$$

Pages 651–653 • CHAPTER 7

1. 40 s : 120 s = 1 : 3 **2.** 400 cm : 250 cm = 8 : 5 **3.** 3000 g : 45 g = 200 : 3 **4.** 2 : 5

5. $18d : 5$ **6.** $16a^2 : 6a = 8a : 3$ **7.** $180 - 55 = 125; 125 : 55 = 25 : 11$

8. $84 - 48 = 36; 48 : 36 = 4 : 3$ **9.** $5x = 45, x = 9; \{9\}$ **10.** $5a = 175, a = 35; \{35\}$

11. $24c = 28, c = \frac{7}{6}; \left\{\frac{7}{6}\right\}$ **12.** $15x = 4, x = \frac{4}{15}; \left\{\frac{4}{15}\right\}$ **13.** $480a = 2880, a = 6; \{6\}$

14. $d = \frac{\overset{3}{\cancel{51}}}{\underset{5}{\cancel{125}}} \cdot \frac{\overset{1}{\cancel{25}}}{\underset{1}{\cancel{17}}}, d = \frac{3}{5}; \left\{\frac{3}{5}\right\}$ **15.** $9(x - 4) = 7x, 9x - 36 = 7x, 2x = 36, x = 18; \{18\}$

16. $21w = 2(10w + 2), 21w = 20w + 4, w = 4; \{4\}$

17. $8(8a - 5) = 13(5a - 4), 64a - 40 = 65a - 52, a = 12; \{12\}$

18. $9\left(\frac{a}{3} - \frac{a}{9}\right) = 2 \cdot 9, 3a - a = 18, 2a = 18, a = 9; \{9\}$

19. $6\left(\frac{2x}{3} - \frac{x}{2}\right) = 12 \cdot 6, 4x - 3x = 72; x = 72; \{72\}$

20. $14\left(\frac{6}{7}x - \frac{1}{2}x\right) = 5 \cdot 14, 12x - 7x = 70, 5x = 70, x = 14; \{14\}$

21. $9\left(\frac{2}{3}x - \frac{5}{9}x\right) = -1 \cdot 9, 6x - 5x = -9, x = -9; \{-9\}$

22. $6\left(\frac{y + 2}{2}\right) = 6\left(\frac{2y}{3}\right), 3y + 6 = 4y, 6 = y; \{6\}$

23. $10\left(\frac{x + 1}{5} - \frac{3}{2}\right) = 10\left(\frac{3x - 6}{10}\right), 2x + 2 - 15 = 3x - 6, -13 = x - 6, x = -7; \{-7\}$

24. $z\left(\frac{12}{z}\right) = z\left(\frac{4 + 4z}{z}\right), 12 = 4 + 4z, 4z = 8, z = 2; \{2\}$

25. $6x\left(\frac{1}{x} + \frac{1}{3}\right) = 6x\left(\frac{1}{2}\right), 6 + 2x = 3x, x = 6; \{6\}$

26. $5y\left(\frac{4}{5y} + \frac{y - 2}{y}\right) = 5y\left(-\frac{1}{5}\right), 4 + 5y - 10 = -y, 6y = 6, y = 1; \{1\}$

27. $5c = 2(c + 3), 5c = 2c + 6, 3c = 6, c = 2; \{2\}$

28. $6m\left(\frac{3m + 5}{6} - \frac{10}{m}\right) = 6m\left(\frac{m}{2}\right), m(3m + 5) - 60 = 3m^2, 3m^2 + 5m - 60 = 3m^2, 5m = 60,$ $m = 12; \{12\}$

29. $(2h + 4)\left(\frac{h}{2h + 4} - \frac{1}{h + 2}\right) = (2h + 4), h - 2 = 2h + 4, -6 = h; \{-6\}$

30. $\frac{80}{100} \cdot 700 = 560$ **31.** $\frac{45}{100} \cdot 450 = 202.5$ **32.** $\frac{3.25}{100} \cdot 48 = 1.56$

33. $\frac{60}{100} \cdot x = 18, x = 18 \cdot \frac{10}{6} = 30$ **34.** $\frac{150}{100}x = 63, x = 63 \cdot \frac{2}{3} = 42$

35. $\frac{45}{180} = \frac{x}{100}$, $180x = 4500, x = 25$; 25% **36.** $\frac{54}{36} = \frac{x}{100}$, $36x = 5400, x = 150$; 150%

37. $12x = 480, x = 40; \{40\}$ **38.** $6z = 1800, z = 300; \{300\}$ **39.** $8y = 6400, y = 800; \{800\}$

40. $4a - 7 = 29, 4a = 36, a = 9; \{9\}$ **41.** $30b + 3b = 9900, 33b = 9900, b = 300; \{300\}$

42. $5c = 660 - 6c, 11c = 660, c = 60; \{60\}$

43. Let z = amount of zinc in the alloy. $z = \frac{28}{100} \cdot 30 = \frac{84}{10} = 8.4$. 8.4 kg zinc are in the alloy.

44. Let x = percent tax rate. $140 \cdot \frac{x}{100} = 6.30; x = \frac{6.30 \cdot 100}{140} = 4.5$. The sales tax rate = 4.5%.

45. $\frac{85}{100} \cdot 150 = 127.5$. The sale price is $127.50.

46.

	Amount ×	% salt =	Amount salt
Old solution	12	30	3.6
Water	x	0	0
New solution	$x + 12$	20	$0.2(x + 12)$

$3.6 + 0 = 0.2(x + 12)$; $36 = 2(x + 12)$; $36 = 2x + 24$; $2x = 12$; $x = 6$. 6 kg water must be added.

47.

	Amount ×	% salt =	Amount salt
Old solution	40	10	4
Water	x	0	0
New solution	$40 - x$	25	$0.25(40 - x)$

$0.25(40 - x) = 4$; $25(40 - x) = 400$; $1000 - 25x = 400$; $25x = 600$; $x = 24$. 24 kg must be evaporated.

48.

	Work rate ×	Time =	Work done
Machine 1	$\frac{1}{15}$	x	$\frac{x}{15}$
Machine 2	$\frac{1}{30}$	x	$\frac{x}{30}$

$\frac{x}{15} + \frac{x}{30} = 1$; $2x + x = 30$; $3x = 30$; $x = 10$. It would take both machines together 10 min.

49.

	Work rate ×	Time =	Work done
AC I	$\frac{1}{20}$	12	$\frac{12}{20} = \frac{3}{5}$
AC II	$\frac{1}{x}$	12	$\frac{12}{x}$

$\frac{3}{5} + \frac{12}{x} = 1$; $3x + 60 = 5x$; $2x = 60$; $x = 30$.

It would take the second air conditioner 30 min to cool the room alone.

50. $\frac{1}{6^2} = \frac{1}{36}$ **51.** $\frac{1}{5^3} = \frac{1}{125}$ **52.** $\frac{1}{7^2} = \frac{1}{49}$ **53.** $\frac{1}{9^3} = \frac{1}{729}$ **54.** $\frac{1}{2^4} \cdot \frac{1}{2^3} = \frac{1}{2^7} = \frac{1}{128}$

55. $\frac{1}{6^{-2}} = 6^2 = 36$ **56.** $3^{-4-(-3)} = 3^{-1} = \frac{1}{3}$ **57.** $8^{-2-(-4)} = 8^2 = 64$ **58.** $\frac{5}{y^2}$

59. $\frac{1}{(9y)^3} = \frac{1}{729y^3}$ **60.** $\frac{x^2}{y^5}$ **61.** $\frac{1}{a^2b^3}$ **62.** $\frac{u}{v^2w}$ **63.** $\frac{e^2}{d^4f^2}$ **64.** $\left(\frac{b^3}{a^2}\right)^2 = \frac{b^6}{a^4}$

65. $\left(\frac{z^3}{x^4y^5}\right)^{-3} = \frac{(x^4y^5)^3}{(z^3)^3} = \frac{x^{12}y^{15}}{z^9}$ **66.** 6.48×10^7 **67.** 1.47×10^8 **68.** 6.43×10^{11}

69. 9.8×10^{-6} **70.** 6×10^{-9} **71.** 1×10^{-11}

Pages 653–656 • CHAPTER 8

1. $3 - 2 \cdot 0 \neq 6$, no. $0 - 2(-3) = 6$, yes. **2.** $3 + 3 \cdot 2 = 9$, yes. $-3 + 3 \cdot 4 = 9$, yes.

3. $2 \cdot 4 - (-1) \neq 5$, no. $2 \cdot 1 - (-7) \neq 5$, no.

4. $2 \cdot 1 + 3 \cdot 2 \neq 7$, no. $2 \cdot 5 + 3(-1) = 7$, yes.

5. $4\left(\frac{3}{2}\right) + 2 \cdot 0 = 6$, yes. $4 \cdot 1 + 2 \cdot 1 = 6$, yes.

6. $-3 \cdot 1 + 4(-1) = -7$, yes. $-3 \cdot 2 + 4\left(\frac{1}{4}\right) \neq -7$, no.

7. $y = \frac{8 - x}{2}$

x	y
0	4
1	$\frac{7}{2}$
2	3
3	$\frac{5}{2}$
4	2
5	$\frac{3}{2}$
6	1
7	$\frac{1}{2}$
8	0
9	neg.

Solutions: (0, 4), (2, 3), (4, 2), (6, 1), (8, 0)

8. $y = 5 - 3x$

x	y
0	5
1	2
2	neg.

Solutions: (0, 5), (1, 2)

9. $y = \frac{12 - 4x}{3}$

x	y
0	4
1	$\frac{8}{3}$
2	$\frac{4}{3}$
3	0
4	neg.

Solutions: (0, 4), (3, 0)

10. $y = \frac{5}{x}$

x	y
1	5
2	$\frac{5}{2}$
3	$\frac{5}{3}$
4	$\frac{5}{4}$
5	1
6	<1

Solutions: (1, 5), (5, 1)

11. $y = \frac{10 - x}{4}$

x	y
0	$\frac{5}{2}$
1	$\frac{9}{4}$
2	2
3	$\frac{7}{4}$
4	$\frac{3}{2}$
5	$\frac{5}{4}$
6	1
7	$\frac{3}{4}$
8	$\frac{1}{2}$
9	$\frac{1}{4}$
10	0
11	neg.

Solutions: (2, 2), (6, 1), (10, 0)

12. $y = \frac{1}{x}$

x	y
1	1
2	<1

Solution: (1, 1)

13.

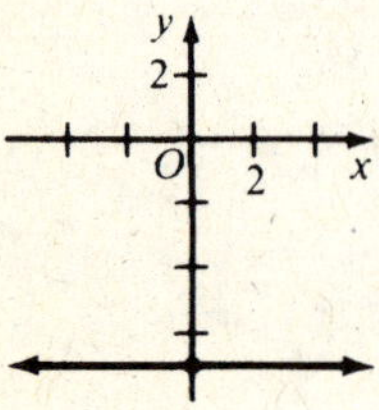

14.

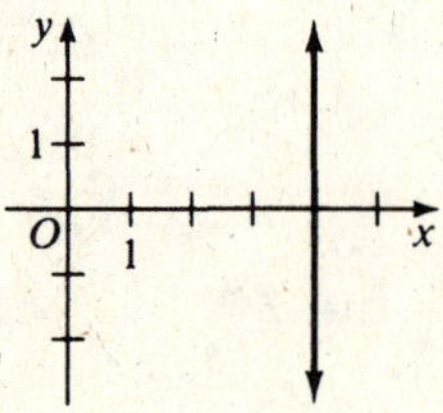

15.

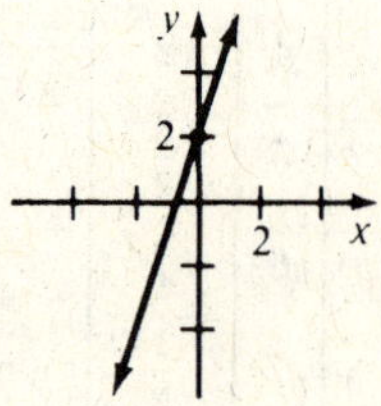

16.

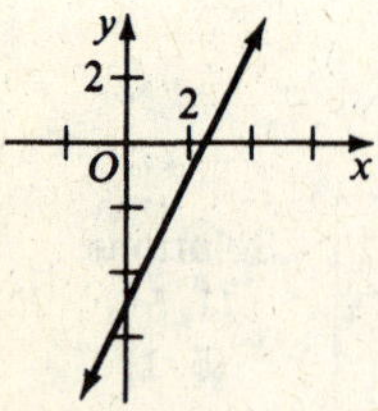

17.

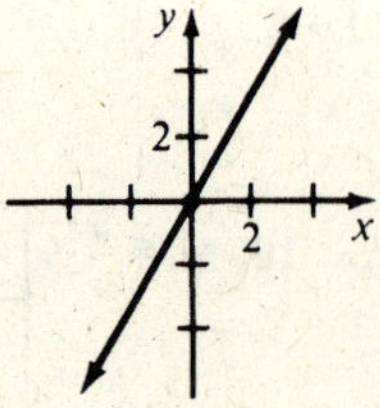

18.

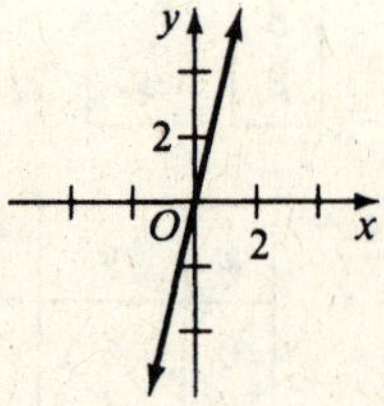

19.

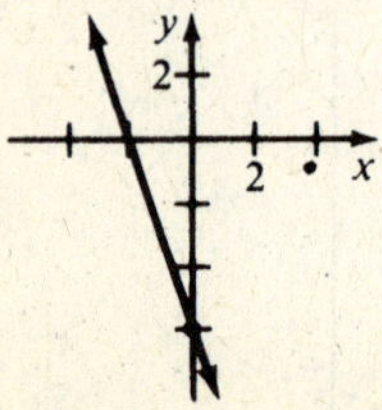

20.

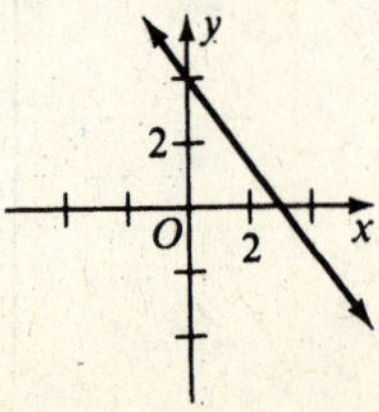

21.

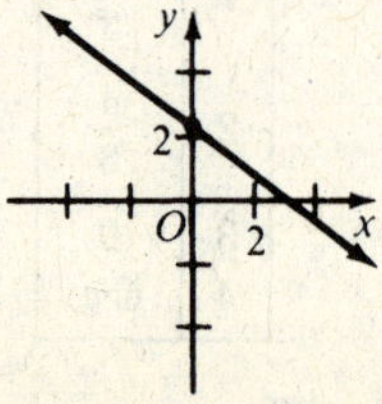

22. $m = \frac{6-2}{4-1} = \frac{4}{3}$ **23.** $m = \frac{2-1}{-1-(-7)} = \frac{1}{6}$ **24.** $m = \frac{6-0}{-1-0} = \frac{6}{-1} = -6$

25. $m = \frac{-3-(-3)}{2-(-4)} = \frac{0}{6} = 0$ **26.** $m = \frac{1-(-2)}{2-8} = \frac{3}{-6} = -\frac{1}{2}$

27. $m = \frac{-7-(-4)}{-7-6} = \frac{-7+4}{-13} = \frac{3}{13}$ **28.** $m = 7$ **29.** no slope

30. $-2y = -3x + 8$, $y = \frac{3}{2}x - 4$; $m = \frac{3}{2}$ **31.** $y = 9$; $m = 0$

32. $4y = -5x + 16$, $y = -\frac{5}{4}x + 4$; $m = -\frac{5}{4}$ **33.** $y = -x + 1$; $m = -1$

34. $\frac{-3-1}{0-2} = \frac{-4}{-2} = 2$; $\frac{5-(-3)}{4-0} = \frac{5+3}{4} = 2$; $\frac{-7-5}{-2-4} = \frac{-12}{-6} = 2$; collinear

35. $\frac{-2-4}{9-0} = \frac{-6}{9} = -\frac{2}{3}$; $\frac{6-(-2)}{-3-9} = \frac{6+2}{-12} = -\frac{8}{12} = -\frac{2}{3}$; $\frac{0-6}{6-(-3)} = \frac{-6}{6+3} = -\frac{6}{9} = -\frac{2}{3}$; collinear

36. $\frac{-4-(-2)}{2-(-3)} = \frac{-4+2}{2+3} = -\frac{2}{5}$; $\frac{-5-(-4)}{6-2} = \frac{-5+4}{4} = -\frac{1}{4}$; not collinear

37. $\frac{3-3}{0-(-5)} = 0$; $\frac{3-3}{5-0} = 0$; $\frac{3-3}{-2-5} = 0$; collinear

38.
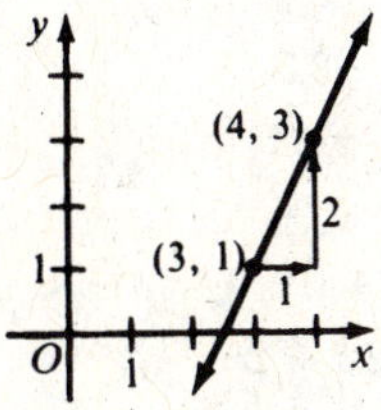

39.
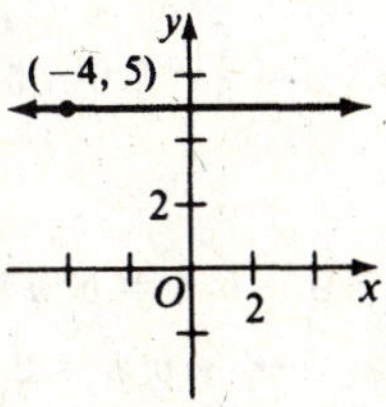

40.
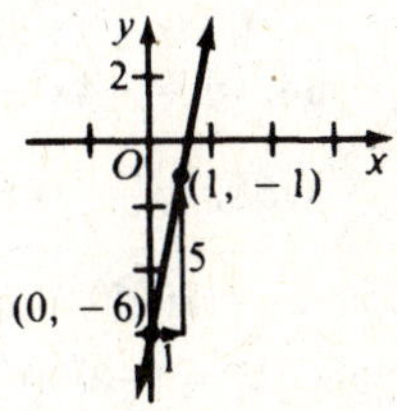

41.
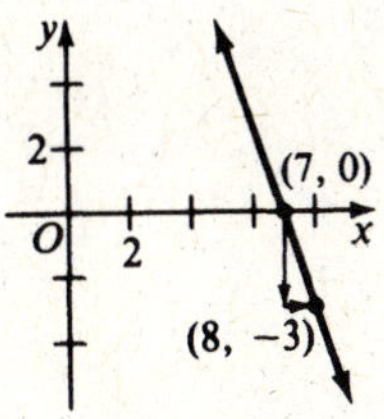

42.
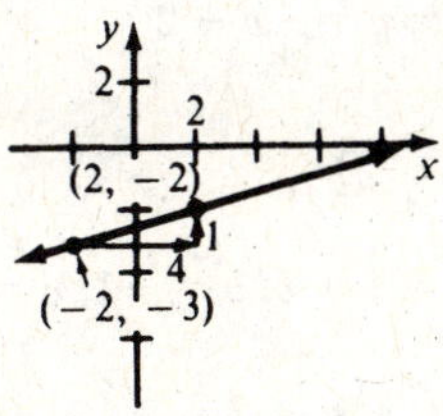

43.
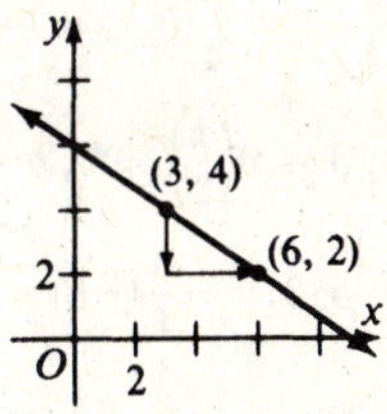

44. $y = -x - 3$

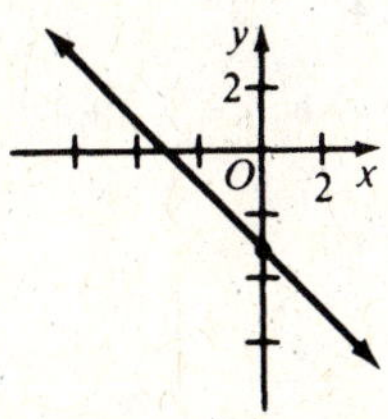

45. $y = \frac{7}{2}x$

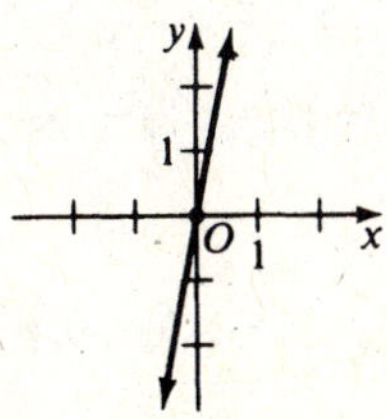

46. $y = 4x - 3$

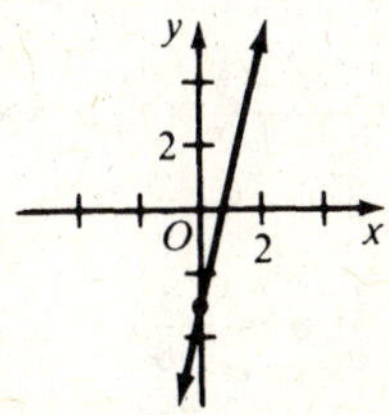

47. $x + y = 3$, $y = -x + 3$

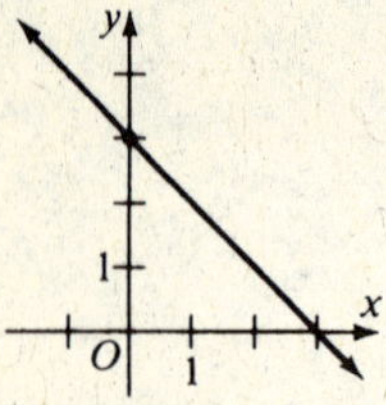

48. $5y = x + 10$, $y = \frac{1}{5}x + 2$

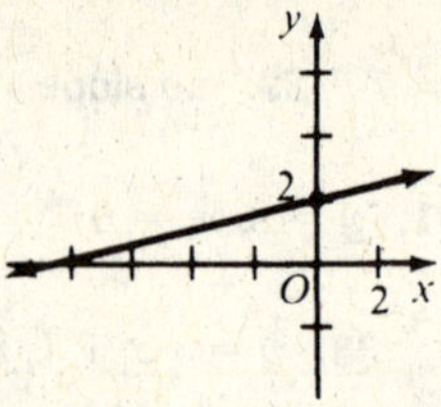

49. $4y = 3x - 5$, $y = \frac{3}{4}x - \frac{5}{4}$

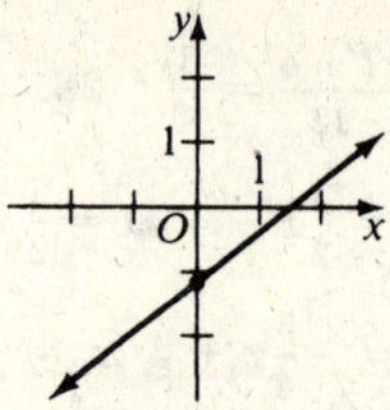

50. $y = x - 2$; $y = x - 3$; both lines have slope of 1 and different y-intercepts, so they are parallel

51. $2y = 3x - 6$, $y = \frac{3}{2}x - 3$; $y = \frac{3}{2}x - 6$; both lines have slope of $\frac{3}{2}$ and different y-intercepts, so they are parallel

52. $5y = -x - 1$, $y = -\frac{1}{5}x - \frac{1}{5}$; $10y = -2x + 2$, $y = -\frac{1}{5}x + \frac{1}{5}$; both lines have slope of $-\frac{1}{5}$ and different y-intercepts, so they are parallel

53. $y = 3x + \frac{1}{2}$ **54.** $y = -4x + \frac{3}{5}$ **55.** $y = \frac{1}{3}x + 6$ **56.** $y = -3.5$

57. $y = -\frac{3}{7}x + \frac{3}{8}$ **58.** $y = -1.5x + 2.7$

59. $y = 3x + b$, $(-5) = 3(-3) + b$, $-5 = -9 + b$, $4 = b$; $y = 3x + 4$

60. $y = -2x + b$, $(-4) = (-2)(3) + b$, $-4 = -6 + b$, $b = 2$; $y = -2x + 2$

61. $y = \frac{3}{4}x + b$, $-2 = \frac{3}{4}(0) + b$, $-2 = b$; $y = \frac{3}{4}x - 2$

62. $y = 0x + b$, $3 = 0\left(\frac{1}{2}\right) + b$, $b = 3$; $y = 3$

63. $y = -\frac{1}{5}x + b$, $0 = -\frac{1}{5}(-5) + b$, $0 = 1 + b$, $b = -1$; $y = -\frac{1}{5}x - 1$

64. $y = \frac{7}{3}x + b$, $(7) = \frac{7}{3}(3) + b$, $7 = 7 + b$, $b = 0$; $y = \frac{7}{3}x$

65. $\frac{4 - 1}{6 - 2} = \frac{3}{4}$; $y = \frac{3}{4}x + b$, $(1) = \frac{3}{4}(2) + b$, $1 = \frac{3}{2} + b$, $b = -\frac{1}{2}$; $y = \frac{3}{4}x - \frac{1}{2}$

66. $\frac{-7 - (-1)}{1 - 2} = \frac{-7 + 1}{-1} = 6$; $y = 6x + b$, $(-1) = 6(2) + b$, $-1 = 12 + b$, $b = -13$; $y = 6x - 13$

67. $\frac{-1 - 0}{6 - 0} = -\frac{1}{6}$; $y = -\frac{1}{6}x + b$, $(0) = -\frac{1}{6}(0) + b$, $b = 0$; $y = -\frac{1}{6}x$

68. $\frac{-4 - 2}{-3 - (-3)} = \frac{-6}{-3 + 3} = \frac{-6}{0}$; no slope; $x = -3$

69. $\frac{2 - 8}{1 - (-2)} = \frac{-6}{1 + 2} = -2$; $y = -2x + b$, $(2) = -2(1) + b$, $2 = -2 + b$, $b = 4$; $y = -2x + 4$

70. $\frac{-7-(-4)}{-7-6} = \frac{-3}{-13} = \frac{3}{13}$; $y = \frac{3}{13}x + b$, $(-7) = \frac{3}{13}(-7) + b$, $-7 = -\frac{21}{13} + b$,

$b = \frac{-91+21}{13} = -\frac{70}{13}$; $y = \frac{3}{13}x - \frac{70}{13}$

71. D = {Mackinac Straits, Humber Estuary, Golden Gate, Ataturk, Verrazano Narrows}
R = {3524 ft., 3800 ft., 4200 ft., 4260 ft., 4626 ft.}

72. D = {1930, 1940, 1950, 1960, 1970}
R = {1782, 2331, 6403, 6881, 11,261}

73.

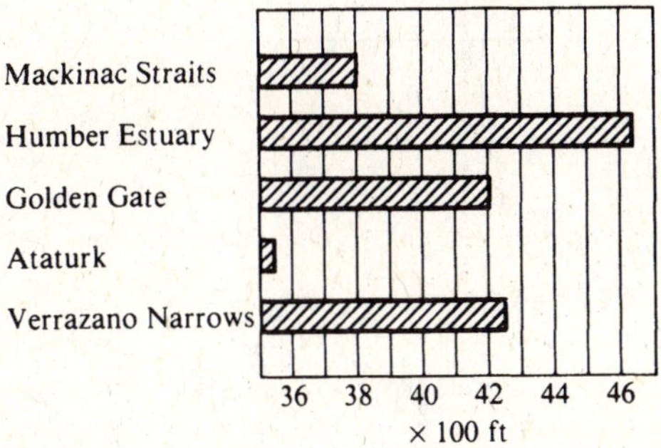

74.

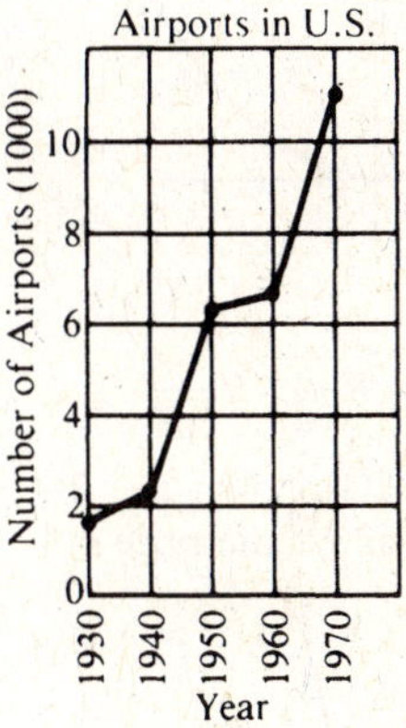

75. $5 - 12 = -7$ **76.** $5 - 3\left(-\frac{1}{3}\right) = 5 + 1 = 6$ **77.** 5 **78.** $5 - 3(-5) = 20$ **79.** 0

80. $(-2)^3 + 2(-2) = -8 - 4 = -12$ **81.** $\frac{1}{8} + 1 = \frac{9}{8}$ **82.** $27 + 6 = 33$

83. $h(1) = 5 - 2 - 1 = 2$; $h(2) = 5 - 4 - 4 = -3$; $h(3) = 5 - 6 - 9 = -10$; $\{2, -3, -10\}$

84. $M(-1) = \frac{6}{-4+2} = \frac{6}{-2} = -3$, $M(0) = \frac{6}{2} = 3$; $M(1) = \frac{6}{4+2} = 1$; $\{-3, 3, 1\}$

85. $r(-2) = -3 + 8 = 5$, $r(-1) = -3 + 4 = 1$; $r(0) = -3$; $R = \{5, 1, -3\}$

86. $N(2) = \frac{10}{-1} = -10$, $N(4) = \frac{10}{1} = 10$, $N(8) = \frac{10}{5} = 2$; $R = \{-10, 10, 2\}$

87. $G(-2) = (-3)(-1) = 3$, $G(0) = (-1)(1) = -1$; $G(2) = (1)(3) = 3$; $R = \{-1, 3\}$

88. $k(3) = 9 - 12 + 2 = -1$, $k(4) = 16 - 16 + 2 = 2$, $k(5) = 25 - 20 + 2 = 7$;
$R = \{-1, 2, 7\}$

89. vertex: $x = -\frac{b}{2a} = -\frac{0}{8} = 0$,
$y = 4(0)^2 = 0$; (0, 0);
axis of symmetry: $x = 0$

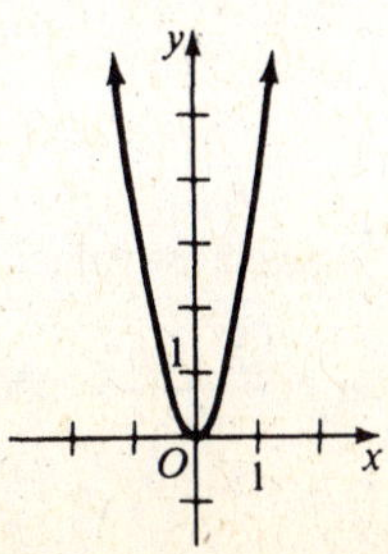

90. vertex: $x = -\frac{b}{2a} = -\frac{0}{-4} = 0$,
$y = -2(0)^2 = 0$; (0, 0);
axis of symmetry: $x = 0$

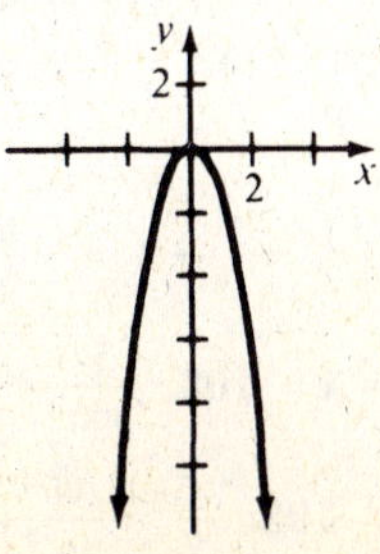

91. vertex: $x = -\frac{b}{2a} = -\frac{0}{\frac{2}{5}} = 0$,

$y = \frac{1}{5}(0)^2 = 0$; (0, 0);

axis of symmetry: $x = 0$

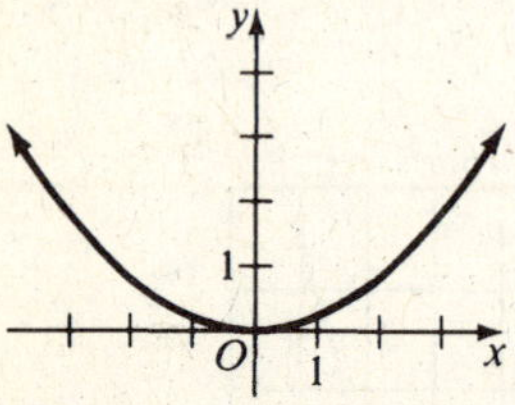

92. vertex: $x = -\frac{b}{2a} = -\frac{3}{-2} = \frac{3}{2}$,

$y = -\left(\frac{3}{2}\right)^2 + 3\left(\frac{3}{2}\right) = -\frac{9}{4} + \frac{9}{2} = \frac{9}{4}$,

$\left(\frac{3}{2}, \frac{9}{4}\right)$; axis of symmetry: $x = \frac{3}{2}$

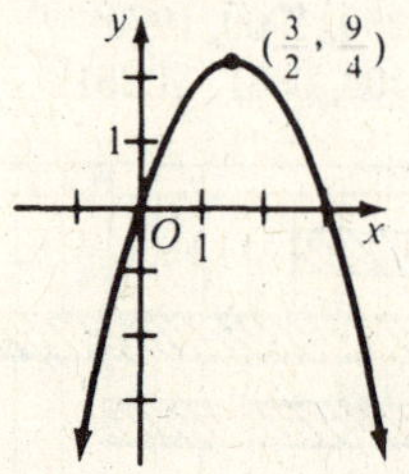

93. vertex: $x = -\frac{b}{2a} = -\frac{-2}{2} = 1$;

$y = (1)^2 - 2(1) + 5 = 1 - 2 + 5 = 4$;
(1, 4); axis of symmetry: $x = 1$

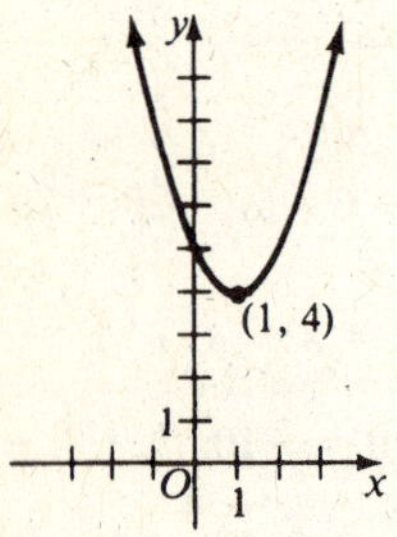

94. vertex: $x = -\frac{b}{2a} = \frac{0}{-1} = 0$;

$y = 4 - \frac{1}{2}(0)^2 = 4$, (0, 4);

axis of symmetry: $x = 0$

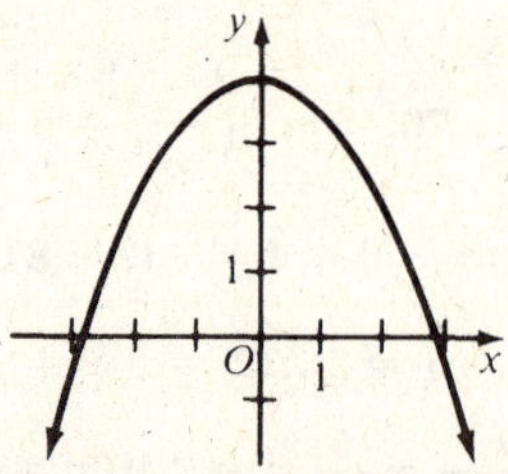

95. Since the parabola opens up, the vertex will be at the least value. $-\frac{b}{2a} = -\frac{49}{4}$;

$f\left(-\frac{7}{2}\right) = \frac{49}{4} - \frac{49}{2} = -\frac{49}{4}$; vertex: $\left(-\frac{7}{2}, -\frac{49}{4}\right)$; least value: $-\frac{25}{4}$

96. $-\frac{b}{2a} = -\frac{-3}{2} = \frac{3}{2}$; $g\left(\frac{3}{2}\right) = \frac{9}{4} - \frac{9}{2} - 4 = -\frac{9}{4} - 4 = -\frac{25}{4}$; vertex: $\left(\frac{3}{2}, -\frac{25}{4}\right)$; least value: $-\frac{25}{4}$

97. $-\frac{b}{2a} = 0$; $h(0) = 0$; vertex: (0, 0); least value: 0

98. Since the parabola opens down, the vertex will be at the greatest value.

$-\frac{b}{2a} = -\frac{1}{-6} = \frac{1}{6}$; $f\left(\frac{1}{6}\right) = \frac{1}{6} - 3\left(\frac{1}{6}\right)^2 = \frac{1}{6} - \frac{1}{12} = \frac{1}{12}$; vertex: $\left(\frac{1}{6}, \frac{1}{12}\right)$; greatest value: $\frac{1}{12}$

99. $-\frac{b}{2a} = -\frac{0}{-\frac{2}{3}} = 0$; $g(0) = 2 - \frac{1}{3}(0)^2 = 2$; vertex: (0, 2); greatest value: 2

100. $-\frac{b}{2a} = -\frac{-1}{-2}$; $h\left(-\frac{1}{2}\right) = -\left(-\frac{1}{2}\right)^2 + \frac{1}{2} - 1 = -\frac{3}{4}$; vertex: $\left(-\frac{1}{2}, -\frac{3}{4}\right)$; greatest value: $-\frac{3}{4}$

101. $\frac{12}{60} = \frac{1}{5} = k$ **102.** $\frac{144}{24} = 6 = k$ **103.** $\frac{5}{300} = \frac{15}{n}$; $5n = 15 \cdot 300$; $n = \frac{15 \cdot 300}{5} = 900$

104. $\frac{28.7}{4.1} = \frac{b}{13}$; $b = \frac{13 \cdot 28.7}{4.1} = 91$ **105.** $\frac{35}{7} = \frac{105}{y_2}$; $y_2 = \frac{7 \cdot 105}{35} = 21$

106. $\frac{5.2}{5} = \frac{x_2}{1}$; $x_2 = \frac{5.2 \times 1}{5} = 1.04$ **107.** $\frac{\frac{3}{8}}{y_1} = \frac{\frac{2}{5}}{\frac{1}{10}}$; $y_1 = \frac{\frac{1}{10} \times \frac{3}{8}}{\frac{2}{5}} = \frac{1}{10} \times \frac{3}{8} \times \frac{5}{2} = \frac{3}{32}$

108. **a.** $C = kd$ **b.** $\frac{C_1}{d_1} = \frac{C_2}{d_2}$ **109.** **a.** $e = km$ **b.** $\frac{e_1}{m_1} = \frac{e_2}{m_2}$

110. **a.** $l = kh$ **b.** $\frac{l_1}{h_1} = \frac{l_2}{h_2}$ **111.** **a.** $rn = k$ **b.** $\frac{n_1}{r_2} = \frac{n_2}{r_1}$ or $r_1n_1 = r_2n_2$

112. **a.** $hb = k$ **b.** $\frac{b_1}{h_2} = \frac{b_2}{h_1}$ or $h_1b_1 = h_2b_2$ **113.** **a.** $ns = k$ **b.** $\frac{s_1}{n_2} = \frac{s_2}{n_1}$ or $n_1s_1 = n_2s_2$

114.

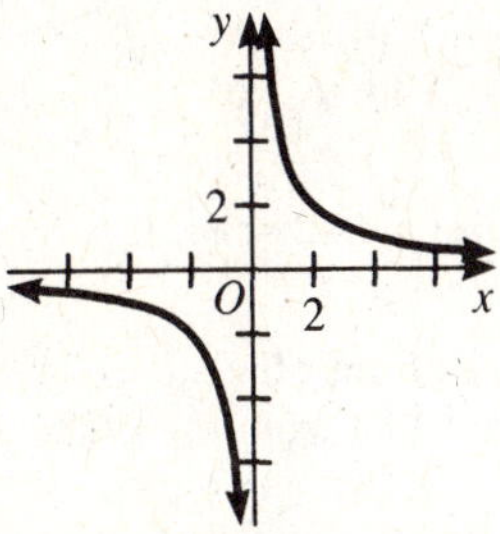

115.

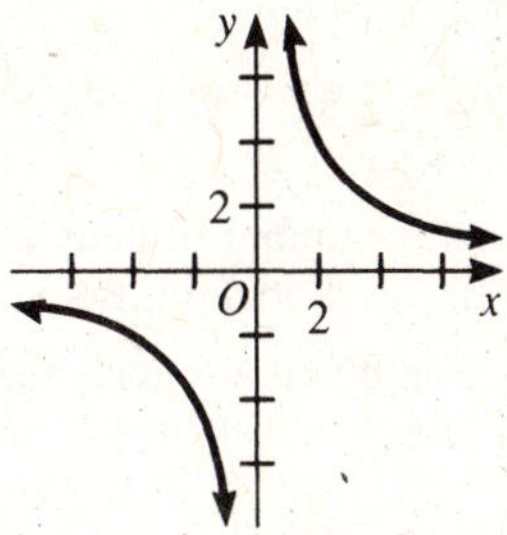

116.

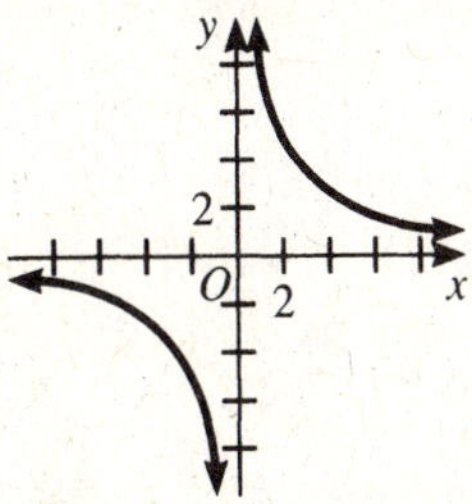

117.

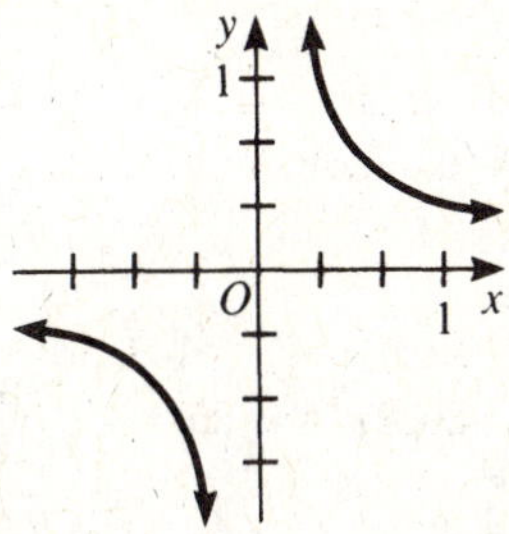

118. $5 \cdot 8 = 4y_2$; $y_2 = \frac{5 \cdot 8}{4} = 10$

119. $(0.6)(1.2) = (0.4)x_2$; $x_2 = \frac{0.6 \times 1.2}{0.4} = 1.8$ **120.** $\frac{1}{4}y_1 = \frac{1}{6} \cdot \frac{1}{2}$; $y_1 = \frac{1}{12} \cdot 4 = \frac{1}{3}$

Pages 656–657 • CHAPTER 9

1.

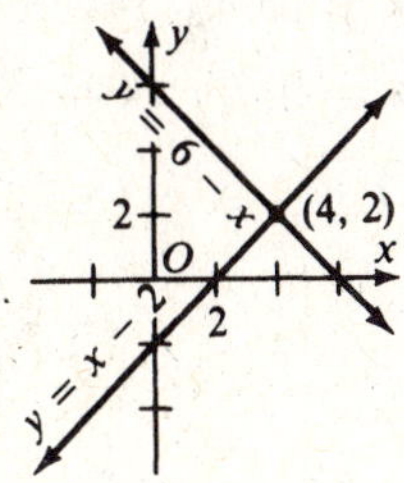

2.

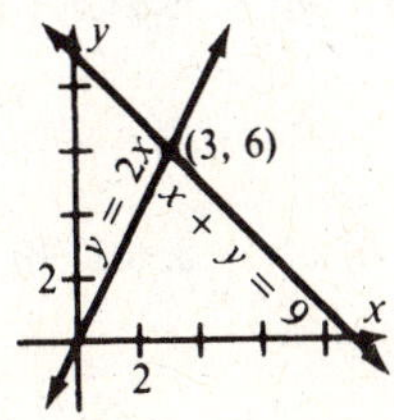

3.

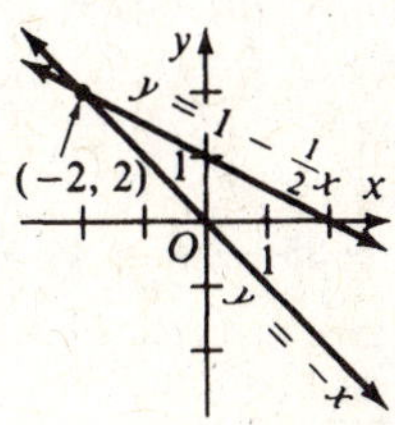

4.

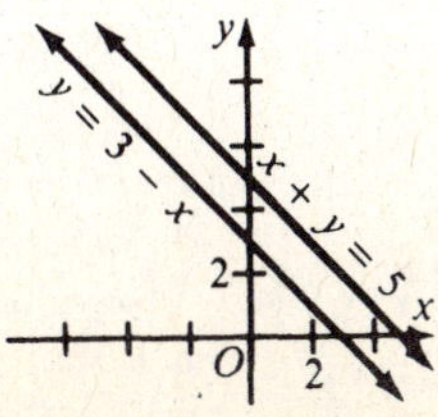

5.

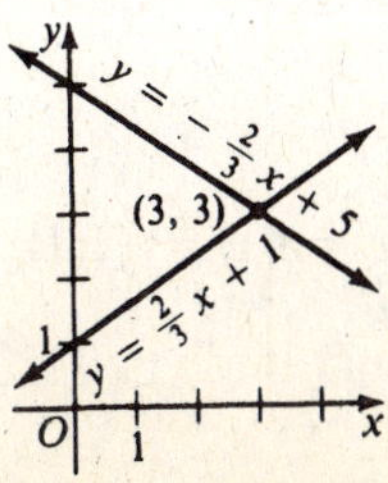

6.

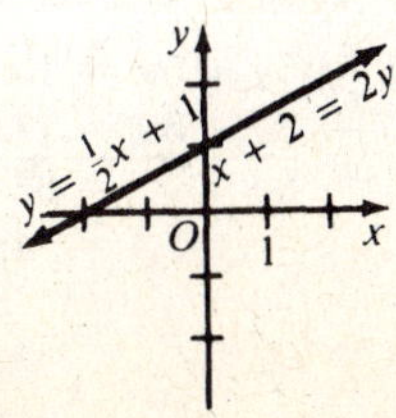

7. $3x + 2x = 5$; $5x = 5$; $x = 1$; $y = 2 \cdot 1$; $y = 2$; $(1, 2)$

8. $m = 3n - 4$; $2(3n - 4) + 6n = 5$; $6n - 8 + 6n = 5$; $12n = 13$; $n = \frac{13}{12}$. $2m + 6\left(\frac{13}{12}\right) = 5$; $2m + \frac{13}{2} = 5$; $2m = -\frac{3}{2}$; $m = -\frac{3}{4}$; $\left(-\frac{3}{4}, \frac{13}{12}\right)$

9. $2a + (1 - a) = 4$; $2a + 1 - a = 4$; $a = 3$. $b = 1 - 3 = -2$; $(3, -2)$

10. $d = 2c - 5$; $4c - 3(2c - 5) = 9$; $4c - 6c + 15 = 9$; $-2c = -6$; $c = 3$. $d = 6 - 5 = 1$; $(3, 1)$

11. $x = 2 - 3y$; $2(2 - 3y) + 3y = 7$; $4 - 6y + 3y = 7$; $4 - 3y = 7$; $-3y = 3$; $y = -1$. $x = 2 + 3 = 5$; $(5, -1)$

12. $x = 15 - 2y$; $3(15 - 2y) - 2y = 5$, $45 - 6y - 2y = 5$, $-8y = -40$, $y = 5$; $x = 15 - 10 = 5$; $(5, 5)$

13. Let w = number of women, m = number of men. $2w - 3 = m$; $w + 2 = m$. $2w - 3 = w + 2$; $w = 5$. There are 7 men on the jury.

14. Let j = distance Janet lives from office, l = distance Lynn lives from office. $l + j = 8$; $2j - 1 = l$. $l = 8 - j$; $2j - 1 = 8 - j$; $3j = 9$; $j = 3$. Janet lives 3 mi from the office; Lynn lives 5 mi from the office.

15.
$$\begin{array}{rcl} r - s &=& -3 \\ \underline{r + s} &=& \underline{9} \\ 2r &=& 6 \\ r &=& 3;\ 3 + s = 9,\ s = 6;\ (3, 6) \end{array}$$

16.
$$\begin{array}{rcl} c + 2n &=& -20 \\ \underline{c - 2n} &=& \underline{30} \\ 2c &=& 10 \\ c &=& 5;\ 5 + 2n = -20,\ 2n = -25,\ n = -\frac{25}{2};\ \left(5, -\frac{25}{2}\right) \end{array}$$

17.
$$\begin{array}{rcl} x - 3y &=& 2 \\ \underline{x + 4y} &=& \underline{16} \\ -7y &=& -14 \\ y &=& 2;\ x - 3(2) = 2,\ x - 6 = 2,\ x = 8;\ (8, 2) \end{array}$$

18.
$$\begin{array}{rcl} 6r + 5y &=& -8 \\ \underline{2r - 5y} &=& \underline{-16} \\ 8r &=& -24 \\ r &=& -3;\ 6(-3) + 5y = -8,\ -18 + 5y = -8,\ 5y = 10,\ y = 2;\ (-3, 2) \end{array}$$

19.
$$\begin{array}{rcl} 12m + 3n &=& 51 \\ \underline{7m - 3n} &=& \underline{44} \\ 19m &=& 95 \\ m &=& 5;\ 35 - 3n = 44;\ -3n = 9;\ n = -3;\ (5, -3) \end{array}$$

20.
$$\begin{array}{rcl} 8g + 7h &=& 26 \\ \underline{8g - 10h} &=& \underline{60} \\ 17h &=& -34 \\ h &=& -2;\ 8g - 14 = 26;\ 8g = 40;\ g = 5;\ (5, -2) \end{array}$$

21.
$$\begin{array}{rcl} v + w = 3 &\rightarrow& 5v + 5w = 15 \\ 3v - 5w = 17 &\rightarrow& \underline{3v - 5w = 17} \\ && 8v = 32 \\ && v = 4;\ 4 + w = 3;\ w = -1;\ (4, -1) \end{array}$$

22. $4a - 3b = -1 \rightarrow 4a - 3b = -1$
$a - b = -1 \rightarrow \underline{4a - 4b = -4}$
$b = 3;\ a - 3 = -1;\ a = 2;\ (2, 3)$

23. $3x - y = 3 \rightarrow 9x - 3y = 9$
$x + 3y = 11 \rightarrow \underline{x + 3y = 11}$
$10x = 20$
$x = 2;\ 2 + 3y = 11;\ 3y = 9;\ y = 3;\ (2, 3)$

24. $3x + 4y = -25 \rightarrow 9x + 12y = -75$
$2x - 3y = 6 \rightarrow \underline{8x - 12y = 24}$
$17x = -51$
$x = -3;\ 3(-3) + 4y = -25;\ 4y = -16;\ y = -4;\ (-3, -4)$

25. $2w - 3z = -1 \rightarrow 8w - 12z = -4$
$3w + 4z = 24 \rightarrow \underline{9w + 12z = 72}$
$17w = 68$
$w = 4;\ 8 - 3z = -1;\ -3z = -9;\ z = 3;\ (4, 3)$

26. $5a - 2b = 0 \rightarrow 15a - 6b = 0$
$2a - 3b = -11 \rightarrow \underline{4a - 6b = -22}$
$11a = 22$
$a = 2;\ 10 - 2b = 0;\ b = 5;\ (2, 5)$

27. Let p = speed of plane, w = speed of wind. $(w + p)\frac{4}{3} = 1120;\ (p - w)\frac{7}{5} = 1120.$

$\frac{4}{3}w + \frac{4}{3}p = 1120 \rightarrow 4w + 4p = 3360 \rightarrow 28w + 28p = 23{,}520$

$-\frac{7}{5}w + \frac{7}{5}p = 1120 \rightarrow -7w + 7p = 5600 \rightarrow \underline{-28w + 28p = 22{,}400}$
$56p = 45{,}920;\ p = 820.$

$\frac{4}{3}w + \frac{3280}{3} = 1120;\ 4w + 3280 = 3360;\ 4w = 80;\ w = 20.$ The plane's speed is 820 km/h; the wind speed is 20 km/h.

28.

	Tens digit	Units digit	Value
Original	t	u	$10t + u$
New	u	t	$10u + t$

$t + u = 7 \rightarrow u + t = 7 \rightarrow 4u + 4t = 28$
$10u + t = 5t \rightarrow 10u - 4t = 0 \rightarrow \underline{10u - 4t = 0}$
$14u = 28;\ u = 2;\ t + 2 = 7;\ t = 5.$ The number is 52.

29.

	3 years ago	Now	5 years from now
Jenny	$x - 8$	$x - 5$	x
Aunt	$y - 8$	$y - 5$	y

$x = \frac{2}{3}y \rightarrow 3x = 2y \rightarrow 3x - 2y = 0 \rightarrow 3x - 2y = 0$

$x - 8 = \frac{1}{2}(y - 5) \rightarrow 2x - 16 = y - 5 \rightarrow 2x - y = 11 \rightarrow \underline{4x - 2y = 22}$
$x = 22$

Jenny is 17; her aunt is 28.

30. Let n = numerator, d = denominator. $n = d - 1$; $\frac{n-1}{d} = \frac{3}{4}$. $n = d - 1$; $4(n - 1) = 3d$;

$$\begin{array}{rcl} n - d = -1 & \rightarrow & 3n - 3d = -3 \\ \underline{4n - 3d = 4} & \rightarrow & \underline{4n - 3d = 4} \\ & & n = 7 \end{array}$$

$n = 7$; $d = 8$. The original fraction is $\frac{7}{8}$.

Pages 657–659 • CHAPTER 10

1. False **2.** True **3.** True **4.** True **5.** False **6.** False **7.** False **8.** True

9. True

10. $5x \leq 15$
$5(-4) \leq 15$ True
$5(-3) \leq 15$ True
$5(-2) \leq 15$ True
$5(-1) \leq 15$ True
$5 \cdot 0 \leq 15$ True
$5 \cdot 1 \leq 15$ True
$5 \cdot 2 \leq 15$ True
$5 \cdot 3 \leq 15$ True
$5 \cdot 4 \leq 15$ False
Solution set: $\{-4, -3, -2, -1, 0, 1, 2, 3\}$

11. $-7x > 14$
$-7(-4) > 14$ True
$-7(-3) > 14$ True
$-7(-2) > 14$ False
$-7(-1) > 14$ False
$-7(0) > 14$ False
$-7(1) > 14$ False
$-7(2) > 14$ False
$-7(3) > 14$ False
$-7(4) > 14$ False
Solution set: $\{-4, -3\}$

12. $-4 - x \geq 0$
$-4 - (-4) \geq 0$ True
$-4 - (-3) \geq 0$ False
$-4 - (-2) \geq 0$ False
$-4 - (-1) \geq 0$ False
$-4 - 0 \geq 0$ False
$-4 - 1 \geq 0$ False
$-4 - 2 \geq 0$ False
$-4 - 3 \geq 0$ False
$-4 - 4 \geq 0$ False
Solution set: $\{-4\}$

13. $x^2 < 10$
$(-4)^2 < 10$ False
$(-3)^2 < 10$ True
$(-2)^2 < 10$ True
$(-1)^2 < 10$ True
$0^2 < 10$ True
$1^2 < 10$ True
$2^2 < 10$ True
$3^2 < 10$ True
$4^2 < 10$ False
Solution set: $\{-3, -2, -1, 0, 1, 2, 3\}$

14.
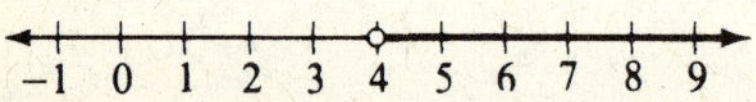

15.
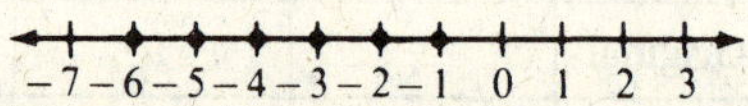

16.
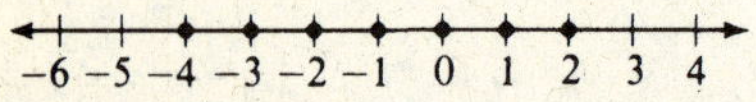

17.
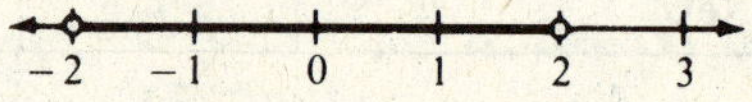

18. $e > 20$; {the real numbers greater than 20}

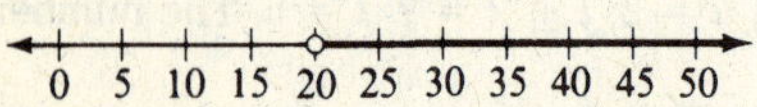

19. $4 > n$; {the real numbers less than 4}

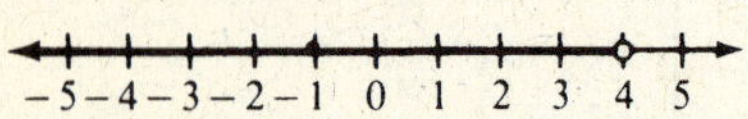

20. $q < -5$; {the real numbers less than -5}

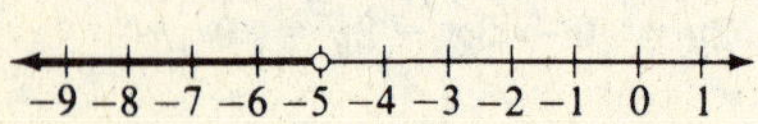

21. $-x < 98$; $x > -98$; {the real numbers greater than -98}

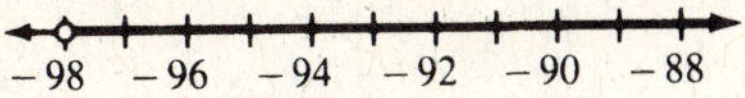

22. $\frac{m}{3} > 3$; $m > 9$; {the real numbers greater than 9}

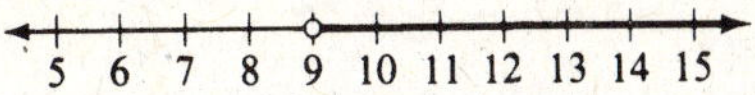

23. $-10 > \frac{4k}{5}$; $-10\left(\frac{5}{4}\right) > k$; $-\frac{25}{2} > k$; $\left\{\text{the real numbers less than } -\frac{25}{2}\right\}$

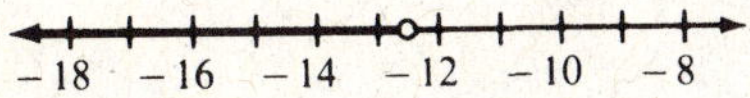

24. $5v > 15$; $v > 3$; {the real numbers greater than 3}

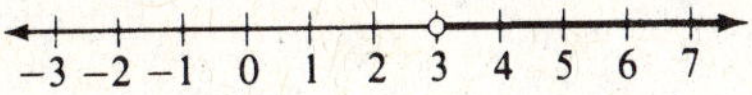

25. $6y > 48$; $y > 8$; {the real numbers greater than 8}

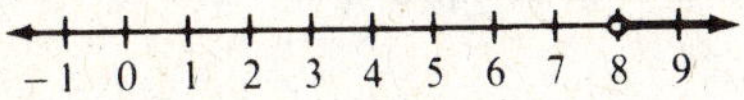

26. $n < 8$; {the real numbers less than 8}

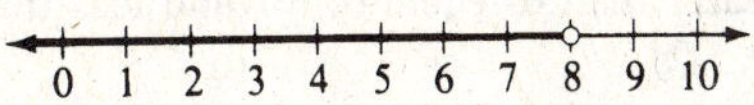

27. $4f > 16$; $f > 4$; {the real numbers greater than 4}

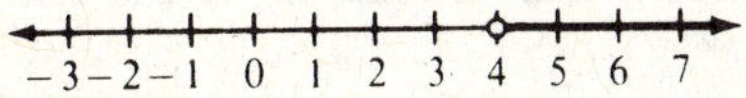

28. $-6v + 18 \leq 42$; $-6v \leq 24$; $v \geq -4$; {the real numbers greater than or equal to -4}

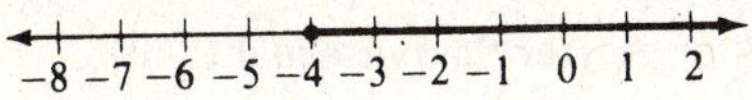

29. $5m + 10 > 4m - 4$; $m > -14$; {the real numbers greater than -14}

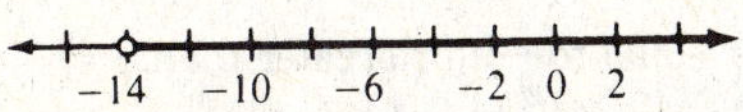

30. $\frac{4}{9}h + 3 \leq \frac{1}{3}$; $4h + 27 \leq 3$; $4h \leq -24$; $h \leq -6$; {the real numbers less than or equal to -6}

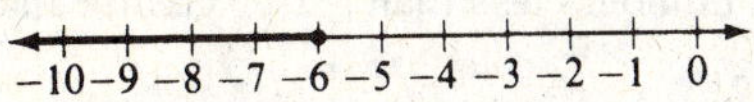

31. $2w - 2 < \frac{3}{2}w$; $4w - 4 < 3w$; $w < 4$; {the real numbers less than 4}

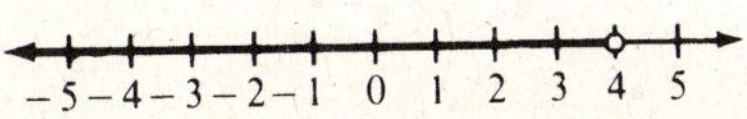

32. $2x - \frac{3}{4}x - 2 > 0$, $\frac{5}{4}x - 2 > 0$, $\frac{5}{4}x > 2$, $x > \frac{8}{5}$; $\left\{\text{the real numbers greater than } \frac{8}{5}\right\}$

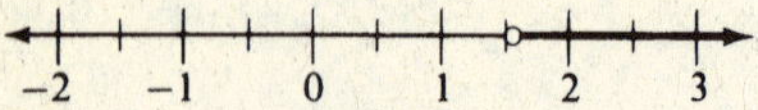

33. **a.** Let x = number Juanita sold. **b.** $2x - 9 \leq 43$

34. **a.** Let s = Sam's age. **b.** $s + 3 < 16$

35. **a.** Let y = distance Roger lives from beach. **b.** $\frac{1}{2}y - 10 \geq 25$

36. **a.** Let w = number who walk or bike. **b.** $\frac{1}{3}w + w \geq 1800$

37. **a.** Let b = Buford's present age. **b.** $b - 6 < \frac{1}{2}b$

38. $-8 < n < 2$; {the real numbers greater than -8 and less than 2}

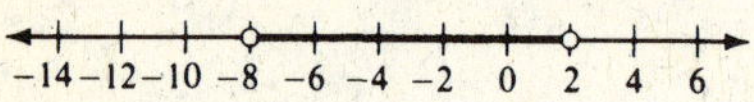

39. $0 < w \leq 8$; {the real numbers greater than 0 and less than or equal to 8}

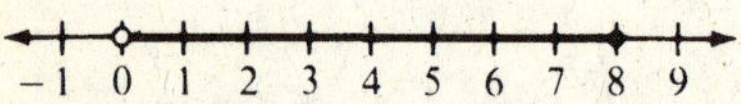

40. $-3 \leq 3a < 6$; $-1 \leq a < 2$; {the real numbers greater than or equal to -1 and less than 2}

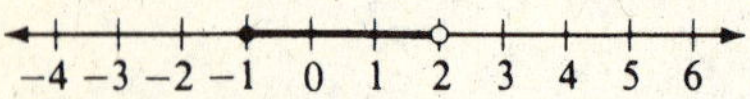

41. $-8 \leq 8m \leq 16$; $-1 \leq m \leq 2$; {the real numbers greater than or equal to -1 and less than or equal to 2}

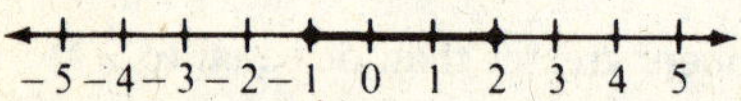

42. $u < -3$ or $u \geq 6$; {the real numbers less than -3 or greater than or equal to 6}

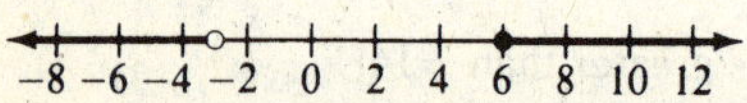

43. $k \leq -9$ or $k > -4$; {the real numbers less than or equal to -9 or greater than -4}

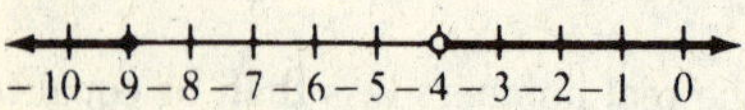

44. $4n < -4$ or $4n > 4$, $n < -1$ or $n > 1$; {the real numbers less than -1 or greater than 1}

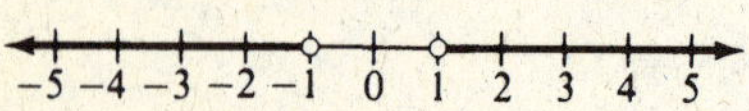

45. $2x \leq -6$ or $10 < 2x$; $x \leq -3$ or $x > 5$; {the real numbers less than or equal to -3 or greater than 5}

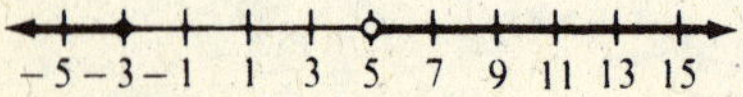

46. $e > -3$ and $3e < -6$; $e > -3$ and $e < -2$; {the real numbers greater than -3 and less than -2}

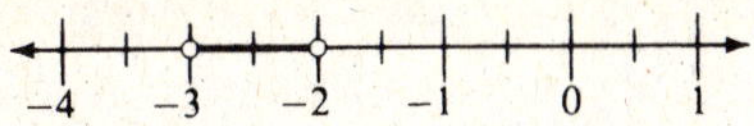

47. $h \geq 6$ or $-h \geq -2$; $h \geq 6$ or $h \leq 2$; {the real numbers greater than or equal to 6 or less than or equal to 2}

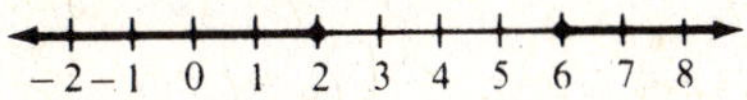

48. $v - 6 = 3$ or $v - 6 = -3$; $v = 9$ or $v = 3$; {3, 9}

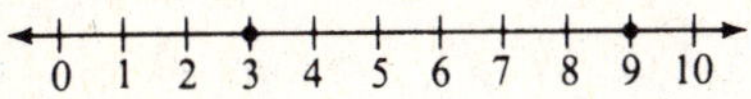

49. $8 - k = 5$ or $8 - k = -5$; $-k = -3$ or $-k = -13$; $k = 3$ or $k = 13$; {3, 13}

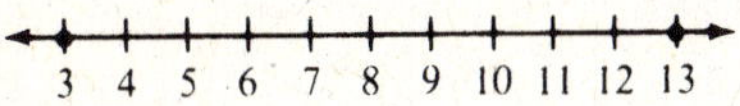

50. $m > 2\frac{1}{2}$ or $m < -2\frac{1}{2}$; $\left\{\text{the real numbers greater than } 2\frac{1}{2} \text{ or less than } -2\frac{1}{2}\right\}$

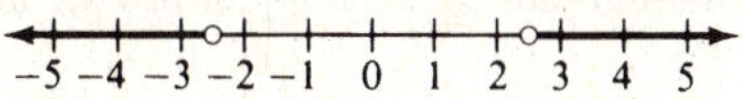

51. $-1.5 \leq x \leq 1.5$; {the real numbers greater than or equal to -1.5 and less than or equal to 1.5}

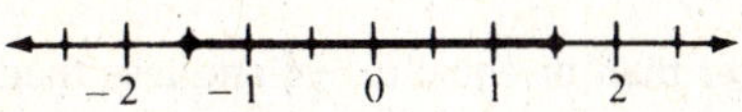

52. $y + 5 > 2$ or $y + 5 < -2$; $y > -3$ or $y < -7$; {the real numbers greater than -3 or less than -7}

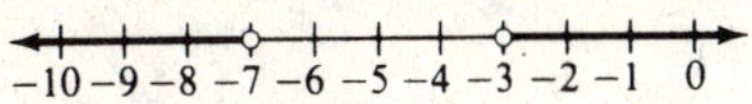

53. $-4 \leq 3 + z \leq 4$; $-7 \leq z \leq 1$; {the real numbers greater than or equal to -7 and less than or equal to 1}

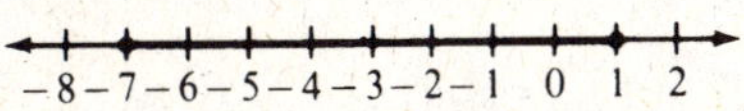

54. $-6 < 7 - f < 6$; $-13 < -f < -1$; $13 > f > 1$; {the real numbers less than 13 and greater than 1}

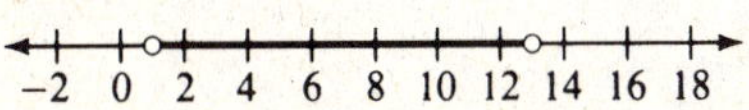

55. $-4 - g > 7$ or $-4 - g < -7$; $-g > 11$ or $-g < -3$; $g < -11$ or $g > 3$; {the real numbers less than -11 or greater than 3}

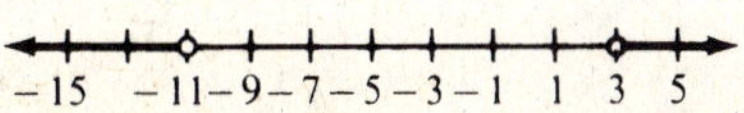

56. $-a - 2 \geq 1$ or $-a - 2 \leq -1$; $-a \geq 3$ or $-a \leq 1$; $a \leq -3$ or $a \geq -1$; {the real numbers less than or equal to -3 or greater than or equal to -1}

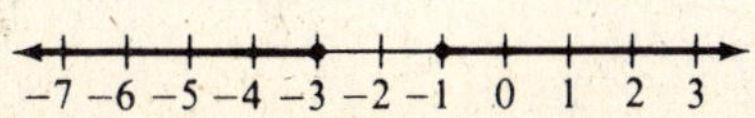

57. $4|s| \le 6$; $|s| \le \frac{6}{4}$; $-\frac{3}{2} \le s \le \frac{3}{2}$; $\left\{\text{the real numbers greater than or equal to } -\frac{3}{2} \text{ and less than or equal to } \frac{3}{2}\right\}$

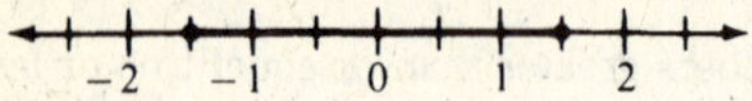

58. $-3|z| < 9$; $|z| > -3$; {all real numbers}

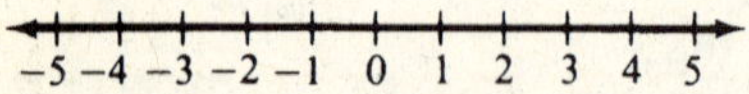

59. $-|3 - b| > -7$; $|3 - b| < 7$; $-7 < 3 - b < 7$; $-10 < -b < 4$; $10 > b > -4$; {the real numbers less than 10 and greater than −4}

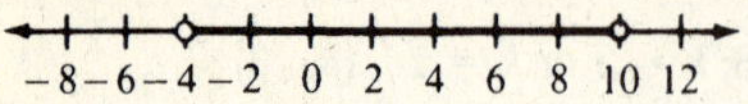

60. $8k = 16$ or $8k = -16$; $k = 2$ or $k = -2$; $\{-2, 2\}$

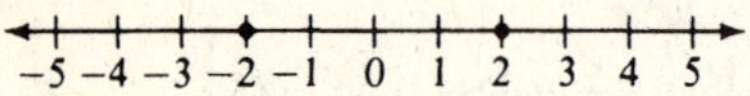

61. $\frac{a}{2} \ge 3$ or $\frac{a}{2} \le -3$; $a \ge 6$ or $a \le -6$; {the real numbers greater than or equal to 6 or less than or equal to −6}

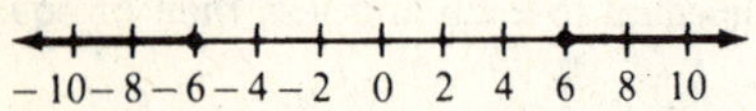

62. $-1 \le \frac{c}{4} \le 1$; $-4 \le c \le 4$; {the real numbers greater than or equal to −4 and less than or equal to 4}

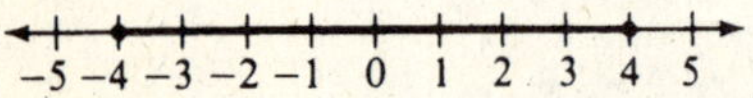

63. $5x - 3 = 17$ or $5x - 3 = -17$; $5x = 20$ or $5x = -14$, $x = 4$ or $x = -\frac{14}{5}$; $\left\{-\frac{14}{5}, 4\right\}$

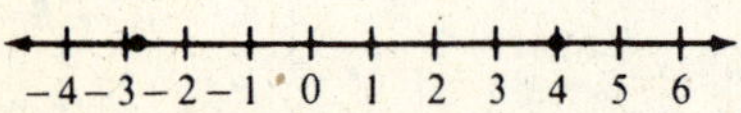

64. $-19 < 7 + 6n < 19$; $-26 < 6n < 12$; $-4\frac{1}{3} < n < 2$; $\left\{\text{the real numbers greater than } -4\frac{1}{3} \text{ and less than } 2\right\}$

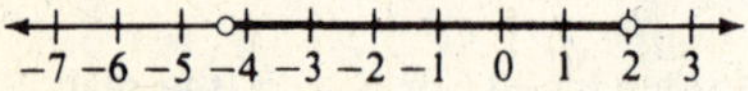

65. $2w - 3 < -5$ or $2w - 3 > 5$, $2w < -2$ or $2w > 8$, $w < -1$ or $w > 4$; {the real numbers less than −1 or greater than 4}

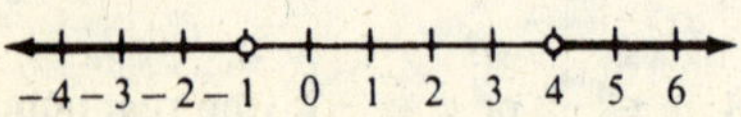

66. $\frac{u}{3} + 4 = 1$ or $\frac{u}{3} + 4 = -1$; $\frac{u}{3} = -3$ or $\frac{u}{3} = -5$; $u = -9$ or $u = -15$; $\{-15, -9\}$

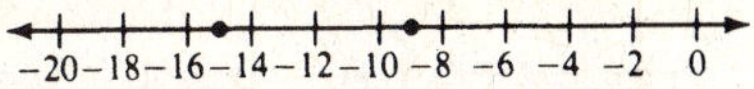

67. $-3 \le \frac{v}{2} - 1 \le 3$, $-2 \le \frac{v}{2} \le 4$, $-4 \le v \le 8$; {the real numbers greater than or equal to -4 and less than or equal to 8}

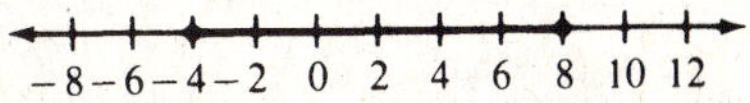

68. $\frac{a}{4} + 2 \ge 1$ or $\frac{a}{4} + 2 \le -1$; $\frac{a}{4} \ge -1$ or $\frac{a}{4} \le -3$; $a \ge -4$ or $a \le -12$; {the real numbers greater than or equal to -4 or less than or equal to -12}

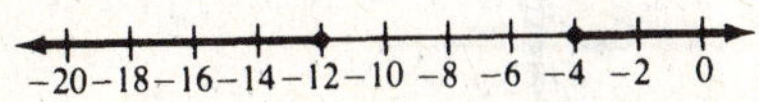

69. $|5 - 3 + 2x| < 6$; $|2 + 2x| < 6$; $-6 < 2 + 2x < 6$; $-8 < 2x < 4$; $-4 < x < 2$; {the real numbers greater than -4 and less than 2}

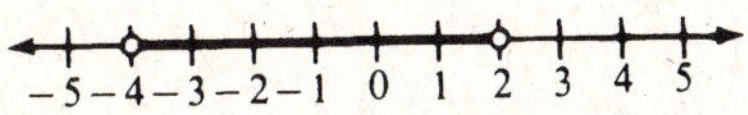

70. $3|2m + 1| = 6$; $|2m + 1| = 2$; $2m + 1 = 2$ or $2m + 1 = -2$; $2m = 1$ or $2m = -3$; $m = \frac{1}{2}$ or $m = -\frac{3}{2}$; $\left\{-\frac{3}{2}, \frac{1}{2}\right\}$

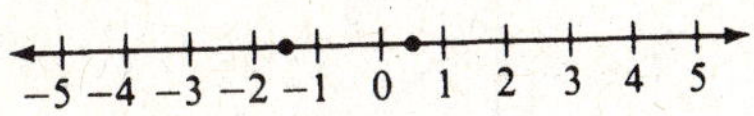

71. $-6|2 - k| \ge 12$; $|2 - k| \le -2$; $\emptyset$

72.
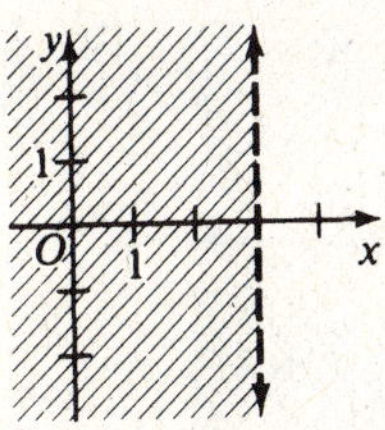

73.
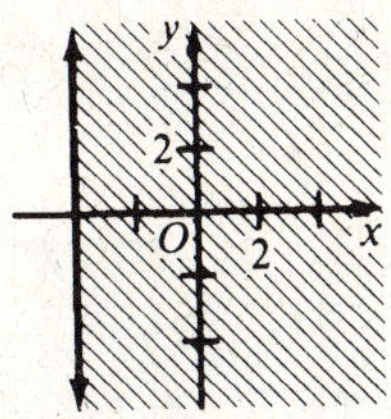

74.
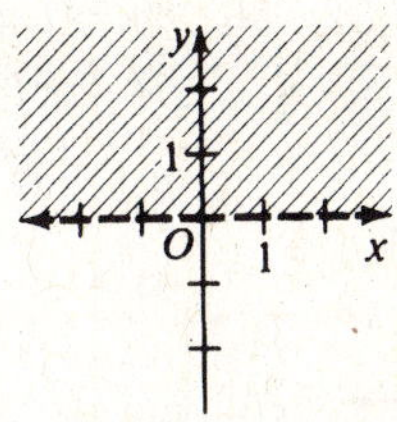

75.
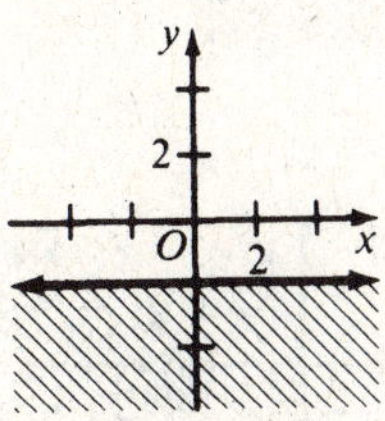

76.
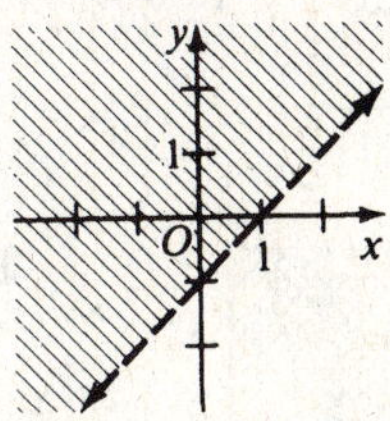

77.
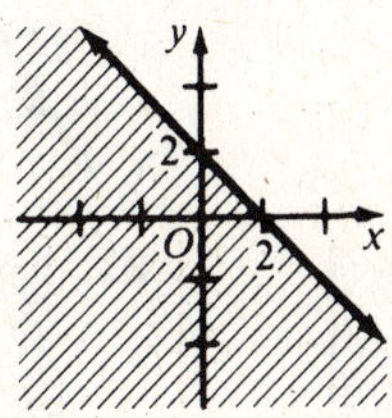

78.

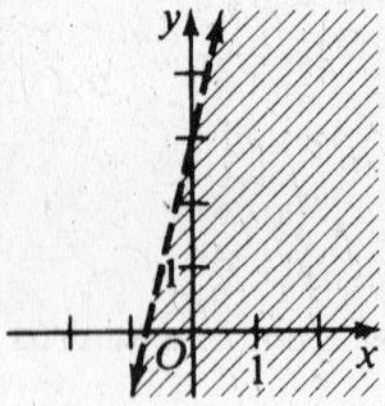

79.

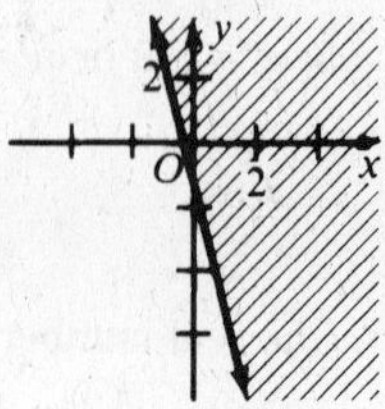

80. $-y \geq 5 - x$; $y \leq x - 5$

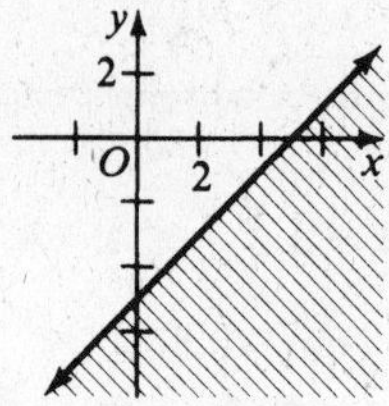

81. $y \leq -4x - 2$

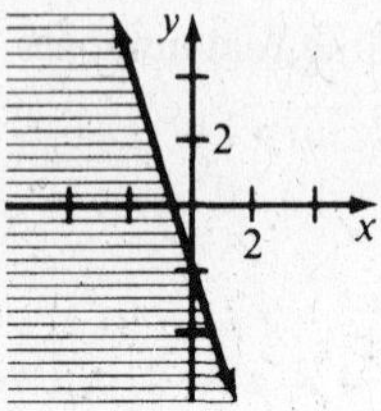

82. $-3y > 6 - x$

$y < \frac{1}{3}x - 2$

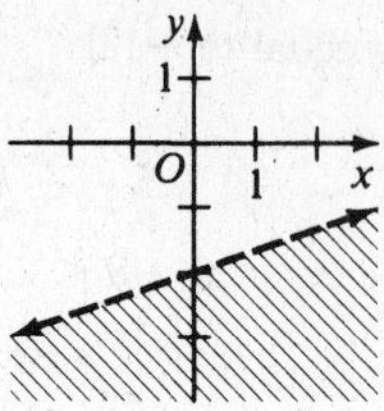

83. $-y < 2 - 6x$

$y > 6x - 2$

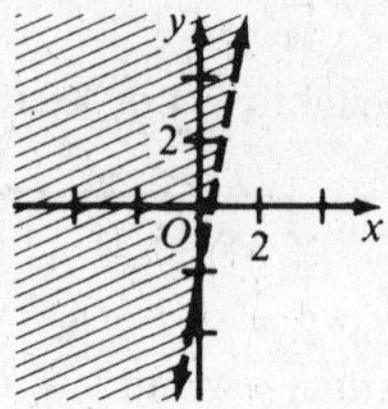

84. $y \geq 5x + 3$

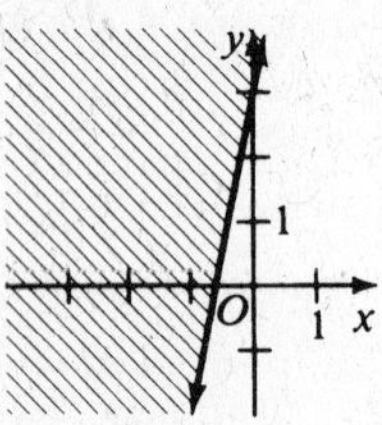

85. $4y < 5x$, $y < \frac{5}{4}x$

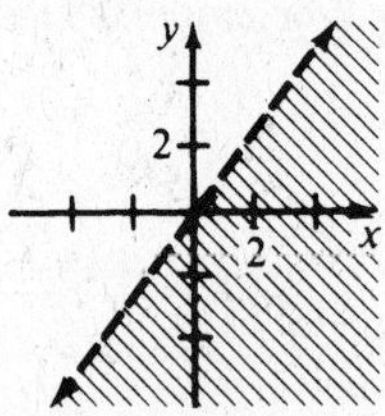

86. $6y \geq -6x - 3$, $y \geq -x - \frac{1}{2}$

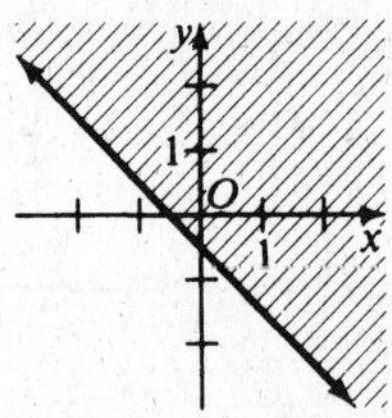

87. $3y > 6x - 2$

$y > 2x - \frac{2}{3}$

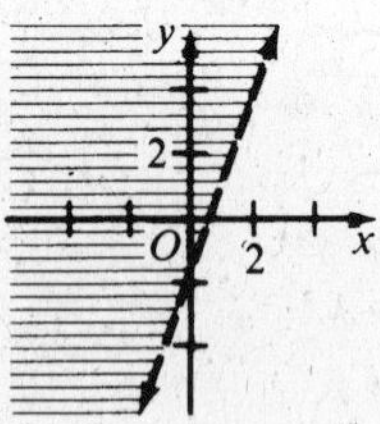

88. $8y - 7 \leq 3x + 6y$

$2y \leq 3x + 7$

$y \leq \frac{3}{2}x + \frac{7}{2}$

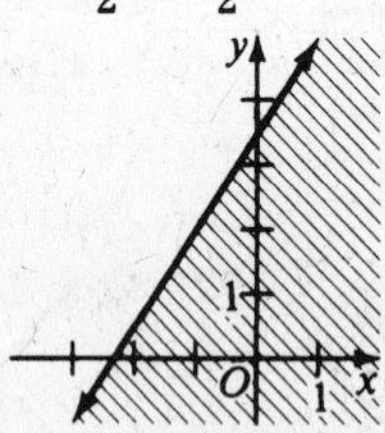

89.

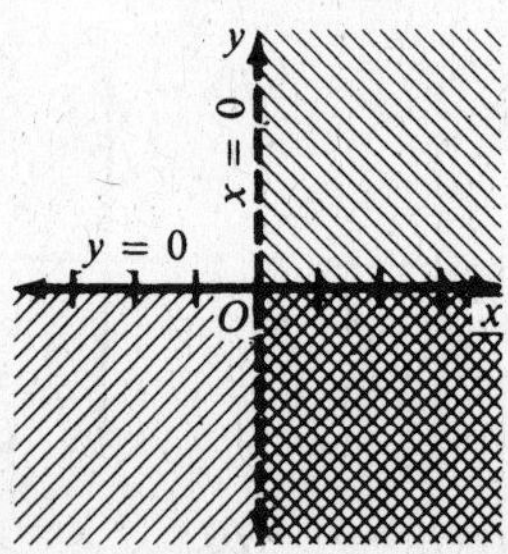

90.

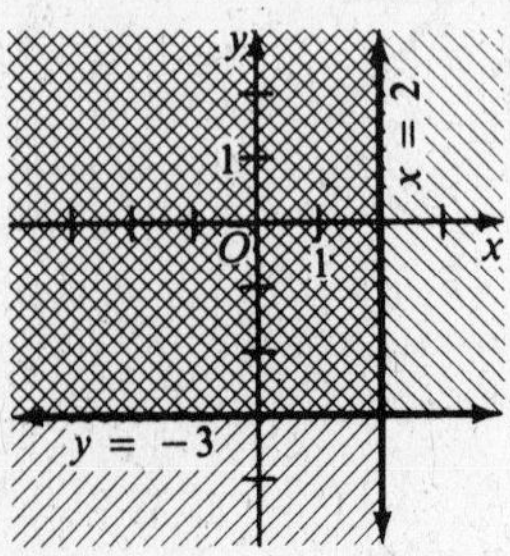

91.

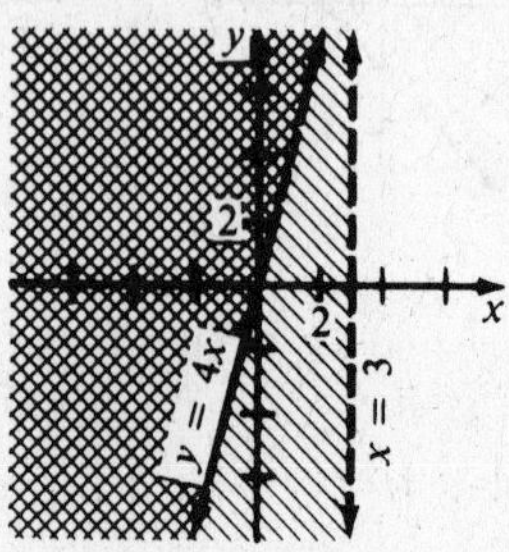

92.

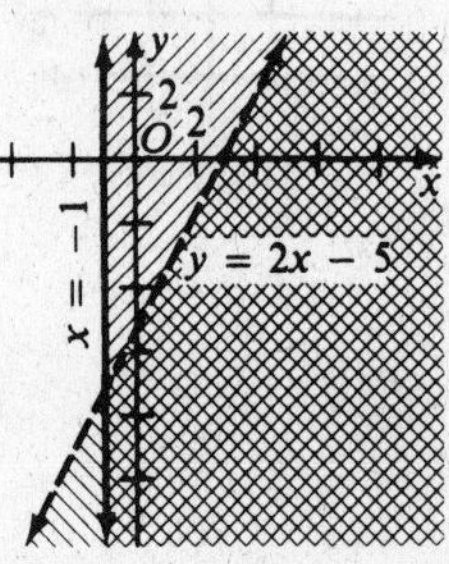

93.

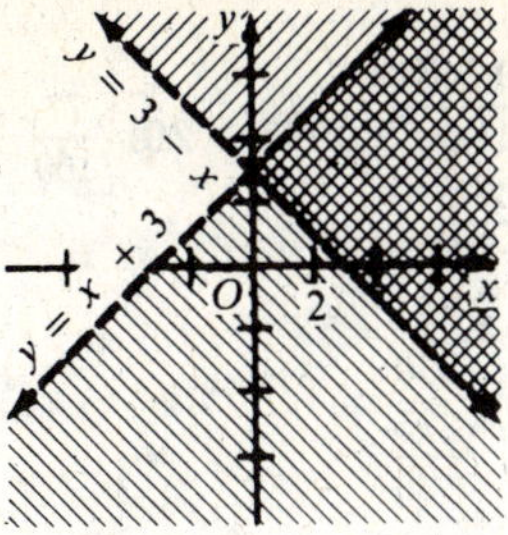

94.

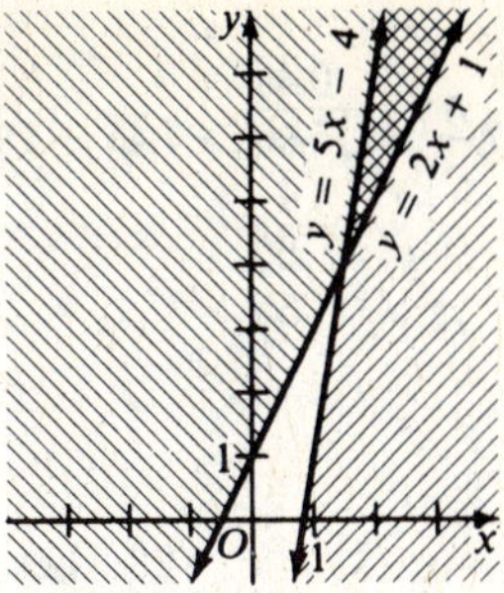

95.

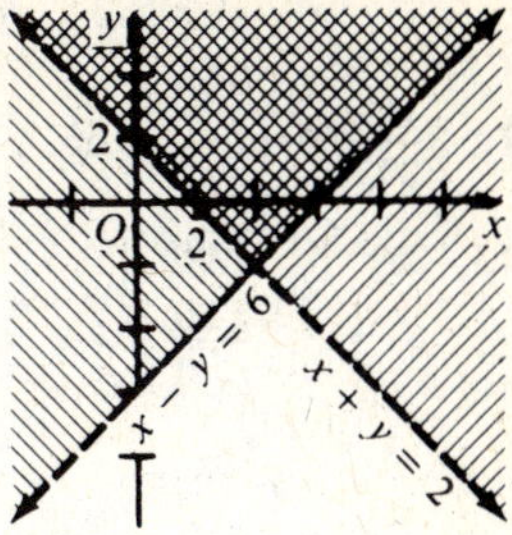

96.

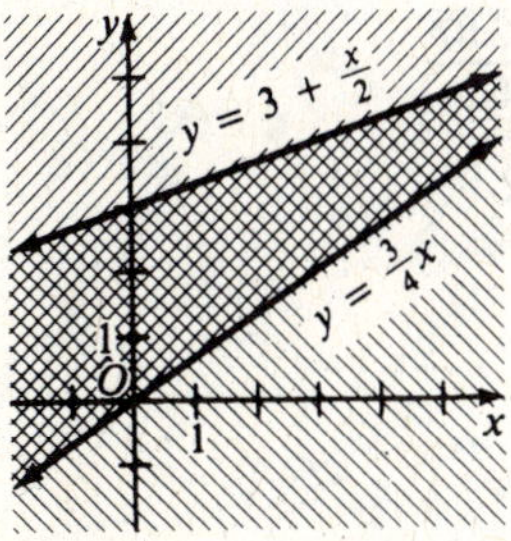

Pages 659–662 • CHAPTER 11

1. $<$ **2.** $=$ **3.** $<$ **4.** $-\frac{39}{8}, -4.7, -\frac{41}{9}$ **5.** $\frac{2}{3}, \frac{12}{17}, \frac{5}{7}, \frac{11}{15}$ **6.** $-\frac{5}{7}, -\frac{5}{8}, -\frac{6}{11}, -\frac{4}{9}$

7. $\dfrac{\frac{27}{41} + \frac{31}{37}}{2} = \dfrac{999 + 1271}{2 \cdot 1517} = \dfrac{2270}{3034} = \dfrac{1135}{1517}$

8. $\dfrac{-\frac{17}{140} + \left(-\frac{11}{32}\right)}{2} = \dfrac{\frac{-136 + (-385)}{1120}}{2} = -\dfrac{521}{1120} \cdot \dfrac{1}{2} = -\dfrac{521}{2240}$

9. $\dfrac{-5\frac{2}{7} + 9\frac{1}{4}}{2} = \dfrac{-\frac{37}{7} + \frac{37}{4}}{2} = \dfrac{\frac{-148 + 259}{28}}{2} = \dfrac{111}{28} \cdot \dfrac{1}{2} = \dfrac{111}{56}$ **10.** $\frac{5}{1}, \frac{5}{2}, \frac{5}{3}, \frac{5}{4}$; decreases

11. $-\frac{3}{7}, -\frac{2}{7}, -\frac{1}{7}, 0$; increases **12.** $\frac{8}{2}, \frac{8}{3}, \frac{8}{4}, \frac{8}{5}$; decreases **13.** $\frac{6}{4}, \frac{5}{4}, \frac{4}{4}, \frac{3}{4}$; decreases

14. $\frac{10}{5}, \frac{10}{7}, \frac{10}{9}, \frac{10}{11}$; decreases **15.** $9\overline{)4.0}$ with quotient $0.\overline{4}$; 36, remainder 4; $= 0.\overline{4}$ **16.** $24\overline{)-29.00}$ with quotient $-1.208\overline{3}$; $-1.208\overline{3}$

$$\begin{array}{r} 24 \\ \hline 50 \\ 48 \\ \hline 200 \\ 192 \\ \hline 80 \\ 72 \\ \hline 8 \end{array}$$

17. $3\frac{11}{20}$; $\begin{array}{r} 0.55 \\ 20\overline{)11.0} \\ \underline{10\ 0} \\ 1\ 00 \\ \underline{1\ 00} \\ 0 \end{array}$; 3.55 **18.** $-7\frac{5}{11}$; $\begin{array}{r} 0.\overline{45} \\ 11\overline{)5.00} \\ \underline{4\ 4} \\ 60 \\ \underline{55} \\ 50 \end{array}$; $-7.\overline{45}$ **19.** $\begin{array}{r} 0.7454 \\ 55\overline{)41.00} \\ \underline{38\ 5} \\ 2\ 50 \\ \underline{2\ 20} \\ 300 \\ \underline{275} \\ 25 \end{array}$; $0.7\overline{45}$ **20.** $\frac{77}{100}$

21. Let $N = 0.\overline{6}$
$10N = 6.\overline{6}$
$-N = -0.\overline{6}$
$9N = 6$
$N = \frac{6}{9} = \frac{2}{3}$

22. Let $N = -0.31\overline{8}$
$10N = -3.18\overline{8}$
$-N = -(-0.31\overline{8})$
$9N = -2.870$
$N = \frac{-2.87}{9} = -\frac{287}{900}$

23. Let $N = 2.\overline{37}$
$100N = 237.\overline{37}$
$-N = -2.\overline{37}$
$99N = 235$
$N = \frac{235}{99}$

24. Let $N = 0.\overline{4135}$
$10{,}000N = 4135.\overline{4135}$
$-N = -0.\overline{4135}$
$9999N = 4135$
$N = \frac{4135}{9999}$

25. $0.625 + \frac{1}{2}(0.63 - 0.625) = 0.625 + \frac{1}{2}(0.005) = 0.625 + 0.0025 = 0.6275$

26. $0.66 + \frac{1}{2}(0.66\overline{6} - 0.66) = 0.66 + \frac{1}{2}(0.00\overline{6}) = 0.66 + 0.00\overline{3} = 0.66\overline{3}$

27. $0.\overline{63} - \frac{1}{2}(0.\overline{63} - 0.6\overline{28}) = 0.\overline{63} - \frac{1}{2}(0.6\overline{36} - 0.6\overline{28}) = 0.\overline{63} - \frac{1}{2}(0.0\overline{08}) =$

$0.6\overline{36} - 0.0\overline{04} = 0.6\overline{32}$

28. $\frac{2}{5} \cdot \frac{85}{100} = \frac{17}{50}$ **29.** $\frac{4}{9} \cdot \frac{2}{3} = \frac{8}{27}$ **30.** $-2\frac{1}{5} \cdot \frac{1}{3} = -\frac{11}{5} \cdot \frac{1}{3} = -\frac{11}{15}$

31. $\sqrt{441} = \sqrt{9 \cdot 49} = 21$ **32.** $\sqrt{784} = \sqrt{16 \cdot 49} = 28$ **33.** $\sqrt{2704} = \sqrt{16 \cdot 169} = 52$

34. $\sqrt{5184} = \sqrt{64 \cdot 81} = 72$ **35.** $\sqrt{10{,}816} = \sqrt{64 \cdot 169} = 8 \cdot 13 = 104$

36. $\sqrt{\frac{4}{100}} = \frac{2}{10} = 0.2$ **37.** $\sqrt{\frac{64}{100}} = \frac{8}{10} = 0.8$ **38.** $\sqrt{\frac{196}{100}} = \frac{14}{10} = 1.4$

39. $\sqrt{\frac{144}{10{,}000}} = \frac{12}{100} = 0.12$ **40.** $\sqrt{\frac{36}{10{,}000}} = \frac{6}{100} = 0.06$ **41.** $\sqrt{\frac{81}{225}} = \frac{9}{15} = \frac{3}{5}$

42. $\sqrt{\frac{1}{289}} = \frac{1}{17}$ **43.** $\sqrt{\frac{324}{1936}} = \sqrt{\frac{81}{484}} = \sqrt{\frac{81}{4 \cdot 121}} = \frac{9}{2 \cdot 11} = \frac{9}{22}$ **44.** $\sqrt{\frac{32}{50}} = \sqrt{\frac{16}{25}} = \frac{4}{5}$

45. $\sqrt{\frac{320}{405}} = \sqrt{\frac{64}{81}} = \frac{8}{9}$ **46.** $\sqrt{9 \cdot 7} = 3\sqrt{7}$ **47.** $\sqrt{16 \cdot 11} = 4\sqrt{11}$

48. $2\sqrt{4 \cdot 13} = 4\sqrt{13}$ **49.** $4\sqrt{9 \cdot 11} = 12\sqrt{11}$ **50.** $5\sqrt{25 \cdot 7} = 25\sqrt{7}$

51. $10\sqrt{81 \cdot 2} = 90\sqrt{2}$ **52.** $\sqrt{64 \cdot 3} = 8\sqrt{3}$ **53.** $\sqrt{672} = \sqrt{16 \cdot 42} = 4\sqrt{42}$

54. $\sqrt{16 \cdot 14} = 4\sqrt{14}$ **55.** $\sqrt{441 \cdot 6} = 21\sqrt{6}$ **56.** 26.8 **57.** -28.3 **58.** 21.0

59. 91.7 **60.** -73.5 **61.** $13|m|$ **62.** $4|a|\sqrt{3}$ **63.** $5x^2\sqrt{5}$

64. $\sqrt{9 \cdot 6 \cdot e^2 \cdot e} = 3|e|\sqrt{6e}$ **65.** $-6|t^3|$ **66.** $7|uv|\sqrt{2}$ **67.** $-12|xy|\sqrt{2x}$

68. $18r^2|s^3|$ **69.** $-\sqrt{\frac{484}{100}w^4} = -\sqrt{\frac{4 \cdot 121}{100}w^4} = -\frac{2 \cdot 11 \cdot w^2}{10} = -\frac{22}{10}w^2 = -2.2w^2$

70. $\sqrt{\frac{576c^6}{100}} = \frac{24|c^3|}{10} = 2.4|c^3|$ **71.** $\frac{a^2|b^3|}{2|c|}\sqrt{\frac{1}{3} \cdot \frac{3}{3}} = \frac{a^2|b^3|\sqrt{3}}{6|c|}, c \neq 0$

72. $\sqrt{\frac{12u^4}{v^2}} = \frac{2u^2\sqrt{3}}{|v|}, u \neq 0, v \neq 0$ **73.** $\frac{12k^4}{16} = \frac{3k^4}{4}$ **74.** $\frac{60}{9m^{18}} = \frac{20}{3m^{18}}, m \neq 0$ **75.** $\frac{15x^{20}}{4}$

76. $|x + 4|$ **77.** $|a - 2|$ **78.** $|9 + k|$ **79.** $g = \pm 7; \{7, -7\}$

80. $h^2 = 64; h = \pm 8; \{8, -8\}$ **81.** $m^2 = \frac{16}{25}; m = \pm\frac{4}{5}; \left\{\frac{4}{5}, -\frac{4}{5}\right\}$

82. $9x^2 = 4; x^2 = \frac{4}{9}; x = \pm\frac{2}{3}; \left\{\frac{2}{3}, -\frac{2}{3}\right\}$ **83.** $6y^2 = 54; y^2 = 9; y = \pm 3; \{3, -3\}$

84. $32t^2 = 27; t^2 = \frac{27}{32}; t = \pm\sqrt{\frac{27}{32}} = \pm\frac{3}{4}\sqrt{\frac{3}{2} \cdot \frac{2}{2}} = \pm\frac{3}{4}\sqrt{\frac{6}{4}} = \pm\frac{3}{8}\sqrt{6}; \left\{\frac{3}{8}\sqrt{6}, -\frac{3}{8}\sqrt{6}\right\}$

85. $a = \pm\sqrt{132} = \pm 2\sqrt{33} = \pm 11.5; \{11.5, -11.5\}$

86. $b^2 = 208; b = \pm\sqrt{208} = \pm 4\sqrt{13} = \pm 14.4; \{14.4, -14.4\}$

87. $c^2 = 43; c = \pm\sqrt{43} = \pm 6.6$ **88.** $\sqrt{3^2 + 4^2} = \sqrt{25} = 5.00$

89. $\sqrt{25 + 64} = \sqrt{89} = 9.43$ **90.** $a = \sqrt{13^2 - 9^2} = \sqrt{169 - 81} = \sqrt{88} = 2\sqrt{22} = 9.38$

91. $a = \sqrt{15^2 - 10^2} = \sqrt{225 - 100} = \sqrt{125} = 5\sqrt{5} = 11.18$

92. $b = \sqrt{16^2 - 8^2} = \sqrt{256 - 64} = \sqrt{192} = \sqrt{64 \cdot 3} = 8\sqrt{3} = 13.86$

93. $b = \sqrt{30^2 - 20^2} = \sqrt{900 - 400} = \sqrt{500} = \sqrt{100 \cdot 5} = 10\sqrt{5} = 22.36$

94. $c = \sqrt{12^2 + \left(\frac{3}{4}a\right)^2} = \sqrt{144 + \left(\frac{3}{4} \cdot 12\right)^2} = \sqrt{144 + 9^2} = \sqrt{144 + 81} = \sqrt{225} = 15.00$

95. $c = \sqrt{\left(\frac{2}{3}b\right)^2 + 15^2} = \sqrt{10^2 + 225} = \sqrt{100 + 225} = \sqrt{325} = 5\sqrt{13} = 18.03$

96. $\sqrt{21^2 + 28^2} = \sqrt{441 + 784} = \sqrt{1225} = 35$; yes **97.** $\sqrt{9^2 + 9^2} = \sqrt{162} \neq 12$; no.

98. $\sqrt{45^2 + 60^2} = \sqrt{2025 + 3600} = \sqrt{5625} = 75$; yes.

99. $\sqrt{31^2 + 41^2} = \sqrt{961 + 1681} = \sqrt{2642} \neq 51$; no.

100. $\sqrt{36a^2 + 64a^2} = \sqrt{100a^2} = 10a$; yes. **101.** $\sqrt{25a^2 + 49a^2} = \sqrt{74a^2} \neq 9a$; no.

102. $4 \cdot 3 = 12$ **103.** $6 \cdot 5 = 30$ **104.** $\sqrt{84} = 2\sqrt{21}$ **105.** $7\sqrt{4} = 7 \cdot 2 = 14$

106. $5\sqrt{6}$ **107.** $8\sqrt{81 \cdot 2} = 8 \cdot 9\sqrt{2} = 72\sqrt{2}$ **108.** $\sqrt{\frac{5}{9} \cdot \frac{9}{5}} = \sqrt{1} = 1$

109. $\sqrt{\frac{7}{5} \cdot \frac{45}{14}} = \sqrt{\frac{9}{2} \cdot \frac{2}{2}} = \frac{3\sqrt{2}}{2}$ **110.** $\sqrt{\frac{35}{6} \cdot \frac{18}{7}} = \sqrt{15}$

111. $\frac{1}{8}\sqrt{\frac{16}{3} \cdot \frac{3}{2}} = \frac{1}{8}\sqrt{4 \cdot 2} = \frac{2}{8}\sqrt{2} = \frac{\sqrt{2}}{4}$ **112.** $\frac{12 \cdot 2\sqrt{5}}{4\sqrt{3}} \cdot \frac{\sqrt{3}}{\sqrt{3}} = \frac{24\sqrt{15}}{12} = 2\sqrt{15}$

113. $\frac{11\sqrt{6}}{7\sqrt{2}} = \frac{11\sqrt{3}}{7}$ **114.** $-15\sqrt{x^2y^2} = -15xy$ **115.** $\sqrt{n^4} + 3\sqrt{n} = n^2 + 3\sqrt{n}$

116. $-140\sqrt{396} = -140\sqrt{36 \cdot 11} = -840\sqrt{11}$ **117.** $4\sqrt{3}$ **118.** $13\sqrt{2}$

119. $9\sqrt{6} - 2\sqrt{6} = 7\sqrt{6}$ **120.** $8\sqrt{7} + 24\sqrt{7} = 32\sqrt{7}$

121. $-30\sqrt{2} - 20\sqrt{2} = -50\sqrt{2}$ **122.** $11\sqrt{2} - 33\sqrt{3}$

123. $2\sqrt{2} - \sqrt{\frac{5 \cdot 6}{6^2}} = 2\sqrt{2} - \frac{\sqrt{30}}{6}$ **124.** $\frac{\sqrt{2}}{\sqrt{3}} - \frac{\sqrt{3}}{\sqrt{2}} = \frac{\sqrt{6}}{3} - \frac{\sqrt{6}}{2} = \frac{2\sqrt{6} - 3\sqrt{6}}{6} = \frac{-\sqrt{6}}{6}$

125. $20\sqrt{\frac{1}{7}\cdot\frac{7}{7}} + \frac{3}{2}\sqrt{\frac{1}{2}\cdot\frac{2}{2}} = \frac{20\sqrt{7}}{7} + \frac{3\sqrt{2}}{4}$, or $\frac{80\sqrt{7} + 21\sqrt{2}}{28}$

126. $9\sqrt{7} + 4\sqrt{7} - \sqrt{35} = 13\sqrt{7} - \sqrt{35}$

127. $2\sqrt{30} - 3\sqrt{30} + 10\sqrt{3} = -\sqrt{30} + 10\sqrt{3}$

128. $2\sqrt{\frac{5\cdot 3}{9}} + \frac{4}{2}\sqrt{\frac{3}{2}\cdot\frac{2}{2}} - \frac{2}{2}\sqrt{17} = \frac{2}{3}\sqrt{15} + \frac{2}{2}\sqrt{6} - \sqrt{17} = \frac{2}{3}\sqrt{15} + \sqrt{6} - \sqrt{17}$

129. $3\sqrt{5\cdot 75} - 6\sqrt{60} = 15\sqrt{15} - 12\sqrt{15} = 3\sqrt{15}$ **130.** $25 - 3 = 22$

131. $7 - 36 = -29$ **132.** $6 - 5 = 1$ **133.** $16 + 8\sqrt{2} + 2 = 18 + 8\sqrt{2}$

134. $25 - 10\sqrt{5} + 5 = 30 - 10\sqrt{5}$ **135.** $18 - 24\sqrt{2} + 16 = 34 - 24\sqrt{2}$

136. $11 + 6\sqrt{77} + 63 = 74 + 6\sqrt{77}$ **137.** $10\sqrt{12} - 8\sqrt{18} = 20\sqrt{3} - 24\sqrt{2}$

138. $8\sqrt{35} - 12\sqrt{7} + 28\sqrt{5} - 42$

139. $15\cdot 14 + 19\sqrt{98} + 6\cdot 7 = 210 + 42 + 19\cdot 7\sqrt{2} = 252 + 133\sqrt{2}$

140. $\frac{5(3 - \sqrt{7})}{(3 + \sqrt{7})(3 - \sqrt{7})} = \frac{15 - 5\sqrt{7}}{9 - 7} = \frac{15 - 5\sqrt{7}}{2}$

141. $\frac{(2 + \sqrt{3})(1 + \sqrt{5})}{(1 - \sqrt{5})(1 + \sqrt{5})} = \frac{2 + \sqrt{3} + 2\sqrt{5} + \sqrt{15}}{1 - 5} = \frac{2 + \sqrt{3} + 2\sqrt{5} + \sqrt{15}}{-4} =$
$-\frac{2 + \sqrt{3} + 2\sqrt{5} + \sqrt{15}}{4}$

142. $m = 49; \{49\}$ **143.** $6x = \frac{9}{4}; x = \frac{9}{4}\cdot\frac{1}{6} = \frac{3}{8}; \left\{\frac{3}{8}\right\}$ **144.** $\sqrt{a} = 9; a = 81; \{81\}$

145. $\sqrt{y} = \frac{4}{5}; y = \frac{16}{25}; \left\{\frac{16}{25}\right\}$ **146.** $\frac{x}{3} = 36; x = 108; \{108\}$ **147.** $n - 2 = 81; n = 83; \{83\}$

148. $\sqrt{5t} = 2; 5t = 4; t = \frac{4}{5}; \left\{\frac{4}{5}\right\}$ **149.** $\sqrt{3z} = 3; 3z = 9; z = 3; \{3\}$

150. $\sqrt{4k - 5} = 7; 4k - 5 = 49; 4k = 54; k = \frac{54}{4} = \frac{27}{2}; \left\{\frac{27}{2}\right\}$

151. $\sqrt{\frac{5u}{2}} = 1; \frac{5u}{2} = 1; 5u = 2; u = \frac{2}{5}; \left\{\frac{2}{5}\right\}$

152. $\frac{4c - 3}{7} = 9; 4c - 3 = 63; 4c = 66; c = \frac{66}{4} = \frac{33}{2}; \left\{\frac{33}{2}\right\}$ **153.** $\sqrt{n} = 3\sqrt{5}; n = 45; \{45\}$

Pages 662–664 • CHAPTER 12

1. $m = \pm\sqrt{\frac{25}{49}} = \pm\frac{5}{7}; \left\{\frac{5}{7}, -\frac{5}{7}\right\}$ **2.** $a^2 = 12; a = \pm\sqrt{12} = \pm 2\sqrt{3}; \{\pm 2\sqrt{3}\}$

3. $w^2 = -52$; no solution **4.** $x^2 = 108; x = \pm\sqrt{108} = \pm 6\sqrt{3}; \{\pm 6\sqrt{3}\}$

5. $7u^2 = 112; u^2 = 16; u = \pm 4; \{4, -4\}$ **6.** $4c^2 = 16; c^2 = 4; c = \pm 2; \{2, -2\}$

7. $3t^2 = 9; t^2 = 3; t = \pm\sqrt{3}; \{\pm\sqrt{3}\}$ **8.** $2n^2 = -5$; no solution

9. $v + 5 = \pm 4; v = 4 - 5$ or $v = -4 - 5; v = -1$ or $v = -9; \{-1, -9\}$

10. $z - 5 = \pm\sqrt{6}; z = 5 + \sqrt{6}$ or $z = 5 - \sqrt{6}; \{5 \pm \sqrt{6}\}$

11. $(k + 4)^2 = 27; k + 4 = \pm\sqrt{27}; k + 4 = \pm 3\sqrt{3}; k = -4 \pm 3\sqrt{3}; \{-4 \pm 3\sqrt{3}\}$

12. $(f - 1)^2 = 15; f - 1 = \pm\sqrt{15}; f = 1 \pm \sqrt{15}; \{1 \pm \sqrt{15}\}$

13. $(h + 7)^2 = 21; h + 7 = \pm\sqrt{21}; h = -7 \pm \sqrt{21}; \{-7 \pm \sqrt{21}\}$

14. $2x + 3 = \pm 10$; $2x = -3 + 10$ or $2x = -3 - 10$; $2x = 7$ or $2x = -13$; $x = \frac{7}{2}$ or $x = -\frac{13}{2}$; $\left\{\frac{7}{2}, -\frac{13}{2}\right\}$

15. $(3y - 1)^2 = 24$; $3y - 1 = \pm\sqrt{24} = \pm 2\sqrt{6}$; $3y = 1 \pm 2\sqrt{6}$; $\left\{\frac{1 \pm 2\sqrt{6}}{3}\right\}$

16. $(e + 3)^2 = 64$; $e + 3 = \pm 8$; $e = -3 \pm 8$; $\{5, -11\}$

17. $(a - 6)^2 = 49$; $a - 6 = \pm 7$; $a = 6 \pm 7$; $\{13, -1\}$

18. $(m + 9)^2 = 36$; $m + 9 = \pm 6$; $m = -9 \pm 6$; $\{-3, -15\}$

19. $x^2 + 16x + 64 = -15 + 64$; $(x + 8)^2 = 49$; $x + 8 = \pm 7$; $x = -8 \pm 7$; $\{-15, -1\}$

20. $y^2 - 8y = -7$; $y^2 - 8y + 16 = -7 + 16$; $(y - 4)^2 = 9$; $y - 4 = \pm 3$; $y = 4 \pm 3$; $\{7, 1\}$

21. $n^2 - 12n = 210$, $n^2 - 12n + 36 = 210 + 36$, $(n - 6)^2 = 246$; $n - 6 = \pm\sqrt{246}$, $n = 6 \pm \sqrt{246}$; $\{6 \pm \sqrt{246}\}$; $\{21.7, -9.7\}$

22. $a^2 + \frac{5}{2}a = 3$; $a^2 + \frac{5a}{2} + \frac{25}{16} = 3 + \frac{25}{16}$; $\left(a + \frac{5}{4}\right)^2 = \frac{73}{16}$; $a + \frac{5}{4} = \pm\sqrt{\frac{73}{16}} = \pm\frac{1}{4}\sqrt{73}$; $a = -\frac{5}{4} \pm \frac{1}{4}\sqrt{73}$; $\left\{\frac{-5 \pm \sqrt{73}}{4}\right\}$; $\{0.9, -3.4\}$

23. $b^2 - 3b = 5$; $b^2 - 3b + \frac{9}{4} = 5 + \frac{9}{4}$; $\left(b - \frac{3}{2}\right)^2 = \frac{29}{4}$; $b - \frac{3}{2} = \pm\frac{1}{2}\sqrt{29}$; $b = \frac{3}{2} \pm \frac{1}{2}\sqrt{29}$; $\left\{\frac{3 \pm \sqrt{29}}{2}\right\}$; $\{4.2, -1.2\}$

24. $c^2 + 2c = 411$; $c^2 + 2c + 1 = 412$; $(c + 1)^2 = 412$; $c + 1 = \sqrt{412} = 2\sqrt{103}$; $c = -1 \pm 2\sqrt{103}$; $\{-1 \pm 2\sqrt{103}\}$; $\{-21.3, 19.3\}$

25. **a.** $e^2 - 10e = -21$; $e^2 - 10e + 25 = -21 + 25$; $(e - 5)^2 = 4$; $e - 5 = \pm 2$; $e = 5 \pm 2$; $\{7, 3\}$

b. $e^2 - 10e + 21 = 0$; $(e - 7)(e - 3) = 0$; $\{7, 3\}$

26. **a.** $f^2 - \frac{9}{2}f = \frac{5}{2}$; $f^2 - \frac{9}{2}f + \frac{81}{16} = \frac{5}{2} + \frac{81}{16}$; $\left(f - \frac{9}{4}\right)^2 = \frac{121}{16}$; $f - \frac{9}{4} = \pm\frac{11}{4}$; $f = \frac{9}{4} \pm \frac{11}{4}$; $\left\{5, -\frac{1}{2}\right\}$

b. $4f^2 - 18f - 10 = 0$; $2(2f + 1)(f - 5) = 0$; $f = 5$ or $f = -\frac{1}{2}$; $\left\{5, -\frac{1}{2}\right\}$

27. **a.** $h^2 + \frac{3}{2}h = 7$; $h^2 + \frac{3}{2}h + \frac{9}{16} = 7 + \frac{9}{16}$; $\left(h + \frac{3}{4}\right)^2 = \frac{121}{16}$; $h + \frac{3}{4} = \pm\frac{11}{4}$; $h = -\frac{3}{4} \pm \frac{11}{4}$; $\left\{2, -\frac{7}{2}\right\}$

b. $6h^2 + 9h - 42 = 0$; $3(h - 2)(2h + 7) = 0$; $h = 2$ or $h = -\frac{7}{2}$; $\left\{2, -\frac{7}{2}\right\}$

28. $m^2 - 6m = 21$; $m^2 - 6m + 9 = 30$; $(m - 3)^2 = 30$; $m - 3 = \pm\sqrt{30}$; $\{3 \pm \sqrt{30}\}$

29. $n^2 + \frac{n}{2} = 5$; $n^2 + \frac{n}{2} + \frac{1}{16} = 5 + \frac{1}{16}$; $\left(n + \frac{1}{4}\right)^2 = \frac{81}{16}$; $n + \frac{1}{4} = \pm\frac{9}{4}$; $n = -\frac{1}{4} \pm \frac{9}{4}$; $n = 2$ or $n = -\frac{5}{2}$; $\left\{2, -\frac{5}{2}\right\}$

30. $x^2 - \frac{x}{2} = 4; x^2 - \frac{x}{2} + \frac{1}{16} = 4 + \frac{1}{16}; \left(x - \frac{1}{4}\right)^2 = \frac{65}{16}; x - \frac{1}{4} = \pm\frac{1}{4}\sqrt{65}; x = \frac{1}{4} \pm \frac{1}{4}\sqrt{65};$ $\left\{\frac{1 \pm \sqrt{65}}{4}\right\}$

31–39. Answers may vary due to rounding.

31. $z = \frac{-7 \pm \sqrt{49 - 4(1)(3)}}{2} = \frac{-7 \pm \sqrt{49 - 12}}{2} = \frac{-7 \pm \sqrt{37}}{2} \approx \frac{-7 \pm 6.08}{2} = -6.5 \text{ or } -0.5;$ $\left\{\frac{-7 \pm \sqrt{37}}{2}\right\}; \{-6.5, -0.5\}$

32. $w = \frac{-8 \pm \sqrt{64 - 4(-4)}}{2} = \frac{-8 \pm \sqrt{64 + 16}}{2} = \frac{-8 \pm \sqrt{80}}{2} = \frac{-8 \pm 4\sqrt{5}}{2} = -4 \pm 2\sqrt{5} \approx$ $-4 \pm 4.48 = -8.5 \text{ or } 0.5; \{-4 \pm 2\sqrt{5}\}; \{-8.5, 0.5\}$

33. $u = \frac{10 \pm \sqrt{100 - 4(2)(-6)}}{4} = \frac{10 \pm \sqrt{100 + 48}}{4} = \frac{10 \pm \sqrt{148}}{4} = \frac{10 \pm 2\sqrt{37}}{4} = \frac{5 \pm \sqrt{37}}{2} \approx$ $\frac{5 \pm 6.08}{2} = 5.5 \text{ or } -0.5; \left\{\frac{5 \pm \sqrt{37}}{2}\right\}; \{5.5, -0.5\}$

34. $5y^2 + 9y + 1 = 0; y = \frac{-9 \pm \sqrt{81 - 4(5)(1)}}{10} = \frac{-9 \pm \sqrt{81 - 20}}{10} = \frac{-9 \pm \sqrt{61}}{10} \approx \frac{-9 \pm 7.81}{10} =$ $-0.1 \text{ or } -1.7; \left\{\frac{-9 \pm \sqrt{61}}{10}\right\}; \{-0.1, -1.7\}$

35. $3k^2 - 5k + 2 = 0; k = \frac{5 \pm \sqrt{25 - 4 \cdot 3 \cdot 2}}{6} = \frac{5 \pm \sqrt{25 - 24}}{6} = \frac{5 \pm 1}{6}; \left\{1, \frac{2}{3}\right\}$

36. $2m^2 + 6m - 3 = 0; m = \frac{-6 \pm \sqrt{36 - 4(2)(-3)}}{4} = \frac{-6 \pm \sqrt{36 + 24}}{4} = \frac{-6 \pm \sqrt{60}}{4} =$ $\frac{-6 \pm 2\sqrt{15}}{4} = \frac{-3 \pm \sqrt{15}}{2} \approx \frac{-3 \pm 3.87}{2} = -3.4 \text{ or } 0.4; \left\{\frac{-3 \pm \sqrt{15}}{2}\right\}; \{-3.4, 0.4\}$

37. $10x^2 + 3x - 2 = 0; x = \frac{-3 \pm \sqrt{9 - 4(10)(-2)}}{20} = \frac{-3 \pm \sqrt{9 + 80}}{20} = \frac{-3 \pm \sqrt{89}}{20} \approx \frac{-3 \pm 9.43}{20} =$ $-0.6 \text{ or } 0.3; \left\{\frac{-3 \pm \sqrt{89}}{20}\right\}; \{-0.6, 0.3\}$

38. $6n^2 + 4n - 3 = 0; n = \frac{-4 \pm \sqrt{16 - 4(6)(-3)}}{12} = \frac{-4 \pm \sqrt{16 + 72}}{12} = \frac{-4 \pm 2\sqrt{22}}{12} =$ $\frac{-2 \pm \sqrt{22}}{6} \approx \frac{-2 \pm 4.69}{6} = -1.1 \text{ or } 0.4; \left\{\frac{-2 \pm \sqrt{22}}{6}\right\}; \{-1.1, 0.4\}$

39. $y^2 - 7y - 2 = 0; y = \frac{7 \pm \sqrt{49 - 4(1)(-2)}}{2} = \frac{7 \pm \sqrt{49 + 8}}{2} = \frac{7 \pm \sqrt{57}}{2} \approx \frac{7 \pm 7.55}{2} = 7.3 \text{ or}$ $-0.3; \left\{\frac{7 \pm \sqrt{57}}{2}\right\}; \{7.3, -0.3\}$

40. discriminant $= 36 - 4 \cdot 1 \cdot 2 = 36 - 8 = 28$; 2 real roots

41. discriminant $= 9 - 4 \cdot 5 \cdot 7 = 9 - 140 = -131$; no real roots

42. discriminant $= 16 - 4(-2)(-2) = 16 - 16 = 0$; one real root

43. discriminant $= 1.44 - 4 \cdot 3 \cdot 1.1 = 1.44 - 13.2 = -11.76$; no real roots

44. discriminant $= 36 - 4 \cdot 3 \cdot 3 = 0$; one real root

45. discriminant $= 1 - (4)\left(\frac{1}{3}\right)(-3) = 1 + 4 = 5$; 2 real roots

46. discriminant $= 4 - 4(3)(-5) = 64$ **a.** 2 x-intercepts **b.** vertex lies below the x-axis

47. discriminant $= 9 - 4(-2)(-6) = -39$ **a.** no x-intercepts **b.** vertex lies below the x-axis

48. discriminant $= 16 - 4(1)(16) = 16 - 64 = 48$ **a.** no x-intercepts
b. vertex lies above the x-axis

49. $(x + 4)(x + 3) = 0; x = -4$ or $-3; \{-4, -3\}$ **50.** $x^2 = 4; x = \pm 2; \{2, -2\}$

51. $x(5x - 9) = 0; x = 0$ or $\frac{9}{5}; \left\{0, \frac{9}{5}\right\}$

52. $3x^2 - 11x - 2 = 0; x = \frac{11 \pm \sqrt{121 - 4 \cdot 3(-2)}}{6} = \frac{11 \pm \sqrt{145}}{6}; \left\{\frac{11 \pm \sqrt{145}}{6}\right\}$

53. $x = \frac{-8 \pm \sqrt{64 - 4 \cdot 1 \cdot 3}}{2} = \frac{-8 \pm \sqrt{64 - 12}}{2} = \frac{-8 \pm \sqrt{52}}{2} = \frac{-8 \pm 2\sqrt{13}}{2} = -4 \pm \sqrt{13};$
$\{-4 \pm \sqrt{13}\}$

54. $x - 3 = \pm\sqrt{6}; x = 3 \pm \sqrt{6}; \{3 \pm \sqrt{6}\}$

55. $6x^2 + 4x - 1 = 0; x = \frac{-4 \pm \sqrt{16 - 4 \cdot 6(-1)}}{12} = \frac{-4 \pm \sqrt{16 + 24}}{12} = \frac{-4 \pm \sqrt{40}}{12} =$
$\frac{-4 \pm 2\sqrt{10}}{12} = \frac{-2 \pm \sqrt{10}}{6}; \left\{\frac{-2 \pm \sqrt{10}}{6}\right\}$

56. $(x + 4)^2 = 24; x + 4 = \pm\sqrt{24}; x = -4 \pm 2\sqrt{6}; \{-4 \pm 2\sqrt{6}\}$

57. $9x^2 - 8x - 12 = 0; x = \frac{8 \pm \sqrt{64 - 4 \cdot 9(-12)}}{18} = \frac{8 \pm \sqrt{64 + 432}}{18} = \frac{8 \pm \sqrt{496}}{18} =$
$\frac{8 \pm 4\sqrt{31}}{18} = \frac{4 \pm 2\sqrt{31}}{9}; \left\{\frac{4 \pm 2\sqrt{31}}{9}\right\}$

58. $3 = 6x^2 - 4x; 6x^2 - 4x - 3 = 0; x = \frac{4 \pm \sqrt{16 - 4 \cdot 6(-3)}}{12} = \frac{4 \pm \sqrt{16 + 72}}{12} = \frac{4 \pm \sqrt{88}}{12} =$
$\frac{4 \pm 2\sqrt{22}}{12} = \frac{2 \pm \sqrt{22}}{6}; \left\{\frac{2 \pm \sqrt{22}}{6}\right\}$

59. $12x^2 - 4x - 2 = 0; x = \frac{4 \pm \sqrt{16 - 4 \cdot 12(-2)}}{24} = \frac{4 \pm \sqrt{16 + 96}}{24} = \frac{4 \pm \sqrt{112}}{24} = \frac{4 \pm 4\sqrt{7}}{24} =$
$\frac{1 \pm \sqrt{7}}{6}; \left\{\frac{1 \pm \sqrt{7}}{6}\right\}$

60. $6x^2 - 7x + 2 = 4x^2 + 7x + 3; 2x^2 - 14x - 1 = 0; x = \frac{14 \pm \sqrt{196 - 4(2)(-1)}}{4} =$
$\frac{14 \pm \sqrt{196 + 8}}{4} = \frac{14 \pm \sqrt{204}}{4} = \frac{14 \pm 2\sqrt{51}}{4} = \frac{7 \pm \sqrt{51}}{2}; \left\{\frac{7 \pm \sqrt{51}}{2}\right\}$

61. $4x^2 - 8x + 3x + 24 = 27 + 5x^2; x^2 + 5x + 3 = 0; x = \frac{-5 \pm \sqrt{25 - 4 \cdot 3}}{2} = \frac{-5 \pm \sqrt{13}}{2};$
$\left\{\frac{-5 \pm \sqrt{13}}{2}\right\}$

62. $x^2 + 12x + 36 + 2x - 2 = 13; x^2 + 14x + 21 = 0; x = \frac{-14 \pm \sqrt{196 - 4 \cdot 21}}{2} =$

$\frac{-14 \pm \sqrt{196 - 84}}{2} = \frac{-14 \pm \sqrt{112}}{2} = \frac{-14 \pm 4\sqrt{7}}{2} = -7 \pm 2\sqrt{7}; \{-7 \pm 2\sqrt{7}\}$

63. Let w = width; then length = $6w$. $6w^2 = 84$; $w^2 = 14$; $w = \pm\sqrt{14} \approx \pm 3.7$. The length is approximately 22.2 cm; the width is approximately 3.7 cm.

64. Let n = number. $n^2 - n = 56$; $n^2 - n - 56 = 0$; $(n - 8)(n + 7) = 0$; $n = 8$ or -7. The number is either 8 or -7.

65. Let b = base; then altitude = $b - 2$. $\frac{1}{2}b(b - 2) = 84$; $b^2 - 2b = 168$; $b^2 - 2b - 168 = 0$;

$(b - 14)(b + 12) = 0$; $b = 14$ or -12. The base is 14 m.

66. Let x = number of inches added. $(30 + x)(40 + x) = 2(30)(40)$, $1200 + 70x + x^2 = 2400$,

$x^2 + 70x - 1200 = 0; x = \frac{-70 \pm \sqrt{4900 - 4(-1200)}}{2} = \frac{-70 \pm \sqrt{9700}}{2} = \frac{-70 \pm 10\sqrt{97}}{2} \approx$

$\frac{-70 \pm 98.5}{2} = 14.2$. The new dimensions will be 44.2 in. by 54.2 in.

67. $hr^2 = k$ **68.** $p = kv^2$ or $\frac{p}{v^2} = k$ **69.** $L = krh$ **70.** $V = khr^2$

71. $rt = kd$ or $\frac{rt}{d} = k$ **72.** $Fr = kv^2$ or $\frac{Fr}{v^2} = k$

Page 665 • CHAPTER 1

(1-7) **1.** *Step 1:* Problem asks for distance train travels. *Step 2:* Let d = distance.

Step 3: $d = 90(2.5)$ *Step 4:* $d = 225$ *Step 5:* $90 \times 2.5 = 225$

The train travels 225 km.

2. *Step 1:* Problem asks for a number. *Step 2:* Let n = number.

Step 3: $n - 27 = 36$ *Step 4:* $n = 36 + 27 = 63$ *Step 5:* $63 - 27 = 36$

The number is 63.

3. *Step 1:* Problem asks for number of games team won.

Step 2: Let l = number of games lost; then $2l$ = number of games won.

Step 3: $l + 2l = 12$ *Step 4:* $3l = 12; l = 4$ *Step 5:* $4 + 8 = 12$

The team won 8 games.

4. *Step 1:* Problem asks for number of records sold first day.

Step 2: Let f = number of records sold first day; then $2f$ = number of records sold second day.

Step 3: $f + 2f = 102$ *Step 4:* $3f = 102; f = 34$ *Step 5:* $34 + 68 = 102$

34 albums were sold on the first day.

5. *Step 1:* Problem asks for a number. *Step 2:* Let n = number.

Step 3: $3n + 11 = 68$ *Step 4:* $3n = 57; n = 19$

Step 5: $3 \cdot 19 + 11 = 57 + 11 = 68$

The number is 19.

6. *Step 1:* Problem asks for amount of money in bank.

Step 2: Let q = number of quarters; then $q + 8$ = number of dimes; $q + 13$ = number of nickels.

Step 3: $q + (q + 8) + (q + 13) = 57$ *Step 4:* $3q + 21 = 57; 3q = 36; q = 12$

Step 5: $12 + (20) + (25) = 57$

There are 12 quarters, 20 dimes, and 25 nickels for a total of $6.25.

Pages 665–666 • CHAPTER 2

(2-3) **1.** $-23 - 15 = -38$. It was 38 yd from the position it was in before the penalty.

2. $26 + 8 - 12 + 4 = 26$. It was on the 26th floor.

3. $250 + 4(10) - 85 = 250 + 40 - 85 = 205$. There was $205 at the end of the month.

4. $500 - (280 + 85 + 124) = 500 - 489 = 11$. It must collect $11 more.

5. $8 + 3 - 1 = 8 + 2 = 10$. It was 10 A.M. **6.** $100 - 5 = 95$ km/h

(2-4) **1.** $-246.09 - (-248.61) = 248.61 - 246.09 = 2.52$. The difference between boiling point and freezing point is 2.52°C.

2. $4418 - (-86) = 4418 + 86 = 4504$. The difference in altitude is 4504 m.

3. $16 - (-12) = 12 + 16 = 28$. She has gone 28 blocks.

4. $88 - (-12) = 100$. The difference is 100°.

5. $8848 - 9245 = -397$. The Dead Sea is 397 m below sea level.

6. $18.3 - 22.7 = -4.4$. The average temperature is −4.4°C.

Pages 666–668 • CHAPTER 3

(3-1) **1.** Let n = number. $n + 13 = -5$; $n = -18$. The number is −18.

2. Let s = cost of sandwich. $70 + s = 250$; $s = 180$. A sandwich costs $1.80.

3. Let n = number. $n - 15 = 43$; $n = 58$. The number is 58.

4. Let p = plane's speed in still air. $970 - p = 145$; $p = 825$. The speed of the plane in still air is 825 km/h.

5. $425 + 642 - 320 = 1067 - 320 = 747$. They had $747 left.

6. $76 + 13 - 7 = 82$. 82 people went.

(3-2) **1.** Let x = number. $-(7x) = 238$; $-7x = 238$; $x = -34$. The number is −34.

2. Let x = number. $\frac{1}{4}x = 73$; $x = 292$. The number is 292.

3. Let c = price per kg. $2.5c = 1.40$; $c = 0.56$. The apples cost $.56 per kg.

4. Let w = number of hours worked per week. $8x = 68$; $x = 8\frac{1}{2}$. He worked $8\frac{1}{2}$ h each week.

5. Let w = width. $2(24) + 2w = 72$; $2w = 72 - 48$; $2w = 24$; $w = 12$. The width is 12 cm.

6. Let x = cost per piece of a 6-piece pizza. $8(0.90) = 6x$; $6x = 7.2$; $x = 1.2$. Each slice would cost $1.20.

(3-4) **1.** Let x = number. $15x - 34 = 146$; $15x = 180$; $x = 12$. The number is 12.

2. Let l = length. $2(35) + 2l = 152$; $70 + 2l = 152$; $2l = 82$; $l = 41$. The length is 41 cm.

3. Let c = cost of desk in cents. $\frac{1}{16}c - 31 + c = 13{,}144$; $\frac{17}{16}c = 13{,}175$;

$c = (13{,}175)\frac{16}{17} = (775) \cdot 16 = 12{,}400$. The desk costs $124.

4. Let t = number of seats in Twin Cinema II; then $t + 150$ = number of seats in Twin Cinema I. $t + t + 150 = 1250$; $2t = 1100$; $t = 550$. There are 550 seats in Twin Cinema II.

5. Let q = number of quarters; then $4 + q$ = number of dimes and $2q$ = number of nickels. $q + 4 + q + 2q = 36$; $4q = 32$; $q = 8$. There are 8 quarters, 12 dimes, and 16 nickels.

6. Let l = length of longest side; then $l - 8$ = length of shortest side and $l - 5$ = length of third side. $l + (l - 8) + (l - 5) = 56$; $l + l - 8 + l - 5 = 56$; $3l - 13 = 56$; $3l = 69$; $l = 23$. The sides are 23 cm, 15 cm, and 18 cm.

(3-5) **1.** Let n, $n + 1$ be the integers. $n + 1 = 2n + 10$; $n = -9$. The integers are −9 and −8.

2. Let n = number. $6n = n + 45$; $5n = 45$; $n = 9$. The number is 9.

3. Let n = number. $5n + 3 = 3n + 27$; $2n = 24$; $n = 12$. The number is 12.

4. Let n and $20 - n$ be the numbers. $2n = 4(20 - n) + 4$; $2n = 80 - 4n + 4$; $6n = 84$; $n = 14$. The numbers are 14 and 6.

5. Let n, $n + 2$, and $n + 4$ = lengths of sides. $n + (n + 2) + (n + 4) = 4n - 1$; $3n + 6 = 4n - 1$; $n = 7$. The sides are 7, 9, and 11.

6. Let s = cost of salad in cents; then $20 + s$ = cost of sandwich. $6(20 + s) = 7s$; $120 + 6s = 7s$; $s = 120$. A salad plate costs \$1.20, a sandwich costs \$1.40.

(3-6) 1.

	Hours worked now	Hours worked then
Kevin	$3x$	$3x + 6$
Karen	x	$x + 6$

$3x + 6 = 2(x + 6)$, $3x + 6 = 2x + 12$, $x = 6$; $3x = 18$; Karen works 6 h and Kevin works 18 h a week.

2. Let a = number of hours Aaron works; then Betsy works $a + 4$ hours and Charita works $2[a + (a + 4)]$ hours. $a + (a + 4) + 2[a + (a + 4)] = 45$;

$a + a + 4 + 2(2a + 4) = 45$; $2a + 4 + 4a + 8 = 45$; $6a = 33$; $a = 5\frac{1}{2}$. Aaron works $5\frac{1}{2}$ h, Betsy works $9\frac{1}{2}$ h and Charita works 30 h.

3. Let q = Zach's first quiz score. $q + (2q - 30) = 150$, $3q - 30 = 150$, $3q = 180$, $q = 60$; Zach's first quiz score was 60.

4.

	Length	Width
First rectangle	$w + 18$	w
Second rectangle	$w + 12$	$w + 3$

$2(w + 12) + 2(w + 3) = 126$; $2w + 24 + 2w + 6 = 126$; $4w = 96$; $w = 24$. The first rectangle is 42 cm by 24 cm; the second is 36 cm by 27 cm.

5. Let b = number of dimes Becky has; then $\frac{1}{3}b$ = number Amy has and $\frac{1}{3}b + 2$ = number Ryan has. $b = \frac{1}{3}b + \frac{1}{3}b + 2$, $b = \frac{2}{3}b + 2$, $\frac{1}{3}b = 2$, $b = 6$; $\frac{1}{3}(6) = 2$; $2 + 2 = 4$; Becky has 6 dimes, Amy has 2 and Ryan has 4.

6. Let w = number of calories in one cup of whole milk; then $\frac{1}{2}w + 10$ = number of calories in one cup of skim milk and $w - 40$ = number of calories in one cup of apple juice. $w + \frac{1}{2}w + 10 + w - 40 = 370$; $2w + w + 20 + 2w - 80 = 740$; $5w = 800$; $w = 160$. There are 160 calories in one cup of whole milk, 120 calories in one cup of apple juice and 90 calories in one cup of skim milk.

(3-7) 1.

	Number	× Value	= Total value
Dimes	$3q - 4$	10	$10(3q - 4)$
Quarters	q	25	$25q$

$10(3q - 4) + 25q = 675$, $30q - 40 + 25q = 675$; $55q = 715$; $q = 13$. There are 13 quarters and 35 dimes.

2.

	Number ×	Cost of ticket =	Total Cost
Students	x	1.50	$1.5(x)$
Adults	$720 - x$	2.50	$2.5(720 - x)$

$1.5x + 2.5(720 - x) = 1220$; $15x + 18{,}000 - 25x = 12{,}200$; $10x = 5800$; $x = 580$; 580 students and 140 adults attended.

3. Let x = number of hours in a regular workday; then $11 - x$ = number of hours overtime. $9x + 13.5(11 - x) = 114.75$; $900x + 1350(11 - x) = 11{,}475$; $900x + 14{,}850 - 1350x = 11{,}475$; $450x = 3375$; $x = 7.5$. The length of a regular workday is $7\frac{1}{2}$ h.

4. Let c = number of kg of carrots; then $9 - c$ = number of kg of potatoes. $0.75c + 0.70(9 - c) = 6.60$; $75c + 70(9 - c) = 660$; $75c + 630 - 70c = 660$; $5c = 30$; $c = 6$. Shopper bought 6 kg of carrots and 3 kg of potatoes.

5.

	Number	× Value =	Total value
Quarters	$102 - (2x + 14)$	25	$25[102 - (2x + 14)]$
Nickels	$x + 14$	5	$5(x + 14)$
Dimes	x	10	$10x$

$25(102 - 2x - 14) + 5(x + 14) + 10x = 1360$;
$2200 - 50x + 5x + 70 + 10x = 1360$; $910 = 35x$; $x = 26$. There are 36 quarters.

Pages 668–669 • CHAPTER 4

(4-8) 1.

	Rate ×	Time =	Distance
Train 1	80	x	$80x$
Train 2	90	x	$90x$

$80x + 90x = 425$; $170x = 425$; $x = 2.5$. They will be 425 km apart in 2.5 h.

2.

	Rate ×	Time =	Distance
Grace	5	$x + \frac{1}{6}$	$5\left(x + \frac{1}{6}\right)$
Will	15	x	$15x$

$5x + \frac{5}{6} = 15x$; $30x + 5 = 90x$; $60x = 5$; $x = \frac{1}{12}$. It will take Will $\frac{1}{12}$ h or 5 min to catch up to Grace.

3.

	Rate ×	Time =	Distance
Out	800	$x + 1$	$800(x + 1)$
Back	1200	x	$1200x$

$800(x + 1) = 1200x$;
$800x + 800 = 1200x$; $400x = 800$; $x = 2$. The distance from Lincoln to Adams is 2400 km.

4.

	Rate	× Time	= Distance
Bike	$x + 10$	$\frac{1}{6}$	$\frac{1}{6}(x + 10)$
Walk	x	$\frac{1}{4}$	$\frac{1}{4}x$

$\frac{1}{6}(x + 10) + \frac{1}{4}x = 2.75$;

$2(x + 10) + 3x = 33$;
$2x + 20 + 3x = 33$; $5x = 13$; $x = 2.6$. It is 0.65 km from his friend's house to the gym.

5.

	Rate	× Time	= Distance
Down	10	$\frac{20}{10}$	20
Back	4	$\frac{20}{4}$	20

Time = 2 + 5 = 7 h. She returned at 7 P.M.

(4-9) **1.** Let w = width; then $w + 4$ = length; $w + 5$ = new width and $w + 9$ = new length. $w(w + 4) + 115 = (w + 5)(w + 9)$; $w^2 + 4w + 115 = w^2 + 14w + 45$; $10w = 70$; $w = 7$. The original dimensions are 7 m by 11 m.

2. Let s = side of square; then $s + 3$ = length of rectangle, $s - 2$ = width of rectangle. $s^2 = (s - 2)(s + 3)$; $s^2 = s^2 + s - 6$; $s = 6$. The square is 6 cm on a side; the rectangle is 9 cm by 4 cm.

3. Let w = width of the pool; then $w + 4$ = length. $(w + 2)(w + 6) - w(w + 4) = 32$; $w^2 + 8w + 12 - w^2 - 4w = 32$; $4w = 20$; $w = 5$. The pool is 5 m by 9 m.

4. Let s = side of square; then $s + 6$ = length of rectangle and $s - 4$ = width. $s^2 = (s + 6)(s - 4)$; $s^2 = s^2 + 2s - 24$; $2s = 24$; $s = 12$. The square is 12 by 12.

5. Let w = the width of the print; then $w + 10$ is the length. $(w + 3)(w + 13) - w(w + 10) = 399$; $w^2 + 16w + 39 - w^2 - 10w = 399$; $6w = 360$, $w = 60$. The print is 60 cm by 70 cm.

(4-10) **1.** Let n, $n + 1$ = the integers. $n + n + 1 = 104$; $2n = 103$; no solution

2.

	Rate	× Time	= Distance
First half	1000	x	$1000x$
Second half	600	$6 - x$	$600(6 - x)$

$1000x = 3600 - 600x$;
$1600x = 3600$; $x = 2.25$. The total distance was 4500 km.

3.

	Quiz Score	Quiz + Bonus
Jill	$x + 12$	$x + 20$
Jack	x	$x + 8$

$x + 20 = 3(x + 8)$, $x + 20 = 3x + 24$,
$2x = -4$, $x = -2$; no solution.

4. Let s = side of smaller square; then $s + 2$ = side of larger square. $(s + 2)^2 = s^2 + 220$; $s^2 + 4s + 4 = s^2 + 220$; $4s = 216$; $s = 54$. The larger square is 56 cm on a side, the smaller, 54 cm.

5. Let n, $n + 1$, $n + 2$ = the integers. $n + n + 1 + n + 2 = 4(n + 2)$; $3n + 3 = 4n + 8$; $n = -5$. The integers are -5, -4, and -3.

Pages 669–670 • CHAPTER 5

(5-13) 1. Let n = the number. $n^2 + n = 132$; $n^2 + n - 132 = 0$; $(n + 12)(n - 11) = 0$; $n = 11$ or $n = -12$. The number is -12 or 11.

2. Let n, $n + 2$ = the integers. $n^2 + (n + 2)^2 = 202$; $n^2 + n^2 + 4n + 4 = 202$; $2n^2 + 4n - 198 = 0$; $n^2 + 2n - 99 = 0$; $(n + 11)(n - 9) = 0$; $n = -11$ or $n = 9$. The integers are 9 and 11.

3. Let w = width, then $8 + w$ = length. $w(8 + w) = 240$; $8w + w^2 = 240$; $w^2 + 8w - 240 = 0$; $(w + 20)(w - 12) = 0$; $w = -20$ or $w = 12$. The rectangle is 12 cm by 20 cm.

4. Let n, $12 - n$ = numbers. $n^2 + (12 - n)^2 = 74$; $n^2 + 144 - 24n + n^2 = 74$; $2n^2 - 24n + 70 = 0$; $n^2 - 12n + 35 = 0$; $(n - 5)(n - 7) = 0$; $n = 5$ or $n = 7$. The numbers are 5 and 7.

5. Let x = width of the strip of grass. $\frac{1}{6}(16 \cdot 12) = (16 - 2x)(12 - 2x)$; $32 = 192 - 56x + 4x^2$; $4x^2 - 56x + 160 = 0$; $x^2 - 14x + 40 = 0$; $(x - 10)(x - 4) = 0$; $x = 4$ or $x = 10$. The strip is 4 m wide; thus, the garden is 4 m by 8 m.

6. Let e = edge of smaller cube; then $e + 4$ = edge of larger cube. $(e + 4)^3 - e^3 = 316$; $(e^2 + 8e + 16)(e + 4) - e^3 = 316$; $e^3 + 12e^2 + 48e + 64 - e^3 = 316$; $12e^2 + 48e - 252 = 0$; $e^2 + 4e - 21 = 0$; $(e + 7)(e - 3) = 0$; $e = 3$ or $e = -7$. Each edge of the first cube is 3 cm; each edge of the second cube is 7 cm.

Pages 670–674 • CHAPTER 7

(7-1) 1. Let $2x$, $3x$ = the numbers. $2x + 3x = 125$; $5x = 125$; $x = 25$. The numbers are 50 and 75.

2. Let $2x$, $3x$ and $5x$ = measures of angles. $2x + 3x + 5x = 180$; $10x = 180$; $x = 18$. The angles measure 36, 54, and 90.

3. Let $2x$, $3x$ and $5x$ = the numbers. $2x + 3x + 5x = 200$; $10x = 200$; $x = 20$. The numbers are 40, 60 and 100.

4. Let $2x$ = number of teachers, x = number of assistants, $9x$ = number of children. $2x + x + 9x = 96$; $12x = 96$; $x = 8$. There are 72 children at the center.

5. Let $5x$ = number of quarters, $3x$ = number of dimes, and $7x$ = number of nickels. $25(5x) + 10(3x) + 5(7x) = 2280$; $125x + 30x + 35x = 2280$; $190x = 2280$; $x = 12$. $5(12) + 3(12) + 7(12) = 180$; There are 180 coins.

6.

	Rate	× Time	= Distance
Train 1	$22x$	2	$44x$
Train 2	$25x$	2	$50x$

$44x + 50x = 376$; $94x = 376$; $x = 4$. The trains travel at 88 km/h and 100 km/h.

(7-2) 1. $\frac{1.5}{5.8} = \frac{2}{x}$, $x = \frac{2(5.8)}{1.5} = \frac{11.6}{1.5} \approx 7.73$; A 2-lb steak costs \$7.73.

2. $\frac{400}{625} = \frac{x}{7500}$; $x = \frac{400 \cdot 7500}{625}$; $x = 4800$. 4800 voters can be expected to vote in favor of Question 1.

3. $\frac{42}{140.70} = \frac{50}{x}$; $x = \frac{50(140.70)}{42} = 167.5$. It would have cost \$167.50 for a group of 50.

4. $\frac{1.44}{24} = \frac{x}{35}$; $x = \frac{35(1.44)}{24} = 2.1$. The tax on a \$35 meal is \$2.10.

5. $\frac{120}{116} = \frac{x}{174}$; $x = \frac{120 \cdot 174}{116} = 180$. It will register 180 lb.

6. $\frac{1}{25} = \frac{25.8}{x}$; $\frac{1}{25} = \frac{18.4}{y}$. $x = 25(25.8) = 645$. $y = 25(18.4) = 460$. The approximate area of Colorado is 645 km × 460 km or 296,700 km^2.

(7-3) 1.

	Amount spent	\$4 less
Juan	$x + 2$	$x - 2$
Sylvia	x	$x - 4$

$x - 4 = \frac{5}{6}(x - 2)$, $6x - 24 = 5x - 10$, $x = 14$; Sylvia spent \$14 and Juan spent \$16.

2. Let n = number. $\frac{3}{5}n + \frac{1}{4}n = 51$; $12n + 5n = 1020$; $17n = 1020$; $n = 60$.

3.

	Now	7 years ago
Bart	$\frac{1}{3}x$	$\frac{1}{3}x - 7$
Mother	x	$x - 7$

$\frac{1}{3}x - 7 = \frac{1}{5}(x - 7)$; $5x - 105 = 3x - 21$; $2x = 84$; $x = 42$. Bart is 14, his mother is 42.

4. Let l = length of rectangle; then $l - 11$ = width. $l = \frac{2}{7}[2l + 2(l - 11)]$; $l = \frac{2}{7}(2l + 2l - 22)$; $l = \frac{2}{7}(4l - 22)$; $7l = 8l - 44$; $l = 44$. The length is 44 cm; the width is 33 cm.

5.

	Number ×	Value =	Total value
Quarters	$\frac{2}{3}x$	25	$\frac{50x}{3}$
Dimes	$\frac{1}{3}x$	10	$\frac{10x}{3}$

$\frac{50x}{3} + \frac{10x}{3} = 1200$; $50x + 10x = 3600$; $60x = 3600$; $x = 60$. There are 20 dimes.

6.

	Rate ×	Time =	Distance
Bus	90	$x + 15.2$	$90(x + 15.2)$
Plane	850	x	$850x$

$850x = 90x + 1368$; $760x = 1368$; $x = 1.8$. The distance from Ardmore to Zepher is 1530 km.

(7-4) 1. Let x = number. $x + \frac{1}{x} = \frac{25}{12}$; $12x^2 + 12 = 25x$; $12x^2 - 25x + 12 = 0$; $(4x - 3)(3x - 4) = 0$; $x = \frac{3}{4}$ or $\frac{4}{3}$. The number is $\frac{3}{4}$ or $\frac{4}{3}$.

2. Let x = number. $x + \frac{1}{x} = \frac{29}{10}$; $10x^2 + 10 = 29x$; $10x^2 - 29x + 10 = 0$; $(5x - 2)(2x - 5) = 0$; $x = \frac{2}{5}$ or $\frac{5}{2}$. The number is $\frac{2}{5}$ or $\frac{5}{2}$.

3.

	Denominator	Numerator
Original fraction	$x + 2$	x
New fraction	$x + 4$	$x + 2$

$\frac{x}{x+2} + \frac{4}{15} = \frac{x+2}{x+4}$; $15x(x + 4) + 4(x + 2)(x + 4) = 15(x + 2)(x + 2)$;

$4x^2 + 24x - 28 = 0$; $x^2 + 6x - 7 = 0$; $(x + 7)(x - 1) = 0$; $x = -7$ or $x = 1$. The original fraction is $\frac{1}{3}$.

4. $\frac{x}{x+2} + \frac{x+2}{x} = \frac{34}{15}$, $15x^2 + 15(x + 2)^2 = 34x(x + 2)$, $15x^2 + 15(x^2 + 4x + 4) = 34x^2 + 68x$, $15x^2 + 15x^2 + 60x + 60 = 34x^2 + 68x$, $4x^2 + 8x - 60 = 0$, $x^2 + 2x - 15 = 0$, $(x + 5)(x - 3) = 0$; $x = -5$ or 3; The fraction is $\frac{3}{5}$.

5.

	Rate	× Time	= Distance
Usual	x	$\frac{100}{x}$	100
New	$x - 10$	$\frac{100}{x-10}$	100

$\frac{100}{x} + \frac{1}{2} = \frac{100}{x-10}$; $200(x - 10) + x(x - 10) = 200x$;

$200x - 2000 + x^2 - 10x = 200x$; $x^2 - 10x - 2000 = 0$; $(x - 50)(x + 40) = 0$; $x = 50$ or $x = -40$. The usual speed limit is 50 km/h.

6.

	Rate	× Time	= Distance
Walking	x	$\frac{2.5}{x}$	2.5
Riding	$x + 14$	$\frac{2.5}{x+14}$	2.5

$\frac{2.5}{x} - \frac{17.5}{60} = \frac{2.5}{x+14}$; $60(x + 14)(2.5) - 17.5x(x + 14) = 2.5(60x)$;

$150x + 2100 - 17.5x^2 - 245x = 150x$; $1500x + 21{,}000 - 175x^2 - 2450x = 1500x$; $175x^2 + 2450x - 21{,}000 = 0$; $x^2 + 14x - 120 = 0$; $(x + 20)(x - 6) = 0$; $x = 6$ or $x = -20$. Walking speed is 6 km/h.

(7-5) **1.** $0.06(175) = 10.5$. Tax is \$10.50.

2. $0.07(182{,}500) = 12{,}775$. The agent makes \$12,775.

3. $0.80x = 32$; $x = \frac{32}{0.8} = \frac{320}{8} = 40$. The original price was \$40.

4. $0.08(84) = 6.72$; $6.72 + 84 = 90.72$. Their new monthly payment will be \$90.72.

5. $0.25(840) = 210$; $840 - 210 = 630$. The final cost is \$630.

6. $0.12(3600) = 432$; $0.08(4200) = 336$; $432 - 336 = 96$. It is \$96 greater.

(7-6) **1.** $56{,}700 - 42{,}000 = 14{,}700; \frac{14{,}700}{42{,}000} = 0.35$. The increase is 35%.

2. $12(2.25) = 27; 27.00 - 21.60 = 5.4; \frac{5.4}{27} = \frac{54}{270} = \frac{2}{10}$. The discount is 20%.

3. $x + 0.16x = 1711; 1.16x = 1711; x = \frac{171{,}100}{116} = 1475$. 1475 people enrolled last year.

4. $0.08(7500) + 0.05(3500) = 600 + 175 = 775$. The total income is \$775.

5. Let x = amount invested. $0.1x = 0.12(x - 1600); 10x = 12(x - 1600);$ $10x = 12x - 19{,}200; 2x = 19{,}200; x = 9600$. They invested \$9600.

6. Let x = amount invested. $0.12(0.75x) + 0.08(0.25x) = 1320;$ $12(0.75x) + 8(0.25x) = 132{,}000; 9x + 2x = 132{,}000; 11x = 132{,}000; x = 12{,}000$. They have invested \$12,000.

CHEMISTRY

(7-7) **1.**

	Amount ×	% acid =	Amount acid
Old solution	20	75%	0.75(20)
Water	x	0	0
New solution	$20 + x$	15%	$0.15(20 + x)$

$0.75(20) = 0.15(20 + x); 75(20) = 15(20 + x); 1500 = 300 + 15x; 1200 = 15x;$ $x = 80$. 80 L of water must be added.

2.

	Amount ×	% acid =	Amount acid
Old solution	4	10%	0.1(4)
Acid	x	100%	x
New solution	$4 + x$	80%	$0.8(4 + x)$

$0.1(4) + x = 0.8(4 + x); 4 + 10x = 8(4 + x); 4 + 10x = 32 + 8x; 2x = 28; x = 14$. 14 L of acid must be added.

3.

	Amount ×	% acid =	Amount acid
1st	16	40%	0.4(16)
2nd	24	16%	0.16(24)
Mixture	40	x	$x(40)$

$0.4(16) + 0.16(24) = 40x; 40(16) + 16(24) = 4000x; 640 + 384 = 4000x;$ $1024 = 4000x; x = 0.256$. The new mixture is 25.6% acid.

4.

	Amount ×	% salt =	Amount salt
Old solution	84	5%	0.05(84)
Water	x	0%	0
New solution	$84 - x$	35%	$0.35(84 - x)$

$0.05(84) = 0.35(84 - x)$; $5(84) = 35(84 - x)$; $420 = 2940 - 35x$; $35x = 2520$; $x = 72$. 72 kg of water must be evaporated.

GROCERY

1.

	Number of liters ×	Price per liter =	Cost
Cranberry	x	0.5	$0.5x$
Apple	$120 - x$	0.35	$0.35(120 - x)$
Mixture	120	0.4	0.4(120)

$0.5x + 0.35(120 - x) = 0.4(120)$; $50x + 35(120 - x) = 40(120)$; $50x + 4200 - 35x = 4800$; $15x = 600$; $x = 40$. They used 40 L of cranberry juice and 80 L of apple juice.

2.

	Number of kg ×	Price per kg =	Cost
Premium	x	17	$17x$
Regular	$36 - x$	7	$7(36 - x)$
Blend	36	11	$11 \cdot 36$

$17x + 7(36 - x) = 396$; $17x + 252 - 7x = 396$; $10x = 144$; $x = 14.4$. The blend includes 14.4 kg of the premium, 21.6 kg of the other.

3.

	Number of lb ×	Price per lb =	Cost
Pork	12	1.25	15
Beef	24	2	48
Mixture	36	x	$36x$

$15 + 48 = 36x$; $36x = 63$; $x = 1.75$. The mixture should be \$1.75 per pound.

4.

	Number of kg ×	Price per kg =	Cost
Cranberries	x	2.10	$2.1x$
Pineapple	10	1.20	12
Relish	$10 + x$	1.35	$1.35(10 + x)$

$2.1x + 12 = 1.35(10 + x)$; $210x + 1200 = 135(10 + x)$; $210x + 1200 = 1350 + 135x$; $75x = 150$; $x = 2$. The grocer should mix 2 kg cranberries with the pineapple.

INVESTMENT AND WAGES

1.

	Number of hours ×	Amount =	Total earnings
Regular	35	x	$35x$
Overtime	10	$\frac{3}{2}x$	$\frac{3}{2}x(10)$

$35x + 15x = 625$; $50x = 625$; $x = 12.5$. The worker's regular hourly wage is \$12.50.

2.

	Amount invested ×	Rate =	Income
Part 1	x	12%	$0.12x$
Part 2	$8000 - x$	8%	$0.08(8000 - x)$

$0.12x + 0.08(8000 - x) = 825$; $12x + 8(8000 - x) = 82{,}500$; $12x + 64{,}000 - 8x = 82{,}500$; $4x = 18{,}500$; $x = 4625$. They invested \$4625 at 12%, \$3375 at 8%.

3.

	Amount invested ×	Rate =	Income
$\frac{2}{3}$	$\frac{2}{3}x$	12.5%	$0.125\left(\frac{2}{3}x\right)$
$\frac{1}{4}$	$\frac{1}{4}x$	8%	$0.08\left(\frac{1}{4}x\right)$
$\frac{1}{12}$	$\frac{1}{12}x$	6%	$0.06\left(\frac{1}{12}x\right)$

$0.125\left(\frac{2}{3}x\right) + 0.08\left(\frac{1}{4}x\right) + 0.06\left(\frac{1}{12}x\right) = 1625$;

$125\left(\frac{2}{3}x\right) + 80\left(\frac{1}{4}x\right) + 60\left(\frac{1}{12}x\right) = 1{,}625{,}000$; $\frac{250}{3}x + 20x + 5x = 1{,}625{,}000$;

$250x + 60x + 15x = 4{,}875{,}000$; $325x = 4{,}875{,}000$; $x = 15{,}000$. They invested \$15,000.

4.

	Amount invested ×	Rate =	Income
Stock 1	x	15%	$0.15x$
Stock 2	$10{,}000 - x$	16%	$0.16(10{,}000 - x)$

$0.15x + 0.16(10{,}000 - x) = 1520$; $15x + 16(10{,}000 - x) = 152{,}000$;
$15x + 160{,}000 - 16x = 152{,}000$; $x = 8000$. \$8000 at 15%, \$2000 at 16% is invested.

(7-8) 1.

	Work rate ×	Time =	Work done
Joe	$\frac{1}{6}$	2	$\frac{2}{6}$
Charlie	$\frac{1}{5}$	2	$\frac{2}{5}$

$x = \frac{2}{6} + \frac{2}{5} = \frac{1}{3} + \frac{2}{5} = \frac{5 + 6}{15} = \frac{11}{15}$.

They can do $\frac{11}{15}$ of the job.

2.

	Work rate × Time = Work done		
Charlotte	$\frac{1}{2}$	$\frac{3}{4}$	$\left(\frac{3}{4}\right)\left(\frac{1}{2}\right)$
Ralph	$\frac{1}{x}$	$\frac{3}{4}$	$\frac{3}{4x}$

$\frac{3}{8} + \frac{3}{4x} = 1$; $3x + 6 = 8x$; $5x = 6$; $x = \frac{6}{5}$. Ralph can do the route in $1\frac{1}{5}$ h, or 1 h 12 min.

3.

	Work rate × Time = Work done		
Crew 1	$\frac{1}{30}$	x	$\frac{x}{30}$
Crew 2	$\frac{1}{24}$	x	$\frac{x}{24}$

$\frac{x}{30} + \frac{x}{24} = 1$; $4x + 5x = 120$; $9x = 120$; $x = 13\frac{1}{3}$. It would take all 5 people $13\frac{1}{3}$ h or 13 h 20 min.

4.

	Work rate × Time = Work done		
Flora	$\frac{1}{4}$	3	$\frac{3}{4}$
Fiona	$\frac{1}{x}$	2	$\frac{2}{x}$

$\frac{3}{4} + \frac{2}{x} = 1$; $3x + 8 = 4x$; $x = 8$. It would take Fiona 8 h.

5.

	Work rate × Time = Work done		
Pipe 1	$\frac{1}{50}$	20	$\frac{20}{50}$
Pipe 2	$\frac{1}{90}$	x	$\frac{x}{90}$

$\frac{20}{50} + \frac{x}{90} = 1$; $\frac{2}{5} + \frac{x}{90} = 1$; $36 + x = 90$; $x = 54$. It would take the second pipe 54 min to finish the job.

6.

	Work rate × Time = Work done		
Machine 1	$\frac{1}{45}$	x	$\frac{x}{45}$
Machine 2	$\frac{1}{60}$	x	$\frac{x}{60}$
Machine 3	$\frac{1}{90}$	x	$\frac{x}{90}$

$\frac{x}{45} + \frac{x}{60} + \frac{x}{90} = 1$;
$4x + 3x + 2x = 180$; $9x = 180$;
$x = 20$.
It would take all 3 machines 20 min.

(7-9) **1.** $P = 10(1.03)^t$ **a.** $t = 0: P = 10(1.03)^0 = 10$ thousand people.

b. $t = -1: P = 10(1.03)^{-1} = \frac{10}{1.03} \approx 9.709$ thousand people.

c. $t = 1: P = 10(1.03)^{+1} = 10.3$ thousand people.

2. **a.** $100(0.5)^{1000/100} = 100(0.5)^{10} = 100\left(\frac{1}{2}\right)^{10} = \frac{100}{1024} = 100(0.0009766) = 0.09766$ g

b. $100(0.5)^{-1000/100} = 100(0.5)^{-10} = 100(2)^{10} = 100 \times 1024 = 102{,}400$ g

3. $1{,}000{,}000(4)^{-14/2} = 1{,}000{,}000(4)^{-7} = \frac{1{,}000{,}000}{4^7} = \frac{1{,}000{,}000}{16{,}384} \approx 61$

4. $10{,}000(1.08)^{-4} = \frac{10{,}000}{(1.08)^4} = \frac{10{,}000}{1.36049} = 7350.29$. Four years ago the amount was \$7,350.29.

5. $12{,}000(1.06)^{-10} = \frac{12{,}000}{(1.06)^{10}} = \frac{12{,}000}{1.7908477} = 6700.73$. Ten years ago the population was 6701 people.

6. $(1.05)^{-5} = \frac{1}{(1.05)^5} = 0.7835258$. It cost \$.78.

(7-10) 1. $\frac{3.84 \times 10^5}{3.00 \times 10^5} = \frac{3.84}{3} = 1.28$. It takes 1.28 s.

2. $4.07 \times 10^5 - 3.56 \times 10^5 = 0.51 \times 10^5 = 5.1 \times 10^4$. The difference is 5.1×10^4 km.

3. $\frac{6.1 \times 10^9}{3 \times 10^5} = 2.03 \times 10^4$. It takes 2.03×10^4 s.

4. a. $5.0 \times 10^2 \times 3.3 \times 10^3 = 16.5 \times 10^5 = 1.65 \times 10^6$. It is 1.65×10^6 light-years.
b. $1.65 \times 10^6 \times 9.5 \times 10^{12} = 15.675 \times 10^{18} = 1.5675 \times 10^{19}$. It is 1.5675×10^{19} km.

5. $6.0 \times 10^3 \times 1.0 \times 10^{-8} = 6 \times 10^{-5}$. It is 6×10^{-5} cm.

Pages 674–675 • CHAPTER 8

(8-9) 1. $\frac{32}{1.6} = \frac{62}{x}; x = \frac{1.6 \cdot 62}{32} = 3.1$. It bends 3.1 cm.

2. $\frac{18}{48} = \frac{x}{104}; x = \frac{18 \cdot 104}{48} = 39$. The baker needs 39 cups of flour.

3. $\frac{22}{54} = \frac{x}{81}; x = \frac{22 \cdot 81}{54} = 33$. 33 kg of the premium nuts are needed.

4. $\frac{6}{4} = \frac{x}{11}; x = \frac{6 \cdot 11}{4} = \frac{33}{2} = 16\frac{1}{2}$. The figure should be $16\frac{1}{2}$ in. tall.

5. $\frac{1}{60} = \frac{7.5}{x}; \frac{1}{60} = \frac{8.4}{y}; x = 60 \cdot 7.5 = 450; y = 60 \cdot 8.4 = 504; 450 \cdot 504 = 226{,}800$. The area of the region is 226,800 km^2.

6. $\frac{1}{5} = \frac{x}{62}; \frac{1}{5} = \frac{y}{30}; \frac{1}{5} = \frac{z}{12}; x = \frac{62}{5} = 12.4; y = \frac{30}{5} = 6; z = \frac{12}{5} = 2.4$. The volume will be $(12.4)(6)(2.4) = 178.56$ cm^3. The volume of the model will be 178.56 cm^3.

(8-10) 1. $7.5(84) = x(90); x = \frac{7.5 \cdot 84}{90} = 7$. It will take 7 h.

2. $36(800) = 24x; x = \frac{36 \cdot 800}{24} = 1200$. Its speed is 1200 rpm.

3. $0.08x = 0.12(1250); x = \frac{0.12 \cdot 1250}{0.08} = 1875$. You would have to invest \$1875.

4. $(1.8)(16) = x(1.92); x = \frac{1.8 \cdot 16}{1.92} = 15$. 15 carrels will fit.

5. $18 \cdot x = 22(100 - x)$; $18x = 2200 - 22x$; $40x = 2200$; $x = 55$. The fulcrum should be placed 55 cm or 0.55 m from the 18 g mass or 45 cm or 0.45 m from the 22 g mass.

6. Let x = distance of 250 g mass from the fulcrum. $400(x - 0.75) = 250x$; $400x - 300 = 250x$; $150x = 300$, $x = 2$. The 250 g mass is 2 m from the fulcrum.

Pages 675–677 • CHAPTER 9

(9-3) **1.** Let q = number of quarters; d = number of dimes. $q + d = 77$; $25q + 10d = 1250$; $d = 77 - q$; $25q + 10(77 - q) = 1250$; $25q + 770 - 10q = 1250$; $15q = 480$; $q = 32$. There are 32 quarters.

2. Let n and m = numbers. $n + m = 32$; $n = m + 4$. $m + 4 + m = 32$; $2m = 28$; $m = 14$. The numbers are 14 and 18.

3. Let l = length, w = width. $l = 2w - 3$; $2l + 2w = 54$. $2w = l + 3$; $2l + l + 3 = 54$; $3l = 51$; $l = 17$. The rectangle is 10 by 17.

4. Let x, y = numbers. $x + y = 66$; $\frac{2}{3}y - x = \frac{1}{3}(y - x)$; $x = 66 - y$;
$\frac{2}{3}y - (66 - y) = \frac{1}{3}y - \frac{1}{3}(66 - y)$; $\frac{2}{3}y - 66 + y = \frac{1}{3}y - 22 + \frac{1}{3}y$.
$2y - 198 + 3y = y - 66 + y$; $3y = 132$; $y = 44$; $x = 22$. The numbers are 22 and 44.

5. Let $\frac{x}{y}$ = fraction. $\frac{x - 1}{y} = \frac{1}{3}$; $\frac{x}{y - 2} = \frac{1}{2}$; $3x - 3 = y$, $2x = y - 2$; $2x = 3x - 3 - 2$;
$x = 5$. $y = 12$. The original fraction is $\frac{5}{12}$.

6. Let $\frac{x}{y}$ = fraction. $\frac{x + 2}{y} = \frac{2}{3}$; $\frac{x}{y - 1} = \frac{1}{2}$; $3x + 6 = 2y$; $2x = y - 1$; $y = 2x + 1$.
$3x + 6 = 2(2x + 1)$; $3x + 6 = 4x + 2$; $x = 4$. $y = 9$. The original fraction is $\frac{4}{9}$.

(9-4) **1.** Let x and y = numbers.

$$\begin{array}{l} x + y = 36 \\ \underline{x - y = \ \ 6} \\ 2x \qquad = 42 \end{array}$$
$x = 21$; $y = 15$. The numbers are 21 and 15.

2. Let x and y = numbers.

$$\begin{array}{lcr} x + y = 73 & \rightarrow & x + y = 73 \\ 2y - x = 50 & \rightarrow & \underline{-x + 2y = 50} \\ & & 3y = 123 \end{array}$$
$y = 41$. $x = 32$. The numbers are 41 and 32.

3. Let b = number of boys, g = number of girls.

$$\begin{array}{lcr} b + g = 158 & \rightarrow & b + g = 158 \\ b = g + 16 & \rightarrow & \underline{b - g = \ \ 16} \\ & & 2b \qquad = 174 \end{array}$$
$b = 87$. There are 87 boys.

4. Let w = rate of walking, b = rate of biking.

$$\begin{array}{lcr} 2w + 1b = 36 & \rightarrow & 2w + b = 36 \\ 2w + 2b = 56 & \rightarrow & \underline{2w + 2b = 56} \\ & & b = 20 \end{array}$$
$w = 8$. She can walk 8 km/h and bike 20 km/h.

5. Let q = number of quarters, d = number of dimes.

$$\begin{array}{lcr} q + d = 38 & \rightarrow & 10q + 10d = 380 \\ 2q(25) + 10d = 1100 & \rightarrow & \underline{50q + 10d = 1100} \\ & & 40q \qquad = 720 \end{array}$$
$q = 18$. $d = 20$. He has 18 quarters, 20 dimes.

6. Let o = amount Olivia has, and c = amount Carl has. $o = c + 30$;
$(o - 7) + (c - 7) = o, o + c - 14 = o, c = 14; o = 14 + 30 = 44$; Olivia has $44 and Carl has $14.

(9-5) 1. Let x and y = numbers.

$$\begin{array}{l} x + y = 51 \\ \underline{x - y = 13} \\ 2x \quad\quad = 64; \end{array}$$

$x = 32$, $y = 19$. The numbers are 32 and 19.

2. Let n = number of nickels, d = number of dimes; $n + d = 27$; $5n + 10d = 195$.
$n = 27 - d$; $5(27 - d) + 10d = 195$; $135 - 5d + 10d = 195$; $5d = 60$; $d = 12$. There are 15 nickels and 12 dimes.

3. Let l = length of lot, w = width of lot. $l = w + 50$; $2l + 2w = 5(4 \cdot 24) + 20$;
$2l + 2w = 480 + 20$; $2l + 2w = 500$; $2(w + 50) + 2w = 500$; $2w + 100 + 2w = 500$;
$4w = 400$; $w = 100$. $l = 150$. The length of the lot is 150 ft.

4. Let a = number of adult tickets, c = number of children's tickets.

$$\begin{array}{rcl} a + c = 1820 & \rightarrow & 2a + 2c = 3640 \\ 5a + 2c = 6100 & \rightarrow & \underline{5a + 2c = 6100} \\ & & 3a \quad\quad = 2460; \end{array}$$

$a = 820$. There were 820 adult tickets and 1000 children's tickets sold.

5. Let r = number right, w = number wrong.

$$\begin{array}{rcl} r + w = 50 & \rightarrow & 8r + 8w = 400 \\ 15r - 8w = 566 & \rightarrow & \underline{15r - 8w = 566} \\ & & 23r \quad\quad = 966; \end{array}$$

$r = 42$. The team answered 42 questions correctly.

6. Let a = cost Brand A, b = cost Brand B.

	Number of kg ×	Cost per kg =	Total cost
Brand A	4	a	$4a$
Brand B	6	b	$6b$
Mixture	10	6.20	62

	Number of kg ×	Cost per kg =	Total cost
Brand A	2	a	$2a$
Brand B	8	b	$8b$
Mixture	10	5.60	56

$$\begin{array}{rcl} 4a + 6b = 62 & \rightarrow & -2a - 3b = -31 \\ 2a + 8b = 56 & \rightarrow & \underline{\;\;2a + 8b = \;\;56} \\ & & 5b = \;\;25; \end{array}$$

$b = 5$; $a = 8$.
Brand A costs $8.00/kg; Brand B costs $5.00/kg.

(9-6) 1. Let b = rate of boat, c = rate of current. $b - c = 16$; $b + c = 30$;

$$\begin{array}{l} b + c = 30 \\ \underline{b - c = 16} \\ 2b \quad\quad = 46; \end{array}$$

$b = 23$; $c = 7$.
The rate of the boat is 23 km/h; the rate of the current is 7 km/h.

2. Let j = rate of jet, w = rate of wind. $j + w = 1100; j - w = 750;$

$$\begin{array}{rl} j + w &= 1100 \\ j - w &= 750 \\ \hline 2j &= 1850; j = 925. \end{array}$$

The rate of the jet is 925 km/h; the rate of the wind is 175 km/h.

3. Let c = rate of current, s = speed of swimmer. 24 min $= \frac{2}{5}$ h; 40 min $= \frac{2}{3}$ h.

$$c + s = 4 \cdot \frac{5}{2}; c + s = 10. s - c = \frac{4}{\frac{2}{3}}; s - c = 6.$$

$$\begin{array}{rl} s + c &= 10 \\ s - c &= 6 \\ \hline 2s &= 16; s = 8. c = 2. \end{array}$$

The swimmer's speed is 8 km/h; the current's rate is 2 km/h.

4. Let p = speed of plane, w = speed of wind. $p - w = \frac{2800}{3.5}; p + w = \frac{2800}{2.5};$

$$p - w = 800, p + w = 1120.$$

$$\begin{array}{rl} p - w &= 800 \\ p + w &= 1120 \\ \hline 2p &= 1920; p = 960, w = 160. \end{array}$$

The speed of the wind is 160 km/h, the speed of the plane is 960 km/h.

5. Let w = wind speed, t = time.

$$840 + w = \frac{3120}{t} \rightarrow t = \frac{3120}{840 + w}$$

$$840 - w = \frac{1920}{t} \rightarrow t = \frac{1920}{840 - w}; \frac{3120}{840 + w} = \frac{1920}{840 - w};$$

$2{,}620{,}800 - 3120w = 1{,}612{,}800 + 1920w; 5040w = 1{,}008{,}000; w = 200$. The wind speed is 200 km/h.

6. Let c = rate of current, r = rate of rowboat.

$$r + c = \frac{66}{3} \rightarrow \quad r + c = 22$$

$$r - c = \frac{33}{3} \rightarrow \quad \begin{array}{rl} r - c &= 11 \\ \hline 2r &= 33; r = 16.5 \text{ km/h}, c = 5.5. \end{array}$$

Page 677 • CHAPTER 10

(10-3) 1. Let n and $n + 1$ = integers. $n + n + 1 < 83; 2n < 82; n < 41$. 40 and 41 are the 2 integers with the greatest sum less than 83.

2. Let d = number of dimes, $2d$ = number of quarters. $10d + 25(2d) > 2000;$

$10d + 50d > 2000; 60d > 2000; d > \frac{100}{3} = 33\frac{1}{3}$. There are at least 34 dimes.

3. $\frac{240 + 180 + 220 + 200 + x}{5} \geq 220; 840 + x \geq 1100; x \geq 260$. The fifth person must bowl at least 260.

4. Let n, $n + 2$, and $n + 4$ = integers. $n + n + 2 + n + 4 < 80; 3n < 74; n < 24\frac{2}{3}$. The greatest such integers are 24, 26, and 28.

5. Let x = current speed limit. $\frac{2x}{(x - 40)} \geq 3.6$; $(x - 40)(3.6) \leq 2x$; $3.6x - 144 \leq 2x$; $1.6x \leq 144$; $x \leq 90$. The current speed limit is at most 90 km/h.

6. Let w = width, $1 + 2w$ = length. $(w + 5)(6 + 2w) \geq w(1 + 2w) + 150$; $2w^2 + 16w + 30 \geq w + 2w^2 + 150$; $15w \geq 120$; $w \geq 8$. The least possible dimensions are 8 cm by 17 cm.

Pages 677–678 • CHAPTER 11

(11-5) 1. Let s = length of side. $s^2 = 184$; $s = \pm\sqrt{184} = \pm 2\sqrt{46} \approx \pm 13.6$. A side is 13.6 cm.

2. Let s = length of a side of square. $s^2 = 25 \cdot 18 = 450$; $s = \pm 15\sqrt{2} \approx \pm 21.2$. The square has side 21.2 m.

3. Let s = side of square. $s^2 = \frac{1}{2} \cdot 8 \cdot 5 = 20$; $s = \pm\sqrt{20} \approx \pm 4.5$. The square has side 4.5 cm.

4. Let r = radius of circle; then $2r$ = side of square. $226.08 = 3.14(r^2)$; $r^2 = 72$; $r = \pm 6\sqrt{2} \approx 8.5$. The square has side 17.0 m.

5. $\pi r^2 = 3.14r^2 = 34.54$; $r^2 = 11$; $r = \pm\sqrt{11} \approx 3.3$. The radius of the pool is 3.3 m.

6. Let R = radius of entire region; r = radius of flower bed. $\pi R^2 = 21.98$; $R^2 = \frac{21.98}{3.14} = 7$; $R = \sqrt{7} \approx 2.6$. $r = R - 1 = 1.6$. The flower bed has a radius of 1.6 m.

(11-6) 1. Let d = diagonal. $d = \sqrt{20^2 + 50^2} = \sqrt{400 + 2500} = \sqrt{2900} = 10\sqrt{29} \approx 53.85$. The road is 53.85 m long.

2. Let x = height of mast. $x = \sqrt{6^2 - 2^2} = \sqrt{36 - 4} = \sqrt{32} = 4\sqrt{2} \approx 5.66$. The mast is 5.66 m high.

3. Let s = length of the second side; then $2s - 1$ = length of the first side and $2s + 1$ = length of hypotenuse. $(2s + 1)^2 = s^2 + (2s - 1)^2$, $4s^2 + 4s + 1 = s^2 + 4s^2 - 4s + 1$, $s^2 - 8s = 0$, $s(s - 8) = 0$; $s = 0$ or 8. The legs are 8 cm and 15 cm.

4. Let x = height of platform. $x = \sqrt{7^2 - 5^2} = \sqrt{49 - 25} = \sqrt{24} = 2\sqrt{6} \approx 4.90$. The platform is 4.90 m high.

5. Let x = length of shorter leg; then $x + 3$ = longer leg and $x + 6$ = hypotenuse. $x^2 + (x + 3)^2 = (x + 6)^2$; $x^2 + x^2 + 6x + 9 = x^2 + 12x + 36$; $x^2 - 6x - 27 = 0$; $(x - 9)(x + 3) = 0$; $x = 9$ or $x = -3$. The legs are 9 cm and 12 cm.

(11-10) 1. Let x = number. $\frac{1}{4}\sqrt{x} = 7$; $\sqrt{x} = 28$; $x = 28^2 = 784$. The number is 784.

2. Let x = number. $\sqrt{3x - 8} = 10$; $3x - 8 = 100$; $3x = 108$; $x = 36$. The number is 36.

3. Let x = number. $4\sqrt{x} = 28$; $\sqrt{x} = 7$; $x = 49$. The number is 49.

4. Let x = number. $\sqrt{2x} - 5 = 9$; $\sqrt{2x} = 14$; $2x = 196$; $x = 98$. The number is 98.

5. Let n and $n + 2$ = integers. $\sqrt{n(n + 2)} = 8\sqrt{15}$; $n(n + 2) = 64 \cdot 15$; $n^2 + 2n - 960 = 0$; $(n - 30)(n + 32) = 0$; $n = -32$ or $n = 30$. The integers are 30 and 32.

6. Let n and $n + 2$ = integers. $\sqrt{n(n + 2)} = 15\sqrt{3}$; $n(n + 2) = 675$; $n^2 + 2n - 675 = 0$; $(n + 27)(n - 25) = 0$; $n = -27$ (reject) or $n = 25$. The integers are 25 and 27.

Pages 679–680 • CHAPTER 12

(12-6) 1. Let n = number. $n + n^2 = 30$; $n^2 + n - 30 = 0$; $(n + 6)(n - 5) = 0$; $n = -6$ or $n = 5$. The number is 5 or -6.

2. Let x = amount of increase. $(13 + x)(7 + x) = 135$; $x^2 + 20x + 91 = 135$; $x^2 + 20x - 44 = 0$; $(x + 22)(x - 2) = 0$; $x = 2$ or $x = -22$ (*reject*). The new dimensions are 15 m by 9 m.

3. Let w = width; then $69 - w$ = length. $w(69 - w) = 540$; $w^2 - 69w + 540 = 0$; $(w - 60)(w - 9) = 0$; $w = 60$ or $w = 9$. The dimensions are 60 m and 9 m.

4. Let n and $n + 2$ = integers. $n^2 + (n + 2)^2 = 340$; $n^2 + n^2 + 4n + 4 = 340$; $2n^2 + 4n - 336 = 0$; $n^2 + 2n - 168 = 0$; $(n + 14)(n - 12) = 0$; $n = 12$ or $n = -14$. The integers are -14 and -12, or 12 and 14.

5.

	Work rate	× Time	= Work done
Crew 1	$\frac{1}{x}$	40	$\frac{40}{x}$
Crew 2	$\frac{1}{x-18}$	40	$\frac{40}{x-18}$

$\frac{40}{x} + \frac{40}{x - 18} = 1$; $40(x - 18) + 40x = x(x - 18)$; $40x - 720 + 40x = x^2 - 18x$; $x^2 - 98x + 720 = 0$; $(x - 90)(x - 8) = 0$; $x = 90$ or $x = 8$. The first crew would take 90 h, the second crew, 72 h.

6. Let n, $3n + 2$ = the numbers. $n^2 + (3n + 2)^2 = 212$; $n^2 + 9n^2 + 12n + 4 = 212$; $10n^2 + 12n - 208 = 0$; $5n^2 + 6n - 104 = 0$; $(5n + 26)(n - 4) = 0$; $n = -\frac{26}{5}$ or $n = 4$. The numbers are 4 and 14, or -5.2 and -13.6.

(12-7) 1. $\frac{112}{64^2} = \frac{x}{56^2}$; $x = \frac{56^2 \cdot 112}{64^2} = 85.75$ m. The stopping distance at 56 km/h is 85.75 m.

2. $\frac{1764}{(1.4)^2} = \frac{x}{(1.7)^2}$; $x = \frac{(1.7)^2 \cdot 1764}{(1.4)^2} = 2601$. Such a diamond would cost \$2601.

3. $4 \cdot 3^2 = h \cdot 6^2$; $h = \frac{4 \cdot 9}{36} = 1$. The new cone base height = 1 unit.

4. $8(3.5)^2 = x(2)^2$; $\frac{98}{4} = x$; $x = 24.5$. It will take the smaller hose 24.5 min.

5. $f_1 4^2 = f_2 12^2$; $16f_1 = 144f_2$; $f_2 = \frac{16f_1}{144} = \frac{f_1}{9}$. The force is $\frac{1}{9}$ of what it was.

6. $\frac{175.5}{6^2} = \frac{487.5}{x^2}$; $x^2 = \frac{6^2 \cdot 487.5}{175.5} = 100$; $x = \pm 10$. It would take 10 s to fall 487.5 m.

(12-8) 1. $\frac{45}{3000 \cdot 2 \cdot 0.075} = \frac{x}{1200 \cdot \frac{2}{3} \cdot 0.075}$; $x = \frac{(45)(800)}{(6000)} = 6$. It would cost 6¢.

2. $\frac{8 \cdot 4}{520} = \frac{x \cdot 3}{585}$; $x = \frac{32 \cdot 585}{520 \cdot 3} = 12$. It will take 12 workers.

3. $\frac{2 \cdot 3}{320} = \frac{3x}{840}$; $x = \frac{840 \cdot 6}{320 \cdot 3} = 5.25$. It will take them 5.25 h.

4. $\frac{840}{2 \cdot 5^2} = \frac{x}{(0.5)(3)^2}$; $x = \frac{840 \cdot 9 \cdot 0.5}{50} = 75.6$; The disc has mass 75.6 g.

Page 685 • EXPLORING DENSITY OF REAL NUMBERS

1. a. Sample answer: 8; $8 \cdot \frac{1}{2} = 4$

b. $4 \cdot \frac{1}{2} = 2$

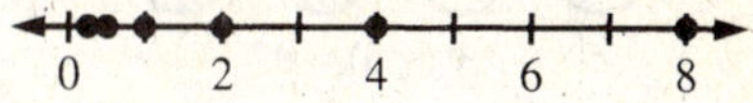

c. $2 \cdot \frac{1}{2} = 1$; $1 \cdot \frac{1}{2} = 0.5$; $0.5 \cdot \frac{1}{2} = 0.25$

d. No. The product approaches, but never reaches, zero.

2. Sample answer: $^-6$; $^-6 \cdot \frac{1}{2} = {}^-3$; $^-3 \cdot \frac{1}{2} = {}^-1.5$; $^-1.5 \cdot \frac{1}{2} = {}^-0.75$;

$^-0.75 \cdot \frac{1}{2} = {}^-0.375$; $^-0.375 \cdot \frac{1}{2} = {}^-0.1875$

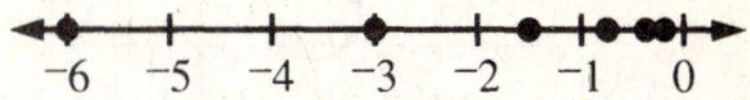

The product approaches, but never reaches, zero.

3. This process of multiplying by $\frac{1}{2}$ produces numbers that approach zero from *either* side of zero on the number line.

4. a. Sample answer: 7, 12; $\frac{7 + 12}{2} = \frac{19}{2} = 9.5$

b. $\frac{9.5 + 12}{2} = \frac{21.5}{2} = 10.75$

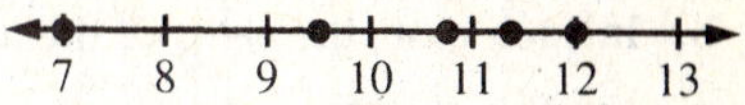

c. $\frac{10.75 + 12}{2} = \frac{22.75}{2} = 11.375$

d. The average approaches, but never reaches, the larger number.

5. a. Sample answer: 7, 13; $\frac{7 + 13}{2} = \frac{20}{2} = 10$

b. $\frac{10 + 7}{2} = \frac{17}{2} = 8.5$; $\frac{8.5 + 7}{2} = \frac{15.5}{2} = 7.75$

The average approaches, but never reaches, the smaller number.

6. a. Multiply x by $\frac{1}{2}$. **b.** Find their average.

7. a. Yes; by averaging you can always find a proper fraction halfway between two proper fractions. For example, the fraction halfway between $\frac{1}{8}$ and $\frac{1}{9}$ is $\frac{17}{144}$. Recall that a proper fraction is a fraction whose numerator is less than its denominator.

b. No; no integer exists between consecutive integers. For example, there is no integer between 7 and 8.

Page 686 • EXPLORING ADDITION OF INTEGERS

1–5. Let one unshaded chip represent +1 and one shaded chip represent −1.

1. a. 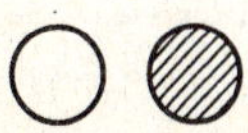

b. 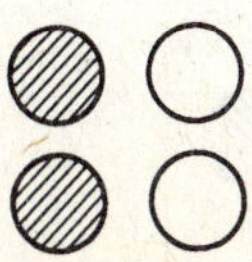

c. 0

d.

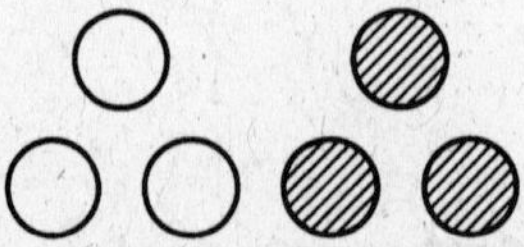

3 + (– 3) = 0

e. Use an equal number of red chips and blue chips.

2. a. 2 + 0 **b.** 2

c.

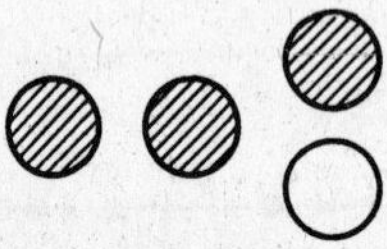

– 2 + 0 = – 2

3. a.

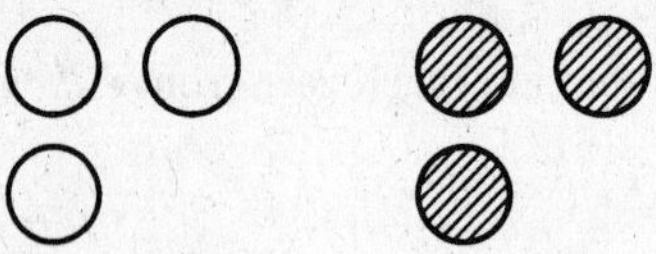

2 + 1 = 3 – 2 + (– 1) = – 3

b. In each case, the signs of the addends and their sum are the same.

c. No; this is not possible.

4. a.

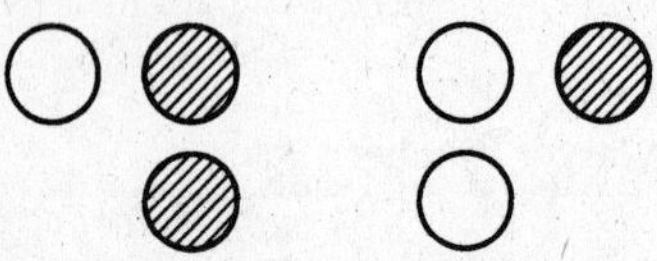

1 + (– 2) = – 1 2 + (– 1) = 1

b. The first is negative, the second positive.

c. The sign of the number with greater absolute value determines the sign of the sum.

5. a.

3 + 5 = 8

b.

– 3 + (– 5) = – 8

c.

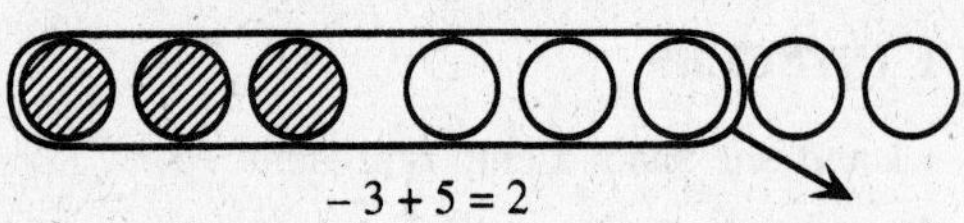

– 3 + 5 = 2

d.

3 + (– 5) = – 2

Page 687 • EXPLORING SUBTRACTION OF INTEGERS

1. c. 1 **d.** $4 - 3 = 1$

2–6. Let one unshaded chip represent $+1$ and one shaded chip represent -1.

2.

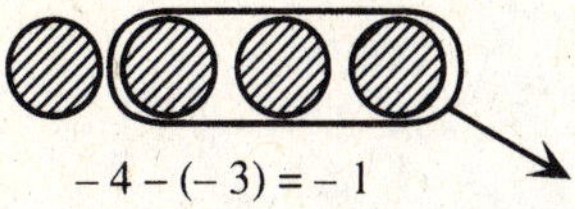

$-4 - (-3) = -1$

3. a. Represent $4 - (-3)$ with chips by removing 3 red chips from 4 blue chips.

b–d.

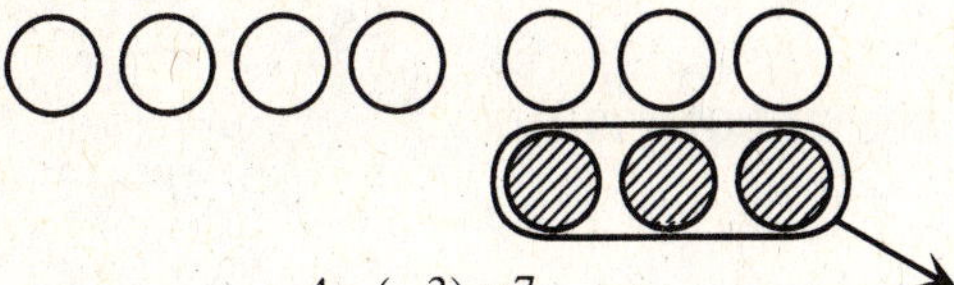

There are 7 blue chips left.

$4 - (-3) = 7$

e. $4 - (-3) = -7$

4. a. 4 red chips **b.** 3 blue chips **c.** 3 pairs of red and blue chips

d.

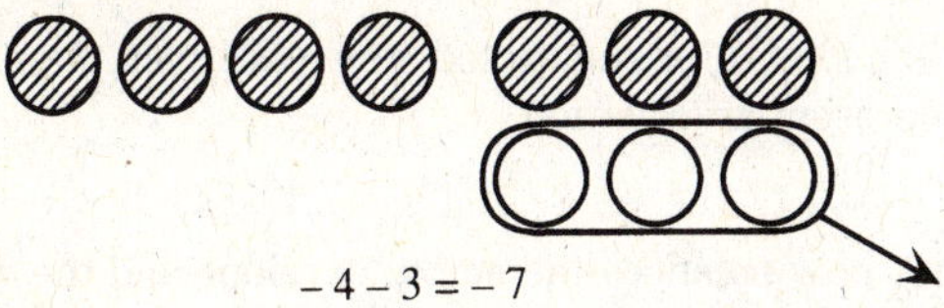

There are 7 red chips left.

$-4 - 3 = -7$

e. $-4 - 3 = -7$

5.

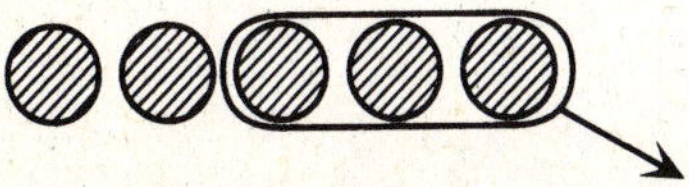

$-5 - (-3) = -2$

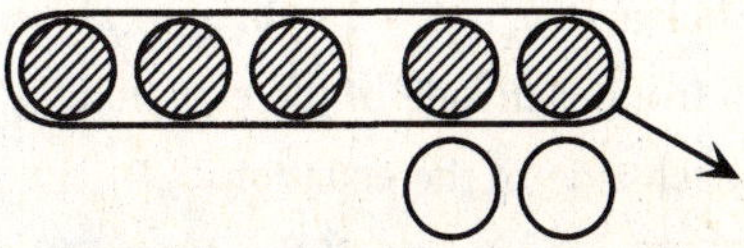

$-3 - (-5) = 2$

The answers are opposites.

6. a.

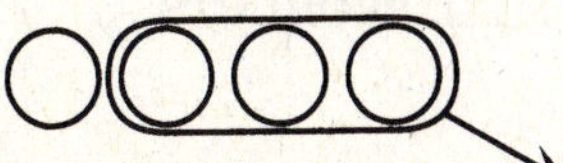

$4 - 3 = 1$

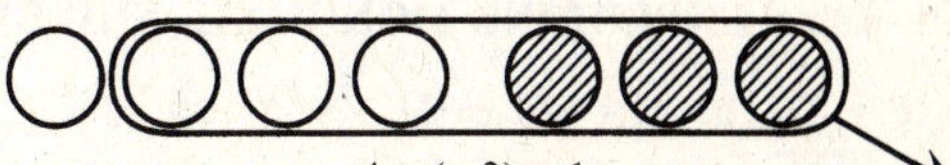

$4 + (-3) = 1$

b.

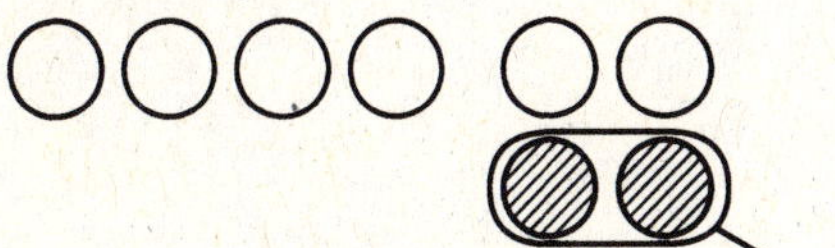

$4 - (-2) = 6$

$4 + 2 = 6$

c.

$4 - 5 = -1$

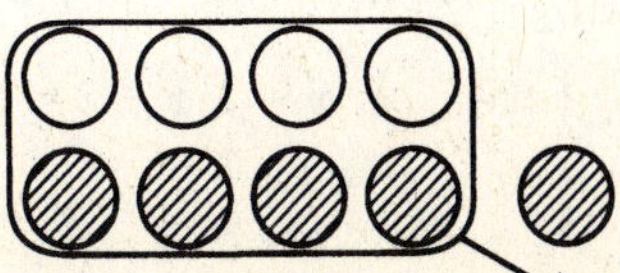

$4 + (-5) = -1$

d. 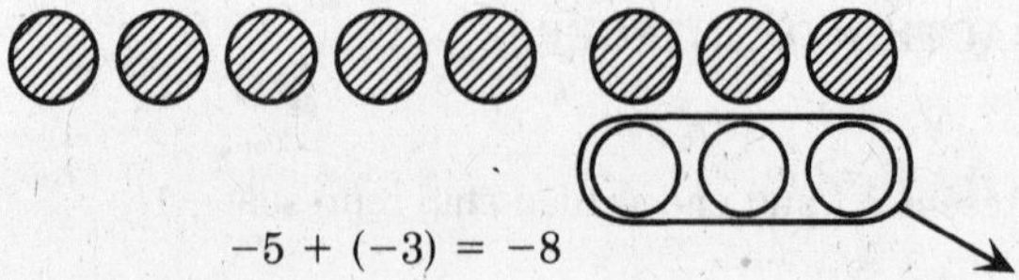

$-5 + (-3) = -8$

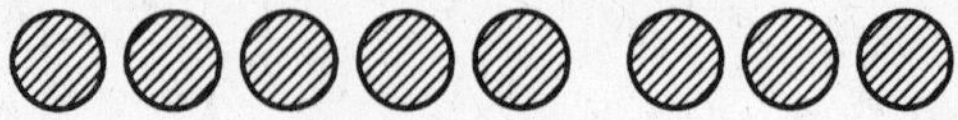

$-5 - 3 = -8$

7. true

Page 688 • EXPLORING WAYS TO SOLVE EQUATIONS

1. a. The pans balance; $6 = 6$

b. The pans balance; 8; $6 + 2 = 6 + 2$, or $8 = 8$

c. The pans balance; 5; $8 - 3 = 8 - 3$, or $5 = 5$

d. The pans do not balance; the left side is lower (heavier) than the right side. $5 - 1 \neq 5 - 2$, or $4 \neq 3$, or $4 > 3$

e. If the same number of washers is added to or subtracted from each pan, the pans stay in balance. Balanced pans can represent an equation.

2. a. The pans balance; $x + 3 = 10$

b. 3 washers; the envelope marked X now balances the empty envelope and the washers on the other pan.

c. 7; 7; $x = 7$

3. a. The pans balance; $y - 4 = 7$ **b.** $y = 11$

4. a. Subtract 6 from each side of the equation.

b. Add 6 to each side of the equation.

c. Subtract 75 from each side of the equation.

d. Add 173 to each side of the equation.

Pages 689–690 • EXPLORING MONOMIAL AND BINOMIAL PRODUCTS

1. a. $x; x + 1$ **b.** $A = x(x + 1)$ **c.** $x^2; x$

d. The area of the rectangle is the sum of the areas of the tiles.

e. $x(x + 1) = x^2 + x$

2. b. 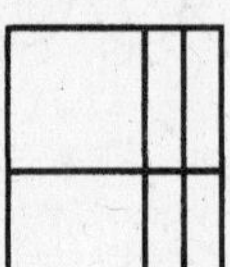

c. $2x, x + 2$

d. $A = 2x(x + 2)$

e. $2x^2 + 4x$

f. $2x(x + 2) = 2x^2 + 4x$

3. b. three 1-by-1 tiles **c.** $x + 1, x + 3$ **d.** $x^2 + 4x + 3$

e. $(x + 1)(x + 3) = x^2 + 4x + 3$

4. a. $x(2x + 3) = 2x^2 + 3x$ **b.** $3x(x + 1) = 3x^2 + 3x$

c. $(x + 2)(x + 2) = x^2 + 4x + 4$

d. $(x + 1)(x + 5) = x^2 + 6x + 5$

5. a. $x(2x + 3) =$
$2x^2 + 3x$

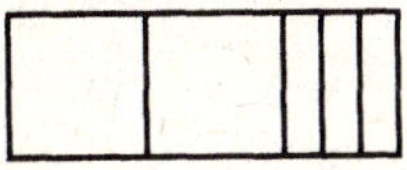

b. $3x(2x + 1) =$
$6x^2 + 3x$

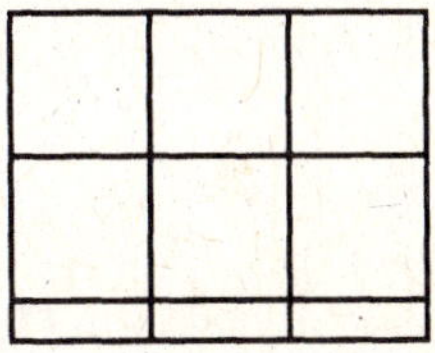

c. $(x + 5)3x =$
$3x^2 + 15x$

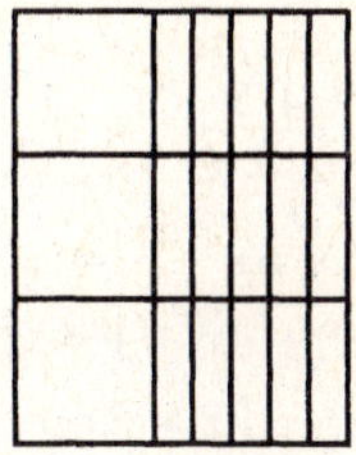

6. a. $(x + 2)(2x + 1) =$
$2x^2 + 5x + 2$

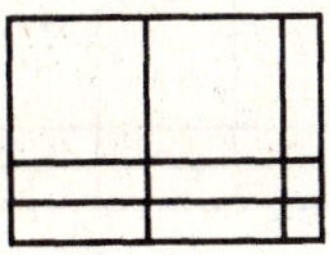

b. $(2x + 3)(3x + 1) =$
$6x^2 + 11x + 3$

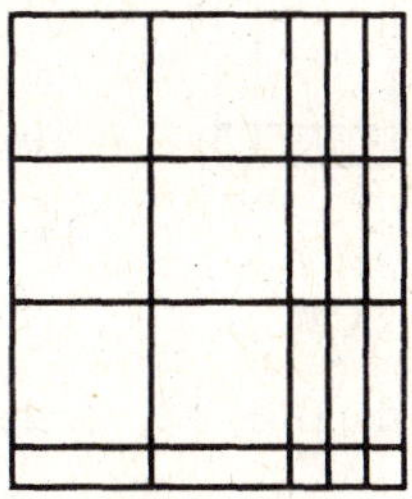

c. $(x + 3)^2 =$
$x^2 + 6x + 9$

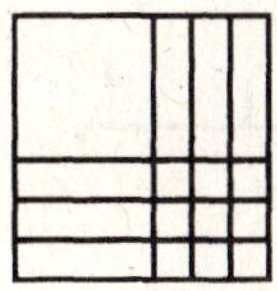

Page 691 • EXPLORING POLYNOMIAL FACTORS

1. a. $A = x^2 + 3x$

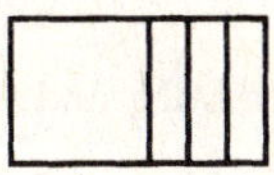

b. $x + 3; x$

c. $A = x(x + 3)$

d. $x^2 + 3x = x(x + 3)$

2. a. Use two x-by-x tiles and four 1-by-x tiles to represent $2x^2 + 4x$.

b. $2x + 4$

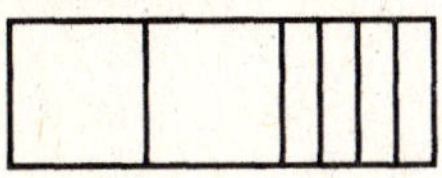

c. $2x^2 + 4x = x(2x + 4)$

d. $2x, x + 2$

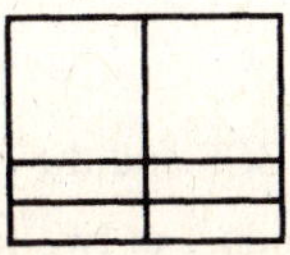

e. $2x^2 + 4x = 2x(x + 2)$

3. a. One x-by-x tile, seven 1-by-x tiles, and twelve 1-by-1 tiles

b.

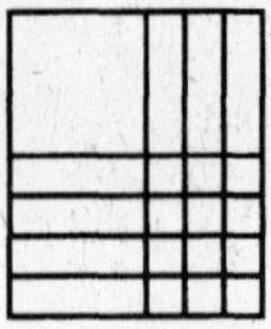

c. $x + 4, x + 3$

d. $x^2 + 7x + 12 = (x + 4)(x + 3)$

4. a. $x^2 + 5x + 6 = (x + 3)(x + 2)$

b. $2x^2 + 3x + 1 = (2x + 1)(x + 1)$

c. $3x^2 + 9x = 3x(x + 3)$

5. a. $2x^2 + x =$
$x(2x + 1)$

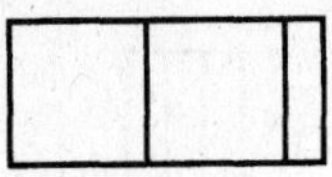

b. $x^2 + 2x + 1 =$
$(x + 1)^2$

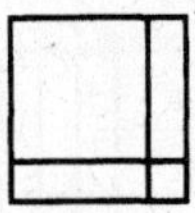

c. $3x^2 + 7x + 2 =$
$(3x + 1)(x + 2)$

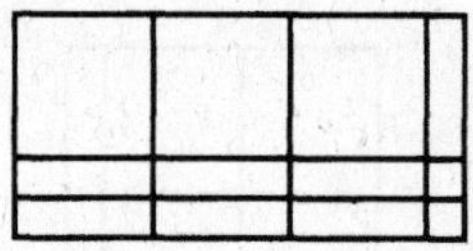

Page 692 • EXPLORING GCF AND LCM

1. a. $18 = 2 \cdot 3^2$ and $24 = 2^3 \cdot 3$, so the GCF $= 2 \cdot 3$, the product of the smaller powers of each common prime factor.

b. 2^3 and 3^2

c. The LCM $= 2^3 \cdot 3^2$, the product of the larger powers of each prime factor.

d. 2 and 3

e. The factors not included in the GCF form the LCM, and those not included in the LCM form the GCF.

2. a. $18 = 2 \cdot 3^2$ and $24 = 2^3 \cdot 3$, so $18 \cdot 24 = 2 \cdot 3^2 \cdot 2^3 \cdot 3 = (2 \cdot 2^3) \cdot (3^2 \cdot 3) = 2^4 \cdot 3^3$

b. The product of the two numbers, $2^4 \cdot 3^3$, is equal to the product of the GCF, $2 \cdot 3$, and the LCM, $2^3 \cdot 3^2$.

3. a. $N = ab^2c^3$ and $P = a^2bc^2d$, so the GCF $= abc^2$, the product of the smaller powers of each common prime factor.

b. $a^2b^2c^3d$

c. The LCM $= a^2b^2c^3d$, the product of the larger powers of each prime factor.

d. abc^2

e. The factors not included in the GCF form the LCM, and those not included in the LCM form the GCF.

4. a. $N \cdot P = ab^2c^3 \cdot a^2bc^2d = (a \cdot a^2)(b^2 \cdot b)(c^3 \cdot c^2)d = a^3b^3c^5d$

b. The product of the two expressions is equal to the product of the GCF and the LCM.

5. $48 = 2^4 \cdot 3$ and $60 = 2^2 \cdot 3 \cdot 5$, so the GCF $= 2^2 \cdot 3 = 12$. Since the product of two numbers equals the product of their GCF and LCM, the LCM equals the product of the two numbers divided by the GCF. Thus, the LCM $= \frac{48 \cdot 60}{12} = 240$.

6. a. $4 = 2^2$ and $6 = 2 \cdot 3$, so the GCF = 2. The LCD $= \frac{4 \cdot 6}{2} = 12.$

b. $12 = 2^2 \cdot 3$ and $18 = 3^2 \cdot 2$, so the GCF $= 2 \cdot 3 = 6$. The LCD $= \frac{12 \cdot 18}{6} = 36.$

c. $25 = 5^2$ and $15 = 3 \cdot 5$, so the GCF = 5. The LCD $= \frac{25 \cdot 15}{5} = 75.$

Page 693 • EXPLORING APPLICATIONS OF PROPORTIONS

1. A sample answer is given.

a. One phone book of 228 pages has 88 entries in one column and 4 columns per page, or about $88 \cdot 4 = 352$ listings per page. So $\frac{352 \text{ listings}}{1 \text{ page}} = \frac{x \text{ listings}}{228 \text{ pages}}$, or $x = 80{,}256$ listings in the entire book.

b. In the same phone book, each page has about 60 entries using only initials for first names.

c. $\frac{60}{352}$ **d.** $\frac{60}{352} \cdot 80{,}256$ listings $= 13{,}680$ listings with initials.

2. a–b. Answers will vary. **c.** Answers will vary (but should be approximately 40).

d. $\frac{200}{1000} = \frac{x}{200}$; $x = \frac{200 \cdot 200}{1000} = 40$ animals

3. $\frac{100}{p} = \frac{8}{100}$; $p = \frac{100 \cdot 100}{8} = 1250$ fish

4. Possible answer: Survey your algebra class to find a ratio, then set up a proportion using the total population of the school. For example, if 2 people in your class of 30 students are in the drama club and there are 680 students in your school, then the number of students, n, you would expect to be in the drama club is determined by $\frac{2}{30} = \frac{n}{680}$; $n = \frac{2 \cdot 680}{30} \approx$ 45 students.

Page 694 • EXPLORING LINEAR EQUATIONS

1. a–c. Sample answer: $m = 1$, $b = 1$; start with $y = x + 1$.

d. The lines all go through the same point, $(0, b)$; the slant, or slope of each line is different; the equations are alike except for the value of m.

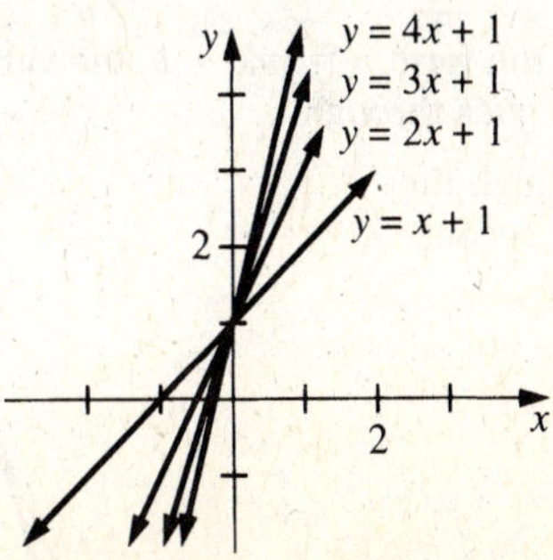

2. a–b. Sample answer: Start with $y = x + 1$.

c. The lines all go through the same point, (0, b); the slant, or slope of each line is different; the equations are alike except for the value of m.

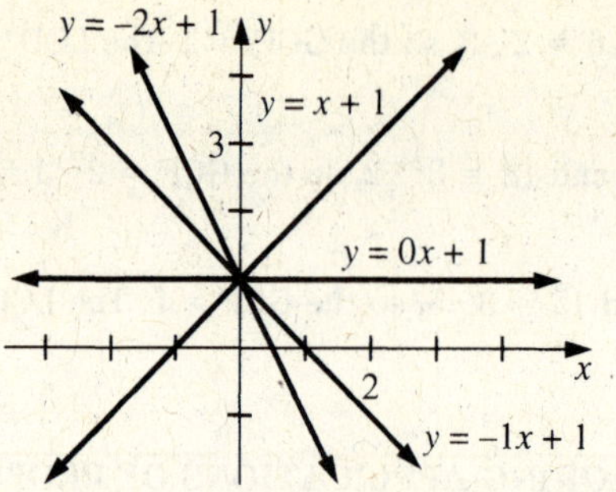

3. For an equation of the form $y = mx + b$, the value of m determines the slope of the line.

4. a–b. Sample answer: Start with $y = x + 1$.

c. The lines are parallel (have the same slope); each line goes through a different point (0, b) on the y-axis; the equations are alike except for the value of b.

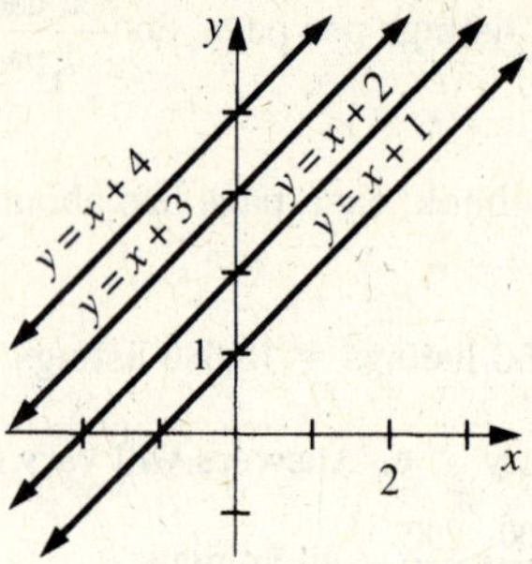

5. a–b. Sample answer: Start with $y = x + 1$.

c. The lines are parallel (have the same slope); each line goes through a different point (0, b) on the y-axis; the equations are alike except for the value of b.

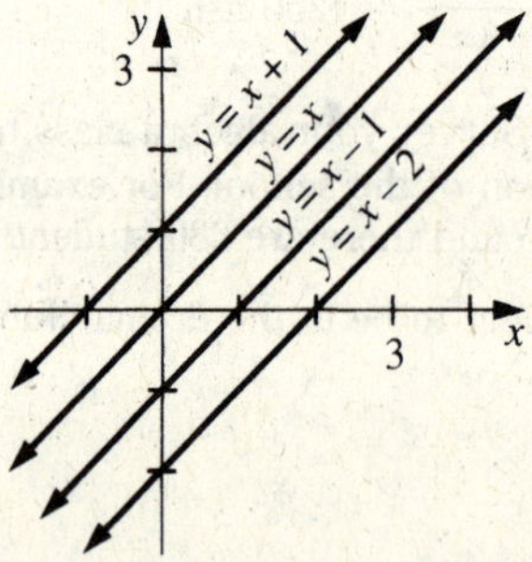

6. For an equation of the form $y = mx + b$, the value of b determines the point (0, b) on the y-axis that the line goes through.

7. They both pass through the point (0, 5), but the second line is steeper.

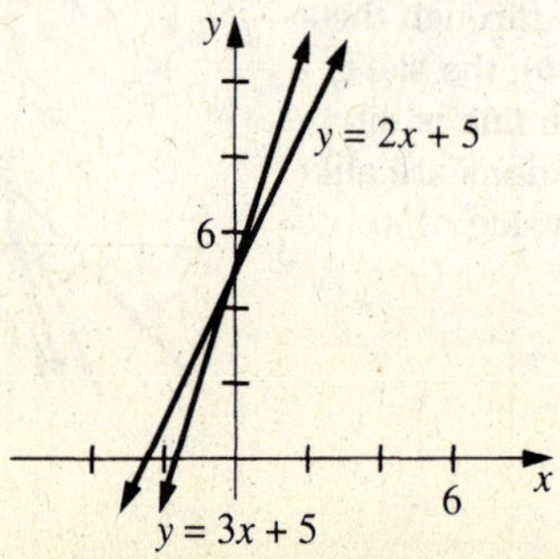

8. They are parallel (have the same slope); the first line intersects the y-axis at (0, 5); the second line intersects the y-axis at (0, 1).

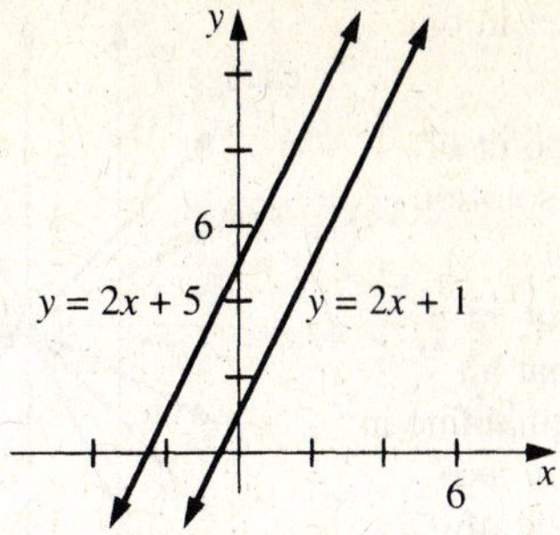

Page 695 • EXPLORING SYSTEMS OF LINEAR EQUATIONS

1. a–c. Start with $x - 2y = \frac{1}{2}$.

b. The lines are parallel.

c. Changing the value of c produces lines that are parallel to the original line.

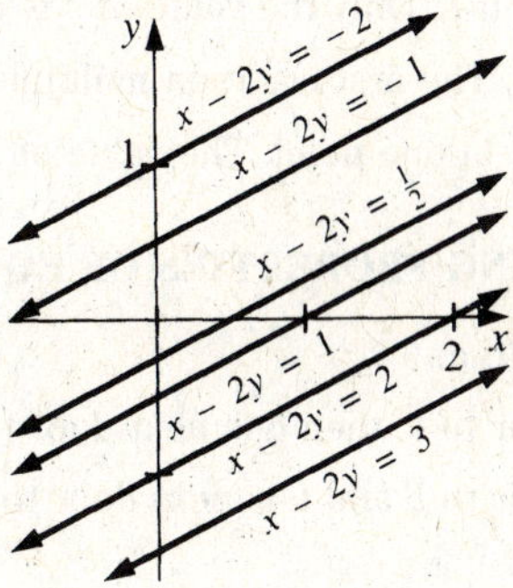

2. a–c. Start with $x - 2y = \frac{1}{2}$.

b. The lines are the same; the graphs coincide.

c. Multiplying by a nonzero constant produces lines that coincide with the original line.

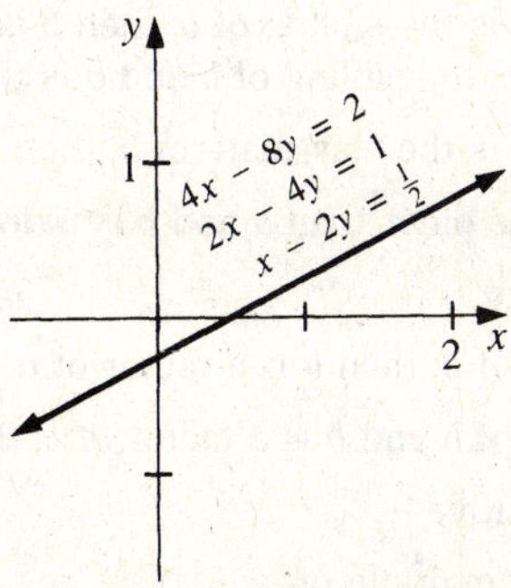

3. a–d. Start with $x - 2y = \frac{1}{2}$.

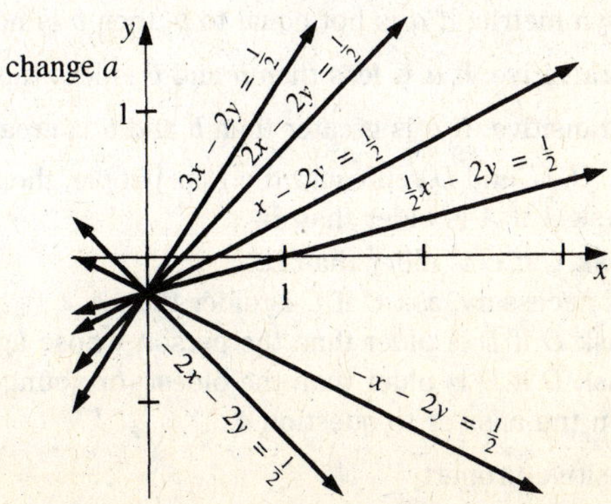

b. The lines intersect in one point.

c. Yes; yes; so the point of intersection is a solution of both equations.

d. Changing the value of a produces lines that all intersect the original line in one point on the y-axis. Changing the value of b produces lines that all intersect the original line in one point on the x-axis.

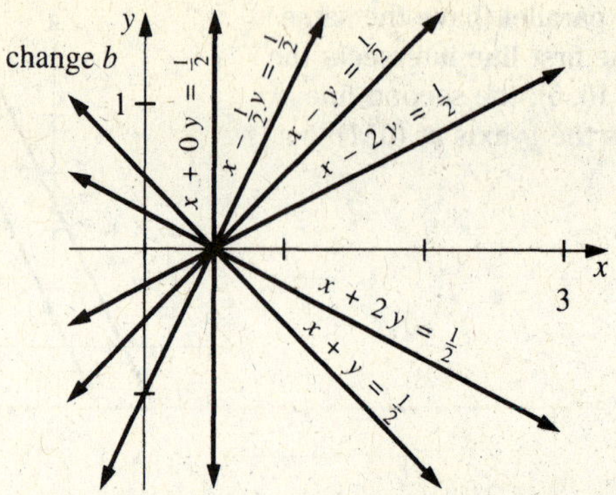

4. They are parallel lines. Only the value of c is different in each.

5. The lines coincide. The first equation multiplied by -2 yields the second equation.

6. The lines intersect in one point. The value of b, but not a, is different.

Page 696 • EXPLORING PROPERTIES OF EQUALITY AND ORDER

1. **a.** a is next door to a.

b. If a is next door to b, then b is next door to a.

c. If a is next door to b and b is next door to c, then a is next door to c.

2. symmetric

3. **a.** transitive: If a is shorter than b and b is shorter than c, then a is shorter than c.

b. symmetric: If a is the sibling of b, then b is the sibling of a.
transitive: If a is the sibling of b and b is the sibling of c, then a is the sibling of c.

c. symmetric: If a is the classmate of b, then b is the classmate of a.

d. transitive: If a is older than b and b is older than c, then a is older than c.

4. **a.** a is a factor of a.

b. If a is a factor of b, then b is a factor of a.

c. If a is a factor of b and b is a factor of c, then a is a factor of c.

5. reflexive and transitive

6. **a.** reflexive; a is a multiple of a.
transitive: If a is a multiple of b and b is a multiple of c, then a is a multiple of c.

b. symmetric: If a is not equal to b, then b is not equal to a.

c. transitive: If a is less than b and b is less than c, then a is less than c.

d. transitive: If a is greater than b and b is greater than c, then a is greater than c.

7. If A, B, C, and D represent the four people, then the minimum of 5 questions are as follows:
1) Ask A if A is older than B.
2) Ask C if C is older than B.
3) If necessary, ask C if C is older than A.
4) Ask D if D is older than the person whose age is between the other two.
5) Ask D if D is older than the oldest (or youngest) person of the other three, depending on the answer to question 4.

8. transitive property

Page 697 • EXPLORING DECIMALS

1. a. 0.5

b.

$\frac{1}{2}$	$\frac{1}{3}$	$\frac{1}{4}$	$\frac{1}{5}$	$\frac{1}{6}$	$\frac{1}{7}$	$\frac{1}{8}$	$\frac{1}{9}$	$\frac{1}{10}$	$\frac{1}{11}$	$\frac{1}{12}$
0.5	$0.\overline{3}$	0.25	0.2	$0.1\overline{6}$	$0.\overline{142857}$	0.125	$0.\overline{1}$	0.1	$0.\overline{09}$	$0.08\overline{3}$

c. $\frac{1}{2}, \frac{1}{4}, \frac{1}{5}, \frac{1}{8}, \frac{1}{10}$ **d.** $\frac{1}{3}, \frac{1}{6}, \frac{1}{7}, \frac{1}{9}, \frac{1}{11}, \frac{1}{12}$

e.

Denominator	2	4	5	8	10
Prime Factorization	2	2^2	5	2^3	$2 \cdot 5$

Denominator	3	6	7	9	11	12
Prime Factorization	3	$2 \cdot 3$	7	3^2	11	$2^2 \cdot 3$

The denominators for **c** (shown in the first table) have only 2 and/or 5 for prime factors. The denominators for **d** (shown in the second table) have prime factors other than 2 or 5.

f. Their decimal forms are terminating.

g. Their decimal forms are repeating.

2. a. repeats, since 13 is prime; $\frac{1}{13} = 0.\overline{076923}$

b. terminates, since the prime factors for $20 = 2^2 \cdot 5$ are 2 and 5; $\frac{1}{20} = 0.05$

c. repeats, since the prime factors for $27 = 3^3$ are not just 2 or 5; $\frac{1}{27} = 0.\overline{037}$

d. terminates, since the prime factors for $200 = 2^3 \cdot 5^2$ are 2 and 5; $\frac{1}{200} = 0.005$

3. a. $0.4 = \frac{4}{10} = \frac{2}{5}$; $0.29 = \frac{29}{100}$; $0.325 = \frac{325}{1000} = \frac{13}{40}$; $0.4791 = \frac{4791}{10{,}000}$

b.

Denominator	5	100	40	10,000
Prime Factorization	5	$2^2 \cdot 5^2$	$2^3 \cdot 5$	$2^4 \cdot 5^4$

c. Their only prime factors are 2 and 5; agree.

4. They are the product of prime factors of 2 and/or 5.

5. They are the product of prime factors that include at least one prime other than 2 or 5.

Pages 698–699 • EXPLORING IRRATIONAL NUMBERS

1. Use the Pythagorean theorem.

a. $1^2 + 1^2 = x^2, x = \sqrt{1^2 + 1^2} = \sqrt{2}.$

b. $1^2 + (\sqrt{2})^2 = x^2, x = \sqrt{1^2 + (\sqrt{2})^2} = \sqrt{1 + 2} = \sqrt{3}.$

c. $1^2 + (\sqrt{3})^2 = x^2, x = \sqrt{1^2 + (\sqrt{3})^2} = \sqrt{1 + 3} = \sqrt{4} = 2.$

d. $1^2 + 2^2 = x^2, x = \sqrt{1^2 + 2^2} = \sqrt{5};$

$1^2 + (\sqrt{5})^2 = x^2, x = \sqrt{1^2 + (\sqrt{5})^2} = \sqrt{1 + 5} = \sqrt{6};$

$1^2 + (\sqrt{6})^2 = x^2, x = \sqrt{1^2 + (\sqrt{6})^2} = \sqrt{1 + 6} = \sqrt{7};$

$1^2 + (\sqrt{7})^2 = x^2, x = \sqrt{1^2 + (\sqrt{7})^2} = \sqrt{1 + 7} = \sqrt{8} = 2\sqrt{2};$

$1^2 + (\sqrt{8})^2 = x^2, x = \sqrt{1^2 + (\sqrt{8})^2} = \sqrt{1 + 8} = \sqrt{9} = 3.$

e. The length of the new hypotenuse is the square root of the number that is one more than the radicand of the longer leg.

2. Use the Pythagorean theorem.

a. $2^2 + 2^2 = x^2, x = \sqrt{2^2 + 2^2} = \sqrt{8} = 2\sqrt{2}.$

b. $2^2 + (2\sqrt{2})^2 = x^2, x = \sqrt{2^2 + (2\sqrt{2})^2} = \sqrt{4 + 8} = \sqrt{12} = 2\sqrt{3}.$

c. $2^2 + (2\sqrt{3})^2 = x^2, x = \sqrt{2^2 + (2\sqrt{3})^2} = \sqrt{4 + 12} = \sqrt{16} = 2\sqrt{4} = 4.$

d. $2^2 + 4^2 = x^2, x = \sqrt{2^2 + 4^2} = \sqrt{4 + 16} = \sqrt{20} = 2\sqrt{5};$

$2^2 + (2\sqrt{5})^2 = x^2, x = \sqrt{2^2 + (2\sqrt{5})^2} = \sqrt{4 + 20} = \sqrt{24} = 2\sqrt{6};$

$2^2 + (2\sqrt{6})^2 = x^2, x = \sqrt{2^2 + (2\sqrt{6})^2} = \sqrt{4 + 24} = \sqrt{28} = 2\sqrt{7};$

$2^2 + (2\sqrt{7})^2 = x^2, x = \sqrt{2^2 + (2\sqrt{7})^2} = \sqrt{4 + 28} = \sqrt{32} = 2\sqrt{8} = 4\sqrt{2};$

$2^2 + (2\sqrt{8})^2 = x^2, x = \sqrt{2^2 + (2\sqrt{8})^2} = \sqrt{4 + 32} = \sqrt{36} = 2\sqrt{9} = 6.$

e. The length of the new hypotenuse is 2 times the square root of the number that is one more than the radicand of the longer leg.

3. The length of the hypotenuse will be $\sqrt{x^2 + (x\sqrt{y})^2} = \sqrt{x^2 + x^2y} = \sqrt{x^2}\sqrt{1 + y} = x\sqrt{y + 1}.$

4. a–f.

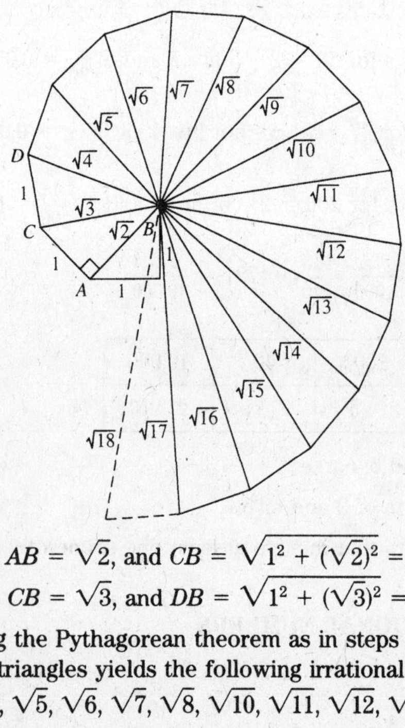

c. $AC = 1, AB = \sqrt{2}$, and $CB = \sqrt{1^2 + (\sqrt{2})^2} = \sqrt{3}$

e. $CD = 1, CB = \sqrt{3}$, and $DB = \sqrt{1^2 + (\sqrt{3})^2} = \sqrt{4} = 2$

g. Applying the Pythagorean theorem as in steps **c** and **e** to find the hypotenuses of successive triangles yields the following irrational lengths:
$\sqrt{2}, \sqrt{3}, \sqrt{5}, \sqrt{6}, \sqrt{7}, \sqrt{8}, \sqrt{10}, \sqrt{11}, \sqrt{12}, \sqrt{13}, \sqrt{14}, \sqrt{15}, \sqrt{17}.$

5. Use the Pythagorean theorem.

a. The length of the hypotenuse is $\sqrt{2^2 + 3^2} = \sqrt{13}.$

b. The length of the hypotenuse is $\sqrt{1^2 + 2^2} = \sqrt{5}$; yes, $\sqrt{5}$ is irrational.

c. Possible answer: 3, 5, $\sqrt{34}$, since $\sqrt{3^2 + 5^2} = \sqrt{34}$, and $\sqrt{34}$ is irrational.

d. The length of the diagonal is $\sqrt{2^2 + 2^2} = \sqrt{8} = 2\sqrt{2}$.

e. The length of the diagonal is $\sqrt{2^2 + 5^2} = \sqrt{29}$.

6. Continue the Wheel of Theodorus until you construct a triangle with hypotenuse of length $\sqrt{17}$. On a geoboard, build a right triangle with legs of lengths 1 and 4, since $\sqrt{1^2 + 4^2} = \sqrt{17}$.

7. Yes; continue the Wheel of Theodorus until you construct a triangle with hypotenuse of length $\sqrt{21}$. No, you cannot construct a segment with length $\sqrt{21}$ on a geoboard, since $\sqrt{21}$ cannot be expressed as $\sqrt{x^2 + y^2}$ where x and y are integers.

8. Begin a new Wheel of Theodorus using a right triangle with legs that measure 5 in. each. Using perpendicular segments of 5 in., draw five more triangles of the wheel. The last hypotenuse yields $5\sqrt{7}$ in.

Page 700 • EXPLORING QUADRATIC EQUATIONS

1. a. For $a = 2$, $b = 4$, and $c = 2$, $y = 2x^2 + 4x + 2$.

b. one

c. Root is -1.

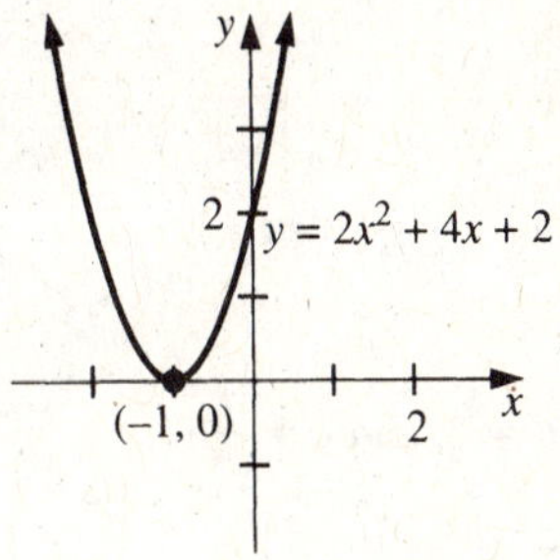

d. Let $a = 1$, $b = -4$, and $c = 4$ ($b^2 - 4ac = (-4)^2 - 4 \cdot 1 \cdot 4 = 16 - 16 = 0$). Then $y = x^2 - 4x + 4$, which yields one root, 2.

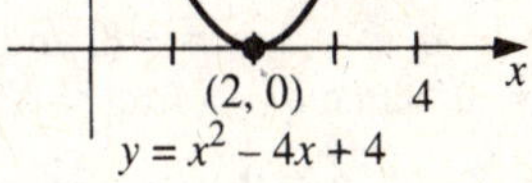

2. a. Let $a = 1$, $b = 3$, and $c = 2$ ($b^2 - 4ac = 3^2 - 4 \cdot 2 \cdot 1 = 1 > 0$). Then $y = x^2 + 3x + 2$.

b. two

c. Roots are -2 and -1.

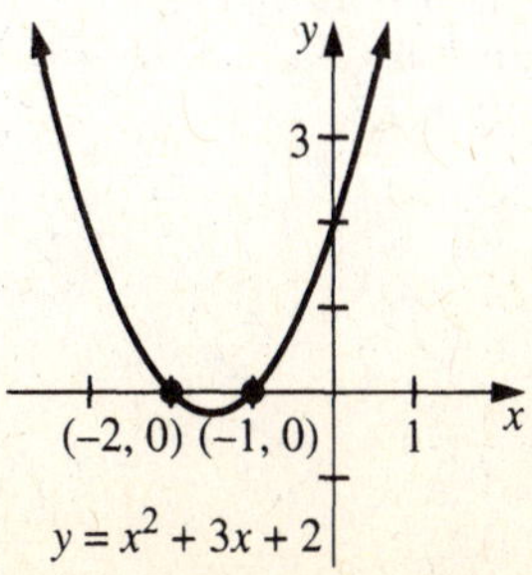

d. Let $a = -2$, $b = -4$, and $c = 1$ $(b^2 - 4ac = (-4)^2 - 4(-2)1 = 16 + 8 = 24 > 0)$. Then $y = -2x^2 - 4x + 1$, which yields two roots at ≈ 0.2 and ≈ -2.2.

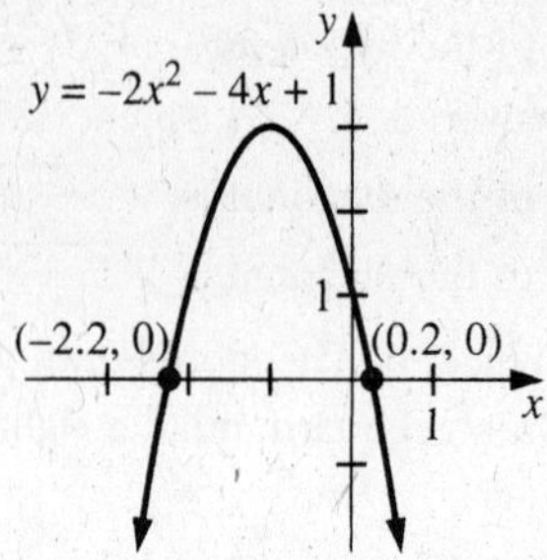

3. a. Let $a = 1$, $b = 2$, and $c = 2$ $(b^2 - 4ac = 2^2 - 4(1)(2) = -4 < 0)$. Then $y = x^2 + 2x + 2$.

b. none

c. There are no real roots.

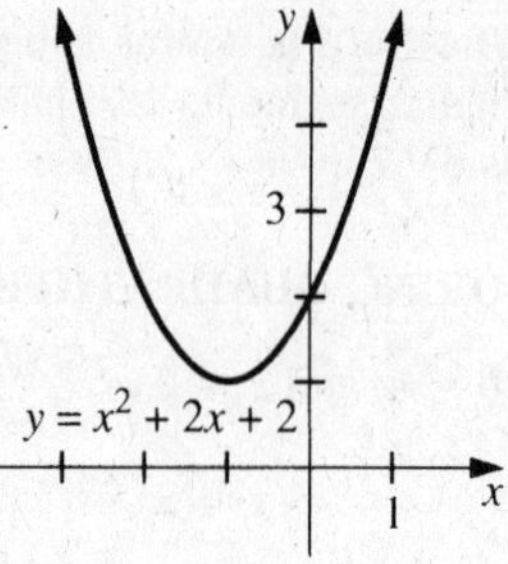

d. Let $a = -\frac{1}{2}$, $b = 3$, and $c = -5$ $\left(b^2 - 4ac = 3^2 - 4\left(-\frac{1}{2}\right)\left(-5\right) = 9 - 10 = -1 < 0\right)$. Then $y = -\frac{1}{2}x^2 + 3x - 5$, which yields no real roots.

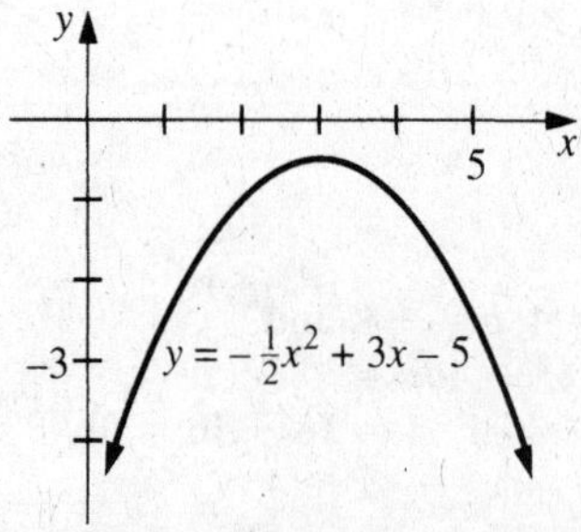

4. For $3x^2 - 8x + 9 = 0$, $a = 3$, $b = -8$, and $c = 9$, so $b^2 - 4ac = (-8)^2 - 4 \cdot 3 \cdot 9 = 64 - 108 = -44 < 0$. Since $b^2 - 4ac < 0$, there are no real roots.

5. For $x^2 - \sqrt{12}x + 3 = 0$, $a = 1$, $b = -\sqrt{12}$, and $c = 3$, so $b^2 - 4ac = (-\sqrt{12})^2 - 4 \cdot 1 \cdot 3 = 12 - 12 = 0$. Since $b^2 - 4ac = 0$, there is one real root; $x \approx 1.7$.

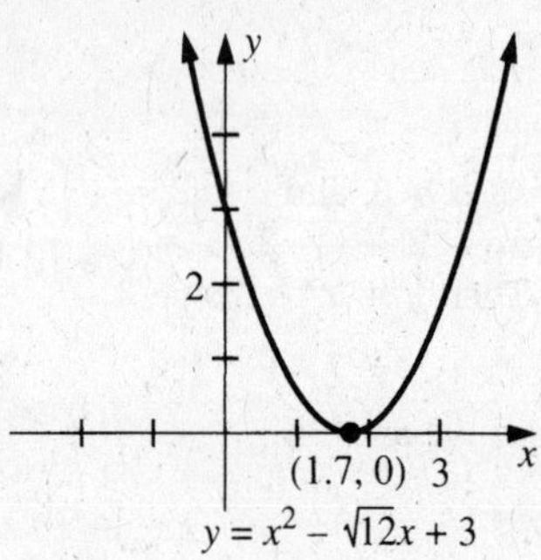

6. For $3x^2 - 6x + 1 = 0$, $a = 3$, $b = -6$, and $c = 1$, so $b^2 - 4ac = (-6)^2 - 4 \cdot 3 \cdot 1 = 36 - 12 = 24 > 0$. Since $b^2 - 4ac > 0$, there are two real roots; $x \approx 1.8$, $x \approx 0.2$.

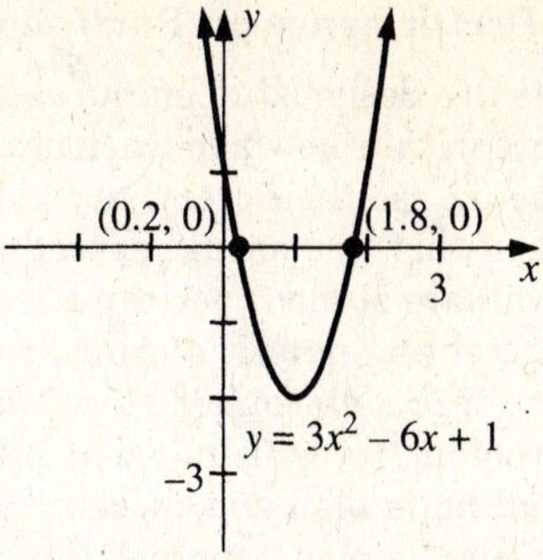

7. Answers will vary. To make up a quadratic equation $ax^2 + bx + c = 0$ that has no real roots, choose values for a, b, and c so that $b^2 - 4ac < 0$. For one root, choose values so that $b^2 - 4ac = 0$. For two roots, choose values so that $b^2 - 4ac > 0$. A graph of each equation showing the number of times the graph crosses the x-axis should verify the number of roots. For example, $x^2 - 2x + 3 = 0$ has no real roots, $x^2 + 6x + 9 = 0$ has one real root, and $x^2 + x - 6$ has two real roots.

Assessing Student Performance on Portfolio Projects

The Portfolio Projects are designed to encourage independent or group exploration of mathematical concepts within new applications. The projects are intended to be idea-starters. Students should use their creativity both in forming the problem and in shaping the solution. Because of this, you should expect a variety of responses from students. Similarly, you can evaluate student performance in a number of ways.

Some teachers prefer an analytic scoring system, assigning a specific number of points to each of a list of predetermined steps (for example, awarding 2 points for understanding the problem, recognizing what information is given and what must be found; 2 points for making a plan and choosing an appropriate problem-solving strategy; 2 points for carrying out the plan appropriately; and 2 points for the presentation of results). Other teachers prefer a holistic approach, assessing each student's work in its entirety rather than as a collection of discrete parts.

Holistic Scoring and Scoring Rubrics

Many teachers who use holistic scoring start with written descriptions, or scoring rubrics, that distinguish among different levels of performance. Rather than assigning so many points per part of an answer, scoring rubrics tend to focus on qualitative distinctions. As you read the student responses the first time through, the rubrics should help you sort the works into four categories. The first group includes responses that show complete understanding and in which the ideas are clearly expressed. The second group includes responses that show some understanding of the basic ideas and concepts but which are incomplete or unclear. The third group includes responses that show little or no understanding of the problem or its solution and suggest the need for remediation. The fourth group includes responses that show no effort. After the initial sorting of all student responses, the responses within each category can be arranged into subcategories as necessary in a second reading.

General Rubrics and Specific Rubrics

Scoring rubrics help clarify the key assessment criteria for yourself and for your students. General scoring rubrics serve as a guideline. This broad framework helps establish a dependable system, familiar to students, colleagues, administrators, and parents. One such general rubric follows:

Complete Understanding
4 points — Complete responses that demonstrate competence
Some Understanding
3 points — Responses show competence, though they may not be complete or entirely accurate
2 points — Responses show limited competence and contain significant errors
Little or No Understanding
1 point — Responses that show effort but little evidence of competence
No Attempt
0 points — Responses that show no effort

This is an example of a 4-point scoring system, with the primary emphasis being the distinction among the strongest responses, the weakest responses, and those in the middle. Which general rubric you use will depend somewhat on your own needs and preferences. Many educators with experience using rubrics, however, recommend using the general rubrics only as a framework from which specific rubrics for each problem or activity are written.

Specific rubrics allow you to highlight the project's distinctive requirements. The process of creating specific scoring rubrics for a project helps define the project's essential elements. Sharing these rubrics with students helps focus their work. Many teachers, in fact, involve the students in writing the project's specific scoring rubrics.

Writing Scoring Rubrics

Scoring rubrics written for a specific project should be general enough to allow for the unexpected (in order to recognize what each student might bring to the problem). At the same time, they need to be specific enough to make qualitative distinctions among the levels of performance. The elements you may want to consider in writing specific rubrics for a project are mathematical content, connections, the problem-solving process, reasoning, and communication.

The project-specific rubrics you develop should reflect your instructional goals and the needs of your students. Consider the elements as you identify the qualities that make up a complete response for a given project, as well as the qualitative differences between top-level and lower-level performance. Each of the Portfolio Projects will call upon students to use their abilities in all five of these areas, but some projects rely more heavily on one of these areas than on another. Your scoring rubrics for a particular project should focus primarily on the key elements for that project.

Suggestions for Specific Scoring Rubrics

ISBNs and Check Digits (Chapter 2)
Complete Understanding
Calculations are carried out correctly. Explanations are clear and understandable, using math equations where appropriate. Student devises and explains own code and check calculation formula.
Some Understanding
Calculations are carried out correctly. Explanations may be fragmented or not use math equations where appropriate. Student-devised code may not work as described.
Little or No Understanding
No show of understanding beyond doing calculations.

A Transportation Problem (Chapter 7)
Complete Understanding
Written report is clear and understandable and uses all necessary facts describing time and cost factors involved. Math equations are used correctly and where appropriate.
Some Understanding
Written report may not use all facts or may not accurately address both time and cost factors. Some math equations may be used incorrectly.
Little or No Understanding
Written report is fragmented, does not use all necessary facts, and makes no use of math equations. Both questions posed in the project may not have been answered.

Using Rubrics to Encourage Excellence

Assessment should not be considered the end of the instructional sequence. Many teachers find holistic scoring rubrics useful in encouraging students toward higher levels of performance. Scoring rubrics can be particularly helpful in a "review-and-revise" cycle within the process of working on the projects. The review and preliminary rating can be done by the teacher, or it can be done within a conference among students. In either case, because holistic scoring rubrics focus on qualitative distinctions among varying levels of performance, this preliminary review often helps the student see more clearly how to move his or her work from mid-range to top-level performance.